Epigenetics in Human Disease

Epigenetics in Human Disease

Edited by

Trygve O. Tollefsbol

Department of Biology, University of Alabama at Birmingham, AL, USA

Amsterdam • Boston • Heidelberg • London • New York • Oxford
Paris • San Diego • San Francisco • Singapore • Sydney • Tokyo

Academic Press is an imprint of Elsevier

ELSEVIER

Academic Press is an imprint of Elsevier
32 Jamestown Road, London NW1 7BY, UK
225 Wyman Street, Waltham, MA 02451, USA
525 B Street, Suite 1800, San Diego, CA 92101-4495, USA

First edition 2012

British Library Cataloguing-in-Publication Data
A catalogue record for this book is available from the British Library

Library of Congress Cataloging-in-Publication Data
A catalog record for this book is available from the Library of Congress

ISBN: 978-0-12-388415-2

For information on all Academic Press publications
visit our website at elsevierdirect.com

Typeset by TNQ Books and Journals Pvt Ltd. www.tnq.co.in

Printed and bound in United States of America

12 13 14 15 16   10 9 8 7 6 5 4 3 2 1

Working together to grow
libraries in developing countries

www.elsevier.com | www.bookaid.org | www.sabre.org

ELSEVIER   BOOK AID International   Sabre Foundation

CONTENTS

CONTENTS

Now that epigenetics has moved to the forefront of biological sciences, this book is focused on practical aspects of this burgeoning field of science and is intended to provide the most recent, pertinent, and comprehensive information with respect to the role of epigenetics in human disease. Epigenetic diseases consist of the increasing number of human diseases that have at least part of their basis in aberrations of epigenetic processes such as DNA methylation, histone modifications, or non-coding RNAs. Unlike most genetic defects as a cause for human disease, epigenetic alterations are potentially reversible. This is perhaps the most important aspect of epigenetic diseases because their reversibility makes these diseases amenable to pharmacological treatment. The goal of this volume is to highlight those diseases or conditions for which we have considerable epigenetic knowledge such as cancer, autoimmune disorders, and aging as well as those that are yielding exciting breakthroughs in epigenetics such as diabetes, neurological disorders, and cardiovascular disease.

Where applicable, attempts are made not only to detail the role of epigenetics in the etiology, progression, diagnosis, and prognosis of these diseases, but also to present novel epigenetic approaches to the treatment of these diseases. The book is designed such that each featured human disease is first described in terms of the underlying role of epigenetics in the disease and, where possible, followed by a chapter describing the most recent advances in epigenetic approaches for treating the disease. This allows basic scientists to readily view how their efforts are currently being translated to the clinic and it also allows clinicians to review in side-by-side chapters the epigenetic basis of the diseases they are treating. In some cases our knowledge of the epigenetics of human diseases is more extensive. Therefore, in those cases, such as cancer, more than one chapter on the underlying epigenetic causes appears. In other cases, such as for neurological disorders, the epigenetic basis of the diseases can vary somewhat due to the varying nature of the disorders. Chapters are also presented on the epigenetics of human imprinting disorders, respiratory diseases, infectious diseases, and gynecological and reproductive diseases, as well as the epigenetics of stem cells, obesity, and allergic diseases. Although aging is not considered to be a human disease per se, there are many age-associated diseases. Moreover, since epigenetics plays a major role in the aging process, advances in the epigenetics of aging are highly relevant to many human diseases. Therefore, this volume closes with chapters on aging epigenetics and breakthroughs that have been made to delay the aging process through epigenetic approaches.

The intended audience for this book is the vast body of students and scientists who are interested in either the underlying basis of human diseases and/or novel means to treat human diseases that are caused by reversible epigenetic processes. This book is tailored for those with interests ranging from basic molecular biology to clinical therapy and who could benefit from a comprehensive analysis of epigenetics as it applies to human diseases.

Trygve O. Tollefsbol

CONTRIBUTORS

Manori Amarasekera
School of Paediatrics and Child Health, University of Western Australia, Roberts Road, Subiaco, WA 6008, Australia

Daisuke Aoki
Department of Obstetrics and Gynecology, Keio University School of Medicine, Shinanomachi 35, Shinjuku-ku, Tokyo 160-8582, Japan

Eri Arai
Division of Molecular Pathology, National Cancer Center Research Institute, Tokyo 104-0045, Japan

Esteban Ballestar
Chromatin and Disease Group, Cancer Epigenetics and Biology Programme (PEBC), Bellvitge Biomedical Research Institute (IDIBELL), L'Hospitalet de Llobregat, Barcelona, Spain

Kouji Banno
Department of Obstetrics and Gynecology, Keio University School of Medicine, Shinanomachi 35, Shinjuku-ku, Tokyo 160-8582, Japan

Peter J. Barnes
National Heart and Lung Institute, Imperial College School of Medicine, Dovehouse St, London SW3 6LY, UK

Christopher G. Bell
Medical Genomics, UCL Cancer Institute, University College London, 72 Huntley Street, London, WC1 6BT, UK

Graham C. Burdge
Academic Unit of Human Development and Health, Faculty of Medicine, University of Southampton, Southampton, UK

J. Richard Chaillet
Department of Microbiology and Molecular Genetics, Magee-Womens Research Institute, University of Pittsburgh School of Medicine, Pittsburgh, PA, USA

Christopher Chang
Division of Allergy, Asthma and Immunology, Thomas Jefferson University, Nemours/A.I. duPont Hospital for Children, 1600 Rockland Road, Wilmington, DE 19803, USA.

Fabio Coppedè
Faculty of Medicine, Section of Medical Genetics, University of Pisa, Pisa, Italy

Lorenzo De La Rica
Chromatin and Disease Group, Cancer Epigenetics and Biology Programme (PEBC), Bellvitge Biomedical Research Institute (IDIBELL), L'Hospitalet de Llobregat, Barcelona, Spain

Andrea Fuso
Dept. of Surgery "P. Valdoni", Sapienza University of Rome, Via Antonio Scarpa, 14 00161, Rome, Italy

Peter D. Gluckman
Liggins Institute, University of Auckland, Auckland, New Zealand

Keith M. Godfrey
Academic Unit of Human Development and Health, Faculty of Medicine, University of Southampton, Southampton, UK; MRC Lifecourse Epidemiology Unit, Southampton, UK; NIHR Nutrition Biomedical Research Unit, Southampton General Hospital, Tremona Road, Southampton SO16 6YD, UK

Steven G. Gray
Department of Clinical Medicine, Trinity Centre for Health Sciences, St James's Hospital, Dublin 8, Ireland

Sun-Wei Guo
Shanghai Obstetrics and Gynecology Hospital, Fudan University, 419 Fangxie Road, Shanghai 200011, China

Mark A. Hanson
Academic Unit of Human Development and Health, Faculty of Medicine, University of Southampton, Southampton, UK

Akira Hirasawa
Department of Obstetrics and Gynecology, Keio University School of Medicine, Shinanomachi 35, Shinjuku-ku, Tokyo 160-8582, Japan

Takae Hirasawa
Department of Epigenetics Medicine, Interdisciplinary Graduate School of Medicine and Engineering, University of Yamanashi, Yamanashi, Japan

Kai Huang
Fortiss gemeinnützige GmbHAn-Institut der TU MünchenGuerickestr. 25, 80805 Munich, Germany

Biola M. Javierre
Chromatin and Disease Group, Cancer Epigenetics and Biology Programme (PEBC), Bellvitge Biomedical Research Institute (IDIBELL), L'Hospitalet de Llobregat, Barcelona, Spain

Yae Kanai
Division of Molecular Pathology, National Cancer Center Research Institute, Tokyo 104-0045, Japan

Iori Kisu
Department of Obstetrics and Gynecology, Keio University School of Medicine, Shinanomachi 35, Shinjuku-ku, Tokyo 160-8582, Japan

Yusuke Kobayashi
Department of Obstetrics and Gynecology, Keio University School of Medicine, Shinanomachi 35, Shinjuku-ku, Tokyo 160-8582, Japan

Takeo Kubota
Department of Epigenetics Medicine, Interdisciplinary Graduate School of Medicine and Engineering, University of Yamanashi, Yamanashi, Japan

Abigail S. Lapham
Academic Unit of Human Development and Health, Faculty of Medicine, University of Southampton, Southampton, UK

Karen A. Lillycrop
Faculty of Natural and Environmental Sciences, University of Southampton, Southampton, UK

Charlotte Ling
Epigenetics and Diabetes Unit, Lund University Diabetes Centre, Department of Clinical Sciences, Lund University, Malmö, Sweden

Niamh Lynam-Lennon
Department of Surgery, Trinity College Dublin, Trinity Centre for Health Sciences, St James's Hospital, Dublin 8, Ireland

Masato Maekawa
Department of Laboratory Medicine, Hamamatsu University School of Medicine, 1-20-1 Handayama, Higashi-ku, Hamamatsu 431-3192, Japan

Stephen G. Maher
Department of Surgery, Trinity College Dublin, Trinity Centre for Health Sciences, St James's Hospital, Dublin 8, Ireland

Andriana Margariti
Cardiovascular Division, King's College London BHF Centre, London, UK

David Martino
Developmental Epigenetics, Murdoch Children's Research Institute, Flemington Road, Parkville, VIC 3052, Australia

Kenta Masuda
Department of Obstetrics and Gynecology, Keio University School of Medicine, Shinanomachi 35, Shinjuku-ku, Tokyo 160-8582, Japan

Ciro Mercurio
DAC-Genextra group, Via Adamello 16, 20100 Milan Italy and European Institute of Oncology, IFOM-IEO campus, Via Adamello 16, 20100 Milan Italy

Janos Minarovits
Microbiological Research Group, National Center for Epidemiology, H-1529 Budapest, Piheno u. 1. Hungary

Saverio Minucci
European Institute of Oncology, IFOM-IEO campus, Via Adamello 16, 20100 Milan Italy and University of Milan, Via Festa del Perdono 7, 20100 Milan Italy

Rajesh C. Miranda
Texas A&M Health Science Ctr., College of Medicine, Dept. Neuroscience and Experimental Therapeutics, Medical Research and Education Building, 8447 State Highway 47, Bryan, TX 77807-3260, USA

Kunio Miyake
Department of Epigenetics Medicine, Interdisciplinary Graduate School of Medicine and Engineering, University of Yamanashi, Yamanashi, Japan

K. Naga Mohan
Centre for Human Genetics, Biotech Park Phase one, Electronic City, Bangalore-560100, India; Department of Biological Sciences, BITS-Pilani, Hyderabad Campus, Shameerpet, Hyderabad-500078, India and Department of Microbiology and Molecular Genetics, Magee-Women's Research Institute, University of Pittsburgh School of Medicine, Pittsburgh, PA USA

Gudrun E. Moore
Clinical and Molecular Genetics Unit, Institute of Child Health, University College London, London, UK

Hans Helmut Niller
Institute for Medical Microbiology and Hygiene at the University of Regensburg, Franz-Josef-Strauß-Allee 11, D-93053 Regensburg, Germany

Hiroyuki Nomura
Department of Obstetrics and Gynecology, Keio University School of Medicine, Shinanomachi 35, Shinjuku-ku, Tokyo 160-8582, Japan

Tatsushi Onaka
Division of Brain and Neurophysiology, Department of Physiology, School of Medicine, Jichi Medical University, Tochigi, Japan

Simon Plyte
Congenia-Genextra group, Via Adamello 16, 20100 Milan Italy and European Institute of Oncology, IFOM-IEO campus, Via Adamello 16, 20100 Milan Italy

Susan Prescott
School of Paediatrics and Child Health, University of Western Australia, Roberts Road, Subiaco, WA 6008, Australia

Derrick E. Rancourt
Department of Biochemistry and Molecular Biology, University of Calgary, 3330 Hospital Drive NW, Calgary, AB, Canada, T2N 4N1

Tina Rönn
Epigenetics and Diabetes Unit, Lund University Diabetes Centre, Department of Clinical Sciences, Lund University, Malmö, Sweden

Richard Saffery
Developmental Epigenetics, Murdoch Children's Research Institute, Flemington Road, Parkville, VIC 3052, Australia

Sabita N. Saldanha
Department of Biology, University of Alabama at Birmingham, 175A Campbell Hall, 1300 University Boulevard, University of Alabama at Birmingham, Birmingham, AL 35294-1170 and Department of Math and Sciences, Alabama State University, P. O. Box, 271, Montgomery, AL 36101-0271, USA

Richard H. Scott
Clinical and Molecular Genetics Unit, Institute of Child Health, University College London, London and Department of Clinical Genetics, Great Ormond Street Hospital, London, UK

Mehdi Shafa
Department of Biochemistry and Molecular Biology, University of Calgary, 3330 Hospital Drive NW, Calgary, AB, Canada, T2N 4N1

Cassandra L. Smith
Molecular Biotechnology Research Laboratory, Boston University, 44 Cummington Street, Boston, MA, USA

Nobuyuki Susumu
Department of Obstetrics and Gynecology, Keio University School of Medicine, Shinanomachi 35, Shinjuku-ku, Tokyo 160-8582, Japan

Trygve O. Tollefsbol
Department of Biology, 175 Campbell Hall, 1300 University Boulevard, Birmingham, AL 35294-1170, USA

Kosuke Tsuji
Department of Obstetrics and Gynecology, Keio University School of Medicine, Shinanomachi 35, Shinjuku-ku, Tokyo 160-8582, Japan

Meri K. Tulic
School of Paediatrics and Child Health, University of Western Australia, Roberts Road, Subiaco, WA 6008, Australia

Arisa Ueki
Department of Obstetrics and Gynecology, Keio University School of Medicine, Shinanomachi 35, Shinjuku-ku, Tokyo 160-8582, Japan

Alexander M. Vaiserman
Laboratory of Epigenetics, Institute of Gerontology, Kiev, Ukraine

Yoshihisa Watanabe
Department of Laboratory Medicine, Hamamatsu University School of Medicine, 1-20-1 Handayama, Higashi-ku, Hamamatsu 431-3192, Japan

Qingbo Xu
Cardiovascular Division, King's College London BHF Centre, London, UK

Hidenori Yamasue
Department of Neuropsychiatry, Graduate School of Medicine, The University of Tokyo, Tokyo, Japan

Megumi Yanokura
Department of Obstetrics and Gynecology, Keio University School of Medicine, Shinanomachi 35, Shinjuku-ku, Tokyo 160-8582, Japan

Boda Zhou
Cardiovascular Division, King's College London BHF Centre, London, UK and Department of Physiology, Peking University, Beijing, China

CHAPTER 1

Epigenetics of Human Disease

Trygve O. Tollefsbol
University of Alabama at Birmingham, Birmingham, AL, USA

1.1 INTRODUCTION

Epigenetics does not involve changes in DNA sequence but is nevertheless able to influence heritable gene expression through a number of processes such as DNA methylation, modifications of chromatin and non-coding RNA. Aberrations in DNA methylation are common contributors to disease. For example, imprinting diseases such as the Angelman, Silver–Russell, Prader–Willi and Beckwith–Wiedemann syndromes are often associated with alterations in DNA methylation [1]. Human diseases attributable to DNA methylation-based imprinting disorders, however, have not been limited to these genetic diseases as diabetes, schizophrenia, autism and cancer have also been associated with aberrations in imprinting. Abnormalities of the enzymes that mediate DNA methylation can also contribute to disease as illustrated by the rare Immunodeficiency–Centromere instability–Facial anomalies (ICF) syndrome caused by mutations in DNA methyltransferase 3B (DNMT3B). Likewise, Rett syndrome, related to mutations in the methyl-binding domain (MBD) protein, MeCP2, leads to dysregulations in gene expression and neurodevelopmental disease [2]. Perhaps most commonly, DNA methylation aberrations can often contribute to cancer either through DNA hypo- or hypermethylation. DNA hypomethylation leads to chromosomal instability and can also contribute

T. Tollefsbol (Ed): Epigenetics in Human Disease. DOI: 10.1016/B978-0-12-388415-2.00001-9

to oncogene activation, both common processes in oncogenesis, and DNA hypermethylation is often associated with tumor suppressor gene inactivation during tumorigenesis.

Histone modifications frequently contribute to disease development and progressions and histone acetylation or deacetylation are the most common histone modifications involved in diseases. Aberrations in histone modifications can significantly disrupt gene regulation, a common factor in disease, and could potentially be transmissible across generations [3]. Histone modifications have in fact been associated with a number of diseases such as cancer and neurological disorders. Collaborations between DNA methylation and histone modifications can occur and either or both of these epigenetic processes may lead to disease development [4].

Non-coding RNAs are an emerging area of epigenetics and alternations in these RNAs, especially microRNAs (miRNAs), contribute to numerous diseases. miRNAs can inhibit translation of mRNA if the miRNA binds to the mRNA, a process that leads to its degradation, or the miRNA may partially bind to the 3′ end of the mRNA and prohibit the actions of transfer RNA [5]. Although miRNAs have been associated with a number of diseases such as Crohn's disease [6], their role in tumorigenesis is now established and is considered to be a frequent epigenetic aberration in cancer.

Collectively, epigenetic processes are now generally accepted to play a key role in human diseases. As the knowledge of epigenetic mechanisms in human diseases expands, it is expected that approaches to disease prevention and therapy using epigenetic interventions will also continue to develop and may eventually become mainstays in disease management.

1.2 EPIGENETIC VARIATION METHODS

Technological advances often serve as a major stimulus for knowledge development and the field of epigenetics is no exception in this regard. Recent advances in epigenetic-based methods have served as major driving forces in the fascinating and ever-expanding epigenetic phenomena that have been revealed especially over the past decade. Although genome-wide maps have been developed, there is still a need for maps of the human methylome and histone modifications in healthy and diseased tissues, as discussed in Chapter 2. Epigenetic variation is especially prominent in human diseases and established techniques such as bisulfite genomic methylation sequencing and chromatin immunoprecipitation (ChIP) analyses are revealing numerous epigenetic aberrations involved in disease processes. However, cutting-edge advances in comparative genomic hybridization (CGH) and microarray analyses as well as quantitative analysis of methylated alleles (QAMA) and many other developing technologies are now facilitating the elucidation of epigenetic alterations in disease that were previously unimagined. Combinations of epigenetic technologies are also emerging that show promise in leading to new advances in understanding the epigenetics of disease.

1.3 CANCER EPIGENETICS

As mentioned above, DNA methylation is often an important factor in cancer development and progression. DNA methylation changes can now be readily assessed from body fluids and applied to cancer diagnosis as well as the prognosis of cancer (Chapter 3). Epigenome reference maps will likely have an impact on our understanding of many different diseases and may lead the way to breakthroughs in the diagnosis, prevention and therapy of human cancers. Histone modifications are frequently altered in many human cancers and the development of a histone modification signature may be developed that will aid in the prognosis and treatment of cancers (Chapter 4). These histone maps may also have potential in guiding therapy of human cancers. MicroRNAs (miRNAs) are central to many cellular functions and they are frequently dysregulated during oncogenesis (Chapter 5). In fact, miRNA expression profiles may be more useful than gene expression profiles for clinical applications since there are fewer mRNA regulatory molecules. These miRNA profiles may be applicable to identifying various cancers or to stratify tumors in addition to serving prognostic or therapeutic roles. Epigenetic therapy for cancer is perhaps one of the most exciting and rapidly developing areas of epigenetics. As discussed in Chapter 6, approaches are available for targeting enzymes such as the DNMTs, histone acetyltransferases (HATs), histone deacetylases (HDACs), histone methyl-transferases (HMTs) and histone demethylases (HDMTs). The development of drug-based inhibitors of these epigenetic-modifying enzymes could be further improved through drug combinations or even natural plant-based products, many of which have been found to harbor properties that can mimic the often more toxic and perhaps less bioavailable epigenetic drugs that are currently in use.

1.4 EPIGENETICS OF NEUROLOGICAL DISEASE

One of the newer areas of epigenetics that has been rapidly expanding is its role in neuro-logical disorders or diseases. These disorders are not limited to the brain as the disease target, but also often involve nutritional and metabolic factors that contribute as well to conditions such as neurobehavioral diseases (Chapter 7). At this point, however, the number of neurodevelopmental disorders that have been associated with epigenetic aberrations is not very extensive. A possible explanation for this is that the pervasive nature of epigenetic processes could serve as a negative selective force against more localized disease such as neurodevelopmental disorders (Chapter 8). In fact, many neurodevelopmental disorders are due to partial loss-of-function mutations or are X-chromosomal mosaics with recessive X-linked mutations. Neurodegenerative diseases such as Alzheimer's disease have been increasingly associated with alternations in epigenetic processes. Environmental factors such as diet and exposure to heavy metals may lead to the epigenetic changes often involved in Alzheimer's disease eventually contributing to increased amyloid β peptide (Chapter 9). These factors may begin early in life and manifest as late-onset forms of Alzheimer's disease. Fortunately, as reviewed in Chapter 10, a number of new approaches are currently being developed that could have translational potential in preventing or treating many of the epigenetic changes that are being revealed as an important component of neurobiological disorders.

1.5 AUTOIMMUNITY AND EPIGENETICS

There is a strong association between environmental factors, age and the development of autoimmune disorders. Epigenetic processes are central to aging and are also an important mediator between the environment and disease and it is thought that these factors may be important in the development and progression of numerous autoimmune diseases. For example, systemic lupus erythematosus (SLE) and rheumatoid arthritis (RA) are autoimmune disorders that have frequently been associated with aberrations in epigenetic mechanisms (Chapter 11). Often the epigenomic and sequence-specific DNA methylation changes found in SLE and RA affect key genes in immune function. Two challenges are to increase the use of high-throughput approaches to these diseases to mine for additional gene aberrations and to translate these epigenetic changes to the clinic through the development of novel approaches for preventing or treating SLE and RA. Fortunately, there is hope for epigenetic therapy of autoimmune disorders as reviewed in Chapter 12. Much of the current research for drug development relevant to autoimmune dysfunction is focused on correcting alterations in DNA methylation and histone acetylation. However, recent exciting advances suggest promising avenues for drug development as applied to miRNAs. For instance, miRNAs or inhibitors of miRNA to impact DNA methylation may have utility in affecting gene transcription in immune cells that often lead to the development of SLE.

3

1.6 HUMAN IMPRINTING DISORDERS

Both DNA methylation and histone modifications can impact imprinting centers that control parent-of-origin-specific expression and lead to human imprinting disorders. These disorders, such as Angelman, Prader–Willi, Silver–Russell and Beckwith–Wiedemann syndromes frequently involve epigenetic changes that contribute to these disorders and they often manifest at a very young age (Chapter 13). However, both epigenetic and genetic factors are often important in human imprinting disorders and the development of epigenetic therapy approaches in this particular area represents a considerable challenge. Advances are being made in understanding the epigenetic basis of human imprinting disorders which may provide breakthroughs in treating these tragic diseases.

1.7 EPIGENETICS OF OBESITY

Rare obesity-associated imprinting disorders have been described and dietary modulation efforts have suggested an epigenetic component may exist in these disorders. In fact, the major role of environmental factors in obesity strongly suggests a role of epigenetic changes such as those involving DNA methylation in obesity (Chapter 14). Early-life environmental factors could be especially important in controlling epigenetic aberrations that may contribute to obesity as reviewed in Chapter 15. It is likely that increased identification of obesity biomarkers and their associated epigenetic factors may lead to new advances in controlling the extant epidemic in childhood obesity in many developed countries. It is highly likely that nutritional or lifestyle interventions either during pregnancy or early in life could impact processes such as DNA methylation and histone modifications that are highly responsive to environmental stimuli and lead to means to control obesity at very early ages.

1.8 DIABETES: THE EPIGENETIC CONNECTION

Similar to obesity, environmental factors are also often important in the development of type 2 diabetes. Non-genetic risk factors such as aging and a sedentary lifestyle have been associated with epigenetic aberrations characteristic of type 2 diabetes (Chapter 16). Since markers such as DNA methylation have been shown to vary in diabetic versus non-diabetic individuals, it is very possible that epigenetic manifestations may have a key role in the pathogenesis of type 2 diabetes. However, multisystem studies are currently needed to further substantiate this concept and additional studies on the prediction and prevention of type 2 diabetes are sorely needed. Histone modifications have also been strongly implicated in diabetes as reviewed in Chapter 17. In fact, HDAC inhibitors may have potential in treating diabetes in the short term. Nutritional compounds that lead to HDAC inhibition may have potential in treating type 2 diabetes as well as the development of miRNA-based therapeutics that would have greater targeting potential.

1.9 EPIGENETICS AND ALLERGIC DISORDERS

Consistent with many other epigenetic diseases, early environmental factors appear to be a critical component to the development of numerous allergic disorders. For example, exposure to specific factors in utero may be associated with epigenetic aberrations that affect gene expression, immune programming and the development of allergic maladies in the offspring (Chapter 18). Additionally, this transgenerational component may allow for the transmittance of epigenetic changes to future generations beyond the offspring leading to allergic disorders. Novel early interventions into epigenetic-modifying factors such as maternal diet may contribute to an eventual decline in allergy-based disorders. Asthma is a common disorder of this nature and there is some evidence that corticosteroids exert their anti-inflammatory effects in part by inducing acetylation of anti-inflammatory genes (Chapter 19). The potential recruitment of HDAC2 to activated inflammatory genes by corticosteroids may be a key mechanism for epigenetic-based therapy of allergic disorders such as asthma. Future efforts are now being directed toward modifiers of other epigenetic processes in allergic disorders such as histone phosphorylation and ubiquitination.

1.10 CARDIOVASCULAR DISEASE AND EPIGENETICS

Atherosclerosis is a major precipitating factor in cardiovascular diseases and the functions of smooth muscle cells (SMCs) and endothelial cells (ECs) are central to the development of atherosclerosis. Mounting evidence has indicated that epigenetic processes such as DNA methylation and histone acetylation have critical functions in modulating SMC and EC homeostasis. The SMC and EC proliferation, migration, apoptosis and differentiation not only contribute to atherosclerosis, but also cardiomyocyte hypertrophy and heart failure, as reviewed in Chapter 20. The role of HDACs in cardiovascular disease such as arteriosclerosis has been showing promise, although concerns surround the tissue-specificity of these agents. Given this concern, the development of highly selective and cell type-specific HDAC inhibitors may have potential in epigenetic-based therapies for cardiovascular diseases of varied types.

1.11 EPIGENETICS OF HUMAN INFECTIOUS DISEASES

A common theme is the environmental impact on the epigenome and its role in epigenetic disease processes. Consistent with this concept, bacterial and viral infections often cause epigenetic changes in host cells that lead to pathology as reviewed in Chapter 21. The consequences of these epigenome-modifying infections are not limited to neoplasia. There are, in fact, many other diseases that have an epigenetic basis induced by infectious agents such as diseases of the oral cavity. Even organisms like protozoa can contribute to host epigenetic dysregulation. Knowledge accumulated regarding epigenetic "invaders" of the genome and their pathological consequences will undoubtedly lead to the development of more sophisticated and novel approaches to controlling and treating epigenetic-based infectious diseases.

1.12 REPRODUCTIVE DISORDERS AND EPIGENETIC ABERRATIONS

Endometriosis, or the presence of functional endometrial-like tissues outside of the uterine cavity, is often secondary to hormonal and immunological aberrations. Most exciting in the context of epigenetics, however, is that many recent studies have indicated that endometriosis may have an important epigenetic component that contributes to its pathological progression (Chapter 22). A number of investigations have indicated HDAC inhibitors may be effective in treating endometriosis. There is also potential for the development of epigenetic biomarkers for endometriosis such as changes in DNA methylation as well as miRNA-based biomarkers. Epigenetic processes are also gaining increasing importance in endometrial cancer (Chapter 23). Damage to the mismatch repair system appears to play a significant role in the development of endometrial cancer through the mechanism of *hMLH1* hypermethylation. These findings may have important epigenetic therapeutic implications for endometrial cancer and could also have potential for the prevention, diagnosis and risk assessment of endometrial cancer.

1.13 STEM CELL EPIGENETICS IN HUMAN DISEASE

Stem cell-based therapeutic approaches could lead to powerful means of treating human diseases and epigenetic regulatory signals play an important role in the maintenance of stem cell potency (Chapter 24). Chromatin modifications and dynamics appear to have an important role in conservation of pluripotency and the differentiation of embryonic stem cells which are central factors in stem cell-based therapeutics. In fact, several epigenetic disorders have been modeled in vitro through the use of induced pluripotent stem cells (iPSCs) from the cells of patients. Understanding the basic epigenetic changes central to these processes may have considerable potential in the treatment of human epigenetic diseases. Non-coding RNAs also participate in stem cell renewal and differentiation (Chapter 25). The role of epigenetics and non-coding RNAs may provide many useful tools for manipulating stem cell programming as applied to therapy of epigenetic-based diseases.

1.14 EPIGENETICS OF AGING AND AGE-ASSOCIATED DISEASES

Few processes are as pervasive as aging which impacts not only the entire physiological fitness of an organism, but also its predisposition to developing age-related diseases which is comprised of an ever-growing list of diseases. It is now apparent that epigenetic processes are major components of aging, which opens many avenues to human diseases (Chapter 26). Although aging is not considered a disease in and of itself, it is perhaps the most frequent contributor to human disease. Therefore, delaying the epigenetic aberrations associated with aging through epigenetic intervention and treating epigenetic-based age-associated diseases could have a tremendous impact on the role of epigenetics in human disease. Although they are on opposite sides of the lifespan spectrum, early developmental processes are likely linked to later life aging and age-associated diseases (Chapter 27). The role of nutrition, hormones and metabolic environment early in life can have effects throughout life, influence epigenetic pathways and markers and manifest in the form of aging and age-related diseases. Considerable interest is now focused on the impact of early life epigenetic impacts and the outcome of these effects on the myriad of age-associated diseases which comprise much of the pathology that forms the basis of human disease.

1.15 CONCLUSION

Epigenetic processes not only take many forms, but they also can readily become expressed as human diseases. These diseases, that can be loosely grouped under the heading of "epigenetic diseases", are vast and the list of diseases that fit into this description is rapidly growing. Elucidation of the epigenetic aberrations in human diseases not only has implications for epigenetic-based therapy, but also for risk assessment, prevention, progression analysis, prognosis and biomarker development. A common theme of many epigenetic-based human diseases is the role of the environment. This may take varied forms, ranging from maternal nutrition to infectious agents. Exciting advances are rapidly developing that are contributing significantly toward the management of human diseases through epigenetic intervention. It is anticipated that epigenetic-based preventive and therapeutic strategies will continue to develop at a rapid pace and may assume a role at the forefront of medicine in the not too distant future.

References

[1] Falls JG, Pulford DJ, Wylie AA, Jirtle RL. Genomic imprinting: implications for human disease. Am J Pathol 1999;154:635−47.

[2] Yasui DH, Peddada S, Bieda MC, Vallero RO, Hogart A, Nagarajan RP, et al. Integrated epigenomic analyses of neuronal MeCP2 reveal a role for long-range interaction with active genes. Proc Natl Acad Sci USA 2007;104:19416−21.

[3] Cavalli G, Paro R. The Drosophila Fab-7 chromosomal element conveys epigenetic inheritance during mitosis and meiosis. Cell 1998;93:505−18.

[4] Esteller M. Cancer epigenomics: DNA methylomes and histone-modification maps. Nat Rev Genet 2007;8:286−98.

[5] Gibney ER, Nolan CM. Epigenetics and gene expression. Heredity (Edinb) 2010;105:4−13.

[6] Wu F, Zhang S, Dassopoulos T, Harris ML, Bayless TM, Meltzer SJ, et al. Identification of microRNAs associated with ileal and colonic Crohn's disease. Inflamm Bowel Dis 2010;16:1729−38.

CHAPTER 2

Methods and Strategies to Determine Epigenetic Variation in Human Disease

Yoshihisa Watanabe, Masato Maekawa
Hamamatsu University School of Medicine, Hamamatsu, Japan

CHAPTER OUTLINE

T. Tollefsbol (Ed): Epigenetics in Human Disease. DOI: 10.1016/B978-0-12-388415-2.00002-0

2.1 INTRODUCTION

Epigenetics is not only one of the most rapidly expanding fields of study in biomedical research but is also one of the most exciting and promising in terms of increasing our understanding of disease etiologies and of developing new treatment strategies. Among the recent landmark events in this field are the characterization of the human DNA methylome at single nucleotide resolution, the discovery of CpG island shores, the identification of new histone variants and modifications, and development of genome-wide maps of nucleosome positions. Much of our increased understanding is the result of technological breakthroughs that have made it feasible to undertake large-scale epigenomic studies. These new methodologies have enabled ever finer mapping of the epigenetic marks, such as DNA methylation, histone modifications and nucleosome positioning, that are critical for regulating the expression of both genes and noncoding RNAs [1]. In turn, we have a growing understanding of the consequences of aberrant patterns of epigenetic marks and of mutations in the epigenetic machinery in the etiology of disease.

However, there are several aspects of the methods used to analyze epigenetic variation associated with disease that present potential problems. First, the tissue used to obtain the DNA. This depends to some extent on the nature of the disease, and can influence the analytical methods that are employed. For example, the DNA of some tissues may have a low incidence of moieties with the diagnostic pattern of methylation, which would limit the choice of analytic methodologies to those with high sensitivity for these molecular signatures. Second, different diseases may require analysis of either regional or genome-wide epigenetic variation, with the choice depending on the predicted variation in the specific disease. The continuing increase in the number of "epigenetic" diseases means that the list of methods that are practical for the different diseases is also increasing. Third, epigenetic variation can be a consequence or a cause of the disease. Therefore, use of strategies that can differentiate the role, or otherwise, of epigenetic variation in the causality of a disease is fundamental. It might, for example, allow determination of whether epigenetic variation is a marker of disease progression, a potential therapeutic target, or a useful marker for assessing the efficiency of a therapy.

Although the new technologies have provided considerable insights into epigenetic aspects of disease, there is still considerably more work that needs to be carried out. In particular, there is a great need for detailed descriptions of human DNA methylomes and for maps of histone modifications and nucleosome positions in healthy and diseased tissues. A number of international projects and initiatives have been established to meet this need: the NIH Roadmap Epigenomics Program, the ENCODE Project, the AHEAD Project, and the Epigenomics NCBI browser, among others [2,3]. The availability of detailed epigenetic maps will be of enormous value to basic and applied research and will enable pharmacological research to focus on the most promising epigenetic targets.

This chapter summarizes some of the contemporary methods used to study epigenetics and highlights new methods and strategies that have considerable potential for future epigenetic and epigenomic studies.

2.2 DNA METHYLATION ANALYSIS

Methylation of cytosine bases in DNA is not only an important epigenetic modification of the genome but is also crucial to the regulation of many cellular processes. DNA methylation is important in many eukaryotes for both normal biology and disease etiology [1]. Therefore, identifying which genomic sites DNA are methylated and determining how this epigenetic mark is maintained or lost is vital to our understanding of epigenetics. In recent years, the technology used for DNA methylation analysis has progressed substantially: previously, analyses were essentially limited to specific loci, but now, they can be performed on a genome-wide scale to characterize the entire "methylome" with single-base-pair resolution [4].

The new wealth of profiling techniques raises the challenge of which is the most appropriate to select for a given experimental purpose. Here, we list different methodologies available for analyzing DNA methylation and briefly compare their relative strengths and limitations [5]. We also discuss important considerations for data analysis.

2.2.1 Methylation-Sensitive Restriction Enzymes

The identification of DNA methylation sites using methylation-sensitive restriction enzymes requires high-molecular-weight DNA and is limited by the target sequence of the chosen enzyme. The use of restriction enzymes that are sensitive to CpG methylation within their cleavage recognition sites [6] is a relatively low-resolution method, but it can be useful when combined with genomic microarrays [7,8].

2.2.2 Bisulfite Conversion of Unmethylated Cytosines, PCR and Sequencing

Conversion of unmethylated sequences with bisulfite followed by PCR amplification and sequencing analyses provides an unbiased and sensitive alternative to the use of restriction enzymes. This approach is therefore generally regarded as the "gold-standard technology" for detection of 5-methyl cytosine as it enables mapping of methylated sites at single-base-pair resolution [9]. The bisulfite method requires a prolonged incubation of the DNA sample with sodium bisulfite; during this period, unmethylated cytosines in the single-stranded DNA are deaminated to uracil. However, the modified nucleoside 5-methyl cytosine is immune to transformation and, therefore, any cytosines that remain following bisulfite treatment must have been methylated. This method is currently one of the most popular approaches to methylation analysis and yields reliable, high-quality data [9,10]. The drawback to the method is that it is labor-intensive and is not suitable for screening large numbers of samples.

2.2.3 Comparative Genomic Hybridization (CGH) and Microarray Analysis

A combination of CGH and microarray analysis can overcome the limitations of the bisulfite method. This combination can enable high-throughput methylation analyses. The various advantages and disadvantages of this approach have been reviewed previously [11–13]. Recent high-throughput studies have used protein affinity to enrich for methylated sequences and then exploited these sequences as probes in genomic microarrays. Methylated DNA fragments can be affinity-purified either with an anti-5-methyl cytosine antibody or by using the DNA-binding domain of a methyl-CpG-binding protein [14,15].

2.2.4 Bisulfite Treatment and PCR Single-Strand Conformation Polymorphism (SSCP) (BiPS)

The combination of bisulfite treatment with PCR-based single-strand DNA conformation polymorphism (SSCP) analysis offers a potentially quantitative assay for methylation [16]. This combination approach, sometimes referred to as BiPS analysis, can be used for the rapid identification of the methylation status of multiple samples, for the quantification of methylation differences, and for the detection of methylation heterogeneity in amplified DNA fragments. This technique has been successfully used to investigate the methylation status of the promoter region of the hMLH1, p16, and HIC1 genes in several cancer cell lines and colorectal cancer tissues [17].

2.2.5 Methylation-Sensitive Single-Nucleotide Primer Extension

Methylation-sensitive single-nucleotide primer extension (MS-SNuPE) is a technique that can be used for rapid quantitation of methylation at individual CpG sites [18,19]. Treatment of genomic DNA with sodium bisulfite is used to convert unmethylated cytosine to uracil while leaving 5-methylcytosine unaltered. Strand-specific PCR is performed to generate a DNA template for quantitative methylation analysis using MS-SNuPE. This protocol can be carried using multiplex reactions, thus enabling the simultaneous quantification of multiple CpG sites in each assay.

2.2.6 Combined Bisulfite and Restriction Analysis

The combined bisulfite and restriction analysis (COBRA) approach involves combining the bisulfite and restriction analysis protocols [20]. It is relatively simple to use while still retaining quantitative accuracy. Although both COBRA and MS-SNuPE are quantitative, they have the restrictions that the former can only analyze a specific sequence because it utilizes restriction enzymes and the latter is somewhat laborious. MS-SnuPE has also been combined with microarray analysis to allow parallel detection of DNA methylation in cancer cells [19].

2.2.7 Quantitative Bisulfite Sequencing using Pyrosequencing Technology

Quantitaive bisulfite sequencing using pyrosequencing technology (QBSUPT) is based on the luminometric detection of pyrophosphate release following nucleotide incorporation [21]. The advantage of QBSUPT is that quantitative DNA methylation data are obtained directly from PCR products, without the need for cloning and sequencing a large number of clones. However, QBSUPT cannot be used to analyze haplotype-specific DNA methylation patterns. Thus, while very sensitive, this assay may be more suited to laboratory diagnosis.

2.2.8 MethyLight Technology

MethyLight technology provides a tool for the quantitative analysis of methylated DNA sequences via fluorescence detection in PCR-amplified samples [22]. This method has two particular advantages: first, the fluorescent probe can be designed to detect specific DNA methylation patterns, not simply to discriminate methylated from unmethylated sequences; second, it has the potential ability to rapidly screen hundreds or even thousands of samples.

2.2.9 Quantitative Analysis of Methylated Alleles (QAMA)

QAMA is a quantitative variation of MethyLight that uses TaqMan probes based on minor groove binder (MGB) technology [23]. QAMA has the main advantage of being simple to set up, making it suitable for high-throughput methylation analyses.

2.2.10 DNA Methylation Analysis by Pyrosequencing

Pyrosequencing is a replication-based sequencing method in which addition of the correct nucleotide to immobilized template DNA is signaled by a photometrically detectable reaction. This method has been adapted to quantify methylation of CpG sites. The template DNA is treated with bisulfite and PCR is used for sequencing; the ratio of T and C residues is then used to quantify methylation. Pyrosequencing offers a high-resolution and quantitatively accurate measurement of methylation of closely positioned CpGs [24].

2.2.11 Matrix-Assisted Laser Desorption Ionization Time-of-Flight Mass Spectrometry

Tost et al. [25] described a method using matrix-assisted laser desorption ionization time-of-flight (MALDI-TOF) for analysis and quantification of methylation at CpGs. Although the method requires gene-specific amplification, and should therefore be considered a candidate gene method, it is amenable to automation as it can make use of the EpiTYPER platform developed by Sequenom. EpiTYPER can be used to determine methylation status following gene-specific amplification of bisulfite-treated DNA followed by in vitro transcription, base-specific RNA cleavage and MALDI-TOF analysis [26]. Although it is not a genome-wide technology, it is quantitative for multiple CpG dinucleotides for large numbers of gene loci and can be reliably applied to pooled DNA samples to obtain group averages for valuable samples.

2.2.12 New Technologies

Several second-generation sequencing platforms became available in 2007 and were further developed with the launch of the first single-molecule DNA sequencer (Helicos Biosciences) in 2008 [27]. These new sequencing tools have been applied to epigenetic research, for example, studies on DNA methylation. Undoubtedly, future developments of these technologies hold the tantalizing prospect of high-throughput sequencing to identify DNA methylation patterns across the whole mammalian genome, possibly even opening up the prospect of genotyping individual cancers to aid the application of custom-designed cancer therapies [28].

2.2.13 Computational Tools

The development of computational tools and resources for DNA methylation analysis is accelerating rapidly [29]. Sequence-based analyses involve alignment to a reference genome, collapsing of clonal reads, read counts or bisulfite-based analysis [30], and further data analysis. Comparison of the relative strengths and weaknesses of the various methods for DNA methylation analysis is hampered by their complexity and diversity. Inevitably, choice of method is based on pragmatic grounds, for example, the number of samples, the quality and quantity of DNA samples, the desired coverage of the genome, and the required resolution.

2.3 HISTONE MODIFICATION ANALYSIS

Histones are abundant, small basic proteins that associate with the DNA in the eukaryotic nucleus to form chromatin. The four core histones (H2A, H2B, H3 and H4) can show substantial modifications of 20–40 N-terminal amino acids that are highly conserved despite playing no structural role. The modifications are thought to constitute a histone code by which the cell encodes various chromatin conformations and controls gene expression states. The analysis of these modified histones can be used as a model for the dissection of complex epigenetic modification patterns and for investigation of their molecular functions. In this section, we review the techniques that have been used to decipher these complex histone modification patterns.

Posttranslational modification (PTM) of proteins plays a key role in regulating the biological function of many polypeptides. Initially, analyses of the modification status were performed using either a specialized gel system or a radioactive precursor molecule followed by complete protein hydrolysis and identification of the labeled amino acid [31–35]. This approach showed that histones could be modified *in vivo* by acetylation, methylation or phosphorylation [31,36,37]. As most of the modifications occurred at the N-terminus of the histone, it was feasible to map the site of some modifications using Edman degradation [38]. However, this is only possible when histones can be purified in sufficient quantities and with a high purity. The purification process is labor-intensive and involves multiple steps; this precludes the possibility of analyzing histone modifications from small numbers of cells or of mapping posttranslational modifications at specific loci.

Mass spectrometry is the method of choice for analyzing PTM in histones [39–42], as each modification adds a defined mass to the molecule. The high resolution of modern mass spectrometers and recent developments in soft ionization techniques have facilitated the mapping of posttranslational modifications. As a result, these high-resolution methods have enabled much faster detection of PTMs and have shown that such modifications are considerably more abundant than expected. The increased complexity of the proteome revealed by these analyses presents major challenges both for investigation and for the processing of the raw data. The mass spectrometry methods currently used to precisely map a modified residue are very elaborate and require enrichment of the peptides that carry particular modifications [43–46]. Different molecules can carry several modifications that localize on a single peptide within a protease digest [47–50]. These short stretches of dense modifications have been termed eukaryotic linear motifs (ELMs) and are thought to play a critical role in regulating the global function of proteins [51]. The high level of sequence conservation within these short ELMs also supports this idea. Many ELMs contain a number of amino acids that can be modified and the position of each modification has to be precisely determined [51]. Identification of each modification at different sites within a highly modified ELM is laborious and also hampered by the fact that some modifications result in similar mass differences.

A variety of different methods are available to study complex histone modification patterns; these range from "bottom-up approaches" to produce detailed and quantitative measurements of particular histone modifications, to "top-down approaches" aimed at elucidating the interactions of different modifications [52]. The use of a range of methods should greatly facilitate analysis of complex modification patterns and provide a greater insight into the biological roles of these histone modifications. Many of the methods used to analyze histone modifications can equally be applied to other types of modified protein that can function as integrators in multiple signaling pathways. The information on epigenomic analyses, including histone modifications using new technology such as next-generation sequencing (NGS), is reviewed below.

2.4 NON-CODING RNA ANALYSIS: MicroRNA

There is increasing evidence that small non-coding RNAs, such as microRNA, and long non-coding RNAs, such as lincRNA, can regulate gene expression. Mature microRNAs (miRNAs) are very small molecules, 19–25 nucleotides (nt), which poses a problem for their quantification. As small RNAs are less efficiently precipitated in ethanol, it is necessary to avoid resuspension in ethanol when using the standard Trizol protocol for RNA isolation. On the other hand, miRNAs appear to be more stable than longer RNAs and, consequently, in degraded samples it is still possible to obtain readable miRNA expression data. miRNAs have been reported to have greater stability than mRNAs in samples obtained from tissues which were fixed with formalin and paraffin embedded [53–55].

However, the intrinsic characteristics of miRNAs make production of miRNA expression profiles very problematic. For example, mature miRNAs lack common sequence features, such as a poly-A tail or 5′ cap, that can be used to drive selective purification. As mentioned above, the mature miRNAs are very small, which reduces the effectiveness of most conventional biological amplification methods. This problem arises because of poor specificity in primer binding. As a consequence, standard real-time PCR methods can only be applied to miRNA precursors. Furthermore, sequence heterogeneity among the miRNAs with respect to GC content, results in a wide range of optimal melting temperatures for these nucleic acid duplexes and hampers the simultaneous detection of multiple miRNAs. An additional problem for the specificity of miRNA detection arises from the close sequence similarity of miRNAs of the same family (mature miRNA, pri-miRNA, and pre-miRNA) and of the target sequence.

Currently, various methodologies have been adapted to detect miRNAs, including Northern blot analysis with radiolabeled probes [56,57], microarray-based [58] and PCR-based analyses [59], single molecule detection in a liquid phase, in situ hybridization [60,61] and high-throughput sequencing [62]. However, all of these methods have inherent limitations and the choice of *method* for miRNA detection depends mainly on specific experimental conditions. Ideally, an miRNA profiling method should fulfill the following requirements: sufficient sensitivity to allow quantitative analysis of miRNA levels, even with small amounts of starting material; sensitive to single-nucleotide differences between miRNAs; highly reproducible; capable of processing many samples at one time; and, easy to perform without the need for expensive reagents or equipment [63].

miRNAs were first identified using Northern blotting [64–66]. Small RNA molecules can be detected with a modified version of the standard protocol for Northern blotting in which high-percentage urea-acrylamide gels are mainly used; this modified approach can detect small RNA molecules that are approximately 100 times smaller than the average coding RNA. There are three main techniques for detecting and quantifying miRNA in tissue samples: cloning of miRNA; PCR-based detection; and, hybridization with selective probes. Initially, cloning was the main approach used as it offers advantages for discovery of new miRNAs not predicted from bioinformatic analysis and for sequencing the miRNAs [65,67,68]. However, cloning is less precise than the other methods for quantifying miRNAs. The PCR-based technique is able to detect low copy numbers with high sensitivity and specificity of both the precursor and mature form of miRNAs [69]. It is relatively inexpensive, can be used for clinical samples, and can work with minute amounts of RNA. Various hybridization techniques can be used on miRNAs, namely, Northern blotting, bead-based flow cytometry, in situ hybridization and microarray [70,71]. Northern blotting using radioactive probes is very sensitive; however, it is very time-consuming, is only practical in large clinical studies for detecting expression of hundreds of miRNAs, and requires large amounts of total RNA from each sample.

Following their initial discovery, the number of miRNAs quickly increased and they were shown to be present in all eukaryotic species [66,67,72]. In order to analyze a large number of miRNAs in many patients, it is essential to have a technique that can simultaneously process multiple miRNAs using the relatively small amounts of RNA that can be obtained from each patient. Designing probes for miRNAs is complicated by their short length and their low abundance. As each miRNA is only 19–25 nt long, the probe is almost exclusively determined by the sequence of the miRNA itself, which necessitates a different annealing temperature for each probe and miRNA interaction.

Microarray technology was developed in 1995 and has been applied to miRNA quantification [71,73]. In brief, microarrays are based on multiple hybridizations in parallel, using a glass or quartz support where probes have either been spotted or synthesized by photochemical synthesis [74–76]. The ability to include a high density of spots on an array enables a high number of genes to be analyzed simultaneously [76,77]. Three approaches are in general use for detecting nucleic acids such as DNA or RNA on an array platform. The first, which is common for custom arrays, uses glass slides and is based on the spotting of unmodified oligonucleotides over the slide [78]. The second also uses glass slides and is based on the deposition of probes on the slide. The distinction is that the 5′ terminus of the probe is cross-linked to the matrix on the glass. This allows the spotting of a much higher number of probes on these slides. In the third method, probes are photochemically synthesized directly on a quartz surface, allowing the number of probes to rise to millions on a small and compact area [75]. Usually, but not always, the first two methods compare two samples on each slide (one used as reference) that are stained in different colors. The third method uses single-color hybridization where each slide is hybridized with only one sample.

Most microarrays use DNA oligo spotting, a few use locked nucleic acid (LNA) that may enable increased affinity between probes and miRNAs, thereby achieving more uniform conditions of hybridization with different probes. Ideally, microarray-based detection of miRNAs should avoid manipulation of the samples, such as enrichment of low-molecular-weight RNA species and amplification of miRNAs. Additionally, it is feasible to develop microarrays able to discriminate the two predominant forms of miRNAs (precursor and mature). The

13

International Human Epigenome Consortium (IHEC) recommended that the identity and abundance of all non-coding RNA species in a cell type should be determined and suggested that this should be accomplished by RNA-seq by next-generation DNA sequencing after isolation of large or small RNA species.

2.5 ANALYSIS OF GENOME DNA REPLICATION PROGRAM BASED ON DNA REPLICATION TIMING

Chromosomal DNA replication is essential for normal cellular division and also has a significant role in the maintenance of genomic integrity. Genomic instability increases when DNA replication errors occur and, thus, mistakes in replication may be an important factor in the etiology of cancers and neuronal disorders. Replication in eukaryotes is initiated from discrete genomic regions, termed origins. The replication program is strict within a cell or tissue type but can vary among tissues and during development. The genetic program that controls activation of replication origins in mammalian cells awaits elucidation. Nevertheless, there is evidence that the specification of replication sites and the timing of replication are responsive to epigenetic modifications. Over the last decade, many new techniques have been developed and applied to analysis of DNA replication timing in the human genome. These techniques have provided significant insights into cell cycle controls, human chromosome structure, and the role of epigenetic changes to the genome with respect to DNA replication. In this section, we describe the methods that are currently employed for determining the spatiotemporal regulation of DNA replication in the human genome (DNA replication timing).

Two approaches are generally used to investigate DNA replication timing, fluorescence in situ hybridization (FISH) or PCR [79–82]. The FISH method is based on the cytogenetic discrimination of replicated (two double signals for autosomal loci; DD) and unreplicated loci (two single signals; SS) using DNA probes that are labeled with a fluorescent dye [79]. By comparing the frequencies of the two types of signal, the relative replication timing of each locus can be determined. However, the method is absolutely dependent on the assumption that replicated loci will provide DD-type FISH signals, that is, the replicated signals created by passage of the replication fork will separate sufficiently to be seen as a DD signal.

The PCR-based method involves labeling cells in exponential growth with BrdU for 60–90 min and then fractionating them by flow cytometry. Typically, this allows discrimination of six cell cycle fractions: G1, four successive S phase stages, and G2/M (mitotic) [80–82]. Samples containing equal numbers of cells from each cell-cycle fraction are collected, and newly replicated DNA labeled with BrdU is extracted and purified from each fraction. Whether or not a locus has commenced or completed replication can be determined by quantitative PCR of the newly replicated DNA. This approach has been exploited to provide replication timings for sequence tagged sites on human chromosomes 11q and 21q [81,82] and identified Mb-sized zones that replicated early or late in S phase (i.e. early/late transition zones). The early zones were found to be more GC-rich and gene-rich than the late zones, and the early/late transitions occurred primarily in genome regions that showed rapid switches in the relative GC content in the chromatin [81,82].

Woodfine et al. [83] performed the first microarray-based analysis to map replication timing in the human genome. They adapted the comparative genomic hybridization technique, which had been developed to assess genomic copy-number differences in cancer cells. Relative replication times can be inferred by measuring the relative amounts of different sequences in a population of S-phase cells compared to a non-replicating G1 genome. In this method, S-phase cells in an asynchronously growing human cell culture are isolated and their DNA extracted. The DNA is color-labeled and then mixed with DNA from G1 phase cells that has been differentially color-labeled. The combined DNA sample is hybridized to an array of genomic sequences and, after normalization of the data, the relative fluorescence intensities of the S-phase DNA at each array spot (the S to G1 replication timing ratio) provide a measure of replication timing. Comparison of the data obtained by this method with those from nascent strand quantitative PCR methods described above [81−84], showed that this new approach provided estimations of replication timing that were consistent with those obtained earlier [82,83].

White et al. [85] modified the experimental approach of Woodfine et al. [83] by comparing the representation of genomic sequences in newly replicated DNA isolated from early S-phase cells with that from late S-phase cells. In this way, they obtained a replication timing ratio for early S to late S. Although their measure has a different basis to the S phase to G1 ratio of Woodfine et al. [83], nevertheless, the results provide a similar description of replication timing. For example, a replication profile for chromosome 22 in a lymphoblastoid cell line obtained by White et al. [85] was consistent with that of Woodfine et al. [83]. To date, high-resolution analyses have shown a positive correlation between replication timing and a range of genomic parameters such as GC content, gene density and transcriptional activity [82,83,85].

DNA replication errors have been implicated in the etiology of many diseases [86−89]. One possible mechanism for this relationship is that disease-related reprogramming of the epigenome might depend on impaired regulation of replication timing patterns [90]. Thus, for example, chromosomal rearrangements in cancers have been reported to be associated with replication timing changes in translocation breakpoints [91,92]. Likewise, peripheral blood cells from prostate cancer patients have an altered pattern of replication accompanied by aneuploidy that distinguishes them from individuals with benign prostate hyperplasia (a common disorder in elderly men). These cellular characteristics have been suggested to be a better marker for prostate cancer than use of the blood marker, prostate-specific antigen (PSA) [93,94]. Analyses of changes in replication timing in the human genome have shown that the tumor suppressor gene *p53* plays a role in its regulation through the control of cell cycle checkpoints [95]. Thus, in cancer cells, the normal order of DNA replication is altered: regions that normally replicate late sometimes replicate early, and vice versa [84,91−93,96−98]. Replication timing has also been shown to change during development, differentiation and tumorigenesis; moreover, the structure of the chromatin may also change. The model illustrated in Figure 2.1 shows a possible mode of interaction of chromatin conformation, replication timing and the expression of genes, including oncogenes, in an early/late-switch region of replication timing (R/G-chromosome band boundary) [99]. For example, the replication timing environment of an oncogene (or a tumor suppressor gene) located in an early/late-switch region of replication timing may change from intermediate replication, between early and late S phase, to early replication timing (or late replication timing) by an increase (or decrease) in the number of early replication origins at the edge of an early replication zone (Figure 2.1B). In addition, the chromatin environment of such an oncogene (or tumor suppressor gene) may also change from that of an R/G-chromosome band boundary to an R band (or from that of an R/G-chromosome band boundary to a G band). Stalling of the replication fork in the vicinity of oncogenes might also induce translocation events, thereby altering the structure or the local environment of the oncogenes and affecting their function (Figures 2.1A, 2.1B) [99]. The interrelationship of these various factors suggests that analysis of replication timing assays as part of an epigenetics investigation might, in future, allow much earlier cancer detection than is possible today [5,99,100].

2.6 STRATEGY FOR EPIGENOMIC INVESTIGATION BASED ON CHROMOSOMAL BAND STRUCTURES

The various methods for genome-wide epigenetic analyses described in the above sections are summarized in Table 2.1. The replication timing of genes along the entire lengths of human chromosomes 11q and 21q has been described previously; these analyses showed that cancer-related genes, including several oncogenes, are concentrated in regions showing transition

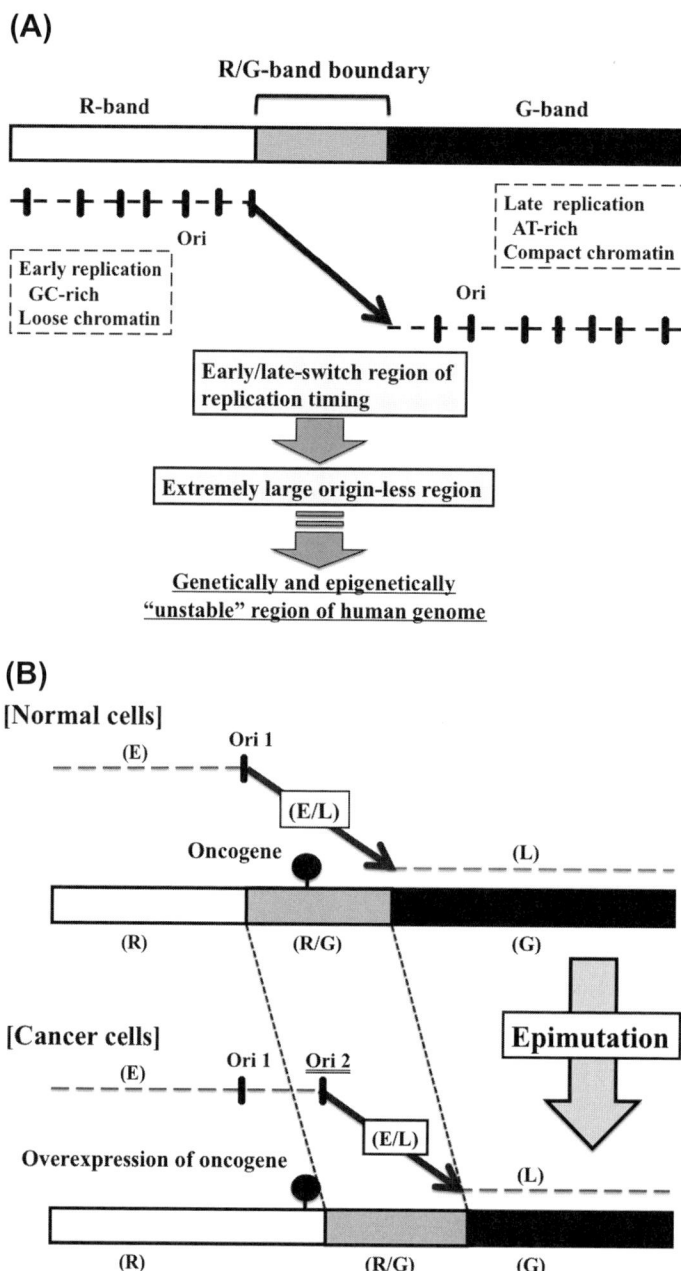

(A)

(B)

FIGURE 2.1

(A) Characterization of genetically and epigenetically "unstable" region of human genome. Replication origins (Ori) generally fire bi-directionally. The transition zone, which is shown by a thick arrow, is a large origin-free region between early and late-replicating domains [134,135]. Only the replication fork that starts at the edge of the early zone is predicted to be able to continue replicating over a period of hours or to pause at specific sites in the replication-transition region until it meets the fork initiated from the adjacent later-replicating zone. A pause during replication is known to increase the risk of DNA breaks and rearrangements [105–107]. Therefore, later replication sites within early/late-switch regions are particularly "unstable" regions of human genome [82]. The possible structure of the R/G-chromosome band boundary is shown above the origin map. Typical characteristics of early and late replicating regions are indicated. R- band, G-band, and R/G-band boundaries are shown in white, black, and gray, respectively. We propose that R/G chromosome band boundaries, which are transition regions for replication timing and GC content, are genetically and epigenetically more "unstable" regions of human genome than other genomic regions [5,82,84,96–98]. **(B)** Model for epigenetic changes to genes, including oncogenes in the R/G-chromosome band boundary. During tumorigenesis, chromatin structures as well as replicon structures may change. For example, the replication timing environment of an oncogene located in an early/late-switch region of replication timing (R/G-chromosome band boundary) may change from intermediate replication, between early and late S phase, to early replication timing by an increase in the number of early replication origins at the edge of an early replication zone. In addition, the chromatin environment of such an oncogene may also change from that of an R/G-chromosome band boundary to an R band. Stalling of the replication fork in the vicinity of oncogenes could also induce chromosome translocations that alter the structure or the local environment of the oncogenes, and thereby affect their function. The position of the oncogene is indicated by the black circle. (E), early replication zone; (L), late replication zone; (E/L), early/late-switch region; (R), R-band; (G), G-band; (R/G), R/G-band boundary; (Ori 1), origin 1; (Ori 2), origin 2.

TABLE 2.1 Methods of genome-wide epigenetics analysis

Method of Analysis	Techniques	Microarray Techniques	Next-Generation Sequencing
DNA methylation analysis	Bisulfite reaction treatment	BiMP	BC-seq, WGSBS
	Concentration of methylated DNA by immunoprecipitation	MeDIP, mDIP, MIRA	MeDIP-seq, MIRA-seq
	Application of methylation-sensitive restriction enzyme	HELP, MIAMI	HELP-seq
Histone modification analysis	Chromatin immunoprecipitation (ChIP)	ChIP-on-chip	ChIP-seq
MicroRNA analysis	Extraction of RNA	MicroRNA-chip	MicroRNA-seq
DNA replication timing analysis	Concentration of newly replicated DNA by immunoprecipitation	ReplicationTiming-chip	ReplicationTiming-seq

BC-seq, bisulfite conversion followed by capture and sequencing; BiMP, bisulfite methylation profiling; ChIP, chromatin immunoprecipitation; -chip, followed by microarray; HELP, HpaII tiny fragment enrichment by ligation-mediated PCR; MeDIP, mDIP, methylated DNA immunoprecipitation; MIAMI, microarray-based integrated analysis of methylation by isoschizomers; MIRA, methylated CpG island recovery assay; WGSBS, whole-genome shotgun bisulfite sequencing.

from early to late replication timing [82,84,96,97]. Scrutiny of the updated replication timing map for human chromosome 11q found that amplicons, gene amplifications associated with cancer, are located in the early/late switch regions of replication timing in human cell lines [84]. These transition regions also contain genes related to neural diseases, such as *APP* associated with familial Alzheimer's disease (AD1), and *SOD1* associated with familial amyotrophic lateral sclerosis (ALS1) [82]. Several neural disease genes are present in chromosomal regions with early/late transitions [82,96]. Interestingly, in metaphase and interphase nuclei, early-replicating zones have a looser chromatin structure, whereas late-replication zones have compact chromatin [101–104]. Therefore, transitions in chromatin compaction coincide with replication transition regions. In terminally differentiated cells, such as neurons, it is expected that the level of chromatin compaction established during the final round of DNA replication will be maintained. Transitions in chromatin compaction within a gene might lead to reduced genomic stability, and may also increase susceptibility to agents that can influence gene expression. Thus, the probability of epimutation, such as instability of chromatin structures and DNA damage (including DNA rearrangements) appears to be greater in replication transition regions than elsewhere in the genome [82,84,96,97,105–107] (Figure 2.1A).

It is likely that transition zones are subject to tight regulation, as changing their positions would affect the replication timing patterns of several flanking replicons. There is strong evidence that transition zones are conserved among different ES cell lines [108]. During development, transition zones may therefore be targets for chromatin-modifying enzymes to facilitate rapid reconfiguration and establishment of new replication timing patterns. Early and late replication zones tend to be located in different regions of the nucleus during S phase; it is possible that transition regions flanking these replication zones might be subject to dynamic reorganization or relocation during replication fork movement. The transition zones for replication timing are known to be associated with genomic instability, which is suspected to be involved in the etiology of human diseases such as cancer. Common fragile sites (CFSs) represent the best-known examples of regions of the human genome that break under replication stress. CFSs are associated with very large genes [109] and are frequently found at R/G band boundaries [110]. The human genome appears to have a large excess of so-called "dormant" or "backup" origins and these may be used to rescue stalled replication forks. Interestingly, "spare" origins appear to be absent from R/G band boundaries [111,112].

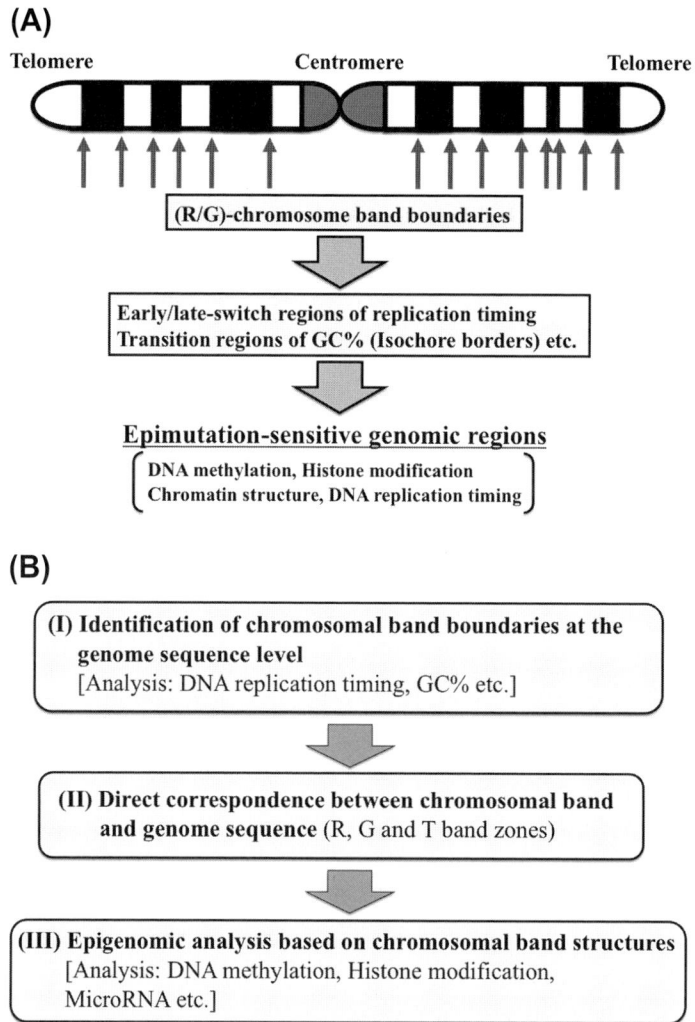

(A)

(B)

FIGURE 2.2

(A) Epimutation-sensitive genomic regions on the human genome associated with chromosomal bands. Chromosomal band boundaries, indicated by gray arrows, are suggested to be "unstable" genomic regions in the human genome, which are more epimutation-sensitive than other genomic regions. (B) Strategy for epigenomic analysis based on chromosomal band structures.

In conclusion, early/late-switch regions of replication timing generally correspond with transitions in relative GC content, are correlated with R/G chromosome band boundaries, and are suspected of being "unstable" genomic regions that have increased susceptibility to epigenetic mutation, as well as DNA damage (Figure 2.2A) [5,99]. There is a clear need for further epigenomic analysis on chromosomal band structures, in particular, to obtain a greater understanding of these epimutation-sensitive regions at the genome sequence level (Figure 2.2B, Table 2.2). Before performing epigenomic analysis using DNA methylation and

TABLE 2.2 Characterization of Chromosome Bands

Chromosome Band	G/Q Band	R Band	T Band (R Subgroup)
Replication timing	Late	Early	Very early
GC%	AT-rich	Medium	GC-rich
Gene density	Low	High	Very high
Chromatin compactness	Compact	Loose	Loose

histone modification, we propose clarifying the direct correspondence between chromosomal band (R-, T-, and G-band) and genome sequence by analyzing DNA replication timing and GC %, etc. (Figures 2.2A, 2.2B). Additionally, we suggest that epigenomic analysis focused on chromosomal band structures (the boundaries of which were identified as epimutation-sensitive genomic regions at the genome sequence level) will provide considerable insights into normal and disease conditions. In the future, this will be a promising strategy for epigenetic analysis.

2.7 OVERVIEW OF RECENT EPIGENETIC GENOME-WIDE OR BIOINFORMATIC STUDIES AND STRATEGIES

Over the last few years, genome-wide association studies (GWASs) have successfully identified loci associated with common diseases; however, the basis of many diseases still remains to be determined. The development of new genomic technologies has opened up the possibility of performing similar genome-wide studies to GWASs but with the aim of identifying epigenetic variations, particularly with respect to DNA methylation, that are associated with disease. Although such epigenome-wide association studies (EWASs) will provide valuable new information, they do pose specific problems that are not inherent in GWASs.

Performance of an EWAS is predicated on the assumption that it will be equally successful as a GWAS for identifying disease-associated variations. However, the differences between the epigenome and the genome influence the nature of the study design. For example, tissue-specific epigenetic modifications or epigenetic changes that occur downstream of the disease initiation step, might be important considerations in an EWAS for determining the cohorts and samples that should be analyzed. Although it is technically feasible to use array- and sequencing-based technologies in an EWAS, the computational and statistical methods required to analyze the data still require further development [113].

Several next-generation sequencing (NGS) platforms harness the power of massively parallel short-read DNA sequencing (MPSS) to analyze genomes with considerable precision. These methods can be applied to genome-wide epigenomic studies and they offer a potentially revolutionary change in nucleic acid analysis. The ability to sequence complete genomes will undoubtedly change the types of question that can be asked in many disciplines of biology. Recent excellent reviews provide a comprehensive description of the chemistry and technology behind the leading NGS platforms [114]. In this section, we discuss the application of NGS in epigenomic research, with a particular focus on chromatin immunoprecipitation combined with Mpss (ChIP sequencing or "ChIP-seq").

ChIP-seq offers the possibility of genome-wide profiling of DNA-binding proteins, histone modifications or nucleosomes. This method has advantages over the longer-established ChIP-chip (chromatin immunoprecipitation combined with microarray) technique [52]. For example, although arrays can be tiled at a high density, they require large numbers of probes and are expensive [115]. The hybridization process also imposes a fundamental limitation in the resolution of the arrays. ChIP-seq does not suffer from the "noise" generated by the hybridization step in ChIP-chip, which is complex and dependent on many factors, including the GC content, length, concentration and secondary structure of the target and probe sequences. Cross-hybridization between imperfectly matched sequences can occur frequently and contribute to the noise. In addition, the intensity signal measured on an array might not be linear over its entire range, and its dynamic range is limited below and above saturation points. A recent study reported that distinct and biologically meaningful peaks seen in ChIP-seq were obscured when the same experiment was conducted with ChIP-chip [116]. Genome coverage using ChIP-seq is not limited by the selection of probe sequences on the array. This is an important constraint in microarray analysis of repetitive regions of the genome, which are often "masked out" on the arrays. As a consequence, investigation of heterochromatin or microsatellites is optimized by use of ChIP-seq. Sequence variations within repeat elements can be identified and used to align the reads in the genome; unique sequences that flank repeats are similarly helpful [117].

The main disadvantages of ChIP-seq are cost and availability. Several groups have successfully developed and applied their own protocols for library construction, which has substantially lowered that part of the cost. For high-resolution profiling of an entire large genome, ChIP-seq can already be less expensive than ChIP-chip; however, this depends on the genome size and the level of sequencing detail required; a ChIP-chip experiment on selected regions using a customized microarray may yield as much biologically meaningful data. The recent decrease in sequencing cost per base-pair has not had as large an effect on ChIP-seq as on other applications, since the decrease has come as much from increased read lengths as from the number of sequenced fragments. The gain in the fraction of reads that can be uniquely aligned to the genome declines rapidly after 25–35 bp and is marginal beyond 70–100 nucleotides [118]. However, as the cost of sequencing decreases and institutional support for sequencing platforms grows, ChIP-seq is likely to become the method of choice for nearly all ChIP experiments in the near future.

ChIP-seq analyses have been performed on multiple transcription factors with their transcriptional co-regulators, boundary elements, numerous types of histone modifications, histone variants, nucleosome occupancy, DNA methylation patterns and gene transcription [119]. The data from these analyses are providing fresh insights into complex transcriptional regulatory networks. Furthermore, "chromatin signatures", characteristic chromatin structures in particular genomic regions, enable genome annotation based on predicting histone modifications and an overall landscape of the epigenome in human cells [52,119]. In addition, identification of the specific chromatin signatures associated with genomic features such as enhancers, insulators, boundary elements and promoters, will provide another means of annotating complex genomes. NGS technologies provide an increasing ability to query multiple genomic features, which were previously too technically challenging and costly; this inevitably has raised expectations and ambitions, as exemplified by the published goals of the International Human Epigenome Consortium.

Histone PTMs influence gene expression patterns and genome function by establishing and orchestrating DNA-based biological processes [120]. PTMs can either directly affect the structure of chromatin or can recruit co-factors that recognize histone marks and thereby adjust local chromatin structures and their behavior. A comprehensive and high-resolution analysis of histone modifications across the human genome will help our understanding of the functional correlation of various PTMs with processes such as transcription, DNA repair and DNA replication [121,122]. Use of modification-specific antibodies in ChIP has revolutionized the ability to ascribe biological functions to histone modifications. ChIP-on-chip has allowed a description of the global distribution and dynamics of various histone modifications [123]. However, prior to NGS, it had not been practical to map multiple modifications in an unbiased genomic fashion.

One of the first applications of ChIP-seq was in the analysis of the genome-wide distribution of histone modifications [119]. This study, and others that followed, exemplified the newfound feasibility and utility of obtaining collections of comprehensive genomic datasets. Twenty histone methylation sites in human T-cells were mapped [124], while five histone methylation patterns in pluripotent and lineage-committed mouse cells were described [125]. Such genome-wide analyses have revealed associations between specific modified histones and gene activity as well as the spatial and combinatorial relationship between different types of histone modifications. Moreover, dynamic changes in histone modification patterns during cellular differentiation and allele-specific histone modifications were revealed [125].

These initial ChIP-seq studies, in combination with more recent analyses examining the distribution of other types of histone modifications, have revealed that specific genomic features are associated with distinct types of chromatin signatures [126,127]. Such genome-wide chromatin landscape maps have subsequently been exploited as a tool for defining and predicting novel transcription units, enhancers, promoters, and most recently ncRNAs in previously unannotated regions of the human genome [128]. In future, the influence and utilization of NGS technologies will undoubtedly find widespread use and relevance in many different areas of biology, far beyond the test-bed of epigenetics.

Recent studies of the epigenome have shown that many promoters and enhancers have distinctive chromatin signatures. These characteristic motifs can be used as to search and map the regulatory elements of the genome. Won et al. [129] used this approach in a supervised learning method involving a trained Hidden Markov model (HMM) based on histone modification data for known promoters and enhancers. They used the trained HMMs to identify promoter or enhancer-like sequences in the human genome [129]. In a somewhat similar manner, Ernst and Kellis [130] sought to identify biologically meaningful combinations of epigenetic combinations in the genome of human T-cells. They defined these genomic regions as having "spatially coherent and biologically meaningful chromatin mark combinations", and applied a multivariate HMM analysis to search for them. Fifty-one distinct chromatin states were identified by the analysis, including those associated with promoters, transcription, active intergenic regions, large-scale repressed regions and repetitive chromatin. Each chromatin state showed specific enrichments for particular sequence motifs, suggesting distinct biological roles. This approach, therefore, provides a means of annotating the human genome with respect to function and describes the locations of regions with diverse classes of epigenetic function across the genome [130].

There is considerable uncertainty regarding the influence of variations in chromatin structure and transcription factor binding on gene expression, and whether such variations underlie or contribute to phenotypic differences. To address this question, McDaniell et al. [131] cataloged variation in chromatin structure and transcription factor binding between individuals and between homologous chromosomes within individuals (allele-specific variation). The analysis was carried out on lymphoblastoid cells from individuals with diverse geographical ancestries. They reported that 10% of active chromatin sites were specific to individuals, and a similar proportion was allele-specific. Both individual-specific and allele-specific sites could be transmitted from parent to child, suggesting that these epigenetic marks are heritable features of the human genome. The study highlights the potential importance of heritable epigenetic variation for phenotypic variation in humans [131].

Ernst et al. [132] extended their earlier chromatin profiling analysis described above by mapping nine chromatin marks in nine different human cell types with the aim of identifying regulatory elements, their cell-type specificities and their functional interactions. By comparing chromatin profiles across a range of cell types they were able to define cell-type-specific patterns of promoters and enhancers affecting chromatin status, gene expression, regulatory motif enrichment and regulator expression. Using the profiles, they linked enhancers to putative target genes and predicted the cell-type-specific activators and repressors with which they interacted [132].

Computational methods for analyzing data from epigenomic studies are being continually developed and becoming ever more sophisticated; they have been used to identify functional genomic elements and to determine gene structures and *cis*-regulatory elements. For example, Hon et al. [133] described a statistical program called ChromaSig with the capacity to identify commonly occurring chromatin signatures from histone modification data. They demonstrated the potential utility of the algorithm in data from HeLa cells by identifying five clusters of chromatin signatures associated with transcriptional promoters and enhancers. Thus, through use of ChromaSig, chromatin signatures associated with specific biological functions were identified.

2.8 GENERAL OVERVIEW AND FUTURE PERSPECTIVE

Over the last decade, the technologies available to study the mechanisms and consequences of epigenetic modifications have increased exponentially. The stimulus for this has been the rapid increase in our understanding and appreciation of the importance of epigenetic changes on phenotypes and in the etiology of diseases. Technological advances now enable large-scale epigenomic analyses. The first whole-genome, high-resolution maps of epigenetic modifications have been produced, but there is clearly much more to do. Detailed maps of the human methylome, histone modifications and nucleosome positions in healthy and diseased tissues are still needed. This review section has attempted to provide an overview of the currently available techniques and to discuss some of the advantages and limitations of each technology. With the rapid growth in interest in understanding the epigenetic regulation of disease development, a variety of new and improved methodologies are certain to emerge in the coming years. These technologies will undoubtedly change the scope of epigenetic studies and will provide valuable new insights into the developmental basis of diseases and into reproductive toxicology. Particularly, in future, the influence and utilization of NGS technologies will find widespread use and relevance in many different areas of biology, far beyond the test-bed of epigenetics.

Here, we outline a promising strategy for epigenome investigation that combines several of the epigenetic methods described above (Figures 2.2A, 2.2B). The early/late-switch regions of replication timing generally correspond to chromosomal zones with transitions in relative GC content; they are also correlated to R/G chromosome band boundaries, and are suspected of being "unstable" genomic regions that have increased susceptibility to epigenetic mutation and DNA damage [5,99]. There is a clear need for further epigenomic analysis on chromosomal band structures, in particular, to obtain a greater understanding of these epimutation-sensitive regions at the genome sequence level. Before performing epigenomic analysis using DNA methylation and histone modification, the direct correspondence between chromosomal band (R-, T-, and G-bands) and genome sequence should be elucidated by analyzing DNA replication timing and GC%, etc. (Figures 2.2A, 2.2B). Finally, we suggest that epigenomic analysis focused on chromosomal band structures, the boundaries of which were identified as epimutation-sensitive genomic regions at the genome sequence level, will provide considerable insights into normal and disease conditions.

References

[1] Portela A, Esteller M. Epigenetic modifications and human disease. Nat Biotechnol 2010;28:1057–68.

[2] Bernstein B. The NIH Roadmap Epigenome Mapping Consortium. Nat Biotechnol 2010;28:1045–8.

[3] Satterlee J. Tackling the epigenome: challenges and opportunities for collaborative efforts. Nat Biotechnol 2010;28:1039–44.

[4] Laird PW. Principles and challenges of genomewide DNA methylation analysis. Nat Rev Genet 2010;11:191–203.

[5] Watanabe Y, Maekawa M. Methylation of DNA in cancer. Adv Clin Chem 2010;52:145–68.

[6] Bird AP. Use of restriction enzymes to study eukaryotic DNA methylation. II: the symmetry of methylated sites supports semi-conservative copying of the methylation pattern. J Mol Biol 1978;118:48–60.

[7] Khulan B, Thompson RF, Ye K, Fazzari MJ, Suzuki M, Stasiek E, et al. Comparative isoschizomer profiling of cytosine methylation: the HELP assay. Genome Res 2006;16:1046–55.

[8] Schumacher A, Kapranov P, Kaminsky Z, Flanagan J, Assadzadeh A, Yau P, et al. Microarray-based DNA methylation profiling: technology and applications. Nucleic Acids Res 2006;34:528–42.

[9] Zhang Y, Rohde C, Tierling S, Stamerjohanns H, Reinhardt R, Walter J, et al. DNA methylation analysis by bisulfite conversion, cloning, and sequencing of individual clones. Methods Mol Biol 2009;507:177–87.

[10] Derks S, Lentjes MH, Hellebrekers DM, de Bruïne AP, Herman JG, van Engeland M. Methylation-specific PCR unraveled. Cell Oncol 2004;26:291–9.

[11] Liu ZJ, Maekawa M. Polymerase chain reaction-based methods of DNA methylation analysis. Anal Biochem 2003;317:259—65.

[12] Fazzai MJ, Greally JM. Epigenomics: beyond CpG islands. Nat Rev Genet 2004;5:446—55.

[13] Hou P, Ji M, Li S, He N, Lu Z. High-throughput method for detecting DNA methylation. J Biochem Biophys Methods 2004;60:139—50.

[14] Cross SH, Charlton JA, Nan X, Bird AP. Purification of CpG islands using a methylated DNA binding column. Nat Genet 1994;6:236—44.

[15] Fraga MF, Ballestar E, Montoya G, Taysavang P, Wade PA, Esteller M. The affinity of different MBD proteins for a specific methylated locus depends on their intrinsic binding properties. Nucleic Acids Res 2003;31:1765—74.

[16] Maekawa M, Sugano K, Kashiwabara H, Ushiama M, Fujita S, Yoshimori M, et al. DNA methylation analysis using bisulfite treatment and PCR-single-strand conformation polymorphism in colorectal cancer showing microsatellite instability. Biochem Biophys Res Commun 1999;262:671—6.

[17] Maekawa M, Sugano K, Ushiama M, Fukayama N, Nomoto K, Kashiwabara H, et al. Heterogeneity of DNA methylation status analyzed by bisulfite-PCR-SSCP and correlation with clinico-pathological characteristics in colorectal cancer. Clin Chem Lab Med 2001;39:121—8.

[18] Gonzalgo ML, Liang G. Methylation-sensitive single-nucleotide primer extension (Ms-SNuPE) for quantitative measurement of DNA methylation. Nat Protoc 2007;2:1931—6.

[19] Wu Z, Luo J, Ge Q, Lu Z. Microarray-based Ms-SNuPE: near-quantitative analysis for a high-throughput DNA methylation. Biosens Bioelectron 2008;23:1333—9.

[20] Xiong Z, Laird PW. COBRA: a sensitive and quantitative DNA methylation assay. Nucleic Acids Res 1997;25:2532—4.

[21] Colella S, Shen L, Baggerly KA, Issa JP, Krahe R. Sensitive and quantitative universal Pyrosequencing methylation analysis of CpG sites. Biotechniques 2003;35:146—50.

[22] Eads CA, Danenberg KD, Kawakami K, Saltz LB, Blake C, Shibata D, et al. MethyLight: a high-throughput assay to measure DNA methylation. Nucleic Acids Res 2000;28:E32.

[23] Zeschnigk M, Böhringer S, Price EA, Onadim Z, Masshöfer L, Lohmann DR. A novel real-time PCR assay for quantitative analysis of methylated alleles (QAMA): analysis of the retinoblastoma locus. Nucleic Acids Res 2004;32:e125.

[24] Tost J, Gut IG. DNA methylation analysis by pyrosequencing. Nat Protoc 2007;2:2265—75.

[25] Tost J, Schatz P, Schuster M, Berlin K, Gut IG. Analysis and accurate quantification of CpG methylation by MALDI mass spectrometry. Nucleic Acids Res 2003;31:e50.

[26] Ehrich M, Nelson MR, Stanssens P, Zabeau M, Liloglou T, Xinarianos G, et al. Quantitative high-throughput analysis of DNA methylation patterns by base-specific cleavage and mass spectrometry. Proc Natl Acad Sci USA 2005;102:15785—90.

[27] Gupta PK. Single-molecule DNA sequencing technologies for future genomics research. Trends Biotechnol 2008;26:602—11.

[28] Zilberman D, Henikoff S. Genome-wide analysis of DNA methylation patterns. Development 2007;134:3959—65.

[29] Bock C, Lengauer T. Computational epigenetics. Bioinformatics 2008;24:1—10.

[30] Lister R, Pelizzola M, Dowen RH, Hawkins RD, Hon G, Tonti-Filippini J, et al. Human DNA methylomes at base resolution show widespread epigenomic differences. Nature 2009;462:315—22.

[31] Allfrey VG, Faulkner R, Mirsky AE. Acetylation and methylation of histones and their possible role in the regulation of RNA synthesis. Proc Natl Acad Sci USA 1964;51:786—94.

[32] Waterborg JH, Fried SR, Matthews HR. Acetylation and methylation sites in histone H4 from *Physarum polycephalum*. Eur J Biochem 1983;136:245—52.

[33] Waterborg JH, Matthews HR. Patterns of histone acetylation in the cell cycle of *Physarum polycephalum*. Biochemistry 1983;22:1489—96.

[34] Waterborg JH. Dynamic methylation of alfalfa histone H3. J Biol Chem 1993;268:4918—21.

[35] Waterborg JH. Histone synthesis and turnover in alfalfa. Fast loss of highly acetylated replacement histone variant H3.2. J Biol Chem 1993;268:4912—7.

[36] Tidwell T, Allfrey VG, Mirsky AE. The methylation of histones during regeneration of the liver. J Biol Chem 1968;243:707—15.

[37] Johnson EM, Vidali G, Littau VC, Allfrey VG. Modulation by exogenous histones of phosphorylation of non-histone nuclear proteins in isolated rat liver nuclei. J Biol Chem 1973;248:7595—600.

[38] Sobel RE, Cook RG, Perry CA, Annunziato AT, Allis CD. Conservation of deposition-related acetylation sites in newly synthesized histones H3 and H4. Proc Natl Acad Sci USA 1995;92:1237—41.

23

[39] Wilkins MR, Gasteiger E, Gooley AA, Herbert BR, Molloy MP, Binz PA, et al. High-throughput mass spectrometric discovery of protein post-translational modifications. J Mol Biol 1989;289:645—57.

[40] Sickmann A, Mreyen M, Meyer HE. Identification of modified proteins by mass spectrometry. IUBMB Life 2002;54:51—7.

[41] Mann M, Jensen ON. Proteomic analysis of post-translational modifications. Nat Biotechnol 2003;21:255—61.

[42] Sickmann A, Mreyen M, Meyer HE. Mass spectrometry: a key technology in proteome research. Adv Biochem Eng Biotechnol 2003;83:141—76.

[43] Posewitz MC, Tempst P. Immobilized gallium(III) affinity chromatography of phosphopeptides. Anal Chem 1999;71:2883—92.

[44] Stensballe A, Andersen S, Jensen ON. Characterization of phosphoproteins from electrophoretic gels by nanoscale Fe(III) affinity chromatography with off-line mass spectrometry analysis. Proteomics 2001;1:207—22.

[45] Zhou H, Watts JD, Aebersold R. A systematic approach to the analysis of protein phosphorylation. Nat Biotechnol 2001;19:375—8.

[46] Schlosser A, Vanselow JT, Kramer A. Mapping of phosphorylation sites by a multi-protease approach with specific phosphopeptide enrichment and NanoLC-MS/MS analysis. Anal Chem 2005;77:5243—50.

[47] Wisniewski JR, Szewczuk Z, Petry I, Schwanbeck R, Renner U. Constitutive phosphorylation of the acidic tails of the high mobility group 1 proteins by casein kinase II alters their conformation, stability, and DNA binding specificity. J Biol Chem 1999;274:20116—22.

[48] Appella E, Anderson CW. Signaling to p53: breaking the posttranslational modification code. Pathol Biol (Paris) 2000;48:227—45.

[49] Puri PL, Sartorelli V. Regulation of muscle regulatory factors by DNA-binding, interacting proteins, and post-transcriptional modifications. J Cell Physiol 2000;185:155—73.

[50] Jenuwein T, Allis CD. Translating the histone code. Science 2001;293:1074—80.

[51] Puntervoll P, Linding R, Gemünd C, Chabanis-Davidson S, Mattingsdal M, Cameron S, et al. ELM server: a new resource for investigating short functional sites in modular eukaryotic proteins. Nucleic Acids Res 2003;31:3625—30.

[52] Schones DE, Zhao K. Genome-wide approaches to studying chromatin modifications. Nat Rev Genet 2008;9:179—91.

[53] Nelson PT, Baldwin DA, Scearce LM, Oberholtzer JC, Tobias JW, Mourelatos Z. Microarray-based, high-throughput gene expression profiling of microRNAs. Nat Methods 2004;1:155—61.

[54] Wang H, Ach RA, Curry B. Direct and sensitive miRNA profiling from low-input total RNA. RNA 2007;13:151—9.

[55] Tricoli JV, Jacobson JW. MicroRNA: Potential for Cancer Detection, Diagnosis, and Prognosis. Cancer Res 2007;67:4553—5.

[56] Sempere LF, Freemantle S, Pitha-Rowe I, Moss E, Dmitrovsky E, Ambros V. Expression profiling of mammalian microRNAs uncovers a subset of brain-expressed microRNAs with possible roles in murine and human neuronal differentiation. Genome Biol 2004;5:R13.

[57] Válóczi A, Hornyik C, Varga N, Burgyán J, Kauppinen S, Havelda Z. Sensitive and specific detection of microRNAs by northern blot analysis using LNA-modified oligonucleotide probes. Nucleic Acids Res 2004;32:e175.

[58] Castoldi M, Benes V, Hentze MW, Muckenthaler MU. miChip: a microarray platform for expression profiling of microRNAs based on locked nucleic acid (LNA) oligonucleotide capture probes. Methods 2007;43:146—52.

[59] Chen C, Ridzon DA, Broomer AJ, Zhou Z, Lee DH, Nguyen JT, et al. Real-time quantification of microRNAs by stem-loop RT-PCR. Nucleic Acids Res 2005;33:e179.

[60] Kloosterman WP, Wienholds E, de Bruijn E, Kauppinen S, Plasterk RH. In situ detection of miRNAs in animal embryos using LNA-modified oligonucleotide probes. Nat Methods 2006;3:27—9.

[61] de Planell-Saguer M, Rodicio MC, Mourelatos Z. Rapid in situ codetection of noncoding RNAs and proteins in cells and formalin-fixed paraffin-embedded tissue sections without protease treatment. Nat Protoc 2010;5:1061—73.

[62] Schulte JH, Marschall T, Martin M, Rosenstiel P, Mestdagh P, Schlierf S, et al. Deep sequencing reveals differential expression of microRNAs in favorable versus unfavorable neuroblastoma. Nucleic Acids Res 2010;38:5919—28.

[63] Takada S, Mano H. Profiling of microRNA expression by mRAP. Nat Protoc 2007;2:3136—45.

[64] Lee RC, Feinbaum RL, Ambros V. The *C. elegans* heterochronic gene lin-4 encodes small RNAs with antisense complementarity to lin-14. Cell 1993;75:843—54.

24

[65] Lau NC, Lim LP, Weinstein EG, Bartel DP. An abundant class of tiny RNAs with probable regulatory roles in *Caenorhabditis elegans*. Science 2001;294:858–62.

[66] Lee RC, Ambros V. An extensive class of small RNAs in *Caenorhabditis elegans*. Science 2001;294:862–4.

[67] Lagos-Quintana M, Rauhut R, Lendeckel W, Tuschl T. Identification of novel genes coding for small expressed RNAs. Science 2001;294:853–8.

[68] Grishok A, Pasquinelli AE, Conte D, Li N, Parrish S, Ha I, et al. Genes and mechanisms related to RNA interference regulate expression of the small temporal RNAs that control *C. elegans* developmental timing. Cell 2001;106:23–34.

[69] Jiang J, Lee EJ, Gusev Y, Schmittgen TD. Real-time expression profiling of microRNA precursors in human cancer cell lines. Nucleic Acids Res 2005;33:5394–403.

[70] Lu J, Getz G, Miska EA, Alvarez-Saavedra E, Lamb J, Peck D, et al. MicroRNA expression profiles classify human cancers. Nature 2005;435:834–8.

[71] Liu CG, Calin GA, Meloon B, Gamliel N, Sevignani C, Ferracin M, et al. An oligonucleotide microchip for genome-wide microRNA profiling in human and mouse tissues. Proc Natl Acad Sci USA 2004;101:9740–4.

[72] Rhoades MW, Reinhart BJ, Lim LP, Burge CB, Bartel B, Bartel DP. Prediction of plant microRNA targets. Cell 2002;110:513–20.

[73] Schena M, Shalon D, Davis RW, Brown PO. Quantitative monitoring of gene expression patterns with a complementary DNA microarray. Science 1995;270:467–70.

[74] Kreil DP, Russell RR, Russell S. Microarray oligonucleotide probes. Methods Enzymol 2006;410:73–98.

[75] Wolber PK, Collins PJ, Lucas AB, De Witte A, Shannon KW. The Agilent in situ-synthesized microarray platform. Methods Enzymol 2006;410:28–57.

[76] Dalma-Weiszhausz DD, Warrington J, Tanimoto EY, Miyada CG. The affymetrix GeneChip platform: an overview. Methods Enzymol 2006;410:3–28.

[77] Gershon D. DNA microarrays: more than gene expression. Nature 2005;437:1195–8.

[78] Hughes TR, Hiley SL, Saltzman AL, Babak T, Blencowe BJ. Microarray analysis of RNA processing and modification. Methods Enzymol 2006;410:300–16.

[79] Selig S, Okumura K, Ward DC, Cedar H. Delineation of DNA replication time zones by fluorescence in situ hybridization. EMBO J 1992;11:1217–25.

[80] Hansen RS, Canfield TK, Lamb MM, Gartler SM, Laird CD. Association of fragile X syndrome with delayed replication of the FMR1 gene. Cell 1993;73:1403–9.

[81] Watanabe Y, Tenzen T, Nagasaka Y, Inoko H, Ikemura T. Replication timing of the human X-inactivation center (XIC) region: correlation with chromosome bands. Gene 2000;252:163–72.

[82] Watanabe Y, Fujiyama A, Ichiba Y, Hattori M, Yada T, Sakaki Y, et al. Chromosome-wide assessment of replication timing for human chromosomes 11q and 21q: disease-related genes in timing-switch regions. Hum Mol Genet 2002;11:13–21.

[83] Woodfine K, Fiegler H, Beare DM, Collins JE, McCann OT, Young BD, et al. Replication timing of the human genome. Hum Mol Genet 2004;13:191–202.

[84] Watanabe Y, Ikemura T, Sugimura H. Amplicons on human chromosome 11q are located in the early/late-switch regions of replication timing. Genomics 2004;84:796–805.

[85] White EJ, Emanuelsson O, Scalzo D, Royce T, Kosak S, Oakeley EJ, et al. DNA replication-timing analysis of human chromosome 22 at high resolution and different developmental states. Proc Natl Acad Sci USA 2004;101:17771–6.

[86] Mirkin EV, Mirkin SM. Replication fork stalling at natural impediments. Microbiol Mol Biol Rev 2007;71:13–35.

[87] Sequeira-Mendes J, Díaz-Uriarte R, Apedaile A, Huntley D, Brockdorff N, Gómez M. Transcription initiation activity sets replication origin efficiency in mammalian cells. PLoS Genet 2009;5:e1000446.

[88] Bacolla A, Wells RD. Non-B DNA conformations as determinants of mutagenesis and human disease. Mol Carcinog 2009;48:273–85.

[89] Hiratani I, Gilbert DM. Replication timing as an epigenetic mark. Epigenetics 2009;4:93–7.

[90] Göndör A, Ohlsson R. Replication timing and epigenetic reprogramming of gene expression: a two-way relationship? Nat Rev Genet 2009;10:269–76.

[91] State MW, Greally JM, Cuker A, Bowers PN, Henegariu O, Morgan TM, et al. Epigenetic abnormalities associated with a chromosome 18(q21–q22) inversion and a Gilles de la Tourette syndrome phenotype. Proc Natl Acad Sci USA 2003;100:4684–9.

[92] D'Antoni S, Mattina T, Di Mare P, Federico C, Motta S, Saccone S. Altered replication timing of the HIRA/Tuple1 locus in the DiGeorge and Velocardiofacial syndromes. Gene 2004;333:111–9.

25

[93] Dotan ZA, Dotan A, Ramon J, Avivi L. Altered mode of allelic replication accompanied by aneuploidy in peripheral blood lymphocytes of prostate cancer patients. Int J Cancer 2004;111:60−6.

[94] Dotan ZA, Dotan A, Ramon J, Avivi L. Aberrant allele-specific replication, independent of parental origin, in blood cells of cancer patients. BMC Cancer 2008;8:390.

[95] Watanabe Y, Shibata K, Sugimura H, Maekawa M. p53-dependent change in replication timing of the human genome. Biochem Biophys Res Commun 2007;364:289−93.

[96] Watanabe Y, Shibata K, Ikemura T, Maekawa M. Replication timing of extremely large genes on human chromosomes 11q and 21q. Gene 2008;421:74−80.

[97] Watanabe Y, Abe T, Ikemura T, Maekawa M. Relationships between replication timing and GC content of cancer-related genes on human chromosomes 11q and 21q. Gene 2009;433:26−31.

[98] Göndör A, Ohlsson R. Replication timing and epigenetic reprogramming of gene expression: a two-way relationship? Nat Rev Genet 2009;10:269−76.

[99] Watanabe Y, Maekawa M. Spatiotemporal regulation of DNA replication in the human genome and its association with genomic instability and disease. Curr Med Chem 2010;17:222−33.

[100] Maekawa M, Watanabe Y. Epigenetics: relations to disease and laboratory findings. Curr Med Chem 2007;14:2642−53.

[101] Holmquist G, Gray M, Porter T, Jordan J. Characterization of Giemsa dark- and light-band DNA. Cell 1982;31:121−9.

[102] Holmquist GP. Evolution of chromosome bands: molecular ecology of noncoding DNA. J Mol Evol 1989;28:469−86.

[103] Bernardi G. The isochore organization of the human genome. Annu Rev Genet 1989;23:637−61.

[104] Craig JM, Bickmore WA. Chromosome bands—flavours to savour. Bioessays 1993;15:349−54.

[105] Bierne H, Michel B. When replication forks stop. Mol Microbiol 1994;13:17−23.

[106] Verbovaia LV, Razin SV. Mapping of replication origins and termination sites in the Duchenne muscular dystrophy gene. Genomics 1997;45:24−30.

[107] Rothstein R, Michel B, Gangloff S. Replication fork pausing and recombination or 'gimme a break'. Genes Dev 2000;14:1−10.

[108] Hiratani I, Ryba T, Itoh M, Yokochi T, Schwaiger M, Chang CW, et al. Global reorganization of replication domains during embryonic stem cell differentiation. PLoS Biol 2008;6:e245.

[109] Helmrich A, Stout-Weider K, Hermann K, Schrock E, Heiden T. Common fragile sites are conserved features of human and mouse chromosomes and relate to large active genes. Genome Res 2006;16:1222−30.

[110] Debatisse M, El Achkar E, Dutrillaux B. Common fragile sites nested at the interfaces of early and late-replicating chromosome bands: cis acting components of the G2/M checkpoint? Cell Cycle 2006;5:578−81.

[111] Ge XQ, Jackson DA, Blow JJ. Dormant origins licensed by excess Mcm2-7 are required for human cells to survive replicative stress. Genes Dev 2007;21:3331−41.

[112] Ibarra A, Schwob E, Méndez J. Excess MCM proteins protect human cells from replicative stress by licensing backup origins of replication. Proc Natl Acad Sci USA 2008;105:8956−61.

[113] Rakyan VK, Down TA, Balding DJ, Beck S. Epigenome-wide association studies for common human diseases. Nat Rev Genet 2011;12:529−41.

[114] Mardis ER. Next-generation DNA sequencing methods. Annu Rev Genomics Hum Genet 2008;9:387−402.

[115] Kim TH, Barrera LO, Zheng M, Qu C, Singer MA, Richmond TA, et al. A high-resolution map of active promoters in the human genome. Nature 2005;436:876−80.

[116] Alekseyenko AA, Peng S, Larschan E, Gorchakov AA, Lee OK, Kharchenko P, et al. A sequence motif within chromatin entry sites directs MSL establishment on the *Drosophila* X chromosome. Cell 2008;134:599−609.

[117] Rozowsky J, Euskirchen G, Auerbach RK, Zhang ZD, Gibson T, Bjornson R, et al. PeakSeq enables systematic scoring of ChIP−seq experiments relative to controls. Nature Biotech 2009;27:66−75.

[118] Whiteford N, Haslam N, Weber G, Prügel-Bennett A, Essex JW, Roach PL, et al. An analysis of the feasibility of short read sequencing. Nucleic Acids Res 2005;33:e171.

[119] Park PJ. ChIP-seq: advantages and challenges of a maturing technology. Nat Rev Genet 2009;10:669−80.

[120] Kouzarides T. Chromatin modifications and their function. Cell 2007;128:693−705.

[121] Li B, Carey M, Workman JL. The role of chromatin during transcription. Cell 2007;128:707−19.

[122] Groth A, Rocha W, Verreault A, Almouzni G. Chromatin challenges during DNA replication and repair. Cell 2007;128:721−33.

[123] Rando OJ. Global patterns of histone modifications. Curr Opin Genet Dev 2007;17:94−9.

[124] Barski A, Cuddapah S, Cui K, Roh TY, Schones DE, Wang Z, et al. High-resolution profiling of histone methylations in the human genome. Cell 2007;129:823−37.

[125] Mikkelsen TS, Ku M, Jaffe DB, Issac B, Lieberman E, Giannoukos G, et al. Genome-wide maps of chromatin state in pluripotent and lineage-committed cells. Nature 2007;448:553−60.

[126] Wang Z, Zang C, Rosenfeld JA, Schones DE, Barski A, Cuddapah S, et al. Combinatorial patterns of histone acetylations and methylations in the human genome. Nat Genet 2008;40:897−903.

[127] Robertson AG, Bilenky M, Tam A, Zhao Y, Zeng T, Thiessen N, et al. Genome-wide relationship between histone H3 lysine 4 mono- and tri-methylation and transcription factor binding. Genome Res 2008;18:1906−17.

[128] Guttman M, Amit I, Garber M, French C, Lin MF, Feldser D, et al. Chromatin signature reveals over a thousand highly conserved large non-coding RNAs in mammals. Nature 2009;458:223−7.

[129] Won KJ, Chepelev I, Ren B, Wang W. Prediction of regulatory elements in mammalian genomes using chromatin signatures. BMC Bioinformatics 2008;9:547.

[130] Ernst J, Kellis M. Discovery and characterization of chromatin states for systematic annotation of the human genome. Nat Biotechnol 2010;28:817−25.

[131] McDaniell R, Lee BK, Song L, Liu Z, Boyle AP, Erdos MR, et al. Heritable individual-specific and allele-specific chromatin signatures in humans. Science 2010;328:235−9.

[132] Ernst J, Kheradpour P, Mikkelsen TS, Shoresh N, Ward LD, Epstein CB, et al. Mapping and analysis of chromatin state dynamics in nine human cell types. Nature 2011;473:43−9.

[133] Hon G, Ren B, Wang W. ChromaSig: a probabilistic approach to finding common chromatin signatures in the human genome. PLoS Comput Biol 2008;4:e1000201.

[134] Ermakova OV, Nguyen LH, Little RD, Chevillard C, Riblet R, Ashouian N, et al. Evidence that a single replication fork proceeds from early to late replicating domains in the Igh locus in a non-B cell line. Mol Cell 1999;3:321−30.

[135] Karnani N, Taylor C, Malhotra A, Dutta A. Pan-S replication patterns and chromosomal domains defined by genome-tiling arrays of ENCODE genomic areas. Genome Res 2007;17:865−76.

CHAPTER 3

DNA Methylation Alterations in Human Cancers

Yae Kanai, Eri Arai
National Cancer Center Research Institute, Tokyo, Japan

CHAPTER OUTLINE

3.1 INTRODUCTION: BIOLOGICAL ROLES OF DNA METHYLATION

Epigenetic processes, i.e. alterations to biological information without changes in the DNA sequences that are mitotically and/or meiotically heritable, go beyond DNA-stored information and are essential for packaging and interpretation of the genome [1]. The modulation of epigenetic profiles contributes significantly to embryonic development, differentiation, and transition from a stem cell to a lineage-committed cell, and underlies responses to environmental signals such as hormones, nutrients and inflammation [2]. DNA methylation is a key element of epigenetic mechanisms that include histone-modification, positioning of histone variants, nucleosome remodeling, and non-coding RNA. DNA methylation is a covalent

T. Tollefsbol (Ed): Epigenetics in Human Disease. DOI: 10.1016/B978-0-12-388415-2.00003-2

chemical modification resulting in addition of a methyl (CH_3) group at the carbon 5 position of the cytosine ring of CpG dinucleotides. CpG sites are concentrated either in repetitive sequences or CpG islands in promoter regions.

The C-terminal catalytic domain of DNA (cytosine-5-)-methyltransferases (DNMTs) transfers methyl groups from S-adenosyl-L-methionine (AdoMet) to cytosines. Dietary folate, vitamins B6 and B12, methionine and choline can critically affect the synthesis of AdoMet [3]. The C-terminal catalytic domain of DNMTs is composed of five conserved amino acid motifs, namely I, IV, VI, IX and X [4,5]. Motifs I and X are filed together to form the binding site for AdoMet. Motif IV contains the prolylcysteinyl dipeptide that provides the thiolate at the active site. Motif VI contains the glutamyl residue that protonates the 3 position of the target cytosine. Motif IX forms the target recognition domain. The N-terminal regulatory domain of DNMT1 contains a PCNA (proliferating cell nuclear antigen)-binding domain, a cysteine-rich ATRX (alpha thalassemia/mental retardation syndrome X-linked) zinc finger DNA-binding motif, and a polybromo homology domain targeting DNMT1 to the replication foci. The preference of DNMT1 for hemimethylated over unmethylated substrates in vitro and its targeting of replication foci are believed to allow copying of the methylation pattern of the parental strand to the newly synthesized daughter DNA strand. Thus, DNMT1 has been recognized as the "maintenance" DNMT, whereas DNMT3A and DNMT3B show de novo DNA methylation activity in vitro [6]. However, since de novo methylation of CpG islands has actually been observed in human fibroblasts overexpressing DNMT1, DNMT1 is capable of de novo DNA methylation activity in vivo as well as having a maintenance function [3]. DNA methylation profiles in vivo may be determined on the basis of cooperation between DNMT1 and the DNMT3 family. DNMT3L lacks conserved motifs of the catalytic domain and cooperates with the DNMT3 family to establish an imprinting pattern [7].

30

DNA methylation plays critical roles in the maintenance of chromatin integrity and regulation of gene expression [8]: (a) repetitive and parasitic sequences, such as retrotransposons and endogenous retroviral elements, are usually repressed due to DNA methylation and (b) methylation of CpGs islands can directly impede the binding of transcription factors to their target sites, thus prohibiting the transcription of specific genes. Moreover, methylation of CpG islands normally promotes a highly condensed heterochromatin structure, where active transcription does not occur. Weber et al. reported that approximately 70% of human genes are linked to promoter CpG islands and about 4% of CpG island promoters are methylated in somatic cells [9]. Methylation of CpG islands naturally takes place during X chromosome inactivation and imprinting, though the majority of CpG islands remain unmethylated during development and differentiation. Extensive changes in DNA methylation during the processes of differentiation are known to take place at CpG island shores, regions of comparatively low CpG density close to CpG islands [10].

On the other hand, DNA demethylation is a process involving removal of a methyl group from a nucleotide in DNA. Although passive demethylation occurs in the absence of methylation of newly synthesized DNA strands by "maintenance" DNMT during replication rounds, active removal of cytosine methylation has long remained a mystery. Recently, it has been proved that 5-methylcytosine can be converted to 5-hydroxymethylcytosine, an intermediate form potentially involved in active demethylation, by the 2-oxoglutarate and Fe (II)-dependent oxygenases TET1, TET2 and TET3 [11,12]. MLL (myeloid/lymphoid or mixed-lineage leukemia)—TET1 translocations have been found in patients with acute lymphoblastic leukemia [13] and deletions or mutations at the TET2 locus have been reported in myelodysplastic syndrome and acute myeloid leukemia [14], indicating that impairment of the conversion of 5-methylcytosine to 5-hydroxymethylcytosine may also participate in tumorigenesis [15]. Further investigation will be needed to elucidate the significance of conversion to 5-hydroxymethylcytosine.

3.2 DNA METHYLATION ALTERATIONS IN HUMAN CANCERS

Heterozygosity of the *Dnmt1* gene, in conjunction with treatment using the DNMT inhibitor 5-aza-deoxycytidine, reduces the average number of intestinal adenomas in ApcMin mice [16]. On the other hand, genomic hypomethylation in p53+/− mice due to the introduction of a hypomorphic allele of Dnmt 1 induces sarcomas at an earlier age in comparison with littermates possessing normal levels of DNMT1 activity [17,18]. Increased loss of heterozygosity (LOH) accompanied by activation of endogenous retroviral elements has been observed in Dnmt1 hypomorphic mice [19]. These observations of genetically engineered animals clearly demonstrate a causal relationship between alterations of DNA methylation and human cancers.

In fact, human cancer cells obtained from clinical tissue specimens frequently show genome-wide DNA hypomethylation and region-specific DNA hypermethylation [20]. DNA hypomethylation induces a higher probability of translocation of parasitic sequences to other genomic regions, and chromosomal rearrangement resulting in chromosomal instability [21]. Furthermore, aberrant DNA hypomethylation can also induce activation of oncogenes and loss of imprinting. However, a more widely recognized epigenetic change in human cancers is DNA hypermethylation at the CpG islands of promoters that silences specific genes, including tumor-suppressor genes [20] such as CDKN2A (cyclin-dependent kinase inhibitor 2A), *CDKN2B* (cyclin-dependent kinase inhibitor 2B), *TP73* (tumor protein p73), *MLH1* (mutL homolog 1), *Apc* (adenomatosis polyposis coli), *BRCA1* (breast cancer 1), *MGMT* (O-6-methylguanine-DNA methyltransferase), *VHL* (von Hippel-Lindau tumor-suppressor), *GSTP1* (glutathione S-transferase pi 1), *CDH1* (cadherin 1) and *DAPK1* (death-associated protein kinase 1). DNA hypermethylation of tumor-suppressor genes frequently becomes the second hit for driver events in accordance with the two-hit theory [22]. Moreover, some tumor-suppressor genes, such as *TIMP3* (tissue inhibitor of metalloproteinase 3), *SFRP1* (secreted frizzled-related protein 1), *SFRP2*, *SFRP4*, *SFRP5* and *RASSF1* (Ras association (RalGDS/AF-6) domain family member 1), are seldom mutated, or their mutations have never been reported in human cancers [23]. Therefore, intensive screening of genes that are methylated in human cancers may be a strategy for identification of tumor-related genes that have potential as therapeutic targets. In some instances, genes can be silenced simultaneously due to a process of long-range epigenetic silencing, and the spreading of silencing seems to affect neighboring unmethylated genes through repressive chromatin [24].

MiRNAs are the best-known class of short non-coding RNAs, which are typically around 21 nucleotides in length, imperfectly aligned with the 3′UTR of target mRNAs, and induce their translational repression. Observations of silencing due to DNA hypermethylation have expanded to tumor-suppressive microRNAs (25), such as miR-34a and 34b/c, miR-124, miR-137, miR-152, miR-193a, miR-200, miR-203, miR-205, miR-218 and miR-345. In addition to their tumor-suppressor function, miRNAs can also serve as oncogenes to promote cancer growth. B-cell integration cluster (BIC)/miR-155 is the first miRNA shown to have such tumor-promoting activity. miR-10b is another oncogene highly associated with cancer metastasis. Transcription of miR-10b is regulated by the transcription factor Twist, and the downstream targets of miR-10b include homeobox D10. Other miRNAs with oncogene function include miR-17 clusters, miR-21, and miR-373 and miR-520c as metastasis-promoting miRNAs [26].

3.3 ABERRANT DNA METHYLATION IN PRECANCEROUS CONDITIONS ASSOCIATED WITH CHRONIC INFLAMMATION, PERSISTENT VIRAL INFECTION AND SMOKING

DNA methylation alterations are frequently observed even in precancerous conditions and early-stage cancers, suggesting that epigenetic alterations may precede the classical transforming events, such as mutations of tumor-suppressor genes, amplification of oncogenes and

31

chromosomal instability. Environmental factors influence health, and epigenetic profiles are known to be responses to environmental signals. Thus, aberrant DNA methylation participates especially in precancerous conditions associated with chronic inflammation, persistent viral infection and smoking [27,28]. For example, in the 1990s, although LOH on chromosome 16 was frequently detected by classical Southern blotting in hepatocellular carcinomas (HCCs) associated with metastasis, the molecular events occurring in non-cancerous liver tissue showing chronic hepatitis or liver cirrhosis, which are widely considered to be precancerous conditions, were unknown. When we examined the DNA methylation status on chromosome 16 using Southern blotting with a DNA methylation-sensitive restriction enzyme, DNA methylation alterations at multiple loci were frequently revealed even in chronic hepatitis or liver cirrhosis, compared with normal liver tissue, indicating that DNA methylation alterations are a very early event during multistage hepatocarcinogenesis [29]. This was one of the earliest reports of DNA methylation alterations at the precancerous stage.

We then examined whether aberrant DNA methylation precedes chromosomal instability during hepatocarcinogenesis. Bisulfite modification, which converts unmethylated cytosine residues to uracil, leaving methylated cytosine residues unchanged, was applied to micro-dissected specimens obtained from lobules, pseudo-lobules or regenerative nodules in non-cancerous liver tissue from patients with HCCs. Although no degree of DNA methylation of any of the examined C-type CpG islands, which are generally methylated in a cancer-specific but not age-dependent manner, was ever detected in normal liver tissue from patients without HCCs, DNA hypermethylation of such islands was frequently found even in microdissected specimens of non-cancerous liver tissue showing no remarkable histological changes obtained from patients with HCCs in which LOH was never detected by PCR using multiple micro-satellite markers. Thus it was directly confirmed that aberrant DNA methylation is an earlier event preceding chromosomal instability during hepatocarcinogenesis [30].

32

DNA hypermethylation around the promoter region of the CDH1 gene at 16q22.1 [31], which encodes a Ca^{2+}-dependent cell–cell adhesion molecule [32], has been detected even in samples of non-cancerous liver tissue showing chronic hepatitis or cirrhosis [33]. Heterogeneous E-cadherin expression in such non-cancerous liver tissue, which is associated with small focal areas of hepatocytes showing only slight E-cadherin immunoreactivity, might be due, at least partly, to DNA hypermethylation [33]. Reduction of E-cadherin expression due to DNA hypermethylation around the promoter region may participate even in the very early stage of hepatocarcinogenesis through loss of intercellular adhesiveness and destruction of tissue morphology.

In addition to the chronic hepatitis and liver cirrhosis stages resulting from infection with hepatitis B virus (HBV) and/or hepatitis C virus (HCV) [30,34,35], DNA methylation alterations are frequently found at the precancerous stage in various organs, especially in association with chronic inflammation and/or persistent infection with viruses. Epstein–Barr virus (EBV) infection in stomach cancers is significantly associated with marked accumulation of DNA hypermethylation of C-type CpG islands [36,37], and viral latent membrane protein 2A up-regulates DNMT1 in cultured cancer cells [38]. Helicobacter pylori infection, another etiologic factor that is believed to be involved in stomach carcinogenesis, has also been reported to strongly promote regional DNA hypermethylation [39]. Cervical intraepithelial neoplasia (CIN) is a precursor lesion for squamous cell carcinoma of the uterine cervix closely associated with human papillomavirus (HPV) infection. DNMT1 protein expression is increased even in low-grade CINs relative to normal squamous epithelia, and is further increased in higher-grade CINs and squamous cell carcinomas of the uterine cervix [40]. HPV-16 E7 protein has been reported to associate directly with DNMT1 and stimulate the methyltransferase activity of DNMT1 in vitro [41].

DNA hypermethylation at the HIC1 (hypermethylated in cancer 1) locus has been observed in non-cancerous lung tissues, which may contain progenitor cells for cancers, obtained from patients with non-small-cell lung cancers, and in the corresponding non-small-cell lung cancers [42]. HIC1 is a growth-regulatory and tumor-repressor gene [43] that was first identified in the commonly methylated chromosomal region in human cancer cells [44]. The incidence of DNA hypermethylation at this locus was significantly associated with poorer differentiation of lung adenocarcinomas. The incidence of DNA hypermethylation in samples of both non-cancerous lung tissue and non-small-cell lung cancer from patients who were current smokers was significantly higher than in patients who had never smoked [42]. The incidence of DNA hypermethylation in non-cancerous lung tissue from patients with non-small-cell lung cancers was significantly correlated with the extent of pulmonary anthracosis, as an index of the cumulative effects of smoking [27]. Cigarette smoking seems to be another background factor associated with alterations of DNA methylation during multistage carcinogenesis.

3.4 ABNORMAL EXPRESSION OF DNMTS IN HUMAN CANCERS

At least a proportion of DNA methylation alterations in human cancers may be attributable to abnormalities of DNMTs. In fact, altered expression of DNMTs has been reported in human cancers. For example, the levels of DNMT1 mRNA expression are significantly higher in samples of non-cancerous liver tissue showing chronic hepatitis or cirrhosis than in normal liver tissue, and are even higher in HCCs [45,46]. The incidence of DNMT1 overexpression in HCCs is significantly correlated with poorer tumor differentiation and portal vein tumor involvement. Moreover, DNMT1 overexpression in tumors is inversely correlated with the recurrence-free and overall survival rates of patients with HCCs [47].

Ductal adenocarcinomas of the pancreas frequently develop after chronic damage due to pancreatitis. At least a proportion of peripheral pancreatic ductal epithelia with an inflammatory background may be at the precancerous stage. The incidence of DNMT1 protein expression increases with progression from peripheral pancreatic ductal epithelia with an inflammatory background, to another precancerous lesion, pancreatic intraductal neoplasia (PanIN), to well-differentiated ductal adenocarcinoma, and finally to poorly differentiated ductal adenocarcinoma, in comparison with normal peripheral pancreatic duct epithelia [48]. DNMT1 overexpression in ductal adenocarcinomas of the pancreas is significantly correlated with the extent of invasion to surrounding tissue, an advanced stage, and poorer patient outcome [48]. The average number of methylated CpG islands of examined tumor-suppressor genes in microdissected specimens of peripheral pancreatic ductal epithelia with an inflammatory background, PanIN and ductal adenocarcinoma was significantly correlated with the level of DNMT1 protein expression demonstrated immunohistochemically in precisely microdissected areas [49].

When the human DNMT3A and DNMT3B genes were first cloned, the expression levels of DNMT1, 3A and 3B were reported in ten paired samples of normal and cancerous tissue obtained from various organs. Robertson et al. observed ≥2-fold overexpression of DNMT3A in five of ten samples, DNMT1 in six of ten samples, and DNMT3B in eight of ten samples, and DNMT3B clearly showing the largest fold increases among the three enzymes [50]. On the other hand, the cancer phenotype associated with accumulation of DNA methylation on C-type CpG islands is defined as the CpG-island methylator phenotype (CIMP) [51], and such accumulation is generally associated with frequent silencing of tumor-related genes due to DNA hypermethylation only, or a two-hit mechanism involving DNA hypermethylation and LOH in human cancers of various organs [52]. Expression levels of DNMT1 mRNA and protein are significantly correlated with poorer differentiation and CIMP in stomach cancers, but no such association has been observed for the expression of DNMT2, DNMT3A or DNMT3B [53]. EBV infection in stomach cancers is significantly associated with marked accumulation of DNA methylation on C-type CpG islands and overexpression of DNMT1 protein, although

33

Helicobacter pylori infection, another etiologic factor strongly promoting regional DNA hypermethylation, was not correlated with DNMT1 expression levels.

Urothelial carcinomas (UCs) of the urinary bladder are clinically remarkable because of their multicentricity and tendency to recur due to a "field effect". Even non-cancerous urothelia showing no remarkable histological changes obtained from patients with UCs can be considered precancerous, because they may have been exposed to carcinogens in the urine. Our immunohistochemical examinations have clearly revealed that the incidence of nuclear DNMT1 immunoreactivity is already higher in non-cancerous urothelia showing no remarkable histological changes obtained from patients with UCs, where the PCNA labeling index had not yet increased, compared to that in normal urothelia from patients without UCs, indicating that DNMT1 overexpression was not a secondary result of increased cell proliferative activity, but in fact preceded such activity [54]. The incidence of nuclear DNMT1 immunoreactivity showed a progressive increase in dysplastic urothelia, and during transition to UCs, being significantly correlated with accumulation of DNA methylation on C-type CpG islands [55].

With respect to the mechanisms regulating the expression levels of DNMTs [56], the members of the miR-29 family, including miR-29a, miR-29b and mir-29c, have been shown to directly target DNMT3A and DNMT3B [57]. Enforced expression of miR-29s in lung cancer cell lines restores the normal patterns of DNA methylation, induces re-expression of methylation-silenced tumor-suppressor genes, and inhibits tumorigenicity in vitro and in vivo [57]. Enforced expression of miR-29b in acute myeloid leukemia cells resulted in markedly reduced expression of DNMT1, DNMT3A, and DNMT3B at both the RNA and protein levels [58]. Although down-regulation of DNMT3A and DNMT3B was the result of direct interaction of miR-29b with the 3'UTRs of these genes, miR-29b down-regulates DNMT1 indirectly by targeting Sp1, a transactivator of the DNMT1 gene [58]. miR-148 has been observed to bind to the coding region, outside the usual 3'UTR, of DNMT3B and to induce splicing alteration of DNMT3B in human cancer cells [59]. DNMT1 may also be directly regulated by miR-148 [60] and miR-126 [61]. Down-regulated miR-152 induces aberrant DNA methylation in HCC cells by targeting DNMT1 [62]. In addition to miRNAs, Hu-antigen R (HuR) proteins bind to target mRNAs and modify their levels of expression by altering their stability. HuR proteins target the 3'UTR of DNMT3B in human colon cancer cells, resulting in DNA hypermethylation of its target genes [63].

3.5 MUTATIONS, POLYMORPHISM AND SPLICING ALTERATIONS OF DNMTS AND HUMAN CANCERS

Even though our previous screening indicated that mutations of DNMT1 are not the major event during carcinogenesis in the liver and stomach [64], recent massively parallel DNA sequencing has identified somatic mutations including missense mutations, frameshifts, splice-site mutations and large deletions, which were predicted to affect DNMT3A translation in acute myeloid leukemia cells [65]. The overall survival of patients showing DNMT3A mutations was significantly shorter than that of patients without such mutations. Mutations of the DNMT3A gene, which reduce its enzymatic activity and alter the DNA methylation profiles, have also been reported in acute monocytic leukemia [66]. These observations add to the evidence for participation of aberrant DNMT activity in the pathogenesis of malignancies.

DNMT3A gene polymorphism can affect transcriptional levels of DNMT3A and susceptibility to cancers. The effect of a single nucleotide polymorphism, A/G, in the DNMT3A promoter region on transcriptional activity has been evaluated using a luciferase assay. Carriage of the A allele conferred significantly higher promoter activity in comparison with the G allele, and AA homozygotes had a six-fold increased risk of gastric cancer [67]. Similarly, a marked association between DNMT3B6 promoter C/T polymorphism and overall survival of patients with head and neck squamous cell carcinoma has been reported [68]: the homozygotes (CC-genotype and TT-genotype) survived significantly longer than the heterozygotes (CT-type). Such polymorphism may affect the gene expression profiles through distinct DNA methylation patterns.

Pericentromeric satellite regions are considered to be one of the specific targets of DNMT3B, since Dnmt3B−/− mice lack DNA methylation in such regions and die in utero [6]. DNA hypomethylation in pericentromeric satellite regions is known to result in centromeric decondensation and enhanced chromosome recombination. In fact, germline mutations of the DNMT3B gene have been reported in patients with immunodeficiency, centromeric instability, and facial anomalies (ICF) syndrome, a rare recessive autosomal disorder characterized by DNA hypomethylation of pericentromeric satellite regions [69]. In HCCs [70] and UCs [71], DNA hypomethylation of these regions is correlated with copy number alterations on chromosomes 1 and 9, respectively, where satellite regions are rich. The major splice variant of DNMT3B in normal liver tissue samples is DNMT3B3, which possesses the conserved catalytic domains. DNMT activity of human DNMT3B3 has been confirmed in vitro [72]. On the other hand, DNMT3B4 lacks the conserved catalytic domains, although it retains the N-terminal domain required for targeting to heterochromatin sites. Samples of normal liver tissue show only a trace level of DNMT3B4 expression. The levels of DNMT3B4 mRNA expression and the ratio of DNMT3B4 mRNA to DNMT3B3 in samples of non-cancerous liver tissue obtained from patients with HCCs, and in HCCs themselves, are significantly correlated with the degree of DNA hypomethylation in pericentromeric satellite regions [73]. DNA demethylation on satellite 2 has been observed in DNMT3B4-transfected human epithelial 293 cells [73]. Since DNMT3B4 lacking DNMT activity competes with DNMT3B3 for targeting to pericentromeric satellite regions, DNMT3B4 overexpression may lead to chromosomal instability through induction of DNA hypomethylation in such regions.

As another molecular mechanism involved in site-specific DNA methylation alterations, interaction between DNMT3A and c-myc has been reported. This interaction promotes the site-specific methylation of CpG dinucleotides localized in c-myc boxes in the promoter regions of the CDKN2a, CCND1 and TIMP2 genes [74]. The invalidation of c-myc reveals that c-myc allows recruitment of DNMT 3A on the c-myc box of c-myc-regulated genes. Monitoring transcription factor arrays have identified transcription factors interacting with DNMT3A and DNMT3B (such as CREB and FOS), those interacting with DNMT 3A (such as AP2alpha and p53) and those interacting with DNMT 3B (such as SP1 and SP4) [74]. Thus, direct interaction between DNMT 3A and/or DNMT 3B and transcription factors provides a rational molecular explanation for the mechanism of targeted DNA methylation.

35

3.6 SIGNAL PATHWAYS AFFECTING DNA METHYLATION STATUS DURING TUMORIGENESIS

Molecular links between the major signaling pathways involved in tumorigenesis and epigenetic events have been reported [75]. For example, correlations between the phosphatidylinositol 3-kinase (PI3K)/AKT pathway and epigenetic events in tumorigenesis and progression have been attracting attention. It has been reported that PTEN methylation becomes progressively higher from benign thyroid adenoma to follicular thyroid cancer and to aggressive anaplastic thyroid cancer, which harbors activating genetic alterations in the PI3K/AKT pathway that correspond to a progressively higher prevalence [76]. An association of PTEN methylation with PIK3CA alterations and ras mutations has been reported in thyroid tumors [76]. Aberrant methylation and hence silencing of the PTEN gene, which coexists with activating genetic alterations of the PI3K/AKT pathway, may enhance the signaling of this pathway and contribute to tumor progression.

With regard to BRAF-MEK signaling, BRAF is highly expressed in neurons. Expression of MAP2, a neuron-specific microtubule-associated protein that binds and stabilizes dendritic microtubules, is expressed in cutaneous primary melanomas and inversely associated with melanoma progression. Ectopic expression of MAP2 in metastatic melanoma cells inhibits cell growth by inducing mitotic spindle defects and apoptosis [77]. Levels of MAP2 promoter activity in melanoma cell lines are correlated with activating mutation in BRAF: hyperactivation of BRAF-MEK signaling activates MAP2 expression in melanoma cells through promoter demethylation or down-regulation of the neuronal transcription repressor HES1 [77]. Thus, BRAF oncogene levels can regulate the neuronal differentiation and tumor progression of melanoma. Genome-wide DNA methylation analysis after shRNA knockdown of BRAF V600E in thyroid cancer cells has revealed numerous methylation targets including hyper- or hypo-methylated genes with metabolic and cellular functions [78]. Among such genes, the HMGB2 gene plays a role in thyroid cancer cell proliferation, and the FDG1 gene in cell invasion [78]. A prominent epigenetic mechanism through which BRAF V600E can promote tumorigenesis is alteration of the expression of numerous important genes through DNA methylation alterations.

The Ras signaling pathway also regulates DNA methylation status. Forced expression of a cDNA encoding human GAP120 (hGAP), a down-modulator of Ras activity, or delta 9-Jun, a transdominant negative mutant of Jun, in adrenocortical tumor Y1 cells causes transformed cells to revert to their original morphology, resulting in a reduced level of DNA methylation through a reduction of both mRNA expression and the enzymatic activity of DNMTs [79]. Introduction of oncogenic Ha-ras into GAP transfectants has been found to increase the levels of DNA methylation and DNMT activity. Moreover, transient transfection CAT assays have demonstrated that the DNMT promoter in Y1 cells is activated by AP-1 and inhibited by down-regulators of Ras signaling [79]. In addition to Y1 cells, it has been reported that over-expression of unmutated Ha-ras in human T cells causes an increase in DNMT expression, and that DNMT is decreased by inhibitory signaling via the ras-MAPK pathway [80].

The apoptosis-promoting protein Par-4 has been shown to be down-regulated in Ras-transformed NIH 3T3 fibroblasts through the Raf/MEK/ERK MAPK pathway. The par-4 promoter is methylated in Ras-transformed cells through a MEK-dependent pathway, and treatment with a DNMT inhibitor restores the levels of both the Par-4 mRNA transcript and protein, suggesting that the Ras-mediated down-regulation of Par-4 occurs through promoter methylation [81]. In fact, it has been revealed that Ras transformation is associated with up-regulation of DNMT1 and DNMT3 expression [81].

3.7 DNA METHYLATION AND HISTONE MODIFICATIONS

DNA methylation determines chromatin configuration and regulates the expression levels of genes in cooperation with histone modifications [82,83]. Covalent histone modifications mark active promoters (methylation of lysine 4 of histone H3 [H3K4] and acetylation of histone H3 lysine 27 [H3K27]), active enhancers (H3K4 methylation, H3K27 acetylation), actively transcribed genes (H3K36 methylation), or heterochromatin regions (H3K9 methylation, H3K27 methylation) [82,83]. When methyl-CpG-binding proteins, such as MeCP2 and MBD2, bind to methylated CpG dinucleotide, their transcriptional repression domain recruits a co-repressor complex containing histone deacetylases (HDACs) [84]. On the other hand, histone methyltransferases, such as G9A and SUV39H1, are required to recruit DNMTs [85].

Transcriptionally repressive chromatin modifications within the promoters of tumor-suppressor genes silenced by DNA methylation are known to resemble the chromatin modifications of these genes in normal embryonic stem cells, e.g. polycomb (PcG) complex binding and H3K27 methylation. These genes also have an active marker, H3K4 methylation, in normal stem cells, and this bivalent state is converted to a primary active or repressive chromatin conformation after differentiation cues have been received [86]. During carcinogenesis, such modifications may render the genes vulnerable to errors, resulting in aberrant DNA methylation. These PcG complexes have been shown to directly interact with DNMTs, and possibly to promote cancer-specific gene silencing. EZH2, the PcG proteins in the polycomb repressive complex 2/3 (PRC2/3) that catalyzes the trimethylation of H3K27, may be a key player [87]. Overexpression of EZH2 is correlated with tumor progression and poorer prognosis in various cancers [88,89]. Depletion of EZH2 in cancer cells leads to growth arrest [90]. CBX7, another PcG protein, is a constituent of PRC1, and has also been shown to read the repressive histone marks, H3K9me3 and H3K27me3 [91]. Similarly to EZH2, CBX7 is able to recruit DNA methylation machinery to gene promoters and facilitate gene silencing during the development of cancers.

It has long been known that individual cancers each consist of heterogeneous cell populations. The recently proposed cancer stem cell hypothesis has emphasized that only certain subpopulations, known as cancer stem cells, cancer-initiating cells or tumor-propagating cells, have tumorigenic potential. These cancer-initiating cells are usually resistant to chemotherapy and radiotherapy, leading to treatment failure. Moreover, they may be capable of forming metastatic foci in distant organs. Despite the existence of such subpopulations, the cancer stem cell hypothesis continues to generate controversy. Since the PcG complex targets similar sets of genes in embryonic stem cells and cancer cells, much effort should be focused on how epigenetic mechanisms participate in the generation of cancer-initiating cells [20,23].

3.8 SUBCLASSIFICATION OF HUMAN CANCERS BASED ON DNA METHYLATION PROFILING

Almost all cancers are heterogeneous diseases composed of distinct clinicopathological subtypes. DNA methylation profiles may, at least partly, represent the molecular basis of each subtype [92,93]. Recently, analysis on a genome-wide scale has become possible using DNA methylation-sensitive restriction enzyme-based or anti-methyl-cytosine antibody affinity techniques that enrich the methylated and unmethylated fractions of genomic DNA [94]. These fractions can then be hybridized to DNA microarrays. Such DNA methylation profiling may provide new insight into disease entities and help to provide more accurate classifications of human cancers [23]. Such subclassification may yield clues for clarification of distinct mechanisms of carcinogenesis in various organs, and identify possible target molecules for prevention and therapy in patients belonging to specific clusters.

For example, progressive accumulation of genetic and epigenetic abnormalities has been best described in colon cancers. Clustering analyses based on either epigenetic (DNA methylation of multiple CpG island promoter regions) profiling or a combination of genetic (mutations of BRAF, KRAS, and p53 and microsatellite instability [MSI]) and epigenetic profiling have revealed distinct molecular signatures. Colon cancers were clustered into CIMP1, CIMP2, and CIMP-negative groups based on DNA methylation data [95]. CIMP1 is characterized by MSI and BRAF mutations and rare KRAS and p53 mutations. CIMP2 is associated with KRAS mutations and rare MSI, BRAF, or p53 mutations. CIMP-negative cases have a high rate of p53 mutation and lower rates of MSI or mutation of BRAF or KRAS. Together, the data show that colon cancers can be grouped into three molecularly distinct disease subclasses [95]. These three groups also differ clinically: CIMP1 and CIMP2 are more often proximal, CIMP1 has a good prognosis because it consists mostly of MSI-high cancers, and CIMP2 has a poor prognosis. Moreover, these groups may have distinct precancerous lesions that can be diagnosed endoscopically, such as serrated adenomas for CIMP1, and villous adenomas for CIMP2.

We focused on renal carcinogenesis and examined the DNA methylation status of C-type CpG islands of multiple tumor-related genes using bisulfite conversion. Even in non-cancerous renal tissue showing no remarkable histological changes obtained from patients with conventional-type clear cell renal cell carcinomas (RCCs), the average number of methylated CpG islands was significantly higher than in normal renal tissue obtained from patients without any primary renal tumor, regardless of patient age [96]. Stepwise accumulation of DNA methylation on CpG islands has been clearly shown to progress from normal renal tissue, to non-cancerous renal tissue showing no remarkable histological changes obtained from patients with RCCs, and to RCCs. Since it has not been possible to observe any histological change in non-cancerous renal tissue obtained from patients with RCCs, and RCCs usually develop from backgrounds without chronic inflammation or persistent viral infection, precancerous conditions in the kidney have been rarely described. However, from the viewpoint of altered DNA methylation, we have shown that it is possible to recognize the presence of precancerous conditions even in the kidney [96]. In other words, regional DNA methylation alterations may participate in the early and precancerous stage of multistage renal carcinogenesis. Surprisingly, the average number of methylated CpG islands in non-cancerous renal tissues obtained from patients with RCCs showing higher histological grades was significantly higher than that in equivalent tissue obtained from patients with low-grade RCCs, suggesting that precancerous conditions showing regional DNA hypermethylation may generate more malignant RCCs [96].

In order to further clarify the significance of DNA methylation alterations during renal carcinogenesis, we performed genome-wide DNA methylation analysis using BAC array-based methylated CpG island amplification (BAMCA), which may be suitable, not for focusing on specific promoter regions or individual CpG sites, but for overviewing the DNA methylation tendency of individual large regions among all chromosomes [92,93], in tissue samples. The average numbers of BAC clones showing DNA hypo- or hypermethylation in non-tumorous renal tissue obtained from patients with chromophobe RCCs and oncocytomas were significantly lower than the average number in non-tumorous renal tissue obtained from patients with clear cell RCCs [97]. In non-tumorous renal tissue from all examined patients with renal tumors (clear cell RCCs, papillary RCCs, chromophobe RCCs and oncocytomas), biphasic accumulation of DNA methylation alterations was evident. Among such patients, the recurrence-free survival rate of patients showing DNA hypo- or hypermethylation on more BAC clones in their non-tumorous renal tissue was significantly lower than that of patients showing DNA hypo- or hypermethylation on fewer BAC clones [97]. Significant DNA methylation profiles determining the histological subtype (chromophobe RCCs and oncocytomas vs clear cell RCCs) of future developing renal tumors and/or patient outcome (favorable outcome vs poorer outcome) may already be established at the precancerous stage.

We performed two-dimensional unsupervised hierarchical clustering analysis based on the genome-wide DNA methylation status (signal ratios obtained by BAMCA) of samples of non-cancerous renal tissue. On the basis of the DNA methylation profiles of these samples, the patients with clear cell RCCs were clustered into two subclasses, Clusters KA_N and KB_N [98]. The corresponding clear cell RCCs of patients in Cluster KB_N showed more frequent macroscopically evident multinodular growth, vascular involvement and renal vein tumor thrombi, and higher pathological tumor-node-metastasis (TNM) stages than those in Cluster KA_N. Our Clusters KA_N and KB_N in precancerous tissue can be considered clinicopathologically valid: the overall survival rate of patients in Cluster KB_N was significantly lower than that of patients in Cluster KA_N. DNA methylation alterations at the precancerous stage may even determine the outcome of patients with clear cell RCCs.

Two-dimensional unsupervised hierarchical clustering analysis based on BAMCA data for clear cell RCCs themselves was able to group patients into two subclasses, Clusters KA_T and KB_T [98]. Clear cell RCCs in Cluster KB_T showed more frequent vascular involvement and renal vein tumor thrombi, and also higher pathological TNM stages than those in Cluster KA_T. The overall survival rate of patients in Cluster KB_T was significantly lower than that of patients in

Cluster KA_T. Multivariate analysis revealed that our clustering was a predictor of recurrence and was independent of histological grade, macroscopic configuration, vascular involvement or presence of renal vein tumor thrombi.

When we compared the DNA methylation profiles of non-cancerous renal tissue and those of the corresponding clear cell RCC, Cluster KB_N was completely included in Cluster KB_T. BAC clones, of which DNA methylation status significantly discriminated Cluster KB_N from Cluster KA_N, also discriminated Cluster KB_T from Cluster KA_T without exception. When we examined each of the representative BAC clones characterizing both Clusters KB_N and KB_T, the BAMCA signal ratio in the non-cancerous renal tissue was at almost the same level as that in the corresponding clear cell RCC developing in each individual patient [98]. Accordingly, we concluded that the genome-wide DNA methylation profiles of non-cancerous renal tissue are basically inherited by each corresponding clear cell RCC [99].

The average number of examined methylated C-type CpG islands was significantly higher in Cluster KB_T than in Cluster KA_T. The frequency of CIMP in Cluster KB_T was significantly higher than that in Cluster KA_T. Genome-wide DNA methylation alterations consisting of both hypo- and hypermethylation of DNA revealed by BAMCA in Cluster KB_T are associated with regional DNA hypermethylation on CpG islands. Moreover, a subclass of Cluster KB_N and KB_T based on BAMCA data showed particularly marked accumulation of copy number alterations [100]: specific DNA methylation profiles at the precancerous stage may be closely related to, or may be prone to, chromosomal instability. DNA methylation alterations in precancerous conditions, which do not occur randomly but are prone to further accumulation of epigenetic and genetic alterations, can generate more malignant cancers and even determine the outcome of individual patients [92] (Figure 3.1).

With respect to urothelial carcinogenesis, unsupervised hierarchical clustering of UCs based on array comparative genomic hybridization (CGH) data clustered UCs into three subclasses, Clusters UA, UB_1, and UB_2 [101] (Figure 3.2). In Cluster UA, copy number alterations, especially chromosomal gains, revealed by array CGH analysis, and DNA

39

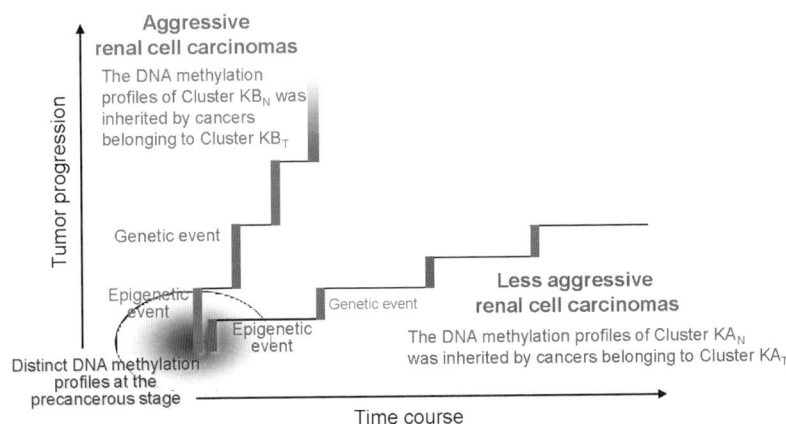

FIGURE 3.1

DNA methylation profiles in precancerous conditions and renal cell carcinomas (RCCs). Two-dimensional unsupervised hierarchical clustering analysis based on BAC array-based methylated CpG island amplification (BAMCA) data for non-cancerous renal tissue samples clustered patients with clear cell RCCs into two subclasses, Clusters KA_N and KB_N [98]. On the basis of the DNA methylation profiles of clear cell RCCs themselves, the patients with clear cell RCCs were divided into Clusters KA_T and KB_T [98]. Patients with more malignant RCCs and showing a poorer outcome were accumulated in Clusters KB_N and KB_T. The DNA methylation profile of Cluster KB_N was inherited by patients with RCCs belonging to Cluster KB_T. Regional DNA hypermethylation of C-type CpG islands and copy number alterations were accumulated in Cluster KB_T. DNA methylation alterations in precancerous conditions, such as the DNA methylation profile corresponding to Cluster KB_N, may be prone to further accumulation of epigenetic and genetic alterations, thus generating more malignant cancers, such as the RCCs in patients belonging to Cluster KB_T. This figure is reproduced in the color plate section.

FIGURE 3.2

Hierarchical clustering analysis of urothelial carcinomas (UCs) based on array comparative genomic hybridization (CGH) data. In Cluster UA, copy number alterations, especially chromosomal gains, revealed by array CGH analysis, and DNA hypomethylation revealed by BAMCA were both accumulated in a genome-wide manner [109]. Cluster UB$_1$ showed accumulation of regional DNA hypermethylation on C-type CpG islands [109]. In Cluster B$_2$, the number of BAC clones showing both DNA hypo- and hypermethylation by BAMCA was rather high, and the number of probes showing loss or gain by array CGH was rather low, in comparison to Cluster UB$_1$ [109]. Genetic and epigenetic events appear to accumulate in a complex manner during the developmental stage of individual tumors. This figure is reproduced in the color plate section.

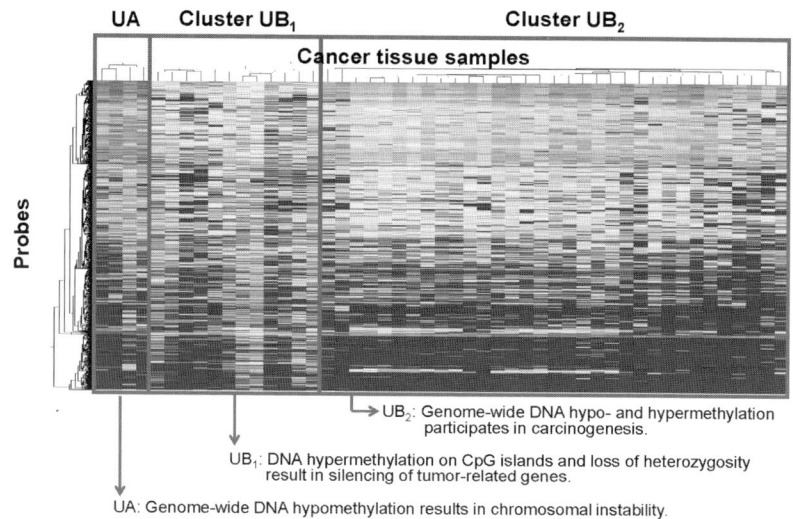

UB$_2$: Genome-wide DNA hypo- and hypermethylation participates in carcinogenesis.

UB$_1$: DNA hypermethylation on CpG islands and loss of heterozygosity result in silencing of tumor-related genes.

UA: Genome-wide DNA hypomethylation results in chromosomal instability.

hypomethylation revealed by BAMCA, were both accumulated in a genome-wide manner, suggesting that DNA hypomethylation may result in chromosomal instability through changes in chromatin configuration and enhancement of chromosomal recombination [101]. Cluster UB1 showed accumulation of regional DNA hypermethylation on C-type CpG islands. Silencing of tumor-related genes due to DNA hypermethylation and chromosomal losses may be critical for the development of UCs belonging to Cluster UB1 [101]. In Cluster UB2, the number of BAC clones shown by BAMCA to have both DNA hypo- and hypermethylation was rather high, and the number of probes shown by array CGH to have loss or gain was rather low, in comparison to Cluster UB1 [101]. In addition to copy number alterations, genome-wide DNA methylation alterations may also participate in the development of UCs belonging to Cluster UB$_2$. Taken together, the data suggest that genetic and epigenetic events accumulate in a complex manner during the developmental stage of individual UCs (Figure 3.2).

3.9 DIAGNOSIS OF CANCERS IN BODY FLUIDS AND BIOPSY SPECIMENS BASED ON DNA METHYLATION PROFILES

The incidence of DNA methylation alterations is generally high in human cancers derived from various organs. Therefore, DNA methylation alterations are applicable as biomarkers for early diagnosis of patients with cancers [102]. Cancer diagnosis based on DNA methylation alterations was initially attempted using body fluids, such as urine, that can be collected non-invasively. For example, DNA hypermethylation of regulatory sequences at the GSTP1 gene locus is present in the majority of primary prostate carcinomas, but not in normal prostatic tissue or other normal tissues. Matched specimens of primary tumor, peripheral blood lymphocytes, and simple voided urine were collected from patients with prostate cancers at various clinical stages, and the DNA methylation status of GSTP1 was examined using methylation-specific PCR [102]. Decoding of the results indicated that urine from prostate cancer patients contained shed cancer cells or debris. Furthermore, there was no case where urine-sediment DNA harbored methylation when the corresponding tumor was negative, suggesting the feasibility of molecular diagnosis using DNA methylation status as an indicator of prostatic cancer cells in urine [103].

Quantitative analysis has been introduced for cancer diagnosis based on DNA methylation alterations. For example, quantitative fluorogenic real-time PCR assay has been used to examine primary tumor DNA and urine sediment DNA from patients with UCs of the urinary bladder for promoter hypermethylation of multiple genes in order to identify potential biomarkers for bladder cancer [104]. The promoter methylation pattern in urine generally matched that in the primary tumors. A selected gene panel including CDKN2A, MGMT, and GSTP1 was validated in urine-sediment DNA samples from an additional validation cohort of patients with UCs of various stages and grades, and from additional age-matched control subjects [104]. Testing of such a gene panel using quantitative methylation-specific PCR assay has been shown to be a powerful non-invasive approach for detection of cancers.

DNA methylation may become an alternative biomarker which can compensate for the demerits of conventional diagnostic techniques. Gastrointestinal endoscopy followed by pathological diagnosis of biopsy specimens is useful for diagnosis of stomach cancers. However, the diagnostic power depends on the technical skill of the endoscopist. Endoscopic biopsy is a topical procedure whereby only a small portion of the lesion is removed. Moreover, gastrointestinal endoscopy is neither comfortable nor risk-free for patients, and is associated with frequent morbidity. Therefore, a method for sensitive and specific detection of early gastric cancer has been established using DNA methylation analysis of gastric washes [105]. This revealed a close correlation between the DNA methylation level of the MINT 25 locus in tumor biopsy specimens and that in gastric washes. MINT25 methylation had high sensitivity, specificity, and area under the receiver operating characteristic curve for tumor cell detection in gastric washes [105]. In addition, even when compared with potential protein or mRNA biomarkers in gastric washes, DNA methylation in such samples may be optimal because of its stability and amplifiability.

In general, pancreatic biopsy yields only a small amount of tissue, and in specimens of pancreatic juice the cellular morphology is not well preserved due to degeneration. We applied the BAMCA method to normal pancreatic tissue obtained from patients without ductal adenocarcinomas, non-cancerous pancreatic tissue obtained from patients with ductal adenocarcinomas, and cancerous tissue. The results of BAMCA for normal pancreatic tissue samples reflected the DNA methylation profiles of normal peripheral pancreatic duct epithelia (the origin of ductal adenocarcinomas), acinar cells and islet cells. In samples of non-cancerous pancreatic tissue obtained from patients with ductal adenocarcinomas, BAMCA revealed DNA hypo- or hypermethylation on many BAC clones in comparison to normal pancreatic tissue samples. Microscopic observation of non-cancerous pancreatic tissue samples obtained from patients with ductal adenocarcinomas revealed lymphocytes and fibroblasts associated with various degrees of chronic pancreatitis, which is considered to be one of the precancerous conditions for ductal adenocarcinomas (Figure 3.3). Our previous studies using microdissection and immunohistochemistry revealed accumulation of DNA hyper-methylation of tumor-related genes associated with DNMT1 overexpression, even in peripheral pancreatic duct epithelia at the precancerous stage [48,49]. Therefore, the results of BAMCA for samples of non-cancerous pancreatic tissue from patients with ductal adenocarcinomas may reflect the DNA methylation profiles of peripheral pancreatic duct epithelia at the precancerous stage, lymphocytes, fibroblasts, acinar cells, and islet cells. In order to diagnose ductal adenocarcinomas in tissue samples, cancer-specific DNA methylation profiles should be discriminated from those of normal or precancerous peripheral pancreatic duct epithelia, lymphocytes, fibroblasts, acinar cells, and islet cells. Therefore, we identified 12 BAC clones whose DNA methylation status was able to discriminate cancerous tissue samples from both normal pancreatic tissue and non-cancerous pancreatic tissue samples obtained from patients with ductal adenocarcinomas in the learning cohort with a specificity of 100%. Using the criteria that combined these 12 BAC clones, cancerous tissues were precisely diagnosed with 100% sensitivity and specificity in both the learning and validation cohorts [106]. Our diagnostic criteria may be advantageous for supporting the histological and cytological assessment of pancreatic cancers (Figure 3.3).

41

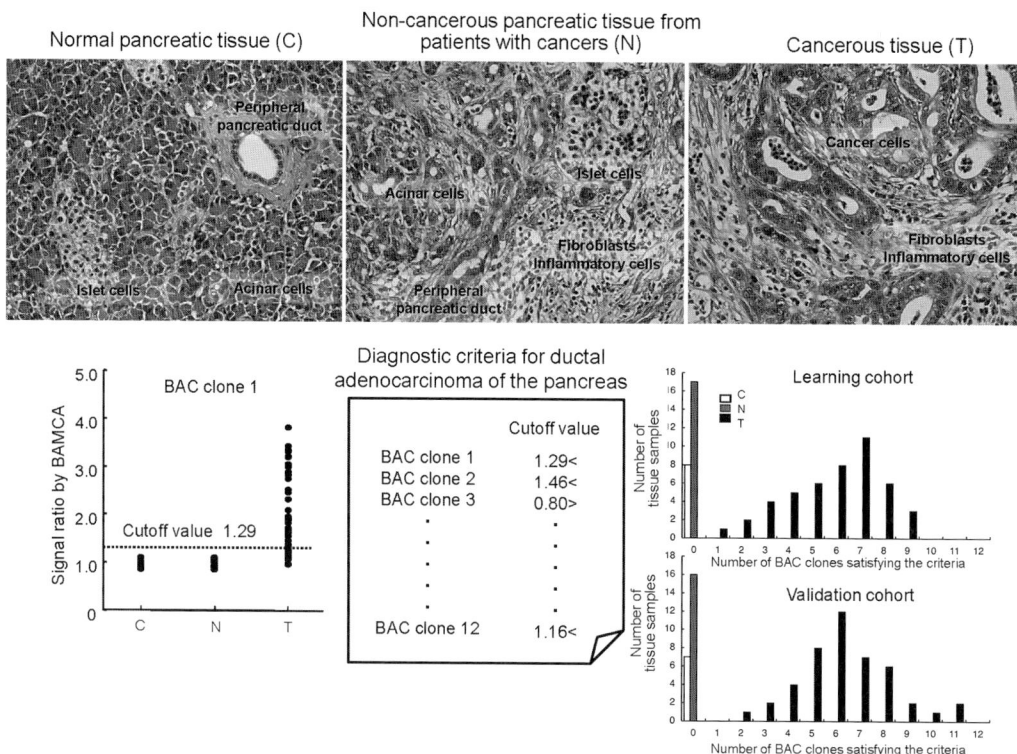

FIGURE 3.3

Diagnostic criteria based on DNA methylation profiles for ductal adenocarcinomas of the pancreas. In order to diagnose ductal adenocarcinomas in tissue samples, cancer-specific DNA methylation profiles should be discriminated from those of normal and precancerous peripheral pancreatic duct epithelia, inflammatory cells, fibroblasts, acinar cells, and islet cells. Therefore, we identified 12 BAC clones for which the DNA methylation status was able to discriminate cancerous tissue (T) samples from both normal pancreatic tissue obtained from patients without pancreatic cancers (C) and non-cancerous pancreatic tissue from patients with cancers (N). Using criteria that combined these 12 BAC clones, cancerous tissue samples were precisely diagnosed with 100% sensitivity and specificity in both the learning and validation cohorts [106]. In general, pancreatic biopsy yields only a small amount of tissue, and in specimens of pancreatic juice the cellular morphology is not well preserved due to degeneration. Our diagnostic criteria may be advantageous for supporting the histological and cytological assessment of such specimens. This figure is reproduced in the color plate section.

3.10 CARCINOGENETIC RISK ESTIMATION BASED ON DNA METHYLATION PROFILES

DNA methylation alterations play a role even in the early and precancerous stage during multistage carcinogenesis. Since even subtle alterations of DNA methylation profiles at the precancerous stage are stably preserved on DNA double strands by covalent bonds, they may be better indicators for risk estimation than mRNA and protein expression profiles, which can be easily affected by the microenvironment of precursor cells. Personalized prevention by elimination of inflammatory conditions, viruses, and other microorganisms, together with prohibition of smoking, which causes DNA methylation alterations, may be applicable to patients with precancerous conditions.

Since HCC usually develops in liver already affected by chronic hepatitis or liver cirrhosis associated with HBV and/or HCV infection, the prognosis of patients with HCC is deemed poor unless the cancer is diagnosed at an early stage. Therefore, surveillance at the precancerous stage becomes a priority. In clinical practice, especially intensive surveillance should be performed for patients at high risk of HCC development, even if the patients are asymptomatic. Therefore, we applied the BAMCA method to samples of liver tissue. Wilcoxon test showed that 25 BAC clones, whose DNA methylation status was inherited by HCCs from non-cancerous liver tissue in patients with HCCs, were able to discriminate such non-cancerous liver tissue from normal liver tissue obtained from patients without HCCs. The criteria for carcinogenetic risk estimation that combined the 25 BAC clones allowed diagnosis of non-cancerous liver tissue from patients with HCCs in the learning cohort as being at high risk of carcinogenesis with 100% sensitivity and specificity [107]. In the validation cohort, these criteria allowed such discrimination with 96% sensitivity and specificity [107]. In patients with HCCs, there were no significant differences in DNA methylation status in these 25 BAC clones between samples of non-cancerous liver tissue showing chronic hepatitis and those showing cirrhosis, indicating that the criteria we employed were not associated with inflammation or fibrosis. In addition, the average number of BAC clones satisfying these criteria was significantly lower in liver tissue from patients with HBV or HCV infection but without HCCs than in non-cancerous liver tissue from patients with HCCs. DNA methylation status in these 25 BAC clones does not simply depend on hepatitis virus infection but may actually reflect the risk of carcinogenesis itself. Therefore, our criteria not only discriminate non-cancerous liver tissue from patients with HCCs from normal liver tissues, but may be capable of discriminating patients who may or may not develop HCCs from among those who are being followed up for HBV or HCV infections, chronic hepatitis, or cirrhosis.

Next, to precisely identify the CpG sites having the largest diagnostic impact on each of the 25 BAC clones and to improve the sensitivity and specificity of carcinogenetic risk estimation, we quantitatively evaluated the DNA methylation status of 203 Sma I sites on these 25 BAC clones using highly quantitative pyrosequencing of tissue specimens. In order to overcome PCR bias, we optimized the PCR conditions for each pyrosequencing primer set. It was revealed that 30 regions including 45 CpG sites had the largest diagnostic impact. Using these 30 regions, we then established criteria revised on the basis of pyrosequencing for estimation of carcinogenetic risk [108]. The revised criteria allowed diagnosis of all samples of non-cancerous liver tissue obtained from HCC patients in the validation cohort as being at high risk of carcinogenesis, with improved sensitivity and specificity [108]. It is feasible that only one CpG site in the promoter region was included in the revised criteria, because DNA methylation status in genomic regions, which do not directly participate in gene silencing, may be altered at the precancerous stage before alterations in the promoter regions themselves occur. Many CpG sites with evident diagnostic impact are located within non-CpG islands, gene bodies, and non-coding regions that have been overlooked as DNA methylation biomarkers. Meticulous examination of such regions may be important for identifying optimal indicators of carcinogenetic risk.

During the surveillance period, in order to clarify the baseline liver histology, liver biopsy is performed in patients with HBV or HCV infection prior to interferon therapy. Therefore, carcinogenetic risk estimation using such liver biopsy specimens will be advantageous for close follow-up of patients who are at high risk of HCC development. We have confirmed that carcinogenetic risk estimation using pyrosequencing is applicable to routine formalin-fixed, paraffin-embedded liver biopsy specimens. Our next step is to validate the reliability of such risk estimation prospectively using liver biopsy specimens obtained prior to interferon therapy from a large cohort of patients with HBV or HCV infection.

As mentioned above, UC is clinically remarkable because of its multicentricity due to the "field effect". Even non-cancerous urothelia showing no remarkable histological changes obtained from patients with UCs can be considered to be at the precancerous stage, because they may have been exposed to carcinogens in the urine. In fact, principal component analysis based on BAMCA data have revealed progression of DNA methylation alterations from normal urothelia to non-cancerous urothelia obtained from patients with UCs, and to UCs themselves. Unsupervised hierarchical clustering analysis of patients with UCs based on the DNA methylation status of their non-cancerous urothelia showed that the DNA methylation profiles of non-cancerous urothelia were significantly correlated with the invasiveness of UCs developing in individual patients, suggesting that DNA methylation alterations at the precancerous stage may generate more malignant cancers [109]. The combination of DNA methylation status on 83 BAC clones was able to completely discriminate between non-cancerous urothelia from patients with UCs and normal urothelia, and allowed diagnosis of non-cancerous urothelia from patients with UCs as having a high risk of carcinogenesis, with 100% sensitivity and specificity [109]. Differences in DNA methylation profiles between muscle-invasive UCs and non-invasive UCs have also been extensively examined: frequent DNA hypermethylation of the HOXB2 [110] and RASSF1A [111] genes is known to be associated with invasiveness of UCs.

3.11 PERSONALIZED MEDICINE BASED ON DNA METHYLATION PROFILES: PROGNOSTICATION OF PATIENTS WITH CANCERS AND PREDICTION OF RESPONSE TO CHEMOTHERAPY

Since DNA methylation alterations frequently correlate with clinicopathological parameters of cancers, they can be used as prognostic indicators in patients with cancers. For example, based on BAMCA data, 41 BAC clones, whose DNA methylation status was able to discriminate HCC patients who survived more than 4 years after hepatectomy from patients who suffered recurrence within 6 months and died within a year after hepatectomy, have been identified [107]. The DNA methylation status of these 41 BAC clones was correlated with the cancer-free survival rate of HCC patients in the validation cohort. Prognostication based on our criteria may be promising for supportive use during follow-up after surgical resection, since multivariate analysis revealed that our criteria are able to predict overall patient outcome independently of parameters observed in hepatectomy specimens, such as the degree of histological differentiation, presence of portal vein tumor thrombi, intrahepatic metastasis and multicentricity, which are already known to have a prognostic impact. Such prognostication using liver biopsy specimens obtained before transarterial embolization, transarterial chemoembolization, and radiofrequency ablation may be advantageous even for patients who undergo such therapies.

Even when surgery is performed with curative intent for patients with pancreatic cancers, the rate of recurrence is very high. Although previous studies have suggested the efficacy of adjuvant chemotherapy, it needs to be carried out carefully, paying close attention to adverse reactions. In order to decide the indications for such adjuvant chemotherapy, prognostic criteria should be explored. We have identified 11 BAC clones whose DNA methylation status was able to discriminate patients showing early relapse from those without relapse in the learning cohort with 100% specificity, and this was correlated with the recurrence-free and overall survival rates in the validation cohort [106]. Multivariate analysis revealed that satisfying the prognostic criteria using these 11 BAC clones was a parameter independent of surgical margin positivity and lymph node metastasis at the time of surgery [106].

The quality of life of patients with urinary bladder cancers is generally poor after total cystectomy. In general, therefore, after therapeutic diagnosis of UC tumors obtained by transurethral resection, patients are followed-up by repeat cystoscopy examinations. In patients showing sudden prominent malignant progression, it is difficult to determine the appropriate timing of total cystectomy. Therefore, prognostic indicators need to be explored. The combination of DNA methylation status on 20 BAC clones selected by Wilcoxon test was able to completely discriminate patients who suffered recurrence after surgery from patients who did not [109]. DNA methylation profiling may thus provide optimal indicators for prognostication in patients with UCs. Other recently published DNA methylation alterations in

TABLE 3.1 DNA methylation alterations in human cancers that are correlated with the outcome of patients and can be used as prognostic indicators

Tumor	Gene	DNA Methylation Status	References
Glioma	MGMT	DNA hypermethylation	PLoS One 2011; 6: e23332
			J Neurooncol 2011; 102: 311–16
	LINE-1	DNA hypomethylation	PLoS One 2011; 6: e23332
	POTEH	DNA hypomethylation	Brain Res 2011; 1391: 125–31
Head and neck cancer	miRNA-137	DNA hypermethylation	Cancer 2011; 117: 1454–62
	ESR1	DNA hypermethylation	Clin Epigenetics 2010; 1: 61–69
	HIC1	DNA hypermethylation	Clin Epigenetics 2010; 1: 61–69
	LATS2	DNA hypomethylation	BMC Cancer 2010; 10: 538
	p16	DNA hypermethylation	Oral Oncol 2010; 46: 734–9
Salivary gland cancer	RUNX3	DNA hypermethylation	Cancer Sci 2011; 102: 492–7
Lung cancer	RASSF1A	DNA hypermethylation	Carcinogenesis 2011; 32: 411–16.
Esophageal cancer	p 14	DNA hypermethylation	J Clin Pathol 2011; 64: 246–51
	p 15	DNA hypermethylation	J Clin Pathol 2011; 64: 246–51
	p 16	DNA hypermethylation	J Clin Pathol 2011; 64: 246–51
	p 21	DNA hypermethylation	J Clin Pathol 2011; 64: 246–51
	p 27	DNA hypermethylation	J Clin Pathol 2011; 64: 246–51
	p 57	DNA hypermethylation	J Clin Pathol 2011; 64: 246–51
	p 73	DNA hypermethylation	J Clin Pathol 2011; 64: 246–51
	PAX6	DNA hypermethylation	Ann Surg Oncol 2011; 18: 1185–94
	ENST00000363328	DNA hypermethylation	Ann Surg Oncol 2011; 18: 1185–94
Stomach cancer	Claudin-4	DNA hypomethylation	Lab Invest 2011; 91: 1652–67
	BNIP3	DNA hypermethylation	Oncol Rep 2011; 25: 513–18
	DAPK	DNA hypermethylation	Oncol Rep 2011; 25: 513–18
	S100A6	DNA hypomethylation	Am J Pathol 2010; 177: 586–97
	EphA1	DNA hypermethylation	Oncol Rep 2010; 24: 1577–84
Colorectal cancer	fibulin-3b	DNA hypermethylation	Neoplasma 2011; 58: 441–8
	p16	DNA hypermethylation	Anticancer Res 2011; 31: 1643–6
			Oncol Rep 2011; 25: 789–94
	LINE-1	DNA hypomethylation	Cancer 2011; 117: 1847–54
	RASSF1A	DNA hypermethylation	J Cell Physiol 2011; 226: 1934–9
	SFRP2	DNA hypermethylation	Clin Invest Med 2011; 34: E88–95
	DSC3	DNA hypermethylation	Br J Cancer 2011; 104: 1013–19
	IGFBP3	DNA hypermethylation	Clin Cancer Res 2011; 17: 1535–45
	EVL	DNA hypermethylation	Clin Cancer Res 2011; 17: 1535–45
	hMLH1	DNA hypermethylation	Oncol Rep 2011; 25: 789–94
	PPARG	DNA hypermethylation	PLoS One 2010; 5: e14229
	MGMT	DNA hypermethylation	Cancer Causes Control 2011; 22: 301–9
	IGF2	DNA hypomethylation	Gastroenterology 2010; 139: 1855–64
	RARβ2	DNA hypermethylation	Tumour Biol 2010; 31: 503–11
Gastrointestinal stromal tumor	REC8	DNA hypermethylation	Gut 2012; 61: 392–401
	PAX3	DNA hypermethylation	Gut 2012; 61: 392–401
	p16	DNA hypermethylation	Gut 2012; 61: 392–401
Hepatocellular carcinoma	RASSF1A	DNA hypermethylation	Asian Pac J Cancer Prev 2010; 11: 1677–81
	CADM1	DNA hypermethylation	Oncol Rep 2011; 25: 1053–62
	WIF-1	DNA hypermethylation	Tumour Biol 2011; 32: 233–40
	RELN	DNA hypermethylation	Ann Surg Oncol 2011; 18: 572–9
Renal cell carcinoma	HOXA5	DNA hypermethylation	Pathol Int 2010; 60: 661–6
	MSH2	DNA hypermethylation	Pathol Int 2010; 60: 661–6
	hsa-miR-9	DNA hypermethylation	Oncogene 2010; 29: 5724–8
Neuroblastoma	CASP8	DNA hypermethylation	Mol Carcinog 2011; 50: 153–62
	TMS1	DNA hypermethylation	Mol Carcinog 2011; 50: 153–62
	APAF1	DNA hypermethylation	Mol Carcinog 2011; 50: 153–62

45

Continued

TABLE 3.1 —continued

Tumor	Gene	DNA Methylation Status	References
Breast cancer	Endoglin	DNA hypermethylation	Oncogene 2011; 30: 1046—58
	RASSF1A	DNA hypermethylation	Breast Cancer Res Treat 2011; 129: 1—9
	CDO1	DNA hypermethylation	BMC Cancer 2010; 10: 247
Cervical cancer	APC1A	DNA hypermethylation	Int J Oncol 2011; 39: 683—8
Endometrioid cancer	CDH1	DNA hypermethylation	Cancer Invest 2011; 29: 86—92
Ovarian cancer	GREB1	DNA hypomethylation	Oncology 2011; 80: 12—20
	TGIF	DNA hypomethylation	Oncology 2011; 80: 12—20
	TOB1	DNA hypomethylation	Oncology 2011; 80: 12—20
	TMCO5	DNA hypermethylation	Oncology 2011; 80: 12—20
	PTPRN	DNA hypermethylation	Oncology 2011; 80: 12—20
	GUCY2C	DNA hypermethylation	Oncology 2011; 80: 12—20
	HERV-K	DNA hypomethylation	Int J Gynecol Cancer 2011; 21: 51—7
Trophoblastic tumor	ASPP1	DNA hypermethylation	Mod Pathol 2011; 24: 522—32
Melanoma	LINE-1	DNA hypomethylation	J Transl Med 2011; 9: 78
Acute myeloid leukemia	CEBPA	DNA hypermethylation	Leukemia 2011; 25: 32—40
Multiple myeloma	p16	DNA hypermethylation	Ann Hematol 2011; 90: 73—9

human cancers that are correlated with patient outcome and can be used as prognostic indicators are summarized in Table 3.1.

In addition, DNA methylation profiles may be predictive indicators of response to chemotherapy. One such example is silencing of the mitotic checkpoint gene CHFR (checkpoint with forkhead and ring finger domains) in gastric cancers. Mitotic checkpoints prevent errors in chromosome segregation that can lead to neoplasia, and it is notable that gastric cancers often show impaired checkpoint function. CHFR expression was silenced by DNA methylation of the 5′ region of the gene in tested gastric cancer cell lines and primary gastric cancers; expression was restored by treatment with 5-aza-2′-deoxycytidine. In addition, histones H3 and H4 were found to be deacetylated in cell lines showing aberrant methylation. Cells not expressing CHFR showed impaired checkpoint function, leading to nuclear localization of cyclin B1 after treatment with microtubule inhibitors such as docetaxel or paclitaxel. Absence of CHFR appears to be associated with the sensitivity of cells to mitotic stress caused by microtubule inhibition, and restoration of CHFR expression by 5-aza-2′-deoxycytidine or adenoviral gene transfer restores the checkpoint. By affecting mitotic checkpoint function, CHFR inactivation likely plays a key role in gastric cancer tumorigenesis [112]. Moreover, aberrant methylation of CHFR appears to be a good molecular marker with which to predict the sensitivity of gastric cancers to microtubule inhibitors.

Another example is MGMT, a DNA repair protein, which reverses the addition of alkyl groups to the guanine base of DNA. Silencing of MGMT due to DNA methylation in glioma is a useful predictor of response to alkylating agents such as carmustine or temozolomide [113]. Similarly, methylation of a mismatch repair gene, hMLH1, in ovarian and colon cancer cell lines confers chemoresistance to many chemotherapeutic agents. Treatment with a DNA demethylating agent, 5-aza-2′-deoxycytidine, can reactivate hMLH1 and reverse the chemoresistance. Likewise, silencing of APAF-1 (apoptotic peptidase activating factor-1), a proapoptotic gene, confers chemoresistance to melanoma and leukemia cells by mediating resistance to cytochrome c-dependent apoptosis [114]. These findings demonstrate the potential clinical utility of DNA methylation markers for individualized therapy of cancer patients.

3.12 NEW TECHNOLOGIES FOR DNA METHYLATION ANALYSIS AND FUTURE DIRECTIONS

Currently available forms of screening technology, such as single-base-pair resolution whole-genome DNA methylation analysis using second-generation sequencers, and international efforts aimed at determining reference epigenome profiles, are now opening new avenues of epigenome therapy for cancer patients. Although broad DNA methylation profiling was initially performed on the basis of two-dimensional gel electrophoresis, adaptation of microarray hybridization techniques used in gene expression and genome studies to the profiling of DNA methylation patterns opened the door to the era of the epigenome. Enzyme-based and affinity enrichment-based DNA methylation analysis techniques have been proved suitable for examination of human tissue samples using hybridization arrays [115]. Currently available high-throughput DNA sequencing technologies using second-generation sequencers are now capable of single-base-pair resolution for whole-genome DNA methylation analysis. Although projects involving analysis of large numbers of human tissue samples will still rely on array-based approaches for several more years, the trend will be towards bisulfite shotgun sequencing [94]. Nanopore sequencing provides single-molecule detection and avoids any bias introduced by differential amplification of methylation-derived states [116]. Moreover, third-generation sequencers for real-time sequencing can directly detect 5-methylcytosine without bisulfite conversion [117]. In addition, genome-wide analysis of histone modification and non-coding RNA is also being robustly performed. Thus, high-throughput mapping of the epigenome, i.e. an overview of DNA methylation, histone modification, non-coding RNA, and chromatin accessibility in normal, precursor and cancer cells, is now highly reproducible and standardized.

Importantly, changes in the epigenome are potentially reversible by drug treatments. This has significant implications for the prevention and therapy of human cancers. Indeed, several inhibitors of chromatin-modifying enzymes, including DNMT inhibitors, as well as HDAC inhibitors, have been approved by the US Food and Drug Administration and the EU, and are now being used in clinical practice [118,119]. However, to maximize the potential of such therapeutic approaches, a more comprehensive characterization of the epigenome changes that occur during normal development and adult cell renewal should be accomplished by international consortia.

Scientists and representatives of major funding agencies have decided to launch the International Human Epigenome Consortium (IHEC) [120]. Just as the Human Genome Project provided a reference "normal" sequence for studying human disease, high-resolution reference epigenome maps consisting of the various epigenetic layers of detailed DNA methylation as well as histone modification, nucleosome occupancy and corresponding coding, and non-coding RNA expression in different normal cell types will be provided by IHEC. Such a reference human epigenome will be available to the worldwide research community. Information on the methods utilized by IHEC members will be useful for producing large epigenomic datasets related to health and diseases in humans. It will become possible to compare profiles of different human populations, thereby helping to evaluate the impact of environment and nutrition on the epigenome. Epigenome reference maps will have an immediate impact on our understanding of cancers as well as diabetes, cardiopulmonary diseases, neuropsychiatric disorders, imprinting disorders, inflammation, and autoimmune diseases, and will hopefully lead to breakthroughs in the prevention, diagnosis, and therapy of human cancers.

47

References

[1] Issa JP, Just W. Epigenetics. FEBS Lett 2011;585:1993.

[2] Skinner MK. Environmental epigenetic transgenerational inheritance and somatic epigenetic mitotic stability. Epigenetics 2011;6:838—42.

[3] Siedlecki P, Zielenkiewicz P. Mammalian DNA methyltransferases. Acta Biochim Pol 2006;53:245—56.

[4] Cheng X, Blumenthal RM. Mammalian DNA methyltransferases: a structural perspective. Structure 2008;16:341—50.

[5] Hermann A, Gowher H, Jeltsch A. Biochemistry and biology of mammalian DNA methyltransferases. Cell Mol Life Sci 2004;61:2571—87.

[6] Okano M, Bell DW, Haber DA, Li E. DNA methyltransferases Dnmt3a and Dnmt3b are essential for de novo methylation and mammalian development. Cell 1999;99:247—57.

[7] Bourc'his D, Xu GL, Lin CS, Bollman B, Bestor TH. Dnmt3L and the establishment of maternal genomic imprints. Science 2001;294:2536—9.

[8] Deaton AM, Bird A. CpG islands and the regulation of transcription. Genes Dev 2011;25:1010—22.

[9] Weber M, Hellmann I, Stadler MB, Ramos L, Pääbo S, Rebhan M, et al. Distribution, silencing potential and evolutionary impact of promoter DNA methylation in the human genome. Nat Genet 2007;39:457—66.

[10] Irizarry RA, Ladd-Acosta C, Wen B, Wu Z, Montano C, Onyango P, et al. The human colon cancer methylome shows similar hypo- and hypermethylation at conserved tissue-specific CpG island shores. Nat Genet 2009;41:178—86.

[11] Tahiliani M, Koh KP, Shen Y, Pastor WA, Bandukwala H, Brudno Y, et al. Conversion of 5-methylcytosine to 5-hydroxymethylcytosine in mammalian DNA by MLL partner TET1. Science 2009;324:930—5.

[12] Veron N, Peters AH. Epigenetics: Tet proteins in the limelight. Nature 2011;473:293—4.

[13] Burmeister T, Meyer C, Schwartz S, Hofmann J, Molkentin M, Kowarz E, et al. The MLL recombinome of adult CD10-negative B-cell precursor acute lymphoblastic leukemia: results from the GMALL study group. Blood 2009;113:4011—5.

[14] Delhommeau F, Dupont S, Della Valle V, James C, Trannoy S, Masse A, et al. Mutation in TET2 in myeloid cancers. N Engl J Med 2009;360:2289—301.

[15] Langemeijer SM, Aslanyan MG, Jansen JH. TET proteins in malignant hematopoiesis. Cell Cycle 2009;8:4044—8.

[16] Laird PW, Jackson-Grusby L, Fazeli A, Dickinson SL, Jung WE, Li E, et al. Suppression of intestinal neoplasia by DNA hypomethylation. Cell 1995;81:197—205.

[17] Eden A, Gaudet F, Waghmare A, Jaenisch R. Chromosomal instability and tumors promoted by DNA hypo-methylation. Science 2003;300:455.

[18] Gaudet F, Hodgson JG, Eden A, Jackson-Grusby L, Dausman J, Gray JW, et al. Induction of tumors in mice by genomic hypomethylation. Science 2003;300:489—92.

[19] Howard G, Eiges R, Gaudet F, Jaenisch R, Eden A. Activation and transposition of endogenous retroviral elements in hypomethylation induced tumors in mice. Oncogene 2008;27:404—8.

[20] Sharma S, Kelly TK, Jones PA. Epigenetics in cancer. Carcinogenesis 2010;31:27—36.

[21] Jones PA, Baylin SB. The epigenomics of cancer. Cell 2007;128:683—92.

[22] Lewandowska J, Bartoszek A. DNA methylation in cancer development, diagnosis and therapy—multiple opportunities for genotoxic agents to act as methylome disruptors or remediators. Mutagenesis 2011;26:475—87.

[23] Tsai HC, Baylin SB. Cancer epigenetics: linking basic biology to clinical medicine. Cell Res 2011;21:502—17.

[24] Clark SJ. Action at a distance: epigenetic silencing of large chromosomal regions in carcinogenesis. Hum Mol Genet 2007;16(Spec No 1):R88—95.

[25] Saito Y, Jones PA. Epigenetic activation of tumor suppressor microRNAs in human cancer cells. Cell Cycle 2006;5:2220—2.

[26] Li M, Marin-Muller C, Bharadwaj U, Chow KH, Yao Q, Chen C. MicroRNAs: control and loss of control in human physiology and disease. World J Surg 2009;33:667—84.

[27] Kanai Y, Hirohashi S. Alterations of DNA methylation associated with abnormalities of DNA methyl-transferases in human cancers during transition from a precancerous to a malignant state. Carcinogenesis 2007;28:2434—42.

[28] Kanai Y. Alterations of DNA methylation and clinicopathological diversity of human cancers. Pathol Int 2008;58:544—58.

[29] Kanai Y, Ushijima S, Tsuda H, Sakamoto M, Sugimura T, Hirohashi S. Aberrant DNA methylation on chro-mosome 16 is an early event in hepatocarcinogenesis. Jpn J Cancer Res 1996;87:1210—7.

[30] Kondo Y, Kanai Y, Sakamoto M, Mizokami M, Ueda R, Hirohashi S. Genetic instability and aberrant DNA methylation in chronic hepatitis and cirrhosis—A comprehensive study of loss of heterozygosity and micro-satellite instability at 39 loci and DNA hypermethylation on 8 CpG islands in microdissected specimens from patients with hepatocellular carcinoma. Hepatology 2000;32:970—9.

[31] Yoshiura K, Kanai Y, Ochiai A, Shimoyama Y, Sugimura T, Hirohashi S. Silencing of the E-cadherin invasion-suppressor gene by CpG methylation in human carcinomas. Proc Natl Acad Sci USA 1995;92:7416—9.

48

[32] Hirohashi S, Kanai Y. Cell adhesion system and human cancer morphogenesis. Cancer Sci 2003;94:575—81.

[33] Kanai Y, Ushijima S, Hui AM, Ochiai A, Tsuda H, Sakamoto M, et al. The E-cadherin gene is silenced by CpG methylation in human hepatocellular carcinomas. Int J Cancer 1997;71:355—9.

[34] Kanai Y, Hui AM, Sun L, Ushijima S, Sakamoto M, Tsuda H, et al. DNA hypermethylation at the D17S5 locus and reduced HIC-1 mRNA expression are associated with hepatocarcinogenesis. Hepatology 1999;29:703—9.

[35] Kanai Y, Ushijima S, Tsuda H, Sakamoto M, Hirohashi S. Aberrant DNA methylation precedes loss of heterozygosity on chromosome 16 in chronic hepatitis and liver cirrhosis. Cancer Lett 2000;148:73—80.

[36] Etoh T, Kanai Y, Ushijima S, Nakagawa T, Nakanishi Y, Sasako M, et al. Increased DNA methyltransferase 1 (DNMT1) protein expression correlates significantly with poorer tumor differentiation and frequent DNA hypermethylation of multiple CpG islands in gastric cancers. Am J Pathol 2004;164:689—99.

[37] Ryan JL, Jones RJ, Kenney SC, Rivenbark AG, Tang W, Knight ER, et al. Epstein—Barr virus-specific methylation of human genes in gastric cancer cells. Infect Agent Cancer 2010;5:27.

[38] Hino R, Uozaki H, Murakami N, Ushiku T, Shinozaki A, Ishikawa S, et al. Activation of DNA methyltransferase 1 by EBV latent membrane protein 2A leads to promoter hypermethylation of PTEN gene in gastric carcinoma. Cancer Res 2009;69:2766—74.

[39] Maekita T, Nakazawa K, Mihara M, Nakajima T, Yanaoka K, Iguchi M, et al. High levels of aberrant DNA methylation in Helicobacter pylori-infected gastric mucosae and its possible association with gastric cancer risk. Clin Cancer Res 2006;12:989—95.

[40] Sawada M, Kanai Y, Arai E, Ushijima S, Ojima H, Hirohashi S. Increased expression of DNA methyltransferase 1 (DNMT1) protein in uterine cervix squamous cell carcinoma and its precursor lesion. Cancer Lett 2007;251:211—9.

[41] Laurson J, Khan S, Chung R, Cross K, Raj K. Epigenetic repression of E-cadherin by human papillomavirus 16 E7 protein. Carcinogenesis 2010;31:918—26.

[42] Eguchi K, Kanai Y, Kobayashi K, Hirohashi S. DNA hypermethylation at the D17S5 locus in non-small cell lung cancers: its association with smoking history. Cancer Res 1997;57:4913—5.

[43] Chen W, Cooper TK, Zahnow CA, Overholtzer M, Zhao Z, Ladanyi M, et al. Epigenetic and genetic loss of Hic1 function accentuates the role of p53 in tumorigenesis. Cancer Cell 2004;6:387—98.

[44] Wales MM, Biel MA, el Deiry W, Nelkin BD, Issa JP, Cavenee WK, et al. p53 activates expression of HIC-1, a new candidate tumour suppressor gene on 17p13.3. Nat Med 1995;1:570—7.

[45] Sun L, Hui AM, Kanai Y, Sakamoto M, Hirohashi S. Increased DNA methyltransferase expression is associated with an early stage of human hepatocarcinogenesis. Jpn J Cancer Res 1997;88:1165—70.

[46] Saito Y, Kanai Y, Sakamoto M, Saito H, Ishii H, Hirohashi S. Expression of mRNA for DNA methyltransferases and methyl-CpG-binding proteins and DNA methylation status on CpG islands and pericentromeric satellite regions during human hepatocarcinogenesis. Hepatology 2001;33:561—8.

[47] Saito Y, Kanai Y, Nakagawa T, Sakamoto M, Saito H, Ishii H, et al. Increased protein expression of DNA methyltransferase (DNMT) 1 is significantly correlated with the malignant potential and poor prognosis of human hepatocellular carcinomas. Int J Cancer 2003;105:527—32.

[48] Peng DF, Kanai Y, Sawada M, Ushijima S, Hiraoka N, Kosuge T, et al. Increased DNA methyltransferase 1 (DNMT1) protein expression in precancerous conditions and ductal carcinomas of the pancreas. Cancer Sci 2005;96:403—8.

[49] Peng DF, Kanai Y, Sawada M, Ushijima S, Hiraoka N, Kitazawa S, et al. DNA methylation of multiple tumor-related genes in association with overexpression of DNA methyltransferase 1 (DNMT1) during multistage carcinogenesis of the pancreas. Carcinogenesis 2006;27:1160—8.

[50] Robertson KD, Uzvolgyi E, Liang G, Talmadge C, Sumegi J, Gonzales FA, et al. The human DNA methyl-transferases (DNMTs) 1, 3a and 3b: coordinate mRNA expression in normal tissues and overexpression in tumors. Nucleic Acids Res 1999;27:2291—8.

[51] Toyota M, Ahuja N, Ohe-Toyota M, Herman JG, Baylin SB, Issa JP. CpG island methylator phenotype in colorectal cancer. Proc Natl Acad Sci USA 1999;96:8681—6.

[52] Issa JP, Shen L, Toyota M. CIMP, at last. Gastroenterology 2005;129:1121—4.

[53] Kanai Y, Ushijima S, Kondo Y, Nakanishi Y, Hirohashi S. DNA methyltransferase expression and DNA methylation of CPG islands and peri-centromeric satellite regions in human colorectal and stomach cancers. Int J Cancer 2001;91:205—12.

[54] Nakagawa T, Kanai Y, Saito Y, Kitamura T, Kakizoe T, Hirohashi S. Increased DNA methyltransferase 1 protein expression in human transitional cell carcinoma of the bladder. J Urol 2003;170:2463—6.

[55] Nakagawa T, Kanai Y, Ushijima S, Kitamura T, Kakizoe T, Hirohashi S. DNA hypermethylation on multiple CpG islands associated with increased DNA methyltransferase DNMT1 protein expression during multistage urothelial carcinogenesis. J Urol 2005;173:1767—71.

[56] Denis H, Ndlovu MN, Fuks F. Regulation of mammalian DNA methyltransferases: a route to new mechanisms. EMBO Rep 2011;12:647—56.

49

[57] Fabbri M, Garzon R, Cimmino A, Liu Z, Zanesi N, Callegari E, et al. MicroRNA-29 family reverts aberrant methylation in lung cancer by targeting DNA methyltransferases 3A and 3B. Proc Natl Acad Sci USA 2007;104:15805—10.

[58] Garzon R, Liu S, Fabbri M, Liu Z, Heaphy CE, Callegari E, et al. MicroRNA-29b induces global DNA hypomethylation and tumor suppressor gene reexpression in acute myeloid leukemia by targeting directly DNMT3A and 3B and indirectly DNMT1. Blood 2009;113:6411—8.

[59] Duursma AM, Kedde M, Schrier M, le Sage C, Agami R. miR-148 targets human DNMT3b protein coding region. Rna 2008;14:872—7.

[60] Braconi C, Huang N, Patel T. MicroRNA-dependent regulation of DNA methyltransferase-1 and tumor suppressor gene expression by interleukin-6 in human malignant cholangiocytes. Hepatology 2010;51:881—90.

[61] Zhao S, Wang Y, Liang Y, Zhao M, Long H, Ding S, et al. MicroRNA-126 regulates DNA methylation in CD4+ T cells and contributes to systemic lupus erythematosus by targeting DNA methyltransferase 1. Arthritis Rheum 2011;63:1376—86.

[62] Huang J, Wang Y, Guo Y, Sun S. Down-regulated microRNA-152 induces aberrant DNA methylation in hepatitis B virus-related hepatocellular carcinoma by targeting DNA methyltransferase 1. Hepatology 2010;52:60—70.

[63] Lopez de Silanes I, Gorospe M, Taniguchi H, Abdelmohsen K, Srikantan S, Alaminos M, et al. The RNA-binding protein HuR regulates DNA methylation through stabilization of DNMT3b mRNA. Nucleic Acids Res 2009;37:2658—71.

[64] Kanai Y, Ushijima S, Nakanishi Y, Sakamoto M, Hirohashi S. Mutation of the DNA methyltransferase (DNMT) 1 gene in human colorectal cancers. Cancer Lett 2003;192:75—82.

[65] Ley TJ, Ding L, Walter MJ, McLellan MD, Lamprecht T, Larson DE, et al. DNMT3A mutations in acute myeloid leukemia. N Engl J Med 2010;363:2424—33.

[66] Yan XJ, Xu J, Gu ZH, Pan CM, Lu G, Shen Y, et al. Exome sequencing identifies somatic mutations of DNA methyltransferase gene DNMT3A in acute monocytic leukemia. Nat Genet 2011;43:309—15.

[67] Fan H, Liu D, Qiu X, Qiao F, Wu Q, Su X, et al. A functional polymorphism in the DNA methyltransferase-3A promoter modifies the susceptibility in gastric cancer but not in esophageal carcinoma. BMC Med 2010;8:12.

[68] Wang L, Rodriguez M, Kim ES, Xu Y, Bekele N, El-Naggar AK, et al. A novel C/T polymorphism in the core promoter of human de novo cytosine DNA methyltransferase 3B6 is associated with prognosis in head and neck cancer. Int J Oncol 2004;25:993—9.

[69] Hansen RS, Wijmenga C, Luo P, Stanek AM, Canfield TK, Weemaes CM, et al. The DNMT3B DNA methyltransferase gene is mutated in the ICF immunodeficiency syndrome. Proc Natl Acad Sci USA 1999;96:14412—7.

[70] Wong N, Lam WC, Lai PB, Pang E, Lau WY, Johnson PJ. Hypomethylation of chromosome 1 heterochromatin DNA correlates with q-arm copy gain in human hepatocellular carcinoma. Am J Pathol 2001;159:465—71.

[71] Nakagawa T, Kanai Y, Ushijima S, Kitamura T, Kakizoe T, Hirohashi S. DNA hypomethylation on pericentromeric satellite regions significantly correlates with loss of heterozygosity on chromosome 9 in urothelial carcinomas. J Urol 2005;173:243—6.

[72] Soejima K, Fang W, Rollins BJ. DNA methyltransferase 3b contributes to oncogenic transformation induced by SV40T antigen and activated Ras. Oncogene 2003;22:4723—33.

[73] Saito Y, Kanai Y, Sakamoto M, Saito H, Ishii H, Hirohashi S. Overexpression of a splice variant of DNA methyltransferase 3b, DNMT3b4, associated with DNA hypomethylation on pericentromeric satellite regions during human hepatocarcinogenesis. Proc Natl Acad Sci USA 2002;99:10060—5.

[74] Hervouet E, Vallette FM, Cartron PF. Dnmt3/transcription factor interactions as crucial players in targeted DNA methylation. Epigenetics 2009;4:487—99.

[75] Simon JA, Lange CA. Roles of the EZH2 histone methyltransferase in cancer epigenetics. Mutat Res 2008;647:21—9.

[76] Hou P, Ji M, Xing M. Association of PTEN gene methylation with genetic alterations in the phosphatidylinositol 3-kinase/AKT signaling pathway in thyroid tumors. Cancer 2008;113:2440—7.

[77] Maddodi N, Bhat KM, Devi S, Zhang SC, Setaluri V. Oncogenic BRAFV600E induces expression of neuronal differentiation marker MAP2 in melanoma cells by promoter demethylation and down-regulation of transcription repressor HES1. J Biol Chem 2010;285:242—54.

[78] Hou P, Liu D, Xing M. Genome-wide alterations in gene methylation by the BRAF V600E mutation in papillary thyroid cancer cells. Endocr Relat Cancer 2011;18:687—97.

[79] MacLeod AR, Rouleau J, Szyf M. Regulation of DNA methylation by the Ras signaling pathway. J Biol Chem 1995;270:11327—37.

[80] Deng C, Yang J, Scott J, Hanash S, Richardson BC. Role of the ras-MAPK signaling pathway in the DNA methyltransferase response to DNA hypomethylation. Biol Chem 1998;379:1113—20.

[81] Pruitt K, Ulkü AS, Frantz K, Rojas RJ, Muniz-Medina VM, Rangnekar VM, et al. Ras-mediated loss of the pro-apoptotic response protein Par-4 is mediated by DNA hypermethylation through Raf-independent and Raf-dependent signaling cascades in epithelial cells. J Biol Chem 2005;280:23363—70.

[82] Young NL, Dimaggio PA, Garcia BA. The significance, development and progress of high-throughput combinatorial histone code analysis. Cell Mol Life Sci 2010;67:3983—4000.

[83] Gardner KE, Allis CD, Strahl BD. Operating on chromatin, a colorful language where context matters. J Mol Biol 2011;409:36—46.

[84] Bogdanovic O, Veenstra GJ. DNA methylation and methyl-CpG binding proteins: developmental requirements and function. Chromosoma 2009;118:549—65.

[85] Miremadi A, Oestergaard MZ, Pharoah PD, Caldas C. Cancer genetics of epigenetic genes. Hum Mol Genet 2007;16(Spec No 1):R28—49.

[86] Ohm JE, McGarvey KM, Yu X, Cheng L, Schuebel KE, Cope L, et al. A stem cell-like chromatin pattern may predispose tumor suppressor genes to DNA hypermethylation and heritable silencing. Nat Genet 2007;39:237—42.

[87] Vire E, Brenner C, Deplus R, Blanchon L, Fraga M, Didelot C, et al. The Polycomb group protein EZH2 directly controls DNA methylation. Nature 2006;439:871—4.

[88] Ding L, Kleer CG. Enhancer of Zeste 2 as a marker of preneoplastic progression in the breast. Cancer Res 2006;66:9352—5.

[89] Bryant RJ, Cross NA, Eaton CL, Hamdy FC, Cunliffe VT. EZH2 promotes proliferation and invasiveness of prostate cancer cells. Prostate 2007;67:547—56.

[90] Fussbroich B, Wagener N, Macher-Goeppinger S, Benner A, Falth M, Sultmann H, et al. EZH2 depletion blocks the proliferation of colon cancer cells. PLoS One 2011;6:e21651.

[91] Mohammad HP, Cai Y, McGarvey KM, Easwaran H, Van Neste L, Ohm JE, et al. Polycomb CBX7 promotes initiation of heritable repression of genes frequently silenced with cancer-specific DNA hypermethylation. Cancer Res 2009;69:6322—30.

[92] Kanai Y. Genome-wide DNA methylation profiles in precancerous conditions and cancers. Cancer Sci 2010;101:36—45.

[93] Arai E, Kanai Y. DNA methylation profiles in precancerous tissue and cancers: carcinogenetic risk estimation and prognostication based on DNA methylation status. Epigenomics 2010;2:467—81.

[94] Laird PW. Principles and challenges of genomewide DNA methylation analysis. Nat Rev Genet 2010;11:191—203.

[95] Shen L, Toyota M, Kondo Y, Lin E, Zhang L, Guo Y, et al. Integrated genetic and epigenetic analysis identifies three different subclasses of colon cancer. Proc Natl Acad Sci USA 2007;104:18654—9.

[96] Arai E, Kanai Y, Ushijima S, Fujimoto H, Mukai K, Hirohashi S. Regional DNA hypermethylation and DNA methyltransferase (DNMT) 1 protein overexpression in both renal tumors and corresponding nontumorous renal tissues. Int J Cancer 2006;119:288—96.

[97] Arai E, Wakai-Ushijima S, Fujimoto H, Hosoda F, Shibata T, Kondo T, et al. Genome-wide DNA methylation profiles in renal tumors of various histological subtypes and non-tumorous renal tissues. Pathobiology 2011;78:1—9.

[98] Arai E, Ushijima S, Fujimoto H, Hosoda F, Shibata T, Kondo T, et al. Genome-wide DNA methylation profiles in both precancerous conditions and clear cell renal cell carcinomas are correlated with malignant potential and patient outcome. Carcinogenesis 2009;30:214—21.

[99] Arai E, Kanai Y. Genetic and epigenetic alterations during renal carcinogenesis. Int J Clin Exp Pathol 2011;4:58—73.

[100] Arai E, Ushijima S, Tsuda H, Fujimoto H, Hosoda F, Shibata T, et al. Genetic clustering of clear cell renal cell carcinoma based on array-comparative genomic hybridization: its association with DNA methylation alteration and patient outcome. Clin Cancer Res 2008;14:5531—9.

[101] Nishiyama N, Arai E, Nagashio R, Fujimoto H, Hosoda F, Shibata T, et al. Copy number alterations in urothelial carcinomas: their clinicopathological significance and correlation with DNA methylation alterations. Carcinogenesis 2011;32:462—9.

[102] Sepulveda AR, Jones D, Ogino S, Samowitz W, Gulley ML, Edwards R, et al. CpG methylation analysis—current status of clinical assays and potential applications in molecular diagnostics: a report of the Association for Molecular Pathology. J Mol Diagn 2009;11:266—78.

[103] Cairns P, Esteller M, Herman JG, Schoenberg M, Jeronimo C, Sanchez-Cespedes M, et al. Molecular detection of prostate cancer in urine by GSTP1 hypermethylation. Clin Cancer Res 2001;7:2727—30.

[104] Hoque MO, Begum S, Topaloglu O, Chatterjee A, Rosenbaum E, Van Criekinge W, et al. Quantitation of promoter methylation of multiple genes in urine DNA and bladder cancer detection. J Natl Cancer Inst 2006;98:996—1004.

[105] Watanabe Y, Kim HS, Castoro RJ, Chung W, Estecio MR, Kondo K, et al. Sensitive and specific detection of early gastric cancer with DNA methylation analysis of gastric washes. Gastroenterology 2009;136:2149–58.

[106] Gotoh M, Arai E, Wakai-Ushijima S, Hiraoka N, Kosuge T, Hosoda F, et al. Diagnosis and prognostication of ductal adenocarcinomas of the pancreas based on genome-wide DNA methylation profiling by bacterial artificial chromosome array-based methylated CpG island amplification. J Biomed Biotechnol 2011;2011:780836.

[107] Arai E, Ushijima S, Gotoh M, Ojima H, Kosuge T, Hosoda F, et al. Genome-wide DNA methylation profiles in liver tissue at the precancerous stage and in hepatocellular carcinoma. Int J Cancer 2009;125:2854–62.

[108] Nagashio R, Arai E, Ojima H, Kosuge T, Kondo Y, Kanai Y. Carcinogenetic risk estimation based on quantification of DNA methylation levels in liver tissue at the precancerous stage. Int J Cancer 2011;129:1170–9.

[109] Nishiyama N, Arai E, Chihara Y, Fujimoto H, Hosoda F, Shibata T, et al. Genome-wide DNA methylation profiles in urothelial carcinomas and urothelia at the precancerous stage. Cancer Sci 2010;101:231–40.

[110] Marsit CJ, Houseman EA, Christensen BC, Gagne L, Wrensch MR, Nelson HH, et al. Identification of methylated genes associated with aggressive bladder cancer. PLoS One 2010;5:e12334.

[111] Jarmalaite S, Jankevicius F, Kurgonaite K, Suziedelis K, Mutanen P, Husgafvel-Pursiainen K. Promoter hypermethylation in tumour suppressor genes shows association with stage, grade and invasiveness of bladder cancer. Oncology 2008;75:145–51.

[112] Satoh A, Toyota M, Itoh F, Sasaki Y, Suzuki H, Ogi K, et al. Epigenetic inactivation of CHFR and sensitivity to microtubule inhibitors in gastric cancer. Cancer Res 2003;63:8606–13.

[113] Hegi ME, Liu L, Herman JG, Stupp R, Wick W, Weller M, et al. Correlation of O6-methylguanine methyltransferase (MGMT) promoter methylation with clinical outcomes in glioblastoma and clinical strategies to modulate MGMT activity. J Clin Oncol 2008;26:4189–99.

[114] Campioni M, Santini D, Tonini G, Murace R, Dragonetti E, Spugnini EP, et al. Role of Apaf-1, a key regulator of apoptosis, in melanoma progression and chemoresistance. Exp Dermatol 2005;14:811–8.

[115] Estecio MR, Issa JP. Tackling the methylome: recent methodological advances in genome-wide methylation profiling. Genome Med 2009;1:106.

[116] Branton D, Deamer DW, Marziali A, Bayley H, Benner SA, Butler T, et al. The potential and challenges of nanopore sequencing. Nat Biotechnol 2008;26:1146–53.

[117] Eid J, Fehr A, Gray J, Luong K, Lyle J, Otto G, et al. Real-time DNA sequencing from single polymerase molecules. Science 2009;323:133–8.

[118] Kelly TK, De Carvalho DD, Jones PA. Epigenetic modifications as therapeutic targets. Nat Biotechnol 2010;28:1069–78.

[119] Boumber Y, Issa JP. Epigenetics in cancer: what's the future? Oncology (Williston Park) 2011;25:220–6, 228.

[120] International Human Epigenome Consortium [homepage on the internet] Cited. Available from: http://www.ihec-epigenomes.org/; February 2012.

CHAPTER 4

Alterations of Histone Modifications in Cancer

Ciro Mercurio[1,3], Simon Plyte[2,3], Saverio Minucci[3,4]
[1]DAC-Genextra Group, Milan, Italy
[2]Congenia-Genextra Group, Milan, Italy
[3]European Institute of Oncology, Milan, Italy
[4]University of Milan, Milan Italy

4.1 INTRODUCTION

Dynamic changes in chromatin structure, which permit local decondensation and remodeling and are necessary for the role of chromatin in processes as gene transcription, DNA replication and repair, are achieved by modification of the chromatin and in particular by post-translational modifications of the histone component [1,2].

The post-translational modifications of histones identified so far include acetylation, phosphorylation, methylation, monoubiquitination, sumoylation, and ADP ribosylation [2]. The reversibility of these modifications is what confers the necessary dynamicity of the chromatin remodeling events and these are tightly controlled by the opposing activity of enzymes responsible for adding or removing the modifications (for example, histone acetyltransferases and histone deacetylates for histone acetylation and histone methyltransferases and histone demethylase for histone methylation). It is worth noting that those enzymes responsible for histone modifications can also modify non-histone proteins: while this observation has enormous implications, we have chosen here to restrict our analysis to the study of histone modifications and to the action of those enzymes on histones. We will first introduce the known molecular and biochemical properties of the different types of histone post-translational modifications, concentrating primarily on acetylation, methylation, phosphorylation, and ubiquitination. Then, we will summarize the current knowledge regarding the relevance of histone modifications in cancer, with a particular emphasis on the description of global changes to the pattern of histone modifications in cancer cells and their potential role as prognostic factors. Finally we will discuss the molecular mechanisms that are potentially involved in the generation of these altered patterns in cancer cells.

The various histone modifications act in a coordinate and ordered manner to control the conformation of chromatin [3]. A further level of complexity is present, due to the interplay between the different histone modifications, DNA methylation, and ATP-dependent chromatin remodeling components [3]. Remarkable progress has been made in recent years in the identification of these histone modifications, their genome-wide distribution and the level of interconnection between them and other relevant events such as DNA methylation. The increased knowledge and interest in the role of epigenetic modifications in cancer has been reinforced by the identification of a deregulated pattern of histone modification in several cancer types. The reversibility of histone modification and the identification of the molecular machinery that governs these modifications have made histone-modifying enzymes attractive new targets for anticancer therapy. In addition, a clear role for the pattern of histone modification as a predictor of prognosis in several cancers has emerged, although the use of such "histone modification signature" as predictor of therapeutic response is still at an initial stage.

54

4.2 CHROMATIN ORGANIZATION

The structural and functional unit of chromatin is the nucleosome, which consists of a disc-shaped octamer composed of two copies of each histone protein (H2A, H2B, H3, and H4), around which 147 base-pairs of DNA are wrapped twice (Figure 4.1) [4–7]. Electron microscopy studies revealed that organization of nucleosomal arrays is constituted by a series of "beads on a string", with the "beads" being the individual nucleosomes and the "string" being the linker DNA [4,5]. Linker histones, such as histone H1 and other non-histone proteins interact with the nucleosomal arrays to further package the nucleosomes to form higher-order chromatin structures [7,8].

Histones are high evolutionarily conserved proteins with flexible N and C terminal domains and a conserved related globular domain which mediates histone–histone interactions within the octamer (Figure 4.1) [9–11]. There are two small domains protruding from the globular domain: an aminoterminal domain constituted by 20–35 residues rich in basic amino acids and a short protease accessible carboxyterminal domain [9–11]. Histone H2A is unique among the histones having an additional 37 amino acids carboxy-terminal domain that protrudes from the nucleosome [11]. Additional histone variants have also been identified and tend to have specialized roles [12]. The N-terminal tail of histones, as well as more recently defined positions in the globular domain, are subject to eight different classes of post-translational modification involving more than 60 distinct modification sites: lysine acetylation, ubiquitination and sumolylation, serine, threonine and thyrosine phosphorylation, lysine and arginine methylation, glutamate poly-ADP ribosylation, arginine deimination, and proline isomerization (Figure 4.1) [2,11]. The combination of these histone modifications, the interplay between them and DNA methylation and ATP chromatin remodeling proteins, dynamically regulates chromatin structure and in so doing,

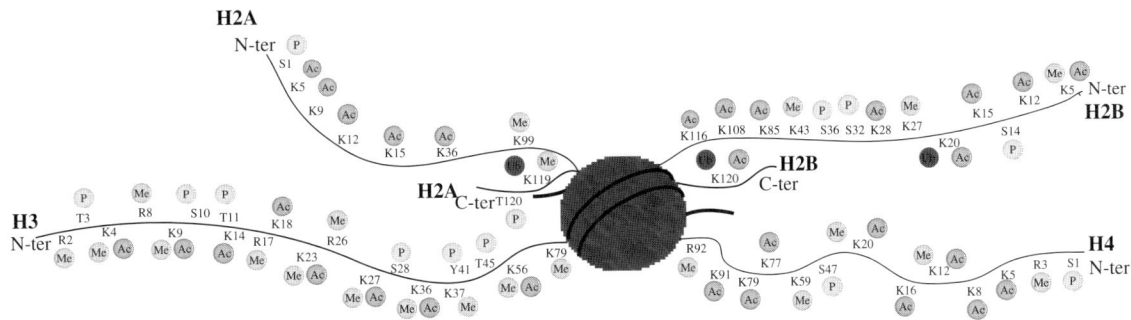

Chromatin Modification	Residues modified	Function regulated
Acetylation	Lysine	Transcription, DNA repair, replication and condensation
Methylation (Lysine)	Lysine me1, me2, me3	Transcription, DNA repair
Methylation (Arginine)	Arginine-me1, Arginine-me2a Arginine-me2s	Transcription
Phosphorylation	Serine, Threonine, Tyrosine	Transcription, DNA repair and condensation
Ubiqutination	Lysine	Transcription, DNA repair
Sumoylation	Lysine	Transcription
ADP ribosylation	Glutamic	Transcription
Deimination	Arginine	Transcription
Proline isomerization	P-cis, P-trans	Transcription

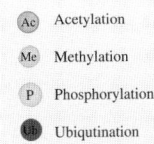

Different classes of histone modifications

FIGURE 4.1

Mammalian core histone modifications. N- and C-terminal histone tails extend from the globular domains of histones H2A, H2B, H3, and H4. DNA is wrapped around the nucleosome octamer made up of two H2A—H2B dimers and a H3—H4 tetramer. Post-translational covalent modifications include acetylation, methylation, phosphorylation, and ubiquitylation. Human histone tail amino acid sequences are shown. Lysine positions 56 and 79 on histone H3 are located within the globular domain of the histone. This figure is reproduced in the color plate section.

coordinates several relevant cellular processes including transcription, DNA replication, DNA repair, and genomic stability [13—16].

4.3 HISTONE MODIFICATIONS

We will now look at the various histone modifications available.

4.3.1 Histone Acetylation

Histone acetylation is a reversible covalent modification, occurring at specific lysines residues in the histone tails. This modification is able to neutralize the positive charge of the targeted lysine, weakening the histone-DNA [2,16,17] or nucleosome—nucleosome interactions and consequently inducing conformational changes resulting in an open chromatin architecture [16]. Furthermore, histone acetylation represents a "histone mark" recognized by specific proteins such as bromo domain-containing proteins, whose interaction with the modified chromatin leads to a cascade of additional modifications often culminating in increased transcriptional activity [2—17]. The steady state level of histone lysine acetylation is determined by the opposing activity of two different types of enzymes: histone acetyltransferases (HATs), which use acetyl CoA to transfer an acetyl group to the ε-amino group of the N terminal of histone tails, and histone deacetylases (HDACs) which reverse this modification (Table 4.1) [18].

HATs are often part of large multiprotein complexes, and can be divided into two main groups: type A and type B based on their cellular localization [19]. The type A HATs are nuclear

TABLE 4.1 Histone Modifying Enzymes: Histone Acetyltransferases and Histone Deacetylases

Category	Gene	Histone Specificity
Histone Acetyltransferases		
GNC5/PCAF family	HAT1	H4
	Gcn5	H3/H4
	PCAF	H3/H4
MYST family	MOF (MYST1)	H4K16
	HBO1 (MYST2)	H4>H3
	MOZ (MYST3) KAT6a	H3
	MORF (MYST4) KAT6b	H3
	Tip60	H4 H2a
P300/CBP family	P300	H2A/H2B/H3/H4
	CBP	H2A/H2B/H3/H4
Histone Deacetylases		
HDAC class I	HDAC1,HDAC2,HDAC3,HDAC8	—
HDAC class IIA	HDAC4,HDAC5,HDAC7,HDAC9	—
HDAC class IIB	HDAC6,HDAC10	—
HDAC class IV	HDAC11	—
HDAC Class III (Sirtuin)		
Sirtuin class I	SIRT1,SIRT2,SIRT3	—
Sirtuin class II	SIRT4	—
Sirtuin class III	SIRT5	—
Sirtuin class IV	SIRT6,SIRT7	—

proteins acetylating both nucleosomal histones and other chromatin-associated proteins, whereas type B HATs are prevalently cytoplasmatic, and acetylate newly synthesized histones not yet deposited into chromatin [19,20].

Type A HATs may be further grouped into at least three different sub-groups based on their sequence homology and conformational structure: GNATs (Gcn5-related N-acetyl transferases) related HATs, MYST-related HATs (MOZ (monocytic leukaemia zinc-finger protein), Ybf2 (yeast binding factor 2)/Sas3 (something about silencing 3), Sas2, Tip60 (Tat interactive protein-60)), and p300/CBP family (CREB (cAMP response element binding protein)-binding protein) [19].

The type B HATs share sequence homology with scHat1 [20], the founding member of this type of HATs and have a role in the deposition of histones into the nucleosome through the acetylation of newly synthesized histone H4 at lysine 5 (K5) and lysine 12 (K12) (as well as certain sites within H3) [20]. HATs generally contain multiple subunits and their catalytic activity depends, in part, on the context of the other subunits in those complexes [19].

HDACs are also part of large multiprotein complexes and comprise a family of 18 genes subdivided into four classes on the basis of their sequence homology to ortholog yeast proteins, subcellular localization and enzymatic activities (Table 4.1) [21,22]. Of these classes, class I, II, and IV have in common a zinc (Zn)-dependent enzymatic activity [23], while class III (or Sirtuins from the homology with the yeast Sir2) constitutes a separate, structurally unrelated, nicotinamide adenine dinucleotide (NAD)-dependent subfamily [24,25]. HDAC class I (HDAC1, 2, 3, and 8) are nuclear proteins with homology to the yeast RPD3 protein and are ubiquitously expressed in various human cell lines and tissues [21]. HDAC class II is constituted by large proteins with homology to the Hda1 yeast protein and shuttle between the cytoplasm and the nucleus. They can be subdivided into class IIa (HDAC4, 5, 7, and 9) and class IIb (HDAC6 and 10) on the basis of the presence of a double deacetylase domain typical of HDAC6 and HDAC10 [26]. Class IV is represented by a single protein: HDAC11 which is characterized by a deacetylase domain sharing homology with both HDAC class I and class II domains [27].

HDACs, in opposition to the enzymatic activity of HATs, reverse the acetylation of lysine and in so doing, restore the positive charge of the lysine residue. In general, HDACs, as well as the HATs, have relatively low substrate specificity by themselves. In addition, the determination of their specificity is complicated by the fact that these enzymes are present in multiple complexes often including other HDAC family members [21]. For instance, HDAC1 is found together with HDAC2 within the NuRD, Sin3a, and Co-REST complexes and this may lead to changes in their activity/specificities [21]. Removal of acetylation tends to stabilize the chromatin structure and leads to a more closed chromatin conformation which suggests a main, but not exclusive, function of HDACs as transcriptional repressors.

Histone acetylation is almost invariably associated with transcriptional activation [19] and although most of the acetylation sites fall within the N-terminal tail of the histones, which are more accessible for modification, acetylation within the core domain of H3 at lysine 56 (H3K56ac) has also been reported [28]. With respect to the specific role of histone acetylation on gene transcription, gene-specific and global effects can be distinguished [16]. Histone acetylation targeted at specific promoters regulates the transcription of specific genes, whereas, histone acetylation over large regions of chromatin, including coding regions and non-promoter regions, affects global gene expression levels. A characteristic enrichment of histone acetylation at enhancer elements, and particularly in gene promoters, where they presumably facilitate the transcription factor access, has been recently reported [29]. However, genome-wide mapping of HATs and HDACs revealed that the above scenario, linking histone acetylation to transcriptional activation and histone deacetylation to transcriptional repression, is much too simplistic [30]. The study conducted by Zhao and colleagues also revealed a surprisingly strong association of HDACs with active genes. This observation suggests that there are two distinct functions for HDACs on active or primed genes. On active genes, by removing the acetyl group added during transcription, HDACs would reset the elongation of the chromatin for subsequent transcription rounds. In contrast, on primed genes, HDACs would exert a repressive action by maintaining low levels of acetylation, which in turn could prevent pol II binding and gene transcription [30]. The suppression of spurious internal transcription initiation is an additional function of HDACs that is thought to be important to assure accurate gene transcription and maintain chromatin integrity following transcription [31−33]. Seminal studies conducted in yeast have demonstrated the relevant role of Rpd3S deacetylase in inhibiting the assembly of transcription factors at inappropriate or "cryptic" sites within genes and in the suppression of cryptic transcription initiation [31−33].

The role of histone acetylation is not limited to the regulation of gene transcription but extended to additional processes such as nucleosomal assembly, chromatin folding, chromosome condensation, genome stability, DNA duplication and repair [16,28,34]. For instance, a specific role of H4K16 acetylation (H4K16ac) in chromosome condensation [34] and a role of H3K56 acetylation (H3K56ac) in DNA repair and genome stability have been reported [16,28]. The correlation between histone tail acetylation and the timing of replication also strongly supports a critical role of this histone modification in DNA replication [35]. This is further supported by the essential role of ING5−HBO1 acetyltransferase complex in DNA replication [36].

4.3.2 Histone Methylation

Histone methylation is a reversible modification mainly occurring on the side chains of both lysines and arginines [37−40]. Methylation is unique among the histone post-translational modifications because up to three methyl groups can be added to a single lysine residue, creating a total of four different methyl states: un-methylated, mono-, di-, or trimethylated states. Furthermore, arginine residues can undergo both monomethylation and dimethylation, with the latter in a symmetric or asymmetrical configuration [37–40]. Histone methylation, in contrast to acetylation and phosphorylation, does not alter the charge of the histone tails but influences the basicity and hydrophobicity and consequently the affinity of certain molecules such as transcription factors toward DNA [38,41,42]. Histone methylation on lysine and arginine is mediated by histone lysine methyltransferases (HMTs) (Table 4.2) [38, 43] and protein arginine methyltransferases (PRMTs) (Table 4.2) [37,44], respectively, and these enzymes transfer a methyl group, from the cofactor S-adenosyl-L-methionine to either the ε-amino group of lysine or to the guanidino group of arginine.

HMTs have been grouped in two main different classes: lysine-specific SET domain containing histone methyltransferases, characterized by a 130-amino-acid catalytic domain known as SET (Su (var), Enhancer of Zeste, and Trithorax), and the non-SET-containing lysine methyltransferases [45]. Remarkably, all the known HMTs have a SET domain harboring the enzymatic activity with the exception of DOT1, that methylates lysine-79 within the core domain of H3 only in nucleosomal substrates and not in free histones [46]. The SET domain containing HMTs catalyze mono-, di-, and trimethylation of their target lysine residue localized on histone tails and are classified into six subfamilies: SET1, SET2, SUV39, EZH, SMYD, and PRDM. Of note, a small number of SET-containing HMTs that do not fall into the above six subfamilies due to an absence of conserved sequences flanking their SET domains include Set8/PR-Set7, SUV4-20H1, and SUV4-20H2, Set7/9, as well as MLL5, RIZ (retino-blastoma protein-interacting zinc-finger) and SMYD3 (SET- and MYND-domain containing protein 3) [43].

HMTs, in contrast to HATs and HDACs, which can be promiscuous in their histone substrate specificity, typically show a high degree of specificity toward their histone targets (Figure 4.2) [38,43] and are able to recognize and modulate different degrees of methylation on the same lysine (i.e. mono- vs. di- and tri-methyaltion) [2,42].

PRMTs proteins, characterized by a common catalytic methyltransferase domain and unique N-terminal and C-terminal regions, are divided into two different classes: type I and type II [37,44]. The type II enzymes (PRMT 5, 7, and 9) catalyze mono- and di-symmetric methyla-tion of the arginines (R), whereas, the formation of mono- and di-asymmetric tails is achieved by type I enzymes (PRMT 1–4, 6, and 8). All the PRMTs transfer methyl group to several substrates but PRMT1, 4, 5, and 6 are the most relevant enzymes with respect to histone arginine methylation [37,44].

In addition, within the family of PR domain-containing proteins (PRDMs) [47], PRDM2 (RIZ1) and PRDM6 have been shown to have intrinsic methyltransferase activity toward lysine residues, such as lysine 9 of histone H3 (H3K9) and lysine 20 of histone H4 (H4K20) [48,49].

Prior to recent identification of numerous proteins involved in histone demethylation, a single mechanism of arginine deimination has been described. This mechanism is catalyzed by peptidylarginine deiminase 4 (PAD4) [50,51] and characterized by the conversion of the arginine to citrulline via a deimination reaction [50,51]. The identification of this mechanism was the first demonstration of the reversibility of the histone methylation mark, although representing deimination (that is, a further modification of the methylated residue) rather than a direct reversion of methylation [51]. Over the last few years several "bona fide" histone demethylases, able to revert both lysine and arginine methylation, have been identified (Table 4.2) [51–54].

Lysine demethylating enzymes have been subdivided into two main families: KDM1 (lysine (K) demethylase 1) family which are FAD-dependent amine oxidases, acting only on mono- and dimethylated lysine [55] ,and the JMJC domain contain proteins which are Fe(II) and 2-oxoglutarate-dependent enzymes able to remove all methylation states (Table 4.2) [54]. The

TABLE 4.2 Histone Modifying Enzymes: Histone Lysine Methyltransferases, Histone Lysine Demethylases and Histone Arginine Methyltransferases

Category	Gene	Histone Specificity
Histone Lysine Methyltransferases		
	SUV39-H1/KMT1A	H3K9me 2,3
	SUV39-H2/KMT1B	H3K9me 2,3
	G9a/KTM1C	H3K9me 1,2
	EuHMTase/GLP/KMT1D	H3K9me 1,2
	ESET/SETDB1/KMT1E	H3K9me 2,3
	CLL8/KMT1F	H3K9me
	MLL1/KMT2A	H3K4me1,2,3
	MLL2/KMT2B	H3K4me1,2,3
	MLL3/KMT2C	H3K4me1,2,3
	MLL4/KMT2D	H3K4me1,2,3
	MLL5/KMT2E	H3K4me
	hSET1A/KMT2F	H3K4me1,2,3
	hSET1B/KMT2G	H3K4me1,2,3
	ASH1/KMT2H	H3K4
	SET2/KMT3A	H3K4
	NSD1/KMT3B	H3K3me6 2,3
	NSD2	H3K36me3
	NSD3	-
	SMYD1	-
	SMYD2/KMT3C	H3K36me 2, H3K4me
	SMYD3	H3K4me 2,3
	DOT1L/KMT4	H3K79
	PR-SET7−8/KMT5A	
	SUV4−20H1/KMT5B	H4K20me 2,3
	SUV4−20H2/KMT5C	H4K20me 2,3
	EZH2/KMT6	H3K27me 2,3
	EZH1	H3K27me 2,3
	SET7−9/KMT7	H3K4 1
	RIZ1/KMT8	H3K9me
Histone Lysine Demethylases		
	LSD1/KDM1A	H3K4me1/2, H3K9me1/2
	LSD2/KDM1B	H3K4me1/2, H3K9me1/2
KDM2 cluster	FBXL11A/JHDM1A/KDM2A	H3K36me1/2
	FBXL10B/JHDM1B/ KDM2B	H3K36me1/2
KDM3 cluster	JMJD1A,JHDM2A/ KDM3A	H3K9me1/2
	JMJD1B/JHDM2B/KDM3B	H3K9me
KDM4 cluster	JMJD2A/JHDM3A/ KDM4A	H3K9/ K36me2/3
	JMJD2B/ KDM4B	H3K9/ K36me2/3
	JMJD2C/GASC1/KDM4C	H3K9/ K3K36me2/3
	JMJD2D/KDM4D	H3K9me2/3
KDM5 cluster	JARID1A/RBP2/KDM5A	H3K4me2/3
	JARID1B/PLU1/KDM5B	H3K4me1/2/3
	JARID1C/SMCX /KDM5C	H3K4me2/3
	JARID1D/SMCY/KDM5D	H3K4me2/3
KDM6 cluster	UTX/KDM6A	H3K27me2/3
	JMJD3/KDM6B	H3K27me2/3
	KDM7	H3K9me1/2, H3K27me1/2
	KDM8	H3K36me2
	PFH8	H3K9me1/2, H4K20me
Arginine Methyltransferases		
PRMT type I	PRMT1, PRMT3, PRMT4, PRMT6, PRMT8	H4 mono di-symmetric
PRMT type II	PRMT5, PRMT7, PRMT9	H4 mono di-symmetric
Not defined	PRMT2, PRMT10, PRMT11	

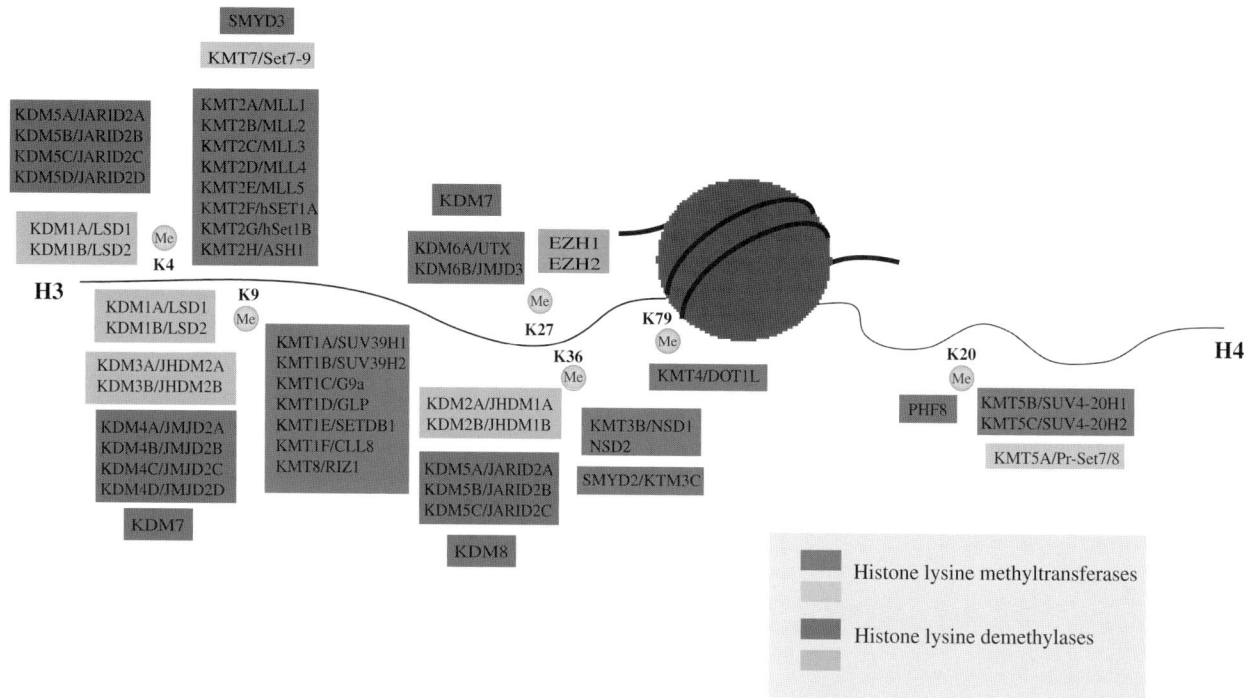

FIGURE 4.2

Histone lysine methylases and demethylases. Histone H3 and H4 tails with known lysine methylases (green box) and demethylases (red box). Lighter shades indicate monomethylation by HMTs or demethylation of mono- and di-methylated lysine residues by HDMs. Darker shade indicates mono-, di-, and tri-methylation by HMTs or demethylation of di- and tri-methylated lysine residues by HDMs. This figure is reproduced in the color plate section.

60

JMJC domain-containing proteins, comprising 25 proteins of which 15 have confirmed demethylase activity, are divided into subfamilies based on sequence similarity: KDM2 (FBXL), KDM3 (JMJD1), KDM4 (JMJD2/JHDM3), KDM5 (Jarid1), KDM6(UTX/JMJD6), and Jumonji clusters [54].

The jumonji protein JMJD6 is, at the moment, the only known enzyme able to demethylate arginine (R) 2 and 3 specifically on histones H3 (H3R2) and H4 (H4R3) [56]. However, due to the low catalytic rate observed for this enzyme, it is currently unclear whether this reaction actually occurs physiologically or whether there are, as yet, uncharacterized arginine demethylases that perform this function in cells. As with the lysine methyltransferases, the histone demethylases possess a high level of substrate specificity with respect to their target lysine and appear sensitive to the degree of lysine methylation (Figure 4.2). The substrate specificity of histone demethylase can be further influenced by the association of additional proteins. For example the specificity of KDM1A for H3K9 when in complex with steroid receptors changes to H3K4 when the protein is in a complex with Co-Rest [57]. Interestingly, this substrate switching also correlates with a switching of the role of KDM1A from activator to repressor of gene transcription [57].

Histone methylation represents a mechanism of "marking" the histone in order to recruit several effector proteins with recognition domains specific for different methylated lysine residues [58]. For instance, plant homeodomain (PHD) of bromodomain-PHD transcription-factor (BPTF) binds to H3K4 trimethylated/dimethylated (H3K4me3/me2) and recruits the nucleosome remodeling factor (NURF) complex to the target gene leading to gene activation [59]. In contrast, the chromodomain of heterochromatin protein 1 (HP1) binds H3K9 trimethylated (H3K9me3) leading to heterochromatin formation and gene silencing [60].

In contrast to acetylation, which is generally associated with transcriptional activation, histone lysine methylation correlates with either an activation or a repression of transcription depending both on the methylated site and the degree of methylation [41,42]. A better understanding of the organization and the complexity of histone methlyation and acetylation has come from the generation of high-resolution, genome-wide maps of the distribution of histone lysine and arginine methylations [61] as well as from the analysis of combinatorial pattern of histone acetylation and methylation [29]. These works [29,61] led to the identification of distinct and combinatorial patterns of histone marks at different genomic regions including promoters, insulators, enhancers, and transcribed regions. More importantly, these analyses have highlighted the cooperative manner in which diverse modifications can act to globally regulate gene expression. For instance, these studies clearly demonstrated the association of methylation of histones H3K9, H3K27, and H4K20 with gene silencing, as well as the link between active gene transcription and H3K4, H3K36, and H3K79 methylation. Interestingly, a strict equilibrium between methylation of H3K4 which activates transcription and methylation of H3K27, which represses transcription, was recently reported to be important in the activity of "stemness" transcription patterns to maintain pluripotency of embryonic stem cells [62].

4.3.3 Histone Phosphorylation

Phosphorylation of histones, like the other histone modifications, is a highly dynamic process specifically characterized by the addition of a phosphate group from ATP to the hydroxyl group of the target amino acid side chain of several and different residues within histone tails. The addition of phosphate and hence negative charge is able to modify the chromatin structure and in so doing, is able to influence interactions between transcription factors and other chromatin components [63,64]. Histone phosphorylation takes place on serine (S), tyrosine (Y), and threonine (T), with the large majority of histone phosphorylation sites being found within the N-terminal tails and only a very few examples, such as H3Y41, are found within the histone core [63,65].

Distinct phosphorylation patterns of histones have been linked to several cellular processes [63,64]. The contribution and interdependency of cross-talk between histone phosphorylation and other histone modifications is important in defining the role of histone phosphorylation. This is clearly seen in the interdependency of histone acetylation and methylation on phosphorylation of histone H3 and vice versa [66]. A well-characterized case of this interdependency is the phosphorylation of H3 at S10 which, to facilitate gene transcription, enhances H3K14 acetylation and H3K4 methylation and simultaneously inhibits H3K9 methylation [66]. Conversely, the methylation of H3K9 interferes with H3S10 phosphorylation [66]. As with acetylation and methylation, histone phosphorylation represents a histone mark recognized by specific ancillary proteins which in this case comprise the 14-3-3 protein family [67,68]. Several distinct histone kinases and histone phosphatases have been identified (Table 4.3) [63,64]. For example, phosphorylation of histone H2AX (a variant of histone H2A), induced by a DNA-damage signaling pathway is dependent on phosphatidylinositol-3-OH kinases (PI3Ks) such as ATM, ATR, and DNA-PK [69]. Histone H2B phosphorylation at S14, catalyzed by Mst1 (mammalian sterile-20-like kinase), has a role in the induction of apoptosis [70]. Phosphorylation of histone H3 at S10 and S28, associated with the seemingly contrasting functions of chromatin condensation and transcriptional activation, is due to the catalytic activity of aurora kinase family and primarily to aurora-B activity [71–73]. Other kinases mediating the phosphorylation of histone H3S10 and regulating gene expression have been identified in the MSK/RSK/Jil-1 family [73]. Haspin has recently been identified as the kinase responsible for the phosphorylation at H3T3, an event required for normal metaphase chromosome alignment [74], while Dlk/ZIP kinase (Dlk: Death-associated protein (DAP)-like kinase, ZIP: Zipper interacting protein kinase) is the enzyme responsible for phosphorylation of H3T11 during mitosis [75]. Recently, Chk1 has been identified as another histone kinase that regulates DNA damage-induced transcriptional repression through the phosphorylation of H3T11 [76]. Ribosomal S6 kinase 2 (RSK2) appears to be the main kinase that catalyzes the phosphorylation of H2BS32 and H2AX [77,78]. JAK2 is the kinase responsible for

TABLE 4.3 Histone-Modifying Enzymes: Kinases, Phosphatases, and Ubiquitilases

Category	Gene	Histone Specificity
Kinases		
	AURORA-B	H3S10,H3S28
	MST1	H2BS14
	ATM, ATR, DNAPK.	H2AX
	MSK-RSK	H3S10
	HASPIN	H3T3
	DLK/ZIP	H3T11
	CHK1	H3T11
	JAK2	H3Y41
Phosphatases		
	PP1	H3S10,H3S28
	PP2A	H2AX
	PPγ	H3T11
Ubiquitilases		
E2 and E3 ligases	hHR6A/hHR6B (E2 ligase)	H2BK120
	RNF20 (E3 ligase)	H2BK120
	RING1B , 2A-HUB	H2AK119
	(E3 ligases)	
	RNF8 (E3 ligase)	H2AK119
Dubs	USP22	H2AK119
	USP16	H2BK120
	MYSM1	H2BK120
	USP3	H2BK120
	USP7	H2AK119
	BRCC36	H2AX

phosporylation of H3Y41 [65] and in doing so determines the release of the transcriptional repressor heterochromatin protein 1α (HP1α) from chromatin [65].

Little is known about the role of protein phosphatases in regulating the dephosphorylation of histones. Protein phosphatase 1 (PP1) appears to be involved in the dephosphorylation of H3S10 and H3S28 [73,74]. Protein phosphatase 2 (PP2A) is responsible for the dephosphorylation of H2AX after DNA repair [79,80] and Protein phosphatase γ (PPγ) is involved into dephosphorylation of H3T11 after DNA damage [81].

4.3.4 Histone Ubiquitination

Histone ubiquitination differs substantially from the other modifications, since it is a very large modification with the ubiquitin moiety comprising of a 76-amino-acid polypeptide [82]. Histone ubiquitination is a reversible modification whose steady state is determined by two enzymatic activities involved in addition and removal of the ubiquitin moiety from histones [39]. Histone ubiquitination, occurring in human mainly on histone H2A at lysine 119 (H2AK119ub1) and histone H2B at lysine 120 (H2BK120ub1), is catalyzed by the formation of an isopeptide bond between the carboxy-terminal glycine of ubiquitin and the ε-group of a lysine residue on carboxyterminal tail of the histones. This bond is the result of the sequential catalytic actions of E1 activating, E2 conjugating, and E3 ligase enzymes with the E3 ligases responsible for specific recognition and ligation of ubiquitin to its substrates [39,82,83]. Substrates can be both poly- and monoubiquitinated. Polyubiquitination creates an irreversible signal for proteosomal-mediated degradation whilst monoubiquitination generates a regulatory signal which can be reversed by the action of ubiquitin-specific proteases called deubiquitinating enzymes [39,82,83]. Histone ubiquitination occurs largely in the mono-ubiqutinated form and correlates with active and open chromatin, although histone ubiq-uitination has been linked with both transcriptional activation and silencing depending on the genomic context [39,84,85]. Furthermore, a role of monoubiquitination at histone H2A linked to DNA repair mechanism has been reported [86]. Interestingly, conjugation of a single ubiquitin moiety to histone H2A results in a significantly different outcome when compared to the addition of ubiquitin to H2B. H2A ubiquitination, being associated predominantly with transcriptional repression, may be considered a repressive mark whilst H2B ubiquitination appears be involved both in transcriptional activation and gene silencing [84–90]. The role of histone ubiquitination in transcriptional control is due to both a direct effect of histone H2B ubiquitination on RNA polymerase II transcription elongation [85–87] and to the cross-talk with other histone modifications [39,84,88–90]. Several data have demonstrated the ability of histone H2B ubiquitination to directly promote accurate and efficient RNA pol II transcription affecting nucleosomal dynamics [85–87]. Further, in the context of trans-histone cross-talk, it is well documented that the monoubiquitination of H2B is required for lysine methylation of histone H3K4 whilst this methylation is inhibited by the monoubiquitination of H2A [39,84,88–90]. The possible molecular mechanisms linking histone ubiquitination to tran-scriptional regulation are at least two. One mechanism envisages that the addition of a large macromolecule, such as ubiquitin, to a histone tail would lead to a modification of the high-order chromatin structure. The other one suggests that ubiquitination represents a signal for successive histone modifications, and/or a signal for recruitment of other proteins to the chromatin. For example, recent evidence obtained using chemically modified histones, demonstrated that H2B monoubiquitination, in contrast to H2A ubiquitination, had a clear impact on chromatin organization by inhibiting both nucleosome-array folding as well as interfiber oligomerization [90].

Table 4.3 lists some of the enzymes involved in this modification. The E2 conjugating enzymes hHR6A/hHR6B and the E3 ligases, RNF20, are responsible of H2B ubiquitination in cells [91–94]. Ring1B and 2A-HUB are the E3 ligase responsible of H2A ubiquitination [95,96], whilst RNF8 is E3 ligase responsible for monoubiquitination of H2A during DNA repair [97,98]. The deubiquitinating enzyme (Dubs) USP22, as a component of hSAGA complex, is involved in the deubiquitination of both H2B and H2A [99–101]. USP22 is required for the transcription of cell cycle genes and target genes regulated by the Myc oncoprotein whereby it is recruited to specific target gene loci [99–101]. In addition, Ubp-M (USP16) and 2A-DUB (MYSM1) have also been implicated in the deubiquitination of H2B [102], whilst recently USP3 [103] and USP7 [104] have been suggested to be involved in the deubiquitination of both H2B and H2A in human cells. BRCA1-containing complex (BRCC36) appears to be the DUB responsible for the deubiquitination of H2AX [105].

4.3.5 Mode of Action of Histone Modifications

The molecular mechanisms underlying the function of each individual histone modification can be generalized into two categories: "cis" and "trans"' mechanisms [106]. The *cis* mechanism, for which histone acetylation and phosphorylation represent the best examples, corresponds to alterations of intra- and internucleosomal contacts via changes of steric or charge interactions, influencing chromatin structure [106]. The *trans* mechanism is characterized by the involvement of non-histone protein "readers" that bind to specific histone modifications giving rise to functional consequences [107]. As anticipated, the *cis* mechanism is responsible for a direct structural perturbation of the chromatin. Explicative of this is the case of acetylation that, by reducing the positive charge of histones, disrupts the electrostatic interaction between histones and DNA, and in so doing influences chromatin organization [2,106].

The enrichment of multiple histone acetylation sites on regions involved in active transcrip-tion, such as gene promoters, represents a striking example of the *cis* mechanism [29]. The

63

fundamental role of the H4K16 acetylation in the control of the chromatin structure demonstrates that a single modification site can also have a strong impact on chromatin organization and reveals that the presence of multiple acetylation sites is not necessary to invoke gross structural changes in chromatin [31]. Similarly, conjugation with ubiquitin can cause direct structural perturbations in chromatin. The ubiquitination of histone H2B has been shown to disrupt compaction of local and higher-order chromatin [90]. In the *trans* mechanism, histone modifications represent a mark for the recruitment of so-called "chromatin readers" [108–110]. These proteins are characterized by recognizing specific histone modifications and include bromodomain proteins that recognize lysine acetylation [111], the "Royal super family" comprising chromodomain, Tudor, PWWP, and MBT domain that recognize histone lysine methylation [108–113], and PHD fingers protein [111]. Additional histone modification "readers" include the 14-3-3 sigma protein which recognized histone phosphorylation [68] and MDC1 which contains tandem BRCT domains that bind to γH2AX, the DSB-induced phosphorylated H2A variant [114].

Interestingly, within the group of methyl lysine binders the same modified site can be recognized by different domains. For example, trimethylation of H3K4 binds both the tandem chromodomains within CHD1 and the tandem Tudor domains within JMJD2A [115,116]. In some cases, the chromatin reader proteins can also simultaneously bind different histone modifications, as in the case of the L3MBTL1 protein simultaneously binds to mono/dimethylated lysine 20 of histone H4 (H4K20me1/2) and mono/dimethylated lysine 26 of histone H1B (H1BK26me1/2), and in so doing compacts nucleosomal arrays bearing the two histone modifications [117]. Not only can histone modifications generate a platform for "reader" recruitment but they can also disrupt interactions between histones and "readers". For example, H3K4me3 can prevent the NuRD complex from binding to the H3 N-terminal tail [118], as well as prevent binding of the PHD finger of DNMT3L to the H3 tail [119].

The functional consequences of histone modifications can be of two different types: establishment of global chromatin environments and orchestration of DNA-based biological tasks such as transcription, chromosome condensation, DNA replication, and repair [106,107]. Histone modifications contribute to the establishment of the global chromatin environment by arranging the genome into distinct domains. For example, euchromatin, where DNA is kept "accessible" for transcription and heterochromatin, where chromatin is maintained "inaccessible" for transcription [106,107,120]. Euchromatin is typically enriched in acetylated histones and methylation of H3K4, H3K36, and H3K79, whilst heterochromatin is characterized by histone hypoacetylation, a high level of methylation on H3K9, H3K27, and H4K20 and association of heterochromatin protein-1 (HP1) [121,122]. Histone modifications coordinate chromatin folding to facilitate the execution of specific functions [106,107]. Generally, for transcription, histone modifications can be divided into those correlating with activation and those correlating with repression. However, under specific conditions, certain types of modification have the potential to either activate or repress transcription [2,106,107]. For example, acetylation, methylation, phosphorylation, and ubiquitination have all been implicated in activation of transcription [123], whereas methylations and ubiquitination have also been implicated in repression [123]. These functions can be both local, as in the case of regulation of specific gene transcription, and genome wide as in the case of DNA replication, and chromosome condensation [2,104].

An important feature is that histone modifications have both short- and long-term functional effects [123]. An example of the short-term effect can be seen by the rapid and cyclic changes in histone modifications associated with transcription in response to external stimulation [124]. In this case, histone modifications on chromatin are the endpoint of a signaling pathway that corresponds to a mechanism through which the genome responds to external stimuli. An example of the longer-term effects of histone modification on genomic function is evident in the definition and maintenance of chromatin structures throughout the cell cycle [125]. Histone modifications having the longest effect are related to modification of heterochromatin. Constitutive heterochromatin is characterized by a specific pattern of histone modifications including an enrichment of trimethylation of H3K9 and H4K20 and a depletion of overall acetylation [121,122]. Similarly, facultative heterochromatin, as observed in the inactive X chromosome of females, is characterized by the loss of H3K4 methylation and an increase in H3K27 methylation [126].

4.3.6 Histone Cross-Talk

Histone modifications do not necessarily act on a stand-alone basis and frequently crosstalking interactions are in place as a mechanism of signaling integration [2,107,127]. Histone cross-talk occurs on single and multiple histone tails and between histones within the same or in different nucleosomes [127−129]. A first level of cross-talk can be identified in the mutually exclusive antagonism between different types of modifications, such as acetylation and methylation, occurring on the same lysine residue. Another level corresponds to the interdependency between different modifications. A good example of histone *trans*-tail cross-talk, reported for the first time in yeast [130,131], is represented by the prerequisite of ubiquitination of H2BK123 for H3K4 methylation, leading to the silencing of genes located near chromosome telomers [130,131]. An additional example of this *trans*-histone tail cross-talk is the dependency of H3K4 methylation by sCOMPASS and H3K79 methylation by scDot1 on the ubiquitylation of H2BK123 by scRad6/Bre1 [132]. Interestingly, this mechanism is conserved in humans [133]. Furthermore, adjacent modifications can influence each other as in the case of HP1 binding to H3K9 methylation being affected by the presence of an adjacent phosphorylated H3S10 [134].

Additionally, the catalytic activity of an enzyme could be influenced by modification of its substrate recognition site, for example the isomerization of H3P38 can influence the ability of Set2 to methylate H3K36 [135].

Similarly, the efficiency of substrate recognition by an enzyme could be different in the context of an additional modification as demonstrated by higher efficiency of H3 recognition by GCN5 acetyltransferase in the presence of H3S10 phosphorylation [136]. A further level involves a cooperative mechanism between some modifications to recruit specific factors, an example of which is the involvement of H3K9 and H3K14 acetylation in the recruitment of PHF8 to methylated H3K4 [110].

Equally important in the fine tuning control of chromatin organization is the interplay between the histone modifications, DNA methylation [137,138] and ATP-dependent chromatin remodeling [139]. The large number of histone modifications and the possible interplay between them led to the proposition of the so-called "histone code hypothesis" in which "multiple histone modifications, acting in a combinatorial or sequential fashion on one or multiple histone tails, specify unique downstream functions" [140,141]. This hypothesis led the scientific community to adopt some metaphors to describe it such that the code is written by some enzymes ("writers"), removed by others ("erasers"), and is readily recognized by proteins ("readers") recruited to modifications through the binding of specific domains.

Independently of the debate on the accuracy of considering the cross-talk between histone modifications as a histone code rather than an epigenetic code or a language of histone crosstalk [142,143], it is evident that the dynamic plasticity of chromatin, necessary for the control of cellular processes, represents the end point of a fine-tuning mechanism that involves the concerted action of the histone modifications together with the actions of DNA and chromatin remodeling proteins.

4.4 HISTONE MODIFICATIONS AND CANCER

Epigenetic alterations have been known to occur in cancer cells for decades. In fact, aberrant DNA methylation was described as an early event in tumorigenesis, although only recently was recognized to play a causal role, and not just an unwanted molecular consequence of the transformation process. With this appreciation, it became clear that other epigenetic modifications, such as histone post-translational changes, are also altered in cancer cells. The first examples of altered patterns of histone modifications in cancer [144,145] were found using different methodologies (liquid chromatography–electrospray mass spectrometry (LC-ES/MS) [146] and immunohistochemical (IHC) analysis [145]). One study showed the global level of trimethylation of H4K20 (H4K20me3) and acetylation of H4K16 (H4K16ac) in several types of cancer cells [144], while another reported the global level of the dimethylation of H3K4 (H3K4me2) and H3R3 (H3R3me2) as well as the level of acetylation of H3K9, H3K18, and H4K12 in primary prostate cancer tissues [145]. An impressive set of data/publications has confirmed and extended those initial studies. A comprehensive analysis of all of the alterations in the histone modification patterns found in cancer cells is prohibitive and beyond the scope of our work. However, we will try to review several cases of well-documented alterations in histone modifications in cancer and discuss their mechanistic implications. For clarity, we will describe separately histone H4 and histone H3 alterations.

4.4.1 Alterations in the Pattern of Histone H4 Modifications

Fraga and colleagues [144] compared histone H4 modifications in normal lymphocytes against several types of cancer cells and reported, for the first time, a globally lower level of both trimethylation of H4K20 (H4K20me3) and acetylation of H4K16 (H4K16ac) in cancer tissues, which was further validated in an analysis of primary tumors versus their normal counterpart [144]. Interestingly, the reduction of both histone modifications correlated with DNA hypomethylation at repetitive DNA sequences and with silencing of the promoter of tumor suppressor genes such as CDKN2A, BRCA1, and MLH1 [144]. In the same study, using a well-recognized model of tumor progression (the mouse multistage skin carcinogenesis model), the authors reported that the loss of those two histone modifications occurred progressively from the first stage of carcinogenesis, represented by benign papilloma, to the most malignant stage [144]. The progressive loss of H4K20me3 has been subsequently observed in additional animal models of carcinogenesis [146], including estradiol-induced mammary carcinogenesis in rats [147], and then reinforced by several studies performed on tissues derived from different cancer patients [148–152]. An aberrant pattern of histone H4 modifications, characterized by hypoacetylation of H4K12/H4K16, a loss of H4K20me3, and hyperacetylation of H4K5/H4K8, has been reported in a study conducted on non-small-cell lung cancer (NSCL) patients [148]. Loss of H4K20me3 in this case also represents an early event in tumorigenesis that was already present in early lesions and that becomes more evident during the sequential progression of disease moving from cell hyperplasia to metaplasia, dysplasia, and then to carcinoma in situ [148]. Reduction of H4K20me3 was more frequent in squamous cell carcinomas (67%) compared to adenocarcinomas (27%), whilst H4K16ac was more homogeneously reduced in the two histological types [148]. In lung adenocarcinomas, the observed down-regulation of H4K20me3 correlated with prognosis and permitted the identification of two populations of stage I tumor samples with distinct clinical outcome where a longer survival was observed in patients having higher levels of H4K20me3 [148]. Interestingly, loss of H4K20me3 correlated with decreased expression of a specific H4K20 trimethyltransferase, Suv4-20h2 [148]. Similar findings were also obtained in an experimental model of hepatocarcinogenesis induced by methyl deficiency in rats, strengthening the link between the two events [153]. In ovarian cancer, a significant correlation between H4K20me3 loss and an increase in malignancy has been observed by an IHC analysis conducted on normal ovarian epithelium, ovarian adenomas, and ovarian epithelial carcinomas [149]. In addition, in breast cancer, reduced levels of both H4K20me3 and H4K16ac correlated well with tumor grade, progression of disease, and with worse prognosis [150]. Similarly, a progressive loss of H4K16ac and H4K20me3 has been reported from low- to high-grade lung neuroendocrine tumors, reflecting both the degree of differentiation and the proliferation rate of the tumors [151]. In bladder cancer [152], IHC analysis of the global expression levels of different histone modifications (H3K4me1, H3K4me3, H4K20me1, H4K20me2, and H4K20me3), generally confirmed a correlation between cancer progression and a progressive decrease of histone modification levels. However, increased H4K20me3 correlated with worse prognosis [152]. Therefore, changes in H4K20 methylation levels appear to be frequently associated with chromatin alterations in cancer cells, but the precise significance of this finding is not necessarily consistent from cancer to cancer, excluding a simple interpretation of this phenomenon.

Other histone H4 modifications have been found associated with cancer cells. As mentioned above, H4K16 hypoacetylation correlates with worse prognosis in breast cancer and medulloblastoma [150,154]. In breast cancer, a study conducted on a very large dataset of patients revealed low or absent acetylation of H4K16 in the majority of analyzed cases and a strong correlation with clinico-histological features such as tumor grade, vascular invasion, and prognosis [150]. In medulloblastoma patients, a concurrent analysis of H4K16 acetylation and of the acetylase responsible for the modification (hMOF), demonstrated that the hypoacetylation of H4K16 in primary medulloblastomas, compared to normal tissues, correlated well with a reduction in hMOF and poor prognosis [154].

H4K12 acetylation (H4K12ac) is another histone H4 modification found altered in cancer [145,148,155,156]. A good correlation between hypoacetylation of H4K12, tumor grade, and cancer recurrence has been reported in prostate cancer patients [145]. In this cancer type, the prognostic value of H4K12ac was independent of tumor stage. If measured together with H3K9 and H3K18 acetylation, H4K12 acetylation permitted the clustering of low-grade prostate cancer cases (Gleason 6 or less) into two prognostically separate groups [145]. This finding highlights another important principle (see also below): it will require an integrated analysis of the different histone modifications to reveal complex histone patterns that will lead to a more consistent "epigenetic" classification of cancer types rather than a single histone modification which will only provide partial information.

A general decrease in H4K12ac has been reported in lung cancer, predominantly in adenocarcinoma patients [148]. In addition, a correlation between H4K12 hypoacetylation and tumor grade has been reported for colorectal cancer [155]. Hypoacetylation of H4K12 is also observed in aggressive breast carcinomas including basal carcinomas and HER-2-positive tumors [150]. Histone H4 and tubulin acetylation has been analyzed in breast cancer patients at different stages with a lower level of H4K12 acetylation being observed in both ductal carcinomas in situ (DCIS) and invasive ductal carcinoma (IDC) relative to synchronous normal breast epithelium [156]. A greater reduction in acetylation was seen in high-grade IDC versus low/intermediate-grade, ER-negative versus ER-positive, and PR-negative versus PR-positive tumors. Unexpectedly, the observed gradual loss of H4K12ac occurred in the context of a parallel reduction in the expression of HDAC1, 2 and HDAC6 [156]. Though this observation does not have an explanation so far, it does underline the difficulties in drawing mechanistical conclusions at this stage (discussed below).

Finally, we note that other technical approaches have been attempted to study histone modifications in cancer cells, and may also provide further insights. Cuomo and colleagues comprehensively analyzed post-translational modifications of histone H4 in a panel of breast cancer cell lines, compared to normal epithelial mammary cells combining high-resolution mass spectrometry analysis of histones with stable isotope labeling with amino acids in cell culture (SILAC) [157]. Besides confirming the presence of known alterations in histone H4 modifications (H4K16 hypoacetylation and loss of H4K20me3), a novel alteration was identified in the levels of H4K20me1 [157]. Although, at this stage, this method cannot be

67

applied to routine clinical samples, it may provide information that can be used for testing with different approaches (such as standard IHC) new candidate markers in patient samples.

4.4.2 Alterations in the Pattern of Histone H3 Modifications

A pioneer study by Seligston and colleagues (discussed above for histone H4) highlighted the possibility of using the study of histone modifications as prognostic predictors of clinical outcome [145]. The work characterized by IHC analysis on primary prostate cancer tissue, the levels of acetylation of H3K9, H3K18, and H4K12, and of the dimethylation of H4R3 and H3K4. This work revealed a clear difference in the pattern of modification on histone H3 in tumor versus normal prostate tissue. While no single histone modification analyzed was predictive per se, a more complex pattern obtained combining global histone modifications at multiple sites was able to define the clinical outcome of the analyzed patients: lower levels of modified histones characterized patients with poorer prognosis and with increased risk of tumor recurrence after removal of primary tumor [145]. These observations have been subsequently confirmed and expanded by a larger study reporting low levels of H3K4 monomethylation (H3K4me1), H3K9 dimethylation (H3K9me2), H3K9 trimethylation (H3K9me3), H3 and H4 acetylation in prostate cancer compared to non-malignant prostate tissue [158]. H3K4 dimethylation (H3K4me2) and H3K18 acetylation (H3K18ac), identified as the most predictive histone modifications in prostate cancer, have been further analyzed and their prognostic power has been confirmed in different cancer types [150,159–166]. Low levels of H3K4me2 and H3K18ac correlate with worse prognosis and survival in lung and kidney cancer [159]. In the same study, H3K9me2 alone predicts a poorer prognosis in prostate and kidney cancer [159]. The findings in kidney cancer have been confirmed by another study, in which the analysis of several histone modifications (H3K9ac, H3K18ac, total H3ac, and H4ac) led to the identification of a strong correlation between global level of H3K18 and H3K9 acetylation with cancer progression and worse prognosis [160]. Low levels of H3K4me2, H3K9me2, or H3K18ac have also been identified as significant and independent predictors of poor survival in pancreatic adenocarcinoma patients [161]. In this patient population, the combination of low levels of H3K4me2 and H3K18ac was identified as the most significant predictor of overall survival [161]. Importantly, in this study there was a significant correlation between low levels of H3K4me2 and/or H3K8me2 and a worse overall survival in the subgroup of patients treated with 5-FU, but not in the subgroup of patients treated with Gemcitabine: though not conclusive, this information poses the relevant question about the link between epigenetic pattern of the tumor and the response to treatment [161,162]. The largest correlative study of several histone modifications (H3K9ac, H3K18ac, H4K12ac, and H4K16ac, H3K4me2, and H4K20me3, and H4R3me2), clinicopathologic features and prognosis has been conducted in breast cancer [150]. From this study, beside the data pertaining to the histone H4 modifications discussed above, an additional finding was the correlation between low levels of H3K9 and H3K18 acetylation with high tumor grade and with biological markers such as the absence of steroid receptor expression [150]. In the study there was a strong correlation between low levels of H3K9 and H3K18 acetylation and breast carcinoma with poorer prognosis including basal carcinoma and Her2-positive tumors [150]. In the case of NSCLC patients of early stage (stages I to III), the most significant observation was a correlation between low levels of acetylation of lysine 5 of histone H2A (H2AK5ac) and worse survival for patients with tumor stage II, as well as a correlation between low levels of H3K4me2 and poor prognosis for patients with stage I [163]. In contrast to the data available for prostate, breast, and pancreatic cancer, where there was a poor prognosis [145,150,151], the study showed a correlation between low levels of H3K9 acetylation and better survival [163]. Further, contrasting data for H3K18 acetylation (H3K18ac) have been reported for esophageal and glioma cancer patients [170,171]. Nonetheless, for glioma, the analysis of several histone modifications (H3K9Ac, H3K18Ac, H3K4me2, H4K12Ac, and H4R3Me) confirmed the predictive power of low levels of H3K4me2 for worse survival [165]. In relation to the modification of additional sites on histone H3, an increased level of H3K56 acetylation correlated with higher grade in many cancer types including thyroid, laryngo-pharynx, colon cancer, and astrocytomas [166]. H3K27 methylation plays an important regulatory role in gene transcription and is found frequently altered in cancer cells compared to normal tissue, although not consistently among the different cancer types [164,167−170].

Low levels of H3K27 trimethylation (H3K27me3) have been reported in breast, ovarian, and pancreatic cancers [167]. In breast cancer, low levels of H3K27me3 correlated with additional prognostic factors such as large tumor size, estrogen receptor-negative and lymph node-positive status, but not with HER-2/Neu status [167]. A similar correlation with high tumor grade was observed in ovarian and pancreatic cancer [167]. Importantly, in all these tumor types, low levels of H3K27me3 correlated with a significantly shorter overall survival time [167]. In contrast with these observations, low levels of H3K27me3 and H3K18ac correlated with an improved prognosis in patients with esophageal squamous cell carcinoma [166,168]. In the case of esophageal squamous cell carcinoma patients treated with chemo/radiotherapy, a higher expression of H3K27me3 correlated positively with expression of the H3K27 methylase EZH2, tumor grade, and worse survival [168]. High levels of H3K27me3 associated with advanced clinical stage and short overall survival have also been reported in nasopharyngeal carcinoma [169] and hepatocellular carcinoma [170].

All of the studies reported above provide a first glimpse of the clinical relevance of the study of altered histone modification patterns in tumors: they are mainly if not exclusively correlative, and have been conducted using methodologies that do not allow a detailed mechanistic analysis of the molecular consequences of the observed alterations. In some cases, a different histone pattern between normal and cancer cell has been reported looking at the level of gene promoter, using chromatin immunoprecipitation-based techniques [171−173]. Interestingly, to our knowledge, little has been done so far to cross these two types of epigenetic analyses (at a more global level and a greater molecular detail) to provide a more refined epigenetic profile of cancer samples; something that is urgently needed.

By applying chromatin immunoprecipitation microarrays, in a comparative study between normal and prostate cancer cells, Kondo and colleagues found that 5% of promoters were enriched with H3K27me3 in cancer cells and that this enrichment corresponded to gene silencing independently of DNA methylation [171]. This effect could be reverted by EZH2 down-regulation [171]. A subsequent genome-wide analysis of H3K4me3 and H3K27me3 in prostate cancer cells and normal epithelial cells confirmed and expanded the observations reported above [172].

As another example of molecular studies, it is worth mentioning a genome-wide chromatin immunoprecipitation study conducted in leukemia patients [173]. The analysis of H3K9me3 revealed a specific H3K9me3 alteration pattern in leukemia patients which was able to distinguish not only normal hematopoietic cells from leukemic blasts, but also acute myeloid leukemia (AML) samples from acute lymphoblastic leukemia (ALL) samples. The analysis led to the identification of a characteristic signature of AML in which a decrease in H3K9me3 levels, occurring mainly at specific promoter regions such as those containing cyclic adenosine monophosphate response elements (CREs), correlated with higher transcriptional activity and most importantly, predicted survival [173].

4.5 MECHANISMS UNDERLYING HISTONE ALTERATIONS IN CANCER

An obvious question stemming from the studies summarized above regards the molecular mechanism(s) leading to the global alterations in the pattern of histone modifications observed in cancer cells. As an initial level of analysis, perhaps not surprisingly, it was found that altered expression levels of enzymes that add, remove, or recognize specific histone

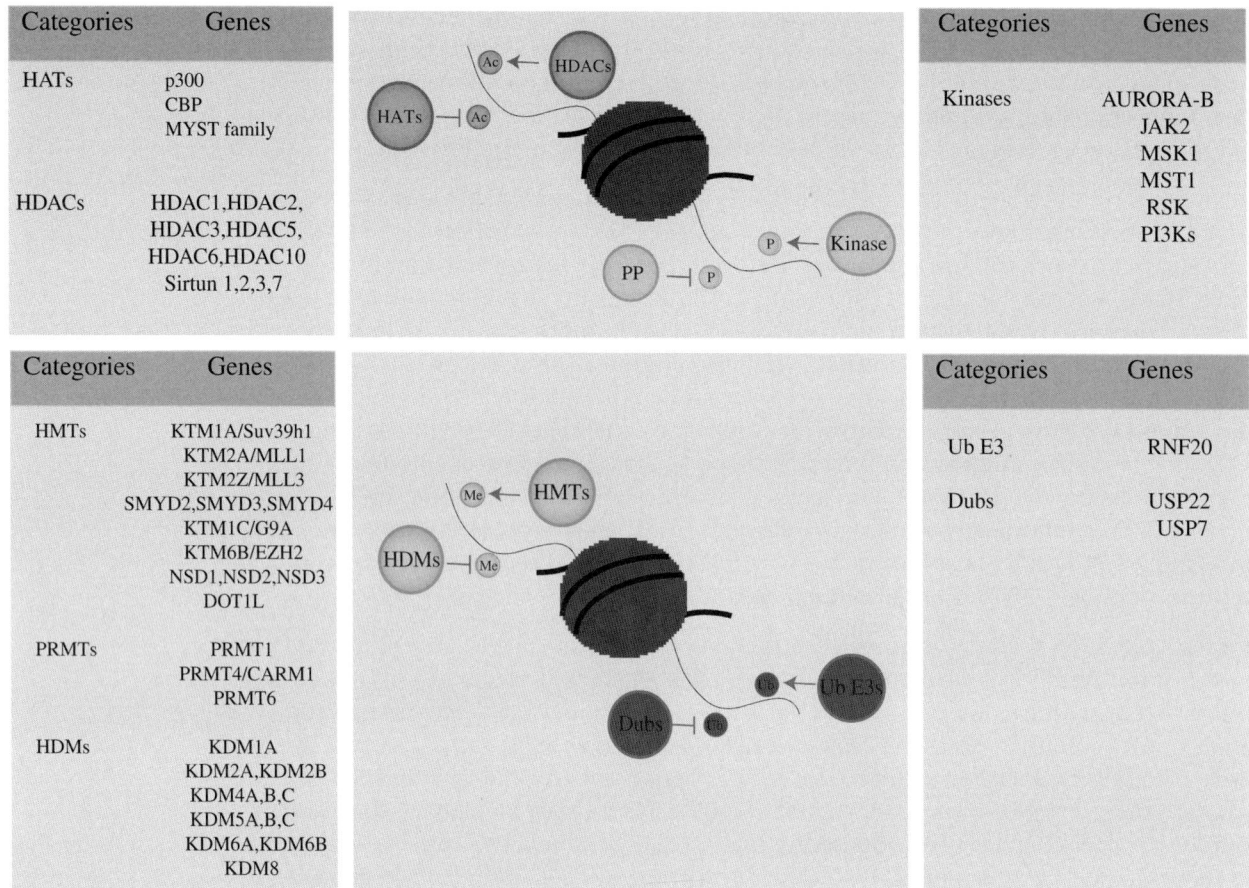

Categories	Genes
HATs	p300
	CBP
	MYST family
HDACs	HDAC1,HDAC2,
	HDAC3,HDAC5,
	HDAC6,HDAC10
	Sirtun 1,2,3,7

Categories	Genes
Kinases	AURORA-B
	JAK2
	MSK1
	MST1
	RSK
	PI3Ks

Categories	Genes
HMTs	KTM1A/Suv39h1
	KTM2A/MLL1
	KTM2Z/MLL3
	SMYD2,SMYD3,SMYD4
	KTM1C/G9A
	KTM6B/EZH2
	NSD1,NSD2,NSD3
	DOT1L
PRMTs	PRMT1
	PRMT4/CARM1
	PRMT6
HDMs	KDM1A
	KDM2A,KDM2B
	KDM4A,B,C
	KDM5A,B,C
	KDM6A,KDM6B
	KDM8

Categories	Genes
Ub E3	RNF20
Dubs	USP22
	USP7

FIGURE 4.3

Histone-modifying enzymes and cancer. Selection of histone modifying enzymes altered in human cancer. This figure is reproduced in the color plate section.

modifications can be measured in cancer cells. At a further level of complexity, in some cases, even in the presence of normal levels of those factors, an aberrant recruitment of histone-modifying enzymes at specific chromatin regions could be considered as a determining molecular event (Figure 4.3).

4.5.1 Alteration of the Histone Acetylation Network (HATs, HDACs and Sirtuins)

Chromosomal translocations, mutations or simply over-expression involving several HATs (such as E1A-binding protein p300 (EP300), cAMP response element-binding protein (CBP), and enzymes of the MYST family) strongly support the involvement of these enzymes in cancer [174–178]. Several inactivating mutations in p300, correlating with lack of enzymatic activity [177], have been detected in primary solid tumors and tumor-derived cell lines of epithelial origin, hinting at a tumor-suppressive role for the enzyme [177,178]. More intriguingly, in support of the view that p300 can act both as a tumor suppressor and an oncogene, down-regulation of p300 leads to growth inhibition and activation of a senescence checkpoint in human melanocytes [179]. For CBP, several inactivating mutations have been identified in epithelial tumors [180,181]. In AMLs, this enzyme has been found translocated and fused to either the HAT monocytic leukemia zinc finger (MOZ) gene or to the histone methylase MLL (mixed lineage leukemia) [182–184]. A translocation between CBP and a member of the MYST family (MORF), resulting in a fusion protein with two functional HAT domains, has also been reported in some cases of AML [185]. Sequence or deletion mutations of CBP, impairing its acetyltransferase activity, have been reported both in relapsed acute lymphoblastic leukemia (ALL) and B-cell lymphoma [186,187]. Several other HATs have also been found as altered in cancer. The MOZ gene has been identified as a common retroviral integration site leading to myeloid/lymphoid tumors [188], while hMOF expression appears frequently down-regulated both in primary breast carcinomas and in medulloblastoma [154]. A significant down-regulation of Tip60 expression in colon and lung carcinomas has been reported [189] as well as a link between Tip60 down-regulation and disease progression in colorectal and gastric cancer [190,191]. Monoallelic loss of the enzyme, with concomitant reduction in mRNA levels, has been reported in human lymphomas, head-and-neck and mammary carcinomas [192]. As for many other cases discussed above, up-regulation of Tip60 has been linked to promotion of epithelial tumorigenesis, suggesting that the enzyme can have both oncosuppressive and oncogenic properties [193,194]. A critical point (valid also for the other cases described below) is that there have been few, if any, attempts to correlate these observations with the altered patterns of histone modifications occurring in cancer cells. It is tempting to speculate that crossing these two levels of molecular analyses will lead to the definition of specific molecular lesions in histone-modifying enzymes (such as HATs discussed here) that generate defined alterations in the pattern of histone modifications in cancer cells.

In the case of HDACs, aberrant recruitment of HDAC1, 2 in acute promyelocytyc leukemia (APL) represents the first and well-characterized example of the contribution of these families of enzymes in cancer and particularly in hematological malignancies [195−197]. Another example of HDAC-dependent aberrant transcriptional repression can be found in the AML subtype M2, characterized by the presence of the fusion protein AML1-ETO, derived from the (t8;21) chromosomal translocation [198]. Similarly to the APL fusion protein, AML1-ETO works as a potent transcription repressor through the interaction with N-CoR and the formation of a complex with N-Cor/Sin3/HDAC1 [199,200]. HDACs are apparently involved in transcriptional repression in the case of leukemia due to inversion of chromosome 16 (8% of AML cases) [201], as well as through the interaction with the over-expressed transcription factor SCL/TAL1 in the development of T-cell acute lymphoblastic leukemia (T-ALL) [202]. An elevated expression of HDAC1, HDAC2, and HDAC6 and a possible correlation between expression levels of HDAC2 and histone H4 acetylation with tumor aggressiveness have been reported for cutaneous T-cell lymphoma patients [203]. Elevated expression of HDACs of class I, except HDAC3, was also reported in patients with primary myelofibrosis as compared to other myeloproliferative diseases or normal volunteers [204]. Moreover, over-expression of HDAC1, HDAC2, and HDAC6 and higher acetylation levels of histone H4 compared to normal lymphoid tissue have been reported in patients with peripheral T-cell lymphoma and diffuse B large cell lymphoma [205].

Over-expression of single HDACs, such as HDAC1, HDAC2, HDAC3, and HDAC6 among others, has also been reported in solid tumors [206]. Furthermore, reduced expression levels of HDAC1 in gastric cancer, and HDAC5 and HDAC10 in lung cancers have been linked with poor prognosis [207,208]. The class III HDAC, SIRT1, has been found up-regulated in murine lung carcinomas, lymphomas, and prostate cancer and in human AMLs, glioblastoma, colorectal, prostate, and skin cancer [208−211].

4.5.2 Alteration of the Histone Methylation Network (HMTs, PRMTs and HDMs)

Similarly to HATs and HDACs, altered expression and/or activity both of HMTs and HDMs, due to genetic mutations or to non-genetic events, has been reported in several cancer types [212,213].

Examining the various HMTs and HDMs, a differential expression profile of the H3K9me3 methyltransferase Suv39h1 (KMT1A) has been found in colorectal cancers versus normal colorectal mucosa [214]. In support of this, double knock-out of Suv39h1 (KMT1A)/Suv39h2

(KMT1B) in mice compromised genomic stability and led to cancer predisposition and lymphoma development as a consequence of a greatly reduced level of methylated histone H3K9 [215]. The most striking example of HMTs involvement in cancer is highlighted by the mixed lineage leukemia gene MLL1 (KMT2A), encoding an H3K4 HMT. Recurrent chromosomal translocations of MLL/KMT2A represent a common and frequent characteristic both in human lymphoid and myeloid leukemia [216,217]. More than 50 functional MLL fusion proteins have been found in approximately 10% of human acute leukemia, which can be defined as AMLs, acute lymphoblastic leukemias (ALLs) or mixed lineage leukemias (MLLs) [216–219]. Furthermore, intragenic mutations of MLL3 (KMT2Z) in colorectal cancer have been reported [220].

Another H3K4 HMT frequently found up-regulated in colorectal and hepatocellular carcinoma is SMYD3 [221, 222]. An increased expression of this enzyme has also been reported in breast cancer in which SMYD3 was found to promote carcinogenesis by directly regulating expression of the proto-oncogene WNT10B [223]. Similarly, SMYD2 has been found over-expressed in hepatocellular carcinoma [224], whilst down-regulation of SMYD4 expression has been reported both in medulloblastoma and breast cancer [225,226]. Increased levels of G9A (KMT1C), a H3K9 HMT, has been described in leukemia, hepatocellular carcinoma, and in prostate and lung cancer [227–230]. Another HMT involved in cellular transformation and oncogenesis is EZH2 (KMT6B), a H3K27 methytrasferase, that is a member of polycomb repressive complex 2 (PRC2) together with EED, SUZ12, and RbAp48 [231]. Increased expression of EZH2 has been found in several solid tumors such as prostate, breast, colon, skin, bladder, and lung cancer [231]. Moreover, tissue microarray analysis demonstrated that EZH2 protein levels strongly correlated with tumor aggressiveness [232], and recently, a polycomb repression signature, consisting of 14 direct targets of EZH2 repression, has been identified as a prognostic factor for the clinical outcomes in metastatic breast and prostate tumors [233]. Additional evidence correlating high expression of EZH2 with a poor prognosis in both breast and bladder cancer have recently been published [234,235]. Interesting data are emerging on the role of EZH2, and particularly of EZH2-mediated epigenetic repression of DNA damage repair, in a mechanism that could promote expansion of breast-tumor-initiating cells [236]. In contrast to the data reported above [231–235], inactivating mutation of EZH2, associated with a loss of function, were identified in myeloid malignancies [237,238]. Additionally, somatic and heterozygous EZH2 Y641 mutations that lead to a loss of enzymatic activity have been found in B-cell lymphomas [239,240]. Surprisingly, B-cell lymphoma cell lines and lymphoma samples harboring heterozygous EZH2 Y641 mutations showed increased levels of H3K27me3 [241]. Elegant in vitro enzyme assays have shed light on these observations, demonstrating that the Y641 mutation causes a concurrent decrease in monomethylation and increase in trimethylation activity of the mutated enzyme relative to its wild-type form [241]. Essentially, all the available evidence points to an oncogenic role for EZH2 and is consistent with the enhanced levels of H3K27Me3 found in tumor cells over-expressing the enzyme.

Additional HMTs found altered in cancer belong to the NSD family: NSD1 (KMT3B), NSD2, and NSD3 are responsible for methylation of H3K36 and at a less extent of H4K20 [242]. The NSD1 locus is involved in translocations with the nucleoporin gene (NUP98) and implicated in the pathogenesis of childhood acute myeloid leukemias and myelodysplastic syndromes [243,244]. Moreover, NSD1 expression has been found abrogated by CpG island-promoter hypermethylation both in neuroblastoma and glioma cells [245]. The translocation t (4;14) (p16;q32) involving the NSD2 locus has been reported in 20% of multiple myeloma cases [246,247] and a significant over-expression of NSD2 has been measured in different tumor types, correlating both with tumor aggressiveness and prognosis [248]. Similarly, NSD3 translocations with NUP98 in leukemia and NSD3 gene amplification in breast cancer cell lines and primary breast carcinoma have been reported [249,250]. A pivotal role of DOT-1, an H3K79 methyltransferase, in MLL fusion protein-mediated leukemogenesis has recently been reported [251,252].

All of the enzymes described above belong to the category of lysine methyltransferases. In the case of arginine methyltransferases, over-expression of CARM1/PRMT4 has been reported in breast, colorectal, and hormone-dependent prostate tumors [255−257]. PRMT1 is also apparently involved in MLL-mediated transformation [257]. Some data regarding the de-regulation of expression of both PRMT1 and PRMT6 in different cancer types have been published [258].

Due to their recent discovery, compared to HMTs, the HDMs have not been extensively characterized as yet; however, several recent results fully support a critical role for them in cancer [213].

The correlation of KDM1A over-expression with adverse clinical outcome has been reported in bladder, lung, prostate, neuroblastoma, and breast cancer [259−263], whereas low levels of both KDM1A and H3K4me2 have been reported in hepatocellular carcinoma [264]. Both KDM2A and KDM2B have been considered as either tumor suppressors or oncogenes depending on the cellular context [213]. Both KDM2A and KDM2B have been found in a retroviral-mediated insertional mutagenesis screening for genes involved in the induction of lymphomas in rats [265,266]. In support of this, both proteins are involved in immortaliza-tion of fibroblasts and oncogene-induced senescence [266,267]. Moreover, KDM2B expression has been found significantly increased both in B- and T-cell acute lymphoblastic leukemias, AMLs and in seminomas cells [268,269]. A relevant role for KDM2B in initiation and main-tenance of AML has recently been described [270]. A significantly decreased expression of KDM2A and KDM2B has been reported in prostate cancer and glioblastoma, respectively [271,272]. Amplification of KDM4C has been reported for esophageal squamous cell carcinomas, breast cancers, medulloblastomas, and metastatic lung sarcomatoid carcinomas [273−277]. Over-expression of KDM4A, KDM4B, and KDM4C has been reported in prostate cancer [278] and in medulloblastoma [279], whilst high expression of KDM4B has also been reported in ER-positive breast cancer [280]. KDM5A has been found over-expressed in gastric cancer [281] and has been involved in the determination of a drug tolerance phenotype in a non-small-cell lung carcinoma cell system [282]. Increased expression of KDM5B has been reported for several cancer types [283−286].

A systematic sequencing of renal carcinoma samples revealed the presence of inactivating mutations of KDM5C [287]. Functionally, a genome-wide siRNA screen led to identify KDM5C as one of the E2-dependent regulators of human papilloma virus oncogene expression [288]. Inactivating mutations in KDM6A have been reported in multiple myeloma, esophageal squamous cell carcinoma and clear cell renal cell carcinoma [287,289]. A reduced expression of KDM6B has been reported in lung and liver carcinoma, and in a subset of lymphomas and leukemias [290]. In contrast, but potentially consistent with a role as tumor suppressor, high expression of KDM6B has been reported in melanocytic nevi [291]. HPV-mediated induction of the expression of both KDM6A and KDM6B in cervical cancer cells has been reported supporting an involvement of both enzymes in HPV-mediated oncogenesis [292]. Interestingly, KDM6B expression could also be induced by Epstein−Barr virus and appears to be higher in Hodgkin's lymphoma [293]. Finally, multiple types of tumors including thyroid, adrenal, bladder, uterine, and liver exhibited a higher expression of the H3K36 me2 demethylase KDM8 protein in comparison to their respective normal tissue controls [294].

4.5.3 Alterations of Histone Kinases, Histone Ubiquitinating and Deubiquitinating Enzymes

A significant amount of data has been generated showing the deregulation of the enzymes responsible for histone phosphorylation in cancer.

Aurora B (the primary kinase responsible for H3S10 phosphorylation) has been found over-expressed in breast, colorectal, kidney, lung, and prostate cancers [295−298]. JAK2, the kinase that phosphorylates H3Y41 in hematopoietic cells, has also been found to be constitutively activated, due to gene mutation and/or rearrangement, in several hematological malignancies [299]. Furthermore, gene amplification of JAK2 and KDM4C has been recently found both in primary mediastinal B-cell lymphoma and Hodgkin's lymphoma [300]. An important role of MSK1 and RSK2, and consequently of histone H3 phosphorylation, in cellular transformation has been documented [301−303], together with a role of MST1 kinase as a tumor-suppressor gene [304]. Moreover, data in support of a possible interdependence between RSK2-H2AX and RSK2-H3 phosphorylation in the control of cell transformation have been published [305].

The histone ubiquitination network has not been fully characterized as yet, but a picture is emerging of its role in cancer [306]. The E3 ligase RNF20, responsible for H2B ubiquitination, has been found commonly silenced via DNA hypermethylation in breast cancers [307]. In support of this, RNF20-depleted cells showed an increased oncogenic potential after measuring the ability of cells to form tumors once transplanted in mice [307]. The deubiquitinating enzyme USP22 has been defined as a putative marker of cancer stem cells [308], and it has also been identified among 11 genes forming a gene signature associated with poor prognosis in multiple tumor types [309]. An elevated expression of USP22, correlating with a poor prognosis, has been reported both for colorectal and breast cancer [310,311], and more recently in melanoma [312]. Similarly, USP7 over-expression has been reported in prostate, bladder, colon, liver, and lung cancers [313], although in a more recent analysis its expression was not found significantly deregulated [312].

Taken together, these data (i.e pertaining to histone-modifying enzymes and secondary proteins involved in the network) conclusively demonstrate that alterations in the histone modification network occur invariably in cancer cells and have a complex modality of action. Although an exhaustive molecular description of the events/links underlying these alterations does not exist, the data generated to date can explain a large part of the epigenetic alterations found in cancer cells.

4.5.4 Other Mechanisms

The abundance and diversity of histone-modifying enzymes, as well as their overlapping and redundant substrate specificities, clearly exclude a simplistic explanation of the causal contribution of each enzyme to the altered pattern of histone modifications found in cancer.

Kurdistani proposed that at the basis of the altered pattern of histone modifications in cancer and in view of the low levels of histone modifications found in several cancer types, there is an anomalous allocation of co-factors as acetyl coenzymeA and S-adenosylmethionine required by HATs and HMTs to modify histones, respectively [314]. Thus, metabolic pathways involved in cell growth and division are also centrally involved in the mechanisms of histone modification. One speculation deriving from this proposal is that the altered metabolism of cancer cells, where there is a high and continuous need for macromolecular biosynthesis, could lead the cancer cells to prevalently divert the use of these co-factors from histone modification pathways to more immediately vital pathways. This diversion of essential cofactors away from histone modification enzymes would have a major impact on the global levels of histone acetylation and methylation. In fact, recent data linking the alteration of metabolic enzymes (IDH1-2, for example) and deregulated expression of histone methylation tend to support the link between altered metabolism, histone modification, and cancer [315].

4.6 CONCLUSIONS

Over the last few years a large number of histone modifications have been identified, information regarding the role of each type of modification has been generated, the enzymes catalyzing these modifications have been characterized and the elucidation of the molecular mechanisms linking these modifications to cellular processes has begun.

Concurrently, it has become evident that the histone post-translational modifications act in an ordered and coordinated manner and should be considered in the context of an "epigenetic code", or better, in the context of a "global cross-talk" network between the different histone modifications. Moreover, in agreement with the relevant role of histone modifications in several key cellular processes, significant data have been published suggesting that a deregulation of the histone modification pattern is linked to different human malignancies and particularly to cancer. In this regard, the most striking example is the loss of H3K16ac and H4K20me3 which represents a well-recognized cancer histone mark. In addition, several data have been generated from different cancer types highlighting the correlation between altered global histone modification patterns and cancer aggressiveness and there is now the possibility to use them as independent prognostic factors.

The reversible nature of epigenetic modifications, including all the histone modifications, has provided the basis for development of "epigenetic therapies". So far this has led to the approval of two DNA methyltransferase inhibitors (Vidaza and Decitabine) for myelodysplastic syndrome and two HDAC inhibitors (Vorinostat and Romidepsin) for cutaneous T-cell lymphoma. Large investments in both time and money for identifying inhibitors of other proteins involved in regulation of epigenetic modifications, including HAT, Sirtuin, HMTs, and HDMs, is ongoing.

It is now important for both basic and applied science to acquire additional knowledge regarding the functional relevance of each histone modification and, more importantly, knowledge pertaining to the interplay between these modifications and the machinery involved in their addition and removal in the context of cancer biology. The different pattern of histone modifications reported between normal and cancer cells, together with the accumulating evidence that these differences can be prognostic factors and potential predictors of therapeutic response, suggest that further research in these fields could open the way for better personalized medicine in both epigenetic and non-epigenetic therapies. Future studies intended to increase our knowledge of epigenetic modifications, histone modifications in particular, could have a striking impact on the relevance of epigenetic events in cancer biology and on the design of more efficient strategies for epigenetic therapies in the treatment of cancer. In this regard, of significant relevance is the advent and use of new technologies in the epigenomic field. The application of techniques such as chromatin immunoprecipitation followed by modern high-density microarrays, next-generation sequencing that permits profiling of large sample series and the accurate determination of the location of different histone modifications at global level are expected to have a major impact in the field.

Moreover, the identification of new methodologies that would permit the technologies described above to be performed on paraffin-embedded tissue samples (PAT-ChIP-seq) [316] would have an enormous impact. Analysis of historical cancer tissue samples would permit the determination of the location and type of different histone modifications on a global scale. Detailed analysis on samples where clinical outcome can be associated with an "epigenetic signature" would represent a real breakthrough in the identification of epigenetic biomarkers for different pathological conditions.

From the clinical perspective, one can envisage a future where a histone modification signature will exist for each type of cancer and that this signature will be correlated with prognosis and more importantly, with the choice of best possible treatment. Last, but by no means least, one could expect to use these histone maps to monitor the efficacy of epigenetic drug treatment at the molecular level.

75

References

[1] Alberts B, Johnson A, Lewis J, Raff M, Roberts K, Walter P. Chromosomal DNA and Its Packaging in the Chromatin Fiber. Mol Biol cell 4th edition 2002.

[2] Kouzarides T. Chromatin modifications and their function. Cell 2007;128:693—705.

[3] Murr R. Interplay between different epigenetic modifications and mechanisms. Adv Genet 2010;70:101—41.

[4] Kornerber RD. Chromatin structure: A repeating unit of histones and DNA. Science 1974;184:868—71.

[5] Oudet P, Gross-Bellard M, Chambon P. Electron microscopic and biochemical evidence that chromatin structure is a repeating unit. Cell 1975;4:281—300.

[6] Luger K, Mader AW, Richmond RK, Sargent DF, Richmond TJ. Crystal structure of the nucleosome core particle at 2.8 A resolution. Nature 1997;389:251—60.

[7] Kornberg RD, Lorch Y. Twenty-five years of the nucleosome, fundamental particle of the eukaryote chromosome. Cell 1999;98:285—94.

[8] Robinson PJ, Fairall L, Huynh VA, Rhodes D. EM measurements define the dimensions of the "30-nm" chromatin fiber: evidence for a compact, interdigitated structure. Proc Natl Acad Sci USA 2006;103:6506—11.

[9] Van Holde KE. Histone modifications. In: Rich A, editor. Chromatin, Springer Series in Molecular Biology. New York: Springer; 1988. p. 111—48.

[10] Arents G, Moudrianakis EN. The histone fold: a ubiquitous architectural motif utilized in DNA compaction and protein dimerization. Proc Natl Acad Sci USA 1995;92:11170—4.

[11] Peterson CL, Laniel MA. Histones and histone modifications. Curr Biol 2004;14:R546—51.

[12] Paul B, Henikoff T, Henikoff S. Histone variants — ancient wrap artists of the epigenomeS. Henikoff, K. Ahmad, Assembly of variant histones into chromatin. Annu Rev Cell Dev Biol 2005;21:133—53.

[13] Cosgrove MS, Boeke JD, Wolberger C. Regulated nucleosome mobility and the histone code. Nat Struct Mol Biol 2004;11:1037—104.

[14] Goll MG, Bestor TH. Histone modification and replacement in chromatin activation. Genes Dev 2002;16:1739—42.

[15] Grant PA. A tale of histone modifications. Genome Biol 2001;2. REVIEWS0003.

[16] Shahbazian MD, Grunstein M. Functions of site-specific histone acetylation and deacetylation. Annu Rev Biochem 2007;76:75—100.

[17] Struhl K. Histone acetylation and transcriptional regulatory mechanisms. Genes Dev 1998;12:599—606.

[18] Yang XJ, Seto E. HATs and HDACs: from structure, function and regulation to novel strategies for therapy and prevention. Oncogene 2007;26:5310—8.

[19] Lee KK, Workman JL. Histone acetyltransferase complexes: One size doesn't fit all. Nat Rev Mol Cell Biol 2007;8:284—95.

[20] Parthun MR. Hat1: the emerging cellular roles of a type B histone acetyltransferase. Oncogene 2007;26:5319—28.

[21] Yang XJ, Seto E. The Rpd3/Hda1 family of lysine deacetylases: from bacteria and yeast to mice and men. Nat Rev Mol Cell Biol 2008;9:206—18.

[22] Thiagalingam S, Cheng KH, Lee HJ, Mineva N, Thiagalingam A, Ponte JF. Histone deacetylases: unique players in shaping the epigenetic histone code. Ann NY Acad Sci 2003;983:84—100.

[23] Mai A. The therapeutic uses of chromatin-modifying agents. Expert Opin Ther Targets 2007;11:835—51.

[24] Finkel T, Deng CX, Mostoslavsky R. Recent progress in the biology and physiology of sirtuins. Nature 2009;460:587—91.

[25] Michan S, Sinclair D. Sirtuins in mammals: insights into their biological function. Biochem J 2007;404:1—13.

[26] Verdin E, Dequiedt F, Kasler HG. Class II histone deacetylases: versatile regulators. Trends Genet 2003;5:286—93.

[27] Gao L, Cueto MA, Asselbergs F, Atadja P. Cloning and functional characterization of HDAC11, a novel member of the human histone deacetylase family. J Biol Chem 2002;277:25748—55.

[28] Tjeertes JV, Miller KM, Jackson SP. Screen for DNA-damage-responsive histone modifications identifies H3K9Ac and H3K56Ac in human cells. EMBO J 2009;28:1878—89.

[29] Wang Z, Zang C, Rosenfeld JA, Schones DE, Barski A, Cuddapah S, et al. Combinatorial patterns of histone acetylations and methylations in the human genome. Nat Genet 2008;40:897—903.

[30] Wang Z, Zang C, Cui K, Schones DE, Barski A, Peng W, et al. Genome-wide mapping of HATs and HDACs reveals distinct functions in active and inactive genes. Cell 2009;138:1019—31.

[31] Keogh MC, Kurdistani SK, Morris SA, Ahn SH, Podolny V, Collins RS, et al. Cotranscriptional Set2 Methylation of Histone H3 Lysine 36 Recruits a Repressive Rpd3 Complex. Cell 2005;123:593—605.

[32] Carrozza MJ, Li B, Florens L, Suganuma T, Swanson SK, Lee KK, et al. Histone H3 Methylation by Set2 Directs Deacetylation of Coding Regions by Rpd3S to Suppress Spurious Intragenic Transcription. Cell 2005;123:581—92.

[33] Li B, Gogol M, Carey M, Pattenden SG, Seidel C, Workman JL. Infrequently transcribed long genes depend on the Set2/Rpd3S pathway for accurate transcription. Genes Dev 2007;21:1422−30.

[34] Shogren-Knaak M, Ishii H, Sun JM, Pazin MJ, Davie JR, Peterson CL. Histone H4-K16 acetylation controls chromatin structure and protein interactions. Science 2006;311:844−7.

[35] Cimbora DM, Schubeler D, Reik A, Hamilton J, Francastel C, Epner EM, et al. Long-distance control of origin choice and replication timing in the human beta-globin locus are independent of the locus control region. Mol Cell Biol 2000;20:5581−91.

[36] Doyon Y, Cayrou C, Ullah M, Landry AJ, Cote V, Selleck W, et al. ING tumor suppressor proteins are critical regulators of chromatin acetylation required for genome expression and perpetuation. Mol Cell 2006;21:51−64.

[37] Bedford MT, Clarke SG. Protein arginine methylation in mammals: who, what, and why. Mol Cell 2009;33:1−13.

[38] Martin C, Zhang Y. The diverse functions of histone lysine methylation. Nat Rev Mol Cell Biol 2005;6:838−49.

[39] Shilatifard A. Chromatin modifications by methylation and ubiquitination: implications in the regulation of gene expression. Annu Rev Biochem 2006;75:243−69.

[40] Ng SS, Yue WW, Oppermann U, Klose RJ. Dynamic protein methylation in chromatin biology. Cell Mol Life Sci 2009;66:407−22.

[41] Zhang Y, Reinberg D. Transcription regulation by histone methylation: interplay between different covalent modifications of the core histone tails. Genes Dev 2001;15:2343−60.

[42] Rice JC, Briggs SD, Ueberheide B, Barber CM, Shabanowitz J, Hunt DF, et al. Histone methyltransferases direct different degrees of methylation to define distinct chromatin domains. Mol Cell 2003;12:1591−8.

[43] Keppler BR, Archer TK. Chromatin-modifying enzymes as therapeutic targets-Part 1. Expert Opin Ther Targets 2008;12:1301−12.

[44] Wolf SS. The protein arginine methyltransferase family: an update about function, new perspectives and the physiological role in humans. Cell Mol Life Sci 2009;66:2109−21.

[45] Allis CD, Berger SL, Cote J, Dent S, Jenuwien T, Kouzarides T, et al. New nomenclature for chromatin-modifying enzymes. Cell 2007;131:633−6.

[46] Feng Q, Wang H, Ng HH, Erdjument-Bromage H, Tempst P, Struhl K, et al. Methylation of H3-lysine 79 is mediated by a new family of HMTases without a SET domain. Curr Biol 2002;12:1052−8.

[47] Fumasoni I, Meani N, Rambaldi D, Scafetta G, Alcalay M, Ciccarelli FD. Family expansion and gene re-arrangements contributed to the functional specialization of PRDM genes in vertebrates. BMC Evol Biol 2007;4(7):187.

[48] Kim KC, Geng L, Huang S. Inactivation of a histone methyltransferase by mutations in human cancers. Cancer Res 2003;63:7619−23.

[49] Funabiki T, Kreider BL, Ihle JN. The carboxyl domain of zinc fingers of the Evi-1 myeloid transforming gene binds a consensus sequence of GAAGATGAG. Oncogene 1994;9:1575−158.

[50] Cuthbert GL, Daujat S, Snowden AW, Erdjument-Bromage H, Hagiwara T, Yamada M, et al. Histone de-imination antagonizes arginine methylation. Cell 2004;118:545−53.

[51] Wang Y, Wysocka J, Sayegh J, Lee YH, Perlin JR, Leonelli L, et al. Human PAD4 regulates histone arginine methylation levels via demethylimination. Science 2004;306:279−83.

[52] Trojer P, Reinberg D. Histone lysine demethylases and their impact on epigenetics. Cell 2006;125:213−7.

[53] Klose RJ, Zhang Y. Regulation of histone methylation by demethylimination and demethylation. Nat Rev Mol Cell Biol 2007;8:307−18.

[54] Mosammaparast N, Shi Y. Reversal of histone methylation: biochemical and molecular mechanisms of histone demethylases. Annu Rev Biochem 2010;79:155−79.

[55] Shi Y, Lan F, Matson C, Mulligan P, Whetstine JR, Cole PA, et al. Histone demethylation mediated by the nuclear amine oxidase homolog LSD1. Cell 2004;119:941−53.

[56] Chang B, Chen Y, Zhao Y, Bruick RK. JMJD6 is a histone arginine demethylase. Science 2007;318:444−7.

[57] Metzger E, Wissmann M, Yin N, Müller JM, Schneider R, Peters AH, et al. LSD1 demethylates repressive histone marks to promote androgen-receptor-dependent transcription. Nature 2005;437:436−9.

[58] Yun M, Wu J, Workman JL, Li B. Readers of histone modifications. Cell Res 2011;21:564−78.

[59] Wysocka J, Swigut T, Xiao H, Milne TA, Kwon SY, Landry J, et al. A PHD finger of NURF couples histone H3 lysine 4 trimethylation with chromatin remodelling. Nature 2006;442:86−90.

[60] Bannister AJ, Zegerman P, Partridge JF, Miska EA, Thomas JO, Allshire RC, et al. Selective recognition of methylated lysine 9 on histone H3 by the HP1 chromo domain. Nature 2001;410:120−4.

[61] Barski A, Cuddapah S, Cui K, Roh TY, Schones DE, Wang Z, et al. High-resolution profiling of histone methylations in the human genome. Cell 2007;129:823−37.

77

[62] Bernstein BE, Mikkelsen TS, Xie X, Kamal M, Huebert DJ, Cuff J, et al. A bivalent chromatin structure marks key developmental genes in embryonic stem cells. Cell 2006;125:315−26.

[63] Oki M, Aihara H, Ito T. Role of histone phosphorylation in chromatin dynamics and its implications in diseases. Subcell Biochem 2007;41:319−36.

[64] Cheung P, Allis CD, Sassone-Corsi P. Signaling to chromatin through histone modifications. Cell 2000;103:263−71.

[65] Dawson MA, Bannister AJ, Göttgens B, Foster SD, Bartke T, Green AR, et al. JAK2 phosphorylates histone H3Y41 and excludes HP1alpha from chromatin. Nature 2009;461:819−22.

[66] Ito T. Role of histone modification in chromatin dynamics. J Biochem 2007;141:609−14.

[67] Macdonald N, Welburn JP, Noble ME, Nguyen A, Yaffe MB, Clynes D, et al. Molecular basis for the recognition of phosphorylated and phosphoacetylated histone h3 by 14-3-3. Mol Cell 2005;20:199−211.

[68] Winter S, Simboeck E, Fischle W, Zupkovitz G, Dohnal I, Mechtler K, et al. 14-3-3 proteins recognize a histone code at histone H3 and are required for transcriptional activation. EMBO Journal 2008;27:88−99.

[69] Foster ER, Downs JA. Histone H2A phosphorylation in DNA double-strand break repair. FEBS J 2005;272:3231−40.

[70] Cheung WL, Ajiro K, Samejima K, Kloc M, Cheung P, Mizzen CA, et al. Apoptotic phosphorylation of histone H2B is mediated by mammalian sterile twenty kinase. Cell 2003;113:507−17.

[71] Nowak SJ, Corces VG. Phosphorylation of histone H3: A balancing act between chromosome condensation and transcriptional activation. Trends Genet 2004;20:214−20.

[72] Sugiyama K, Sugiura K, Hara T, Sugimoto K, Shima H, Honda K, et al. Aurora-B associated protein phosphatases as negative regulators of kinase activation. Oncogene 2002;21:3103−11.

[73] Goto H, Yasui Y, Nigg EA, Inagaki M. Aurora-B phosphorylates histone H3 at serine28 with regard to the mitotic chromosome condensation. Genes Cells 2002;7:11−7.

[74] Dai J, Sultan S, Taylor SS, Higgins JM. The kinase haspin is required for mitotic histone H3 Thr 3 phosphorylation and normal metaphase chromosome alignment. Genes Dev 2005;19:472−88.

[75] Preuss U, Landsberg G, Scheidtmann KH. Novel mitosis-specific phosphorylation of histone H3 at Thr11 mediated by Dlk/ZIP kinase. Nucleic Acids Res 2003;31:878−85.

[76] Shimada M, Niida H, Zineldeen DH, Tagami H, Tanaka M, Saito H, et al. Chk1 is a histone H3 threonine 11 kinase that regulates DNA damage-induced transcriptional repression. Cell 2008;132:221−32.

[77] Lau AT, Lee SY, Xu YM, Zheng D, Cho YY, Zhu F, et al. Phosphorylation of Histone H2B Serine 32 Is Linked to Cell Transformation. J Biol Chem 2011;286:26628−37.

[78] Zhu F, Zykova TA, Peng C, Zhang J, Cho YY, Zheng D, et al. Phosphorylation of H2AX at Ser139 and a new phosphorylation site Ser16 by RSK2 decreases H2AX ubiquitination and inhibits cell transformation. Cancer Res 2011;71:393−403.

[79] Chowdhury D, Keogh MC, Ishii H, Peterson CL, Buratowski S, Lieberman J. γ-H2AX dephosphorylation by protein phosphatase 2A facilitates DNA double-strand break repair. Mol Cell 2005;20:801−9.

[80] Chowdhury D, Xu X, Zhong X, Ahmed F, Zhong J, Liao J, et al. A PP4-phosphatase complex dephosphorylates γ-H2AX generated during DNA replication. Mol Cell 2008;31:33−46.

[81] Shimada M, Haruta M, Niida H, Sawamoto K, Nakanishi M. Protein phosphatase 1γ is responsible for dephosphorylation of histone H3 at Thr 11 after DNA damage. EMBO Rep 2010;11:883−9.

[82] Pickart CM, Eddins MJ. Ubiquitin: Structures, functions, mechanisms. Biochim Biophy Acta 2004;1695:55−72.

[83] Hershko A, Ciechanover A. The ubiquitin system. Annu Rev Biochem 1998;67:425−79.

[84] Zhang Y. Transcriptional regulation by histone ubiquitination and deubiquitination. Genes Dev 2003;17:2733−40.

[85] Tanny JC, Erdjument-Bromage H, Tempst P, Allis CD. Ubiquitylation of histone H2B controls RNA polymerase II transcription elongation independently of histone H3 methylation. Genes Dev 2007;21:835−47.

[86] Fleming AB, Kao CF, Hillyer C, Pikaart M, Osley MA. H2B Ubiquitylation Plays a Role in Nucleosome Dynamics during Transcription Elongation. Mol Cell 2008;31:57−66.

[87] Shukla A, Bhaumik SR. H2B-K123 ubiquitination stimulates RNAPII elongation independent of H3-K4 methylation. B.B.R.C 2007;59:214−20.

[88] Weake VM, Workman JL. Histone Ubiquitination: Triggering Gene Activity. Mol Cell 2008;29:653−63.

[89] Osley MA. Regulation of histone H2A and H2B ubiquitylation. Brief Funct Genomic Proteomic 2006;5: 179−18.

[90] Fierz B, Chatterjee C, McGinty RK, Bar-Dagan M, Raleigh DP, Muir TW. Histone H2B ubiquitylation disrupts local and higher-order chromatin compaction. Nat Chem Biol 2011;7:113−9.

78

[91] Koken MH, Reynolds P, Jaspers-Dekker I, Prakash L, Prakash S, Bootsma D, et al. Structural and functional conservation of two human homologs of the yeast DNA repair gene RAD6. Proc Natl Acad Sci USA 1991;88:8865—9.

[92] Roest HP, van Klaveren J, de Wit J, van Gurp CG, Koken MH, Vermey M, et al. Inactivation of the HR6B ubiquitin-conjugating DNA repair enzyme n mice causes male sterility associated with chromatin modification. Cell 1996;86:799—810.

[93] Kim J, Hake SB, Roeder RG. The human homolog of yeast BRE1 functions as a transcriptional coactivator through direct activator interactions. Mol Cell 2005;20:759—70.

[94] Zhu B, Zheng Y, Pham AD, Mandal SS, Erdjument-Bromage H, Tempst P, et al. Monoubiquitination of human histone H2B: the factors involved and their roles in HOX gene regulation. Mol Cell 2005;20:601—11.

[95] Cao R, Tsukada Y, Zhang Y. Role of Bmi-1 and Ring1A in H2A ubiquitylation and Hox gene silencing. Mol Cell 2005;20:845—54.

[96] Zhou W, Zhu P, Wang J, Pascual G, Ohgi KA, Lozach J, et al. Histone H2A monoubiquitination represses transcription by inhibiting RNA polymerase II transcriptional elongation. Mol Cell 2008;29:69—80.

[97] Berginck S, Salomons FA, Hoogstraten D, Groothuis TA, de Waard H, Wu J, et al. DNA damage triggers nucleotide excision repair-dependent monoubiquitylation of histone H2A. Genes Dev 2006;20:1343—52.

[98] Mailand N, Bekker-Jensen S, Faustrup H, Melander F, Bartek J, Lukas C, et al. RNF8 ubiquitylates histones at DNA double-strand breaks and promotes assembly of repair proteins. Cell 2007;131:887—900.

[99] Zhang XY, Varthi M, Sykes SM, Phillips C, Warzecha C, Zhu W, et al. The putative cancer stem cell marker USP22 Is a subunit of the human SAGA complex required for activated transcription and cell-cycle progression. Mol Cell 2008;29:102—11.

[100] Zhao Y, Lang G, Ito S, Bonnet J, Metzger E, Sawatsubashi S, et al. A TFTC/ STAGA module mediates histone H2A and H2B deubiquitination, coactivates nuclear receptors, and counteracts heterochromatin silencing. Mol Cell 2008;29:92—101.

[101] Zhang XY, Pfeiffer HK, Thorne AW, McMahon SB. USP22, an hSAGA subunit and potential cancer stem cell marker, reverses the polycomb-catalyzed ubiquitylation of histone H2A. Cell Cycle 2008; 7(1):522—4.

[102] Zhu P, Zhou W, Wang J, Puc J, Ohgi KA, Erdjument-Bromage H, et al. A histone H2A deubiquitinase complex coordinating histone acetylation and H1 dissociation in transcriptional regulation. Mol Cell 2007;27:609—21.

[103] Nicassio F, Corrado N, Vissers JH, Areces LB, Berginck S, Marteijn JA, et al. Human USP3 is a chromatin modifier required for S phase progression and genome stability. Curr Biol 2007;17:1972—7.

[104] Van der Knaap JA, Kumar BR, Moshkin YM, Langenberg K, Krijgsveld J, Heck AJ, et al. GMP synthetase stimulates histone H2B deubiquitylation by the epigenetic silencer USP7. Mol Cell 2005;17:695—707.

[105] Feng L, Wang J, Chen J. The Lys 63-specific deubiquitinating enzyme BRCC36 is regulated by two scaffold proteins localizing in different subcellular compartments. J Biol Chem 2010;285:30982—8.

[106] Wang GG, Allis CD, Chi P. Chromatin remodeling and cancer, Part I: Covalent histone modifications. Trends Mol Med 2007;13:363—72.

[107] Bannister AJ, Kouzarides T. Regulation of chromatin by histone modifications. Cell Res 2011;221:381—95.

[108] Taverna SD, Li H, Ruthenburg AJ, Allis CD, Patel DJ. How chromatin-binding modules interpret histone modifications: lessons from professional pocket pickers. Nat Struct Mol Biol 2007;11:1025—40.

[109] Vermeulen M, Eberl HC, Matarese F, Marks H, Denissov S, Butter F, et al. Quantitative interaction proteomics and genome-wide profiling of epigenetic histone marks and their readers. Cell 2010;142:967—80.

[110] Bartke T, Vermeulen M, Xhemalce B, Robson SC, Mann M, Kouzarides T. Nucleosome-interacting proteins regulated by DNA and histone methylation. Cell 2010;43:470—84.

[111] Mujtaba S, Zeng L, Zhou MM. Structure and acetyl-lysine recognition of the bromodomain. Oncogene 2007;26:5521—7.

[112] Maurer-Stroh S, Dickens NJ, Hughes-Davies L, Kouzarides T, Eisenhaber F, Ponting CP. The Tudor domain 'Royal Family': Tudor, plant Agenet, Chromo, PWWP and MBT domains. Trends Biochem Sci 2003;28:69—74.

[113] Champagne KS, Kutateladze TG. Structural insight into histone recognition by the ING PHD fingers. Curr Drug Targets 2009;10:432—41.

[114] Stucki M, Clapperton JA, Mohammad D, Yaffe MB, Smerdon SJ, Jackson SP. MDC1 directly binds phosphorylated histone H2AX to regulate cellular responses to DNA doublestrand breaks. Cell 2005;123:1213—26.

[115] Sims RJ, Chen CF, Santos-Rosa H, Kouzarides T, Patel SS, Reinberg D. Human but not yeast CHD1 binds directly and selectively to histone H3 methylated at lysine 4 via its tandem chromodomains. J Biol Chem 2005;280:41789—92.

[116] Huang Y, Fang J, Bedford MT, Zhang Y, Xu RM. Recognition of histone H3 lysine-4 methylation by the double Tudor domain of JMJD2A. Science 2006;312:748–51.

[117] Trojer P, Li G, Sims 3rd RJ, Vaquero A, Kalakonda N, Boccuni P, et al. L3MBTL1, a histonemethylation-dependent chromatin lock. Cell 2007;129:915–28.

[118] Zegerman P, Canas B, Pappin D, Kouzarides T. Histone H3 lysine 4 methylation disrupts binding of nucleosome remodeling and deacetylase (NuRD) repressor complex. J Biol Chem 2002;277:11621–4.

[119] Adams-Cioaba MA, Min J. Structure and function of histone methylation binding proteins. Biochem Cell Biol 2009;87:93–105.

[120] Huisinga KL, Brower-Toland B, Elgin SCR. The contradictory definitions of heterochromatin: Transcription and silencing. Chromosoma 2006;115:110–22.

[121] Peters AH, Kubicek S, Mechtler K, O'Sullivan RJ, Derijck AA, Perez-Burgos L, et al. Partitioning and plasticity of repressive histone methylation states in mammalian chromatin. Mol Cell 2003;12:1577–89.

[122] Schotta G, Lachner M, Sarma K, Ebert A, Sengupta R, Reuter G, et al. A silencing pathway to induce H3-K9 and H4-K20 trimethylation at constitutive heterochromatin. Genes Dev 2004;18:1251–62.

[123] Sawan C, Herceg Z. Histone modification and cancer. Adv Genet 2010;70:57–85.

[124] Turner BM. Defining an epigenetic code. Nat Cell Biol 2007;9:2–6.

[125] Hazzalin CA, Mahadevan LC. Dynamic acetylation of all lysine 4-methylated histone H3 in the mouse nucleus: analysis at c-fos and c-jun. PLoS Biol 2005;3:e393.

[126] Kohlmaier A, Savarese F, Lachner M, Martens J, Jenuwein T, Wutz A. A chromosomal memory triggered by Xist regulates histone methylation in X inactivation. PLoS Biol Jul 2004;2(7):e171.

[127] Lee JS, Smith E, Shilatifard A. The language of histone crosstalk. Cell 2010;142:682–5.

[128] Latham JA, Dent SY. Cross-regulation of histone modifications. Nat Struct Mol Biol 2007;14:1017–24.

[129] Shukla A, Chaurasia P, Bhaumik SR. Histone methylation and ubiquitination with their cross-talk and roles in gene expression and stability. Cell Mol Life Sci 2009;66:1419–33.

[130] Sun ZW, Allis CD. Ubiquitination of histone H2B regulates H3 methylation and gene silencing in yeast. Nature 2002;418:104–8.

[131] Dover J, Schneider J, Tawiah-Boateng MA, Wood A, Dean K, Johnston M, et al. Methylation of histone H3 by COMPASS requires ubiquitination of histone H2B by Rad6. J Biol Chem 2002;277:28368–71.

[132] Lee JS, Shukla A, Schneider J, Swanson SK, Washburn MP, Florens L, et al. Histone crosstalk between H2B monoubiquitination and H3 methylation mediated by COMPASS. Cell 2007;131:1084–96.

[133] Kim J, Guermah M, McGinty RK, Lee JS, Tang Z, Milne TA, et al. RAD6-Mediated transcription-coupled H2B ubiquitylation directly stimulates H3K4 methylation in human cells. Cell 2009;137:459–71.

[134] Fischle W, Tseng BS, Dormann HL, Ueberheide BM, Garcia BA, Shabanowitz J, et al. Regulation of HP1-chromatin binding by histone H3 methylation and phosphorylation. Nature 2005;438:1116–22.

[135] Nelson CJ, Santos-Rosa H, Kouzarides T. Proline isomerization of histone H3 regulates lysine methylation and gene expression. Cell 2006;126:905–16.

[136] Clements A, Poux AN, Lo WS, Pillus L, Berger SL, Marmorstein R. Structural basis for histone and phosphohistone binding by the GCN5 histone acetyltransferase. Mol Cell 2003;12:461–73.

[137] Cedar H, Bergman Y. Linking DNA methylation and histone modification: patterns and paradigms. Nat RevGenet 2009;10:295–304.

[138] Bartke T, Vermeulen M, Xhemalce B, Robson SC, Mann M, Kouzarides T. Nucleosome-interacting proteins regulated by DNA and histone methylation. Cell 2010;143:470–84.

[139] Santos-Rosa H, Caldas C. Chromatin modifier enzymes, the histone code and cancer. Eur J Cancer 2005;41:2381–402.

[140] Strahl BD, Allis CD. The language of covalent histone modifications. Nature 2000;403:41–5.

[141] Jenuwein T, Allis CD. Translating the histone code. Science 2001;293:1074–80.

[142] Gardner KE, Allis CD, Strahl BD. Operating on chromatin, a colorful language where context matters. J Mol Biol 2011;409:36–46.

[143] Henikoff S, Shilatifard A. Histone modification: cause or cog? Trends Genet 2011;10:389–96.

[144] Fraga MF, Ballestar E, Villar-Garea A, Boix-Chornet M, Espada J, Schotta G, et al. Loss of acetylation at Lys16 and trimethylation at Lys20 of histone H4 is a common hallmark of human cancer. Nat Genet 2005;37:391–400.

[145] Seligson DB, Horvath S, Shi T, Yu H, Tze S, Grunstein M, et al. Global histone modification patterns predict risk of prostate cancer recurrence. Nature 2005;435:1262–6.

[146] Bagnyukova TV, Tryndyak VP, Montgomery B, Churchwell MI, Karpf AR, James SR, et al. Genetic and epigenetic changes in rat preneoplastic liver tissue induced by 2-acetylaminofluorene. Carcinogenesis 2008;29:638–46.

[147] Kovalchuk O, Tryndyak VP, Montgomery B, Boyko A, Kutanzi K, Zemp F, et al. Estrogen-induced rat breast carcinogenesis is characterized by alterations in DNA methylation, histone modifications and aberrant microRNA expression. Cell Cycle 2007;6:2010—8.

[148] Van Den Broeck A, Brambilla E, Moro-Sibilot D, Lantuejoul S, Brambilla C, Eymin B, et al. Loss of histone H4K20 trimethylation occurs in preneoplasia and influences prognosis of non-small cell lung cancer. Clin Cancer Res 2008;14:7237—45.

[149] Zhen L, Gui-lan L, Ping Y, Jin H, Ya-li W. The expression of H3K9Ac, H3K14Ac, and H4K20TriMe in epithelial ovarian tumors and the clinical significance. Int J Gynecol Cancer 2010;20:82—6.

[150] Elsheikh SE, Green AR, Rakha EA, Powe DG, Ahmed RA, Collins HM, et al. Global histone modifications in breast cancer correlate with tumor phenotypes, prognostic factors, and patient outcome. Cancer Res 2009;69:3802—9.

[151] Li F, Ye B, Hong L, Xu H, Fishbein MC. Epigenetic Modifications of Histone H4 in Lung Neuroendocrine Tumors. Appl Immunohistochem Mol Morphol 2011;5:389—94.

[152] Schneider AC, Heukamp LC, Rogenhofer S, Fechner G, Bastian PJ, von Ruecker A, et al. Global histone H4K20 trimethylation predicts cancer-specific survival in patients with muscleinvasive bladder cancer BJU Int; 2011;8(pt2):E290—6.

[153] Pogribny IP, Ross SA, Tryndyak VP, Pogribna M, Poirier LA, Karpinets TV. Histone H3 lysine 9 and H4 lysine 20 trimethylation and the expression of Suv4-20h2 and Suv-39h1 histone methyltransferases in hepato-carcinogenesis induced by methyl deficiency in rats. Carcinogenesis 2006;27:1180—6.

[154] Pfister S, Rea S, Taipale M, Mendrzyk F, Straub B, Ittrich C, et al. The histone acetyltransferase hMOF is frequently downregulated in primary breast carcinoma and medulloblastoma and constitutes a biomarker for clinical outcome in medulloblastoma. Int J Cancer 2008;122:1207—13.

[155] Ashktorab H, Belgrave K, Hosseinkhah F, Brim H, Nouraie M, Takkikto, et al. Global histone H4 acetylation and HDAC2 expression in colon adenoma and carcinoma. Dig Dis Sci 2009;54:2109—1.

[156] Suzuki J, Chen YY, Scott GK, Devries S, Chin K, Benz CC, et al. Protein acetylation and histone deacetylase expression associated with malignant breast cancer progression Clin Cancer Res 2009;15:3163—71.

[157] Cuomo A, Moretti S, Minucci S, Bonaldi T. SILAC-based proteomic analysis to dissect the "histone modification signature" of human breast cancer cells. Amino Acids 2011;41:387—99.

[158] Ellinger J, Kahl P, von der Gathen J, Rogenhofer S, Heukamp LC, Gütgemann I, et al. Global levels of histone modifications predict prostate cancer recurrence. Prostate 2010;70:61—9.

[159] Seligson DB, Horvath S, McBrian MA, Mah V, Yu H, Tze S, et al. Global levels of histone modifications predict prognosis in different cancers. Am J Pathol 2009;4:1619—28.

[160] Mosashvilli D, Kahl P, Mertens C, Holzapfel S, Rogenhofer S, Hauser S, et al. Global histone acetylation levels: prognostic relevance in patients with renal cell carcinoma. Cancer Sci 2010;101:2664—9.

[161] Manuyakorn A, Paulus R, Farrell J, Dawson NA, Tze S, Cheung-Lau G, et al. Cellular histone modification patterns predict prognosis and treatment response in resectable pancreatic adenocarcinoma: results from RTOG 9704. J Clin Oncol 2010;28:1358—65.

[162] Altucci L, Minucci S. Epigenetic therapies in haematological malignancies: searching for true targets. Eur J Cancer 2009;45:1137—45.

[163] Barlési F, Giaccone G, Gallegos-Ruiz MI, Loundou A, Span SW, Lefesvre P, et al. Global histone modifications predict prognosis of resected non small-cell lung cancer. J Clin Oncol 2007;25:4358—64.

[164] Tzao C, Tung HJ, Jin JS, Sun GH, Hsu HS, Chen BH, et al. Prognostic significance of global histone modifications in resected squamous cell carcinoma of the esophagus. Mod Pathol 2009;22:252—60.

[165] Liu BL, Cheng JX, Zhang X, Wang R, Zhang W, Lin H, et al. Global histone modification patterns as prognostic markers to classify glioma patients. Cancer Epidemiol Biomarkers Prev 2010;19:2888—96.

[166] Das C, Lucia MS, Hansen KC, Tyler JK. CBP/p300-mediated acetylation of histone H3 on lysine 56. Nature 2009;459:113—7.

[167] Wei Y, Xia W, Zhang Z, Liu J, Wang H, Adsay NV, et al. Loss of trimethylation at lysine 27 of histone H3 is a predictor of poor outcome in breast, ovarian, and pancreatic cancers. Mol Carcinog 2008;47:701—6.

[168] He LR, Liu MZ, Li BK, Rao HL, Liao YJ, Guan XY, et al. Prognostic impact of H3K27me3 expression on locoregional progression after chemoradiotherapy in esophageal squamous cell carcinoma. B M C Cancer 2009;22(9):461.

[169] Cai MY, Tong ZT, Zhu W, Wen ZZ, Rao HL, Kong LL, et al. H3K27me3 Protein Is a Promising Predictive Biomarker of Patients' Survival and Chemoradioresistance in Human Nasopharyngeal Carcinoma. Mol Med 2011. doi: 10.2119/molmed.2011.00054. [Epub ahead of print].

[170] Cai MY, Hou JH, Rao HL, Luo RZ, Li M, Pei XQ, et al. High expression of H3K27me3 in human hepato-cellular carcinomas correlates closely with vascular invasion and predicts worse prognosis in patients. Mol Med 2011;(1—2):12—20.

81

[171] Kondo Y, Shen L, Cheng AS, Ahmed S, Boumber Y, Charo C, et al. Gene silencing in cancer by histone H3 lysine 27 trimethylation independent of promoter DNA methylation. Nat Genet 2008;40:741—50.

[172] Ke XS, Qu Y, Rostad K, Li WC, Lin B, Halvorsen OJ, et al. Genome-wide profiling of histone h3 lysine 4 and lysine 27 trimethylation reveals an epigenetic signature in prostate carcinogenesis. PLoS One 2009;4(3):e4687.

[173] Muller-Tidow C, Ulrich Klein H, Hascher A, Isken F, Tickenbrock L, Thoennissen N, et al. Profiling of histone H3 lysine 9 trimethylation levels predicts transcription factor activity and survival in acute myeloid leukemia. Blood 2010;18:3564—71.

[174] Dekker FJ, Haisma HJ. Histone acetyl transferases as emerging drug targets. Drug Discov. Today 2009;14:942—8.

[175] Avvakumov N, Cote J. The MYST family of histone acetyltransferases and their intimate links to cancer. Oncogene 2007;26:5395—407.

[176] Yang XJ. The diverse superfamily of lysine acetyltransferases and their roles in leukemia and other diseases. Nucleic Acids Res 2004;32:959—76.

[177] Muraoka M, Konishi M, Kikuchi-Yanoshita R, Tanaka K, Shitara N, Chong JM, et al. p300 gene alterations in colorectal and gastric carcinomas. Oncogene 1996;12:1565—9.

[178] Gayther SA, Batley SJ, Linger L, Bannister A, Thorpe K, Chin SF, et al. Mutations truncating the EP300 acetylase in human cancers. Nat Genet 2000;24:300—3.

[179] Bandyopadhyay D, Okan NA, Bales E, Nascimento L, Cole PA, Medrano EE. Down-regulation of p300/CBP histone acetyltransferase activates a senescence checkpoint in human melanocytes. Cancer Res 2002;62:6231—9.

[180] Iyer NG, Ozdag H, Caldas C. p300/CBP and cancer. Oncogene 2004;23:4225—31.

[181] Kishimoto M, Kohno T, Okudela K, Otsuka A, Sasaki H, Tanabe C, et al. Mutations and deletions of the CBP gene in human lung cancer. Clin Cancer Res 2005;11:512—9.

[182] Crowley JA, Wang Y, Rapoport AP, Ning Y. Detection of MOZ-CBP fusion in acute myeloid leukemia with 8;16 translocation. Leukemia 2005;19:2344—5.

[183] Giles RH, Dauwerse JG, Higgins C, Petrij F, Wessels JW, Beverstock G, et al. Detection of CBP rearrangements in acute myelogenous leukemia with t (8; 16). Leukemia 1997;11:2087—96.

[184] Taki T, Sako M, Tsuchida M, Hayashi Y. The t (11; 16) (q23; p13) translocation in myelodysplastic syndrome fuses the MLL gene to the CBP gene. Blood 1997;89:3945—50.

[185] Mullighan CG, Zhang J, Kasper LH, Lerach S, Payne-Turner D, Phillips LA, et al. CREBBP mutations in relapsed acute lymphoblastic leukaemia. Nature 2011;471:235—9.

[186] Pasqualucci L, Dominguez-Sola D, Chiarenza A, Fabbri G, Grunn A, Trifonov V, et al. Inactivating mutations of acetyltransferase genes in B-cell lymphoma. Nature 2011:471,189—471,195.

[187] Panagopoulos I, Fioretos T, Isaksson M, Samuelsson U, Billström R, Strömbeck B, et al. Fusion of the MORF and CBP genes in acute myeloid leukemia with the t (10;16)(q22;p13). Hum Mol Genet 2001;10:395—404.

[188] Lund AH, Turner G, Trubetskoy A, Verhoeven E, Wientjens E, Hulsman D, et al. Genome-wide retroviral insertional tagging of genes involved in cancer in Cdkn2a-deficient mice. Nat Genet 2002;32:160—5.

[189] Leonart ME, Vidal F, Gallardo D, Diaz-Fuertes M, Rojo F, Cuatrecasas M, et al. New p53 related genes in human tumors: significant downregulation in colon and lung carcinomas. Oncol Rep 2006;16:603—8.

[190] Sakuraba K, Yasuda T, Sakata M, Kitamura YH, Shirahata A, Goto T, et al. Down-regulation of Tip60 gene as a potential marker for the malignancy of colorectal cancer. Anticancer Res 2009;10:3953—5.

[191] Sakuraba K, Yokomizo K, Shirahata A, Goto T, Saito M, Ishibashi K, et al. TIP60 as a potential marker for the malignancy of gastric cancer. Anticancer Res 2011;31:77—9.

[192] Gorrini C, Squatrito M, Luise C, Syed N, Perna D, Wark L, et al. Tip60 is a haplo-insufficient tumour suppressor required for an oncogene-induced DNA damage response. Nature 2007;448:1063—7:

[193] Hobbs CA, Wei G, Defeo K, Paul B, Hayes CS, Gilmour SK. Tip60 protein isoforms and altered function in skin and tumors that over-express ornithine decarboxylase. Cancer Res 2006;66:8116—812.

[194] Shiota M, Yokomizo A, Masubuchi D, Tada Y, Inokuchi J, Eto M, et al. Tip60 promotes prostate cancer cell proliferation by translocation of androgen receptor into the nucleus. Prostate 2010;70:540—54.

[195] Minucci S, Nervi C, Lo Coco F, Pelicci PG. Histone deacetylases: A common molecular target for differentiation treatment of acute myeloid leukemias? Oncogene 2001;20:3110—5.

[196] Peterson LF, Zhang DE. The 8; 21 translocation in leukemogenesis. Oncogene 2004;23:4255—62.

[197] Mercurio C, Minucci S, Pelicci PG. Histone deacetylases and epigenetic therapies of hematological malignancies. Pharmacol Res 2010;62:18—34.

[198] Wang J, Hoshino T, Redner RL, Kajigaya S, Liu JM. ETO, fusion partner in t (8; 21) acute myeloid leukemia, represses transcription by interaction with the human N-CoR/mSin3/HDAC1 complex. Proc Natl Acad Sci USA 1998;95:10860—5.

82

[199] Lutterbach B, Westendorf JJ, Linggi B, Patten A, Moniwa M, Davie JR, et al. ETO, a target of t (8; 21) in acute leukemia, interacts with the N-CoR and mSin3 corepressors. Mol Cell Biol 1998;18:7176—84.

[200] Gelmetti V, Zhang J, Fanelli M, Minucci S, Pelicci PG, Lazar MA. Aberrant recruitment of the nuclear receptor corepressor-histone deacetylase complex by the acute myeloid leukemia fusion partner ETO. Mol Cell Biol 1998;18:7185—91.

[201] Shigesada K, van de Sluis B, Liu PP. Mechanism of leukemogenesis by the inv (16) chimeric gene CBFB/PEBP2B-MHY11. Oncogene 2004;23:4297—307.

[202] Ferrando AA, Herblot S, Palomero T, Hansen M, Hoang T, Fox EA, et al. Biallelic transcriptional activation of oncogenic transcription factors in T-cell acute lymphoblastic leukemia. Blood 2004;103:1909—11.

[203] Marquard L, Gjerdrum LM, Christensen IJ, Jensen PB, Sehested M, Ralfkiaer E. Prognostic significance of the therapeutic targets histone deacetylase 1, 2, 6 and acetylated histone H4 in cutaneous T-cell lymphoma. Histopathology 2008;53:267—77.

[204] Wang JC, Chen C, Dumlao T, Naik S, Chang T, Xiao YY, et al. Enhanced histone deacetylase enzyme activity in primary myelofibrosis. Leuk Lymphoma 2008;49:2321—7.

[205] Marquard L, Poulsen CB, Gjerdrum LM, de Nully Brown P, Christensen IJ, Jensen PB. Histone deacetylase 1, 2, 6 and acetylated histone H4 in B- and T-cell lymphomas. Histopathology 2009;54:688—98.

[206] Bolden JE, Peart MJ, Johnstone RW. Anticancer activities of histone deacetylase inhibitors. Nat Rev Drug Discov 2006;5:769—84.

[207] Osada H, Tatematsu Y, Saito H, Yatabe Y, Mitsudomi T, Takahashi T. Reduced expression of class II histone deacetylase genes is associated with poor prognosis in lung cancer patients. Int J Cancer 2004;112:26—32.

[208] Huffman DM, Grizzle WE, Bamman MM, Kim JS, Eltoum IA, Elgavish A, et al. SIRT1 is significantly elevated in mouse and human prostate cancer. Cancer Res 2007;67:6612—8.

[209] Liu G, Yuan X, Zeng Z, Tunici P, Ng H, Abdulkadir IR, et al. Analysis of gene expression and chemoresistance of CD133þ cancer stem cells in glioblastoma. Mol Cancer 2006;5:67.

[210] Ozdag H, Teschendorff AE, Ahmed AA, Hyland SJ, Blenkiron C, Bobrow L, et al. Differential expression of selected histone modifier genes in human solid cancers. B M C Genomics Apr 2006;25(7):90.

[211] Biancotto C, Frigè G, Minucci S. Histone modification therapy of cancer. Adv Genet 2010;70:341—86.

[212] Albert M, Helin K. Histone methyltransferases in cancer. Seminars in Cell & Developmental Biology 2010;21:209—20.

[213] Varier RA, Timmers HT. Histone lysine methylation and demethylation pathways in cancer. Biochimica et Biophysica Acta 2011;1815:75—89.

[214] Ka MY, Lee BB, Kim YH, Chang DK, Kyu Park S, Chun HK, et al. Association of the SUV39H1 histone methyltransferase with the DNA methyltransferase 1 at mRNA expression level in primary colorectal cancer. Int J Cancer 2007;121:2—7.

[215] Peters AH, O'Carroll D, Scherthan H, Mechtler K, Sauer S, Schöfer C, et al. Loss of the Suv39h histone methyltransferases impairs mammalian heterochromatin and genome stability. Cell 2001;107:323—37.

[216] Hess JL. MLL: a histone methyltransferase disrupted in leukemia. Trends Mol Med 2004;10:500—7.

[217] Krivtsov AV, Armstrong SA. MLL translocations, histone modifications and leukaemia stem-cell development. Nat Rev Cancer 2007;7:823—33.

[218] Shih LY, Liang DC, Fu JF, Wu JH, Wang PN, Lin TL, et al. Characterization of fusion partner genes in 114 patients with de novo acute myeloid leukemia and MLL rearrangement. Leukemia 2006;20. 218-2.

[219] Slany RK. The molecular biology of mixed lineage leukemia. Haematologica 2009;94:984—93.

[220] Sjöblom T, Jones S, Wood LD, Parsons DW, Lin J, Barber TD, et al. The consensus coding sequences of human breast and colorectal cancers. Science 2006;314:268—74.

[221] Hamamoto R, Furukawa Y, Morita M, Iimura Y, Silva FP, Li M, et al. SMYD3 encodes a histone methyltransferase involved in the proliferation of cancer cells. Nat Cell Biol 2004;6:731—40.

[222] Luo XG, Xi T, Guo S, Liu ZP, Wang N, Jiang Y, et al. Effects of SMYD3 over-expression on transformation, serum dependence, and apoptosis sensitivity in NIH3T3 cells. IUBMB Life 2009;61:679—84.

[223] Hamamoto R, Silva FP, Tsuge M, Nishidate T, Katagiri T, Nakamura Y, et al. Enhanced SMYD3 expression is essential for the growth of breast cancer cells. Cancer Sci 2006;97:113—8.

[224] Skawran B, Steinemann D, Weigmann A, Flemming P, Becker T, Flik J, et al. Gene expression profiling in hepatocellular carcinoma: upregulation of genes in amplified chromosome regions. Mod Pathol 2008;21:505—16.

[225] Northcott PA, Nakahara Y, Wu X, Feuk L, Ellison DW, Croul S, et al. Multiple recurrent genetic events converge on control of histone lysine methylation in medulloblastoma. Nat Genet 2009;41:465—72.

[226] Hu L, Zhu YT, Qi C, Zhu YJ. Identification of Smyd4 as a potential tumor suppressor gene involved in breast cancer development. Cancer Res 2009;69:4067—72.

83

[227] Huang J, Dorsey J, Chuikov S, Pérez-Burgos L, Zhang X, Jenuwein T, et al. G9A and GLP methylate lysine 373 in the tumor suppressor p53. J Biol Chem 2010;285:9636—41.

[228] Kondo Y, Shen L, Ahmed S, Boumber Y, Sekido Y, Haddad BR, et al. Downregulation of histone H3 lysine 9 methyltransferase G9a induces centrosome disruption and chromosome instability in cancer cells. PLoS ONE 2008;3:e2037.

[229] Kondo Y, Shen L, Suzuki S, Kurokawa T, Masuko K, Tanaka Y, et al. Alterations of DNA methylation and histone modifications contribute to gene silencing in hepatocellular carcinomas. Hepatol Res 2007;37:974—83.

[230] Watanabe H, Soejima K, Yasuda H, Kawada I, Nakachi I, Yoda S, et al. Deregulation of histone lysine methyltransferases contributes to oncogenic transformation of human bronchoepithelial cells. Cancer Cell Int 2008;8:15.

[231] Chase A, Cross NC. Aberrations of EZH2 in cancer. Clin Cancer Res 2011;17:2613—8.

[232] Kleer CG, Cao Q, Varambally S, Shen R, Ota I, Tomlins SA, et al. EZH2 is a marker of aggressive breast cancer and promotes neoplastic transformation of breast epithelial cells. Proc Natl Acad Sci USA 2003;100:11606—11.

[233] Yu J, Yu J, Rhodes DR, Tomlins SA, Cao X, Chen G, et al. A polycomb repression signature in metastatic prostate cancer predicts cancer outcome. Cancer Res 2007;67:10657—63.

[234] Alford SH, Toy K, Merajver SD, Kleer CG. Increased risk for distant metastasis in patients with familial early-stage breast cancer and high EZH2 expression. Breast Cancer Res Treat 2011. 2011 May 26. [Epub ahead of print].

[235] Takawa M, Masuda K, Kunizaki M, Daigo Y, Takagi K, Iwai Y, et al. Validation of the histone methyltransferase EZH2 as a therapeutic target for various types of human cancer and as a prognostic marker. Cancer Sci 2011;102:1298—305.

[236] Chang CJ, Yang JY, Xia W, Chen CT, Xie X, Chao CH, et al. EZH2 promotes expansion of breast tumor initiating cells through activation of RAF1-beta-catenin signaling. Cancer Cell 2011;19:86—100.

[237] Ernst T, Chase AJ, Score J, Hidalgo-Curtis CE, Bryant C, Jones AV, et al. Inactivating mutations of the histone methyltransferase gene EZH2 in myeloid disorders. Nat Genet 2010;42:722—6.

[238] Nikoloski G, Langemeijer SM, Kuiper RP, Knops R, Massop M, Tönnissen ER, et al. Somatic mutations of the histone methyltransferase gene EZH2 in myelodysplastic syndromes. Nat Genet 2010;42:665—7.

[239] Makishima H, Jankowska AM, Tiu RV, Szpurka H, Sugimoto Y, Hu Z, et al. Novel homo- and hemizygous mutations in EZH2 in myeloid malignancies. Leukemia 2010;24:1799—804.

[240] Morin RD, Johnson NA, Severson TM, Mungall AJ, An J, Goya R, et al. Somatic mutations altering EZH2 (Tyr641) in follicular and diffuse large B-cell lymphomas of germinal-center origin. Nat Genet 2010;42:181—18.

[241] Yap DB, Chu J, Berg T, Schapira M, Cheng SW, Moradian A, et al. Somatic mutations at EZH2 Y641 act dominantly through a mechanism of selectively altered PRC2 catalytic activity, to increase H3K27 trime-thylation. Blood 2011;117:2451—9.

[242] Morishita M, di Luccio E. Cancers and the NSD family of histone lysine methyltransferases. Biochim Biophys Acta 2011;1816:158—63.

[243] Wang GG, Caim L, Pasillas MP, Kamps MP. NUP98-NSD1 links H3K36 methylation to Hox-A gene activation and leukaemogenesis. Nat Cell Biol 2007;9:804—12.

[244] Jaju RJ, Fidler C, Haas OA, Strickson AJ, Watkins F, Clark K, et al. A novel gene, NSD1, is fused to NUP98 in the t(5;11)(q35;p15.5) in de novo childhood acute myeloid leukemia,. Blood 2001;98:1264—12.

[245] Berdasco M, Ropero S, Setien F, Fraga MF, Lapunzina P, Losson R, et al. Epigenetic inactivation of the Sotos overgrowth syndrome gene histone methyltransferase NSD1 in human neuroblastoma and glioma. Proc Natl Acad Sci USA 2009;106:21830—5.

[246] Martinez-Garcia E, Popovic R, Min DJ, Sweet SM, Thomas PM, Zamdborg L, et al. The MMSET histone methyl transferase switches global histone methylation and alters gene expression in t (4;14) multiple myeloma cells. Blood 2011;117:211—20.

[247] Brito JL, Walker B, Jenner M, Dickens NJ, Brown NJ, Ross FM, et al. MMSET deregulation affects cell cycle progression and adhesion regulons in t (4; 14) myeloma plasma cells. Haematologica 2009;94:78—86.

[248] Kassambara A, Klein B, Moreaux J. MMSET is over-expressed in cancers: link with tumor aggressiveness. Biochem. Biophys Res Comm 2009;379:840—5.

[249] Rosati R, La Starza R, Veronese A, Aventin A, Schwienbacher C, Vallespi T, et al. NUP98 is fused to the NSD3 gene in acute myeloid leukemia associated with t (8;11)(p11.2;p15). Blood 2002;99:3857—60.

[250] Angrand PO, Apiou F, Stewart AF, Dutrillaux B, Losson R, Chambon P. NSD3, a new SET domain-containing gene, maps to 8p12 and is amplified in human breast cancer cell lines. Genomics 2001;74:79—88.

[251] Bernt KM, Zhu N, Sinha AU, Vempati S, Faber J, Krivtsov AV, et al. MLL-Rearranged Leukemia Is Dependent on Aberrant H3K79 Methylation by DOT1L. Cancer Cell 2011;20:66—78.

[252] Nguyen AT, Taranova O, He J, Zhang Y. DOT1L, the H3K79 methyltransferase, is required for MLL-AF9-mediated leukemogenesis. Blood 2011;117:6912—22.

[253] Chang MJ, Wu H, Achille NJ, Reisenauer MR, Chou CW, Zeleznik-Le NJ. Histone H3 lysine 79 methyl-transferase Dot1 is required for immortalization by MLL oncogenes. Cell 2005;121:167—78.

[254] Barry ER, Corry GN, Rasmussen TP. Targeting DOT1L action and interactions in leukemia: the role of DOT1L in transformation and development. Expert Opin Ther Targets 2010;14:405—18.

[255] Majumder S, Liu Y, Ford 3rd OH, Mohler JL, Whang YE, Involvement of arginine methyltransferase. CARM1 in androgen receptor function and prostate cancer cell viability. Prostate 2006;66:1292—301.

[256] Cheung N, Chan LC, Thompson A, Cleary ML, Eric So CW. Protein argininemethyltransferase- dependent oncogenesis. Nat Cell Biol 2007;9:1208—15.

[257] Pal S, Vishwanath SN, Erdjument-Bromage H, Tempst P, Sif S. Human SWI/SNFassociated PRMT5 methylates histone H3 arginine 8 and negatively regulates expression of ST7 and NM23 tumor suppressor genes. Mol Cell Biol 2004;24:9630—45.

[258] Yoshimatsu M, Toyokawa G, Hayami S, Unoki M, Tsunoda T, et al. Dysregulation of PRMT1 and PRMT6, Type I arginine methyltransferases, is involved in various types of human cancers. Int J Cancer 2011;128:562—73.

[259] Kahl P, Gullotti L, Heukamp LC, Wolf S, Friedrichs N, Vorreuther R, et al. Androgen receptor coactivators lysine-specific histone demethylase 1 and four and a half LIM domain protein 2 predict risk of prostate cancer recurrence. Cancer Res 2006;66:11341—1134.

[260] Schulte JH, Lim S, Schramm A, Friedrichs N, Koster J, Versteeg R, et al. Lysine-specific demethylase 1 is strongly expressed in poorly differentiated neuroblastoma: implications for therapy. Cancer Res 2009;69:2065—71.

[261] Lim S, Janzer A, Becker A, Zimmer A, Schule R, Buettner R, et al. Lysine specific demethylase 1 (LSD1) is highly expressed in ER-negative breast cancers and a biomarker predicting aggressive biology. Carcinogenesis 2010;31:512—5.

[262] Hayami S, Kelly JD, Cho HS, Yoshimatsu M, Unoki M, Tsunoda T, et al. Over-expression of LSD1 contributes to human carcinogenesis through chromatin regulation in various cancers. Int J Cancer 2011;128:574—86.

[263] Kauffman EC, Robinson BD, Downes MJ, Powell LG, Lee MM, Scherr DS, et al. Role of androgen receptor and associated lysine-demethylase coregulators, LSD1 and JMJD2A, in localized and advanced human bladder cancer. Mol Carcinog 2011. 2011 Mar 11. doi: 10.1002/mc.20758. Epub ahead of print.

[264] Magerl C, Ellinger J, Braunschweig T, Kremmer E, Koch LK, Höller T, et al. H3K4 dimethylation in hepato-cellular carcinoma is rare compared with other hepatobiliary and gastrointestinal carcinomas and correlates with expression of the methylase Ash2 and the demethylase LSD1. Hum Pathol 2010;41:181—9.

[265] Suzuki T, Minehata K, Akagi K, Jenkins NA, Copeland NG. Tumor suppressor gene identification using retroviral insertional mutagenesis in Blm-deficient mice. EMBO J 2006;25:3422—31.

[266] Pfau R, Tzatsos A, Kampranis SC, Serebrennikova OB, Bear SE, Tsichlis PN. Members of a family of JmjC domain-containing oncoproteins immortalize embryonic fibroblasts via a JmjC domain dependent process. Proc Natl Acad Sci USA 2008;105:1907—12.

[267] Tzatsos A, Pfau R, Kampranis SC, Tsichlis PN. Ndy1/KDM2B immortalizes mouse embryonic fibroblasts by repressing the Ink4a/Arf locus. Proc Natl Acad Sci USA 2009;106:2641—6.

[268] Andersson A, Ritz C, Lindgren D, Edén P, Lassen C, Heldrup J, et al. Microarray-based classification of a consecutive series of 121 childhood acute leukemias: prediction of leukemic and genetic subtype as well as of minimal residual disease status. Leukemia 2007;21:1198—203.

[269] Sperger JM, Chen X, Draper JS, Antosiewicz JE, Chon CH, Jones SB, et al. Gene expression patterns in human embryonic stem cells and human pluripotent germ cell tumors. Proc Natl Acad Sci USA 2003;100:13350—5.

[270] He J, Nguyen AT, Zhang Y. KDM2b/JHDM1b, an H3K36me2-specific demethylase, is required for initiation and maintenance of acute myeloid leukemia. Blood 2001;117:3869—80.

[271] Frescas D, Guardavaccaro D, Kuchay SM, Kato H, Poleshko A, Basrur V, et al. KDM2A represses transcription of centromeric satellite repeats and maintains the heterochromatic state. Cell Cycle 2008;7:3539—47.

[272] Frescas D, Guardavaccaro D, Bassermann F, Koyama-Nasu R, Pagano M. JHDM1B/FBXL10 is a nucleolar protein that represses transcription of ribosomal RNA genes. Nature 2007;450:309—13.

[273] Yang ZQ, Imoto I, Fukuda Y, Pimkhaokham A, Shimada Y, Imamura M, et al. Identification of a novel gene, GASC1, within an amplicon at 9p23-24 frequently detected in esophageal cancer cell lines. Cancer Res 2000;60:4735—9.

[274] Vinatzer U, Gollinger M, Mullauer L, Raderer M, Chott A, Streubel B. Mucosa associated lymphoid tissue lymphoma: novel translocations including rearrangements of ODZ2, JMJD2C, and CNN3. Clin Cancer Res 2008;14:6426—31.

[275] Italiano A, Attias R, Aurias A, Pérot G, Burel-Vandenbos F, Otto J, et al. Molecular cytogenetic characterization of a metastatic lung sarcomatoid carcinoma: 9p23 neocentromere and 9p23-p24 amplification including JAK2 and JMJD2C. Cancer Genet Cytogenet 2006;167:122−30.

[276] Liu G, Bollig-Fischer A, Kreike B, van de Vijver MJ, Abrams J, Ethier SP, et al. Genomic amplification and oncogenic properties of the GASC1 histone demethylase gene in breast cancer. Oncogene 2009;28:4491−500.

[277] Ehrbrecht A, Müller U, Wolter M, Hoischen A, Koch A, Radlwimmer B, et al. Comprehensive genomic analysis of desmoplastic medulloblastomas: identification of novel amplified genes and separate evaluation of the different histological components. J Pathol 2006;208:554−63.

[278] Cloos PA, Christensen J, Agger K, Maiolica A, Rappsilber J, Antal T, et al. The putative oncogene GASC1 demethylates tri- and dimethylated lysine 9 on histone H3. Nature 2006;442:307−11.

[279] Northcott PA, Nakahara Y, Wu X, Feuk L, Ellison DW, Croul S, et al. Multiple recurrent genetic events converge on control of histone lysine methylation in medulloblastoma. Nat Gen 2009;41:465−72.

[280] Yang J, Jubb AM, Pike L, Buffa FM, Turley H, Baban D, et al. The histone demethylase JMJD2B is regulated by estrogen receptor alpha and hypoxia, and is a key mediator of estrogen induced growth. Cancer Res 2010;70:6456−66.

[281] Zeng KJ, Ge Z, Wang L, Li Q, Wang N, Bjorkholm M, et al. The histone demethylase RBP2 Is over-expressed in gastric cancer and its inhibition triggers senescence of cancer cells. Gastroenterology 2010;138:981−92.

[282] Sharma SV, Lee DY, Li B, Quinlan MP, Takahashi F, Maheswaran S, et al. A chromatin-mediated reversible drug-tolerant state in cancer cell subpopulations. Cell 2010;141:69−80.

[283] Xiang Y, Zhu Z, Han G, Ye X, Xu B, Peng Z, et al. JARID1B is a histone H3 lysine 4 demethylase up-regulated in prostate cancer. Proc Natl Acad Sci USA 2007;104:19226−31.

[284] Lu PJ, Sundquist K, Baeckstrom D, Poulsom R, Hanby A, Meier-Ewert S, et al. A novel gene (PLU-1) containing highly conserved putative DNA/chromatin binding motifs is specifically up-regulated in breast cancer. J Biol Chem 1999;274:15633−45.

[285] Yamane K, Tateishi K, Klose RJ, Fang J, Fabrizio LA, Erdjument-Bromage H, et al. PLU-1 is an H3K4 demethylase involved in transcriptional repression and breast cancer cell proliferation. Mol Cell 2007;25:801−12.

[286] Hayami S, Yoshimatsu M, Veerakumarasivam A, Unoki M, Iwai Y, Tsunoda T, et al. Over-expression of the JmjC histone demethylase KDM5B in human carcinogenesis: involvement in the proliferation of cancer cells through the E2F/RB pathway. Mol Cancer 2010;13:9−59.

[287] Dalgliesh GL, Furge K, Greenman C, Chen L, Bignell G, Butler A, et al. Systematic sequencing of renal carcinoma reveals inactivation of histone modifying genes. Nature 2010;463:360−3.

[288] Smith JA, White EA, Sowa ME, Powell ML, Ottinger M, Harper JW, et al. Genome-wide siRNA screen identifies SMCX, EP400, and Brd4 as E2-dependent regulators of human papillomavirus oncogene expression. Proc Natl Acad Sci USA 2010;107:3752−7.

[289] van Haaften G, Dalgliesh GL, Davies H, Chen L, Bignell G, Greenman C, et al. Somatic mutations of the histone H3K27 demethylase gene UTX in human cancer. Nat Genet 2009;41:521−3.

[290] Agger K, Cloos PAC, Rudkjær L, Williams K, Andersen G, Christensen J, et al. The H3K27me3 demethylase JMJD3 contributes to the activation of the INK4A−ARF locus in response to oncogene- and stress-induced senescence. Genes Dev 2009;23:1171−6.

[291] Barradas M, Anderton E, Acosta JC, Li S, Banito A, Rodriguez-Niedenführ M, et al. Histone demethylase JMJD3 contributes to epigenetic control of INK4a/ARF by oncogenic RAS. Genes Dev 2009;23:1177−82.

[292] McLaughlin-Drubin ME, Crum CP, Münger K. Human papillomavirus E7 oncoprotein induces KDM6A and KDM6B histone demethylase expression and causes epigenetic reprogramming. Proc Natl Acad Sci USA 2011;2108:2130−5.

[293] Anderton JA, Bose S, Vockerodt M, Vrzalikova K, Wei W, Kuo M, et al. The H3K27me3 demethylase, KDM6B, is induced by Epstein−Barr virus and over-expressed in Hodgkin's Lymphoma. Oncogene 2011;30:2037−43.

[294] Hsia DA, Tepper CG, Pochampalli MR, Hsia EY, Izumiya C, Huerta SB, et al. KDM8, a H3K36me2 histone demethylase that acts in the cyclin A1 coding region to regulate cancer cell proliferation. Proc Natl Acad Sci USA 2010;107:9671−6.

[295] Vader G, Lens SM. The Aurora kinase family in cell division and cancer. Biochim Biophys Acta 2008;1786:60−72.

[296] Carvajal RD, Tse A, Schwartz GK. Aurora kinases: new targets for cancer therapy. Clin Cancer Res 2006;12:6869−75.

[297] Tatsuka M, Katayama H, Ota T, Tanaka T, Odashima S, Suzuki F, et al. Multinuclearity and increased ploidy caused by over-expression of the aurora- and Ipl1-like midbody-associated protein mitotic kinase in human cancer cells. Cancer Res 1998;58:4811−6.

[298] Katayama H, Ota T, Jisaki F, Ueda Y, Tanaka T, Odashima S, et al. Mitotic kinase expression and colorectal cancer progression. J Natl Cancer Inst 1999;91:1160−2.

[299] Levine RL, Gilliland DG. JAK-2 mutations and their relevance to myeloproliferative disease. Curr Opin Hematol 2007;14:43−7.

[300] Rui L, Emre NC, Kruhlak MJ, Chung HJ, Steidl C, Slack G, et al. Cooperative epigenetic modulation by cancer amplicon genes. Cancer Cell 2010;18:590−605.

[301] Odgerel T, Kikuchi J, Wada T, Shimizu R, Kano Y, Furukawa Y. MSK1 activation in acute myeloid leukemia cells with FLT3 mutations. Leukemia 2010;24:1087−90.

[302] Kim HG, Lee KW, Cho YY, Kang NJ, Oh SM, Bode AM, et al. Mitogen- and stress-activated kinase 1-mediated histone H3 phosphorylation is crucial for cell transformation. Cancer Res 2008;68:2538−47.

[303] Cho YY, Yao K, Kim HG, Kang BS, Zheng D, Bode AM, et al. Ribosomal S6 kinase 2 is a key regulator in tumor promoter induced cell transformation. Cancer Res 2007;67:8104−12.

[304] Song H, Mak KK, Topol L, Yun K, Hu J, Garrett L, et al. Mammalian Mst1 and Mst2 kinases play essential roles in organ size control and tumor suppression. Proc Natl Acad Sci USA 2010;107:1431−6.

[305] Zhu F, Zykova TA, Peng C, Zhang J, Cho YY, Zheng D, et al. Phosphorylation of H2AX at Ser139 and a new phosphorylation site Ser16 by RSK2 decreases H2AX ubiquitination and inhibits cell transformation. Cancer Res 2011;71:393−403.

[306] Espinosa JM. Histone H2B ubiquitination: the cancer connection. Genes Dev 2008;22:2743−9.

[307] Shema E, Tirosh I, Aylon Y, Huang J, Ye C, Moskovits N, et al. The histone H2B-specific ubiquitin ligase RNF20/hBRE1 acts as a putative tumor suppressor through selective regulation of gene expression. Genes Dev 2008;22:2664−76.

[308] Glinsky GV. Genomic models of metastatic cancer: functional analysis of death from- cancer signature genes reveals aneuploid, anoikis-resistant, metastasis-enabling phenotype with altered cell cycle control and activated Polycomb Group (PcG) protein chromatin silencing pathway. Cell Cycle 2006;5:1208−16.

[309] Glinsky GV, Berezovska O, Glinskii AB. Microarray analysis identifies a death-from cancer signature predicting therapy failure in patients with multiple types of cancer. J Clin Invest 2005;115:1503−21.

[310] Liu YL, Yang YM, Xu H, Dong XS. Aberrant expression of USP22 is associated with liver metastasis and poor prognosis of colorectal cancer. J Surg Oncol 2011;103:283−9.

[311] Zhang Y, Yao L, Zhang X, Ji H, Wang L, Sun S, et al. Elevated expression of USP22 in correlation with poor prognosis in patients with invasive breast cancer. J Cancer Res Clin Oncol 2011;137:1245−53.

[312] Luise C, Capra M, Donzelli M, Mazzarol G, Jodice MG, Nuciforo P, et al. An Atlas of Altered Expression of Deubiquitinating Enzymes in Human Cancer. PLoS One 2011;6(1):e15891.

[313] Hussain S, Zhang Y, Galardy PJ. DUBs and cancer: the role of deubiquitinating enzymes as oncogenes, nononcogenes and tumor suppressors. Cell Cycle 2009;8:1688−97.

[314] Kurdistani SK. Histone modifications in cancer biology and prognosis. Prog Drug Res 2011;67:91−106.

[315] Xu W, Yang H, Liu Y, Yang Y, Wang P, Kim SH, et al. Oncometabolite 2-hydroxyglutarate is a competitive inhibitor of α-ketoglutarate-dependent dioxygenases. Cancer Cell 2011;19:17−30.

[316] Fanelli M, Amatori S, Barozzi I, Soncini M, Zuffo Dal, Bucci R, et al. Pathology tissue-chromatin immuno-precipitation, coupled with high-throughput sequencing, allows the epigenetic profiling of patient samples. Proc Natl Acad Sci USA 2010;107:21535−40.

CHAPTER 5

MicroRNA in Oncogenesis

Niamh Lynam-Lennon, Steven G. Gray, Stephen G. Maher
Trinity College Dublin, Trinity Centre for Health Sciences, Dublin, Ireland

5.1 INTRODUCTION

Identified in 2001, microRNA (miRNA) compose a large part of a family of small non-coding RNA, which includes small nuclear RNA (snRNA), small nucleolar RNA (snoRNA), small interfering RNA (siRNA), and piwi-interacting RNA (piRNA). Within cells, miRNA function as endogenous gene silencers, repressing target mRNA at a translational level. miRNA are highly evolutionarily conserved, being present in the genomes of animals, plants, and viruses. It is now considered that miRNA may represent anywhere in the region of 1–3% of the entire human genome [1,2] and estimates of the number of miRNA targets indicate that they may play a role in regulating as many as 30% of mammalian genes [3]. Consequently, miRNA have been shown to play central roles in developmental timing, hematopoietic cell differentiation, programmed cell death, and oncogenesis [4–7].

5.2 miRNA BIOGENESIS

miRNA are synthesized in the nucleus as long primary transcripts (pri-miRNA) up to 1000 nucleotides (nt) in length, which are characterized by imperfect hairpin structures. While the majority of transcripts are less than 1 kb, longer primary transcripts have been documented, for example, pri-miR-21 is 3433 nt [8]. For the majority of miRNA, transcription is mediated by RNA polymerase II (Pol II) [9], although a small subset are transcribed by Pol III [10]. The pri-miRNA hairpin structures have a 5′-cap and a 3′-polyA tail. Most pri-miRNA arise from

89

T. Tollefsbol (Ed): Epigenetics in Human Disease. DOI: 10.1016/B978-0-12-388415-2.00005-6

intergenic spaces, or are in antisense orientation to known genes, indicating independent transcription units. Other genes for miRNA are found in intronic regions and could be transcribed as part of the primary transcript for the corresponding gene. Many miRNA also form genomic clusters, and can be transcribed as a single polycistronic transcript [11,12]. A double-stranded RNA (dsRNA)-specific endonuclease RNase III, Drosha, in conjunction with a dsRNA-binding protein DGCR8, processes the pri-miRNA into hairpin RNAs 70–100-nt long, called precursor miRNA (pre-miRNA). These pre-miRNA molecules are then transported from the nucleus to the cytoplasm, a process mediated by the nuclear transport receptor exportin-5 and the nuclear protein Ran-GTP. In the cytoplasm, GTP is hydrolyzed to guanosine diphosphate (GDP), and the pre-miRNA is released from the transporting complex [13,14]. Subsequently, the RNAse III Dicer enzyme cooperates with a dsRNA-binding partner, transactivating response RNA-binding protein (TRBP), and the protein activator of protein

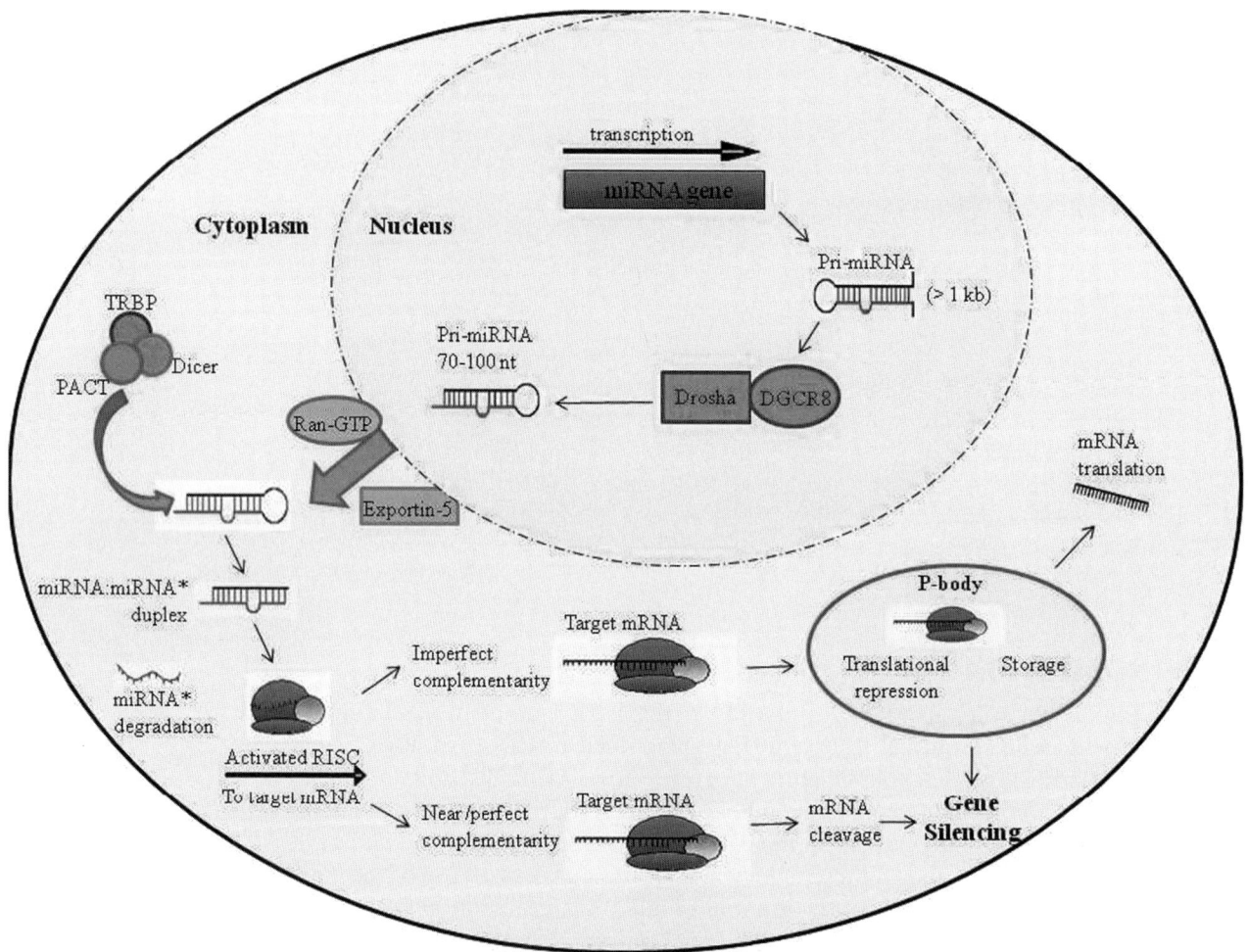

FIGURE 5.1

miRNA biogenesis and the mechanism of gene silencing. The miRNA is transcribed by Pol II in the nucleus. The resulting pri-miRNA, is then cleaved by Drosha and DGCR8, producing a pre-miRNA molecule, approximately 70–100 nucleotides in length. The pre-miRNA is transported to the cytoplasm by Exportin-5 and Ran-GTP. Here, the pre-miRNA undergoes its final processing step, which, facilitated by TRBP and PACT involves cleavage by Dicer below the stem-loop. This produces a duplex molecule, containing the single-stranded mature miRNA molecule and a miRNA* fragment. The miRNA:miRNA* complex is then incorporated into the RNA-induced silencing complex (RISC), which is activated upon unwinding of the miRNA:miRNA* duplex. Preferentially, the miRNA* fragment is degraded, with the mature miRNA molecule guiding the RISC to the target mRNA. miRNA regulate target genes by base pairing to sites of complementarity in the 3'-UTR and coding sequence of target mRNA. The degree of complementarity between the seed region of the miRNA and the binding sites in the target mRNA determines the mechanism of regulation. If there is sufficient (near perfect) complementarity, regulation is carried out by RNA interference, whereby the RISC cleaves the target mRNA. If there is insufficient complementarity, regulation is carried out by repression of translation and/or degradation. Degradation involves deadenylation of target mRNA followed by decapping and degradation in P-bodies. Translationally repressed targets can also be stored in P-bodies. This figure is reproduced in the color plate section.

kinase R (PACT) to process the pre-miRNA into a mature form, by cutting the structure at the base of the stem-loop. The remaining dsRNA duplex structure contains the mature miRNA as well as a complementary fragment termed miRNA*. The mature miRNA is single-stranded and is loaded into an RNA-induced silencing complex (RISC). It is the interaction of the miRNA/RISC and its target mRNA that results in gene regulation (Figure 5.1).

5.3 miRNA-MEDIATED REGULATION OF TARGETS

Within the RISC, miRNA interact with target genes via base-pairing. The interaction between a miRNA and its target mRNA is restricted to the 5′ end of the miRNA. Sequence complementarity between nucleotides 2–8, also known as the "seed region", is vital for target sequence recognition [15], although exceptions to this rule have been demonstrated [16,17]. Most commonly, miRNA binding sites are present in the 3′-untranslated region (UTR) of target mRNAs, usually in multiple copies [18]. However, miRNA have also been demonstrated to target the 5′-UTR and coding regions of mRNA [19,20]. A study by Tay et al. demonstrated that a network of miRNA can bind to multiple sites within the coding and 3′-UTR of a single mRNA target, adding to the complexity of miRNA-mediated target regulation [21]. The degree of complementarity between the seed region of the miRNA and the binding site in the target mRNA determines the mechanism by which the miRNA regulates the target [22]. If the miRNA bares sufficient sequence complementarity (near perfect) to the target mRNA, then regulation is carried out by a process called RNA interference, whereby the RISC is directed to cleave the target mRNA [22]. If there is insufficient complementarity, which is generally the case in mammals [15], regulation is achieved by repression of translation and/or destabilization of the mRNA [23].

The core components of the RISC are the Argonaute (Ago) family of proteins, which play a key role in its function [24]. All four mammalian Ago proteins (Ago1–Ago4) can direct the translational repression of a target mRNA, however, only Ago2 possesses "slicer" activity, and is responsible for cleaving target mRNA [25]. The exact mechanism(s) of miRNA-mediated translational repression of target genes is still uncertain [26–28]. Several studies have provided evidence that translational repression occurs pre-initiation of translation [27–29]. However, other studies suggest that repression occurs post-initiation of translation [26,30,31]. It was initially thought that miRNA-mediated repression of target genes was predominantly reflected at the protein level, with no or minimal effect on mRNA levels. However, it has now been demonstrated that miRNA-mediated repression of target genes is frequently associated with destabilization of mRNA, although it is not known if this is a secondary effect of translational repression. miRNA-mediated degradation of mRNA targets involves deadenylation (removal of the Poly A tail), followed by decapping and exonucleolytic digestion [32–34]. In addition, processing bodies (P-bodies), cytoplasmic structures involved in the storage and degradation of mRNA, are also thought to play a role in miRNA regulation [35,36]. miRNA are thought to guide target mRNA and associated RISC proteins to these storage structures, which are enriched for mRNA degradation and translational repression factors [27,36]. The mechanisms dictating whether a target mRNA follows the degradation or translational repression pathway are presently unknown. Adding to the complexity of miRNA-mediated regulation are the recent discoveries that under different stress conditions miRNA-induced repression of targets can be reversed [37] and that miRNA can activate translation of target mRNA [38].

miRNA-mediated regulation appears to be an extremely dynamic process, its complexity is increased by the fact that perfect complimentarity to the target is not required for regulation. This indicates that a single miRNA has the potential to regulate multiple target genes. In addition, a network of miRNA can function simultaneously to regulate a single mRNA. This ultimately makes in silico identification of target genes and the elucidation of miRNA function much more difficult.

The seed region, located at positions 2–7 from the 5′ end of the miRNA, is employed by the RISC as a nucleation signal for recognizing target mRNA [15,39,40]. On the mRNA the corresponding sites are referred to as "seed sites". There are a number of stringencies associated with target seed site recognition and binding [41]. A stringent seed site has perfect Watson—Crick binding and can be divided into four "seed" types: 8mer, 7mer-m8, 7mer-A1, and 6mer [41]. Each of these types differs depending on the combination of the nucleotide of position 1 and pairing at position 8. 8mer has both an adenine at position 1 of the mRNA target site and base-pairing at position 8. An adenine on the target site corresponding to position 1 of a miRNA is known to increase efficiency of target recognition [3]. 7mer-A1 has an adenine at position 1 only, while 7mer-m8 has base-pairing at position 8 only. By contrast, 6mer has neither an adenine at position 1 nor base-pairing at position 8 [42].

In addition to stringent seed recognition, moderately stringent recognition is also possible, as the RISC can tolerate small mismatches or wobble pairing within the seed region. The thermodynamic stability of a wobble pairing (such as G:U) is comparable to that of a Watson—Crick pairing [41,43].

Watson—Crick pairing in the 3′ part of the miRNA molecule is known to enhance the site recognition efficacy in miRNA targets that have seed pairing [42]. The preferable nucleotide number of matches in the 3′ part differs between the site that has stringent-seed pairing and the one that has moderate-stringent-seed pairing [41]. Stringent seeds require 3—4 matches in the positions 13—16, whereas moderate-stringent-seeds require 4—5 matches in positions 13—19. Sites with this additional 3′ pairing are called 3-supplementary and 3′ compensatory sites [44].

It has been extensively demonstrated that the vast majority of miRNA target recognition sequences are found in the 3′-UTR of the target gene, even though miRNA-loaded RISC can in theory bind any segment of mRNA. Target genes generally have longer 3′ UTR, whereas certain ubiquitous genes, such as house-keeping genes, tend to have short 3′ UTR, potentially to avoid being regulated by miRNA [45]. Target sites are not evenly distributed with 3′ UTR. They are located near both ends on long 3′ UTR (generally ≥2000 nt). For shorter 3′ UTR, target sites tend to be ∼15—20 nt away from the stop codon [42,43].

While it is generally considered that functional miRNA sites are preferentially located in 3′ UTR, seed sites in the coding sequence and 5′ UTR regions can also promote mRNA down-regulation [46,47]. The basis for preferential miRNA binding in the 3′ UTR may have a number of explanations. For example, the RISC may need to compete with other protein complexes, such as ribosomes, binding to the coding sequence and translation initiation complexes in the 5′ UTR. As such the 3′ UTR might simply be more accessible for long-term binding than the other two sites [41,48].

5.4 miRNA AND CANCER

miRNA are involved in the regulation of important cellular pathways, such as proliferation [49], cell death [50], angiogenesis [51], invasion and metastasis [52], the deregulation of which are all hallmarks of cancer [53]. Thus, it is not surprising that aberrant miRNA expression has been demonstrated in many different cancers.

The link between miRNA and cancer was first highlighted by Calin and colleagues [54], when it was discovered that two miRNA, miR-15a and miR-16-1, are located in a region on chromosome 13 that is deleted in over 65% of chronic lymphocytic leukemia (CLL) patients. Despite extensive profiling, no cancer-associated genes had been identified within this region, suggesting that miR-15a and miR-16-1 were the genomic targets of this frequent deletion. Interestingly, this deletion is present in almost all cases of indolent CLL, suggesting that down-regulation of miR-15a and miR-16-1 is an early event in the pathogenesis of this disease. Supporting this, a subsequent study identified the proto-oncogene Bcl2, which is overexpressed in CLL, as a target of miR-15a and miR-16-1 [55]. This suggests that the down-regulation of these two miRNA in CLL provides a mechanism for oncogenesis via the increased expression of Bcl2. Further supporting a tumor suppressor role for miR-15a and miR-16-1, ectopic expression of both miRNA was demonstrated to prevent tumor growth in a xenograft model of leukemia [56].

A role for miRNA in cancer was further supported by several studies that performed genome-wide profiling of miRNA expression in multiple primary tumors and cell lines. Volinia and colleagues [57] profiled miRNA expression in over 500 samples of both normal and tumor tissue taken from breast, colon, lung, pancreatic, prostate, and stomach. miRNA expression was demonstrated to be significantly dysregulated in malignant tissue, demonstrating a common miRNA expression profile in solid cancer. A study by Lu et al., profiled the expression of 217 miRNA in 334 samples comprised of normal non-cancer tissue, primary tumors, and tumor cell-lines, from 20 different cancer types. miRNA profiles were demonstrated to accurately distinguish normal from malignant tissue, separate cancer type, categorize differentiation state and cluster samples according to their embryonic lineage. Importantly, miRNA profiles were demonstrated to be more accurate at classifying poorly differentiated tumors than mRNA [58], suggesting a potential role as diagnostic biomarkers. Unique miRNA profiles have now been demonstrated for all cancers studied to date, these include breast [59], esophageal [60], colo-rectal [61], lung [62], prostate [63], and gastric [64], supporting a role for miRNA in the development of cancer and suggesting their potential use as diagnostic biomarkers.

In addition to their role in the development of cancer, evidence also suggests a role for miRNA in cancer progression. Metastasis, the process by which cancer cells disseminate from the primary tumor site and establish secondary tumors at distant sites, is the predominant cause of cancer-related deaths. Dysregulation of miRNA has been demonstrated in metastasis, with both pro- and antimetastatic miRNA identified. miR-10b is highly expressed in metastatic breast cancer. Overexpression of miR-10b increases cell motility and invasiveness, effects that are reversed upon inhibition of expression, both in vitro [65] and in vivo [66], suggesting a prometastatic role for miR-10b. Conversely, the miR-200 family negatively regulates epithelial–mesenchymal transition (EMT), which is thought to facilitate metastasis via increased cell motility. The miR-200 family, which is repressed during EMT, was demonstrated to target the E-cadherin repressors ZEB1 and ZEB2, resulting in increased E-cadherin expression and inhibition of EMT [67], suggesting an antimetastatic function for this miRNA family.

In addition, specific miRNA profiles have also been demonstrated to predict progression and outcome in a number of cancers. In breast cancer, dysregulation of miR-145 and miR-21 was associated with tumor progression, whilst reduced *let-7* expression was associated with increased lymph node metastasis and proliferation capacity [59]. A 9-miRNA signature was associated with time to progression in CLL [68]. In colorectal cancer, miR-21 expression is associated with tumor stage, invasion, and prognosis [69]. Whilst in lung cancer, reduced *let-7* was indicative of a poor prognosis in two independent studies [62,70]. The demonstrated role of miRNA in cancer progression highlights their potential as both novel therapeutics for the treatment of cancer and as prognostic markers.

5.5 OncomiRs

Whilst miRNA profiles in cancer are generally tumor-specific, several miRNA are dysregulated across multiple cancers, suggesting a common role in tumorigenesis. miRNA that have been shown to be down-regulated in cancers, such as miR-15a, miR-16-1, and the *let-7* family, have been proposed to be tumor suppressors, whilst up-regulated miRNA, such as miR-21 and the miR-17-92 cluster, have been classified as oncogenes.

5.5.1 miRNA as Tumor Suppressors

In addition to their tumor suppressor role in CLL, a potential anticancer role for miR-15a and miR-16-1 has also been demonstrated in several other cancers such as lung [71], pituitary [72], and ovarian [73]. In prostate cancer [74], miR-15a and miR-16-1 were demonstrated to be down-regulated in 80% of tumors. Inhibition of miR-15a and miR-16-1 promoted survival, growth, invasiveness, and tumorigenicity in untransformed prostate cells, suggesting down-regulation of miR-15a and miR-16-1 as an early event in the pathogenesis of prostate cancer. Conversely, overexpression of miR-15a and miR-16-1 reduced growth, induced apoptosis and tumor regression in a prostate cancer xenograft model. This study identified the oncogenes CCND1 and WNT3A as direct targets, all of which promote cell growth, survival, and invasiveness, suggesting a mechanism for tumor development in prostate tumors deficient in miR-15a and miR-16-1.

The *let-7* family, which consists of 12 closely related miRNA, is also a potential tumor suppressor miRNA that is frequently down-regulated in cancer. A study by Takamizawa and colleagues was the first to identify a potential tumor suppressor role for *let-7* [70], demonstrating the down-regulation of *let-7* in lung cancer, with low expression also associated with a shorter postoperative survival time. Overexpression of *let-7* significantly inhibited growth in lung cancer cells, suggesting a functional role for *let-7* in modulating tumorigenesis. This is supported by several studies that have demonstrated the antiproliferative effects of *let-7* in lung [4], thyroid [75], and prostate [76] cancer. Furthermore, *let-7* has been demonstrated to inhibit tumor growth in vivo [77,78], further supporting a tumor suppressor role. The effects of *let-7* on cell growth may be explained by its negative regulation of several oncogenes, such as Ras [5], myc [79], HMGA2 [80], and cell cycle regulators, including Cdc25A, CDK6, and Cyclin D2 [4]. This suggests that the frequent down-regulation of *let-7* seen in multiple cancers provides a mechanism for tumor development and progression via increased expression of these gene targets.

5.5.2 miRNA as Oncogenes

miR-21 was one of the first miRNA identified in humans [8]. Overexpression of miR-21 has been demonstrated in multiple cancers, such as glioblastoma [50], breast [59], esophageal [81], and CLL [82], suggesting an oncogenic role for this miRNA. Several studies have provided evidence for miR-21 as an antiapoptotic factor. In glioblastoma and breast cancer cells, knockdown of miR-21 inhibited cell growth and induced apoptosis [50,83], suggesting that overexpression of miR-21 may promote tumor development via the negative regulation of proapoptotic factors. The oncogenic properties of miR-21 are mediated, at least in part, via its negative regulation of several important tumor suppressor genes. Several studies have demonstrated miR-21 regulation of the tumor suppressor proteins programmed cell death 4 (PDCD4) [84−86] phosphatase and tensin homolog (PTEN) [87] and tropomyosin 1 (TMP1) [88]. In addition, miR-21-mediated regulation of all three tumor suppressors is associated with increased invasion and metastasis, suggesting a role for miR-21 in cancer progression. miR-21 has also been implicated in the resistance of cancer cells to various chemotherapeutics. miR-21 was demonstrated to modulate sensitivity to the chemotherapeutic agent gemcitabine in cholangiocarcinoma [89], doxorubicin in bladder cancer [90], and 5-fluorouracil in colorectal cancer [91], suggesting an additional role for miR-21 in oncogenesis. The role of miR-21 in both the development and progression of cancer, in addition to the response to anticancer treatment, highlights the potential of miR-21 as a novel therapeutic target.

The polycistronic cluster miR-17-92, is a family of homologous miRNA that are transcribed as a single pri-miRNA and then processed to produce seven mature miRNA molecules (miR-17-5p, miR-17-3p, miR-18, miR-19a, miR-20, miR-19b-1, and miR-92-1) [92]. The cluster is located on chromosome 13, at a region commonly amplified in a number of hematopoietic malignancies and solid tumors [93−95]. The oncogenic potential of miR-17-92 was first highlighted in a mouse model of B-cell lymphoma [94]. miR-17-92 was demonstrated to accelerate c-myc-induced lymphoma development, producing tumors with a more aggressive phenotype. These tumors also demonstrated lower apoptosis, suggesting that the role of miR-17-92 in tumor development may be via antiapoptotic mechanisms. This was supported by the discovery that the proapoptotic factors PTEN and Bim are both direct targets of miR-17-92 [96]. miR-17-92 is also overexpressed in lung cancer where it was demonstrated to enhance cellular proliferation, whilst inhibition of miR-17-92 induced apoptosis [93,97], further supporting an antiapoptotic role. The negative regulation of the cell cycle regulator p21 by miR-17-92 [98] may explain the effects of this polycistron on proliferation. Interestingly, miR-17-92 has been demonstrated to be directly induced by the oncogene c-myc [99], which is frequently up-regulated in cancer, further supporting an oncogenic role for miR-19-72.

Surprisingly, loss of the genomic region encoding the miR-17-92 cluster has also been linked to malignancy in hepatocellular cancer [100]. This potential tumor suppressor role of miR-17-92 may be explained by its negative regulation of the transcription factor E2F1 [99]. E2F1 is induced by c-myc and promotes cell cycle progression [101], it also forms a positive feedback loop by inducing c-myc expression [102]. Thus, the repression of E2F1 by miR-17-92 provides a mechanism for inhibition of c-myc-mediated growth. The down-regulation of miR-17-92 in hepatocellular cancer may therefore provide a mechanism for tumor development. This suggests that miR-17-92 may have dual tumor suppressor and oncogenic roles in a tissue/tumor-dependent manner.

5.6 MECHANISMS OF miRNA DEREGULATION

Several mechanisms are thought to contribute either alone or in combination to the dysregulation of miRNA in human cancer. These include chromosomal alterations, dysregulated transcriptional activation, epigenetic modifications, and alterations in biogenesis. However, the exact mechanism(s) directing the tumor/tissue-specific dysregulation of miRNA in carcinogenesis is still unknown.

5.6.1 Chromosomal Aberrations

Following the initial study which demonstrated the deletion of miR-15a and miR-16-1 in CLL, Calin and colleagues [6] used a computational approach to identify the genomic location of 186 miRNA. Over 50% of the miRNA investigated were demonstrated to be located at fragile regions, regions that commonly undergo deletion, amplification or translocation events or genomic regions associated with human cancers, such as viral DNA integration sites, suggesting a role for altered miRNA in the development of cancer. Interestingly, profiling of several of these sites of cancer-associated chromosomal alterations had previously failed to identify protein-coding tumor suppressors or oncogenes implicated in the development of the disease, suggesting that genomic alterations of miRNA may be a causal factor in tumorigenesis.

Several studies have since experimentally validated the non-random genomic location of miRNA in cancer, demonstrating significant chromosomal alterations at miRNA-encoded regions across multiple cancer types. A study by Zhang et al. [103] used array-based comparative genomic hybridization to profile the genomic loci of 283 miRNA in over 200 samples of breast, ovarian, and melanoma primary tumors and cancer cell lines. A high proportion of miRNA gene copy number alterations were demonstrated in all three cancer types, with both distinct and common alterations observed. Furthermore, a strong correlation between miRNA copy number alterations and miRNA transcript levels was demonstrated both in cancer cell lines and a previously published independent set of breast tumor samples [59], further supporting genomic alterations as a mechanism for miRNA dysregulation. Another study profiling miRNA expression in over 200 normal tissues and the NCI-60 panel of cancer cell lines, identified potential tumor suppressor and oncogenic miRNA, which are encoded at known cancer-associated genomic regions [104].

Both genomic loss and gain have been demonstrated to result in altered expression of miRNA. The region encoding miR-15a and miR-16-1 is deleted in more than 50% of B-cell chronic lymphocytic leukemia cases [54]. In contrast, the miR-17-92 cluster is encoded in an 800-bp region of the non-protein coding gene C13, which is frequently amplified in B-cell lymphoma and lung cancer [93,94]. Increased gene copy number has been demonstrated to correlate with overexpression of the cluster in both cancer types, supporting chromosomal amplifications as a mechanism for dysregulation of miR-17-92. Chromosomal translocations also contribute to miRNA dysregulation. A t(8;17) translocation that juxtaposes the oncogene c-myc to chromosome 17 resulting in its overexpression, is associated with an aggressive form of B-cell leukemia [105]. The underlying mechanism of c-myc up-regulation was unknown until it was demonstrated that this translocation positions c-myc at the promoter region of miR-142, which is encoded 50 nt from this chromosomal break [6]. This suggests the involvement of miR-142 regulatory elements in the overexpression of c-myc. A translocation at chromosome 12 disrupts the HMGA2 gene, resulting in the generation of a truncated version of HMGA2, which is overexpressed in lung cancer [106]. The up-regulation of HMGA2 has been attributed to loss of *let-7* binding sites within the 3'-UTR of HMGA2, caused by the translocation. Indeed, disruption of *let-7*-mediated regulation of HMGA2 has been demonstrated to promote oncogenic transformation [80,107], supporting chromosomal translocation as a mechanism for dysregulation of important tumor suppressor miRNA.

Both distinct and common miRNA copy alterations have been demonstrated in oncogenesis, suggesting that genomic alterations of miRNA play a functional role in malignancy, rather than being a mere by-product. Whilst the miR-17-92 cluster loci is frequently amplified in lymphoma and lung cancer, it has been demonstrated to be deleted in ovarian, breast, and melanoma tumors [103], suggesting tumor-specific copy number alterations. In contrast, overexpression of miR-21, which has been demonstrated in multiple malignancies, is thought to be due to amplification, suggesting a common mechanism of dysregulation.

5.6.2 Dysregulation of Transcription Factors

As the majority of miRNA are transcribed by Pol II, transcriptional control is an important mechanism regulating miRNA expression. Many transcription factors are dysregulated in cancer, resulting in the altered expression of miRNA, which may promote tumorigenesis.

The transcription factor c-myc is involved in the regulation of approximately 15% of human genes, regulating cell death, proliferation, and differentiation [108] via both positive and negative regulation of gene expression. c-myc is frequently overexpressed in human malignancies, with almost all cases of Burkitt's lymphoma caused by the juxtaposition of c-myc with immune regulatory elements, which results in hyperactivation [109]. A study by O'Donnell and colleagues [99] demonstrated that the oncogenic miR-17-92 miRNA cluster located on chromosome 13 is a transcriptional target of c-myc. c-myc binds directly to the genomic locus of the miR-17-92 cluster, activating transcription. Co-expression of c-myc and the miR-17-92 polycistron accelerated oncogenesis in a murine model of B-cell lymphoma [94], supporting c-myc-mediated activation of miR-17-92 as a mechanism for tumorigenesis. Adding to the complexity of c-myc-mediated miRNA deregulation, induction of c-myc has also been demonstrated to result in the widespread repression of miRNA expression in B-cell lymphoma [110]. Many of these c-myc-repressed miRNA (including miR-15a, miR-16a, miR-34a, *let-7a-1*, and *let-7f-1*), have been demonstrated to be down-regulated in human cancers and have known tumor suppressor activity, suggesting c-myc-mediated repression as a mechanism for oncogenic transformation. Similar to the direct activation of miR-17-92, c-myc was demonstrated to bind directly to the promoters of several of these down-regulated miRNA, suggesting direct repression of transcription. Additionally, c-myc has been demonstrated to modulate miRNA expression via indirect mechanisms, as in the case of the *let-7* family, in which c-myc inhibits biogenesis via activation of the RNA-binding protein Lin-28B [111].

The tumor suppressor p53 regulates the expression of a complex network of genes via both transcriptional activation and repression. In response to cellular stresses, such as DNA damage, hypoxia, mitotic spindle damage, and activation of oncogenes, p53 becomes stabilized and orchestrates cell cycle arrest, apoptosis or senescence, thereby maintaining genomic integrity. Consequently, dysfunction of p53 is considered to be an early event in tumorigenesis. This is highlighted by the fact that p53 is mutated in over 50% of human cancers [112]. Several studies have demonstrated a miRNA component to the p53 network, with the miR-34 family, which consist of miR-34a, miR-34b, and miR-34c, identified as direct transcriptional targets of p53 [113]. p53 binds to responsive elements in the promoter region, inducing expression. Ectopic expression of all three miR-34 members has been demonstrated to have antiproliferative effects, inducing cell cycle arrest and apoptosis. Consistent with the tumor suppressor effects of miR-34 is the finding that miR-34 is down-regulated in a number of human cancers. Thus, p53-mediated activation of miR-34 provides a novel mechanism for tumor suppressor activity, and suggests that dysregulation of this pathway may provide a mechanism for oncogenic transformation. Several other p53-regulated miRNA have since been identified, these include, miR-107 [114], miR-145 [115], miR-192, and miR-215 [116]. In addition, p53 may also be involved in the regulation of miRNA processing [117], suggesting an additional mechanism for miRNA dysregulation in oncogenesis. Interestingly, both p53 and myc have been demonstrated to be direct targets of miRNA [118,119], adding to the complexity of the transcriptional regulation of miRNA.

5.6.3 Epigenetic Alterations

Epigenetic modifications describe heritable and reversible changes in chromatin that do not alter the original DNA sequence. The most frequently studied epigenetic changes in cancer are differential methylation of DNA and histone modifications. Methylation of CpG islands, which are associated with the promoter regions of genes, results in transcriptional silencing. Aberrant methylation that results in silencing of tumor suppressor genes is thought to contribute to tumorigenesis. Conversely, a reduction in global methylation levels (hypo-methylation) is also associated with cancer. Similarly, post-translational modifications of histones, such as acetylation, methylation, and phosphorylation play an important role in the regulation of gene expression. Consequently, alterations in histone modification patterns, often in combination with dysregulated hypermethylation, are commonly demonstrated in carcinogenesis.

Similar to protein-coding genes, miRNA have also been demonstrated to be subject to epigenetic regulation. Indeed, over 50% of miRNA genes are associated with CpG islands, suggesting the potential for epigenetic modifications [120]. Saito and colleagues [121] demonstrated that treatment of bladder cancer cells with both the demethyltransferase inhibitor 5-aza-2'-deoxycytidine and the histone deacetylase inhibitor 4-phenylbutyric acid, resulted in the reactivation of expression of 17 miRNA, suggesting that these miRNA were under epigenetic control. One of these miRNA, miR-127, which is located within a CpG island, was demonstrated to be down-regulated in tumor tissue and cell lines across multiple cancer types, when compared to normal tissue, suggesting that epigenetic regulation may provide a mechanism for repression of miR-127 in tumorigenesis. Interestingly, both methylation and histone deacteylation were demonstrated to be involved in transcriptional repression of miR-127, as induction of expression was only evident following both demethylating and deace-tylase inhibition treatment. This induction of miR-127 expression was concomitant with a decrease in expression of the miR-127-predicted target, the proto-oncogene BCL6, suggesting a tumor suppressor role for miR-127. Since then, a plethora of studies have demonstrated epigenetic control of a number of miRNA using several different methods. A study by Scott et al. [122] demonstrated significant alterations in miRNA expression in breast cancer cells following histone deacetylase inhibition, suggesting that histone modifications alone can regulate miRNA expression. Lujambio and colleagues [123] used a DNA methyltransferase-deficient cell line model of colon cancer to investigate epigenetically silenced miRNA, demonstrating hypermethylation of miR-124a. Silencing of miR-124a was associated with the increased expression of target gene CDK6, which plays a demonstrated role in oncogenesis, supporting a tumor suppressor role for miR-124a. Interestingly, hypermethylation of miR-124a was demonstrated across multiple tumor and cancer cell lines including breast, lung, and lymphoma, suggesting a common mechanism of dysregulation in tumorigenesis. This was supported in acute lymphoblastic leukemia (ALL), where methylation of miR-124a was demonstrated in 59% of patients, and was significantly associated with a poor prognosis [124]. The epigenetic silencing of miR-124a in ALL was associated with increased cell growth, which was mediated by the miR-124a target CDK6 [124], suggesting methylation-induced silencing of tumor suppressor miRNA as a mechanism for tumorigenesis.

In contrast, up-regulation of miRNA via hypomethylation has also been demonstrated. Demethylating treatment of ovarian cancer cell lines induced expression of miR-21, miR-203, and miR-205, all of which were demonstrated to be up-regulated in ovarian tumors, when compared to normal tissue [125]. Given the known oncogenic properties of miR-21 and its up-regulation in multiple cancers, these data suggest hypomethylation as a potential mechanism for overexpression of oncogenic miRNA. Whilst epigenetic regulation of miRNA does appear to be a mechanism for dysregulation of miRNA in cancer, evidence does suggest that epigenetic control of miRNA expression is a cell-specific effect. For example, whilst demethylation and histone deacetylase inhibition reactivated miR-127 in colon, cervical, and lymphoma cell lines [121], expression was not induced in MCF-7 breast cancer or CALU-1 lung cancer cell lines.

Taken together these data suggest epigenetic regulation as a major mechanism of altered miRNA expression in cancer. The additional discovery that several miRNA negatively regulate components of the epigenetic regulation machinery [126], only adds to the complexity of the epigenetic-miRNA network.

98

5.6.4 Altered Processing

pri-miRNA molecules undergo a co-ordinated multistep process to produce a mature functional miRNA. Alterations in the processing of miRNA that affect either miRNA maturation and/or the interaction of miRNA and gene targets are therefore likely to play a role in the dysregulation of miRNA in cancer.

A potential role for altered miRNA processing in cancer was first highlighted by Kumar and colleagues [127], who demonstrated that knockdown of Drosha, Dicer, and DGCR8 resulted in a global decrease in miRNA levels, which enhanced proliferation and transformation in cancer cells. Several other studies have since demonstrated altered expression of miRNA processing machinery in multiple cancer types. A study in prostate cancer demonstrated the up-regulation of Dicer and several other components of the miRNA processing pathway [128], suggesting a mechanism for the up-regulation of miRNA previously identified in prostate tumors [57]. The up-regulation of Dicer was also demonstrated in colorectal cancer [129], where increased expression was significantly associated with disease progression and poor survival, suggesting a role for altered miRNA processing in tumor progression. Conversely, reduced expression of Dicer was demonstrated in a subset of non-small-cell lung cancers with a poor prognosis [130]. This down-regulation was significantly associated with the reduced expression of let-7, suggesting altered processing as a mechanism for let-7 dysregulation. Interestingly, the downregulation of Dicer was not due to methylation of the promoter, suggesting alternative mechanisms of dysregulation. This was also supported in ovarian cancer where reduced expression of Dicer was associated with advanced tumor stage [131]. The inconsistencies in these studies may suggest that the role of altered miRNA processing in cancer is tissue/tumor-specific, and may be dependent on the expression of cofactors involved in biogenesis.

A common chromosomal amplification in cervical cancer was demonstrated to result in up-regulation of Drosha in advanced squamous cell carcinoma (SCC) tumors [132]. This over-expression resulted in an altered miRNA expression profile, suggesting that Drosha-mediated dysregulation of miRNA may play a role in the development and progression of cervical cancer. This was supported in esophageal SCC [133], where overexpression of Drosha was associated with poor prognosis and increased cell growth. In neuroblastoma [134], low expression of Dicer and Drosha was associated with wide-scale down-regulation of miRNA in advanced tumors with a poor prognosis, supporting dysregulation of miRNA processing machinery as a mechanism for altered miRNA expression. Interestingly, silencing of Drosha and Dicer promoted transformation and cell growth, suggesting a mechanism for tumor development. The altered expression of Dicer and Drosha in multiple cancers strongly suggests that dys-regulation of miRNA processing proteins plays a role in cancer development and progression via altered expression of miRNA.

In addition to altered levels of miRNA processing proteins, alterations in post-transcriptional regulation of miRNA can also contribute to their dysregulation. During biogenesis, Drosha cleaves the pri-miRNA to produce a pre-miRNA molecule. By doing so, Drosha also determines one end of the mature miRNA molecule. It has been demonstrated that variation in the site at which Drosha cleavage occurs can produce multiple pre-miRNA molecules from a single pri-miRNA [135]. These pre-miRNA isoforms differ slightly in their 5′ sequence, which may alter interaction with potential gene targets. Interestingly, this shift in Drosha cleavage was non-random and only occurred during the biogenesis of specific miRNA, suggesting that this alteration may be functionally important.

In addition, evidence also suggests that miRNA are subject to RNA editing, such as the conversion of adenosine residues into inosine (A-to-I) at specific editing sites [136]. Approxi-mately 16% of pri-miRNAs undergo A-to-I editing [137], which has been demonstrated to interfere with both biogenesis and function of miRNA. A study by Yang and colleagues [138], demonstrated that A-to-I editing of pri-miR-142 interfered with the Drosha processing step, resulting in degradation of the edited pri-miRNA molecule and consequently, reduced expres-sion of mature miR-142. Similarly, A-to-I editing of pri-miR-151 inhibited the Dicer processing step, which resulted in an accumulation of pre-miR-151 [139], suggesting this post-transcriptional regulation as a mechanism for altered miRNA expression levels. In addition, A-to-I editing has also been demonstrated to provide a mechanism for dysregulation of miRNA function. Editing within the miRNA seed region results in the generation of novel edited mature miRNA isoforms, with altered target specificity [140]. Whilst the role this plays in cancer remains to be elucidated, it does suggest that alterations in post-transcriptional processing of miRNA may play a role in tumorigenesis via dysregulation of miRNA expression and/or function.

5.7 miRNA AND TREATMENT RESISTANCE

Whilst the role of miRNA in the initiation, progression, and prognosis of cancer is well documented, the role of miRNA in the response to cancer therapy is less well known. Given the role of miRNA in regulating pathways involved in the cellular response to chemotherapy and radiation, such as cell cycle [141], apoptosis [55], survival [142], oxidative stress [143], and DNA repair [144], it is likely that miRNA are involved in the tumor response to anticancer therapy.

5.7.1 Role of miRNA in Chemoresistance

It is now becoming increasingly clear that miRNAs may also play important roles in resistance to chemotherapeutic drugs. With regard to cisplatin resistance, in the non-small-cell lung cancer (NSCLC) cell line A549, miR-181a, miR-181b, and miR-630 have been shown to be involved with cellular responses to cisplatin [145,146]. miR-181a enhances cisplatin-mediated cell death, via the induction of apoptosis through Bax oligomerization, mitochondrial transmembrane potential dissipation, and proteolytic maturation of caspase-9 and caspase-3 [145]. miR-181b was found to be down-regulated in an isogenic model of cisplatin resistance in A549 cells (A549/CDDP), and overexpression of this miRNA decreased levels of Bcl2, enhancing sensitivity to cisplatin-induced cell death [146]. Of note, miR-181a has also been found to be both significantly down-regulated and associated with poor survival in primary NSCLC tissues [147,148], while in head and neck SCC cells, miR-181a was found to be down-regulated in response to cisplatin treatment [149].

miR-497 has been shown to be downregulated in a cisplatin-resistant NSCLC cell line (A549/CDDP). This was associated with increased expression of Bcl-2, and overexpression of exogenous miR-497 reduced Bcl2 protein levels and sensitized the A549/CDDP cells to cisplatin [150].

In the study by Kroemer and colleagues miR-630 was found to be up-regulated in A549 cells in response to cisplatin. This miRNA was subsequently demonstrated to attenuate the DNA damage response (phosphorylation of ATM, histone H2AX, and p53), with concomitant induction of p27(Kip1), decreased cell proliferation, and G0−G1 phase cell cycle arrest as opposed to the late G2−M cell cycle arrest normally mediated by cisplatin [145].

ZEB1 is a master regulator of the epithelial−mesenchymal transition (EMT). Reports have demonstrated that ZEB1 is important for this process in lung cancer through its regulation of many EMT genes, including E-cadherin [151−153]. Knockdown of ZEB1 results in the suppression of anchorage-independent cell growth of lung cancer cells [154]. Of note, the miR-200 family has been shown to target ZEB1 [67,155,156], and one of its members, miR-200c, has been shown to be down-regulated in NSCLC, resulting in an aggressive, invasive, and chemoresistant phenotype in NSCLC [157]. The loss of miR-200c expression occurs as a consequence of DNA CpG methylation, and re-expression was shown to restore the sensitivity of a drug-resistant cell line to cisplatin and cetuximab [157].

One of the well-established mechanisms involving cisplatin resistance concerns the over-expression of ERCC1. This DNA repair gene is involved in the repair of DNA adducts and stalled DNA replication forks, and its expression levels can predict both survival and cisplatin-based therapeutic benefit in patients with resected NSCLC [158,159]. In a cohort of patients for which ERCC1 expression and response to cisplatin therapy was known, Friboulet et al. identified miR-375 as significantly underexpressed in ERCC1-positive tumors [160]. Similar to miR-200c, this miRNA has also been shown to be epigenetically regulated by DNA CpG methylation [161].

It is becoming clear that miRNA may play important roles in tumor cell responses to chemotherapeutics, such as cisplatin. Many miRNA appear to be epigenetically silenced by DNA CpG methylation, and as such it may be possible to resensitize patients to cisplatin-based chemotherapy through the use of demethylating agents, such as Vidaza® or Decitabine®. Indeed reactivation of genes silenced by DNA CpG methylation can result in re-sensitivity to cisplatin in cell line models [162], and as such this strategy may work for miRNA.

5.7.2 Role of miRNA in Radioresistance

Radiation has been demonstrated to modulate miRNA expression in lung cancer [163], lymphoblastoma [164], colon cancer [165] and glioma [166] cell lines. Moreover, this radiation-mediated modulation of miRNA has been demonstrated to be dose-dependent [143], suggesting that miRNA play a functional role in the cellular response to radiation. In addition, altered basal miRNA expression has also been associated with radioresistance [167], suggesting a role for miRNA in determining the initial response to radiation.

Several studies have identified a direct role for miRNA in regulating pathways involved in the cellular response to radiation. Simone and colleagues demonstrated a role for miRNA in the response to oxidative stress and DNA damage. Radiation significantly altered the expression of 17 miRNA in human fibroblast cells [143]. Interestingly, expression of all 17 miRNA were also modulated by treatment with the chemotherapeutic etoposide, which mimics the effect of radiation by inducing double-strand breaks and H_2O_2, which induces oxidative stress, suggesting a common role for these miRNA in the cellular response to genotoxic stress. Interestingly, the radiation-induced modulation of miRNA expression was inhibited by the addition of cysteine, a free radical scavenger that abrogates the effects of ionizing radiation (IR), suggesting a role for miRNA in the cellular response to oxidative stress. In addition, the study by He et al., which identified the p53-mediated regulation of the miR-34 family, highlighted a role for miR-34 in the DNA damage response to IR [113], supporting a role for miRNA in the regulation of pathways involved in the radioresponse.

Furthermore, miRNA have been demonstrated to play a role in modulating the cellular response to radiation. Inhibition of miR-221/222 sensitized glioma cells to radiation, which was mediated by abrogation of miR-221/222-mediated regulation of the cyclin-dependent kinase inhibitor p27(kip1) [168]. p27 is an important negative regulator of cell cycle progression, specifically the G1 phase arrest [169]. Loss of p27 is associated with a more a aggressive cancer phenotype [170] and reduced survival in patients undergoing radiation therapy and surgery in prostate cancer [171]. Reduced expression is associated with a poor response to neoadjuvant chemoradiotherapy (CRT) in rectal cancer [172]. Thus, alterations in miR-221/222 expression may modulate the cellular response to radiation via regulation of p27. Another study demonstrated the miR-181a-mediated modulation of radiosensitivity in glioma cells [173]. miR-181a was down-regulated in response to radiation, however, ectopic expression significantly sensitized cells to the cytotoxic effects of radiation. Overexpression of miR-181a also resulted in down-regulation of the antiapoptotic Bcl2, indicating Bcl2 as a potential target of miR-181a. Bcl2 expression is associated with resistance to radiation in numerous cancers [174–176]. This suggests that down-regulation of miR-181a in glioma cells following exposure to radiation, provides a mechanism for radioresistance via abrogation of miR-181a-mediated regulation of Bcl2. A study by Kato and colleagues, demonstrated the miR-34-mediated modulation of radiosensitivity in breast cancer cells, and highlighted a role for miR-34 in the in vivo response to IR in *C. elegans* [177]. miRNA-mediated modulation of radiosensitivity has also been demonstrated in prostate cancer. miR-521 was significantly down-regulated in response to radiation in two prostate cancer cell lines, suggesting a role in the radiation response [178]. Modulation of miR-521 expression altered sensitivity to radiation, with overexpression inducing sensitivity, whilst inhibition induced resistance. The DNA repair protein CSA and the antioxidant enzyme MnSOD were identified as potential targets of miR-521, suggesting a mechanism for modulation of the radioresponse [178]. A study by Weidhaas and colleagues, demonstrated a role for *let-7* in determining the sensitivity to radiation in lung cancer [179]. *let-7* has been demonstrated to regulate the oncogene Ras, which is commonly overexpressed in cancer and has been shown to be critical for protection from radiation-induced cell death [180]. The authors identified a common radiation-induced pattern of miRNA expression in both normal and tumor lung cells, with seven members of the *let-7* family significantly down-regulated, suggesting a common global miRNA response to radiation. Ectopic expression of *let-7a* and *let-7b* sensitized lung cancer cells to radiation whilst inhibition induced a radioprotective effect, suggesting a functional role for *let-7* in the response to radiation in lung cancer. *let-7* overexpression also induced sensitivity to radiation in *C. elegans*, which was indicated to be regulated, at least in part, via *let-7*-mediated regulation of Ras and other DNA damage response genes such as Rad51, Rad21, Cdc25, and Fancd2 [179]. This is supported by Oh et al., who demonstrated that overexpression of *let-7a* sensitized lung cancer cells to radiation, via regulation of Ras [181].

A role for miRNA in the in vivo radiation response has also been demonstrated. Svoboda et al. identified the up-regulation of miR-137 and miR-125b in rectal tumor biopsies 2 weeks after the initiation of neoadjuvant capecitabine CRT, suggesting a role for these miRNA in the

101

tumor response to CRT. Furthermore, increased expression of both miRNA was associated with a poor response to CRT [182]. A study by Wang and colleagues demonstrated significantly altered expression of just 12 miRNA in resected lung tissue of patients who were resistant and sensitive to adjuvant radiation therapy. Additionally, miR-126, which was up-regulated in radiosensitive tumor tissue, was demonstrated to inhibit proliferation and induce IR-mediated apoptosis via the PI3K-AKT pathway in vitro. miR-126 also enhanced radiosensitivity in vitro [183]. Therefore, the down-regulation of miR-126 in tumor tissue may provide a mechanism for resistance to radiation therapy in vivo, via enhanced proliferation and evasion of IR-induced apoptosis.

Taken together these studies strongly suggest a role for altered miRNA expression as a mechanism for resistance to radiation. Furthermore, they strongly suggest a role for miRNA as both predictive biomarkers of response to radiation, and potential novel therapeutics with which to enhance the efficacy of radiation therapy.

5.8 CLINICAL APPLICATIONS

miRNA have been shown to be intimately involved in the oncogenic process, and as such they may also be exploited in the clinical setting for diagnostic, prognostic, and therapeutic applications. miRNA array profiling has revealed unique, tissue-specific miRNA profiles that may be useful in the clinical situation. Previously, gene expression profiling has been used in a diagnostic and prognostic capacity, as well as in predicting treatment outcome, but these approaches have not translated well into a routine clinical setting for numerous reasons. For example, most of the techniques require fresh tumor material, or have issues with reproducibility, have complicated bioinformatics due to large data sets, and/or are not cost-effective. Presently, in an evolving field, employing miRNA in diagnostics may be of more efficient clinical utility. The number of miRNA for profiling is far fewer than the number of genes. miRNA are also very short in length, and as such are highly stable, hence they are not subject to the same degradation problems as mRNA. Many investigators now agree that given the direct involvement of miRNA in the regulation of protein expression, miRNA expression profiles may be superior to gene expression profiles for clinical applications, since only a small number of mRNA are regulatory molecules [184].

miRNA profiles can be used to discriminate between normal and malignant tissue, in cancers such as lung [62], colorectal [185], breast, pancreatic [186,187], hepatocellular [188], and CLL [189], among others. Additionally, it is possible to delineate and stratify tumors of the same organ of origin, but that have different histologies, for example pulmonary adenocarcinoma and squamous cell carcinoma [62] and endocrine and acinar pancreatic tumors [186]. In a similar fashion, miRNA profiles may also be employed as prognostic indicators for factors such as therapeutic outcome and overall survival, in cancers such as lung [190], esophageal [191], gastric [192], osteosarcoma [193], and breast [194,195]. Many studies have identified gene expression profiles that are predictive of therapeutic benefit, in a variety of cancer types, including breast [196,197], esophageal [198], and colon [199,200]. For breast cancer, these gene signatures have led to the development of clinical diagnostics, such as the Mammaprint [201] and Oncotype DX [202] signature arrays. Perhaps, the future may provide a miRNA-based diagnostic for prognostication in cancer patients.

A significant advantage of miRNA from a clinical perspective is the recent finding that they can be found free in circulation and can be purified from serum and plasma [203–205]. Due to their small size, miRNA are highly stable molecules in serum, being protected from RNases, and may be ideal candidates for the development of relatively non-invasive screening tests.

A company called *mirna therapeutics* have developed a new type of anticancer miRNA technology, which involves using chemically modified synthetic miRNA mimics and a liposomal-based delivery system to reintroduce a down regulated miRNA back into tumors in vivo

[206,207]. This strategy may be used to reintroduce miRNA that are important for tumor cell responsiveness to other anticancer therapeutics. It is favorable to a gene therapy approach, as rather than altering a single gene, which will have limited cellular impact, altering a single miRNA will have multiple downstream effects on tumor cells in terms of signaling pathways and effector molecules. Indeed a similar strategy has been employed in the treatment of the hepatitis C virus (HCV). It is known that miR-122 is a liver-specific miRNA essential for HCV replication [208]. It has been shown that HCV replication in liver can be dramatically reduced using oligonucleotide inhibitors of miR-122 [209]. This anti-miR-122 (miravirsen) HCV treatment strategy is now in phase II clinical trials. A possibility for the future lies in miRNA replacement therapy for cancer.

References

[1] Sontheimer EJ, Carthew RW. Silence from within: endogenous siRNAs and miRNAs. Cell 2005;122:9–12.

[2] Bentwich I, Avniel A, Karov Y, Aharonov R, Gilad S, Barad O, et al. Identification of hundreds of conserved and nonconserved human microRNAs. Nat Genet 2005;37:766–70.

[3] Lewis BP, Burge CB, Bartel DP. Conserved seed pairing, often flanked by adenosines, indicates that thousands of human genes are microRNA targets. Cell 2005;120:15–20.

[4] Johnson CD, Esquela-Kerscher A, Stefani G, Byrom M, Kelnar K, Ovcharenko D, et al. The let-7 microRNA represses cell proliferation pathways in human cells. Cancer Res 2007;67:7713–22.

[5] Johnson SM, Grosshans H, Shingara J, Byrom M, Jarvis R, Cheng A, et al. RAS is regulated by the let-7 microRNA family. Cell 2005;120:635–47.

[6] Calin GA, Sevignani C, Dumitru CD, Hyslop T, Noch E, Yendamuri S, et al. Human microRNA genes are frequently located at fragile sites and genomic regions involved in cancers. Proc Natl Acad Sci USA 2004;101:2999–3004.

[7] Ambros V. The functions of animal microRNAs. Nature 2004;431:350–5.

[8] Cai X, Hagedorn CH, Cullen BR. Human microRNAs are processed from capped, polyadenylated transcripts that can also function as mRNAs. RNA 2004;10:1957–66.

[9] Lee Y, Kim M, Han J, Yeom KH, Lee S, Baek SH, et al. MicroRNA genes are transcribed by RNA polymerase II. EMBO J 2004;23:4051–60.

[10] Borchert GM, Lanier W, Davidson BL. RNA polymerase III transcribes human microRNAs. Nat Struct Mol Biol 2006;13:1097–101.

[11] Lau NC, Lim LP, Weinstein EG, Bartel DP. An abundant class of tiny RNAs with probable regulatory roles in Caenorhabditis elegans. Science 2001;294:858–62.

[12] Lee Y, Jeon K, Lee JT, Kim S, Kim VN. MicroRNA maturation: stepwise processing and subcellular localization. EMBO J 2002;21:4663–70.

[13] Bohnsack MT, Czaplinski K, Gorlich D. Exportin 5 is a RanGTP-dependent dsRNA-binding protein that mediates nuclear export of pre-miRNAs. RNA 2004;10:185–91.

[14] Lund E, Guttinger S, Calado A, Dahlberg JE, Kutay U. Nuclear export of microRNA precursors. Science 2004;303:95–8.

[15] Lewis BP, Shih IH, Jones-Rhoades MW, Bartel DP, Burge CB. Prediction of mammalian microRNA targets. Cell 2003;115:787–98.

[16] Vella MC, Choi EY, Lin SY, Reinert K, Slack FJ. The C. elegans microRNA let-7 binds to imperfect let-7 complementary sites from the lin-41 3′UTR. Genes Dev 2004;18:132–7.

[17] Didiano D, Hobert O. Perfect seed pairing is not a generally reliable predictor for miRNA-target interactions. Nat Struct Mol Biol 2006;13:849–51.

[18] Brennecke J, Stark A, Russell RB, Cohen SM. Principles of microRNA-target recognition. PLoS Biol 2005;3:e85.

[19] Duursma AM, Kedde M, Schrier M, le Sage C, Agami R. miR-148 targets human DNMT3b protein coding region. RNA 2008;14:872–7.

[20] Orom UA, Nielsen FC, Lund AH. MicroRNA-10a binds the 5′UTR of ribosomal protein mRNAs and enhances their translation. Mol Cell 2008;30:460–71.

[21] Tay Y, Zhang J, Thomson AM, Lim B, Rigoutsos I. MicroRNAs to Nanog, Oct4 and Sox2 coding regions modulate embryonic stem cell differentiation. Nature 2008;455:1124–8.

[22] Hutvagner G, Zamore PD. A microRNA in a multiple-turnover RNAi enzyme complex. Science 2002;297:2056–60.

[23] Valencia-Sanchez MA, Liu J, Hannon GJ, Parker R. Control of translation and mRNA degradation by miRNAs and siRNAs. Genes Dev 2006;20:515—24.

[24] Filipowicz W. RNAi: the nuts and bolts of the RISC machine. Cell 2005;122:17—20.

[25] Liu J, Carmell MA, Rivas FV, Marsden CG, Thomson JM, Song JJ, et al. Argonaute2 is the catalytic engine of mammalian RNAi. Science 2004;305:1437—41.

[26] Petersen CP, Bordeleau ME, Pelletier J, Sharp PA. Short RNAs repress translation after initiation in mammalian cells. Mol Cell 2006;21:533—42.

[27] Pillai RS, Bhattacharyya SN, Artus CG, Zoller T, Cougot N, Basyuk E, et al. Inhibition of translational initiation by Let-7 MicroRNA in human cells. Science 2005;309:1573—6.

[28] Wang B, Love TM, Call ME, Doench JG, Novina CD. Recapitulation of short RNA-directed translational gene silencing in vitro. Mol Cell 2006;22:553—60.

[29] Humphreys DT, Westman BJ, Martin DI, Preiss T. MicroRNAs control translation initiation by inhibiting eukaryotic initiation factor 4E/cap and poly(A) tail function. Proc Natl Acad Sci USA 2005;102:16961—6.

[30] Maroney PA, Yu Y, Fisher J, Nilsen TW. Evidence that microRNAs are associated with translating messenger RNAs in human cells. Nat Struct Mol Biol 2006;13:1102—7.

[31] Nottrott S, Simard MJ, Richter JD. Human let-7a miRNA blocks protein production on actively translating polyribosomes. Nat Struct Mol Biol 2006;13:1108—14.

[32] Behm-Ansmant I, Rehwinkel J, Doerks T, Stark A, Bork P, Izaurralde E. mRNA degradation by miRNAs and GW182 requires both CCR4:NOT deadenylase and DCP1:DCP2 decapping complexes. Genes Dev 2006;20:1885—98.

[33] Giraldez AJ, Mishima Y, Rihel J, Grocock RJ, Van Dongen S, Inoue K, et al. Zebrafish MiR-430 promotes deadenylation and clearance of maternal mRNAs. Science 2006;312:75—9.

[34] Wu L, Fan J, Belasco JG. MicroRNAs direct rapid deadenylation of mRNA. Proc Natl Acad Sci USA 2006;103:4034—9.

[35] Engels BM, Hutvagner G. Principles and effects of microRNA-mediated post-transcriptional gene regulation. Oncogene 2006;25:6163—9.

[36] Liu J, Valencia-Sanchez MA, Hannon GJ, Parker R. MicroRNA-dependent localization of targeted mRNAs to mammalian P-bodies. Nat Cell Biol 2005;7:719—23.

[37] Bhattacharyya SN, Habermacher R, Martine U, Closs EI, Filipowicz W. Relief of microRNA-mediated translational repression in human cells subjected to stress. Cell 2006;125:1111—24.

[38] Vasudevan S, Tong Y, Steitz JA. Switching from repression to activation: microRNAs can up-regulate translation. Science 2007;318:1931—4.

[39] Stark A, Brennecke J, Russell RB, Cohen SM. Identification of Drosophila MicroRNA targets. PLoS Biol 2003;1:e60.

[40] Rajewsky N, Socci ND. Computational identification of microRNA targets. Dev Biol 2004;267:529—35.

[41] Saito T, Saetrom P. MicroRNAs—targeting and target prediction. N Biotechnol 2010;27:243—9.

[42] Grimson A, Farh KK, Johnston WK, Garrett-Engele P, Lim LP, Bartel DP. MicroRNA targeting specificity in mammals: determinants beyond seed pairing. Mol Cell 2007;27:91—105.

[43] Gaidatzis D, van Nimwegen E, Hausser J, Zavolan M. Inference of miRNA targets using evolutionary conservation and pathway analysis. BMC Bioinformatics 2007;8:69.

[44] Bartel DP. MicroRNAs: target recognition and regulatory functions. Cell 2009;136:215—33.

[45] Stark A, Brennecke J, Bushati N, Russell RB, Cohen SM. Animal MicroRNAs confer robustness to gene expression and have a significant impact on 3′UTR evolution. Cell 2005;123:1133—46.

[46] Lytle JR, Yario TA, Steitz JA. Target mRNAs are repressed as efficiently by microRNA-binding sites in the 5′ UTR as in the 3′ UTR. Proc Natl Acad Sci USA 2007;104:9667—72.

[47] Kloosterman WP, Wienholds E, Ketting RF, Plasterk RH. Substrate requirements for let-7 function in the developing zebrafish embryo. Nucleic Acids Res 2004;32:6284—91.

[48] Bartel DP. MicroRNAs: genomics, biogenesis, mechanism, and function. Cell 2004;116:281—97.

[49] Cheng AM, Byrom MW, Shelton J, Ford LP. Antisense inhibition of human miRNAs and indications for an involvement of miRNA in cell growth and apoptosis. Nucleic Acids Res 2005;33:1290—7.

[50] Chan JA, Krichevsky AM, Kosik KS. MicroRNA-21 is an antiapoptotic factor in human glioblastoma cells. Cancer Res 2005;65:6029—33.

[51] Hua Z, Lv Q, Ye W, Wong CK, Cai G, Gu D, et al. MiRNA-directed regulation of VEGF and other angiogenic factors under hypoxia. PLoS One 2006;1:e116.

[52] Ma L, Teruya-Feldstein J, Weinberg RA. Tumour invasion and metastasis initiated by microRNA-10b in breast cancer. Nature 2007;449:682—8.

[53] Hanahan D, Weinberg RA. The hallmarks of cancer. Cell 2000;100:57—70.

[54] Calin GA, Dumitru CD, Shimizu M, Bichi R, Zupo S, Noch E, et al. Frequent deletions and down-regulation of micro- RNA genes miR15 and miR16 at 13q14 in chronic lymphocytic leukemia. Proc Natl Acad Sci USA 2002;99:15524−9.

[55] Cimmino A, Calin GA, Fabbri M, Iorio MV, Ferracin M, Shimizu, et al. miR-15 and miR-16 induce apoptosis by targeting BCL2. Proc Natl Acad Sci USA 2005;102:13944−9.

[56] Calin GA, Cimmino A, Fabbri M, Ferracin M, Wojcik SE, Shimizu M, et al. MiR-15a and miR-16-1 cluster functions in human leukemia. Proc Natl Acad Sci USA 2008;105:5166−71.

[57] Volinia S, Calin GA, Liu CG, Ambs S, Cimmino A, Petrocca F. A microRNA expression signature of human solid tumors defines cancer gene targets. Proc Natl Acad Sci USA 2006;103:2257−61.

[58] Lu J, Getz G, Miska EA, Alvarez-Saavedra E, Lamb J, Peck D, et al. MicroRNA expression profiles classify human cancers. Nature 2005;435:834−8.

[59] Iorio MV, Ferracin M, Liu CG, Veronese A, Spizzo R, Sabbioni, et al. MicroRNA gene expression deregulation in human breast cancer. Cancer Res 2005;65:7065−70.

[60] Guo Y, Chen Z, Zhang L, Zhou F, Shi S, Feng X, et al. Distinctive microRNA profiles relating to patient survival in esophageal squamous cell carcinoma. Cancer Res 2008;68:26−33.

[61] Michael MZ, SM OC, van Holst Pellekaan NG, Young GP, James RJ. Reduced accumulation of specific microRNAs in colorectal neoplasia. Mol Cancer Res 2003;1:882−91.

[62] Yanaihara N, Caplen N, Bowman E, Seike M, Kumamoto K, Yi M, et al. Unique microRNA molecular profiles in lung cancer diagnosis and prognosis. Cancer Cell 2006;9:189−98.

[63] Ambs S, Prueitt RL, Yi M, Hudson RS, Howe TM, Petrocca F, et al. Genomic profiling of microRNA and messenger RNA reveals deregulated microRNA expression in prostate cancer. Cancer Res 2008;68:6162−70.

[64] Guo J, Miao Y, Xiao B, Huan R, Jiang Z, Meng D, et al. Differential expression of microRNA species in human gastric cancer versus non-tumorous tissues. J Gastroenterol Hepatol 2009;24:652−7.

[65] Ma L. Role of miR-10b in breast cancer metastasis. Breast Cancer Res 2010;12:210.

[66] Ma L, Reinhardt F, Pan E, Soutschek J, Bhat B, Marcusson EG, et al. Therapeutic silencing of miR-10b inhibits metastasis in a mouse mammary tumor model. Nat Biotechnol 2010;28:341−7.

[67] Korpal M, Lee ES, Hu G, Kang Y. The miR-200 family inhibits epithelial-mesenchymal transition and cancer cell migration by direct targeting of E-cadherin transcriptional repressors ZEB1 and ZEB2. J Biol Chem 2008;283:14910−4.

[68] Calin GA, Ferracin M, Cimmino A, Di Leva G, Shimizu M, Wojcik SE, et al. A MicroRNA signature associated with prognosis and progression in chronic lymphocytic leukemia. N Engl J Med 2005;353:1793−801.

[69] Schetter AJ, Leung SY, Sohn JJ, Zanetti KA, Bowman ED, Yanaihara N, et al. MicroRNA expression profiles associated with prognosis and therapeutic outcome in colon adenocarcinoma. JAMA 2008;299:425−36.

[70] Takamizawa J, Konishi H, Yanagisawa K, Tomida S, Osada H, Endoh H, et al. Reduced expression of the let-7 microRNAs in human lung cancers in association with shortened postoperative survival. Cancer Res 2004;64:3753−6.

[71] Bandi N, Zbinden S, Gugger M, Arnold M, Kocher V, Hasan L, et al. miR-15a and miR-16 are implicated in cell cycle regulation in a Rb-dependent manner and are frequently deleted or down-regulated in non-small cell lung cancer. Cancer Res 2009;69:5553−9.

[72] Bottoni A, Piccin D, Tagliati F, Luchin A, Zatelli MC, degli Uberti EC. miR-15a and miR-16-1 down-regulation in pituitary adenomas. J Cell Physiol 2005;204:280−5.

[73] Bhattacharya R, Nicoloso M, Arvizo R, Wang E, Cortez A, Rossi S, et al. MiR-15a and MiR-16 control Bmi-1 expression in ovarian cancer. Cancer Res 2009;69:9090−5.

[74] Bonci D, Coppola V, Musumeci M, Addario A, Giuffrida R, Memeo L, et al. The miR-15a-miR-16-1 cluster controls prostate cancer by targeting multiple oncogenic activities. Nat Med 2008;14:1271−7.

[75] Ricarte-Filho JC, Fuziwara CS, Yamashita AS, Rezende E, da-Silva MJ, Kimura ET. Effects of let-7 microRNA on Cell Growth and Differentiation of Papillary Thyroid Cancer. Transl Oncol 2009;2:236−41.

[76] Dong Q, Meng P, Wang T, Qin W, Wang F, Yuan J, et al. MicroRNA let-7a inhibits proliferation of human prostate cancer cells in vitro and in vivo by targeting E2F2 and CCND2. PLoS One 2010;5:e10147.

[77] Esquela-Kerscher A, Trang P, Wiggins JF, Patrawala L, Cheng A, Ford L, et al. The let-7 microRNA reduces tumor growth in mouse models of lung cancer. Cell Cycle 2008;7:759−64.

[78] Trang P, Medina PP, Wiggins JF, Ruffino L, Kelnar K, Omotola M, et al. Regression of murine lung tumors by the let-7 microRNA. Oncogene 2010;29:1580−7.

[79] Wong TS, Man OY, Tsang CM, Tsao SW, Tsang RK, Chan JY, et al. MicroRNA let-7 suppresses nasopharyngeal carcinoma cells proliferation through downregulating c-Myc expression. J Cancer Res Clin Oncol 2010.

[80] Lee YS, Dutta A. The tumor suppressor microRNA let-7 represses the HMGA2 oncogene. Genes Dev 2007;21:1025−30.

[81] Feber A, Xi L, Luketich JD, Pennathur A, Landreneau RJ, Wu M, et al. MicroRNA expression profiles of esophageal cancer. J Thorac Cardiovasc Surg 2008;135:255−60. discussion 260.

[82] Fulci V, Chiaretti S, Goldoni M, Azzalin G, Carucci N, Tavolaro S, et al. Quantitative technologies establish a novel microRNA profile of chronic lymphocytic leukemia. Blood 2007;109:4944−51.

[83] Si ML, Zhu S, Wu H, Lu Z, Wu F, Mo YY. miR-21-mediated tumor growth. Oncogene 2007;26:2799−803.

[84] Asangani IA, Rasheed SA, Nikolova DA, Leupold JH, Colburn NH, Post S, et al. MicroRNA-21 (miR-21) post-transcriptionally downregulates tumor suppressor Pdcd4 and stimulates invasion, intravasation and metastasis in colorectal cancer. Oncogene 2008;27:2128−36.

[85] Frankel LB, Christoffersen NR, Jacobsen A, Lindow M, Krogh A, Lund AH. Programmed cell death 4 (PDCD4) is an important functional target of the microRNA miR-21 in breast cancer cells. J Biol Chem 2008;283:1026−33.

[86] Lu Z, Liu M, Stribinskis V, Klinge CM, Ramos KS, Colburn NH, et al. MicroRNA-21 promotes cell transformation by targeting the programmed cell death 4 gene. Oncogene 2008.

[87] Meng F, Henson R, Wehbe-Janek H, Ghoshal K, Jacob ST, Patel T. MicroRNA-21 regulates expression of the PTEN tumor suppressor gene in human hepatocellular cancer. Gastroenterology 2007;133:647−58.

[88] Zhu S, Si ML, Wu H, Mo YY. MicroRNA-21 targets the tumor suppressor gene tropomyosin 1 (TPM1). J Biol Chem 2007;282:14328−36.

[89] Meng F, Henson R, Lang M, Wehbe H, Maheshwari S, Mendell JT, et al. Involvement of human micro-RNA in growth and response to chemotherapy in human cholangiocarcinoma cell lines. Gastroenterology 2006;130:2113−29.

[90] Tao J, Lu Q, Wu D, Li P, Xu B, Qing W, et al. microRNA-21 modulates cell proliferation and sensitivity to doxorubicin in bladder cancer cells. Oncol Rep 2011;25:1721−9.

[91] Valeri N, Gasparini P, Braconi C, Paone A, Lovat F, Fabbri M, et al. MicroRNA-21 induces resistance to 5-fluorouracil by down-regulating human DNA MutS homolog 2 (hMSH2). Proc Natl Acad Sci USA 2010;107:21098−103.

[92] Tanzer A, Stadler PF. Molecular evolution of a microRNA cluster. J Mol Biol 2004;339:327−35.

[93] Hayashita Y, Osada H, Tatematsu Y, Yamada H, Yanagisawa K, Tomida S, et al. A polycistronic microRNA cluster, miR-17-92, is overexpressed in human lung cancers and enhances cell proliferation. Cancer Res 2005;65:9628−32.

[94] He L, Thomson JM, Hemann MT, Hernando-Monge E, Mu D, Goodson S, et al. A microRNA polycistron as a potential human oncogene. Nature 2005;435:828−33.

[95] Ota A, Tagawa H, Karnan S, Tsuzuki S, Karpas A, Kira S, et al. Identification and characterization of a novel gene, C13orf25, as a target for 13q31-q32 amplification in malignant lymphoma. Cancer Res 2004;64:3087−95.

[96] Xiao C, Srinivasan L, Calado DP, Patterson HC, Zhang B, Wang J, et al. Lymphoproliferative disease and autoimmunity in mice with increased miR-17-92 expression in lymphocytes. Nat Immunol 2008;9:405−14.

[97] Matsubara H, Takeuchi T, Nishikawa E, Yanagisawa K, Hayashita Y, Ebi H, et al. Apoptosis induction by antisense oligonucleotides against miR-17-5p and miR-20a in lung cancers overexpressing miR-17-92. Oncogene 2007;26:6099−105.

[98] Wong P, Iwasaki M, Somervaille TC, Ficara F, Carico C, Arnold C, et al. The miR-17-92 microRNA polycistron regulates MLL leukemia stem cell potential by modulating p21 expression. Cancer Res 2010;70:3833−42.

[99] O'Donnell KA, Wentzel EA, Zeller KI, Dang CV, Mendell JT. c-Myc-regulated microRNAs modulate E2F1 expression. Nature 2005;435:839−43.

[100] Lin YW, Sheu JC, Liu LY, Chen CH, Lee HS, Huang GT, et al. Loss of heterozygosity at chromosome 13q in hepatocellular carcinoma: identification of three independent regions. Eur J Cancer 1999;35:1730−4.

[101] Bracken AP, Ciro M, Cocito A, Helin K. E2F target genes: unraveling the biology. Trends Biochem Sci 2004;29:409−17.

[102] Matsumura I, Tanaka H, Kanakura Y. E2F1 and c-Myc in cell growth and death. Cell Cycle 2003;2:333−8.

[103] Zhang L, Huang J, Yang N, Greshock J, Megraw MS, Giannakakis A, et al. microRNAs exhibit high frequency genomic alterations in human cancer. Proc Natl Acad Sci USA 2006;103:9136−41.

[104] Gaur A, Jewell DA, Liang Y, Ridzon D, Moore JH, Chen C, et al. Characterization of microRNA expression levels and their biological correlates in human cancer cell lines. Cancer Res 2007;67:2456−68.

[105] Gauwerky CE, Huebner K, Isobe M, Nowell PC, Croce CM. Activation of MYC in a masked t(8;17) translocation results in an aggressive B-cell leukemia. Proc Natl Acad Sci USA 1989;86:8867−71.

[106] Sarhadi VK, Wikman H, Salmenkivi K, Kuosma E, Sioris T, Salo J, et al. Increased expression of high mobility group A proteins in lung cancer. J Pathol 2006;209:206−12.

[107] Mayr C, Hemann MT, Bartel DP. Disrupting the pairing between let-7 and Hmga2 enhances oncogenic transformation. Science 2007;315:1576−9.

106

[108] Grandori C, Cowley SM, James LP, Eisenman RN. The Myc/Max/Mad network and the transcriptional control of cell behavior. Annu Rev Cell Dev Biol 2000;16:653—99.

[109] Dalla-Favera R, Bregni M, Erikson J, Patterson D, Gallo RC, Croce CM. Human c-myc onc gene is located on the region of chromosome 8 that is translocated in Burkitt lymphoma cells. Proc Natl Acad Sci USA 1982;79:7824—7.

[110] Chang TC, Yu D, Lee YS, Wentzel EA, Arking DE, West KM, et al. Widespread microRNA repression by Myc contributes to tumorigenesis. Nat Genet 2008;40:43—50.

[111] Chang TC, Zeitels LR, Hwang HW, Chivukula RR, Wentzel EA, Dews M, et al. Lin-28B transactivation is necessary for Myc-mediated let-7 repression and proliferation. Proc Natl Acad Sci USA 2009;106:3384—9.

[112] Hollstein M, Sidransky D, Vogelstein B, Harris CC. p53 mutations in human cancers. Science 1991;253:49—53.

[113] He L, He X, Lim LP, de Stanchina E, Xuan Z, Liang Y, et al. A microRNA component of the p53 tumour suppressor network. Nature 2007;447:1130—4.

[114] Yamakuchi M, Lotterman CD, Bao C, Hruban RH, Karim B, Mendell JT, et al. P53-induced microRNA-107 inhibits HIF-1 and tumor angiogenesis. Proc Natl Acad Sci USA 2010;107:6334—9.

[115] Sachdeva M, Zhu S, Wu F, Wu H, Walia V, Kumar S, et al. p53 represses c-Myc through induction of the tumor suppressor miR-145. Proc Natl Acad Sci USA 2009;106:3207—12.

[116] Georges SA, Biery MC, Kim SY, Schelter JM, Guo J, Chang AN, et al. Coordinated regulation of cell cycle transcripts by p53-Inducible microRNAs, miR-192 and miR-215. Cancer Res 2008;68:10105—12.

[117] Suzuki HI, Yamagata K, Sugimoto K, Iwamoto T, Kato S, Miyazono K. Modulation of microRNA processing by p53. Nature 2009;460:529—33.

[118] Sampson VB, Rong NH, Han J, Yang Q, Aris V, Soteropoulos P, et al. MicroRNA let-7a down-regulates MYC and reverts MYC-induced growth in Burkitt lymphoma cells. Cancer Res 2007;67:9762—70.

[119] Le MT, Teh C, Shyh-Chang N, Xie H, Zhou B, Korzh V, et al. MicroRNA-125b is a novel negative regulator of p53. Genes Dev 2009;23:862—76.

[120] Weber B, Stresemann C, Brueckner B, Lyko F. Methylation of human microRNA genes in normal and neoplastic cells. Cell Cycle 2007;6:1001—5.

[121] Saito Y, Liang G, Egger G, Friedman JM, Chuang JC, Coetzee GA, et al. Specific activation of microRNA-127 with downregulation of the proto-oncogene BCL6 by chromatin-modifying drugs in human cancer cells. Cancer Cell 2006;9:435—43.

[122] Scott GK, Mattie MD, Berger CE, Benz SC, Benz CC. Rapid alteration of microRNA levels by histone deacetylase inhibition. Cancer Res 2006;66:1277—81.

[123] Lujambio A, Ropero S, Ballestar E, Fraga MF, Cerrato C, Setien F, et al. Genetic unmasking of an epigenetically silenced microRNA in human cancer cells. Cancer Res 2007;67:1424—9.

[124] Agirre X, Vilas-Zornoza A, Jimenez-Velasco A, Martin-Subero JI, Cordeu L, Garate L, et al. Epigenetic silencing of the tumor suppressor microRNA Hsa-miR-124a regulates CDK6 expression and confers a poor prognosis in acute lymphoblastic leukemia. Cancer Res 2009;69:4443—53.

[125] Iorio MV, Visone R, Di Leva G, Donati V, Petrocca F, Casalini P, et al. MicroRNA signatures in human ovarian cancer. Cancer Res 2007;67:8699—707.

[126] Fabbri M, Garzon R, Cimmino A, Liu Z, Zanesi N, Callegari E, et al. MicroRNA-29 family reverts aberrant methylation in lung cancer by targeting DNA methyltransferases 3A and 3B. Proc Natl Acad Sci USA 2007;104:15805—10.

[127] Kumar MS, Lu J, Mercer KL, Golub TR, Jacks T. Impaired microRNA processing enhances cellular transformation and tumorigenesis. Nat Genet 2007;39:673—7.

[128] Chiosea S, Jelezcova E, Chandran U, Acquafondata M, McHale T, Sobol RW, et al. Up-regulation of dicer, a component of the MicroRNA machinery, in prostate adenocarcinoma. Am J Pathol 2006;169:1812—20.

[129] Faber C, Horst D, Hlubek F, Kirchner T. Overexpression of Dicer predicts poor survival in colorectal cancer. Eur J Cancer 2011;47:1414—9.

[130] Karube Y, Tanaka H, Osada H, Tomida S, Tatematsu Y, Yanagisawa K, et al. Reduced expression of Dicer associated with poor prognosis in lung cancer patients. Cancer Sci 2005;96:111—5.

[131] Merritt WM, Lin YG, Han LY, Kamat AA, Spannuth WA, Schmandt R, et al. Dicer, Drosha, and outcomes in patients with ovarian cancer. N Engl J Med 2008;359:2641—50.

[132] Muralidhar B, Goldstein LD, Ng G, Winder DM, Palmer RD, Gooding EL, et al. Global microRNA profiles in cervical squamous cell carcinoma depend on Drosha expression levels. J Pathol 2007;212:368—77.

[133] Sugito N, Ishiguro H, Kuwabara Y, Kimura M, Mitsui A, Kurehara H, et al. RNASEN regulates cell proliferation and affects survival in esophageal cancer patients. Clin Cancer Res 2006;12:7322—8.

[134] Lin RJ, Lin YC, Chen J, Kuo HH, Chen YY, Diccianni MB, et al. microRNA signature and expression of Dicer and Drosha can predict prognosis and delineate risk groups in neuroblastoma. Cancer Res 2010;70:7841—50.

107

[135] Wu H, Ye C, Ramirez D, Manjunath N. Alternative processing of primary microRNA transcripts by Drosha generates 5′ end variation of mature microRNA. PLoS One 2009;4:e7566.

[136] Luciano DJ, Mirsky H, Vendetti NJ, Maas S. RNA editing of a miRNA precursor. RNA 2004;10:1174−7.

[137] Kawahara Y, Megraw M, Kreider E, Iizasa H, Valente L, Hatzigeorgiou AG, et al. Frequency and fate of microRNA editing in human brain. Nucleic Acids Res 2008;36:5270−80.

[138] Yang W, Chendrimada TP, Wang Q, Higuchi M, Seeburg PH, Shiekhattar R, et al. Modulation of microRNA processing and expression through RNA editing by ADAR deaminases. Nat Struct Mol Biol 2006;13:13−21.

[139] Kawahara Y, Zinshteyn B, Chendrimada TP, Shiekhattar R, Nishikura K. RNA editing of the microRNA-151 precursor blocks cleavage by the Dicer-TRBP complex. EMBO Rep 2007;8:763−9.

[140] Kawahara Y, Zinshteyn B, Sethupathy P, Iizasa H, Hatzigeorgiou AG, Nishikura K. Redirection of silencing targets by adenosine-to-inosine editing of miRNAs. Science 2007;315:1137−40.

[141] Linsley PS, Schelter J, Burchard J, Kibukawa M, Martin MM, Bartz SR, et al. Transcripts targeted by the microRNA-16 family cooperatively regulate cell cycle progression. Mol Cell Biol 2007;27:2240−52.

[142] Meng F, Henson R, Wehbe-Janek H, Smith H, Ueno Y, Patel T. The MicroRNA let-7a modulates interleukin-6-dependent STAT-3 survival signaling in malignant human cholangiocytes. J Biol Chem 2007;282:8256−64.

[143] Simone NL, Soule BP, Ly D, Saleh AD, Savage JE, Degraff W, et al. Ionizing radiation-induced oxidative stress alters miRNA expression. PLoS One 2009;4:e6377.

[144] Crosby ME, Kulshreshtha R, Ivan M, Glazer PM. MicroRNA regulation of DNA repair gene expression in hypoxic stress. Cancer Res 2009;69:1221−9.

[145] Galluzzi L, Morselli E, Vitale I, Kepp O, Senovilla L, Criollo A, et al. miR-181a and miR-630 regulate cisplatin-induced cancer cell death. Cancer Res 2010;70:1793−803.

[146] Zhu W, Shan X, Wang T, Shu Y, Liu P. miR-181b modulates multidrug resistance by targeting BCL2 in human cancer cell lines. Int J Cancer 2010;127:2520−9.

[147] Gao W, Shen H, Liu L, Xu J, Shu Y. MiR-21 overexpression in human primary squamous cell lung carcinoma is associated with poor patient prognosis. J Cancer Res Clin Oncol 2011;137:557−66.

[148] Gao W, Yu Y, Cao H, Shen H, Li X, Pan S, et al. Deregulated expression of miR-21, miR-143 and miR-181a in non small cell lung cancer is related to clinicopathologic characteristics or patient prognosis. Biomed Pharmacother 2010;64:399−408.

[149] Huang Y, Chuang A, Hao H, Talbot C, Sen T, Trink B, et al. Phospho-DeltaNp63alpha is a key regulator of the cisplatin-induced microRNAome in cancer cells. Cell Death Differ 2011;18:1220−30.

[150] Zhu W, Zhu D, Lu S, Wang T, Wang J, Jiang B, et al. miR-497 modulates multidrug resistance of human cancer cell lines by targeting BCL2. Med Oncol 2011.

[151] Dohadwala M, Yang SC, Luo J, Sharma S, Batra RK, Huang M, et al. Cyclooxygenase-2-dependent regulation of E-cadherin: prostaglandin E(2) induces transcriptional repressors ZEB1 and snail in non-small cell lung cancer. Cancer Res 2006;66:5338−45.

[152] Gemmill RM, Roche J, Potiron VA, Nasarre P, Mitas M, Coldren CD, et al. ZEB1-responsive genes in non-small cell lung cancer. Cancer Lett 2011;300:66−78.

[153] Witta SE, Gemmill RM, Hirsch FR, Coldren CD, Hedman K, Ravdel L, et al. Restoring E-cadherin expression increases sensitivity to epidermal growth factor receptor inhibitors in lung cancer cell lines. Cancer Res 2006;66:944−50.

[154] Takeyama Y, Sato M, Horio M, Hase T, Yoshida K, Yokoyama T, et al. Knockdown of ZEB1, a master epithelial-to-mesenchymal transition (EMT) gene, suppresses anchorage-independent cell growth of lung cancer cells. Cancer Lett 2010;296:216−24.

[155] Gregory PA, Bert AG, Paterson EL, Barry SC, Tsykin A, Farshid G, et al. The miR-200 family and miR-205 regulate epithelial to mesenchymal transition by targeting ZEB1 and SIP1. Nat Cell Biol 2008;10: 593−601.

[156] Park SM, Gaur AB, Lengyel E, Peter ME. The miR-200 family determines the epithelial phenotype of cancer cells by targeting the E-cadherin repressors ZEB1 and ZEB2. Genes Dev 2008;22:894−907.

[157] Ceppi P, Mudduluru G, Kumarswamy R, Rapa I, Scagliotti GV, Papotti M, et al. Loss of miR-200c expression induces an aggressive, invasive, and chemoresistant phenotype in non-small cell lung cancer. Mol Cancer Res 2010;8:1207−16.

[158] Cobo M, Isla D, Massuti B, Montes A, Sanchez JM, Provencio M, et al. Customizing cisplatin based on quantitative excision repair cross-complementing 1 mRNA expression: a phase III trial in non-small-cell lung cancer. J Clin Oncol 2007;25:2747−54.

[159] Allingham-Hawkins D, Lea A, Levine S. ERCC1 Expression Analysis to Guide Therapy in Non-Small Cell Lung Cancer. PLoS Curr 2010;2:RRN1202.

[160] Friboulet L, Barrios-Gonzales D, Commo F, Olaussen KA, Vagner S, Adam J, et al. Molecular characteristics of ERCC1 negative vs. ERCC1 positive tumors in resected NSCLC. Clin Cancer Res 2011.

[161] Mazar J, Deblasio D, Govindarajan SS, Zhang S, Perera RJ. Epigenetic regulation of microRNA-375 and its role in melanoma development in humans. FEBS Lett 2011.

[162] Chang X, Monitto CL, Demokan S, Kim MS, Chang SS, Zhong X, et al. Identification of hypermethylated genes associated with cisplatin resistance in human cancers. Cancer Res 2010;70:2870—9.

[163] Shin S, Cha HJ, Lee EM, Lee SJ, Seo SK, Jin HO, et al. Alteration of miRNA profiles by ionizing radiation in A549 human non-small cell lung cancer cells. Int J Oncol 2009;35:81—6.

[164] Cha HJ, Shin S, Yoo H, Lee EM, Bae S, Yang KH, et al. Identification of ionizing radiation-responsive microRNAs in the IM9 human B lymphoblastic cell line. Int J Oncol 2009;34:1661—8.

[165] Shin S, Cha HJ, Lee EM, Jung JH, Lee SJ, Park IC, et al. MicroRNAs are significantly influenced by p53 and radiation in HCT116 human colon carcinoma cells. Int J Oncol 2009;34:1645—52.

[166] Chaudhry MA, Sachdeva H, Omaruddin RA. Radiation-induced micro-RNA modulation in glioblastoma cells differing in dna-repair pathways. DNA Cell Biol 2010;29:553—61.

[167] Chaudhry MA, Kreger B, Omaruddin RA. Transcriptional modulation of micro-RNA in human cells differing in radiation sensitivity. Int J Radiat Biol 2010;86:569—83.

[168] Zhang C, Wang G, Kang C, Du Y, Pu P. [Up-regulation of p27(kip1) by miR-221/222 antisense oligonu-cleotides enhances the radiosensitivity of U251 glioblastoma]. Zhonghua Yi Xue Yi Chuan Xue Za Zhi 2009;26:634—8.

[169] Polyak K, Lee MH, Erdjument-Bromage H, Koff A, Roberts JM, Tempst P, et al. Cloning of p27Kip1, a cyclin-dependent kinase inhibitor and a potential mediator of extracellular antimitogenic signals. Cell 1994;78:59—66.

[170] Cheville JC, Lloyd RV, Sebo TJ, Cheng L, Erickson L, Bostwick, et al. Expression of p27kip1 in prostatic adenocarcinoma. Mod Pathol 1998;11:324—8.

[171] Cheng L, Lloyd RV, Weaver AL, Pisansky TM, Cheville JC, Ramnani DM, et al. The cell cycle inhibitors p21WAF1 and p27KIP1 are associated with survival in patients treated by salvage prostatectomy after radi-ation therapy. Clin Cancer Res 2000;6:1896—9.

[172] Esposito G, Pucciarelli S, Alaggio R, Giacomelli L, Marchiori E, Iaderosa, et al. P27kip1 expression is asso-ciated with tumor response to preoperative chemoradiotherapy in rectal cancer. Ann Surg Oncol 2001;8:311—8.

[173] Chen G, Zhu W, Shi D, Lv L, Zhang C, Liu P, et al. MicroRNA-181a sensitizes human malignant glioma U87MG cells to radiation by targeting Bcl-2. Oncol Rep 2010;23:997—1003.

[174] Condon LT, Ashman JN, Ell SR, Stafford ND, Greenman J, Cawkwell L. Overexpression of Bcl-2 in squamous cell carcinoma of the larynx: a marker of radioresistance. Int J Cancer 2002;100:472—5.

[175] An J, Chervin AS, Nie A, Ducoff HS, Huang Z. Overcoming the radioresistance of prostate cancer cells with a novel Bcl-2 inhibitor. Oncogene 2007;26:652—61.

[176] Streffer JR, Rimner A, Rieger J, Naumann U, Rodemann HP, Weller M. BCL-2 family proteins modulate radiosensitivity in human malignant glioma cells. J Neurooncol 2002;56:43—9.

[177] Kato M, Paranjape T, Muller RU, Nallur S, Gillespie E, Keane K, et al. The mir-34 microRNA is required for the DNA damage response in vivo in C. elegans and in vitro in human breast cancer cells. Oncogene 2009;28:2419—24.

[178] Josson S, Sung SY, Lao K, Chung LW, Johnstone PA. Radiation modulation of microRNA in prostate cancer cell lines. Prostate 2008;68:1599—606.

[179] Weidhaas JB, Babar I, Nallur SM, Trang P, Roush S, Boehm M, et al. MicroRNAs as potential agents to alter resistance to cytotoxic anticancer therapy. Cancer Res 2007;67:11111—6.

[180] Weidhaas JB, Eisenmann DM, Holub JM, Nallur SV. A conserved RAS/mitogen-activated protein kinase pathway regulates DNA damage-induced cell death postirradiation in Radelegans. Cancer Res 2006;66:10434—8.

[181] Oh JS, Kim JJ, Byun JY, Kim IA. Lin28-let7 modulates radiosensitivity of human cancer cells with activation of K-Ras. Int J Radiat Oncol Biol Phys 2010;76:5—8.

[182] Svoboda M, Izakovicova Holla L, Sefr R, Vrtkova I, Kocakova I, Tichy B, et al. Micro-RNAs miR125b and miR137 are frequently upregulated in response to capecitabine chemoradiotherapy of rectal cancer. Int J Oncol 2008;33:541—7.

[183] Wang XC, Du LQ, Tian LL, Wu HL, Jiang XY, Zhang H, et al. Expression and function of miRNA in post-operative radiotherapy sensitive and resistant patients of non-small cell lung cancer. Lung Cancer 2011;72:92—9.

[184] Cowland JB, Hother C, Gronbaek K. MicroRNAs and cancer. APMIS 2007;115:1090—106.

[185] Bandres E, Cubedo E, Agirre X, Malumbres R, Zarate R, Ramirez N, et al. Identification by Real-time PCR of 13 mature microRNAs differentially expressed in colorectal cancer and non-tumoral tissues. Mol Cancer 2006;5:29.

[186] Roldo C, Missiaglia E, Hagan JP, Falconi M, Capelli P, Bersani S, et al. MicroRNA expression abnormalities in pancreatic endocrine and acinar tumors are associated with distinctive pathologic features and clinical behavior. J Clin Oncol 2006;24:4677—84.

[187] Lee EJ, Gusev Y, Jiang J, Nuovo GJ, Lerner MR, Frankel WL, et al. Expression profiling identifies microRNA signature in pancreatic cancer. Int J Cancer 2007;120:1046—54.

[188] Murakami Y, Yasuda T, Saigo K, Urashima T, Toyoda H, Okanoue T, et al. Comprehensive analysis of microRNA expression patterns in hepatocellular carcinoma and non-tumorous tissues. Oncogene 2006;25:2537—45.

[189] Calin GA, Liu CG, Sevignani C, Ferracin M, Felli N, Dumitru CD, et al. MicroRNA profiling reveals distinct signatures in B cell chronic lymphocytic leukemias. Proc Natl Acad Sci USA 2004;101:11755—60.

[190] Bianchi F, Nicassio F, Marzi M, Belloni E, Dall'olio V, Bernard L, et al. A serum circulating miRNA diagnostic test to identify asymptomatic high-risk individuals with early stage lung cancer. EMBO Mol Med 2011;3:495—503.

[191] Feber A, Xi L, Pennathur A, Gooding WE, Bandla S, Wu M, et al. MicroRNA prognostic signature for nodal metastases and survival in esophageal adenocarcinoma. Ann Thorac Surg 2011;91:1523—30.

[192] Liu R, Zhang C, Hu Z, Li G, Wang C, Yang C, et al. A five-microRNA signature identified from genome-wide serum microRNA expression profiling serves as a fingerprint for gastric cancer diagnosis. Eur J Cancer 2011;47:784—91.

[193] Gougelet A, Pissaloux D, Besse A, Perez J, Duc A, Dutour A, et al. Micro-RNA profiles in osteosarcoma as a predictive tool for ifosfamide response. Int J Cancer 2011;129:680—90.

[194] Janssen EA, Slewa A, Gudlaugsson E, Jonsdottir K, Skaland I, Soiland H, et al. Biologic profiling of lymph node negative breast cancers by means of microRNA expression. Mod Pathol 2010;23:1567—76.

[195] Iorio MV, Casalini P, Tagliabue E, Menard S, Croce CM. MicroRNA profiling as a tool to understand prognosis, therapy response and resistance in breast cancer. Eur J Cancer 2008;44:2753—9.

[196] van 't Veer LJ, Dai H, van de Vijver MJ, He YD, Hart AA, Mao M, et al. Gene expression profiling predicts clinical outcome of breast cancer. Nature 2002;415:530—6.

[197] Knauer M, Mook S, Rutgers EJ, Bender RA, Hauptmann M, van de Vijver MJ, et al. The predictive value of the 70-gene signature for adjuvant chemotherapy in early breast cancer. Breast Cancer Res Treat 2010;120:655—61.

[198] Maher SG, Gillham CM, Duggan SP, Smyth PC, Miller N, Muldoon C, et al. Gene expression analysis of diagnostic biopsies predicts pathological response to neoadjuvant chemoradiotherapy of esophageal cancer. Ann Surg 2009;250:729—37.

[199] Del Rio M, Molina F, Bascoul-Mollevi C, Copois V, Bibeau F, Chalbos P, et al. Gene expression signature in advanced colorectal cancer patients select drugs and response for the use of leucovorin, fluorouracil, and irinotecan. J Clin Oncol 2007;25:773—80.

[200] Yi JM, Dhir M, Van Neste L, Downing SR, Jeschke J, Glockner SC, et al. Genomic and epigenomic integration identifies a prognostic signature in colon cancer. Clin Cancer Res 2011;17:1535—45.

[201] Mook S, Van't Veer LJ, Rutgers EJ, Piccart-Gebhart MJ, Cardoso F. Individualization of therapy using Mammaprint: from development to the MINDACT Trial. Cancer Genomics Proteomics 2007;4:147—55.

[202] Cobleigh MA, Tabesh B, Bitterman P, Baker J, Cronin M, Liu ML, et al. Tumor gene expression and prognosis in breast cancer patients with 10 or more positive lymph nodes. Clin Cancer Res 2005;11:8623—31.

[203] Chen X, Ba Y, Ma L, Cai X, Yin Y, Wang K, et al. Characterization of microRNAs in serum: a novel class of biomarkers for diagnosis of cancer and other diseases. Cell Res 2008;18:997—1006.

[204] Lawrie CH, Gal S, Dunlop HM, Pushkaran B, Liggins AP, Pulford K, et al. Detection of elevated levels of tumour-associated microRNAs in serum of patients with diffuse large B-cell lymphoma. Br J Haematol 2008;141:672—5.

[205] Mitchell PS, Parkin RK, Kroh EM, Fritz BR, Wyman SK, Pogosova-Agadjanyan EL, et al. Circulating microRNAs as stable blood-based markers for cancer detection. Proc Natl Acad Sci USA 2008;105:10513—8.

[206] Trang P, Wiggins JF, Daige CL, Cho C, Omotola M, Brown D, et al. Systemic delivery of tumor suppressor microRNA mimics using a neutral lipid emulsion inhibits lung tumors in mice. Mol Ther 2011;19:1116—22.

[207] Wiggins JF, Ruffino L, Kelnar K, Omotola M, Patrawala L, Brown D, et al. Development of a lung cancer therapeutic based on the tumor suppressor microRNA-34. Cancer Res 2010;70:5923—30.

[208] Roberts AP, Lewis AP, Jopling CL. miR-122 activates hepatitis C virus translation by a specialized mechanism requiring particular RNA components. Nucleic Acids Res 2011.

[209] Young DD, Connelly CM, Grohmann C, Deiters A. Small molecule modifiers of microRNA miR-122 function for the treatment of hepatitis C virus infection and hepatocellular carcinoma. J Am Chem Soc 2010;132:7976—81.

CHAPTER 6

Epigenetic Approaches to Cancer Therapy

Sabita N. Saldanha[1,2], Trygve O. Tollefsbol[1]
[1]University of Alabama at Birmingham, Birmingham, AL, USA
[2]Alabama State University, Montgomery, AL, USA

6.1 INTRODUCTION

Condensation of DNA is achieved by the interaction of basic proteins called histones that encircle 147 bp of DNA forming a structure called the nucleosome [1,2]. The histones are arranged as dimers of each subunit; H2A, H2B, H3, and H4 in the octet [2]. Histone H1 is independent of the octet but helps tether the nucleosome complex [2]. The octect complex with the DNA is so arranged that certain amino acid residues of the histones extend out serving as regulatory substrates for nucleosomal stability [1]. These substrates establish the condensed and decondensed states of the chromatin [1]. Condensation of the chromatin prevents the transcriptome machinery from binding and consequently inhibits gene expression. However, when these projected tails are modified through enzymatic transformations such as acetylation, methylation, phosphorylation, sumoylation, and ubiquitination, the accessibility of DNA changes based on the residue modified [3]. Interestingly, current research has emphasized the roles of these modifications in the transformation process of a normal cell to a tumorigenic phenotype by creating imbalances in net expression of tumor suppressor versus oncogenes or overall genomic imbalances [4]. These covalent modifications are reversible and therefore can have profound impacts on the cellular phenotype when the activities of the enzymes that mediate these modifications are altered. Intense interest has been directed toward the mechanistic pathways of these modifications in carcinogenesis. However, substrate specificity and residue-specific alterations still need to be ascertained.

In addition to histone modifications, CpG dinucleotides can be subjected to epigenetic changes by the methylation of cytosine residues [5,6]. These methylation patterns are heritable and are governed by four isoforms of DNA methyltransferases; DNMT1, DNMT3a, DNMT3b,

T. Tollefsbol (Ed): Epigenetics in Human Disease. DOI: 10.1016/B978-0-12-388415-2.00006-8

and DNMT3L [6]. Another area of epigenetics that still requires further exploration and can potentially compound the effects of chromatin epigenomics in a neoplastic cell is the epigenetic regulation of non-histone proteins. Epigenetic regulations of non-histone proteins can drastically affect pathways within the cell, the cell cyclical controls, and cellular pheno-types. For example, acetylation of key residues of p53 stabilizes the protein and thus the cell cyclical function with which it is associated [7,8]. This chapter discusses the current treatments that are designed to target epigenetic enzymes with the hope of reversing the epigenome of cancerous cells. Non-histone protein modifications are also important in cancerous cells and therefore the current approaches to therapy aimed at targeting non-histone proteins will also be discussed.

6.2 HISTONE ACETYLATION

Positively charged amino acids such as lysine (K) and arginine (R) located at amino ($-NH_2$) terminal ends of histones are variously modified (Figure 6.1) [3]. Histones are preferentially methylated or phosphorylated at arginine residues and acetylated at lysine residues [3,9]. Acetylation of lysines initiates active gene expression. Acetylation of histone residues not only establishes euchromatin states but has crucial roles in nucleosome assembly and maintenance of chromatin states that affect various phases of the cell, including DNA repair [9–11]. Currently, no mathematical models are available that can determine the exact pattern of epigenetic marks which alter sets of genes in cancer tissues. Another hindrance in determining these marks is that these chemical transformations are dynamic and affect the genome globally

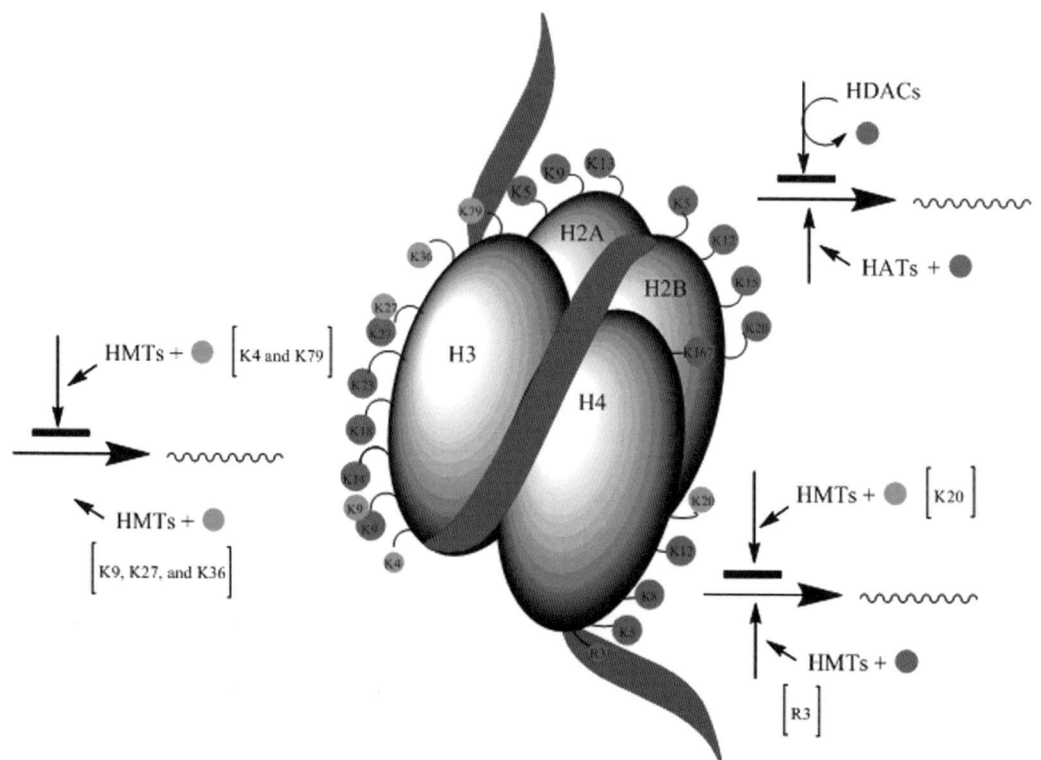

FIGURE 6.1
Effects of acetylation and methylation on histone residues. The red circle represents acetyl groups, the yellow circle symbolizes methylation and the green symbolizes methylation of arginine residue. Acetylation of lysine residues is associated with gene expression whereas methylation-mediated expression is dependent on the residue methylated and the position. These reversible processes exerted by epigenetic enzymes HATs, HDACs, and HMTs largely affect genomic stabilities, local gene expression, and factors governing cell fate or phenotype. This figure is reproduced in the color plate section.

rather than at specific targets. However, certain histone-lysine residues are specifically acetylated or deacetylated at key positions. Histone acetyltransferases (HATs) are enzymes that orchestrate the acetylation of histones and are placed in three superfamilies based on homologies to the yeast class of HATs; GNAT (Gcn5-related N-acetyltransferase), MYST (MOZ, Ybf2/Sas3, Sas2 and TIP60), and p300/CBP145 [4]. In a few cancers, dysregulation of HAT activities by mutations in the HAT genes or the dysfunction of the gene by translocations account for tumor promotion [4]. In breast cancer cells, NCOA3 (AIB1) is overexpressed and in certain leukemias, p300, CBP, and MYST3 (MOZ) translocations are observed [4]. Targeting HAT activity can affect the acetylation patterns and possibly control the expression of oncogenes that are overexpressed.

6.3 HISTONE DEACETYLASES

The dynamic equilibrium of chromatin architecture is finely regulated by the activity of HATs and HDACs. In most cancers, HATs are mutated and include chromosomal translocations of the respective HAT, but HDACs are frequently overexpressed [12]. Certain cofactors exhibit intrinsic HAT or HDAC activity and in most instances the effects are conglomerative with other complexes. Therefore aberrant recruitment of HDACs to transcription factors that affect genes such as oncogenes or tumor suppressor genes or their expression may facilitate a switch from normal to abnormal phenotype. A total of 18 HDACs have been identified and are classified into four major classes [4]. The classification of HDACs is based on the homology of the catalytic site [13]. Class IV HDACs exhibit homologies similar to class I and II HDACs [4]. The Sir2 HDACs have prominent roles in DNA repair and different Sir2s have varied nuclear functional roles [14,15]. Chromatin organization is a well-studied area; however, the actual roles of HDACs at the gene level and their roles in specific cancers still require further elucidation. In vitro analysis of the effects of HDAC inhibitors has demonstrated their profound effects on inhibiting cell proliferation, and inducing cell differentiation and apoptosis [4]. HDAC inhibitors are considered important tools in cancer therapeutics and are currently being evaluated for their therapeutic efficacies in vitro and in clinical trials. HDAC inhibitors tested to date fall under four categories, short-chain fatty acids (SCFA), hydroxamic acid derivatives, benzamindes, and cyclic tertrapeptides [4] (Table 6.1). Inhibition of HDACs increases the acetylation levels of specific histone residues and in some instances increases stabilities of non-histone proteins, both of which are essential to gene regulatory functions (Figure 6.2). These observations have been supported by gene expression profiling studies [16]. Of about 1750 proteins that have been identified to be acetylated at lysine residues, 200 of these become modified in the presence of HDAC inhibitors and represent a significant number that may contribute to changes in gene expression and probably initiate antitumor activity [16]. What needs to be determined however is the substrate specificities for each HDAC and what drugs are specific for each HDAC.

TABLE 6.1 HDAC Inhibitors

Class Type	Type	Compound	Class of HDAC Inhibited	Reference
I	Short-chain fatty acid (SCFA)	Sodium butyrate Valproate Phenylbutyrate	Class I and II	[4]
II	Hydroxamic acid derivatives	Vorinostat Belinostat (PXD-101) Panobinostat (LBH-589)	Class II	[4]
III	Benzamindes	Entinostat (SNDX-275)	Class I	[4]
IV	Cyclic tetrapeptides	Romidepsin (FK-228)	Not determined	[4]

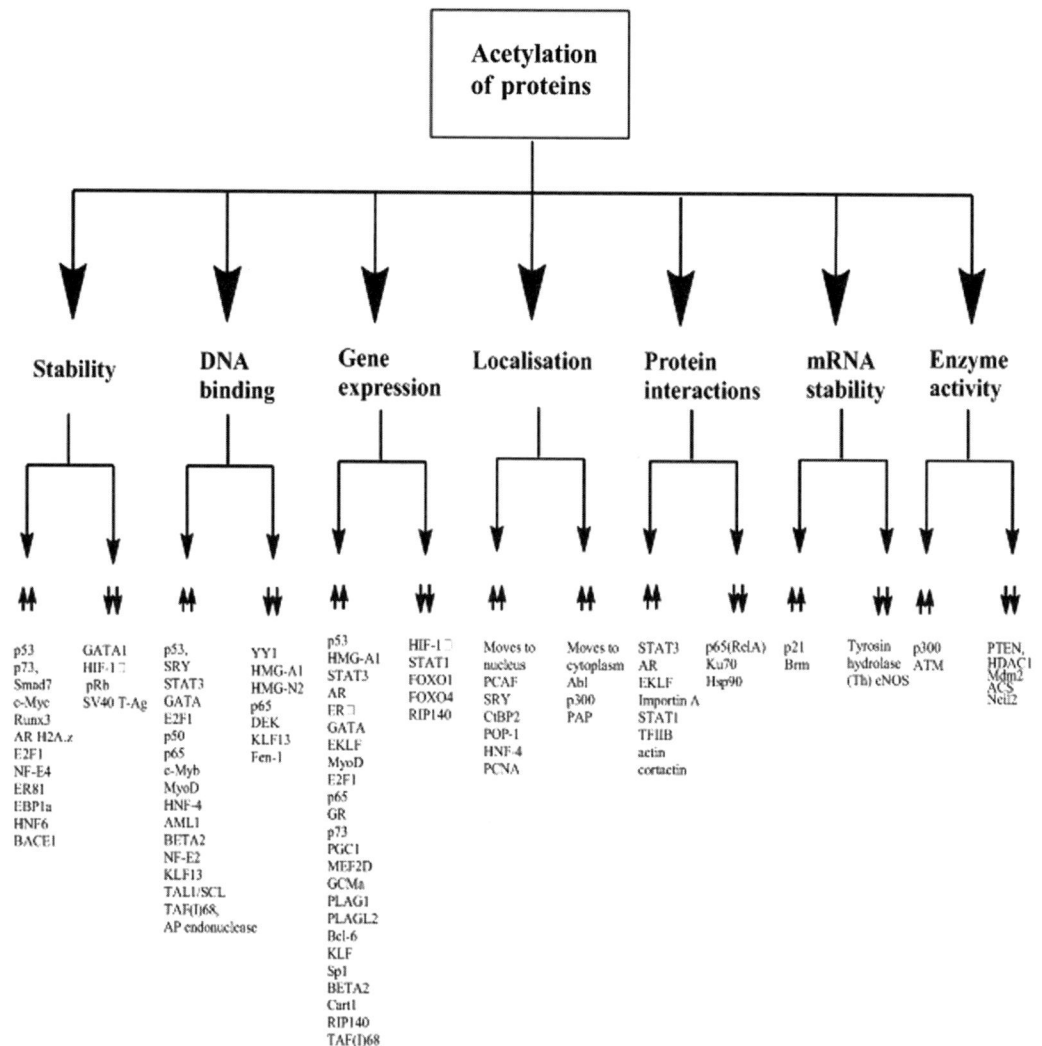

Acetylation of proteins

Stability		DNA binding		Gene expression		Localisation		Protein interactions		mRNA stability		Enzyme activity
↑↑	↓↓	↑↑	↓↓	↑↑	↓↓	↑↑	↑↑	↑↑	↓↓	↑↑	↓↓ ↑↑	↓↓

Stability ↑↑	Stability ↓↓	DNA binding ↑↑	DNA binding ↓↓	Gene expression ↑↑	Gene expression ↓↓	Localisation (to nucleus)	Localisation (to cytoplasm)	Protein interactions ↑↑	Protein interactions ↓↓	mRNA stability ↑↑	Enzyme activity
p53	GATA1	p53,	YY1	p53	HIF-1	Moves to	Moves to	STAT3	p65(RelA)	p21	Tyrosin p300
p73,	HIF-1	SRY	HMG-A1	HMG-A1	STAT1	nucleus	cytoplasm	AR	Ku70	Brm	hydrolase ATM
Smad7	pRb	STAT3	HMG-N2	STAT3	FOXO1	PCAF	Abl	EKLF	Hsp90		(Th) eNOS
c-Myc	SV40 T-Ag	GATA	p65	AR	FOXO4	SRY	p300	Importin A			
Runx3		E2F1	DEK	ER	RIP140	CtBP2	PAP	STAT1			PTEN,
AR H2A.z		p50	KLF13	GATA		POP-1		TFIIB			HDAC1
E2F1		p65	Fen-1	EKLF		HNF-4		actin			Mdm2
NF-E4		c-Myb		MyoD		PCNA		cortactin			ACS
ER81		MyoD		E2F1							Neil2
EBP1a		HNF-4		p65							
HNF6		AML1		GR							
BACE1		BETA2		p73							
		NF-E2		PGC1							
		KLF13		MEF2D							
		TAL1/SCL		GCMa							
		TAF(I)68,		PLAG1							
		AP endonuclease		PLAGL2							
				Bcl-6							
				KLF							
				Sp1							
				BETA2							
				Cart1							
				RIP140							
				TAF(I)68							

FIGURE 6.2
Effects of acetylation on protein functions. Acetylation of proteins affects many different functions, some of which are listed. The double up-arrows indicate increase and the double down-arrows indicate decrease with respect to the particular function. Some of the genes affected by acetylation under specific protein functions are listed [60].

6.4 HISTONE METHYLATION AND DEMETHYLATION

Histone acetylation is key to promoting gene expression. Acetylated lysines are always associated with gene expression but their methylation status contributes to varied gene expression and primarily depends on the position and form of the methylated lysine residue (Figure 6.1). Lysines exists in mono-, di-, and tri-methylated forms and in some instances the same lysine can be acetylated or methylated, for example, K4 and K9 residues of histone H3 [17]. However, in this instance the acetylated lysine status will not govern the methylated status of the same lysine in the histone [17]. In other cases, either acetylation or methylation will influence the covalent modified status of the neighboring lysine residues and the summation of these effects will determine the outcome. A commonly found histone pattern in many cancers is the loss of H4K16 acetylation and H4K20 tri-methylation [18].

Patterns affect the histone residues globally or histones of gene-specific loci can independently influence cancer outcomes (Table 6.2). This implies that when tumor suppressor genes are down-regulated by hypermethylation, oncogenes may be stimulated by acetylation or

TABLE 6.2 Histone Methylation Marks in Cancer Development

Histone	Residue	Change in Histone Pattern	Type of Cancer	Reference
H3	K16	Loss of acetylation	Cancer	[18]
	K20	Loss of trimethylation		
H3	K9	Hypomethylation	Medulloblastoma	[19]
H3	K4	Global decrease in methylation	Poor prognosis or increased risk of recurrence in prostate, breast, kidney, lung, and ovarian cancer	[20,21]
	K18	Global decrease in acetylation		
	K9	Global decrease in methylation		
	K27	Global decrease in tri-methylation		
H3	K9	Hypermethylation	Silences tumor suppressor genes in colorectal, breast and prostate cancer	[22,23]
	K27	Hypermethylation		

hypomethylation. For example, hypermethylation of H3K79 promotes leukemogenesis [24]. Tumor-specific epigenetic abnormalities can stem from altered modifications of the histone residues, and/or altered expression of the enzymes that catalyze the modifications. These changes are driven by mutations or chromosomal rearrangement of genes that code for epigenetic enzymes regardless of their epigenetic modification. As shown in Table 6.2, medulloblastoma arises from the hypomethylation of H3K9, and the loss of H3K9 methyltransferase; amplification of demethylases or acetyltransferases could trigger the outcome observed. Hyperacetylation of H3K9 could inhibit the methylation of its residue.

Like DNA, histone lysine residues are methylated by the activity of methyltransferases and utilize S-adenosyl methionine (SAM) in catalyzing the transfer of the methyl group to specific histone residues [6]. The methyltransferases are specific based on the residues they target. Protein lysine methyltransferases (PKMTs) and arginine methyltransferases (PRMTs) are specific for lysine and arginine residues respectively, and mediate mono-, di-, and tri-methylation. PRMTs primarily catalyze mono- and di-methylation of histone arginine residues 2, 8, 17, and 26 of H3 and arginine residue 3 of H4 [25]. PKMTs have a conserved SET domain that is required for the methyltransferase activity and several of these have been implicated in cancers [25]. H3K27 methylation is mediated by a PKMT called EZH2 [26,27], but this enzyme is over-expressed in many tumors and appears to have major roles in cancer aggressiveness as seen in breast and prostate tissues [26,27]. In another case, leukemogenesis is promoted by the aberrant recruitment of H3K79 non-SET domain DOT1L [28]. Most of the tumor-related effects of HMTs are associated with the over-expression, amplification, and translocation of the genes coding the enzymes. Some of these include SMYD3 [29], CARM1 [30], and PRMT1 [31]. Drugs that can inhibit the activities of these enzymes are currently being investigated and a certain few are showing great promise in clinical trials. 3-Deazaneplanocin (DZNep) is a compound that targets HMTs, including EZH2, and is instrumental in inducing global hypomethylation at several lysine residues and includes H3K27 and H4K20 [32]. As seen with HDACs, in vitro analysis of DZNep treatment of tumor cell lines induced apoptosis through the activation of key target genes [32]. A combination of a HMT and HDAC inhibitors may profoundly affect the synergistic induction of apoptosis and has been demonstrated in colon cancer cells [33,34].

Like with most chemical compounds, non-specific and indirect mechanisms of action may limit their clinical applications. Certain generic compounds may inhibit HMT activity directly with low to no specificity (generic analogs of SAM such as S-adenosyl homocysteine (SAH) and sinefungin) [35]. However, more selective compounds have been identified through current screening methods. A fungal mycotoxin, chaetocin, is a potent inhibitor of H3K9 HMTases SUV39H1 and G9a (EHMT2) at IC_{50} concentrations of 0.8 mM and 2.5 mM, respectively [36,37]. This compound specifically depletes H3K9 di- and tri-methylation levels. However the compound exhibits cytotoxic effects independent of its inhibitory activity. Another non-SAM-competitive inhibitor of HMT G9a, BIX-01294 has been effective at 1.7 mM and the inhibition is selective toward HMTs with SET domains such as SETDB1 (ESET), and SETD7 (SET7/9) [38,39]. BIX-01294 was found after screening 12 500 compounds and is specific in action towards H3K9me2, reducing the di-methylated levels in mammalian cells. Structural analysis of the SET domain reveals that the compound binds to the H3 substrate-binding groove [38]. Both natural and chemical analogs with similar affinities to SET domains can serve as suitable inhibitors in cancer treatments. Once the mechanistic action is determined, the compounds can be modified to improve concentration efficacies and minimize non-specific or cytotoxic effects. Similar inhibitors towards PRMTs, have been found [40]. However, more robust inhibitors are yet to be discovered as the current PRMT inhibitors are non-specific and have low selectivity and activity. Pyrazole-containing CARM1 inhibitors with lower but highly potent IC_{50} concentrations have been reported [41]. It is encouraging that such molecules targeting essential epigenetic enzymes can potentially reverse epigenetic-mediated cancerous phenotypes and that further optimizations and discoveries of effective yet non-cytotoxic drugs need to be identified for clinical testing.

Histone demethylases (HDMs) promote hypomethylation of their target residues and the gene output is determined by the residue, or position demethylated and/or gene-specific target that is demethylated. Therapeutically, targeting HDMs can be of significant importance as onco-genes that are normally expressed through histone hypomethylation of their promoters can be switched off by hypermethylation. Certainly, smaller molecules with effective catalytic inhibitory activity toward HDMs are valuable and screening for such compounds is crucial. HDMs catalyze the removal of methyl groups from lysine residues in two ways. First by the amine oxidation reaction which is specific for mono- and di-methylated residues and second by the hydroxylation of methylated residues creating an unstable intermediate that degrades to release formaldehyde [42]. The second process is specific towards mono-, di-, and tri-methylated residues. The first process requires FAD and the second uses alpha-ketoglutarate and iron as cofactors [42]. The only known HDM to date is lysine-specific demethylase 1 (LSD1) which mediates its demethylating action through the amine oxidation process [43]. The family of Jumonji (JmjC) domain-containing proteins demethylate by the hydroxylation of the methyl groups and includes JARID1B (PLU-1) and Jumonji C (JMJC) 4 domain-containing protein (JMJD2C) [42]. These HDMs have been implicated in tumor progression as well. LSD1 has varied roles in terms of the residues it catalyzes. H3K4 mono- or di-methylated residues are demethylated by LSD1 in conjunction with corepressor RE1-silencing transcription factor (CoREST) [44]. However, H3K9 mono- and di- methylated marks are demethylated by LSD1, in which it interacts with androgen receptor (AR) as a coactivator to enhance the demethylating function [45]. This enzyme is over-expressed in certain cancers and has been reported to be associated with aggressive prostate cancer and poorly differentiated neuroblastomas.

Since LSD1 is homologous to monoamine oxidases (MAO) [46], molecules that are effective against MAO can inhibit the activity of LSD1. One such inhibitor is the tranylcypromine. When used, this inhibitor increases the di-methylated levels of H3K4 both in vitro and in vivo and inhibits the neuroblastoma tumor growth [47]. LSD1 has been found to be inhibited by polyamine compounds, a few of which have shown remarkable positive outcomes in colon cancer cells [47]. These compounds increase the mono- and di-methylated H3K4 levels and reexpresse many silenced genes important in colon cancer development. In breast and prostate cancer, the HMD, JARID1B, is overexpressed and demethylates H3K4me3 that induces cell proliferation, inhibits tumor suppressor functions, and results in AR coactivation [48]. JMJD2C, a H3K9me2/3 demethylase, is amplified and overexpressed in esophageal squamous carcinoma and targeting this enzyme has been proven effective in inhibiting esophageal squamous cell growth in vitro [49]. JMJD2C in conjunction with LSD1 enhances AR-dependent gene expression in prostate cancer where this enzyme is overexpressed [50].

Targeting the enzyme or cofactor can effectively inhibit the activity of HDMs such as JmjC. For example, derivatives of noggin (NOG), analog of alpha-ketoglutarate, are effective against members of the Jumonji-C family of HDMs and a dimethylester of NOG (DMOG) showing cellular activity has been reported [51]. However, these molecules that are being used as HDM inhibitors are effective at much higher concentrations. Further screening of potential drugs with higher efficacies and potencies at lower concentrations is still required and the knowledge of the structural configurations of LSD1 and JMJD2C can assist in the find.

6.5 DNA METHYLATION

The architectural configuration of the nucleosome is strictly governed by histones and their covalent modifications. However, the DNA encompassed by the histone octet dictates cellular functions and stability. Apart from the normally associated transcription factors with promoters, methylation of CpG residues is another important mechanism regulating gene expression. Aberrant expression of this predominant epigenetic modification has been reported to play significant roles in a variety of diseases, including cancers. Fortunately, CpG methylation can be reversed and therefore this heritable change when exposed to demethylating compounds or compounds that inhibit the catalytic function of the enzyme itself (DNMTs), presents as potential cancer therapeutic tools. DNMTs are required for CpG methylation and small molecules that can target these enzymes are being tested in vitro as well as in clinical trials. The very first DNMT inhibitor was Vidaza (5-azacytidine) and was approved for use by the US Food and Drug Administration (FDA) as treatment against myelodysplastic syndrome (MDS) [52,53]. Another molecule, Dacogen (5-aza-2′-deoxycytidine, or decitabine) developed by MGI Pharma Inc. (Bloomington, MN, USA) has also been used to treat MDS [52,53]. These compounds facilitate their action through both methylation-dependent and -independent pathways and in some cases direct proteasomal degradation of the enzyme has been reported [53]. In theory, the use of these inhibitors is aimed at reversing the expression of methylation-silenced critical gene expression. Clinical trials using these molecules have shown great potential as therapeutic agents against leukemia, including MDS, acute myeloid leukemia, chronic myelogenous leukemia, and chronic myelomonocytic leukemia [53]. So far, the therapeutic improvements are seen against leukemias, although this is not the case with solid tumors since it is likely that a multitude of factors govern the growth of the mass. Cellular toxicity is also a major concern and the use of these molecules triggers cell cycle arrest by their integration into the DNA molecule itself. Therefore, it is imperative to develop or screen for drugs that have less cytotoxicity and more efficacy at lower concentrations. Another concern is the relatively low stability of these compounds in vivo and therefore modifications that enhance their stability are to be considered when selecting the drug as a chemotherapeutic agent. It is imperative to design and develop drugs that are relatively stable, mediate the degradations of DNMTs without incorporation into DNA, and facilitate gene expression crucial to cell differentiation and apoptotic pathways.

In mammalian cells, DNMTs exist as four active forms, DNMT1, DNMT3a, DNMT3b, and DNMT3L and either singly or in combination catalyze the methylome patterning crucial to gametogenesis, embryogenesis, development, and carcinogenesis [54]. Structural analysis of these enzymes shows that the catalytic domain resides in the C-terminal region of the protein with the N-terminal essential for DNA recognition states, hemimethylated versus unmethylated. DNMT3a is ubiquitously expressed and DNMT3b is present at very low levels with the exception of a few tissues [55]. These levels change in tumor cells and global hypomethylation and regional hypermethylation of specific genes becomes an apparent pattern and has been reported to be the case in cervical, prostate, and metastatic hepatocellular carcinoma [55]. Most preclinical studies focus on the hypermethylation of key genes, in some cases tumor suppressor genes and others oncogenes and the correlation of DNMT levels, methylation patterns of the promoters, and gene expression. There appears to be some level of correlation between DNMT levels and hypermethylation; however, regression analysis does not seem to support this one-to-one correlation indicative of a much more complex regulatory mechanism in vivo [55]. Some of the commonly reported hypermethylated genes include RAR, RASSF1A, CDNK2A, CHD13, APC, p15, and p16 [56]. p15 is used as a marker to determine leukemia transformation, and in some other tumors the levels of hypermethylation of tumor suppressor genes determine the stage of the tumor [57]. Inhibitors of DNMTs are grouped under two categories, those that interfere with the methylation of cytosine residues by chelating into the DNA complex and second, the non-nucleosides that target DNMT activity or stability (Table 6.3).

Results from both non-nucleoside and nucleoside DNMT inhibitors (DNMTi) in preclinical and clinical settings are encouraging but many of these have drawbacks that need to be revamped. Certainly the need to screen for and design small molecules with potent DNMTi activity, less cytotoxicity, and improved specificities is very evident and significant progress has been made in this direction. Combination treatments of two nucleosides and two non-nucleosides are in progress and the data generated from in vitro studies have shown remarkable synergistic DNA-hypermethylating activity. These have been well described in a review by Jiang S-W [55].

6.6 ACETYLATION OF NON-HISTONE PROTEINS

Histones are the likely targets for reversible modifications and much focus has been on understanding the roles of these modifications in cellular processes and the enzymes that catalyze these chemistries. However, gene products, primarily proteins, are in many ways subjected to similar regulations, of which acetylation and phosphorylation are key modifications. Many cellular processes are governed by the activity of proteins and are involved in cell signaling, transcription, and even protein degradation (Figure 6.2). Reviews on the acetylation of non-histone proteins are limited, yet this protein modification in conjunction with histone modifications is very relevant to cancer epigenomics. Acetylation of proteins can affect many aspects of protein function as shown in Figure 6.2. Especially pertinent are the effects of acetylation on p53 (tumor suppressor), nuclear factor-κB (NF-κB) (metastatic gene), and myelocytomatosis oncogene (c-Myc) (oncogene) and the plausible roles of indirect effects of HAT/HDAC inhibitors on these proteins in restoring normal cellular phenotypes.

Protein p53 is essential to many cell regulatory functions, and in particular, is important to rescue a cell from DNA damage and maintain normal cell division [58]. As seen in many cancers, p53 is mutated and the protein is dysfunctional, contributing to a tumorigenic phenotype. Activation of p53 in cells requires the phosphorylation of the protein that promotes the acetylation of key lysine residues 120, 164, 320, 370, 372, 373, 381, 382, and 386 mediated by different acetyltransferases [59,60]. Although controversial, findings strongly support that acetylation of p53 at the C-terminus end favorably enhances the DNA binding ability of the protein to its target genes [60]. K120 and K164 along with the C-terminus are required for p53 activity, and single-site mutational losses can be rescued by the acetylation of key functional residues [60]. Acetylation of p53 K120 by specific HATs, such as Tip60 and hMOF, induces genes of the apoptotic pathway [60]. When K382 is acetylated, p53 recruits CREB binding protein (CBP) that further enhances the transcription of genes, suggestive of coactivator functions [60]. In all, acetylation of p53 improves the stability and binding ability of p53, allowing for recruitment of coactivators to the transcriptional binding sites in the promoters of genes crucial to cell cycle regulatory functions, such as for p21. Therefore, HDAC inhibitors that increase acetylated levels can contribute to a much more stable p53 even when it is mutated and can enhance its DNA-binding abilities, which otherwise would be ineffective. Further studies are necessary to validate the roles of key acetylated residues that improve the DNA-binding ability of p53 in the mutated form.

TABLE 6.3 Nucleoside and Non-Nucleoside DNMT Inhibitors (DNMTi)

Type	DNMTi	Structure	Mode of Action	Advantages	Disadvantages	Reference
Nucleoside inhibitors	Azacytidine (5-azacytidine; 5-aza-CR)		Incorporates into RNA affecting nuclear and cytoplasmic RNA metabolism and related functions including protein synthesis Incorporates into DNA as 5-Aza-dCTP trapping DNMT and rendering it susceptible to proteasomal degradation	Is effective at lower concentration. Acts on the S-phase of the cell cycle and therefore is very effective against highly proliferative cells	Cytotoxic and when treatment is withdrawn remethylation is observed. Low stability. Not effective against solid tumors	[55]
	deoxycytidine (5-aza-2'-deoxycytidine; 5-aza-CdR),		Interferes with the intermediate of DNMT-DNA covalent reaction; Traps and inactivates DNMTs	Advantage similar to azacytidine	Disadvantage similar to azacytidine	[55]
	Zebularine (1- (β-D-ribofuranosyl)-2(1H)-pyrimidinone)		Forms a covalent complex with DNMT and cytidine deaminase	Less cytotoxic and can be given for extended periods of time. Higher stability in vivo and therefore can be administered orally. Enhances chemo- and radiosensitivity of the tumor cell. Has angiostatic and antimitogenic properties	Unknown	[55]

Continued

TABLE 6.3 Nucleoside and Non-Nucleoside DNMT Inhibitors (DNMTi)—continued

Type	DNMTi	Structure	Mode of Action	Advantages	Disadvantages	Reference
	5-Fluoro-2'-deoxycytidine (FdCyd, NSC 48006)		Inhibits cytidine deaminase. Prevents the transfer of the methyl moiety at the B-elimination step mediated by DNMT. Traps DNMT in the covalent complex inhibiting its function	Unknown	Generation of 5-fluorodeoxyuridine as a metabolite in vivo which is potentially toxic	[55]
Non-nucleoside inhibitors	Procaine		Demethylates CpG-rich residues by binding to them preventing DNMTs from binding. Reduces cell viability and causes growth inhibition	Effective at very high concentrations (µM) and is cell-type specific	Pyrrolidine (procaine derivative) highly effective in demethylating and is considered a potential compound to be pursued as a DNMTi	[55]
	Procainamide		Specifically inhibits DNMT1 activity and induces hypomethylation. Blocks enzyme catalytic activity	Reactivates tumor suppressor genes. Effectively reactivates WIF-1 and inhibits Wnt pathway. Very promising in the treatment of lung cancers	Unknown	[55]

Type	DNMTi	Structure	Mode of Action	Advantages	Disadvantages	Reference
	Hydralazine		Specifically inhibits DNMT1 activity and induces hypomethylation. Blocks enzyme catalytic activity	Combination studies with other epigenetic inhibitors make it well tolerable and effective against solid tumors. The efficacy of this compound has also been tested in cervical cancers using oral doses indicative of higher stability	Unknown	[55]
	RG108		Exact mechanism not fully understood. Direct interaction with DNMTs or binding to CpG-rich sites are plausible explanations	Studies have shown that the satellite DNA are left intact and hypomethylation is observed at tumor suppressor genes. Conserving the methylation patterns of satellite DNA ensures that chromosomal stability is achieved	Unknown	[55]

121

NF-κB is an important gene of the immune system and is observed to be important in the inflammatory process. This gene is also essential to cell survival, differentiation, and proliferation and is reported to be overexpressed in many tumors that are aggressive and metastatic [60]. Nuclear activation of NF-κB target genes occur only when the protein is acetylated but otherwise exists as an inactive complex tethered by Iκ-B inhibitor [60]. The ubiquitination-mediated degradation of Iκ-B initiated by its phosphorylation, frees the NF-κB that translocates to the nucleus and binds to its target sequence. The NF-κB is a complex of p50, p52, p65(RelA), c-Rel, and RelB [60]. In mammals, the p50/p65 heterodimer is the most commonly found complex [60]. Modifications of these dimers are essential to many of its downstream functions. Phosphorylation of p65 initiates its acetylation at multiple sites mediated by the recruitment of p300/CBP. NF-κB target gene expression is enhanced greatly when its K221 and K310 residues are acetylated and the full activity of the protein is dependent on this modification [61]. This observation is supported by the fact that SIRT1 deacetylation of p65 K310, HDAC1, and HDAC3 deacetylation of K221 or K310 inhibits transcription of its target genes [61]. Acetylation of NF-κB supports many functions and the acetylation of K122 and K123 enhances the export of the protein and re-association with IκB forming an inactive complex [62]. When p50 subunit is acetylated at positions K431, K440, and K441 the protein molecules bind with a higher affinity to its target gene sequences [63,64]. Thus the role of acetylation in NF-κB is many-fold. Since this gene is deregulated in many diseases including cancers a further investigation into the application of HDACi or HATi in regulating NF-κB functions is warranted.

Overexpression of c-Myc has been documented in many cancers. c-Myc binds and activates target genes as a complex with Max [64]. However, regulatory functions of the complex are soley through the transcription activation domain (TAD) of c-Myc located at the N-terminus [64]. This domain interacts with HATs, such as GCN5 and Tip60, forming coactivator complexes [64]. The c-Myc interacts with p300 via its TAD region that acetylates lysines at several positions between the TAD and DNA-binding regions of the protein, enhancing Myc turnover [64]. Thus, HAT-specific interactions with c-Myc dictate its stability and turnover in mammalian cells, and molecules that target these interactions by the induction of deacetylated levels are promising strategies in cancer therapy.

6.7 FUTURE DIRECTIONS

Epigenetic phenomena affect histone and non-histone proteins and molecular compounds that target enzymes influencing these roles are important to further develop. Drawbacks of compounds such as those with high cytotoxicity, low specificity, and low stability all have to be considered when selecting an antiepigenetic compound promoting antitumor activity. One way of improving drug design or compound efficacy is by a dual approach which has been proven to be much more efficient as seen with 5 azacytidine and other chemical compounds. Further work in these areas is therefore warranted. In addition to drug-based compounds, natural plant-based products with similar characteristics need to be screened and tested.

References

[1] Bernstein BE, Humphrey EL, Erlich RL, Schneider R, Bouman P, Liu JS, et al. Methylation of histone H3 Lys 4 in coding regions of active genes. Proc Natl Acad Sci USA 2002;99:8695—700.

[2] Wong CA, Recktenwald AJ, Jones ML, Waterman BM, Bollini ML, Dunagan WC. The cost of serious fall-related injuries at three Midwestern hospitals. Jt Comm J Qual Patient Saf 2011;37:81—7.

[3] Cohen I, Poreba E, Kamieniarz K, Schneider R. Histone modifiers in cancer: friends or foes? Genes Cancer 2011;2:631—47.

[4] Ma X, Ezzeldin HH, Diasio RB. Histone deacetylase inhibitors: current status and overview of recent clinical trials. Drugs 2009;69:1911—34.

[5] Deaton AM, Bird A. CpG islands and the regulation of transcription. Genes Dev 2011;25:1010—22.

[6] Li Y, Tollefsbol TO. Impact on DNA methylation in cancer prevention and therapy by bioactive dietary components. Curr Med Chem 2010;17:2141—51.

[7] Kim JY, Lee KS, Seol JE, Yu K, Chakravarti D, Seo SB. Inhibition of p53 acetylation by INHAT subunit SET/TAF-I {beta} represses p53 activity. Nucleic Acids Res 2011.

[8] Seo SK, Jin HO, Woo SH, Kim YS, An S, Lee JH, et al. Histone deacetylase inhibitors sensitize human non-small cell lung cancer cells to ionizing radiation through acetyl p53-mediated c-myc down-regulation. J Thorac Oncol 2011;6:1313—9.

[9] Torres-Padilla ME, Parfitt DE, Kouzarides T, Zernicka-Goetz M. Histone arginine methylation regulates pluripotency in the early mouse embryo. Nature 2007;445:214—8.

[10] Kouzarides T. SnapShot: Histone-modifying enzymes. Cell 2007;128:802.

[11] Kouzarides T. Chromatin modifications and their function. Cell 2007;128:693—705.

[12] Marks PA, Richon VM, Miller T, Kelly WK. Histone deacetylase inhibitors. Adv Cancer Res 2004;91:137—68.

[13] Yang XJ, Seto E. The Rpd3/Hda1 family of lysine deacetylases: from bacteria and yeast to mice and men. Nat Rev Mol Cell Biol 2008;9:206—18.

[14] Horio Y, Hayashi T, Kuno A, Kunimoto R. Cellular and molecular effects of sirtuins in health and disease. Clin Sci (Lond) 2011;121:191—203.

[15] McGuinness D, McGuinness DH, McCaul JA, Shiels PG. Sirtuins, bioageing, and cancer. J Aging Res 2011;235754.

[16] Choudhary C, Kumar C, Gnad F, Nielsen ML, Rehman M, Walther TC, et al. Lysine acetylation targets protein complexes and co-regulates major cellular functions. Science 2009;325:834—40.

[17] Rice JC, Allis CD. Histone methylation versus histone acetylation: new insights into epigenetic regulation. Curr Opin Cell Biol 2001;13:263—73.

[18] Varier RA, Timmers HT. Histone lysine methylation and demethylation pathways in cancer. Biochim Biophys Acta 2011;1815:75—89.

[19] Lindsey JC, Lusher ME, Anderton JA, Bailey S, Gilbertson RJ, Pearson AD, et al. Identification of tumour-specific epigenetic events in medulloblastoma development by hypermethylation profiling. Carcinogenesis 2004;25:661—8.

[20] Seligson DB, Horvath S, McBrian MA, Mah V, Yu H, Tze S, et al. Global levels of histone modifications predict prognosis in different cancers. Am J Pathol 2009;174:1619—28.

[21] Seligson DB, Horvath S, Shi T, Yu H, Tze S, Grunstein M, et al. Global histone modification patterns predict risk of prostate cancer recurrence. Nature 2005;435:1262—6.

[22] Kondo Y, Shen L, Cheng AS, Ahmed S, Boumber Y, Charo C, et al. Gene silencing in cancer by histone H3 lysine 27 trimethylation independent of promoter DNA methylation. Nat Genet 2008;40:741—50.

[23] Kondo Y, Shen L, Issa JP. Critical role of histone methylation in tumor suppressor gene silencing in colorectal cancer. Mol Cell Biol 2003;23:206—15.

[24] Muller CI, Ruter B, Koeffler HP, Lubbert M. DNA hypermethylation of myeloid cells, a novel therapeutic target in MDS and AML. Curr Pharm Biotechnol 2006;7:315—21.

[25] Yost JM, Korboukh I, Liu F, Gao C, Jin J. Targets in epigenetics: inhibiting the methyl writers of the histone code. Curr Chem Genomics 2011;5:72—84.

[26] Chase A, Cross NC. Aberrations of EZH2 in cancer. Clin Cancer Res 2011;17:2613—8.

[27] Chang CJ, Yang JY, Xia W, Chen CT, Xie X, Chao CH, et al. EZH2 promotes expansion of breast tumor initiating cells through activation of RAF1-beta-catenin signaling. Cancer Cell 2011;19:86—100.

[28] Okada Y, Feng Q, Lin Y, Jiang Q, Li Y, Coffield VM, et al. hDOT1L links histone methylation to leukemogenesis. Cell 2005;121:167—78.

[29] Hamamoto R, Furukawa Y, Morita M, Iimura Y, Silva FP, Li M, et al. SMYD3 encodes a histone methyltransferase involved in the proliferation of cancer cells. Nat Cell Biol 2004;6:731—40.

[30] Majumder S, Liu Y, Ford 3rd OH, Mohler JL, Whang YE. Involvement of arginine methyltransferase CARM1 in androgen receptor function and prostate cancer cell viability. Prostate 2006;66:1292—301.

[31] Cheung N, Chan LC, Thompson A, Cleary ML, So CW. Protein arginine-methyltransferase-dependent oncogenesis. Nat Cell Biol 2007;9:1208—15.

[32] Tan J, Yang X, Zhuang L, Jiang X, Chen W, Lee PL, et al. Pharmacologic disruption of Polycomb-repressive complex 2-mediated gene repression selectively induces apoptosis in cancer cells. Genes Dev 2007;21:1050—63.

[33] Yadav S, Singhal J, Singhal SS, Awasthi S. hSET1: a novel approach for colon cancer therapy. Biochem Pharmacol 2009;77:1635—41.

123

[34] Lea MA. Recently identified and potential targets for colon cancer treatment. Future Oncol 2010;6:993–1002.

[35] Saavedra OM, Isakovic L, Llewellyn DB, Zhan L, Bernstein N, Claridge S, et al. SAR around (l)-S-adenosyl-l-homocysteine, an inhibitor of human DNA methyltransferase (DNMT) enzymes. Bioorg Med Chem Lett 2009;19:2747–51.

[36] Iwasa E, Hamashima Y, Fujishiro S, Higuchi E, Ito A, Yoshida M, et al. Total synthesis of (+)-chaetocin and its analogues: their histone methyltransferase G9a inhibitory activity. J Am Chem Soc 2010;132:4078–9.

[37] Greiner D, Bonaldi T, Eskeland R, Roemer E, Imhof A. Identification of a specific inhibitor of the histone methyltransferase SU(VAR)3-9. Nat Chem Biol 2005;1:143–5.

[38] Chang Y, Zhang X, Horton JR, Upadhyay AK, Spannhoff A, Liu J, et al. Structural basis for G9a-like protein lysine methyltransferase inhibition by BIX-01294. Nat Struct Mol Biol 2009;16:312–7.

[39] Kubicek S, O'Sullivan RJ, August EM, Hickey ER, Zhang Q, Teodoro ML, et al. Reversal of H3K9me2 by a small-molecule inhibitor for the G9a histone methyltransferase. Mol Cell 2007;25:473–81.

[40] Lakowski TM, Hart P, Ahern CA, Martin NI, Frankel A. Neta-substituted arginyl peptide inhibitors of protein arginine N-methyltransferases. ACS Chem Biol 2010;5:1053–63.

[41] Sack JS, Thieffine S, Bandiera T, Fasolini M, Duke GJ, Jayaraman L, et al. Structural basis for CARM1 inhibition by indole and pyrazole inhibitors. Biochem J 2011;436:331–9.

[42] Schneider J, Shilatifard A. Histone demethylation by hydroxylation: chemistry in action. ACS Chem Biol 2006;1:75–81.

[43] Shi Y, Lan F, Matson C, Mulligan P, Whetstine JR, Cole PA, et al. Histone demethylation mediated by the nuclear amine oxidase homolog LSD1. Cell 2004;119:941–53.

[44] Yang M, Gocke CB, Luo X, Borek D, Tomchick DR, Machius M, et al. Structural basis for CoREST-dependent demethylation of nucleosomes by the human LSD1 histone demethylase. Mol Cell 2006;23:377–87.

[45] Yu V, Fisch T, Long AM, Tang J, Lee JH, Hierl M, et al. High-Throughput TR-FRET Assays for Identifying Inhibitors of LSD1 and JMJD2C Histone Lysine Demethylases. J Biomol Screen 2011.

[46] Mao B, Zhao G, Lv X, Chen HZ, Xue Z, Yang B, et al. Sirt1 deacetylates c-Myc and promotes c-Myc/Max association. Int J Biochem Cell Biol 2011;43:1573–81.

[47] Singh MM, Manton CA, Bhat KP, Tsai WW, Aldape K, Barton MC, et al. Inhibition of LSD1 sensitizes glioblastoma cells to histone deacetylase inhibitors. Neuro Oncol 2011;13:894–903.

[48] Xiang Y, Zhu Z, Han G, Ye X, Xu B, Peng Z, et al. JARID1B is a histone H3 lysine 4 demethylase up-regulated in prostate cancer. Proc Natl Acad Sci USA 2007;104:19226–31.

[49] Hamada S, Suzuki T, Mino K, Koseki K, Oehme F, Flamme I, et al. Design, synthesis, enzyme-inhibitory activity, and effect on human cancer cells of a novel series of jumonji domain-containing protein 2 histone demethylase inhibitors. J Med Chem 2010;53:5629–38.

[50] Wissmann M, Yin N, Muller JM, Greschik H, Fodor BD, Jenuwein T, et al. Cooperative demethylation by JMJD2C and LSD1 promotes androgen receptor-dependent gene expression. Nat Cell Biol 2007;9:347–53.

[51] Spannhoff A, Hauser AT, Heinke R, Sippl W, Jung M. The emerging therapeutic potential of histone methyltransferase and demethylase inhibitors. ChemMedChem 2009;4:1568–82.

[52] Vigna E, Recchia AG, Madeo A, Gentile M, Bossio S, Mazzone C, et al. Epigenetic regulation in myelodysplastic syndromes: implications for therapy. Expert Opin Investig Drugs 2011;20:465–93.

[53] Ghoshal K, Bai S. DNA methyltransferases as targets for cancer therapy. Drugs Today (Barc) 2007;43:395–422.

[54] Liu L, Li Y, Tollefsbol TO. Gene-environment interactions and epigenetic basis of human diseases. Curr Issues Mol Biol 2008;10:25–36.

[55] Ren J, Singh BN, Huang Q, Li Z, Gao Y, Mishra P, et al. DNA hypermethylation as a chemotherapy target. Cell Signal 2011;23:1082–93.

[56] Tsou JA, Hagen JA, Carpenter CL, Laird-Offringa IA. DNA methylation analysis: a powerful new tool for lung cancer diagnosis. Oncogene 2002;21:5450–61.

[57] Abd El-Hamid TM, Mossallam GI, Sherisher MA. The Clinical Implications of Methylated p15 and p73 Genes in Adult Acute Lymphoblastic Leukemia. J Egypt Natl Canc Inst 2010;22:175–84.

[58] Benkirane M, Sardet C, Coux O. Lessons from interconnected ubiquitylation and acetylation of p53: think metastable networks. Biochem Soc Trans 2010;38:98–103.

[59] Das C, Kundu TK. Transcriptional regulation by the acetylation of nonhistone proteins in humans – a new target for therapeutics. IUBMB Life 2005;57:137–49.

[60] Spange S, Wagner T, Heinzel T, Kramer OH. Acetylation of non-histone proteins modulates cellular signalling at multiple levels. Int J Biochem Cell Biol 2009;41:185–98.

[61] Yeung F, Hoberg JE, Ramsey CS, Keller MD, Jones DR, Frye RA, et al. Modulation of NF-kappaB-dependent transcription and cell survival by the SIRT1 deacetylase. EMBO J 2004;23:2369–80.

[62] Kiernan R, Bres V, Ng RW, Coudart MP, El Messaoudi S, Sardet C, et al. Post-activation turn-off of NF-kappa B-dependent transcription is regulated by acetylation of p65. J Biol Chem 2003;278:2758—66.

[63] Deng WG, Wu KK. Regulation of inducible nitric oxide synthase expression by p300 and p50 acetylation. J Immunol 2003;171:6581—8.

[64] Deng WG, Zhu Y, Wu KK. Up-regulation of p300 binding and p50 acetylation in tumor necrosis factor-alpha-induced cyclooxygenase-2 promoter activation. J Biol Chem 2003;278:4770—7.

CHAPTER 7

Epigenomics in Neurobehavioral Diseases

Cassandra L. Smith[1], Kai Huang[2]
[1]Boston University, Boston, MA, USA
[2]Fortiss gemeinnützige GmbHAn-Institut der TU, Munich, Germany

127

7.1 INTRODUCTION

The spectrum of neurobehavioral diseases, recently named brain diseases by the United States National Institute of Mental Health (NIMH) perhaps to counteract stigma associated with these diseases, includes rare single gene diseases like Huntington's disease (HD) along with common multifactorial and multilevel diseases with genetic and environmental factors, such as schizophrenia (SZ), autism spectrum disorder (ASD), and bipolar disease (BD) (formerly manic depressive disease). Common neurodegenerative diseases associated with aging are dementia, Alzheimer's and Parkinson's diseases (discussed in Chapter 9). Alzheimer's disease and prion diseases are epigenomic "templating" diseases that involve the formation of pathogenic proteins [1].

Neurodevelopmental disorders such as Rett (RTT) and Fragile X (FRAX) syndromes are discussed briefly here and in Chapter 8. These diseases arise from specific defects in epigenomic processes at specific genetic loci. Another class of neurobehavioral diseases is substance abuse. These diseases have well-defined environmental triggers (e.g. alcohol or cocaine) that perturb neurobehavioral processes. Here, our emphasis is on SZ, but included are observations on BP and ASD because of their shared features. These disorders have many characteristics in common with rare neurobehavioral disorders with well-defined genetic causes, but like other common (cancer and cardiovascular) diseases have strong epigenomic and environmental components.

T. Tollefsbol (Ed): Epigenetics in Human Disease. DOI: 10.1016/B978-0-12-388415-2.00007-X

An issue with neurobehavioral diseases is that the targets are fuzzy. Here, we have attempted to cover issues that are important in understanding epigenomic (another fuzzy term) aspects of these diseases. Not surprisingly a complete picture is not possible, instead we show where links are known to exist. Other links may not exist or may not have been studied. A glossary of terms is presented at the end of the chapter to aid the reader.

7.2 WHAT IS THE EPIGENOME?

Although the term "epigenetics" was first used by Waddington in the 1940s, this field of science has multiple origins. As originally defined by Waddington, epigenetics is: "All those events which lead to the unfolding of the genetic program for development." This definition was non-specific because although Waddington did not propose or know of any mechanisms to connect genetics with development, importantly he linked these two aspects of biology. The idea was that development of a multicellular organism involving differentiation of primordial totipotent cells into specific cells and tissues was like a ball moving down a landscape of branching valleys to its final destination (Figure 7.1). Independently, the term epigenetics was invoked to explain a variety of "bizarre" phenotypic phenomena in different organisms that could not be explained by simple Mendelian genetics.

Epigenetics evolved to mean modifications to DNA that affected gene expression but do not involve base changes (aka mutations). The term "epi", meaning "on top of genetics" (with genetic = DNA), certainly encompasses changes to DNA and to DNA packaging by histones.

The best-studied epigenetic changes are DNA and histone (i.e. chromatin) modifications that modulate access of macromolecules that regulate gene expression. In addition, modifications impact chromosome localization within the nucleus; hence, impacting potential interactions between chromatin regions [2,3].

We prefer the broad term "epigenomics" that not only refers to changes to DNA and histones, but also encompasses other non-linear aspects of the information transfer from genotype to phenotype. Some of these changes are well known. For instance, we include somatic DNA base changes, and chromosomal localization in the epigenomic landscape. At the RNA level, epigenomics changes include alternative splicing, editing, capping, methylation, and other modifications, polyA tailing and non-coding RNA (ncRNA) regulation. At the protein level, epigenomics changes include the many post-translational modifications to protein (e.g. phosphorylation, methylation, acetylation, palmitoylation, ubiquination, etc.).

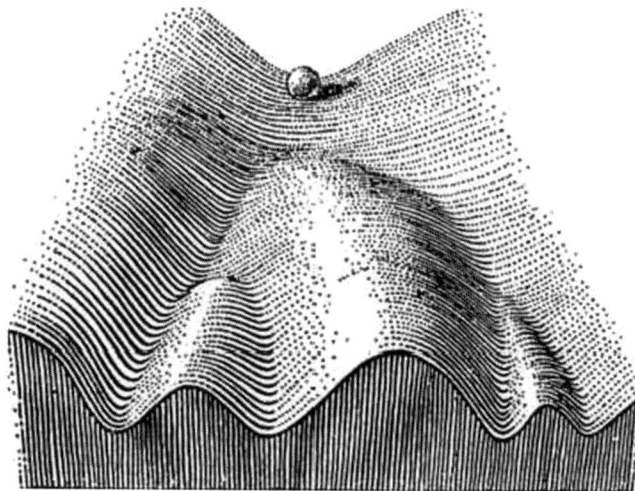

FIGURE 7.1
Waddington's epigenetic landscape.

Some RNA and protein modifications have been studied for a long time and may not be generally thought of as part of the epigenomic landscape. However, within our viewpoint, all these changes represent non-linear information transfer and fall under the umbrella of epigenomics. Today, many epigenomic processes have been identified, some studied in some detail, but many others remain to be discovered.

In brief, the term "epigenomics" describes non-linear transfer of genotype to phenotype, and is a more global and inclusive term than epigenetics. In the future, an important issue will be how to integrate and understand the multitude of genetic, environmental, and epigenomic factors that contribute to disease phenotypes. Epigenomic programming is specific for individual genes for cell type, time, environment, and history.

The amount of epigenomic variation is greater than genetic variation. Besides the amount of variation, a second difficulty in this research area is how to collect the information, and a third is how to integrate and analyze the multilevel data. These issues are similar but greater than those faced with sequencing the human genome. Although, there are a large number of reviews on epigenomic programming much of the needed data does not exist today. This review will focus on epigenomic changes to chromatin.

7.3 EPIGENOMIC MODULATION OF CHROMATIN

DNA methylation is the most studied and best-understood epigenomic process with many reviews published yearly. DNA methyl transferases (DNMT) transfers methyl ($-CH_3$) groups from S-adenosyl methionine (SAM) to the 5 carbon atom of cytosine residues within the dinucleotide sequence, 5'CpG3' (Figure 7.2). The maintenance DNA methylase, DNMT1, acts soon after DNA replication to copy parental strand methylation to the newly synthesized DNA strand, thus conserving the DNA methylation pattern through mitosis. DNMT 3a and 3b are linked to *de novo* DNA methylation. Passive demethylation may occur when DNA is replicated but parental strand methylation is not transferred to the newly synthesized strand.

129

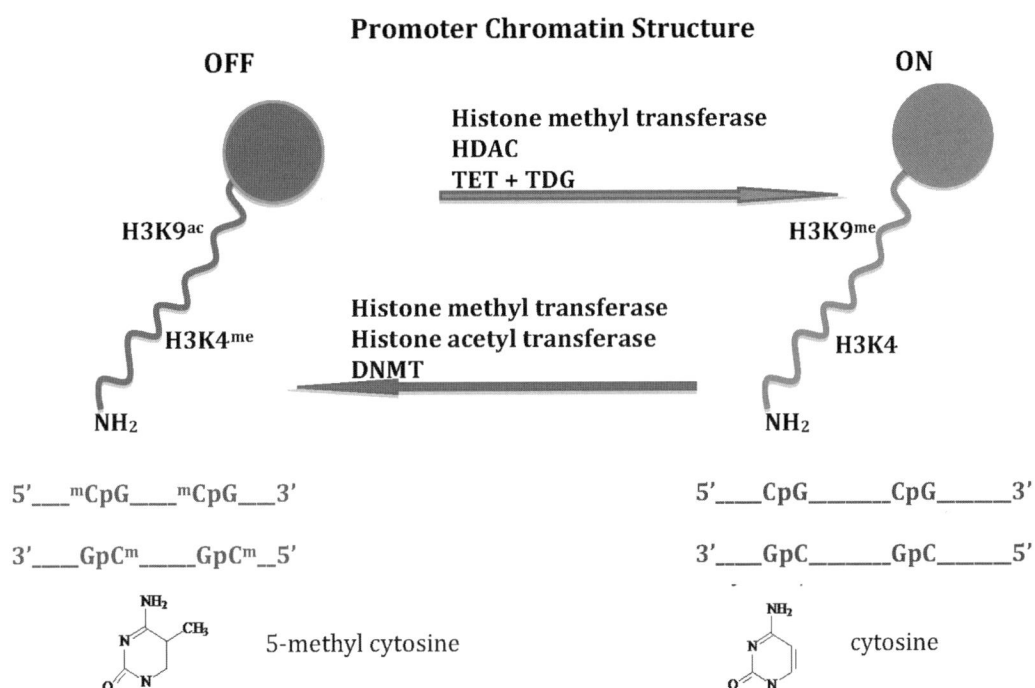

FIGURE 7.2
Epigenomic programming of chromatin. This figure is reproduced in the color plate section.

Usually, the absence of DNA methylation at a promoter region and the presence of DNA methylation in singly occurring gene body 5'CpG3' sites is associated with gene expression [4,5]. The rate of CpG methylation is determined by the availability of DNMT enzymes and their affinity for a given CpG, the number of CpGs and the non-CpG DNA sequence within a site, as well as other factors like the level of the methyl donor, and with DNMT1, whether the double-stranded DNA is hemimethylated. Cytosine methylation can mediate silencing of gene expression through the binding actions of proteins such as methylated CpG binding proteins (MeCP1 and 2), and the methyl CpG binding domain proteins MBD1, MBD2, MBD3, and MBD4 (e.g. [6]).

A recent exciting caveat to DNA methylation studies is the discovery of modification to the 5 ring position carbon of cytosine in addition to methylation. The modifications include 5-hydroxymethyl cytosine, 5-formyl cytosine, and 5-carboxyl cytosine created by the reiterative oxidation by the 10—11 translocation family of dioxygenases, TET1, TET2, and TET3 [7—9]. In an active demethylation process, the modified cytidines are de-aminated producing a mismatched base pair (T-G) in a dinucleotide sequence 5'TpG3'/3'GpC5'. Thymine DNA glycosylase (TDG) removes the mismatched thymine, which is replaced by an unmethylated cytosine residue during base excision repair [10,11]. TDG interacts with DNA methylases, histone acetyl transferases, and transcription factors.

Hydroxymethylcytosine distribution varies in a tissue and cell-type-specific manner with high levels in mouse and human embryonic stem cells [7,12,13]. Hydroxymethyl cytosine is enriched at promoter regions and within gene bodies but the former does not correlate with gene expression, and the latter is more positively correlated with gene expression than 5-methyl cytosine [14].

These recent discoveries raise many interesting questions besides revealing an active mechanism for DNA de-methylation. For instance, most past studies on DNA methylation did not distinguished between these modifications; hence, almost all past studies need to be revisited in order to clarify what modifications are present.

Histones condense genomic DNA to fit within the nucleus of a cell. Modifications to histones are linked to different levels of compaction referred to as "open" and "closed" chromatin that allow or prevent access and expression of genes. Histones can be methylated, phosphorylated, acetylated, and/or ubiquinated (for a recent review see [15]).

The nucleosome is composed of an octamer of histones H2A, H2B, H3, and H4. Most histone modifications occur in free amino terminal histone tails rather than in the globular core portions of the nucleosome. The H3 tail has the most sites for modification, 36 residues, with some sites having multiple modifications simultaneously (e.g. mono-, di-, or tri-methylations). The potential variation in H3 variation is $\sim 2^{40}$ potentially providing $\sim 10^{14}$ variants. For the most part, the "histone code" for any process is not known and certainly the relationship with the newly discovered cytosine codes.

Histone H3 lysine 4 methylation (H3K4^{me}) and histone H3 lysine 9 acetylation (H3K9^{ac}) are generally associated with open chromatin and gene expression. Demethylated H3K4 and deacetylated H3K9^{me} are associated with closed chromatin and lack of gene expression. Generally, these modification states correlate with modification of DNA in promoter regions and with gene expression (Figure 7.2). One mechanism for coordinating DNA and histone modification involves a complex that binds to methylated CpG sites and includes the methylated CpG binding protein 2 (MeCP2), a transcriptional repression domain (TRD) protein, co-repressor SIN3 homolog A (SIN3A), and histone deacetylase (HDAC) [16].

Valproic acid (VPA), a HDAC inhibitor, is used to treat BP as an alternative to lithium, to treat epilepsy, and less commonly migraines, major depression, and schizophrenia [17—19]. In utero, exposure to VPA increased the risk for autism and neural tube defects [20] and likely other neurobehavioral diseases like SZ and ASD linked to folate deficiencies (see below).

Epigenomic programming is used to regulate gene expression as a function of parental origin. Genes that are programmed by parental origin are called imprinted genes (http://www. imprint.com; http://www.igc.otago.ac.nz). However, chromatin codes and epigenomic codes are dynamic and may change in response to environmental and developmental cues.

There are approximately 80 genes proven and another 200 thought-to-be imprinted in the human genome. Imprinting appears to be regional, with 20 proven regions, and 50 addition putative regions. In some regions, differential imprinting occurs. Imprinting defects at specific chromosomal regions is associated with neurodevelopmental disorders (see Chapter 8). ASD, and in some cases SZ, is co-morbid with several imprinted diseases including Prader–Willi/Angelman syndrome, CHARGE, fragile X disease and Rett (RTT) syndromes (see below) [21,22].

In meiosis, most paternal histones in sperm are replaced with protamines [23]. Hence, paternal epigenomic programming of chromatin appears to be through DNA modification. In contrast, maternal chromatin modifications to DNA and histone can be preserved through gamete production.

Epigenomic complexities include other differences between males and females. Overall DNA methylation patterns are different in male and female cells [24]. For instance, Shimabukuro et al. [25] found global methylation differences in leukocytes from males and females, and reported a slight decrease overall in samples from males but not female SZ patients, especially in younger individuals.

Females have an extra layer of epigenomic complexity. One X chromosome is inactivated through epigenomic programming early in development (before the 32-cell stage)(for review see [26]). This ensures that genes on the X chromosome are expressed at the same level in female (XX) and male (XY) cells. Usually inactivation is random; hence, females are functional mosaics. The inactive X chromosome can be reactivated with age, and in some individuals inactivation is not random.

An excess of chromosome X aneuploids have been detected in neurobehavioral disorders for some time including schizophrenic patients [27,28]. Recently, systematic re-sequencing of synaptic genes on chromosome X revealed a higher than expected level of rare damaging variants in SZ and ASD patients [29]. Such mutations may account for the increased severity seen in male versus female patients because males have single copies of chromosome X. X chromosome inactivation (especially if not random, as found for SZ and BP) may account for disease discordance in twins [30]. Both X and autosomal chromosome number abnormalities in neurons are detected in patients with different neurobehavioral disorders including SZ, BP and ASD (for review see [31]).

In 2009, Lister et al. [32] reported on methylated sites across the genome. As expected, low levels of methylation were found at CpG islands, promoter methylation level was inversely linked gene expression. Gene body and intergenic regions had high levels of methylation (>70%) and were called highly methylated domains (HMDS). Other regions were partially methylated (<70%) and called partially methylated domains (PMD). Subsequent studies by Schroeder et al. [33] revealed genes linked to ASD were enriched in PMD regions.

Mutations in genes involved in epigenomic programming are directly linked to neurobehavioral disorder. DNMT1 expression, already at unusually high levels in cortical GABAergic neurons, is increased in SZ and BP [19,34,35]. In these neurons, hypermethylation of the promoter regions of GABAnergic genes glutamic acid decaboxylase$_{67}$ (GAD$_{67}$) and reelin (RELN) and decreased gene expression is linked to psychotic aspects of SZ and BD diseases. In mice, combined DNMT1 and DNMT3a deficiencies in post-mitotic neurons leads to defects in learning and memory [36].

RTT syndrome, an ASD disease, is usually due to mutations in the methylated cytosine binding protein, MECP (an X linked gene), although some cases are due to known mutations in other genes (for review see [37]). In RTT syndrome, usually a de novo mutation occurs in the proband, although some cases are due to de novo mutations in the paternally inherited allele. RTT syndrome is more common in females because a mutation in the single copy of the gene present in males leads to embryonic lethality. MECP2 protein binds to methylated DNA and inhibits transcription factor binding; hence influencing the expression level of many genes.

DNA, RNA, and histone epigenomic changes can be analyzed by the same approaches used in the past to investigate DNA sequence variation. These approaches have been converted to second-generation ultra-high-throughput methods for collecting large amounts of data. Hence, it is likely that much of the missing epigenomic data on DNA, RNA, and histones will be uncovered in the near future. Here, we focus on chromatin because these are the best-characterized epigenomic changes corrected to neurobehavioral disease. Other epigenomic changes are now being collected.

7.4 ISSUES UNIQUE TO NEUROBEHAVIORAL DISEASES

There are unique major issues related to neurobehavioral disorders. For instance, lack of funding at least by the US National Institute of Health (NIMH) prevents progress on these diseases and discourages innovative approaches. Another issue is that almost any positive observation on these diseases is accompanied by a contradictory negative finding. Other issues include the subjective diagnosis and the unknown cause(s) of the common neurobehavioral disease.

Many of the severe neurobehavioral diseases have overlapping symptoms and can be viewed as points on a continuum of phenotypes that share characteristics. And there are so many changes linked to neuropsychiatric diseases that it is difficult to distinguish between cause and consequence. Our approach has been to view these seemingly disparate observations as windows into a disrupted fundamental cellular process such as that described below.

Neurobehavioral diseases are diagnosed from subjective behavioral reporting by afflicted individuals and trained observers because objective criterion is not established. SZ is diagnosed more frequently in males and BP in females. Males tend to be diagnosed in their early twenties for SZ and BP while women are diagnosed more frequently in their late twenties. Attempts to standardize subjective criteria reach back in time to Kraepelin and Bleuler in the early 1900s. Generally, two major categories of neurobehavioral disease are recognized: SZ and BP based on symptom groupings, course, and outcome.

Today, the classification of neurobehavioral disorders is based on criteria developed by the American Psychiatric Association and published as "Diagnostic and Statistical Manual of Mental Disorders (DSM)" and the ICD (International Criteria of Disease; published by WHO). These classification schemes undergo periodic revision with current versions DMS-IV and ICD-10, respectively. Generally, classification is based on qualitative behavioral characteristics rather than quantitative objective criteria.

Symptoms for SZ are divided into positive, negative, and cognitive (Table 7.1). The most common positive symptom is hallucination. Hallucinations are usually auditory but can be visual, tactile, olfactory, or gustatory. Positive symptoms reflect traits added to the personality. Negative symptoms such as depression represent behavioral deficits such as flat or blunted affect, alogia (poverty of speech), anhedonia (inability to experience pleasure), and asociality. The negative symptoms contribute to poor quality of life, functional disability, and lack of motivation, and have been linked to folate deficiencies (see below). Cognitive symptoms, including deficits in working memory and executive function, are related to the ability to function in society.

TABLE 7.1 Positive, Negative, and Executive Symptoms in SZ

I. Positive Symptoms (Reality Distortion)

1. Unusual thought
2. Delusions
3. Grandiosity
4. Suspiciousness/persecution
5. Hallucinatory behavior

II. Negative Symptoms (Poverty Syndrome)

1. Emotional withdrawal
2. Passive, apathetic social withdrawal
3. Lack of spontaneity and flow of conversation
4. Poor rapport
5. Blunted affect
6. Motor retardation
7. Disturbance of volition

III. Disorganized Symptoms (Formal Thought Disorder, Odd Behaviors, Cognition)

1. Conceptual disorganization – circumstantial speech, loose, tangential, illogical associations
2. Incoherent speech – incomprehensive sentences
3. Poverty of speech content – fluent speech with little information content
4. Inappropriate affect – bizarre behavior

A SZ diagnosis using the DSMIV requires three components be present. (1) The occurrence of two or more of the following symptoms for 1 month or longer: delusions, hallucinations, disorganized speech, grossly disorganized behavior (e.g. dressing inappropriately, crying frequently) or catatonic behavior, negative symptoms (blunted affect (lack or decline in emotional response), alogia (lack or decline in speech), or avolition (lack or decline in motivation)). Occurrence of hallucinations with a single voice in a running commentary of patient's activity, or two or more voices or voices that are "bizarre" can be used singly for diagnosis. (2) Social or occupational dysfunction (work, interpersonal relationship, or self-care) that is significantly lower since the onset of symptoms. (3) Disturbances present for a significant amount of time, i.e. continuous signs for 6 months, with at least 1 month of active symptoms or less if symptoms disappear with treatment.

Today, there is movement to develop and implement objective criteria to patient classification based on "endophenotypes" or subphenotypes that can be measured by behavioral, physiological, and DNA testing. Endophenotypes are: (1) heritable characteristics that co-segregate with disease in a family; (2) disease state independent (i.e. present whether disease is active or inactive); and (3) present in a higher than expected frequency in families segregating disease [38–40]. Endophenotypes for SZ are listed in Box 7.1. Today, several companies claim to diagnose SZ with ~85% accuracy based on genetic and/or endophenotypic testing. This accuracy level is inadequate given the devastating consequences of receiving a SZ diagnosis.

BP-related mood disorders includes episodes of severe and mild to moderate mania, and severe and mild to moderate depression; some episodes may be mixed episodes. In a manic period heightened energy, creativity, and euphoria are common, accompanied by hyper-activity, little sleep, and feelings of being all-powerful, invincible, and destined for greatness. During this period the individual may spiral out of control and engage in reckless self-damaging decisions involving gambling, other financial activity, sexual activity, etc.

ASD is a grouping of diseases in the DMS-IV- text revision (DMS-IV-TR) called pervasive developmental disorder (PDD) [42]. Besides autism, PDD syndromes include: Asperger syndrome, Rett syndrome, childhood disintegrative disorder, and pervasive developmental

133

BOX 7.1 BEHAVIORAL ENDOPHENOTYPES IN SZ [41]

Prepulse inhibition
P50 suppression
Antisaccade
Continuous performance
Letter–number span
Verbal learning
Abstraction
Face memory
Spatial processing
Sensorimotor dexterity
Emotion recognition

disorder not otherwise specified. The three behaviors used for diagnosis of ASD include defects in (1) social interactions, (2) communication and imaginative play, and (3) interests and activities. The age of onset is <3 years, and there appears to be evidence that the incidence of ASD is increasing. Other symptoms include brain malformations, seizures, defective immunological responses, inflammation (especially in the gut [43]), and oxidative stress [44]. Today, treatment includes intensive education along with behavioral and occupational therapies, but there are no standard medical interventions. Instead, most parents have pursued non-conventional treatments (e.g. nutritional) with anecdotal but not proven efficacies (see below).

Brain dysfunction may be due to a disparate range of environmental and/or genetics factors. For instance, response to medication or infectious disease can present with behavioral symptoms. In 1988, Templer and Cappelletty [45] proposed separating primary and second schizophrenia into different categories based on features of disease as shown in Table 7.2. This view is gaining more importance as genetic studies have failed to reveal clear-cut causes of neurobehavioral disease. Recently, Sachdev and Keshavan published a comprehensive text entitled "Secondary Schizophrenia" [46] that brings together for the first time information on primary disease that may present symptoms of neurobehavioral dysfunction.

In some cases, the primary disease involves damage to the brain, especially in the temporal lobe, that goes undetected because appropriate screening is not done. Severe closed head

TABLE 7.2 Proposed Features Separating Primary and Secondary Schizophrenias [45]

Primary	Secondary
Process	Reactive
Brain atrophy	No brain atrophy
Normal brain asymmetry	Abnormal brain asymmetry
Corpus callosum not thickened	Corpus callosum thickened
Sensory gating Deficit	More normal sensory gating
"Typical" schizophrenia symptoms	"Atypical" schizophrenia
Affective flatness	+ symptoms
Lower IQ	Affective component to illness
Less seasonality of birth	Higher IQ
No head trauma	Head trauma
No epilepsy	Epilepsy
More genetic involvement	Less genetic involvement
Insidious onset	Rapid onset
Poor premorbid function	Pre-morbid functioning

trauma, subarachnoid hemorrhage, cerebral tumors, cerebral infection (e.g. syphilis, HIV, etc.), demyelinating diseases, and Wilson's disease (where excess copper accumulates in the brain and other tissues) are linked to SZ. Traumatic head injury increases risk for SZ by 2–5-fold, although disease may emerge up to 20 years later. Vitamin B12 (cobalamin) deficiency and thyroid disease are associated usually with depressive disorder but also with SZ (see below). Other rare genetic diseases linked to secondary SZ are HD, Friedreich's ataxia (a trinucleotide repeat disease, see below), and Prader–Willi syndrome. Psychosis is linked to of stimulants like amphetamine, phencyclidine (PCP), lysergic acid diethylamine (LSD), or 3-4-methylenedioxymethamphetamine (ecstasy) (see below).

An estimated 5–15% of SZ patients have underlying undiagnosed illness [47]. When conventional medical testing of individuals suspected or given a psychiatric diagnosis is not done, medical conditions in these patients go untreated. The push for routine health assessment and care for patients with neurobehavioral disorders is increasing (for instance, see [48]). Medical testing is critical for helping patients with underlying pathologies that impact symptoms, or that impair the quality of life, and for research purposes where disease and symptoms need to be clearly defined.

SZ is a systemic disease linked to loss of taste and/or smell [49–52], palate and/teeth abnormalities [53–55], minor dysmorphic features [56,57], and increased risk for metabolic syndrome including diabetes [58]. BP has not been linked to such systemic abnormalities, whereas immunological and gut abnormalities have been detected in some ASD patients.

The possible outcomes of SZ are varied (Figure 7.3). Some disease presentations are episodic and others have spontaneous remission. These observations cannot be accounted for by genetics alone. Few studies have rigorously examined the environmental factors that can lead to remission, although recently consensus standards for remission and recovery are being developed to facilitate treatment and research [59].

Besides the paucity of systematic long-term studies on neurobehavioral patients and few, if any, genetic studies have attempted to group observations into primary and secondary SZs. This is

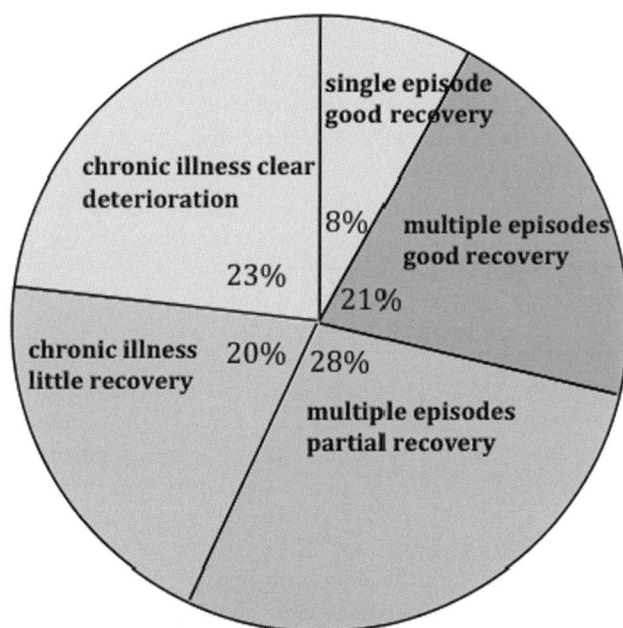

FIGURE 7.3
Outcome of schizophrenia diagnosis (adapted from http://www.science.org.au). This figure is reproduced in the color plate section.

unfortunate because patients without severe brain damage may present opportunities for reversal of disease, perhaps through epigenomic manipulation. Clearly, the development of new and effective treatment modalities will require further dissection of neurobehavioral disease types.

7.5 GENETICS

Genetic dissection of rare single gene diseases with high penetrance is effective, provided families and funds are available. Penetrance describes the level of phenotypic expression of a genetic trait in individuals with a causal mutation. For instance, HD, with <1:100000 prevalence and 100% penetrance, is caused by a dominant mutation in the Huntingtin gene on the small arm of chromosome 4 [60]. HD is characterized as a brain disease with motor, behavioral, and cognitive symptoms, but in fact widespread pathology is present [61]. Even HD is complicated by the identification of modifying genes (e.g. [62]) and a chromosomal haplotype that predisposes carriers to acquire an HD mutation [63].

Although the discovery of the Huntingtin gene opened up new areas of research, the neural cause of HD is unclear and no treatments have been developed from the gene discovery in 1993. Admittedly, treatment for a disease caused by a dominant mutation is more difficult than a disease due to a recessive mutation because the mutant allele or its product must be eliminated. HD is an example of a trinucleotide disease rooted in genomic instability and epigenomics (see below).

Individual genetic studies link a large number of genes to SZ, BP, and ASD with rare alleles having greater affects than common alleles [64,65]. Some genes are link to more than one neurobehavioral disease. However, the results are complex. Some genes and even some single nucleotide polymorphisms (SNPs) appear to have both positive and negative effects on disease occurrence and presentation in different studies. Today, no single, or small number of genes is sufficient and/or necessary determinants of SZ, with similar observations found for BP and ASD. The vasoactive intestinal peptide receptor gene (VIPR2) within the 7q36.3 rare copy number variant (CNV) region is linked to 1/300 cases of SZ but required screening of >8000 patients and >7000 controls. Most meta-analyses have not been this successful, nor been this extensive.

The causes of disappointing genetic results can be true false positives from small sample sizes and inadequate statistical criteria; true heterogeneity in disease and/or inadequate gene testing. In the past, gene SNP testing focused on known "functional" SNPs with detected phenotypic consequences. However, any variation in a gene is likely to have some phenotypic consequence. Perhaps the application of second-generation ultra-high-throughput sequencing for complete genome and/or complete gene sequencing will clarify gene association results.

Genome side association studies (GWAS) studies analyze SNPs across the genome both within and outside of genes. These studies are generally used to look for linkage disequilibrium between genes, i.e. haplotypes and phenotype. A surprise of GWAS studies was the detection of a slight elevation in the occurrence of CNVs in SZ, BP, and ASD (for review see [64]). The CNVs include small and large regions of the genome, and some are common across the multiple disorders (e.g. [66]). De novo mutations are detected as CNVs [67] and within exons [68], and support the notion proposed by us in 2003 [69] that genome instability is a characteristic of SZ (see below). The absence of strong genetic links in SZ is similar to findings on other common illnesses like cardiovascular disease and cancer.

The results of these genetic studies have led to several generalities. The risk for developing SZ, BP, or ASD is generally proportional to the degree of genetic relationship with an afflicted individual. These diseases have high heritability but low penetrance. Low penetrance is attributed to several factors including: (a) epistasis (interaction between genes, i.e. polygenic disease); (2) de novo confounding germline or somatic mutations; and (c) epigenomic programming changes including gene–environment interactions.

Epistasis is commonly invoked to explain the large number of genes linked to neurobehavioral illnesses. There are many possible types of epistatic interactions with different outcomes. Interacting genes may code for proteins or non-coding RNAs (ncRNAs) within the same, a redundant, or alternative pathway. Other mutations affecting epistasis may decrease or increase protein activity and affect which pathway step is rate-limiting. Mutation may lead to a loss, gain, or change of function, each with different consequences.

Despite the spectrum of possible epistatic affect, epistasis does not account for some genetic observation on neurobehavioral illness. For instance, epistasis cannot explain the discordance rates of disease in monozygotic twins. Discordance in monozygotic twins has traditionally been used to distinguish between genetic and environmental components of disease especially when twins are raised apart. However, although monozygotic twins share the highest percentage of their genome, their genomes are not identical because of somatic and epigenomic changes.

Monozygotic twins that develop from the same fertilized egg are discordant for neurobehavioral disease at levels that are significantly below 100%, i.e. arguably ~50% (with studies ranging from ~35% to ~80% for SZ and BP, and ~65% for ASD [70]). Variation in concordance rates may be due to several factors including differences in ascertainment of disease and/or twinship. More recently studies have used standardized methods for disease diagnosis and quantitative, rather than qualitative, assessment of twinship. However, conclusions from quantitative assessment results can vary also (e.g. [71]) and concordant illness in monozygotic twins may not have the same presentation and course.

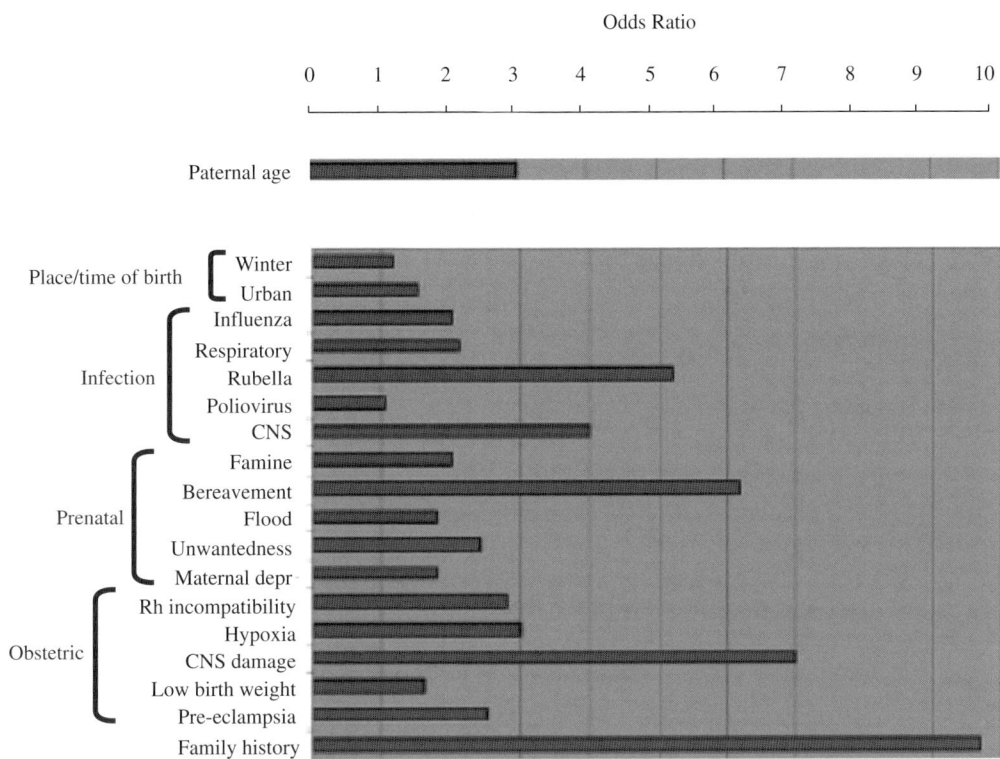

DOI: 10.1371/Journal.pmed.0020212.g001

FIGURE 7.4
Odds ratio for schizophrenia as a function of genetic and environmental risk factors (adapted from [73]). This figure is reproduced in the color plate section.

The prevalence of SZ varies from 0.8—2.0% in different countries [72]. Some patients come from families where related individuals have the same or a related neurobehavioral disorder (or an associated endophenotype); others have, arguably, no family history (aka spontaneous forms of illness). Despite the high heritability (Figure 7.4) of SZ, preliminary modeling experiments indicate the rate of spontaneous forms of illness is responsible for maintaining the level of the disease in the population over many generations. This result underscores the importance of prevention through environmental interventions.

7.6 ENVIRONMENT

Environmental factors linked to SZ and BP occur early in development (for recent reviews see [74,75]) whereas the age of disease onset for SZ and BP is typically young adulthood (see above). Early life environmental impacts must persist over many years in order to impact disease phenotype (see Figures 7.4 and 7.5). Figure 7.4 provides the odds ratio of well-studied factors for SZ. Figure 7.5 provides developmental time windows where environmental components appear to impact subsequent disease development. Environmental factors early in life are linked to other common diseases. Early life periods have uniquely high levels of DNA replication and epigenomic programming.

Studies going back to 1958 link elder fathers as a risk factor for SZ in progeny with a 3-fold increase associated with fathers >50 years of age (for review see [76]). The increase is linear with paternal age. Similar observations are found for BP77 and ASD78. In fact, paternal age is generally linked to negative neurobehavioral outcomes [79]. Increased paternal age is associated with increases in other diseases such as Down syndrome; neural tube defects (see discussion of folate below); congenital cataracts; and reduction defects of the upper limb [80,81], with the greatest affect seen in autosomal dominant mutations [82—84]. All paternal age affects are likely DNA-mediated and due to the high levels of DNA replication and the associated epigenomic programming (see below).

Undernutrition (general caloric or protein deficiency) and malnutrition (deficiencies in specific elements, e.g. folic acid, zinc, copper, etc.) occur worldwide and are the most common diseases of childhood and prenatal life. Moderate to severe undernutrition occurring prior to 2 years of age is associated with persistent behavioral and cognitive deficits that resist nutritional rehabilitation [85]. Pregnant mothers exposed to famine or malnourished (especially for folate), notably in the second trimester (e.g. [86]), have an increased risk for children with SZ (Figure 7.4). Prenatal nutrition is linked to problems in dopamine-mediated behaviors and dopamine receptor binding in adults [87].

Perhaps some environmental factors shown in Figures 7.4 and 7.5 and linked to disease are bystander markers for poor nutrition. The immune response to an infection agent, or a response to oxidative conditions imposes additional nutritional requirement on a cell and organism (see below). Measles (rubella infections), associated with severe immune responses, early in pregnancy increases the risk for SZ. Other infectious agents linked to SZ include influenza, herpes simplex virus type 2 (HSV-2), *Toxoplasma gondii*, as well as elevated elevation of inflammatory markers link interleukin-8 (IL-8) early in pregnancy, and tumor necrosis factor (TNF) alpha at birth (for review see [88]). Famine, infection, hypoxia, and brain damage might be classified as secondary SZs (see above). The simultaneous occurrence of multiple risk factors (genetic and/or environmental) may be necessary when individual risk factors have mild effects.

Nutrition may underlie some maternal and rearing environment stress factors. Stress may lead to poor nutrition, as well as increases in cytokines. Chronic "defeat stress" is linked to decreased levels of HDAC-5 in the nucleus accumbens, long-lasted decreases in brain-derived neurotrophic factor (BDNF) transcription, and aberrant epigenomic changes [74]. Although early life factors appear to be paradoxical, all have the ability to influence mutation rates and

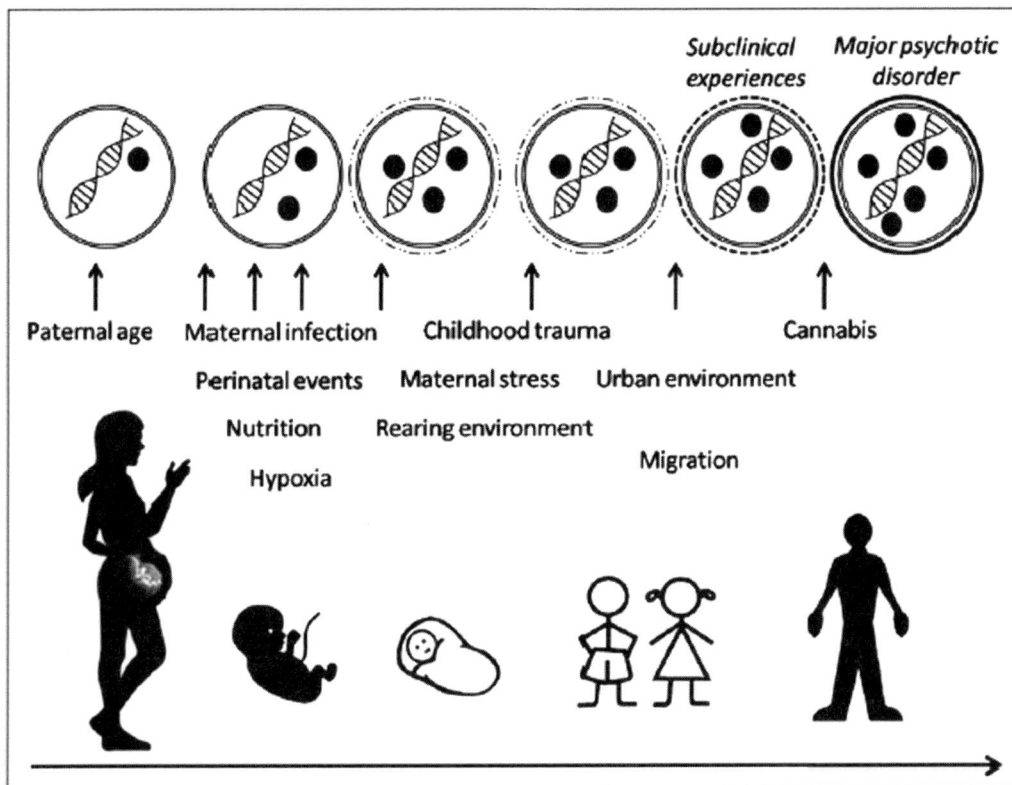

FIGURE 7.5
Developmental periods affected by environmental factors linked to SZ, BP, and ASD that impact the epigenome *(adapted from [74])*.

epigenomic programming in germline and somatic cells, and be impacted by nutrition (see below).

Generally, nutrition is an under-considered factor for phenotype (and in medicine). Few studies have examined the affect of nutrition on other risk factors for disease. Further, the importance of nutrition in preventing or reducing symptoms in neurobehavioral disorders is under-appreciated and poorly translated to the clinic although evidence that nutrition is a critical component for brain development and function is well documented [89].

Cannabis and drugs are clear risk factors for SZ and other neurobehavioral disorders. Evidence suggests the cannabis risk is limited to individuals with the VAL158 allele of the enzyme catechol-O-methyl transferase (COMT) [74]. Tetrahydrocannabinol (THC), the active ingredient of cannabis, induces expression of HDAC3 [90]. Abuse of stimulants like cocaine, amphetamine, and phenycyclidine are linked to neurobehavioral disorders (positive and cognitive deficits), and long-term effects of drug use lead to brain changes similar to those seen in neurobehavioral disorders, and a sensitization to low dose exposures.

7.7 GENOMIC INSTABILITY

At the molecular level, severe neurobehavioral disorders most resemble cancer. Both disorders are linked to DNA instability (i.e. replication and repair) and aberrant chromatin modifications throughout the genome. These processes are closely linked at both the macromolecular level and at the metabolic level (see below).

The first neurobehavioral diseases linked to genomic instability were the trinucleotide repeats diseases (for recent review see [91]; Table 7.3). The disease mutations involved increase ("expansions") in trinucleotide repeating sequences, usually $(CCG)_n$ or $(CAG)_n$ located

TABLE 7.3 Trinucleotide Repeat Diseases

Disease[1]	Sequence	Gene[2]	Parent of Origin[3]	Repeat Length (n)			Somatic Instability
				Normal	Disease	Permutation	
Diseases with Coding Trinucleotide Repeats							
DRLPA	CAG	ATN1(exon5)	P	6-35	35-48	49-88	Yes
HD	CAG	HTT (exon1)	P	6-29	29-37	38-180	Yes
OPMD	CGN	PABPN1 (exon 1)	P & M	10	12-17	>11 None is tissue tested (hypothala mus)	
SCA1	CAG	ATXN1 (exon 8)	P	6-39	40	41-83	Yes
SCA2	CAG	ATXN2 (exon 1)	P	<31	31-32	32-200	Unknown
SCA3 (Machado–Joseph Disease)	CAG	ATXN3 (exon 8)	P	12-40	41-85	52-86	Unknown
SCA6	CAG	CACNA1A (exon 47)	P	<18	10	20-33	None found
SCA7	CAG	ATXN7 (exon 3)		p	4-17	28-33	>36 - >460
SCA17	CAG	TBP (exon3)	P>M	25-42	43-48	45-77	YES
SMBA	CAG	AR (exon 1)		P	13-31	32-39	40
Diseases with Non-Coding Trinucleotides Repeats							
DM1	CTG	DMPK (3' UTR)		M	5-37	37-50	>50
DM2	CCTG	CNBP (intron 1)	Unknown	<30	31-74	75-11,000	Yes
FRAX-E	GCC	aff2 (5' UTR)		M	4-39	40-200	>200
FRDA	GAA	FXN (intron 1)	Recessive	5-30	31-100	70-1,000	Yes
FXS	CGG	FMR1 (5' UTR)		M	6-50	55-200	>200
HDL2	CTG	JPH# (exon2A)		M	6-27	29-35	36-57
SCA8	CTG	ATXN*OS (3' UTR)		M	15-34	34-89	89-250
SCA10	CATTCT	ATXN10 (intron 9)	M & P but smaller changes with M	10-299	29-400	400-4,500	Yes
SCA12	CAG	PPP2R2B (5'UTR) but more unstable with P	M&P	7-28	28-56	66-78	None found

outside or within coding regions, respectively. Expansions outside of coding regions are usually larger than those within coding regions, most likely because the latter are constrained by protein function.

The first described and best studied repeat disease is FRAX disease — the most frequent cause of hereditary mental retardation (for review see [92]). FRAX disease, caused by a CGG trinucleotide repeat in the 3′ promoter region of the FMR1 gene, is the most common genetic cause of autism, and is associated with SZ. Disease occurs when the repeat number becomes greater than 230 but can range up to 2000 copies. Between 200 and 230 repeats, disease occurrence is associated with hypermethylation of the promoter DNA region. FRAX disease is diagnosed by detection of the FRAXA site, cytologically, on metaphase chromosomes at Xp27 from cells grown in culture medium deficient in folate (see below). The site appears as broken chromosomal regions on metaphase chromosomes. Usually, the chromosome is not broken but the DNA within the region of site is unreplicated.

HD is an example of a trinucleotide repeat disease with a $(CAG)_n$ expansion within a coding region of a specific gene. CAG repeating sequences within exons usually code for the amino acid glutamine.

SZ is associated with specific fragile sites in some families [28,69]. In addition, a higher than expected frequency of genes or regions linked to SZ co-localize with fragile site regions. In addition, the distribution of repeat sizes in multiple genes are skewed towards larger but not expanded size in SZ [69,93].

Fragile sites are unusual heritable elements. The sites must be induced in growing cells under conditions that interfere with DNA replication. When induced, the sites appear only in a small number of chromosomes, i.e. <30%. There are ~120 fragile sites in the human genome, most induced by folate deficiency, others induced following treatment with agents that interfere with DNA replication. Sites are classified as common or rare depending on their frequency in the populations (i.e. occurring in >5% and <5% of the population, respectively). Some rare sites (like FRAGX) are associated with specific neurobehavioral disorders (and/or cancers), while globally rare fragile sites distributed throughout the genome are linked to SZ and cancer. Fragile sites are regions of the genome that are unstable and linked to transpositions, deletions/insertions, mutations, and epigenetic programming changes.

In 2003, our research [69] uncovered a connection between genome instability (encompassing possible base and DNA methylation changes) and SZ. We detected a larger than expected number of DNA differences around anonymous $(CAG)_n$ repeating sequences between monozygotic twins discordantly afflicted by schizophrenia. In contrast, monozygotic twins concordant for, or unaffected by, SZ had the expected low level of DNA differences. The DNA differences in the discordant twin pairs appeared to be due to DNA base changes, although in a minor number of loci, differences may have been due to DNA methylation differences.

Some researchers have found DNA methylation differences between monozygotic twins and proposed these epigenetic differences could account for discordance in twins [94–96]. As DNA stability and epigenomic programming are closely linked (see below), there is no strong reason to believe one will occur in the absence of the other.

Our twin research began as a search to explain why all progeny from monozygotic twins discordant for SZ have the same elevated probability of becoming ill with SZ [97]. This observation argued that the predisposition toward disease is passed on, but that genes alone did not determine disease occurrence or phenotype. Our goal was to find DNA differences that might account for discordance in disease phenotype. Furthermore, given that SZ is a syndrome affecting many biological processes and that changes in patient lymphocyte DNA was detected by others, we reasoned that DNA changes in lymphocytes would be detected.

An increasing number of genetic and epigenetic changes in lymphocytes are detected in SZ, BP, and ASD, as more laboratories use lymphocytes as sentinels of disease. For instance, Uranova et al. [98] found ultrastructure abnormalities that included an increased frequency in large activated lymphocytes and small mitochondria in samples from schizophrenic patients. Cytogenetic studies on fragile site induction have revealed an increased sensitivity to folate deprivation, here-to-fore unknown fragile sites in psychiatric, SZ, and BP patients [99–102]. Aberrant DNA damage signaling response reflected as abnormal histone phosphorylation levels, with unchanged repair rates were detected by Catts et al. [103]. Decondensed chromatin changes [104], lower levels of DNA methylation overall and in male schizophrenic patients [25] an increased expression of high mobility group A protein 1A (HMGA1), were well-documented in transcription regulator interacting with chromatin [105].

Several groups are examining the utility of using whole gene expression in blood cells for diagnosing SZ (for review see [106]). Changes in gene expression of specific genes linked to SZ linked to have been detected, e.g. RELN receptors [107,108]. These cells are also used for functional studies. James et al. [109] reported abnormalities in the response of lymphoblastoid cells from autistic patient to oxidative stress metabolites (see below). The results are exciting because they may provide quantitative diagnostic tests, as well as means of studying epigenomic changes relevant to brain disease in a readily available tissue.

7.8 TRANSCRIPTIONAL DYSREGULATION

Recent studies from several groups, including us, have linked aberrant chromatin changes and gene expression to different neurobehavioral diseases. Costa and colleagues have focused on the GABAergic neurons (see above). Our own work focused on developing a multi-level understanding of dopamine metabolism in the synaptic cleft.

We analyzed samples from the prefrontal cortex of individuals with SZ and BP versus unaffected controls. In a series of manuscripts (e.g. [110–112]), we reported on: the promoter methylation status; transcript level and presence of SNPs of genes important in dopamine metabolism including membrane-bound catechol-O-methyl transferase (MB-COMT) and RELN previously found to be hypermethylated in SZ [113]; and the dopamine receptors (DRD) 1 and 2.

Our results revealed that patient samples had significantly lower levels of MB-COMT promoter methylation than controls, particularly in the left frontal lobe, and that the level of methylation inversely correlated with mRNA expression [111]. The VAL158MET polymorphism within the MB-COMT gene is correlated with protein activity, with the VAL allele codes for a hyperactive protein that leads to excess dopamine degradation [114]. The valine (VAL) allele was present in more SZ patients, and the VAL/VAL genotype more common in both SZ and BP patients than in well controls, although other studies report mixed results. Most individuals with MB-COMT promoter hypomethylation and the VAL/VAL genotype were in the patient groups (SZ and BD), while most cases with methylated MB-COMT promoters and MET/MET genotypes were in the control group.

In the same samples, hypomethylation of the MB-COMT and DRD2 promoters almost always correlated. This implies that synaptic dopamine deficiency resulting from overexpression of MB-COMT was usually compensated for by DRD2 overexpression. However, DRD2 hypomethylation was significantly lower in the patient samples than in controls, possibly reflecting a reduced ability to compensate for low levels of dopamine in the synaptic cleft. Correlations were detected with RELN expression and promoter methylation status [111,115].

Understanding even a relatively simple function like dopamine metabolism requires multi-level data on DNA sequences, chromatin modifications, RNA expression and structure, protein expression and structure, metabolite levels, and phenotype. As discussed above, our own studies have begun to dissect multilevel dopamine metabolism regulation. Integration and understanding of any process will require complex mathematical modeling of function such as those pioneered by Reed and colleagues [116] for folate, and by Voit and colleagues [117] for dopamine metabolism.

7.9 RNA EPIGENOMICS

Other epigenomic modulations known to occur in the brain and associated with neurobehavioral disorders are at the RNA and protein level. Of the known modifications, the most advanced studies are focused on alternative splicing. The BDNF gene, linked to dopamine metabolism (see above) and many neurobehavioral diseases, is a striking example of the amount of variation that is likely present in genes expressed in the brain. The BDNF gene has 11 exons, nine promoters, >30 splice variants, and antisense regulation [118]. Some variation in BDNF RNA processing is likely linked to neurobehavioral diseases. Disrupted in schizophrenia 1(DISC1) mRNA splice variants are up-regulated and associated with risk polymorphisms in SZ [119]. Splice variations are detected in protein regulator of G protein signaling 4 (RFS4) in SZ but not BP [120].

RNA editing and alternative splicing changes are linked to the serotonin 2C receptor in SZ [121]. Other RNA editing will be uncovered given that the adenosine deaminases (ADARs) are mainly expressed in the nervous system (for review see [122]). microRNA (miRNA) dysregulation occurs in SZ, BP, and ASD (for reviews see [123,124]). Although the function of most non-coding RNAs (ncRNAs) is not known yet, their link to chromatin modification, genomic instability imprinting, X-chromosome, and gene expression argues for their importance in a number of processes linked to disease [125]. Today we are seeing the beginning of tantalizing level of here-to-fore unknown RNA complexity.

7.10 METABOLISM

A key intracellular metabolic hub important for genomic stability and epigenomic programming is the crossroads (Figure 7.6) of the folate—methionine—transulfuration—dopamine (FMTD hub) pathways (for review see [28]). The FMTD hub directly links the synthesis of DNA and RNA precursor and metabolites used for energy transduction (i.e. ATP and GTP) to metabolites important for epigenomic programming, cellular response to oxidative stress, and dopamine metabolism.

DNA and RNA precursor syntheses require folate as a methyl donor. Specifically, the formation of TTP from UTP by thymidylate synthase requires folate. Further, two steps in the de novo formation of inosine monophosphate (IMP) utilizes folate as a methyl donor. IMP is a precursor for the synthesis all other purines. Given all the data that point towards abnormalities in the FMTD hub it is not surprising that homeostatic purine imbalances are present in naive first-episode SZ patients [126].

SAM is produced in the methionine cycle from exogenous methionine and ATP. SAM is the major intracellular methyl donor that acts as a cofactor for over 100 methyl transferases, including enzymes that participate in epigenomic regulation through modification of DNA and RNA, proteins, including histones, and in other catabolic and anabolic reactions such as those that degrade dopamine and nicotine. Clearly, interference in SAM production will have a major impact on a cell.

In the methionine cycle, demethylated SAM is broken down into the key metabolites s-adenosyl homocysteine (SAH) and homocysteine (HCY). HCY is re-methylated by the enzyme betaine homocysteine methyl transferase (BHMT) or the enzyme methionine synthase (MS), to reform methionine. The enzyme BHMT transfers a methyl group from betaine to HCY. Betaine or its precursor choline is obtained from the diet. Choline can also be obtained

143

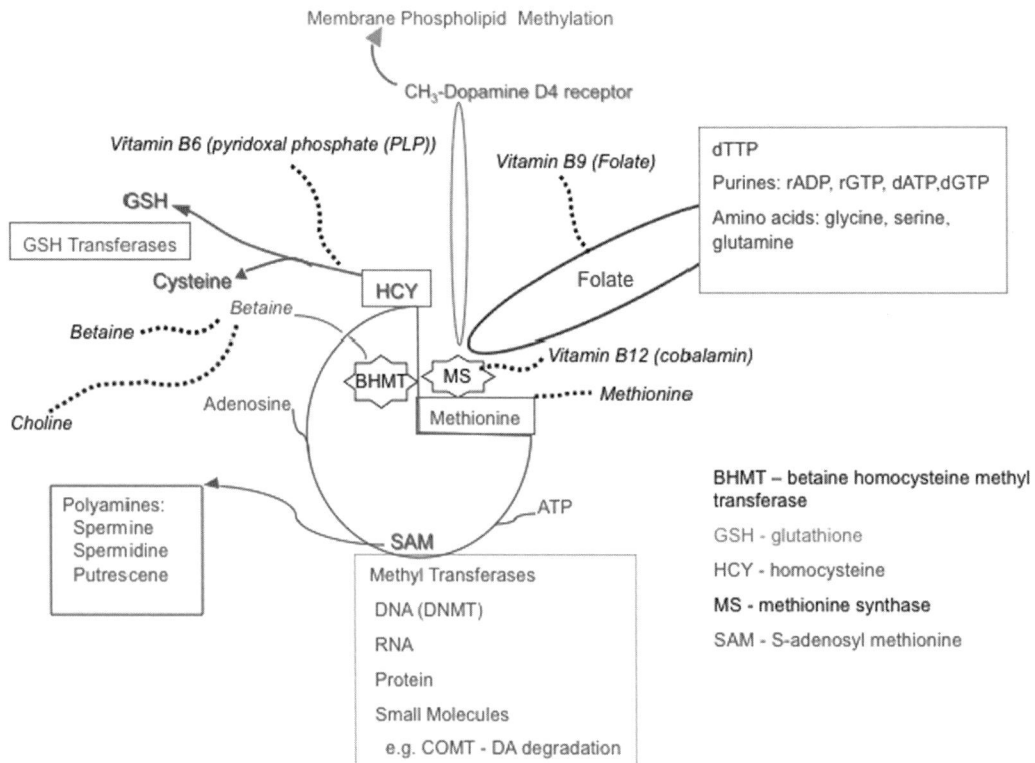

FIGURE 7.6

The FMTD hub: Metabolic link between DNA and RNA synthesis, energy production, epigenomic programming, oxidative stress responses, inflammation, and dopamine metabolism (not all metabolic steps are shown). This figure is reproduced in the color plate section.

from degradation of the cell membrane. Besides being used to reform methionine, betaine is used for the synthesis of phospholipids.

The MS enzyme is a cobalamin (vitamin B12)-dependent enzyme sensitive to oxidative conditions, transfers a methyl groups from a folate derivative, 5-methyl tetrahydrofolate formed in the folate cycle. Curiously, demethylated (MET313HCY) dopamine receptor D4 (DRD4) competes with HCY as a methyl acceptor from MS. Methylated DRD4 acts as a methyl transferase by methylating membrane phospholipids and causing a change in membrane permeability. Our recent unpublished studies revealed that the DRD4 promoter is hypo-methylated and the gene overexpressed in brain samples from SZ and BP patients.

HCY is the input metabolite into the transulfuration pathway. The transulfuration pathway not only synthesizes glutathione (GSH), the primary intracellular antioxidant, but also the amino acid cysteine in mammals. Two rate-limiting enzyme steps require another essential vitamin, B6, in the transulfuration cycle. Inhibition of MS activity by oxidative stress directs HCY towards the transulfuration pathway and away from the production of SAM.

Methionine re-cycling through the methionine cycle depends on the amino acid's availability, and metabolic need. Reed and colleagues have pioneered mathematical modeling of this hub (e.g. [127–130]) and applied this model to understanding disease [128,131,132] like SNPs linked to autism [133], cellular processes like DNA methylation [129], and toxic arsenic exposures [134]. These kinds of efforts are needed to understand the multitude of factors linked to neuropsychiatric disease.

In summary, the FMTD hub depends on five nutrients (methionine, vitamins B9, B12, and B6, and betaine (or choline)) that must be obtained from the diet, or from the intestinal microbiome. Multiple evidence points towards abnormalities in this pathway in neuro-behavioral disorders.

Dopamine metabolism is linked to these pathways in three manners: (1) degradation of dopamine by COMT; (2) methylation regulation of genes linked to dopamine metabolism (research by others and us, see above); and (3) DRD4 methylation by the enzyme MS. Multiple genetic variants within the folate cycle contribute to negative symptoms in SZ with low serum folate [135]. SNPs within this hub are linked to SZ, BP, and ASD patients (see above and [109,136,137]), and even to parents of autistic patients [133]. However, deficits in the FMTD hub are linked to many diseases. Hence, it is not clear whether a particular abnormality is linked to or affected by disease occurrence or presentation. This distinction may not be important. It is likely that normalization of the FMTD hub dynamics in patients will have positive outcomes.

Clearly, nutrition has the potential to impact a large number of cellular processes that require energy, for instance, DNA replication, RNA synthesis, epigenomic programming, and the reduction of oxidative stress. Likewise, processes that require DNA replication and/or RNA synthesis, or cause oxidative stress like ATP generation in the mitochondria (the primary source of oxidative stress) or inflammation, will generally impact multiple cellular processes linked to this metabolic hub including epigenomics programming.

7.11 NUTRITIONAL AND DRUG INTERVENTIONS

Nutritional and drug interventions used to treat neurobehavioral disorders impact the epigenome. In many cases, the epigenomic link was detected after disease efficacy was discovered. In reality, most, if not all, nutrients and drugs impact the epigenome. The important issue is to distinguish between primary and secondary affects.

A number of required and intermediate metabolites in the FMTD hub impact neurobehavioral behaviors. That folic acid is useful for treating refractory major depression and SZ has been known for some time, while SAM and methionine are known to increase psychotic symptoms (see [112] for a review).

Drugs used to treat neurobehavioral disorders impact the epigenome directly. Haloperidol, an older anti-psychotic drug used to treat schizophrenia, alters DNA methylation in a tissue- and in a sex-specific manner in rats [138]. Clozapine and sulpiride (but not haloperidol or olanzapine) reduced promoter methylation of the GAD_{67} and RELN [139] previously shown to be epigenomically remodeled in a methionine model of schizophrenia [140]. These effects were enhanced by VPA. VPA, an inhibitory of HDAC, is commonly used to treat BP in place of lithium (see above). Fluoxetine, a new generation selective serotonin-reuptake inhibitor used to treat depression, increased MeCP2, MBD1, and HDAC2 in rat GABAergic neurons after a 10-day treatment regimen [141]. These changes were accompanied by a decrease in acetylation at histone 3 lysine 9 (H3K9) and histone 3 lysine 14 (K3K14). Similar results were found following administration of cocaine (an inhibitor of the serotonin, dopamine, and norepinephrine transporters) but not of nortriptyline (an inhibitor of the serotonin and dopamine transporters). These results suggest that some drugs that target the serotoninergic system may exert their effects through epigenetic mechanisms.

Chronic, but not acute, cocaine dosing induces histone H3 hyperacetylation at the *BDNF* and *Cdk5* promoters that is accompanied by DNA demethylation [142] _ENREF_112. Sodium butyrate, a non-specific HDAC inhibitor, augments the cocaine-induced histone modifications in striatum and the cocaine-induced association of phosphoacetylated histone H3 within the *cFos* gene promoter.

The list of drugs used for the treatment or induction of disease that impact the epigenome is growing. In fact, some modes of action may be mediated through changes in the epigenome

145

rather than through, for instance, direct binding of a dopamine, serotonin, or nicotinamide receptor. Today we have tools that allow the exploration for novel interactions that can be important for understanding complex diseases.

7.12 PUTTING IT ALL TOGETHER

Research and treatment of severe neurobehavioral diseases has a history of many failed claims of success or ways to succeed. The severity of illness argues for optimism but not over-optimism. Recent disappointment in the outcome of genetic approaches for common diseases is an example of this phenomenon. Unfortunately, such (over)enthusiasm impairs research on neurobehavioral diseases more severely than other diseases because of the unique deficit in research funds.

Today, there is an increased interest in epigenomics and environmental components of neuropsychiatric disease. The development of new high-throughput techniques, and computational analytical tools, for example in system biology and complexity science, has renewed interest in integrating complex, multilevel disease models.

Our own view is that the seemingly disparate observations on these diseases provide windows onto different and/or overlapping disease processes that lead to similar behavioral symptoms. Further, although the brain is the major disease target, metabolic and nutritional factors are clearly linked to these diseases and need to be considered along with other risk factors. Nutritional interventions provide exciting possibilities but single supplements as drugs cannot be considered in isolation of their impact of multiple metabolic processes.

GLOSSARY

Note: See tables for additional gene names and abbreviations
ADAR adenosine deaminase RNA
ASD autism spectrum disorder
BDNF brain-derived neurotrophic factor
BHMT betainehomocysteine methyl transferase
BP bipolar disease
CNV copy number variants
COMT catechol-O-methyl transferase
DISC1 Some variation in BDNF RNA processing is likely linked to neurobehavioral diseases. Disrupted in schizophrenia 1
DNMT DNA methyl transferase
DRD dopamine receptor
DSM Diagnostic and Statistical Manual of Mental Disorders
FRAX Fragile X
FMTD folate-methionine-transulfuration-dopamine hub
GAD glutamate decarboxylase
GSH glutathione
GWAS genome-wide association studies
H2A histone 2A
H2b histone 2B
H3 histone 3
H4 histone 4
H3K4^{me} histone 3 lysine residue 4 methylatedHDAC—histone deacetylase
H3K9^{ac} histone H3 lysine 9 acetylation
H3K9^{me} histone H3 lysine 9 methylation
HCY homocysteine
HD Huntington disease
HDAC histone deacetylase
HMD highly methylated domains
ICD International Criteria of Disease
IL-8 interleukin-8
IMP inosine monophosphate
MB-COMT membrane-bound catechol-O-methyl transferase (COMT)

MBD methylated CpG dinging domain
MECP methylated CpG binding protein
MS methionine synthase
mSin3A co-repressor SIN3 homolog A, transcription regulator
ncRNA non-coding RNA
NIMH National Institute of Mental Health (United States)
PDD pervasive developmental disorders
PMD partially methylated domains
RFS4 regulator of G protein signaling 4
RELN reelin
RTT Rett syndrome
SAH S-adenosyl homocysteine
SAM S-adenosyl methionine
SIN3 SIN3 homolog A, transcription regulator
SZ schizophrenia
TDG thymine DNA clygosylase
TET ten eleven-translocation family of dioxygenses
THC tetrahydrocannabinol
TNF tumor necrosis factor
TRD transcriptional repression domain
VIPR2 vasoactive intestinal peptide receptor
VPA valproic acid

Acknowledgments

The authors would like to thank Shiva Singh, Andrew Bolton, and Myron Lewis for comments and encouragement.

References

[1] Jucker M, Walker LC. Pathogenic protein seeding in alzheimer disease and other neurodegenerative disorders. Ann Neurol 2011;70:532—40.

[2] Schneider R, Grosschedl R. Dynamics and interplay of nuclear architecture, genome organization, and gene expression. Genes Dev 2007;21:3027—43.

[3] Misteli T, Soutoglou E. The emerging role of nuclear architecture in DNA repair and genome maintenance. Nat Rev Mol Cell Biol 2009;10:243—54.

[4] Deaton AM, Bird A. CpG islands and the regulation of transcription. Genes Dev 2011;25:1010—22.

[5] Ball MP, Li JB, Gao Y, Lee JH, LeProust EM, Park IH, et al. Targeted and genome-scale strategies reveal gene-body methylation signatures in human cells. Nat Biotechnol 2009;27:361—8.

[6] Chatagnon A, Perriaud L, Nazaret N, Croze S, Benhattar J, Lachuer J, et al. Preferential binding of the methyl-CpG binding domain protein 2 at methylated transcriptional start site regions. Epigenetics 2011;6.

[7] Tahiliani M, Koh KP, Shen Y, Pastor WA, Bandukwala H, Brudno Y, et al. Conversion of 5-methylcytosine to 5-hydroxymethylcytosine in mammalian DNA by MLL partner TET1. Science 2009;324:930—5.

[8] Ito S, Shen L, Dai Q, Wu SC, Collins LB, Swenberg JA, et al. Tet proteins can convert 5-methylcytosine to 5-formylcytosine and 5-carboxylcytosine. Science 2011;333:1300—3.

[9] He YF, Li BZ, Li Z, Liu P, Wang Y, Tang Q, et al. Tet-mediated formation of 5-carboxylcytosine and its excision by TDG in mammalian DNA. Science 2011;333:1303—7.

[10] Cortellino S, Xu J, Sannai M, Moore R, Caretti E, Cigliano A, et al. Thymine DNA glycosylase is essential for active DNA demethylation by linked deamination-base excision repair. Cell 2011;146:67—79.

[11] Li YQ, Zhou PZ, Zheng XD, Walsh CP, Xu GL. Association of Dnmt3a and thymine DNA glycosylase links DNA methylation with base-excision repair. Nucleic Acids Res 2007;35:390—400.

[12] Kriaucionis S, Heintz N. The nuclear DNA base 5-hydroxymethylcytosine is present in Purkinje neurons and the brain. Science 2009;324:929—30.

[13] Li W, Liu M. Distribution of 5-hydroxymethylcytosine in different human tissues. J Nucleic Acids 2011;2011:870726.

[14] Jin SG, Wu X, Li AX, Pfeifer GP. Genomic mapping of 5-hydroxymethylcytosine in the human brain. Nucleic Acids Res 2011;39:5015—24.

[15] Bannister AJ, Kouzarides T. Regulation of chromatin by histone modifications. Cell Res 2011;21:381—95.

[16] Nan X, Ng HH, Johnson CA, Laherty CD, Turner BM, Eisenman RN, et al. Transcriptional repression by the methyl-CpG-binding protein MeCP2 involves a histone deacetylase complex. Nature 1998;393:386—9.

[17] Muskiet FA, Kemperman RF. Folate and long-chain polyunsaturated fatty acids in psychiatric disease. J Nutr Biochem 2006;17:717−27.

[18] Rosenberg G. The mechanisms of action of valproate in neuropsychiatric disorders: can we see the forest for the trees? Cell Mol Life Sci 2007;64:2090−103.

[19] Guidotti A, Auta J, Chen Y, Davis JM, Dong E, Gavin DP, et al. Epigenetic GABAergic targets in schizophrenia and bipolar disorder. Neuropharmacology 2011;60:1007−16.

[20] Ornoy A. Valproic acid in pregnancy: how much are we endangering the embryo and fetus? Reprod Toxicol 2009;28:1−10.

[21] Grafodatskaya D, Chung B, Szatmari P, Weksberg R. Autism spectrum disorders and epigenetics. J Am Acad Child Adolesc Psychiatry 2010;49:794−809.

[22] Robertson KD, Wolffe AP. DNA methylation in health and disease. Nat Rev Genet 2000;1:11−9.

[23] Zini A, Gabriel MS, Zhang X. The histone to protamine ratio in human spermatozoa: comparative study of whole and processed semen. Fertil Steril 2007;87:217−9.

[24] Fuke C, Shimabukuro M, Petronis A, Sugimoto J, Oda T, Miura K, et al. Age related changes in 5-methylcytosine content in human peripheral leukocytes and placentas: an HPLC-based study. Ann Hum Genet 2004;68:196−204.

[25] Shimabukuro M, Sasaki T, Imamura A, Tsujita T, Fuke C, Umekage T, et al. Global hypomethylation of peripheral leukocyte DNA in male patients with schizophrenia: a potential link between epigenetics and schizophrenia. J Psychiatr Res 2007;41:1042−6.

[26] Wutz A. Gene silencing in X-chromosome inactivation: advances in understanding facultative heterochromatin formation. Nat Rev Genet 2011;12:542−53.

[27] van Rijn S, Swaab H, Aleman A, Kahn RS. X Chromosomal effects on social cognitive processing and emotion regulation: A study with Klinefelter men (47, XXY). Schizophr Res 2006;84:194−203.

[28] Smith CL, Bolton A, Nguyen G. Genomic and epigenomic instability, fragile sites, schizophrenia and autism. Curr Genomics 2011;11:447−69.

[29] Piton A, Gauthier J, Hamdan FF, Lafreniere RG, Yang Y, Henrion E, et al. Systematic resequencing of X-chromosome synaptic genes in autism spectrum disorder and schizophrenia. Mol Psychiatry 2011;16:867−80.

[30] Rosa A, Picchioni MM, Kalidindi S, Loat CS, Knight J, Toulopoulou T, et al. Differential methylation of the X-chromosome is a possible source of discordance for bipolar disorder female monozygotic twins. Am J Med Genet B Neuropsychiatr Genet 2008;147B:459−62.

[31] Iourov IY, Vorsanova SG, Yurov YB. Somatic genome variations in health and disease. Curr Genomics 2010;11:387−96.

[32] Lister R, Ecker JR. Finding the fifth base: genome-wide sequencing of cytosine methylation. Genome Res 2009;19:959−66.

[33] Schroeder DI, Lott P, Korf I, LaSalle JM. Large-scale methylation domains mark a functional subset of neuronally expressed genes. Genome Res 2011;21:1583−91.

[34] Costa E, Dong E, Grayson DR, Guidotti A, Ruzicka W, Veldic M. Reviewing the role of DNA (cytosine-5) methyltransferase overexpression in the cortical GABAergic dysfunction associated with psychosis vulnerability. Epigenetics 2007;2:29−36.

[35] Veldic M, Caruncho HJ, Liu WS, Davis J, Satta R, Grayson DR, et al. DNA-methyltransferase 1 mRNA is selectively overexpressed in telencephalic GABAergic interneurons of schizophrenia brains. Proc Natl Acad Sci USA 2004;101:348−53.

[36] Feng J, Zhou Y, Campbell SL, Le T, Li E, Sweatt JD, et al. Dnmt1 and Dnmt3a maintain DNA methylation and regulate synaptic function in adult forebrain neurons. Nat Neurosci 2010;13:423−30.

[37] Moretti P, Zoghbi HY. MeCP2 dysfunction in Rett syndrome and related disorders. Curr Opin Genet Dev 2006;16:276−81.

[38] Gershon ES, Goldin LR. Clinical methods in psychiatric genetics. I. Robustness of genetic marker investigative strategies. Acta Psychiatr Scand 1986;74:113−8.

[39] Gottesman II , Gould TD. The endophenotype concept in psychiatry: etymology and strategic intentions. Am J Psychiatry 2003;160:636−45.

[40] Leboyer M, Bellivier F, Nosten-Bertrand M, Jouvent R, Pauls D, Mallet J. Psychiatric genetics: search for phenotypes. Trends Neurosci 1998;21:102−5.

[41] Greenwood TA, Lazzeroni LC, Murray SS, Cadenhead KS, Calkins ME, Dobie DJ, et al. Analysis of 94 candidate genes and 12 endophenotypes for schizophrenia from the Consortium on the Genetics of Schizophrenia. Am J Psychiatry 2011;168:930−46.

[42] Manning-Courtney P, Brown J, Molloy CA, Reinhold J, Murray D, Sorensen-Burnworth R, et al. Diagnosis and treatment of autism spectrum disorders. Curr Probl Pediatr Adolesc Health Care 2003;33:283−304.

[43] Cohly HH, Panja A. Immunological findings in autism. Int Rev Neurobiol 2005;71:317–41.

[44] Melnyk S, Fuchs GJ, Schulz E, Lopez M, Kahler SG, Fussell JJ, et al. Metabolic Imbalance Associated with Methylation Dysregulation and Oxidative Damage in Children with Autism. J Autism Dev Disord 2011.

[45] Templer DI, Cappelletty GG, Kauffman I. Schizophrenia and multiple sclerosis. Distribution in Italy. Br J Psychiatry 1988;153:389–90.

[46] Sachdev PS, Keshavan MS. Secondary Schizophrenia. Cambridge, UK: Cambridge University Press; 2010.

[47] Lewis SW. Secondary Schizophrenia. In: Hirsh SR, Weinberger DR, editors. Schizophrenia. Oxfod, UK: Blackwell Science; 1995. p. 324–40.

[48] Montejo AL. The need for routine physical health care in schizophrenia. Eur Psychiatry 2010;25(Suppl. 2): S3–5.

[49] Rupp CI. Olfactory function and schizophrenia: an update. Curr Opin Psychiatry 2010;23:97–102.

[50] Stevenson RJ, Langdon R, McGuire J. Olfactory hallucinations in schizophrenia and schizoaffective disorder: a phenomenological survey. Psychiatry Res 2011;185:321–7.

[51] Javitt DC. Sensory processing in schizophrenia: neither simple nor intact. Schizophr Bull 2009;35:1059–64.

[52] Turetsky BI, Hahn CG, Arnold SE, Moberg PJ. Olfactory receptor neuron dysfunction in schizophrenia. Neuropsychopharmacology 2009;34:767–74.

[53] Velasco-Ortega E, Monsalve-Guil L, Velasco-Ponferrada C, Medel-Soteras R, Segura-Egea JJ. Temporomandibular disorders among schizophrenic patients. A case-control study. Med Oral Patol Oral Cir Bucal 2005;10:315–22.

[54] Kirkpatrick B, Hack GD, Higginbottom E, Hoffacker D, Fernandez-Egea E. Palate and dentition in schizophrenia. Schizophr Res 2007;91:187–91.

[55] Moraes N, Moraes E, Cunha Marques HH. Dental anomalies in mental patients. Bull Pan Am Health Organ 1975;9:325–8.

[56] Xu T, Chan RC, Compton MT. Minor physical anomalies in patients with schizophrenia, unaffected first-degree relatives, and healthy controls: a meta-analysis. PLoS One 2011;6:e24129.

[57] Weinberg SM, Jenkins EA, Marazita ML, Maher BS. Minor physical anomalies in schizophrenia: a meta-analysis. Schizophr Res 2007;89:72–85.

[58] Meyer JM, Stahl SM. The metabolic syndrome and schizophrenia. Acta Psychiatr Scand 2009;119:4–14.

[59] Emsley R, Chiliza B, Asmal L, Lehloenya K. The concepts of remission and recovery in schizophrenia. Curr Opin Psychiatry 2011;24:114–21.

[60] Group, T. H. s. D. C. R. A novel gene containing a trinucleotide repeat that is expanded and unstable on Huntington's disease chromosomes. Cell 1993;72:971–83.

[61] van der Burg JM, Bjorkqvist M, Brundin P. Beyond the brain: widespread pathology in Huntington's disease. Lancet Neurol 2009;8:765–74.

[62] Djousse L, Knowlton B, Hayden MR, Almqvist EW, Brinkman RR, Ross CA, et al. Evidence for a modifier of onset age in Huntington disease linked to the HD gene in 4p16. Neurogenetics 2004;5:109–14.

[63] Andrew SE, Hayden MR. Origins and evolution of Huntington disease chromosomes. Neurodegeneration 1995;4:239–44.

[64] Gejman PV, Sanders AR, Kendler KS. Genetics of schizophrenia: new findings and challenges. Annu Rev Genomics Hum Genet 2011;12:121–44.

[65] Aldinger KA, Plummer JT, Qiu S, Levitt P. SnapShot: Genetics of Autism. Neuron 2011;72. 418–418 e1.

[66] Brunetti-Pierri N, Berg JS, Scaglia F, Belmont J, Bacino CA, Sahoo T, et al. Recurrent reciprocal 1q21.1 deletions and duplications associated with microcephaly or macrocephaly and developmental and behavioral abnormalities. Nat Genet 2008;40:1466–71.

[67] Rees E, Moskvina V, Owen MJ, O'Donovan MC, Kirov G. De Novo Rates and Selection of Schizophrenia-Associated Copy Number Variants. Biol Psychiatry 2011.

[68] Xu B, Roos JL, Dexheimer P, Boone B, Plummer B, Levy S, et al. Exome sequencing supports a de novo mutational paradigm for schizophrenia. Nat Genet 2011;43:864–8.

[69] Nguyen GH, Bouchard J, Boselli MG, Tolstoi LG, Keith L, Baldwin C, et al. DNA stability and schizophrenia in twins. Am J Med Genet B Neuropsychiatr Genet 2003;120B:1–10.

[70] Le Couteur A, Bailey A, Goode S, Pickles A, Robertson S, Gottesman I, et al. A broader phenotype of autism: the clinical spectrum in twins. J Child Psychol Psychiatry 1996;37:785–801.

[71] St Clair DM, St Clair JB, Swainson CP, Bamforth F, Machin GA. Twin zygosity testing for medical purposes. Am J Med Genet 1998;77:412–4.

[72] McGrath JJ. Variations in the incidence of schizophrenia: data versus dogma. Schizophr Bull 2006;32:195–7.

[73] Sullivan PF. The genetics of schizophrenia. PLoS Med 2005;2:e212.

149

[74] Rutten BP, Mill J. Epigenetic mediation of environmental influences in major psychotic disorders. Schizophr Bull 2009;35:1045−56.

[75] Brown AS. The environment and susceptibility to schizophrenia. Prog Neurobiol 2011;93:23−58.

[76] Malaspina D, Perrin M, Kleinhaus KR, Opler M, Harlap S. Growth and schizophrenia: aetiology, epigemiology and epigenetics. Novartis Foundation Symposium 2008;289:196−203.

[77] Frans EM, Sandin S, Reichenberg A, Lichtenstein P, Langstrom N, Hultman CM. Advancing paternal age and bipolar disorder. Arch Gen Psychiatry 2008;65:1034−40.

[78] Reichenberg A, Gross R, Weiser M, Bresnahan M, Silverman J, Harlap S, et al. Advancing paternal age and autism. Arch Gen Psychiatry 2006;63:1026−32.

[79] Saha S, Barnett AG, Foldi C, Burne TH, Eyles DW, Buka SL, et al. Advanced paternal age is associated with impaired neurocognitive outcomes during infancy and childhood. PLoS Med 2009;6:e40.

[80] Fisch H, Hyun G, Golden R, Hensle TW, Olsson CA, Liberson GL. The influence of paternal age on down syndrome. J Urol 2003;169:2275−8.

[81] McIntosh GC, Olshan AF, Baird PA. Paternal age and the risk of birth defects in offspring. Epidemiology 1995;6:282−8.

[82] Risch N, Reich EW, Wishnick MM, McCarthy JG. Spontaneous mutation and parental age in humans. Am J Hum Genet 1987;41:218−48.

[83] Glaser RL, Jabs EW. Dear old dad. Sci Aging Knowledge Environ 2004;2004. re1.

[84] Tiemann-Boege I, Navidi W, Grewal R, Cohn D, Eskenazi B, Wyrobek AJ, et al. The observed human sperm mutation frequency cannot explain the achondroplasia paternal age effect. Proc Natl Acad Sci USA 2002;99:14952−7.

[85] Galler JR, Shumsky JS, Morgane PJ. Malnutrition and brain development. In: Walker WA, Watkins JB, editors. Nutrition in Pediatrics: Basic Science and Clinical Application Second edit. Neuilly-sur-Seine, France: J. B. Decker Europe, Inc; 1996. p. 196−212.

[86] Brown AS, Susser ES. Prenatal nutritional deficiency and risk of adult schizophrenia. Schizophr Bull 2008;34:1054−63.

[87] Palmer AA, Brown AS, Keegan D, Siska LD, Susser E, Rotrosen J, et al. Prenatal protein deprivation alters dopamine-mediated behaviors and dopaminergic and glutamatergic receptor binding. Brain Res 2008;1237:62−74.

[88] Brown AS, Derkits EJ. Prenatal infection and schizophrenia: a review of epidemiologic and translational studies. Am J Psychiatry 2010;167:261−80.

[89] Hawkins DR, Pauling L. Orthomolecular psychiatry: treatment of schizophrenia. San Francisco: W. H. Freeman; 1973.

[90] Khare M, Taylor AH, Konje JC, Bell SC. Delta9-tetrahydrocannabinol inhibits cytotrophoblast cell proliferation and modulates gene transcription. Mol Hum Reprod 2006;12:321−33.

[91] McMurray CT. Mechanisms of trinucleotide repeat instability during human development. Nat Rev Genet 2010;11:786−99.

[92] Santoro MR, Bray SM, Warren ST. Molecular Mechanisms of Fragile X Syndrome: A Twenty-Year Perspective. Annu Rev Pathol 2011.

[93] Culjkovic B, Stojkovic O, Savic D, Zamurovic N, Nesic M, Major T, et al. Comparison of the number of triplets in SCA1, MJD/SCA3, HD, SBMA, DRPLA, MD, FRAXA and FRDA genes in schizophrenic patients and a healthy population. Am J Med Genet 2000;96:884−7.

[94] Kaminsky ZA, Tang T, Wang SC, Ptak C, Oh GH, Wong AH, et al. DNA methylation profiles in monozygotic and dizygotic twins. Nat Genet 2009;41:240−5.

[95] Singh SM, Murphy B, O'Reilly R. Epigenetic contributors to the discordance of monozygotic twins. Clin Genet 2002;62:97−103.

[96] Singh SM, O'Reilly R. (Epi)genomics and neurodevelopment in schizophrenia: monozygotic twins discordant for schizophrenia augment the search for disease-related (epi)genomic alterations. Genome 2009;52:8−19.

[97] Gottesman I, Bertelsen A. Confirming unexpressed genotypes for schizophrenia. Risks in the offspring of Fischer's Danish identical and fraternal discordant twins. Arch Gen Psychiatry 1989;46:867−72.

[98] Uranova NA, V, V, Vikhreva OV, Zimina IS, Kolomeets NS, Orlovskaya DD. The role of oligodendrocyte pathology in schizophrenia. Int J Neuropsychopharmenia 2007;10:537−45.

[99] Tastemir D, Demirhan O, Sertdemir Y. Chromosomal fragile site expression in Turkish psychiatric patients. Psychiatry Res 2006;144:197−203.

[100] Demirhan O, Tastemir D. Chromosome aberrations in a schizophrenia population. Schizophr Res 2003;65:1−7.

[101] Demirhan O, Tastemir D, Sertdemir Y. Chromosomal fragile sites in schizophrenic patients. Genetika 2006;42:985−92.

[102] Demirhan O, Tastemir D, Sertdemir Y. The expression of folate sensitive fragile sites in patients with bipolar disorder. Yonsei Med J 2009;50:137—41.

[103] Catts VS, Catts SV, Jablensky A, Chandler D, Weickert CS, Lavin MF. Evidence of aberrant DNA damage response signalling but normal rates of DNA repair in dividing lymphoblasts from patients with schizophrenia. World J Biol Psychiatry 2011.

[104] Kloukina-Pantazidou I, Havaki S, Chrysanthou-Piterou M, Kontaxakis VP, Papadimitriou GN, Issidorides MR. Chromatin alterations in leukocytes of first-episode schizophrenic patients. Ultrastruct Pathol 2010;34:106—16.

[105] Morikawa T, Manabe T, Ito Y, Yamada S, Yoshimi A, Nagai T, et al. The expression of HMGA1a is increased in lymphoblastoid cell lines from schizophrenia patients. Neurochem Int 2010;56:736—9.

[106] Woelk CH, Singhania A, Perez-Santiago J, Glatt SJ, Tsuang MT. The utility of gene expression in blood cells for diagnosing neuropsychiatric disorders. Int Rev Neurobiol 2011;101:41—63.

[107] Suzuki K, Nakamura K, Iwata Y, Sekine Y, Kawai M, Sugihara G, et al. Decreased expression of reelin receptor VLDLR in peripheral lymphocytes of drug-naive schizophrenic patients. Schizophr Res 2008;98:148—56.

[108] Glatt SJ, Everall IP, Kremen WS, Corbeil J, Sasik R, Khanlou N, et al. Comparative gene expression analysis of blood and brain provides concurrent validation of SELENBP1 up-regulation in schizophrenia. Proc Natl Acad Sci USA 2005;102:15533—8.

[109] James SJ, Rose S, Melnyk S, Jernigan S, Blossom S, Pavliv O, et al. Cellular and mitochondrial glutathione redox imbalance in lymphoblastoid cells derived from children with autism. FASEB J 2009;23:2374—83.

[110] Abdolmaleky HM, Cheng KH, Russo A, Smith CL, Faraone SV, Wilcox M, et al. Hypermethylation of the reelin (RELN) promoter in the brain of schizophrenic patients: a preliminary report. Am J Med Genet B Neuropsychiatr Genet 2005;134B:60—6.

[111] Abdolmaleky HM, Cheng KH, Faraone SV, Wilcox M, Glatt SJ, Gao F, et al. Hypomethylation of MB-COMT promoter is a major risk factor for schizophrenia and bipolar disorder. Hum Mol Genet 2006;15:3132—45.

[112] Abdolmaleky HM, Zhou JR, Thiagalingam S, Smith CL. Epigenetic and pharmacoepigenomic studies of major psychoses and potentials for therapeutics. Pharmacogenomics 2008;9:1809—23.

[113] Guidotti A, Auta J, Davis JM, Di-Giorgi-Gerevini V, Dwivedi Y, Grayson DR, et al. Decrease in reelin and glutamic acid decarboxylase67 (GAD67) expression in schizophrenia and bipolar disorder: a postmortem brain study. Arch Gen Psychiatry 2000;57:1061—9.

[114] Lachman HM, Papolos DF, Saito T, Yu YM, Szumlanski CL, Weinshilboum RM. Human catechol-O-methyltransferase pharmacogenetics: description of a functional polymorphism and its potential application to neuropsychiatric disorders. Pharmacogenetics 1996;6:243—50.

[115] Martinowich K, Hattori D, Wu H, Fouse S, He F, Hu Y, et al. DNA methylation-related chromatin remodeling in activity-dependent BDNF gene regulation. Science 2003;302:890—3.

[116] Best JA, Nijhout HF, Reed MC. Homeostatic mechanisms in dopamine synthesis and release: a mathematical model. Theor Biol Med Model 2009;6:21.

[117] Qi Z, Miller GW, Voit EO. Computational systems analysis of dopamine metabolism. PLoS One 2008;3:e2444.

[118] Pruunsild P, Kazantseva A, Aid T, Palm K, Timmusk T. Dissecting the human BDNF locus: bidirectional transcription, complex splicing, and multiple promoters. Genomics 2007;90:397—406.

[119] Nakata K, Lipska BK, Hyde TM, Ye T, Newburn EN, Morita Y, et al. DISC1 splice variants are upregulated in schizophrenia and associated with risk polymorphisms. Proc Natl Acad Sci USA 2009;106:15873—8.

[120] Ding L, Hegde AN. Expression of RGS4 splice variants in dorsolateral prefrontal cortex of schizophrenic and bipolar disorder patients. Biol Psychiatry 2009;65:541—5.

[121] Dracheva S, Elhakem SL, Marcus SM, Siever LJ, McGurk SR, Haroutunian V. RNA editing and alternative splicing of human serotonin 2C receptor in schizophrenia. J Neurochem 2003;87:1402—12.

[122] Hogg M, Paro S, Keegan LP, O'Connell MA. RNA editing by mammalian ADARs. Adv Genet 2011;73:87—120.

[123] Miller BH, Wahlestedt C. MicroRNA dysregulation in psychiatric disease. Brain Res 2010;1338:89—99.

[124] Xu B, Karayiorgou M, Gogos JA. MicroRNAs in psychiatric and neurodevelopmental disorders. Brain Res 2010;1338:78—88.

[125] Wapinski O, Chang HY. Long noncoding RNAs and human disease. Trends Cell Biol 2011;21:354—61.

[126] Yao JK, Dougherty Jr GG, Reddy RD, Keshavan MS, Montrose DM, Matson WR, et al. Homeostatic imbalance of purine catabolism in first-episode neuroleptic-naive patients with schizophrenia. PLoS One 2010;5:e9508.

[127] Nijhout HF, Reed MC, Ulrich CM. Mathematical models of folate-mediated one-carbon metabolism. Vitam Horm 2008;79:45—82.

[128] Nijhout HF, Gregory JF, Fitzpatrick C, Cho E, Lamers KY, Ulrich CM, et al. A mathematical model gives insights into the effects of vitamin B-6 deficiency on 1-carbon and glutathione metabolism. J Nutr 2009;139:784—91.

[129] Ulrich CM, Neuhouser M, Liu AY, Boynton A, Gregory 3rd JF, Shane B, et al. Mathematical modeling of folate metabolism: predicted effects of genetic polymorphisms on mechanisms and biomarkers relevant to carcinogenesis. Cancer Epidemiol Biomarkers Prev 2008;17:1822—31.

151

[130] Reed MC, Nijhout HF, Neuhouser ML, Gregory 3rd JF, Shane B, James SJ, et al. A mathematical model gives insights into nutritional and genetic aspects of folate-mediated one-carbon metabolism. J Nutr 2006;136:2653—61.

[131] Ulrich CM, Reed MC, Nijhout HF. Modeling folate, one-carbon metabolism, and DNA methylation. Nutr Rev 2008;66(Suppl. 1):S27—30.

[132] Boyles AL, Billups AV, Deak KL, Siegel DG, Mehltretter L, Slifer SH, et al. Neural tube defects and folate pathway genes: family-based association tests of gene-gene and gene-environment interactions. Environ Health Perspect 2006;114:1547—52.

[133] James SJ, Melnyk S, Jernigan S, Hubanks A, Rose S, Gaylor DW. Abnormal Transmethylation/transsulfuration Metabolism and DNA Hypomethylation Among Parents of Children with Autism. J Autism Dev Disord 2008;38:1976.

[134] Lawley SD, Cinderella M, Hall MN, Gamble MV, Nijhout HF, Reed MC. Mathematical model insights into arsenic detoxification. Theor Biol Med Model 2011;8:31.

[135] Roffman JL, Brohawn DG, Nitenson AZ, Macklin EA, Smoller JW, Goff DC. Genetic Variation Throughout the Folate Metabolic Pathway Influences Negative Symptom Severity in Schizophrenia. Schizophr Bull 2011.

[136] LaSalle JM. A genomic point-of-view on environmental factors influencing the human brain methylome. Epigenetics 2011;6:862—9.

[137] Gilbody S, Lewis S, Lightfoot T. Methylenetetrahydrofolate reductase (MTHFR) genetic polymorphisms and psychiatric disorders: a HuGE review. Am J Epidemiol 2007;165:1—13.

[138] Shimabukuro M, Jinno Y, Fuke C, Okazaki Y. Haloperidol treatment induces tissue- and sex-specific changes in DNA methylation: a control study using rats. Behav Brain Funct 2006;2:37.

[139] Dong E, Nelson M, Grayson DR, Costa E, Guidotti A. Clozapine and sulpiride but not haloperidol or olanzapine activate brain DNA demethylation. Proc Natl Acad Sci USA 2008;105:13614—9.

[140] Dong E, Agis-Balboa RC, Simonini MV, Grayson DR, Costa E, Guidotti A. Reelin and glutamic acid decarboxylase67 promoter remodeling in an epigenetic methionine-induced mouse model of schizophrenia. Proc Natl Acad Sci USA 2005;102:12578—83.

[141] Cassel S, Carouge D, Gensburger C, Anglard P, Burgun C, Dietrich JB, et al. Fluoxetine and cocaine induce the epigenetic factors MeCP2 and MBD1 in adult rat brain. Mol Pharmacol 2006;70:487—92.

[142] Kumar A, Choi KH, Renthal W, Tsankova NM, Theobald DE, Truong HT, et al. Chromatin remodeling is a key mechanism underlying cocaine-induced plasticity in striatum. Neuron 2005;48:303—14.

Emerging Role of Epigenetics in Human Neurodevelopmental Disorders

K. Naga Mohan[1,2]**, J. Richard Chaillet**[2]
[1]Centre for Human Genetics, Bangalore-560100, India; Department of Biological Sciences, Hyderabad, India
[2]University of Pittsburgh School of Medicine, Pittsburgh, PA, USA

8.1 INTRODUCTION: FEW NEURODEVELOPMENTAL DISORDERS HAVE EPIGENETIC DEFECTS

Considering the complexity of the processes of epigenetic inheritance and epigenetic regulation of transcription, the list of neurodevelopmental disorders associated with mutations in genes

T. Tollefsbol (Ed): Epigenetics in Human Disease. DOI: 10.1016/B978-0-12-388415-2.00008-1

encoding epigenetic regulators is remarkably short. This is particularly noteworthy because of the numerous proteins involved and the likely profound involvement of epigenetic processes in mammalian development. Why are there so few neurodevelopmental disorders with epigenetic etiologies? The most likely explanation is that epigenetic regulatory processes act broadly across the genome, target many genes, function throughout gametogenesis and embryogenesis, and function in all cells. The ubiquitous role of epigenetic processes likely means humans carrying loss-of-function mutations in genes encoding epigenetic machinery will not be viable due to adverse effects on cell function and organismal development. Because of this, we might expect that viable outcomes of mutations in genes governing epigenetic processes would be rare, and found only when the mutations affect a lesser number of target genes, are confined to specific developmental stages, or affect only a small percentage of cells. As will be evident in the following description of relevant neurodevelopmental syndromes with aberrant epigenetic modifications, the effects of the causative mutations are usually muted (partial loss-of-function mutations) through a variety of different genetic and/or epigenetic processes. The two general categories of epigenetic defects we describe here are either mutations involving components of the epigenetic machinery or deal with aberrant epigenetic patterns due to genetic defects.

8.2 COMPONENTS OF THE EPIGENETIC MACHINERY

The epigenetic machinery regulating genes and controlling cell and organismal physiology is complex. It establishes and maintains the large variety of chromatin states that coat the entire genome and change in programmed ways during development and cellular differentiation. As mentioned above, the two main targets of covalent modifications in mammalian genomes are DNA and the core histones. Two types of DNA modifications have been described so far, methylation and hydroxylmethylation of the C-5 position of cytosine residues [1]. Of these two modifications, the molecular and developmental events whereby cytosine methylation is established and maintained are better understood. DNA methylation in mammals is localized to cytosine bases in the context of CpG dinucleotides. Typically, in non-dividing cells or in dividing cells outside of S phase of the cell cycle, both cytosine bases of complementary base-paired CpG dinucleotides are methylated (so-called fully methylated DNA). Figure 8.1A shows how fully methylated DNA can be used as a form of heritable genomic information. The biochemical mechanism of inheritance of DNA methylation is a well-established two-step process, in which fully methylated DNA replicates to form two short-lived hemimethylated double helices, which are rapidly acted on by the DNA cytosine methyltransferase 1 (DNMT1) enzyme to regenerate two identical fully methylated double helices [2]. The epigenetic information that is inherited is typically in the form of patterns of DNA methylation, in which a methylation pattern is defined as the positions of methylated CpG dinucleotides among interspersed unmethylated CpG dinucleotides. The density of CpG dinucleotides varies widely in the genome, ranging from a high of 5–10 CpG dinucleotides per 100 base-pairs in CpG islands to nearly zero CpGs per 100 base-pairs in intronic or intergenic regions. Methylation that can be inherited on CpG islands, and the normal very low degree of CpG methylation on most CpG islands is related to high CpG methylation in the adjacent intronic and intergenic regions [3].

The details of the process of inherited DNA methylation were established in the mouse, and the most enlightening examples of this process are parental alleles of imprinted genes (Figure 8.1B). In primordial germ cells, unmethylated alleles of imprinted genes become fully methylated as the cells differentiate into more mature germ cells through a de novo methylation process involving the DNMT3A cytosine methyltransferase enzyme (Figures 8.1A and 8.1B). The biochemical mechanism to generate fully methylated alleles functions in both male and female germ cells. For some imprinted genes, the paternal allele becomes fully methylated during spermatogenesis, and for the remainder, the maternal allele becomes fully methylated during oogenesis. Following fertilization, through repetitive rounds of biochemical maintenance methylation occurring coincidentally with every cell cycle, the germ-line methylated allele

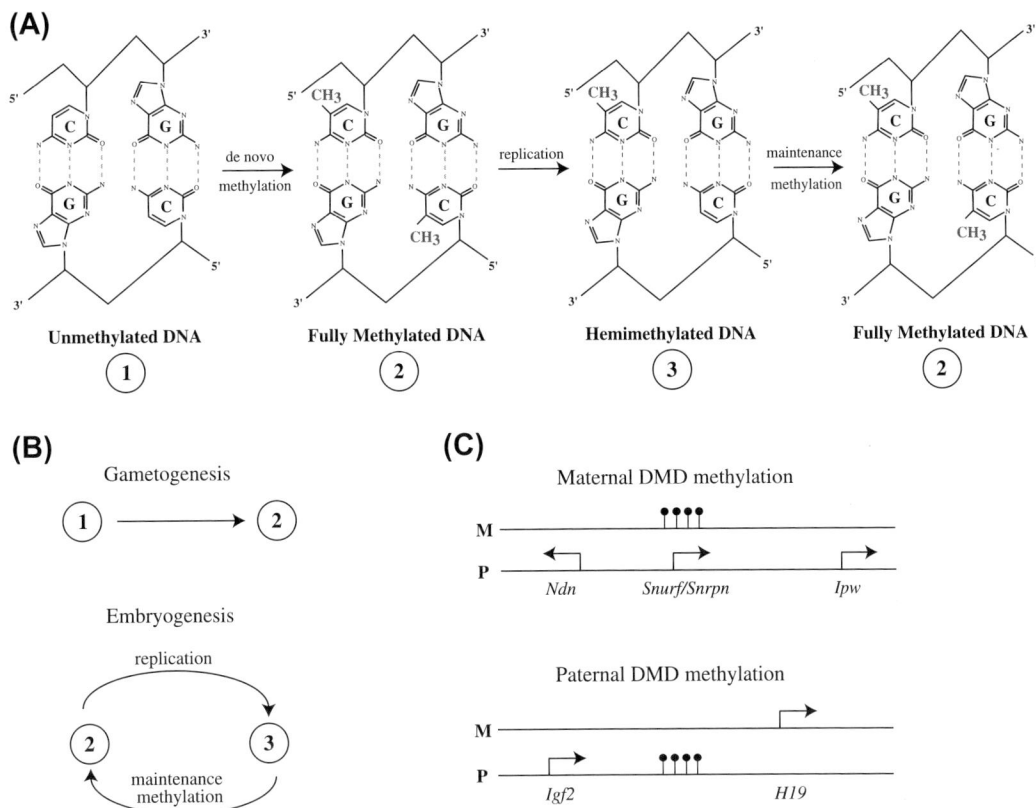

FIGURE 8.1

Establishing and maintaining DNA methylation in mammals. (A) Stepwise progression from unmethylated based-paired CpG dinucleotides (1) to fully methylated DNA (2), and its maintenance following DNA replication. Transition between (1) and (2) is de novo methylation. Maintenance requires an obligate hemimethylated intermediate (3). (B) De novo methylation of DNA occurs largely in the early stages of gametogenesis and maintenance methylation is primarily a post-fertilization process. (C) Differentially methylated domains (DMDs) of imprinted genes are created through a process in which only one parental allele undergoes de novo methylation in the germ lineage, and following its maintenance in the embryo, a difference in methylation of the two parental alleles is seen. *Snurf/Snrpn* is a maternally methylated DMD and *Igf2/H19* is a paternally methylated DMD. This figure is reproduced in the color plate section.

maintains its methylation in all somatic cells, whereas the opposite unmethylated parental allele remains unmethylated. This maintenance methylation process is an efficient and accurate mechanism of epigenetic inheritance using the DNMT1 cytosine methyltransferase enzyme, which acts immediately following DNA replication and maintains fully methylated DNA patterns [4]. During this process, hemimethylated DNA is recognized by DNMT1 and converted to fully methylated DNA. Importantly, this essential aspect of epigenetic inheritance functions after fertilization and depends on the prior establishment of fully methylated DNA patterns during gametogenesis [2]. The parent-specific allelic differences in DNA methylation in embryonic and adult mammals are found in just a few regions of the genome called differentially methylated domains (DMDs). Two examples of DMDs, one with a maternally methylated allele and the other with a paternally methylated allele, are shown in Figure 8.1C.

CpG methylation patterns are not inert covalent modifications of genomic DNA. Rather, they are "interpreted" in various ways to change genome function, typically by altering transcription efficiency in the vicinity of a methylation pattern. A main mechanism of interpretation is via binding of methyl-binding proteins (MBPs) that contain methyl-binding domains (MBDs). Methyl CpG binding protein 2 (MeCP2) is the prominent example of a MBP that binds to methylated DNA, typically in promoters of genes (Figure 8.2). The cardinal consequence of this binding is the recruitment of other protein complexes that lead to transcriptional repression [5]. In the absence of promoter methylation (e.g. in *Dnmt*-null cell lines), there may be more transcription, and in the absence of MeCP2, there may also be more transcription (Figure 8.2).

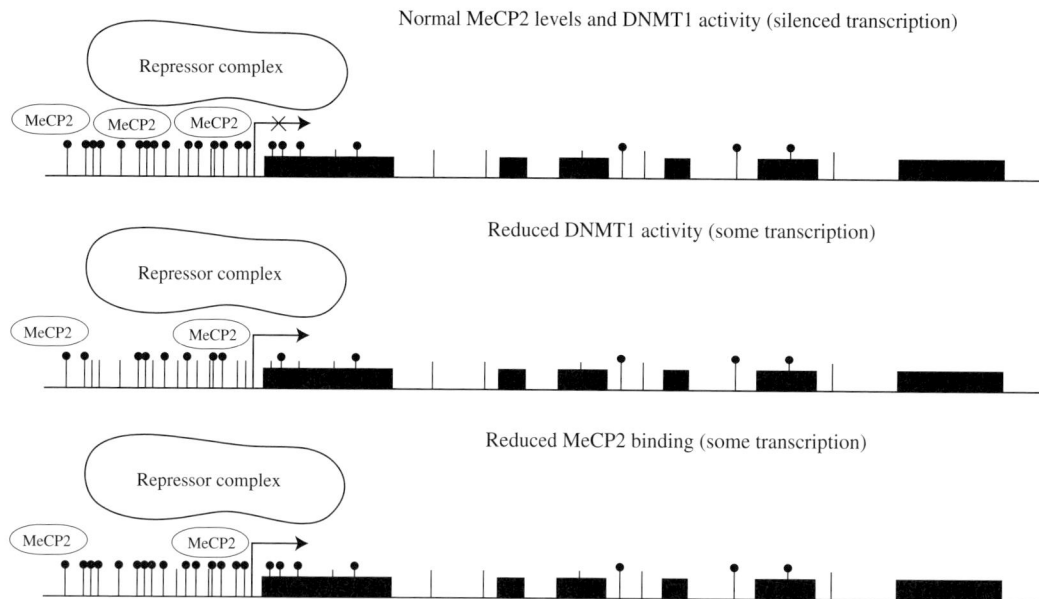

FIGURE 8.2

Gene repression by DNA methylation occurs via MBD-containing protein binding and recruitment of a repressor complex (top schematic). Reduced DNMT1 activity (middle schematic) or reduced MeCP2 (bottom schematic) lead to promoter activity and some transcription.

DNA in the nucleus of cells does not exist alone, but is organized into chromatin by a variety of DNA-interacting proteins. The basic unit of organized chromatin is DNA complexed with basic histones in the form of nucleosomes. An individual nucleosome is comprised of a histone octomer core with 146 base-pairs of DNA wrapped around the outside. Methylation and acetylation of lysines in the N-terminal tails are the two main core histone modifications that affect gene expression. Histone acetyltransferases (HATs) and methyltransferases (HMTs) are responsible for these modifications, whereas histone deacetylases (HDACs) remove acetyl groups from the lysines so that the deacetylated lysines are methylated by HMTs [6]. As in the case of DNA methyltransferases, a variety of transcriptional coactivators and corepressors influence gene expression via local histone modifications involving any of the three different classes of histone-modifying enzymes.

Ordered states of chromatin are not static, but change in a dynamic and highly regulated way. The main role of the multiple regulated chromatin states is the control of gene transcription, either at the level of transcription initiation or elongation. Two types of chromatin regulation are relevant to the neurodevelopmental disorders discussed in this chapter. The position of nucleosomes along a stretch of genomic DNA can be adjusted to change transcription efficiency [6]. For example, the nucleosome-remodeling complex SWI/SNF (SWItch defective/Sucrose NonFermentor) can be recruited to sites rich in histone acetylation, and mediate the repositioning or clearance of nucleosomes (Figure 8.3A). This clearance occurs in gene promoters, and leads to DNA sequence-specific binding of transcriptional activators and an increase in gene transcription. A different chromatin organization can be mediated by cohesin [7]. In addition to its role in pairing of sister chromatids during interphase of the cell cycle, at this and other cell-cycle stages cohesin can mediate functional interactions between distant DNA sequence elements, which in turn govern gene expression (Figure 8.3B).

8.3 NEURODEVELOPMENTAL DISORDERS DUE TO DEFECTS IN EPIGENETIC MACHINERY

We will now discuss some neurodevelopmental disorders that are due to defects in epigenetic machinery.

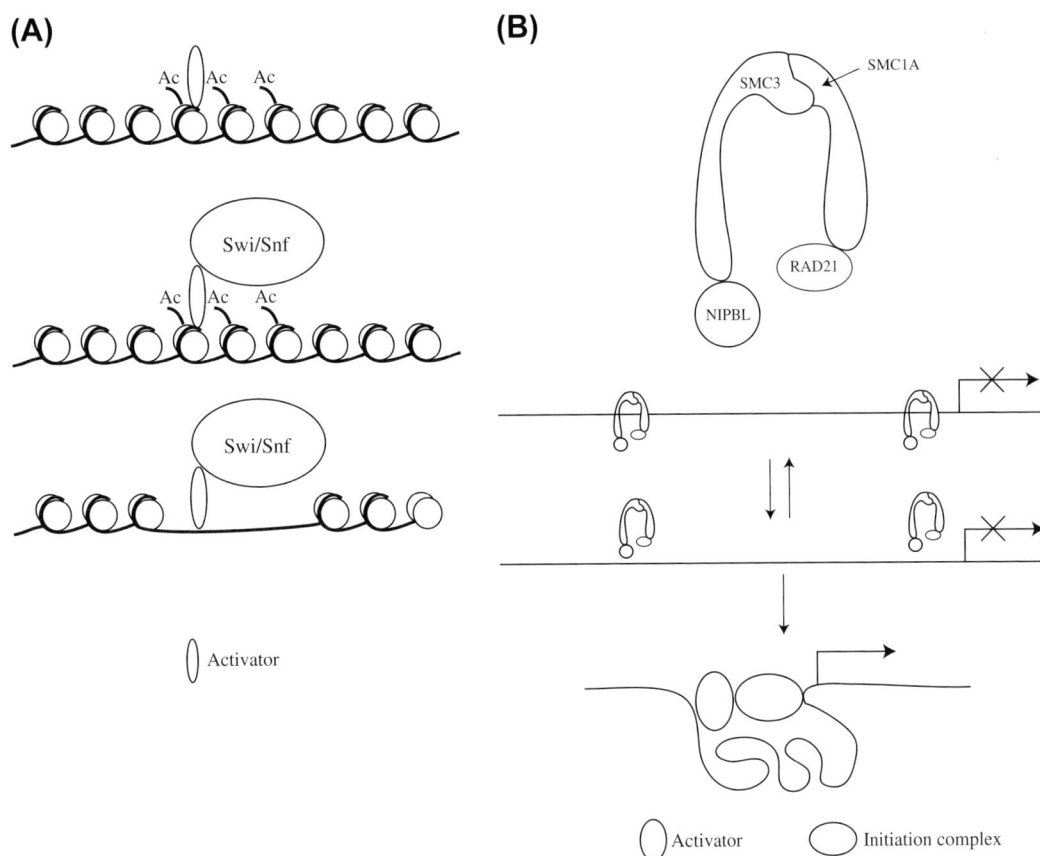

FIGURE 8.3
Higher-order chromatin organization relevant to neurodevelopmental syndromes. (A) Chromatin remodeling complexes with ATPase activity such as SWI/SNF are recruited by concentrated acetylated histones, and repositioned nucleosomes, thus leading to activator binding and gene transcription. (B) Positioned cohesin complexes comprised of structural (SMC3 and SMC1A) and regulatory (NIPBL and RAD21) proteins are thought to be regulated in an ON—OFF fashion to control gene expression via long-range interactions between different bound DNA *cis*-regulatory elements, shown here as activator and initiation complexes.

8.3.1 DNMT1 Levels and Neuronal Development

Experiments in mice using conditional mutant alleles of both *Dnmt1* and *Dnmt3a* suggest that these enzymes have different effects in different regions of the mouse brain. Loss of DNMT1 in postmitotic neurons does not affect DNA methylation or survival of neurons [8]. However, in premitotic neurons, absence of DNMT1 results in neuronal cell death. Interestingly, a combined loss of DNMT1 and DNMT3A results in smaller-sized postmitotic hippocampal neurons with hypomethylation and dysregulation of genes involved in long-term synaptic plasticity and defects in learning and memory [9]. We can conclude from these studies that proper levels of DNMT1 and probably DNMT3A are clearly important for the development and function of the mammalian nervous system, and that the vast majority of mutations in these enzymes will lead to embryonic death and therefore the absence of recoverable syndromes.

8.3.2 Schizophrenia and Bipolar Disorders

Evidence suggesting the importance of de novo and maintenance DNA methyltransferase levels in neuronal function and neurological behavior comes from a series of reports. Elevated levels of DNMT1 were found in GABAergic neurons from layers I—IV of prefrontal cortex of schizophrenic and bipolar patients [10]. In these cell types, the promoters of GAD_{67} and *RELN* were hypermethylated and associated with low transcript levels of the two genes, leading to the suggestion that elevated DNMT1 levels are responsible for their down-regulation. Reduced levels of GAD_{67} have been implicated in lowering production of the inhibitory transmitter GABA and trophic REELIN protein from the cortical neurons resulting in reduced dendritic spines and neuropil hypoplasticity in the pyramidal neurons. In turn, reduced dendritic spine plasticity has been suggested to be responsible for cognitive dysfunctions observed in psychotic patients. It is also possible that other genes involved in neuronal function are altered by elevated DNMT1 levels. Recently, it was shown that in telencephalic GABAergic neurons, both DNMT1 and DNMT3A are elevated in layers I and II of BA10 cortical neurons [11]. These findings may help in understanding the mechanism of de novo hypermethylation and maintenance methylation of GAD_{67} and *RELN* in these neurons. Taken together, these results suggest that abnormal levels of de novo and maintenance methyltransferases may at least in part explain abnormal neuronal morphology and function in psychotic patients.

8.3.3 ES Cell Models

The near absence of human syndromes with mutations in the molecular machinery controlling the establishment and maintenance of DNA methylation indicates the essential nature of DNA methylation and the difficulty in determining the function of DNA methylation in nervous system development through the recognition and study of human congenital syndromes. Rather, investigation of the role of DNA methylation in development of the nervous system requires experimental models of mammalian development such as genetically modified mice and pluripotent mouse ES cells. Mouse studies are important because of the limited human phenotypes associated with mutations in the epigenetic machinery, making the discipline of human genetics less informative in regards to function of methylation mechanisms and epigenetics in general. It is now clear that mutations in components of epigenetic processes produce pleiotropic effects, probably leading most commonly to embryonic lethality rather than a syndrome. Broadly pleiotropic effects can be mitigated by either mosaicism or dominant heterozygous effects. The mouse offers additional advantages because of the ability to spatially confine the mutational effect using conditional mutant alleles.

Overexpression of DNMT1 protein in early embryogenesis leads to embryonic death [12], suggesting that the concentration of DNMT1 protein regulates epigenetic features of the mouse genome that affect its function, most likely in aspects of gene transcription. DNMT1 activity can be increased in later embryonic development and in the adult mouse without obvious adverse effects, suggesting that the concentration of DNMT1 protein is critically important during an early period of embryogenesis [13]. Afterwards, above normal activity is tolerated by the mouse. Interestingly, overexpression of DNMT1 protein in ES cells has no apparent effect on the phenotype of pluripotent ES cells [14]. When these overexpressing cells are subjected to differentiation stimuli however, the resulting neurons showed abnormal dendritic arborization, branching and elevated levels of functional N-methyl D-aspartate receptor (NMDAR), a feature also reported in some neurological and neurodegenerative disorders. In contrast to human cortical neurons of schizophrenic patients [10], the promoters of *Gad67* and *Reln* were not hypermethylated in these ES cell-derived mouse neurons. These findings suggest that overexpression of DNMT1 may alter the epigenetic composition of the ES cell genome, which would involve dysregulation of genes other than *Gad67* and *Reln* (such as genes encoding components of NMDAR) leading to abnormal neuronal differentiation.

8.3.4 HSAN1

Hereditary sensory neuropathy with dementia and hearing loss (HSAN1) is a neurodevelopmental syndrome recently reported to be caused by mutations in the human *DNMT1* gene [15]. HSAN1 is one of just a handful of human syndromes in which abnormalities in genomic methylation are associated with neurodevelopmental disorders. Interestingly, HSAN1 is inherited in an autosomal dominant manner, and is due to mutations in a region of DNMT1 that targets it to nuclear replication foci. The mutations lead to premature degradation of mutant proteins, reduced enzyme activity, and decreased heterochromatin binding during the G2 cell cycle phase resulting in genome-wide hypomethylation and local hypermethylation. The autosomal dominant feature is particularly interesting; the wild-type protein presumably remains in cells, but in the presence of the mutant protein is unable to maintain methylation. To our knowledge, this is the first example of a heterozygous *DNMT1* mutation in either mouse or human giving rise to a mutant developmental phenotype. Possible explanations for the molecular abnormalities in heterozygous HSAN1 individuals include haploinsufficiency and dominant-negative effects. Because HSAN1 mutations fall within a DNMT1 domain implicated in dimerization [15,16], we can speculate that heterodimers between wild-type and mutant DNMT1 proteins readily form and, through a dominant-negative mechanism, such as instability and impaired heterochromatin binding of the heterodimer, lead to insufficient and/or inaccurate maintenance methyltransferase activity.

8.3.5 ICF Syndrome

DNMT3B mutations are found in patients with immunodeficiency-centromeric instability-facial anomalies (ICF) syndrome [17]. Facial dysmorphism, mental retardation, recurrent and prolonged respiratory infections, infections of the skin and digestive system, and variable immune deficiency with a constant decrease of IgA are the most common features observed in the ICF syndrome patients. Cytogenetic abnormalities in these patients include chromosome alterations affecting the heterochromatic regions of chromosomes 1, 9, and 16 with despiralization, chromatid and chromosome breaks, somatic pairing, and interchanges between homologous and non-homologous chromosomes. The pericentromeric heterochromatin of these chromosomes always becomes decondensed and fuses to produce multiradial configurations in ICF syndrome. At molecular level, ICF syndrome patients show undermethylation of classical satellite DNA located in the pericentromeric regions of chromosomes 1, 9, and 16, and on the distal long arm of the Y chromosome. Alpha satellite DNA is largely unaffected. Studies using cells from ICF syndrome patients showed extensive hypomethylation, advanced replication time, nuclease hypersensitivity, and a variable escape from silencing for genes on the Y and the inactive X chromosomes. Global expression profiling of lymphoblastoid cell lines from three ICF syndrome patients and five normal controls revealed significant changes in expression levels of genes involved in immune function, signal transduction, mRNA transcription, development, and neurogenesis. Abnormal epigenetic modifications in ICF cells include loss of methylation at promoter regions of several genes such as *LHX2*, loss of histone H3K27 trimethylation (repressive mark), and gain of H3K9 acetylation and H3K4 trimethylation (activation marks). Several of the derepressed genes include homeobox genes critical for immune system, brain, and craniofacial development. These genes showed a consistent loss of binding of the SUZ12 component of the PRC2 polycomb repression complex. These molecular and cytogenetic findings in ICF syndrome cells suggests that DNMT3B mutations are pleiotropic and affect the expression of several genes including those related to the development of the nervous system.

8.3.6 Rett Syndrome

A major functional consequence of DNA methylation, whether heritable or not, is its negative effect on gene transcription (Figure 8.4). As exemplified by the Rett syndrome, MBD proteins play major roles in this process. Rett syndrome is the prototypical neurodevelopmental syndrome with a heritable defect in the epigenetic machinery. It is an X-linked neurological disease caused by genetic defects in MeCP2 that affects one girl in 10 000–15 000 live births. Girls with Rett syndrome develop normally, but between 6 and 18 months of age, there is the onset of an autistic stage, characterized by loss of cognitive, motor, and social skills. During this stage, they develop characteristic neurological symptoms, such as stereotypical hand-wringing

159

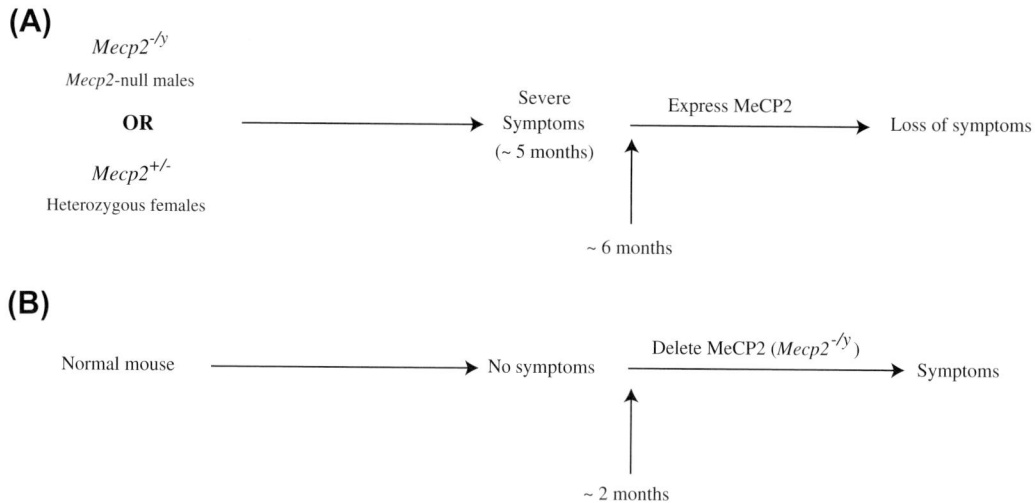

(A)

$Mecp2^{-/y}$

Mecp2-null males

OR $\longrightarrow$ Severe Symptoms (~ 5 months) $\xrightarrow{\text{Express MeCP2}}$ Loss of symptoms

$Mecp2^{+/-}$

Heterozygous females

~ 6 months

(B)

Normal mouse $\longrightarrow$ No symptoms $\xrightarrow{\text{Delete MeCP2 } (Mecp2^{-/y})}$ Symptoms

~ 2 months

FIGURE 8.4

Rett syndrome features require ongoing loss of MeCP2 function. (A) Summary of experimental scheme showing that expression of MeCP2 in *Mecp2*-null adult male mice or in adult female mice heterozygous for an *Mecp2*-null allele beginning at approximately 6 months of age will reverse the Rett syndrome-like symptoms. Scheme is based on [33]. (B) Summary of experimental scheme showing that deletion of the single wild-type *Mecp2* allele in normal adult male mice at approximately 2 months of age leads to Rett syndrome-like symptoms. Scheme is based on [34].

and clapping, indicative of loss of purposeful motor control. The disease is progressive, and after a period in which the clinical manifestations appear stable, additional deterioration occurs, leading to severe mental retardation and motor impairments, including ataxia, apraxia, and tremors. A few boys with Rett syndrome have been described, and they typically have more severe disease progression [18]. These observations are consistent with an X-linked mosaicism in females, with normally functioning cells (X chromosome with mutant *MeCp2* allele is inactive) intermixed with dysfunctional cells (wild-type *MeCp2* allele is inactive), each cell functioning autonomously with respect to its MeCP2 protein activity. Males are more severely affected because they have a single, mutant *Mecp2* allele. The specific *Mecp2* mutation has some effect on the phenotypic manifestation; mutations in the amino-terminus of the protein correlated with a more severe clinical presentation compared with mutations near the carboxyl-terminus [19]. However, it is the pattern of X inactivation that is most closely associated with disease severity; those individuals whose neurons show inactivation of X chromosomes containing the mutant *MeCp2* allele have milder Rett syndrome phenotypes. Even a few healthy female carriers have been identified because a very skewed pattern of X inactivation resulted in a high percentage of cells with an inactive mutant *MeCp2* allele [20].

Skewed X inactivation is also associated, in a poorly understood way, with familial cases of Rett syndrome. In four out of five families with Rett syndrome, no *MECP2* mutations were found [21]. In these familial cases, all the mothers and six out of eight affected girls had a highly skewed pattern of X inactivation and it was the paternal X chromosome that was active. Analysis using polymorphic DNA markers suggested that the two traits, highly skewed X inactivation and Rett syndrome, are not linked. Based on these results, Villard et al. [21] proposed that familial Rett syndrome is due to the inheritance of two traits: an X-linked locus abnormally escaping X inactivation, and the presence of highly skewed X inactivation in carrier women.

The precise molecular mechanism of MeCP2 function has been debated since MeCP2 was identified as a methyl-cytosine binding protein, prior to the realization of *MeCp2* mutations in Rett syndrome. Studies of mutations in Rett syndrome, males with duplications of the *Mecp2* gene, and mouse models of Rett syndrome have been highly instructive in advancing our knowledge of MeCP2's function. Mutations in *MeCp2* leading to Rett syndrome are found in highly conserved regions, and virtually all of these are loss-of-function or partial loss-of-function mutations [22]. This conclusion is consistent with the aforementioned strong correlation of Rett syndrome phenotype with skewing of X inactivation rather than with specific mutation. Interestingly, males with *MeCp2* gene duplications also exhibit neuro-developmental and mental retardation that overlaps with phenotypes seen in traditional Rett syndrome [23], indicating that the proper allelic dose of *MeCp2* is important for normal neurodevelopment and nervous system function and that loss or gain of *MeCP2* activity lead to similar nervous system defects.

Observations that both *MeCP2*-null mice and mice with brain-specific deletions of *MeCP2* recapitulate a number of key features of Rett syndrome suggest that the site of dose-dependent *MeCP2* defects is the central nervous system [24]. Furthermore, a mouse model expressing *MeCP2* under the *Tau* promoter, which is primarily neuron-specific, rescued *MeCp2*-null mice, and transgenic mice overexpressing *MeCP2* had phenotypes similar to the human *MeCP2* duplications [25]. The finding that *MeCP2* is very abundant in neuronal nuclei and that MeCP2 deficiency results in global changes in neuronal chromatin structure further supports the notion that MeCP2 primarily functions in neurons of the CNS and that Rett-syndrome neurodevelopmental abnormalities are a consequence of disturbances in *MeCP2's* neuronal functions [26]. From these observations, Skene et al. [26] postulated that *MeCP2* may not act as a gene-specific transcriptional repressor in neurons, but might instead dampen transcriptional noise genome-wide in a DNA methylation-dependent manner. This unifying idea may help explain the apparent contradiction that MeCP2 functions as a transcriptional repressor in complex with the co-repressor Sin3a and HDACs, and also binds to promoters of activated genes in association with the transcriptional activator CREB1 (cAMP response element binding protein 1) [27]. Recently, *MeCP2* expression in astrocytes and microglial cells has been described, and loss of glial expression of MeCP2 in mouse models of Rett syndrome was found to contribute to behavioral phenotypes [28–31]. Collectively, these findings suggest that both cell-autonomous neuronal deficits and interactions between MeCP2-deficient non-neuronal cells and neurons contribute to the pathologies of Rett syndrome.

Detailed histomorphological analyses of Rett syndrome brains revealed abnormal neuronal morphology but not neuronal death [32]. This observation, together with highly informative mouse studies, has helped to clarify the role of MeCP2 in nervous system development and function. Guy et al. [33] showed that the neurological symptoms of Rett syndrome can be reversed by reactivating MeCP2 in an adult model where the disease was already established. Specifically, reactivation of MeCP2 expression in *Mecp2*-null male mice beginning 4 weeks after birth (before the development of symptoms) prevented their certain death by 11 weeks of age [33]. In both MeCP2-null males and heterozygous mutant females that developed clear neurological symptoms, the symptoms were reversed upon restoration of MeCP2 expression. Importantly, these studies showed that the developmental absence of MeCP2 does not irreversibly damage neurons, and they are consistent with the notion that MeCP2 is required to stabilize and maintain the mature neuronal state. An explanation for this remarkable reversal of phenotype when MeCP2 expression is restored is that DNA methylation patterns are laid down normally in the absence of MeCP2 and restored MeCP2 will distribute according to the genomic patterns of CpG methylation. That is, essential MeCP2 target sites in neuronal genomes are encoded solely by patterns of DNA methylation that are established and maintained normally in cells lacking the protein. This work provided a proof of concept that symptoms of the disorder may be reversible in humans.

Recently, it has been shown that switching off MeCP2 in adult mice results in the appearance of behaviors typically seen in Rett syndrome [34]. This finding argues that early expression of the gene does not protect against the development of symptoms if the disease gene is later inactivated. Thus brain cells must have MeCP2 at all times to function normally, yet loss of MeCP2 does not cause irreversible cellular damage.

8.3.7 ATR-X Syndrome

MeCP2 interacts with several other proteins [35] including other chromatin proteins, and mutations in some of the genes encoding these proteins are also associated with neuro-developmental disorders. MeCP2 physically interacts with alpha thalassemia/mental retardation syndrome X-linked homolog protein (ATRX), which belongs to the sucrose nonfermenting 2 (SNF2) family of chromatin remodeling proteins. These proteins use energy from hydrolysis of ATP to disrupt nucleosome stability. ATR-X syndrome is an X-linked disorder with mutations in the *ATRX* gene. The syndrome is confined to males. Female carriers are intellectually normal and no consistent physical manifestations have been recognized. The syndrome is characterized by severe mental retardation, associated with unique physical anomalies, varying degrees of urogenital abnormalities, and alpha thalassemia. Phenotypic abnormalities are not confined to the nervous system, as there are non-neurological abnormalities, such as craniofacial defects, skeletal abnormalities, and lung, kidney, and digestive problems. Mild cerebral atrophy may be seen, and in two cases partial or complete agenesis of the corpus callosum was reported [36].

Studies of the pattern of X inactivation in carriers for ATR-X syndrome showed that in most of them, the abnormal X chromosome is predominantly inactivated in cells from a variety of tissues. ATRX is found in the nucleus in association with pericentric heterochromatin [37]. ATRX protein was significantly reduced in a group of patients with a wide variety of *ATRX* mutations, and in some patients the protein was virtually undetectable. These findings are consistent with the view that the significant effect of these mutations is a decrease in normal cellular ATRX activity. How loss of ATRX activity leads to ATR-X syndrome is not clear. ATRX influences effects on gene expression through interactions with MeCP2. *ATRX* mutations in humans result in aberrant DNA methylation patterns in several repetitive elements, including ribosomal DNA repeats, subtelomeric repeats, and Y-specific satellite repeats [36]. Moreover, the consistent clinical features of ATR-X syndrome suggest that ATRX regulates expression of a discrete subset of genes, alpha-globin being one well-defined example. This notion is supported by the observation that human *ATRX* mutations cause alpha-thalassemia by down-regulating alpha-globin but not beta-globin expression.

The importance of the MeCP2-ATRX interaction in neurodevelopment is highlighted in the phenotypes of patients harboring *MeCP2* variants that are not Rett syndrome mutations. In particular, the A140V mutant MeCP2 protein has been reported in hemizygous males of several independent X-linked mental retardation (XLMR) families [38]. Because these XLMR males survive past birth, the A140V mutation is therefore less severe than *MeCP2* mutations associated with Rett syndrome. The only measurably altered property of the A140V mutant is loss of the ability to interact with ATRX, whereas MeCP2 mutant proteins in Rett syndrome patients have additional abnormalities in addition to reduced ATRX binding. Thus, inappropriate targeting of ATRX likely contributes to transcriptional defects, altered neuronal phenotypes, and the neurological abnormalities seen in both Rett syndrome patients and *Mecp2*-null mice.

8.3.8 Cornelia de Lange Syndrome

MeCP2 also interacts with structural components of the cohesin complex, and MeCP2, ATRX and cohesin co-occupy genomic imprinting control regions in cells of the mouse forebrain [39]. Because of these interactions, it is not surprising that mutations in genes encoding cohesin structural and regulatory proteins are associated with neurodevelopmental abnormalities. Cornelia de Lange syndrome (CdLS) is a dominantly inherited disorder with a characteristic facial appearance, malformations of the upper extremities, gastroesophageal dysfunction, growth retardation, and abnormalities in cognitive development. Mental retardation in CdLS patients, although typically moderate to severe, displays a wide range of variability. The majority of CdLS cases are due to mutations in the cohesin regulatory subunit NIPBL (nipped B-like) or structural subunits SMC1A and SMC3 [7]. Mutations in NIPBL, the vertebrate homolog of the yeast Scc2 protein and a regulator of cohesin loading and unloading, comprise ~50% of CdLS cases. The predicted outcomes of *NIPBL* mutations are truncated or untranslated proteins, suggesting that NIPBL haploinsufficiency results in CdLS phenotypes [40]. That haploinsufficiency is a mechanism in CdLS is confirmed by the child with a large deletion of the NIPBL region and severe manifestations of CdLS [41].

Mutations in genes encoding other components of cohesin are associated with CdLS or the related Roberts and SC phocomelia syndromes (RS/SC). Mutations in SMC1A and SMC3, which are structural cohesin subunits, contribute to ~5% of CdLS cases and result in consistently milder CdLS phenotypes with absence of major anatomical anomalies [42,43]. *Smc1L1*, which encodes the SMC1A cohesin subunit, is on the X chromosome and the affected individuals are male. This indicates that the mutations are unlikely to be strong loss-of-function alleles, which would likely be lethal. Notably, *Smc1L1* escapes X-inactivation, and of the three female carriers that passed on a *Smc1L1* gene mutation, one displayed very mild characteristics consistent with CdLS. This additional clinical and cytogenetic information suggests that human *Smc1L1* mutations are probably partial loss-of-function mutations, and the retention of some function in mutant SMC1A proteins is a very likely explanation for their association with viable, yet neurodevelopmentally abnormal, CdLS patients. In this regard analysis of CdLS-associated mutant SMC1A and SMC3 proteins indicates that they are likely to be stable proteins *in vivo* [43]. As stable, expressed proteins, it is very likely that they contribute to cohesin complexes with altered (partial) function. The Roberts and SC phocomelia syndromes (RS/SC) are due to mutations in the cohesin regulator ESCO2 [44]. ESCO2 is a member of a conserved family that is required for the establishment of sister chromatid cohesion during S phase and may have acetyltransferase activity (Figure 8.3B).

As evidenced by near absence of sister chromatid cohesion defects among CdLS patients, sister chromatid cohesion seems to be minimally affected or not affected at all by dosage changes of cohesin and its associated proteins. Thus, the clinical phenotypes of CdLS and RS/SC are more likely the result of cohesin-mediated gene dysregulation during embryonic development. Studies in *Drosophila* have shown that the sister chromatid cohesion complex controls gene activation [45,46]. Such studies suggest a molecular etiology of abnormal phenotypes in CdLS patients, because effects on gene expression during human embryogenesis could lead to the development anomalies seen in CdLS patients. The *Drosophila Nipped-B* gene, an ortholog of the budding yeast gene *Scc2*, is required for cohesin to bind to chromosomes. Nipped-B regulates the loading of the structural subunits of cohesin onto chromosomes. In *Drosophila*, Nipped-B-mediated cohesin-loading is required for long-range activation of two homeobox protein genes that play critical roles in development. Specifically, reducing cohesin concentrations boosts *cut* gene expression, whereas reducing Nipped-B levels diminishes *cut* gene expression in the emergent wing margin [45]. The interpretation of these findings is that cohesin acts as an insulator that blocks enhancer–promoter communication, similar to the manner in which cohesin blocks spreading of SIR protein complexes at the HMR locus in budding yeast [47]. How cohesin might hinder enhancer–promoter communication in this model is not known. One possibility is similar to that proposed for the *gypsy* transposon insulator, where the insulator blocks a spread of homeoprotein binding between the enhancer and the promoter [48]. Nipped-B might facilitate gene activation by regulating the dynamic equilibrium between bound and unbound cohesin. Reduced Nipped-B activity, in principle, would slow the kinetics of cohesin removal and therefore block timely gene activation during *Drosophila* development. It is important to note that although effects on sister chromatid cohesion are seen in homozygous *Nipped-B* mutants, they were not observed in heterozygous mutants. However significant effects on gene expression were observed in heterozygous *Nipped-B* mutants [46]. A schematic of the structure of cohesin and how it may control the expression of genes is shown in Figure 8.3B.

163

Human *Smc1L1* mutations cause similar developmental problems as *NIPBL* mutations. An intriguing possibility is that the identified *Smc1L1* mutations are not null alleles, but slow down cohesin-binding dynamics as is proposed above in the case of reduced NIPBL levels. Similar effects may also occur in RS/SC syndrome patients because there are common features in Cornelia de Lange and RS/SC syndrome patients. Nevertheless, the detailed molecular etiology of RS/SC syndromes is probably distinct from the proposed effects of CdLS mutations on long-range gene activity. For example, key features of RS/SC patients not found in CdLS patients are the presence of chromosomal and mitotic defects. In addition, RS/SC patients show effects on cohesion specifically in heterochromatic but not euchromatic regions. In conclusion, the chromosomal and mitotic defects in RS/SC syndrome, taken together with the similar abnormal developmental phenotypes of Cornelia de Lange syndrome suggest an overlap in the cohesin-mediated functions of chromatid cohesion and control of gene expression.

8.3.9 MeCP2-ATRX-Cohesin Gene Regulation

That most human neurodevelopmental disorders are associated with mutations in the interacting chromatin proteins MeCP2, ATRX, and cohesin is interesting and raises the issue of why mutations in genes encoding these proteins are compatible with embryonic and postnatal development, albeit abnormal development? Although we do not know the answer to this question, the modes of inheritance of the associated neurodevelopmental disorders provide some insight into this issue. *ATRX* is an X-linked gene and the ATR-X syndrome is found in males carrying mutations that are likely loss-of-function mutations, indicating that complete loss of ATRX is tolerated and probably required to produce the syndrome. MeCP2 is more essential for development than ATRX because male carriers have a much more severe disease progression than females, which presumably leads to an early embryonic death. Moreover, females with Rett syndrome are protected from severe early consequences because they are X-chromosome mosaics. Likewise, CdLS is inherited as an autosomal dominant trait and more rarely as an X-linked trait, suggesting that only a partial loss of function is tolerated (reviewed in [49]). All in all, based on our discussion so far, there are just a handful of neurodevelopmental disorders due to mutations in epigenetic genes, and most of these involve partial loss-of-function of proteins (or loss of just mutant-cell function in an X-chromosome mosaic background) associated with higher-order chromatin structure and function.

How might dysregulation of MECP2, ATRX, and cohesin interactions lead to neuro-developmental disorders? In in vivo and cell culture experiments, using mice and their imprinted genes and DMDs as a model, Kernohan et al. [39] showed that ATRX, cohesin, and MeCP2 co-occupy imprinting control regions in the mouse forebrain. Furthermore, loss of ATRX led to reduced occupancy of cohesin, CTCF, and MeCP2 at the *Gtl2* and other imprinting DMDs. The fact that ATRX is essential to achieve full occupancy of cohesin and CTCF at target imprinted domains potentially implicates ATRX in the regulation of higher-order chromatin conformation, insulator functions, or monoallelic gene expression. In neurons however, ATRX is not likely to govern all occupancy and effects of MeCP2 because the loss of MeCP2 activity in mice and in males with *MeCP2* mutations leads to more severe effects than loss of ATRX activity. The investigators concluded from their studies that ATRX partners with cohesin and MeCP2 and contributes to developmental silencing of imprinted genes in the brain. Mechanistically, MeCP2 and cohesin function in chromosomal looping, and the presence of ATRX at target genes with cohesin and MeCP2 suggests that ATRX may also modulate chromatin loop formation by promoting specific long-range interactions. In regards to the ATR-X and Rett syndromes, the interaction between ATRX and MeCP2 might also control the expression of certain MeCP2-bound non-imprinted genes involved in cognitive function through their epigenetic regulation.

The finding that ATRX is essential to achieve full occupancy of cohesin and CTCF at target imprinted domains potentially implicates ATRX in the regulation of higher-order chromatin conformation, insulator functions, or monoallelic gene regulation. It is possible that CTCF and cohesin perform their insulator functions in different tissues in collaboration with various members of the SNF2 family of chromatin remodeling proteins, including ATRX. The interplay among ATRX, MeCP2, CTCF, and cohesin proteins at imprinting control regions, and probably at other sites in the neuronal genome, raises the intriguing possibilities that neuro-developmental syndromes may also be associated with mutations in the CTCF gene or in one or more of the genes encoding cohesin subunits. No human mutations are known in CTCF and there are also no mouse knockout models to study inactivation of *Ctcf* gene in neuro-development. Limited information available based on conditional knockouts of *Ctcf* or depletion of CTCF in oocytes suggests that CTCF is critical for cell cycle and development. For example, lack of CTCF in the female germline results in abnormal meiosis and mitosis, associated with lethality at 4-cell stage of preimplantation development [50]. By extension of the studies using conditional *Ctcf* alleles, it is possible that CTCF may be essential for survival of neurons as well. As discussed above, mutations in genes encoding cohesin structural and regulatory subunits are associated with CdLS, a neurodevelopmental syndrome.

8.3.10 Rubenstein—Taybi Syndrome

The Rubenstein—Taybi syndrome is characterized by a range of phenotypic abnormalities. Many organs are affected, not just the nervous system. It is an autosomal dominant disorder characterized primarily by mental retardation. There are also physical abnormalities, including postnatal growth deficiency often followed by excessive weight gain in later childhood or puberty, abnormal craniofacial features and an increased risk of cancer. In approximately 55% of cases, the syndrome is associated with de novo mutations in the CREB binding protein (CBP). Genetic cause in about 42% of the cases is not known. CBP is a histone acetyl-transferase (HAT) closely related to the p300 family of proteins, and a transcriptional co-activator that is shown to interact with more than 300 transcription factors and the general transcription machinery [51]. CBP and p300 have been shown to regulate hematopoietic stem cells raising the possibility of their involvement in stem cell biology [52]. Consistent with the autosomal inheritance of the syndrome, mouse studies showed that haploinsufficiency of CBP results in cognitive dysfunction associated with mature circuit abnormalities [53]. In regard to its role in neurological development, knockdown of CBP protein levels in utero results in significant reduction in the efficiency of neurogenesis and gliogenesis from cortical precursor cells [54]. CBP has been shown to bind to neural promoters and mediate histone acetylation, and therefore is thought to be essential for the expression of the neural lineage-specific genes during differentiation of neurons. Taken together these results are in line with our earlier suggestion that neurological disorders are only the outcomes of "less severe effects" of mildly aberrant epigenetic machinery.

8.3.11 Coffin—Lowry Syndrome

Coffin—Lowry syndrome is another of the handful of rare neurodevelopmental syndromes linked to defects in chromatin remodeling and maintenance of chromatin architecture. It is an X-linked disorder associated with a broad set of severe systemic abnormalities in male patients, including neurodevelopmental abnormalities. Phenotypic features of male patients include growth and psychomotor retardation, general hypotonia, and skeletal abnormalities. The mental retardation is usually severe. Facial features in affected males are very characteristic, and include a prominent forehead, down-slanting palpebral fissures, orbital hypertelorism, thick lips, a thick nasal septum with anteverted nares, and irregular or missing teeth. In heterozygous "carrier" females, there is a wide range of milder phenotypes, such as minor facial coarsening and obesity.

At the molecular level, Coffin—Lowry syndrome is caused by loss-of-function mutations in RSK2 (also known as RPS6KA3), which encodes a serine/threonine protein kinase belonging to the RSK family of kinases (RSK1 to RSK4) in humans. Mice lacking RSK2 survive to birth, but have defects in osteoblast differentiation resulting in skeletal abnormalities. They also show impaired spatial learning and reduced control of exploratory behavior [55]. In utero mouse experiments involving *RSK2* knockdowns indicated that loss of RSK2 results in a significant decrease in neurogenesis, an increase in the radial precursor cells, defective dopaminergic receptor function, impaired learning and coordination [56]. Coffin—Lowry is an epigenetic disease because RSK2 normally influences chromatin structure through two different mechanisms: direct phosphorylation of histones and by interacting with CBP, a histone acetyltransferase. Even though phosphorylation of H3 in response to epidermal growth factor is defective in cell lines derived from patients with Coffin—Lowry syndrome, whether dysregulation of transcription through this mechanism contributes to cognitive impairment in patients with Coffin—Lowry syndrome is not clear. A possible reason behind survival of $Rsk^{-/-}$ mice with developing skeletal and neurological abnormalities is that RSK2 plays an important role in osteogenesis and neurogenesis, whereas in development of other tissues, other members of the RSK family play an important role.

8.3.12 Kabuki Syndrome

Kabuki syndrome is characterized by mental retardation, distinctive facial appearance, developmental delay, short stature, and urogenital tract anomalies. A recent study identified mutations in the autosomal *MLL2* gene in approximately half of Kabuki syndrome patients [57,58]. Most of these cases are due to dominant de novo *MLL2* mutations. Mutations were found throughout the gene and included nonsense mutations, splice-site mutations, small deletions or insertions and missense mutations. *MLL2* encodes a large 5262-residue protein that is part of the SET family of proteins, of which Trithorax, the *Drosophila* homolog of MLL, is the best characterized. The SET domain of MLL2 confers strong histone 3 lysine 4 methyl-transferase activity and is important in the epigenetic control of active chromatin states. Most of the *MLL2* variants identified in individuals with Kabuki syndrome are predicted to truncate the polypeptide chain before translation of the SET domain. The syndrome more likely results from haploinsufficiency than a gain of function because the few pathogenic missense variants are located in region of *MLL2* that encode C-terminal domains. Therefore, a human syndrome with surviving affected individuals with *MLL2* mutations is probably only evident through partial loss of function.

8.4 NEURODEVELOPMENTAL DISORDERS DUE TO ABERRANT EPIGENETIC PATTERNS

We will now look at some neurodevelopmental disorders that have arisen due to aberrant epigenetic patterns.

8.4.1 Fragile X Spectrum Disorders

The *FMR1* locus on the X chromosome is associated with three distinct fragile X spectrum disorders: fragile X syndrome (FXS), fragile X-associated tremor/ataxia syndrome (FXTAS), and fragile X-associated primary ovarian insufficiency (FXPOI). FXS is the leading cause of inherited cognitive disability, in which affected males have mild to severe cognitive disability, macro-orchidism and connective tissue dysplasia due to an expansion of a CGG trinucleotide (triplet) repeat located in the 5′ untranslated region (UTR) of the *FMR1* gene. The repeat length is variable in the normal human population, ranging from six to 55 repeats [59]. Upon maternal transmission, a repeat can become unstable, resulting in an expansion in the offspring. Expansions into the range of 55—200 repeats generate premutation *FMR1* alleles, and expansions above 200 repeats from a premutation carrier female generate full-mutation alleles. The latter expansions initiate a series of molecular events leading to transcriptional silencing of the *FMR1* gene. The order of events appears to be histone deacetylation and H3-K9 methylation, followed by DNA methylation and H3-K4 demethylation [60]. Repeat binding proteins that attract HDACs, HMTs, DNMTs, and heterochromatin protein 1 (HP1) are implicated in this cascade of silencing events (reviewed in [61]. The absence of *FMR1* transcription and consequentially FMR1 protein (FMRP) in males with full-mutation alleles is the cause of FXS, whereas high expression of CGG-containing transcripts from premutation alleles initiates a cascade of events that culminate in central nervous system degeneration and FXTAS, probably via alterations in the availability of RNA-binding proteins. Female carriers of premutation alleles are at risk of developing FXPOI [62].

Recent studies using neuronal and non-neuronal cell lines from normal, premutation, and FXS individuals showed that the transcription of *FMR1* gene is quite complex in these individuals. Exon 1 of *FMR1* contains a sense promoter with three transcription start sites (sites 1−3) and an antisense promoter with multiple start sites. In addition there is another antisense promoter located >10 kb downstream to the sense promoter. The activity of these promoters and usage of the transcription start sites was shown to be dependent on the repeat length. In cell lines with normal repeat length, the distal-most start site in the sense promoter (site 3) is mostly used. As the repeat number increases, the proximal sites (sites 1 and 2) are used more heavily [63]. This promoter usage may explain the observation that in female carriers, the *FMR1* transcript levels are higher. In these carriers, the antisense transcripts (ASFMR1) from the distal promoter also predominate. However, in full-mutation cells, the antisense transcript is not detected. It has been suggested that both FMR1 and ASFMR1 may contribute to the variable phenotypes associated with the repeat expansion [64]. On the basis of the observations of the antisense transcripts in FXS and other patients with triplet nucleotide expansions, Kumari and Usdin [61] proposed an RNA-based model for formation of heterochromatin in the full-mutation cases. In this model, the affected gene generates a region of double-stranded RNA (dsRNA) that is a Dicer substrate. Irrespective of the source of dsRNA, the dicer products load onto the RNA-induced transcriptional silencing (RITS) complex and target the complex to the affected gene. RITS complex then facilitates hypoacetylation of histones H3 and H4, methylation of H3-K9, and DNA methylation.

The involvement of epigenetic mechanisms in fragile X spectrum disorders extends beyond transcriptional silencing of full-mutation *FMR1* alleles by DNA methylation. In rare cases of males with unmethylated full mutations, histone H3/H4 acetylation and H3 lysine 9 methylation patterns were similar to those found in FXS cell lines, suggesting that these post-translational changes alone are not sufficient for full-mutation allele inactivation [65]. Rather, it is likely that the acquisition of stable DNA methylation in the repeat-expanded promoter region is required for complete transcription repression. Consistent with this notion are recent observations in iPS cells from FXS individuals of the permanence of embryonically derived methylation on the full-mutation repeats [66]. In another example of epigenetic regulation of repeat-expanded *FMR1* alleles by mechanisms other than DNA methylation, overexpression of *FMR1* mRNA and the accompanying neurodegeneration in a *Drosophila* model of premutation FXTAS can be effectively suppressed by the application of histone acetyltransferase inhibitors [67].

Recently, much has been learned about the normal function of FMRP, which in turn has provided insight into pathogenesis of FXS, in which FMRP is not expressed or expressed at very low levels. FMRP is a cytoplasmic RNA-binding protein found in synapses in CNS neurons, and known to repress translation of mRNAs [68]. One hypothesis on the cause of FXS postulates that, in the absence of FMRP, loss of repression of metabotropic glutamate receptor (mGluR)-mediated pathways leads to the distinct cognitive and behavioral features of FXS [69]. More specifically, FMRP may be indirectly involved in the temporal and spatial controls of translation by controlling microRNA-mediated translational inhibition in response to mGluR signals [70,71]. This connection between FMRP and mGluR signaling may be the reason for the mitigation of behavioral abnormalities in individuals with FXS by the administration of an inhibitor of metabotropic glutamate receptor 5 (mGluR5)-mediated signaling [72].

8.4.2 Imprinting Disorders

Genomic imprinting is an epigenetic process that distinguishes alleles based on their parental origin. The main consequences of the imprinting processes are imprinted genes in diploid cells of the developing mammalian organism in which one allele is transcribed and the opposite allele is transcriptionally silent (monoallelic expression). There are approximately 100 known imprinted genes. There are a number of different ways in which genome-wide imprinting (set of monoallelically expressed genes) can be experimentally disrupted in the developing mouse embryo. These methods include failure to inherit the complete maternal set of imprints by blocking their formation in $Dnmt3L^{-/-}$ oocytes, absence of paternal set of imprints (parthenogenotes) and absence of maternal set of imprints (androgenotes). All of these methods lead to a combination of loss of expression of some imprinted genes and biallelic expression of other imprinted genes and embryonic death. The approximately 100 imprinted genes are organized into 16 distinct clusters, which are distributed among approximately one-half of the autosomes. Imprinting can also be lost locally by disrupting the genetic or epigenetic integrity of a single imprinted gene or cluster of imprinted genes. This can occur spontaneously or it can be experimentally induced, and is usually associated with viable fetal outcomes.

Mutations or epigenetic changes within one cluster of imprinted genes has been particularly revealing about the role of some imprinted genes in neurodevelopment. The Angelman and Prader–Willi syndromes are associated with mutations in a cluster of imprinted genes on chromosome 15. Angelman syndrome is a neurodevelopmental disorder characterized by intellectual and developmental delays, sleep disturbance, seizures, jerky movement, frequent laughter or smiling, and usually a happy demeanor. It is due to a loss of expression of the Ube3A gene, which is normally imprinted and expressed in the central nervous system from just the maternal allele [73]. This gene encodes for an E6-AP ubiquitin ligase. The frequency is approximately 1 in 20 000 live births, and it is most commonly due to de novo deletions in an upstream imprinting control element for Ube3A expression. Besides the behavioral and neurological symptoms, there are also pathognomonic neurophysiological findings. Angelman syndrome is not a neurodegenerative syndrome, and individuals with Angelman's syndrome have a near normal lifespan. Prader–Willi syndrome is a neurodevelopmental disorder distinct from Angelman syndrome that occurs at a frequency of 1/25 000 to 1/10 000 births. IQ is low, but there is much variability. Children show an unusual cognitive profile, and have behavioral problems, primarily hyperphagia with the development of morbid obesity. These are recognizable, viable syndromes with distinct clinical characteristics and neurodevelopmental abnormalities because of the localized nature of the mutations or epigenetic abnormality, rather than a complete disruption of genomic imprinting at all DMDs.

Separate disruptions within the same cluster of imprinted genes on chromosome 15 lead to the two clinically distinct syndromes. Important studies of Angelman and Prader–Willi syndrome patients with deletions within the chromosome-15 cluster of imprinted genes identified a region deleted in all Prader–Willi patients and a separate region deleted in all Angelman syndrome patients with Chromosome-15 deletions. These are designated SRO (shared region of overlap; [74], reviewed in [75]). Studies of the Prader–Willi SRO revealed important insights into the molecular and developmental etiology of the syndrome. Within the SRO is a differentially methylated domain (DMD). Approximately one-half of Prader–Willi syndrome cases are due to microdeletions of paternal chromosome 15, within the SNURF/SNRPN cluster, removing the paternal allele of the DMD and usually adjacent sequences/genes. Deletion of the unmethylated paternal *SNURF/SNRPN* DMD allele leads to transcriptional repression of the imprinted genes controlled by the DMD. Most of the remaining PWS cases are associated with maternal heteroisodisomy and paternal nullisomy of chromosome 15. In these cases, genes regulated by the DMD sequences are transcriptionally repressed because both (maternal) alleles of the DMD are methylated. Importantly, the other DMDs in the genome function normally. In a similar vein, most cases of Russell—Silver syndrome (not a neurodevelopmental disorder) are caused by loss of paternal *IGF2/H19* DMD methylation, leading to a loss of *IGF2* expression in the developing fetus, which accounts for the markedly reduced weight at birth. Most cases of Beckwith—Weidemann syndrome (another imprinting disorder) are due to de novo epigenetic events in which methylation on the maternal KvDMR (*KCNQ1* DMD) is never established during oogenesis or, if established, not maintained in the period immediately following fertilization. Again, as with PWS and RSS, only a single DMD functions abnormally, either because of a mutation or because of an abnormality in its normal epigenetic integrity.

The developmental etiologies of the two syndromes can be best understood with the realization that the ICs for the two syndromes are sites of critical parent-specific epigenetic modifications. The imprinting process creates a few DMDs in the genome (16 in the mouse and probably a similar number in the human), which are unique landmarks of inherited parent-specific epigenetic information. For each DMD, one parental allele is methylated, having acquired its methylation in the parent's germline and maintained (inherited) this methylation after fertilization. The opposite parental DMD allele does not acquire methylation during gametogenesis, and thus maintains the unmethylated epigenetic state following fertilization in the embryo. How do DMDs appear in the genome? DNA methylation is an epigenetic process that can be logically divided into two distinct purposes, the inheritance of information and the control of transcription. Imprinting is primarily an issue of the inheritance of epigenetic information and secondarily an issue of regulation of transcription. Thus, certain imprinting disorders are neurodevelopmental disorders associated with disruption of critically important inherited methylation patterns.

Another reason for the existence of imprinting disorders is that the majority of imprinted genes are expressed either in the placenta or in the developing embryonic nervous system, possibly at later stages of development, resulting in survivors with the syndrome. The spatial and temporal expression of imprinted genes is even more constrained when referring to genes within a single cluster.

8.5 MATERNAL DUPLICATIONS VERSUS PATERNAL DUPLICATIONS IN PWS/AS REGION

Familial analyses of individuals affected with neuropsychiatric disorders suggest that whereas maternal duplications in the human 15q11.2—15q13 regions are associated with autism, paternal duplications are associated with milder phenotypes with reduced penetrance and even benign in some cases [76]. A recent study shows that maternal duplications are more frequently associated with schizophrenia and schizoaffective disorders than paternal duplications [77]. Although the precise molecular basis for pathogenicity is not established, it is clear from these studies that altered gene dosage is the underlying mechanism. Consistent with this possibility, maternal isodicentric chromosomes significantly increase the risk for autism spectrum disorder (ASD). Expression analysis of cells from patients carrying the maternal duplications suggested involvement of altered epigenetic patterns with complex changes in expression of the genes in the 15q11.2—15q13 region. This region also includes three non-imprinted GABA receptors and therefore changes in the copy number/gene dosage involving GABA receptors may lead to ASD phenotypes. Consistent with this possibility, altered expression patterns of *GABRB3*,

GABRA5, and *GABRG3* was observed in some of the postmortem cerebral cortex samples from patients with ASD [78]. Taken together, these reports suggest that mechanism of pathogenesis due to changes in the copy number of genes in the 15q11.2−15q13 is not yet clear and can be fairly complex.

8.5.1 Possibilities of Epimutations

Epimutation can be described as a heritable change in gene expression that does not affect the actual base-pair sequence of DNA [79] and should be able to explain the phenotype observed in a patient. From the discussion above, epimutations include both DNA methylation and histone modification changes. Epimutations, by definition should involve one of the two alleles of a candidate gene that are equivalent to loss of function through silencing of an essential gene or inappropriate activation of a gene which is otherwise silent in that particular tissue. At present there are only two cases implicating epigenetic modifications of candidate genes for neuropsychiatric disorders. One of these two examples includes *FMR1* as described above and the second one is *RELN*. A polymorphic form of *RELN* containing longer GGC repeats was shown to be preferentially transmitted to ASD patients and was associated with a reduced expression of RELN [80]. The mechanism of silencing could involve altered epigenetic modifications as described in the case of *FMR1*. It may be noted however that epimutations (without involving a change in the DNA sequence) of a vast majority of candidate genes for neuropsychiatric disorders are yet to be discovered.

8.6 CONCLUSIONS

All of the many different cell types of mammalian species use aspects of the epigenetic machinery for regulation of gene expression or control the inheritance of epigenetic modifications of DNA sequences. Very few epigenetic processes, most notably the influence of MeCP2 on neuronal gene expression, are functionally confined to the nervous system. Because of this, autosomal recessive or X-linked mutations in genes encoding components of epigenetic processes most likely affect the development and function of many organs and consequently lead to embryonic death in humans. Few neurodevelopmental syndromes associated with genetic mutations in a component of an epigenetic process have been recognized. Most individuals with neurodevelopmental syndromes due to mutations in genes controlling epigenetic processes are X-chromosomal mosaics (with recessive X-linked mutations), have autosomal dominant mutations, or have partial loss-of-function mutations.

References

[1] Tahiliani M, Koh KP, Shen Y, Pastor WA, Bandukwala H, Brudno Y, et al. Conversion of 5-methylcytosine to 5-hydroxymethylcytosine in mammalian DNA by MLL partner TET1. Science 2009;324:930−5.

[2] Reinhart B, Chaillet JR, Genomic imprinting: cis-acting sequences and regional control. Int Rev Cytol 2005;243:173−213.

[3] Ji H, Ehrlich LI, Seita J, Murakami P, Doi A, Lindau P, et al. Comprehensive methylome map of lineage commitment from haematopoietic progenitors. Nature 2010;467:338−42.

[4] Bestor TH. The DNA methyltransferases of mammals. Hum Mol Genet 2000;9:2395−402.

[5] Nan X, Campoy FJ, Bird A. MeCP2 is a transcriptional repressor with abundant binding sites in genomic chromatin. Cell 1997;88:471−81.

[6] Suganuma T, Workman JL. Signals and combinatorial functions of histone modifications. Annu Rev Biochem 2011;80:473−99.

[7] Liu J, Krantz ID. Cohesin and human Disease. Annu Rev Genom Human Genetics 2008;9:303−20.

[8] Fan G, Beard C, Chen RZ, Csankovszki G, Sun Y, Siniaia M, et al. DNA hypomethylation perturbs the function and survival of CNS neurons in postnatal animals. J Neurosci 2001;21:788−97.

[9] Feng J, Zhou Y, Campbell SL, Le T, Li E, Sweatt JD, et al. Dnmt1 and Dnmt13a maintain DNA methylation and regulate synaptic function in adult forebrain neurons. Nat Neurosci 2010;13:423−30.

[10] Veldic M, Caruncho HJ, Liu WS, Davis J, Satta R, Grayson DR, et al. DNA-methyltransferase 1 mRNA is selectively overexpressed in telencephalic GABAergic interneurons of schizophrenia brains. Proc Natl Acad Sci USA 2004;101:348−53.

[11] Zhubi A, Veldic M, Puri NV, Kadriu B, Caruncho H, Loza I, et al. An upregulation of DNA-methyltransferase 1 and 3a expressed in telencephalic GABAergic neurons of schizophrenia patients is also detected in peripheral blood lymphocytes. Schizophr Res 2009;111:115−22.

[12] Biniszkiewicz D, Gribnau J, Ramsahoye B, Gaudet F, Eggan K, Humpherys D, et al. Dnmt1 overexpression causes genomic hypermethylation, loss of imprinting, and embryonic lethality. Mol Cell Biol 2002;22:2124−35.

[13] Ding F, Chaillet JR. In vivo stabilization of the Dnmt1 (cytosine-5)-methyltransferase protein. Proc Natl Acad Sci USA 2002;99:14861−6.

[14] D'Aiuto LD, Di Maio R, Mohan KN, Minervini C, Saporiti F, Soreca I, et al. Mouse ES cells overexpressing DNMT1 produce abnormal neurons with upregulated NMDA/NR1 subunit. Differentiation 2011;82:9−17.

[15] Klein CJ, Botuyan MV, Wu Y, Ward CJ, Nicholson GA, Hammans S, et al. Mutations in DNT1 cause hereditary sensory neuropathy with dementia and hearing loss. Nat Genet 2011;43:595−600.

[16] Fellinger K, Rothbauer U, Felle M, Langst G, Leonhardt H. Dimerization of DNA methyltransferase 1 is mediated by its regulatory domain. J Cell Biochem 2009;106:521−8.

[17] Xu GL, Bestor TH, Bourc'his D, Hsieh CL, Tommerup N, Bugge M, et al. Chromosome instability and immunodeficiency syndrome caused by mutations in a DNA methyltransferase gene. Nature 1999;402:187−91.

[18] Urdinguio RG, Sanchez-Mut JV, Esteller M. Epigenetic mechanisms in neurological diseases: genes, syndromes, and therapies. Lancet Neurol 2009;8:1056−72.

[19] Amir RE, Zoghbi HY. Rett syndrome: Methyl-CpG binding protein 2 mutations and phenotype-genotype correlations. Am J Med Genet 2000;97:147−52.

[20] Ishii T, Makita Y, Ogawa A, Amamiya S, Yamamoto M, Miyamoto A, et al. The role of different X-inactivation pattern on the variable clinical phenotype with Rett syndrome. Brain Dev 2001;23:S161−4.

[21] Villard L, Levy N, Xiang F, Kpebe A, Labelle V, Chevillard C, et al. Segregation of a totally skewed pattern of X chromosome inactivation in four familial cases of Rett syndrome without MECP2 mutation: implications for the disease. J Med Genet 2001;38:435−42.

[22] Amir RE, Van den Veyver IB, Wan M, Tran CQ, Francke U, Zoghbi HY. Rett syndrome is caused by mutations in X-linked MWCP2, encoding methyl-CpG-binding protein 2. Nat Genet 1999;23:185−8.

[23] Del Gaudio D, Fang P, Scaglia F, Ward PA, Craigen WJ, Glaze DG, et al. Increased MECP2 gene copy number as the result of genomic duplication in neurodevelopmentally delayed males. Genet Med 2006;8:784−92.

[24] Chen RZ, Akbarian S, Tudor M, Jaenisch R. Deficiency of methyl-CpG binding protein-2 in CNS neurons results in a Rett-like phenotype in mice. Nat Genet 2001;27:327−31.

[25] Collins AL, Levenson JM, Vilaythong AP, Richman R, Armstrong DL, Noebels JL, et al. Mild overexpression of MeCP2 causes a progressive neurological disorder in mice. Hum Molec Genet 2004;13:2679−89.

[26] Skene PJ, Illingworth RS, Webb S, Kerr ARW, James KD, Turner DJ, et al. Neuronal MeCP2 is expressed at near histone-octomer levels and globally alters the chromatin state. Mol Cell 2010;37:457−68.

[27] Chahrour M, Jung SY, Shaw C, Zhou X, Wong STC, Qin J, et al. MeCP2, a key contributor to neurological disease, activates and represses transcription. Science 2008;320:1224−9.

[28] Ballas N, Lioy DT, Grunseich C, Mandel G. Non-cell autonomous influence of MeCP2-deficient glia on neuronal dendritic morphology. Nat Neurosci 2009;12:311−7.

[29] Maezawa I, Swanberg S, Harvey D, LaSalle JM, Jin LW. Rett syndrome astrocytes are abnormal and spread MeCP2 deficiency through gap junctions. J Neurosci 2009;29:5051−61.

[30] Maezawa I, Jin LW. Rett syndrome microglia damage dendrites and synapses by the elevated release of glutamate. J Neurosci 2010;30:5346−56.

[31] Lioy DT, Garg SK, Monaghan CE, Raber J, Foust KD, Kaspar BK, et al. A role for glia in the progression of Rett's syndrome. Nature 2011;475:497−500.

[32] Armstrong D, Dunn JK, Antalffy B, Trivedi R. Selective dendritic alterations in the cortex of Rett syndrome. J. Neuropathol Exp Neurol 1995;54:195−201.

[33] Guy J, Gan J, Selfridge J, Cobb S, Bird A. Reversal of neurological defects in a mouse model of Rett syndrome. Science 2007;315:1143−7.

[34] McGraw CM, Samaco RC, Zoghbi HY. Adult neural function requires MeCP2. Science 2011;333:186.

[35] Nan X, Hou J, Maclean A, Nasir J, Lafuente MJ, Shu X, et al. Interaction between chromatin proteins MECP2 and ATRX is disrupted by mutations that cause inherited mental retardation. Proc Natl Acad Sci USA 2007;104:2709−14.

[36] Gibbons RJ, Higgs DR. Molecular-clinical spectrum of the ATR-X syndrome. Am J Med Genet 2000;97:204−12.

171

[37] McDowell TL, Gibbons RJ, Sutherland H, O'Rourke DM, Bickmore WA, Pombo A, et al. Localization of a putative transcriptional regulator (ATRX) at pericentromeric heterochromatin and the short arms of acrocentric chromosomes. Proc Natl Acad Sci USA 1999;96:13983–8.

[38] Winnepenninckx B, Errijgers V, Hayez-Delatte F, Reyniers E, Kooy RF. Identification of a family with nonspecific mental retardation (MRX79) with the A140V mutation in the MECP2 gene: Is there a need for routine screening? Hum Mutat 2002;20:249–52.

[39] Kernohan KD, Jiang Y, Tremblay DC, Bonvissuto AC, Eubanks JH, Mann MRW, et al. ATRX partners with cohesin and MeCP2 and contributes to developmental silencing of imprinted genes in the brain. Dev Cell 2010;18:191–202.

[40] Krantz ID, McCallum J, DeScipio C, Kaur M, Gillis LA, Yaeger D, et al. Cornelia de Lange syndrome is caused by mutations in NIPBL, the human homolog of Drosophila melanogaster Nipped-B. Nat Genet 2004;36:631–5.

[41] Hulinsky R, Byrne JLB, Lowchik A, Viskochil DH. Fetus with interstitial del(5)(p13.1p14.2) diagnosed postnatally with Cornelia de Lange Syndrome. Am J Med Genet 2005;137A:336–8.

[42] Musio A, Selicorni A, Focarelli ML, Gervasini C, Milani D, Russo S, et al. X-linked Cornelia de Lange syndrome owing to SMC1L1 mutations. Nat Genet 2005;38:528–30.

[43] Deardorff MA, Kaur M, Yaeger D, Rampuria A, Korolev S, Pie J, et al. Mutations in cohesin complex members SMC3 and SMC1A cause a mild variant of Cornelia de Lange syndrome with predominant mental retardation. Am J Hum Genet 2007;80:485–94.

[44] Vega H, Waisfisz Q, Gordillo M, Sakai N, Yanagihara I, Yamada M, et al. Roberts syndrome is caused by mutations in ESCO2, a human homolg of yeast ECO1 that is essential for the establishment of sister chromatid cohesin. Nat Genet 2005;37:468–70.

[45] Rollins RA, Morcillo P, Dorsett D. Nipped-B, a Drosophila homologue of chromosomal adherins, participates in activation by remote enhancers in the cut and Ultrabithorax genes. Genetics 1999;152:577–93.

[46] Rollins RA, Korom M, Aulnr N, Martens A, Dorsett D. Drosophila Nipped-B protein supports sister chromatid cohesin and opposes the Stromalin/Scc3 cohesion factor to facilitate long-range activation of the cut gene. Mol Cell Biol 2004;24:3100–11.

[47] Chang CR, Wu CS, Hom Y, Gartenberg MR. Targeting of cohesin by transcriptionally silent chromatin. Genes Development 2005;19:3031–42.

[48] Dorsett D. Distant liaisons: long range enhancer-promoter interactions in Drosophila. Curr Opin Genet Dev 1999;9:505–14.

[49] Dorsett D. Roles of the sister chromatid cohesin apparatus in gene expression, development, and human syndromes. Chromosoma 2007;116:1–13.

[50] Wan LB, Pan H, Hannenhalli S, Cheng Y, Ma J, Fedoriw A, et al. Maternal depletion of CTCF reveals multiple functions during oocyte and preimplantation embryo development. Development 2008;135:2729–38.

[51] Lee J, Hagerty S, Cormier KA, Kim J, Kung AL, Ferrante RJ, et al. Monoallelic deletion of CBP leads to pericentromeric heterochromatin condensation through ESET expression and histone H3 (K9) methylation. Hum Mol Genet 2008;17:1774–82.

[52] Rebel VI, Kung AL, Tanner EA, Yang H, Bronson RT, Livingston DM. Distinct roles for CREB-binding protein and p300 in hematopoietic stem cell self-renewal. Proc Natl Acad Sci USA 2002;99:14789–94.

[53] Josselyn SA. What's right with my mouse model? New insights into the molecular and cellular basis of cognition from mouse models of Rubinstein-Taybi syndrome. Learn Mem 2005;12:80–3.

[54] Wang J, Weaver IC, Gauthier-Fisher A, Wang H, He L, Yeomans J, et al. CBP histone acetyltransferase activity regulates embryonic neural differentiation in the normal and Rubinstein-Taybi syndrome brain. Dev Cell 2010;18:114–25.

[55] Poirier R, Jacquot S, Vaillend C, Southiphong AA, Libbey M, Davis S, et al. Deletion of the Coffin–Lowry syndrome gene Rsk2 in mice is associated with impaired spatial learning and reduced control of exploratory behavior. Behav Genet 2007;37:31–50.

[56] Dugani CB, Paquin A, Kaplan DR, Miller FD. Coffin–Lowry syndrome: a role for RSK2 in mammalian neurogenesis. Dev Biol 2010;347:348–59.

[57] Ng SB, Bigham AW, Buckingham KJ, Hannibal MC, McMillin MJ, Gildersleeve HI, et al. Exome sequencing identifies MLL2 mutations as a cause of Kabuki syndrome. Nat Genet 2010;42:790–3.

[58] Li Y, Bogershausen N, Alanay Y, Kiper POS, Plume N, Keupp K, et al. A mutation screen in patients with Kabuki syndrome. Hum Genet 2011;130:715–24.

[59] Willemsen R, Levenga J, Oostra BA. CGG repeat in the FMR1 gene: size matters. Clin Genet 2011;80:214–25.

[60] Pietrobono R, Tabolacci E, Zalfa F, Zito I, Terracciano A, Moscato U, et al. Molecular dissection of the events leading to inactivation of the FMR1 gene. Hum Mol Genet 2005;14:267–77.

[61] Kumari D, Usdin K. Chromatin remodeling in the noncoding repeat expansion diseases. J Biol Chem 2009;284:7413–7.

[62] Allingham-Hawkins DJ, Babul-Hirji R, Chitayat D, Holden JJA, Yang KT, Lee C, et al. Fragile X premutation is a significant risk factor for premature ovarian failure: The international collaborative POF in fragile X study—preliminary data. J Med Genet 1999;83:322—5.

[63] Ladd PD, Smith LE, Rabaia NA, Moore JM, Georges SA, Hansen RS, et al. An antisense transcript spanning the CGG repeat region of FMR1 is upregulated in premutation carriers but silenced in full mutation individuals. Hum Mol Genet 2007;24:3174—87.

[64] Beilina A, Tassone F, Schwartz PH, Sahota P, Hagerman PJ. Redistribution of transcription start sites within the FMR1 promoter region with expansion of the downstream CGG-repeat element. Hum Mol Genet 2004;13:543—9.

[65] Tabolacci E, Moscato U, Zalfa F, Bagni C, Chiurazzi P, Neri G. Epigenetic analysis reveals a euchromatic configuration in the FMR1 unmethylated full mutations. Eur J Hum Genet 2008;16:1487—98.

[66] Urbach A, BarNur O, Daley GQ, Bevenistry N. Differential modeling of fragile X syndrome by human embryonic stem cells and induced pluripotent stem cells. Cell Stem Cell 2010;6:407—11.

[67] Todd PK, Oh SY, Krans A, Pandey UB, Di Prospero NA, Min K-T, et al. Histone deacetylases suppress CGG repeat-induced neurodegeneration via transcriptional silencing in models of fragile X tremor ataxia syndrome. PLoS Genet 2010;6:12.e1001240.

[68] Zalfa F, Giorgi M, Primerano B, Moro A, D Penta A, et al. The fragile X syndrome protein FMRP associates with BC1 RNA and regulates the translation of specific mRNAs at synapses. Cell 2003;112:317—27.

[69] Bear MF, Huber KM, Warren ST. The mGluR theory of fragile X mental retardation. Trends Neurosci 2004;27:370—7.

[70] Jin P, Zarnescu DC, Ceman S, Nakamoto M, Mowrey J, Jongens TA, et al. Biochemical and genetic interaction between the fragile X mental retardation protein and the microRNA pathway. Nat Neurosci 2004;2:113—7.

[71] Muddashetty RS, Nalavadi VC, Gross C, Yao X, Xing L, Laur O, et al. Reversible inhibition of PSD-95 mRNA translation by mir-125a, FMRP phosphorylation, and mGluR signaling. Mol Cell 2011;42:673—88.

[72] Jacquemont S, Curie A, des Portes V, Torrioli MG, Berry-Kravis E, Hagerman RJ, et al. Epigenetic modification of he FMR1 gene in fragile X syndrome is associated with differential response to the mGluR antagonist AFQ056. Sci Trans Med 2011;3:1—9.

[73] Kishino T, Lalande M, Wagstaff J. UBE3A/E6-AP mutations cause Angelman syndrome. Nat Genet 1997;15:70—3.

[74] Ohta T, Gray TA, Rogan PK, Buiting K, Gabriel JM, Saitoh S, et al. Imprinting-mutation mechanisms in Prader—Willi syndrome. Am J Hum Genet 1999;64:397—413.

[75] Horsthemke B, Wagstaff J. Mechanisms of imprinting of the Prader—Willi/Angelman region. Am J Med Genet 2008;146A:2041—52.

[76] Cook Jr EH, Lindgren V, Leventhal BL, Courchesne R, Lincoln A, Shulman C, et al. Autism or atypical autism in maternally but not paternally derived proximal 15q duplication. Am J Hum Genet 1997;60:928—34.

[77] Ingason A, Kirov G, Giegling I, Hansen T, Isles AR, Jakobsen KD, et al. Maternally derived microduplications at 15q11-q13: implication of imprinted genes in psychotic illness. Am J Psychiatry 2011;168:408—17.

[78] Hogart A, Nagarajan RP, Patzel KA, Yasui DH, Lasalle JM. 15q11 to 15q13 GABAA receptor genes are normally biallelically expressed in brain yet are subject to epigenetic dysregulation in autism-spectrum disorders. Hum Mol Genet 2007;16:691—703.

[79] Jablonka E, Lamb MJ. Evolution in Four Dimensions: Genetic, Epigenetic, Behavioral, and Symbolic Variation in the History of Life. Cambridge: MIT Press; 2005.

[80] Persico AM, Levitt P, Pimenta AF. Polymorphic GGC repeat differentially regulates human reelin gene expression levels. J Neural Transm 2006;113:1373—82.

173

CHAPTER 9

The Epigenetics of Alzheimer's Disease

Fabio Coppedè
University of Pisa, Pisa, Italy

9.1 ALZHEIMER'S DISEASE

Alzheimer's disease (AD) is an age-related neurodegenerative disorder and the most common form of dementia in the elderly. The disease is clinically characterized by a progressive neurodegeneration in selected brain regions, including the temporal and parietal lobes and restricted regions within the frontal cortex and the cingulate gyrus, leading to memory loss accompanied by changes in behavior and personality severe enough to affect daily life. It was reported that worldwide 26.6 million individuals suffered from AD in 2006, and this number will quadruple by 2050 following the global aging of the world's population [1]. Therefore, AD is becoming a serious health concern in both developed and developing countries, particularly because there is actually no cure for the disease.

Affected brain regions in AD are characterized by the occurrence of extracellular amyloid deposits or senile plaques (SP) and by the presence of neurofibrillary tangles (NFT) composed of intraneuronal aggregates of hyperphosphorylated tau protein [2]. The primary component of SP is an approximately 40-residue-long peptide, known as amyloid β (Aβ) peptide, resulting

T. Tollefsbol (Ed): Epigenetics in Human Disease. DOI: 10.1016/B978-0-12-388415-2.00009-3

from the proteolytic processing of its precursor, the amyloid precursor protein (APP). APP can be processed either by α-secretase and γ-secretase (a protein complex composed by presenilins and other proteins) producing non-amyloidogenic peptides, or by β-secretase (β-site APP cleaving enzyme 1, BACE1) and γ-secretase-producing Aβ peptides. Therefore the balance between different secretase activities is very important in the maintenance of the physiological levels of non-amyloidogenic and amyloidogenic fragments. The two major forms of Aβ that are produced by APP processing under normal conditions are 40 and 42 residues in length (Aβ$_{40}$ and Aβ$_{42}$, respectively). Aβ$_{42}$ is the major component of SP. In a normal individual the majority of Aβ produced is of the shorter variety, Aβ$_{40}$; whereas mutations causing familial AD lead to increased Aβ$_{42}$/Aβ$_{40}$ ratio [3,4].

AD is a complex multifactorial disorder. Rare mutations in *APP*, presenilin-1 (*PSEN1*), and presenilin-2 (*PSEN2*) genes cause early-onset (< 65 years) familial forms of the disease accounting for less than 1% of the total AD cases [5]. AD-causative mutations lead to altered APP production and/or processing and the disease is transmitted in families following a Mendelian inheritance pattern [3,4]. Interestingly, 50% or more of early-onset AD cases are not explained by the known *APP*, *PSEN1*, and *PSEN2* mutations, suggesting the existence of yet unknown genetic factors [5]. Most of AD (90−95%) are however sporadic forms diagnosed in people over 65 years of age and referred to as late-onset (LOAD) forms, likely resulting from the interaction among genetic, epigenetic, environmental, and stochastic factors super-imposed on the physiological decline of cognitive functions with age [6]. Hundreds of genes have been investigated in genetic association studies as possible AD susceptibility or modifier genes, and more-recent genome-wide association studies (GWAS) are revealing novel poly-morphisms that could account for increased LOAD risk. The ALZGene database (www.alzgene.org) is a continuously updated database containing meta-analyses of published studies, including GWAS, in order to provide a list of the genetic variants that most likely affect disease risk [7]. Accessed on July 2011, the database contained data on 1395 genetic association studies covering 695 genes and 2973 polymorphisms. In addition, 320 meta-analyses were available for those genetic polymorphisms which had been investigated in at least four independent research studies. According to these meta-analyses the top ten genes that most likely contribute to LOAD risk are the following: (1) apolipoprotein E (*APOE ε4* allele, OR = 3.68), (2) bridging integrator 1 (*BIN1* rs744373, OR = 1.17), (3) clusterin (*CLU* rs11136000, OR = 0.88), (4) ATP-binding cassette sub-family A member 7 (*ABCA7* rs3764650, OR = 1.23), (5) comple-ment receptor 1 (*CR1* rs3818361, OR = 1.17), (6) phosphatidylinositol binding clathrin assembly protein (*PICALM* rs3851179, OR = 0.88), (7) membrane-spanning 4-domains, subfamily A, member 6A (*MS4A6A* rs610932, OR = 0.90), (8) myeloid cell surface antigen CD33 (*CD33* rs3865444, OR = 0.89), (9) membrane-spanning 4-domains, subfamily A, member 4E (*MS4A4E* rs670139, OR = 1.08), and (10) CD2-associated protein (*CD2AP* rs9349407, OR = 1.12) (www.alzgeme.org). Additional genes linked to LOAD risk include those involved in sorting mechanisms, such as the sortilin-related receptor (SORL1, several polymorphisms, OR = 1.10−1.20) and its homolog, the sortilin-related VPS10 domain containing receptor 1 (SORCS1, rs600879, OR = 1.24), both causing APP, BACE1 and γ-secretases to localize in the same compartment that is critical in the regulation of Aβ production (www.alzgene.org). Overall, with the exception of the APOE *ε4* variant which gives an odds ratio of 3.7, all the other variants only confer a modest LOAD risk with ORs ranging from 0.9 to 1.2, suggesting a multifactorial nature of the disease and complex gene−gene and gene−environment interactions yet to be clarified [7].

Alongside with genetic variants, also several environmental agents have been investigated as possible LOAD risk factors, often with conflicting or inconclusive results [6]. Among them metals, pesticides, brain injuries, and the resulting inflammation and oxidative stress, have been suggested to increase LOAD risk; by contrast, Mediterranean diet, the consumption of fruit, vegetables and fish, dietary antioxidants, regular physical activity, and brain stimulation until late in life are considered to be among LOAD protective factors [6]. However, despite active research in the field the etiology of sporadic AD is still uncertain. Interestingly, a recent hypothesis linking environmental exposure to AD risk suggests that several environmental factors could epigenetically modify the expression of AD-related genes, such as *APP*, *PSEN1*, and many others [6]. This chapter describes the current evidence of AD-related epigenetic modifications. Epigenetics has also a role in aging, cognitive functions, and several other age-related diseases than AD [8,9]. All these points are nicely addressed in Chapter 26.

9.2 ONE-CARBON METABOLISM AND DNA METHYLATION IN ALZHEIMER'S DISEASE

Folates are essential nutrients required for one-carbon biosynthetic and epigenetic processes. They are derived entirely from dietary sources, mainly from the consumption of green vegetables, fruits, cereals, and meat. After intestinal absorption, folate metabolism requires reduction and methylation into the liver to form 5-methyltetrahydrofolate (5-MTHF), release into the blood and cellular uptake; then it can be used for the synthesis of DNA and RNA precursors or for the conversion of homocysteine (Hcy) to methionine, which is then used to form S-adenosylmethionine (SAM), the main DNA-methylating agent. Several enzymes and cofactors, such as vitamin B12 and vitamin B6, participate in one-carbon metabolism (Figure 9.1), and common polymorphisms of genes involved in folate uptake and metabolism have been often associated with increased AD risk, though results are still conflicting or inconclusive for most of them [10]. Folic acid is the synthetic form added to foods and found in dietary supplements, it is converted to a natural biological form of the vitamin as it passes through the intestinal wall, with enzymatic reduction and methylation resulting in the circulating form of the vitamin, 5-methylTHF [10]. Given its role in the DNA methylation pathway, the contribution of folate metabolism (also referred to as one-carbon metabolism) to epigenetic modifications of AD-related genes is currently investigated by several research groups (see below).

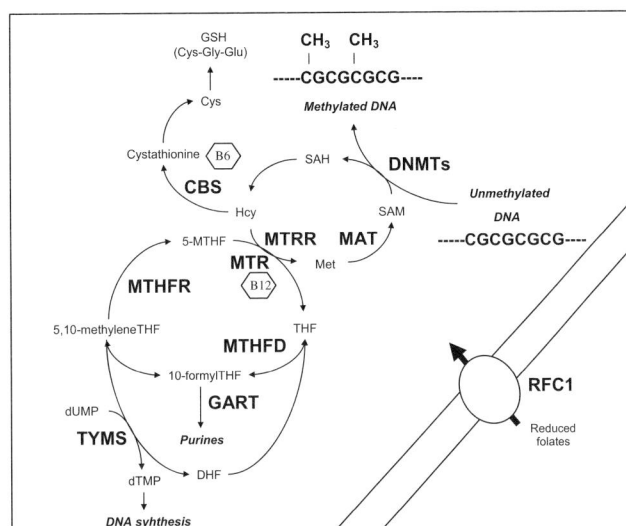

FIGURE 9.1
Overview of the folate metabolic pathway, adapted from Coppedè et al. [86]: *Metabolites:* Cys, cysteine; dTMP, deoxythymidine monophosphate; dUMP, deoxyuridine monophosphate; DHF, dihydrofolate; 10-formyl-THF, 10-formyl-tetrahydrofolate; GSH, glutathione; Hcy, homocysteine; Met, methionine; 5-MTHF, 5- methyltetrahydrofolate; 5,10-MTHF, 5,10-methylenetetrahydrofolate; SAH, S-adenosylhomocysteine; SAM, S-adenosylmethionine; THF, tetrahydrofolate. *Enzymes:* CBS, cystathionine β-synthase; DNMTs, DNA methyltransferases; GART, phosphoribosylglycinamide transformylase; MAT, methionine adenosyltransferase; MTHFD, methylenetetrahydrofolate dehydrogenase; MTHFR, methylenetetrahydrofolate reductase; MTR, methionine synthase; MTRR, methionine synthase reductase; RFC1, reduced folate carrier; TYMS, thymidylate synthase. *Cofactors:* B6, vitamin B6; B12, vitamin B12.

9.2.1 An Overview of One-Carbon Metabolism

As illustrated in Figure 9.1 cellular folates can be used either for DNA methylation processes or for the synthesis of nucleic acid precursors (Figure 9.1). Folates are highly hydrophilic molecules that do not cross biological membranes by diffusion alone, so it is not surprising that sophisticated membrane transport systems have evolved for facilitating their uptake by mammalian cells and tissues, the most ubiquitous and best characterized being the reduced folate carrier (RFC1). Methylenetetrahydrofolate reductase (MTHFR) is the first enzyme in the DNA methylation pathway and reduces 5,10-methylentetrahydrofolate (5,10-MTHF) to 5-methylTHF. Subsequently, methionine synthase (MTR) transfers a methyl group from 5-methylTHF to homocysteine forming methionine and tetrahydrofolate (THF). Methionine is then converted to SAM in a reaction catalyzed by methionine adenosyltransferase (MAT). Most of the SAM generated is used in transmethylation reactions, whereby SAM is converted to S-adenosylhomocysteine (SAH) by transferring the methyl group to diverse biological acceptors, including proteins and DNA. Vitamin B12 (or cobalamin) is a cofactor of MTR, and methionine synthase reductase (MTRR) is required for the maintenance of MTR in its active state. If not converted into methionine, Hcy can be condensed with serine to form cystathionine in a reaction catalyzed by cystathionine β-synthase (CBS), which requires vitamin B6 as a cofactor. Cystathionine can be then utilized to form the antioxidant compound glutathione (GSH). Another important function of folates is in the de novo synthesis of DNA and RNA precursors, required during nucleic acid synthesis and for DNA repair processes. Therefore, depending on cellular demands 5,10-MTHF can be used for the synthesis of SAM or for the synthesis of nucleic acids precursors. Thymidylate synthase (TYMS) converts deoxyuridine monophosphate (dUMP) and 5,10-MTHF to deoxythymine monophospate (dTMP) and dihydrofolate (DHF) in the de novo synthesis of pyrimidines (Figure 9.1). Methylenetetrahydrofolate dehydrogenase (MTHFD1) consists of three activities: 5,10-MTHF dehydrogenase, 5,10-methenyltetrahydrofolate cyclohydrolase, and 10-formyltetrahydrofolate synthetase, respectively, and catalyzes three sequential reactions in the interconversion of THF derivatives. Once generated by the 10-formyltetrahydrofolate synthetase activity, 10-formyltetrahydrofolate can be utilized for the production of purines by the phosphoribosylglycineamide transformylase (GART) enzyme. Overall, the folate metabolic pathway involves several enzymes and is tightly regulated by intracellular levels of metabolites and cofactors [10].

9.2.2 One-Carbon Metabolism in Alzheimer's Disease

The author recently reviewed the literature dealing with one-carbon metabolism in AD, and most of the published studies agree that AD individuals are characterized by decreased folate values, as well as increased plasma homocysteine levels (hyperhomocysteinemia) [10]. The hypothesis that Hcy is a risk factor for AD was initially prompted by the observation that patients with histologically confirmed AD had higher plasma levels of Hcy, than age-matched controls. Although most evidence accumulated so far implicates hyperhomocysteinemia as an AD risk factor, there are also conflicting results [11]. For example, a recent meta-analysis of relevant studies suggests that individuals with AD and vascular dementia have higher plasma Hcy levels than controls, but a causal relationship between hyperhomocysteinemia and the risk of developing dementia was not observed [12]. Therefore, the role of hyperhomocysteinemia as a risk factor for dementia is still controversial.

Alongside data concerning plasma Hcy values, there is also some "although controversial" indication that Hcy levels are increased in the cerebrospinal fluid (CSF) of AD patients with respect to controls [13–15]. A recent study including 70 AD patients, 33 patients with another type of dementia, and 30 age-matched control subjects, revealed that folates in CSF were significantly different between groups, but not Hcy. In addition, the average folate in CSF was lower in AD patients compared with controls, and in AD CSF there was a significant inverse correlation between Hcy and folate, supporting the hypothesis of a possible role of folate in the onset or worsening of AD [16].

DNA methylation is closely dependent on the DNA methylation potential, which is referred to as the ratio between SAM and SAH levels. There is an indication of reduced DNA methylation potential in AD brains [17], and it has been suggested that increased SAH concentrations in the brains of AD patients might inhibit DNA methyltransferases (DNMTs) [18]. A recent study showed that Hcy plasma levels in the highest quartile were more frequent in AD patients than in controls. In addition, AD patients had significantly lower CSF levels of the methyl group donor SAM. Accordingly, the SAM/SAH ratio, i.e. the methylation capacity, was significantly lower in the CSF of the AD patients. Further, explorative analysis of all subjects showed that CSF SAM levels were lower in carriers of the *APOE ε4* allele compared with non-carriers [19]. Studies in AD animal models also revealed altered SAM levels in the brain. However, the analysis of SAM and SAH levels in the aging brain of APP/PS1 Alzheimer mice during aging, revealed that SAM levels decreased in the brains of wild-type mice, whereas SAH levels diminished in both wild-type and APP/PS1 mice. In contrast to wild-type mice, SAM levels in APP/PS1 mice are not decreased during aging, probably related to less demand due to neurodegeneration [20].

DNMTs are the key enzymes for DNA methylation and catalyze the transfer of a methyl group from SAM to cytosine, thus forming 5-methyl-cytosine and SAH (Figure 9.2). There are multiple families of DNMTs in mammals. DNMT1 is primarily involved in the maintenance of DNA

179

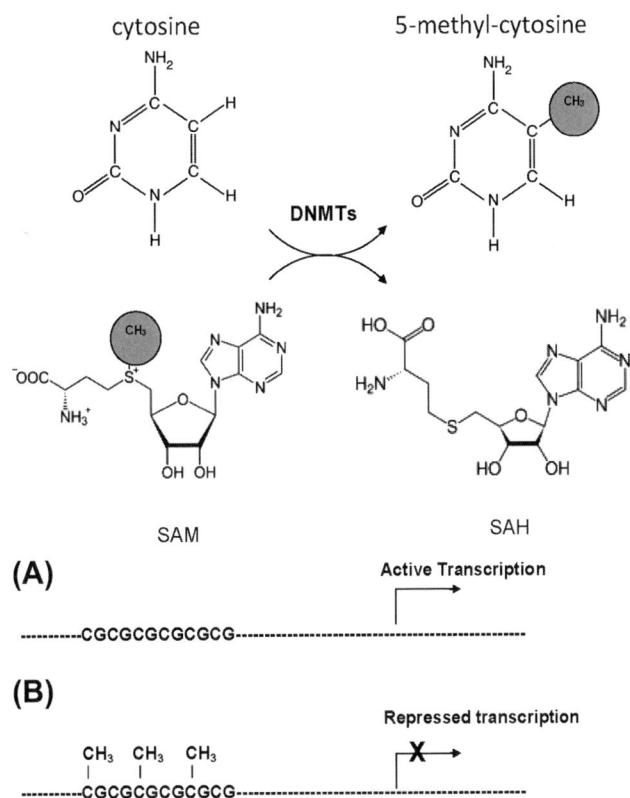

FIGURE 9.2
The reaction catalyzed by DNA methyltransferases (DNMTs). DNMTs are the key enzymes for DNA methylation and catalyze the transfer of a methyl group from SAM to cytosine, thus forming 5-methyl-cytosine and SAH. Methylation of CpG sequences might induce chromatin conformational modifications and inhibit the access of the transcriptional machinery to gene promoter regions, thus altering gene expression levels. Therefore, promoter rmethylation of CpG islands is commonly associated with gene silencing and promoter demethylation with gene expression, though several exceptions to this rule are known.

methylation patterns during development and cell division, whereas DNMT3a and DNMT3b are the de novo methyltransferases and establish DNA methylation patterns during early development. DNMT3L induces de novo DNA methylation by recruitment or activation of DNMT3a, whilst DNMT2 is primarily involved in the methylation of transfer RNA molecules [21,22].

Following the observation of impaired one-carbon metabolism and methylation potential in AD, some investigators raised the possibility of a link between DNA methylation and AD risk, and several studies are now available (Table 9.1), mainly from animal disease models and neuronal cell cultures, suggesting that dietary factors, but also other environmental factors, can lead to epigenetic modifications and deregulated expression of key AD genes [23]. Little is, however, still known in humans.

9.2.3 Studies in Cell Cultures and Animal Models

One of the most exciting hypothesis linking one-carbon metabolism to AD risk suggests that impaired folate/Hcy metabolism and subsequent reduction of SAM levels might result in epigenetic modifications of the promoters of AD-related genes leading to increased Aβ peptide production [24,25]. DNA methylation represents one of the most important epigenetic

TABLE 9.1 Evidence of DNA Methylation in AD and AD Model Systems

Experimental Model	Observation	References
Human neuroblastoma SK-N-SH or SK-N-BE cells	Folate and vitamin B12 deprivation induced epigenetic modifications in the promoter of *PSEN1*, resulting in up-regulation of gene expression and increased production of presenilin 1, BACE1, and APP	[25]
	B vitamin deficiency induced decreased de novo DNA methylation activity	[32]
	SAM administration resulted in down-regulation of *PSEN1* expression	[24]
	SAM administration was able to modulate the expression of seven out of 588 genes of the nervous system	[35]
BV-2 mouse microglial cells	SAH administration increased the production of Aβ peptide likely through induction of hypomethylation of *APP* and *PSEN1* gene promoters	[31]
Murine cerebral endothelial cells	Aβ reduces global DNA demethylation whilst increasing DNA methylation of the gene encoding neprilysin	[36]
Rodents	B vitamin deprivation altered the methylation potential and induced hypomethylation of the promoter of *PSEN1*, resulting in increased gene expression	[27,28]
	B vitamin deficiency induced decreased *de novo* DNA methylation activity	[32]
Rodents and monkeys	Early life exposure to Pb resulted in inhibition of DNA-methyltransferase, hypomethylation of the promoter of *APP* and delayed up-regulation of gene expression later in life	[37,38]
Post-mortem human brains	The analysis of DNA methylation in selected regions of *MAPT*, *APP*, and *PSEN1* genes revealed no difference between control samples and AD samples	[42]
	Only two (*SORBS3* and *S100A2*) out of 50 analyzed loci showed AD-related methylation changes, which appeared to reflect an acceleration of age-related changes	[43]
	AD brains showed an increased epigenetic drift with unusual methylation patterns, and inter-individual variations concerning *PSEN1*, *APOE*, *MTHFR*, and *DNMT1* genes	[44]
Blood DNA	AD brains showed a marked reduction of DNA methylation	[40,41]
	Higher levels of LINE-1 methylation in AD subjects with respect to controls	[45]

processes, alongside histone tail modifications and mechanisms involving small RNA molecules. Methylation of CpG sequences might induce chromatin conformational modifications and inhibit the access of the transcriptional machinery to gene promoter regions, thus altering gene expression levels. Therefore, promoter hypermethylation is commonly associated with gene silencing and promoter demethylation with gene expression, though several exceptions to this rule have been reported [26].

APP and *PSEN*1 promoter methylation was analyzed in human neuroblastoma SK-N-SH or SK-N-BE cell lines under conditions of folate and vitamin B12 deprivation from the media. The study revealed demethylation of a CpG site in the promoter of *PSEN1*, with a subsequent increased production of presenilin1, BACE1, and APP proteins [25]. By contrast, SAM administration to human neuroblastoma SK-N-SH cells resulted in down-regulation of *PSEN1* gene expression and reduced Aβ peptide production [24]. Similarly, a combination of dietary folate, vitamin B12, and vitamin B6 deprivation (B vitamin deprivation) resulted in hyper-homocysteinemia, increased brain SAH levels, depletion of brain SAM, hypomethylation of specific CpG moieties in the 5′-flanking region of *PSEN1* and enhancement of presenilin 1 and BACE1 expression and Aβ deposition in mice [27,28]. Dietary deficiency of folate and vitamin E, in condition of oxidative stress (the diet contained iron as a pro-oxidant), increased presenilin 1 expression, BACE1 activity, and Aβ levels in normal adult mice. These increases were particularly evident in mice lacking apolipoprotein E. On the contrary, dietary SAM supplementation attenuated these deleterious consequences [29]. A similar experiment was performed in mice expressing the human *APOE* gene. Mice expressing human apolipoprotein ε4, apolipoprotein ε3, or apolipoprotein ε2, were subjected to a diet lacking folate and vitamin E, and containing iron as a pro-oxidant. The study revealed that presenilin 1 and gamma-secretase were over-expressed in ε3 mice to the same extent as in ε4 mice even under a complete diet, and were not alleviated by SAM supplementation. Aβ increased only in ε4 mice maintained under the complete diet, and was alleviated by SAM supplementation [30]. Lin and co-workers showed that SAH increases the production of Aβ in BV-2 mouse microglial cells possibly by an increased expression of APP and induction of hypomethylation of *APP* and *PSEN1* gene promoters [31].

A recent study demonstrated that B vitamin deficiency induces DNMT3a and DNMT3b protein down-regulation in both neuroblastoma cell cultures and mice brain regions. Moreover, the de novo DNA methylation activity was modulated in both neuroblastoma cells and mice, and resulted decreased under B vitamin deficiency conditions and increased by SAM supplementation, whilst the activity of a putative DNA demethylase (MBD2) showed an opposite tendency [32]. It is however worth mentioning that little is still known concerning DNA demethylation in the adult mammalian brain. Guo and co-workers [33] have recently demonstrated that the 5-methylcytosine hydroxylase TET1, by converting 5-methylcytosine to 5-hydroxymethylcytosines (5hmCs), promotes DNA demethylation in mammalian cells through a process that requires the DNA base excision repair pathway. Demethylation of 5hmCs is promoted by the AID (activation-induced deaminase)/APOBEC (apolipoprotein B mRNA-editing enzyme complex) family of cytidine deaminases. Furthermore, Tet1 and Apobec1 are involved in neuronal activity-induced, region-specific, active DNA demethylation and subsequent gene expression in the dentate gyrus of the adult mouse brain in vivo [33]. Nutritional B vitamin restriction was also used to study the variation of protein expression profile in mice brain regions. A group of proteins mainly involved in neuronal plasticity and mitochondrial functions was identified as modulated by one-carbon metabolism [34]. The same group also analyzed 588 genes of the central nervous system in SK-N-BE neuroblastoma cells, observing that only seven genes were modulated by SAM treatment (and therefore by DNA methylation); three were up-regulated and four down-regulated [35]. Others observed that the Aβ peptide induces global DNA hypo-methylation in murine cerebral endothelial cells, but specifically leads to hypermethylation of the gene encoding neprilysin (*NEP*), one of the enzymes responsible for Aβ degradation, thus suppressing *NEP* expression in mRNA and protein levels [36].

Alongside with the dietary factors (dietary B vitamins), metal exposure has also been suggested to be able to epigenetically modulate the expression of AD-related genes. Particularly, the analysis of rodents and monkeys exposed to lead (Pb) during embryogenesis and/or early postnatal period revealed that early life exposure to this metal could result in epigenetic modifications of AD-related genes, such as *APP*, and subsequent deregulated expression later in life [37,38]. Particularly, it was observed that developmental exposure of rats to lead resulted in a delayed overexpression (20 months later) of the amyloid precursor protein and its amyloidogenic Aβ product. Similarly, aged monkeys exposed to lead as infants also responded in the same way [37,38]. These data indicate that environmental influences occurring during brain development predetermine the expression and regulation of APP later in life, potentially influencing the course of amyloidogenesis, and suggesting that early life neurogenesis is likely to be a critical step for environmentally induced epigenetic modifications [39].

9.2.4 Studies in Humans

Despite evince of possible epigenetic modifications of AD-related genes obtained in neuronal cell cultures as well as in rodents and primates, epigenetic studies in AD patients are still scarce, likely depending on the difficulty to obtain human post-mortem brain specimens. Mastroeni and co-workers examined global DNA methylation in post-mortem brain regions of mono-zygotic twins discordant for AD, observing significantly reduced levels of DNA methylation in the temporal neocortex of the AD twin [40]. The same group analyzed brain tissues from 20 AD patients and 20 age-matched controls, and used immunocytochemistry for two markers of DNA methylation (5-methylcytosine and 5-methylcytidine) and eight methylation maintenance factors in the entorhinal cortex layer II, again observing significant decrements in AD cases [41]. However, Barrachina and Ferrer [42] analyzed DNA methylation in selected regions of *MAPT* (the gene encoding for microtubule-associated tau protein), *APP*, and *PSEN1* in post-mortem frontal cortex and hippocampus of healthy controls and various stages of AD progression, observing no differences in the percentage of CpG methylation in any region analyzed [42]. Siegmund and co-workers examined the DNA methylation status at 50 loci, encompassing primarily 5′ CpG islands of genes related to CNS growth and development, in the temporal neocortex of 125 subjects ranging in age from 17 weeks of gestation to 104 years old, and including a cohort of AD patients [43]. Only two loci showed AD-related changes (*SORBS3* and *S100A2*, encoding a cell adhesion molecule and a calcium-binding protein, respectively), which appeared to reflect an acceleration of age-related changes [43]. Wang and collaborators [44] analyzed 12 potential AD susceptibility loci in peripheral lymphocytes and post-mortem brain samples of LOAD patients and matched controls. From the 12 analyzed CpG-rich regions, only one was hypermethylated, and was the only region analyzed outside of a promoter region (DNA methylation can also occur in CpG sites outside the promoter region "gene-body methylation", and is usually associated with transcriptional activation). Overall, the authors failed to find significant differences in DNA methylation between LOAD cases and controls [44]. However, four of the studied genes (*PSEN1*, *APOE*, *MTHFR*, and *DNMT1*) showed a significant interindividual epigenetic variability, and LOAD brain samples showed an "age-specific epigenetic drift" associated with unusual methylation patterns [44]. More recently, Bollati and co-workers analyzed the methylation pattern of repetitive elements (i.e. Alu, LINE-1, and α-satellite DNA) in the DNA obtained from blood samples of 43 AD patients and 38 matched controls, observing higher levels of LINE-1 methylation in AD subjects with respect to controls [45].

9.2.5 Linking the Methylation Potential to Tau Phosphorylation in Alzheimer's Disease

The results of prospective cohort studies suggest that increased serum Hcy levels might predispose to AD [46–48], and there is also an indication that higher folate intake is related to lower AD risk in the elderly [48,49]. Folate deficiency fosters a decline in SAM levels, thus decreasing DNA methylation during aging and AD. Moreover, folate deficiency and the resultant SAM depletion lead to increased levels of Hcy [46]. Hcy is a critical branch point metabolite that can influence cellular levels of SAM and SAH, which in turn regulate the activity of methyltransferases during DNA methylation and post-translational modification of proteins [50]. There is an indication that elevated Hcy causes tau hyperphosphorylation, NFT formation, and SP formation via inhibition of methyltransferases and reduced methylation of protein phosphatase 2A (PP2A) [51,52]. Tau phosphorylation is regulated by the equilibrium between protein kinases and phosphatases. PP2A is the major protein phosphatase that dephosphorylates tau. Methylation of the PP2A catalytic subunit can potently activate PP2A. Therefore, demethylation of PP2A resulting from inhibition of methyltransferases could result in increased production of hyperphosphorylated tau protein [51,52]. Moreover, B vitamin deficiency was shown to reduce PP2A activity and increase that of glycogen synthase kinase 3beta (GSK3beta) one of the most important protein kinase involved in tau phosphorylation [53]. In addition, a B vitamin-deficient diet increased phosphorylated tau levels in *APOE ε4* but not in *APOE ε3* mice, and tau phosphorylation was prevented by SAM supplementation [54]. Overall, there is an indication that deficiency in one-carbon metabolism, reduced SAM levels, and SAH-mediated inhibition of methyltransferase proteins could also contribute to NFT formation during AD pathogenesis.

9.3 HISTONE TAIL MODIFICATIONS AND ALZHEIMER'S DISEASE

Chromatin can exist in a condensate inactive state (heterochromatin) or in a decondensed and transcriptionally active state (euchromatin). Conformational changes in histone proteins or modifications of the way in which DNA wraps around the histone octamer in nucleosomes may either alter or facilitate the access of the transcriptional machinery to the promoter region of some genes, leading to gene silencing or activation, respectively. Histone tail modifications include acetylation, methylation, phosphorylation, ubiquitylation, sumoylation, and other post-translational modifications (Figure 9.3). Histone tail acetylation represents one of the most studied modifications and is associated with chromatin relaxation and transcriptional activation, while deacethylation is related to a more condensed chromatin state and transcriptional repression [55]. Acetylation occurs at lysine residues on the amino-terminal tails of the histones, thereby neutralizing the positive charge of the histone tails and decreasing their affinity for DNA. As a consequence, histone acetylation alters nucleosomal conformation, which can increase the accessibility of transcriptional regulatory proteins to chromatin templates [55]. Histone acetyltransferases (HATs) catalyze the acetylation of lysine residues in histone tails, whereas histone deacetylation is mediated by histone deacetylases (HDACs). Another frequently studied modification of histone tails is methylation on either lysine or arginine residues. Methylation of histone tails can be associated with either condensation or relaxation of the chromatin structure, since several sites for methylation are present on each tail thus allowing several combinations [56]. A number of lysine methylation sites of H3 and H4 have been well characterized. Methylation at H3K4, H3K36, and H3K79 has been linked to actively transcribed genes, whereas di- and tri-methylated H3K9, methylated H3K27 and H4K20 are considered repressive marks [57]. Protein lysine methyltransferases (PKMTs) and protein arginine methyltransferases (PRMTs) are two of the writers responsible for adding the methyl marks to histones. Moreover, these writers only methylate the residue to a specific methylation level, which is known as product specificity. For example, lysine residues can be mono-, di-, or trimethylated, while arginines can be mono- or dimethylated in either a symmetric (one methyl group on each of the two N terminal atoms) or asymmetric (both methyl groups on the same N terminal atom) manner. Both PKMTs and PRMTs require SAM as the methyl donor for histone tail methylation reactions [58]. G9a and GLP are the primary enzymes for mono- and dimethylation at Lys 9 of histone H3 (H3K9me1 and H3K9me2), and exist predominantly as a G9a−GLP heteromeric complex that appears to be a functional H3K9 methyltransferase in vivo. It was shown that in mammals G9a/GLP suppresses transcription by independently inducing both H3K9 and DNA methylation [59].

183

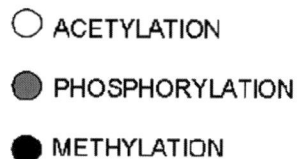

FIGURE 9.3
Histone tail modifications. The figure shows some examples of histone tail modifications. Acetylation of lysine residues is associated with chromatin relaxation and transcriptional activation, while deacethylation is related to a more condensed chromatin state and transcriptional repression. DNA methylation can be associated with either transcriptional repression (for example the H3K9 methylation, shown in the figure) or with transcriptional activation (see the text, for further details on the histone code).

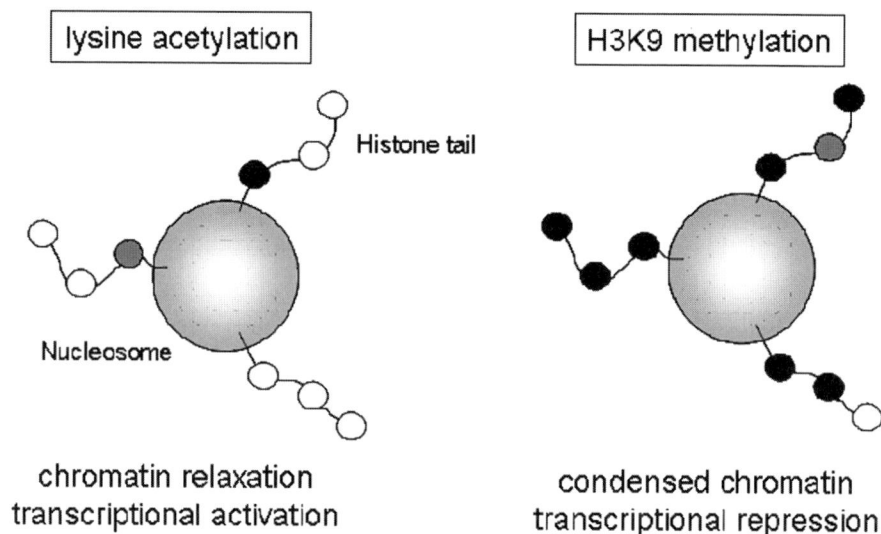

9.3.1 Histone Modifications in Alzheimer's Disease Brains

Several authors reported histone modifications in AD (Table 9.2). For example Ogawa et al. [60] observed increased phosphorylation of histone 3 (H3) in AD hippocampal neurons. An increased phosphorylation of histone H2AX was observed in hippocampal astrocytes of AD individuals [61]. Concerning HDACs, elevated levels of HDAC6 have been observed in post-mortem brain regions of AD subjects, and it was proposed that HDAC6 is a tau-interacting protein and a potential modulator of tau phosphorylation and accumulation [62].

9.3.2 HDAC Inhibitors and Memory Function in AD Animal Models

Several investigators have manipulated histone acetylation with HDAC inhibitors (HDACi) in AD animal models, often observing prevention of cognitive deficits and memory recovery. For example, Francis and co-workers have undertaken preclinical studies in the APP/PS1 mouse model of AD to determine whether there are differences in histone acetylation levels during associative memory formation. After fear-conditioning training, levels of hippocampal acetylated histone 4 (H4) in APP/PS1 mice were about 50% lower than in wild-type littermates [63]. Interestingly, an acute treatment with the HDACi trichostatin A prior to training rescued both acetylated H4 levels and contextual freezing performance to wild-type values [63]. It was also observed that inhibitors of class 1 histone deacetylases (sodium valproate, sodium butyrate, or vorinostat) reverse contextual memory deficits in a mouse model of AD that showed pronounced contextual memory impairments beginning at 6 months of age [64]. Clusterin is a secreted molecular chaperone, also called apolipoprotein J, and a significant susceptibility gene for LOAD (www.alzgene.org). Clusterin shares several properties with

TABLE 9.2 Evidence of Histone Modifications in AD and AD Model Systems

Experimental Model	Observation	Refs.
AD hippocampal neurons	Increased phosphorylation of histone H3	[60]
AD hippocampal astrocytes	Increased phosphorylation of histone H2AX	[61]
Post-mortem human brains	Increased levels of HDAC6, primarily involved in cytoplasmic acetylation, and likely involved in tau phosphorylation and accumulation	[62]
AD animal models	HDAC inhibitors improved cognition and memory functions	[63–67]

apolipoprotein E, since they bind to Aβ peptides and are present in neuritic plaques, enhance the clearance of Aβ peptides in brain, and are included in lipid particles and thus regulate cholesterol traffic. It was shown that HDACi, such as valproic acid and vorinostat, stimulated the expression of clusterin in human astrocytes, and might therefore be able to prevent Aβ aggregation in AD [65]. A recent paper by Govindarajan et al. showed that severe amyloid pathology correlates with a pronounced deregulation of histone acetylation in the forebrain of APP/PS1-21 mice, and that a prolonged treatment with the pan-HDAC inhibitor sodium butyrate improved associative memory in APP/PS1-21 mice even when administered at a very advanced stage of pathology. The recovery of memory function correlated with elevated hippocampal histone acetylation and increased expression of genes implicated in associative learning [66]. Others observed that systemic administration of the HDAC inhibitor 4-phenylbutyrate (PBA) reinstated fear learning in the Tg2576 mouse model of AD. Memory reinstatement by PBA was observed independently of the disease stage: both in young mice at the onset of the first symptoms, but also in aged mice, when amyloid plaque deposition and major synaptic loss has occurred [67]. In this section the author has discussed the most recent findings concerning HDACi and memory function in AD animal models. Several additional examples can be found in the literature, overall suggesting that the administration of HDACi was able to improve cognition in various animal models of AD.

9.4 RNA-MEDIATED MECHANISMS AND ALZHEIMER'S DISEASE

MicroRNAs (miRNAs) are a group of small non-coding RNAs that bind to the 3′ untranslated region (3′-UTR) of target mRNAs and mediate their post-transcriptional regulation leading to either degradation or translational inhibition (Figure 9.4), depending on the degree of sequence complementarity. MiRNA-mediated mechanisms are therefore commonly considered among epigenetic regulators, and J. Satoh (68) has recently reviewed all the studies performed in recent years and reporting an aberrant expression of miRNAs in AD brains (Table 9.3). For most of them the target mRNA has not yet been clarified, however miR-107, miR29a/b-1, miR-15a, miR-9, and miR-19b have as a target BACE1 mRNA, and were found to be down-regulated in the temporal cortex of AD patients, likely resulting in increased BACE1 expression and production of Aβ peptides [69,70]. Similarly miR-15a, miR-101, let-7i, and miR-106b, target the APP mRNA and were down-regulated in the anterior temporal cortex of AD patients [70,71]. Additionally, miR-132 was down-regulated in cerebellum, hippocampus, and medial frontal gyrus of AD patients [72]. MiR132 targets the ARHGAP32 mRNA encoding for the rho GTPase activating protein 32, a member of a family of proteins that have been implicated in regulating multiple processes in the morphological development of neurons, including axonal growth and guidance, dendritic elaboration, and formation of synapses [73]. Non-coding ribonucleic acids (ncRNAs) are transcribed from introns and intergenic regions. They are also implicated in epigenetic regulation and might be involved in site-specific recruitment of chromatin-modifying complexes [74]. Faghihi et al. [75] identified a conserved non-coding antisense transcript for BACE1 (BACE1-AS) that regulates BACE1 mRNA and

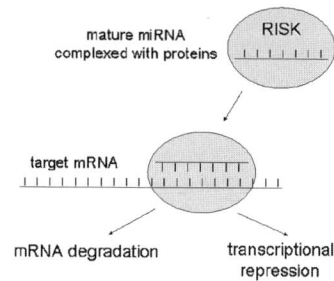

FIGURE 9.4

MicroRNAs (miRNAs) mediated gene regulation. miRNAs are a group of small non-coding RNAs that bind to the 3′ untranslated region (3′-UTR) of target mRNAs and mediate their post-transcriptional regulation. A mature single-strand miRNA complexed with proteins forms the RISC—miRNA complex that binds to the target mRNA leading to either degradation or translational inhibition, depending on the degree of sequence complementarity.

TABLE 9.3 Some Examples of Altered miRNAs in AD Brains

Experimental Model	Observation	Refs.
Post-mortem AD brains	Down-regulation of miR-107, miR29a/b-1, miR-15a, miR-9, and miR-19b that target BACE1 mRNA	[69,70]
	Down-regulation of miR-15a, miR-101, let-7i, and miR-106b that target APP mRNA	[70,71]
	Down-regulation of miR-132 that targets ARHGAP32 mRNA	[72]

protein expression in vitro and in vivo. BACE1-AS concentrations were elevated in subjects with AD and in APP transgenic mice [75]. Others observed that a subset of mRNA-like sense-antisense transcript pairs are co-expressed near synapses of the adult mouse forebrain [76]. Several of these pairs involve mRNAs that have been implicated in synaptic functions and in AD pathways, including BACE1, reticulon 3 (a protein that binds BACE1 and inhibits APP cleavage); APP binding protein 2 (which binds the cytoplasmic tail of APP); rab6 (a protein that is elevated in AD cortex); sirtuin-3 (a member of a class of proteins thought to be neuroprotective in AD); and integrin-linked kinase (which phosphorylates GSK3beta to inhibit its activity) [76].

9.5 DISCUSSION AND CONCLUSIONS

Epigenetic mechanisms have been linked to aging, age-related diseases, and memory formation [8,9,77,78]. Overall, there is consensus indicating that DNA methylation is dynamically regulated in the adult central nervous system (CNS) in response to experience, and that this cellular mechanism is a crucial step in memory formation, that DNMT activity is necessary for memory, and that DNA methylation may work in concert with histone modifications in memory formation [78]. The analysis of Dnmt3a and Dnmt3b proteins during the development of the mouse CNS revealed that whereas Dnmt3b is important for the early phase of neurogenesis, Dnmt3a likely plays a dual role in regulating neurogenesis prenatally and CNS maturation and function postnatally. Particularly, Dnmt3b is specifically expressed in progenitor cells during neurogenesis, suggesting an important role in the initial steps of progenitor cell differentiation. Dnmt3a is expressed in post-mitotic young neurons following the Dnmt3b expression and may be required for the establishment of tissue-specific methylation patterns of the genome [79,80]. Dnmt1 and Dnmt3a are expressed in post-mitotic neurons and required for synaptic plasticity, learning and memory through their overlapping roles in maintaining DNA methylation and modulating neuronal gene expression in adult CNS rodent neurons [81]. The study by Siegmund et al. [43] clearly showed that DNMT3a is expressed in human neurons across all ages, and that DNA methylation is dynamically regulated in the human cerebral cortex throughout the lifespan, and involves differentiated neurons [43]. The studies performed in mice and monkeys exposed to Pb as infants have nicely demonstrated that early life might be particularly vulnerable to epigenetic modifications that could predispose to AD pathogenesis in adulthood [37,38]. The same group has recently shown that Dnmt1 and Dnmt3a proteins, methyl-CpG binding proteins, and proteins involved in histone acetylation and methylation were decreased in the brain of 23-year-old monkeys exposed to Pb as infants, with respect to non-exposed monkeys [82]. Even if DNA demethylation processes are less clear than methylation ones, base excision repair is likely to be involved [32,78]. The studies by Fuso and colleagues suggest that dietary manipulation in rodents is able to methylate/demethylate the promoter of the *PSEN1* gene, with subsequent regulation of gene expression [27,28], and that the same dietary manipulation can modulate the expression of de novo DNMTs [32]. Similarly, all the following studies [63−67] indicate that HDAC inhibitors improved cognition and memory functions in AD animal models. Although promising, the limit of all these papers is that most of the experiments have been performed in disease animal models, and little is still known in humans. Severe alterations of the one-carbon metabolic pathway in humans are associated with disease, for example *DNMT3B* mutations cause a rare chromosome breakage disease characterized by aberrant DNA methylation and called the immunodeficiency, centromeric region instability, and facial anomalies syndrome (ICF syndrome) [83]. Complex interactions between polymorphisms in one-carbon metabolic genes and plasma Hcy or serum folate and vitamin B12 levels are known in LOAD subjects [84], however we still do not know to what extent dietary "one-carbon nutrients" and their metabolism can modulate and/or reverse AD-related epigenetic marks in the human brain. Therefore, some caution should be taken prior to recommend "one-carbon nutrients" or epigenetic drugs, such as HDACi, as preventative strategies for age-related neurodegeneration. Indeed, many issues still need to be clarified before compounds with epigenetic properties could be effectively used to treat patients with a cognitive decline, the most important being the possible short- and long- term side effects. If a diet rich in antioxidants and in methyl donor compounds, such as fruit and vegetables, is likely to protect against the risk of late-onset degenerative diseases, a manipulation of our genome by means of a widespread usage of HDACi or other epigenetic compounds could have several side-effects in different organs and tissues of the body, such as, for example, cancer promotion [10,56,85]. Additional studies are therefore required to clarify the interplay between dietary methyl-donor compounds, polymorphisms of metabolic genes and epigenetic modifications of key AD genes, as well as the interconnections between DNA methylation/demethylation reactions, histone tail modifications and RNA-mediated mechanisms in the regulation of AD-related genes [10]. Only a better understanding of the networks and the interplay of these three epigenetic mechanisms in AD pathophysiology could help the design of preventative strategies to delay disease onset and/or progression.

References

[1] Brookmeyer R, Johnson E, Ziegler-Graham K, Arrighi HM. Forecasting the global burden of Alzheimer's disease. Alzheimer's Dement 2007;307:186−91.

[2] Wenk GL. Neuropathologic changes in Alzheimer's disease. J Clin Psychiatry 2003;64:7−10.

[3] Findeis MA. The role of amyloid beta peptide 42 in Alzheimer's disease. Pharmacol Ther 2007;116:266−86.

[4] Bentahir M, Nyabi O, Verhamme J, Tolia A, Horré K, Wiltfang J, et al. Presenilin clinical mutations can affect gamma secretase activity by different mechanisms. J Neurochem 2006;96:732−42.

[5] Ertekin-Taner N. Genetics of Alzheimer's disease: a centennial review. Neurol Clin 2007;25:611−7.

[6] Migliore L, Coppedè F. Genetics, environmental factors and the emerging role of epigenetics in neurodegenerative diseases. Mutat Res 2009;667:82−97.

[7] Bertram L, McQueen MB, Mullin K, Blacker D, Tanzi RE. Systematic meta-analyses of Alzheimer disease genetic association studies: the AlzGene database. Nat Genet 2007;39:17—23.

[8] Rodríguez-Rodero S, Fernández-Morera JL, Fernandez AF, Menéndez-Torre E, Fraga MF. Epigenetic regulation of aging. Discov Med 2010;10:225—33.

[9] Reichenberg A, Mill J, MacCabe JH. Epigenetics, genomic mutations and cognitive function. Cogn Neuro-psychiatry 2009;14:377—90.

[10] Coppedè F. One-carbon metabolism and Alzheimer's disease: focus on epigenetics. Curr Genomics 2010;11:246—60.

[11] Zhuo JM, Wang H, Praticò D. Is hyperhomocysteinemia an Alzheimer's disease (AD) risk factor, an AD marker, or neither? Trends Pharmacol Sci 2011. In press.

[12] Ho RC, Cheung MW, Fu E, Win HH, Zaw MH, Ng A, et al. Is high homocysteine level a risk factor for cognitive decline in elderly? A systematic review, meta-analysis, and meta-regression. Am J Geriatr Psychiatry 2011;19:607—17.

[13] Selley ML, Close DR, Stern SE. The effect of increased concentrations of homocysteine on the concentration of (E)-4-hydroxy-2-nonenal in the plasma and cerebrospinal fluid of patients with Alzheimer's disease. Neuro-biol Aging 2002;23:383—8.

[14] Hasegawa T, Ukai W, Jo DG, Xu X, Mattson MP, Nakagawa M, et al. Homocysteic acid induces intraneuronal accumulation of neurotoxic Abeta42: implications for the pathogenesis of Alzheimer's disease. J Neurosci Res 2005;80:869—76.

[15] Isobe C, Murata T, Sato C, Terayama Y. Increase of total homocysteine concentration in cerebrospinal fluid in patients with Alzheimer's disease and Parkinson's disease. Life Sci 2005;77:1836—43.

[16] Smach MA, Jacob N, Golmard JL, Charfeddine B, Lammouchi T, Ben Othman L, et al. Folate and homo-cysteine in the cerebrospinal fluid of patients with Alzheimer's disease or dementia: a case control study. Eur Neurol 2011;65:270—8.

[17] Morrison LD, Smith DD, Kish SJ. Brain S-adenosylmethionine levels are severely decreased in Alzheimer's disease. J Neurochem 1996;67:1328—31.

[18] Kennedy BP, Bottiglieri T, Arning E, Ziegler MG, Hansen LA, Masliah E. Elevated S-adenosyhomocysteine in Alzheimer brain: influence on methyltransferases and cognitive function. J Neural Transm 2004;111:547—67.

[19] Linnebank M, Popp J, Smulders Y, Smith D, Semmler A, Farkas M, et al. S-adenosylmethionine is decreased in the cerebrospinal fluid of patients with Alzheimer's disease. Neurodegener Dis 2010;7:373—8.

[20] Hooijmans CR, Blom HJ, Oppenraaij-Emmerzaal D, Ritskes-Hoitinga M, Kiliaan AJ. The analysis of S-adenosylmethionine and S-adenosylhomocysteine levels in the aging brain of APP/PS1 Alzheimer mice. Neurol Sci 2009;30:439—45.

[21] Goll MG, Bestor TH. Eukaryotic cytosine methyltransferases. Annu Rev Biochem 2005;74:481—514.

[22] Goll MG, Kirpekar F, Maggert KA, Yoder JA, Hsieh CL, Zhang X, et al. Methylation of tRNAAsp by the DNA methyltransferase homolog Dnmt2. Science 2006;311:395—8.

[23] Mastroeni D, Grover A, Delvaux E, Whiteside C, Coleman PD, Rogers J. Epigenetic mechanisms in Alzheimer's disease. Neurobiol Aging 2011. In Press.

[24] Scarpa S, Fuso A, D'Anselmi F, Cavallaro RA. Presenilin 1 gene silencing by S-adenosylmethionine: a treatment for Alzheimer disease? FEBS Lett 2003;541:145—8.

[25] Fuso A, Seminara L, Cavallaro RA, D'Anselmi F, Scarpa S. S-adenosylmethionine/homocysteine cycle alter-ations modify DNA methylation status with consequent deregulation of PS1 and BACE and beta-amyloid production. Mol Cell Neurosci 2005;28:195—204.

[26] Gius D, Cui H, Bradbury CM, Smart DK, Zhao S, Young L, et al. Distinct effects on gene expression of chemical and genetic manipulation of the cancer epigenome revealed by a multimodality approach. Cancer Cell 2004;6:361—71.

[27] Fuso A, Nicolia V, Cavallaro RA, Ricceri L, D'Anselmi F, Coluccia P, et al. B-vitamin deprivation induces hyperhomocysteinemia and brain S-adenosylhomocysteine, depletes brain S-adenosylmethionine, and enhances PS1 and BACE expression and amyloid-beta deposition in mice. Mol Cell Neurosci 2008;37:731—46.

[28] Fuso A, Nicolia V, Pasqualato A, Fiorenza MT, Cavallaro RA, Scarpa S. Changes in Presenilin 1 gene methylation pattern in diet-induced B vitamin deficiency. Neurobiol Aging 2011;32:187—99.

[29] Chan A, Shea TB. Folate deprivation increases presenilin expression, gamma-secretase activity, and Abeta levels in murine brain: potentiation by ApoE deficiency and alleviation by dietary S-adenosyl methionine. J Neurochem 2007;102:753—60.

[30] Chan A, Tchantchou F, Rogers EJ, Shea TB. Dietary deficiency increases presenilin expression, gamma-secretase activity, and Abeta levels: potentiation by ApoE genotype and alleviation by S-adenosyl methionine. J Neurochem 2009;110:831—6.

[31] Lin HC, Hsieh HM, Chen YH, Hu ML. S-Adenosylhomocysteine increases beta-amyloid formation in BV-2 microglial cells by increased expressions of beta-amyloid precursor protein and presenilin 1 and by hypomethylation of these gene promoters. Neurotoxicology 2009;30:622−7.

[32] Fuso A, Nicolia V, Cavallaro RA, Scarpa S. DNA methylase and demethylase activities are modulated by one-carbon metabolism in Alzheimer's disease models. J Nutr Biochem 2011;22:242−51.

[33] Guo JU, Su Y, Zhong C, Ming GL, Song H. Hydroxylation of 5-methylcytosine by TET1 promotes active DNA demethylation in the adult brain. Cell 2011;145:423−34.

[34] Borro M, Cavallaro RA, Gentile G, Nicolia V, Fuso A, Simmaco M, et al. One-carbon metabolism alteration affects brain proteome profile in a mouse model of Alzheimer's disease. J Alzheimer's Dis 2010;22:1257−68.

[35] Cavallaro RA, Fuso A, D'Anselmi F, Seminara L, Scarpa S. The effect of S-adenosylmethionine on CNS gene expression studied by cDNA microarray analysis. J Alzheimer's Dis 2006;9:415−9.

[36] Chen KL, Wang SS, Yang YY, Yuan RY, Chen RM, Hu CJ. The epigenetic effects of amyloid-beta(1-40) on global DNA and neprilysin genes in murine cerebral endothelial cells. Biochem Biophys Res Commun 2008;378:57−61.

[37] Basha MR, Wei W, Bakheet SA, Benitez N, Siddiqi HK, Ge YW, et al. The fetal basis of amyloidogenesis: exposure to lead and latent overexpression of amyloid precursor protein and beta-amyloid in the aging brain. J Neurosci 2005;25:823−9.

[38] Wu J, Basha MR, Brock B, Cox DP, Cardozo-Pelaez F, McPherson CA, et al. Alzheimer's disease (AD)-like pathology in aged monkeys after infantile exposure to environmental metal lead (Pb): evidence for a developmental origin and environmental link for AD. J Neurosci 2008;28:3−9.

[39] Lahiri DK, Zawia NH, Greig NH, Sambamurti K, Maloney B. Early-life events may trigger biochemical pathways for Alzheimer's disease: the "LEARn" model. Biogerontology 2008;9:375−9.

[40] Mastroeni D, McKee A, Grover A, Rogers J, Coleman PD. Epigenetic differences in cortical neurons from a pair of monozygotic twins discordant for Alzheimer's disease. PLoS One 2009;4:e6617.

[41] Mastroeni D, Grover A, Delvaux E, Whiteside C, Coleman PD, Rogers J. Epigenetic changes in Alzheimer's disease: Decrements in DNA methylation. Neurobiol Aging 2010;31:2025−37.

[42] Barrachina M, Ferrer I. DNA methylation of Alzheimer disease and tauopathy-related genes in postmortem brain. J Neuropathol Exp Neurol 2009;68:880−91.

[43] Siegmund KD, Connor CM, Campan M, Long TI, Weisenberger DJ, Biniszkiewicz D, et al. DNA methylation in the human cerebral cortex is dynamically regulated throughout the life span and involves differentiated neurons. PLoS One 2007;2:e895.

[44] Wang SC, Oelze B, Schumacher A. Age-specific epigenetic drift in late-onset Alzheimer's disease. PLoS One 2008;3:e2698.

[45] Bollati V, Galimberti D, Pergoli L, Dalla Valle E, Barretta F, Cortini F, et al. DNA methylation in repetitive elements and Alzheimer disease. Brain Behav Immun 2011. In press.

[46] Seshadri S, Beiser A, Selhub J, Jacques PF, Rosenberg IH, D'Agostino RB, et al. Plasma homocysteine as a risk factor for dementia and Alzheimer's disease. N Engl J Med 2002;346:476−83.

[47] Luchsinger JA, Tang MX, Shea S, Miller J, Green R, Mayeux R. Plasma homocysteine levels and risk of Alzheimer disease. Neurology 2004;62:1972−6.

[48] Ravaglia G, Forti P, Maioli F, Martelli M, Servadei L, Brunetti N, et al. Homocysteine and folate as risk factors for dementia and Alzheimer disease. Am J Clin Nutr 2005;82:636−43.

[49] Luchsinger JA, Tang MX, Miller J, Green R, Mayeux R. Relation of higher folate intake to lower risk of Alzheimer disease in the elderly. Arch Neurol 2007;64:86−92.

[50] Fowler B. Homocysteine: overview of biochemistry, molecular biology, and role in disease processes. Semin Vasc Med 2005;5:77−86.

[51] Vafai SB, Stock JB. Protein phosphatase 2A methylation: a link between elevated plasma homocysteine and Alzheimer's Disease. FEBS Lett 2002;518:1−4.

[52] Sontag E, Nunbhakdi-Craig V, Sontag JM, Diaz-Arrastia R, Ogris E, Dayal S, et al. Protein phosphatase 2A methyltransferase links homocysteine metabolism with tau and amyloid precursor protein regulation. J Neurosci 2007;27:2751−9.

[53] Nicolia V, Fuso A, Cavallaro RA, Di Luzio A, Scarpa S. B vitamin deficiency promotes tau phosphorylation through regulation of GSK3beta and PP2A. J Alzheimer's disease 2010;19:895−907.

[54] Chan A, Rogers E, Shea TB. Dietary Deficiency in Folate and Vitamin E Under Conditions of Oxidative Stress Increases Phospho-Tau Levels: Potentiation by ApoE4 and Alleviation by S-Adenosylmethionine. J Alzheimer's Dis 2009;17:483−7.

[55] Berger SL. The complex language of chromatin regulation during transcription. Nature 2007;447:407−12.

[56] Chouliaras L, Rutten BP, Kenis G, Peerbooms O, Visser PJ, Verhey F, et al. Epigenetic regulation in the pathophysiology of Alzheimer's disease. Prog Neurobiol 2010;90:498−510.

189

[57] Martin C, Zhang Y. The diverse functions of histone lysine methylation. Nat Rev Mol Cell Biol 2005;6:838−49.

[58] Yost JM, Korboukh I, Liu F, Gao C, Jin J. Targets in epigenetics: inhibiting the methyl writers of the histone code. Curr Chem Genomics 2011;5:72−84.

[59] Tachibana M, Matsumura Y, Fukuda M, Kimura H, Shinkai Y. G9a/GLP complexes independently mediate H3K9 and DNA methylation to silence transcription. EMBO J 2008;27:2681−90.

[60] Ogawa O, Zhu X, Lee HG, Raina A, Obrenovich ME, Bowser R, et al. Ectopic localization of phosphorylated histone H3 in Alzheimer's disease: a mitotic catastrophe? Acta Neuropathol 2003;105:524−8.

[61] Myung NH, Zhu X, Kruman II, Castellani RJ, Petersen RB, Siedlak SL, et al. Evidence of DNA damage in Alzheimer disease: phosphorylation of histone H2AX in astrocytes. Age (Dordr) 2008;30:209−15.

[62] Ding H, Dolan PJ, Johnson GV. Histone deacetylase 6 interacts with the microtubule-associated protein tau. J Neurochem 2008;106:2119−30.

[63] Francis YI, Fà M, Ashraf H, Zhang H, Staniszewski A, Latchman DS, et al. Dysregulation of histone acetylation in the APP/PS1 mouse model of Alzheimer's disease. J Alzheimer's Dis 2009;18:131−9.

[64] Kilgore M, Miller CA, Fass DM, Hennig KM, Haggarty SJ, Sweatt JD, et al. Inhibitors of class 1 histone deacetylases reverse contextual memory deficits in a mouse model of Alzheimer's disease. Neuro-psychopharmacology 2009;35:870−80.

[65] Nuutinen T, Suuronen T, Kauppinen A, Salminen A. Valproic acid stimulates clusterin expression in human astrocytes: Implications for Alzheimer's disease. Neurosci Lett 2010;475:64−8.

[66] Govindarajan N, Agis-Balboa RC, Walter J, Sananbenesi F, Fischer A. Sodium Butyrate Improves Memory Function in an Alzheimer's Disease Mouse Model When Administered at an Advanced Stage of Disease Progression. J Alzheimer's Dis 2011. In Press.

[67] Ricobaraza A, Cuadrado-Tejedor M, Marco S, Pérez-Otaño I, García-Osta A. Phenylbutyrate rescues dendritic spine loss associated with memory deficits in a mouse model of Alzheimer disease. Hippocampus 2010. In Press.

[68] Satoh J. MicroRNAs and their therapeutic potential for human diseases: aberrant microRNA expression in Alzheimer's disease brains. J Pharmacol Sci 2010;114:269−75.

[69] Wang WX, Rajeev BW, Stromberg AJ, Ren N, Tang G, Huang Q, et al. The expression of microRNA miR-107 decreases early in Alzheimer's disease and may accelerate disease progression through regulation of beta-site amyloid precursor protein-cleaving enzyme 1. J Neurosci 2008;28:1213−23.

[70] Hébert SS, Horré K, Nicolaï L, Papadopoulou AS, Mandemakers W, Silahtaroglu AN, et al. Loss of microRNA cluster miR-29a/b-1 in sporadic Alzheimer's disease correlates with increased BACE1/beta-secretase expression. Proc Natl Acad Sci USA 2008;105:6415−20.

[71] Hébert SS, Horré K, Nicolaï L, Bergmans B, Papadopoulou AS, Delacourte A, et al. MicroRNA regulation of Alzheimer's Amyloid precursor protein expression. Neurobiol Dis 2009;33:422−8.

[72] Cogswell JP, Ward J, Taylor IA, Waters M, Shi Y, Cannon B, et al. Identification of miRNA changes in Alzheimer's disease brain and CSF yields putative biomarkers and insights into disease pathways. J Alzheimer's Dis 2008;14:27−41.

[73] Nasu-Nishimura Y, Hayashi T, Ohishi T, Okabe T, Ohwada S, Hasegawa Y, et al. Role of the Rho GTPase-activating protein RICS in neurite outgrowth. Genes Cells 2006;11:607−14.

[74] Costa FF. Non-coding RNAs: Meet thy masters. Bioessays 2010;32:599−608.

[75] Faghihi MA, Modarresi F, Khalil AM, Wood DE, Sahagan BG, Morgan TE, et al. Expression of a noncoding RNA is elevated in Alzheimer's disease and drives rapid feed-forward regulation of beta-secretase. Nat Med 2008;14:723−30.

[76] Smalheiser NR, Lugli G, Torvik VI, Mise N, Ikeda R, Abe K. Natural antisense transcripts are co-expressed with sense mRNAs in synaptoneurosomes of adult mouse forebrain. Neurosci Res. 2008;62:236−9.

[77] Lubin FD. Epigenetic gene regulation in the adult mammalian brain: Multiple roles in memory formation. Neurobiol Learn Mem 2011;96:68−78.

[78] Day JJ, Sweatt JD. DNA methylation and Memory Formation. Nat Neurosci 2010;13:1319−23.

[79] Feng J, Chang H, Li E, Fan G. Dynamic expression of de novo DNA methyltransferases Dnmt3a and Dnmt3b in the central nervous system. J Neurosci Res 2005;79:734−46.

[80] Watanabe D, Uchiyama K, Hanaoka K. Transition of mouse de novo methyltransferases expression from Dnmt3b to Dnmt3a during neural progenitor cell development. Neuroscience 2006;142:727−37.

[81] Feng J, Zhou Y, Campbell SL, Le T, Li E, Sweatt JD, et al. Dnmt1 and Dnmt3a maintain DNA methylation and regulate synaptic function in adult forebrain neurons. Nat Neurosci 2010;13:423−30.

[82] Bihaqi SW, Huang H, Wu J, Zawia NH. Infant Exposure to Lead (Pb) and Epigenetic Modifications in the Aging Primate Brain: Implications for Alzheimer's Disease. J Alzheimer's Dis 2011. In press.

[83] Ehrlich M. The ICF syndrome, a DNA methyltransferase 3B deficiency and immunodeficiency disease. Clin Immunol 2003;109:17−28.

[84] Coppedè F, Tannorella P, Pezzini I, Migheli F, Ricci G, Caldarazzo-Ienco E, et al. Folate, homocysteine, vitamin B12 and polymorphisms of genes participating in one-carbon metabolism in late onset Alzheimer's disease patients and healthy controls. Antioxid Redox Signal 2011. In press.

[85] Mason JB. Unraveling the complex relationship between folate and cancer risk. Biofactors 2011;37:253—60.

[86] Coppedè F, Grossi E, Migheli F, Migliore L. Polymorphisms in folate-metabolizing genes, chromosome damage, and risk of Down syndrome in Italian women: identification of key factors using artificial neural networks. BMC Med Genomics 2010;3:42.

CHAPTER 10

Epigenetic Modulation of Human Neurobiological Disorders

Takeo Kubota[1], Kunio Miyake[1], Takae Hirasawa[1], Tatsushi Onaka[2], Hidenori Yamasue[3]
[1]University of Yamanashi, Yamanashi, Japan
[2]Jichi Medical University, Tochigi, Japan
[3]The University of Tokyo, Tokyo, Japan

193

10.1 INTRODUCTION

Neurons are sensitive to the dosage of genes, and a subset of neurobiological disorders can be caused by either an underexpression or an overexpression of a molecule in the neuron. Examples of such neurodevelopmental disorders are Pelizaeus-Merzbacher disease that is associated with a deletion, mutation, or duplication of the PLP1 gene [1], adult-onset neuromuscular disease, Charcot–Marie–Tooth disease associated with a mutation or duplication of PMP22 [2], and several types of mental retardation that are associated with deletion or duplication of the neuronal migration factor LIS1 [3,4]. Also, it was recently demonstrated that mutation or duplication (multiplication) of the α-synuclein gene is associated with Parkinson's disease [5]. These clinical findings suggest that the brain is extremely sensitive to perturbations in gene-regulation, and further indicate that the brain is an organ that requires a proper control system for gene expression.

"Epigenetic" mechanism is one such system of control of gene expression in higher vertebrates. The term "epigenetics" was first used by Conrad Waddington in 1939 to describe "the causal interactions between genes and their products, which bring the phenotype into being" [6]. The current definition is "the study of heritable changes in gene expression that occur independent

T. Tollefsbol (Ed): Epigenetics in Human Disease. DOI: 10.1016/B978-0-12-388415-2.00010-X

of changes in the primary DNA sequence" [7]. Waddington's definition initially referred to the role of epigenetics in embryonic development, in which cells develop distinct identities despite having the same genetic information; however, the definition of epigenetics has evolved over time as it is implicated in a wide variety of biological processes. The mechanism is essential for normal development during embryogenesis [8] and neural cell differentiation [9] as well as differentiation of other types of cells [10]. Therefore, epigenetics, which includes DNA methylation, histone modifications, and regulation by microRNAs, may be a key concept to understanding of the pathogenesis of neurobiological disorders [11–14].

Precise understanding of neurobiological disorders is important, because the number of children with autism and related disorders (autistic spectrum disorder; ASD) have increased 30-fold (prevalence from 1/2500 to 1/86) in the last 50 years in England [15]. Affected children are presently numbered approximately 100 (between 34 and 264) per 10 000 children worldwide [15–20]. Autism is one of the most common neurobiological disorders, which is characterized by three specific criteria: (1) abnormal reciprocal social interactions: reduced interest in peers and difficulty maintaining social interaction and failure to use eye gaze and facial expressions to communicate efficiently, (2) impaired communication: language delays, deficits in language comprehension and response to voices, stereotyped or literal use of words and phrases, poor pragmatics and lack of prosody, resulting in monotone or exaggerated speech patterns, and (3) repetitive behaviors: motor stereotypies, repetitive use of objects, compulsions and rituals, insistence on sameness, upset to change and unusual or very narrow restricted interests [21]. However, there have so far not been appropriate biological markers for autism, and thus, diagnosis is made based purely on behavioral criteria, such as DSM-IV99, the diagnostic manual of the American Psychiatric Association, and ICD-10100, the diagnostic manual of the World Health Organization [22,23].

While various environmental factors are thought to contribute to the pathogenesis of autism [24], recent genetic studies have revealed rare mutations in more than 20 genes in a subset of autistic children [25]. Many of the genes encode proteins that are associated with synaptic function, including synaptic scaffolding proteins, receptors, transporters on synapses, and neuronal cell adhesion molecules [25]. These findings suggest that autism may be a disorder of the synapse [26] (Figure 10.1).

However, the increase in the incidence of autism cannot be solely attributed to genetic factors, because it is unlikely that mutation rates suddenly increased in recent years. Therefore,

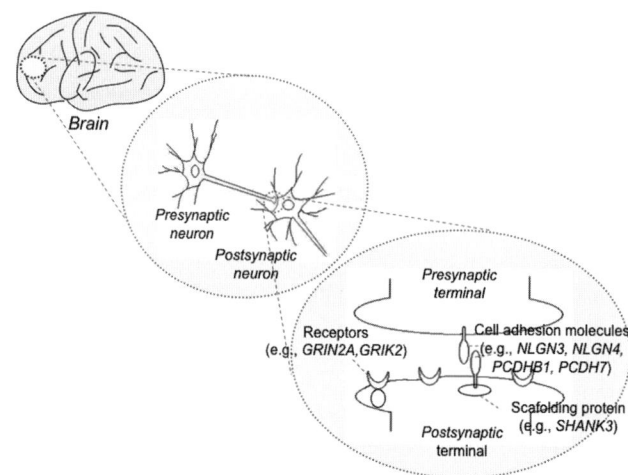

FIGURE 10.1

Location of the molecules in the synapse, which are associated with the pathogenesis of autism (references [25] and [45]). This figure is reproduced in the color plate section.

environmental factors are very likely to be involved in this increase. This is partly supported by a study in twins that revealed that environmental factors contribute to the occurrence of autism [27,28]. Furthermore epigenetic mechanisms are affected by environmental factors [11], and environment-induced epigenetics changed in early life can persist through adulthood and can be transmitted across generations resulting in abnormal behavior traits in the offspring [29]. Consideration of the foregoing led us to propose a hypothesis that various environmental factors change the epigenetic status of brain-specific genes leading to alterations in the expression of a number of neuronal genes associated with the synapse, resulting in abnormal brain function (aberrant synaptic function) in certain neurobiological disorders (namely autism and ASD).

In this chapter, we show various examples of neurobiological disorders associated with epigenetics, environmental factors that affect epigenetic gene regulation, and discuss future directions in medicine for neurobiological disorders based on recent epigenetic understandings.

10.2 EPIGENETIC MECHANISM ASSOCIATED WITH CONGENITAL NEUROBIOLOGICAL DISORDERS

Epigenetic control of gene expression is an intrinsic mechanism for normal brain development [30] and abnormalities in the molecules associated with this mechanism are associated with congenital neurodevelopmental disorders including autistic disorders [31−40]. Genomic imprinting is the epigenetic phenomenon initially discovered in human disorders. In an imprinted gene, one out of the two parental alleles is active and the other allele is inactive due to epigenetic mechanisms such as DNA methylation. Therefore, mutations in the active allele or deletion of the active allele of the imprinted gene results in no expression. This has been found in autistic disorders, Angelman syndrome, and Prader−Willi syndrome [31].

Since there are more genes in the two X chromosomes than in the X and Y chromosomes, females (XX) have more genes than males (XY). To minimize this sex imbalance, one of the two X chromosomes in females is inactivated by an epigenetic mechanism [32]. If X-inactivation does not properly occur in a female, such a female is believed to die in utero and abort. This hypothesis is supported by the recent findings in cloned animals produced by somatic nuclear transfer in which failure of X-chromosome inactivation induces embryonic abortion [33,34]. Even if one of the X chromosomes is extremely small due to a large terminal deletion, making the overdosage effect of X-linked genes small, the female shows a severe congenital neuro-developmental delay [35], indicating that proper epigenetic control of gene expression is essential for normal brain development.

DNA methylation and histone modifications are a fundamental step in epigenetic gene control. DNA methylation is achieved by the addition of a methyl group (CH_3) to CpG dinucleotides in a reaction that is mediated by DNA methyltransferases (e.g. DNMT1, 3A, 3B). Defect in a methyltransferase DNMT3B activity causes a syndrome characterized by immunodeficiency, centromere instability, facial anomalies, and mild mental retardation (ICF syndrome) [36−38].

Methyl-CpG binding proteins, which bind to the methylated DNA region of genes, are also important molecules for control of gene expression. Mutations in one of the methyl-CpG binding proteins, named methyl CpG binding protein 2 (MeCP2), cause Rett syndrome, which is characterized by seizures, ataxic gait, language dysfunction, and autistic behavior [39,40]. Thus, it is thought that the mutation inMeCP2 results in a dysfunctional protein and mis-binding of MeCP2 to the methylated regions of genes, which leads to malsuppression of gene expression in the brain leading to manifestation of Rett syndrome including autism. Recent studies have shown that MeCP2 controls the expression of some neuronal genes, such as brain-derived neurotrophic factor (*BDNF*), distal-less homeobox 5 (*DLX5*), insulin-like growth factor binding protein 3 (*IFGBP3*), and protocadherins *PCDHB1* and *PCDH7* (neuronal cell adhesion molecules) [41–45]. These findings suggest that not only mutations [25], but also epigenetic dysregulation of genes that encode synaptic molecules is possibly associated with autism (Figure 10.2). In fact, either deficiency of MeCP2 (i.e. Rett syndrome) or excessive production of MeCP2 (duplication of the *MECP2* genomic region) can lead to autistic features [46].

10.3 EPIGENETIC MECHANISM UNDERLYING ALTERATION OF THE BRAIN FUNCTION BY ENVIRONMENTAL FACTORS

In autism, both environmental factors (e.g. toxins and infections) and genetic factors (e.g. mutations in synaptic molecules) have historically been implicated [19,47,48]. However, a biological mechanism that links these two groups of factors has not been identified. Epigenetics may bridge these two groups of factors contributing to disease development [11]. Besides the *intrinsic* (congenital) epigenetic defects (associated with the diseases described above), several lines of evidence suggest that *extrinsic* (environmental) factors, such as malnutrition, drugs, mental stress, maternal care, and neuronal stimulation, alter the epigenetic status of genes thereby affecting brain function [49–63]. Therefore, it is intriguing to think that acquired neurodevelopmental disorders, including autistic disorders, may be the result of epigenetic dysregulation caused by environmental factors (Figure 10.3).

The epigenetic mechanism is also likely to be relevant in drug addiction. Gene expression in the dopaminergic and glutamatergic systems is mediated by an epigenetic mechanism, and cocaine and alcohol alter the epigenetic state. For example, cocaine induces either hyperacetylation or hypoacetylation of histones H3 and H4 in the nucleus accumbens of mice, and alcohol induces hyperacetylation of histones H3 and H4 in the frontal cortex and the nucleus accumbens of adolescent rats; such alterations may be associated with permanent behavioral consequences [64,65].

The above studies were mainly performed in animals, and there is little evidence for such changes in humans. However, the fact that epigenomic differences are greater in older monozygotic twins than in younger twins suggests that epigenetic status may be altered during aging by environmental factors in humans [66].

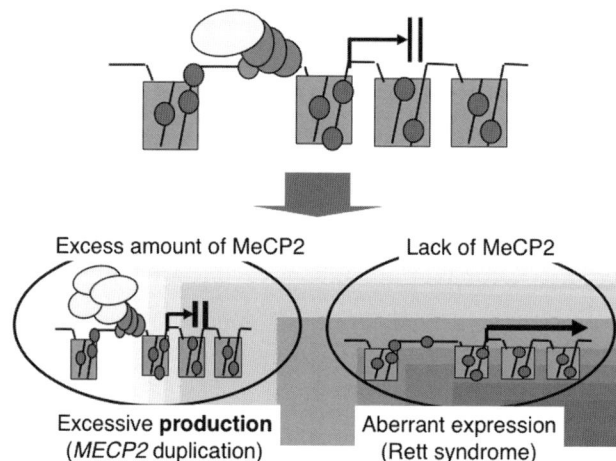

FIGURE 10.2

Schematic representation of epigenetic control of the gene expression. Either an excessive amount or deficiency of MeCP2 leads to aberrant expression pattern of its target genes, leading to neurobiological disorders that exhibit autistic features. Yellow circle: MeCP2, orange circle: protein associated with histone modification, red circle: methyl-residue, green square: chromosomal histone protein, arrow: transcription start site. This figure is reproduced in the color plate section.

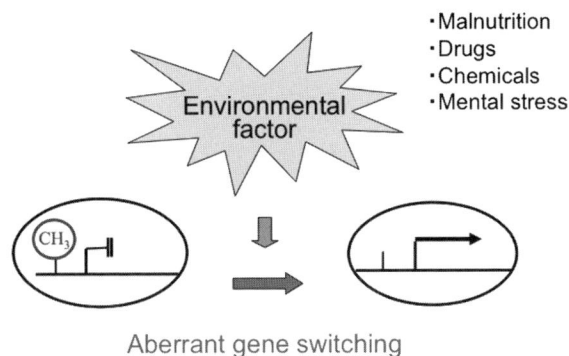

- Malnutrition
- Drugs
- Chemicals
- Mental stress

Aberrant gene switching

FIGURE 10.3
DNA methylation mechanisms can be affected by environmental factors. This figure is reproduced in the color plate section.

10.4 TRANSGENERATIONAL EPIGENETIC INHERITANCE (NON-MENDELIAN DISEASE INHERITANCE)

In the current understanding of biology, one's acquired character is not inherited in the next generation. We call this way of thinking Darwinian inheritance. Therefore, we can rest easy, because this leads us to believe that the bad habits obtained during one's lifetime will not be inherited by our progeny. However, recent advances in epigenetics have revealed that such undesirable acquired traits might be transmitted into the next generation.

Epigenetic marks, e.g. DNA methylation or histone modifications, allow the transmission of gene activity states from one cell to its daughter cells. A fundamental question in epigenetics is whether these marks can also be transmitted through the germline. If so, an aberrant epigenetic mark acquired in one generation could be inherited by the next generation. In general, epigenetic marks should be erased by demethylating factors such as the cytidine deaminases (e.g. AID, APOBEC1) [67] and re-established in each generation, but there have been reports that this erasure is incomplete at some loci in the genome of several model organisms, possibly due to deficiency of demethylating factors (e.g. AID) [67]. "Transgenerational epigenetic inheritance" refers to the germline transmission of an epigenetic mark [68,69], which may provide direct biological proof for an unexplained hypothesis, Lamarckism, which is the idea of heritability of acquired characteristics.

Transgenerational inheritance of epigenetic marks was first demonstrated in a specific mouse strain. The methylation status at the *Axin (Fu)* locus in mature sperm reflects the methylation state of the allele in the somatic tissue of the animal. This epigenetic status is linked to the shape of tail of the animals, and it does not undergo epigenetic reprogramming during gametogenesis [70]. This observation was recently confirmed in *Drosophila*, in which an aberrant epigenetic mark (defective chromatin state) acquired in one generation induced by environmental stress (e.g. heat shock) was inherited by the next generation [71].

It has also been demonstrated that an aberrant epigenetic mark acquired in one generation by mental stress (maternal separation in early life) can be inherited by the next generation [29]. Chronic maternal separation alters behaviors as well as the profile of DNA methylation in the promoter of several candidate genes in both germline of the separated mice and the brains of the offspring with altered gene expression (e.g. decreases in the expression of the corticotropin releasing factor receptor 2 in the amygdala and the hypothalamus) [72]. These findings may provide the biological evidence for the current social issue that traumatic experiences in early life are risk factors for the development of behavioral and emotional disorders.

Based on the evidences described above, the readers of this chapter might be led to believe that epigenetics is a scientific field that portends adverse news for society. However, if we could create an environment conducive to good human health, we will be able to sever the

197

deleterious environment-induced epigenetic patterns across the generations. Even if epigenetic markings provide a "memory" of past experiences and the markings persist across the lifespan of an individual and then be transmitted to the offspring via epigenetic inheritance, future epigenetic research can possibly establish restorative methods taking advantage of the reversibility of stress-induced epigenetic modifications. It can also show us the appropriate environment for keeping a healthy physical and mental condition [73,74].

10.5 EPIGENETIC MEDICINE FOR NEUROBIOLOGICAL DISORDERS

Environmental factors via epigenetic mechanisms are not always harmful. Imipramine, a major antidepressant, was recently found to have the effect of restoring a depressive state by alteration of the epigenetic state (increasing histone H3 and H4 acetylation at the Bdnf P3 and P4 promoters and histone H3-K4 dimethylation at Bdnf P3 promoter), leading to up-regulation of Bdnf (brain-derived neurotrophic factor) in the hippocampus [53]. It was recently shown that other drugs used for mental illness also have an epigenetic restoring effect; these include valproic acid [51,54], clozapine [55], sulpiride [55], and lithium [56]. Oxytocin, which is a hormone associated with social interactions [57–59], is tested for the treatment for autism and ASD, since previous studies have revealed the effects of this neuropeptide on autistic social cognition in adults with autism [60,61]. Since aberrant hypermethylation of the promoter region of the oxytocin receptor gene is found in the DNA from brain and blood samples of some autistic children [62], oxytocin as well as other drugs for autism may have epigenetic restoring effects. Based on these findings, chemicals that alter epigenetic gene expression are candidates of new drugs for a subset of the patients with mental and neuro-biological disorders [65].

Recent research has shown that not only drugs, but also general nutrition alters the epigenetic status of DNA. The best example is folic acid. Folic acid is the substrate that supplies methyl-residues during methylation of cytosine in DNA. Therefore, in order to maintain DNA methylation, proper intake of folic acid is essential. In Japan, the number of young women who do not take a sufficient amount of folic acid during pregnancy is increasing, and this increases the risk of having babies with neural tube defects [75]. In rats, inappropriate supply of nutrients from the mother to the fetus also increases the susceptibility of the fetus to develop diabetes mellitus through epigenetic effects [76]. These hypotheses are supported by observation from a rat study in which protein restriction during pregnancy induced a state of malnutrition and hyperlipidemia in the fetus. However, supplementation of the maternal protein-restricted diet with folic acid during pregnancy relieved these abnormalities. The effect was achieved by an increase in DNA methylation of the promoter regions of the PPAR-alpha and glucocorticoid receptor genes in the liver by folic acid, which led to proper suppression of the genes [49]. These findings indicate that specific nutrient intakes may alter the phenotype of the offspring through epigenetic changes. Protein restriction during pregnancy also reduced the expression of Mecp2 gene in the liver of the fetus [50]. This observation implies that malnutrition during pregnancy may contribute to the development of neurobiological disorders, although the effect of malnutrition on Mecp2 has not been investigated in the brain.

Royal jelly, known to change the phenotype from a genetically identical female honeybee to a fertile queen, also has epigenetic effects. A recent study revealed that royal jelly has the effect of erasing global DNA methylation, because silencing the expression of Dnmt3, a DNA methyltransferase, had a similar effect on the larval development [77].

Since the 1980s, folic acid is empirically used for the treatment of autistic patients, and these studies have shown that it is effective only for a subset of the patients [78–80]. However, it has not been completely proven that the effect of folic acid on autistic patients is based on epigenetic effects. Therefore, it is important to confirm the genomic regions (genes) where DNA methylation is altered by treatment with folic acid. Improved methods for the genome-

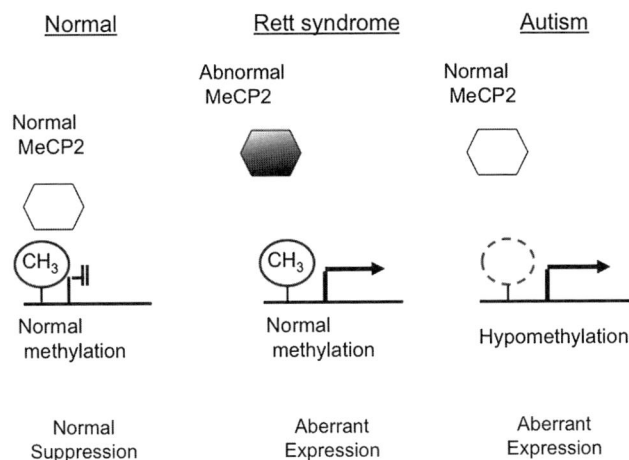

FIGURE 10.4
Pathogenesis of Rett syndrome and putative pathogenesis of autism. This figure is reproduced in the color plate section.

wide methylation studies [81,82] may allow us to identify folate-responsive genes where DNA methylation is altered by administration of folic acid in autistic children. Such gene, if identified, can be a therapeutic marker for folic acid treatment, and the folic acid responsive (treatable) autistic patients can be distinguished from the non-responsive (non-treatable) patients (Figure 10.4).

Folic acid-based treatment may be safe, since it is a nutrient, and it is expected to have global effects, but not on individual genes. Besides these nutrition-based treatments, several alternative epigenetic-based treatments are currently developed, which is gene-specific and able to restore the epigenetic status that has been changed by an environmental factor. Examples are inhibitors of either DNA methylation or histone deacetylases are bound to pyrrole-imidazole (PI) polyamides, small synthetic molecules that recognize and attach to the minor groove of DNA, which can be designed for DNA sequences of any genes and can regulate gene transcription by binding to DNA [83]. In fact, molecules that consist of an epigenetic inhibitor and a PI polyamide have been delivered to a target gene and have altered its expression [84].

It has recently been discovered that the DNA sequence is not the same in each neuron [85], and that epigenetic change underlies the somatic change of the DNA sequence [86]. This is the phenomenon that is referred to as retrotransposition, in which a repetitive LINE-1 element is inserted into various genomic regions when it is hypomethylated and can change the expression of adjacent genes. The retrotransposition is accelerated by deficiency of MeCP2 [87]. Interestingly, in mice, the retrotransposition is also activated by voluntary exercise (running) [88], suggesting that exercise may alter the DNA methylation status in neurons.

Studies using a Rett syndrome mouse model (*Mecp2* knockout mice) show that environmental enrichment (e.g. availability of stimulating toys) during early postnatal development produces effects on neural development and ameliorates the neurological phenotypes associated with Rett syndrome [89,90]. This suggests that DNA methylation status may be corrected by an appropriate environmental stimulus, compensating for the insufficient MeCP2 function. The reversibility of this epigenetic change is supported by a different study in which activation of *Mecp2* expression after birth leads to relief of neurological symptoms in *Mecp2* knockout mice [91].

10.6 CONCLUSION

Various environmental factors potentially rewrite epigenetic codes. However, since epigenetic code is reversible, being different from genetic codes, it is potentially treatable, and preventable

199

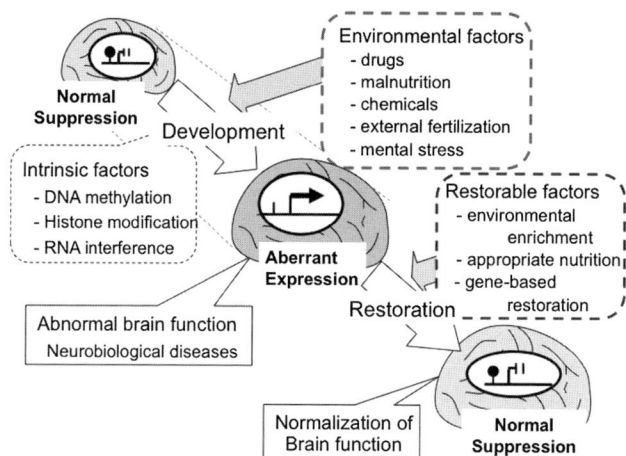

FIGURE 10.5

Overview of the DNA methylation changes and environmental factors. Note: This diagram depicts aberrant expression of normally suppressed gene by promoter hypermethylation; however, suppression of normally expressed genes by promoter hypermethylation is equally important for pathogenesis of human neurobiological disorders. This figure is reproduced in the color plate section.

when we further understand the mechanism(s) of how environmental factors induce epigenetic changes (Figure 10.5). In this context, better understanding of epigenetics is now important for development of new therapies for neurobiological disorders.

References

[1] Inoue K, Kanai M, Tanabe Y, Kubota T, Kashork CD, Wakui K, Fukushima Y, Lupski JR, Shaffer LG. Prenatal interphase FISH diagnosis of PLP1 duplication associated with Pelizaeus-Merzbacher disease. Prenat Diagn 2001;21:1133–6.

[2] Online Mendelian Inheritance in Man (OMIM): #118220 http://www.ncbi.nlm.nih.gov/entrez/

[3] Reiner O, Carrozzo R, Shen Y, Wehnert M, Faustinella F, Dobyns WB, Caskey CT, Ledbetter DH. Isolation of a Miller-Dieker lissencephaly gene containing G protein beta-subunit-like repeats. Nature 1993;364:717–21.

[4] Bi W, Sapir T, Shchelochkov OA, Zhang F, Withers MA, Hunter JV. Increased LIS1 expression affects human and mouse brain development. Nat Genet 2009;41:168–77.

[5] Obi T, Nishioka K, Ross OA, Terada T, Yamazaki K, Sugiura A, Takanashi M, Mizoguchi K, Mori H, Mizuno Y, Hattori N. Clinicopathologic study of a SNCA gene duplication patient with Parkinson disease and dementia. Neurology 2008;70:238–41.

[6] Waddington CH. Epigenotype. Endeavour 1942;1:18–20.

[7] Sharma S, Kelly TK, Jones PA. Epigenetics in cancer. Carcinogenesis 2010;31:27–36.

[8] Li E, Beard C, Jaenisch R. Role for DNA methylation in genomic imprinting. Nature 1993;366:362–5.

[9] Takizawa T, Nakashima K, Namihira M, Ochiai W, Uemura A, Yanagisawa M, et al. DNA methylation is a critical cell-intrinsic determinant of astrocyte differentiation in the fetal brain. Dev Cell 2001;1:749–58.

[10] Sakashita K, Koike K, Kinoshita T, Shiohara M, Kamijo T, Taniguchi S, Kubota T. Dynamic DNA methylation change in the CpG island region of p15 during human myeloid development. J Clin Invest 2001;108:1195–204.

[11] Qiu J. Epigenetics: unfinished symphony. Nature 2006;441:143–5.

[12] Abel T, Zukin RS. Epigenetic targets of HDAC inhibition in neurodegenerative and psychiatric disorders. Curr Opin Pharmacol 2008;8:57–64.

[13] Urdinguio RG, Sanchez-Mut JV, Esteller M. Epigenetic mechanisms in neurological diseases: genes, syndromes, and therapies. Lancet Neurol 2009;8:1056–72.

[14] Wu H, Tao J, Chen PJ, Shahab A, Ge W, Hart RP, Ruan X, Ruan Y, Sun YE. Genome-wide analysis reveals methyl-CpG-binding protein 2-dependent regulation of microRNAs in a mouse model of Rett syndrome. Proc Natl Acad Sci USA 2010;107:18161–6.

[15] Baird G, Simonoff E, Pickles A, Chandler S, Loucas T, Meldrum D, Charman T. Prevalence of disorders of the autism spectrum in a population cohort of children in South Thames: the Special Needs and Autism Project (SNAP). Lancet 2006;368:210–5.

[16] Yeargin-Allsopp M, Rice C, Karapurkar T, Doernberg N, Boyle C, Murphy C. Prevalence of Autism in a US Metropolitan Area. JAMA 2003;289:49–55.

[17] Honda H, Shimizu Y, Imai M, Nitto Y. Cumulative incidence of childhood autism: a total population study of better accuracy and precision. Dev Med Child Neurol 2005;47:10–8.

[18] Holden C. Autism Now. Science 2009;323:565.

[19] Fombonne E. Epidemiology of pervasive developmental disorders. Pediatr Res 2009;65:591–8.

[20] Kim YS, Leventhal BL, Koh YJ, Fombonne E, Laska E, Lim EC, et al. Prevalence of Autism Spectrum Disorders in a Total Population Sample. Am J Psychiatry 2011;168:904–12.

[21] Silverman JL, Yang M, Lord C, Crawley JN. Behavioural phenotyping assays for mouse models of autism. Nat Rev Nerusosci 2010;490:490–502.

[22] American Psychiatric Association. Diagnostic and Statistical Manual of Mental Disorders. 4th ed. Washington D.C: APA; 1994.

[23] World Health Organization. The ICD-10 Classification of Mental and Behavioural Disorders, Geneva, Switzerland; 1992

[24] Herbert MR. Contributions of the environment and environmentally vulnerable physiology to autism spectrum disorders. Curr Opin Neurol 2010;23:103–10.

[25] Persico AM, Bourgeron T. Searching for ways out of the autism maze: genetic, epigenetic and environmental clues. Trends Neurosci 2006;29:349–58.

[26] Zoghbi HY. Postnatal neurodevelopmental disorders: meeting at the synapse? Science 2003;302:826–30.

[27] Hoekstra RA, Bartels M, Hudziak JJ, Van Beijsterveldt TC, Boomsma DI. Genetics and environmental covariation between autistic traits and behavioral problems. Twin Res Hum Genet 2007;10:853–86.

[28] Hallmayer J, Cleveland S, Torres A, Phillips J, Cohen B, Torigoe T, et al. Genetic heritability and shared environmental factors among twin pairs with autism. Arch Gen Psychiatry 2011;68:1095–102.

[29] Franklin TB, Russig H, Weiss IC, Gräff J, Linder N, Michalon A, Vizi S, Mansuy IM. Epigenetic transmission of the impact of early stress across generations. Biol Psychiatry 2010;68:408–15.

[30] Guy J, Hendrich B, Holmes M, Martin JE, Bird A. A mouse Mecp2-null mutation causes neurological symptoms that mimic Rett syndrome. Nat Genet 2001;27:322–6.

[31] Kubota T, Das S, Christian SL, Baylin SB, Herman JG, Ledbetter DH. Methylation-specific PCR symplifies imprinting analysis. Nat Genet 1997;16:16–7.

[32] Kubota T, Nonoyama S, Tonoki H, Masuno M, Imaizumi K, Kojima M, Wakui K, Shimadzu M, Fukushima Y. A new assay for the analysis of X-chromosome inactivation based on methylation-specific PCR. Hum Genet 1999;104:49–55.

[33] Xue F, Tian XC, Du F, Kubota C, Taneja M, Dinnyes A, Dai Y, Levine H, Pereira LV, Yang X. Aberrant patterns of X chromosome inactivation in bovine clones. Nat Genet 2002;31:216–20.

[34] Nolen LD, Gao S, Han Z, Mann MR, Gie Chung Y, Otte AP, Bartolomei MS, Latham KE. X chromosome reactivation and regulation in cloned embryos. Dev Biol 2005;279:525–40.

[35] Kubota T, Wakui K, Nakamura T, Ohashi H, Watanabe Y, Yoshino M, et al. Proportion of the cells with functional X disomy is associated with the severity of mental retardation in mosaic ring X Turner syndrome females. Cytogenet Genome Res 2002;99:276–84.

[36] Okano M, Bell DW, Haber DA, Li E. DNA methyltransferases Dnmt3a and Dnmt3b are essential for de novo methylation and mammalian development. Cell 1999;99:247–57.

[37] Shirohzu H, Kubota T, Kumazawa A, Sado T, Chijiwa T, Inagaki K, Suetake I, Tajima S, Wakui K, Miki Y, Hayashi M, Fukushima Y, Sasaki H. Three novel DNMT3B mutations in Japanese patients with ICF syndrome. Am J Med Genet 2002;112:31–7.

[38] Kubota T, Furuumi H, Kamoda T, Iwasaki N, Tobita N, Fujiwara N, Goto Y, Matsui A, Sasaki H, Kajii T. ICF syndrome in a girl with DNA hypomethylation but without detectable DNMT3B mutation. Am J Med Genet A 2004;129:290–3.

[39] Amir RE, Van den Veyver IB, Wan M, Tran CQ, Francke U, Zoghbi HY. Rett syndrome is caused by mutations in X-linked MECP2, encoding methyl-CpG-binding protein 2. Nat Genet 1999;23:185–8.

[40] Chunshu Y, Endoh K, Soutome M, Kawamura R, Kubota T. A patient with classic Rett syndrome with a novel mutation in MECP2 exon 1. Clin Genet 2006;70:530–1.

[41] Chen WG, Chang Q, Lin Y, Meissner A, West AE, Griffith EC, Jaenisch R, Greenberg ME. Derepression of BDNF Transcription Involves Calcium-Dependent Phosphorylation of MeCP2. Science 2003;302:885–9.

[42] Martinowich K, Hattori D, Wu H, Fouse S, He F, Hu Y, Fan G, Sun YE. DNA methylation-related chromatin remodeling in activity-dependent BDNF gene regulation. Science 2003;302:890–3.

[43] Horike S, Cai S, Miyano M, Cheng JF, Kohwi-Shigematsu T. Loss of silent-chromatin looping and impaired imprinting of DLX5 in Rett syndrome. Nat Genet 2005;37:31–40.

[44] Itoh M, Ide S, Takashima S, Kudo S, Nomura Y, Segawa M, et al. Methyl CpG-binding protein 2, whose mutation causes Rett syndrome, directly regulates Insulin-like Growth Factor Binding Protein 3 in mouse and human Brains. J Neuropathol Exp Neurol 2007;66:117—23.

[45] Miyake K, Hirasawa T, Soutome M, Itoh M, Goto Y, Endoh K, et al. The protocadherins, PCDHB1 and PCDH7, are regulated by MeCP2 in neuronal cells and brain tissues: implication for pathogenesis of Rett syndrome. BMC Neurosci 2011 (in press).

[46] Ramocki MB, Peters SU, Tavyev YJ, Zhang F, Carvalho CM, Schaaf CP, et al. Autism and other neuropsychiatric symptoms are prevalent in individuals with MeCP2 duplication syndrome. Ann Neurol 2009;66:771—82.

[47] Bailey A, Le Couteur A, Gottesman I, Bolton P, Simonoff E, Yuzda E, Rutter M. Autism as a strongly genetic disorder: evidence from a British twin study. Psychol Med 1995;25:63—77.

[48] Zafeiriou DI, Ververi A, Vargiami E. Childhood autism and associated comorbidities. Brain Dev 2007;29:257—72.

[49] Burdge GC, Lillycrop KA, Phillips ES, Slater-Jefferies JL, Jackson AA, Hanson MA. Folic Acid Supplementation during the Juvenile-Pubertal Period in Rats Modifies the Phenotype and Epigenotype Induced by Prenatal Nutrition. J Nutr 2009;139:1054—60.

[50] Lillycrop KA, Slater-Jefferies JL, Hanson MA, Godfrey KM, Jackson AA, Burdge GC. Induction of altered epigenetic regulation of the hepatic glucocorticoid receptor in the offspring of rats fed a protein-restricted diet during pregnancy suggests that reduced DNA methyltransferase-1 expression is involved in impaired DNA methylation and changes in histone modifications. Br J Nutr 97:1064—73.

[51] Jessberger S, Nakashima K, Clemenson Jr GD, Mejia E, Mathews E, Ure K, Ogawa S, Sinton CM, Gage FH, Hsieh J. Epigenetic Modulation of Seizure-Induced Neurogenesis and Cognitive Decline. J Neurosci 2007;27:5967—75.

[52] Weaver IC, Cervoni N, Champagne FA, D'Alessio AC, Sharma S, Seckl JR, Dymov S, Szyf M, Meaney MJ. Epigentic programming by maternal behavior. Nat Neurosci 2004;9:847—54.

[53] Tsankova NM, Berton O, Renthal W, Kumar A, Neve RL, Nestler EJ. Sustained hippocampal chromatin regulation in a mouse model of depression and antidepressant action. Nat Neurosci 2006;9:519—25.

[54] Dong E, Chen Y, Gavin DP, Grayson DR, Guidotti A. Valproate induces DNA demethylation in nuclear extracts from adult mouse brain. Epigenetics 2010;5:730—5.

[55] Dong E, Nelson M, Grayson DR, Costa E, Guidotti A. Clozapine and sulpiride but not haloperidol or olanzapine activate brain DNA demethylation. Proc Natl Acad Sci USA 2008;105:13614—9.

[56] Wang Q, Xu X, Li J, Liu J, Gu H, Zhang R, et al. Lithium, an anti-psychotic drug, greatly enhances the generation of induced pluripotent stem cells. Cell Res 2011;21:1424—35.

[57] Takayanagi Y, Yoshida M, Bielsky IF, Ross HE, Kawamata M, Onaka T, et al. Pervasive social deficits, but normal parturition, in oxytocin receptor-deficient mice. Proc Natl Acad Sci USA 2005;102:16096—101.

[58] Young LJ. Being human: love: neuroscience reveals all. Nature 2009;457:148.

[59] Bartz JA, Zaki J, Bolger N, Ochsner KN. Social effects of oxytocin in humans: context and person matter. Trends Cogn Sci 2011;15:301—9.

[60] Yamasue H, Kuwabara H, Kawakubo Y, Kasai K. Oxytocin, sexually dimorphic features of the social brain, and autism. Psychiatry Clin Neurosci 2010;63:129—40.

[61] Andari E, Duhamel JR, Zalla T, Herbrecht E, Leboyer M, Sirigu A. Promoting social behavior with oxytocin in high-functioning autism spectrum disorders. Proc Natl Acad Sci USA 2010;107:4389—94.

[62] Gregory SG, Connelly JJ, Towers AJ, Johnson J, Biscocho D, Markunas CA, et al. Genomic and epigenetic evidence for oxytocin receptor deficiency in autism. BMC Med 2009;(7):62. 2009 Oct 22.

[63] Ma DK, Jang MH, Guo JU, Kitabatake Y, Chang ML, Pow-Anpongkul N, Flavell RA, Lu B, Ming GL, Song H. Neuronal Activity-Induced Gadd45b Promotes Epigenetic DNA Demethylation and Adult Neurogenesis. Science 2009;323:1074—7.

[64] Renthal W, Kumar A, Xiao G, Wilkinson M, Covington 3rd HE, Maze I, et al. Genome-wide analysis of chromatin regulation by cocaine reveals a role for sirtuins. Neuron 2009;62:335—48.

[65] Pascual M, Boix J, Felipo V, Guerri C. Repeated alcohol administration during adolescence causes changes in the mesolimbic dopaminergic and glutamatergic systems and promotes alcohol intake in the adult rat. J Neurochem 2009;108:920—31.

[66] Fraga MF, Ballestar E, Paz MF, Ropero S, Setien F, Ballestar ML, et al. Epigenetic differences arise during the lifetime of monozygotic twins. Proc Natl Acad Sci USA 2005;102:10604—9.

[67] Popp C, Dean W, Feng S, Cokus SJ, Andrews S, Pellegrini M, Jacobsen SE, Reik W. Genome-wide erasure of DNA methylation in mouse primordial germ cells is affected by AID deficiency. Nature 2011;463:1101—5.

[68] Horsthemke B. Heritable germline epimutations in humans. Nat Genet 2007;39:573—4.

[69] Daxinger L, Whitelaw E. Transgenerational epigenetic inheritance: more questions than answers. Genome Res 2010;20:1623—8.

[70] Rakyan VK, Chong S, Champ ME, Cuthbert PC, Morgan HD, Luu KV, Whitelaw E. Transgenerational inheritance of epigenetic states at the murine Axin(Fu) allele occurs after maternal and paternal transmission. Proc Natl Acad Sci USA 2003;100:2538—43.

[71] Seong KH, Li D, Shimizu H, Nakamura R, Ishii S. Inheritance of Stress-Induced, ATF-2-Dependent Epigenetic Change. Cell 2011;145:1049—61.

[72] Weiss IC, Franklin TB, Vizi S, Mansuy IM. Inheritable effect of unpredictable maternal separation on behavioral responses in mice. Front Behav Neurosci. Front Behav Neurosci 2011;5:3.

[73] Thayer ZM, Kuzawa CW. Biological memories of past environments: Epigenetic pathways to health disparities. Epigenetics 2011;6:798—803.

[74] Arai JA, Feig LA. Long-lasting and transgenerational effects of an environmental enrichment on memory formation. Brain Res Bull 2011;85:30—5.

[75] Watanabe H, Fukuoka H, Sugiyama T, Nagai Y, Ogasawara K, Yoshiike N. Dietary folate intake during pregnancy and birth weight in Japan. Eur J Nutr 2008;47:341—7.

[76] Park JH, Stoffers DA, Nicholls RD, Simmons RA. Development of type 2 diabetes following intrauterine growth retardation in rats is associated with progressive epigenetic silencing of Pdx1. J Clin Invest 2008;118:2316—24.

[77] Kucharski R, Maleszka J, Foret S, Maleszka R. Nutritional control of reproductive status in haneybees via DNA methylation. Science 2008;319:1827—30.

[78] Rimland B. Controversies in the Treatment of Autistic Children: Vitamin and Drug Therapy. J Child Neurol 1998;3:68—72.

[79] James SJ, Cutler P, Melnyk S, Jernigan S, Janak L, Gaylor DW, Neubrander JA. Metabolic biomarkers of increased oxidative stress and impaired methylation capacity in children with autism. Am J Clin Nutr 2004;20:1611—7.

[80] Moretti P, Sahoo T, Hyland K, Bottiglieri T, Peters S, del Gaudio D, et al. Cerebral folate deficiency with developmental delay, autism, and response to folinic acid. Neurology 2005;64:1088—90.

[81] Laird PW. Principles and challenges of genomewide DNA methylation analysis. Nat Rev Genet 2005;11:191—203.

[82] Varley KE, Mitra RD. Bisulfite Patch PCR enables multiplexed sequencing of promoter methylation across cancer samples. Genome Res 2010;20:1279—87.

[83] Ohtsuki A, Kimura MT, Minoshima M, Suzuki T, Ikeda M, Bando T, Nagase H, Shinohara K, Sugiyama H. Synthesis and properties of PI polyamide—SAHA conjugate. Tetrahedron Lett 2009;50:7288—92.

[84] Matsuda H, Fukuda N, Ueno T, Katakawa M, Wang X, Watanabe T, et al. Transcriptional inhibition of progressive renal disease by gene silencing pyrrole-imidazole polyamide targeting of the transforming growth factor-β1 promoter. Kidney Int 2011;79:46—56.

[85] Coufal NG, Garcia-Perez JL, Peng GE, Yeo GW, Mu Y, Lovci MT, et al. L1 Retrotransposition in Human Neural Progenitor Cells. Nature 2009;460:1127—31.

[86] Muotri AR, Chu VT, Marchetto MC, Deng W, Moran JV, Gage FH. Somatic mosaicism in neuronal precursor cells mediated by L1 retrotransposition. Nature 2005;435:903—10.

[87] Muotri AR, Marchetto MC, Coufal NG, Oefner R, Yeo G, Nakashima K, Gage FH. L1 retrotransposition in neurons is modulated by MeCP2. Nature 2010;468:903—10.

[88] Muotri AR, Zhao C, Marchetto MC, Gage FH. Environmental Influence on L1 Retrotransposons in the Adult Hippocampus. Hypoocumpus 2009;19:1002—7.

[89] Lonetti G, Angelucci A, Morando L, Boggio EM, Giustetto M, Pizzorusso T. Early environmental enrichment moderates the behavioral and synaptic phenotype of MeCP2 null mice. Biol Psychiatry 2010;67:657—65.

[90] Kerr B, Silva PA, Walz K, Young JI. Unconventional transcriptional response to environmental enrichment in a mouse model of Rett syndrome. PLoS One 2010;5:e11534.

[91] Guy, J, Gan J, Selfridge J, Cobb S, Bird A. Reversal of neurological defects in a mouse model of Rett syndrome. Science 2007;315:1143—7.

203

CHAPTER 11

Epigenetic Basis of Autoimmune Disorders in Humans

Biola M. Javierre, Lorenzo De La Rica, Esteban Ballestar
Bellvitge Biomedical Research Institute (IDIBELL), L'Hospitalet de Llobregat, Barcelona, Spain

11.1 IMMUNITY AND AUTOIMMUNITY

The term immunity refers to the group of mechanisms developed by multicellular organisms to defend themselves against potentially harmful agents. The development of an immune system occurred during the evolution from unicellular to multicellular organisms. Autoimmune reactions emerged in parallel with the increasing complexity of the immune system to recognize and eliminate pathogenic elements.

T. Tollefsbol (Ed): Epigenetics in Human Disease. DOI: 10.1016/B978-0-12-388415-2.00011-1

To achieve specific recognition of harmful agents, it is essential first to identify their own organic components as something innocuous; in other words, the immune system must be autotolerant. When the fine balance between recognition of self-components and defense against foreign agents is broken the immune system can react against the body's own components, inducing cell destruction. If the immune system recognizes a particular component located in a specific organ as harmful, and reacts solely against this, it gives rise to a process of organ-specific autoimmunity. Examples include diabetes mellitus type 1, primary biliary cirrhosis, Hashimoto's thyroiditis, Grave's disease, and celiac disease (Table 11.1). Conversely, if the reaction affects various organs and systems it is classified as systemic autoimmunity. This group of disorders includes systemic lupus erythematosus, rheumatoid arthritis, Sjögren's syndrome, scleroderma, and psoriasis among others (Table 11.1).

11.2 EPIGENETIC DEREGULATION IN AUTOIMMUNITY

Epigenetics is one of the most rapidly developing areas in the fields of molecular biology and biomedicine. This discipline has contributed to our understanding of key aspects of cell biology and the pathogenesis of a variety of diseases, among which cancer is the most widely investigated [1].

Epigenetics is generally defined as the scientific discipline that studies the reversible and potentially heritable changes in gene expression that do not affect the DNA sequence [2]. Epigenetic mechanisms can also be described as those that register, signal, or perpetuate gene activity states and involve the chemical modification of chromatin. Epigenetics mainly focuses on two groups of chemical modifications: DNA methylation and histone post-translational modifications. Both groups of modifications have direct and/or indirect effects on gene expression and nuclear structure, and ultimately, determine cell identity.

There is a wide spectrum of critical biological phenomena regulated by epigenetic mechanisms. Epigenetic modifications are implicated in cell differentiation, cell cycle, apoptosis and signaling processes and are able to couple external signals to fine regulation of gene expression.

TABLE 11.1 Autoimmune Disorders: Types and Main Affected Organs

Type	Disease	Main Affected Organ(s)	References
Systemic	Systemic lupus erythematosus	Skin, joints, heart, brain, kidneys, other	[4,5]
	Rheumatoid arthritis	Joints, skin, lungs, heart and blood vessels, other	[37]
	Sjögren's syndrome	Salivary glands, tear glands, joints	[85]
	Scleroderma	Skin, intestine, lung	[93]
	Psoriasis	Skin, joints	[97]
	Crohn's disease	Gastrointestinal tract, skin, joints, eye	[113]
	Ulcerative colitis	Gastrointestinal tract, eye, muscle, skin	[113]
	Ankylosing spondylitis	Spinal and sacroiliac joints, eye, lung, heart	[137]
Organ-specific	Multiple sclerosis	Brain, spinal cord (CNS)	[75]
	Vitiligo	Skin	[88]
	Primary biliary cirrhosis	Liver	[103]
	Hashimoto's thyroiditis	Thyroid	[110]
	Grave's disease	Thyroid	[111]
	Diabetes mellitus type 1	Pancreatic islet cells	[135]
	Celiac disease	Gastrointestinal tract	[136]

Given the wide spectrum of processes where epigenetic regulation participates, deregulation is associated with a wide variety of diseases, including cancer and genetically complex diseases like autoimmune disorders.

In recent years, some laboratories have begun to elucidate the relevance of epigenetic alterations in autoimmune disorders. Research in epigenetic changes has mainly focused on autoimmune rheumatic disorders, primarily in systemic lupus erythematosus (SLE) and rheumatoid arthritis (RA). Results from these studies highlight the importance of investing further efforts to identify the full range of epigenetic alterations, different cell types involved and to explore the potential of these changes as targets for therapy. This chapter will summarize the most important aspects of our knowledge about the influence of epigenetic deregulation in human autoimmune disorders (Table 11.2) with special emphasis in SLE and RA.

11.2.1 Epigenetic Deregulation in SLE

SLE is a systemic and chronic autoimmune disorder characterized by the production of non-organ-specific autoantibodies, multisystem inflammation and damage to multiple organs [3]. SLE patients experience multiple and systemic clinical manifestations and unpredictable

TABLE 11.2 Concordance Rates in Monozygotic Twins for Different Autoimmune Disorders as well as Some Examples of Epigenetic Deregulation Events in these Disorders

Disease	MZ CR*	Examples of Epigenetic Deregulation Events
Systemic lupus erythematosus	25% [6]	— DNMT1 down-regulation [21] — Global DNA hypomethyaltion [20] — Focal hypomethylation (*PRF1, CSF3R, TNFSF5, IFNGR2*) [25,29,31] — Global histone H3 & H4 hypoacetylation [35] — Focal hypoacetylation (*IL-10*) [33]
Rheumatoid arthritis	15% [6]	— Global DNA hypomethylation [44] — Focal hypomethylation (LINE-1, *MAPK13, MET*, hsa-mir-203) [45,46] — Focal hypermethylation (*DR3*) [49]
Multiple sclerosis Sjögren's syndrome Scleroderma	25% [76]	— Focal hypomethylation (*PAD2*) [81] — Focal hypermethylation (*BP230*) [86] — DNMT1 overexpression [94] — Global DNA hypomethyaltion [19] — Focal hypermethylation (*FL1*) [95]
Psoriasis	67% [97]	— Focal hypomethylation (*SHP1, p15, p16, p21*) [98,99,101] — Focal hypermethylation (*p14, p16*) [100,102]
Primary biliary cirrhosis	60% [6]	
Autoimmune thyroid diseases	30–50% [110,111]	— Skewed X chromosome inactivation [112]
Crohn's disease	60% [113]	
Ulcerative colitis	6% [113]	— Global hypomethylation [114] — Focal hypermethylation (*p14, p16, PAR2, MDR1, CDH1*) [115,116,119–121]
Diabetes mellitus type 1	21–70% [6]	
Celiac disease	75% [6]	
Ankylosing spondylitis	50% [6]	

MZ, monozygotic; CR, concordance rate.

exacerbations and relapses. These mainly affect the skin and joints, although there are also cardiac, pulmonary, renal, neuropsychiatric, hematological, and reproductive alterations [4,5]. The mechanisms responsible for the breakdown of immune tolerance to self-components and autoantibody production remain unknown, however alterations in apoptosis, cytokine levels, signaling pathways, and immune cell behavior have been characterized in SLE patients [4].

SLE is characterized by a complex etiopathology, including not only genetic but also environmental factors, such as changes in hormone levels, viral infection or exposure to chemicals. Recently, it has been proposed that epigenetic alterations induced by the environment, and therefore mysregulation of associated genes, could trigger autoimmunity when occurring in specific genetic backgrounds [3,4]. The observed concordance rates for monozygotic and dizygotic twins (around 25% and 2%, respectively), and the high heritability of SLE (more than 66%), underscores the key influence of genetic factors in this autoimmune disorder [6,7]. Currently, a variety of susceptibility genes for SLE, some of which are common to other autoimmune disorders, have been identified [3,8,9]. In addition, different environmental factors are also key to the onset and the progression of SLE disease. Much evidence supports this contribution, including partial concordance between monozygotic twins and the induction of lupus-like syndromes following administration of certain drugs. More than one hundred chemical substances, many of which influence the activity of epigenetic modifiers, have the ability to induce SLE symptoms after long-term exposure. One clear example is that of 5-azacytidine and related compounds. 5-Azacytidine is a chemical analog of cytidine and is incorporated into the DNA as cytosine during replication. Its demethylating properties are based on the formation of stable complexes between DNA methyltransferases and DNA, where the analog is incorporated [10]. 5-Azacytidine, also known under the commercial name of Vidaza®, has been used therapeutically since 2004 to treat myelodysplastic syndrome and patients receiving such treatment have frequently reported to develop lupus-like syndrome [11]. This autoimmune response has also been studied in mice. In both cases, lupus-like symptoms disappear when the treatment is discontinued. On the other hand, the in vitro treatment of human or murine CD4+ T lymphocytes with 5-azacytidine, or other demethylating agents, generates autoreactivity and the direct injection of these treated cells into healthy mice induces an autoimmune response with characteristics similar to those of SLE [12−18].

DNA METHYLATION PROFILE ALTERATION IN SYSTEMIC LUPUS ERYTHEMATOSUS

Epigenetic dysregulation in SLE has been observed at the DNA methylation and histone modification levels. In the particular case of DNA methylation, $CD4^+$ T lymphocytes of SLE patients are characterized by a significant reduction in the total content of 5-methylcytosine and the symptomatology is directly correlated with the loss of this epigenetic mark [19,20]. DNA methyltransferases were early reported to be down-regulated T cells from lupus patients [21]. Alterations in DNA methylation levels have also been described in SLE mouse models. For example, T cells obtained from the murine model MRL/lpr are also characterized by lower levels of DNA methylation and reduced DNMT1 expression than control mice and this is directly correlated with aging and SLE progression [22,23].

Global hypomethylation can potentially have different consequences, including altered gene expression, erasure of imprinting signature, and reactivation of endoparasitic sequences, ultimately contributing to the loss of autotolerance as well as to SLE development. Alterations in DNA methylation occur at repetitive sequences and at gene promoters (Figure 11.1). With respect to the repetitive elements, a global decrease in the content of 5-methylcytosine suggests hypomethylation in repetitive elements that are the major contributors of CpG dinucleotides to the genome 24. However, to date there is little information on the specific repetitive elements that are affected. Recently, it has been demonstrated that the 18S and 28S regions of the ribosomal RNA genes that are present in the genome in several hundred copies, undergo

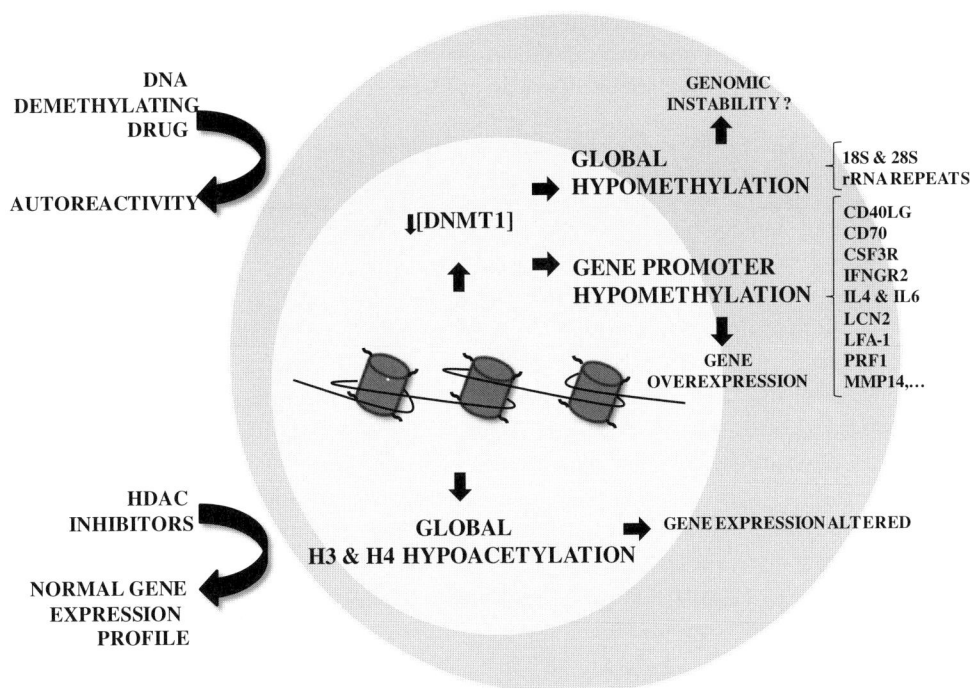

FIGURE 11.1
A scheme depicting some of the best-known epigenetic deregulation events in blood cells from SLE patients: changes in DNA methylation and histone modifications, as well as the effects of epigenetic drugs.

hypomethylation in white blood cells of SLE patients [25]. Hypomethylation at the ribosomal RNA gene repeats correlate with the overexpression of the pre-RNA and the 18S RNA. This finding could potentially be associated with increased production of ribosomal particles in SLE, perhaps associated with the production of the autoantibodies against these particles that are frequently detected in serum of SLE patients [26, 27].

With respect to promoters, various studies have reported the occurrence of hypomethylation at the regulatory region of several genes in SLE. The majority of these studies have addressed this issue by using candidate gene analysis. More recently, high-throughput analysis of monozygotic twins discordant for SLE has led to the identification of a larger set of genes, as well as dissect the contribution of DNA methylation changes to this disease [25].

Among the identified gene promoters, several of them are implicated in immune processes such as cellular death (*PRF1*), B-cell co-stimulation (*TNFSF7* and *TNFSF5*), immune synapse (*ITGAL* and *CSF3R*) and interleukin signaling (*IL4* and *IL6*). One example is represented by *PRF1*, the gene that encodes perforin, involved in cellular death through generation of a pore by insertion into the cytoplasmic membrane [28]. In CD4$^+$ T lymphocytes of SLE patients, the promoter region of the PRF1 gene is hypomethylated, the gene overexpressed and there is a consequent increase in monocyte and macrophage death [29,30]. The treatment of healthy human CD4$^+$ cells with DNA demethylating drugs induces autoreactivity in vitro and monocyte killing ability. On the other hand, the use of concanamycin A, a perforin inhibitor, reduces monocyte killing in SLE [29]. Another example is the B-cell costimulating protein CD40LG, also called CD154 or TNFSF5. TNFSF5 is encoded on the X-chromosome and the two alleles are methylated in healthy women, whereas in men the single copy is unmethylated. In CD4+ lymphocytes of women with SLE, the TNFSF5 promoter tends to become hypomethylated and, for this reason, the costimulating molecule is overexpressed [31,32]. This specific epigenetic dysregulation could partially explain the higher incidence of SLE in women than in men. These are just some examples of genes characterized by DNA methylation alteration in SLE. Undoubtedly, many other genes remain to be discovered that develop epigenetic alterations in SLE. In-depth studies based on fractionated cell populations and more powerful high-throughput technologies will provide us with a better knowledge of the role of DNA methylation changes in SLE.

HISTONE MODIFICATION PROFILE ALTERATION IN SYSTEMIC LUPUS ERYTHEMATOSUS

Our knowledge about the alterations in the histone modification patterns in SLE is also far from complete. Most of the evidence about the role of changes in histone modifications in SLE comes from the use of epigenetic drugs. Specifically, the use of histone deacetylase (HDAC) inhibitors suggests that deacetylation is involved in the skewed expression of certain genes that are associated with the disease. For instance, SLE T-helper cells exhibit increased and prolonged expression of cell-surface CD40 ligand (CD154), spontaneously overproduce interleukin-10 (IL-10), but underproduce interferon-gamma (IFN-γ). The histone deacetylase inhibitor trichostatin A (TSA) significantly reverses this skewed expression of these gene products [33] (Figure 11.1). It is likely that this reversion is the result of modification of the histone acetylation status, although alteration of the acetylation levels of regulatory proteins cannot be discarded. This result not only suggests that histone acetylation might account for this aberrant expression but also that this pharmacological agent may be a candidate for the treatment of this autoimmune disease.

Little information on the histone acetylation changes associated with SLE is available. Most of the information comes from the MRL/lpr mouse model, where histone hypoacetylation has been demonstrated [34]. In CD4$^+$ T lymphocytes of SLE patients with active disease are also characterized by global histone H3 and H4 hypoacetylation [35] (Figure 11.1). In contrast, monocytes from SLE patients have been reported to undergo histone H4 hyperacetylation at gene promoters and this alteration associates with aberrant overexpression of associated genes. Interestingly, the 179 identified genes are characterized by the presence of potential IRF1 binding sites within the 5 Kb upstream region and the IFNα treatment increases gene expression and histone acetylation [36]. The use of high-throughput approaches and analysis of specific cell types and more histone modification marks will surely lead to detailed information about genomic sites that undergo histone modification changes in SLE.

11.2.2 Epigenetic Deregulation in RA

RA is a systemic autoimmune disease characterized by synovial hyperplasia and joint inflammation and progressive destruction [37]. In the affected joint, there is an inflammatory microenvironment involving many immune cells [37]. These include synovial cells, which are hyperactivated and hyper-reactive due to high concentrations of proinflammatory cytokines. In addition, there is bone and cartilage destruction by the main effector cell types, osteoclasts and rheumatoid arthritis synovial fibroblasts (RASF), respectively [38].

RASFs constitute one of the cell types involved in RA pathogenesis as well as one of the more extensively studied. RASFs are more aggressive than their normal counterparts [39]. They overexpress metalloproteinases (MMPs) and cytokines [40] and show tumoral behavior (invasiveness [41], resistance to apoptosis [42], and anchorage-independent growth [43]).

DNA METHYLATION ALTERATIONS IN RHEUMATOID ARTHRITIS

Synovial fibroblasts are the best-characterized cells for epigenetic alterations in RA. In the early 1990s, global DNA hypomethylation was reported to occur in blood, synovial mononuclear cells, and synovial tissue of RA individuals [44] (Figure 11.2). In RASFs, hypomethylation was associated with aberrant overexpression of retrotransposable Line 1, affecting the expression of other genes [45]. More recently, met proto-oncogene (*MET*), p38delta MAP kinase (*MAPK13*), and galectin 3 binding protein (*LGALS3BP*) have been shown to be overexpressed, contributing to the aggressive phenotype of these effector cell

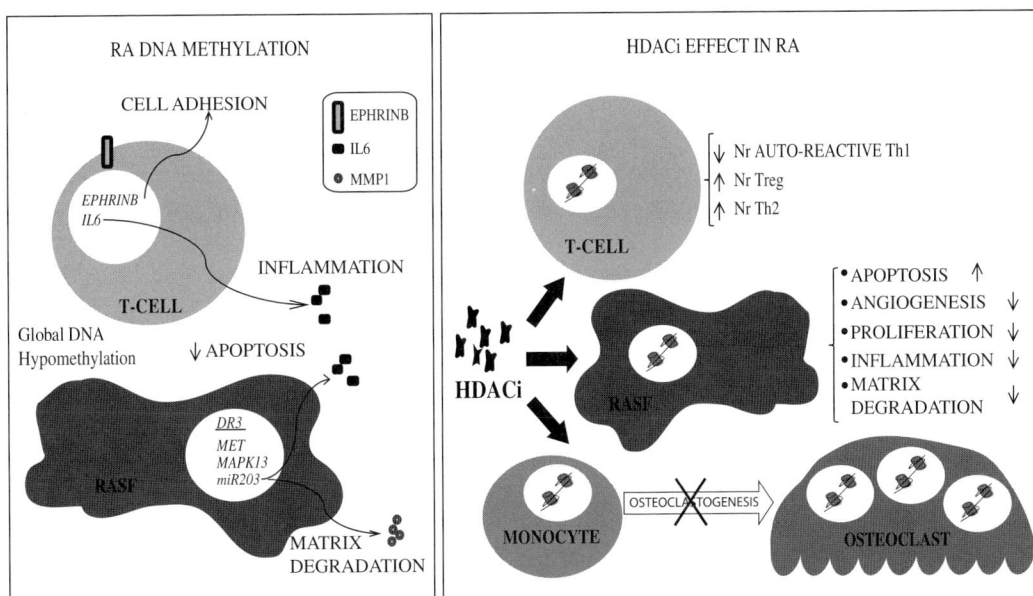

FIGURE 11.2

Epigenetic changes in RA. There is global DNA hypomethylation in the context of T-cells [20] and RASFs [44]. Examples of gene promoters hypomethylated and overexpressed in T-cells are *EPHRINB* [51] (cell adhesion molecule) and *IL6* [52] (proinflammatory cytokine). LINE1 repetitive sequences [46] and hsa-mir-203 [50] are hypomethylated in RASFs. Hsa-miR-203 deregulation leads to MMP1 (matrix degradating enzyme) and IL6 overexpression, while LINE1 deregulation causes MET and MAPK13 up-regulation. There is also specific promoter hypermethylation, as it happens in DR3 (underlined) [49], and is related to higher resistance to apoptosis. A proinflammatory microenviroment is developed within the affected joint. HDACi have shown several inhibitory effects in RA disease effector cell types. They are able to decrease the number of autoreactive Th1 while increasing Treg and Th2, balancing T cells to a more protective status [68—71]. In RASFs, HDACi are able to induce apoptosis [60,61], inhibit proliferation, and decrease secretion of inflammatory cytokines, matrix degrading proteins [56] as well as angiogenic factors [67]. It also achieves lower bone degradation by inhibiting osteoclastogenesis [74]. This figure is reproduced in the color plate section.

211

type [46]. Neidhart and Gay's team has identified that RASF display lower levels of DNMT1 than their healthy counterparts. They have also shown that these alterations in DNMT1 and global DNA methylation levels associate with the aggressive phenotype of RASFs. Moreover, synovial fibroblasts obtained from healthy donors treated with DNA hypomethylating drugs, like 5-aza-2-desoxicytidine, become as aggressive as RASFs [47].

RASFs also display specific promoter hypermethylation, like cancer cells [48,49] (Figure 11.2). For instance, the promoter of the tumor necrosis factor receptor superfamily, member 25 gene (*TNFRSF25*), also known as death receptor 3 (*DR3*), is hypermethylated, and consequently down-regulated. This could explain the increased resistance to apoptosis that this cell type exhibits [49]. Some microRNAs are also deregulated in RASF by aberrant methylation changes at their promoters. As an example, hsa-miR-203 is up-regulated in RASFs relative to OASFs, and leads to an increase in the production of MMP-1 and IL-6 [50].

Blood cells are also reported to exhibit DNA methylation changes in RA. One example is Ephrin B1 (*EFNB1*), which codes for a membrane protein involved in cell adhesion and inflammation process signaling. EFNB1 mRNA and protein levels are up-regulated in blood and synovial T lymphocytes owing to the hypomethylation of its promoter [51] (Figure 11.2). Another example is the multifunctional interleukin-6 (IL-6). Loss of methylation of a single CpG at the promoter of IL-6 results in overexpression, that probably results in immune cell hyperactivation [52].

In RA CD4$^+$ T cells, as well as in RASFs, there is global DNA hypomethylation and lower activity levels of DNMTs [20].

There is a subset of "senescent" CD4$^+$ CD28$^-$ T cells that are more frequent in the elderly and RA patients, and that, among other abnormalities, are autoreactive. In this subset, ERK and

JNK pathways are decreased and DNMT1 and DNMT3 down-regulated. These decreased levels of these DNMTs are associated with demethylation of some gene promoters and subsequent protein overexpression. Some examples are *CD70*, perforin (*PRF1*), and *KIR2DL4*, which could contribute to the inflammatory phenotype [53].

It has been speculated that there is an imbalance of helper lymphocyte differentiation in RA. In this disease, peripheral CD4$^+$ T cells are characterized by *FOXP3* promoter hypermethylation. On the other hand, synovial CD4$^+$ T cells experience *FOXP3* and *IFNG* promoter demethylation, possibly indicating a higher representation of regulatory T and T helper 1 cells in the affected joints [54].

HISTONE MODIFICATION PROFILE ALTERATION IN RHEUMATOID ARTHRITIS

In RA, there is also evidence of a potential role of histone modifications. Although no specific histone modification alterations have been described in this autoimmune disorder, the important role of this epigenetic mechanism is demonstrated by research based on HDAC inhibitors (Figure 11.2).

In RA, synovial fibroblasts show higher levels of HDAC1, and these are positively regulated by the concentration of TNF-alpha [55]. The depletion of HDAC1 and HDAC2 results in decreased cell proliferation and inhibition of certain TNF-alpha cytokines such as MMP1 [56]. Nucleosomes located along the MMP1 promoter are hyperacetylated in RASFs, where this gene is overexpressed [57].

The first observation that HDACi could modulate the expression of RA-related genes was made in 2003 in an animal model of this autoimmune disorder. Adjuvant arthritis rats treated with phenylbutyrate and trichostatin A (TSA) showed RA pathology suppression and the isolated RASFs from these treated rats did not proliferate. This effect was due to the up-regulation of two CDK inhibitors (p16^{INK4a} and p21$^{WAF1/Cip1}$) and inhibition of *TNF-alpha*, *IL-1*, and *IL-6* in the affected joints [58]. In 2004, systemic administration of the HDACi FK228 succeeded in ameliorating synovial proliferation and joint destruction when administered to autoantibody-mediated arthritis mice by up-regulating the same CDK inhibitors [59]. It has also been shown that treatment of RASFs with HDACi induces apoptosis via Fas receptor and TRAIL [60,61]. Other HDACi, such as MS-275 or suberoylanilide hydroxamic acid (SAHA), have also demonstrated growth arrest activity, inhibition of proinflammatory cytokines (TNF-alpha, IL-1beta) as well as down-regulation of angiogenesis and MMP expression in synovial fibroblasts in several RA animal models and patients [62–65]. The inhibition of HDACs, and thus, of MMPs, produces decreased cartilage resorption [66].

HDACi show strong anti-inflammatory effects in vivo that cannot only be explained by their effects on cell cycle and cell proliferation. These drugs are also involved in the inhibition of angiogenesis in the affected joint by down-regulating the hypoxia-induced factors HIF-1α and VEGF, which play central roles in the angiogenic process [67].

Apart from RASFs, the antiarthritic effects of HDACi have also been tested in T cells. In RA patients, regulatory T cells fail to suppress CD4$^+$ effector cells. TSA and valproic acid are able to increase the function and number of *FOXP3* expressing CD25$^+$ CD4$^+$ regulatory T lymphocytes, which are able to inhibit the proliferative response of effector T cells [68], leading to better clinical features [69].

TSA also has the ability to suppress arthritis severity in collagen-induced arthritis mice. This is achieved by inhibiting autoantigen-specific T helper 1 cell proliferation and apoptosis induction and also gives rise to less IFN-gamma release. On the other hand, it enhances T helper 2 response by acetylating IL-4 promoter and thus its overexpression [70]. In conclusion, TSA is able to balance the relationship between T helper 1 and T helper 2 to produce a protective status. Moreover, another HDACi, LAQ824, has previously been characterized by the ability to regulate this balance in human healthy T lymphocytes [71].

HDACi have also been shown to induce osteoblast proliferation and maturation as well as accelerating matrix mineralization by regulating the expression of several genes, such as those coding for growth factors and Wnt receptors [72,73].

On the other hand, the HDACi FR901228 can inhibit osteoclastogenesis. It inhibits nuclear translocation of NFATc1 (a crucial factor for osteoclastogenic differentiation) and boosts production of IFN-beta (an inhibitor of osteoclastogenesis). As a result, bone destruction is suppressed, and the clinical features of RA are improved [74].

11.2.3 Epigenetic Deregulation in Multiple Sclerosis

Multiple sclerosis (MS) is a multisystemic inflammatory and chronic autoimmune disorder characterized by demyelination and neurodegeneration of brain and spinal cord [75]. It is a genetically complex disease, but its etiology depends on environmental factors including Epstein—Barr virus infection. It is also thought that epigenetic dysregulation also participates in the onset of the disease driven by environmental factors. In fact, concordance rates of around 25% and 5% in monozygotic and dizygotic twins have been estimated [76]. Sunlight deficiency, low vitamin D ingestion, radioactivity, geomagnetism, atmospheric contaminants, toxins, and nicotine poisoning have been identified as environmental triggers of epigenetic deregulation in MS [77]. Various studies have demonstrated that some of these factors have the ability to modify the epigenetic profile directly or indirectly [78,79].

Although the participation of epigenetic deregulation events in MS development has been proposed, there is little experimental evidence. One key protein in MS is myelin basic protein (MBP), a major component of the myelin sheath of Schwann cells and oligodendrocytes in the nervous system. This protein is citrullated by peptide arginine deiminase 2 (PAD2), generating loss of myelin instability. In MS, the *PAD2* promoter is hypomethylated and consequently PAD2 is overproduced and the myelin is more easily degraded [80,81]. Moreover, MBP citrullation has the ability to induce autocleavage, generating new potential autoantigens and allowing molecular mimicry [82,83].

Recently, Baranzini and collaborators studied the genome, the epigenome, and the transcriptome of CD4$^+$ T cells from three pairs of monozygotic twins discordant for MS. Although, they did not detect any reproducible and significant difference at the DNA methylation profile between co-siblings, this study represented the first high-throughput approach to investigate DNA methylation alterations in MS [84]. It is expected that future high-throughput studies with larger sets of samples and perhaps using other sets of cell types will help to dissect epigenetic profiles specific to MS.

11.2.4 Epigenetic Deregulation in Sjögren's Syndrome

Sjögren's syndrome (SjS) is a systemic autoimmune disease characterized by chronic inflammation of the exocrine glands that produce saliva and tears [85].

As in the previously described autoimmune disorders, SjS is a multifactorial disease in which genetic predisposition and environmental factors both feature. One report has described *BP230* promoter hypermethylation in labial salivary glands of SjS patients. *BP230* is the epithelial splice isoform of the dystonin gene, a component of the plakin protein family of adhesion junction plaque molecules and it is involved in basal lamina anchorage. This hypermethylation correlates with mRNA down-regulation, although paradoxically the protein level is increased [86]. The *BP230* epigenetic deregulation could explain the anchorage alterations of salivary gland cells that characterize SjS patients. Also, in SjS patients' salivary glands, up-regulation of two miRNAs has been detected and have been shown to be useful as markers of disease progression [87].

11.2.5 Epigenetic Deregulation in Vitiligo

Vitiligo is a depigmenting autoimmune disorder characterized by melanocyte destruction and patchy loss of skin pigmentation. Unlike other autoimmune disorders, it equally affects both sexes and all races, although it usually appears between the ages of 10 and 30 years [88]. Interestingly, vitiligo occurs in association with other autoimmune disorders including Addison's disease, autoimmune thyroiditis, SLE, RA, and psoriasis among others, highlighting that mechanisms of deregulation are shared between all these conditions [89,90]. The etiology of vitiligo is a matter of a complex interaction between genetic predisposition and epigenetic alteration [91]. Environmental factors are key regulators of the onset of vitiligo. In fact, sunburn and traumas are associated with the manifestation of this autoimmune disorder. There are few data about epigenetic deregulation events in this autoimmune disorder, although the use of the Smyth line (SL) chicken, an animal model for human vitiligo, has yielded some information. Specifically, in vivo 5-azacytidine treatment of these chickens induces the production of antibodies against the melanocyte-specific protein TRP-1, skin melanocyte depletion and consequent depigmentation [92]. This study represents a clear example of how the administration of a DNA hypomethylating drug can induce vitiligo disease in genetically predisposed individuals, reinforcing the role of genetic and epigenetic influences on this autoimmune disorder.

11.2.6 Epigenetic Deregulation in Progressive Systemic Sclerosis

Progressive systemic sclerosis (PSS), or scleroderma, is a rare disease characterized by excessive collagen deposition, mainly in skin, but also in other organs, and progressive vasculopathy. It is considered an autoimmune disease because of the presence of autoantibodies, several of which act against nuclear epitopes. A greater incidence in women and frequent autoimmune comorbidities have been observed [93].

PSS is characterized by aberrant fibroblast activation in which there is greater deposition of collagen and lower secretion of MMPs, resulting in an elastic skin. These fibroblasts maintain their activated phenotype even when cultured in vitro, suggesting that gene expression is epigenetically regulated. Specifically, PSS fibroblasts produce higher levels of some cytokines and growth factors as well as lower levels of some MMPs. A correlation has been reported between the scleroderma fibroblast phenotype and higher detected levels of DNMT1, emphasizing the key role of epigenetic deregulation in PSS [94]. For example, FL1, a transcription factor that inhibits collagen production, undergoes hypermethylation in PSS fibroblasts and thus silencing, resulting in increased production of collagen. This property can be reverted by treatment with epigenetic modifiers, including HDACi and DNA methylation inhibitors. These observations provide a bona fide example of how epigenetic mechanisms orchestrate the pathogenic phenotype of a cell involved in an autoimmune disorder [95].

T lymphocytes are another cell type involved in the pathogenesis of this disease. In particular, this cell population displays lower levels of 5-methylcytosine and down-regulation of the methyl-CpG binding domain protein (MBD4) [19].

On the other hand, the reported important role of skewed X-chromosome inactivation mosaicism in the pathogenesis of scleroderma could explain the great prevalence of PSS in females [96]. This gene–dosage equilibration mechanism is also altered in autoimmune thyroid diseases (AITDs), as is explained in more detail below.

11.2.7 Epigenetic Deregulation in Psoriasis

Psoriasis is a chronic systemic skin disease characterized by precocious keratinocyte differentiation and hyperproliferation, triggered by autoimmune reactions.

Psoriasis etiopathology consists of a complex combination of genetic risk and epigenetic deregulation, which is reflected by the concordance rates for monozygotic and dizygotic twins for this autoimmune disorder (around 67% and 15%, respectively) [97].

With respect to the identification of epigenetic alterations in psoriasis, *SHP-1* (also called *PTPN6*) promoter 2 is hypomethylated in psoriatic skin samples and the isoform II expression levels are increased. Conversely, this promoter is hypermethylated in normal skin, repressing gene expression. This observation contrasts with the silencing of this gene observed in other hyperproliferative syndromes, such as leukemias and lymphomas [98].

Abnormal proliferative activity has been detected in the hematopoietic cells of patients suffering psoriasis. The p16^{INK4A} tumor suppressor is a negative regulator of CDK4, a protein that accelerates cell cycle progression. In hematopoietic cells obtained from psoriatic patients, p16 gene promoter is hypomethylated and p16 mRNA level is increased. This overproduction results in a reduced ability to form colonies in hematopoietic cells from psoriasis patients, showing that bone marrow, and not only immunocytes, is dysfunctional in psoriatic patients [99]. Conversely, p16 is hypermethylated in epidermal samples from psoriatic patients [100]. Consistent with this line of evidence, other research shows a lower proliferative potential of hematopoietic cells in psoriatic patients, and is correlated with lower promoter methylation of p15 and p21 genes, and thus higher expression levels [101].

An increase in DNA methylation has been reported in PBMCs from psoriatic patients as well as in psoriatic skin lesions. This methylation is correlated positively with PASI (Psoriasis Area and Severity Index) scores. Specifically, DNMT1 is up-regulated while the MBD2 and MeCP2 are down-produced, indicating that the methylation machinery of psoriatic patients is altered. Another example is that of the p14 gene promoter. This gene, a homologues of p16^{INK4A}, is hypermethylated and its expression is consequently downregulated [102].

11.2.8 Epigenetic Deregulation in Primary Biliary Cirrhosis

Primary biliary cirrhosis (PBC) is a chronic and progressive organic-specific autoimmune disease of the liver that it is characterized by the destruction of small-to-medium bile ducts, cholestasis fibrosis, and cirrhosis [103].

The etiology remains unknown, although it is clearly related to a combination of genetic predisposition and environmental stimulation, as indicated by the 60% concordance for monozygotic twins and the striking predominance in women (female:male ratio of 10:1) [104]. Various studies have demonstrated that familiar incidence of PBC, recurrent urinary tract infections, use of nail polish, and hormone treatment are risk factors for PBC development [105]. Some of these external factors have the ability to induce epigenetic changes. Despite the lack of studies, it is likely that dysregulation of epigenetic factors is implicated in PBC. For example, two X-chromosome genes are known to have lower expression in peripheral blood cells of PBC twins in comparison with their healthy counterparts [106]. These genes are *CLIC2*, which encodes a chloride channel involved in stabilization of cell membrane potential, transepithelial transport, maintenance of intracellular pH, and regulation of cell volume [107,108], and *PIN4*, a member of the parvulin subfamily of the peptidyl-prolyl *cis/trans* isomerases [109]. Interestingly, promoter methylation analysis revealed some dysregulation events but not always correlated with the expression behavior [106]. This study emphasizes the relationship between sex, epigenetics, and PBC.

11.2.9 Epigenetic Deregulation in Thyroid Diseases

Autoimmune thyroid diseases (AITDs) are the group of conditions in which the autoimmune attack of the thyroid takes place by infiltration of lymphocytes of the glandule. It is an organ-specific autoimmune disease that affects mainly women. This attack results in two opposite clinical outcomes: Hashimoto's thyroiditis (HT) and Grave's disease (GD). HT is a hypothyroidism in which there is apoptosis of thyroid cells, while GD involves hyperactivation of the thyroid due to TSH receptor-stimulating antibodies causing hyperthyroidism.

Many susceptibility loci have been identified including HLA-DR, CTLA-4, CD40, FOXP3, CD25, TSHR, and others. However, as with many other autoimmune diseases, genetics cannot

215

explain all the factors involved in AITDs. The concordance rates between monozygotic twins are around 30–50% [110,111], implying that other mechanisms such as epigenetic changes could trigger the onset of the disease.

A mechanism of skewed X chromosome inactivation (XCI) has been proposed in AITD. Under physiological conditions, females have in their cells either the maternal or paternal X-chromosome active, at random and at a ratio of 50:50. When this ratio is modified (for example 20:80) there is a process of skewed XCI. Much epigenetic machinery is involved in the process of XCI in order to methylate CpG on DNA, lysines of histones, and deacetylate the silenced region.

Mechanistically, the expression of antigens may not be sufficiently tolerated in some cells. Some self-antigens on one X-chromosome may not be present in the thymus or in other tolerance-related tissues, but may be highly expressed in other tissues, triggering immune responses [112].

11.2.10 Epigenetic Deregulation in Inflammatory Bowel Diseases: Crohn's Disease and Ulcerative Colitis

Inflammatory bowel diseases (IBD) are a group of pathologies that affect the digestive tract, including the small intestine and colon, and are characterized by inflammation and tissue destruction. There are two main forms: Crohn's disease (CD) and ulcerative colitis (UC). CD, also known as regional enteritis or ileitis, is an organ-specific autoimmune disorder that compromises the gastrointestinal epithelia, from mouth to anus. It has a concordance rate among monozygotic twins of around 60% [113], showing a stronger genetic component than UC. UC is a type of inflammatory bowel disease in which only the large intestine or colon is affected by the autoimmune pathological mechanism. UC patients present ulcers and open sores in their gastric mucosa that cause abdominal pain, diarrhea, fever, and weight loss, among other symptoms. The concordance rate for monozygotic twins is as low as 6% [113], displaying the slight effect that genetic fingerprints have on this disease. The causes triggering this condition are still unknown, but there are some lines of evidence that point to the contribution of epigenetic mechanisms to UC pathogenesis.

With respect to DNA methylation, it has been reported that global hypomethylation correlates with higher proliferative activity in the affected mucosa of the rectum from patients with longstanding UC in comparison with healthy individuals [114]. Moreover, CpG island hypermethylation at specific gene promoters has been described, and this epigenetic signature is involved in UC-associated colorectal carcinogenesis [115]. Specific promoter hypermethylation has been assessed for many key sequences. Among these, CDKN2A at both p16^{INK4a} [115] and p14^{ARF} [116] stand out, since the promoters of both are commonly silenced by methylation in sporadic colorectal cancer [117,118].

The methylation status of other genes, including PAR2, MDR1, CDH1, and GDNF, has been correlated with the severity of the disease. In particular, PAR2 promoter methylation correlates with severe clinical phenotypes of UC [119] and the chronic continuous type and earlier onset of UC are associated with MDR1 promoter hypermethylation [120]. In the parts of the gastric mucosa with active methylation, there is specific methylation of CDH1 and GDNF [121]. On the other hand, patients suffering UC have an increased incidence of colorectal cancer (UC-associated colorectal carcinoma) and there are some genes that could be used as biomarkers for early detection of cancer or dysplasia in UC. Some examples of these epigenetic biomarkers are CDH1, HPP1, EYA4, SFRP2, ESR1, RUNX3, MINT1, COX-2, and DAPK. Interestingly, the methylation level of these regulatory sequences directly increases with malignancy [122–128]. It is remarkable that in a subset of IBD-associated colorectal carcinoma there is microsatellite instability due to the hypermethylation and silencing of MLH1 gene and mutation in TGFbetaRII, whose proteins are involved in the DNA mismatch repair gene pathway [129,130].

It has been speculated that UC is a disease of accelerated aging of the colon, an idea for which there is evidence from the short length of the telomeres and the level of DNA damage detected in the colonocytes of these people [131]. In fact, in the mucosa of UC patients, the promoter of some genes, such as *ER*, *MYOD*, *p16*, and *CSPG2* have accelerated age-related methylation [132]. These findings could help explain the higher risk of UC-associated colorectal carcinogenesis associated with this autoimmune disorder.

On the other hand, there are some polymorphisms, affecting XRCC1 and GST genes among others, that have a protective effect against methylation in UC. This could be useful to classify patients with high or low risk of developing cancerous lesions in ulcerative colitis [133].

UC and CD share epigenetic dysregulation features, as shown in a genome-wide methylation study in which many genes related to immune system were found to be differentially methylated in both diseases. Some examples of these commonly altered genes are *STAT5A*, *TNFTSF1A*, PECAM1, FABP3, FGF, and TNFSF8, among others [134].

11.2.11 Epigenetic Deregulation in Other Organ-Specific Autoimmune Disorders

Type 1 diabetes mellitus (T1DM), also known as juvenile diabetes, is a type of diabetes mellitus generated by the organ-specific autoimmune destruction of the insulin-producing beta cells of the pancreas [135]. As in the diseases described above, T1DM is mainly determined by epigenetic deregulation against a background of genetic susceptibility, data corroborated by the concordance rate of 21% to 70% in monozygotic twins and around 10% in dizygotic twins [6]. Although there is no doubt about the contribution of the epigenetic component in this autoimmune disorder, no specific examples of epigenetic dysregulation have yet been described.

Celiac disease, also called celiac sprue or gluten-sensitive enteropathy, is another example of organ-specific autoimmune disorder determined by the environment and genetic susceptibility. It involves chronic inflammation and destruction of the proximal small intestine and consequently the alteration of nutrient absorption. Patients with this condition cannot tolerate gluten. In fact, a transglutaminase enzyme modifies gliadin, a gluten component, generating an immune system crossreaction and destruction of the tissue of the small bowel [136]. In this case, the concordance rates for monozygotic and dizygotic twins are around 75% and 11%, respectively [6].

Ankylosing spondylitis (AS), also known as Bekhterev syndrome or Marie-Strümpell disease, is an organ-specific autoimmune disease. This disorder is one of the spondyloarthropathies which mainly occurs in young men. This chronic arthritis affects the spinal and sacroiliac joints, generating inflammation, joint damage and, in severe cases, bone formation and consequent spine fusion [137]. Despite its genetic contribution, environmentally driven epigenetic changes play an important role in AS etiology, a fact reflected in the high percentage of concordance between monozygotic and dizygotic siblings (50% and 20%, respectively) [6].

11.3 CONCLUSIONS

The study of epigenetic alterations in autoimmune diseases still needs greater efforts to determine its relevance and potential as targets for therapeutic approaches. Further studies are therefore needed to understand the epigenetic contributions to the pathogenesis of autoimmune diseases. Many of these disorders share clinical and genetic features and are influenced by similar environmental factors. Epigenetic modifications are influenced by environmental factors and are known to directly determine gene function, therefore constituting a relevant target to investigate its participation in the etiology of these diseases. Most efforts to identify the epigenetic alterations that occur in autoimmune disease have focused on SLE and RA and have served to identify both global and sequence-specific hypomethylation and overexpression of key genes in immune function. Several issues are now key to address: to make extensive use of high-throughput approaches, to systematically analyze all potential specific cell types relevant to disease pathogenesis, and to find the best way of using this information in a clinical setting.

References

[1] Portela A, Esteller M. Epigenetic modifications and their relevance to disease. Nat Biotech 2010.

[2] Berger SL, Kouzarides T, Shiekhattar R, Shilatifard A. An operational definition of epigenetics. Genes Dev 2009;23:781−3.

[3] Rahman A, Isenberg DA. Systemic lupus erythematosus. N Engl J Med 2008;358:929−39.

[4] D'Cruz DP, Khamashta MA, Hughes GR. Systemic lupus erythematosus. Lancet 2007;369:587−96.

[5] Rothfield N, Sontheimer RD, Bernstein M. Lupus erythematosus: systemic and cutaneous manifestations. Clin Dermatol 2006;24:348−62.

[6] Ballestar E. Epigenetics lessons from twins: prospects for autoimmune disease. Clin Rev Allergy Immunol 2010;39:30−41.

[7] Sullivan KE. Genetics of systemic lupus erythematosus. Clinical implications. Rheum Dis Clin North Am 2000;26:229−56. v-vi.

[8] Moser KL, Kelly JA, Lessard CJ, Harley JB. Recent insights into the genetic basis of systemic lupus erythematosus. Genes Immun 2009;10:373−9.

[9] Rhodes B, Vyse TJ. The genetics of SLE: an update in the light of genome-wide association studies. Rheumatology (Oxford) 2008;47:1603−11.

[10] Cihak A. Biological effects of 5-azacytidine in eukaryotes. Oncology 1974;30:405−22.

[11] Kaminskas E, Farrell A, Abraham S, Baird A, Hsieh LS, Lee SL, et al. Approval summary: azacitidine for treatment of myelodysplastic syndrome subtypes. Clin Cancer Res 2005;11:3604−8.

[12] Richardson B. Effect of an inhibitor of DNA methylation on T cells. II. 5-Azacytidine induces self-reactivity in antigen-specific T4+ cells. Hum Immunol 1986;17:456−70.

[13] Yung R, Powers D, Johnson K, Amento E, Carr D, Laing T, et al. Mechanisms of drug-induced lupus. II. T cells overexpressing lymphocyte function-associated antigen 1 become autoreactive and cause a lupuslike disease in syngeneic mice. J Clin Invest 1996;97:2866−71.

[14] Yung RL, Quddus J, Chrisp CE, Johnson KJ, Richardson BC. Mechanism of drug-induced lupus. I. Cloned Th2 cells modified with DNA methylation inhibitors in vitro cause autoimmunity in vivo. J Immunol 1995;154:3025−35.

[15] Yung R, Chang S, Hemati N, Johnson K, Richardson B. Mechanisms of drug-induced lupus. IV. Comparison of procainamide and hydralazine with analogs in vitro and in vivo. Arthritis Rheum 1997;40:1436−43.

[16] Richardson BC, Strahler JR, Pivirotto TS, Quddus J, Bayliss GE, Gross LA, et al. Phenotypic and functional similarities between 5-azacytidine-treated T cells and a T cell subset in patients with active systemic lupus erythematosus. Arthritis Rheum 1992;35:647−62.

[17] Cornacchia E, Golbus J, Maybaum J, Strahler J, Hanash S, Richardson B. Hydralazine and procainamide inhibit T cell DNA methylation and induce autoreactivity. J Immunol 1988;140:2197−200.

[18] Quddus J, Johnson KJ, Gavalchin J, Amento EP, Chrisp CE, Yung RL, et al. Treating activated CD4+ T cells with either of two distinct DNA methyltransferase inhibitors, 5-azacytidine or procainamide, is sufficient to cause a lupus-like disease in syngeneic mice. J Clin Invest 1993;92:38−53.

[19] Lei W, Luo Y, Yan K, Zhao S, Li Y, Qiu X, et al. Abnormal DNA methylation in CD4+ T cells from patients with systemic lupus erythematosus, systemic sclerosis, and dermatomyositis. Scand J Rheumatol 2009:1−6.

[20] Richardson B, Scheinbart L, Strahler J, Gross L, Hanash S, Johnson M. Evidence for impaired T cell DNA methylation in systemic lupus erythematosus and rheumatoid arthritis. Arthritis Rheum 1990;33:1665−73.

[21] Deng C, Kaplan MJ, Yang J, Ray D, Zhang Z, McCune WJ, et al. Decreased Ras-mitogen-activated protein kinase signaling may cause DNA hypomethylation in T lymphocytes from lupus patients. Arthritis Rheum 2001;44:397−407.

[22] Wu J, Zhou T, He J, Mountz JD. Autoimmune disease in mice due to integration of an endogenous retrovirus in an apoptosis gene. J Exp Med 1993;178:461−8.

[23] Mizugaki M, Yamaguchi T, Ishiwata S, Shindo H, Hishinuma T, Nozaki S, et al. Alteration of DNA methylation levels in MRL lupus mice. Clin Exp Immunol 1997;110:265−9.

[24] Wilson AS, Power BE, Molloy PL. DNA hypomethylation and human diseases. Biochim Biophys Acta 2007;1775:138−62.

[25] Javierre BM, Fernandez AF, Richter J, Al-Shahrour F, Martin-Subero JI, Rodriguez-Ubreva J, et al. Changes in the pattern of DNA methylation associate with twin discordance in systemic lupus erythematosus. Genome Res 2010;20:170−9.

[26] Isenberg DA, Manson JJ, Ehrenstein MR, Rahman A. Fifty years of anti-ds DNA antibodies: are we approaching journey's end? Rheumatology (Oxford) 2007;46:1052−6.

[27] Sawalha AH, Harley JB. Antinuclear autoantibodies in systemic lupus erythematosus. Curr Opin Rheumatol 2004;16:534−40.

[28] van den Broek MF, Hengartner H. The role of perforin in infections and tumour surveillance. Exp Physiol 2000;85:681—5.

[29] Kaplan MJ, Lu Q, Wu A, Attwood J, Richardson B. Demethylation of promoter regulatory elements contributes to perforin overexpression in CD4$^+$ lupus T cells. J Immunol 2004;172:3652—61.

[30] Luo Y, Zhang X, Zhao M, Lu Q. DNA demethylation of the perforin promoter in CD4(+) T cells from patients with subacute cutaneous lupus erythematosus. J Dermatol Sci 2009;56:33—6.

[31] Lu Q, Wu A, Tesmer L, Ray D, Yousif N, Richardson B. Demethylation of CD40LG on the inactive X in T cells from women with lupus. J Immunol 2007;179:6352—8.

[32] Zhou Y, Yuan J, Pan Y, Fei Y, Qiu X, Hu N, et al. T cell CD40LG gene expression and the production of IgG by autologous B cells in systemic lupus erythematosus. Clin Immunol 2009;132:362—70.

[33] Hu N, Qiu X, Luo Y, Yuan J, Li Y, Lei W, et al. Abnormal histone modification patterns in lupus CD4$^+$ T cells. J Rheumatol 2008;35:804—10.

[34] Garcia BA, Busby SA, Shabanowitz J, Hunt DF, Mishra N. Resetting the epigenetic histone code in the MRL-lpr/lpr mouse model of lupus by histone deacetylase inhibition. J Proteome Res 2005;4:2032—42.

[35] Mishra N, Brown DR, Olorenshaw IM, Kammer GM. Trichostatin A reverses skewed expression of CD154, interleukin-10, and interferon-gamma gene and protein expression in lupus T cells. Proc Natl Acad Sci USA 2001;98:2628—33.

[36] Zhang Z, Song L, Maurer K, Petri MA, Sullivan KE. Global H4 acetylation analysis by ChIP-chip in systemic lupus erythematosus monocytes. Genes Immun 2009.

[37] Duke O, Panayi GS, Janossy G, Poulter LW. An immunohistological analysis of lymphocyte subpopulations and their microenvironment in the synovial membranes of patients with rheumatoid arthritis using monoclonal antibodies. Clin Exp Immunol 1982;49:22—30.

[38] Neumann E, Lefevre S, Zimmermann B, Gay S, Muller-Ladner U. Rheumatoid arthritis progression mediated by activated synovial fibroblasts. Trends Mol Med 2010;16:458—68.

[39] Tolboom TC, van der Helm-Van Mil AH, Nelissen RG, Breedveld FC, Toes RE, Huizinga TW. Invasiveness of fibroblast-like synoviocytes is an individual patient characteristic associated with the rate of joint destruction in patients with rheumatoid arthritis. Arthritis Rheum 2005;52:1999—2002.

[40] Distler JH, Jungel A, Huber LC, Seemayer CA, Reich 3rd CF, Gay RE, et al. The induction of matrix metalloproteinase and cytokine expression in synovial fibroblasts stimulated with immune cell microparticles. Proc Natl Acad Sci USA 2005;102:2892—7.

[41] Muller-Ladner U, Kriegsmann J, Franklin BN, Matsumoto S, Geiler T, Gay RE, et al. Synovial fibroblasts of patients with rheumatoid arthritis attach to and invade normal human cartilage when engrafted into SCID mice. Am J Pathol 1996;149:1607—15.

[42] Baier A, Meineckel I, Gay S, Pap T. Apoptosis in rheumatoid arthritis. Curr Opin Rheumatol 2003;15:274—9.

[43] Lafyatis R, Remmers EF, Roberts AB, Yocum DE, Sporn MB, Wilder RL. Anchorage-independent growth of synoviocytes from arthritic and normal joints. Stimulation by exogenous platelet-derived growth factor and inhibition by transforming growth factor-beta and retinoids. J Clin Invest 1989;83:1267—76.

[44] Corvetta A, Della Bitta R, Luchetti MM, Pomponio G. 5-Methylcytosine content of DNA in blood, synovial mononuclear cells and synovial tissue from patients affected by autoimmune rheumatic diseases. J Chromatogr 1991;566:481—91.

[45] Neidhart M, Rethage J, Kuchen S, Kunzler P, Crowl RM, Billingham ME, et al. Retrotransposable L1 elements expressed in rheumatoid arthritis synovial tissue: association with genomic DNA hypomethylation and influence on gene expression. Arthritis Rheum 2000;43:2634—47.

[46] Kuchen S, Seemayer CA, Rethage J, von Knoch R, Kuenzler P, Beat AM, et al. The L1 retroelement-related p40 protein induces p38delta MAP kinase. Autoimmunity 2004;37:57—65.

[47] Karouzakis E, Gay RE, Michel BA, Gay S, Neidhart M. DNA hypomethylation in rheumatoid arthritis synovial fibroblasts. Arthritis Rheum 2009;60:3613—22.

[48] Esteller M. Epigenetics in cancer. N Engl J Med 2008;358:1148—59.

[49] Takami N, Osawa K, Miura Y, Komai K, Taniguchi M, Shiraishi M, et al. Hypermethylated promoter region of DR3, the death receptor 3 gene, in rheumatoid arthritis synovial cells. Arthritis Rheum 2006;54:779—87.

[50] Stanczyk J, Ospelt C, Karouzakis E, Filer A, Raza K, Kolling C, et al. Altered expression of microRNA-203 in rheumatoid arthritis synovial fibroblasts and its role in fibroblast activation. Arthritis Rheum 2011;63:373—81.

[51] Kitamura T, Kabuyama Y, Kamataki A, Homma MK, Kobayashi H, Aota S, et al. Enhancement of lymphocyte migration and cytokine production by ephrinB1 system in rheumatoid arthritis. Am J Physiol Cell Physiol 2008;294:C189—96.

[52] Nile CJ, Read RC, Akil M, Duff GW, Wilson AG. Methylation status of a single CpG site in the IL6 promoter is related to IL6 messenger RNA levels and rheumatoid arthritis. Arthritis Rheum 2008;58:2686—93.

[53] Chen Y, Gorelik GJ, Strickland FM, Richardson BC. Decreased ERK and JNK signaling contribute to gene overexpression in "senescent" CD4+CD28- T cells through epigenetic mechanisms. J Leukoc Biol 2010;87:137–45.

[54] Janson PC, Linton LB, Bergman EA, Marits P, Eberhardson M, Piehl F, et al. Profiling of CD4$^+$ T cells with epigenetic immune lineage analysis. J Immunol 2011;186:92–102.

[55] Kawabata T, Nishida K, Takasugi K, Ogawa H, Sada K, Kadota Y, et al. Increased activity and expression of histone deacetylase 1 in relation to tumor necrosis factor-alpha in synovial tissue of rheumatoid arthritis. Arthritis Res Ther 2010;12:R133.

[56] Horiuchi M, Morinobu A, Chin T, Sakai Y, Kurosaka M, Kumagai S. Expression and function of histone deacetylases in rheumatoid arthritis synovial fibroblasts. J Rheumatol 2009;36:1580–9.

[57] Maciejewska-Rodrigues H, Karouzakis E, Strietholt S, Hemmatazad H, Neidhart M, Ospelt C, et al. Epigenetics and rheumatoid arthritis: the role of SENP1 in the regulation of MMP-1 expression. J Autoimmun 2010;35:15–22.

[58] Chung YL, Lee MY, Wang AJ, Yao LF. A therapeutic strategy uses histone deacetylase inhibitors to modulate the expression of genes involved in the pathogenesis of rheumatoid arthritis. Mol Ther 2003;8:707–17.

[59] Nishida K, Komiyama T, Miyazawa S, Shen ZN, Furumatsu T, Doi H, et al. Histone deacetylase inhibitor suppression of autoantibody-mediated arthritis in mice via regulation of p16INK4a and p21(WAF1/Cip1) expression. Arthritis Rheum 2004;50:3365–76.

[60] Jungel A, Baresova V, Ospelt C, Simmen BR, Michel BA, Gay RE, et al. Trichostatin A sensitises rheumatoid arthritis synovial fibroblasts for TRAIL-induced apoptosis. Ann Rheum Dis 2006;65:910–2.

[61] Morinobu A, Wang B, Liu J, Yoshiya S, Kurosaka M, Kumagai S. Trichostatin A cooperates with Fas-mediated signal to induce apoptosis in rheumatoid arthritis synovial fibroblasts. J Rheumatol 2006;33:1052–60.

[62] Choo QY, Ho PC, Tanaka Y, Lin HS. Histone deacetylase inhibitors MS-275 and SAHA induced growth arrest and suppressed lipopolysaccharide-stimulated NF-kappaB p65 nuclear accumulation in human rheumatoid arthritis synovial fibroblastic E11 cells. Rheumatology (Oxford) 2010;49:1447–60.

[63] Nasu Y, Nishida K, Miyazawa S, Komiyama T, Kadota Y, Abe N, et al. Trichostatin A, a histone deacetylase inhibitor, suppresses synovial inflammation and subsequent cartilage destruction in a collagen antibody-induced arthritis mouse model. Osteoarthritis Cartilage 2008;16:723–32.

[64] Lin HS, Hu CY, Chan HY, Liew YY, Huang HP, Lepescheux L, et al. Anti-rheumatic activities of histone deacetylase (HDAC) inhibitors in vivo in collagen-induced arthritis in rodents. Br J Pharmacol 2007;150:862–72.

[65] Carta S, Tassi S, Semino C, Fossati G, Mascagni P, Dinarello CA, et al. Histone deacetylase inhibitors prevent exocytosis of interleukin-1beta-containing secretory lysosomes: role of microtubules. Blood 2006;108:1618–26.

[66] Young DA, Lakey RL, Pennington CJ, Jones D, Kevorkian L, Edwards DR, et al. Histone deacetylase inhibitors modulate metalloproteinase gene expression in chondrocytes and block cartilage resorption. Arthritis Res Ther 2005;7:R503–12.

[67] Manabe H, Nasu Y, Komiyama T, Furumatsu T, Kitamura A, Miyazawa S, et al. Inhibition of histone deacetylase down-regulates the expression of hypoxia-induced vascular endothelial growth factor by rheumatoid synovial fibroblasts. Inflamm Res 2008;57:4–10.

[68] Tao R, de Zoeten EF, Ozkaynak E, Chen C, Wang L, Porrett PM, et al. Deacetylase inhibition promotes the generation and function of regulatory T cells. Nat Med 2007;13:1299–307.

[69] Saouaf SJ, Li B, Zhang G, Shen Y, Furuuchi N, Hancock WW, et al. Deacetylase inhibition increases regulatory T cell function and decreases incidence and severity of collagen-induced arthritis. Exp Mol Pathol 2009;87:99–104.

[70] Zhou X, Hua X, Ding X, Bian Y, Wang X. Trichostatin Differentially Regulates Th1 and Th2 Responses and Alleviates Rheumatoid Arthritis in Mice. J Clin Immunol 2011.

[71] Brogdon JL, Xu Y, Szabo SJ, An S, Buxton F, Cohen D, et al. Histone deacetylase activities are required for innate immune cell control of Th1 but not Th2 effector cell function. Blood 2007;109:1123–30.

[72] Schroeder TM, Westendorf JJ. Histone deacetylase inhibitors promote osteoblast maturation. J Bone Miner Res 2005;20:2254–63.

[73] Schroeder TM, Nair AK, Staggs R, Lamblin AF, Westendorf JJ. Gene profile analysis of osteoblast genes differentially regulated by histone deacetylase inhibitors. BMC Genomics 2007;8:362.

[74] Nakamura T, Kukita T, Shobuike T, Nagata K, Wu Z, Ogawa K, et al. Inhibition of histone deacetylase suppresses osteoclastogenesis and bone destruction by inducing IFN-beta production. J Immunol 2005;175:5809–16.

[75] Compston A, Coles A. Multiple sclerosis. Lancet 2008;372:1502–17.

[76] Ebers GC, Sadovnick AD. The role of genetic factors in multiple sclerosis susceptibility. J Neuroimmunol 1994;54:1–17.

[77] Kragt J, van Amerongen B, Killestein J, Dijkstra C, Uitdehaag B, Polman C, et al. Higher levels of 25-hydroxyvitamin D are associated with a lower incidence of multiple sclerosis only in women. Mult Scler 2009;15:9−15.

[78] Mikaeloff Y, Caridade G, Tardieu M, Suissa S. Parental smoking at home and the risk of childhood-onset multiple sclerosis in children. Brain 2007;130:2589−95.

[79] Marrie RA. Environmental risk factors in multiple sclerosis aetiology. Lancet Neurol 2004;3:709−18.

[80] Moscarello MA, Mastronardi FG, Wood DD. The role of citrullinated proteins suggests a novel mechanism in the pathogenesis of multiple sclerosis. Neurochem Res 2007;32:251−6.

[81] Mastronardi FG, Noor A, Wood DD, Paton T, Moscarello MA. Peptidyl argininedeiminase 2 CpG island in multiple sclerosis white matter is hypomethylated. J Neurosci Res 2007;85:2006−16.

[82] D'Souza CA, Wood DD, She YM, Moscarello MA. Autocatalytic cleavage of myelin basic protein: an alternative to molecular mimicry. Biochemistry 2005;44:12905−13.

[83] Musse AA, Boggs JM, Harauz G. Deimination of membrane-bound myelin basic protein in multiple sclerosis exposes an immunodominant epitope. Proc Natl Acad Sci USA 2006;103:4422−7.

[84] Baranzini SE, Mudge J, van Velkinburgh JC, Khankhanian P, Khrebtukova I, Miller NA, et al. Genome, epigenome and RNA sequences of monozygotic twins discordant for multiple sclerosis. Nature 2010;464:1351−6.

[85] Mariette X, Gottenberg JE. Pathogenesis of Sjogren's syndrome and therapeutic consequences. Curr Opin Rheumatol 2010;22:471−7.

[86] Gonzalez S, Aguilera S, Alliende C, Urzua U, Quest AF, Herrera L, Molina C, Hermoso M, Ewert P, Brito M, Romo R, Leyton C, Perez P, Gonzalez MJ. Alterations in type I hemidesmosome components suggestive of epigenetic control in the salivary glands of patients with Sjogren's syndrome. Arthritis Rheum 63:1106−15.

[87] Lu Q, Renaudineau Y, Cha S, Ilei G, Brooks WH, Selmi C, Tzioufas A, Pers JO, Bombardieri S, Gershwin ME, Gay S, Youinou P. Epigenetics in autoimmune disorders: highlights of the 10th Sjogren's syndrome symposium. Autoimmun Rev 9:627−30.

[88] Das SK, Majumder PP, Chakraborty R, Majumdar TK, Haldar B. Studies on vitiligo. I. Epidemiological profile in Calcutta, India. Genet Epidemiol 1985;2:71−8.

[89] Laberge G, Mailloux CM, Gowan K, Holland P, Bennett DC, Fain PR, et al. Early disease onset and increased risk of other autoimmune diseases in familial generalized vitiligo. Pigment Cell Res 2005;18:300−5.

[90] Alkhateeb A, Fain PR, Thody A, Bennett DC, Spritz RA. Epidemiology of vitiligo and associated autoimmune diseases in Caucasian probands and their families. Pigment Cell Res 2003;16:208−14.

[91] Spritz RA, The genetics of generalized vitiligo: autoimmune pathways and an inverse relationship with malignant melanoma. Genome Med 2:78.

[92] Sreekumar GP, Erf GF, Smyth Jr JR. 5-azacytidine treatment induces autoimmune vitiligo in parental control strains of the Smyth line chicken model for autoimmune vitiligo. Clin Immunol Immunopathol 1996;81:136−44.

[93] Arora-Singh RK, Assassi S, del Junco DJ, Arnett FC, Perry M, Irfan U, et al. Autoimmune diseases and autoantibodies in the first degree relatives of patients with systemic sclerosis. J Autoimmun 2010;35:52−7.

[94] Qi Q, Guo Q, Tan G, Mao Y, Tang H, Zhou C, et al. Predictors of the scleroderma phenotype in fibroblasts from systemic sclerosis patients. J Eur Acad Dermatol Venereol 2009;23:160−8.

[95] Wang Y, Fan PS, Kahaleh B. Association between enhanced type I collagen expression and epigenetic repression of the FLI1 gene in scleroderma fibroblasts. Arthritis Rheum 2006;54:2271−9.

[96] Ozbalkan Z, Bagislar S, Kiraz S, Akyerli CB, Ozer HT, Yavuz S, et al. Skewed X chromosome inactivation in blood cells of women with scleroderma. Arthritis Rheum 2005;52:1564−70.

[97] Raval K, Lofland JH, Waters H, Piech CT. Disease and treatment burden of psoriasis: examining the impact of biologics. J Drugs Dermatol 10:189−96.

[98] Ruchusatsawat K, Wongpiyabovorn J, Shuangshoti S, Hirankarn N, Mutirangura A. SHP-1 promoter 2 methylation in normal epithelial tissues and demethylation in psoriasis. J Mol Med 2006;84:175−82.

[99] Zhang K, Zhang R, Li X, Yin G, Niu X, Hou R. The mRNA expression and promoter methylation status of the p16 gene in colony-forming cells with high proliferative potential in patients with psoriasis. Clin Exp Dermatol 2007;32:702−8.

[100] Chen M, Chen ZQ, Cui PG, Yao X, Li YM, Li AS, et al. The methylation pattern of p16INK4a gene promoter in psoriatic epidermis and its clinical significance. Br J Dermatol 2008;158:987−93.

[101] Zhang K, Zhang R, Li X, Yin G, Niu X. Promoter methylation status of p15 and p21 genes in HPP-CFCs of bone marrow of patients with psoriasis. Eur J Dermatol 2009;19:141−6.

[102] Zhang P, Su Y, Chen H, Zhao M, Lu Q. Abnormal DNA methylation in skin lesions and PBMCs of patients with psoriasis vulgaris. J Dermatol Sci 2010;60:40−2.

[103] Poupon R. Primary biliary cirrhosis: a. J Hepatol 2010;52:745—58.

[104] Selmi C, Invernizzi P, Miozzo M, Podda M, Gershwin ME. Primary biliary cirrhosis: does X mark the spot? Autoimmun Rev 2004;3:493—9.

[105] Gershwin ME, Selmi C, Worman HJ, Gold EB, Watnik M, Utts J, et al. Risk factors and comorbidities in primary biliary cirrhosis: a controlled interview-based study of 1032 patients. Hepatology 2005;42: 1194—202.

[106] Mitchell MM, Lleo A, Zammataro L, Mayo MJ, Invernizzi P, Bach N, Shimoda S, Gordon S, Podda M, Gershwin ME, Selmi C, LaSalle JM. Epigenetic investigation of variably X chromosome inactivated genes in monozygotic female twins discordant for primary biliary cirrhosis. Epigenetics 6:95—102.

[107] Thiemann A, Grunder S, Pusch M, Jentsch TJ. A chloride channel widely expressed in epithelial and non-epithelial cells. Nature 1992;356:57—60.

[108] Rogner UC, Heiss NS, Kioschis P, Wiemann S, Korn B, Poustka A. Transcriptional analysis of the candidate region for incontinentia pigmenti (IP2) in Xq28. Genome Res 1996;6:922—34.

[109] Uchida T, Fujimori F, Tradler T, Fischer G, Rahfeld JU. Identification and characterization of a 14 kDa human protein as a novel parvulin-like peptidyl prolyl cis/trans isomerase. FEBS Lett 1999;446:278—82.

[110] Brix TH, Kyvik KO, Hegedus L. A population-based study of chronic autoimmune hypothyroidism in Danish twins. J Clin Endocrinol Metab 2000;85:536—9.

[111] Brix TH, Kyvik KO, Christensen K, Hegedus L. Evidence for a major role of heredity in Graves' disease: a population-based study of two Danish twin cohorts. J Clin Endocrinol Metab 2001;86:930—4.

[112] Brix TH, Knudsen GP, Kristiansen M, Kyvik KO, Orstavik KH, Hegedus L. High frequency of skewed X-chromosome inactivation in females with autoimmune thyroid disease: a possible explanation for the female predisposition to thyroid autoimmunity. J Clin Endocrinol Metab 2005;90:5949—53.

[113] Tysk C, Lindberg E, Jarnerot G, Floderus-Myrhed B. Ulcerative colitis and Crohn's disease in an unselected population of monozygotic and dizygotic twins. A study of heritability and the influence of smoking. Gut 1988;29:990—6.

[114] Gloria L, Cravo M, Pinto A, de Sousa LS, Chaves P, et al. DNA hypomethylation and proliferative activity are increased in the rectal mucosa of patients with long-standing ulcerative colitis. Cancer 1996;78: 2300—6.

[115] Hsieh CJ, Klump B, Holzmann K, Borchard F, Gregor M, Porschen R. Hypermethylation of the p16INK4a promoter in colectomy specimens of patients with long-standing and extensive ulcerative colitis. Cancer Res 1998;58:3942—5.

[116] Sato F, Harpaz N, Shibata D, Xu Y, Yin J, Mori Y, et al. Hypermethylation of the p14(ARF) gene in ulcerative colitis-associated colorectal carcinogenesis. Cancer Res 2002;62:1148—51.

[117] Gonzalez-Zulueta M, Bender CM, Yang AS, Nguyen T, Beart RW, Van Tornout JM, et al. Methylation of the 5' CpG island of the p16/CDKN2 tumor suppressor gene in normal and transformed human tissues correlates with gene silencing. Cancer Res 1995;55:4531—5.

[118] Esteller M, Tortola S, Toyota M, Capella G, Peinado MA, Baylin SB, et al. Hypermethylation-associated inactivation of p14(ARF) is independent of p16(INK4a) methylation and p53 mutational status. Cancer Res 2000;60:129—33.

[119] Tahara T, Shibata T, Nakamura M, Yamashita H, Yoshioka D, Okubo M, et al. Promoter methylation of protease-activated receptor (PAR2) is associated with severe clinical phenotypes of ulcerative colitis (UC). Clin Exp Med 2009;9:125—30.

[120] Tahara T, Shibata T, Nakamura M, Yamashita H, Yoshioka D, Okubo M, et al. Effect of MDR1 gene promoter methylation in patients with ulcerative colitis. Int J Mol Med 2009;23:521—7.

[121] Saito S, Kato J, Hiraoka S, Horii J, Suzuki H, Higashi R, Kaji E, Kondo Y, Yamamoto K. DNA methylation of colon mucosa in ulcerative colitis patients: Correlation with inflammatory status. Inflamm Bowel Dis 2011;17(9):1955—65.

[122] Wheeler JM, Kim HC, Efstathiou JA, Ilyas M, Mortensen NJ, Bodmer WF. Hypermethylation of the promoter region of the E-cadherin gene (CDH1) in sporadic and ulcerative colitis associated colorectal cancer. Gut 2001;48:367—71.

[123] Sato F, Shibata D, Harpaz N, Xu Y, Yin J, Mori Y, et al. Aberrant methylation of the HPP1 gene in ulcerative colitis-associated colorectal carcinoma. Cancer Res 2002;62:6820—2.

[124] Osborn NK, Zou H, Molina JR, Lesche R, Lewin J, Lofton-Day C, et al. Aberrant methylation of the eyes absent 4 gene in ulcerative colitis-associated dysplasia. Clin Gastroenterol Hepatol 2006;4:212—8.

[125] Huang Z, Li L, Wang J. Hypermethylation of SFRP2 as a potential marker for stool-based detection of colorectal cancer and precancerous lesions. Dig Dis Sci 2007;52:2287—91.

[126] Fujii S, Tominaga K, Kitajima K, Takeda J, Kusaka T, Fujita M, et al. Methylation of the oestrogen receptor gene in non-neoplastic epithelium as a marker of colorectal neoplasia risk in longstanding and extensive ulcerative colitis. Gut 2005;54:1287—92.

222

[127] Garrity-Park MM, Loftus Jr EV, Sandborn WJ, Bryant SC, Smyrk TC. Methylation status of genes in non-neoplastic mucosa from patients with ulcerative colitis-associated colorectal cancer. Am J Gastroenterol 2010;105:1610−9.

[128] Kuester D, Guenther T, Biesold S, Hartmann A, Bataille F, Ruemmele P, Peters B, Meyer F, Schubert D, Bohr UR, Malfertheiner P, Lippert H, Silver AR, Roessner A,Schneider-Stock, R. Aberrant methylation of DAPK in long-standing ulcerative colitis and ulcerative colitis-associated carcinoma. Pathol Res Pract 206:616−24.

[129] Fleisher AS, Esteller M, Harpaz N, Leytin A, Rashid A, Xu Y, et al. Microsatellite instability in inflammatory bowel disease-associated neoplastic lesions is associated with hypermethylation and diminished expression of the DNA mismatch repair gene, hMLH1. Cancer Res 2000;60:4864−8.

[130] Fujiwara I, Yashiro M, Kubo N, Maeda K, Hirakawa K. Ulcerative colitis-associated colorectal cancer is frequently associated with the microsatellite instability pathway. Dis Colon Rectum 2008;51:1387−94.

[131] Risques RA, Lai LA, Brentnall TA, Li L, Feng Z, Gallaher J, et al. Ulcerative colitis is a disease of accelerated colon aging: evidence from telomere attrition and DNA damage. Gastroenterology 2008;135:410−8.

[132] Issa JP, Ahuja N, Toyota M, Bronner MP, Brentnall TA. Accelerated age-related CpG island methylation in ulcerative colitis. Cancer Res 2001;61:3573−7.

[133] Tahara T, Shibata T, Nakamura M, Okubo M, Yamashita H, Yoshioka D, Yonemura J, Hirata I, Arisawa T. Association between polymorphisms in the XRCC1 and GST genes, and CpG island methylation status in colonic mucosa in ulcerative colitis. Virchows Arch 458:205−11.

[134] Lin Z, Hegarty J, Cappel J, Yu W, Chen X, Faber P, et al. Identification of disease-associated DNA methylation in intestinal tissues from patients with inflammatory bowel disease. Clin Genet 2011;80:59−67.

[135] van Belle TL, Coppieters KT, von Herrath MG. Type 1 diabetes: etiology, immunology, and therapeutic strategies. Physiol Rev 2011;91:79−118.

[136] Moore JK, West SR, Robins G. Advances in celiac disease. Curr Opin Gastroenterol 2011;27:112−8.

[137] Dougados M, Baeten D. Spondyloarthritis. Lancet 2011;377:2127−37.

CHAPTER 12

Approaches to Autoimmune Diseases Using Epigenetic Therapy

Christopher Chang
Thomas Jefferson University, Wilmington, DE, USA

12.1 INTRODUCTION

Autoimmune diseases affect more than 23.5 million people in the United States and about an estimated 5% of the world's population. The incidence of pediatric systemic lupus erythematosus ranges from 0.36 to 2.5 per 100 000 per year [1]. Rates vary in different parts of the world, reflecting a genetic basis for many of these conditions. There are "hotspots" where incidence rates may increase, and this is believed to be influenced by environmental or other

T. Tollefsbol (Ed): Epigenetics in Human Disease. DOI: 10.1016/B978-0-12-388415-2.00012-3

exposures. There is also a role for gender in autoimmunity, with the prevalence in women being significantly higher than in men [2]. In some conditions, such as scleroderma and autoimmune thyroiditis, this gender bias has been traced to an imbalance in X-chromosome inactivation, known as the X chromosome inactivation skew theory [3]. This is evidence that not only genetics, but epigenetics may play a role in the pathogenesis of autoimmune diseases. Phenotypic variation within each of the autoimmune diseases may indeed be a function of epigenetic influences on a baseline level of gene expression [4–6]. Because epigenetic modifications are reversible [7], this also opens the door for potential treatments to be developed that will reverse the epigenetic changes that contribute to the pathogenesis of the disease.

The treatment of autoimmune diseases has undergone several very significant paradigm changes over the past century. With a better understanding of the mechanisms of this group of diseases have come newer and more innovative modes of therapy. The discovery of cortisone, initially called "Compound E" in the 1940s was hailed as a wonder drug after the successful treatment of a woman with rheumatoid arthritis at the Mayo Clinic. This would ultimately lead to a Noble Prize for Hench, Kendall, and Reichstein. Kendall, in 1964, proclaimed that it would be "…highly improbable that any product will ever be found which can be used in place of cortisone…" [8]. The steroids and non-steroidal inflammatory agents that were initially used in the treatment of these diseases was followed by the discovery and subsequent development of disease-modifying antirheumatic drugs (DMARDs). The difference is that the former group would be effective in treating symptoms but unlike DMARDs, would not slow progression of the disease. DMARD drugs were initially used to treat rheumatoid arthritis, hence the name, but their use was then extended to include other autoimmune diseases including systemic lupus erythematosus (SLE), myasthenia gravis, immune thrombocytopenic purpure (ITP), Crohn's disease, and many others. The earlier DMARDs consisted of traditional drugs that were of low molecular weight, but more recently another new class of DMARDs has emerged. These are the biological agents, which are synthesized by genetic engineering and have proven to be extremely effective in the control of these diseases. The earliest biological agent to treat rheumatoid arthritis was rituximab, introduced in 1986. Other biologics used to treat autoimmune diseases such as Crohn's disease include the tumor necrosis factor alpha inhibitors. The first anti-TNF drug was infliximab, introduced in 1998. Although generally considered safer than chronic corticosteroid use, the potential for serious side effects can occur.

More recently, a new strategy towards the treatment of autoimmune disease has been introduced. This strategy is based on observations that epigenetics may play a role in the development of autoimmunity. The bulk of experience in the use of the epigenetic drugs has so far been in the treatment of cancer (Box 12.1). This experience has led to a great deal of promise for a similar application in the treatment of autoimmunity. Interestingly, the use of corticosteroids in the treatment of these illnesses may be intertwined with the development of epigenetic drugs because of the impact of epigenetic drugs on the glucocorticoid receptor [9,10]. Epigenetic drugs may also play a role in treatment of other inflammatory diseases states such as asthma [11,12] as well as other classes of disease, including neurologic [13] or psychiatric [13,14] disorders. The challenges may be different, since the target genes and cells that have gone awry may be different depending on disease states, but the principles that lead to the development of epigenetic drugs are similar. A historical and current timeline for the development of drugs for autoimmune diseases is shown in Figure 12.1.

12.2 PATHOPHYSIOLOGIC BASIS FOR THE DEVELOPMENT OF EPIGENETIC TREATMENTS IN AUTOIMMUNITY

Like many diseases, the pathophysiology of autoimmune diseases may include both genetic and environmental factors. Epigenetics describes changes in gene expression which are stable and heritable, but reversible. One of the key features of epigenetics that makes it an attractive target for development of new drugs is that the changes in gene expression are occasionally

---

BOX 12.1 CATEGORIES OF EPIGENETIC DRUGS

HDAC Inhibitors

Phenylbutyrate

Trichostatin A

Suberoylanilide hydroxamic acid

MS275

FK228

Valproic acid

DNA Methylating Agents

5-azacytidine

Decitabine

Zebularine

Procainamide

Procaine

Hydralazine

Epigallocatechin-3-gallate (EGCG) [126]

DNA methyltransferase 1 antisense oligonucleotides (DNMT1 ASO) [127]

MicroRNA

---

limited to certain cell types [15], although genome-wide epigenetic changes in disease do exist [16]. On the other hand, the knowledge that we need to devise ways to specifically target the gene or cell responsible for the disease is still not available. While global, non-specific epigenetic changes may be easier to induce by epigenetic drugs, and may yet prove to be

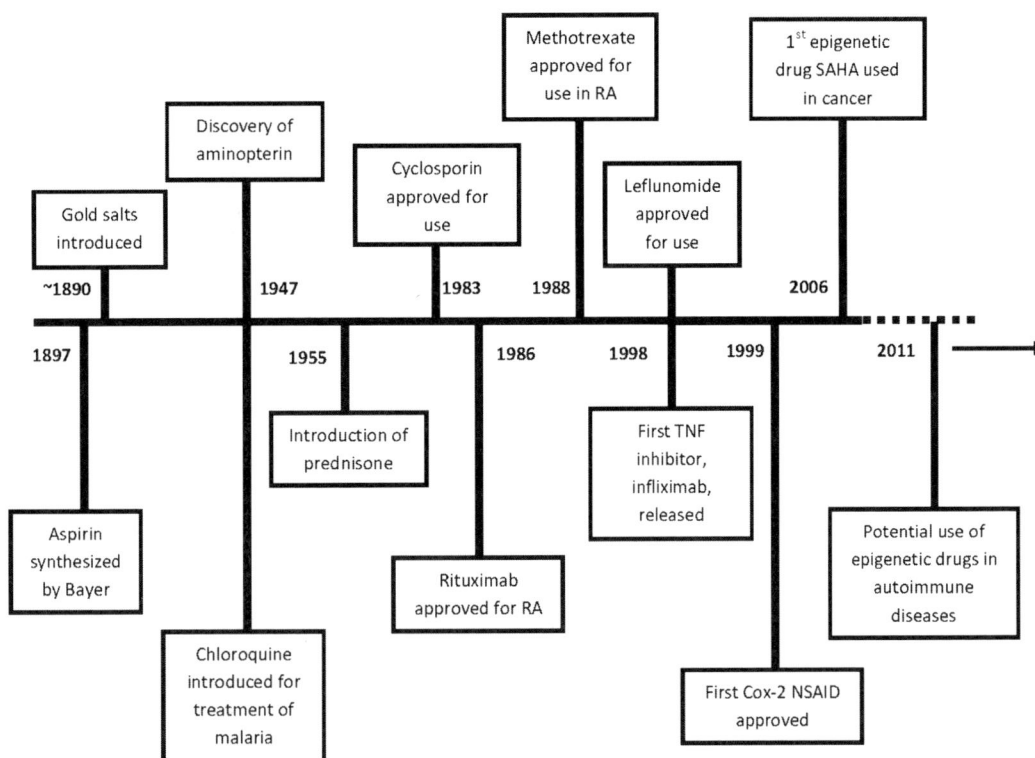

FIGURE 12.1

Timeline outlining significant discoveries in autoimmunity drugs.

clinically valuable in treating autoimmune diseases, a greater success would arise from the ability to target the effect of epigenetic drugs directly to the cells in which dysregulation of transcription occurs. The successful targeting of the control of a single gene or cell type may be associated with a lower risk of side effects, since genes irrelevant to the disease will be spared. The fact that epigenetic changes are believed to be reversible indicates that drugs known to affect gene transcription may be used to restore normal transcription and lead to resolution of clinical symptoms.

The two target areas by which epigenetic modification influences gene expression are histone modification and DNA methylation [17]. In autoimmune diseases such as SLE, it is suspected that DNA methylation plays a role in the disease process. Studies of identical twins with SLE have suggested that there is a significant role for non-genetic factors as the highest concordance rate among identical twins fails to reach 60% [18]. The observation of a global hypo-methylation of DNA in the T cells of patients with SLE [19] has indicated epigenetic modifications may play a role in disease pathogenesis and progression. In general, DNA methylation occurs at the 5′ position of cytosine in CpG dinucleotides located in the promoter regions and leads to repression of transcription.

The existence of a role of chromatin and histone modification in the regulation of gene expression is a common phenomenon of many cell types and genes. Epigenetic modification is involved in the regulation of various proinflammatory cascades responsible for many disease states, including infection, cancer, and autoimmune diseases. The central event in this regulatory network is the activation of nuclear factor kappa light chain enhancer of activated B cells (NFκB). NFκB is a heterodimeric protein, present in almost all mammalian cell types, that regulates DNA transcription. It is at the core of most inflammatory processes and its activation is closely linked to a number of histone acetyltransferases. Histone modification enzymes include histone acetyltransferases (HAT) or histone deacetylases (HDAC). Histone deacetylases remove acetyl groups from lysine residues forming compact and condensed chromatin which is transcriptionally silenced. Conversely, HAT enzymes acetylate lysine residues and render chromatin less compact and more transcriptionally active. Histone deacetylase inhibitors (HDACi) block the action of HDAC, and lead to increased acetylation of nucleosome core histones (see Figure 12.2). In some cases this leads to transcriptional activation, but many genes are repressed by HDACi as well. It should also be noted that HDAC activity is not limited to histone proteins, and other proteins can be deacetylated at the lysine residue as well.

DNA methylation and histone modification are processes that work in concert with each other to determine transcriptional activity [20]. The hallmark of these processes is reversibility, although early on it was not believed to be so. Besides acetylation, histone proteins can undergo other reactions to render the chromatin open or closed, and these include phosphorylation, methylation, ribosylation, sumoylation, and ubiquitination. The reactions are not mutually exclusive. The primary site of action is at the histone tail, which is near the amino terminus of the protein. In general, opening the chromatin, as occurs through acetylation is associated with increased gene expression.

There are at least 18 known HDACs distributed among four classes of histone deacetylases, I through IV. They act on a variety of cells and signaling pathways to regulate chromatin architecture and immunologic function [21]. The class I, II, and IV HDAC enzyme activity is dependent on Zn^{2+} ion. Class III HDACs are the sirtuins, which also possess ability to influence immune function [22]. The class I HDAC enzymes are the most widely studied. These are generally found in the nucleus and regulate the production of inflammatory cytokines. Class II HDAC enzymes translocate from the nucleus to the cytoplasm in response to external stimuli. Their primary effect is in the regulation of lymphocyte differentiation and activation [23].

HDAC enzymes have been found to be able to regulate innate immunity by virtue of their effects on Toll-like receptor (TLR) signaling. In addition, HDAC can increase the expression of

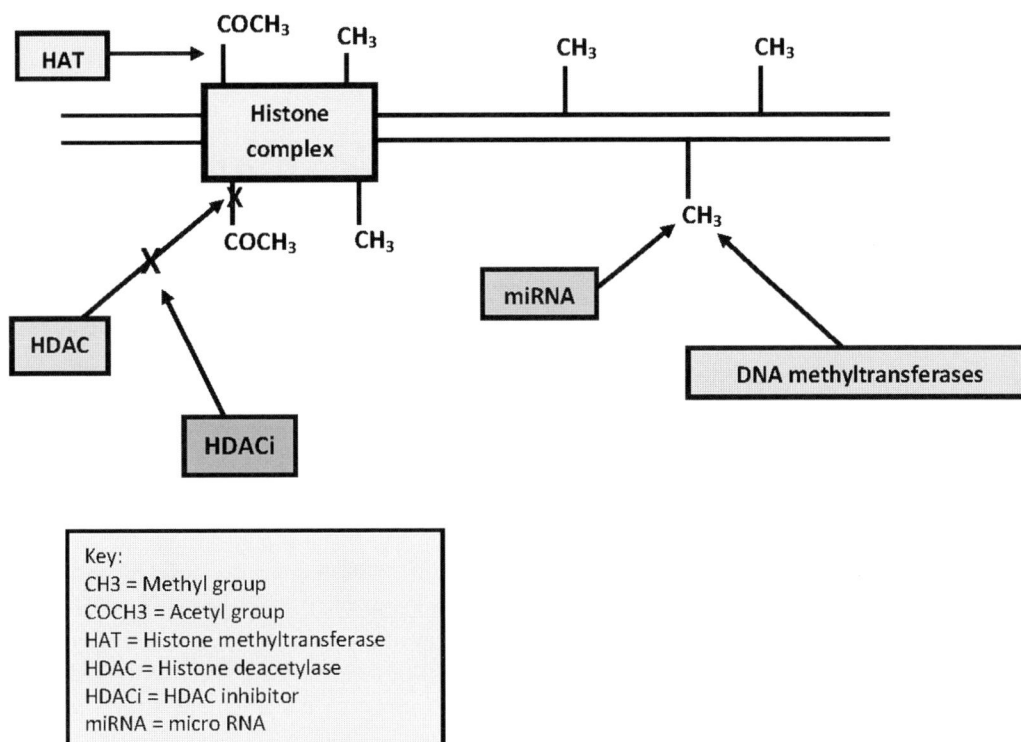

FIGURE 12.2

Molecular targets for epigenetic treatment of autoimmune diseases. This figure is reproduced in the color plate section.

proinflammatory mediators through suppression of inhibitors of NFκB such as IKK-β (inhibitor of nuclear factor kappa-B kinase subunit beta). It has been shown that HDACs can induce type I interferon production. HDACs play a role in regulatory T cell (Treg) homeostasis through cooperative modulation of forkhead box P3 (Foxp3) transcription factor. HDAC7, HDAC9, Tip60, and Foxp3 coexist in a complex and together regulate function and stability of Treg cells [24]. HDACs appear to suppress Treg cell function while acetylation of Foxp3 seems to augment it. On the other hand, HDACs appear to also suppress cytokine production from activated T cells. In a mouse allergy model, deletion of *Hdac1* led to exacerbation of airway inflammation, mucus hypersecretion, increased airway resistance, and parenchymal lung inflammation [25]. Th2 cells stimulated in vitro had higher IL-4 production and eosinophil recruitment was observed as well. Clearly the interaction between histone acetylation and immune function is highly complex, with opposing forces acting to maintain balance in immune homeostasis. Abnormalities in HDAC activity may lead to either proinflammatory cell activation or may lead to immune suppression. HDAC inhibitors may therefore play a role in reversing these epigenetic changes induced by HDACs.

In cancer, where most of the experience in the efficacy of HDAC inhibitors exists, HDAC inhibitors have been shown to induce cell cycle arrest, cell differentiation, and apoptotic cell death of "transformed cells". If one views synovial hyperplasia in rheumatoid arthritis in the context of "tumor-like" synovial cells that have gone out of control, then many of the applications of HDAC inhibitors in cancer therapy may be extended to autoimmune diseases.

12.3 PATHOLOGY OF AUTOIMMUNE DISEASE AND POTENTIAL TARGETS FOR EPIGENETIC DRUGS

The pathophysiology of autoimmune diseases is complex and the immunological pathways that are impacted vary depending on the specific autoimmune disease. However, there are common features that may lend it to strategic targeting of epigenetic pharmacotherapeutics. Autoimmune diseases arise as a result of an imbalance in the immune system that leads to loss of tolerance to self antigens. The presence of autoreactive T cells and autoantibodies plays a role in the disease pathogenesis. Pathologic events that lead to the disease phenotype may include defects in important signaling pathways for apoptosis, inflammatory cell activation, mediator release, immune tolerance, and regulatory cell function. T helper lymphocytes, B lymphocytes, dendritic cells, T regulatory cells, and other cell types play a role in the pathogenesis of various autoimmune diseases. For each of the diseases, specific cell types may play a role, for example synoviocytes in rheumatoid arthritis, lymphocytes in systemic lupus erythematosus, or neurons in multiple sclerosis.

The cytokine profile, which is intricately linked to the selective activation of various cell types, is also important. Cytokines that have been identified to play a role in the development of autoimmune diseases include IL-6, TGF-β, IL-17, IL-12, IL-23, IL-2, TNFα, interferon γ and the chemokines CXCL9, CXCL10, and CXCL11, which are all Th1-attracting chemokines [26]. Recently Th1 and Th17 cells also have been found to play a potential role in autoimmune disease pathogenesis [27,28]. While there may also be many known and as yet unknown pathways, cell lines and humoral factors involved in the pathogenesis of autoimmune diseases, the above illustrates the numerous potential points of attack for epigenetic drugs. All of the factors listed above are regulated in some way at the DNA level, and control of gene expression with epigenetic drugs may allow restoration of normal immune homeostasis and reverse the aberrant function seen in patients with autoimmune diseases. The sheer extent of the involvement of multiple pathways, cells, regulatory factors, mediators, and signaling molecules indicates that there are probably numerous redundant pathways in the disease process. Box 12.2 shows potential targets for epigenetic drugs in autoimmunity.

At a different level, the field of biological modulators also targets the same factors, thus the development of drugs which antagonize the function of such proinflammatory mediators such as TNFα. Epigenetics, by targeting identical pathways, provides a means to regulate the generation of these mediators at the DNA transcription level, rather than blocking already excessive production of the mediator. Epigenetics may also have the potential to regulate the expression of more than one inflammatory mediator at a time, thus helping to account for redundancy of the immune system.

An example of the potential utility of epigenetic drugs in treating autoimmune diseases can be illustrated by first appreciating the role of regulator T cells in the pathogenesis of autoimmune diseases. The development and function of regulatory T cells in the human is under the control of a critical transcription factor known as foxp3. While the precise role of Treg cells in the pathogenesis of autoimmune diseases, such as SLE and rheumatoid arthritis, is still incompletely understood, there is some evidence that the levels of certain phenotypes of Treg cells such as CD4+CD25+Foxp3+ T reg cells correlate inversely with disease activity [29,30], suggesting a suppressive effect of these cells. Histone deacetylases have been shown to control the functions of Treg cells by altering transcription factors of the foxp3 gene. Acetylation leads to improved DNA binding and histone acetyltransferases (HATs) such as Tip60 have the opposite effect; acetylation would lead to increased expression of Foxp3 [31] and resistance to protease degradation, as would the use of HDAC inhibitors. Thus, HDAC inhibitors may potentially augment Treg cell function, and potentially reverse disease progression in autoimmune disorders. Figure 12.3 illustrates how an epigenetic treatment using HDAC inhibitors may regulate proinflammatory cytokines and autoimmune disease activity.

12.4 HDAC INHIBITORS

As mentioned above, the clinical data regarding HDAC inhibitors is derived primarily from cancer research. Some previously known drugs have HDAC inhibition activity, including

BOX 12.2 POSSIBLE TARGETS FOR EPIGENETIC REGULATION IN LUPUS AND OTHER AUTOIMMUNE DISEASES

Pathways
Inflammatory signaling pathways
Apoptosis [128]
B-cell stimulatory pathways [129]
T-cell stimulatory pathways [130]
NFκB pathway [119]
Glucocorticoid receptor activation [131]
Complement system
Toll-like receptor

Gene Clusters/Systems
MHC [132]
FCγRs
TCR ξ-chain gene

Cytokines and Signaling Molecules
CTLA-4 [133]
PCD-1 (programmed cell death-1)
IL-2 [55]
CD40 ligand [134]
MMPs [81]
IFN-γ
IL-10
CD70 [72]
LFA-1 [72]
IL-12, IL-23 [35]
IL-6 [135]
IL-17 [136]
TGF-β
TNFα [137]
IL-18 [138]
Cyclin-dependent kinase inhibitor p21 [139]

Chemokines and Chemokine Receptors
CXCL9 [35]
CXCL10 [35]
CXCL11 [35]
CXCR12 [80]

valproic acid used in the treatment of seizures and trichostatin A used in the treatment of fungal infections. Other compounds with HDAC inhibition activity include butyrates such as phenylbutyrate, benzamides, and cyclic peptides such as depsipeptides and apicidin [32]. Newer HDAC inhibitors such as suberoylanilide hydroxamic acid (SAHA) are available for the treatment of cancers, but no indication has been established for the use of these drugs in treating autoimmune diseases. However, as illustrated above, the role of HDACs in immune function is significant. In fact, multiple studies have documented effects of HDAC inhibitors on immune response. HDAC inhibitors have been shown to play a role in the reduction of expression of costimulatory molecules in antigen presentation by dendritic cells [33]. Other effects of HDAC inhibitors include a reduction in IL12 and IL6 by dendritic cells and macrophages [34]. Expression of Th cell chemokines CXCL9 and CXCL10 is also reduced. HDAC inhibitors have also been found to impair differentiation of Th1 and Th17 cells [35].

Suberoylanilide hydroxamic acid (SAHA) is a HDAC inhibitor that has been used successfully in the treatment of gliomas, cutaneous T-cell lymphoma, Sezary syndrome, and other solid

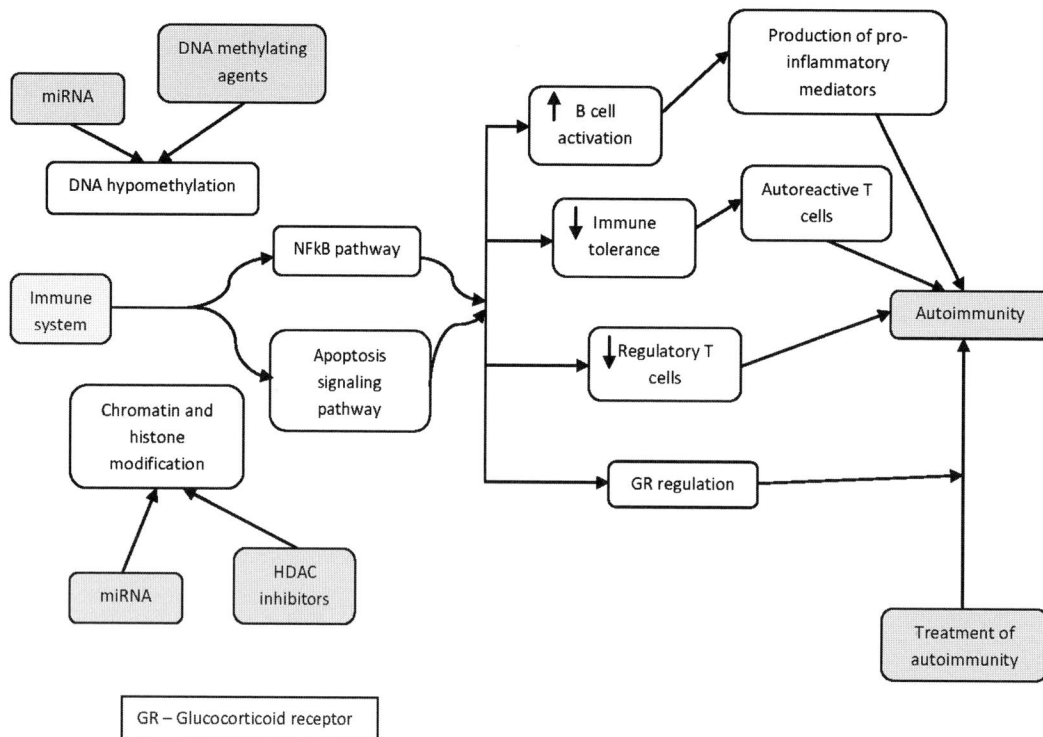

FIGURE 12.3
Potential effects of epigenetic changes in autoimmune diseases. This figure is reproduced in the color plate section.

cancers [36]. A recent study also suggests a role in the treatment of lung fibrosis, indicating that SAHA may play a role in regulation of transcription of genes encoding proinflammatory cytokines [37]. Both SAHA and MS275 have been shown to have antirheumatic properties in a rat model of collagen-induced arthritis [38]. MS275, an HDAC1 selective inhibitor, was found to have the capability to delay onset of disease in rats and mice, and to prevent bone erosion [39].

Trichostatin A is an HDAC inhibitor that has been studied as a potential anticancer drug [40]. Trichostatin A inhibits HDACs 1, 3, 4, 6, and 10. From a global hypomethylation standpoint, patients with SLE display an altered level of gene expression than healthy controls. Differences in DNA methylation are tied to activity of histone acetylation and methylation, DNA methylation, gene positioning, and their interactions. The histone code, representing post-translational modification and positioning of histones, is one of the master regulators of DNA methylation and gene expression [41,42]. It has been demonstrated in a MRL-lpr/lpr mouse model of SLE that trichostatin A can alter site-specific hypoacetylation of histones H3 and H4, and this is associated with an improvement in disease phenotype [43]. The authors interpreted this as a resetting of an aberrant "histone code" that was present in diseased mice. It is important to emphasize that while these observations provide optimism for the potential use of this agent in treating SLE, much is still unknown, including how a global hypomethylation affects suppression of anti-lupus genes and stimulation of pro-lupus genes.

Trichostatin A has also been found to be able to stimulate overexpression of the genes encoding for cycloxygenase-2 (Cox-2) and CXCL12 in mouse macrophages. On the other hand, it suppresses expression of genes coding for other immunostimulatory molecules such as TNFα, IL-6, IL-12p40, CCL2/MCP1, CCL17, and Endothelin-1 [44,45]. Along with SAHA, it was also found to reduce costimulatory molecule expression in dendritic cells. Bosisio showed that TSA could block the production of Th17-polarizing cytokines [35] and Tao showed that TSA promotes the production and activity of regulatory T cells [23,46]. These two observations suggest that TSA acts on different cell lines to suppress inflammation and may play a role in hyperactive immune states such as autoimmune diseases.

The pathophysiology of rheumatoid arthritis ultimately leads to tissue inflammation and pannus, or synovial hyperplasia. This leads to destruction of cartilage and bone. The pathogenesis is mediated through antigen-dependent T-cell activation, followed by infiltration of T and B lymphocytes into synovium and also release of proinflammatory mediators into synovium by macrophages and fibroblasts. These proinflammatory cytokines lead to proliferation of synovial cells. Historically, the treatment of rheumatoid arthritis has targeted the inflammatory response at various levels. Non-steroidal anti-inflammatory agents, steroids, DMARDs, and immunosuppressives are all non-selective medications which through their individual mechanisms of action, act to suppress this inflammatory response. Because of this, they each have their own set of adverse side effects, and some are serious, including gastrointestinal bleeding, hepatorenal toxicity, bone marrow suppression, etc. Will the use of epigenetic agents be any different?

The unrelated HDAC inhibitors trichostatin A (TSA) and phenylbutazone were assessed for their ability to modulate gene expression in rheumatoid arthritis and for their clinical efficacy in an adjuvant arthritis rat model for rheumatoid arthritis. The goal was to investigate the mechanism of action of these agents and to determine if disease progression can be suppressed. Of interest was the role of regulation of cell cycle inhibitors genes $p16^{INK4}$ and $p21^{Cip1}$ by epigenetic modification. Experiments on gene therapy had already shown that injection of an adenovirus that carries $p16^{INK4}$ and $p21^{Cip1}$ can inhibit synovial fibroblast proliferation. The technique was deemed impractical because it must be administered into individual joints and rheumatoid arthritis typically affects multiple joints. Because of the role of HDAC inhibitors as chromatin modifiers in cancer, and because of the role of histone acetylation in up-regulating cell cycle inhibitors, down-regulating proinflammatory cytokines such as IL-6, IL-1, TNF-α, and IL8, and because of the parallels between tumorigenesis and synovial hyperplasia as resulting from a failure of gene regulation, it was believed that HDAC inhibition may play a role in the treatment of rheumatoid arthritis.

In the study by Chung et al., HDAC inhibitors were found to be able to up-regulate $p16^{INK4}$ and $p21^{Cip1}$. Levels of acetylated H3 and H4 were also increased when cultured synovial fibroblasts were incubated with TSA or phenylbutyrate. The effect for $p21^{Cip1}$ was reversible, but the increased expression of $p16^{INK4}$ in synovial fibroblasts from rats with adjuvant arthritis was sustained even when HDAC inhibitors were removed. Topically applied phenylbutyrate and TSA also induced expression of $p21^{Cip1}$ and $p16^{INK4}$ in the synovium of rats with adjuvant arthritis, but not in normal synovial cells. In vivo studies showed that joint swelling was improved in rats who were administered TSA or PB topically. Other findings of this study included an ability of 10% PB cream or 1% Trichostatin A ointment to suppress paw swelling and pannus formation, and to promote wound healing. The proinflammatory cytokine TNF-α was also suppressed in affected tissues in the rats with adjuvant arthritis who were treated with both TSA and phenylbutyrate topically [47]. A summary of anti-inflammatory effects of HDAC inhibitors and other epigenetic drugs in rheumatoid arthritis is shown in Box 12.3.

12.4.1 HDAC Inhibitors in Bone Homeostasis

The role of HDAC inhibitors in the pathogenesis of bone loss that is frequently seen in chronic diseases, including autoimmune diseases, is currently being investigated. Osteoimmunology is a relatively new field, but recent evidence has demonstrated that cell surface receptors may play a complex and common role in bone homeostasis and immunity. The maintenance of healthy bone is a balance between bone formation and bone resorption, which is a dynamic process that leads to replacement of about 10% of our bone matrix every year. The regulatory factors that govern this balance include growth factors such as Receptor Activator for Nuclear Factor κB Ligand (RANKL) and Colony Stimulating Factor 1 (CSF-1)[48]. RANKL is necessary for the

233

BOX 12.3 EPIGENETIC AGENT-INDUCED ANTI-INFLAMMATORY EFFECTS IN RHEUMATOID ARTHRITIS*

HDAC Inhibitors [140]
1. Suppression of IL-6 and TNFα [47]
2. Suppression of cytokine gene expression [44]
3. Blocking of cytokine production from synovial macrophages [140]
4. Induction of apoptosis in human macrophages [140]
5. Induction of apoptosis in rheumatoid arthritis synovial macrophages [140]
6. Suppression of macrophage expression of Bfl-1 protein [140]

DNA Methylation Agents
1. Increased expression of CXCL12 [141]
2. Potential inhibition of CXCL12-induced expression of matrix metalloproteinases [81]
3. Potential reduction in CXCL12-induced collagenase activity and bone destruction [81]

*Other effects seen in lupus cells are not shown here.

formation of osteoclasts and is thus a promoter of bone resorption. The first monoclonal antibody against RANKL, denosumab, is FDA-approved to treat osteoporosis. RANKL is also present on T cells and is believed to be a key factor in dendritic cell maturation.

RANKL has been shown to cause substantial bone resorption when administered locally to mouse calvariae, but TSA, when coadministered, was found to inhibit this effect by promoting p21-WAF-dependent apoptosis of osteoclasts [49]. TSA and sodium butyrate were also found to inhibit in vitro differentiation of murine bone marrow cultures into osteoclast-like cells [50]. SAHA also inhibited RANKL-mediated osteoclastogenesis and potentiated apoptosis by suppressing NF-κB activation in a RAW264 mouse macrophage system. In vitro, SAHA was also able to suppress class I and II histone deacetylase activity in human osteoblasts [51]. Not all HDAC inhibitors were found to suppress osteoclast activity. Interestingly, MS275, which is a class I HDAC inhibitor, and 2664.12, a class II HDAC inhibitor, failed to show osteoclast activity suppression by themselves, but when combined, were able to show marked levels of inhibition. This effect was similar to that seen in the broad-spectrum HDAC inhibitor 1179.4b, suggesting that both HDAC class I and HDAC class II enzymes need to be suppressed in order for bone resorption to be inhibited.

In addition to the effects of HDAC inhibitors on osteoclast activity, there is also evidence that they may impact osteoblasts as well. The HDAC inhibitors valproic acid, sodium butyrate, and MS275 all promoted osteoblast maturation. TSA was shown to enhance the function of mature osteoblasts and enhance the expression of the osteoblast genes, type I collagen, osteopontin, osteocalcin, and bone sialoprotein in the calvarial-derived primary osteoblast cell line MC3T3-E1[52]. Runx2 transcription was also increased, and TSA accelerated matrix mineralization and alkaline phosphatase production in MC3T3-E1 cells.

These data suggest that HDAC inhibitors may play a role in treating osteoporosis related to autoimmune and other diseases. In particular, bone loss in rheumatoid arthritis may be an area where HDAC inhibitors may provide the most benefit because of the duplicity of action in also reining in the hyperactive immune response in this disease. Much research still needs to be done to elucidate the efficacy and safety of these agents in the treatment of autoimmune disease and their related morbidities.

12.4.2 HDAC Inhibitors in the Treatment of Inflammatory Bowel Disease

The HDAC inhibitor butyrate was found to inhibit NF-κB-mediated inflammation in peripheral blood mononuclear cell (PBMC) cultures from intestinal biopsy specimens of patients with Crohn's disease. Butyrate inhibited the production of TNF and LPS induced mRNA expression of proinflammatory cytokines, and also blocked translocation of NFκB from the cytoplasm to the nucleus. This was believed to be mediated by an inhibition of degradation of the NFκB inhibitory protein IκB, as the decrease in IκB normally seen in LPS-induced NFκB activation was not seen in the presence of HDAC inhibitor. An additional finding was that in vivo administration of butyrate to a trinitrobenzene sulfonic acid (TNBS) experimental colitis model in rats reduced inflammation [53]. The authors were not certain this effect was due to HDAC inhibitor activity because the same effect was not seen with TSA.

12.4.3 HDAC Inhibitors in the Treatment of Multiple Sclerosis

The efficacy of HDAC inhibitors in treating multiple sclerosis may be related to several different mechanisms. The existence of the experimental autoimmune encephalomyelitis (EAE) animal model for MS is particularly useful in studying various novel treatments. It has been demonstrated that interferon-γ plays a role in pathogenesis by virtue of its ability to induce antigen-presenting cells to secrete IL-12, which facilitates the differentiation of naïve Th0 cells to Th1 cells [54]. As mentioned above, Th1 cells may play a role in autoimmune diseases, and they are known to secrete IL-2, a potent proinflammatory cytokine. HDAC inhibitors, particularly TSA, inhibit gene expression of IL-2 [55,56]. HDAC inhibitors have also been found to increase expression of IL-10 in Th2 cells in EAE, but the effect was opposite in SLE cells. When NFκB is associated with HAT, gene expression increases, but it decreases when associated with HDACs [57]. NFκB is normally bound to IκB in the cytoplasm, and upon degradation of the inhibitory protein, NFκB is released to translocate to the nucleus. HDAC inhibitors such as butyrate interfere with degradation of IκB, leading to sequestration of NFκB in the cytoplasm and inhibition of NF-κB-mediated inflammatory responses [53,58,59].

12.4.4 Epigenetic Regulation of the Glucocorticoid Receptor and Related Immune Regulation

Glucocorticoids have been used since the 1950s in the treatment of many diseases. Glucocorticoids have been shown to have anti-inflammatory effects in the treatment of asthma, autoimmune diseases, neoplastic diseases, adverse effects of transplantation and many others. The mechanism of action is through reversal of histone acetylation of activated inflammatory genes. The effects of glucocorticoids on the glucocorticoid receptor (GR) are complex and dose-dependent. The anti-inflammatory effect of glucocorticoids is mediated through binding of liganded GRs to coactivators and by recruitment of the class I HDAC1 and HDAC2, to the activated transcription complex. The result is activation of GR which translocates to the nucleus, and in the nucleus, the GR can either stimulate or inhibit gene transcription. Studies have shown that at low concentrations, GR reduces gene transcription of NFκB and AP-1 associated inflammatory genes. GR can also bind to specific DNA elements known as glucocorticoid response elements (GREs), which then results in inhibition of proinflammatory genes such as IL-6 [60].

This mechanism has been shown to play a role in oxidative stress-mediated resistance to the anti-inflammatory effects of glucocorticoids, such as in smoking. Because glucocorticoids act on the NFκB pathway to mediate their anti-inflammatory effects through HDAC2 recruitment, the possibility of using HDAC2 to restore glucocorticoid sensitivity in conditions of glucocorticoid resistance may prove to be a way of limiting glucocorticoid side effects.

The question remains as to whether or not these epigenetic drugs can be used safely as anti-inflammatory agents in the treatment of autoimmune and other immunological diseases. Trichostatin A has been found to attenuate airway inflammation in a mouse model for asthma [61], and SAHA antirheumatic effects further suggest that this is possible, but again, the actual likelihood of side effects of these agents is completely unknown at this time.

The world of HDAC inhibitors is a rapidly changing one, with new agents continually being developed. Most clinical trials currently being conducted are focused on the efficacy and safety of these drugs in treating cancer. Some of the HDAC inhibitors undergoing laboratory investigation or clinical trials include romidepsin and panobinostat in the treatment of cytotoxic T-cell lymphomas [62,63], CI-994 in the treatment of cancers, MS-275 in the treatment of myelodysplastic syndromes [64], BML-210 and NVP-LAQ824, both antileukemic agents, M344, a benzamide with potential application for the treatment of spinal muscular atrophy, Mocetinostat in solid tumors and hematologic cancers [65], and PXD101, a hydroxamate type HDAC inhibitor undergoing clinical trials in the treatment of ovarian and other solid tumors [66]. This list is by no means all-inclusive. The development of HDAC inhibitors in the treatment of autoimmune diseases is in its early phases, and lags behind progress made in the oncologic and hematologic diseases.

12.5 DNA METHYLATION AND DNA METHYLTRANSFERASES

The other potential target for epigenetic drugs is at the level of DNA methylation. In the genome of most vertebrates, including humans, most CG di-nucleotide sequences, the primary target of DNA methylation, are methylated. DNA methylation is associated in vertebrates with chromatin structure. In general, hypomethylation is associated with active chromatin [67]. The methylation of genes has been shown to be an important mechanism by which genes are regulated or silenced.

DNA methylating agents have been used successfully in the treatment of cancer. These include 5-aza-cytidine, decitabine, zebularine, procainamide, procaine, EGCG, and DNMT1 ASO [68]. DNA methylation is generally thought to be associated with transcription repression, while hypomethylation usually has a stimulatory role and leads to activation of genes. The consequence of this gene activation in the context of autoimmune diseases is that this may lead to autoreactivity. But not all genes that are overexpressed as a result of DNA hypomethylation are immunostimulatory in nature.

The methylating enzymes that maintain global methylation of DNA include the DNA(Cytosine-5) methyltransferases DNMT1, DNMT2, and DNMT3A and 3B. DNMT1 is the one that is most well studied. It is constitutively expressed and it has been shown that its activity is decreased in the CD4+ cells of patients with SLE. Corresponding low levels of DNMT mRNA are also seen in SLE patients [69]. Inhibition of DNA methylation can lead to overexpression of genes that may be involved in the pathogenesis of lupus, including ITGAL [70], perforin [71] and CD70 [72], perhaps via a mechanism that involves decreased T-cell ERK pathway signaling [73]. It is presumed that while the demethylation seen in SLE patients is a global phenomenon, it is genes that play a role in the pathogenesis of SLE that are of interest when developing methylating drugs as pharmacotherapeutic agents.

Besides DNA methyltransferases, other proteins may play a role in the active methylation of DNA. These include the methyl-CpG-binding proteins MBD2 and MBD4. It has been suggested that these "demethylating" proteins may in fact be the main reason for the DNA hypomethylation in SLE patients, as it has been observed that MBD2 mRNA transcript levels were elevated in SLE patients [74]. MBD4 transcripts were observed to be elevated in SLE patients. Thus, MBD2 and MBD4 may be the primary regulators of DNA hypomethylation in patients with SLE [75,76].

Certain drugs can cause a condition known as drug-induced lupus. The mechanism by which DNA methylation occurs in this group of conditions may vary from drug to drug [77,78]. Procainamide, one of the earlier drugs associated with lupus, is a competitive DNA methyltransferase (DNMT) inhibitor. Another demethylating drug is hydralazine, though via a different mechanism. Hydralazine inhibits ERK pathway signaling and thereby inhibits DNMT activity, leading to the DNA hypomethylation that is also seen in idiopathic lupus [78].

The DNA hypomethylation resulting from both pathways leads to increased transcription of genes coding for inflammatory proteins, including LFA-1 and CD70, which in turn leads to increased T-cell autoreactivity and a breakdown in peripheral tolerance.

In mice, 5-azacytidine can induce lupus. Azacytidine is a cytosine analog that is incorporated into DNA during synthesis, and treatment of mice with this drug leads to a genome-wide hypomethylation. Karouzakis has shown that 5-azacytidine increases expression of CXCL12 and decreased methylation of CpG nucleotides in the promoter region of CXCL12 in rheumatoid arthritis synovial fibroblasts. CXCL12, or stromal-derived factor-1 (SDF-1) plays a role in the trafficking of progenitor cells and is activated in response to tissue damage. It is a mediator of chronic inflammation by attraction of lymphocytes and monocytes into joints in patients with rheumatoid arthritis [79,80]. It also appears to increase expression of matrix metalloproteinases, which induce collagenase activity and may be significant mediators of joint destruction in rheumatoid arthritis [81].

miRNAs, which will be discussed later in the article, may be intricately linked with DNA methylation patterns as well. miRNAs have been linked to regulation of DNA in the cells of patients with lupus. Two microRNAs which show an ability to down-regulate protein levels of DNMT1. This leads to relatively hypomethylated CD4$^+$ T cells. Since these two miRNAs, miRNA-21 and miRNA-148 are both found in abundance in the CD4$^+$ T cells of animals and humans with lupus, the hypomethylated state may explain the increase in autoimmune-related cellular markers such as CD70 and LFA-1. In this study, inhibition of miRNA-148 and miRNA-21 resulted in reversal of the hypomethylation in CD4$^+$ T cells [82,83].

This observation that drugs can lead to the global hypomethylation seen in drug-induced lupus has driven research into developing drugs that can do the opposite, i.e. deactivate transcription of those genes that lead to autoreactivity. But the epigenetic dysregulation of DNA methylation in the T cells of patients with SLE is indeed a complex process. Many genes can be affected, some with more clinical significance than others.

237

12.6 MicroRNA

MicroRNA is another avenue of research with a potential impact on epigenetic therapeutics. While microRNAs are not recognized universally as an epigenetic phenomenon, it has been shown that miRNAs can regulate DNA methylation. Autoimmune diseases involve a dysregulation of the immune system leading to the production of antibodies to self-antigens. These antibodies, along with other changes in cellular immunity generate an aberrant inflammatory response. The result of this response is the destruction or abnormal function of tissues and organs to which these antibodies are directed. MicroRNAs are non-coding RNAs about 21–23 nucleotides long that function as master regulators or post-transcriptional regulators of gene expression. These extremely important molecules were discovered in the early 1990s and play a role in the regulation of approximately half of all protein-coding genes in mammals [84]. miRNAs exert their action by binding to partially complementary nucleotides in the 3'-untranslated region of messenger RNA, and exert their action by inhibition of translation or degradation of RNA [82]. In order to facilitate this binding, "mature" miRNAs are loaded onto miRNA-induced silencing complexes (miRISCs), which will then find a target mRNA complementary to the "seed sequence" on the mature RNA. This sequence generally recognizes complementary RNA at the 3'UTR region of the messenger RNA.

miRNAs may be degraded after exerting their effects on gene expression, but it has been recently reported that miRNAs may be stably expressed [85]. Because they are involved in the reversible modification of gene expression, they may be considered an epigenetic phenomenon. Since their discovery and the recognition of their importance in gene expression, the targets of miRNA have been hotly sought after. The number of miRNAs identified has grown to over 800, and one mRNA may have multiple gene targets. They are now known to play a role in the regulation of every facet of cellular activity, affecting all cell types and tissues. Their role in autoimmune diseases has been supported by the finding that there is aberrant expression of miRNAs in many autoimmune diseases. The normal up-regulation or down-regulation of the expression of several specific miRNAs appears to be critical for the maintenance of immune homeostasis. Defects in balancing elements of our immune system can either lead to a hyperactive or hypoactive immune system, the former leading to allergies or autoimmunity, and the latter leading to defects in immune surveillance and in protection against infection.

A defect in a particular miRNA may be responsible for a target gene coding for proteins that are involved in the pathogenesis of multiple autoimmune diseases, thus this may be a potential reason for the "overlap" syndrome, whereby individuals with one autoimmune disease are susceptible to others as well. Indeed, certain miRNAs have been identified to be directed against targets that code for mediators and signaling molecules responsible for the pathogenesis of autoimmune diseases. The computer age has ushered in new methodologies and technologies to identify miRNA targets. The development of algorithms that are based on evolutionary conservation and seed sequences, leads to the "narrowing down" of predicted targets for a particular miRNA [86]. But this still leaves hundreds or thousands of possibilities. The identification of biologically relevant proteins can further be facilitated by additional computational approaches that utilize known information on gene ontology and study of potential target genes [87]. This can be followed by laboratory studies that utilize microarray analysis techniques in conjunction with miRNA knockdown or overexpression to identify function. An example would be a study by Zhao et al., who investigated the relationship between miRNA-126, DNA methylation, and immune activity [88]. They were able to show that in $CD4^+$ T cells from patients with systemic lupus erythematosus, miRNA-126 is highly overexpressed, and this correlated inversely with DNMT1 protein levels. Lower DNMT protein levels leads to DNA hypomethylation and increased gene transcription of immune-related genes LFA-1 and CD20. The authors showed that in vitro knockdown of miRNA-126 in $CD4^+$ cells from lupus patients resulted in decreased autoimmune activity, and decreased IgG production from B cells.

Other methods of identifying miRNA targets include proteomic analysis. The technique involves stable isotope labeling with amino acids in cell culture (SILAC) and analysis of the nuclear proteome after overexpression of a particular miRNA. Matching of changes in miRNA expression with protein levels can provide clues to the target genes of miRNAs. Other computational and laboratory techniques exist to correlate miRNA with gene targets. These include gene network analysis, identification of miRISC-associated miRNA targets, identification of miRISC-bound MREs and software algorithms utilizing gene ontology and interactome analysis. Further discussion of these techniques is beyond the scope of this chapter.

An understanding of the factors that play a role in disease pathogenesis is important to the development of treatment strategies. Since it is now likely that miRNAs may play a role in disease pathogenesis of autoimmune diseases, one might consider treatments that can either reverse the disease process or replenish missing regulatory factors. The administration of disease-relevant miRNAs to treat patients with autoimmune disease has been investigated in animal models. In mice with autoantibody-mediated arthritis, up-regulation of Bcl-2 interferes with normal apoptosis. It was shown that administration of intrarticular miRNA-15a into the synovium leads to induction of apoptosis. In this study, miRNA was found to be able to reverse Bcl-2 induced suppression of normal apoptosis [89]. The clinical utility of this methodology is still under investigation.

A number of miRNAs play an important role in autoimmune diseases. MiRNA-181 plays a role in the development of immune tolerance by setting the threshold for TCR signaling during T cell maturation in the thymus [90]. Positive and negative selection in the thymus dictates the subsequent T-cell reactivity towards that antigen. If the threshold is set low, as seen with high miRNA-181 expression, then the T-cell response will be enhanced. Conversely low miRNA-181 expression leads to tolerance. miRNA-181 also has effects on B cells, its expression leading to an increased fraction of B lineage cells in vitro in tissue culture differentiation systems and in vivo in a mouse model [91]. Other miRNAs involved in immune function include miRNA-101, miRNA-146a, and miRNA-155. miRNA-146a is a master regulator of cytokine production, including TNF-α, which is elevated in SLE. miRNA-155 is an important factor in autoimmunity, and plays a role in promoting the development of T helper cells Th1 and Th17. MiRNA-155 increases the presence of CD4$^+$CD25$^+$Foxp3$^+$ T reg cells in a mouse model (MRL/lpr) of systemic lupus erythematosus. These Treg cells have reduced suppressive ability, and one of the targets of miRNA-155 is CD62L. The effects of miRNA-155 in RA patients are discussed below.

12.6.1 miRNAs in Systemic Lupus Erythematosus (SLE)

A study identifying abnormal expression in CD4$^+$ T cells in SLE revealed 11 miRNA candidates, some of which were up-regulated and some of which were down-regulated. MiRNA-126 in particular was overexpressed. The overexpression correlated with decreased DNMT1 levels. This was demonstrated to be a direct effect. miRNA-126 was found to inhibit DNMT1 translation. In CD4+ T cells from patients with SLE, up-regulation of miRNA-126 was associated with a reduced activity of DNMT1. Reduced activity of DNMT1 leads to reduced methylation of the immune-related genes TNFSF7 and ITGAL, which code for CD70 and CD11a respectively. This leads to increased expression of CD70 and CD11a and subsequently to increased B- and T-cell autoreactivity [70]. CD70 is the ligand for CD27, which induces B-cell costimulatory function and results in an increase in IgG. CD11a is also known as lymphocyte function-associated antigen and plays a role in adhesion of T cells to other immune cells [92].

More interestingly, inhibition of the overexpression of miRNA-126 in SLE patients led to restoration of DNMT1 levels and increased methylation of the ITGAL and TNFSF7 promoter. The consequence of this is a reduction in the generation of CD11a and CD70 promoter activity, decreasing T- and B-cell autoreactivity and a potential therapy for SLE [93]. While the early results in these in vitro studies are promising, and demonstrate that DNMT1 is a direct target of miRNA-126, it is still a long way from using miRNA-126 or an inhibitor of miRNA-126 for clinical applications. The effects on DNA hypomethylation are not limited to microRNA-126. Other miRNA can also directly induce hypomethylation, including miRNA-21 and miRNA-148a [94].

Another miRNA that is involved in SLE is miRNA-146a, which is down-regulated in SLE and which targets TNF receptor-associated factor 6 (TRAF6) and IL-1 receptor associated kinase 1 (IRAK1), both of which are involved in the activation of NF-κB. Type-1 interferons (IFN) have been shown to play a role in the pathogenesis of SLE [95]. An IFN score is formulated by the activity of IFN signature genes, MX1, OAS1, and LY6E [96]. Tang et al. showed a correlation between low miR-146a and high interferon expression, which are both seen in patients with high SLE disease activity, especially with involvement of renal disease [96]. miRNA-146a is also involved in many other immune regulatory effects, playing a role in Treg suppressor functions. miRNA146a is also a suppressor of apoptosis in T cells, and regulates levels of proinflammatory cytokines such as IL-2 [97].

miRNA-125a has also been found to be relevant in the pathogenesis of SLE. Levels of miRNA-125 are reduced in the mononuclear cells in the blood of patients with SLE, resulting in elevated expression of KLF13 and subsequent overexpression of CCL5. This chemokine is associated with inflammatory activity in SLE, including nephritis and arthritis [98].

12.6.2 miRNAs in Rheumatoid Arthritis (RA)

One of the miRNAs with the greatest potential for evolving into a therapeutic for the treatment of rheumatoid arthritis is miRNA-155 [99]. The immune effects attributed to miRNA-155 include a lack of developing collagen-induced arthritis when miRNA-155 is knocked out in mice. MiRNA-155 knockout mice in a K/BxN serum transfer arthritis model exhibited a reduction in protein expression of IL-6, IL-17, and IL-22, along with a reduction in pathogenic T and B cells [100]. Decreased numbers of osteoclasts result in less bone destruction. This observation suggests that inhibition of miRNA-155 may play a potential role in the treatment of RA.

miRNAs that have been shown to be up-regulated in RA include miRNA-223 and miRNA-346. Up-regulation of miRNA-146a is also observed in RA, but the targets of this miRNA, TRAF6 and IRAK1, are not overexpressed in patients compared with controls. On the other hand, TNF-α regulation appears to be impaired in these patients [101]. It has been noted also that up-regulation of miRNA146a/b is associated with an overexpression of IL-17 [102]. Other miRNAs that may play a role in RA include miRNA-124a and miRNA-203.

12.6.3 miRNAs in Multiple Sclerosis, Scleroderma, Sjögren's Syndrome, Polymyositis, Dermatomyositis, and Primary Biliary Cirrhosis

The pathogenesis of scleroderma involves the replacement of normal tissue with collagen-rich extracellular matrix by aberrant functioning of fibroblasts. Important cytokines that lead to increased proliferation and activation of fibroblasts include IL-4, platelet-derived growth factor (PDGF)-β, and transforming growth factor (TGF)- β[103]. The activity of these cytokines correlates inversely with levels of miR-29a. miRNAs found to be important in polymyositis or dermatomyositis include miRNA-146b, miRNA-155, miRNA-214, miRNA-221, and miRNA 222 [104].

Relevant miRNAs in Sjögren's syndrome include the miRNA-17-92 cluster. MiRNAs are already being used as biomarkers in Sjögren's syndrome [105]. miRNA-146a is also important in Sjögren's syndrome in that levels are increased in Sjögren's syndrome compared with health controls. The effects of miRNA-146a have already been discussed, but functional assays revealed that miRNA-146a significantly up-regulated phagocytic activity in human monocytic THP-1 cells and negatively regulated production of TNF-α, IL-1β, MIP-1α, IP-10, and IL-6 [106].

While some might consider that miRNAs are not strictly epigenetic agents, they merit discussion here as epigenetic approaches to treatment of autoimmune diseases because of the significant role they play in autoimmunity, and because of recent evidence showing that variations in miRNA expression are directly related to DNMT expression and alterations in DNA methylation. The mechanism for this association is still not entirely clear.

In multiple sclerosis, miRNA-326 is up-regulated in patients with relapsing—remitting disease, and studies using the experimental autoimmune encephalomyelitis (EAE) mouse model showed that up-regulation of miRNA-326 is linked to increased Th-17 cell number, and clinically to increased severity of EAE symptoms [107]. Other miRNAs that are up-regulated in active multiple sclerosis lesions include miRNA-34a and miRNA-155, which is also up-regulated in a number of other autoimmune diseases. A recent report suggests that these miRNAs function by inducing CD47 to reduce inhibitory control signaling of macrophages [108].

In end-stage PBC, 35 different miRNAs were expressed in the liver. Down-regulation of miRNA-122a and miRNA-26a was detected, as was up-regulation of miRNA-328 and miRNA-299-5p. These miRNAs act upon targets that control cell proliferation, inflammation, and apoptosis [109]. Table 12.1 illustrates the effect of miRNAs in various autoimmune disease states.

12.7 ANTAGOMIRS

Antagomirs are synthetic antagonists of miRNAs which were first developed in 2005 as silencing agents of miRNAs. One of the earlier studies of the use of antagomirs to silence miRNA involved miRNA-122 in mice [110]. Antagomir-122 was an effective miRNA silencing

TABLE 12.1 miRNAs Involved in Autoimmune Diseases

	RA	SLE	SS	SC	MS	PM/DM	PBC	Psoriasis	IBD	ITP
17-92										
16	↑									
17					↓					
17-5p					↑					
18b					↑					
20					↓					
21		↑							↑	
26a							↓			
29										
34a					↑					
96					↑					
122a							↓			
124a	↓									
125a		↓								
126	↑	↑								
132										
145					↑					
146	↑							↑		
146a	↑	↓								
148a		↑								
155	↑				↑					
184		↓								
192									↓	
198		↑								
203	↑							↑		
206										↓
207										↓
214										↓
221										
222										
223	↑									
296										↓
299-5p							↑			
326					↑					
328							↑			
346										
371-5p		↑								
375									↑	
379										↓
423-5p		↑								
513										↓
599					↑					
1224-3p		↓								

PM, polymyositis; DM, dermatomyositis; SLE, systemic lupus erythematosus; RA, rheumatoid arthritis; MS, multiple sclerosis; PBC, primary biliary cirrhosis; IBD, inflammatory bowel disease; ITP, idiopathic thrombocytopenic purpura.

agent that was able to reduce miRNA levels which are normally abundant in the livers of mice. While many functions of miRNAs themselves are still not known, miRNA-122 is thought to be involved in cholesterol biosynthesis, and indeed, antagomir-122-treated mice demonstrated lower plasma cholesterol levels.

A study comparing the efficacy of an antagomir to miRNA-145 and glucocorticoid therapy in the treatment of mice with allergy to dust mite showed that the antagomir-treated group

TABLE 12.2 A comparison of Various forms of Current and Future Therapy in RA

Therapy	Conventional (Non-Steroidal)	Conventional (Steroids)	DMARDS (Non-Biologics)	Biologics	Gene Therapy	Epigenetic Therapy
Mechanism	Variable	Regulation of inflammatory pathways	Variable	Blocking of action of mediators	Induction of immunological tolerance	Regulation of gene expression, regulation of adaptive and innate immunity
Goals and expectations	Treating symptoms	Treating symptoms	Treating symptoms and slowing/ceasing progression of disease	Treating symptoms and slowing/ceasing progression of disease	Disease modification	Disease modification
Currently Available	Yes	Yes	Yes	Yes	Yes	Yes
Approved indication for autoimmune diseases	Yes	Yes	Yes	Yes	Yes	No
Side effects (known or potential)	Gastrointestinal bleeding, asthma	Immunosuppression, Cushing's syndrome, hypertension, diabetes, osteonecrosis, osteoporosis	Bone marrow suppression, hypersensitivity reactions, immunosuppression	Anaphylaxis, infection, hypersensitivity reactions, immunosuppression	Cancer, infection, immunosuppression	Cancer, infection, immunosuppression
Benefit/risk ratio	+/−	+/−	+	+++	Unknown	Unknown

had similar reductions in airway hyper-responsiveness, mucous secretion, eosinophilic inflammation, and Th2 cytokine levels [111,112]. miRNA-126 is another miRNA which has immunologically related regulatory function, and an antagomir to miRNA-126 has been shown to reduce Th2 cell function when administered to mice with house dust mite allergy [105].

The impact of antagomirs on immunologic diseases such as allergic rhinitis portends the rapid development of new innovative treatment for allergic and autoimmune diseases. Given the widespread and constantly increasing association between various miRNAs and autoimmune disease, it is only a matter of time before antagomirs will be used in clinical trials to treat diseases such as SLE, rheumatoid arthritis, scleroderma, etc.

12.8 TECHNIQUES TO MEASURE EPIGENETIC ALTERATIONS — APPLICATION OF EPIGENETICS AS BIOMARKERS

Since the observation of increased DNA hypomethylation seen in SLE patients, the possibility of using epigenetic modifications or activity as biomarkers has been proposed. Measurement of epigenetic alteration techniques has been developed. Differential methylation hybridization takes advantage of microarray technology and oligonucleotide probes to measure changes in gene expression levels using picogram levels of DNA attached to a solid surface [113]. This technique has been utilized as a potential screening technique to identify biomarkers in lung cancer [113]. Chromatin immunoprecipitation assays coupled to hybridization of genomic microarrays (ChIP-on-chip) is a method to investigate the interaction between DNA and protein. Because Chip-on-chip deals with chromatin, one might use this techniques to preferentially identify genes with a regulatory component, including replication factors, promoter regions, repressors and silencing elements, enhancers, and other functional elements [114]. Another method of quantifying global DNA methylation is fluorescence correlation spectroscopy, which takes advantage of the high affinity of MBD for methylated DNA [115].

miRNAs can also be used for disease monitoring and prognostication. This has been demonstrated in cancer, for example, where miRNA29 targets DNMT3A and DNMT3B, and is down-regulated in lung cancer tissue [116]. But expression levels of other miRNAs such as miRNA-155 and miRNA 101, which are implicated in the pathogenesis of SLE, may potentially be developed as a diagnostic marker or used to monitor disease activity. Other miRNAs that may someday be utilized as diagnostic markers include miRNA-181a, miRNA-146, miRNA-223, and the miRNA17-92 gene cluster [117].

12.9 POTENTIAL SIDE EFFECTS OF TREATMENT WITH EPIGENETIC DRUGS IN AUTOIMMUNE DISEASES

Our knowledge of the effects of reversing epigenetic changes that have been shown to be present in autoimmune disease is incomplete. Changing the histone code, or changing levels of global demethylation, may provide relief from the disease by altering protein synthesis of inflammatory mediators that are responsible for the disease characteristics, but may affect other cell lines and genes that can be crucial to normal function elsewhere in the organism. The fact that about 50% of promoter regions are located in CpG islands [118] may provide an innovative means for the design of drug or delivery systems to provide improved specificity, but whether this will yield the desired outcomes is unknown.

The use of miRNA provides slightly more specificity as to what is targeted, but even then, a single miRNA may target multiple genes and vice versa. This lack of targeting directly to the gene of interest on the cell type of interest suggests that there will be as yet unidentified side effects. These side effects could potentially be short term or long term, leading to significant drug-induced morbidity. The field of epigenetic therapy in treating autoimmune diseases is still

---

BOX 12.4 POTENTIAL SIDE EFFECTS OF EPIGENETIC DRUGS

1. Unexpected activation or repression of bystander genes
2. Bacterial infection
3. Reactivation of viral infections
4. Embryogenic side effects — teratogenicity
5. Cytokine storm
6. Hypersensitivity to the drug
7. Cancer — reactivation of cancer genes, inhibition of tumor suppressive genes
8. Autoimmune diseases
9. Interaction with other concurrent medications or treatments

---

in its infancy, and a search of "epigenetic therapy" and "side effects" on Pubmed revealed no articles discussing side effects.

A further consideration is that epigenetic changes are, by nature, hereditary. When one considers the use of epigenetic modification in the treatment of autoimmune diseases, one should also consider what effects such changes may have on offspring. It is not known if DNA methylation levels, or methylation indices change during an individual's lifetime, or even if they are different embryonically. Will the use of these drugs lead to teratogenicity? Genes are turned on and off constantly during development in response to host and environmental stimuli. Will epigenetic drugs disrupt the normal regulation during development? These are all important considerations in the development of this potentially very effective treatment of autoimmunity and other diseases. Specificity is another important issue. There is always the possibility that use of HDAC inhibitors, DNA methylating enzymes and microRNA to treat autoimmune disease can also lead to effects of gene expression of constituent and necessary genes. In administering any treatment to a patient, the risk–benefit ratio must be considered carefully and conveyed to the patient. In the case of epigenetics, knowledge or risk is limited. Box 12.4 shows potential side effects of epigenetic drugs.

12.10 BALANCING CONVENTIONAL THERAPY AND EPIGENETIC THERAPY

As we move well into the 21st century, new approaches and new drugs are being developed for the treatment of autoimmune diseases. Older drugs may become obsolete, or may still be valuable tools for treatment. Avenues for combined use of conventional and innovative therapies will undoubtedly be a topic of debate as new drugs appear. A comparison of the various therapies for autoimmune diseases is outlined in Table 12.2.

The effects of HDAC on the glucocorticoid receptor has been discussed earlier, and this must be taken into consideration when combining HDAC inhibitors and glucocorticoids to treat autoimmune diseases. Other examples of possible conflicts of drugs exist. Methotrexate, for example is a dihydrofolate reductase inhibitor and limits the action of DNA methyltransferases [119]. Another example of how two therapeutic modalities may affect each other involves IL-6 and HDAC inhibitors. IL-6 is a cytokine that is involved in DNA methylation [120]. Tocilizumab is a monoclonal antibody against IL-6, and therefore it may have epigenetic properties itself, but whether or not it can be combined with HDAC inhibitors as a treatment for autoimmune diseases has not been studied.

Combination of DNA methyltransferases with HDIC inhibitors has been attempted in cancer but not autoimmune diseases. 5-aza-CdR has been used in conjunction with valproic acid and retinoic acid, two HDAC inhibitors, and reversed silencing of regulatory genes in breast cancer cells [121]. This approach has also been successful in the treatment of myeloplastic syndromes [122]. Another strategy may be to use HDAC inhibitors or DNA methylation agents to restore expression of certain receptors in order to enhance the response to drugs that target those receptors. In breast cancer studies, 5-aza-CdR and TSA have been used in combination to increase expression of estrogen receptors to render the conventional drug, tamoxifen, more effective [123].

12.11 WHERE DO WE GO FROM HERE?

The future holds great promise for the utilization of epigenetic drugs in the treatment of autoimmune diseases. There is also potential for great danger. In an article in 2008, Ballas wrote that the last two decades of the 20th century would be known for the development of biologics and was a venture into a "brave new world" [124]. Perhaps epigenetic drug development will similarly be the highlight of the early- to mid-21st century and should be labeled a "braver new world", for the simple reason that with epigenetic drugs, side effects may extend to future generations. Some considerations in the development of epigenetic drugs must include the following:

1. How to target specific genes?
2. Will the alteration of global DNA hypomethylation in autoimmune diseases shut off necessary genes?
3. How reversible may side effects of the next generation of epigenetic drugs be?
4. What is the cost of development of epigenetic drugs?
5. Are miRNAs the answer?
6. What is the specificity of miRNAs used as therapeutic drugs?
7. What are the potential adverse effects of epigenetic treatment of autoimmune diseases?
8. Are there other ways than pharmaceutical development to utilize epigenetics in the treatment of autoimmune diseases?
9. What is the most effective and safest delivery method for administration of these agents to the patient?
10. Will epigenetic modification have an effect in pregnancy or future offspring?

Our current goals should be focused on answering the above questions. Drummond has studied the delivery of HDAC inhibitors by packaging in liposomes [125]. The use of siRNAs or miRNAs in the treatment of diseases is being investigated by several commercial entities. One of the immediate goals is to identify all of the target genes for the known miRNAs. In cases where a specific miRNA overexpression is pathologic, the development of antagomirs [110], synthetic oligonucleotides that target specific miRNAs, may be helpful as a method of enhancing the specificity of epigenetic drugs in the treatment of autoimmunity.

12.12 DISCUSSION

As of today, there is no Food and Drug Administration (FDA)-approved epigenetic drug for the treatment of autoimmune diseases. Several drugs are approved for treatment of cancer and myeloplastic syndromes (MDS), and are therefore available commercially. The target genes involved in the epigenetic treatment of cancer are typically tumor-suppressive genes. In autoimmunity, the target genes, proteins, and cell types are involved in normal immune homeostasis as well, rendering the treatment with epigenetics much more complex. Well-known epigenetic targets include DNA methylation and histone/chromatin structure. Our knowledge of epigenetics is in its infancy. We are just beginning to characterize the mechanisms of action of potential epigenetic drugs. Even less information is available about unexpected or unwanted side effects associated with the use of these drugs. It is important to consider these important issues when developing new targets for the treatment of autoimmunity.

MicroRNAs can impact epigenetics by virtue of their effects on DNA methylation in autoimmune diseases. MicroRNAs can be up-regulated or down-regulated in autoimmune diseases.

The use of microRNAs or inhibitors of miRNA to impact DNA methylation is a method of targeting post-transcriptional gene transcription changes in CD4+ T cells in SLE patients, and thus may be a new method of influencing epigenetics to treat autoimmune diseases. Micro-RNAs are directed against specific genes and thus may be associated with a higher specificity of action than global demethylating agents or HDAC inhibitors. Some of the relevant microRNAs in various autoimmune diseases such as SLE or RA have been identified and include miRNAs that target proinflammatory genes, genes that regulate T-cell function including Treg function, and genes that regulate apoptosis or the cell cycle.

Besides pharmaceutical development, epigenetics may have other uses as potential biomarkers in monitoring the effectiveness of therapy. In SLE, disease activity is associated with levels of methylation and of expression of genes such as TNFSF7 and ITGAL. While our treatment may as yet not involve the use of epigenetic manipulation, levels of gene expression can be potentially used to monitor the success of other forms of therapy. For example, miRNAs can be used to assist in the diagnosis and determine prognosis and therapeutic outcome as well. This can be accomplished with the help of computer analysis of the activity of multiple miRNAs along with levels of hypomethylation and histone deacetylase activity.

In summary, epigenetics is an area of research with tremendous potential. As our knowledge increases, we will learn how to control expression of the critical factors that lead to autoimmune disease, and how to do it in a selective manner than does not endanger the patient. Treatments of the past may be abandoned in favor of these more effective and potentially safer therapeutic methods. Morbidity and mortality will decrease, and patients with these disorders will be able to enjoy a higher quality of life.

References

[1] Pineles D, Valente A, Warren B, Peterson MG, Lehman TJ, Moorthy LN. Worldwide incidence and prevalence of pediatric onset systemic lupus erythematosus. Lupus 2011;20:1187—92.

[2] Cooper GS, Miller FW, Germolec DR. Occupational exposures and autoimmune diseases. Int Immunopharmacol 2002;2:303—13.

[3] Morey C, Avner P. Genetics and epigenetics of the X chromosome. Ann N Y Acad Sci 2010;1214:E18—33.

[4] Crow MK, Kirou KA, Wohlgemuth J. Microarray analysis of interferon-regulated genes in SLE. Autoimmunity 2003;36:481—90.

[5] Banchereau J, Pascual V, Palucka AK. Autoimmunity through cytokine-induced dendritic cell activation. Immunity 2004;20:539—50.

[6] Baechler EC, Batliwalla FM, Karypis G, et al. Interferon-inducible gene expression signature in peripheral blood cells of patients with severe lupus. Proc Natl Acad Sci USA 2003;100:2610—5.

[7] Sakao S, Tatsumi K. The importance of epigenetics in the development of chronic obstructive pulmonary disease. Respirology 2011;16:1056—63.

[8] Kendall E. The development of cortisone as a therapeutic agent. Amsterdam: Elsevier Publishing Company; 1964.

[9] Barnes PJ, Adcock IM, Ito K. Histone acetylation and deacetylation: importance in inflammatory lung diseases. Eur Respir J 2005;25:552—63.

[10] Barnes PJ. Targeting histone deacetylase 2 in chronic obstructive pulmonary disease treatment. Expert Opin Ther Targets 2005;9:1111—21.

[11] Koppelman GH, Nawijn MC. Recent advances in the epigenetics and genomics of asthma. Curr Opin Allergy Clin Immunol 2011;11:414—9.

[12] Durham AL, Wiegman C, Adcock IM. Epigenetics of asthma. Biochim Biophys Acta 2011;1810:1103—9.

[13] Abel T, Zukin RS. Epigenetic targets of HDAC inhibition in neurodegenerative and psychiatric disorders. Curr Opin Pharmacol 2008;8:57—64.

[14] Ptak C, Petronis A. Epigenetic approaches to psychiatric disorders. Dialogues Clin Neurosci 2010;12:25—35.

[15] Bell CG, Beck S. The epigenomic interface between genome and environment in common complex diseases. Brief Funct Genomics 2010;9:477—85.

[16] Park YJ, Claus R, Weichenhan D, Plass C. Genome-wide epigenetic modifications in cancer. Prog Drug Res 2011;67:25—49.

[17] Janson PC, Winqvist O. Epigenetics - the key to understand immune responses in health and disease. Am J Reprod Immunol 2011;66(Suppl. 1):72—4.

[18] Meda F, Folci M, Baccarelli A, Selmi C. The epigenetics of autoimmunity. Cell Mol Immunol 2011;8:226—36.

[19] Renaudineau Y, Youinou P. Epigenetics and autoimmunity, with special emphasis on methylation. Keio J Med 2011;60:10—6.

[20] Ooi SK, Qiu C, Bernstein E, et al. DNMT3L connects unmethylated lysine 4 of histone H3 to de novo methylation of DNA. Nature 2007;448:714—7.

[21] de Ruijter AJ, van Gennip AH, Caron HN, Kemp S, van Kuilenburg AB. Histone deacetylases (HDACs): characterization of the classical HDAC family. Biochem J 2003;370:737—49.

[22] Stunkel W, Campbell RM, Sirtuin 1 (SIRT1). The Misunderstood HDAC. J Biomol Screen 2011;16:1153—69.

[23] Wang L, Tao R, Hancock WW. Using histone deacetylase inhibitors to enhance Foxp3(+) regulatory T-cell function and induce allograft tolerance. Immunol Cell Biol 2009;87:195—202.

[24] Beier UH, Akimova T, Liu Y, Wang L, Hancock WW. Histone/protein deacetylases control Foxp3 expression and the heat shock response of T-regulatory cells. Curr Opin Immunol 2011;23(5):670—8.

[25] Grausenburger R, Bilic I, Boucheron N, et al. Conditional deletion of histone deacetylase 1 in T cells leads to enhanced airway inflammation and increased Th2 cytokine production. J Immunol 2010;185:3489—97.

[26] Sozzani S. Dendritic cell trafficking: more than just chemokines. Cytokine Growth Factor Rev 2005;16:581—92.

[27] Annunziato F, Cosmi L, Santarlasci V, et al. Phenotypic and functional features of human Th17 cells. J Exp Med 2007;204:1849—61.

[28] Yen D, Cheung J, Scheerens H, et al. IL-23 is essential for T cell-mediated colitis and promotes inflammation via IL-17 and IL-6. J Clin Invest 2006;116:1310—6.

[29] Valencia X, Yarboro C, Illei G, Lipsky PE. Deficient CD4+CD25high T regulatory cell function in patients with active systemic lupus erythematosus. J Immunol 2007;178:2579—88.

[30] Lyssuk EY, Torgashina AV, Soloviev SK, Nassonov EL, Bykovskaia SN. Reduced number and function of CD4+CD25highFoxP3+ regulatory T cells in patients with systemic lupus erythematosus. Adv Exp Med Biol 2007;601:113—9.

[31] Li B, Samanta A, Song X, et al. FOXP3 interactions with histone acetyltransferase and class II histone deacetylases are required for repression. Proc Natl Acad Sci USA 2007;104:4571—6.

[32] Halili MA, Andrews MR, Sweet MJ, Fairlie DP. Histone deacetylase inhibitors in inflammatory disease. Curr Top Med Chem 2009;9:309—19.

[33] Nencioni A, Beck J, Werth D, et al. Histone deacetylase inhibitors affect dendritic cell differentiation and immunogenicity. Clin Cancer Res 2007;13:3933—41.

[34] Brogdon JL, Xu Y, Szabo SJ, et al. Histone deacetylase activities are required for innate immune cell control of Th1 but not Th2 effector cell function. Blood 2007;109:1123—30.

[35] Bosisio D, Vulcano M, Del Prete A, et al. Blocking TH17-polarizing cytokines by histone deacetylase inhibitors in vitro and in vivo. J Leukoc Biol 2008;84:1540—8.

[36] Kelly WK, Marks PA, Drug insight. Histone deacetylase inhibitors—development of the new targeted anti-cancer agent suberoylanilide hydroxamic acid. Nat Clin Pract Oncol 2005;2:150—7.

[37] Wang Z, Chen C, Finger SN, et al. Suberoylanilide hydroxamic acid: a potential epigenetic therapeutic agent for lung fibrosis? Eur Respir J 2009;34:145—55.

[38] Lin T, Chen H, Koustova E, et al. Histone deacetylase as therapeutic target in a rodent model of hemorrhagic shock: effect of different resuscitation strategies on lung and liver. Surgery 2007;141:784—94.

[39] Lin HS, Hu CY, Chan HY, et al. Anti-rheumatic activities of histone deacetylase (HDAC) inhibitors in vivo in collagen-induced arthritis in rodents. Br J Pharmacol 2007;150:862—72.

[40] Drummond DC, Noble CO, Kirpotin DB, Guo Z, Scott GK, Benz CC. Clinical development of histone deacetylase inhibitors as anticancer agents. Annu Rev Pharmacol Toxicol 2005;45:495—528.

[41] Morley M, Molony CM, Weber TM, et al. Genetic analysis of genome-wide variation in human gene expression. Nature 2004;430:743—7.

[42] Sims 3rd RJ, Reinberg D. From chromatin to cancer: a new histone lysine methyltransferase enters the mix. Nat Cell Biol 2004;6:685—7.

[43] Garcia BA, Busby SA, Shabanowitz J, Hunt DF, Mishra N. Resetting the epigenetic histone code in the MRL-lpr/lpr mouse model of lupus by histone deacetylase inhibition. J Proteome Res 2005;4:2032—42.

[44] Aung HT, Schroder K, Himes SR, et al. LPS regulates proinflammatory gene expression in macrophages by altering histone deacetylase expression. FASEB J 2006;20:1315—27.

[45] Bode KA, Schroder K, Hume DA, et al. Histone deacetylase inhibitors decrease Toll-like receptor-mediated activation of proinflammatory gene expression by impairing transcription factor recruitment. Immunology 2007;122:596—606.

[46] Tao R, de Zoeten EF, Ozkaynak E, et al. Deacetylase inhibition promotes the generation and function of regulatory T cells. Nat Med 2007;13:1299−307.

[47] Chung YL, Lee MY, Wang AJ, Yao LF. A therapeutic strategy uses histone deacetylase inhibitors to modulate the expression of genes involved in the pathogenesis of rheumatoid arthritis. Mol Ther 2003;8:707−17.

[48] Blair HC, Zaidi M. Osteoclastic differentiation and function regulated by old and new pathways. Rev Endocr Metab Disord 2006;7:23−32.

[49] Yi T, Baek JH, Kim HJ, et al. Trichostatin A-mediated upregulation of p21(WAF1) contributes to osteoclast apoptosis. Exp Mol Med 2007;39:213−21.

[50] Rahman MM, Kukita A, Kukita T, Shobuike T, Nakamura T, Kohashi O. Two histone deacetylase inhibitors, trichostatin A and sodium butyrate, suppress differentiation into osteoclasts but not into macrophages. Blood 2003;101:3451−9.

[51] Cantley MD, Fairlie DP, Bartold PM, et al. Inhibitors of histone deacetylases in class I and class II suppress human osteoclasts in vitro. J Cell Physiol 2011;226:3233−41.

[52] Schroeder TM, Westendorf JJ. Histone deacetylase inhibitors promote osteoblast maturation. J Bone Miner Res 2005;20:2254−63.

[53] Segain JP, Raingeard de la Bletiere D, Bourreille A, et al. Butyrate inhibits inflammatory responses through NFkappaB inhibition: implications for Crohn's disease. Gut 2000;47:397−403.

[54] Bright JJ, Musuro BF, Du C, Sriram S. Expression of IL-12 in CNS and lymphoid organs of mice with experimental allergic encephalitis. J Neuroimmunol 1998;82:22−30.

[55] Takahashi I, Miyaji H, Yoshida T, Sato S, Mizukami T. Selective inhibition of IL-2 gene expression by trichostatin A, a potent inhibitor of mammalian histone deacetylase. J Antibiot (Tokyo) 1996;49:453−7.

[56] Camelo S, Iglesias AH, Hwang D, et al. Transcriptional therapy with the histone deacetylase inhibitor trichostatin A ameliorates experimental autoimmune encephalomyelitis. J Neuroimmunol 2005;164:10−21.

[57] Gray SG, De Meyts P. Role of histone and transcription factor acetylation in diabetes pathogenesis. Diabetes Metab Res Rev 2005;21:416−33.

[58] Luhrs H, Gerke T, Boxberger F, et al. Butyrate inhibits interleukin-1-mediated nuclear factor-kappa B activation in human epithelial cells. Dig Dis Sci 2001;46:1968−73.

[59] Yin L, Laevsky G, Giardina C. Butyrate suppression of colonocyte NF-kappa B activation and cellular proteasome activity. J Biol Chem 2001;276:44641−6.

[60] Ray A, Prefontaine KE, Ray P. Down-modulation of interleukin-6 gene expression by 17 beta-estradiol in the absence of high affinity DNA binding by the estrogen receptor. J Biol Chem 1994;269:12940−6.

[61] Choi JH, Oh SW, Kang MS, Kwon HJ, Oh GT, Kim DY. Trichostatin A attenuates airway inflammation in mouse asthma model. Clin Exp Allergy 2005;35:89−96.

[62] Whittaker SJ, Demierre MF, Kim EJ, et al. Final results from a multicenter, international, pivotal study of romidepsin in refractory cutaneous T-cell lymphoma. J Clin Oncol 2010;28:4485−91.

[63] Shao W, Growney JD, Feng Y, et al. Activity of deacetylase inhibitor panobinostat (LBH589) in cutaneous T-cell lymphoma models: Defining molecular mechanisms of resistance. Int J Cancer 2010;127:2199−208.

[64] Knipstein J, Gore L. Entinostat for treatment of solid tumors and hematologic malignancies. Expert Opin Investig Drugs 2011;20:1455−67.

[65] Boumber Y, Younes A, Garcia-Manero G. Mocetinostat (MGCD0103): a review of an isotype-specific histone deacetylase inhibitor. Expert Opin Investig Drugs 2011;20:823−9.

[66] Ramalingam SS, Belani CP, Ruel C, et al. Phase II study of belinostat (PXD101), a histone deacetylase inhibitor, for second line therapy of advanced malignant pleural mesothelioma. J Thorac Oncol 2009;4:97−101.

[67] Razin A, Cedar H. Distribution of 5-methylcytosine in chromatin. Proc Natl Acad Sci USA 1977;74:2725−8.

[68] Peedicayil J. Epigenetic therapy—a new development in pharmacology. Indian J Med Res 2006;123:17−24.

[69] Deng C, Kaplan MJ, Yang J, et al. Decreased Ras-mitogen-activated protein kinase signaling may cause DNA hypomethylation in T lymphocytes from lupus patients. Arthritis Rheum 2001;44:397−407.

[70] Lu Q, Kaplan M, Ray D, Zacharek S, Gutsch D, Richardson B. Demethylation of ITGAL (CD11a) regulatory sequences in systemic lupus erythematosus. Arthritis Rheum 2002;46:1282−91.

[71] Kaplan MJ, Lu Q, Wu A, Attwood J, Richardson B. Demethylation of promoter regulatory elements contributes to perforin overexpression in CD4+ lupus T cells. J Immunol 2004;172:3652−61.

[72] Oelke K, Lu Q, Richardson D, et al. Overexpression of CD70 and overstimulation of IgG synthesis by lupus T cells and T cells treated with DNA methylation inhibitors. Arthritis Rheum 2004;50:1850−60.

[73] Oelke K, Richardson B. Decreased T cell ERK pathway signaling may contribute to the development of lupus through effects on DNA methylation and gene expression. Int Rev Immunol 2004;23:315−31.

[74] Detich N, Theberge J, Szyf M. Promoter-specific activation and demethylation by MBD2/demethylase. J Biol Chem 2002;277:35791—4.

[75] Zhu B, Zheng Y, Angliker H, et al. 5-Methylcytosine DNA glycosylase activity is also present in the human MBD4 (G/T mismatch glycosylase) and in a related avian sequence. Nucleic Acids Res 2000;28:4157—65.

[76] Vilain A, Vogt N, Dutrillaux B, Malfoy B. DNA methylation and chromosome instability in breast cancer cell lines. FEBS Lett 1999;460:231—4.

[77] Chang C, Gershwin ME. Drug-induced lupus erythematosus: incidence, management and prevention. Drug Saf 2011;34:357—74.

[78] Chang C, Gershwin ME. Drugs and autoimmunity—a contemporary review and mechanistic approach. J Autoimmun 2010;34:J266—75.

[79] Nanki T, Hayashida K, El-Gabalawy HS, et al. Stromal cell-derived factor-1-CXC chemokine receptor 4 interactions play a central role in CD4+ T cell accumulation in rheumatoid arthritis synovium. J Immunol 2000;165:6590—8.

[80] Blades MC, Ingegnoli F, Wheller SK, et al. Stromal cell-derived factor 1 (CXCL12) induces monocyte migration into human synovium transplanted onto SCID Mice. Arthritis Rheum 2002;46:824—36.

[81] Konttinen YT, Ainola M, Valleala H, et al. Analysis of 16 different matrix metalloproteinases (MMP-1 to MMP-20) in the synovial membrane: different profiles in trauma and rheumatoid arthritis. Ann Rheum Dis 1999;58:691—7.

[82] Ceribelli A, Yao B, Dominguez-Gutierrez PR, Chan EK. Lupus T cells switched on by DNA hypomethylation via microRNA? Arthritis Rheum 2011;63:1177—81.

[83] Pan W, Zhu S, Yuan M, et al. MicroRNA-21 and microRNA-148a contribute to DNA hypomethylation in lupus CD4+ T cells by directly and indirectly targeting DNA methyltransferase 1. J Immunol 2010;184:6773—81.

[84] Krol J, Loedige I, Filipowicz W. The widespread regulation of microRNA biogenesis, function and decay. Nat Rev Genet 2010;11:597—610.

[85] Fabian MR, Sonenberg N, Filipowicz W. Regulation of mRNA translation and stability by microRNAs. Annu Rev Biochem 2010;79:351—79.

[86] Miranda KC, Huynh T, Tay Y, et al. A pattern-based method for the identification of MicroRNA binding sites and their corresponding heteroduplexes. Cell 2006;126:1203—17.

[87] John B, Enright AJ, Aravin A, Tuschl T, Sander C, Marks DS. Human MicroRNA targets. PLoS Biol 2004;2. e363.

[88] Zhao S, Wang Y, Liang Y, et al. MicroRNA-126 regulates DNA methylation in CD4+ T cells and contributes to systemic lupus erythematosus by targeting DNA methyltransferase 1. Arthritis Rheum 2011;63:1376—86.

[89] Nagata Y, Nakasa T, Mochizuki Y, et al. Induction of apoptosis in the synovium of mice with autoantibody-mediated arthritis by the intraarticular injection of double-stranded MicroRNA-15a. Arthritis Rheum 2009;60:2677—83.

[90] Li QJ, Chau J, Ebert PJ, et al. miR-181a is an intrinsic modulator of T cell sensitivity and selection. Cell 2007;129:147—61.

[91] Chen CZ, Li L, Lodish HF, Bartel DP. MicroRNAs modulate hematopoietic lineage differentiation. Science 2004;303:83—6.

[92] Hogg N, Leitinger B. Shape and shift changes related to the function of leukocyte integrins LFA-1 and Mac-1. J Leukoc Biol 2001;69:893—8.

[93] Zhao S, Wang Y, Liang Y, et al. MicroRNA-126 regulates DNA methylation in CD4+ T cells and contributes to systemic lupus erythematosus by targeting DNA methyltransferase 1. Arthritis Rheum 2010;63:1376—86.

[94] Pan W, Zhu S, Yuan M, et al. MicroRNA-21 and microRNA-148a contribute to DNA hypomethylation in lupus CD4+ T cells by directly and indirectly targeting DNA methyltransferase 1. J Immunol 184:6773—81.

[95] Higgs BW, Liu Z, White B, et al. Patients with systemic lupus erythematosus, myositis, rheumatoid arthritis and scleroderma share activation of a common type I interferon pathway. Ann Rheum Dis 2011;70:2029—36.

[96] Tang Y, Luo X, Cui H, et al. MicroRNA-146A contributes to abnormal activation of the type I interferon pathway in human lupus by targeting the key signaling proteins. Arthritis Rheum 2009;60:1065—75.

[97] Curtale G, Citarella F, Carissimi C, et al. An emerging player in the adaptive immune response: microRNA-146a is a modulator of IL-2 expression and activation-induced cell death in T lymphocytes. Blood 2010;115:265—73.

[98] Zhao X, Tang Y, Qu B, et al. MicroRNA-125a contributes to elevated inflammatory chemokine RANTES levels via targeting KLF13 in systemic lupus erythematosus. Arthritis Rheum 62:3425—35.

[99] Stanczyk J, Pedrioli DM, Brentano F, et al. Altered expression of MicroRNA in synovial fibroblasts and synovial tissue in rheumatoid arthritis. Arthritis Rheum 2008;58:1001—9.

[100] Leng RX, Pan HF, Qin WZ, Chen GM, Ye DQ. Role of microRNA-155 in autoimmunity. Cytokine Growth Factor Rev 2011;22:141—7.

[101] Pauley KM, Satoh M, Chan AL, Bubb MR, Reeves WH, Chan EK. Upregulated miR-146a expression in peripheral blood mononuclear cells from rheumatoid arthritis patients. Arthritis Res Ther 2008;10. R101.

[102] Niimoto T, Nakasa T, Ishikawa M, et al. MicroRNA-146a expresses in interleukin-17 producing T cells in rheumatoid arthritis patients. BMC Musculoskelet Disord 2010;11:209.

[103] Maurer B, Stanczyk J, Jungel A, et al. MicroRNA-29, a key regulator of collagen expression in systemic sclerosis. Arthritis Rheum 2010;62:1733—43.

[104] Eisenberg I, Eran A, Nishino I, et al. Distinctive patterns of microRNA expression in primary muscular disorders. Proc Natl Acad Sci USA 2007;104:17016—21.

[105] Michael A, Bajracharya SD, Yuen PS, et al. Exosomes from human saliva as a source of microRNA biomarkers. Oral Dis 2010;16:34—8.

[106] Pauley KM, Stewart CM, Gauna AE, et al. Altered miR-146a expression in Sjogren's syndrome and its functional role in innate immunity. Eur J Immunol 2011;41:2029—39.

[107] Du C, Liu C, Kang J, et al. MicroRNA miR-326 regulates TH-17 differentiation and is associated with the pathogenesis of multiple sclerosis. Nat Immunol 2009;10:1252—9.

[108] Junker A, Krumbholz M, Eisele S, et al. MicroRNA profiling of multiple sclerosis lesions identifies modulators of the regulatory protein CD47. Brain 2009;132:3342—52.

[109] Padgett KA, Lan RY, Leung PC, et al. Primary biliary cirrhosis is associated with altered hepatic microRNA expression. J Autoimmun 2009;32:246—53.

[110] Krutzfeldt J, Rajewsky N, Braich R, et al. Silencing of microRNAs in vivo with 'antagomirs'. Nature 2005;438:685—9.

[111] Collison A, Mattes J, Plank M, Foster PS. Inhibition of house dust mite-induced allergic airways disease by antagonism of microRNA-145 is comparable to glucocorticoid treatment. J Allergy Clin Immunol 2011;128:160—7:e4.

[112] Mattes J, Collison A, Plank M, Phipps S, Foster PS. Antagonism of microRNA-126 suppresses the effector function of TH2 cells and the development of allergic airways disease. Proc Natl Acad Sci USA 2009;106:18704—9.

[113] Son JW, Jeong KJ, Jean WS, et al. Genome-wide combination profiling of DNA copy number and methylation for deciphering biomarkers in non-small cell lung cancer patients. Cancer Lett 2011;311:29—37.

[114] Ballestar E, Paz MF, Valle L, et al. Methyl-CpG binding proteins identify novel sites of epigenetic inactivation in human cancer. EMBO J 2003;22:6335—45.

[115] Umezu T, Ohyashiki K, Ohyashiki JH. Detection method for quantifying global DNA methylation by fluorescence correlation spectroscopy. Anal Biochem 2011;415:145—50.

[116] Dai Y, Sui W, Lan H, Yan Q, Huang H, Huang Y. Comprehensive analysis of microRNA expression patterns in renal biopsies of lupus nephritis patients. Rheumatol Int 2009;29:749—54.

[117] Zhao S, Long H, Lu Q. Epigenetic perspectives in systemic lupus erythematosus: pathogenesis, biomarkers, and therapeutic potentials. Clin Rev Allergy Immunol 2010;39:3—9.

[118] Bird AP. CpG-rich islands and the function of DNA methylation. Nature 1986;321:209—13.

[119] Wu J, Wood GS. Reduction of Fas/CD95 promoter methylation, upregulation of Fas protein, and enhancement of sensitivity to apoptosis in cutaneous T-cell lymphoma. Arch Dermatol 2011;147:443—9.

[120] Gasche JA, Hoffmann J, Boland CR, Goel A. Interleukin-6 promotes tumorigenesis by altering DNA methylation in oral cancer cells. Int J Cancer 2011;129:1053—63.

[121] Mongan NP, Gudas LJ. Valproic acid, in combination with all-trans retinoic acid and 5-aza-2'-deoxycytidine, restores expression of silenced RARbeta2 in breast cancer cells. Mol Cancer Ther 2005;4:477—86.

[122] Soriano AO, Yang H, Faderl S, et al. Safety and clinical activity of the combination of 5-azacytidine, valproic acid, and all-trans retinoic acid in acute myeloid leukemia and myelodysplastic syndrome. Blood 2007;110:2302—8.

[123] Fan J, Yin WJ, Lu JS, et al. ER alpha negative breast cancer cells restore response to endocrine therapy by combination treatment with both HDAC inhibitor and DNMT inhibitor. J Cancer Res Clin Oncol 2008;134:883—90.

[124] Ballas ZK. Immunomodulators: a brave new world. J Allergy Clin Immunol 2008;121:331—3.

[125] Drummond DC, Marx C, Guo Z, et al. Enhanced pharmacodynamic and antitumor properties of a histone deacetylase inhibitor encapsulated in liposomes or ErbB2-targeted immunoliposomes. Clin Cancer Res 2005;11:3392—401.

[126] Nandakumar V, Vaid M, Katiyar SK. (-)-Epigallocatechin-3-gallate reactivates silenced tumor suppressor genes, Cip1/p21 and p16INK4a, by reducing DNA methylation and increasing histones acetylation in human skin cancer cells. Carcinogenesis 2011;32:537—44.

[127] Yan L, Nass SJ, Smith D, Nelson WG, Herman JG, Davidson NE. Specific inhibition of DNMT1 by antisense oligonucleotides induces re-expression of estrogen receptor-alpha (ER) in ER-negative human breast cancer cell lines. Cancer Biol Ther 2003;2:552—6.

250

[128] Kaplan MJ. Apoptosis in systemic lupus erythematosus. Clin Immunol 2004;112:210—8.

[129] Loh C, Pau E, Chang NH, Wither JE. An intrinsic B-cell defect supports autoimmunity in New Zealand black chromosome 13 congenic mice. Eur J Immunol 2011;41:527—36.

[130] Church LD, Filer AD, Hidalgo E, et al. Rheumatoid synovial fluid interleukin-17-producing CD4 T cells have abundant tumor necrosis factor-alpha co-expression, but little interleukin-22 and interleukin-23R expression. Arthritis Res Ther 2010;12:R184.

[131] Usmani OS, Ito K, Maneechotesuwan K, et al. Glucocorticoid receptor nuclear translocation in airway cells after inhaled combination therapy. Am J Respir Crit Care Med 2005;172:704—12.

[132] Kitamura H, Torigoe T, Asanuma H, Honma I, Sato N, Tsukamoto T. Down-regulation of HLA class I antigens in prostate cancer tissues and up-regulation by histone deacetylase inhibition. J Urol 2007;178:692—6.

[133] Yang HX, Zhang W, Zhao LD, et al. Are CD4+CD25-Foxp3+ cells in untreated new-onset lupus patients regulatory T cells? Arthritis Res Ther 2009;11:R153.

[134] Pau E, Chang NH, Loh C, Lajoie G, Wither JE. Abrogation of pathogenic IgG autoantibody production in CD40L gene-deleted lupus-prone New Zealand Black mice. Clin Immunol 2011;139:215—27.

[135] Gillespie J, Savic S, Wong C, et al. Histone deacetylases are dysregulated in rheumatoid arthritis and a novel HDAC3-selective inhibitor reduces IL-6 production by PBMC of RA patients. Arthritis Rheum 2012;64:418—22.

[136] Jie G, Jiang Q, Rui Z, Yifei Y. Expression of interleukin-17 in autoimmune dacryoadenitis in MRL/lpr mice. Curr Eye Res 2010;35:865—71.

[137] Cui LF, Guo XJ, Wei J, et al. Overexpression of TNF-alpha and TNFRII in invasive micropapillary carcinoma of the breast: clinicopathological correlations. Histopathology 2008;53:381—8.

[138] Majumdar G, Rooney RJ, Johnson IM, Raghow R. Pan-histone deacetylase inhibitors inhibit pro-inflammatory signaling pathways to ameliorate interleukin-18-induced cardiac hypertrophy. Physiol Genomics 2011;16:1319—33.

[139] Zupkovitz G, Grausenburger R, Brunmeir R, et al. The cyclin-dependent kinase inhibitor p21 is a crucial target for histone deacetylase 1 as a regulator of cellular proliferation. Mol Cell Biol 2010;30:1171—81.

[140] Grabiec AM, Krausz S, de Jager W, et al. Histone deacetylase inhibitors suppress inflammatory activation of rheumatoid arthritis patient synovial macrophages and tissue. J Immunol 2010;184:2718—28.

[141] Karouzakis E, Rengel Y, Jungel A, et al. DNA methylation regulates the expression of CXCL12 in rheumatoid arthritis synovial fibroblasts. Genes Immun 2011;12:643—52.

CHAPTER 13

Epigenetic Mechanisms of Human Imprinting Disorders

Richard H. Scott[1,2], Gudrun E. Moore[1]
[1]Institute of Child Health, University College London, London, UK
[2]Great Ormond Street Hospital, London, UK

CHAPTER OUTLINE

T. Tollefsbol (Ed): Epigenetics in Human Disease. DOI: 10.1016/B978-0-12-388415-2.00013-5

13.1 INTRODUCTION

Epigenetics literally means "above genetics" and refers to the biological mechanisms other than alterations in DNA sequence that influence gene expression and that are stable through cell division [1,2]. The word epigenetics was originally coined by Waddington in 1942 as a portmanteau of "epigenesis" and "genetics" to describe the process by which the genotypes give rise to phenotypes during development [3]. Nowadays, Waddington's definition would be considered to apply to the field of developmental biology in general whereas the meaning of the word epigenetics has narrowed to specifically refer to non-genetic factors that influence gene expression.

There are three widely accepted and closely interacting epigenetic mechanisms: (1) DNA methylation; (2) histone modifications; (3) DNA binding of Polycomb/Trithorax proteins (Table 13.1). Many other specific factors as well as general alterations in chromatin structure also correlate with different states of gene activity but are not considered primary epigenetic modifications as they are not stable through cell division independent of their initial trigger. This caveat eliminates, for example, the DNA binding of transcription factors from consideration as truly epigenetic. However, it also leads some to question the use of the term for systems widely referred to as epigenetic such as histone modification, whose independent heritability through cell division is uncertain.

13.2 CHROMATIN STRUCTURE REFLECTS EPIGENETIC MODIFICATIONS

Chromatin is the complex of DNA, histones, and other DNA-binding proteins and RNA that together make up chromosomes. Differences in chromatin structure are seen between genes in active and inactive states and reflect underlying epigenetic modifications. At its most extreme, chromatin can be considered to be either in an open, active conformation (euchromatin) or

TABLE 13.1 Features of Transcriptionally Active and Inactive Chromatin

Feature	Active	Inactive
Chromatin structure	Open, extended	Closed, condensed
DNA methylation at promoter	No	Yes
Histone methylation		
— H3K4 mono-/trimethylation	Yes	No
— H3K4 dimethylation	No	Yes
— H3K9 monomethylation	Yes	No
— H3K9 trimethylation	No	Yes
— H3K27 monomethylation	Yes	No
— H3K27 trimethylation	No	Yes
Histone acetylation		
— H3K9 acetylation	Yes	No
— H3K14 acetylation	Yes	No
Polycomb complex binding	No	Yes

H3K4, Histone 3 lysine 4; H3K9, Histone 3 lysine 9; H3K14, Histone 3 lysine 14; H3K27, Histone 3 lysine 27.

a closed, inactive conformation (heterochromatin). Other alterations observed include changes in large-scale chromatin conformation and physical interactions between normally distant regions of chromatin.

13.2.1 DNA Methylation

DNA methylation was the first epigenetic modification to be identified and is perhaps the best studied. In mammals, it is well established to have a mitotically stable silencing effect on genes when present at CpG dense promoter sequences [4]. In mammals, DNA methylation occurs at cytosine residues to form 5-methylcytosine and almost exclusively affects CpG dinucleotides [4]. DNA methylation affects the large majority of CpG dinucleotides in the genome and is found broadly across inter- and intragenic sequences including the gene bodies of active genes. Unmethylated domains account for only 1−2% of the genome, the majority of which are CpG islands, short CG-rich stretches of sequence found preferentially at gene promoter regions. Genetic knock-out experiments have demonstrated that DNA methylation is essential for embryonic development, genomic imprinting and X-inactivation and may be involved in the silencing of transposons [5−10].

13.3 DNA METHYLATION AND TRANSCRIPTIONAL SILENCING

DNA methylation can directly reduce binding of transcription factors, but its principal means of transcriptional repression is thought to be via the recruitment of methyl-CpG binding domain (MBD) proteins which effect alterations in chromatin conformation, for example MBD1 and MECP2, which both result in histone modification [11].

The silencing effect of DNA methylation is well established when it is present in CpG dense promoter regions. However, the large majority of silent genes do not have a methylated CpG island at their promoter, indicating that other means of epigenetic control must exist. The effect of methylation at promoters with low CpG density is not established. Furthermore, in most tissue types, DNA methylation is normally stably present through cell division and relatively uniform between most cell types. The contribution of dynamic/tissue-specific changes in methylation in the control of gene expression remains unclear [12,13].

13.4 MAINTENANCE AND ESTABLISHMENT OF DNA METHYLATION DURING DEVELOPMENT

DNA methylation is maintained through cell division by the DNA methyltransferase, DNMT1. The symmetry of the CpG sequence means that both strands of DNA have a CpG dinucleotide. The two strands typically share the same methylation status and this is crucial to the maintenance of stable DNA methylation through mitotic division. Following DNA replication, the two daughter double-stranded DNA molecules are hemimethylated (i.e. methylated on one strand only). Methylation of the new strand of each daughter molecule is then performed by DNMT1 [14].

In contrast to its relative stability in differentiated cells, dramatic changes in DNA methylation occur during mammalian development. This epigenetic reprogramming occurs in two stages; (1) reprogramming of germ cells; and (2) reprogramming of early embryonic cells (Figure 13.1). Each stage involves a round of demethylation and a round of de novo methylation. De novo methylation is carried out by a variety of DNA methyltransferases including DNMT1, DNMT3A, and DNMT3B, some of which have germ cell and sex-specific isoforms [15].

Primordial germ cells undergo genome-wide demethylation early in development, like other post-zygotic cell types and are largely demethylated until gonadal differentiation. After gonadal differentiation, de novo methylation occurs and leads to substantial methylation in both sperm and eggs, principally targeting transposons and repeat sequences but also

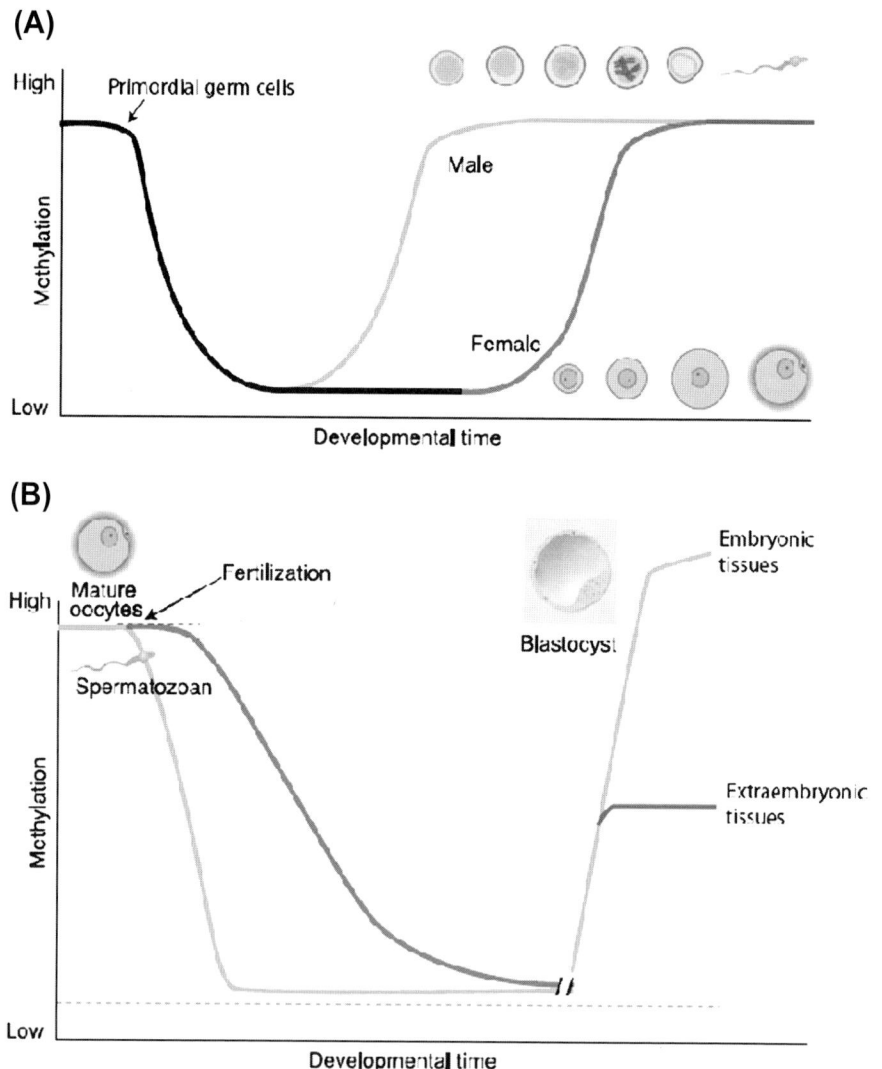

FIGURE 13.1

Changes in the overall level of DNA methylation during mammalian development. (A) Changes in DNA methylation in germ cells. (B) Changes in DNA methylation following fertilization. In both panels the level of DNA methylation is shown on the vertical axis and developmental time on the horizontal axis. Adapted from [16]. This figure is reproduced in the color plate section.

imprinted loci [15,16]. The overall level of methylation is somewhat higher in sperm than eggs and sex-specific differences occur at imprinted loci. The timing of de novo methylation also differs between the sexes. It occurs before meiosis in male germ cells and during meiotic prophase arrest I in female germ cells.

After fertilization a further round of genome-wide demethylation then occurs. Shortly before gastrulation, de novo methylation occurs. Following this, somatic embryonic cells show the high level of methylation at sequences other than CpG islands that are seen in maturity. Trophoblast cells undergo de novo methylation but remain relatively less methylated. Primordial germ cells remain largely unmethylated until after gastrulation [16].

13.4.1 Histone Modifications

Histones are the chief protein component of chromatin and are assembled as octameric particles made up of two copies of each of the four classes of core histone molecule, H2A, H2B, H3, and H4. One hundred and forty-six base-pairs of DNA are wound around each histone octamer forming the basic building block of chromatin — the nucleosome. Many post-translational modifications involve histones, often in combination with one another, and exert epigenetic control on gene expression. Foremost amongst these are the methylation and acetylation of lysine residues in the N-terminal tails of histones H3 and H4 (Table 13.1).

Modifications most commonly associated with active gene expression include mono-methylation at H3K9 and H3K27 (i.e. H3 lysine 9 and H3 lysine 27), trimethylation of H3K4, and acetylation at H3K9 and H3K14. Repressive modifications include dimethylation of H3K4 and trimethylation of H3K9, H3K27. Control of these modifications is exerted by a wide variety of histone methyltransferase, histone demethylase, histone acetylase, and histone deacetylase enzymes [17,18].

Several models have been proposed to explain the heritability of histone modifications through cell division, but none is proven (reviewed in [19]). Indeed it is unclear whether histone modifications themselves are transmitted to daughter chromatin strands following DNA replication or if this is transmitted via a separate system, for example DNA methylation or the binding of non-histone proteins. While this uncertainty remains, some authors argue that histone modifications should not be regarded as true epigenetic modifications.

13.4.2 Non-Histone DNA-Binding Proteins

A variety of non-histone proteins also bind to DNA to affect chromatin structure and exert epigenetic control on gene expression. The best established of these are the Polycomb and Trithorax group proteins which promote transcriptional repression and activation respectively, and both of which act stably through cell division. The two systems interact closely with one another and with other epigenetic systems and have been implicated in the regulation of genes in early development and stem cell renewal. They are also involved in X-inactivation and genomic imprinting [20,21].

The Polycomb system consists of two protein complexes. The Polycomb repressor complex 2, the so-called initiation complex, binds to target DNA sequences and, through the action of its component protein EZH2, results in the repressive histone modification H3K27 trimethylation (Table 13.1). The Polycomb repressor complex 1, the so-called maintenance complex, recognizes this repressive mark and is crucial in the resultant transcriptional repression. The mechanism by which the complex causes repression is unknown [21]. The Trithorax system also acts through histone modification with the Trithorax protein MLL causing the activating histone modification H3K4 methylation. Polycomb/Trithorax-induced epigenetic states are stable through cell division, and the Polycomb repressor complex 1 has recently been shown to remain bound to DNA during DNA replication [22].

13.4.3 Non-Coding RNAs

Non-coding RNAs have been shown to contribute to epigenetic control of gene expression, for example the *Xist* non-coding RNA which is central to X-inactivation [23,24].

13.4.4 Cross-Talk Between Epigenetic Systems

There is extensive correlation between activating and repressive modifications in different epigenetic systems so that they often appear to work in a concerted manner rather than as separate systems (Table 13.1). In addition to simple correlations, an increasing number of specific interactions are being identified between the systems, for example between DNA methylation and histone deacetylation through the recruitment of methyl cytosine binding proteins such as MECP2 and MBD1 and through the inhibition of binding of DNMT3A and its co-factor DNMT3L to H3 by trimethylation at H3K4. A detailed exposition of these interactions is beyond the scope of this chapter. They are reviewed by Reik [25] and Cedar and Bergman [11].

13.5 GENOMIC IMPRINTING

Genomic imprinting is an epigenetic phenomenon observed in mammals, seeded plants, and some insects in which certain genes show parent-of-origin-specific patterns of expression.

257

Around 60 genes have been shown to be consistently imprinted in man (Geneimprint database 2008; Catalogue of Parent of Origin Effects 2009) [74,75]. Some are imprinted in all cell types examined, while others show tissue-specific imprinting, or are only imprinted at certain stages of development. Imprinted genes are often arranged in clusters, each cluster spanning up to several megabases (Table 13.2).

One theory that perhaps best explains the evolution of imprinting is that of parental genome conflict. This theory suggests that there is a conflict of evolutionary advantage between the paternal genome with the maternal genome which is as a result of the mother carrying the offspring in utero. The maternal genome must preserve herself and resources for future offspring so limits supplies to the baby, whereas the paternal genome only needs to consider the baby and encourages growth. Proponents of the theory point to the existence of a number of imprinted genes that regulate growth and the tendency for paternally expressed genes to promote growth and for maternally expressed genes to suppress growth [26,27].

13.5.1 Imprinted Loci

Formal demonstration of imprinting requires demonstration of parent-of-origin-specific gene expression. This can be technically challenging as human tissues are difficult to obtain, limiting systematic expression analysis of the human genes. Instead, known imprinted loci have often been identified following the observation of features suggestive of imprinting, including:

1. Phenotypic abnormality resulting from uniparental disomy (UPD)
2. Parent-of-origin-specific effects of mutation, copy number abnormality or chromosomal rearrangement
3. Parent-of-origin effects of allelic loss in cancers
4. Parent-of-origin-specific epigenetic modifications in the region (for example differential methylation)
5. DNA sequence characteristics similar to other imprinted loci
6. Evidence of imprinting in other species, e.g. mouse.

Major clusters of imprinted genes have been identified at 7q21.3, 7q32.2, 11p15, 15q11.2, 19q13.4, and 20q13.32 (Table 13.2).

13.5.2 Imprinting Centers Control Imprinting in *cis*

Imprinting is controlled by epigenetic modifications at *cis*-acting regulatory sequence elements termed imprinting centers. The mechanism of control has been elucidated only in a subset of imprinted loci. Even in this small number of loci, the variety of different mechanisms operating is striking. A shared feature is the presence of areas of parent-of-origin-specific DNA methylation termed differentially methylated regions (DMRs). These DMRs are CpG-rich sequences located either at the promoter of an imprinted gene or in a more distant regulatory sequence called an imprinting center. This differential methylation is associated with maintenance of differential (i.e. imprinted) expression of genes in the region. A single differentially methylated imprinting center often appears to control imprinting of multiple genes in a cluster. This is sometimes termed a primary DMR. Differential methylation is sometimes seen at other nearby "secondary" DMRs which are established as a result of the action of the imprinting center/primary DMR and which in some cases have been shown to be involved in the maintenance of imprinting in the region [28].

At some loci, differences in other epigenetic modifications have been demonstrated between the two parental alleles including differential histone modification, Polycomb complex binding as well as the differential binding of the CCCTC-binding factor CTCF and differences in high-order chromatin structure such as looping of chromatin to allow access of genes to distant enhancer elements. A further feature shared by a number of loci is the presence of multiple overlapping, often untranslated, transcripts that may play a regulatory function.

TABLE 13.2 Genes/Transcripts Reported to Show Imprinted Expression in Man

Gene	Location	Expressed Allele
TP73	1p36.3	Maternal
DIRAS3	1p31.3	Paternal
LRRTM1	2p12	Paternal
NAP1L5	4q22.1	Paternal
PRIM2	6p11.2	Maternal
PLAGL1	6q24.2	Paternal
HYMAI	6q24.2	Paternal
IGF2R	6q25.3	Maternal
SLC22A2	6q25.3	Maternal
SLC22A3	6q25.3	Maternal
DDC	7p12.2	Isoform dependent
GRB10	7p12.2	Isoform dependent
CALCR	7q21.3	Maternal
TFPI2	7q21.3	Maternal
SGCE	7q21.3	Paternal
PEG10	7q21.3	Paternal
PPP1R9A	7q21.3	Maternal
DLX5	7q21.3	Maternal
CPA4	7q32.2	Maternal
MEST	7q32.2	Paternal
MESTIT1	7q32.2	Paternal
COPG2	7q32.2	Paternal
COPG2IT1	7q32.2	Paternal
KLF14	7q32.2	Maternal
DLGAP2	8p23.3	Paternal
KCNK9	8q24.3	Maternal
INPP5F V2	10q26.11	Paternal
H19	11p15.5	Maternal
IGF2	11p15.5	Paternal
IGF2AS	11p15.5	Paternal
INS	11p15.5	Paternal
KCNQ1	11p15.5	Maternal
KCNQ1OT1	11p15.5	Paternal
KCNQ1DN	11p15.5	Maternal
CDKN1C	11p15.5	Maternal
SLC22A18AS	11p15.5	Maternal
SLC22A18	11p15.5	Maternal
PHLDA2	11p15.5	Maternal
OSBPL5	11p15.4	Maternal
ZNF215	11p15.4	Maternal
AWT1	11p13	Paternal
WT1-AS	11p13	Paternal
WIF1	12q14.3	Paternal
DLK1	14q32.2	Paternal
MEG3	14q32.2	Maternal
RTL1	14q32.2	Paternal
RTLas	14q32.2	Maternal
MKRN3	15q11.2	Paternal
MAGEL2	15q11.2	Paternal
NDN	15q11.2	Paternal
W89101	15q11.2	Paternal
SNRPN	15q11.2	Paternal
SNURF	15q11.2	Paternal
SNORD107	15q11.2	Paternal
SNORD64	15q11.2	Paternal
SNORD108	15q11.2	Paternal

Continued

259

TABLE 13.2	Genes/Transcripts Reported to Show Imprinted Expression in Man—continued	
Gene	**Location**	**Expressed Allele**
SNORD109A	15q11.2	Paternal
PWCR1	15q11.2	Paternal
SNORD115 cluster	15q11.2	Paternal
SNORD116	15q11.2	Paternal
SNORD109B	15q11.2	Paternal
UBE3A	15q11.2	Maternal
ATP10A	15q11.2	Maternal
H73492	15q13	Paternal
ZNF597	16p13.3	Maternal
TCEB3C	18q21.1	Maternal
ZNF331	19q13.41	Maternal
ZIM2	19q13.43	Paternal
PEG3	19q13.43	Paternal
ITUP1	19q13.43	Paternal
ZNF264	19q13.43	Maternal
PSIMCT-1	20q11.21	Paternal
NNAT	20q11.23	Paternal
BLCAP	20q11.2	Isoform dependent
L3MBTL	20q13.12	Paternal
GNAS	20q13.32	Maternal
GNAS Exon A/B	20q13.32	Paternal
GNASXL	20q13.32	Paternal
NESP	20q13.32	Maternal
NESPAS	20q13.32	Paternal

260

13.5.3 Establishment of DNA Methylation at Imprinted Loci During Development

The establishment of imprinting also involves DNA methylation, histone modification, and/or Polycomb gene binding, although it should be noted that many of the data in this area derive from analyses in the mouse and are assumed to apply also in man. As with other sequences, DNA methylation at imprinted loci is reprogrammed during germ cell development. However, the pattern of methylation differs between the male and female germ cells, resulting in the establishment of primary DMRs and therefore of imprinting. A further difference from reprogramming of non-imprinted loci is that DMRs escape the second round of demethylation and de novo remethylation in early embryonic development.

Imprinted regions undergo DNA demethylation similar to that at non-imprinted regions in primordial germ cells and are almost entirely demethylated when they enter the gonads shortly after gastrulation. Methylation of imprinted genes in the male germline occurs at only a small number of loci, the best-studied being the *H19* DMR at 11p15. As with non-imprinted loci, methylation at paternally methylated DMRs such as *H19* occurs before meiosis [29]. Methylation at a larger number of imprinted loci occurs in the female germline. As with methylation at other loci, this occurs after meiosis I [30]. Unlike the majority of other sequences, imprinted loci appear to escape the genome-wide demethylation that occurs after fertilization, allowing them to retain the differential methylation of the paternal and maternal alleles established during germ cell development.

13.5.4 The 11p15 Imprinted Region

The detailed description of the structure and control mechanisms of each of the known imprinted regions is beyond the scope of this chapter. However, in order to illustrate a number of the shared features of imprinted regions, we set out the normal structure and function of the 11p15 imprinted region. The region has been studied extensively in man and is disrupted in the human disorders Beckwith—Wiedemann syndrome and Silver—Russell syndrome.

In the simplest terms, the paternal 11p15 allele promotes growth through the expression of growth-promoting genes and the silencing of growth-suppressing genes and the maternal 11p15 allele suppresses growth through the expression of growth-suppressing genes and the silencing of growth-promoting genes. In fact, the 11p15 Growth Regulatory Region (GRR) consists of two, apparently independent, imprinted domains each controlled in *cis* by its own differentially methylated imprinting center (Figure 13.2). Each domain contains a cluster of imprinted genes which include growth promoters and growth suppressors.

13.5.5 Imprinted Domain 1

Imprinted domain 1 is the more telomeric of the two domains and contains the paternally expressed growth promoter *IGF2* (*insulin-like growth factor 2*) and the maternally expressed non-coding *H19* (Figure 13.2) [31]. Imprinting at *H19* and *IGF2* is observed in both placental and embryonic tissues and is maintained in maturity in many differentiated tissues. Imprinting in the domain is controlled by a paternally methylated imprinting center immediately upstream of *H19*, known as the *H19* DMR. It is also referred to as imprinting center 1 or imprinting control region 1 and is thought to act as a physical insulator, controlling access of

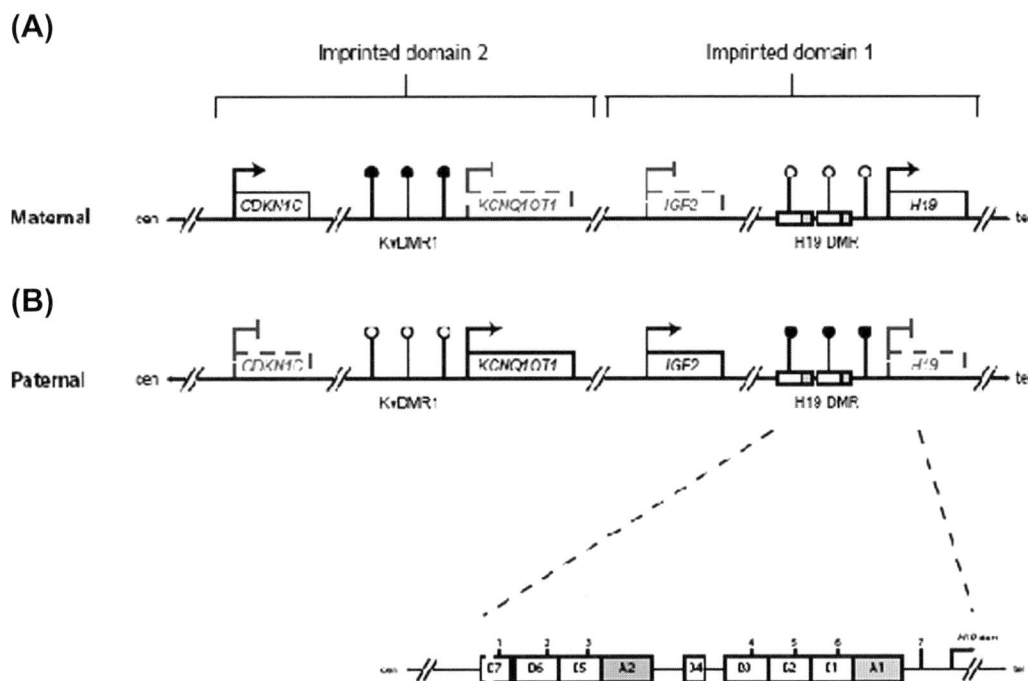

261

FIGURE 13.2

Schematic diagram of the normal 11p15 GRR. (A) The maternal and paternal alleles at the normal 11p15 growth-regulatory region. The region is arranged in two imprinted domains, the more telomeric imprinted domain 1 and the more centromeric imprinted domain 2. Each imprinted domain is controlled by an imprinting center containing a differentially methylated region (DMR). Imprinted domain 1 is controlled by the *H19* DMR. This is methylated on the paternal allele (filled lollipops) and unmethylated on the paternal allele (open lollipops). Imprinted domain 2 is controlled by KvDMR1. This is methylated on the maternal allele and unmethylated on the paternal allele. At imprinted domain 1, *IGF2* is expressed (solid outline and arrow at 5′ end) from the paternal allele but silent (gray dashed outline and bar at 5′ end) on the maternal allele while the non-coding *H19* transcript is expressed from the maternal allele and silent on the paternal allele. At imprinted domain 2, a number of genes including *CDKN1C* are expressed from the maternal allele but silent form the paternal allele, while *KCNQ1OT1* is expressed from the paternal allele but silent from the maternal allele. (B) Detailed view of the structure of the *H19* DMR, which is arranged in two repeat blocks each containing one A- and three or four B-repeat elements. Six of the B-repeat elements containing target sites for the CTCF zinc-finger protein (numbered above the repeat blocks) and a seventh CTCF target site lies between the repeat blocks and the *H19* transcription start site.

the competing *H19* and *IGF2* genes to telomeric enhancer elements [32]. On the normal maternal allele, the *H19* DMR is unmethylated allowing *H19* access to these enhancers. This results in expression of *H19* and silencing of *IGF2*. On the normal paternal allele, the *H19* DMR is methylated, allowing *IGF2* access to the enhancers. This results in expression of *IGF2* and silencing of *H19*.

H19 DMR CONTROLS IMPRINTED DOMAIN 1

In man, differential methylation at the H19 DMR extends for at least 5.5 kb upstream from the 5′ end of the *H19* gene. At its core is a, 3.8-kb span of repeated sequence elements which contain multiple target sites for CTCF (Figure 13.2) [33,34]. These are arranged in two repeat blocks containing A- and B-repeat elements. Six CTCF target sites are present within B-repeat elements and a seventh lies between the repeat blocks and the *H19* transcription start site.

Binding of CTCF at these sites occurs preferentially to the unmethylated maternal allele and may protect it from abnormal methylation and/or mediate its function as a chromatin insulator [35]. The functional importance of these CTCF target sites is supported by the *H19* hypermethylation and *IGF2* loss of imprinting seen with their deletion or disruption of the maternal allele in model organisms and in man [31,34−37].

CHROMATIN LOOPING IN THE CONTROL OF IMPRINTED DOMAIN 1

In the mouse, parent-of-origin-specific chromatin loops have been found at imprinted domain 1 and are thought to be important in the control of imprinting. These physical alterations in chromatin conformation mediate what is often referred to as the chromatin insulator function of the *H19* DMR. The chromatin loops are formed by the physical interaction of the *H19* DMR with differentially methylated regions at *IGF2* and the different loops formed on each parental allele probably result in imprinting at *H19* and *IGF2* by altering their access to enhancer elements [38].

The unmethylated maternal *H19* DMR is bound by CTCF and physically interacts with *IGF2* DMR1. This creates two chromatin domains, with *H19* in an active domain with its enhancers and *IGF2* in an inactive domain away from the enhancers. The methylated paternal *H19* DMR is not bound by CTCF and interacts with *IGF2* DMR2. This results in *IGF2* lying in the active domain, with access to the enhancers telomeric of *H19*. *H19* is in the active domain but silenced, probably by the presence DNA methylation at its promoter (which is within the *H19* DMR). It is presumed that a similar mechanism operates in man, though this is yet to be demonstrated and the role of the human *IGF2* DMRs is uncertain.

THE *IGF2* DIFFERENTIALLY METHYLATED REGIONS

Three differentially methylated regions at *IGF2* have been identified in the mouse (*IGF2* DMR0, DMR1, and DMR2). As described above, *IGF2* DMR1 and DMR2 have been demonstrated to be involved in the formation of parent-of-origin-specific chromatin loops. In man, only two *IGF2* DMRs have been identified, the paternally methylated *IGF2* DMR0 and *IGF2* DMR2 [38]. Their roles are not known and understanding of the region was hindered by initial reports which incorrectly assigned the parent-of-origin of the methylated allele at *IGF2* DMR0 [39,40].

13.5.6 Imprinted Domain 2

Imprinted domain 2 contains the paternally expressed non-coding *KCNQ1OT1* (*KCNQ1-overlapping transcript 1*) and a number of maternally expressed genes including the growth suppressor *CDKN1C* (*Cyclin dependent kinase inhibitor 1C*; Figure 13.2) [64]. *KCNQ1OT1* and *CDKN1C* maintain imprinted expression in embryonic and many differentiated tissues. By contrast a number of maternally expressed genes in the region other than *CDKN1C* are imprinted only in the placenta or in a subset of embryonic or differentiated tissues [41].

Imprinting in the domain is controlled in cis by an imprinting center at the promoter of *KCNQ1OT1* known as KvDMR1, which is also referred to as *KCNQ1OT1*, *LIT1*, imprinting center 2, or imprinting control region 2. On the normal maternal allele KvDMR1 is methylated, resulting in silencing of *KCNQ1OT1* and expression of *CDKN1C*. On the normal paternal allele KvDMR1 is unmethylated, resulting in expression of *KCNQ1OT1* and silencing of *CDKN1C*.

KVDMR1 CONTROLS IMPRINTING AT IMPRINTED DOMAIN 2

KvDMR1 may control imprinting in domain 2 through more than one mechanism. Two main mechanisms have been proposed: (1) a non-coding RNA mediated mechanism; and (2) an enhancer competition-mediated mechanism.

On the paternal allele, many of the maternally expressed genes show imprinted expression in the placenta in the absence of differential DNA methylation at their promoters [42]. This is similar to that seen in the process of X inactivation and it has been proposed that silencing of these genes on the paternal allele occurs by a similar process to that seen on the inactive X: through repressive histone H3K27 methylation mediated by Polycomb group proteins. This repressive histone modification is observed in placental tissues across the domain and may be targeted to the paternal allele by coating of the region in cis by the paternally expressed non-coding RNA *KCNQ1OT1*, a mechanism parallel to that mediated by the *Xist* transcript on the inactive X [43].

The enhancer-competition mechanism similar to that seen at imprinted domain 1 has also been proposed to explain the maintenance of imprinting in the domain that may be central to the maintenance of imprinting at *KCNQ1OT1*, *CDKN1C*, and other genes that are imprinted in embryonic tissues [44]. There is currently limited evidence to provide mechanistic understanding of this model [45].

13.5.7 Establishment of Imprinting at the 11p15 GRR

Differential methylation is observed in the germline at the imprinting centers in both domains: at the *H19* DMR in imprinted domain 1 and at KvDMR1 in imprinted domain 2 [42,44]. It is thought that this germline differential methylation is the driver of the establishment of post-zygotic imprinting at each domain. At imprinted domain 1 this model predicts that differential methylation at the *H19* DMR is the first event in a cascade that involves differential binding of CTCF and possibly other factors, differential methylation at the *IGF2* DMRs, establishment of parent-of-origin-specific chromatin loops, and therefore the differential access to enhancers that mediate imprinting in the region. At imprinted domain 2, the model predicts that differential methylation at KvDMR1 is the first event in a cascade that involves imprinted expression of the non-coding *KCNQ1OT1* transcript, which in turn is crucial to the establishment of parent-of-origin-specific histone modifications and other epigenetic modifications that mediate imprinting at other genes.

13.5.8 Abnormalities at Imprinted Loci

A variety of classes of molecular defects can result in disruption at imprinted loci. These include both epigenetic and genetic defects.

13.6 UNIPARENTAL DISOMY

Uniparental disomy (UPD) results when both chromosomes of a pair are inherited from the same parent. When UPD encompasses an imprinted locus, both alleles show the characteristics of the retained allele. For example, in a region of paternal UPD (pUPD), paternally expressed genes are expressed from both alleles and maternally expressed genes are silenced. Uniparental disomy can occur by a variety of mechanisms, either prezygotic (usually errors of meiosis) or postzygotic (errors of mitosis) and can affect whole chromosomes or be segmental [46].

13.7 EPIMUTATIONS

Epimutations are isolated epigenetic defects that result in disruption of the normal pattern of expression. This can result in the silencing of the normally active allele or expression of a normally silent allele. At imprinted loci, the activation of a normally silent allele is termed loss of imprinting and results in biallelic expression of a normally monoallelically expressed gene. Disrupted expression is frequently associated with disruption of differential DNA methylation, resulting in hypomethylation or hypermethylation at a DMR. The primary defect underlying epimutations is not known. One possibility is that they result from the stochastic loss or gain of DNA methylation at key CpGs within the relevant DMR.

13.8 IMPRINTING CENTER MUTATIONS

Imprinting center mutations are genetic mutations at imprinting control regions that result in epigenetic disruption of expression of the genes under their control (in cis). They are often associated with disruption of methylation at DMRs and, as with other mutations at imprinted loci, they show parent-of-origin-specific pathogenicity. Imprinting center mutations identified to date have largely been microdeletions spanning several kilobases and in some cases megabases [34,47–51]. The 2.9-kb microinsertion identified at the 11p15 H19 DMR represents a further class of imprinting center mutation and some balanced chromosome rearrangements with breakpoints at imprinted loci may be further examples. The mechanism of pathogenicity of imprinting center mutations is often obscure.

13.9 MUTATIONS IN IMPRINTED GENES

Mutations in a number of imprinted genes have been reported. They are typically only of consequence when inherited on the active allele, that is they show parent-of-origin-dependent pathogenicity [52].

13.10 COPY NUMBER ABNORMALITIES ENCOMPASSING IMPRINTED GENES

Large copy number defects encompassing imprinted genes have been reported at a number of loci, either as a result of interstitial disruptions or unbalanced chromosome translocations. As with mutations in imprinted genes, they have parent-of-origin-specific effects: they would only be expected to alter the expression of genes which are active on the disrupted allele.

13.11 MUTATIONS IN IMPRINTING ESTABLISHMENT OR MAINTENANCE MACHINERY

Recently, biallelic mutations in ZFP57, a gene important in the establishment of DNA methylation at maternally methylated DMRs, have been found to cause hypomethylation at multiple imprinted loci [56]. This represents an example of a distinct mechanism of imprinting disruption: mutation of a component of the machinery of the establishment or maintenance of imprinting. Other examples include biallelic mutations in the genes NALP7 and C6ORF221, which cause recurrent hydatidiform mole [53,54]. Interestingly, mutations in the related gene NALP2 have been reported in one family with Beckwith–Wiedemann syndrome [55].

13.11.1 Phenotypes Associated with Constitutional Abnormalities at Imprinted Loci

Constitutional abnormalities at imprinted loci underlie a number of congenital syndromes in man (Table 13.3). The study of these disorders and the molecular abnormalities that underlie them resulted in the identification of many of the known imprinted loci and has been central to much of our understanding of normal and abnormal imprinting. In a number of cases these constitutional molecular abnormalities, despite being present soma-wide, are mosaic.

13.12 CHROMOSOME 6Q24

Disruption of the imprinted locus at 6q24 results in transient neonatal diabetes mellitus [57]. In addition to diabetes mellitus from the neonatal period that can last until 18 months of age, other features of the condition include intrauterine growth retardation, macroglossia, and umbilical hernia. Abnormalities at 6q24 account for approximately 90% of cases of transient neonatal diabetes mellitus and result in increased expression of the paternally expressed genes *PLAGL1* (*Pleomorphic adenoma gene-like 1*) and the *HYMAI* (*Hydatidiform mole-associated and imprinted*) transcript. Reported abnormalities include paternal UPD, epimutation at the DMRs at *PLAGL1* and *HYMAI* (hypomethylation of the maternal allele) and paternal duplications encompassing both genes [57–59].

13.13 CHROMOSOME 7

Silver–Russell syndrome is a growth-restriction disorder associated with pre- and postnatal growth restriction, relative macrocephaly, growth asymmetry, fifth finger clinodactyly, and a characteristic facial appearance. Maternal uniparental disomy for chromosome 7 is found in approximately 10% of cases. Uniparental disomy usually affects the whole of chromosome 7 but maternal segmental abnormalities have also been reported, providing insights into the likely critical region [60,61]. Extensive work has identified a number of imprinted genes on chromosome 7 (Table 13.2) [62,63]. As discussed below, abnormalities at the 11p15 growth regulatory region account for a further 25–40% of cases of Silver–Russell syndrome.

13.14 CHROMOSOME 11P15

As set out earlier in this chapter, the 11p15 growth-regulatory region, is arranged in two adjacent but independent imprinted domains (Figure 13.2). Two opposing groups of abnormalities in the region result in overgrowth (most characteristically Beckwith–Wiedemann syndrome) and growth restriction (most characteristically Silver–Russell syndrome) [64,65].

Overgrowth is caused by abnormalities that result in increased expression of paternally expressed growth promoters such as *IGF2* or decreased expression of maternally expressed growth suppressors such as *CDKN1C*. Growth-promoting abnormalities at 11p15 are found in approximately 80% of cases of Beckwith–Wiedemann syndrome and include: paternal uniparental disomy 11p15; epimutations at imprinted domain 1 (hypermethylation of the maternal *H19* DMR) or imprinted domain 2 (hypomethylation of the maternal KvDMR1); imprinting center mutations at imprinted domain 1 that result in *H19* hypermethylation (microdeletion or microinsertion within the *H19* imprinting control region); paternal duplications; maternal deletion of *KCNQ1OT1* and KvDMR1; and maternal mutations in *CDKN1C* [64].

Conversely, abnormalities that result in a net decrease in expression of growth promoters such as *IGF2* result in growth restriction. Growth-suppressing abnormalities at 11p15 are found in 25–40% of cases of Silver–Russell syndrome and include epimutations at imprinted domain 1 (hypomethylation of the paternal *H19* DMR) and maternal duplications of the region [65].

TABLE 13.3 Phenotypes Associated with Constitutional Abnormalities at Imprinted Loci

Locus/Phenotype	Class of Abnormality	Abnormality
6q24		
Transient neonatal diabetes mellitus	UPD	Paternal UPD 6q24
	Epimutation	Hypomethylation maternal *PLAGL1* and *HYMAI* DMRs
	Duplication imprinted genes	Duplication paternal 6q24
Chromosome 7		
Growth retardation/Silver–Russell syndrome	UPD	Maternal UPD 7, usually affecting the whole chromosome
11p15		
Overgrowth/Beckwith–Wiedemann syndrome	UPD	Paternal UPD 11p15
	Epimutation	Hypermethylation maternal *H19* DMR
	Epimutation	Hypomethylation maternal KvDMR1 (referred to as KvDMR1 loss of methylation)
	Imprinting centre mutation	Microdeletion/microinsertion maternal *H19* DMR causing *H19* hypermethylation
	Mutation imprinted genes	Mutation maternal *CDKN1C*
	Deletion imprinted genes	Deletion encompassing maternal *KCNQ1OT1* and KvDMR1
	Duplication imprinted genes	Duplication paternal 11p15
Growth retardation/Silver–Russell syndrome	Epimutation	Hypomethylation paternal *H19* DMR
	Duplication imprinted genes	Duplication maternal 11p15
14q32.2		
Maternal UPD 14-like phenotype	UPD	Maternal UPD 14q32.2
	Deletion imprinted genes	Deletion encompassing paternal *DLK1*
Paternal UPD 14-like phenotype	UPD	Paternal UPD 14q32.2
	Epimutation	Hypermethylation maternal IG-DMR and *MEG3*-DMR
	Deletion imprinted genes	Deletion encompassing maternal *MEG3*
15q11-q13		
Prader–Willi syndrome	UPD	Maternal UPD 15q11-q13
	Epimutation	Hypermethylation paternal *SNRPN* DMR
	Imprinting center mutation	Microdeletion paternal *SNRPN* region causing hypermethylation
	Deletion imprinted genes	Deletion paternal 15q11.2 encompassing snoRNAs in *SNRPN* region

TABLE 13.3 Phenotypes Associated with Constitutional Abnormalities at Imprinted Loci—continued

Locus/Phenotype	Class of Abnormality	Abnormality
Angelman syndrome	UPD	Paternal UPD 15q11-13
	Epimutation	Hypomethylation maternal *SNRPN* DMR
	Imprinting centre mutation	Microdeletion/inversion upstream of maternal *SNRPN* causing hypomethylation
	Mutation imprinted gene	Mutation maternal *UBE3A*
	Deletion imprinted genes	Deletion maternal 15q11.2 encompassing *UBE3A*
20q13.32		
Pseudohypoparathyroidism type 1a	Mutation imprinted gene	Mutation maternal *GNAS*
Pseudohypoparathyroidsim type 1b	Epimutation	Hypomethylation maternal *GNAS* Exon A/B DMR +/− *GNASXL* and *NESPAS* DMRs
	Imprinting center mutation	Deletion maternal STX16 exon 4–6 causing hypomethylation *GNAS* exon A/B DMR
	Imprinting center mutation	Deletion maternal *NESP55* DMR causing hypomethylation *GNAS* exon A/B, *GNASXL* and *NESPAS* DMRs
Multiple loci Hypomethylation multiple imprinted loci	Epimutation/unknown	Hypomethylation maternally methylated DMRs including: *PLAGL1*, *GRB10*, *KvDMR1* and *NESPAS*
	Mutation imprinting machinery	Biallelic mutations *ZFP57* causing hypomethylation at multiple maternally methylated *DMRs* including *PLAGL1*, *GRB10*, *KCNQ1OT1* and *NESPAS*

UPD, uniparental disomy; DMR, differentially methylated region.

13.15 CHROMOSOME 14Q32.2

Two opposing groups of abnormalities occur at 14q32.2 causing reciprocal abnormalities at imprinted genes in the region and resulting in two distinct phenotypes. The first, referred to as the paternal UPD 14-like phenotype, is associated with developmental delay, a bell-shaped thorax, abdominal wall defects, and distinctive facial appearance. Causative abnormalities result in decreased expression of the maternally expressed genes such as *MEG3* (*Maternally expressed gene 3*) and *RTL1as* (*Retrotransposon-like gene 1 antisense*) and/or increased expression of paternally expressed genes such as *DLK1* (*Delta, drosophila, homolog-like 1*) and *RTL1* (*Retrotransposon-like gene 1*). Reported abnormalities include paternal UPD 14 and maternal 14q32.2 deletions encompassing the *MEG3* DMR and/or the IG-DMR [50,51].

The second phenotype, referred to as the maternal UPD 14-like phenotype, is associated with pre- and postnatal growth restriction, developmental delay, and early puberty. Causative abnormalities result in increased expression of the maternally expressed genes such as *MEG3* and *RTL1as* and/or decreased expression of paternally expressed genes such as *DLK1* and *RTL1*. Reported abnormalities include maternal UPD 14; epimutations at the IG-DMR and *MEG3*-DMR (hypomethylation at the maternal allele); and paternal 14q32.2 deletions encompassing*DLK1* [50].

13.16 CHROMOSOME 15Q11-Q13

Similar to the 11p15 and 14q32.2 imprinted loci, two opposing groups of abnormalities occur at 15q11.2 and cause distinct phenotypes. Prader—Willi syndrome is characterized by moderate developmental delay, neonatal hypotonia, hyperphagia, and hypogonadism. It is caused by a variety of abnormalities at 15q11.2 that reduce expression of paternally expressed genes including *SNRPN* (*Small nucleolar ribonucleoprotein polypeptide N*), the *SNURF* (*SNRPN upstream reading frame*) and/or the nearby small nucleolar RNAs (snoRNAs). Reported abnormalities include maternal uniparental disomy 15q11.2; epimutations at the *SNRPN* DMR (hypermethylation of the paternal allele); imprinting center mutations that result in *SNRPN* DMR hypermethylation (microdeletions affecting a critical region that includes *SNRPN* exon 1); and paternal deletions that include the snoRNAs adjacent to *SNRPN* [66—68]. This last abnormality is the most frequent cause of the condition and often encompasses the whole of 15q11.2.

Angelman syndrome is characterized by developmental delay with absent or nearly absent speech, an ataxic gait, seizures, and microcephaly. Causative abnormalities reduce expression of the maternally expressed *UBE3A* (*Ubiquitin-protein ligase E3A*) and include paternal uniparental disomy 15q11.2; epimutation at the *SNRPN* DMR (hypomethylation of the maternal allele); imprinting center mutations that result in *SNRPN* DMR hypomethylation (microdeletions or chromosomal inversion affecting a critical region upstream of *SNRPN*); maternal deletions at 15q11.2 encompassing *UBE3A*; and maternal mutations in *UBE3A* [48,68,69].

13.17 CHROMOSOME 20Q13.32

Abnormalities at the imprinted 20q13.32 locus that disturb expression of *GNAS* (*Guanine nucleotide binding protein alpha-stimulating activity polypeptide*) and its surrounding transcripts result in a group of disorders associated with parathyroid hormone resistance (pseudohypo-parathyroidism). Pseudoparathryoidism type 1a is characterized by Albright's hereditary osteodystrophy and resistance to numerous hormones typically including thyroid-stimulating hormone and gonadotrophins in addition to parathyroid hormone. It is caused by mutations on the maternal *GNAS* allele [70,71].

Pseudohypoparathroidism type 1b is characterized by resistance to parathyroid hormone and in some cases thyroid-stimulating hormone without features of Albright's hereditary osteo-dystrophy. Causative epimutations are frequently found and result in hypomethylation of the maternal *GNAS Exon A/B* DMR and in some cases the *GNASXL* and *NESPAS* DMRs [71]. In addition, two distinct types of imprinting center mutations have been reported: deletions of the maternal *STX16* (*Syntaxin 16*) exons 4—6 that result in hypomethylation at the *GNAS Exon A/B* DMR; and deletions of the maternal *NESP55* DMR that cause hypomethylation at the maternal *GNAS Exon A/B*, *GNASXL*, and *NESPAS* DMRs.

13.18 HYPOMETHYLATION AT MULTIPLE IMPRINTED LOCI

A number of individuals have recently been reported with hypomethylation at multiple maternally methylated loci including *PLAGL1*, *GRB10*, *KvDMR1*, and *NESPAS*. This pattern has been identified in individuals originally diagnosed with transient neonatal diabetes mellitus and Beckwith—Wiedemann syndrome [72,73]. However, the spectrum of phenotypes with which hypomethylation at multiple imprinted loci manifests and the determinants of the phenotypic features remain to be identified. Some cases presenting with features of transient neonatal diabetes mellitus are caused by biallelic mutations in *ZFP57* [56]. The underlying cause in others is unknown.

13.19 CONCLUSION

A variety of molecular mechanisms can disrupt imprinted loci and cause a number of human syndromes. Their study has proved valuable in our understanding of these disorders themselves. It has also led to important advances in our understanding of the mechanisms by which imprinting is established and maintained and by which it can be abrogated. Despite the considerable advances that have been made in these areas over the last few decades, much remains to be understood.

References

[1] Bird A. Perceptions of epigenetics. Nature 2007;447:396—8.

[2] Ledford H. Language: Disputed definitions. Nature 2008;455:1023—8.

[3] Waddington CH. The Strategy of the Genes. London: Allen & Unwin; 1957.

[4] Suzuki MM, Bird A. DNA methylation landscapes: provocative insights from epigenomics. Nat Rev Genet 2008;9:465—76.

[5] Li E, Bestor TH, Jaenisch R. Targeted mutation of the DNA methyltransferase gene results in embryonic lethality. Cell 1992;69:915—26.

[6] Walsh CP, Chaillet JR, Bestor TH. Transcription of IAP endogenous retroviruses is constrained by cytosine methylation. Nat Genet 1998;20:116—7.

[7] Okano M, Bell DW, Haber DA, Li E. DNA methyltransferases Dnmt3a and Dnmt3b are essential for de novo methylation and mammalian development. Cell 1999;99:247—57.

[8] Chen T, Li E. Structure and function of eukaryotic DNA methyltransferases. Curr Top Dev Biol 2004;60:55—89.

[9] Karpf AR, Matsui S. Genetic disruption of cytosine DNA methyltransferase enzymes induces chromosomal instability in human cancer cells. Cancer Res 2005;65:8635—9.

[10] Dodge JE, Okano M, Dick F, Tsujimoto N, Chen T, Wang S, et al. Inactivation of Dnmt3b in mouse embryonic fibroblasts results in DNA hypomethylation, chromosomal instability, and spontaneous immortalization. J Biol Chem 2005;280:17986—91.

[11] Cedar H, Bergman Y. Linking DNA methylation and histone modification: patterns and paradigms. Nat Rev Genet 2009;10:295—304.

[12] Weber M, Hellmann I, Stadler MB, Ramos L, Paabo S, Rebhan M, et al. Distribution, silencing potential and evolutionary impact of promoter DNA methylation in the human genome. Nat Genet 2007;39:457—66.

[13] Metivier R, Gallais R, Tiffoche C, Le PC, Jurkowska RZ, Carmouche RP, et al. Cyclical DNA methylation of a transcriptionally active promoter. Nature 2008;452:45—50.

[14] Gruenbaum Y, Cedar H, Razin A. Substrate and sequence specificity of a eukaryotic DNA methylase. Nature 1982;295:620—2.

[15] Schaefer CB, Ooi SK, Bestor TH, Bourc'his D. Epigenetic decisions in mammalian germ cells. Science 2007;316:398—9.

[16] Reik W, Dean W, Walter J. Epigenetic reprogramming in mammalian development. Science 2001;293:1089—93.

[17] Shi Y. Histone lysine demethylases: emerging roles in development, physiology and disease. Nat Rev Genet 2007;8:829—33.

[18] Haberland M, Montgomery RL, a ndOlson EN. The many roles of histone deacetylases in development and physiology: implications for disease and therapy. Nat Rev Genet 2009;10:32—42.

[19] Probst AV, Dunleavy E, Almouzni G. Epigenetic inheritance during the cell cycle. Nat Rev Mol Cell Biol 2009;10:192—206.

[20] Wang J, Mager J, Chen Y, Schneider E, Cross JC, Nagy A, et al. Imprinted X inactivation maintained by a mouse Polycomb group gene. Nat Genet 2001;28:371—5.

[21] Schwartz YB, Pirrotta V. Polycomb complexes and epigenetic states. Curr Opin Cell Biol 2008;20:266—73.

[22] Francis NJ, Follmer NE, Simon MD, Aghia G, Butler JD. Polycomb proteins remain bound to chromatin and DNA during DNA replication in vitro. Cell 2009;137:110—22.

[23] Herzing LB, Romer JT, Horn JM, Ashworth A. Xist has properties of the X-chromosome inactivation centre. Nature 1997;386:272—5.

[24] Sleutels F, Zwart R, Barlow DP. The non-coding Air RNA is required for silencing autosomal imprinted genes. Nature 2002;415:810—3.

[25] Reik W. Stability and flexibility of epigenetic gene regulation in mammalian development. Nature 2007;447:425—32.

[26] Moore T, Haig D. Genomic imprinting in mammalian development: a parental tug-of-war. Trends Genet 1991;7:45—9.

[27] Haig D. Genomic imprinting and kinship: how good is the evidence? Annu Rev Genet 2004;38:553—85.

[28] Murrell A, Heeson S, Reik W. Interaction between differentially methylated regions partitions the imprinted genes Igf2 and H19 into parent-specific chromatin loops. Nat Genet 2004;36:889—93.

[29] Davis TL, Yang GJ, McCarrey JR, Bartolomei MS. The H19 methylation imprint is erased and re-established differentially on the parental alleles during male germ cell development. Hum Mol Genet 2000;9:2885—94.

[30] Lucifero D, Mertineit C, Clarke HJ, Bestor TH, Trasler JM. Methylation dynamics of imprinted genes in mouse germ cells. Genomics 2002;79:530—8.

[31] Scott RH, Douglas J, Baskcomb L, Huxter N, Barker K, Hanks S, et al. Constitutional 11p15 abnormalities, including heritable imprinting center mutations, cause nonsyndromic Wilms tumor. Nat Genet 2008a;40:1329—34.

[32] Bell AC, Felsenfeld G. Methylation of a CTCF-dependent boundary controls imprinted expression of the Igf2 gene. Nature 2000;405:482—5.

[33] Cui H, Niemitz EL, Ravenel JD, Onyango P, Brandenburg SA, Lobanenkov VV, et al. Loss of imprinting of insulin-like growth factor-II in Wilms' tumor commonly involves altered methylation but not mutations of CTCF or its binding site. Cancer Res 2001;61:4947—50.

[34] Sparago A, Cerrato F, Vernucci M, Ferrero GB, Silengo MC, Riccio A. Microdeletions in the human H19 DMR result in loss of IGF2 imprinting and Beckwith—Wiedemann syndrome. Nat Genet 2004;36:958—60.

[35] Hark AT, Schoenherr CJ, Katz DJ, Ingram RS, Levorse JM, Tilghman SM. CTCF mediates methylation-sensitive enhancer-blocking activity at the H19/Igf2 locus. Nature 2000;405:486—9.

[36] Pant V, Kurukuti S, Pugacheva E, Shamsuddin S, Mariano P, Renkawitz R, et al. Mutation of a single CTCF target site within the H19 imprinting control region leads to loss of Igf2 imprinting and complex patterns of de novo methylation upon maternal inheritance. Mol Cell Biol 2004;24:3497—504.

[37] Szabo PE, Tang SH, Silva FJ, Tsark WM, Mann JR. Role of CTCF binding sites in the Igf2/H19 imprinting control region. Mol Cell Biol 2004;24:4791—800.

[38] Murrell A, Ito Y, Verde G, Huddleston J, Woodfine K, Silengo MC, et al. Distinct methylation changes at the IGF2-H19 locus in congenital growth disorders and cancer. PLoS ONE 2008;3:e1849.

[39] Cui H, Onyango P, Brandenburg S, Wu Y, Hsieh CL, Feinberg AP. Loss of imprinting in colorectal cancer linked to hypomethylation of H19 and IGF2. Cancer Res 2002;62:6442—6.

[40] Sullivan MJ, Taniguchi T, Jhee A, Kerr N, Reeve AE. Relaxation of IGF2 imprinting in Wilms tumours associated with specific changes in IGF2 methylation. Oncogene 1999;18:7527—34.

[41] Apostolidou S, bu-Amero S, O'Donoghue K, Frost J, Olafsdottir O, Chavele KM, et al. Elevated placental expression of the imprinted PHLDA2 gene is associated with low birth weight. J Mol Med 2007;85:379—87.

[42] Lewis A, Mitsuya K, Umlauf D, Smith P, Dean W, Walter J, et al. Imprinting on distal chromosome 7 in the placenta involves repressive histone methylation independent of DNA methylation. Nat Genet 2004;36:1291—5.

[43] Redrup L, Branco MR, Perdeaux ER, Krueger C, Lewis A, Santos F, et al. The long noncoding RNA Kcnq1ot1 organises a lineage-specific nuclear domain for epigenetic gene silencing. Development 2009;136:525—30.

[44] Lewis A, Reik W. How imprinting centres work. Cytogenet Genome Res 2006;113:81—9.

[45] Shin JY, Fitzpatrick GV, Higgins MJ. Two distinct mechanisms of silencing by the KvDMR1 imprinting control region. EMBO J 2008;27:168—78.

[46] Kotzot D. Complex and segmental uniparental disomy updated. J Med Genet 2008;45:545—56.

[47] Buiting K, Saitoh S, Gross S, Dittrich B, Schwartz S, Nicholls RD, et al. Inherited microdeletions in the Angelman and Prader—Willi syndromes define an imprinting centre on human chromosome 15. Nat Genet 1995;9:395—400.

[48] Buiting K, Barnicoat A, Lich C, Pembrey M, Malcolm S, Horsthemke B. Disruption of the bipartite imprinting center in a family with Angelman syndrome. Am J Hum Genet 2001;68:1290—4.

[49] Bastepe M, Frohlich LF, Linglart A, Abu-Zahra HS, Tojo K, Ward LM, et al. Deletion of the NESP55 differentially methylated region causes loss of maternal GNAS imprints and pseudohypoparathyroidism type Ib. Nat Genet 2005;37:25—7.

[50] Kagami M, Sekita Y, Nishimura G, Irie M, Kato F, Okada M, et al. Deletions and epimutations affecting the human 14q32.2 imprinted region in individuals with paternal and maternal upd(14)-like phenotypes. Nat Genet 2008;40:237—42.

[51] Kagami M, O'Sullivan MJ, Green AJ, Watabe Y, Arisaka O, Masawa N, et al. The IG-DMR and the MEG3-DMR at human chromosome 14q32.2: hierarchical interaction and distinct functional properties as imprinting control centers. PLoS Genet 2010;6:e1000992.

[52] Hatada I, Ohashi H, Fukushima Y, Kaneko Y, Inoue M, Komoto Y, et al. An imprinted gene p57KIP2 is mutated in Beckwith–Wiedemann syndrome. Nat Genet 1996;14:171–3.

[53] Murdoch S, Djuric U, Mazhar B, Seoud M, Khan R, Kuick R, et al. Mutations in NALP7 cause recurrent hydatidiform moles and reproductive wastage in humans. Nat Genet 2006;38:300–2.

[54] Parry DA, Logan CV, Hayward BE, Shires M, Landolsi H, Diggle C, et al. Mutations causing familial biparental hydatidiform mole implicate c6orf221 as a possible regulator of genomic imprinting in the human oocyte. Am J Hum Genet 2011;89:451–8.

[55] Meyer E, Lim D, Pasha S, Tee LJ, Rahman F, Yates JR, et al. Germline mutation in NLRP2 (NALP2) in a familial imprinting disorder (Beckwith–Wiedemann Syndrome). PLoS Genet 2009;5:e1000423.

[56] Mackay DJ, Callaway JL, Marks SM, White HE, Acerini CL, Boonen SE, et al. Hypomethylation of multiple imprinted loci in individuals with transient neonatal diabetes is associated with mutations in ZFP57. Nat Genet 2008;40:949–51.

[57] Gardner RJ, Mackay DJ, Mungall AJ, Polychronakos C, Siebert R, Shield JP, et al. An imprinted locus associated with transient neonatal diabetes mellitus. Hum Mol Genet 2000;9:589–96.

[58] Temple IK, Gardner RJ, Mackay DJ, Barber JC, Robinson DO, Shield JP. Transient neonatal diabetes: widening the understanding of the etiopathogenesis of diabetes. Diabetes 2000;49:1359–66.

[59] Temple IK, Shrubb V, Lever M, Bullman H, Mackay DJ. Isolated imprinting mutation of the DLK1/GTL2 locus associated with a clinical presentation of maternal uniparental disomy of chromosome 14. J.Med.Genet 2007;44:637–40.

[60] Monk D, Bentley L, Hitchins M, Myler RA, Clayton-Smith J, Ismail S, et al. Chromosome 7p disruptions in Silver–Russell syndrome: delineating an imprinted candidate gene region. Hum Genet 2002;111:376–87.

[61] Monk D, Wakeling EL, Proud V, Hitchins M, bu-Amero SN, Stanier P, et al. Duplication of 7p11.2-p13, including GRB10, in Silver–Russell syndrome. Am J Hum Genet 2000;66:36–46.

[62] Hitchins MP, Monk D, Bell GM, Ali Z, Preece MA, Stanier P, et al. Maternal repression of the human GRB10 gene in the developing central nervous system; evaluation of the role for GRB10 in Silver–Russell syndrome. Eur J Hum Genet 2001;9:82–90.

[63] Bentley L, Nakabayashi K, Monk D, Beechey C, Peters J, Birjandi Z, et al. The imprinted region on human chromosome 7q32 extends to the carboxypeptidase A gene cluster: an imprinted candidate for Silver–Russell syndrome. J Med Genet 2003;40:249–56.

[64] Weksberg R, Shuman C, Smith AC. Beckwith–Wiedemann syndrome. Am J Med Genet C.Semin.Med Genet 2005;137:12–23.

[65] Abu-Amero S, Monk D, Frost J, Preece M, Stanier P, Moore GE. The genetic aetiology of Silver–Russell syndrome. J Med Genet 2008;45:193–9.

[66] Ohta T, Gray TA, Rogan PK, Buiting K, Gabriel JM, Saitoh S, et al. Imprinting-mutation mechanisms in Prader–Willi syndrome. Am J Hum Genet 1999;64:397–413.

[67] Sahoo T, del GD, German JR, Shinawi M, Peters SU, Person RE, et al. Prader–Willi phenotype caused by paternal deficiency for the HBII-85 C/D box small nucleolar RNA cluster. Nat Genet 2008;40:719–21.

[68] Buiting K, Gross S, Lich C, Gillessen-Kaesbach G, el-Maarri O, Horsthemke B. Epimutations in Prader–Willi and Angelman syndromes: a molecular study of 136 patients with an imprinting defect. Am J Hum Genet 2003;72:571–7.

[69] Kishino T, Lalande M, Wagstaff J. UBE3A/E6-AP mutations cause Angelman syndrome. Nat Genet 1997;15:70–3.

[70] Patten JL, Johns DR, Valle D, Eil C, Gruppuso PA, Steele G, et al. Mutation in the gene encoding the stimulatory G protein of adenylate cyclase in Albright's hereditary osteodystrophy. N Engl J Med 1990;322:1412–9.

[71] Bastepe M, Juppner H. GNAS locus and pseudohypoparathyroidism. Horm Res 2005;63:65–74.

[72] Mackay DJ, Hahnemann JM, Boonen SE, Poerksen S, Bunyan DJ, White HE, et al. Epimutation of the TNDM locus and the Beckwith–Wiedemann syndrome centromeric locus in individuals with transient neonatal diabetes mellitus. Hum Genet 2006;119:179–84.

[73] Bliek J, Verde G, Callaway J, Maas SM, De CA, Sparago A, et al. Hypomethylation at multiple maternally methylated imprinted regions including PLAGL1 and GNAS loci in Beckwith–Wiedemann syndrome. Eur J Hum Genet 2009;17:611–9.

[74] Imprinted Gene Database. 2008. Available at http://www.geneimprint.com/site/genes-by-species.

[75] Catalogue of Imprinted Genes and Parent of Origin Effects. 2009. Available at http://www.otago.ac.nz/IGC.

CHAPTER 14

Epigenomic Factors in Human Obesity

Christopher G. Bell
University College London, London, UK

273

14.1 INTRODUCTION

The detrimental effects of obesity on health can be starkly brought into focus by a recent publication from the United States, stating that the escalating rates of this condition will now negate all the positive public health gains made by the reduction in tobacco consumption over the past 50 years [1]. This will led, after the steady rise in life expectancy during the last century, to a decline in lifespan for those children born today [2]. The high prevalence of obesity is a result of the current "obesogenic" environment widespread throughout the Western world; a coupling of reduced energy expenditure both at work and leisure, with increasingly easy access to high-calorific foods [3]. A major driver in the energy intake overload has been documented as simply the increase in both portion sizes and eating opportunities [4]. Additionally this high obesity rate is now being swiftly caught up to by those in developing countries, as they are increasingly removed from a rural existence and rapidly adopt modern

T. Tollefsbol (Ed): Epigenetics in Human Disease. DOI: 10.1016/B978-0-12-388415-2.00014-7

urban life [5,6]. Obesity increases the risk of type 2 diabetes (T2D), coronary vascular disease, hypertension, and some forms of cancer [7]. This obesity-driven increase in T2D alone is putting a considerable strain on health care provision because of its chronic nature and multisystemic complications [8]. Furthermore, estimates placed obesity as causative in ≈ 14% and ≈ 20% of all cancer deaths in the US in men and women, respectively [9].

Although on a population-wide scale this environmental and nutritional influence is extremely pervasive, it does not affect all individuals equally. There is considerable variance between those most susceptible to weight gain to those least at risk. Obesity, as a complex polygenic trait, is the result of environmental and genomic effects and there is substantial genetic variation in individual response to this "obesogenic" pressure to put on weight [3]. Due to this genetic susceptibility certain ethnic groups have been found to be at even greater risk when they encounter this environment [10]. However, the emergence of this dramatically increasing rate over the time span of a few generations is too fast for the appearance of new obesity-promoting mutant alleles, therefore cannot be attributed to a pure genetic effect alone.

The discovery of Mendelian childhood-onset extreme obesity syndromes, caused by single gene mutations, revealed the critical role of neurons within the hypothalamus, particularly those residing in the arcuate nucleus [11,12] (see Figure 14.1). These neurons are involved in

FIGURE 14.1

Arcuate nucleus control of central energy balance between food intake and energy expenditure. Peripheral tissue hormonal signals are summated by the neuropeptide Y (NPY)/agouti-related protein (AGRP) neurons (orexigenic-promoting food intake), and the pro-opiomelanocortin (POMC)/cocaine and amphetamine-related transcript (CART) neurons (with opposite anorexigenic effect) in the arcuate nucleus of the hypothalamus. These hormones include PYY$_{3-36}$ from the distal gastrointestinal tract via Y2 receptors (Y2R), Leptin from adipose tissue via the leptin receptor (LEPR), and Ghrelin from the stomach and duodenum via the growth hormone secretagog receptors (GHSRs) and insulin from the pancreas. The NPY/AGRP neurons also have a direct inhibitory effect on the POMC/CART neurons through gamma-aminobutyric acid (GABA) release. Second-order downstream effector neurons are influenced by the NPY/AGRP and POMC/CART neurons, and also receive modifying inputs from dopamine, serotonin, and endocannabinoid signals, with receptors that include the Y1 receptor (Y1R) and the melanocortin 4 receptor (MC4R). This figure is reproduced in the color plate section. Source: From [56], © Nature Publishing Group.

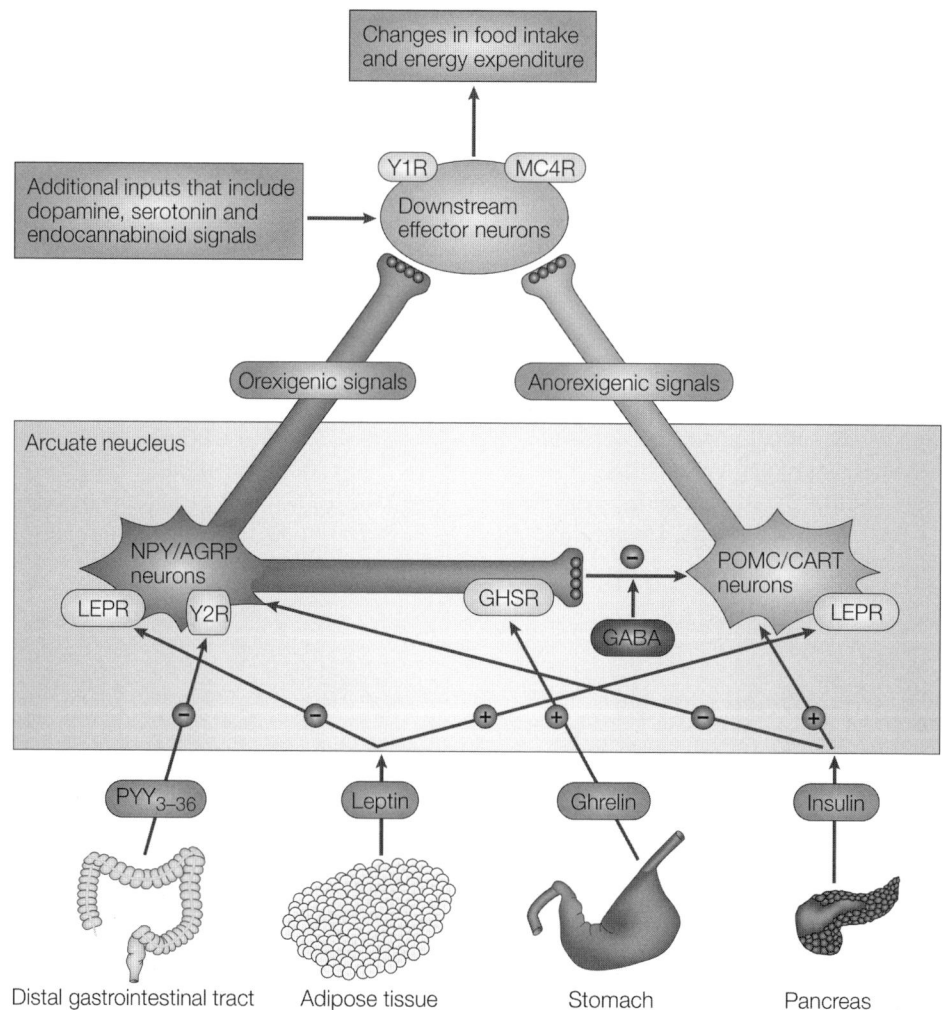

central appetite control and energy balance and have highlighted the importance of the leptin—melanocortin pathway in this peripheral tissue to brain control feedback loop [19]. However these monogenic causes are rare, the most frequent, *MC4R* mutations, accounting for only ~6% of extreme obesity in children [14].

Considerable success has come in gene discovery for all common diseases in the last 5 years, with the advent of high-throughput single nucleotide polymorphism (SNP)-chips facilitating genome-wide association studies (GWAS). These population-based case-control studies led to the identification of common genetic variants associated with common obesity, which currently stands at 32 loci influencing body mass index (BMI) [15,16]. Further, rarer but stronger causal variants have also begun to be found [17,18] and more will be able to be unearthed by whole-genome sequencing studies currently underway. These common variants have modest effect size, the largest within the *FTO* locus leading to a 1.67-fold higher rate of obesity [19], although the potential to gain new insights to the pathophysiology from these replicated associations is considerable. Work, current and on-going in the coming years, will enable the unravelling of the etiological significance of these genomic loci, with likely therapeutic advances.

The combined effect of these common SNP variants known to date is only 2—3% of the estimated 40—70% inherited fraction of obesity risk [20]. Additional power to detect genetic involvement, however, may come by detecting: the actual causal variant that the association SNP may only be an adequate proxy for; further common pathogenic variants within these loci; rare variants within these [21] or yet to be discovered loci; and finally the possible role of non-sequence-based or epigenetic factors in susceptibility [22]. The ability to detect these epigenetic alterations through genetic means alone though, will range from obligatory to stochastic, depending on their direct association with the underlying genome [23]. Furthermore, epigenetic elements are influenced by the environment, as well as the genetic sequence, so therefore act at this crucial interface with the genome [24], hence being a plausible mechanism for the obesogenic environment. These marks can lead to longstanding metabolic changes [25]; consequently epigenetic factors may make up some of the "missing heritability" in the complex disease of obesity [26—28]. However this obviously can be in only those epigenomic marks that bear some obligatory or facultative relationship with genome sequence or they will not contribute to heritability measures.

14.2 EPIGENOMIC MARKS

Epigenetic marks include modifications of DNA, such as DNA methylation or hydroxymethylation, post-translational modifications of histone tails, including acetylation, phosphorylation, sumoylation, ubiquitination, and methylation, and some non-coding RNA species. The ability to self-propagate and be conserved through mitotic division limits this definition currently to DNA methylation, and perhaps a few histone modifications [29]. This maintenance permits lineage-specific epigenomes to be preserved, to allow tissue-specific roles to be performed, in genome-identical somatic cells [23].

DNA methylation is the addition of a methyl group to the 5' carbon of cytosine, principally in the context of the CpG dinucleotide, a C followed by G in the genomic sequence, in differentiated cells. This stable mark is also the most well-studied, so will be the major factor discussed herein. As was the case with the genetic causes of human obesity, the initial insights to the epigenetic influence in this disorder of adiposity have come from rare childhood-onset human syndromes, in this instance the imprinting disorders such as Prader—Willi syndrome (PWS), and murine models, e.g. the Agouti-viable mouse (A^{vy}/a).

14.3 A ROLE FOR IMPRINTING ABNORMALITIES IN OBESITY

Genomic imprinting allows only one of the allelic pair of genes to be expressed, dependent on whether it is of paternal, or maternal, origin [30]. This parent-of-origin conditional

TABLE 14.1 Imprinting Disorders with an Obesity-Associated Phenotype

Syndrome	OMIM	Locus	Parental Imprint	Obesity
Prader–Willi	#176270	15q11.2	Maternal	Severe
Pseudohypoparathyroidism 1a	#103580	20q13.3	Maternal	Moderate
Maternal Uniparental Disomy 14	—	14	—	Truncal
Angelman	#105830	15q11.2	Paternal	Late-onset

Source: Imprinting syndromes with an obesity-associated phenotype [53,191].

allele-specific expression (ASE) is driven by parent-of-origin conditional allele-specific DNA methylation (ASM) [31]. The attribution of which parental allele, paternal or maternal, is imprinted is by convention referred to the allele on which the precise DNA methylation occurs, therefore the allele silenced not that which is expressed. Imprinted genes commonly are found in clusters with reciprocally imprinted genes, i.e. both maternal and paternal imprinted genes are colocated [32]. Imprinted loci are coordinately regulated, via imprinting control regions, various species of non-coding RNA and methylation-sensitive boundary elements, leading to considerable regulatory complexity [33,34]. Furthermore, the imprinting of individual genes may not occur body-wide, but may be present, or escape this marking, in only one or more particular organs. This tissue-specific imprinting is very prevalent in the cells of the placenta and brain [30]. Placental imprinting is key in influencing this vital interface of resource allocation between the fetus and mother, and this has been proposed as a major driver in its evolution in eutherian (placental) mammals [35]. Genomic imprinting and placentation evolved around a similar time period in mammals [36]. With regards to brain imprinting, the potential complexity of orchestrated imprinting variation amongst the numerous regions of the central nervous system and furthermore throughout the process of brain development has only just begun to be explored [37]. However, intriguing evidence for the intricate interplay of imprinting and imprinting loss for correct neurogenesis in the mouse has recently been published [38].

Obesity is often observed as part of the spectrum of an imprinting abnormality phenotype (see Table 14.1). Speculation that dysregulation of the imprinted genes in placental resource allocation pathways, or imprinted control of brain development genes involved in the leptin–melanocortin hypothalamic pathway, has been proposed in metabolic disease [39,40]. Understanding why these imprinting mechanisms have evolved may aid in unravelling how they may be causative in obesity.

14.4 CONFLICT THEORY OF IMPRINTING

Eutherian mammals facilitate nutrient uptake via a placenta during fetal growth. The "kinship" or "gene conflict" theorem proposes that this evolved to control the allocation of resources from mother to offspring – in terms of the "interest" of the genes from the two parents. Maternal genes see equal benefit in all of their progeny, due to an equal genetic contribution to each, so therefore will hope to provide for all in an equal fashion. However paternal genes are divided between offspring sired by possibly differing fathers [41]. Genetic influences in higher mammals that help them acquire maternal resources, as well as those within parents that aid this process, have an evolutionary advantage in that they increase the survival and propagation of those genes [36,42]. Therefore the parental conflict theory of gene imprinting proposes that growth-promoting genes, that increase direct investment in the currently developing fetus, will be under paternal control, whereas the opposite effect will be maternally controlled, as the mother tries to equally distribute resources between all of her offspring [36,43]. The extreme manifestation of this drive can be seen in the pathological case of the hydatidiform mole, formed by the fusion of two paternal gametes, which therefore lacks any maternal restraining imprints and leads to an extreme, tumor-like growth, which can even invade in rare cases [44].

Knockout mouse models show that imprinted genes play a role in the blood nutrient transfer interface of the placenta [45]. Imprinted placental genes control nutrition supply and fetal imprinted genes control demand by growth regulation [36]. The paternally expressed *Igf2* is shown to be critical in placental growth [46]. Deletion of the placenta-specific promoter of this gene leads to decreased permeability, by reducing the exchange barrier and increasing the thickness of the placenta, thereby influencing nutrient supply with subsequent effects on the growth of the developing fetus in early gestation [45]. However, the placenta is capable of functionally adapting to this interference, enabling it to respond to fetal need requirements, by the up-regulation of gene pathways of the placental supply genes in the system A amino acid transport systems, including *Slc38a4*, which is also paternally expressed [47]. This counter-balance is not sufficient to maintain this correction in the late gestation period though. It can be hypothesized that fetal and maternal blood flow and the transportation of nutrients may also be under imprinted genetic regulation [36]. It is possible this "conflict" may also sway maternal postnatal resource allocation via breast milk and the control of suckling. Indeed disruption of the paternally expressed transcript of the imprinted *Gnas* locus, isoform *XLαS*, in a mouse model led to reduced growth and survival due to effects including a poor suckling response [48]. Imprinted genes additionally influence cognitive and social behavior which may also be used to gain resources [49].

14.5 RARE IMPRINTED ABNORMALITIES WITH OBESITY-RELATED PHENOTYPES

Prader—Willi syndrome (PWS) is the classic example of an imprinting-related syndrome and this was first recognized as being the pathogenic mechanism in 1989 [50]. The phenotype includes dysmorphic features, short stature, low lean body mass, muscular hypotonia, mild mental retardation, and behavioral abnormalities [51]. The region of the complex locus associated with PWS is maternally imprinted, so only the paternal copy is normally expressed. The inverse paternally imprinted disorder leds to the non-obese Angelman syndrome (AS) child [51]. However, AS individuals are at risk of moderate later-onset adiposity [52]. The obesity observed in PWS is so severe that it is life-threatening and develops within the first 6 years of childhood. PWS is estimated to occur approximately in 0.5—1 individuals per 10 000 births with no racial bias, although it is clinically diagnosed and reported more frequently in Caucasians, and affects approximately 350 000—400 000 individuals worldwide [53]. Diagnosis is via abnormal DNA methylation in 99% of PWS patients, enabling the detection of deletions, uniparental disomy (UPD) of chromosome 15, or an imprinting control region (ICR) defect [54].

The behavioral and cognitive impairment in these individuals leads to an extreme and uncontrolled appetite, due to a proposed inability to experience the normal satiety response after eating [55]. As mentioned above, energy balance is controlled and regulated within the hypothalamus, with the arcuate nucleus performing a major role in this coordination (Figure 14.1). Within the arcuate there is crosstalk and interchange between the orexigenic or appetite-promoting NPY/AGRP neurons and the opposing anorexigenic POMC/CART neurons [56]. Therefore a mechanism that disrupts this finely tuned apparatus has been suggested to explain this extreme behavior [55]. The stomach-secreted hormone Ghrelin, is associated with inducing satiation following feeding, and is found to be increased in PWS individuals [57]. Ghrelin stimulates the NPY/AGRP neurons via growth hormone secretagog receptors, though a definite role in the pathogenesis of PWS is not yet confirmed. *SNRPN* had long been fingered as causative in PWS, but patients with smaller microdeletions have enabled the critical region to be narrowed to a locus containing non-coding HBII-85 snoRNAs [58]. PWS babies are often born small-for-gestation-age with poor feeding responses [59]. This could be interpreted in a reductionist way by the "conflict theory" as indicative of the lack of fetal paternal gene expression, with the later onset of extreme appetite driving the response due

277

to abnormal neurotrophic central pathway formation in the brain during this restricted development.

The *GNAS* locus is another complex imprinted region with an associated abnormal phenotype that includes obesity. Isoforms of the *GNAS* (guanine nucleotide binding protein, α-stimulating) gene, the α-stimulatory subunit of the G-protein, are created from multiple promoters and splicing variations and an antisense non-coding RNA *GNAS-AS1* (GNAS antisense RNA 1 (non-protein coding)) is also transcribed [60]. Biallelic, paternally and maternally expressed genes are positioned within this locus. *GNAS* itself is derived predominately from both alleles, with the exception of maternal-only expression in the tissues of the renal proximal tubules, thyroid, gonads, and pituitary. Loss of the maternal allele is associated with greater weight gain [48,61]. *XLAS* is a large isoform variant of *GNAS* that is paternally expressed within the nervous system and neuroendocrine tissues. Within this cluster another *GNAS* isoform, previously termed *Nesp* in mouse, is maternally expressed [62]. Whilst maternal nonsense genetic mutations of *GNAS* are causative of pseudohypoparathyroidism type 1A (MIM #103580, also known as Albright's hereditary osteodystrophy), the loss of maternal-specific exon 1A DNA methylation leads to pseudohypoparathyroidism type 1B (MIM #603233) [63]. The associated multiple hormone resistance is proposed to be due to these molecules utilizing signaling pathways through G-protein-coupled receptors and the associated obesity is thought similarly to be due to abnormalities of these G-protein receptors centrally [64]. Allelic variation in the Histone H3K4 state of the *GNAS* exon 1A promoter region within proximal renal tubules has also been identified, illustrating the coordinated nature of these post-translational histone modifications with DNA methylation [65].

The human overgrowth disorder, Beckwith–Wiedemann syndrome (BWS) (MIM #130650), is caused by mutation or deletion of imprinted genes within the chromosome 11p15.5 region. Growth is of an increased rate during the latter half of pregnancy and in the first few years of life, and whilst not leading to obesity per se, also displays the phenotypic potential of imprinting in terms of DNA methylation abnormalities and growth control throughout development. It has also been proposed that in vitro assisted reproductive technology may affect the imprinting process and with some increased levels of BWS in those conceived via this technique and confirmatory molecular studies recognizing BWS-specific epigenetic alterations [66]. This manipulation of germ cells and embryos that occurs in vitro, especially at such an early crucial point in development, therefore demonstrates the fragility of the epigenome compared to the genome, exemplified by this specific abnormality at an imprinted locus.

There is considerable variation in estimates of the actual level of imprinted regions in the mammalian, including the human, genome. Recent evidence has hinted at the possibility of high levels of particular brain tissue-specific imprinting in a mouse model [67]. So it is plausible that there is still an underestimation of imprinted loci, particularly with the inclusion of tissue-specific and developmental-stage-specific variation. Current confirmed estimates stand at around 50 genes in human [68,69]. DNA sequence characteristics and recombination rates at imprinted loci have led to predictions of 150 human genes, however without confirmation of differentially methylated regions (DMRs) [70] and further studies have also indicated there are other undiscovered imprinted loci [71,72]. Moreover whilst there is as yet no definitive set of human or other eutherian mammalian imprinted genes, there does appear to be significant dissimilarities between the species. These differences could be reconciled with the "paternal conflict" theory, for instance, as being driven by variation in litter size between mouse and human [73].

Mouse models have indicated a role of imprinted genes in adipocyte metabolism. This includes the paternally expressed genes *Dlk1* (delta-like 1 homolog, *Drosophila*), *Mest* (mesoderm specific transcript) (also known as *Peg1*) and *Ndn* (Necdin) [74]. Cheverud et al. calculated the effects that imprinting may have on complex traits in mice by estimating the variation body size between reciprocal heterozygote, i.e. Aa compared to aA [75]. There was

a small but discernible difference of 0.25 standard deviation units or approximately 1–4% of the variance, which included body weight, reproductive fat pad, and various organ weights. Further murine work looking specifically at obesity quantitative trait loci (QTL) estimated the possible effect of imprinting on diet-dependent obesity was high, with 61% of the QTLs identified with a strong parental influence [76]. These sites were predominately not in known imprinted regions.

Two further paternally expressed genes *Mest(Peg1)* and *Peg3* are involved not only in fetal and postnatal growth, but also can affect maternal nurturing success [77,78]. These imprinted genes are strongly expressed in hypothalamus, preoptic area, and septum, therefore they are excellent candidates for neuronal programming [39].

14.6 DIETARY INFLUENCE ON DNA METHYLATION IN MURINE MODELS

The epigenetic modulation of the expression of the non-imprinted Agouti gene in the mouse viable yellow (A^{vy}/a) allele is the most extensively studied "metastable epiallele" and has become the archetypical model of dietary modulation of DNA methylation with subsequent phenotypic effects (see Figure 14.2). Metastable epialleles are so termed as these loci of epigenetically variability are established very early in embryogenesis and subsequently remain stable whilst permeating through all ensuing developmental stages and germ layers [79]. In the wild-type mouse the Agouti gene encodes a signaling molecule that produces either black eumelanin (a) or yellow phaeomelanin (A). Transcription is normally initiated from a hair-specific promoter in exon 2, with transient expression of the A allele leading to the mottled brown fur. In the agouti mutant mouse the insertion upstream of the agouti gene of a single intracisternal A particle sequence (IAP) (endogenous retrovirus-like element), creates a cryptic promoter leading to an increase of pheomelainin over eumelanin [80]. This overaction results in a lightening of the coat color as ectopic expression of the inverse agonist at melanocortin receptors, agouti, antagonizes the action of melanin [79]. The "viable yellow" heterozygote (A^{vy}/a) mouse has a shortened live span with yellow fur, obesity, and an increased suscep-tibility to neoplasia [81]. The associated obesity phenotype is due to the ability of the agouti protein to centrally mimic the Agouti-Related Protein (AGRP) and to stimulate appetite in the hypothalamus via the NPY/AGRP orexigenic neurons.

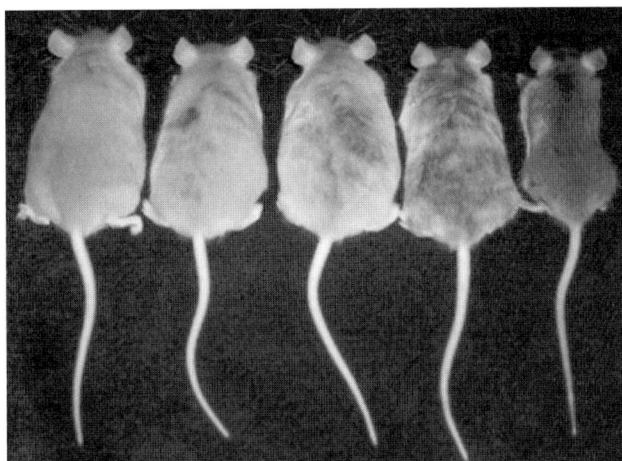

FIGURE 14.2

Agouti mouse showing epigenetic dietary effects with graduation from obese Agouti to normal-sized Pseudoagouti (from [192]. *Reproduced with permission from Environmental Health Perspectives.*

There is a spectrum of expression of agouti depending on the epigenetic state of the IAP element leading to a continuum from the agouti yellow, mottled to the brown "pseudoagouti". The methylation of the IAP results in epigenetic suppression of this novel promoter and the pseudoagouti mouse is by comparison phenotypically lean with a normal lifespan. The crucial functional methylation variability occurs at six CpG sites that reside in the 5' long terminal repeat (LTR) of the IAP element [82]. A dietary influence on the methylation of this upstream IAP element was identified, whereby the use of methyl-supplemented diet or other nutritional modulations, including Genistein, in the pregnant black (a/a) mother led to an alteration in the portion of heterozygote (A^{vy}/a) sired and resultant pups with a skew towards the darkened pseudoagouti phenotype [83,84] (see Figure 14.2). Histone modifications accompany the hypomethylation of the IAP element, with H3K and H4K diacetylation, whilst with DNA methylation of the IAP, the repressive histone H4K20me3 mark is displayed (82). This phenotypic modulation of an individual locus by maternal food intake has led to hypotheses in regards to how the diet affects the developing embryo particularly by direct DNA methylation epiallele variation across the entire genome [79]. Additionally, this visible dietary phenotypic effect is at a non-imprinted locus, further expanding the nutritional possibilities of epigenomic modulation, particularly in a developmental setting, as an example of modulation of gene expression with potential influence on disease susceptibility.

Dietary impact on imprinted genes nevertheless has been documented, in the imprinted *Igf2* locus in a mouse model [85]. Instalment of a methyl-donor-deficient diet post-weaning led to loss of imprinting at this locus with subsequent modification of expression.

The inbred mouse strain C57Bl/6J is documented as being highly susceptible to diet-induced obesity, but furthermore has also been observed to show a wide range of variability in this weight gain when fed a high-calorie diet [86]. Phenotype divergence into those who would become high weight-gainers versus low was even evident in measures before commencing an obesity-promoting diet. After this regimen was introduced this led to a four-fold difference in obesity. Additionally these dissimilarities persisted even when the mice were switched back to a calorie-restricted diet. Koza et al. also detected gene expression differences in these isogenic mice, between the high- and low-weight gainers, prior to the high-calorie diet introduction, leading to the hypothesis that pre-existing epigenetic factors were involved in these observations [86].

Overfeeding in rats, induced by limiting the litter size, led to an obese phenotype [87]. Investigation of hypothalamic tissue showed hypermethylation of the promoter of *POMC*, the anorexigenic neurohormone, proopiomelanocortin, involved in appetite repression [87]. Leptin and insulin stimulate this pathway via two Sp1-related binding sequences within this promoter. The hyperleptinemia and hyperinsulinemia present in these rats therefore did not lead to an up-regulation of *POMC* expression. The DNA methylation levels within these binding sites were inversely correlated to the quotients of *POMC* expression/leptin and *POMC* expression/insulin, demonstrating the function of these acquired epigenomic alterations in modifying the "set point" parameters of this promoter critical for bodyweight regulation by overfeeding: an example of obesity epigenetic reprogramming.

Recently in humans, obese women, who were separated into high and low responders to dietary restriction, were found to have measurable differences in subcutaneous adipose tissue DNA methylation [88]. In another investigation of human muscle cells, a reversible effect on DNA methylation in the key metabolic regulator Peroxisome Proliferator-Activated Receptor γ, coactivator 1 (PPARGC1A) with overfeeding was seen [89].

14.7 OBESOGENIC ENVIRONMENT EFFECTS ON COMMON HUMAN OBESITY

Over and above the cumulative genetic susceptibility risks recently identified from GWAS, are the major environmental elements that are the product of urban and socioeconomic transitions over the last 60 years and the formation of the aforementioned "obesogenic" environment. This can be perceptibly displayed by the observation that the common genetic susceptibilities towards the trait are not acted upon, unless certain compounding causes are encountered. Lifestyle contributors such as diet and exercise are central in liability to metabolic disease and also have a substantial aggregate effect [90]. It has been proposed that the etiology of common diseases are under both genetic and epigenetic influence and these disease-related epigenetic factors could be environmentally induced with subsequent modulatory effects on genetic susceptibility [91]. Epigenetic marks such as DNA methylation provide a mechanism by which these environmental influences can modulate genetic susceptibility to disease, as evidence from murine models and humans has alluded to. Nutrition, behavior, stress, toxins, and stochastic action can all have an impact on DNA methylation leading to variation in epialleles [92]. Epigenetic effect on gene expression by the modification of the target cell epigenome [93], thereby changes metabolic risk [79,94]. Combined gene by environmental interactions have been observed with known genetic susceptibility factors for obesity, as seen with accentuated adiposity effect of the *FTO* susceptibility variant in individuals through low physical activity [95]. The significant role of epigenetics in the pathogenesis of cancer is well established [96] but has also been seen in other diseases such as in the etiology of atherosclerotic plaques [97], and evidence is accumulating in the metabolic syndrome.

14.8 AGING EFFECT ON DNA METHYLATION

In demonstrating the non-rigid state of DNA methylation, alteration over time has been an initial target. An accumulation of DNA variation over a lifetime was found by comparing divergence between monozygotic twins by age [98]. This may be caused by accrued environmental effect and/or "epigenetic drift" due to defective transmission through multiple mitotic replications. DNA methylation over time devolves from its fixed bimodal extremes; with hypomethylated CpGs within islands gaining methylation and hypermethylated CpGs losing methylation [99]. Therefore it has been speculated these shifts may modify metabolic pathways, becoming gradually suboptimal, leading to slow late-onset weight gain [100]. Now that higher-resolution DNA methylome analysis has become available more subtle signatures of aging have begun to be identified, such as hypermethylation at the promoters of the target genes of the repressive Polycomb Group proteins [101].

14.9 DEVELOPMENTAL EPIGENOMIC DYSREGULATION

Whilst DNA methylation is a comparatively stable repressive mark, it is however required to be removed from specific genomic regions, as well as genome-wide, through the process of development. The major stages are at postfertilization and at germ cell differentiation in males and females. Two global waves of demethylation are undertaken with the embryo being most susceptible to environmental perturbation during the re-establishment of DNA methylation postfertilization but preimplantation [92]. Experimental evidence in mice shows that the preimplantation embryo is sensitive to epigenetic modifications [102]. Furthermore, the entire process of embryogenesis involves exact DNA methylation configuration in order that normal tissue development occurs and so is a window of possible influence [92].

Therefore, these adverse environmental effects within specific critical junctures, such as when the tissue-specific epigenome is being cast in developing cells, have been proposed to have significant lasting effects on metabolism into adult life with a concurrent risk of chronic diseases [92,103,104]. Direct evidence of dietary modulation during these time-points, and the latent ability to affect long-term risk of chronic metabolic disease health has been attempted using murine models, through from the periconceptual period to postweaning [105]. The epigenetic state of the transcription factor *Hnf4α* was investigated in the pancreas of rats that had been subjected to poor maternal diet and controls [106]. *HNF4A* is implicated in T2D pathogenesis and mutants are associated with a monogenic form of T2D

(MODY, type I, MIM #125850). Poor maternal diet during critical periods of development, as well as aging, was shown to down-regulate an islet-specific promoter and the interaction between the promoter and an enhancer was also down-regulated. A small study in humans also identified methylation changes in *HNF4A* in cord blood following intrauterine growth retardation (IUGR) [107].

Further rat models of IUGR investigated the DNA methylation and gene expression of pancreatic islets [108]. Cellular memory in the pancreatic cells of the developmental intrauterine environment was sought by the investigation of approximately 1 million CpG sites in the rat methylome of these cells at the later age of 7 weeks. This study pinpointed changes in methylation state of approximately 1400 CpGs consistent with undergoing IUGR. These alterations were preferably found in highly conserved non-coding intergenic genomic loci, which may be *cis*-regulatory sites, and frequently near genes involved in the deregulation of islets following IUGR, including insulin secretion, vascularization, cell proliferation, and cell death. Also they were additionally associated with mRNA expression levels preceding the development of diabetes.

14.10 FETAL PLASTICITY

The dramatic and startling increase in obesity over such a short historical period has led to the concept that fetal programming or plasticity may be involved. This proposes that the periconceptual, in utero, and postnatal developmental environment can impact on long-term risk for adult-onset disease by set point adaptive changes [109—111]. This suggests that "poor" nutrition at critical growth stages increases the chance of developing the metabolic syndrome (insulin resistance, obesity, dyslipidemia, and hypertension) in later life and these developmental origins of adult disease put forward epigenetic inheritance as the possible mechanism in this programming [92].

Although ischemic heart disease increases in a population as it becomes wealthier, it was noted that those in the poorest regions suffered disproportionately [112]. An early epidemiological study in England and Wales identified a connection between poor nutrition in early life and increased susceptibility to ischemic heart disease, that was suggested to be due to a more detrimental effect of an adiposity-promoting diet in these individuals [112]. Furthermore, impaired glucose tolerance in men aged 64 was correlated to low weight at birth and at 1 year [113]. The "thrifty phenotype" hypothesis proposed by Barker encapsulates this prospective effect of fetal programming and the reprogramming of the leptin—melanocortin hypothalamic axis, pituitary, adrenal, or islet development, or insulin-signaling pathways may be mechanisms in these long-term effects [114].

Documented historical famines furthermore allow these questions to be proposed to age cohorts, and the Dutch Hunger Winter of November 1944 to May 1945 has been a classic example in the literature. Upon reaching the army draft age of 19 years the obesity rates of 300 000 men were compared [115]. The outcomes were dependent on when during their development the severe wartime famine had occurred. Those exposed in the last trimester or the first months of life were less likely to be obese, whilst those whose mothers had been in early or mid-pregnancy during this severe wartime famine were twice as likely to be obese at army draft. This would imply early changes may differ in programming from later changes, fitting with hypotheses of developmental programming and critical epigenetic windows. Epigenetic examination of DNA methylation in these Dutch Hunger Winter individuals 60 years after this famine displayed variation at the imprinted DMR of *IGF2* that was found to be less methylated among periconceptionally exposed individuals [116]. Individuals conceived during the famine were compared against a sibling of the same sex to attempt to reduce environmental and genetic modulators. Follow-up work investigated a further 15 loci with known involvement in growth and metabolism [117]. In those with periconceptual famine,

DNA methylation was increased in the promoter and imprinted DMRs of *GNAS-AS1* and *MEG3* (maternally expressed 3 (non-protein coding)), as well as proximal promoters of *IL10* (interleukin 10), *ABCA1* (ATP-binding cassette, sub-family A (ABC1), member 1), and *LEP* (Leptin). Comparative decreased DNA methylation was also recognized in the promoter of imprinted *INS-IGF2* (INS-IGF2 read-through) which is also part of the *INS* (insulin) promoter.

These studies have highlighted the periconceptual period as a volatile spatiotemporal window in epigenomic development with additional support for this coming from animal models [118]. Manipulation of dietary vitamin B and methionine during this periconceptual period in sheep led to heavier, fatter, and insulin-resistant animals [119].

In a large population-based obesity study, the maternal impact on this trait was shown to be greater [120], possibly due to maternally imprinted genes. As these imprinted loci are strongly interconnected with resource allocation, energy balance, and feeding behavior, early environmental effects may have long-reaching consequences. This makes these parental-specific marks obviously strong initial candidates in any "fetal programming" or "plasticity" influence on chronic disease. It may be that these imprinting resource allocation tools are subverted by the "fetal programming" mechanism, enabling a shift of resource regulation over the course of a lifetime and subsequent risk for adult-onset disease [39]. This metabolic plasticity enables a non-changed genome to produce a range of phenotypes in response to variation in environment, specifically early nutritional status [121,122]. Subtle modulation of imprinting pathways could be a mechanism, or it may be that they are too critical for an adaptive process and changes are only seen in severe disease manifestations [123].

Two independent, but sometimes overlapping, pathways showing how the nutritional state through development can lead to increased susceptibility to later-onset obesity have been proposed [123]. First, this may be a "mismatch" pathway, which can be either severe or predictive. These are developmental plasticity modifications of the genetically driven phenotype cued by prenatal undernutrition or possibly stress, which then may not be correctly geared, if the postnatal environment is obesogenic. Second, a pathway due to the risks caused by maternal obesity and the "hypernutrition" experienced by the fetus in this situation.

14.11 POSTNATAL ENVIRONMENTAL MISMATCH

Survival-induced modifications are due to severe maternal ill health or placental abnormality and lead to growth reduction in order to survive. Fetal growth retardation leads to changes in gene expression driven by epigenetic changes [114]. Recent work has shown that via the imprinted *Peg3* placental sacrifice will occur in order to protect brain development [124]. Optimization for survival favors some organs against overall growth, but may lead to persistent metabolic and endocrinological changes that over time become detrimental when subsequently encountering an obesogenic environment [114]. For instance the development of insulin resistance in order to survive in utero undernutrition [103].

Significant correlations between low birth weight and adult-onset disease in studies across Europe and in the USA have been identified, even after corrections for possible confounders such as socioeconomic status, physical activity, smoking, etc. [114]. There is a disproportionate loss of subcutaneous compared to visceral fat in undernourished babies, which leads to a relative visceral adiposity and additionally evidence of specific further gain in visceral fat in IUGR children [125]. Those born with the smallest birth weight (<2.5 kg) were seven times more likely to develop T2D or glucose intolerance [113]. A French study, by Jaquet et al., showed higher levels of insulin resistance in IUGR infants at age 25 years and the reduced glucose uptake was concurrent with a lesser degree of free fatty acid suppression in adipose tissue, implicating this tissue in the early stages of insulin resistance [126] . A genetic explanation for this has also been proposed in that any genetic influences on insulin resistance restrict growth in utero and then subsequently lead to insulin resistance in adulthood [127].

Restricted infant growth and fast weight gain in childhood intensify the effects of impaired prenatal growth [128]. Evidence from many animals points to the additive effect of prenatal diet restriction and then subsequent hypercalorific diet in the developing infant leading to stronger risk of weight gain than the latter alone [129]. Mouse evidence shows high levels of "catch-up" growth after in utero growth restriction leads to even further increased susceptibility to the adverse effects on lifespan of a postweaning obesogenic diet [130]. Further evidence for probable programming of chronic disease by impaired fetal nutrition was established in an Indian population study, where small-for-gestational-age babies with a high "catch up" growth and therefore high fat mass at ages 2 through to 12, later had the highest levels of insulin resistance [5]. Some studies have, however, found inconclusive evidence of early undernutrition increasing risk of later obesity, but did support a role for overnutrition [131].

Predictive responses via phenotypic plasticity mechanisms are in order to improve or optimize fitness at a later stage of development [132]. These are an evolutionary beneficial ability to enable adaptation to probable future environmental likelihoods, but are not advantageous at the time [133]. These modifications allow subtle modification where this is possible. This response is made within the variation of normal range in development and is a non-pathological environmental prediction adjustment.

Hypothalamic neuroendocrine gene expression effects were identified in a rat model of this response [134]. Prenatal undernutrition and postnatal high-fat diet with subsequent assessment at 24 weeks revealed specific alterations in the hypothalamic gene expression of *POMC*, *NPY*, *AGRP*, and the Ob-Rb isoform of *LEPR*, as well as the observation of circulating hyperinsulinemia and hyperleptinemia. These finding were supportive therefore of a predictive response having influenced the neurogenesis of the hypothalamic pathways that had subsequently been incorrect for the later environment encountered.

Exploration in humans for epigenetic evidence investigated 68 CpGs 5′ of five candidate genes from neonatal umbilical cord tissue DNA and ascertained an epiallele association with maternal pregnancy diet and childhood fat mass at age 9 years including an individual CpG 5′ of *RXRA* (Retinoid X receptor-α) and *NOS3* (nitric oxide synthase 3 (endothelial cell)) previously known as eNOS [135]. *RXRA* also associated with lower maternal carbohydrate input and additionally only the *RXRA* adiposity result replicated in a second group. If this result can be corroborated by other researchers this may represent likely "low-hanging fruit" or changes that are strong enough to be exposed by candidate studies. If so this will bode well for more powerful metabolic epigenome-wide association studies (EWAS) currently being proposed. Changes include increases as well as decreases in methylation, therefore diet restriction of methyl donors is a too simplistic explanation of this observation, rather an adaptive response is hypothesized [103,110].

14.12 HYPERNUTRITION

Secondly, the direct effect on the intrauterine environment of maternal diabetes and obesity, as well as infant overfeeding of high-calorific foods, lead to greater risk of adult obesity. Increasing numbers of women now are overweight when pregnant [39]. Hyperinsulinemia is causative of macrosomia in diabetic mothers with the greatest effect on fat mass and subsequently greater risk of childhood obesity [136].

In a Danish study the adult children of women who suffered from gestational diabetes during pregnancy had a 2.5 times increased rate of metabolic syndrome and a twofold risk of overweight compared to a selected control population [137]. Central appetite regulatory circuitry in the hypothalamus is active within the fetus and affected by the nutritional state in developmental stages, therefore it may lead to long-term effects in the gearing of this system due to this chronic hypernutrition [138]. This neural programming may be variable with malleability of this system at critical species-specific windows dependent on restricted or excess diet [138]. High levels of white fat in human infants compared to other mammals is proposed as an energy store buffer to protect brain development post-weaning [139], so this level of preloading will not be drawn upon and reduce with excess feeding. Increasing affluence in developing countries leads to an increase in availability of refined foods which are cheaper but have less nutritional value than fresh foods [6]. Also, early feeding of differing diet, breast or formula, may affect absorption and gut microbiotic changes have been found to have a role in obesity [140]. One study on intrauterine evidence for effect on obesity, showed an association with maternal and paternal BMI, implying that shared environmental influence also cannot be discounted [141].

14.13 EPIGENETIC ANALYSIS OF LEPTIN

Adipose tissue is an endocrine organ, key in energy homeostasis, which secretes adipokines, the major player being leptin. This supplies vital feedback to the hypothalamus about fat mass. The promoter of leptin moves from a highly methylated to a low methylation state in the differentiation from pre-adipocyte to adipocyte, thus facilitating expression of this endocrine hormone in mature fat cells [142]. Leptin additionally has been shown to have a role in neurogenesis and specifically within the critical hypothalamic pathways [143,144]. Visualization of NPY and POMC neurons, by expression of fluorescent proteins, displayed a leptin-mediated modulation of synaptic density onto these cells. Leptin's neurotrophic role in the hypothalamus is furthermore illustrated by the lack of neuronal projection pathways from the arcuate nucleus in the leptin-deficient mouse model. This neuroanatomical deficiency cannot be reversed with adulthood leptin administration, but can, if delivered during the neonatal period [145].

Rats fed a high-fat diet become obese and hyperleptinemic, but also increase methylation in the promoter of leptin and this was collated with a comparative reduction in expression of leptin [146]. Further in vivo analysis of the leptin promoter in human and mouse identified its tissue-specific variability in methylation (t-DMR). A critical C/EBPα (CCAAT/enhancer-binding protein alpha) [147] transcription binding site is located within the CpG Island and generally has overall lower methylation in adipose compared to peripheral blood tissue [148]. Interestingly the mouse CpG Island possessed a higher level of intermediate methylation and is smaller ($^1/_3$ size) and a less dense CpG Island, therefore it is speculated to have lost CpGs via deamination, suggesting it may have higher levels of germline methylation [148]. No DNA methylation differences in the comparison of T2D cases versus controls in the leptin promoter via MeDIP-chip of peripheral blood were however identified [149] [see Figure 14.3] and no changes were found in a study post weigh loss via bariatric surgery [150].

14.14 HISTONE EPIGENOMIC MODIFIERS — MASTER METABOLIC REGULATORS

Recent work in murine models has illuminated the potential role of modifiers of the epigenetic histone state, such as methyltransferases and demethylases, in metabolism and energy homeostasis gene pathways. Particularly this has highlighting the critical role of Kdm3a (lysine (κ)-specific demethylase 3a), also known as Jhdm2a, in metabolic regulation [151]. This acts as a H3K9-specific demethylase, catalyzing the removal of mono- and dimethylation from H3K9, therefore has a de-repressive function. Abrogation of action was shown to lead to significant multigenic effects on metabolic pathways resulting in a maturity-onset obesity, hypercholesterolemia, hyperinsulinemia, hypertriglycerolemia, and hyperlipidemia phenotype. This involved reduction within the skeletal muscle of fat oxidation and the release of glycerol and in brown fat cell disruption of oxygen consumption and the β-adrenergic stimulated secretion of the glycerol. The expression of this demethylase is stimulated by the β-adrenergic mechanisms and it plays a significant role in expression of the transcription factor *Ppara* (peroxisome

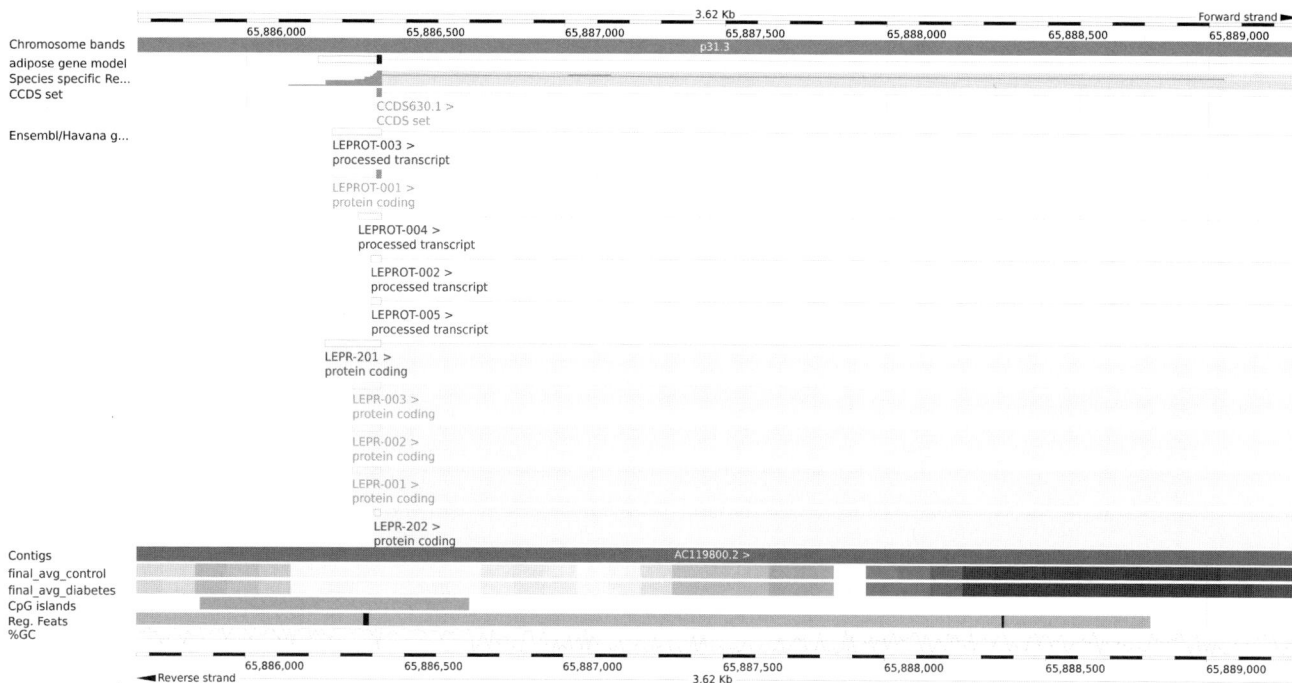

FIGURE 14.3

Methylation DNA immunoprecipitation (MeDIP) graphical result for the Leptin Promoter CpG Island. Sliding scale for methylation level: yellow = low, green = mid, blue = high. Hypomethylation seen over the CpG Island and no significant difference seen between T2D cases (final_avg_diabetes) versus controls (final_ave_control). This figure is reproduced in the color plate section.

proliferator-activated receptor alpha), essential in fatty acid metabolism, and *Ucp1* (Uncoupling protein 1) involved in decoupling energy creation to generate heat in the mitochondria of brown adipose tissue. These abnormalities in adipose storage and energy balance were also displayed in the knockout mouse by hypothermia and reduced energy production via fat utilization [152].

14.15 METASTABLE ALLELES IN HUMAN ASSOCIATED WITH OBESITY

Genetic influence on the methylation state, or allele-specific variation in methylation levels [153,154] by polymorphism such as CpG–SNPs, has been acknowledged as having a considerable bearing on region methylation levels [155,156]. Furthermore, DNA methylation levels driven by CpG gain and loss leading to higher variance in local methylation have been proposed as a major driver in evolution as well as common disease susceptibility [157] and epigenetic and other concepts of non-DNA inheritance are now beginning to be incorporated into evolutionary theory [158]. Transgenerational transmission possibilities are still unclear with current evidence, with some reports of incomplete epigenetic erasure [159], however it is very difficult to completely exclude genetic effects (for review see [160]). Paternal inheritance effects have been proposed in two recent papers [159,160].

Epigenetic variability, or epialleles, can vary purely without genetic influence, or play a facilitative or obligatory role conferred by genetic variants. Furthermore, they are generally intrinsically tissue-specific, however a subset of these epialleles is determined very early in development and subsequently propagate through all differentiating cell lineages as illustrated by the Agouti mouse. Human "metastable epialleles" with correlations across tissues representing all three developing germ layers have also been observed [162]. These loci were shown to have highly different values, between monogenic twins, indicating that they are likely set in a stochastic fashion. Furthermore, DNA methylation at these epialleles was shown to be significantly associated with the experienced periconceptual nutritional environment in a set of rural Gambians. A higher level of methylation was identified in those who had been conceived during the nutrition-reduced rainy season [162]. Metastable epialleles of imprinted genes have also been hypothesized to play a major part in adaption and evolution [163].

Feinberg et al. catalogued 227 genomic regions that showed a wide level of DNA methylation variability between 638 Icelandic individuals [164]. These regions were enriched for developmental and morphogenesis genes. The individuals were sampled at two time periods approximately 11 years apart. Approximately half of these, or 119 regions, termed variable methylation regions (VMRs), were found to be stable over this time course within individuals [164]. These VMRs therefore could create an individual epigenomic signature or fingerprint. Four of these stable VMRs correlated with BMI at both timepoint visits: these loci resided in or near the genes *PM20D1*, *MMP9*, *PRKG1*, and *RFC5*. Up-regulation of *MMP9* has been seen in obese individuals [165], and *PRKG1* has been associated with energy balance and food behavior in a number of organisms [166]. These highly variable regions were proposed to reside within loci that could possibly be highly susceptible to environmental modulation and could be investigated for strong environmental influences such as toxins, smoking, dietary variation, etc. However, a caveat stated that this variability may be contributed to by the mixed cell type of peripheral blood that was examined. The intraindividual time-stable VMRs could be expected to have a strong genetic component, such as possible CpG-SNP or haplotype-specific methylation effect [167].

14.16 PARENT-OF-ORIGIN GENETIC EFFECTS

Genome-wide linkage has been used to try to find imprinted obesity-related genomic loci [168]. Whilst highly successful in discovering monogenic disease genes, this technique has had low success in polygenic traits. GWAS have achieved far more in identifying variants strongly associated with these common diseases, including obesity risk and BMI modulation [15,169]. Furthermore, success has been found in looking for parent-of-origin effects in GWAS association SNPs, with significant results in other complex disease traits, such as type 1 diabetes [170].

Analysis from deCODE Genetics, also of GWAS SNP data, revealed a number of parent-of-origin effects [69]. By utilizing the extensive known genealogies of these Icelandic individuals, as well as long-range phasing of haplotypes, parental state of inherited alleles was determined. They focused on those disease-associated SNPs within 500 kb of known imprinted regions (approximately 1% of the genome), which informed that SNPs within two known clusters on chromosome 11p15 (containing *H19/IGF2* and *KCNQ1* imprinted clusters) and 7q32 (including the maternally expressed genes, *CPA4* and *KLF14* surrounding paternally expressed genes *MEST* (*PEG1*) and *MESTIT1*) were present. Whilst not obesity SNP associations, T2D disease-associated traits were located here. Five parent-of-origin SNP associations were able to be made, including three with T2D. The maternally expressed *KCNQ1* contained two SNPs, rs2237892 and rs231362, each with allele C significantly associated when transmitted maternally. Both these SNPs are at CpG-SNP sites, with the C variant of the polymorphism facilitating methylation by the creation of a CpG.

A third SNP, allele C of rs4731702, again via maternal inheritance at 7q32, was significantly associated and additionally this risk allele correlated with lower expression of *KLF14* in adipose but not blood tissue and again only when inherited from the maternal side. Subsequent work by Small et al. discovered this *cis*-modulatory SNP rs4731702 of this imprinted *KLF14* transcription factor gene to be a significant *trans* regulator in gene expression studies in subcutaneous adipose tissue biopsies from a cohort of 776 healthy female twins [171]. The influence of this factor was enriched for metabolic syndrome phenotypes, including BMI, with evidence from *APH1B*, *ARSD*, *C8orf82*, *GNB1*, *SLC7A10*, and *TPMT* expression levels and associated the control of this *trans* expression network by *KLF14* with risk of metabolic disease. Kong et al. also established a novel association at SNP rs2334499, with susceptibility via the paternal, and a protective effect via maternal lineage, and there was correlation with a decreased methylation state of a nearby CTCF-binding region, which are known genomic insulators, and this susceptibility variant [69].

These parent-of-origin associations with metabolic syndrome traits hint at the role of subtle genetic variation, and potentially epiallele variation, in these imprinted loci influencing these diseases [172]. This is particularly the case when these genetic variants may be either directly impacting on methylation ability, in the case of CpG-SNPs, or additionally influencing neighboring CpGs methylation state, or the density of CpGs, or are perhaps in linkage disequilibrium with such variants. These disparities in monoallelic imprinting may be via slight allele-specific expression balance changes through to complete loss of imprinting.

14.17 EPIGENOMIC-WIDE ASSOCIATION STUDIES (EWAS) IN HUMAN OBESITY

The potential to investigate the "methylome" or "chromatinome" with the same power as that of GWAS may enable commonly perturbed epigenomic pathways in common disease to be discovered. As the genome is controlled by epigenetic mechanisms that inform development, but also respond and are affected by the environment, investigation of this "epigenome" in disease promises major insights into both cause and effect [24,173]. DNA methylation, being the most stable of these marks, is currently the most amenable to the high-throughput analysis required in order to examine enough individuals to make these studies sufficiently powered to identify replicable associations. Recent technological advances have now made it possible for these DNA methylation epigenome-wide association studies (EWAS) to be considered [174,175]. By utilizing a chemical modification of the genome, bisulfite conversion, to create a pseudo-SNP at variably methylated cytosine sites, SNP array technology has been adapted for this task. The recent availability of the next generation of array platform (Illumina 450k — Infinium HumanMethylation450 BeadChip performed on the iScan), superseding previous 27-k arrays, allows greater than a magnitude increase in coverage and importantly includes investigation of more dynamic regions of DNA methylation change, such as CpG Island shores [176].

EWAS design needs to carefully consider the principal facets that differentiate the epigenome from the genome. First, that the epigenome is tissue-specific; second, that epigenetic marks can be influenced to varying degrees by the underlying genome; and third, that it can be modified by environmental factors. Regarding tissue-specificity when investigating the obesity pheno-type, adipose, muscle and more difficult to acquire hypothalamic tissue are initial strong etiological candidate epigenomes. Mixture of cell type may impede signal detection, therefore pure cell isolation is a major advantage. Inflammatory processes in obesity may also make inflammatory cells an interesting target [177,178]. A population-based design will encounter significant genetic heterogeneity with subsequent impact on DNA methylation. This will lead to a loss of power in comparison to a disease-discordant monozygotic twin design analysis, although these cohorts are very difficult to resource for any, but particularly this, phenotype. Furthermore the genetically induced variation may be more complex than can be dissected from array genotype data alone. Finally, cause and effect cannot be separated with regards to any DNA methylation changes seen. In obesity therefore any epigenetic association may be induced by the adiposity state itself or confounding factors related to obesity such as diet or physical activity [104].

There is obviously added complexity interpreting the functional significance of small changes in individual CpG, versus a consistent result over an entire CpG island or shore region comprising a number of CpGs commonly seen in cancer methylation results. Caution will be required in the interpretation of EWAS and the lessons from genetic association studies need to be remembered. Without proper correction for multiple hypothesis testing all EWAS will be in danger of overinterpretation due to the high type 1 error [104]. This will make validation by replication in multiple subject groups vital, as is required in step-wise GWAS.

It is currently unclear whether GWAS-type population studies will be adequately powered and highly successful in the EWAS setting, though the increase in coverage of the newer arrays may help to answer this question. Other study designs however may be more amenable, such as the longitudinal. The extreme phenotype has been a powerful and successful strategy in the search for obesity susceptibility genetic alleles [179] and may also be useful in epiallelic hunting. Longitudinal studies investigating epigenomic tissue-specific changes in adipose or muscle tissue, before and after dramatic physiological adjustment, such as gastric bypass operation, in severe adult morbid obesity cases, would remove genetic variability. This EWAS temporal analysis of extreme intervention, leading to dramatic reduction in BMI and glucose control improvement, may reveal epigenetic deviations that can be attempted to be validated in less extreme lifestyle or weight-loss program subjects. Follow-up validation in larger sample sets by focusing on a smaller number of CpG, using for instance quantitative methylation analysis by Qiagen Pyromark Bisulphite conversion Pyrosequencing (Pyro Q-CpG) would facilitate rapid target validation. Subsequent correlation with expression in these tissues of these genes with their methylation changes would also be strongly supportive. Successful EWAS epiallele associations to environmental influences have recently begun to be published, with a single CpG, cg03636183, being associated and replicated with lower methylation in DNA derived from peripheral blood in smokers [180].

14.18 FUTURE PROSPECTS

Large amounts of genomic data are now available for common complex diseases, including obesity, therefore an integrative appropriate utilizing this gathered information would be complementary and also theoretically more powerful [181]. Initial pilot work into the integration of DNA methylation data with GWAS T2D loci has shown the ability to detect DNA methylation variation with respect to risk haplotype status [149]. This method was used to show higher levels of methylation were present on the *FTO* obesity-associated risk haplotype [167]. Chromatin state integration recognized allelic dissimilarity in the open/closed chromatin structure in the pancreas, dependent on the T2D-associated SNP in *TCF7L2* [182] and these approaches can aid formulation of further novel functional hypotheses.

Full integration of genomic, epigenomic, and transcriptomic data, or an "Omni-seq" approach, will have the potential to mine out subtle disease association variation that is not possible by one modality only [183,184]. Furthermore, the advent of third-generation sequencing [185,196] will remove the necessity of chemical alteration enabling direct reading of DNA modifications and will greatly improve the ability to interpret disease significance of DNA methylation as well as newer modifications, such as hydroxymethylation [187], formylcytosine, and carboxylcytosine [188]. This will also be useful in the extremely complex area of the multitude of histone tail modifications and variants, and non-coding RNAs and the possiblity of RNA epigenetic change. The ability for environmental influences to modulate the epigenetic state make sequence-specific epidrugs to correct these epimutations an attractive concept, thereby including pharmacoepigenomics in the future of personalized medicine for the metabolic disease [189,190].

References

[1] Stewart ST, Cutler DM, Rosen AB. Forecasting the effects of obesity and smoking on U.S. life expectancy. N Engl J Med 2009;361:2252−60.

[2] Olshansky SJ, Passaro DJ, Hershow RC, Layden J, Carnes BA, Brody J, et al. A potential decline in life expectancy in the United States in the 21st century. N Engl J Med 2005;352:1138–45.

[3] Friedman JM. Obesity: Causes and control of excess body fat. Nature 2009;459:340–2.

[4] Duffey KJ, Popkin BM. Energy density, portion size, and eating occasions: contributions to increased energy intake in the United States, 1977-2006. PLoS medicine 2011;8:e1001050.

[5] Bhargava SK, Sachdev HS, Fall CH, Osmond C, Lakshmy R, Barker DJ, et al. Relation of serial changes in childhood body-mass index to impaired glucose tolerance in young adulthood. N Engl J Med 2004;350:865–75.

[6] Gluckman PD, Hanson M, Zimmet P, Forrester T. Losing the war against obesity: the need for a developmental perspective. Sci Trans Med 2011;3:93cm19.

[7] Kopelman PG. Obesity as a medical problem. Nature 2000;404:635–43.

[8] Zimmet P, Alberti KG, Shaw J. Global and societal implications of the diabetes epidemic. Nature 2001;414:782–7.

[9] Calle EE, Rodriguez C, Walker-Thurmond K, Thun MJ. Overweight, obesity, and mortality from cancer in a prospectively studied cohort of U.S. adults. N Engl J Med 2003;348:1625–38.

[10] Ogden CL, Carroll MD, Curtin LR, McDowell MA, Tabak CJ, Flegal KM. Prevalence of overweight and obesity in the United States, 1999–2004. JAMA: the journal of the American Medical Association 2006;295:1549–55.

[11] Choquet H, Meyre D. Genomic insights into early-onset obesity. Genome medicine 2010;2:36.

[12] Blakemore AI, Froguel P. Investigation of Mendelian forms of obesity holds out the prospect of personalized medicine. Ann NY Acad Sci 2010;1214:180–9.

[13] O'Rahilly S. Human genetics illuminates the paths to metabolic disease. Nature 2009;462:307–14.

[14] Farooqi IS, Keogh JM, Yeo GS, Lank EJ, Cheetham T, O'Rahilly S. Clinical spectrum of obesity and mutations in the melanocortin 4 receptor gene. N Engl J Med 2003;348:1085–95.

[15] Walley AJ, Asher JE, Froguel P. The genetic contribution to non-syndromic human obesity. Nat Rev Genet 2009;10:431–42.

[16] Kilpelainen TO, Zillikens MC, Stancakova A, Finucane FM, Ried, et al. Genetic variation near IRS1 associates with reduced adiposity and an impaired metabolic profile. Nat Genet 2011;43:753–60.

[17] Walters RG, Jacquemont S, Valsesia A, de Smith AJ, Martinet D, et al. A new highly penetrant form of obesity due to deletions on chromosome 16p11.2. Nature 2010;463:671–5.

[18] Bochukova EG, Huang N, Keogh J, Henning E, Purmann C, Blaszczyk K, et al. Large, rare chromosomal deletions associated with severe early-onset obesity. Nature 2010;463:666–70.

[19] Frayling TM, Timpson NJ, Weedon MN, Zeggini E, Freathy RM, Lindgren CM, et al. A common variant in the FTO gene is associated with body mass index and predisposes to childhood and adult obesity. Science 2007;316:889–94.

[20] Marti A, Ordovas J. Epigenetics lights up the obesity field. Obesity facts 2011;4:187–90.

[21] Nejentsev S, Walker N, Riches D, Egholm M, Todd JA. Rare variants of IFIH1, a gene implicated in antiviral responses, protect against type 1 diabetes. Science 2009;324:387–9.

[22] McCarthy MI, Hirschhorn JN. Genome-wide association studies: potential next steps on a genetic journey. Hum Mol Genet 2008;17:R156–65.

[23] Richards EJ. Inherited epigenetic variation—revisiting soft inheritance. Nat Rev Genet 2006;7:395–401.

[24] Bell CG, Beck S. The epigenomic interface between genome and environment in common complex diseases. Briefings funct genomics 2010;9:477–85.

[25] Barres R, Zierath JR. DNA methylation in metabolic disorders. Am J Clin Nutr 2011;93:897S–900.

[26] Bjornsson HT, Fallin MD, Feinberg AP. An integrated epigenetic and genetic approach to common human disease. Trends Genet 2004;20:350–8.

[27] Manolio TA, Collins FS, Cox NJ, Goldstein DB, Hindorff LA, Hunter DJ, et al. Finding the missing heritability of complex diseases. Nature 2009;461:747–53.

[28] Herrera BM, Keildson S, Lindgren CM. Genetics and epigenetics of obesity. Maturitas 2011;69:41–9.

[29] Berger SL, Kouzarides T, Shiekhattar R, Shilatifard A. An operational definition of epigenetics. Genes Dev 2009;23:781–3.

[30] Ferguson-Smith AC. Genomic imprinting: the emergence of an epigenetic paradigm. Nat Rev Genet 2011;12:565–75.

[31] Reik W, Walter J. Genomic imprinting: parental influence on the genome. Nat Rev Genet 2001;2:21–32.

[32] Edwards CA, Ferguson-Smith AC. Mechanisms regulating imprinted genes in clusters. Curr Opin Cell Biol 2007;19:281–9.

[33] Sleutels F, Zwart R, Barlow DP. The non-coding Air RNA is required for silencing autosomal imprinted genes. Nature 2002;415:810–3.

[34] Ponting CP, Oliver PL, Reik W. Evolution and functions of long noncoding RNAs. Cell 2009;136:629–41.

[35] Reik W, Lewis A. Co-evolution of X-chromosome inactivation and imprinting in mammals. Nat Rev Genet 2005;6:403–10.

[36] Constancia M, Kelsey G, Reik W. Resourceful imprinting. Nature 2004;432:53–7.

[37] Wilkinson LS, Davies W, Isles AR. Genomic imprinting effects on brain development and function. Nat Rev Neurosci 2007;8:832–43.

[38] Ferron SR, Charalambous M, Radford E, McEwen K, Wildner H, Hind E, et al. Postnatal loss of Dlk1 imprinting in stem cells and niche astrocytes regulates neurogenesis. Nature 2011;475:381–5.

[39] Gallou-Kabani C, Junien C. Nutritional epigenomics of metabolic syndrome: new perspective against the epidemic. Diabetes 2005;54:1899–906.

[40] Waterland RA, Garza C. Potential mechanisms of metabolic imprinting that lead to chronic disease. Am J Clin Nutr 1999;69:179–97.

[41] Moore T, Haig D. Genomic imprinting in mammalian development: a parental tug-of-war. Trends Genet 1991;7:45–9.

[42] Dawkins R. The selfish gene. Oxford: Oxford University Press; 1976.

[43] Wilkins JF, Haig D. What good is genomic imprinting: the function of parent-specific gene expression. Nat Rev Genet 2003;4:359–68.

[44] Berkowitz RS, Goldstein DP. Clinical practice. Molar pregnancy. N Engl J Med 2009;360:1639–45.

[45] Sibley CP, Coan PM, Ferguson-Smith AC, Dean W, Hughes J, Smith P, et al. Placental-specific insulin-like growth factor 2 (Igf2) regulates the diffusional exchange characteristics of the mouse placenta. Proc Natl Acad Sci USA 2004;101:8204–8.

[46] Constancia M, Hemberger M, Hughes J, Dean W, Ferguson-Smith A, Fundele R, et al. Placental-specific IGF-II is a major modulator of placental and fetal growth. Nature 2002;417:945–8.

[47] Constancia M, Angiolini E, Sandovici I, Smith P, Smith R, Kelsey G, et al. Adaptation of nutrient supply to fetal demand in the mouse involves interaction between the Igf2 gene and placental transporter systems. Proc Natl Acad Sci USA 2005;102:19219–24.

[48] Plagge A, Gordon E, Dean W, Boiani R, Cinti S, Peters J, et al. The imprinted signaling protein XL alpha s is required for postnatal adaptation to feeding. Nat Genet 2004;36:818–26.

[49] Isles AR, Davies W, Wilkinson LS. Genomic imprinting and the social brain. Philosophical transactions of the Royal Society of London. B Biol Sci 2006;361:2229–37.

[50] Nicholls RD, Knoll JH, Butler MG, Karam S, Lalande M. Genetic imprinting suggested by maternal hetero-disomy in nondeletion Prader–Willi syndrome. Nature 1989;342:281–5.

[51] Horsthemke B, Wagstaff J. Mechanisms of imprinting of the Prader–Willi/Angelman region. American journal of medical genetics. A 2008;146A:2041–52.

[52] Williams CA, Beaudet AL, Clayton-Smith J, Knoll JH, Kyllerman M, Laan LA, et al. Angelman syndrome 2005: updated consensus for diagnostic criteria. Am J Med Genet A 2006;140:413–8.

[53] Butler MG. Genomic imprinting disorders in humans: a mini-review. J Assist Reprod Genet 2009;26:477–86.

[54] Boyle J, Hawkins M, Barton DE, Meaney K, Guitart M, O'Grady A, et al. Establishment of the first WHO international genetic reference panel for Prader Willi and Angelman syndromes. Eur J Hum Genet 2011;19:857–64.

[55] Stoger R. Epigenetics and obesity. Pharmacogenomics 2008;9:1851–60.

[56] Bell CG, Walley AJ, Froguel P. The genetics of human obesity. Nature reviews. Genetics 2005;6:221–34.

[57] Cummings DE, Clement K, Purnell JQ, Vaisse C, Foster KE, Frayo RS, et al. Elevated plasma ghrelin levels in Prader Willi syndrome. Nat Med 2002;8:643–4.

[58] Sahoo T, del Gaudio D, German JR, Shinawi M, Peters SU, Person RE, et al. Prader–Willi phenotype caused by paternal deficiency for the HBII-85 C/D box small nucleolar RNA cluster. Nat Genet 2008;40:719–21.

[59] Goldstone AP. Prader–Willi syndrome: advances in genetics, pathophysiology and treatment. Trends Endocrinol Metab 2004;15:12–20.

[60] Williamson CM, Turner MD, Ball ST, Nottingham WT, Glenister P, Fray M, et al. Identification of an imprinting control region affecting the expression of all transcripts in the Gnas cluster. Nat Genet 2006;38:350–5.

[61] Chen M, Gavrilova O, Liu J, Xie T, Deng C, Nguyen AT, et al. Alternative Gnas gene products have opposite effects on glucose and lipid metabolism. Proc Natl Acad Sci USA 2005;102:7386–91.

[62] Bastepe M, Frohlich LF, Linglart A, Abu-Zahra HS, Tojo K, Ward LM, et al. Deletion of the NESP55 differentially methylated region causes loss of maternal GNAS imprints and pseudohypoparathyroidism type Ib. Nat Genet 2005;37:25−7.

[63] Mariot V, Maupetit-Mehouas S, Sinding C, Kottler ML, Linglart A. A maternal epimutation of GNAS leads to Albright osteodystrophy and parathyroid hormone resistance. J Clin Endocrinol Metab 2008;93:661−5.

[64] Spiegel AM, Weinstein LS. Inherited diseases involving g proteins and g protein-coupled receptors. Annu Rev Med 2004;55:27−39.

[65] Sakamoto A, Liu J, Greene A, Chen M, Weinstein LS. Tissue-specific imprinting of the G protein Gsalpha is associated with tissue-specific differences in histone methylation. Hum Mol Genet 2004;13:819−28.

[66] DeBaun MR, Niemitz EL, Feinberg AP. Association of in vitro fertilization with Beckwith−Wiedemann syndrome and epigenetic alterations of LIT1 and H19. Am J Hum Genet 2003;72:156−60.

[67] Gregg C, Zhang J, Weissbourd B, Luo S, Schroth GP, Haig D, et al. High-resolution analysis of parent-of-origin allelic expression in the mouse brain. Science 2010;329:643−8.

[68] Morison IM, Ramsay JP, Spencer HG. A census of mammalian imprinting. Trends Genet 2005;21:457−65.

[69] Kong A, Steinthorsdottir V, Masson G, Thorleifsson G, Sulem P, Besenbacher S, et al. Parental origin of sequence variants associated with complex diseases. Nature 2009;462:868−74.

[70] Luedi PP, Dietrich FS, Weidman JR, Bosko JM, Jirtle RL, Hartemink AJ. Computational and experimental identification of novel human imprinted genes. Genome Res 2007;17:1723−30.

[71] Wen B, Wu H, Bjornsson H, Green RD, Irizarry R, Feinberg AP. Overlapping euchromatin/heterochromatin-associated marks are enriched in imprinted gene regions and predict allele-specific modification. Genome Res 2008;18:1806−13.

[72] Choufani S, Shapiro JS, Susiarjo M, Butcher DT, Grafodatskaya D, Lou Y, et al. A novel approach identifies new differentially methylated regions (DMRs) associated with imprinted genes. Genome Res 2011;21:465−76.

[73] Monk D, Arnaud P, Apostolidou S, Hills FA, Kelsey G, Stanier P, et al. Limited evolutionary conservation of imprinting in the human placenta. Proc Natl Acad Sci USA 2006;103:6623−8.

[74] Frontera M, Dickins B, Plagge A, Kelsey G. Imprinted genes, postnatal adaptations and enduring effects on energy homeostasis. Adv Exp Med Biol 2008;626:41−61.

[75] Cheverud JM, Hager R, Roseman C, Fawcett G, Wang B, Wolf JB. Genomic imprinting effects on adult body composition in mice. Proc Natl Acad Sci USA 2008;105:4253−8.

[76] Cheverud JM, Lawson HA, Fawcett GL, Wang B, Pletscher LS, A, R. F., et al. Diet-dependent genetic and genomic imprinting effects on obesity in mice. Obesity 2011;19:160−70.

[77] Curley JP, Barton S, Surani A, Keverne EB. Coadaptation in mother and infant regulated by a paternally expressed imprinted gene. Proc Bio Sci Royal Soc 2004;271:1303−9.

[78] Lefebvre L, Viville S, Barton SC, Ishino F, Keverne EB, Surani MA. Abnormal maternal behaviour and growth retardation associated with loss of the imprinted gene Mest. Nat Genet 1998;20:163−9.

[79] Jirtle RL, Skinner MK. Environmental epigenomics and disease susceptibility. Nat Rev Genet 2007;8:253−62.

[80] Morgan HD, Sutherland HG, Martin DI, Whitelaw E. Epigenetic inheritance at the agouti locus in the mouse. Nat Genet 1999;23:314−8.

[81] Duhl DM, Vrieling H, Miller KA, Wolff GL, Barsh GS. Neomorphic agouti mutations in obese yellow mice. Nat Genet 1994;8:59−65.

[82] Dolinoy DC, Weinhouse C, Jones TR, Rozek LS, Jirtle RL. Variable histone modifications at the A(vy) metastable epiallele. Epig Official J DNA Methy Soc 2010;5:637−44.

[83] Wolff GL, Kodell RL, Moore SR, Cooney CA. Maternal epigenetics and methyl supplements affect agouti gene expression in Avy/a mice. FASEB J Official Publ Fed Am Soc for Exper Bio 1998;12:949−57.

[84] Waterland RA, Jirtle RL. Transposable elements: targets for early nutritional effects on epigenetic gene regulation. Mol Cell Bio 2003;23:5293−300.

[85] Waterland RA, Lin JR, Smith CA, Jirtle RL. Post-weaning diet affects genomic imprinting at the insulin-like growth factor 2 (Igf2) locus. Hum Mol Genet 2006;15:705−16.

[86] Koza RA, Nikonova L, Hogan J, Rim JS, Mendoza T, Faulk C, et al. Changes in gene expression foreshadow diet-induced obesity in genetically identical mice. PLoS genetics 2006;2:e81.

[87] Plagemann A, Harder T, Brunn M, Harder A, Roepke K, Wittrock-Staar M, et al. Hypothalamic proopiomelanocortin promoter methylation becomes altered by early overfeeding: an epigenetic model of obesity and the metabolic syndrome. J Physiol 2009;587:4963−76.

[88] Bouchard L, Rabasa-Lhoret R, Faraj M, Lavoie ME, Mill J, Perusse L, et al. Differential epigenomic and transcriptomic responses in subcutaneous adipose tissue between low and high responders to caloric restriction. Am J Clin Nut 2010;91:309−20.

[89] Brons C, Jacobsen S, Nilsson E, Ronn T, Jensen CB, Storgaard H, et al. Deoxyribonucleic acid methylation and gene expression of PPARGC1A in human muscle is influenced by high-fat overfeeding in a birth-weight-dependent manner. J Clin Endocrinol Metab 2010;95:3048−56.

[90] Mozaffarian D, Hao T, Rimm EB, Willett WC, Hu FB. Changes in diet and lifestyle and long-term weight gain in women and men. N Engl J Med 2011;364:2392−404.

[91] Feinberg AP. Phenotypic plasticity and the epigenetics of human disease. Nature 2007;447:433−40.

[92] Faulk C, Dolinoy DC. Timing is everything: The when and how of environmentally induced changes in the epigenome of animals. Epigenetics 2011;6.

[93] Esau C, Kang X, Peralta E, Hanson E, Marcusson EG, Ravichandran LV, et al. MicroRNA-143 regulates adipocyte differentiation. J Bio Chem 2004;279:52361−5.

[94] Jaenisch R, Bird A. Epigenetic regulation of gene expression: how the genome integrates intrinsic and environmental signals. Nat Genet 2003;33(Suppl):245−54.

[95] Andreasen CH, Stender-Petersen KL, Mogensen MS, Torekov SS, Wegner L, Andersen G, et al. Low physical activity accentuates the effect of the FTO rs9939609 polymorphism on body fat accumulation. Diabetes 2008;57:95−101.

[96] Esteller M. Aberrant DNA methylation as a cancer-inducing mechanism. Annu Rev Pharmacol Toxicol 2005;45:629−56.

[97] Ying AK, Hassanain HH, Roos CM, Smiraglia DJ, Issa JJ, Michler RE, et al. Methylation of the estrogen receptor-alpha gene promoter is selectively increased in proliferating human aortic smooth muscle cells. Cardio Res 2000;46:172−9.

[98] Fraga MF, Ballestar E, Paz MF, Ropero S, Setien F, Ballestar ML, et al. Epigenetic differences arise during the lifetime of monozygotic twins. Proc Natl Acad Sci USA 2005;102:10604−9.

[99] Christensen BC, Houseman EA, Marsit CJ, Zheng S, Wrensch MR, Wiemels JL, et al. Aging and environmental exposures alter tissue-specific DNA methylation dependent upon CpG island context. PLoS Genet 2009;5:e1000602.

[100] Issa JP. Epigenetic variation and human disease. J Nut 2002;132:2388S−92S.

[101] Teschendorff AE, Menon U, Gentry-Maharaj A, Ramus SJ, Weisenberger DJ, Shen H, et al. Age-dependent DNA methylation of genes that are suppressed in stem cells is a hallmark of cancer. Genome Res 2010;20:440−6.

[102] Reik W, Romer I, Barton SC, Surani MA, Howlett SK, Klose J. Adult phenotype in the mouse can be affected by epigenetic events in the early embryo. Develop 1993;119:933−42.

[103] Hales CN, Barker DJ. The thrifty phenotype hypothesis. Br Med Bull 2001;60:5−20.

[104] Franks PW, Ling C. Epigenetics and obesity: the devil is in the details. BMC Med 2010;8:88.

[105] McMillen IC, Robinson JS. Developmental origins of the metabolic syndrome: prediction, plasticity, and programming. Physiol Rev 2005;85:571−633.

[106] Sandovici I, Smith NH, Nitert MD, Ackers-Johnson M, Uribe-Lewis S, Ito Y, et al. Maternal diet and aging alter the epigenetic control of a promoter-enhancer interaction at the Hnf4a gene in rat pancreatic islets. Proc Natl Acad Sci USA 2011;108:5449−54.

[107] Einstein F, Thompson RF, Bhagat TD, Fazzari MJ, Verma A, Barzilai N, et al. Cytosine methylation dysregulation in neonates following intrauterine growth restriction. PLoS One 2010;5:e8887.

[108] Thompson RF, Fazzari MJ, Niu H, Barzilai N, Simmons RA, Greally JM. Experimental intrauterine growth restriction induces alterations in DNA methylation and gene expression in pancreatic islets of rats. J Biol Chem 2010;285:15111−8.

[109] Barker DJ, Eriksson JG, Forsen T, Osmond C. Fetal origins of adult disease: strength of effects and biological basis. Int J Epidemiol 2002;31:1235−9.

[110] Gluckman PD, Hanson MA, Cooper C, Thornburg KL. Effect of in utero and early-life conditions on adult health and disease. N Engl J Med 2008;359:61−73.

[111] Godfrey KM, Gluckman PD, Hanson MA. Developmental origins of metabolic disease: life course and intergenerational perspectives. Trends Endocrinol Metab 2010;21:199−205.

[112] Barker DJ, Osmond C. Infant mortality, childhood nutrition, and ischaemic heart disease in England and Wales. Lancet 1986;1:1077−81.

[113] Hales CN, Barker DJ, Clark PM, Cox LJ, Fall C, Osmond C, et al. Fetal and infant growth and impaired glucose tolerance at age 64. BMJ 1991;303:1019−22.

[114] Simmons R. Developmental origins of adult metabolic disease: concepts and controversies. Trends in endocrinology and metabolism: TEM 2005;16:390−4.

[115] Ravelli GP, Stein ZA, Susser MW. Obesity in young men after famine exposure in utero and early infancy. N Engl J Med 1976;295:349−53.

[116] Heijmans BT, Tobi EW, Stein AD, Putter H, Blauw GJ, Susser ES, et al. Persistent epigenetic differences associated with prenatal exposure to famine in humans. Proc Natl Acad Sci USA 2008;105:17046—9.

[117] Tobi EW, Lumey LH, Talens RP, Kremer D, Putter H, Stein AD, et al. DNA methylation differences after exposure to prenatal famine are common and timing- and sex-specific. Hum Mol Genet 2009;18:4046—53.

[118] Reik W. Stability and flexibility of epigenetic gene regulation in mammalian development. Nature 2007;447:425—32.

[119] Sinclair KD, Allegrucci C, Singh R, Gardner DS, Sebastian S, Bispham J, et al. DNA methylation, insulin resistance, and blood pressure in offspring determined by maternal periconceptional B vitamin and methionine status. Proc Natl Acad Sci USA 2007;104:19351—6.

[120] Whitaker KL, Jarvis MJ, Beeken RJ, Boniface D, Wardle J. Comparing maternal and paternal intergenerational transmission of obesity risk in a large population-based sample. Am J Clin Nutr 2010;91:1560—7.

[121] Radford EJ, Ferron SR, Ferguson-Smith AC. Genomic imprinting as an adaptive model of developmental plasticity. FEBS letters 2011;585:2059—66.

[122] Gluckman PD, Lillycrop KA, Vickers MH, Pleasants AB, Phillips ES, Beedle AS, et al. Metabolic plasticity during mammalian development is directionally dependent on early nutritional status. Proc Natl Acad Sci USA 2007;104:12796—800.

[123] Gluckman PD, Hanson MA. Developmental and epigenetic pathways to obesity: an evolutionary-developmental perspective. International journal of obesity 2008;32(Suppl. 7):S62—71.

[124] Broad KD, Keverne EB. Placental protection of the fetal brain during short-term food deprivation. Proc Natl Acad Sci USA 2011;108:15237—41.

[125] Ibanez L, Suarez L, Lopez-Bermejo A, Diaz M, Valls C, de Zegher F. Early development of visceral fat excess after spontaneous catch-up growth in children with low birth weight. J Clin Endocrinol Metab 2008;93:925—8.

[126] Jaquet D, Gaboriau A, Czernichow P, Levy-Marchal C. Insulin resistance early in adulthood in subjects born with intrauterine growth retardation. J Clin Endocrinol Metab 2000;85:1401—6.

[127] Hattersley AT, Tooke JE. The fetal insulin hypothesis: an alternative explanation of the association of low birthweight with diabetes and vascular disease. Lancet 1999;353:1789—92.

[128] Barker DJ. The developmental origins of adult disease. J Am Coll Nutr 2004;23:588S—95S.

[129] Vickers MH, Breier BH, Cutfield WS, Hofman PL, Gluckman PD. Fetal origins of hyperphagia, obesity, and hypertension and postnatal amplification by hypercaloric nutrition. Am J Physiol Endocrinal Metab 2000;279:E83—7.

[130] Ozanne SE, Hales CN. Lifespan: catch-up growth and obesity in male mice. Nature 2004;427:411—2.

[131] Martorell R, Stein AD, Schroeder DG. Early nutrition and later adiposity. J Nutr 2001;131:874S—80S.

[132] Gluckman PD, Hanson MA, Spencer HG. Predictive adaptive responses and human evolution. Trends Eco Evol 2005;20:527—33.

[133] Bateson P, Barker D, Clutton-Brock T, Deb D, D'Udine B, Foley RA, et al. Developmental plasticity and human health. Nature 2004;430:419—21.

[134] Ikenasio-Thorpe BA, Breier BH, Vickers MH, Fraser M. Prenatal influences on susceptibility to diet-induced obesity are mediated by altered neuroendocrine gene expression. J Endocrinol 2007;193:31—7.

[135] Godfrey KM, Sheppard A, Gluckman PD, Lillycrop KA, Burdge GC, McLean C, et al. Epigenetic Gene Promoter Methylation at Birth Is Associated With Child's Later Adiposity. Diabetes 2011;60:1528—34.

[136] Hillier TA, Pedula KL, Schmidt MM, Mullen JA, Charles MA, Pettitt DJ. Childhood obesity and metabolic imprinting: the ongoing effects of maternal hyperglycemia. Diabetes care 2007;30:2287—92.

[137] Clausen TD, Mathiesen ER, Hansen T, Pedersen O, Jensen DM, Lauenborg J, et al. Overweight and the metabolic syndrome in adult offspring of women with diet-treated gestational diabetes mellitus or type 1 diabetes. J Clin Endocrinol Metab 2009;94:2464—70.

[138] Muhlhausler BS. Programming of the appetite-regulating neural network: a link between maternal over-nutrition and the programming of obesity? J neuroendocrinol 2007;19:67—72.

[139] Kuzawa CW. Adipose tissue in human infancy and childhood: an evolutionary perspective. Am J Phys Anthropol 1998;(Suppl 27):177—209.

[140] Ley RE, Turnbaugh PJ, Klein S, Gordon JI. Microbial ecology: human gut microbes associated with obesity. Nature 2006;444:1022—3.

[141] Davey Smith G, Steer C, Leary S, Ness A. Is there an intrauterine influence on obesity? Evidence from parent child associations in the Avon Longitudinal Study of Parents and Children (ALSPAC). Arch Dis Child 2007;92:876—80.

[142] Melzner I, Scott V, Dorsch K, Fischer P, Wabitsch M, Bruderlein S, et al. Leptin gene expression in human preadipocytes is switched on by maturation-induced demethylation of distinct CpGs in its proximal promoter. J Bio Chem 2002;277:45420—7.

[143] Pinto S, Roseberry AG, Liu H, Diano S, Shanabrough M, Cai X, et al. Rapid rewiring of arcuate nucleus feeding circuits by leptin. Science 2004;304:110–5.

[144] Bouret SG, Simerly RB. Minireview: Leptin and development of hypothalamic feeding circuits. Endocrinol 2004;145:2621–6.

[145] Bouret SG, Draper SJ, Simerly RB. Trophic action of leptin on hypothalamic neurons that regulate feeding. Science 2004;304:108–10.

[146] Milagro FI, Campion J, Garcia-Diaz DF, Goyenechea E, Paternain L, Martinez JA. High fat diet-induced obesity modifies the methylation pattern of leptin promoter in rats. J Phys Biochem 2009;65:1–9.

[147] Rosen ED, Hsu CH, Wang X, Sakai S, Freeman MW, Gonzalez FJ, et al. C/EBPalpha induces adipogenesis through PPARgamma: a unified pathway. Genes Dev 2002;16:22–6.

[148] Stoger R. In vivo methylation patterns of the leptin promoter in human and mouse. Epigenetics 2006;1:155–62.

[149] Bell CG, Finer S, Lindgren CM, Wilson GA, Rakyan VK, Teschendorff AE, et al. Integrated Genetic and Epigenetic Analysis Identifies Haplotype-Specific Methylation in the FTO Type 2 Diabetes and Obesity Susceptibility Locus. PLoS One 2010;5:e14040.

[150] Marchi M, Lisi S, Curcio M, Barbuti S, Piaggi P, Ceccarini G, et al. Human leptin tissue distribution, but not weight loss-dependent change in expression, is associated with methylation of its promoter. Epi official J DNA Methy Soc 2011;6.

[151] Inagaki T, Tachibana M, Magoori K, Kudo H, Tanaka T, Okamura M, et al. Obesity and metabolic syndrome in histone demethylase JHDM2a-deficient mice. Genes cells 2009;14:991–1001.

[152] Tateishi K, Okada Y, Kallin EM, Zhang Y. Role of Jhdm2a in regulating metabolic gene expression and obesity resistance. Nature 2009;458:757–61.

[153] Schalkwyk LC, Meaburn EL, Smith R, Dempster EL, Jeffries AR, Davies MN, et al. Allelic skewing of DNA methylation is widespread across the genome. Am J Hum Genet 2010;86:196–212.

[154] Kerkel K, Spadola A, Yuan E, Kosek J, Jiang L, Hod E, et al. Genomic surveys by methylation-sensitive SNP analysis identify sequence-dependent allele-specific DNA methylation. Nat Genet 2008;40:904–8.

[155] Shoemaker R, Deng J, Wang W, Zhang K. Allele-specific methylation is prevalent and is contributed by CpG-SNPs in the human genome. Genome Res 2010;20:883–9.

[156] Gertz J, Varley KE, Reddy TE, Bowling KM, Pauli F, Parker SL, et al. Analysis of DNA Methylation in a Three-Generation Family Reveals Widespread Genetic Influence on Epigenetic Regulation. PLoS Genet 2011;7:e1002228.

[157] Feinberg AP, Irizarry RA. Evolution in health and medicine Sackler colloquium: Stochastic epigenetic variation as a driving force of development, evolutionary adaptation, and disease. Proc Natl Acad Sci USA 2010;107(Suppl. 1):1757–64.

[158] Danchin E, Charmantier A, Champagne FA, Mesoudi A, Pujol B, Blanchet S. Beyond DNA: integrating inclusive inheritance into an extended theory of evolution. Nature reviews. Genetics 2011;12:475–86.

[159] Carone BR, Fauquier L, Habib N, Shea JM, Hart CE, Li R, et al. Paternally induced transgenerational environmental reprogramming of metabolic gene expression in mammals. Cell 2010;143:1084–96.

[160] Daxinger L, Whitelaw E. Transgenerational epigenetic inheritance: more questions than answers. Genome Res 2010;20:1623–8.

[161] Ng SF, Lin RC, Laybutt DR, Barres R, Owens JA, Morris MJ. Chronic high-fat diet in fathers programs beta-cell dysfunction in female rat offspring. Nature 2010;467:963–6.

[162] Waterland RA, Kellermayer R, Laritsky E, Rayco-Solon P, Harris RA, Travisano M, et al. Season of Conception in Rural Gambia Affects DNA Methylation at Putative Human Metastable Epialleles. PLoS Genet 2010;6:e1001252.

[163] Dolinoy DC, Das R, Weidman JR, Jirtle RL. Metastable epialleles, imprinting, and the fetal origins of adult diseases. Pediatric Res 2007;61:30R–7R.

[164] Feinberg AP, Irizarry RA, Fradin D, Aryee MJ, Murakami P, Aspelund T, et al. Personalized epigenomic signatures that are stable over time and covary with body mass index. Sci Transl Med 2010;2:49ra67.

[165] Nair S, Lee YH, Rousseau E, Cam M, Tataranni PA, Baier LJ, et al. Increased expression of inflammation-related genes in cultured preadipocytes/stromal vascular cells from obese compared with non-obese Pima Indians. Diabetologia 2005;48:1784–8.

[166] Kaun KR, Sokolowski MB. cGMP-dependent protein kinase: linking foraging to energy homeostasis. Genome 2009;52:1–7.

[167] Bell CG. Integration of genomic and epigenomic DNA methylation data in common complex diseases by haplotype-specific methylation analysis. Personalized Medicine 2011;8:243.

[168] Dong C, Li WD, Geller F, Lei L, Li D, Gorlova OY, et al. Possible genomic imprinting of three human obesity-related genetic loci. Am J Hum Genet 2005;76:427–37.

[169] Hindorff LA, Sethupathy P, Junkins HA, Ramos EM, Mehta JP, Collins FS, et al. Potential etiologic and functional implications of genome-wide association loci for human diseases and traits. Proc Natl Acad Sci USA 2009;106:9362−7.

[170] Wallace C, Smyth DJ, Maisuria-Armer M, Walker NM, Todd JA, Clayton DG. The imprinted DLK1-MEG3 gene region on chromosome 14q32.2 alters susceptibility to type 1 diabetes. Nat Genet 2010;42:68−71.

[171] Small KS, Hedman AK, Grundberg E, Nica AC, Thorleifsson G, Kong A, et al. Identification of an imprinted master trans regulator at the KLF14 locus related to multiple metabolic phenotypes. Nat Genet 2011;43:561−4.

[172] Heijmans BT, Kremer D, Tobi EW, Boomsma DI, Slagboom PE. Heritable rather than age-related environmental and stochastic factors dominate variation in DNA methylation of the human IGF2/H19 locus. Hum Mol Genet 2007;16:547−54.

[173] Relton CL, Davey Smith G. Epigenetic epidemiology of common complex disease: prospects for prediction, prevention, and treatment. PLoS Med 2010;7:e1000356.

[174] Bell CG, Teschendorff AE, Rakyan VK, Maxwell AP, Beck S, Savage DA. Genome-wide DNA methylation analysis for diabetic nephropathy in type 1 diabetes mellitus. BMC Med Genomics 2010;3:33.

[175] Rakyan VK, Down TA, Balding DJ, Beck S. Epigenome-wide association studies for common human diseases. Nat Rev Genet 2011;12:529−41.

[176] Irizarry RA, Ladd-Acosta C, Wen B, Wu Z, Montano C, Onyango P, et al. The human colon cancer methylome shows similar hypo- and hypermethylation at conserved tissue-specific CpG island shores. Nat Genet 2009;41:178−86.

[177] Vandanmagsar B, Youm YH, Ravussin A, Galgani JE, Stadler K, Mynatt RL, et al. The NLRP3 inflammasome instigates obesity-induced inflammation and insulin resistance. Nat Med 2011.

[178] Pietilainen KH, Rog T, Seppanen-Laakso T, Virtue S, Gopalacharyulu P, Tang J, et al. Association of lipidome remodeling in the adipocyte membrane with acquired obesity in humans. PLoS biology 2011; 9: e1000623.

[179] Froguel P, Blakemore AI. The power of the extreme in elucidating obesity. N Engl J Med 2008;359:891−3.

[180] Breitling LP, Yang R, Korn B, Burwinkel B, Brenner H. Tobacco-smoking-related differential DNA methylation: 27K discovery and replication. Am J Hum Genet 2011;88:450−7.

[181] Birney E. Chromatin and heritability: how epigenetic studies can complement genetic approaches. Trends Genet 2011;27:172−6.

[182] Gaulton KJ, Nammo T, Pasquali L, Simon JM, Giresi PG, Fogarty MP, et al. A map of open chromatin in human pancreatic islets. Nat Genet 2010;42:255−9.

[183] Bell CG, Beck S. Advances in the identification and analysis of allele-specific expression. Genome Med 2009;1:56.

[184] Cedar H, Bergman Y. Linking DNA methylation and histone modification: patterns and paradigms. Nat Rev Genet 2009;10:295−304.

[185] Flusberg BA, Webster DR, Lee JH, Travers KJ, Olivares EC, Clark TA, et al. Direct detection of DNA methylation during single-molecule, real-time sequencing. Nat Methods 2010;7:461−5.

[186] Clarke J, Wu HC, Jayasinghe L, Patel A, Reid S, Bayley H. Continuous base identification for single-molecule nanopore DNA sequencing. Nat Nanotechnol 2009;4:265−70.

[187] Kriaucionis S, Heintz N. The nuclear DNA base 5-hydroxymethylcytosine is present in Purkinje neurons and the brain. Science 2009;324:929−30.

[188] Ito S, Shen L, Dai Q, Wu SC, Collins LB, Swenberg JA, et al. Tet Proteins Can Convert 5-Methylcytosine to 5-Formylcytosine and 5-Carboxylcytosine. Science 2011.

[189] Ingelman-Sundberg M, Gomez A. The past, present and future of pharmacoepigenomics. Pharmacogenomics 2010;11:625−7.

[190] Egger G, Liang G, Aparicio A, Jones PA. Epigenetics in human disease and prospects for epigenetic therapy. Nature 2004;429:457−63.

[191] Delrue MA, Michaud JL. Fat chance: genetic syndromes with obesity. Clin Genet 2004;66:83−93.

[192] Dolinoy DC, Weidman JR, Waterland RA, Jirtle RL. Maternal genistein alters coat color and protects Avy mouse offspring from obesity by modifying the fetal epigenome. Environ Health perspect 2006;114:567−72.

CHAPTER 15

Epigenetic Approaches to Control Obesity

Abigail S. Lapham[1], Karen A. Lillycrop[1], Graham C. Burdge[1], Peter D. Gluckman[2], Mark A. Hanson[1], Keith M. Godfrey[1,3,4]
[1]University of Southampton, Southampton, UK
[2]University of Auckland, Auckland, New Zealand
[3]MRC Lifecourse Epidemiology Unit, Southampton, UK
[4]Southampton NIHR Nutrition Biomedical Research Centre, Southampton, UK

15.1 THE CHANGING EPIDEMIOLOGY OF OBESITY

In the developed world, obesity, diabetes, cardiovascular disease, and non-alcoholic fatty liver disease (NAFLD) are all increasing at alarming rates. These non-communicable diseases (NCDs) now account for 60% of deaths globally [1]. It is predicted that by 2030 there will be 2.16 billion overweight and 1.12 billion obese adults worldwide [2]. Obesity in childhood is of particular concern, with recent estimates that as many as 10% of school-aged children are either overweight or obese, although the prevalence is higher in economically developed regions [3]. A recent statement released by the World Watch Institute revealed that for the first time in human history the number of overweight people rivals the number of underweight [4].

T. Tollefsbol (Ed): Epigenetics in Human Disease. DOI: 10.1016/B978-0-12-388415-2.00015-9

They found that while the world's underfed population has declined slightly since 1980 to 1.1 billion, the number of overweight has surged to in excess of this figure.

In the developing world, obesity is also increasingly becoming as significant a problem as underfeeding. The number of overweight people in China has risen from less than 10% to over 15% in a period of 3 years. In Brazil and Colombia the numbers of overweight individuals are comparable to those seen in a number of European countries, at around 40% of adults. Even in sub-Saharan Africa, a region home to the largest proportion of the world's hungry, an increase in obesity has been observed. This increase has been most marked in women living in urban areas [5].

The large and increasing numbers of overweight and obese people presents a huge clinical and public health burden. For example, in the UK alone, annual direct costs are estimated to be £4.2 billion and Foresight (an in-depth study carried out by the Department of Business and Innovation Skills), which is a UK ministerial department, have predicted that this will more than double by 2050 if we continue as we are. There are also costs to society and the economy more broadly — for example, sickness absence reduces productivity. Foresight estimated that weight problems already cost the wider UK economy in the region of £16 billion, and that this will rise to £50 billion per year by 2050 if left unchecked [6].

The number of overweight children is increasing so rapidly that there is an urgent need to identify risk factors for obesity in order to prevent further increases and to identify possible intervention strategies. Apart from the likelihood that these children will remain overweight throughout adolescence and their entire adult life, the consequences of childhood obesity are now beginning to be fully understood. Being overweight has a negative effect on the psychological wellbeing of the child and studies have shown that overweight children have a lower health-related quality of life [7], as well as poorer educational and social outcomes as compared to children of normal weight [8]. Direct health consequences of being an overweight child include an increased risk of type 2 diabetes, which is now being seen in adolescents due to the pediatric obesity epidemic [9]. Studies have also linked being overweight in childhood with increased risk of impaired glucose tolerance and cardiovascular disease in later life [10].

Although it is well established that the risk of an individual developing obesity is dependent upon the interaction between their genotype and lifestyle factors such as an energy-rich diet and sedentary behavior, it is becoming clear that these are not the sole causes of the obesity epidemic. Whilst there is a genetic component related to the ways that genes can favor fat accumulation in a given environment (Table 15.1 shows a list of 54 genes associated with obesity phenotypes), there is now substantial evidence that the fetal and early postnatal environment strongly influences the risk of developing obesity and that altered epigenetic regulation is central to this process.

15.2 DEVELOPMENTAL ORIGINS OF OBESITY

The association between the quality of the early life environment and subsequent risk of cardio-metabolic disease has been described in a series of epidemiological studies. These showed a strong geographical relationship between infant mortality and risk of cardiovascular disease (CVD) disease 50–60 years later [11]. Subsequent retrospective studies in cohorts in developed and developing nations including the UK, North America, India, and China have shown consistently that lower birth weight within the normal range is associated with an increased risk in later life of CVD and the metabolic syndrome (hypertension, insulin resistance, type 2 diabetes, dyslipidemia, and obesity).

The Dutch Hunger Winter provides an example of how the timing of nutritional constraint during pregnancy is important in determining the future risk of disease. This short-term famine during the winter of 1944–1945 resulted in 18 000 deaths with adult rations in cities

TABLE 15.1 The 54 Loci Associated with Anthropomorphic Obesity Phenotypes

Closest Gene(s)	Chromosomal Location	Phenotype	Associated Lead SNP(s)	Proposed Molecular or Cellular Function	Additional Phenotypes
TBX15–WARS2	1p12	WHR	rs984222	Transcription factor involved in adipocyte and specific adipose depot development	Implicated in Cousin syndrome
PTBP2	1p21.3	BMI	rs1555543	—	
NEGR1	1p31	BMI	rs2815752, rs3101336, rs2568958	Neuronal outgrowth	
TNNI3K	1p31.1	BMI	rs1514175	—	
DNM3–PIGC	1q24.3	WHR	rs1011731	Dominant, negative mutations in DNM enzymes promote GLUT6 and GLUT8 transporters to adipocyte cell surface in rats.	
SEC16B, RASAL2	1q25	BMI	rs10913469	—	
LYPLAL1; ZC3H11B	1q41	WHR	rs2605100	Encodes protein thought to act as triglyceride lipase and is upregulated in subcutaneous adipose tissue in obese patients	
SDCCAG8	1q43–q44	BMI	rs12145833	—	
FANCL	2p16.1	BMI	rs887912	—	
RBJ–ADCY3–POMC	2p23.3	BMI	rs713586	—	Rare POMC mutations cause human obesity
TMEM18	2p25	BMI	rs6548238, rs2867125, rs4854344, rs7561317, rs11127485	Neural development	Associated with T2D
ZNRF3–KREMEN1	2q12.1	WHR	rs4823006	—	Kremen1 protein forms a complex with LDL receptor-related protein 6
LRP1B	2q22.2	BMI	rs2890652	—	LRP1B deletions seen in several types of human cancers
GRB14	2q24.3	WHR	rs10195252	—	Associated with triglyceride and insulin levels. GRB14-deficient mice exhibit increased body
ADAMTS9	3p14.1	WHR	rs6795735	Important for spatial distribution of cells in embryonic development	Associated with T2D
NISCH–STAB1	3p21.1	WHR	rs6784615	Interacts with insulin receptor substrate	
CADM2	3p21.1	BMI	rs13078807	—	
ETV5 (locus with three genes, strongest association in ETV5)	3q27	BMI	rs7647305	—	

Continued

TABLE 15.1 The 54 Loci Associated with Anthropomorphic Obesity Phenotypes—continued

Closest Gene(s)	Chromosomal Location	Phenotype	Associated Lead SNP(s)	Proposed Molecular or Cellular Function	Additional Phenotypes
Gene desert; GNPDA2 is one of three genes nearby	4p13	BMI	rs10938397	—	Associated with T2D
SLC39A8	4q24	BMI	rs13107325	—	
FLJ35779	5q13.3	BMI	rs2112347	—	
ZNF608	5q23.2	BMI	rs4836133	—	
CPEB4	5q35.2	WHR	rs6861681	Regulates polyadenylation elongation	
TFAP2B	6p12	WC, BMI	rs987237	—	Associated with weight, not BMI
Locus containing NCR3, AIF1 and BAT2	6p21	BMI	rs2844479, rs2260000, rs1077393	—	
VEGFA	6p21.1	WHR	rs6905288	Involved in vascular development. Key mediator of adipogenesis	VEGFA variants nominally associated with T2D
NUDT3—HMGA1	6p21.31	BMI	rs206936	—	
PRL	6p22.2—p21.3	BMI	rs4712652	—	
LY86	6p25.1	WHR	rs1294421	Plays a role in recognition of lipopolysaccharide	Associated with asthma
RSPOS	6q22.33	WHR	rs9491696	Promotes angiogenesis and vascular development	Oncogene in mouse mammary epithelial cells
NFE2L3	7p15.2	WHR	rs1055144	—	
MSRA	8p23.1	WC, BMI	rs7826222, rs17150703	—	
LRRN6C	9p21.3	BMI	rs10968576	—	
PTER	10p12	BMI	rs10508503	—	
MTCH2 (locus with 14 genes)	11p11.2	BMI	rs10838738	Cellular apoptosis	
BDNF (locus with four genes, strongest association near BDNF)	11p14	BMI	rs4074134, rs4923461, rs925946, rs10501087, rs6265	BDNF expression is regulated by nutritional state and MC4R signalling	Associated with T2D. Individuals with WAGR syndrome with BDNF deletion have BMI >95th
RPL27A	11p15.4	BMI	rs4929949	—	
ITPR2—SSPN	12p21.1	WHR	rs718314	—	Mice lacking ITPR2 and ITPR3 exhibited hypoglycaemia and lean body type
HOXC13	12q13.13	WHR	rs1443512	Transcription factor important in cell spatial distribution in embryonic development	

Gene	Locus	Trait	SNP	Function	Notes
FAIM2 (locus also contains BCDIN3D)	12q13	BMI	rs7138803	Adipocyte apoptosis	
C12orf51	12q24	WHR	rs2074356	—	
MTIF3–GTF3A	13q12.2	BMI	rs4771122	—	
PRKD1	14q12	BMI	rs11847697	—	
NRXN3	14q31	WC, BMI	rs10146997	—	
MAP2K5	15q23	BMI	rs2241423	—	
SH2B1 (locus with 19–25 genes)	16p11.2	BMI	rs7498665, rs8049439, rs4788102, rs7498665	Neuronal role in energy homeostasis	Sh2b1-null mice are obese and diabetic
GPRC5B	16p12.3	BMI	rs12444979	—	
MAF	16q22–q23	BMI	rs1424233	Transcription factor involved in adipogenesis and insulin–glucagon regulation	
FTO	16q22.2	BMI	rs9939609, rs6499640, rs8050136, rs3751812, rs7190492, rs8044769, rs1558902	Neuronal function associated with control of appetite	Associated with T2D
NPC1	18q11.2	BMI	rs1805081	Intracellular lipid transport	NPC1-null mice show late-onset weight loss and poor food intake. NPC1 interferes with function
MC4R	18q22	BMI	rs17782313, rs12970134, rs17700144	Hypothalamic signalling	Haplo-insufficiency in humans is associated with morbid obesity. MC4R-deficient mice show
KCTD15	19q13.11	BMI	rs11084753, rs29941	—	
QPTCL-GIPR	19q13.32	BMI	rs2287019	Encodes incretin receptor	Associated with fasting and 2-h glucose
TMEM160	19q13.32	BMI	rs3810291	—	

(adapted from Herrera (143))

such as Amsterdam dropping to below 1000 kilocalories (4200 kilojoules) a day by the end of November 1944 and to 580 kilocalories in the west by the end of February 1945. The Dutch Famine Birth Cohort Study found that the children of pregnant women exposed to famine in early gestation were more susceptible to CVD and obesity while those exposed in later pregnancy were more susceptible to hypertension and insulin resistance, diabetes, obesity, microalbuminuria, CVD, and other health problems [12].

Suboptimal intrauterine environments, such as in intrauterine growth restriction (IUGR), may lead to fetal adaptations to assist short-term survival but may be detrimental in the long term [13]. Small babies who were born at term and undergo early catch-up growth, characterized by a greater accumulation of fat mass relative to lean body mass, have a particularly increased risk of becoming obese in later life compared to those born at higher birth weights [14]. Early catch-up growth in infants born preterm and who were fed formula milk is also associated with an increased cardio-metabolic risk in later life [15], including obesity. A number of studies have shown a greater incidence of obesity in adults who were formula-fed as opposed to breast-fed during infancy. These findings highlight the likely role of postnatal feeding.

Overnutrition in early life also increases susceptibility to future obesity. It is believed that high maternal plasma concentrations of glucose, free fatty acids, and amino acids can result in lifelong changes in appetite control, neuroendocrine function, or energy metabolism in the developing fetus which lead to obesity later in life. Fetal overnutrition is more likely in mothers who have a greater BMI during pregnancy [16] as factors such as insulin resistance, glucose intolerance leading to higher plasma concentrations of glucose, and free fatty acids correlate positively with BMI. Dorner and Plagemann [17] have reported that children of obese women are themselves more likely to become overweight and develop insulin resistance in later life. Gestational weight gain irrespective of prepregnancy weight is positively associated with greater childhood adiposity [18] and even moderate weight gain between successive pregnancies has been shown to result in an increase in large-for-gestational-age births [19]. However, maternal weight loss through bariatric surgery prevents transmission of obesity to children compared with the offspring of mothers who did not undergo the surgery and remained obese [20]. European and Indian studies have shown that high maternal weight and adiposity are associated with coronary heart disease, insulin deficiency, and type II diabetes in the offspring. These data suggest that even within a relatively normal dietary range, modest alterations can affect the development of the fetus [21].

However, it is possible that these correlations may not be due to an intrauterine effect but result from shared socioeconomic lifestyle factors between the mother and offspring or the transmission of genetic factors. Studies taking into account paternal BMI have been inconsistent [22–24], with some finding a strong maternal effect, some finding a stronger paternal effect, and others finding similar paternal and maternal relations. However, these studies were all relatively small and may have lacked sufficient power.

The thrifty phenotype hypothesis proposes that reduced fetal growth is associated with a number of chronic conditions in later life [25]. These conditions include coronary heart disease, stroke, diabetes, and hypertension. This increased susceptibility is proposed to result from adaptations made by the fetus in utero due to its limited supply of nutrients. The hypothesis is that poor nutrient supply in utero results in fetal adaptations such that the infant will be prepared for survival in an environment in which resources are likely to be limited, resulting in a thrifty phenotype.

Those with a thrifty phenotype who actually develop in an affluent environment may be more prone to metabolic disorders, such as obesity and type 2 diabetes, whereas those who have received a good nutrient supply in utero will be adapted to good conditions and therefore better able to cope with rich diets. This idea is now widely accepted and is a source of concern for societies such as those in the developing world where rapid socioeconomic improvement is underway resulting in a transition from sparse to adequate or good nutrition [26].

A study of IUGR infants born at full term [27] showed that although these infants were born with reduced subcutaneous fat, visceral fat depots were preserved creating an imbalance between central and peripheral fat deposits. In IUGR infants accelerated postnatal growth is known to exacerbate the negative effects of being born small-for-gestational-age, increasing the risk of developing impaired glucose tolerance [28], insulin resistance [29], obesity [30] and type 2 diabetes [31].

15.3 ANIMAL STUDIES OF EARLY DEVELOPMENT AND METABOLIC PROGRAMMING

The processes by which environmental cues induce altered adult phenotypes and an increased risk of obesity in the offspring are not yet fully understood, but increasing evidence points to the importance of epigenetic processes. Animal models have been useful in understanding the effects on adult phenotypes resulting from perturbations in the developmental environment.

15.3.1 The Maternal Protein-Restricted Diet

The best-studied and most-characterized animal model of nutritional induction of an altered phenotype is feeding pregnant rodents a protein-restricted (PR) diet. Feeding the PR diet during pregnancy results in impaired glucose homeostasis [32], vascular dysfunction [33], impaired immunity [34], increased susceptibility to oxidative stress [35], increased fat deposition, and altered feeding behavior [36,37]. The induction during early life of persistent changes to the phenotype of the offspring by perturbations in maternal diet implies stable alteration of gene transcription which, in turn, results in the altered activities of metabolic pathways and homeostatic control processes. Initially using a candidate gene approach many groups reported long-term changes in the expression of key metabolic genes in response to variations in maternal diet. For example, feeding a PR diet to pregnant rats increased glucocorticoid receptor (GR) expression and reduced expression of 11β-hydroxysteroid dehydrogenase type 2 (11βHSD)-2, the enzyme which inactivates corticosteroids, in liver, lung, kidney, and brain in the offspring [38]. In the liver, increased GR activity up-regulates phosphoenolpyruvate carboxykinase (PEPCK) expression and activity and so increases capacity for gluconeogenesis. This may contribute to the induction of insulin resistance in this model [39]. Altered expression of GR has also been reported in the lung, liver, adrenals, and kidneys of the offspring of sheep fed a restricted diet during pregnancy [40–42]. Feeding a PR diet to pregnant rats up-regulates glucokinase expression in the liver of the offspring, which implies increased capacity for glucose uptake [43]. The expression of genes involved in lipid homeostasis is also altered by maternal PR. Peroxisomal proliferator-activated receptor (PPAR)-α expression was increased in the liver of the offspring of rats fed a PR diet during pregnancy and was accompanied by up-regulation of its target gene acyl-CoA oxidase (AOX) [44]. The expression of acetyl-CoA carboxylase and fatty acid synthase have been reported to be increased in the liver of the offspring of rats fed a PR diet during pregnancy and lactation [45].

More recently genome-wide approaches have been used to determine which genes are altered in response to diet. Transcriptome-wide analysis of adult liver from PR offspring revealed approximately 1.3% of genes within the genome are changed in response to maternal protein restriction. This change in a relatively small subset of genes suggests that these may represent an orchestrated response to the nutritional challenge and be part of an adaptive response [46]. The pathways changed in the liver in response to maternal PR were those involved in developmental processes, ion transport, and hormonal and stress responses, which is consistent with the phenotypic changes observed in PR offspring.

The alterations in offspring metabolism and physiology induced by maternal protein restriction are dependent upon the timing of the nutritional challenge. Bertram et al. has shown in the guinea pig model that female offspring born to dams fed a PR diet in the first half of

303

pregnancy (1–35 days) have raised mean arterial blood pressure which was associated with an increased intraventricular septum and anterior left ventricle wall thickness. They did not exhibit growth restriction at any time; in contrast, the offspring from dams fed a PR diet in late gestation (36–70 days) were growth restricted but did not display alterations in blood pressure or left ventricular structure [47].

Animal studies have also shown a clear interaction between the pre- and postnatal environments [48,49], with variations in the diet fed after weaning exacerbating the effects of maternal undernutrition on the phenotype of the offspring. For example, dyslipidemia and impaired glucose homeostasis induced by feeding dams a PR diet during pregnancy were exacerbated in adult male and female rats fed a diet containing 10% (w/w) fat after weaning compared to a 4% (w/w) fat postweaning diet [50].

15.3.2 Global Dietary Restriction

A number of groups have also used global dietary restriction during pregnancy to investigate how maternal diet can influence disease susceptibility in later life. Woodall et al. used global nutrient restriction, feeding rats 30% of an ad libitum intake throughout gestation, which results in a model of IUGR [51]. Offspring born to dams fed this diet during pregnancy are significantly smaller at birth than control offspring. They also exhibit higher systolic blood pressure, hyperinsulinemia, hyperleptinemia, hyperphagia, reduced locomotion, and obesity. These metabolic alterations are all augmented by feeding a high-fat postnatal diet [52]. However, even modest global nutrient restriction during pregnancy induces alterations in metabolism and the HPA axis. In guinea pigs fed 85% of an ad libitum diet throughout gestation, alterations in postnatal cholesterol homeostasis were observed in the male offspring [53]. In the sheep, a 15% global nutrient restriction during the first half of pregnancy led to reduced adrenocorticotrophin hormone (ACTH) and cortisol responses to exogenous corticotropin-releasing hormone and arginine vasopressin administration, and a blunted cortisol response to ACTH [54]. Long-term changes in gene expression have also been reported in adult offspring of dams fed a global undernutrition diet during pregnancy. Gluckman et al., have showed that expression of PPARα and GR in the rat liver are both down-regulated in adult offspring born to dams fed a global nutrient-restricted diet of 30% ad libitum during pregnancy [55].

15.3.3 High-Fat Diet During Pregnancy

With recent concerns about the levels of obesity in the Western world, a number of new animal models of overnutrition during pregnancy have also been developed. Feeding an obesogenic diet to female rats from before mating and through lactation leads to maternal obesity as well as hyperphagia, increased adiposity, decreased muscle mass, reduced locomotion, and accelerated puberty in the offspring [56]. Samuelsson et al. have also shown that offspring from pregnant rats fed a "junk food diet" of 16% fat, 33% sugar throughout pregnancy and lactation exhibited higher blood pressure, greater adiposity, and insulin resistance in comparison to control offspring [57]. The type of fat may also be important as when dams were fed diets with different ratios of n-6/n-3 fatty acids insulin sensitivity and weight gain varied according to the relative amounts of these fatty acids in the maternal diet [58]. Maternal carbohydrate intake has also been shown to have an effect with high-carbohydrate diets found to result in offspring that remained lighter in weight and had increased responsiveness to the neuromodulator, neuropeptide Y (NPY) [59]. Persistent alterations in the expression of PPAR γ2, 11βHSD-1 and the β2 and β3 adrenoreceptors in adipose tissue [57] were seen, which may lead to increased adipogenesis and decreased lipolysis in these rats. Interestingly, however, studies by Ng et al. have shown that not only can variations in maternal diet affect subsequent phenotype but that paternal diet is also important in determining future disease risk. Paternal high-fat-diet (HFD) exposure induces increased body weight, adiposity, impaired glucose tolerance, and insulin sensitivity in female offspring [60]. Paternal HFD altered the expression of 642 pancreatic islet genes in adult female offspring; these genes included those involved in cation and ATP binding, cytoskeleton, and intracellular transport.

In rodents there is increasing evidence that the period of susceptibility extends into postnatal life as the suckling period has been shown to be critical in the developmental induction of metabolic disease. Studies of rats in cross-fostering experiments show that high-fat feeding in the suckling period leads to an increase in adiposity, hyperleptinemia, and hypertension in the adult offspring fed a normal diet after weaning [61—63]. Schmidt et al. have also shown that overfeeding rats during the suckling period by rearing them in small litters produces hyper-phagia and obesity in adulthood [64]. There is growing evidence that overnutrition during prenatal and/or early postnatal life alters the maturation of the appetite and energy-regulating neural network in the hypothalamus. Overfeeding rat pups by rearing them in small litters leads to an increased food intake in the perinatal period and this was also associated with a persistent increase in appetite drive in later life [65,66]. In rodents, exposure to a diabetic environment before birth or exposure in early postnatal life results in significant changes in the architecture of the hypothalamus, a reduced sensitivity of central hypothalamic neuropeptides to signals of increased nutrition, and a central resistance to peripheral signals of satiety signals [62,67—69]. The effect of overnutrition on hypothalamic function has been observed not only in rodents where appetite circuits are not fully mature until postnatal day 16 [67] but also in sheep where the neural network is relatively mature at birth as in humans [70].

15.4 DEVELOPMENTAL PLASTICITY

Development is a period of rapid change in which environmental cues may induce persistent changes in phenotype in order to prepare the offspring for the predicted future environment; allowing the organism to adapt much more rapidly than could be achieved by mutation. For example, the duration of daylight to which meadow voles (*Microtus pennsylvanicus*) are exposed to prior to conception influences the thickness of their coat in anticipation of either winter or summer temperatures [71]. Work by Gluckman and Hanson argues that the developmental environment can produce a range of effects with both immediate and later-life consequences. These effects do not confer any immediate advantage to the offspring but give a later fitness advantage in later life when in an environment as predicted by the developmental experience. This type of response has been termed a predictive adaptive response (PAR) [72].

PARs rely on the environment remaining relatively constant throughout the life of the offspring. For these PARSs to confer a fitness advantage it is not necessary for the fidelity of predictions to be particularly high. If the predicted environment does not fluctuate significantly over many generations then the favorable trait may become assimilated, whereby it is fixed or genetically encoded [73]. It is also important to consider that we now live much longer than our ancestors. Therefore, mechanisms that enhanced fitness in early evolution may no longer have an advantage, or may be advantageous for the young only. Epigenetic/non-genomic inheritance that may have previously conferred a survival advantage may now exacerbate a risk for successive generations. This may play a role in the current epidemic of obesity and CVD.

PARs are only adaptive when the developmental environment of the offspring is within the predicted range. If the environment differs significantly from that which was predicted the individual is said to be "mismatched", that is having a phenotype that is not appropriate for the environment [72]. This mismatch does not have to be as a result of an extreme pre- or postnatal environment, simply a phenotype being induced during development which is not suitable for responding to the postnatal environment. This mismatch can affect the offspring in a range of ways, including abdominal fat deposition [74]. Mismatch can be due to a range of circumstances such as poorer environmental conditions during development followed by richer conditions later in life or vice versa, or due to exposure to a postnatal environment, which is evolutionarily novel and as such outside of the predictive capabilities of the fetus. Maternal disease, unbalanced diet or body composition can lead to mismatch even if the offspring goes on to have a balanced healthy diet; conversely an increase in energy-dense foods and limited physical activity in the offspring (the Western lifestyle) will increase the degree of mismatch if the intrauterine environment was poor. Changes in lifestyle factors between generations are of particular significance for countries in which rapid socioeconomic transition is underway as contemporary westernized diets and lifestyles constitute novel environments, thus compounding the mismatch [75]. The elevated risk of obesity is due to the degree of mismatch between the pre- and postnatal environment rather than any absolute levels in the postnatal environment. This concept is supported by a number of animal studies in which the maternal pre- and postnatal diets were manipulated, as described later in this chapter.

Both fetal and neonatal life are characterized by a high degree of plasticity (the potential of an organism to alter it phenotype) which provides the potential for organisms to respond rapidly and effectively to environmental change. Phenotypic plasticity is usually defined as a property of individual genotypes to produce different phenotypes when exposed to different environmental conditions [76]. This plasticity can be expressed at a number of levels including behavioral, biochemical, physiological, or developmental. Not all phenotypic plasticity is adaptive and it does not necessarily always serve to improve the individual's survival. Some traits are plastic due to unavoidable constraints in the biochemistry or physiology of the organism. Developmental plasticity forms a component of phenotypic plasticity and in contrast to biochemical and physiological responses, which can be reversed over short timescales, developmental plasticity tends to be irreversible or take longer to be reversed.

Recently there have been advances in understanding of epigenetic effects during development and the key role which they can play in plastic processes. However, studies in animal models have proved to be key in showing how the developmental environment, including the mother's diet, alters epigenetic processes [44,77,78]. In animal studies the effects of epigenetic changes induced experimentally during development have been shown to produce lifelong physiological changes of relevance to human disease such as metabolic alterations known to influence obesity [79].

15.5 EPIGENETICS AND DEVELOPMENTAL PROGRAMMING BY THE EARLY LIFE ENVIRONMENT

Most studies examining the effect of early-life nutrition and its role in gene regulation have focused on DNA methylation, rather than other epigenetic processes such as histone modification and non-coding RNAs. DNA methylation can induce transcriptional silencing by blocking the binding of transcription factors and/or through promoting the binding of the methyl CpG-binding protein (MeCP2). The latter binds to methylated cytosines and, in turn, recruits histone-modifying complexes to the DNA. MeCP2 recruits both histone deacetylases (HDACs), which remove acetyl groups from the histones, a signal of transcriptionally active chromatin, and histone methyl transferases (HMTs), such as Suv39H, which methylates lysine 9 on H3, resulting in a closed chromatin structure and transcriptional silencing. Methylation of CpG dinucleotides de novo is catalyzed by DNA methyltransferases (Dnmt) 3a and 3b, and is maintained through mitosis by gene-specific methylation of hemimethylated DNA by Dnmt1. DNA methylation is important for asymmetrical silencing of imprinted genes, X chromosome inactivation, and silencing of retrotransposons. DNA methylation is also critical for cell differentiation by silencing the expression of specific genes during the development and differentiation of individual tissues [80,81]. Methylation of CpGs is largely established during embryogenesis or in early postnatal life. Following fertilization, maternal and paternal genomes undergo extensive demethylation followed by global methylation de novo just prior to blastocyst implantation [82], during which 70% of CpGs are methylated, mainly in repressive heterochromatin regions and in repetitive sequences such as retrotransposable elements [83]. Lineage-specific methylation of tissue-specific genes occurs throughout prenatal development and early postnatal life and determines developmental fates of differentiating cells. Epigenetic marks are essentially maintained throughout life. However, environmental perturbations during periods when methylation patterns are induced may impair the program of gene silencing or activation with potential long-term adverse consequences.

Epigenetic marks induced during development were thought to persist into adulthood. However, there is now much evidence that aging is associated with tissue-specific epigenetic drift. The first large-scale twin study examined 20 3-year-old and 50-year-old Spanish MZ twin pairs. These studies revealed that while the twins had very similar epigenetic profiles, indicative of a high level of epigenetic heritability, there was a degree of epigenetic variability which increased with age across a range of tissues. It is of interest that the greatest differences were in twins who differed most in lifestyle [84]. This study was cross-sectional rather than longitudinal and for this reason it was not possible to investigate observed individual variability. A more recent study [85] measured the DNA methylation status of three genes (DRD4-Dopamine receptor 4 gene/serotonin transporter gene-SERT/X-linked monoamine oxidase A gene) in 46 monozygotic and 45 dizygotic twins sampled at 5 and 10 years of age. This study revealed that DNA methylation differences were apparent even at 5 years in genetically identical individuals. Analysis at the later timepoint suggested that these differences were not stable over time. The study suggests that environmental influences are important for determining methylation patterns found in the individual and highlights the importance of longitudinal research designs for epigenetic studies.

Studies on isolated embryos first supported the hypothesis that variations in nutrient availability can alter the methylation of genes within the embryo [86]. In these studies decreased expression of both H19 and IGF2 was observed, coupled with increased DNA methylation at the H19 imprinting control region in embryos cultured in the presence of fetal calf serum as compared to controls. Data from these studies demonstrate that early nutrition can cause epigenetic changes which are maintained in later developmental stages, at least in the case of imprinted genes.

Manipulation of human embryos in vitro can induce similar imprinting alterations to those seen in mice. Angelman's syndrome is a human neurogenetic disorder caused by loss of function of the maternal allele (paternally imprinted and therefore maternally expressed) of UBE3A. This is usually as the result of a genetic mutation or rarely as a result of a sporadic imprinting error [87]. However, there have been cases reported where Angelman's syndrome has been found in children conceived using intracytoplasmic sperm injection [87,88]. In vitro fertilization (IVF) has also been linked to the increased incidence of another imprinting disorder, Beckwith—Wiedemann syndrome. This congenital disorder is caused by a loss of imprinting of a group of genes (which includes H19 and IGF2) on human chromosome 11p15 [89]. These studies provide evidence that the early environment can cause epigenetic alterations at imprinted loci, leading to human disorders that include obesity as a clinical characteristic.

A number of factors during early life alter the epigenome of the fetus, producing long-term changes in gene expression. In an elegant study of the effect of maternal behavior during suckling on the development of stress response in the offspring, Weaver et al. showed that pups raised by rat dams which showed poorer nurturing had an increased stress response [90]. The effect was due to hypermethylation of specific CpG dinucleotides within the promoter of the GR gene in the hippocampus of the offspring. These changes were reversed in the brains of the adults by intracranial administration of the histone deacetylase inhibitor Trichostatin A and L-methionine [91].

15.6 EPIGENETICS AND EARLY-LIFE NUTRITION

Differences in the maternal intake of nutrients have been shown to alter the methylation of non-imprinted genes, resulting in subtle effects on fetal and offspring development. In the agouti mouse variations in the maternal intake during pregnancy of nutrients involved in 1-carbon metabolism induces differences in the coat color of the offspring. In agouti mice A^{vy} there is an insertion of an IAP retrotransposon in the 5′ end of the *Agouti* gene which acts as a cryptic promoter directing expression of the agouti gene encoding a paracrine signaling molecule which produces black eumelanin or yellow phaeomelanin. The methylation status of the IAP element produces a range of coat colors between yellow (unmethylated) and brown (methylated) [77,92]. Supplementation of the mother's diet with methyl donors such as betaine, choline, folic acid, and vitamin B_{12} shifted the distribution of coat color of the offspring from yellow (agouti) to brown (pseudo-agouti) [93]. This shift is due to increased methylation of the IAP element [77]. Thus maternal intake of nutrients involved in 1-carbon metabolism can induce graded changes to DNA methylation and gene expression in the offspring which persist into adulthood.

Feeding pregnant rats a PR diet induced hypomethylation of the GR and PPARα promoters in the livers of juvenile and adult offspring. Hypomethylation of the GR promoter was associated with histone modifications which facilitate transcription; acetylation of histones H3 and H4 and methylation of histone H3 at lysine K4, while those that suppress gene expression were reduced or unchanged [94]. Although functionally consistent, the mechanistic relationship between GR hypomethylation and the associated histone changes is not known. These studies showed for the first time that, in contrast to modifying the maternal intake of nutrients directly involved 1-carbon metabolism [44], stable changes to the epigenetic regulation of the expression of transcription factors can be induced in the offspring by modest changes to maternal macronutrient balance during pregnancy. Expression of PPARα and GR, and of their respective target genes, acyl-CoA oxidase and carnitine palmitoyl-transferase-1, and PEPCK was increased in juvenile and adult offspring [44,94—96]. This is consistent with raised plasma β-hydroxybutyrate and glucose concentrations in the fasting offspring [97]. Sequencing analysis of the PPARα promoter showed that four specific CpGs were hypomethylated, and that two CpGs located within transcription factor response elements predicted the level of the transcript [95]. Thus the effects of the maternal PR diet on the offspring are targeted to specific CpGs. The mechanisms involved are not known but by regulating effects of transcription factors on expression they may have important effects on phenotype.

Together, these results indicate that modest dietary protein restriction during pregnancy induces an altered phenotype through epigenetic changes in specific genes. Methylation of the GR and PPARα promoters was also reduced in the heart of the offspring [98] and the PPARα promoter was hypomethylated in the whole umbilical cord [99]. These findings are consistent with increased GR mRNA expression in a range of tissues from the offspring of rats fed a PR diet during pregnancy [100]. However, PPARα methylation does not differ between control and PR offspring in skeletal muscle, spleen, and adipose tissue, indicating that the effects of the maternal diet are tissue specific [Lillycrop and Burdge, unpublished]. Hypomethylation of the GR promoter has also been found in the offspring of mice fed a PR diet during pregnancy [99], which suggests that the effect of the PR diet may not be specific to one species.

The fundamental role of changes in the epigenetic regulation of transcription factor expression in altering the activity of pathways controlled by their target genes is underlined by the observation that although glucokinase expression was increased in the liver of the PR offspring, this was not accompanied by changes in the methylation status of the glucokinase promoter [43]. Since GR activity increases glucokinase expression through enhancement of insulin action [101], greater glucokinase expression in the PR offspring may have been due to increased GR activity as a result of hypomethylation of the GR promoter rather than a direct effect of prenatal undernutrition on glucokinase.

In contrast to the effects of maternal PR diet on the epigenetic regulation of hepatic genes in the offspring, a 70% reduction of total food intake during pregnancy in rats induced hypermethylation and lower PPARα and GR expression in the liver of 170-day-old offspring [55]. One explanation may lie in the differences in severity of nutritional restriction between these two dietary regimens. If the induction of altered phenotypes is predictive, then it may be anticipated that induced changes in the epigenome would differ according to dietary regimen, in order to match the phenotype to the predicted future environment. Thus the maternal PR diet could be regarded as a moderate nutrient constraint which induces in the offspring increased capacity for using nutrient reserves for energy production. In contrast more severe global undernutrition induces conservation of energy substrates. These interpretations are consistent with the phenotypes induced in the offspring [52,55,95].

There is also evidence that an excessive early nutritional environment can alter the epigenetic regulation of genes. In the hypothalamus leptin triggers specific neuronal subpopulations such as pro-opiomelanocortin (POMC) and neuropeptide Y (NPY), and activates several intracellular signaling events which result in decreased food intake and/or increased energy expenditure. The importance of POMC in food intake and energy balance can be demonstrated by mutations in the POMC gene which result in an obese phenotype [102]. Targeted disruption of the POMC gene in a mouse model was found to result in hyperphagia and lower oxygen consumption leading to increased fat mass and obesity [103].

Plagemann et al. [104] showed that neonatal overfeeding induced by raising rat pups in small litters induces the hypermethylation of two CpG dinucleotides within the POMC promoter which are essential for POMC induction by leptin and insulin. Consequently POMC expression is not up-regulated in these rats despite hyperinsulinemia and hyperleptinemia [104]. This suggests that overfeeding during early postnatal life when the appetite circuitry within the hypothalamus is still developing can alter the methylation of genes critical for bodyweight regulation, resulting in the altered programming of this system and an increased tendency towards obesity in later life. Ng et al. have also shown that Il13ra2 was hypomethylated in female offspring after high-fat feeding of the fathers [60,60].

15.6.1 Transgenerational Effects

Emerging evidence from small animal models suggests that induced phenotypes can pass to more than one generation by a non-genomic mechanism. In rats, feeding a PR diet to the F_0 generation during pregnancy results in elevated blood pressure, endothelial dysfunction, and insulin resistance in the F_1 and F_2 generations [105–107] despite adequate nutrition during pregnancy in the F_1 generation. The adverse effects on glucose homeostasis of feeding a PR during pregnancy in the F_0 generation have been found in the offspring up to F_3 generation [108]. The administration of dexamethasone to dams in late pregnancy induced increased expression of the glucocorticoid receptor (GR) and its target gene phosphoenolpyruvate carboxykinase (PEPCK) in the liver of the F_1 and F_2, but not F_3, offspring [109]. These findings raise the important issue that assessment of true non-genomic transmission between generations requires studies which continue to at least the F_3 generation [110].

There is substantial evidence for transgenerational epigenetic inheritance in non-mammalian species and its role in evolutionary biology has been reviewed [111,112]. Although epidemiological and experimental studies have shown transmission of induced phenotypes between generations, to date only one study has reported transmission of nutritionally induced epigenetic marks between generations [96]. The tendency towards obesity in A^{vy} mice is exacerbated thorough successive generations [113]. Transmission of the obese phenotype was prevented by supplementation of females with a methyl donors and cofactors, although this was not associated with a change in the methylation status of the A^{vy} locus.

The mechanism by which induced epigenetic marks are transmitted to subsequent generations is not known, although studies have begun to unpick the mechanisms involved [114]. When the transmission is only to the F_2 generation, a direct effect of the diet fed to the F_0 dams on germ cells which gave rise to the F$_2$ offspring cannot be ruled out. Sequential transmission from F$_1$ to F$_2$, and possibly beyond, would involve induction in the germline of altered epigenetic marks and such changes in DNA methylation would have to be preserved during genome-wide demethylation during fertilization, possibly by a similar mechanism to that which preserves the methylation of imprinted genes and/or by targeted preservation of nucleosome structure as occurs for specific developmental genes during spermatogenesis [115]. An alternative possibility is that prenatal nutritional constraint induces physical or physiological changes in the female which, in turn, restrict the intrauterine environment in which her offspring develop. In this case, transmission of an altered phenotype between generations would involve induction of changes in gene methylation de novo in each generation. If so, the magnitude of the induced effect, epigenetic or phenotypic, might differ between generations.

15.6.2 Mechanism of Epigenetic Change

The processes by which environmental cues induce altered epigenetic regulation in the embryo remain unknown. Studies in liver from juvenile offspring have, however, provided some insights. Feeding a PR diet to pregnant rats induced lower Dnmt1 expression and reduced binding of Dnmt1 at the GR promoter [94]. However, the expression of Dnmt3a, Dnmt3b, and MBD-2, and binding of Dnmt3a at the GR promoter were unaltered [94]. This suggests that hypomethylation of the hepatic GR promoter in the offspring, and probably other genes including PPARα, is induced by reduced capacity to maintain patterns of cytosine methylation during mitosis rather than failure of methylation de novo or active demethylation [94,99]. This is consistent with lower MeCP2 binding and increased histone modifications which facilitate transcription at the GR promoter. Reduced Dnmt1 activity might be expected to result in global demethylation. However, studies in vitro show loss of Dnmt1-induced demethylation of only a subset of genes [116,117]. This indicates that Dnmt1 is targeted to specific genes, consistent with selective hypomethylation in the liver in the PR offspring [44]. Dnmt1 activity is also required for progression through mitosis [118] and its expression is substantially reduced in non-proliferating cells [119]. Thus, suppression of Dnmt1 activity in the preimplantation period could also account for the changes in the number of cell types during early embryonic development in this model [120]. However, recent studies have shown that 5-methylcytosine (5mC) is not the only epigenetic mark of DNA itself, and the role of TET (Ten-eleven-translocation) proteins must also be considered. Tet1, is an enzyme which catalyzes the conversion of 5-methylcytosine (5mC) to 5-hydroxymethylcytosine [121,122] and has therefore been considered as a promising candidate for demethylation. The discovery of 5hmC in the mouse cerebellum [123] and in embryonic stem cells (ESCs) [124] has led to the hypothesis that 5hmC may be an intermediate in the removal of 5mC . Studies have shown that 5hmC levels across the genome are low, consistent with the hypothesis that these may be short-lived. Alternatively, 5hmC may be an epigenetic modification in its own right, attracting its own chromatin or transcriptional modifications. The mark is significantly enriched in CpG dinucleotides within genes, particularly at exons and this has been found to be associated with gene expression as well as polycomb-mediated silencing [125]. Genome-wide profiling methods have also shown that the distribution of 5hmC is distinct to that of 5mC [125]. In ESCs Tet1 is the primary Tet enzyme and is repressed following differentiation into embryoid bodies, which correlates with a reduction in 5hmC levels [124]. High levels of Tet1 in primordial germ cells have also been observed [126] suggesting that Tet1 is associated with the pluripotent state.

15.6.3 Animal Models of Maternal Nutrition and Epigenetic Alterations

It is clear that some individuals are more predisposed than others to obesity-associated diseases. It is difficult to identify those individuals most at risk and those who would most benefit from individualized monitoring and care. It is important to remember that all fat is not equal and that the site of fat accumulation can have important implications [127]. In the worst instances preferential accumulation of fat occurs in visceral adipose tissue and ectopic fat deposition in insulin-sensitive tissues such as muscle, liver, and pancreas, which correlates strongly with severe generalized insulin resistance due to the development of a chronic inflammatory state partly due to infiltration of adipose tissue by macrophages.

Supplementation of the maternal PR diet with glycine [128] or folic acid [33] has been shown to prevent the induction of hypertension and endothelial dysfunction in the offspring. Folic acid supplementation of the PR diet was also shown [50] to prevent dyslipidemia in the adult offspring, in contrast to the same supplementation of the control diet, which induced impaired endothelial dysfunction and dyslipidemia in the offspring [33,50]. Earlier studies have shown that this supplementation prevents hypomethylation of the PPARα and GR promoters in the liver of offspring. A more detailed analysis of the promoters of these genes showed that an increase in maternal folic acid intake induced subtle changes in gene regulation and altered the methylation of individual CpGs dependent on the supplementation given [95]. Animal studies have also identified puberty as a key time for intervention. Folic acid supplementation of the diet of rats during their juvenile-pubertal period [129] was found to induce impaired lipid homeostasis in addition to increased weight gain. These effects were seen irrespective of the maternal diet given and were associated with altered methylation status of specific genes in the liver.

These observations are supportive of the view that puberty is a time of increased instability of the epigenome. In this particular instance the phenotypic changes were undesirable. However, this study highlights the ability to alter effects of prenatal nutrition with interventions during puberty. Studies carried out by Waterland and colleagues on a mouse model of obesity [113] were also able to demonstrate that obesity in offspring could be prevented by appropriate supplementation of the maternal diet. The mouse A^{vy} allele results from a transposition of a murine intracisternal A particle retrotransposon upstream of the agouti gene. The agouti signaling molecule induces yellow pigmentation in the hair follicles as well as antagonizing satiety signaling at the melanocortin 4 receptor in the hypothalamus; as a result the mice have yellow coats and are prone to hyperphagic obesity. In these studies the altered $A^{v/y}$ allele was passed through three successive generations of A^{vy}/a females and a cumulative effect on coat color and obesity was observed. This study was carried out on two populations of mice in parallel; one fed a standard diet, the other a methyl-supplemented diet that induces DNA hypermethylation during development. The work found that maternal obesity could cause transgenerational amplification of increased body weight and that a methyl-supplemented diet was able to prevent this effect. This confirms that epigenetic mechanisms such as methylation play a role in the transgenerational increases in mammalian obesity, but also provides evidence that dietary intervention during pregnancy to prevent obesity is possible. These initial studies point to the need for further work to determine whether increased adiposity occurs as a result of increased energy intake, decreased energy expenditure, or both. For this information to be of clinical value further studies will be required to elucidate a causal relationship between DNA methylation and obesity. Having an understanding of the epigenetic mechanisms which underlie the observed increase in obesity presents the opportunity to prevent or reverse further increases in obesity.

Among the best-characterized of the animal models of intervention is neonatal leptin treatment. Leptin is produced by white adipose tissue and plays a key role in maintaining body weight homeostasis [130]. Leptin was initially proposed as an antiobesity therapy to reduce food intake. However, measurement of serum leptin levels in obese subjects showed that circulating leptin levels were in most cases elevated, in keeping with a state of leptin resistance. More recent studies have shown that leptin has a broader range of functions than first thought and that it is particularly important during growth and development. Leptin measurements in the serum of mice show that leptin levels drop during intrauterine and early postnatal life before increasing 5–10-fold at postnatal days 5–10. Breast milk contains leptin and it is thought that this may contribute to the circulating levels in the neonate. However, the source of this leptin surge is controversial with work in rodents suggesting that it is derived entirely from the developing neonate [131]. Cord blood leptin levels reflect neonatal fat mass and low cord blood leptin levels are associated with rapid postnatal weight gain in small-for-gestational-age infants. Studies carried out by Vickers et al. in New Zealand [132] examined neonatal leptin treatment of rats in whom maternal undernutrition during pregnancy results in offspring obesity, hyperinsulinemia, and hyperleptinemia, especially in the presence of a high-fat postnatal diet. Treatment with leptin from postnatal days 3–13 resulted in a slowing of neonatal weight gain, particularly in programmed offspring, and normalized calorie intake, locomoter activity, body weight, fat mass, insulin and leptin concentrations. This was in contrast to the control animals that were given a saline substitute, which were observed to develop all of the features listed above. This study was able to demonstrate that the effects of developmental programming are potentially reversible if intervention is made during a period of developmental plasticity, in this instance the neonatal period.

15.6.4 Human Studies of Maternal Nutrition and Epigenetic Alterations

At present there are sparse data linking maternal nutrition to epigenetic changes in human offspring. A study of genomic DNA prepared from the blood of adults who were in utero during the Dutch Hunger Winter has revealed differences in the methylation of the differentially methylated region of the imprinted insulin-like growth factor-2 gene (IGF2 DMR). This study has revealed that adults who were in utero during the famine have this region of the gene hypomethylated. Comparisons made using same-sex siblings whose gestation was unaffected by the famine reveal that the mean level of methylation of this region was 52% in exposed individuals as compared to 49% in those who were unexposed [133]. Further studies [134] of people exposed to the Dutch Hunger Winter famine have provided evidence that periconceptional famine exposure may alter methylation of the promoter region of imprinted and non-imprinted genes which are implicated in growth and metabolic disease such as the imprinted genes INSIGN and MEG3 (which were hypo- and hypermethylated, respectively). Some evidence of altered methylation in non-imprinted genes was observed. However, differences between unexposed and exposed subjects were very small and within the range of error for the technique used to measure methylation. A recent study looking at the whole blood methylation levels of the IGF2 DMR in the children of mothers who took a 400 μg supplement of folic acid during pregnancy was found to be 49.5% as compared to 47.4% in those mothers not taking the supplement. This study provides further evidence that in humans, maternal nutrition can have an effect on the epigenetic process and levels of methylation in the fetus [135]. Studies of patients with hyperhomocysteinemia have also been supportive of the notion that folate therapy can alter methylation status of specific genes. Hyperhomocysteinemia (defined as a blood homocysteine concentration above 15 μmol/l) is associated with increased risk of thrombosis, myocardial infarction, and stroke and is known to occur in patients with several genetically determined disorders as well as being highly prevalent in patients with uremia. Comparison of methylation levels in leukocytes of normal versus diseased patients revealed that total DNA hypomethylation was increased in hyperhomocysteinemia; however, treatment of patients using folate therapy was found to restore methylation to normal levels in addition to correcting patterns of gene expression [136].

15.7 IDENTIFICATION OF PREDICTIVE EPIGENETIC MARKERS OF FUTURE OBESITY

The evidence to date in both animal models and in humans suggests that the early-life environment, particularly variations in nutrition, can induce epigenetic alterations in the fetus which then persist throughout the lifecourse, leading to long-term changes in gene expression

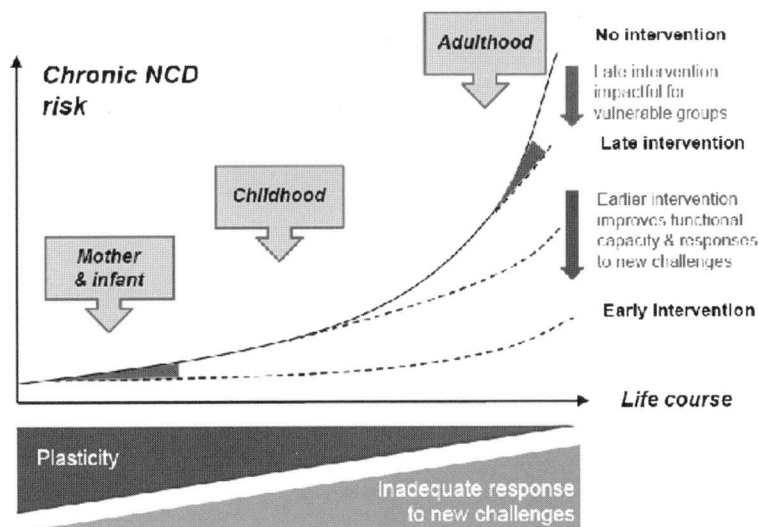

FIGURE 15.1

Non-communicable diseases (NCDs) do not fit the medical model in which an individual is healthy until they contract the disease. Risk increases throughout the lifecourse as a result of declining plasticity (green triangle) and the resulting accumulative effects of inadequate responses to new challenges (brown triangle). However, although the greatest increase occurs in adult life, the trajectory is set much earlier, being influenced by factors such as the mother's diet and body composition before and during pregnancy, and fetal, infant, and childhood nutrition and development. Adopting a lifecourse perspective allows identification of phenotype and markers of risk early, with the possibility of nutritional and other lifestyle interventions. Timely, relatively modest interventions in early life (red area) can have a large effect on disease risk later (red arrow), while later intervention (pink area) can remain impactful for vulnerable groups (pink arrow). Early-life preventive measures require a long-term investment, but are more likely to be effective than population screening programs that identify the early stages of disease or treatments initiated after the disease is manifest. This figure is reproduced in the color plate section.

and phenotype and an altered susceptibility to disease. Characterization of such altered epigenetic marks in early life may allow the identification of individuals at risk of later obesity, enabling early intervention and the development of new therapies. Such concepts are fundamental to current lifecourse strategies to the prevention and treatment of NCDs, including obesity (Figure 15.1).

Proof of concept for a potential role of epigenetic biomarkers in such a lifecourse approach has recently been published. Measurements of the epigenetic profile of a number of genes in umbilical cord tissue at birth were found [137] to predict phenotypic outcomes in childhood independent of birthweight. Greater methylation of a single CpG within the retinoid X receptor alpha (RXRA) promoter measured in umbilical cord was strongly associated with greater adiposity in later childhood. Perinatal measurements of DNA methylation explained >25% of the variance in childhood adiposity. These findings were replicated in a second independent cohort. Given the complexities of retinoid receptor biology, increased RXRA methylation might be acting through a variety of pathways, but an association between increased RXRA methylation and adiposity is consistent with the observation of strongly diminished RXRA expression in visceral white adipose tissue from obese mice [138]. Moreover, a role for retinoid receptor methylation in developmental influences on later adiposity is supported by recent experimental data showing an influence of maternal diet during pregnancy on methylation of LXRA, a heterodimeric partner of RXRA [139].

In the above human studies, associations were also observed between levels of RXRA methylation and the mothers' carbohydrate intake in early pregnancy, supportive of the concept that nutritional conditions in early pregnancy can affect a child's adiposity in later life. The work has provided novel evidence for the importance of the developmental contribution to later adiposity and was able to clearly show that specific components of the epigenetic state at birth could be used to predict adiposity in later childhood. The associations between the methylation of this CpG and both maternal diet and child's phenotype are supportive of the notion that epigenetic processes are able to exert a fine control on developmental outcomes and therefore these epigenetic measurements taken at birth could have prognostic value. It is not known whether methylation in readily available tissue such as blood, buccal, or in this case umbilical cord

313

reflects the levels of methylation in other tissue. In a recent study by Heijmans and collegues in the Netherlands [140] comparison of methylation of candidate loci in blood and buccal cells found that in half of the loci tested DNA methylation measured in blood was a marker for that in buccal cells, despite the fact that these cell types stem from different germ layers (mesoderm and ectoderm, respectively). In other studies DNA methylation in blood has been shown to serve as a marker for methylation measured in colon tissue [141,142]. However there are clearly well-documented tissue-specific differences in gene methylation observed and much more research is required to determine whether methylation levels in blood, buccal, or cord may provide useful proxy markers of methylation in more metabolically relevant tissues and whether such marks can then be used as predictive markers of future disease risk.

15.8 CONCLUSIONS

There is now a considerable body of evidence to suggest that variations in the quality of the early life environment induce a differential risk of obesity and subsequent non-communicable diseases in later life. Furthermore, elements of the heritable or familial component of disease susceptibility may be transmitted by non-genomic means. Studies have shown that the mechanisms include altered methylation of DNA and covalent modification of histones. This non-genomic tuning of the phenotype through developmental plasticity has adaptive value because it attempts to match the individual's responses to the predicted future environment based on cues received during development. When the responses are mismatched the risk of subsequent obesity rises. Epigenetic processes such as those induced by the mother's diet and body composition before and during pregnancy and during the early life of the child set the trajectory for obesity during the lifecourse of the individual. The demonstration of a role for altered epigenetic regulation of genes in the developmental induction of obesity in early life and the identification of obesity biomarkers points to the possibility of nutritional or lifestyle interventions or perhaps pharmacological interventions which could modify long-term obesity risk and reverse the current epidemic of obesity.

References

[1] World Health Organization. 2008−2013 Action Plan for the Global Strategy for the Prevention and Control of Noncommunicable Diseases. World Health Organ 2011:2008.

[2] Kelly T, Yang W, Chen CS, Reynolds K, He J. Global burden of obesity in 2005 and projections to 2030. Int J Obes (Lond) 2008;32:1431−7.

[3] Lobstein T, Baur L, Uauy R. Obesity in children and young people: a crisis in public health. Obes Rev 2004;5(Suppl 1):4−104.

[4] Garder G, Halweil B. Overfed and Underfed. The Global Epidemic of Malnutrition 2011.

[5] International Obesity Taskforce. http, www.iaso.org/iotf/obesity; 2011.

[6] Foresight Project, via the Department for Business, Innovation and Skills. http://www.bis.gov.uk/foresight; 2011.

[7] Williams J, Wake M, Hesketh K, Maher E, Waters E. Health-related quality of life of overweight and obese children. JAMA 2005;293:70−6.

[8] Gortmaker SL, Must A, Perrin JM, Sobol AM, Dietz WH. Social and economic consequences of overweight in adolescence and young adulthood. N Engl J Med 1993;329:1008−12.

[9] Ebbeling CB, Pawlak DB, Ludwig DS. Childhood obesity: public-health crisis, common sense cure. Lancet 2002;360:473−82.

[10] Bhargava SK, Sachdev HS, Fall CH, Osmond C, Lakshmy R, Barker DJ, et al. Relation of serial changes in childhood body-mass index to impaired glucose tolerance in young adulthood. N Engl J Med 2004;350:865−75.

[11] Barker DJ. Intrauterine programming of coronary heart disease and stroke. Acta Paediatr Suppl 1997;423:178−82.

[12] Schulz LC. The Dutch Hunger Winter and the developmental origins of health and disease. Proc Natl Acad Sci USA 2010;107:16757−8.

[13] Barker DJ, Bull AR, Osmond C, Simmonds SJ. Fetal and placental size and risk of hypertension in adult life. BMJ 1990;301:259—62.

[14] Ong KK, Ahmed ML, Emmett PM, Preece MA, Dunger DB. Association between postnatal catch-up growth and obesity in childhood: prospective cohort study. BMJ 2000;320:967—71.

[15] Fall CH, Barker DJ, Osmond C, Winter PD, Clark PM, Hales CN. Relation of infant feeding to adult serum cholesterol concentration and death from ischaemic heart disease. BMJ 1992;304:801—5.

[16] Oken E, Gillman MW. Fetal origins of obesity. Obes Res 2003;11:496—506.

[17] Dorner G, Plagemann A. Perinatal hyperinsulinism as possible predisposing factor for diabetes mellitus, obesity and enhanced cardiovascular risk in later life. Horm Metab Res 1994;26:213—21.

[18] Fraser A, Tilling K, Macdonald-Wallis C, Sattar N, Brion MJ, Benfield L, et al. Association of maternal weight gain in pregnancy with offspring obesity and metabolic and vascular traits in childhood. Circulation 2010;121:2557—64.

[19] Villamor E, Cnattingius S. Interpregnancy weight change and risk of adverse pregnancy outcomes: a population-based study. Lancet 2006;368:1164—70.

[20] Smith J, Cianflone K, Biron S, Hould FS, Lebel S, Marceau S, et al. Effects of maternal surgical weight loss in mothers on intergenerational transmission of obesity. J Clin Endocrinol Metab 2009;94:4275—83.

[21] Guilloteau P, Zabielski R, Hammon HM, Metges CC. Adverse effects of nutritional programming during prenatal and early postnatal life, some aspects of regulation and potential prevention and treatments. J Physiol Pharmacol 2009;60(Suppl 3):17—35.

[22] Davey SG, Steer C, Leary S, Ness A. Is there an intrauterine influence on obesity? Evidence from parent child associations in the Avon Longitudinal Study of Parents and Children (ALSPAC). Arch Dis Child 2007;92:876—80.

[23] Lawlor DA, Timpson NJ, Harbord RM, Leary S, Ness A, McCarthy MI, et al. Exploring the developmental overnutrition hypothesis using parental-offspring associations and FTO as an instrumental variable. PLoS Med 2008;5:e33.

[24] Li L, Law C, Lo CR, Power C. Intergenerational influences on childhood body mass index: the effect of parental body mass index trajectories. Am J Clin Nutr 2009;89:551—7.

[25] Hales CN, Barker DJ. Type 2 (non-insulin-dependent) diabetes mellitus: the thrifty phenotype hypothesis. Diabetologia 1992;35:595—601.

[26] Robinson R. The fetal origins of adult disease. BMJ 2001;322:375—6.

[27] Modi N, Thomas EL, Harrington TA, Uthaya S, Dore CJ, Bell JD. Determinants of adiposity during preweaning postnatal growth in appropriately grown and growth-restricted term infants. Pediatr Res 2006;60:345—8.

[28] Crowther NJ, Cameron N, Trusler J, Gray IP. Association between poor glucose tolerance and rapid post natal weight gain in seven-year-old children. Diabetologia 1998;41:1163—7.

[29] Iniguez G, Ong K, Bazaes R, Avila A, Salazar T, Dunger D, et al. Longitudinal changes in insulin-like growth factor-I, insulin sensitivity, and secretion from birth to age three years in small-for-gestational-age children. J Clin Endocrinol Metab 2006;91:4645—9.

[30] Ong KK, Loos RJ. Rapid infancy weight gain and subsequent obesity: systematic reviews and hopeful suggestions. Acta Paediatr 2006;95:904—8.

[31] Yajnik CS. Early life origins of insulin resistance and type 2 diabetes in India and other Asian countries. J Nutr 2004;134:205—10.

[32] Fernandez-Twinn DS, Wayman A, Ekizoglou S, Martin MS, Hales CN, Ozanne SE. Maternal protein restriction leads to hyperinsulinemia and reduced insulin-signaling protein expression in 21-mo-old female rat offspring. Am J Physiol Regul Integr Comp Physiol 2005;288:R368—73.

[33] Torrens C, Brawley L, Anthony FW, Dance CS, Dunn R, Jackson AA, et al. Folate supplementation during pregnancy improves offspring cardiovascular dysfunction induced by protein restriction. Hypertension 2006;47:982—7.

[34] Calder PC, Yaqoob P. The level of protein and type of fat in the diet of pregnant rats both affect lymphocyte function in the offspring. Nutr Res 2000;20:995—1005.

[35] Langley-Evans SC, Sculley DV. Programming of hepatic antioxidant capacity and oxidative injury in the ageing rat. Mech Ageing Dev 2005;126:804—12.

[36] Bellinger L, Lilley C, Langley-Evans SC. Prenatal exposure to a maternal low-protein diet programmes a preference for high-fat foods in the young adult rat. Br J Nutr 2004;92:513—20.

[37] Bellinger L, Sculley DV, Langley-Evans SC. Exposure to undernutrition in fetal life determines fat distribution, locomotor activity and food intake in ageing rats. Int J Obes (Lond) 2006;30:729—38.

[38] Bertram CE, Hanson MA. Animal models and programming of the metabolic syndrome. Br Med Bull 2001;60:103—21.

315

[39] Burns SP, Desai M, Cohen RD, Hales CN, Iles RA, Germain JP, et al. Gluconeogenesis, glucose handling, and structural changes in livers of the adult offspring of rats partially deprived of protein during pregnancy and lactation. J Clin Invest 1997;100:1768–74.

[40] Whorwood CB, Firth KM, Budge H, Symonds ME. Maternal undernutrition during early to midgestation programs tissue-specific alterations in the expression of the glucocorticoid receptor, 11beta-hydroxysteroid dehydrogenase isoforms, and type 1 angiotensin ii receptor in neonatal sheep. Endocrinol 2001;142:2854–64.

[41] Brennan KA, Gopalakrishnan GS, Kurlak L, Rhind SM, Kyle CE, Brooks AN, et al. Impact of maternal undernutrition and fetal number on glucocorticoid, growth hormone and insulin-like growth factor receptor mRNA abundance in the ovine fetal kidney. Reprod 2005;129:151–9.

[42] Gnanalingham MG, Mostyn A, Dandrea J, Yakubu DP, Symonds ME, Stephenson T. Ontogeny and nutritional programming of uncoupling protein-2 and glucocorticoid receptor mRNA in the ovine lung. J Physiol 2005;565:159–69.

[43] Bogdarina I, Murphy HC, Burns SP, Clark AJ. Investigation of the role of epigenetic modification of the rat glucokinase gene in fetal programming. Life Sci 2004;74:1407–15.

[44] Lillycrop KA, Phillips ES, Jackson AA, Hanson MA, Burdge GC. Dietary protein restriction of pregnant rats induces and folic acid supplementation prevents epigenetic modification of hepatic gene expression in the offspring. J Nutr 2005;135:1382–6.

[45] Maloney CA, Gosby AK, Phuyal JL, Denyer GS, Bryson JM, Caterson ID. Site-specific changes in the expression of fat-partitioning genes in weanling rats exposed to a low-protein diet in utero. Obes Res 2003;11:461–8.

[46] Lillycrop KA, Rodford J, Garratt ES, Slater-Jefferies JL, Godfrey KM, Gluckman PD, et al. Maternal protein restriction with or without folic acid supplementation during pregnancy alters the hepatic transcriptome in adult male rats. Br J Nutr 2010;103:1711–9.

[47] Bertram C, Khan O, Ohri S, Phillips DI, Matthews SG, Hanson MA. Transgenerational effects of prenatal nutrient restriction on cardiovascular and hypothalamic-pituitary-adrenal function. J Physiol 2008;586:2217–29.

[48] Ozanne SE, Hales CN. Lifespan: catch-up growth and obesity in male mice. Nature 2004;427:411–2.

[49] Zambrano E, Bautista CJ, Deas M, Martinez-Samayoa PM, Gonzalez-Zamorano M, Ledesma H, et al. A low maternal protein diet during pregnancy and lactation has sex- and window of exposure-specific effects on offspring growth and food intake, glucose metabolism and serum leptin in the rat. J Physiol 2006;571:221–30.

[50] Burdge GC, Lillycrop KA, Jackson AA, Gluckman PD, Hanson MA. The nature of the growth pattern and of the metabolic response to fasting in the rat are dependent upon the dietary protein and folic acid intakes of their pregnant dams and post-weaning fat consumption. Br J Nutr 2008;99:540–9.

[51] Woodall SM, Johnston BM, Breier BH, Gluckman PD. Chronic maternal undernutrition in the rat leads to delayed postnatal growth and elevated blood pressure of offspring. Pediatr Res 1996;40:438–43.

[52] Vickers MH, Breier BH, Cutfield WS, Hofman PL, Gluckman PD. Fetal origins of hyperphagia, obesity, and hypertension and postnatal amplification by hypercaloric nutrition. Am J Physiol Endocrinol Metab 2000;279:E83–7.

[53] Kind KL, Clifton PM, Katsman AI, Tsiounis M, Robinson JS, Owens JA. Restricted fetal growth and the response to dietary cholesterol in the guinea pig. Am J Physiol 1999;277:R1675–82.

[54] Hawkins P, Steyn C, McGarrigle HH, Saito T, Ozaki T, Stratford LL, et al. Effect of maternal nutrient restriction in early gestation on responses of the hypothalamic-pituitary-adrenal axis to acute isocapnic hypoxaemia in late gestation fetal sheep. Exp Physiol 2000;85:85–96.

[55] Gluckman PD, Lillycrop KA, Vickers MH, Pleasants AB, Phillips ES, Beedle AS, et al. Metabolic plasticity during mammalian development is directionally dependent on early nutritional status. Proc Natl Acad Sci USA 2007;104:12796–800.

[56] Guo F, Jen KL. High-fat feeding during pregnancy and lactation affects offspring metabolism in rats. Physiol Behav 1995;57:681–6.

[57] Samuelsson AM, Matthews PA, Argenton M, Christie MR, McConnell JM, Jansen EH, et al. Diet-induced obesity in female mice leads to offspring hyperphagia, adiposity, hypertension, and insulin resistance: a novel murine model of developmental programming. Hypertension 2008;51:383–92.

[58] Korotkova M, Gabrielsson BG, Holmang A, Larsson BM, Hanson LA, Strandvik B. Gender-related long-term effects in adult rats by perinatal dietary ratio of n-6/n-3 fatty acids. Am J Physiol Regul Integr Comp Physiol 2005;288:R575–9.

[59] Kozak R, Burlet A, Burlet C, Beck B. Dietary composition during fetal and neonatal life affects neuropeptide Y functioning in adult offspring. Brain Res Dev Brain Res 2000;125:75–82.

[60] Ng SF, Lin RC, Laybutt DR, Barres R, Owens JA, Morris MJ. Chronic high-fat diet in fathers programs β-cell dysfunction in female rat offspring. Nature 2010;467:963–6.

[61] Khan IY, Dekou V, Douglas G, Jensen R, Hanson MA, Poston L, et al. A high-fat diet during rat pregnancy or suckling induces cardiovascular dysfunction in adult offspring. Am J Physiol Regul Integr Comp Physiol 2005;288:R127−33.

[62] Plagemann A, Harder T, Rake A, Melchior K, Rittel F, Rohde W, et al. Hypothalamic insulin and neuropeptide Y in the offspring of gestational diabetic mother rats. Neuroreport 1998;9:4069−73.

[63] Plagemann A, Heidrich I, Gotz F, Rohde W, Dorner G. Obesity and enhanced diabetes and cardiovascular risk in adult rats due to early postnatal overfeeding. Exp Clin Endocrinol 1992;99:154−8.

[64] Schmidt I, Fritz A, Scholch C, Schneider D, Simon E, Plagemann A. The effect of leptin treatment on the development of obesity in overfed suckling Wistar rats. Int J Obes Relat Metab Disord 2001;25:1168−74.

[65] Davidowa H, Plagemann A. Decreased inhibition by leptin of hypothalamic arcuate neurons in neonatally overfed young rats. Neuroreport 2000;11:2795−8.

[66] Plagemann A, Harder T, Melchior K, Rake A, Rohde W, Dorner G. Elevation of hypothalamic neuropeptide Y-neurons in adult offspring of diabetic mother rats. Neuroreport 1999;10:3211−6.

[67] Plagemann A, Harder T, Rake A, Melchior K, Rohde W, Dorner G. Increased number of galanin-neurons in the paraventricular hypothalamic nucleus of neonatally overfed weanling rats. Brain Res 1999;818:160−3.

[68] Plagemann A, Harder T, Janert U, Rake A, Rittel F, Rohde W, et al. Malformations of hypothalamic nuclei in hyperinsulinemic offspring of rats with gestational diabetes. Dev Neurosci 1999;21:58−67.

[69] Plagemann A, Harder T, Rake A, Melchior K, Rittel F, Rohde W, et al. Hypothalamic insulin and neuropeptide Y in the offspring of gestational diabetic mother rats. Neuroreport 1998;9:4069−73.

[70] Muhlhausler BS. Programming of the appetite-regulating neural network: a link between maternal over-nutrition and the programming of obesity? J Neuroendocrinol 2007;19:67−72.

[71] Lee TM, Zucker I. Vole infant development is influenced perinatally by maternal photoperiodic history. Am J Physiol 1988;255:R831−8.

[72] Gluckman PD, Hanson MA, Spencer HG. Predictive adaptive responses and human evolution. Trends Ecol Evol 2005;20:527−33.

[73] West-Eberhard MJ. Alternative adaptations, speciation, and phylogeny (A Review). Proc Natl Acad Sci USA 1986;83:1388−92.

[74] Gluckman PD, Hanson MA. The Fetal Matrix: Evolution, Developmental and Disease. Cambridge, UK: Cambridge University Press; 2005.

[75] Godfrey KM, Inskip HM, Hanson MA. The long-term effects of prenatal development on growth and metabolism. Semin Reprod Med 2011;29:257−65.

[76] Pigliucci M. Evolution of phenotypic plasticity: where are we going now? Trends Ecol Evol 2005;20:481−6.

[77] Waterland RA, Jirtle RL. Transposable elements: targets for early nutritional effects on epigenetic gene regulation. Mol Cell Biol 2003;23:5293−300.

[78] Gluckman PD, Hanson MA, Beedle AS. Early life events and their consequences for later disease: a life history and evolutionary perspective. Am J Hum Biol 2007;19:1−19.

[79] Gluckman PD, Hanson MA, Beedle AS, Spencer HG. Predictive adaptive responses in perspective. Trends Endocrinol Metab 2008;19:109−10.

[80] Bird A. DNA methylation patterns and epigenetic memory. Genes Dev 2002;16:6−21.

[81] Bird A. Molecular biology: Methylation talk between histones and DNA. Science 2001;294:2113−5.

[82] Morgan HD, Santos F, Green K, Dean W, Reik W. Epigenetic reprogramming in mammals. Hum Mol Genet 2005;14 Spec No 1:R47−58.

[83] Reik W, Dean W, Walter J. Epigenetic reprogramming in mammalian development. Science 2001;293:1089−93.

[84] Fraga MF, Ballestar E, Paz MF, Ropero S, Setien F, Ballestar ML, et al. Epigenetic differences arise during the lifetime of monozygotic twins. Proc Natl Acad Sci USA 2005;102:10604−9.

[85] Wong CC, Caspi A, Williams B, Craig IW, Houts R, Ambler A, et al. A longitudinal study of epigenetic variation in twins. Epigenetics 2010;5:516−26.

[86] Khosla S, Dean W, Reik W, Feil R. Culture of preimplantation embryos and its long-term effects on gene expression and phenotype. Hum Reprod Update 2001;7:419−27.

[87] Cox GF, Burger J, Lip V, Mau UA, Sperling K, Wu BL, et al. Intracytoplasmic sperm injection may increase the risk of imprinting defects. Am J Hum Genet 2002;71:162−4.

[88] Orstavik KH, Eiklid K, van der Hagen CB, Spetalen S, Kierulf K, Skjeldal O, et al. Another case of imprinting defect in a girl with Angelman syndrome who was conceived by intracytoplasmic semen injection. Am J Hum Genet 2003;72:218−9.

[89] DeBaun MR, Niemitz EL, Feinberg AP. Association of in vitro fertilization with Beckwith−Wiedemann syndrome and epigenetic alterations of LIT1 and H19. Am J Hum Genet 2003;72:156−60.

317

[90] Weaver IC, Cervoni N, Champagne FA, D'Alessio AC, Sharma S, Seckl JR, et al. Epigenetic programming by maternal behavior. Nat Neurosci 2004;7:847—54.

[91] Weaver IC, Champagne FA, Brown SE, Dymov S, Sharma S, Meaney MJ, et al. Reversal of maternal programming of stress responses in adult offspring through methyl supplementation: altering epigenetic marking later in life. J Neurosci 2005;25:11045—54.

[92] Wolff GL, Kodell RL, Moore SR, Cooney CA. Maternal epigenetics and methyl supplements affect agouti gene expression in Avy/a mice. FASEB J 1998;12:949—57.

[93] Bird AP, Wolffe AP. Methylation-induced repression—belts, braces, and chromatin. Cell 1999;99:451—4.

[94] Lillycrop KA, Slater-Jefferies JL, Hanson MA, Godfrey KM, Jackson AA, Burdge GC. Induction of altered epigenetic regulation of the hepatic glucocorticoid receptor in the offspring of rats fed a protein-restricted diet during pregnancy suggests that reduced DNA methyltransferase-1 expression is involved in impaired DNA methylation and changes in histone modifications. Br J Nutr 2007;97:1064—73.

[95] Lillycrop KA, Phillips ES, Torrens C, Hanson MA, Jackson AA, Burdge GC. Feeding pregnant rats a protein-restricted diet persistently alters the methylation of specific cytosines in the hepatic PPAR alpha promoter of the offspring. Br J Nutr 2008;100:278—82.

[96] Burdge GC, Slater-Jefferies J, Torrens C, Phillips ES, Hanson MA, Lillycrop KA. Dietary protein restriction of pregnant rats in the F0 generation induces altered methylation of hepatic gene promoters in the adult male offspring in the F1 and F2 generations. Br J Nutr 2007;97:435—9.

[97] Burdge GC, Slater-Jefferies JL, Grant RA, Chung WS, West AL, Lillycrop KA, et al. Sex, but not maternal protein or folic acid intake, determines the fatty acid composition of hepatic phospholipids, but not of triacylglycerol, in adult rats. Prostaglandins Leukot Essent Fatty Acids 2008;78:73—9.

[98] Lillycrop KA, Phillips ES, Jackson AA, Hanson MA, Burdge GC. Dietary protein restriction in the pregnant rat induces altered epigenetic regulation of the glucocorticoid receptor and peroxisomal proliferator-activated receptor alpha in the heart of the offspring which is prevented by folic acid. Proc Nutr Soc 2006;65:65A.

[99] Burdge GC, Hanson MA, Slater-Jefferies JL, Lillycrop KA. Epigenetic regulation of transcription: a mechanism for inducing variations in phenotype (fetal programming) by differences in nutrition during early life? Br J Nutr 2007;97:1036—46.

[100] Bertram C, Trowern AR, Copin N, Jackson AA, Whorwood CB. The maternal diet during pregnancy programs altered expression of the glucocorticoid receptor and type 2 11beta-hydroxysteroid dehydrogenase: potential molecular mechanisms underlying the programming of hypertension in utero. Endocrinol 2001;142:2841—53.

[101] Printz MP, Jirout M, Jaworski R, Alemayehu A, Kren V. Genetic Models in Applied Physiology. HXB/BXH rat recombinant inbred strain platform: a newly enhanced tool for cardiovascular, behavioral, and developmental genetics and genomics. J Appl Physiol 2003;94:2510—22.

[102] Krude H, Biebermann H, Gruters A. Mutations in the human proopiomelanocortin gene. Ann N Y Acad Sci 2003;994:233—9.

[103] Challis BG, Coll AP, Yeo GS, Pinnock SB, Dickson SL, Thresher RR, et al. Mice lacking pro-opiomelanocortin are sensitive to high-fat feeding but respond normally to the acute anorectic effects of peptide-YY(3-36). Proc Natl Acad Sci USA 2004;101:4695—700.

[104] Plagemann A, Harder T, Brunn M, Harder A, Roepke K, Wittrock-Staar M, et al. Hypothalamic proopiomelanocortin promoter methylation becomes altered by early overfeeding: an epigenetic model of obesity and the metabolic syndrome. J Physiol 2009;587:4963—76.

[105] Torrens C, Poston L, Hanson MA. Transmission of raised blood pressure and endothelial dysfunction to the F2 generation induced by maternal protein restriction in the F0, in the absence of dietary challenge in the F1 generation. Br J Nutr 2008;100:760—6.

[106] Martin JF, Johnston CS, Han CT, Benyshek DC. Nutritional origins of insulin resistance: a rat model for diabetes-prone human populations. J Nutr 2000;130:741—4.

[107] Zambrano E, Martinez-Samayoa PM, Bautista CJ, Deas M, Guillen L, Rodriguez-Gonzalez GL, et al. Sex differences in transgenerational alterations of growth and metabolism in progeny (F2) of female offspring (F1) of rats fed a low protein diet during pregnancy and lactation. J Physiol (Lond) 2005;566:225—36.

[108] Benyshek DC, Johnston CS, Martin JF. Glucose metabolism is altered in the adequately-nourished grand-offspring (F3 generation) of rats malnourished during gestation and perinatal life. Diabetologia 2006;49:1117—9.

[109] Drake AJ, Walker BR, Seckl JR. Intergenerational consequences of fetal programming by in utero exposure to glucocorticoids in rats. Am J Physiol Regul.Integr. Comp Physiol 2005;288:R34—8.

[110] Skinner MK. What is an epigenetic transgenerational phenotype? F3 or F2. Reprod Toxicol 2008;25:2—6.

[111] Anway MD, Skinner MK. Epigenetic programming of the germ line: effects of endocrine disruptors on the development of transgenerational disease. Reprod Biomed Online 2008;16:23—5.

[112] Jablonka E. Inheritance systems and the evolution of new levels of individuality. J Theor Biol 1994;170:301—9.

318

[113] Waterland RA, Travisano M, Tahiliani KG, Rached MT, Mirza S. Methyl donor supplementation prevents transgenerational amplification of obesity. Int J Obes (Lond) 2008;32:1373—9.

[114] Hoile SP, Lillycrop KA, Thomas NA, Hanson MA, Burdge GC. Dietary protein restriction during F0 pregnancy in rats induces transgenerational changes in the hepatic transcriptome in female offspring. PLoS ONE 2011;6. e21668.

[115] Hammoud SS, Nix DA, Zhang H, Purwar J, Carrell DT, Cairns BR. Distinctive chromatin in human sperm packages genes for embryo development. Nature 2009;460:473—8.

[116] Jackson-Grusby L, Beard C, Possemato R, Tudor M, Fambrough D, Csankovszki G, et al. Loss of genomic methylation causes p53-dependent apoptosis and epigenetic deregulation. Nat Genet 2001;27:31—9.

[117] Rhee I, Jair KW, Yen RWC, Lengauer C, Herman JG, Kinzler KW, et al. CpG methylation is maintained in human cancer cells lacking DNMT1. Nature 2000;404:1003—7.

[118] Milutinovic S, Zhuang Q, Niveleau A, Szyf M. Epigenomic stress response. Knockdown of DNA methyltransferase 1 triggers an intra-S-phase arrest of DNA replication and induction of stress response genes. J Biol Chem 2003;278:14985—95.

[119] Suetake I, Shi LH, Watanabe D, Nakamura M, Tajima S. Proliferation stage-dependent expression of DNA methyltransferase (Dnmt1) in mouse small intestine. Cell Struct Funct 2001;26:79—86.

[120] Kwong WY, Wild AE, Roberts P, Willis AC, Fleming TP. Maternal undernutrition during the preimplantation period of rat development causes blastocyst abnormalities and programming of postnatal hypertension. Development 2000;127:4195—202.

[121] Ito S, D'Alessio AC, Taranova OV, Hong K, Sowers LC, Zhang Y. Role of Tet proteins in 5mC to 5hmC conversion, ES-cell self-renewal and inner cell mass specification. Nature 2010;466:1129—33.

[122] Wu H, D'Alessio AC, Ito S, Xia K, Wang Z, Cui K, et al. Dual functions of Tet1 in transcriptional regulation in mouse embryonic stem cells. Nature 2011;473:389—93.

[123] Kriaucionis S, Heintz N. The nuclear DNA base 5-hydroxymethylcytosine is present in Purkinje neurons and the brain. Science 2009;324:929—30.

[124] Tahiliani M, Koh KP, Shen Y, Pastor WA, Bandukwala H, Brudno Y, et al. Conversion of 5-methylcytosine to 5-hydroxymethylcytosine in mammalian DNA by MLL partner TET1. Science 2009;324:930—5.

[125] Wu H, Zhang Y. Tet1 and 5-hydroxymethylation: a genome-wide view in mouse embryonic stem cells. Cell Cycle 2011;10:2428—36.

[126] Hajkova P, Jeffries SJ, Lee C, Miller N, Jackson SP, Surani MA. Genome-wide reprogramming in the mouse germ line entails the base excision repair pathway. Science 2010;329:78—82.

[127] Walley AJ, Blakemore AI, Froguel P. Genetics of obesity and the prediction of risk for health. Hum Mol Genet 2006;15 Spec No 2:R124—30.

[128] Jackson AA, Dunn RL, Marchand MC, Langley-Evans SC. Increased systolic blood pressure in rats induced by a maternal low-protein diet is reversed by dietary supplementation with glycine. Clin Sci (Lond) 2002;103:633—9.

[129] Burdge GC, Lillycrop KA, Phillips ES, Slater-Jefferies JL, Jackson AA, Hanson MA. Folic acid supplementation during the juvenile-pubertal period in rats modifies the phenotype and epigenotype induced by prenatal nutrition. J Nutr 2009;139:1054—60.

[130] Zhang Y, Proenca R, Maffei M, Barone M, Leopold L, Friedman JM. Positional cloning of the mouse obese gene and its human homologue. Nature 1994;372:425—32.

[131] Cottrell EC, Mercer JG, Ozanne SE. Postnatal development of hypothalamic leptin receptors. Vitam Horm 2010;82:201—17.

[132] Vickers MH, Gluckman PD, Coveny AH, Hofman PL, Cutfield WS, Gertler A, et al. Neonatal leptin treatment reverses developmental programming. Endocrinol 2005;146:4211—6.

[133] Heijmans BT, Tobi EW, Stein AD, Putter H, Blauw GJ, Susser ES, et al. Persistent epigenetic differences associated with prenatal exposure to famine in humans. Proc Natl Acad Sci USA 2008;105:17046—9.

[134] Tobi EW, Lumey LH, Talens RP, Kremer D, Putter H, Stein AD, et al. DNA methylation differences after exposure to prenatal famine are common and timing- and sex-specific. Hum Mol Genet 2009;18:4046—53.

[135] Steegers-Theunissen RP, Obermann-Borst SA, Kremer D, Lindemans J, Siebel C, Steegers EA, et al. Periconceptional maternal folic acid use of 400 microg per day is related to increased methylation of the IGF2 gene in the very young child. PLoS ONE 2009;4. e7845.

[136] Ingrosso D, Cimmino A, Perna AF, Masella L, De Santo NG, et al. Folate treatment and unbalanced methylation and changes of allelic expression induced by hyperhomocysteinaemia in patients with uraemia. Lancet 2003;361:1693—9.

[137] Godfrey KM, Sheppard A, Gluckman PD, Lillycrop KA, Burdge GC, McLean C, et al. Epigenetic gene promoter methylation at birth is associated with child's later adiposity. Diabetes 2011;60:1528—34.

319

[138] Lefebvre B, Benomar Y, Guedin A, Langlois A, Hennuyer N, Dumont J, et al. Proteasomal degradation of retinoid X receptor alpha reprograms transcriptional activity of PPARgamma in obese mice and humans. J Clin Invest 2010;120:1454–68.

[139] van Straten EM, Bloks VW, Huijkman NC, Baller JF, van MH, et al. The liver X-receptor gene promoter is hypermethylated in a mouse model of prenatal protein restriction. Am J Physiol Regul Integr Comp Physiol 2010;298:R275–82.

[140] Talens RP, Boomsma DI, Tobi EW, Kremer D, Jukema JW, Willemsen G, et al. Variation, patterns, and temporal stability of DNA methylation: considerations for epigenetic epidemiology. FASEB J 2010;24:3135–44.

[141] Ally MS, Al-Ghnaniem R, Pufulete M. The relationship between gene-specific DNA methylation in leukocytes and normal colorectal mucosa in subjects with and without colorectal tumors. Cancer Epidemiol Biomarkers Prev 2009;18:922–8.

[142] Cui H, Cruz-Correa M, Giardiello FM, Hutcheon DF, Kafonek DR, Brandenburg S, et al. Loss of IGF2 imprinting: a potential marker of colorectal cancer risk. Science 2003;299:1753–5.

[143] Herrera BM, Keildson S, Lindgren CM. Genetics and epigenetics of obesity. Maturitas 2011;69:41–9.

CHAPTER 16

Epigenetics of Diabetes in Humans

Charlotte Ling, Tina Rönn
Lund University, Malmö, Sweden

321

16.1 INTRODUCTION

Diabetes is characterized by chronic hyperglycemia and, according to the World Health Organization (WHO), the definition of chronic hyperglycemia is a fasting plasma glucose concentration ≥ 7.0 mmol/l and/or plasma glucose ≥ 11.1 mmol/l 2 hours after a 75-g oral glucose load measured on two occasions [1]. There are different subtypes of diabetes, the most well-characterized being type 1 diabetes, which is an autoimmune disease and type 2 diabetes, which is a polygenic multifactorial disease that develops due to impaired insulin secretion from pancreatic beta cells in combination with impaired insulin action in target tissues, maturity-onset diabetes of the young (MODY), which is a group of monogenic forms of diabetes, latent autoimmune diabetes in adults (LADA) and maternally inherited diabetes and deafness (MIDD). The prevalence of diabetes is increasing worldwide, particularly in developing countries, and the disease has reached epidemic proportions [2]. Due to a sedentary lifestyle and an increasing age of many populations, type 2 diabetes is not only the most common form of diabetes but also the subtype increasing the most. This chapter will mainly focus on epigenetic mechanisms influencing the development of type 2 diabetes.

It is well established that combinations of non-genetic and genetic risk factors influence the susceptibility for type 2 diabetes. A sedentary lifestyle, including high-calorie food intake, obesity, and physical inactivity, as well as aging, represent non-genetic risk factors for type 2 diabetes. The genetic contribution has been confirmed by family and twin studies [3]. Moreover, recent genome-wide association studies have identified more than 40 polymorphisms associated with an increased risk for the disease [4–13]. Although most of the initial genome-wide analyses were performed in Europeans, many of the identified polymorphisms do also confer risk of type 2 diabetes in other populations [14,15]. There is further a growing body of research suggesting that epigenetic mechanisms may affect the pathogenesis of type 2 diabetes and this chapter will provide some insights into the role of epigenetics in type 2 diabetes.

T. Tollefsbol (Ed): Epigenetics in Human Disease. DOI: 10.1016/B978-0-12-388415-2.00016-0

16.2 EPIGENETIC MECHANISMS

The epigenome, including DNA methylation and histone modifications, plays a key role in controlling the function of the genome. Even though different mammalian cell types in a body contain the same genomic DNA, they have a cell-specific gene expression pattern influencing their phenotype. Epigenetic modifications include chemical modifications of both the DNA sequence itself and of the proteins it is wrapped around, the histones, and are mechanisms used to control this cell-specific gene expression [16]. Cells do also use epigenetic modifications for parental imprinting, X-chromosome inactivation, to regulate cell differentiation and to silence non-coding DNA regions. Epigenetics can be described as heritable changes in gene function that occur without a change in the nucleotide sequence and epigenetic modifications can be transferred between one cell generation and the next (mitotic inheritance) and between generations (meiotic inheritance) [17]. However, although it is well-established that epigenetic modifications can be inherited between generations in plants, there are only a limited number of studies suggesting that this is also the case in mammals [18–21]. In differentiated mammalian cells, DNA methylation mainly takes place on cytosine residues in CG dinucleotides [22]. An increased degree of DNA methylation has been associated with transcriptional silencing through either repressing the binding of transcription factors to the promoter regions or by the recruitment of proteins that specifically bind to methylated CGs, methyl-CG-binding proteins, which can further recruit proteins that close down the chromatin structure. Two groups of methyltransferases are responsible for DNA methylation: DNMT1, which maintains methylation during replication through copying the DNA methylation pattern between cell generations, and DNMT3a and DNMT3b, which are responsible for de novo methylation. Histone modifications can be associated with either an active or repressive state, depending on where the modifications take place [23]. Moreover, numerous different enzymes are responsible for generating these histone modifications [24]. While emerging data demonstrate that the epigenome is dynamic and may change in response to environmental exposures, including risk factors for type 2 diabetes, it is further possible that the epigenetic changes induced by today's sedentary lifestyle may be inherited by future generations [25].

16.3 EPIGENETICS, INSULIN SECRETION, AND DIABETES

Insulin is a key hormone regulating metabolic homeostasis. The secretion of insulin is controlled by fuel metabolism in pancreatic beta cells. When blood glucose levels rise, glucose is transported into the beta cells where it is metabolized to generate elevated ATP/ADP levels, which triggers exocytosis of insulin [26]. Pancreatic insulin secretion is determined both by the total beta cell mass and the function of each individual cell. Type 2 diabetes develops when the insulin secretion is not sufficient to maintain normoglycemia. Both genetic and non-genetic risk factors are known to affect insulin secretion [11–13,27–30]. Moreover, recent studies from our group and others have examined whether changes in DNA methylation and histone modification may play in role in the regulation of insulin secretion. Kuroda et al. hypothesized that since the DNA region surrounding the insulin gene is imprinted, the gene encoding insulin may be regulated by DNA methylation [31]. They demonstrated that the insulin promoter is demethylated in the insulin-producing beta cells compared with other cell types not expressing insulin. Moreover, using a reporter gene construct they showed that increased DNA methylation reduces the expression of the insulin gene. In particular, DNA methylation of a CpG site 182 bp upstream of the insulin promoter, which is part of a cyclic adenosine monophosphate (cAMP) responsive element (CRE), influences the expression of this gene. Mutskov et al. have further shown that the insulin gene is part of a large open chromatin domain in human islets and it displays high levels of histone modifications (hyperacetylation of H4 and dimethylation of H3 lysine 4) typical of active genes [32,33]. To examine if the epigenetic regulation of the insulin gene plays a role in patients with type 2 diabetes, we studied pancreatic islets from diabetic and non-diabetic donors as well as FACS-sorted beta and alpha cells from human islets [34]. We found that four CpG sites, located 234,

180, and 102 bp upstream and 63 bp downstream of the transcription start site respectively, showed increased DNA methylation in islets from type 2 diabetic compared with non-diabetic individuals. Moreover glucose-stimulated insulin secretion, insulin content and insulin expression were reduced in pancreatic islets from the patients with type 2 diabetes [34]. We further showed that DNA methylation of the insulin promoter was increased in alpha compared with beta cells from human donors, proposing that DNA methylation may play an important role in regulating cell-specific insulin gene expression. In agreement with the functional luciferase experiments performed by Kuroda et al., demonstrating a negative effect of DNA methylation on insulin expression, we found a strong negative correlation between the level of methylation and insulin gene expression in human pancreatic islets. As type 2 diabetes is characterized by hyperglycemia and patients with the disease often have elevated HbA1c levels, we also investigated the relation between insulin DNA methylation and HbA1c. Indeed, the level of insulin promoter DNA methylation in the human islets correlated positively with HbA1c levels, suggesting that high levels of glucose may increase methylation of the insulin gene. This hypothesis was further confirmed in clonal beta cells, where cells exposed to high levels of glucose for 48 hours showed increased insulin promoter DNA methylation compared with cells cultured in normal levels of glucose [34].

Insulin secretion is increased in response to nutrient stimulation and metabolism, which leads to elevated ATP/ADP levels in the pancreatic beta cells. Since this process requires a sufficient mitochondrial metabolism and oxidative phosphorylation, it is of no surprise that mitochondrial dysfunctions have been implicated in impaired insulin secretion and type 2 diabetes [11,26,29,35–37]. While common genetic variation near genes influencing oxidative phosphorylation is associated with impaired insulin secretion, we have also shown that epigenetic variation may reduce the expression of genes involved in oxidative phosphorylation (OXPHOS genes) in human pancreatic islets [11,29,36,37]. The transcriptional coactivator PGC-1α (encoded by *PPARGC1A*) regulates the expression of multiple genes with key functions within the mitochondria and it may hence affect ATP production and possibly insulin secretion in pancreatic beta cells. Indeed, while the expression of PGC-1α correlates positively with insulin secretion, it is reduced in pancreatic islets from patients with type 2 diabetes [37]. We further found that DNA methylation of the *PPARGC1A* gene is increased in pancreatic islets from diabetic patients, proposing that epigenetic modifications may affect gene expression and subsequently insulin secretion. In a different study, it was shown that the expression of OXPHOS genes was reduced in islets from diabetic patients [36]. DNA methylation was analyzed in the promoter region of a subset of the OXPHOS genes showing reduced expression, however, there were no differences in methylation of the studied genes due to diabetes. Nevertheless, a negative correlation was found between the level of expression and DNA methylation for one OXPHOS gene, *COX11*. Future studies are needed to further dissect the role for DNA methylation in the regulation of gene expression and insulin secretion in islets from patients with type 2 diabetes.

Two recent studies have performed genome-wide analyses of histone modifications in human pancreatic islets [38,39]. Bhandare et al. analyzed two histone marks associated with gene activation (H3K4me1 and H3K4me2) and one histone mark associated with gene repression (H3K27me3) in human islets and identified relationships between these histone modifications and gene expression. They further identified 18 polymorphisms, previously associated with type 2 diabetes, located within 500 bp of an H3K4me1 GLITR region and suggest that since H3K4me1 is frequently associated with enhancer regions, these SNPs have the potential to affect gene expression [38]. Stitzel et al. did also analyze two histone marks associated with gene activation (H3K4me1 and H3K4me3) and one histone mark associated with gene repression (H3K79me2) in human pancreatic islets and they propose to use their data as a snapshot of the epigenome in human islets [39]. A different method to analyze the chromatin structure is through the use of a formaldehyde-assisted isolation of regulatory elements (FAIRE) assay [40]. Using this approach, approximately 80 000 open chromatin sites in human pancreatic islets have been identified. Interestingly, one polymorphism previously associated with type 2 diabetes and located close to the TCF7L2 gene was found to be located in one of these islet-selective open chromatin regions.

In humans, an adverse intrauterine environment has been associated with an increased risk for diabetes and metabolic disease in postnatal life [41−48]. Inadequate nutrition may lead to chronic alterations in the body's ability to maintain metabolism, hormone levels, and the cell number of important organs [49]. Intrauterine growth restriction could be due to maternal, placental, or genetic factors, and this perturbed environment in early life is thought to affect physiological and cellular adaptive responses in key organs, as summarized in Figure 16.1. The association with susceptibility of metabolic disease in adult life suggests permanent alterations, a cell memory, potentially mediated by epigenetic mechanisms taking place in utero [50,51]. Supportively, rodents exposed to an adverse intrauterine environment show impaired insulin secretion and develop diabetes in adult life due to epigenetic modifications that take place during embryonic development [52−55]. Pancreatic duodenal homeobox 1 (PDX-1) is a homeodomain-containing transcription factor that plays a key role in pancreas development and function and patients with certain mutations in PDX-1 develop a monogenic form of diabetes (MODY4) [56]. Knockout animals lacking Pdx-1 expression in beta cells develop diabetes due to impaired insulin secretion [57]. Moreover, intrauterine growth retardation in rodents results in a decline in islet Pdx-1 expression and diabetes of the offspring [52]. This decline in Pdx-1 expression is also associated with progressive changes in epigenetic modifications including histone modification and DNA methylation at the Pdx-1 locus. A recent study showed that when these animals were treated with Exendin-4, a long-acting glucagon-like peptide-1 (GLP-1) analog, in the newborn period, the development of diabetes was prevented due to increased islet Pdx-1 expression and epigenetic changes [55]. A genome-wide analysis of DNA methylation in islets from rodents exposed to an intrauterine growth retardation did further show epigenetic changes of genes regulating beta cell proliferation, insulin secretion, and vascularization [54]. An additional candidate gene for type 2 diabetes and MODY is HNF4A. A maternal low-protein diet was associated with reduced Hnf4a expression and epigenetic changes in islets of rodent offspring [53]. This phenotype was further associated with impaired insulin secretion and diabetes. In addition, this study showed that epigenetic modifications of Hnf4a controlled the expression from tissue-specific promoters in both rodent and human islets.

Overall, these studies demonstrate that epigenetic changes in pancreatic islets and beta cells may affect the expression of candidate genes for type 2 diabetes and hence insulin secretion and risk for disease.

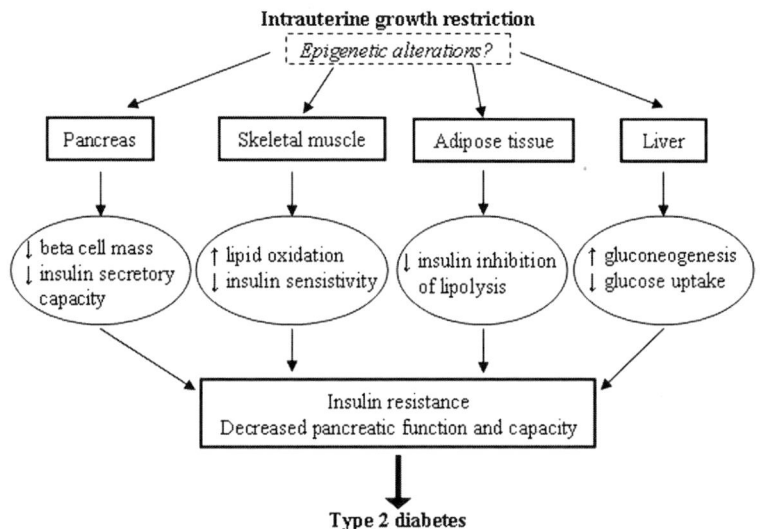

FIGURE 16.1
Intrauterine environment, epigenetic programming, and the development of type 2 diabetes.

16.4 EPIGENETICS, INSULIN RESISTANCE, AND DIABETES

Insulin resistance is a condition when cells in the body become unable to respond to normal amounts of insulin. This results in impaired glucose clearance from the blood and failure to suppress hepatic glucose production, both of which contribute to hyperglycemia. The major target organs for insulin are skeletal muscle and the liver, as these are the sites where the major glucose uptake occurs. Adipose tissue only accounts for a small proportion of glucose clearance, but is still important in maintaining normoglycemia as insulin resistance in fat cells results in increased hydrolysis of triglycerides, which may further increase insulin resistance.

Reduced oxidative capacity of the mitochondria in skeletal muscle has been suggested to contribute to insulin resistance and type 2 diabetes [58]. Moreover, the expression of genes influencing oxidative phosphorylation (OXPHOS genes) has been shown to be down-regulated in skeletal muscle from patients with type 2 diabetes [59–62]. One of these OXPHOS genes, *COX7A1*, was selected as a candidate to investigate the impact of genetic, epigenetic, and non-genetic factors on the gene expression in human skeletal muscle [63]. This study observed an age-related increase in DNA methylation of the *COX7A1* promoter as well as a decrease in mRNA expression of this gene in human skeletal muscle. Additionally, there was a positive correlation between *COX7A1* mRNA expression and insulin-stimulated glucose uptake and also a genetic component influencing gene expression. This proposes a mechanism for regulation of *COX7A1* in human skeletal muscle involving age, epigenetic, and genetic variation, which in turn could affect in vivo metabolism [63].

NDUFB6 is another example of a gene, which encodes for a protein of the respiratory chain, with decreased gene expression in muscle from diabetic patients [59]. To determine whether this defect is inherited or acquired, our group investigated the association of genetic, epigenetic, and non-genetic factors with the mRNA expression of *NDUFB6* [64]. We found a polymorphism (rs629566) in the *NDUFB6* promoter region that was associated with a decline in muscle *NDUFB6* expression with age. Young subjects carrying the rs629566 G/G genotype exhibited higher muscle *NDUFB6* expression, but this genotype was associated with reduced expression in elderly subjects. This was explained by the finding of increased DNA methylation in the promoter of elderly, but not young subjects, carrying the G/G genotype. Moreover, this polymorphism introduced a CG dinucleotide introducing a possible DNA methylation site. The degree of DNA methylation correlated negatively with muscle *NDUFB6* expression, which in turn was associated with insulin sensitivity. This provides an example of how genetic and epigenetic factors may interact to increase age-dependent susceptibility to insulin resistance [64].

PPARGC1A is a key regulator of the OXPHOS genes, and it seems to be differentially methylated not only in pancreatic islets, having a role in insulin secretion [37], but also in human skeletal muscle, possibly contributing to insulin resistance [65,66]. The study by Brons et al. showed that when individuals born with a low birth weight are challenged with a high-fat diet they develop insulin resistance and reduced expression of both *PPARGC1A* and down-stream OXPHOS genes. A novel finding in this study was that individuals with a low birth weight had a constitutive increase in *PPARGC1A* promoter DNA methylation, whereas in contrast, those born with a normal birth weight were able to reverse the changes in DNA methylation induced by the high-fat diet. Supportively, a genome-wide promoter screening of DNA methylation demonstrated epigenetic modifications of the *PPARGC1A* promoter in skeletal muscle from patients with type 2 diabetes [65]. Their methylation levels correlated negatively with *PPARGC1A* mRNA and mitochondrial DNA, and interestingly the highest proportion of cytosine methylation was found in non-CpG nucleotides.

An adverse intrauterine environment could affect not only insulin secretion, but also insulin resistance (Figure 16.1). One potential explanation for this is that epigenetic mechanisms may drive a state that is beneficial for the fetus, for example insulin resistance, which in adult life facilitates the development of type 2 diabetes and the metabolic syndrome when exposed to an obesogenic environment [67]. Human studies in this area are still sparse, but recent animal studies show promising support of this idea. For example Lillycrop et al. have shown a model in rats where a maternal low-protein diet induces persistent, gene-specific epigenetic changes that alter mRNA expression in the liver. These changes can be reversed by folic acid supplementation during pregnancy, suggesting that changes in DNA methylation may reflect an impaired supply of folic acid from the mother [68—70]. Dietary protein restriction of pregnant rats induces, and folic acid supplementation prevents, epigenetic modification of hepatic gene expression in the offspring. In humans, the Dutch Hunger Winter provides an example where the offspring to pregnant women exposed to famine show increases in insulin levels, suggesting an association with insulin resistance [41]. Furthermore, in those individuals exposed to famine, changes in DNA methylation were present six decades later, including decreased DNA methylation in the promoters of *IGF2* and *INSIGF*, and increased DNA methylation in the promoters of *IL10*, *LEP*, *ABCA1*, *GNASAS*, and *MEG3* [50,51].

16.5 PROSPECTIVE

Although there has been substantial progress in the field of epigenetics in human diabetes, additional genome-wide and functional studies are needed to fully dissect and understand the impact of epigenetic modifications in the pathogenesis of the disease. Future studies need to be carried out in several organs under different environmental conditions, since there are multiple environmental risk factors for type 2 diabetes that target different organs. It is possible that these studies will generate information that can be used in the prediction and prevention of type 2 diabetes. Moreover, in the future it is possible that new drugs targeting epigenetic factors can be developed for patients with type 2 diabetes.

References

[1] Alberti KG, Zimmet PZ. Definition, diagnosis and classification of diabetes mellitus and its complications. Part 1: diagnosis and classification of diabetes mellitus provisional report of a WHO consultation. Diabet Med 1998;15:539—53.

[2] Zimmet P, Alberti KG, Shaw J. Global and societal implications of the diabetes epidemic. Nature 2001;414:782—7.

[3] Kobberling J, Tillil H. Genetic and nutritional factors in the etiology and pathogenesis of diabetes mellitus. World Rev Nutr Diet 1990;63:102—15.

[4] McCarthy M. Genomics, Type 2 Diabetes, and Obesity. The New England Journal of Medicine 2010;363:2339—50.

[5] Saxena R, Voight BF, Lyssenko V, Burtt NP, de Bakker PI, Chen H, et al. Genome-wide association analysis identifies loci for type 2 diabetes and triglyceride levels. Science 2007;316:1331—6.

[6] Scott LJ, Mohlke KL, Bonnycastle LL, Willer CJ, Li Y, Duren WL, et al. A genome-wide association study of type 2 diabetes in Finns detects multiple susceptibility variants. Science 2007;316:1341—5.

[7] Sladek R, Rocheleau G, Rung J, Dina C, Shen L, Serre D, et al. A genome-wide association study identifies novel risk loci for type 2 diabetes. Nature 2007;445:881—5.

[8] Zeggini E, Weedon MN, Lindgren CM, Frayling TM, Elliott KS, Lango H, et al. Replication of genome-wide association signals in UK samples reveals risk loci for type 2 diabetes. Science 2007;316:1336—41.

[9] Zeggini E, Scott LJ, Saxena R, Voight BF, Marchini JL, Hu T, et al. Meta-analysis of genome-wide association data and large-scale replication identifies additional susceptibility loci for type 2 diabetes. Nat Genet 2008;40:638—45.

[10] Voight BF, Scott LJ, Steinthorsdottir V, Morris AP, Dina C, Welch RP, et al. Twelve type 2 diabetes susceptibility loci identified through large-scale association analysis. Nat Genet 2010;42:579—89.

[11] Koeck T, Olsson AH, Nitert MD, Sharoyko VV, Ladenvall C, Kotova O, et al. A common variant in TFB1M is associated with reduced insulin secretion and increased future risk of type 2 diabetes. Cell Metab 2011;13:80—91.

[12] Lyssenko V, Nagorny CL, Erdos MR, Wierup N, Jonsson A, Spegel P, et al. Common variant in MTNR1B associated with increased risk of type 2 diabetes and impaired early insulin secretion. Nat Genet 2009;41:82—8.

[13] Lyssenko V, Jonsson A, Almgren P, Pulizzi N, Isomaa B, Tuomi T, et al. Clinical risk factors, DNA variants, and the development of type 2 diabetes. N Engl J Med 2008;359:2220—32.

[14] Ronn T, Wen J, Yang Z, Lu B, Du Y, Groop L, et al. A common variant in MTNR1B, encoding melatonin receptor 1B, is associated with type 2 diabetes and fasting plasma glucose in Han Chinese individuals. Diabetologia 2009.

[15] Wen J, Ronn T, Olsson A, Yang Z, Lu B, Du Y, et al. Investigation of type 2 diabetes risk alleles support CDKN2A/B, CDKAL1, and TCF7L2 as susceptibility genes in a Han Chinese cohort. PLoS ONE 2010;5: e9153.

[16] Gibney ER, Nolan CM. Epigenetics and gene expression. Heredity 2010;105:4—13.

[17] Bird A. Perceptions of epigenetics. Nature 2007;447:396—8.

[18] Cubas P, Vincent C, Coen E. An epigenetic mutation responsible for natural variation in floral symmetry. Nature 1999;401:157—61.

[19] Chong S, Whitelaw E. Epigenetic germline inheritance. Curr Opin Genet Dev 2004;14:692—6.

[20] Anway MD, Cupp AS, Uzumcu M, Skinner MK. Epigenetic transgenerational actions of endocrine disruptors and male fertility. Science 2005;308:1466—9.

[21] Ng SF, Lin RC, Laybutt DR, Barres R, Owens JA, Morris MJ. Chronic high-fat diet in fathers programs beta-cell dysfunction in female rat offspring. Nature 2010;467:963—6.

[22] Lister R, Pelizzola M, Dowen RH, Hawkins RD, Hon G, Tonti-Filippini J, et al. Human DNA methylomes at base resolution show widespread epigenomic differences. Nature 2009;462(7271):315—22.

[23] Lennartsson A, Ekwall K. Histone modification patterns and epigenetic codes. Biochimica et Biophysica Acta 2009;1790(9):863—8.

[24] Turner B M. Environmental sensing by chromatin: an epigenetic contribution to evolutionary change. FEBS Lett 585, 2032-2040

[25] Ling C, Groop L. Epigenetics: a molecular link between environmental factors and type 2 diabetes. Diabetes 2009;58:2718—25.

[26] Mulder H, Ling C. Mitochondrial dysfunction in pancreatic beta-cells in Type 2 diabetes. Mol Cell Endocrinol 2009;297:34—40.

[27] Lyssenko V, Lupi R, Marchetti P, Del Guerra S, Orho-Melander M, Almgren P, et al. Mechanisms by which common variants in the TCF7L2 gene increase risk of type 2 diabetes. J Clin Invest 2007;117: 2155—63.

[28] Lyssenko V, Almgren P, Anevski D, Perfekt R, Lahti K, Nissen M, et al. Predictors of and longitudinal changes in insulin sensitivity and secretion preceding onset of type 2 diabetes. Diabetes 2005;54:166—74.

[29] Olsson AH, Ronn T, Ladenvall C, Parikh H, Isomaa B, Groop L, et al. Two common genetic variants near nuclear encoded OXPHOS genes are associated with insulin secretion in vivo. Eur J Endocrinol 2011.

[30] Ling C, Groop L, Guerra SD, Lupi R. Calpain-10 expression is elevated in pancreatic islets from patients with type 2 diabetes. PLoS ONE 2009;4: e6558.

[31] Kuroda A, Rauch TA, Todorov I, Ku HT, Al-Abdullah IH, Kandeel F, et al. Insulin gene expression is regulated by DNA methylation. PLoS ONE 2009;4: e6953.

[32] Mutskov V, Raaka BM, Felsenfeld G, Gershengorn MC. The human insulin gene displays transcriptionally active epigenetic marks in islet-derived mesenchymal precursor cells in the absence of insulin expression. Stem cells (Dayton, Ohio) 2007;25:3223—33.

[33] Mutskov V, Felsenfeld G. The human insulin gene is part of a large open chromatin domain specific for human islets. Proc Natl Acad Sci USA 2009;106:17419—24.

[34] Cimen H, Han MJ, Yang Y, Tong Q, Koc H, Koc EC. Regulation of succinate dehydrogenase activity by SIRT3 in mammalian mitochondria. Biochem 2010;49:304—11.

[35] Henquin JC. Regulation of insulin secretion: a matter of phase control and amplitude modulation. Diabetologia 2009;52:739—51.

[36] Olsson AH, Yang BT, Hall E, Taneera J, Salehi A, Dekker Nitert M, et al. Decreased expression of genes involved in oxidative phosphorylation in human pancreatic islets from patients with type 2 diabetes. Eur J Endocrinol 2011.

[37] Ling C, Del Guerra S, Lupi R, Ronn T, Granhall C, Luthman H, et al. Epigenetic regulation of PPARGC1A in human type 2 diabetic islets and effect on insulin secretion. Diabetologia 2008;51:615—22.

[38] Bhandare R, Schug J, Le Lay J, Fox A, Smirnova O, Liu C, et al. Genome-wide analysis of histone modifications in human pancreatic islets. Genome Res 2010;20:428—33.

327

[39] Stitzel ML, Sethupathy P, Pearson DS, Chines PS, Song L, Erdos MR, et al. Global epigenomic analysis of primary human pancreatic islets provides insights into type 2 diabetes susceptibility loci. Cell Metabol 2010;12:443—55.

[40] Gaulton KJ, Nammo T, Pasquali L, Simon JM, Giresi PG, Fogarty MP, et al. A map of open chromatin in human pancreatic islets. Nat Genet 2010;42:255—9.

[41] Ravelli AC, van der Meulen JH, Michels RP, Osmond C, Barker DJ, Hales CN, et al. Glucose tolerance in adults after prenatal exposure to famine. Lancet 1998;351:173—7.

[42] Jensen CB, Storgaard H, Dela F, Holst JJ, Madsbad S, Vaag AA. Early differential defects of insulin secretion and action in 19-year-old caucasian men who had low birth weight. Diabetes 2002;51:1271—80.

[43] Jensen CB, Storgaard H, Madsbad S, Richter EA, Vaag AA. Altered skeletal muscle fiber composition and size precede whole-body insulin resistance in young men with low birth weight. J Clin Endocrinol Metab 2007;92:1530—4.

[44] Barker DJ. The intrauterine environment and adult cardiovascular disease. Ciba Foundation symposium 1991;156:3—10. discussion 10—16.

[45] Ozanne SE, Jensen CB, Tingey KJ, Storgaard H, Madsbad S, Vaag AA. Low birthweight is associated with specific changes in muscle insulin-signalling protein expression. Diabetologia 2005;48:547—52.

[46] Ozanne SE, Jensen CB, Tingey KJ, Martin-Gronert MS, Grunnet L, Brons C, et al. Decreased protein levels of key insulin signalling molecules in adipose tissue from young men with a low birthweight: potential link to increased risk of diabetes? Diabetologia 2006;49:2993—9.

[47] Poulsen P, Vaag AA, Kyvik KO, Moller Jensen D, Beck-Nielsen H. Low birth weight is associated with NIDDM in discordant monozygotic and dizygotic twin pairs. Diabetologia 1997;40:439—46.

[48] Poulsen P, Vaag A. The intrauterine environment as reflected by birth size and twin and zygosity status influences insulin action and intracellular glucose metabolism in an age- or time-dependent manner. Diabetes 2006;55:1819—25.

[49] Barker DJ. The developmental origins of insulin resistance. Hormone Res 2005;64(Suppl. 3):2—7.

[50] Tobi EW, Lumey LH, Talens RP, Kremer D, Putter H, Stein AD, et al. DNA Methylation differences after exposure to prenatal famine are common and timing- and sex-specific. Hum Mol Genet 2009.

[51] Heijmans BT, Tobi EW, Stein AD, Putter H, Blauw GJ, Susser ES, et al. Persistent epigenetic differences associated with prenatal exposure to famine in humans. Proc Natl Acad Sci USA 2008;105:17046—9.

[52] Park JH, Stoffers DA, Nicholls RD, Simmons RA. Development of type 2 diabetes following intrauterine growth retardation in rats is associated with progressive epigenetic silencing of Pdx1. J Clin Invest 2008;118:2316—24.

[53] Sandovici IS, NH. Dekker-Nitert M, Ackers-Johnson M, Jones RH, O'Neill LP, Marquez VE, et al. Dynamic epigenetic regulation by early-diet and aging of the type 2 diabetes susceptibility gene Hnf4a in pancreatic islets. Proc Natl Acad Sci USA 2011.

[54] Thompson RF, Fazzari MJ, Niu H, Barzilai N, Simmons RA, Greally JM. Experimental intrauterine growth restriction induces alterations in DNA methylation and gene expression in pancreatic islets of rats. J Biol Chem 2010;285:15111—8.

[55] Pinney SE, Jaeckle Santos LJ, Han Y, Stoffers DA, Simmons RA. Exendin-4 increases histone acetylase activity and reverses epigenetic modifications that silence Pdx1 in the intrauterine growth retarded rat. Diabetologia 2011;54:2606—14.

[56] Stoffers DA, Ferrer J, Clarke WL, Habener JF. Early-onset type-II diabetes mellitus (MODY4) linked to IPF1. Nat Genet 1997;17:138—9.

[57] Ahlgren U, Jonsson J, Jonsson L, Simu K, Edlund H. beta-cell-specific inactivation of the mouse Ipf1/Pdx1 gene results in loss of the beta-cell phenotype and maturity onset diabetes. Gene Dev 1998;12:1763—8.

[58] Kelley DE, He J, Menshikova EV, Ritov VB. Dysfunction of mitochondria in human skeletal muscle in type 2 diabetes. Diabetes 2002;51:2944—50.

[59] Mootha VK, Lindgren CM, Eriksson KF, Subramanian A, Sihag S, Lehar J, et al. PGC-1alpha-responsive genes involved in oxidative phosphorylation are coordinately downregulated in human diabetes. Nat Genet 2003;34:267—73.

[60] Huang X, Eriksson KF, Vaag A, Lehtovirta M, Hansson M, Laurila E, et al. Insulin-regulated mitochondrial gene expression is associated with glucose flux in human skeletal muscle. Diabetes 1999;48:1508—14.

[61] Patti ME, Butte AJ, Crunkhorn S, Cusi K, Berria R, Kashyap S, et al. Coordinated reduction of genes of oxidative metabolism in humans with insulin resistance and diabetes: Potential role of PGC1 and NRF1. Proc Natl Acad Sci USA 2003;100:8466—71.

[62] Sreekumar R, Halvatsiotis P, Schimke JC, Nair KS. Gene expression profile in skeletal muscle of type 2 diabetes and the effect of insulin treatment. Diabetes 2002;51:1913—20.

[63] Ronn T, Poulsen P, Hansson O, Holmkvist J, Almgren P, Nilsson P, et al. Age influences DNA methylation and gene expression of COX7A1 in human skeletal muscle. Diabetologia 2008;51:1159—68.

328

[64] Ling C, Poulsen P, Simonsson S, Ronn T, Holmkvist J, Almgren P, et al. Genetic and epigenetic factors are associated with expression of respiratory chain component NDUFB6 in human skeletal muscle. J Clin Invest 2007;117:3427—35.

[65] Barres R, Osler ME, Yan J, Rune A, Fritz T, Caidahl K, et al. Non-CpG methylation of the PGC-1alpha promoter through DNMT3B controls mitochondrial density. Cell Metab 2009;10:189—98.

[66] Brons C, Jacobsen S, Nilsson E, Ronn T, Jensen CB, Storgaard H, et al. Deoxyribonucleic acid methylation and gene expression of PPARGC1A in human muscle is influenced by high-fat overfeeding in a birth-weight-dependent manner. J Clin Endocrinol Metab 2010;95:3048—56.

[67] Ozanne SE, Constancia M. Mechanisms of disease: the developmental origins of disease and the role of the epigenotype. Nat Clin Pract 2007;3:539—46.

[68] Lillycrop KA, Phillips ES, Jackson AA, Hanson MA, Burdge GC. Dietary protein restriction of pregnant rats induces and folic acid supplementation prevents epigenetic modification of hepatic gene expression in the offspring. J Nutr 2005;135:1382—6.

[69] Lillycrop KA, Slater-Jefferies JL, Hanson MA, Godfrey KM, Jackson AA, Burdge GC. Induction of altered epigenetic regulation of the hepatic glucocorticoid receptor in the offspring of rats fed a protein-restricted diet during pregnancy suggests that reduced DNA methyltransferase-1 expression is involved in impaired DNA methylation and changes in histone modifications. Br J Nutr 2007;97:1064—73.

[70] Lillycrop KA, Phillips ES, Torrens C, Hanson MA, Jackson AA, Burdge GC. Feeding pregnant rats a protein-restricted diet persistently alters the methylation of specific cytosines in the hepatic PPAR alpha promoter of the offspring. Br J Nutr 2008;100:278—82.

CHAPTER 17

The Potential of Epigenetic Compounds in Treating Diabetes

Steven G. Gray
St James's Hospital, Dublin, Ireland

331

T. Tollefsbol (Ed): Epigenetics in Human Disease. DOI: 10.1016/B978-0-12-388415-2.00017-2

17.1 THE PROBLEM OF DIABETES

Diabetes is a lifelong (chronic) disease in which high levels of the sugar glucose are present in the blood. The body produces a hormone, insulin, which regulates blood sugar levels by moving glucose from the bloodstream into muscle, fat, and liver cells, to be used as fuel. Diabetes is essentially a consequence of the body's failure to regulate blood sugar caused primarily by having (a) too little insulin, (b) developing resistance to insulin, or (c) both. Complications associated with diabetes include kidney failure, non-traumatic lower-limb amputations, blindness and diabetes is a major cause of heart disease and stroke [1,2].

Globally, diabetes (and in particular type 2 diabetes) represents a major challenge to world health. According to the WHO more than 220 million people worldwide have diabetes, and it projects that diabetes deaths will double between 2005 and 2030 [3].

In 2010 for the USA alone, diabetes was calculated to affect 25.8 million people, representing approximately 8.3% of the population [4]. Estimated global healthcare expenditures to treat and prevent diabetes and its complications are expected to total at least US Dollar (USD) 376 billion in 2010. By 2030, this number is projected to exceed some USD490 billion [5]. Economically, the cost of diabetes to nations has significant consequences. For example it is estimated that in the period 2006–2015, China will lose $558 billion in foregone national income due to heart disease, stroke, and diabetes alone [3].

Diabetes is a complex syndrome of dysregulation of carbohydrate and lipid metabolism due primarily to beta cell dysfunction associated with a variable degree of insulin resistance. It is clear that a complex interplay between environmental, nutritional, and genetic factors play a role in diabetes pathogenesis. Nevertheless, it is my contention that a common thread, that of histone and transcription factor/protein acetylation links many of the currently identified pathways known to be involved with diabetes pathogenesis. Over the following sections, I will attempt to link these diverse observations into a single unified concept, on the therapeutic potential of targeting HDACs for the treatment of diabetes.

17.2 EPIGENETICS

Epigenetic regulation of gene expression involves stable and heritable changes in gene expression which are not due to changes in the primary DNA sequence. Current known epigenetic mechanisms involve the following: DNA CpG methylation, histone post-translational modifications (PTMs), gene imprinting, and non-coding RNA (ncRNA) (Figure 17.1).

17.2.1 DNA CpG Methylation

DNA CpG methylation was the first epigenetic mechanism identified. In mammalian cells, DNA methylation occurs mainly at the C5 position of CpG dinucleotides and is carried out by

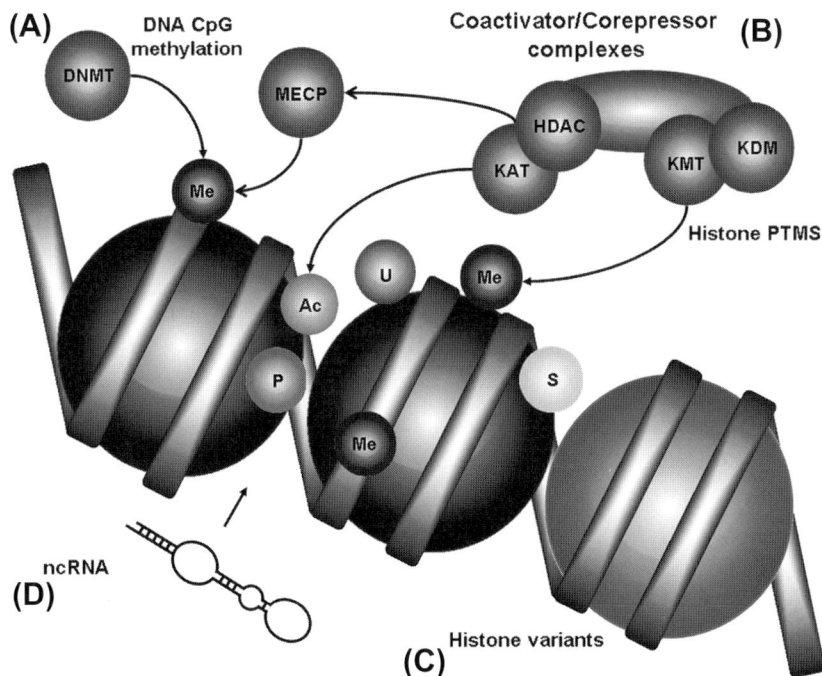

FIGURE 17.1

Overview of epigenetic mechanisms. Four main mechanisms for epigenetic regulation of gene expression have been characterized. (A) DNA methylation, (B) covalent histone modifications, (C) non-covalent mechanisms (e.g.; incorporation of histone variants) and (D) non-coding RNAs (ncRNAs) including microRNAs (miRNAs). DNMT = DNA methyltransferase; KAT = lysine acetyltransferase; KDM = lysine demethylase; KMT = lysine methyltransferase; MeCP = methyl CpG binding protein; PTMS = post-translational modifications; S = sumoylation; U = ubiquitination; Ac = acetylation; Me = methylation. This figure is reproduced in the color plate section.

two general classes of enzymatic activities — maintenance methylation and de novo methylation. The enzymes which transfer these methyl groups are the DNA methyltransferases (DNMTs) [6]. The enzymes and mechanisms for demethylation remain to be elucidated, with base excision repair emerging as the leading candidate [7].

17.2.2 Histone PTMs

The "histone code" is a well-established hypothesis describing the idea that specific patterns of post-translational modifications to histones act like a molecular "code" recognized and used by non-histone proteins to regulate specific chromatin functions [8–10]. This initial concept may however be too simplistic, and it has now been suggested that this "code" should be considered to be more like a "language", whereby PTMs provide more "nuanced" or "context-dependent" effects [11].

These modifications include acetylation, methylation, phosphorylation, sumoylation, and ubiquitination, and various families of proteins have been identified which function to place or remove these PTMs. The best studied of these families are the lysine acetyltransferases or K-acetyltransferases (KATs), histone deacetylases (HDACs), K-methyltransferases (KMTs), and K-demethylases (KDMs) [12].

K-acetyltransferases (KATs), also known as histone acetyltransferases [12], function to covalently add acetyl groups to lysine residues on proteins, and histone deacetylases (HDACs) function to remove acetyl groups [13]. K-methyltransferases (KMTs) add mono-, di-, or trimethyl groups to lysine residues [14], while HDACs and K-demethylases (KDMs) remove these respective modifications [12]. These PTMs play important roles on many proteins in addition to histones and may in fact involve a "protein code" exemplified by the proteins p53 [15] and NFκB [16,17]. The importance of these "non-epigenetic" modifications in the regulation of cellular processes can be exemplified by a recent study that found 3600 acetylation sites on 1750 proteins. To put these in context only 61 acetylation events were found on histones [18]. Nevertheless, if one considers all of the possible combinatorial possibilities for histone modifications, the known modifications on histone H3 alone could produce over one million distinct post-translational "signatures" [19].

17.2.3 microRNAs/Epi-microRNAs

microRNAs (miRNAs) are specialized forms of non-coding RNA (ncRNA). They consist of small, approximately 22 nucleotide single-stranded RNA molecules that regulate gene expression in cells by directly binding to and either degrading or translationally repressing targets and are emerging as key regulators of most, if not all, physiological processes. Altered miRNA expression is implicated in both diabetes [20] and obesity [21]. Some of these miRNAs have been shown to target epigenetic regulators which has led to this subset of miRNAs being described as "epi-miRNAs" [22].

17.3 ABERRANT EPIGENETIC REGULATION OF GENE EXPRESSION OR PROTEIN FUNCTION AS A CAUSE OF DIABETES

Previously the author has hypothesized that aberrant gene expression may be a fundamental issue in the development of diabetes [23]. Three central mechanisms exist (depicted in Figure 17.1) whereby aberrant gene expression could affect the expression of critical "diabetogenes". (a) At a basic level, the activities of these enzmes can directly affect chromatin at target genes, and any alterations or disruptions of their activities could consequently lead to aberrant transcription of diabetogenes. (b) Alternatively, several proteins, which have been identified as the causative factors in monogenic autosomal dominant forms of type 2 diabetes (MODY), have also been shown to associate with KATs/HDACs [23]. Indeed some of the mutations identified in these proteins result in loss of association with KATs/HDACs, or lead to a loss in KAT enzymatic activity [23] (Table 17.1). (c) Alterations to the activities of KATs/HDACs may functionally result in the aberrant regulation of transcription of sets of target genes essential for normal cellular homeostasis/development in diabetes pathogenesis.

17.4 ABERRANT EPIGENETICS WITHIN THE DIABETIC SETTING

The following sections will discuss some of the evidence linking aberrant epigenetics with diabetes pathogenesis.

17.4.1 Aberrant DNA CpG Methylation and Diabetes

DNA CpG methylation was essentially the first epigenetic regulatory mechanism identified, and aberrant DNA CpG methylation was initially associated with transient neonatal diabetes [24,25]. Exciting new genome-wide methylation analyses have now revealed the presence of DNA methylation variations that predispose to both type 1 and type 2 diabetes [26–29], raising the possibility of epigenetic targeting of DNA methyltransferases. The transcriptional coactivator peroxisome proliferator-activated receptor gamma coactivator-1 alpha (PGC-1α) is a critical regulator of cellular metabolic control [30]. Research has shown that levels of PGC-1α are decreased in the muscle of patients with type 2 diabetes [31]. Additional studies have now shown that PGC-1α is also down-regulated in pancreatic islets of diabetics and that furthermore DNA methylation is involved with this down-regulation [32]. Intriguingly while Ling and colleagues found that this was due to DNA CpG methylation [32], non-CpG methylation has also been shown to be essential for the down-regulation of PGC-1α mRNA in myotubes isolated from diabetic patients [33]. As such targeting DNMTs may be a potential therapeutic modality in the treatment of patients with diabetes.

17.4.2 Aberrant Chromatin Regulatory Machinery and Diabetes

Strong evidence now links aberrant expression/regulation of gene expression by chromatin-modifying proteins involved with histone/protein PTMs. For instance, genome-wide association studies (GWAS) identified chromosomal region 6q21 as being linked to both type 1 and type 2 diabetes [34–37], and it is of interest that HDAC2 maps to this region. In support of this observation, among six HDACs tested (HDAC-1 through -5 and HDAC-8) in two diabetes

TABLE 17.1 Linking HATs/HDACs to MODY Forms of Diabetes

Condition	Evidence to link HATs/HDACs to this form	References
MODY1 (HNF-4α)	HNF-4α is directly acetylated by KAT3A (CBP) HNF-4α target gene promoters are regulated by histone acetylation The HNF-4α promoter is regulated by histone acetylation	[23]
MODY2 (GK)	The GK2 promoter is regulated by histone acetylation	[23]
MODY3 (HNF-1α)	Physically interacts with lysine acetyltransferases HNF-1α dominant negative mutants have stronger association with KATs, but this causes decreased KAT activity Point mutations in HNF-1α have decreased affinity for KAT3B (p300) HNF-1α directs hyperacetylation at promoters in pancreatic islets	[23]
MODY4 (PDX-1)	PDX-1 activity depends upon the particular chromatin environment it interacts with Insulin gene activiation is mediated through PDX-1/KAT3B (p300) interactions The PDX-1 promoter is regulated by histone acetylation Activation of PDX-1 function is directly enhanced by KAT3B (p300) Under conditions of low glucose PDX-1 associates with HDAC1 and HDAC2 Under conditions of high glucose PDX-1 associates with KAT3B (p300) In a terminally differentiated glucagonoma cell line, PDX-1 modulates histone H4 acetylation to activate insulin gene expression	[23]
MODY5 (HNF-1β)	Mutations of HNF-1β result in loss of association with KAT3A (CBP) and KAT2B (P/CAF) HNF-1β/KAT2B or KAT3A complexes regulate target promoters	[23]
MODY6 (NEUROD1)	In diabetics mutations of NEUROD1 affect its ability to associate with KAT3A and KAT3B Promoter regions associated with NEUROD1 binding colocalise with KAT3A/KAT3B NEUROD1 associates with KAT3A to drive insulin transcription Direct acetylation of NEUROD1 is required for insulin transcription NEUROD1 expression by INSM1 is regulated via recruitment of HDACs	[23]
MODY7 (KLF11)	A novel KLF-KAT3B-regulated pathway for insulin biosynthesis is disrupted by the c.-331 INS mutation found in neonatal diabetes mellitus KLF11, is a novel KAT3B-dependent regulator of Pdx-1 (MODY4) transcription in pancreatic islet beta cells	[65] [64]

models (STZ-induced diabetic rats and db/db mice), only HDAC-2 activity was significantly increased and was associated with renal injury in these models, reflecting the diabetic nephropathy observed in human diabetes patients [38].

Indeed, one of the first indicators that histone-modifying enzymes may be a central element in diabetes came from a loss-of-function mouse model of KAT3A (CBP). In this model mice which were heterozygous for the mutant displayed increased insulin sensitivity and glucose tolerance, even though they present with a marked lipidystrophy of white adipose tissue [39]. Furthermore, knockout of another k-acetyltransferase KAT13B (SRC-3), prevents the development of obesity in mice, and improves insulin sensitivity by reducing levels of acetylation of PGC-1α [40]. KAT13B would appear to be an important regulator in diabetes pathogenesis as a knock-in model of KAT13B with various mutations displaying a phenotype with increased bodyweight and adiposity, coupled with reduced peripheral insulin sensitivity [41].

As previously stated above, elevated levels of HDAC-2 are associated with diabetic renal injury. The HDAC family comprises 18 members separated into four classes (classes I–IV) based on sequence identity and domain organization [13,42]. The "classical" HDACs comprise classes I, II, and IV and utilize Zn for their functional activity. The class III histone deacetylase are also known as Sirtuins or SIRTs. These HDACs, of which there are seven members, are nicotinamide adenine dinucleotide (NAD)-dependent [43], and strong evidence for their roles in diabetes pathogenesis is emerging. For example, mice which have been bred to overexpress SIRT1 in their forebrain (especially females), exhibit increased fat accumulation accompanied by a significant up-regulation of adipogenic genes in white adipose tissue. Furthermore, glucose tolerance in these mice is impaired and is coupled with a decrease in Glut4 mRNA in their muscle [44]. In addition, specific overexpression of SIRT1 in the liver ameliorates systemic insulin resistance in mice via enhanced liver insulin receptor signaling, leading to decreased hepatic gluconeogenesis and improved glucose tolerance [45]. Furthermore, in livers of mice fed on an HFD (high-fat diet) compared with controls, hyperacetylation of proteins involved in gluconeogenesis, mitochondrial oxidative metabolism, liver injury, and the ER (endoplasmic reticulum) stress response, are observed which was found to be due to reduced SIRT3 activity [46]. SIRT6-null mice die from hypoglycemia, whereas in contrast, neural-specific deletion of Sirt6 in mice leads to obesity [47,48].

Lysine demethylases (KDMs) and lysine methyltransferases (KMTs) may also play functional roles in diabetes pathogenesis. For instance, prior hyperglycemia ("hyperglycemic memory") causes increased recruitment of KDM1 (also known as LSD1) to the NFκB p65 subunit promoter [49]. Furthermore levels of KDM1 have been shown to be significantly decreased in the vascular smooth muscle cells of diabetic mice [50].

KMT7 (also known as SET7/9) has been shown to be a coactivator of NFκB, and in diabetic mice macrophages show increased recruitment of KMT7 at inflammatory genes [51], and the expression of KMT7 is enriched in pancreatic beta cells [52,53].

Taken together, it is clear that many of the enzymes/proteins involved with regulating histone/protein PTMs may play important roles in the pathogenesis of diabetes, and in particular those proteins which regulate histone/protein acetylation. In this regard, targeting HDACs may therefore be important for therapeutically targeting diabetes. In the next section we will discuss some of the results obtained for targeting these proteins within the diabetic setting.

EFFECTS OF HDACı ON DIABETOGENE EXPRESSION

The potential epigenetic basis for diabetes pathogenesis has been discussed in the previous chapter. The importance of histone acetylation in the regulation of genes central to diabetes pathogenesis can be highlighted by a recent genome-wide profiling analysis of human mesenchymal (bone marrow) stem cell-derived adipocytes. Using this technique Fraenkel and colleagues examined H3K56 acetylation (mediated by KAT3A (CBP) and KAT3B (p300), and deacetylated by SIRT1, SIRT2 and SIRT3) in adipocytes [54,55]. Critically, while this modification could be found across half the genome, the highest levels of H3K56 acetylation were associated with transcription factors and proteins in the adipokine signaling and type 2 diabetes pathways [54]. A full discussion on the effects of histone deacetylase inhibitors on these genes/pathways central to diabetes pathogenesis is beyond the remit of this chapter, and for the purposes of the following sections I will limit discussion of the effects of HDACi to a small sample of genes from various classes associated with diabetes and insulin sensitivity.

CLASS I — HORMONES
Insulin

Early studies identified chromatin remodeling involving histone acetylation as a critical regulatory mechanism of insulin biosynthesis in resoponse to glucose levels [56—60]. Further functional analyses have shown that other important regulators of insulin expression including pancreatic and duodenal homeobox-1 (PDX-1) and neurogenic differentiation factor 1/beta-cell E-box transactivator 2 (NeuroD/BETA2) either associate with or require the activities of chromatin-modifying enzymes to regulate insulin expression [61,62]. Hypothalamic expression of KAT3A (CBP) and CBP-binding partner Special AT-rich sequence binding protein 1 (SATB-1) has been found to be highly correlated with lifespan across five strains of mice, and the expression of these genes decreases with both age and diabetes in mice [63].

The transcription factor KLF11 has also been shown to functionally associate with KAT3A (p300) in pancreatic beta cells to mediate the activation of Pdx-1, a master regulator of islet insulin-producing activity [64]. In addition, a novel KLF-KAT3A (p300) pathway for the regulation of insulin biosynthesis has recently been elucidated from studies of the homozygous c.-331 mutation within the insulin promoter associated with neonatal diabetes mellitus [65].

A study examining the histone modifications found at the insulin gene in freshly isolated islets from multiple human donors found that in contrast to most genes where activating modifications tend to be concentrated within 1 kb around the transcription start site; these marks were in fact distributed over the entire coding region of the insulin gene. In addition almost uniformly elevated levels of histone acetylation and H3K4 dimethylation was found at the insulin gene and was accompanied by islet-specific coordinate expression of insulin with two neighboring genes, tyrosine hydroxylase (TH) and insulin-like growth factor 2 (IGF2), confirming the essential role of histone acetylation in the regulation of insulin expression in pancreatic beta cells [66].

A link between the proinflammatory cytokine IL-1β, histone acetylation and insulin expression has emerged. IL-1β has been shown to be an important mediator of pancreatic beta cell loss [67]. In a study examining the impact of hyperglycemia on histone acetylation at the insulin gene promoter, histone acetylation, KAT3A (p300), and RNA polymerase II (pol II) binding were still observed after 4 h in 16mM glucose, but could no longer be detected if IL-1β was also present [68], indicating that IL-1β was preventing appropriate regulation of insulin gene expression and that targeting IL-1β may be a potential means to both prevent pancreatic beta cell loss, and improve insulin expression from pancreatic beta cells. Indeed, the epigenetic targeting drug vorinostat can ablate IL-1β expression [69] and this has been shown to be within the diabetic setting to prevent against beta cell loss [70,71]. As such, epigenetic therapy with HDACi may have a threefold benefit by (a) preventing pancreatic beta cell loss, (b) improving insulin expression, and (c) reducing proinflammatory cues by decreasing IL-1β.

Adiponectin

Adiponectin (Adpn), the most abundant protein secreted by white adipose tissue, regulates energy homeostasis and glucose and lipid metabolism, and functions as an antidiabetic adipokine with insulin-sensitizing, anti-inflammatory, antiatherosclerotic, and cardioprotective properties in obesity-related disorders such as insulin resistance and type 2 diabetes [72].

Chromatin modifications including histone lysine acetylation (H3Ac) and methylation (H3K4me2) are involved with the coordinated regulation of Adpn in adipoctye differentiation [73].

During the development of insulin resistance in NIH-3T3 adipocytes decreased acetylation at lysine 9 of histone H3 (H3K9) is seen at the promoter of this gene [74], indicating that potentially HDACi may be able to alleviate this. As a caveat however, treatment of mice and cells with VPA was actually found to suppress adiponectin expression in mature adipocytes, but adiponectin promoter-driven luciferase expression in fibroblasts [75], but more studies will be required to determine if adiponectin can be successfully induced in the diabetic setting by HDACi.

CLASS II — TRANSCRIPTION FACTORS
PPARγ/PGC-1α

The nuclear hormone receptor peroxisome proliferator-activated receptor gamma (PPARγ) plays central roles in both metabolism and adipogenesis. Ligands such as thiazolidinediones (TZDs) that bind to this receptor (primarily on adipocytes) exert insulin-sensitizing and anti-inflammatory effects, and are widely used to treat metabolic syndrome, especially type 2 diabetes. A number of PPARγ interacting partners have been identified, many of which are known epigenetic regulators, including enzymes for histone acetylation/deacetylation and histone methylation/demethylation and has recently been extensively reviewed by Sugii and Evans [76]. PPARγ regulates the expression of many important diabetogenes. One recent example involves PPARγ and histone acetylation in the regulation of glucose-dependent insulinotropic polypeptide receptor (GIPR), which has important roles with insulinotropic growth and survival of pancreatic β-cells and adipocyte metabolism [77].

Peroxisome proliferator-activated receptor gamma coactivator alpha (PGC-1α) is a critical protein that regulates many pathways linked to energy homeostasis and metabolism [78], through its associations with KATs/HDACs [23]. Given that levels of PGC-1α have been shown to be reduced in the muscles of patients with type 2 diabetes [79], and that the HDACi TSA can up-regulate expression of PGC-1α [80], it will be important to see whether other HDACi can affect PGC-1α in the diabetic setting.

PGC-1α has been shown to associate with the transcription factor Forkhead box protein O (FOXO) and hepatocyte nuclear factor 4 alpha (HNF4α) to activate gluconeogenic gene expression in the liver and regulate glucose homeostasis [81]. In addition acetylation of FOXO itself modulates its promoter specificity and transcriptional activity, and is regulated via the class III deacetylase SIRT1 [82,83]. In a recent development, it has now been demonstrated that the class II HDACs (HDAC4, 5, and 7) also play roles (along with the class I HDAC3) in regulating FOXO transcriptional activity by acetylation, and that depleting class II HDACs in mouse models of type 2 diabetes ameliorates this hyperglycemia [84,85]. From this latest series of developments it would appear that targeting HDACs may be a unique mechanism capable of regulating hepatic glucose homeostasis.

CLASS III — TRANSPORTERS
Glut4

Glut4 is an insulin-dependent glucose transporter first described in rat skeletal muscle by Birnbaum [86]. It is generally only expressed in muscle and adipose tissue, and is typically stored in intracellular lipid rafts in these cells, and rapidly translocates to the plasma membrane in response to insulin signaling [87]. The first evidence that Glut4 may be epigenetically regulated came from studies showing that constitutive localization of HDAC5 into the nucleus in cardiac tissue resulted in a significant (~threefold) decrease in Glut4 expression [88]. Down-regulation of Glut4 expression in the muscle tissue of intrauterine growth-restricted (IUGR) mice was found to involve HDAC1 and HDAC4 [89]. Further studies have shown that the repression of Glut-4 transcription also involves complexes involving the Glut4 enhancer factor (GEF), myocyte enhancer factor 2 (MEF2), and HDAC5 [90,91]. GLUT4 gene expression in preadipocytes is differentiation-dependent, with full expression delayed until late in the differentiation program, and three class II HDACs (HDAC4, HDAC5, and HDAC9) have been shown to regulate this repression [92], and we have confirmed that Glut4 expression can be induced in human SGBS preadipocytes using HDACi, and also show that the CTCF transcription factor may be involved with regulating Glut4 [93]. As CTCF regulates gene transcription through the formation of higher-order chromatin structures [94], this would clearly indicate an important role for chromatin modifications and higher-order structures in the regulation of Glut4.

The potential utility for the use of HDACi in targeting Glut4 came from an initial study which showed that increased Glut4 protein translocation and insulin-induced uptake occurred in muscle cells treated with a HDACi [95]. As shown in Figure 17.2, we have demonstrated that Glut4 expression can be induced in many tissue types including, lung, liver, kidney, muscle, and preadipocytes [93]. Furthermore, levels of Glut4 mRNA transcript could be sustained

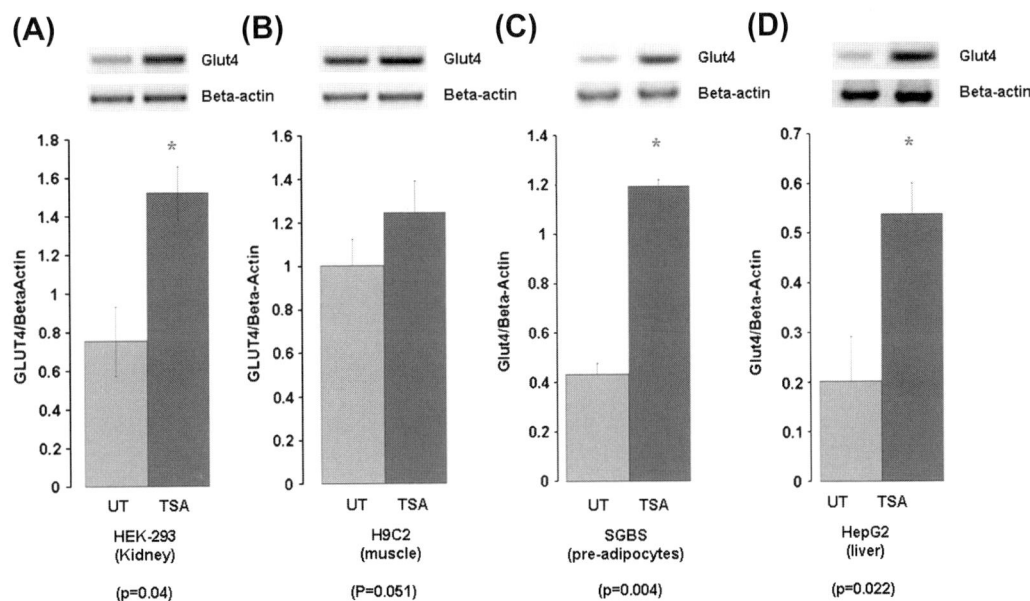

FIGURE 11.2
Effect of HDACi on Glut4. Effects of histone deacetylase inhibitors (HDACi) on Glut-4 in cell lines derived from insulin target tissues. Various cell lines (A) HEK-293 (Kidney), (B) H9C2 (muscle), (C) SGBS (preadipocyte), and (D) HepG2 (liver) were treated with or without the histone deacetylase inhibitor Trichostatin A (TSA), and mRNA expression of Glut-4 was examined by RT-PCR. Densitometric analysis of Glut-4 expression with Beta-actin levels used for normalization purposes was carried out for each. Data are expressed as mean ± SEM. Statistical analysis was performed using a Student's T-test. (UT, untreated; TSA, Trichostatin A). *Previously published in* [93].

following removal of the HDACi [93], indicating that the use of HDACi may provide long-term elevation of Glut4 in insulin target tissues.

17.4.3 Aberrant microRNAs/Epi-microRNAs and Diabetes

Specific miRNAs have now been shown to control the expression of some of the known epigenetic regulators. mIR-34a targets the NAD-dependent histone deacetylase SIRT-1 [96], and is also critical for mediating lipotoxicity induced beta-cell dysfunction [97]. Critically, miR-34a levels are elevated, and SIRT1 protein levels are reduced in diet-induced obese mice [98], indicating that miRNAs that modulate epigenetic regulators may be important in obesity. Another miRNA, mIR-217 has also been shown to target SIRT1 [96,99]. This miRNA is essentially specific to pancreatic cells [100], and as such may be a critical link in diabetes pathogenesis. Another miRNA, miR-132 has been shown to modulate SIRT-1 levels in response to nutrient deprivation [101]. Furthermore, mIR-132 is regulated by hyperglycemia [102], shows dysregulated expression in obese patients [103], and its other known targets include the chromatin modifiers Mecp2, KAT3B (p300), pRB, and KDM5A (Jarid1a) [104,105]. In pancreatic beta-cell islets, the expression of SIRT-1 and mIR-9 are also intimately linked. When mir-9 expression is high during glucose-dependent insulin secretion, levels of SIRT-1 protein are reduced [106].

Other miRNAs directly target the epigenetic regulatory machinery. A family of miRNAs, the miR-29 family, has been shown to target the DNA methyltransferases DNMT3A and DNMT3B [107,108]. In addition to being overexpressed in diabetic rats, overexpression of this miRNA leads to insulin resistance in 3T3-L1 adipocytes, largely through repression of insulin-stimulated glucose uptake [109]. miR-29c has also recently been shown to be a key microRNA in diabetic nephropathy under high glucose conditions where it targets Sprouty homolog 1, and in vivo knockdown of this miRNA prevents progression of diabetic nephropathy [110]. Furthermore, miR-143 has also been shown to target DNMT3A [111], and has been implicated in adipocyte differentiation [112]. This result was confirmed by Lodish and colleagues who further noted that miR-143 was a miRNA whose expression was significantly decreased in the obese state [21]. Two other microRNAs, miR-148a and miR-152, have now been shown to regulate DNMT1 [113]. Intriguingly, in a comparison of miRNA profiles of human mesenchymal stem cells derived from breast, face, and abdominal adipose tissues, mIR-152 was one of a number of differentially expressed microRNAs which separated the breast and abdominal from the facial stem cells [114]. mIR-194 is a further miRNA which has been hypothesized to target DNMT3A [115]. More recently, this miRNA and another miRNA, miR-132, have in fact been shown to regulate the expression of MeCP2 [116], a methyl-binding protein which has been shown to associate with HDACs to regulate transcription.

miR-449a has been shown to target both histone deacetylase 1 (HDAC1) [117], E2F1 [118], and SIRT-1 [119], but has yet to be associated with diabetes pathogenesis.

In another insulin target tissue, miR-1 targets HDAC4 specifically to regulate skeletal muscle proliferation and differentiation [120]. This miRNA may also play a critical role in diabetic cardiomyopathy as under high-glucose conditions, this miRNA is significantly elevated in cardiomyocytes, and accelerates apoptosis by depleting the cells of Hsp60, an important component of the cell's defense mechanisms against diabetic myocardial injury [121].

mIR-27b plays important roles in adipocyte differentiation and has been shown to specifically target PPARγ, a protein known to associate with HDACs [122]. miR-27a has also been shown to target PPARγ, and ectopic expression of miR-27a in 3T3-L1 pre-adipocytes repressed adipocyte differentiation through reduction of PPARγ expression [123]. Another miRNA that regulates PPARγ, miR-519d, was found to be overexpressed in subcutaneous adipose tissue (SAT) from nondiabetic severely obese (n = 20) compared to non-obese adults (n = 8) [124].

The microRNA mIR-101 has been shown to target the lysine methyltransferase KMT6 (aka EZH2) [125]. KMT6 has now been shown to regulate both pancreatic beta cell expression and regeneration in diabetes [126], and also to facilitate adipogenesis [127].

Other miRNAs have been identified which regulate other important pathways on several levels in diabetes and obesity. For instance miRNA-223 has been shown to directly regulate Glut-4 expression [128], while miRNA-133 has been shown to indirectly alter Glut4 expression by regulating KLF15 (MODY7), an essential regulator of Glut-4 [129], and insulin biosynthesis. mIR-696 has now been shown to regulate the transcriptional coactivator PGC-1α in skeletal muscle in response to physical activity [130]. This would indicate that miRNAs may be altered in response to exercise. Intriguingly, levels of critical miRNAs such as mIR-1 (target HDAC4) have been shown to alter in skeletal muscle as an adaptation to acute endurance exercise [131].

Transcription factors strongly associated with diabetes, PDX-1 and NeuroD1, have been shown to regulate the miRNA miR-375 [132]. This microRNA has now been shown to regulate 3'-phosphoinositide-dependent protein kinase-1 and as such glucose-induced biological responses in pancreatic β-cells [133]. mIR-696 has now been shown to regulate the transcriptional coactivator PGC-1α in skeletal muscle in response to physical activity [130].

mIR-103 and mIR-107 have also been shown to be critical regulators of insulin sensitivity via their regulation of caveolin-1. Loss of these miRNAs results in improved glucose homeostasis and insulin sensitivity, while their overexpression in either liver or fat is sufficient to induce impaired glucose homeostasis [134]. In a study of glucose-regulated miRNAs from pancreatic-β cells, Özcan and colleagues demonstrated that in MIN6 pancreatic-β cells, expression of 61 of 108 detectable miRNAs (56%) was altered in response to glucose [135].

Finally, critical roles for miRNAs have now been shown for pancreatic development. miRNA-375 was originally indicated as being a critical regulator of pancreatic islet development [136]. Further studies have now revealed that four different islet-specific microRNAs (miR-7, miR-9, miR-375, and miR-376) are expressed at high levels during human pancreatic islet development [137].

EFFECTS OF INSULIN ON miRNAs

Insulin can also modulate miRNA expression. A comparative microRNA (miRNA) expression profile of human skeletal muscle biopsies before and after a 3-h euglycemic-hyperinsulinemic clamp found that insulin down-regulated the expressions of 39 distinct miRNAs in human skeletal muscle [138].

miRNAS AND DIABETES

One study examining gene and miRNA expression changes in skeletal muscle insulin resistance in patients with type 2 diabetes (average age ~54 years) found that while the muscle mRNA transcriptome is invariant with respect to insulin or glucose homeostasis, a third of miRNAs detected in muscle were altered in disease (n = 62), with many changing prior to the onset of clinical diabetes [139].

EFFECTS OF OBESITY ON miRNAs

In a study profiling approximately 220 miRNAs in pancreatic islets, adipose tissue, and liver from diabetes-resistant (B6) and diabetes-susceptible (BTBR) mice, more than half of the miRNAs profiled were expressed in all three tissues. In addition, many miRNAs in each tissue showed significant changes in response to genetic obesity. In liver there were approximately 40 miRNAs that were down-regulated in response to obesity in B6 but not BTBR mice, indicating that genetic differences between the mouse strains play a critical role in miRNA regulation, and that genetics may play critical roles in miRNA expression [140].

In a separate study of global miRNA expression in human adipocytes during differentiation and in subcutaneous fat samples from non-obese (n = 6) and obese with (n = 9) and without (n = 13) type 2 diabetes mellitus (DM-2) women, approximately 50 miRNAs (6.2%) significantly differed between fat cells from lean and obese subjects. Seventy miRNAs (8.8%) were highly and significantly up- or down-regulated in mature adipocytes as compared to preadipocytes. Seventeen miRNAs (2.1%) were correlated with anthropometrical (BMI) and/or metabolic (fasting glucose and/or triglycerides) parameters, while 11 miRNAs (1.4%) were significantly deregulated in subcutaneous fat from obese subjects with and without DM-2 [141].

In various mouse models of obesity, miR-143 has been shown to impair insulin-stimulated AKT activation and glucose homeostasis, and if mice are made deficient for this miRNA, they are protected from the development of obesity-associated insulin resistance [142]. Furthermore, in preadipocytes overexpression of this miRNA has been shown to increase adipogenesis [21].

EFFECTS OF EPIGENETIC INHIBITORS ON miRNAs

miRNAs themselves have been shown to be epigenetically regulated [143]. Indeed the potential to epigenetically target or modulate miRNAs, which may play a role in diabetes pathogenesis, came from a study on the effects of the histone deacetylase inhibitor (HDACi) trichostatin A (TSA) on miRNA expression in in vivo hepatocyte cultures of rat hepatocytes. In this study mIR-143 was shown to be the most down-regulated miRNA after 7 days of treatment with TSA [144], demonstrating the potential to functionally modulate critical miRNAs in diabetes pathogenesis through HDACi. Another miRNA (mIR-449a) is epigenetically regulated via histone H3 Lys27 trimethylation (H3K27me3), and can be dramatically induced following treatments with a combination of a histone methyltransferase inhibitor (DZNep) and a HDACi (TSA) [118], and while this miRNA has yet to be implicated in diabetes pathogenesis, this result demonstrates the possibility that targeting miRNAs associated with diabetes pathogenesis could be achieved via epigenetic therapies.

17.4.4 Effects of Insulin Resistance/Metabolic Syndrome on Epigenetic Modifying Enzymes

Using peripheral blood mononuclear cells (PBMCs) as an *ex vivo* surrogate model system Avogaro and colleagues examined SIRT1–SIRT7 gene and protein expression in PBMCs of 54 subjects (41 with normal glucose tolerance and 13 with metabolic syndrome). Their analysis showed that SIRT1 levels (both mRNA and protein) were down-regulated in the patients with metabolic syndrome/insulin resistance [145].

17.5 NON-EPIGENETIC EFFECTS OF HISTONE MODIFIER PROTEINS WITH DIABETES PATHOGENESIS

Whilst epigenetics in its purist form relates to the regulation of gene expression through direct modifications of DNA or chromatin, the enzymes involved with this epigenetic regulation also play important roles in regulating various other cellular responses through post-translational modification of non-histone proteins [146]. As such, a fourth mechanism can be envisaged involving a "protein code or language" where the activities of KATs/HDACs and indeed other epigenetic modifying enzymes such as lysine methyltransferases, on such proteins may affect their functions such that the aberrant regulation of target genes in diabetes pathogenesis may also occur [51,147]. This has important implications for using drugs such as HDACi, as in addition to their effects on chromatin, they also elicit important additional effects on other pathways. In the next sections I will detail the potential utility of HDACi to target "non-epigenetic" events on the following important elements of diabetes pathogenesis namely inflammation, and ER stress/chaperone activity.

17.5.1 Inflammation in Diabetes Pathogenesis

Inflammation is a critical component traditionally associated with the destruction of the pancreatic beta cell in type 1 diabetes, but increasingly linked to type 2 diabetes pathogenesis [148]. The major inflammation mechanism associated with diabetes pathogenesis involves the NFκB pathway [23]. The NFκB-Rel family consists of five subunits, but NFκB typically consists of a heterodimeric protein comprised of a p50 and a p65(RelA) subunit. Protein acetylation via lysine acetylases would appear to be a central regulator of NFκB-mediated responses [17]. The lysine acetyltransferases KAT3A (p300) and KAT3B (CBP) act as key coactivators in regulating NFκB-driven gene expression via interactions with the p65(RelA) subunit [149–151]. Likewise, KAT13A (SRC-1) potentiates NFκB transactivation through interactions with the other subunit p50 [152]. Sirt1 also regulates NFκB transactivation by physically interacting with p65(RelA) and inhibiting transcription by deacetylating a critical lysine at position 310 [153].

It has also been shown that NFκB transcription requires IKKα to phosphorylate SMRT which stimulates the exchange of corepressor for coactivator complexes. In this regard, loss of HDAC3 is observed following this phosphorylation event [154].

HDACi have been shown to inhibit NFκB-mediated responses [155]. Importantly, in relation to diabetes pathogenesis HDACi have also been found to dampen NFκB transactivation in studies involving pancreatic beta cells (INS-1) [156].

Two proinflammatory genes associated with diabetes are IL-6 and Il-8 [157,158]. NFκB has also been shown to utilize the lysine acetyltransferase activity of KAT3A/KAT3B (CBP/p300) to stimulate transcription of these genes [159] (Figure 17.3).

Initial observations had linked the inhibitor kappa B kinase complex (IKK) to the development of insulin resistance through phosphorylation of IRS-1 by IKKβ leading to impaired downstream signaling of PI3kinase pathways [160]. Subsequently, two IKKβ-mediated inflammation models have now directly linked IKKβ-mediated inflammatory responses to the

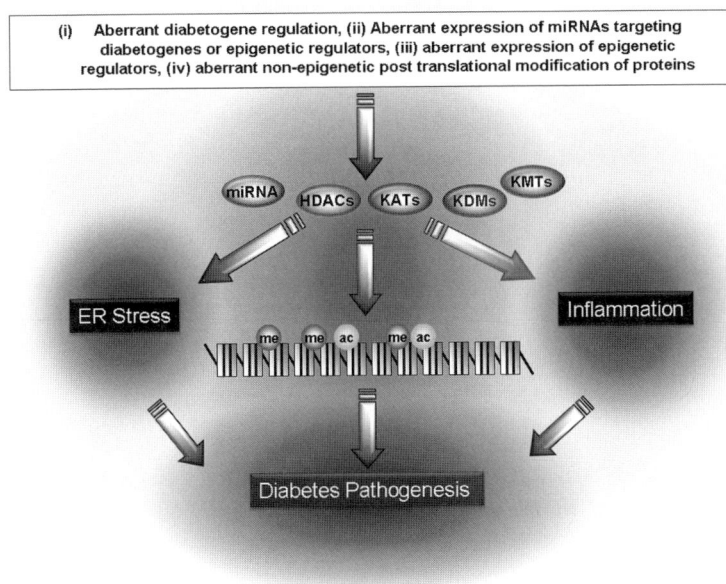

(i) Aberrant diabetogene regulation, (ii) Aberrant expression of miRNAs targeting diabetogenes or epigenetic regulators, (iii) aberrant expression of epigenetic regulators, (iv) aberrant non-epigenetic post translational modification of proteins

FIGURE 17.3

Simplified overview of how HDACi could target diabetes. Diagram summarizing the available evidence linking aberrant epigenetics such as (i) aberrant diabetogene regulation, (ii) aberrant expression of miRNAs, (iii) aberrant expression of epigenetic regulators, (iv) aberrant non-epigenetic post-translational modification of proteins, resulting in inappropriate gene expression, cellular and tissue dysfunction, and the development of diabetes. This figure is reproduced in the color plate section.

development of obesity-induced insulin resistance [161,162]. Other evidence linking the NFκB pathway to HATs/HDACs have come from studies which demonstrate that the sequestration of the p65 subunit of NFκB by the I-kappa-B alpha (IKBA) protein also resulted in the cytoplasmic translocation of the nuclear corepressors N-CoR and SMRT [163], and from data demonstrating that diabetes-induced extracellular matrix protein expression was associated both with increased histone acetyltransferase activity and NFκB activation [164].

The proinflammatory cytokine IL-1β has been described as the gatekeeper of inflammation [165]. Indeed this cytokine mediates important events during type 1 diabetes pathogenesis both in vivo and in vitro including the induction of beta-cell islet apoptosis where it reduced the number of docked insulin granules in live pancreas beta cells by 60% [67]. It has also been demonstrated that IL-1β utilizes NFκB to induce a sustained decrease of specific beta-cell proteins like insulin, GLUT-2 and PDX-1, and regulates the entrance of islet beta-cells into the cell death program [166]. It has now been shown that in pancreatic beta cells and isolated rat islets proinflammatory cytokines (including IL-1β) altered the expression of HDAC- 1, -2, -6, and -11, and that furthermore this regulation was reduced upon the use of HDACi [167].

It is clear from the above that NFκB-mediated inflammation is central to pancreatic beta cell loss, and that epigenetic modifications may be essential to this process.

17.5.2 ER Stress/Chaperones and Diabetes Pathogenesis

Chaperones are important proteins which function to assist the non-covalent folding/unfolding and the assembly/disassembly of other macromolecular structures, but do not occur in these structures when the latter are performing their normal biological functions. The first identified chaperones assist in the correct assembly of nucleosomes, but we now know that such chaperones also contribute to (i) the complex balance between nucleosome assembly and reassembly during transcription, and (ii) may equally be involved as much in histone eviction as in chromatin assembly [168]. Chaperones also act to prevent newly synthesized polypeptide chains and assembled subunits from aggregating into non-functional structures. Many chaperones are heat shock proteins (HSP), which are proteins expressed in response to elevated temperatures or other cellular stresses. Two important families of these chaperones are HSP70 [169], and HSP90 [170,171]. Many of these chaperones can be found in the endoplasmic reticulum (ER), a large organelle consisting of a network of interconnected, closed membrane-bound vesicles. The ER is the site for the synthesis, folding, and modification of secretory and cell-surface proteins and provides many essential cellular functions, including the production of cell membrane components, proteins, lipids, and sterols [172]. Only correctly folded proteins are transported out of the ER, and incompletely folded proteins are retained within the ER to either complete the folding process or to be targeted for destruction [173]. Due to its important roles this organelle is vital to cellular homeostasis.

Intriguingly some HSP proteins have also been described as glucose inducible. These are Glucose-Regulated Protein, 78-KD, GRP78 (also known as HSPA5) [174]; Glucose-Regulated Protein, 94-KD; GRP94 (also known as HSP90 Beta 1) [175]. This may have important implications in the pathogenesis of diabetes, where ER stress and glucose homeostasis play important roles [176–178].

The maintenance of cellular homeostasis requires that a cell be able to sense, and respond to circumvent, any stress elicited. There are many ways by which endogenous or exogenous stress can occur in a cell; these include pathogenic infection, chemical insult, genetic mutation, nutrient deprivation, excess nutrients, and even normal differentiation. The process of mutant protein folding is particularly sensitive to such insults. Adaptive programs exist in the cellular compartments responsible for proteins procession and folding, which enable the detection and correction of folding errors [179].

However, many conditions can interfere with normal ER function leading to a situation called ER stress. Thus, ER stress can arise from a disturbance in protein folding which results in an accumulation of unfolded or misfolded proteins within the organelle [180]. If this occurs, the ER has evolved specialized mechanisms that promote proper folding of aberrant protein, thus preventing its aggregation. Simply put, when ER stress occurs, the ER responds by inducing the expression of specific genes in an attempt to restore normal function and to maintain cellular homeostasis [181]. The principal mechanisms of endoplasmic reticulum (ER) stress responses involve: (1) ER-Associated protein Degradation (ERAD), (2) Endoplasmic overload response (EOR), (3) Unfolded protein response (UPR), and (4) Cellular Death pathway [182–185]. This four-stage model defines the role of ER stress in the onset of clinical manifestations. Two ER stress-induced signal transduction pathways have been described: the UPR [183] and the EOR [186] which attempt to re-establish normal ER function [187]. However, excessive or prolonged ER stress may overwhelm the cell and subsequently initiate cell death via apoptosis or autophagy [188]. Critically, ER stress has been implicated as a major element in diabetes pathogenesis [177,189–192].

LINKING HATs/HDAC ACTIVITIES TO CHAPERONE ACTIVITIES

Proteins with chaperone activity have now been shown to functionally associate with complexes containing HATs/HDACs. Early coimmunoprecipitation studies linked the class I HDACs (HDACs1-3) with the chaperone HSP70, while HSP60 was found to only be associated with HDAC1, and HSP90 was not found associated with any of the class I HDACs [193]. Recently specific chaperones have been identified in a multifunctional KAT–chaperone complex that both acetylates histone H3 and deposits histones H3 and H4 onto DNA, thus functionally linking chaperone activity, histone modification, and nucleosome assembly during replication, chromosome positioning, and double-strand break (DSB) DNA repair [55,168,194–213], and that lysine acetyltransferase activity is important in stabilizing the chaperone/histone H3/H4 complex [214]. The Jun dimerization protein-2 (JDP2) has been shown to have both intrinisic chaperone activity and the ability to inhibit KAT3B (p300)-mediated acetylation of core histones in vitro and in vivo. This factor may therefore control transcription via direct regulation of the modification of histones and the assembly of chromatin [215]. Another protein with histone chaperone activity nucleophosphin, has been shown to be acetylated by KATs both in vitro and in vivo, and this acetylation enhances its ability to regulate chromatin transcription [216], and is in agreement with the previous observations by Turner and colleagues that the abilities of HDACs to remodel chromatin are enhanced by chaperones [193].

Furthermore, regulation of various receptors and signaling pathways often affected in diabetes requires coordinated acetylation of histone chaperones including androgen receptor signaling [217], aryl hydrocarbon receptor (AhR) signaling [218], estrogen receptor (ER) signaling [219,220], glucocorticoid receptor (GR) signaling [221], and the AKT/PI3Kinase signaling pathway [222].

LINKING KATs/HDAC ACTIVITIES TO ER STRESS PATHWAYS

Aggresomes are a cellular response to misfolded protein accumulation as a consequence of ER stress. Emerging evidence links both the activities of KATs/HDACs to ER stress, and further evidence indicates that agents which target these proteins may have utility in alleviating ER stress (Box 17.1). For instance, during cytokeratin aggresome (Mallory bodies) formation in hepatocytes, decreased lysine acetyltransferase and increased histone deacetylase activity was observed [223], while in a model of oxidative stress-induced inclusion formation, treatment with the HDACi 4-phenylbutyrate was found to alleviate the formation of these inclusions [224].

Further direct physical evidence for the association of KATs and HDACs with critical regulatory elements within the ER stress pathway is emerging. CHOP (C/EBP Homologous Protein) is an ER stress-inducible protein which plays a critical role in regulating programmed cell death. Regulation of the CHOP promoter involves a complex containing JDP2 and HDAC3 [225]. In addition, CHOP itself directly associates with the histone acetyltransferase KAT3B (p300) to regulate gene expression responses in response to ER stress [226], and importantly, inhibition of HDACs prevents the degradation of the CHOP protein [226].

BOX 17.1 PATHWAYS LINKING KATs/HDACS TO DIABETES

Direct Associations
GWAS associations (6q21 harbors HDAC2) linked to type 1 and type 2 diabetes
MODY forms of diabetes linked directly to chromatin remodeling activities (see Table 17.1)
Mouse mutant models of chromatin remodeling enzymes have increased levels of adiposity and markers related to diabetes pathogenesis

Pancreas Development
A large body of data links pancreatic development with chromatin modifiers such as HDACs

Stem Cell Differentiation
A significant body of evidence suggests HDACi are useful in differentiating stem cells

Inflammation
Many roles for KATs/HDACs have been identified in the regulation of inflammatory processes including the regulation of NFκB

Diabetogenes
Many genes associated with diabetes pathogenesis are regulated by HDACs

Insulin Resistance/Sensitivity
Genes such as Glut4 can be induced via HDACi. HDACi can enhance insulin-induced glucose uptake

ER Stress/Chaperone
HDACi such as phenylbutyrate can act as chemical chaperones to alleviate ER stress in the pancreas

In the context of diabetes one of the central regulators of ER stress responses is a glucose-responsive gene, Glucose Regulated Protein 78 (GRP78 or BiP). This prosurvival ER chaperone is induced under conditions of ER stress, and has been shown to be regulated via histone PTMs and chromatin remodeling through the activities of protein arginine methyltransferases (PRMT1) and K-acetyltransferases (KAT3B) [227]. Similar results have been observed for other ER-stress responsive promoters, whereby increased histone H4 hyperacetylation was observed following induction of ER stress [228]. The B lymphocyte-induced maturation protein-1 (BLIMP-1) is another protein associated with cellular stress. In some cellular models BLIMP-1 is rapidly up-regulated during the UPR [229]. This repressor protein has been shown to directly associate with histone deacetylases to repress transcription [230], and this may indicate that BLIMP-1 may utilize histone deacetylases to repress important genes during ER stress.

Hepatic regulation of lipid and glucose homeostasis has been shown to involve the activities of an ER stress inducer, activating transcription factor 6 (ATF6). In response to ER stress caused by glucose starvation, ATF6 translocates from the ER to the golgi, where it is cleaved by S1P and S2P proteases. It then translocates to the nucleus whereupon it abrogates SREBP2-mediated sterogenesis and lipogenesis by forming a repressive complex with SREBP and HDAC1 [231].

As previously stated the GR signaling pathway has been shown to be affected by the activities of HDACs and chaperones. GR signaling is critical for maintaining glucose homeostasis in stress, inflammation, and during fasting [232], and a multiprotein HSP90/HSP70-based chaperone complex has been shown to be critical for regulating the steroid binding, trafficking, and turnover of the GR [233]. Recent studies have shown that the acetylation status of HSP90 is critical to GR function. In cells deficient for HDAC6, HSP90-dependent maturation of the glucocorticoid receptor (GR) is compromised, resulting in GR defective in ligand binding, nuclear translocation, and transcriptional activation [221]. Additional studies using an siRNA "knock-down" approach found that depletion of HDAC6 resulted in a markedly decreased ability to assemble stable GR/HSP90 heterocomplexes [234].

In addition, the GR has been shown to functionally associate with HDACs to mediate repression of gene expression [235–237]. This has important implications in preadipocyte differentiation where glucocorticoid treatment strongly potentiates differentiation, by stimulating the titration of the corepressor histone deacetylase 1 (HDAC1) from the C/EBPβ promoter [238,239].

Clearly, the activities of histone/protein-modifying enzymes such as lysine acetyltransferases and histone deacetylases play important roles in ER stress-mediated responses, and may functionally contribute to diabetes pathogenesis. It is also clear that targeting these enzymes may also have some potential in alleviating the ER stress associated with diabetes pathogenesis. The following section details how targeting HDACi in particular may be of benefit in the treatment/management of diabetes.

17.6 POTENTIAL FOR THE USE OF HDACi TO AMELIORATE OR TREAT SYMPTOMS OF DIABETES PATHOGENESIS

HDACi has the potential to be used to ameliorate or treat symptoms of diabetes pathogenesis. We will now discuss this further.

17.6.1 Currently Developed Drugs

Due to their highly conserved active domain, histone deacetylases have been extensively studied for the development of inhibitors. Most of the currently designed inhibitors fall into four broad classes, short-chain fatty acids (SCFAs), hydroxamates, benzamates, and cyclic tetrapeptides, which for the most part target the class I and class II HDACs [240]. Several inhibitors of the class III HDACs (Sirtuins) have now been synthesized and also show therapeutic potential [241—246]. Additionally natural prodrugs which target histone de-acetylases have also been isolated and include sulforaphane (SFN), diallyl disulfide (DADS), and resveratrol [247—249].

17.6.2 Therapeutic Implications

Many histone deacetylase inhibitors have entered clinical trials primarily for the treatment of cancer. Generally, most of the current inhibitors are well tolerated within the clinical setting [23,250,251]. Several old drugs which have recently been discovered to target histone deace-tylases already have FDA approval (e.g. valproic acid (VPA) [252]), and recent clinical trials of this drug targeting solid tumors both alone and in combination with other agents demon-strated reasonable toxicities [253—255]. One of the caveats for the use of VPA in the treatment of diabetes is that known side effects of this drug are increased obesity and insulin resistance [75,256—262]. This may be due to VPAs effects on leptin. In adipocytes it was found that VPA reduced leptin mRNA levels while TSA did not, suggesting that VPA therapy may be associated with altered leptin homeostasis contributing to weight gain in vivo, and that other HDAC inhibitors may not cause similar effects in relation to obesity and insulin resistance [263]. VPA was also found to decrease the expression of adiponectin in differentiated 3T3-L1 adipocytes and this may therefore explain the increased insulin resistance in patients treated with VPA [75]. It has also been suggested that the hyperinsulinemia associated with VPA can be attributed to inhibition of insulin metabolism in the liver [262].

Another therapeutic approach to diabetes involves using salicylates to reduce both insulin resistance [264,265], and protect pancreatic beta cells from apoptosis and impaired function [266]. Salicylates are traditionally thought of as inhibitors of cyclooxygenases, however an early report also indicates that these drugs may also affect histone acetylation [267], more recently confirmed in studies examining embryo development where treatments with sodium salicylate were found to inhibit HDACs activity [268]. In a recent study on the efficacy of aspirin in treating a rat model of diabetes (STZ), the authors found that aspirin treatments resulted in a significant reduction of hepatic NFκB activation and serum TNF-α levels with improved insulin resistance compared to the diabetic group [269].

HDACi have been shown to have effects on adipocyte differentiation. Treatment of 3T3-L1 cells with HDAC inhibitors (apicidin, TSA, or SAHA), led to a dramatic reduction in preadi-pocyte differentiation into adipocytes, however, in contrast, sodium butyrate (NaB) treatment increased adipocyte differentiation [270,271].

Two recent studies have demonstrated that HDACi also have the capacity to ameliorate dia-betic nephropathy. In one study, long-term administration of vorinostat decreased albumin-uria, mesangial collagen IV deposition, and oxidative-nitrosative stress in STZ diabetic mice, and attenuated renal injury [272], while the other study examining one of the earliest features of diabetic nephropathy (renal enlargement), found that vorinostat was able to blunt renal growth and glomerular hypertrophy in STZ-induced diabetes [273].

17.6.3 Pancreatic Islet Development, and Protection Using Histone Deacetylase Inhibitors, or NFκB Blockade

Several recent publications have demonstrated the ability of histone deacetylase inhibitors to protect pancreatic beta cell apoptosis. In various diabetes models, nicotinamide has frequently been observed to both ameliorate and/or accelerate the reversal of diabetes and prevent irre-versible B-cell damage [274—277]. However, the ENDIT clinical trial assessing whether the pretreatment with nicotinamide of non-diabetic individuals predisposed to the development of diabetes could prevent or delay clinical onset of diabetes was ineffective at the dose used [278], but did however reduce high secretion of IFN-γ in high-risk individuals [279]. Insulin-secreting cells exposed long term to either nicotinamide or sodium butyrate were found to have reduced viability and insulin sensitivity, yet enhanced insulin secretory responsiveness to a wide range of beta cell stimulators [280]. Both trichostatin A (TSA) and suberoylanilide hydroxamic acid (SAHA) were shown to prevent cytokine-induced toxicity in pancreatic beta cells [156], while another HDACi (THS-78-5) has been shown to prevent IL-1β-induced metabolic dysfunction in pancreatic beta-cells [71]. Most recently, the HDACi ITF2357 was shown to normalize streptozotocin (STZ)-induced hyperglycemia at the clinically relevant doses of 1.25−2.5 mg/kg, and protects pancreatic β cells in vivo and in vitro [70].

Islet transplantation is a widely pursued potential therapy for the treatment of patients with diabetes. However, currently this is limited by difficulties in their isolation from donors and maintenance in culture. A novel mechanism to enhance this has recently been elicited whereby HDACi (TSA and VPA) were found to rejuvenate isolated islets, enhancing the expression of key genes (including insulin 1, insulin 2), with increased glucose-stimulated insulin secretion and engraftment efficacy in xenogeneic transplantation model [281].

17.6.4 Modulation of Th17-Mediated Autoimmunity

Effector T-cells show great plasticity. Th17 cells are acknowledged to be instrumental in the response against microbial infection, but are also associated with autoimmune inflammatory processes particularly with type 1 diabetes development, and now also with type 2 diabetes [282−287]. The HDACi TSA dramatically reduced the emergence of IL-17-producing cells from Tregs [288], implying (a) that Treg differentiation into IL-17-producing cells depends on histone/protein deacetylase activity and (b) that HDACi may be useful in dampening down Th17-mediated autoimmunity in diabetes pathogenesis. In agreement with this, treatment of a NOD mouse model of autoimmune diabetes during the transition from prediabetic to diabetic stage with a HDACi (TSA) was found to effectively reduce the incidence of diabetes [289].

17.6.5 Stem Cells

A currently hotly pursued therapeutic avenue for diabetes centers on embryonic stem (ES) cell technology [290]. Evidence is emerging indicating that histone deacetylases may be an important consideration in the development of this technology. Indeed histone deacetylase activity has been shown to be required for ES cell differentiation [291]. The importance of HDAC inhibitors in differentiating embryonic stem cells in general has been reviewed recently elsewhere and the reader is directed to the following reviews [292,293].

Nicotinamide, a SIRT-specific inhibitor, was also used to differentiate embryonic stem cells into structures resembling pancreatic islets and which secreted insulin [294]. Indeed a role for SIRT1 is the differentiation of stem cells. Levels of SIRT1 are precisely down-regulated during human embryonic stem cell differentiation at both mRNA and protein levels by a pathway at the arginine methyltransferase coactivator-associated arginine methyltransferase 1 (CARM1), and it is this down-regulation that leads to the reactivation of key developmental genes [295]. Under appropriate culture conditions, bone marrow stem cells (BMSC) cultured in the presence of the histone deacetylase inhibitor trichostatin A can be differentiated into islet-like clusters similar to the cells of the islets of the pancreas and capable of secreting insulin [296]. Using BMSCs derived from diabetic patients, similar results were obtained using nicotinamide as one of the final steps in the differentiation process [297].

Two recent articles have utilized the histone deacetylase inhibitor sodium butyrate to (a) stimulate early pancreatic development in embryonic stem cells [298], and (b) generate islet-like clusters from human embryonic stem cells grown under feeder-free conditions [299].

Human islet-derived precursor cells (hIPCs), are mesenchymal stem cells derived in vitro from adult pancreas,and because they can be expanded prior to differentiation they offer the potential to be a major source of transplantable islet cells if they can be differentiated successfully into insulin-producing cells. However, an issue with most hIPCs is that during proliferative expansion, insulin mRNA transcript becomes undetectable and cannot be induced, a phenomenon consistent with epigenetic silencing of the insulin gene [300,301].

In an analysis of the chromatin from hIPCs, patterns of histone modifications over the insulin gene in human islets and hIPCs were compared against HeLa and human bone marrow-derived mesenchymal stem cells (hBM-MSCs), neither of which expresses insulin. The insulin gene in islets were found to display high levels of histone modifications (H4 hyperacetylation and dimethylation of H3 lysine 4) typical of active genes which were not present in HeLa and hBM-MSCs, which instead had elevated levels of H3 lysine 9 dimethylation (H3K9me2), a mark of inactive genes. hIPCs also showed significant levels of active chromatin modifications, as much as half those seen in islets, and show no measurable H3 K9me2. If cells were expanded from a minor population of mesenchymal stromal cells found in islets, these cells exhibited the same histone modifications as established hIPCs. The authors therefore conclude that hIPCs, which do not express the insulin gene, nonetheless uniquely exhibit epigenetic marks that could poise them for activation of insulin expression [300]. Even with the absence of insulin expression transplanted hIPCs in mice matured into functional cells that secreted human C-peptide in response to glucose, and transcripts for insulin, glucagon, and somatostatin in recovered grafts increased with time in vivo, reaching levels approximately 1% of those in adult islets [302].

When taken together, it is clear that studies with epigenetic inhibitors such as HDACi are warranted in hIPCs and other stem cells as they may be of great utility in enhancing differentiation of patient stem cells into islet cells, which could be then used for transplantation.

17.6.6 The ability of phenylbutyrate to act as a chaperone and alleviate ER stress

A significant body of work has shown that phenylbutyrate and valproate can both function as "chemical chaperones" to alleviate ER stress. For example, 4-PBA relieves ER stress in cell line models of cystic fibrosis [303,304], alpha-1 antitrypsin Z protein (alpha1-ATZ) liver disease [305], and may act as a chemical chaperone to relieve ER stress induced in models of ischemia [306,307], cataract formation [308], and retinitis pigmentosa [309]. PB also prevents ER stress-mediated aggregate formation in a model of hereditary haemochromatosis (HH) [310], and alleviates ER stress in Parkinson's disease [311]. VPA also alleviates ER stress, as cells treated with VPA have up-regulated expression of Grp78/BiP, Grp94, and calreticulin [312,313] resulting in neuroprotection [314].

In a cell line model of misfolded low-density lipoprotein receptors, PB was shown to restore receptor functionality, and shuttle them to the cell surface. The authors concluded from these results that phenylbutyrate did not just solely mediate this response by its ability to induce gene expression of proteins involved in intracellular transport, but could also mediate this effect via a direct chemical chaperone activity [315].

Phenylbutyrate has also been shown to alleviate ER stress within the diabetic setting. For instance, obesity is a major factor in the development of diabetes [316], and the adipose-derived hormone, leptin, is a critical regulator which functions to inhibit food intake and prevent body weight gain [317,318]. However, the use of leptin as a therapy for the treatment of obesity has been hampered by the fact that the majority of obese patients demonstrate "leptin resistance", leading to the notion that leptin resistance may be one of the main causes of obesity [318]. Several reports have now shown that ER stress is a major factor underpinning leptin resistance, and treatment with 4-phenylbutyrate has been shown to alleviate the ER stress and restore leptin sensitivity [319—321]. Increased glycemia and reduced melatonin (Mel) levels have been recently shown to coexist in diabetic patients at the end of the night period. In a rat model mimicking this situation, the absence of melatonin induced night-time hepatic insulin resistance and increased gluconeogenesis due to stimulation of

349

nocturnal unfolded protein response, which could be alleviated using phenylbutyrate [322]. Furthermore, 4-phenybutyrate has also been shown to both relieve ER stress and restore glucose homeostasis by the restoration of systemic insulin sensitivity, resolution of fatty liver disease, and enhancement of insulin action in liver, muscle, and adipose tissues in a mouse model of type 2 diabetes [323]. This has led to the instigation of a phase IV clinical trial (NCT00533559) by the University Health Network, Toronto, to examine the effect of Buphenyl® (phenylbutyrate) on fatty acid-induced impairment of glucose-stimulated insulin secretion, which has recently been completed and this small study involving eight patients found that Buphenyl was able to partially ameliorate both insulin resistance and β-cell dysfunction in these patients, demonstrating the potential utility of this HDACi/chemical chaperone in the treatment of diabetes [324]. Finally treatment of experimentally induced diabetic mice that had undergone islet transplants with phenylbutyrate postoperatively was found to enhance islet engraftment with a higher cumulative cure rate of diabetes (p<0.001) than the control group [325]. As such phenylbutyrate would appear to be a good candidate HDACi with potential clinical efficacy in the treatment of diabetes. One orally available form of this compound (triButyrate®) has been used since the 1980s for the treatment of children with inborn errors of urea synthesis/urea cycle disorders, with a recommended dosage of 450−500 mg/kg bodyweight per day.

17.6.7 Is HDACi Specificity an Issue?

One potential problem with the currently available HDACi concerns the degree of specificity they have to the various HDACs, particularly given that they are not redundant in their biological activity. For instance Vorinostat is known to preferentially inhibit HDACs -1, -2, -3, and 6. In contrast, MGCD0103 (Mocetinostat) primarily targets human HDAC1 and HDAC2 enzymes with weaker inhibition of HDAC3 and HDAC11 and does not inhibit HDAC4, HDAC5, HDAC6, HDAC7, or HDAC8 [326]. MS-275 (Entinostat) is another HDACi that specifically targets the class I HDACs, preferentially inhibiting HDAC 1 compared with HDAC 3 and has little or no effect against HDACs 6 and 8. A HDAC8-specific compound has been developed PCI-34051 [327], but there is currently no evidence linking this HDAC isoform to diabetes pathogenesis. A specific SIRT1 inhibitor compound (CHIC-35) was identified [328], but it is generally perceived that activators of SIRT1 may have utility in the treatment of diabetes [329].

17.6.8 Compounds Which Target Lysine Methylases/Demethylases

Various compounds currently in development are those which are targeting KMTs and KDMs. One such compound is DZNep (3-deazaneplanocin A), a global lysine methyltransferase inhibitor [330]. One of the targets of DZnep is KMT6 (also known as EZH2), and this lysine methyltransferase has been shown to be critical for pancreatic beta-cell regeneration [126]. Therefore, targeting KMTs may not be a suitable therapeutic approach in the treatment of diabetes. As previously discussed in earlier sections, KDMs also may be suitable targets for therapeutic intervention in diabetes. Several inhibitors have been generated which target KDMs [331]. However studies would perhaps indicate that agonists of KDMs as opposed to inhibitors may have more utility in the management of diabetes [332].

Overall, while there is evidence that targeting KDMs and KMTs may have potential benefit more studies will essentially be required to determine whether or not agents which target these proteins will have a role to play in the management of diabetes.

17.6.9 Nutrition-Based Natural Compounds as Therapeutic Agents

As many of the features of type 2 diabetes can be alleviated with good diet and exercise, it must also be noted that dietary histone deacetylase inhibitors exist. Therefore nutrition-based intervention may be an excellent mechanism to epigenetically target patients at risk of developing diabetes, or treat patients for whom diabetes-related comorbidities or complications prevent or reduce their ability to exercise. A large body of evidence exists showing that many natural compounds can inhibit the epigenetic machinery, and include well-known compounds which can act as HDACi [333—336].

SULFORAPHANE

Sulforaphane is an organic isothiocyanate (ITC) found in cruciferous plants such as broccoli, and within the diabetic setting, sulforaphane has been shown to (a) protect β-cell damage by suppressing NFκB pathways [337], and (b) in studies of islet transplantation, pre-treatment with sulforaphane 24 hours prior to transplantation was found to lead to improved long-term islet function in vivo [338]. Some individuals caution against the use of such dietary compounds as these may not be specific enough and may tend to target proteins more essential to an organism than specific disease genes [339]. Nevertheless a nutrition-based approach for delivery of specific targeting therapies may have potential benefit in patients suffering from diabetes. Indeed a recent study found that sulforaphane when used at nutritional levels protected mesenchymal stem cells from apoptosis and senescence and promoted their proliferation [340].

CURCUMIN

One of the most intensively studied natural compounds for its potential role in treating diabetes is the compound curcumin. Derived from turmeric, this compound shows many pleiotropic effects, one of which is to inhibit histone deacetylases [341,342]. Conversely it has also been shown to inhibit lysine acetyltransferases [342—344].

Curcumin has antidiabetic effects in adipocytes, pancreatic cells, hepatic stellate cells, macrophages, and muscle cells, reversing insulin resistance, hyperglycemia, and hyperlipidemia, by suppressing the proinflammatory transcription factors such as NFκB and activates PPARγ leading to the down-regulation of adipokines, and the up-regulation of adiponectin and other gene products [345].

Indeed one of the first case studies of the use of this compound in patients was in a diabetic patient almost four decades ago [346]. In a more recent small patient study, ingestion of 6 g of a curcumin preparation increased postprandial serum insulin levels, but had no significant effect on overall glucose response as measured by an oral glucose tolerance test [347]. One of the ongoing issues with the use of curcumin in humans concerns its poor absorption and the large quantity needed to be effective. However, it has been shown that curcumin has increased bioavailability when combined with phospholipids. When complexed with soy lecitihin curcumin (Meriva®) has 29-fold increased absorption in human patients [348]. Other technologies such as nanoparticles have also been shown to improve the delivery of curcumin [349]. A phase I clinical trial involving nano-curcumin has been initiated in patients with advanced malignancies to identify the maximum dose limits (Clinicaltrial.gov NCT01201694).

Taken together further studies with curcumin are warranted to determine whether a nutritional-based compound such as curcumin may have either chemopreventative benefit in patients at risk of developing diabetes, or may have a role in the management of patients with prediabetes or type 2 diabetes.

RESVERATROL

Resveratrol is another natural compound which has been extensively studied for its potential utility in the management of diabetes. This compound is thought to be an activator of Sirt1 [248], but this has since been called into question [350]. Potential beneficial effects of resveratrol include protection of β cells in diabetes, reduction of insulin secretion, inhibition of cytokine action, and attenuation of the oxidative damage of the pancreatic tissue by resveratrol, and in studies of animals with insulin resistance resveratrol may also improve insulin action involving reduced adiposity, changes in gene expression, and changes in the activities of some enzymes [351].

Recent studies have also evaluated safety and potential mechanisms of activity following multiple dosing, and have found resveratrol to be safe and reasonably well-tolerated at doses of up to 5 g/day, although it is anticipated that the doses used in future trials will be significantly less than this amount. In 2008, an unpublished phase 1b clinical trial of a reformulated version of resveratrol (SRT501) in patients with type 2 diabetes, testing either 1.25 or 2.5 grams of SRT501 given twice daily, found that the patient group receiving 2.5 grams twice a day had significantly lower blood glucose levels as determined through an oral glucose tolerance test (OGTT) at the test's 2-hour timepoint, as compared with the placebo group. Several clinical trials are currently ongoing or have been completed using resveratrol or SRT501 in the diabetic setting (Clinicaltrials.gov identifiers NCT01158417, NCT01375959, NCT01354977, NCT01150955, NCT00823381, NCT00998504, NCT01038089), but to my knowledge the results of these trials have as yet to be published. It must also be noted that the clinical development of SRT501 by GlaxoSmithKline PLC has been discontinued due to the development of kidney complications in malignant myeloma patients undertaking a small clinical trial of this drug. As such, care must be taken in evaluating the potential benefits of resveratrol supplements in patients with diabetes, particularly given that resveratrol is also available as supplement pills and liquids, in which it is sometimes combined with vitamins and/or other ingredients. The supplements are generally labeled as containing from 20 to 500 mg per tablet or capsule. However, the purity of these products is unknown, and because dietary supplements are loosely regulated, it should not be assumed that the labeled dosage is accurate.

Clearly, further work will be required to establish whether resveratrol may be an effective treatment for diabetes.

GlaxoSmithKline PLC have instead focused on the development of new small-molecule chemical entities that activate SIRT1 that are structurally distinct from resveratrol. Three compounds have been identified which have entered various clinical trials (SRT2104, SRT2379, and SRT3025). In preclinical studies, SRT2104 demonstrated a moderate lowering of blood glucose, reduced weight gain, increased energy expenditure, improved exercise tolerance, and improved insulin sensitivity in the diet-induced obesity model of type 2 diabetes. Two phase II clinical trials of this compound have been completed in patients with diabetes (ClinicalTrials.gov identifiers NCT01018017, NCT01031108), but to my knowledge the results of these have yet to be published.

17.6.10 miRNA-/siRNA-Based Therapeutics

Given the emerging body of evidence linking aberrant miRNA expression with diabetes pathogenesis, it is becoming increasingly clear that these ncRNAs may have utility in either the management or treatment of this disease. The possibility of using miRNA and/or siRNA to target diabetes has recently been extensively reviewed [352,353]. For example, during the differentiation of human embryonic stem cells to hepatocytes, the cells are induced into definitive endoderm (DE) cells before being further differentiated to hepatocytes. At the DE stage it has been shown that levels of mIR-9 are significantly elevated [354]. As the pancreas is an endodermal tissue and human embryonic stem cells can be induced to form pancreatic endoderm [355], it may be possible to induce hESCs to form pancreatic endoderm through overexpression of this miRNA.

Clearly, this is an exciting area of epigenetic regulation of gene expression which may have great utility in the treatment of diabetes, and will require further studies. One area which will warrant further studies will be to examine whether epi-miRNAs (those miRNAs which target the epigenetic machinery) can be used to epigenetically target diabetes. As stable miRNAs are found circulating in patient serum either freely or contained in exosomes [356], the nature and role of these in diabetes pathogenesis and the potential to utilize these to treat diabetes pathogenesis is increasingly becoming important. Furthermore, those miRNAs which are aberrantly overexpressed in diabetes could also be targeted using antagomirs [357].

17.7 CONCLUSIONS

It is clear that epigenetic regulation is a central tenet of critical pathways in diabetes pathogenesis.

From the significant body of work presented in this review, it is clear that further studies will be required to examine the therapeutic potential of compounds which target the epigenetic machinery in the treatment/management of diabetes pathogenesis. In the short term, HDAC inhibitors would appear to be the current best candidates. However, as these inhibitors affect many genes it is hoped that the balance of genes altered by such treatments would be tipped from unfavorable to favorable diabetogenes. Indeed the next generation of these inhibitors may have better specificity and efficacy. Currently miRNA/siRNA-based therapeutics are in their infancy, but show great promise for the future. A greater understanding of miRNA target identification/validation, their roles in diabetes pathogenesis, and mechanisms of specific delivery will be required before they can be evaluated as a therapeutic modality. Finally, as nutritional compounds have been shown to act as HDACi, future studies will be required to test the possibility of incorporating nutritional-based interventions as an adjuvant or chemopreventative option in the management and treatment of diabetes.

References

[1] American Diabetes Association. Standards of medical care in diabetes − 2011. Diabetes Care 2011;34(Suppl. 1): S11−61.

[2] Encyclopedia ADAMM. Diabetes. Atlanta (GA): A.D.A.M., Inc; 2011.

[3] WHO. Diabetes Fact sheet N°312. World Health Organization; 2011.

[4] CDC. National diabetes fact sheet: national estimates and general information on diabetes and prediabetes in the United States, 2011. In: Prevention CfDCa, editor. Department of Health and Human Services, Centers for Disease Control and Prevention. Atlanta, GA: U.S; 2011.

[5] IDF. Diabetes Atlas. International Diabetes Federation; 2009.

[6] Denis H, Ndlovu MN, Fuks F. Regulation of mammalian DNA methyltransferases: a route to new mechanisms. EMBO Rep 2011;12:647−56.

[7] Chen ZX, Riggs AD. DNA methylation and demethylation in mammals. J Biol Chem 2011;286:18347−53.

[8] Jenuwein T, Allis CD. Translating the histone code. Science 2001;293:1074−80.

[9] Strahl BD, Allis CD. The language of covalent histone modifications. Nature 2000;403:41−5.

[10] Turner BM. Defining an epigenetic code. Nat Cell Biol 2007;9:2−6.

[11] Lee JS, Smith E, Shilatifard A. The language of histone crosstalk. Cell 2010;142:682−5.

[12] Allis CD, Berger SL, Cote J, Dent S, Jenuwein T, Kouzarides T, et al. New nomenclature for chromatin-modifying enzymes. Cell 2007;131:633−6.

[13] Gray SG, Ekström TJ. The Human Histone Deacetylase Family. Exp Cell Res 2001;262:75−83.

[14] Albert M, Helin K. Histone methyltransferases in cancer. Semin Cell Dev Biol 2010;21:209−20.

[15] Sims 3rd RJ, Reinberg D. Is there a code embedded in proteins that is based on post-translational modifications? Nat Rev Mol Cell Biol 2008;9:815−20.

[16] Calao M, Burny A, Quivy V, Dekoninck A, Van Lint C. A pervasive role of histone acetyltransferases and deacetylases in an NF-kappaB-signaling code. Trends Biochem Sci 2008;33:339−49.

[17] Mankan AK, Lawless MW, Gray SG, Kelleher D, McManus R. NF-kappaB regulation: the nuclear response. J Cell Mol Med 2009;13:631−43.

[18] Choudhary C, Kumar C, Gnad F, Nielsen ML, Rehman M, Walther TC, et al. Lysine acetylation targets protein complexes and co-regulates major cellular functions. Science 2009;325:834−40.

[19] Cyr AR, Domann FE. The redox basis of epigenetic modifications: from mechanisms to functional consequences. Antioxid Redox Signal 2011;15:551−89.

[20] Kolfschoten IG, Roggli E, Nesca V, Regazzi R. Role and therapeutic potential of microRNAs in diabetes. Diabetes Obes Metab 2009;11(Suppl. 4):118—29.

[21] Xie H, Lim B, Lodish HF. MicroRNAs induced during adipogenesis that accelerate fat cell development are downregulated in obesity. Diabetes 2009;58:1050—7.

[22] Fabbri M, Calin GA. Epigenetics and miRNAs in human cancer. Adv Genet 2010;70:87—99.

[23] Gray SG, De Meyts P. Role of histone and transcription factor acetylation in diabetes pathogenesis. Diabetes Metab Res Rev 2005;21:416—33.

[24] Temple IK, Shield JP. Transient neonatal diabetes, a disorder of imprinting. J Med Genet 2002;39:872—5.

[25] Mackay DJ, Temple IK. Transient neonatal diabetes mellitus type 1. Am J Med Genet C Semin Med Genet 2010;154C:335—42.

[26] Li Y, Zhao M, Hou C, Liang G, Yang L, Tan Y, et al. Abnormal DNA methylation in CD4(+) T cells from patients with latent autoimmune diabetes in adults. Diabetes Res Clin Pract 2011.

[27] Pirola L, Balcerczyk A, Tothill RW, Haviv I, Kaspi A, Lunke S, et al. Genome-wide analysis distinguishes hyperglycemia regulated epigenetic signatures of primary vascular cells. Genome Res 2011;21:1601—15.

[28] Rakyan VK, Beyan H, Down TA, Hawa MI, Maslau S, Aden D, et al. Identification of type 1 diabetes-associated DNA methylation variable positions that precede disease diagnosis. PLoS Genet 2011;7:e1002300.

[29] Toperoff G, Aran D, Kark JD, Rosenberg M, Dubnikov T, Nissan B, et al. Genome-wide survey reveals predisposing diabetes type 2-related DNA methylation variations in human peripheral blood. Hum Mol Genet 2011.

[30] Jeninga EH, Schoonjans K, Auwerx J. Reversible acetylation of PGC-1: connecting energy sensors and effectors to guarantee metabolic flexibility. Oncogene 2010;29:4617—24.

[31] Mootha VK, Lindgren CM, Eriksson KF, Subramanian A, Sihag S, Lehar J, et al. PGC-1alpha-responsive genes involved in oxidative phosphorylation are coordinately downregulated in human diabetes. Nat Genet 2003;34:267—73.

[32] Ling C, Del Guerra S, Lupi R, Ronn T, Granhall C, Luthman H, et al. Epigenetic regulation of PPARGC1A in human type 2 diabetic islets and effect on insulin secretion. Diabetologia 2008;51:615—22.

[33] Barres R, Osler ME, Yan J, Rune A, Fritz T, Caidahl K, et al. Non-CpG methylation of the PGC-1alpha promoter through DNMT3B controls mitochondrial density. Cell Metab 2009;10:189—98.

[34] Cox NJ, Wapelhorst B, Morrison VA, Johnson L, Pinchuk L, Spielman RS, et al. Seven regions of the genome show evidence of linkage to type 1 diabetes in a consensus analysis of 767 multiplex families. Am J Hum Genet 2001;69:820—30.

[35] Nerup J, Pociot F. A genomewide scan for type 1-diabetes susceptibility in Scandinavian families: identification of new loci with evidence of interactions. Am J Hum Genet 2001;69:1301—13.

[36] Xiang K, Wang Y, Zheng T, Jia W, Li J, Chen L, et al. Genome-wide search for type 2 diabetes/impaired glucose homeostasis susceptibility genes in the Chinese: significant linkage to chromosome 6q21-q23 and chromosome 1q21-q24. Diabetes 2004;53:228—34.

[37] Concannon P, Erlich HA, Julier C, Morahan G, Nerup J, Pociot F, et al. Type 1 diabetes: evidence for susceptibility loci from four genome-wide linkage scans in 1,435 multiplex families. Diabetes 2005;54:2995—3001.

[38] Noh H, Oh EY, Seo JY, Yu MR, Kim YO, Ha H, et al. Histone deacetylase-2 is a key regulator of diabetes- and transforming growth factor-beta1-induced renal injury. Am J Physiol Renal Physiol 2009;297:F729—739.

[39] Yamauchi T, Oike Y, Kamon J, Waki H, Komeda K, Tsuchida A, et al. Increased insulin sensitivity despite lipodystrophy in Crebbp heterozygous mice. Nat Genet 2002;30:221—6.

[40] Coste A, Louet JF, Lagouge M, Lerin C, Antal MC, Meziane H, et al. The genetic ablation of SRC-3 protects against obesity and improves insulin sensitivity by reducing the acetylation of PGC-1{alpha}. Proc Natl Acad Sci USA 2008;105:17187—92.

[41] York B, Yu C, Sagen JV, Liu Z, Nikolai BC, Wu RC, et al. Reprogramming the posttranslational code of SRC-3 confers a switch in mammalian systems biology. Proc Natl Acad Sci USA 2010;107:11122—7.

[42] Witt O, Deubzer HE, Milde T, Oehme I. HDAC family: What are the cancer relevant targets? Cancer Lett 2009;277:8—21.

[43] Guarente L. Franklin H. Epstein Lecture: Sirtuins, aging, and medicine. N Engl J Med 2011;364:2235—44.

[44] Wu D, Qiu Y, Gao X, Yuan XB, Zhai Q. Overexpression of SIRT1 in Mouse Forebrain Impairs Lipid/Glucose Metabolism and Motor Function. PLoS One 2011;6:e21759.

[45] Li Y, Xu S, Giles A, Nakamura K, Lee JW, Hou X, et al. Hepatic overexpression of SIRT1 in mice attenuates endoplasmic reticulum stress and insulin resistance in the liver. Faseb J 2011;25:1664—79.

[46] Kendrick AA, Choudhury M, Rahman SM, McCurdy CE, Friederich M, Hove Van, et al. Fatty liver is associated with reduced SIRT3 activity and mitochondrial protein hyperacetylation. Biochem J 2011;433:505—14.

[47] Schwer B, Schumacher B, Lombard DB, Xiao C, Kurtev MV, Gao J, et al. Neural sirtuin 6 (Sirt6) ablation attenuates somatic growth and causes obesity. Proc Natl Acad Sci USA 2010;107:21790—4.

354

[48] Xiao C, Kim HS, Lahusen T, Wang RH, Xu X, Gavrilova O, et al. SIRT6 deficiency results in severe hypoglycemia by enhancing both basal and insulin-stimulated glucose uptake in mice. J Biol Chem 2010;285:36776—84.

[49] Brasacchio D, Okabe J, Tikellis C, Balcerczyk A, George P, Baker EK, et al. Hyperglycemia induces a dynamic cooperativity of histone methylase and demethylase enzymes associated with gene-activating epigenetic marks that coexist on the lysine tail. Diabetes 2009;58:1229—36.

[50] Reddy MA, Villeneuve LM, Wang M, Lanting L, Natarajan R. Role of the lysine-specific demethylase 1 in the proinflammatory phenotype of vascular smooth muscle cells of diabetic mice. Circ Res 2008;103:615—23.

[51] Li Y, Reddy MA, Miao F, Shanmugam N, Yee JK, Hawkins D, et al. Role of the histone H3 lysine 4 methyl-transferase, SET7/9, in the regulation of NF-kappaB-dependent inflammatory genes. Relevance to diabetes and inflammation. J Biol Chem 2008;283:26771—81.

[52] Deering TG, Ogihara T, Trace AP, Maier B, Mirmira RG. Methyltransferase Set7/9 maintains transcription and euchromatin structure at islet-enriched genes. Diabetes 2009;58:185—93.

[53] Ogihara T, Vanderford NL, Maier B, Stein RW, Mirmira RG. Expression and function of Set7/9 in pancreatic islets. Islets 2009;1:269—72.

[54] Lo KA, Bauchmann MK, Baumann AP, Donahue CJ, Thiede MA, Hayes LS, et al. Genome-Wide Profiling of H3K56 Acetylation and Transcription Factor Binding Sites in Human Adipocytes. PLoS One 2011;6:e19778.

[55] Vempati RK, Jayani RS, Notani D, Sengupta A, Galande S, Haldar D. p300-mediated acetylation of histone H3 lysine 56 functions in DNA damage response in mammals. J Biol Chem 2010;285:28553—64.

[56] Evans-Molina C, Garmey JC, Ketchum R, Brayman KL, Deng S, Mirmira RG. Glucose regulation of insulin gene transcription and pre-mRNA processing in human islets. Diabetes 2007;56:827—35.

[57] Mosley AL, Corbett JA, Ozcan S. Glucose regulation of insulin gene expression requires the recruitment of p300 by the beta-cell-specific transcription factor Pdx-1. Mol Endocrinol 2004;18:2279—90.

[58] Mosley AL, Ozcan S. Glucose regulates insulin gene transcription by hyperacetylation of histone h4. J Biol Chem 2003;278:19660—6.

[59] Qiu Y, Guo M, Huang S, Stein R. Insulin gene transcription is mediated by interactions between the p300 coactivator and PDX-1, BETA2, and E47. Mol Cell Biol 2002;22:412—20.

[60] Qiu Y, Sharma A, Stein R. p300 mediates transcriptional stimulation by the basic helix-loop-helix activators of the insulin gene. Mol Cell Biol 1998;18:2957—64.

[61] Mosley AL, Ozcan S. The pancreatic duodenal homeobox-1 protein (Pdx-1) interacts with histone deacetylases Hdac-1 and Hdac-2 on low levels of glucose. J Biol Chem 2004;279:54241—7.

[62] Qiu Y, Guo M, Huang S, Stein R. Acetylation of the BETA2 Transcription Factor by p300-associated Factor Is Important in Insulin Gene Expression. J Biol Chem 2004;279:9796—802.

[63] Zhang M, Poplawski M, Yen K, Cheng H, Bloss E, Zhu X, et al. Role of CBP and SATB-1 in aging, dietary restriction, and insulin-like signaling. PLoS Biol 2009;7:e1000245.

[64] Fernandez-Zapico ME, van Velkinburgh JC, Gutierrez-Aguilar R, Neve B, Froguel P, Urrutia R, et al. MODY7 gene, KLF11, is a novel p300-dependent regulator of Pdx-1 (MODY4) transcription in pancreatic islet beta cells. J Biol Chem 2009;284:36482—90.

[65] Bonnefond A, Lomberk G, Buttar N, Busiah K, Vaillant E, Lobbens S, et al. Disruption of a novel KLF-p300-regulated pathway for insulin biosynthesis revealed by studies of the c.-331 INS mutation found in neonatal diabetes mellitus. J Biol Chem 2011.

[66] Mutskov V, Felsenfeld G. The human insulin gene is part of a large open chromatin domain specific for human islets. Proc Natl Acad Sci USA 2009;106:17419—24.

[67] Ohara-Imaizumi M, Cardozo AK, Kikuta T, Eizirik DL, Nagamatsu S. The cytokine interleukin-1beta reduces the docking and fusion of insulin granules in pancreatic beta-cells, preferentially decreasing the first phase of exocytosis. J Biol Chem 2004;279:41271—4.

[68] Lawrence MC, Shao C, McGlynn K, Naziruddin B, Levy MF, Cobb MH. Multiple chromatin-bound protein kinases assemble factors that regulate insulin gene transcription. Proc Natl Acad Sci USA 2009;106:22181—6.

[69] Leoni F, Zaliani A, Bertolini G, Porro G, Pagani P, Pozzi P, et al. The antitumor histone deacetylase inhibitor suberoylanilide hydroxamic acid exhibits antiinflammatory properties via suppression of cytokines. Proc Natl Acad Sci USA 2002;99:2995—3000.

[70] Lewis EC, Blaabjerg L, Storling J, Ronn SG, Mascagni P, Dinarello CA, et al. The Oral Histone Deacetylase Inhibitor ITF2357 Reduces Cytokines and Protects Islet beta Cells In Vivo and In Vitro. Mol Med 2011;17:369—77.

[71] Susick L, Senanayake T, Veluthakal R, Woster PM, Kowluru A. A novel histone deacetylase inhibitor prevents IL-1beta induced metabolic dysfunction in pancreatic beta-cells. J Cell Mol Med 2009;13:1877—85.

[72] Brochu-Gaudreau K, Rehfeldt C, Blouin R, Bordignon V, Murphy BD, Palin MF. Adiponectin action from head to toe. Endocrine 2010;37:11—32.

355

[73] Musri MM, Corominola H, Casamitjana R, Gomis R, Parrizas M. Histone H3 lysine 4 dimethylation signals the transcriptional competence of the adiponectin promoter in preadipocytes. J Biol Chem 2006;281:17180—8.

[74] Sakurai N, Mochizuki K, Goda T. Modifications of histone H3 at lysine 9 on the adiponectin gene in 3T3-L1 adipocytes. J Nutr Sci Vitaminol (Tokyo) 2009;55:131—8.

[75] Qiao L, Schaack J, Shao J. Suppression of adiponectin gene expression by histone deacetylase inhibitor valproic acid. Endocrinology 2006;147:865—74.

[76] Sugii S, Evans RM. Epigenetic codes of PPARgamma in metabolic disease. FEBS Lett 2011;585:2121—8.

[77] Kim SJ, Nian C, McIntosh CH. Adipocyte expression of the glucose-dependent insulinotropic polypeptide receptor involves gene regulation by PPARgamma and histone acetylation. J Lipid Res 2011;52:759—70.

[78] Handschin C, Spiegelman BM. Peroxisome proliferator-activated receptor gamma coactivator 1 coactivators, energy homeostasis, and metabolism. Endocr Rev 2006;27:728—35.

[79] Patti ME, Butte AJ, Crunkhorn S, Cusi K, Berria R, Kashyap S, et al. Coordinated reduction of genes of oxidative metabolism in humans with insulin resistance and diabetes: Potential role of PGC1 and NRF1. Proc Natl Acad Sci USA 2003;100:8466—71.

[80] Crunkhorn S, Dearie F, Mantzoros C, Gami H, da Silva WS, Espinoza D, et al. Peroxisome proliferator activator receptor gamma coactivator-1 expression is reduced in obesity: potential pathogenic role of saturated fatty acids and p38 mitogen-activated protein kinase activation. J Biol Chem 2007;282:15439—50.

[81] Karpac J, Jasper H. Metabolic homeostasis: HDACs take center stage. Cell 2011;145:497—9.

[82] Calnan DR, Brunet A. The FoxO code. Oncogene 2008;27:2276—88.

[83] Haigis MC, Sinclair DA. Mammalian sirtuins: biological insights and disease relevance. Annu Rev Pathol 2010;5:253—95.

[84] Mihaylova MM, Vasquez DS, Ravnskjaer K, Denechaud PD, Yu RT, Alvarez JG, et al. Class IIa histone deacetylases are hormone-activated regulators of FOXO and mammalian glucose homeostasis. Cell 2011;145:607—21.

[85] Wang B, Moya N, Niessen S, Hoover H, Mihaylova MM, Shaw RJ, et al. A hormone-dependent module regulating energy balance. Cell 2011;145:596—606.

[86] Birnbaum MJ. Identification of a novel gene encoding an insulin-responsive glucose transporter protein. Cell 1989;57:305—15.

[87] Garvey WT, Maianu L, Zhu JH, Brechtel-Hook G, Wallace P, Baron AD. Evidence for defects in the trafficking and translocation of GLUT4 glucose transporters in skeletal muscle as a cause of human insulin resistance. J Clin Invest 1998;101:2377—86.

[88] Czubryt MP, McAnally J, Fishman GI, Olson EN. Regulation of peroxisome proliferator-activated receptor gamma coactivator 1 alpha (PGC-1 alpha) and mitochondrial function by MEF2 and HDAC5. Proc Natl Acad Sci USA 2003;100:1711—6.

[89] Raychaudhuri N, Raychaudhuri S, Thamotharan M, Devaskar SU. Histone code modifications repress glucose transporter 4 expression in the intrauterine growth-restricted offspring. J Biol Chem 2008;283:13611—26.

[90] Sparling DP, Griesel BA, Weems J, Olson AL. GLUT4 enhancer factor (GEF) interacts with MEF2A and HDAC5 to regulate the GLUT4 promoter in adipocytes. J Biol Chem 2008;283:7429—37.

[91] McGee SL, van Denderen BJ, Howlett KF, Mollica J, Schertzer JD, Kemp BE, et al. AMP-activated protein kinase regulates GLUT4 transcription by phosphorylating histone deacetylase 5. Diabetes 2008;57:860—7.

[92] Weems J, Olson AL. Class II histone deacetylases limit GLUT4 gene expression during adipocyte differentiation. J Biol Chem 2011;286:460—8.

[93] O'Byrne KJ, Baird A-M, Kilmartin L, Leonard J, Sacevich C, Gray SG. Epigenetic Regulation of Glucose Transporters in Non-Small Cell Lung Cancer. Cancers 2011;3:1550—65.

[94] Phillips JE, Corces VG. CTCF: master weaver of the genome. Cell 2009;137:1194—211.

[95] Takigawa-Imamura H, Sekine T, Murata M, Takayama K, Nakazawa K, Nakagawa J. Stimulation of glucose uptake in muscle cells by prolonged treatment with scriptide, a histone deacetylase inhibitor. Biosci Biotechnol Biochem 2003;67:1499—506.

[96] Yamakuchi M, Ferlito M, Lowenstein CJ. miR-34a repression of SIRT1 regulates apoptosis. Proc Natl Acad Sci USA 2008;105:13421—6.

[97] Lovis P, Roggli E, Laybutt DR, Gattesco S, Yang JY, Widmann C, et al. Alterations in microRNA expression contribute to fatty acid-induced pancreatic beta-cell dysfunction. Diabetes 2008;57:2728—36.

[98] Lee J, Padhye A, Sharma A, Song G, Miao J, Mo YY, et al. A pathway involving farnesoid X receptor and small heterodimer partner positively regulates hepatic sirtuin 1 levels via microRNA-34a inhibition. J Biol Chem 2010;285:12604—11.

[99] Menghini R, Casagrande V, Cardellini M, Martelli E, Terrinoni A, Amati F, et al. MicroRNA 217 modulates endothelial cell senescence via silent information regulator 1. Circulation 2009;120:1524—32.

[100] Szafranska AE, Davison TS, John J, Cannon T, Sipos B, Maghnouj A, et al. MicroRNA expression alterations are linked to tumorigenesis and non-neoplastic processes in pancreatic ductal adenocarcinoma. Oncogene 2007;26:4442−52.

[101] Strum JC, Johnson JH, Ward J, Xie H, Feild J, Hester A, et al. MicroRNA 132 regulates nutritional stress-induced chemokine production through repression of SirT1. Mol Endocrinol 2009;23:1876−84.

[102] Esguerra JL, Bolmeson C, Cilio CM, Eliasson L. Differential glucose-regulation of microRNAs in pancreatic islets of non-obese type 2 diabetes model Goto-Kakizaki rat. PLoS One 2011;6:e18613.

[103] Heneghan HM, Miller N, McAnena OJ, O'Brien T, Kerin MJ. Differential miRNA expression in omental adipose tissue and in the circulation of obese patients identifies novel metabolic biomarkers. J Clin Endocrinol Metab 2011;96:E846−850.

[104] Park JK, Henry JC, Jiang J, Esau C, Gusev Y, Lerner MR, et al. miR-132 and miR-212 are increased in pancreatic cancer and target the retinoblastoma tumor suppressor. Biochem Biophys Res Commun 2011;406:518−23.

[105] Alvarez-Saavedra M, Antoun G, Yanagiya A, Oliva-Hernandez R, Cornejo-Palma D, Perez-Iratxeta C, et al. miRNA-132 orchestrates chromatin remodeling and translational control of the circadian clock. Hum Mol Genet 2011;20:731−51.

[106] Ramachandran D, Roy U, Garg S, Ghosh S, Pathak S, Kolthur-Seetharam U. Sirt1 and mir-9 expression is regulated during glucose-stimulated insulin secretion in pancreatic beta-islets. Febs J 2011;278:1167−74.

[107] Fabbri M, Garzon R, Cimmino A, Liu Z, Zanesi N, Callegari E, et al. MicroRNA-29 family reverts aberrant methylation in lung cancer by targeting DNA methyltransferases 3A and 3B. Proc Natl Acad Sci USA 2007;104:15805−10.

[108] Pass HI, Goparaju C, Ivanov S, Donington J, Carbone M, Hoshen M, et al. hsa-miR-29c* is linked to the prognosis of malignant pleural mesothelioma. Cancer Res 2010;70:1916−24.

[109] He A, Zhu L, Gupta N, Chang Y, Fang F. Overexpression of micro ribonucleic acid 29, highly up-regulated in diabetic rats, leads to insulin resistance in 3T3-L1 adipocytes. Mol Endocrinol 2007;21:2785−94.

[110] Long J, Wang Y, Wang W, Chang BH, Danesh FR. MicroRNA-29c is a signature microRNA under high glucose conditions that targets Sprouty homolog 1, and its in vivo knockdown prevents progression of diabetic nephropathy. J Biol Chem 2011;286:11837−48.

[111] Ng EK, Tsang WP, Ng SS, Jin HC, Yu J, Li JJ, et al. MicroRNA-143 targets DNA methyltransferases 3A in colorectal cancer. Br J Cancer 2009;101:699−706.

[112] Esau C, Kang X, Peralta E, Hanson E, Marcusson EG, Ravichandran LV, et al. MicroRNA-143 regulates adipocyte differentiation. J Biol Chem 2004;279:52361−5.

[113] Braconi C, Huang N, Patel T. MicroRNA-dependent regulation of DNA methyltransferase-1 and tumor suppressor gene expression by interleukin-6 in human malignant cholangiocytes. Hepatology 2010;51:881−90.

[114] Wang KH, Kao AP, Singh S, Yu SL, Kao LP, Tsai ZY, et al. Comparative expression profiles of mRNAs and microRNAs among human mesenchymal stem cells derived from breast, face, and abdominal adipose tissues. Kaohsiung J Med Sci 2010;26:113−22.

[115] Koturbash I, Boyko A, Rodriguez-Juarez R, McDonald RJ, Tryndyak VP, Kovalchuk I, et al. Role of epigenetic effectors in maintenance of the long-term persistent bystander effect in spleen in vivo. Carcinogenesis 2007;28:1831−8.

[116] Klein ME, Lioy DT, Ma L, Impey S, Mandel G, Goodman RH. Homeostatic regulation of MeCP2 expression by a CREB-induced microRNA. Nat Neurosci 2007;10:1513−4.

[117] Noonan EJ, Place RF, Pookot D, Basak S, Whitson JM, Hirata H, et al. miR-449a targets HDAC-1 and induces growth arrest in prostate cancer. Oncogene 2009;28:1714−24.

[118] Yang X, Feng M, Jiang X, Wu Z, Li Z, Aau M, et al. miR-449a and miR-449b are direct transcriptional targets of E2F1 and negatively regulate pRb-E2F1 activity through a feedback loop by targeting CDK6 and CDC25A. Genes Dev 2009;23:2388−93.

[119] Bou Kheir T, Futoma-Kazmierczak E, Jacobsen A, Krogh A, Bardram L, Hother C, et al. miR-449 inhibits cell proliferation and is down-regulated in gastric cancer. Mol Cancer 2011;10:29.

[120] Chen JF, Mandel EM, Thomson JM, Wu Q, Callis TE, Hammond SM, et al. The role of microRNA-1 and microRNA-133 in skeletal muscle proliferation and differentiation. Nat Genet 2006;38:228−33.

[121] Shan ZX, Lin QX, Deng CY, Zhu JN, Mai LP, Liu JL, et al. miR-1/miR-206 regulate Hsp60 expression contributing to glucose-mediated apoptosis in cardiomyocytes. FEBS Lett 2010;584:3592−600.

[122] Karbiener M, Fischer C, Nowitsch S, Opriessnig P, Papak C, Ailhaud G, et al. microRNA miR-27b impairs human adipocyte differentiation and targets PPARgamma. Biochem Biophys Res Commun 2009;390:247−51.

[123] Kim SY, Kim AY, Lee HW, Son YH, Lee GY, Lee JW, et al. miR-27a is a negative regulator of adipocyte differentiation via suppressing PPARgamma expression. Biochem Biophys Res Commun 2010;392:323−8.

[124] Martinelli R, Nardelli C, Pilone V, Buonomo T, Liguori R, Castano I, et al. miR-519d Overexpression Is Associated With Human Obesity. Silver Spring: Obesity; 2010.

[125] Varambally S, Cao Q, Mani RS, Shankar S, Wang X, Ateeq B, et al. Genomic loss of microRNA-101 leads to overexpression of histone methyltransferase EZH2 in cancer. Science 2008;322:1695–9.

[126] Chen H, Gu X, Su IH, Bottino R, Contreras JL, Tarakhovsky A, et al. Polycomb protein Ezh2 regulates pancreatic beta-cell Ink4a/Arf expression and regeneration in diabetes mellitus. Genes Dev 2009;23:975–85.

[127] Wang L, Jin Q, Lee JE, Su IH, Ge K. Histone H3K27 methyltransferase Ezh2 represses Wnt genes to facilitate adipogenesis. Proc Natl Acad Sci USA 2010;107:7317–22.

[128] Lu H, Buchan RJ, Cook SA. MicroRNA-223 regulates Glut4 expression and cardiomyocyte glucose metabolism. Cardiovasc Res 2010;86:410–20.

[129] Horie T, Ono K, Nishi H, Iwanaga Y, Nagao K, Kinoshita M, et al. MicroRNA-133 regulates the expression of GLUT4 by targeting KLF15 and is involved in metabolic control in cardiac myocytes. Biochem Biophys Res Commun 2009;389:315–20.

[130] Aoi W, Naito Y, Mizushima K, Takanami Y, Kawai Y, Ichikawa H, et al. The microRNA miR-696 regulates PGC-1{alpha} in mouse skeletal muscle in response to physical activity. Am J Physiol Endocrinol Metab 2010;298:E799–806.

[131] Safdar A, Abadi A, Akhtar M, Hettinga BP, Tarnopolsky MA. miRNA in the regulation of skeletal muscle adaptation to acute endurance exercise in C57Bl/6J male mice. PLoS One 2009;4:e5610.

[132] Keller DM, McWeeney S, Arsenlis A, Drouin J, Wright CV, Wang H, et al. Characterization of pancreatic transcription factor Pdx-1 binding sites using promoter microarray and serial analysis of chromatin occupancy. J Biol Chem 2007;282:32084–92.

[133] El Ouaamari A, Baroukh N, Martens GA, Lebrun P, Pipeleers D, van Obberghen E. miR-375 targets 3'-phosphoinositide-dependent protein kinase-1 and regulates glucose-induced biological responses in pancreatic beta-cells. Diabetes 2008;57:2708–17.

[134] Trajkovski M, Hausser J, Soutschek J, Bhat B, Akin A, Zavolan M, et al. MicroRNAs 103 and 107 regulate insulin sensitivity. Nature 2011;474:649–53.

[135] Tang X, Muniappan L, Tang G, Ozcan S. Identification of glucose-regulated miRNAs from pancreatic {beta} cells reveals a role for miR-30d in insulin transcription. Rna 2009;15:287–93.

[136] Kloosterman WP, Lagendijk AK, Ketting RF, Moulton JD, Plasterk RH. Targeted inhibition of miRNA maturation with morpholinos reveals a role for miR-375 in pancreatic islet development. PLoS Biol 2007;5:e203.

[137] Joglekar MV, Joglekar VM, Hardikar AA. Expression of islet-specific microRNAs during human pancreatic development. Gene Expr Patterns 2009;9:109–13.

[138] Granjon A, Gustin MP, Rieusset J, Lefai E, Meugnier E, Guller I, et al. The microRNA signature in response to insulin reveals its implication in the transcriptional action of insulin in human skeletal muscle and the role of a sterol regulatory element-binding protein-1c/myocyte enhancer factor 2C pathway. Diabetes 2009;58:2555–64.

[139] Gallagher IJ, Scheele C, Keller P, Nielsen AR, Remenyi J, Fischer CP, et al. Integration of microRNA changes in vivo identifies novel molecular features of muscle insulin resistance in type 2 diabetes. Genome Med 2010;2:9.

[140] Zhao E, Keller MP, Rabaglia ME, Oler AT, Stapleton DS, Schueler KL, et al. Obesity and genetics regulate microRNAs in islets, liver, and adipose of diabetic mice. Mamm Genome 2009;20:476–85.

[141] Ortega FJ, Moreno-Navarrete JM, Pardo G, Sabater M, Hummel M, Ferrer A, et al. MiRNA expression profile of human subcutaneous adipose and during adipocyte differentiation. PLoS One 2010;5:e9022.

[142] Jordan SD, Kruger M, Willmes DM, Redemann N, Wunderlich FT, Bronneke HS, et al. Obesity-induced overexpression of miRNA-143 inhibits insulin-stimulated AKT activation and impairs glucose metabolism. Nat Cell Biol 2011;13:434–46.

[143] Iorio MV, Piovan C, Croce CM. Interplay between microRNAs and the epigenetic machinery: An intricate network. Biochim Biophys Acta 2010.

[144] Bolleyn J, Fraczek J, Vinken M, Lizarraga D, Gaj S, van Delft JH, et al. Effect of Trichostatin A on miRNA expression in cultures of primary rat hepatocytes. Toxicol In Vitro 2011.

[145] de Kreutzenberg SV, Ceolotto G, Papparella I, Bortoluzzi A, Semplicini A, Dalla Man C, et al. Downregulation of the longevity-associated protein sirtuin 1 in insulin resistance and metabolic syndrome: potential biochemical mechanisms. Diabetes 2010;59:1006–15.

[146] Arif M, Senapati P, Shandilya J, Kundu TK. Protein lysine acetylation in cellular function and its role in cancer manifestation. Biochim Biophys Acta 2010;1799:702–16.

[147] Lee J, Saha PK, Yang QH, Lee S, Park JY, Suh Y, et al. Targeted inactivation of MLL3 histone H3-Lys-4 methyltransferase activity in the mouse reveals vital roles for MLL3 in adipogenesis. Proc Natl Acad Sci USA 2008;105:19229–34.

[148] Donath MY, Shoelson SE. Type 2 diabetes as an inflammatory disease. Nat Rev Immunol 2011;11:98–107.

[149] Gerritsen ME, Williams AJ, Neish AS, Moore S, Shi Y, Collins T. CREB-binding protein/p300 are transcriptional coactivators of p65. Proc Natl Acad Sci USA 1997;94:2927–32.

[150] Perkins ND, Felzien LK, Betts JC, Leung K, Beach DH, Nabel GJ. Regulation of NF-kappaB by cyclin-dependent kinases associated with the p300 coactivator. Science 1997;275:523–7.

[151] Wadgaonkar R, Phelps KM, Haque Z, Williams AJ, Silverman ES, Collins T. CREB-binding protein is a nuclear integrator of nuclear factor-kappaB and p53 signaling. J Biol Chem 1999;274:1879–82.

[152] Na SY, Lee SK, Han SJ, Choi HS, Im SY, Lee JW. Steroid receptor coactivator-1 interacts with the p50 subunit and coactivates nuclear factor kappaB-mediated transactivations. J Biol Chem 1998;273:10831–4.

[153] Yeung F, Hoberg JE, Ramsey CS, Keller MD, Jones DR, Frye RA, et al. Modulation of NF-kappaB-dependent transcription and cell survival by the SIRT1 deacetylase. Embo J 2004;23:2369–80.

[154] Hoberg JE, Yeung F, Mayo MW. SMRT derepression by the IkappaB kinase alpha: a prerequisite to NF-kappaB transcription and survival. Mol Cell 2004;16:245–55.

[155] Fabre C, Grosjean J, Tailler M, Boehrer S, Ades L, Perfettini JL, et al. A novel effect of DNA methyltransferase and histone deacetylase inhibitors: NFkappaB inhibition in malignant myeloblasts. Cell Cycle 2008;7:2139–45.

[156] Larsen L, Tonnesen M, Ronn SG, Storling J, Jorgensen S, Mascagni P, et al. Inhibition of histone deacetylases prevents cytokine-induced toxicity in beta cells. Diabetologia 2007;50:779–89.

[157] Dandona P, Aljada A, Bandyopadhyay A. Inflammation: the link between insulin resistance, obesity and diabetes. Trends Immunol 2004;25:4–7.

[158] Esposito K, Nappo F, Giugliano F, Di Palo C, Ciotola M, Barbieri M, et al. Cytokine milieu tends toward inflammation in type 2 diabetes. Diabetes Care 2003;26:1647.

[159] Vanden Berghe W, De Bosscher K, Boone E, Plaisance S, Haegeman G. The nuclear factor-kappaB engages CBP/p300 and histone acetyltransferase activity for transcriptional activation of the interleukin-6 gene promoter. J Biol Chem 1999;274:32091–8.

[160] Gao Z, Hwang D, Bataille F, Lefevre M, York D, Quon MJ, et al. Serine phosphorylation of insulin receptor substrate 1 by inhibitor kappa B kinase complex. J Biol Chem 2002;277:48115–21.

[161] Arkan MC, Hevener AL, Greten FR, Maeda S, Li ZW, Long JM, et al. IKK-beta links inflammation to obesity-induced insulin resistance. Nat Med 2005;11:191–8.

[162] Cai D, Yuan M, Frantz DF, Melendez PA, Hansen L, Lee J, et al. Local and systemic insulin resistance resulting from hepatic activation of IKK-beta and NF-kappaB. Nat Med 2005;11:183–90.

[163] Espinosa L, Ingles-Esteve J, Robert-Moreno A, Bigas A. IkappaBalpha and p65 regulate the cytoplasmic shuttling of nuclear corepressors: cross-talk between Notch and NFkappaB pathways. Mol Biol Cell 2003;14:491–502.

[164] Kaur H, Chen S, Xin X, Chiu J, Khan ZA, Chakrabarti S. Diabetes-induced extracellular matrix protein expression is mediated by transcription coactivator p300. Diabetes 2006;55:3104–11.

[165] Dinarello CA. A clinical perspective of IL-1beta as the gatekeeper of inflammation. Eur J Immunol 2011;41:1203–17.

[166] Papaccio G, Graziano A, d'Aquino R, Valiante S, Naro F. A biphasic role of nuclear transcription factor (NF)-kappaB in the islet beta-cell apoptosis induced by interleukin (IL)-1beta. J Cell Physiol 2004.

[167] Lundh M, Christensen DP, Rasmussen DN, Mascagni P, Dinarello CA, Billestrup N, et al. Lysine deacetylases are produced in pancreatic beta cells and are differentially regulated by proinflammatory cytokines. Diabetologia 2010;53:2569–78.

[168] Park YJ, Luger K. Histone chaperones in nucleosome eviction and histone exchange. Curr Opin Struct Biol 2008;18:282–9.

[169] Meimaridou E, Gooljar SB, Chapple JP. From hatching to dispatching: the multiple cellular roles of the Hsp70 molecular chaperone machinery. J Mol Endocrinol 2009;42:1–9.

[170] Barginear MF, Van Poznak C, Rosen N, Modi S, Hudis CA, Budman DR. The heat shock protein 90 chaperone complex: an evolving therapeutic target. Curr Cancer Drug Targets 2008;8:522–32.

[171] Wandinger SK, Richter K, Buchner J. The Hsp90 chaperone machinery. J Biol Chem 2008;283:18473–7.

[172] Hebert DN, Molinari M. In and out of the ER: protein folding, quality control, degradation, and related human diseases. Physiol Rev 2007;87:1377–408.

[173] Rapoport TA. Protein translocation across the eukaryotic endoplasmic reticulum and bacterial plasma membranes. Nature 2007;450:663–9.

[174] Quinones QJ, de Ridder GG, Pizzo SV. GRP78: a chaperone with diverse roles beyond the endoplasmic reticulum. Histol Histopathol 2008;23:1409–16.

[175] Paris S, Denis H, Delaive E, Dieu M, Dumont V, Ninane N, et al. Up-regulation of 94-kDa glucose-regulated protein by hypoxia-inducible factor-1 in human endothelial cells in response to hypoxia. FEBS Lett 2005;579:105–14.

359

[176] Cryer PE. The barrier of hypoglycemia in diabetes. Diabetes 2008;57:3169—76.

[177] Eizirik DL, Cardozo AK, Cnop M. The role for endoplasmic reticulum stress in diabetes mellitus. Endocr Rev 2008;29:42—61.

[178] Scheuner D, Kaufman RJ. The unfolded protein response: a pathway that links insulin demand with beta-cell failure and diabetes. Endocr Rev 2008;29:317—33.

[179] Rutkowski DT, Kaufman RJ. A trip to the ER: coping with stress. Trends Cell Biol 2004;14:20—8.

[180] Lai E, Teodoro T, Volchuk A. Endoplasmic reticulum stress: signaling the unfolded protein response. Physiology (Bethesda) 2007;22:193—201.

[181] Caruso ME, Chevet E. Systems biology of the endoplasmic reticulum stress response. Subcell Biochem 2007;43:277—98.

[182] Kim I, Xu W, Reed JC. Cell death and endoplasmic reticulum stress: disease relevance and therapeutic opportunities. Nat Rev Drug Discov 2008;7:1013—30.

[183] Ron D, Walter P. Signal integration in the endoplasmic reticulum unfolded protein response. Nat Rev Mol Cell Biol 2007;8:519—29.

[184] Tabas I, Ron D. Integrating the mechanisms of apoptosis induced by endoplasmic reticulum stress. Nat Cell Biol 2011;13:184—90.

[185] Vembar SS, Brodsky JL. One step at a time: endoplasmic reticulum-associated degradation. Nat Rev Mol Cell Biol 2008;9:944—57.

[186] Pahl HL, Baeuerle PA. The ER-overload response: activation of NF-kappa B. Trends Biochem Sci 1997;22:63—7.

[187] Xu C, Bailly-Maitre B, Reed JC. Endoplasmic reticulum stress: cell life and death decisions. J Clin Invest 2005;115:2656—64.

[188] Heath-Engel HM, Chang NC, Shore GC. The endoplasmic reticulum in apoptosis and autophagy: role of the BCL-2 protein family. Oncogene 2008;27:6419—33.

[189] Kaufman RJ, Back SH, Song B, Han J, Hassler J. The unfolded protein response is required to maintain the integrity of the endoplasmic reticulum, prevent oxidative stress and preserve differentiation in beta-cells. Diabetes Obes Metab 12(Suppl. 2):99—107.

[190] Lei X, Zhang S, Emani B, Barbour SE, Ramanadham S. A link between endoplasmic reticulum stress-induced beta-cell apoptosis and the group VIA Ca2+-independent phospholipase A2 (iPLA2beta). Diabetes Obes Metab 12(Suppl. 2):93—98

[191] Leibowitz G, Bachar E, Shaked M, Sinai A, Ketzinel-Gilad M, Cerasi E, et-al. Glucose regulation of beta-cell stress in type 2 diabetes. Diabetes Obes Metab 12(Suppl. 2):66—75

[192] Oslowski CM, Urano FA. switch from life to death in endoplasmic reticulum stressed beta-cells. Diabetes Obes Metab 12(Suppl. 2):58—65

[193] Johnson CA, White DA, Lavender JS, O'Neill LP, Turner BM. Human Class I Histone Deacetylase Complexes Show Enhanced Catalytic Activity in the Presence of ATP and Co-immunoprecipitate with the ATP-dependent Chaperone Protein Hsp70. J Biol Chem 2002;277:9590—7.

[194] Battu A, Ray A, Wani AA. ASF1A and ATM regulate H3K56-mediated cell-cycle checkpoint recovery in response to UV irradiation. Nucleic Acids Res 2011.

[195] Minard LV, Lin LJ, Schultz MC. SWI/SNF and Asf1 Independently Promote Derepression of the DNA Damage Response Genes under Conditions of Replication Stress. PLoS One 2011;6:e21633.

[196] Das C, Lucia MS, Hansen KC, Tyler JK. CBP/p300-mediated acetylation of histone H3 on lysine 56. Nature 2009;459:113—7.

[197] Lin LJ, Minard LV, Johnston GC, Singer RA, Schultz MC. Asf1 can promote trimethylation of H3 K36 by Set2. Mol Cell Biol 2010;30:1116—29.

[198] Berndsen CE, Tsubota T, Lindner SE, Lee S, Holton JM, Kaufman PD, et al. Molecular functions of the histone acetyltransferase chaperone complex Rtt109-Vps75. Nat Struct Mol Biol 2008;15:948—56.

[199] Chen CC, Carson JJ, Feser J, Tamburini B, Zabaronick S, Linger J, et al. Acetylated lysine 56 on histone H3 drives chromatin assembly after repair and signals for the completion of repair. Cell 2008;134:231—43.

[200] Hiraga S, Botsios S, Donaldson AD. Histone H3 lysine 56 acetylation by Rtt109 is crucial for chromosome positioning. J Cell Biol 2008;183:641—51.

[201] Kaplan T, Liu CL, Erkmann JA, Holik J, Grunstein M, Kaufman PD, et al. Cell cycle- and chaperone-mediated regulation of H3K56ac incorporation in yeast. PLoS Genet 2008;4:e1000270.

[202] Li Q, Zhou H, Wurtele H, Davies B, Horazdovsky B, Verreault A, et al. Acetylation of histone H3 lysine 56 regulates replication-coupled nucleosome assembly. Cell 2008;134:244—55.

[203] Miller A, Yang B, Foster T, Kirchmaier AL. Proliferating cell nuclear antigen and ASF1 modulate silent chromatin in Saccharomyces cerevisiae via lysine 56 on histone H3. Genetics 2008;179:793—809.

[204] Tang Y, Meeth K, Jiang E, Luo C, Marmorstein R. Structure of Vps75 and implications for histone chaperone function. Proc Natl Acad Sci USA 2008;105:12206—11.

[205] Williams SK, Truong D, Tyler JK. Acetylation in the globular core of histone H3 on lysine-56 promotes chromatin disassembly during transcriptional activation. Proc Natl Acad Sci USA 2008;105:9000—5.

[206] Adkins MW, Carson JJ, English CM, Ramey CJ, Tyler JK. The histone chaperone anti-silencing function 1 stimulates the acetylation of newly synthesized histone H3 in S-phase. J Biol Chem 2007;282:1334—40.

[207] Fillingham J, Recht J, Silva AC, Suter B, Emili A, Stagljar I, et al. Chaperone control of the activity and specificity of the histone H3 acetyltransferase Rtt109. Mol Cell Biol 2008;28:4342—53.

[208] Han J, Zhou H, Li Z, Xu RM, Zhang Z. Acetylation of lysine 56 of histone H3 catalyzed by RTT109 and regulated by ASF1 is required for replisome integrity. J Biol Chem 2007;282:28587—96.

[209] LeRoy G, Rickards B, Flint SJ. The double bromodomain proteins Brd2 and Brd3 couple histone acetylation to transcription. Mol Cell 2008;30:51—60.

[210] Munemasa Y, Suzuki T, Aizawa K, Miyamoto S, Imai Y, Matsumura T, et al. Promoter region-specific histone incorporation by the novel histone chaperone ANP32B and DNA-binding factor KLF5. Mol Cell Biol 2008;28:1171—81.

[211] Recht J, Tsubota T, Tanny JC, Diaz RL, Berger JM, Zhang X, et al. Histone chaperone Asf1 is required for histone H3 lysine 56 acetylation, a modification associated with S phase in mitosis and meiosis. Proc Natl Acad Sci USA 2006;103:6988—93.

[212] Rufiange A, Jacques PE, Bhat W, Robert F, Nourani A. Genome-wide replication-independent histone H3 exchange occurs predominantly at promoters and implicates H3 K56 acetylation and Asf1. Mol Cell 2007;27:393—405.

[213] Schneider J, Bajwa P, Johnson FC, Bhaumik SR, Shilatifard A. Rtt109 is required for proper H3K56 acetylation: a chromatin mark associated with the elongating RNA polymerase II. J Biol Chem 2006;281:37270—4.

[214] Barman HK, Takami Y, Nishijima H, Shibahara K, Sanematsu F, Nakayama T. Histone acetyltransferase-1 regulates integrity of cytosolic histone H3-H4 containing complex. Biochem Biophys Res Commun 2008;373:624—30.

[215] Jin C, Kato K, Chimura T, Yamasaki T, Nakade K, Murata T, et al. Regulation of histone acetylation and nucleosome assembly by transcription factor JDP2. Nat Struct Mol Biol 2006;13:331—8.

[216] Swaminathan V, Kishore AH, Febitha KK, Kundu TK. Human histone chaperone nucleophosmin enhances acetylation-dependent chromatin transcription. Mol Cell Biol 2005;25:7534—45.

[217] Thomas M, Dadgar N, Aphale A, Harrell JM, Kunkel R, Pratt WB, et al. Androgen receptor acetylation site mutations cause trafficking defects, misfolding, and aggregation similar to expanded glutamine tracts. J Biol Chem 2004;279:8389—95.

[218] Kekatpure VD, Dannenberg AJ, Subbaramaiah K. HDAC6 modulates HSP90 chaperone activity and regulates activation of aryl hydrocarbon receptor signaling. J Biol Chem 2009;284:7436—45.

[219] Fiskus W, Ren Y, Mohapatra A, Bali P, Mandawat A, Rao R, et al. Hydroxamic acid analogue histone deacetylase inhibitors attenuate estrogen receptor-alpha levels and transcriptional activity: a result of hyperacetylation and inhibition of chaperone function of heat shock protein 90. Clin Cancer Res 2007;13:4882—90.

[220] Suuronen T, Ojala J, Hyttinen JM, Kaarniranta K, Thornell A, Kyrylenko S, et al. Regulation of ER alpha signaling pathway in neuronal HN10 cells: role of protein acetylation and Hsp90. Neurochem Res 2008;33:1768—75.

[221] Kovacs JJ, Murphy PJ, Gaillard S, Zhao X, Wu JT, Nicchitta CV, et al. HDAC6 regulates Hsp90 acetylation and chaperone-dependent activation of glucocorticoid receptor. Mol Cell 2005;18:601—7.

[222] Giustiniani J, Daire V, Cantaloube I, Durand G, Pous C, Perdiz D, et al. Tubulin acetylation favors Hsp90 recruitment to microtubules and stimulates the signaling function of the Hsp90 clients Akt/PKB and p53. Cell Signal 2008;21:529—39.

[223] Bardag-Gorce F, Dedes J, French BA, Oliva JV, Li J, French SW. Mallory body formation is associated with epigenetic phenotypic change in hepatocytes in vivo. Exp Mol Pathol 2007;83:160—8.

[224] Hanada S, Harada M, Kumemura H, Bishr Omary M, Koga H, Kawaguchi T, et al. Oxidative stress induces the endoplasmic reticulum stress and facilitates inclusion formation in cultured cells. J Hepatol 2007;47:93—102.

[225] Cherasse Y, Chaveroux C, Jousse C, Maurin AC, Carraro V, Parry L, et al. Role of the repressor JDP2 in the amino acid-regulated transcription of CHOP. FEBS Lett 2008;582:1537—41.

[226] Ohoka N, Hattori T, Kitagawa M, Onozaki K, Hayashi H. Critical and Functional Regulation of CHOP (C/EBP Homologous Protein) through the N-terminal Portion. J Biol Chem 2007;282:35687—94.

361

[227] Baumeister P, Luo S, Skarnes WC, Sui G, Seto E, Shi Y, et al. Endoplasmic reticulum stress induction of the Grp78/BiP promoter: activating mechanisms mediated by YY1 and its interactive chromatin modifiers. Mol Cell Biol 2005;25:4529—40.

[228] Donati G, Imbriano C, Mantovani R. Dynamic recruitment of transcription factors and epigenetic changes on the ER stress response gene promoters. Nucleic Acids Res 2006;34:3116—27.

[229] Doody GM, Stephenson S, Tooze RM. BLIMP-1 is a target of cellular stress and downstream of the unfolded protein response. Eur J Immunol 2006;36:1572—82.

[230] Yu J, Angelin-Duclos C, Greenwood J, Liao J, Calame K. Transcriptional repression by blimp-1 (PRDI-BF1) involves recruitment of histone deacetylase. Mol Cell Biol 2000;20:2592—603.

[231] Zeng L, Lu M, Mori K, Luo S, Lee AS, Zhu Y, et al. ATF6 modulates SREBP2-mediated lipogenesis. Embo J 2004;23:950—8.

[232] Sugden MC, Holness MJ. Role of nuclear receptors in the modulation of insulin secretion in lipid-induced insulin resistance. Biochem Soc Trans 2008;36:891—900.

[233] Pratt WB, Morishima Y, Murphy M, Harrell M. Chaperoning of glucocorticoid receptors. Handb Exp Pharmacol 2006:111—38.

[234] Murphy PJ, Morishima Y, Kovacs JJ, Yao TP, Pratt WB. Regulation of the dynamics of hsp90 action on the glucocorticoid receptor by acetylation/deacetylation of the chaperone. J Biol Chem 2005;280:33792—9.

[235] Amat R, Solanes G, Giralt M, Villarroya F. SIRT1 is involved in glucocorticoid-mediated control of uncoupling protein-3 gene transcription. J Biol Chem 2007;282:34066—76.

[236] Ito K, Barnes PJ, Adcock IM. Glucocorticoid receptor recruitment of histone deacetylase 2 inhibits interleukin-1beta-induced histone H4 acetylation on lysines 8 and 12. Mol Cell Biol 2000;20:6891—903.

[237] Kelly A, Bowen H, Jee YK, Mahfiche N, Soh C, Lee T, et al. The glucocorticoid receptor beta isoform can mediate transcriptional repression by recruiting histone deacetylases. J Allergy Clin Immunol 2008;121:203—8. e201.

[238] Wiper-Bergeron N, Salem HA, Tomlinson JJ, Wu D, Hache RJ. Glucocorticoid-stimulated preadipocyte differentiation is mediated through acetylation of C/EBPbeta by GCN5. Proc Natl Acad Sci USA 2007;104:2703—8.

[239] Wiper-Bergeron N, Wu D, Pope L, Schild-Poulter C, Hache RJ. Stimulation of preadipocyte differentiation by steroid through targeting of an HDAC1 complex. EMBO J 2003;22:2135—45.

[240] Bolden JE, Peart MJ, Johnstone RW. Anticancer activities of histone deacetylase inhibitors. Nat Rev Drug Discov 2006;5:769—84.

[241] Outeiro TF, Kontopoulos E, Altman S, Kufareva I, Strathearn KE, Amore AM, et al. Sirtuin 2 Inhibitors Rescue α-Synuclein-Mediated Toxicity in Models of Parkinson's Disease. Science 2007;317:516—9.

[242] Schuetz A, Min J, Antoshenko T, Wang CL, Allali-Hassani A, Dong A, et al. Structural basis of inhibition of the human NAD+-dependent deacetylase SIRT5 by suramin. Structure 2007;15:377—89.

[243] Kiviranta PH, Leppanen J, Kyrylenko S, Salo HS, Lahtela-Kakkonen M, Tervo AJ, et al. N, N′-Bisbenzylidenebenzene-1,4-diamines and N, N′-Bisbenzylidenenaphthalene-1,4-diamines as Sirtuin Type 2 (SIRT2). Inhibitors. J Med Chem 2006;49:7907—11.

[244] Tervo AJ, Suuronen T, Kyrylenko S, Kuusisto E, Kiviranta PH, Salminen A, et al. Discovering inhibitors of human sirtuin type 2: novel structural scaffolds. J Med Chem 2006;49:7239—41.

[245] Olaharski AJ, Rine J, Marshall BL, Babiarz J, Zhang L, Verdin E, et al. The flavoring agent dihydrocoumarin reverses epigenetic silencing and inhibits sirtuin deacetylases. PLoS Genet 2005;1:e77.

[246] Mai A, Massa S, Lavu S, Pezzi R, Simeoni S, Ragno R, et al. Design, synthesis, and biological evaluation of sirtinol analogues as class III histone/protein deacetylase (Sirtuin) inhibitors. J Med Chem 2005;48:7789—95.

[247] Dashwood RH, Ho E. Dietary histone deacetylase inhibitors: From cells to mice to man. Semin Cancer Biol 2007;17:363—9.

[248] Howitz KT, Bitterman KJ, Cohen HY, Lamming DW, Lavu S, Wood JG, et al. Small molecule activators of sirtuins extend Saccharomyces cerevisiae lifespan. Nature 2003;425:191—6.

[249] Kaeberlein M, McDonagh T, Heltweg B, Hixon J, Westman EA, Caldwell SD, et al. Substrate-specific activation of sirtuins by resveratrol. J Biol Chem 2005;280:17038—45.

[250] Gray SG. Histone acetyltransferases and histone deacetylases in gene regulation and as drug targets. In: DevitaJr VT, Hellman S, Rosenberg SA, editors. Progress in Oncology. Sudbury: Jones and Bartlett; 2004. p. 99—128.

[251] Tambaro F, Dell'Aversana C, Carafa V, Nebbioso A, Radic B, Ferrara F, et al. Histone deacetylase inhibitors: clinical implications for hematological malignancies. Clin Epigenet 2010:25—44.

362

[252] Blaheta RA, Cinatl Jr J. Anti-tumor mechanisms of valproate: a novel role for an old drug. Med Res Rev 2002;22:492—511.

[253] Soriano AO, Yang H, Faderl S, Estrov Z, Giles F, Ravandi F, et al. Safety and clinical activity of the combination of 5-azacytidine, valproic acid and all-trans retinoic acid in acute myeloid leukemia and myelodysplastic syndrome. Blood 2007;110:2302—8.

[254] Atmaca A, Al-Batran SE, Maurer A, Neumann A, Heinzel T, Hentsch B, et al. Valproic acid (VPA) in patients with refractory advanced cancer: a dose escalating phase I clinical trial. Br J Cancer 2007;97:177—82.

[255] Münster P, Marchion D, Bicaku E, Schmitt M, Lee JH, DeConti R, et al. Phase I trial of histone deacetylase inhibition by valproic acid followed by the topoisomerase II inhibitor epirubicin in advanced solid tumors: a clinical and translational study. J Clin Oncol 2007;25:1979—85.

[256] Greco R, Latini G, Chiarelli F, Iannetti P, Verrotti A. Leptin, ghrelin, and adiponectin in epileptic patients treated with valproic acid. Neurology 2005;65:1808—9.

[257] Pylvanen V, Knip M, Pakarinen AJ, Turkka J, Kotila M, Rattya J, et al. Fasting serum insulin and lipid levels in men with epilepsy. Neurology 2003;60:571—4.

[258] Pylvanen V, Pakarinen A, Knip M, Isojarvi J. Characterization of insulin secretion in Valproate-treated patients with epilepsy. Epilepsia 2006;47:1460—4.

[259] Pylvanen V, Pakarinen A, Knip M, Isojarvi J. Insulin-related metabolic changes during treatment with valproate in patients with epilepsy. Epilepsy Behav 2006;8:643—8.

[260] Tan H, Orbak Z, Kantarci M, Kocak N, Karaca L. Valproate-induced insulin resistance in prepubertal girls with epilepsy. J Pediatr Endocrinol Metab 2005;18:985—9.

[261] Verrotti A, di Corcia G, Salladini C, Trotta D, Morgese G, Chiarelli F. Serum insulin and leptin levels and valproate-associated obesity. Epilepsia 2003;44:1606.

[262] Verrotti A, D'Egidio C, Mohn A, Coppola G, Chiarelli F. Weight gain following treatment with valproic acid: pathogenetic mechanisms and clinical implications. Obes Rev 2011;12:e32—43.

[263] Lagace DC, McLeod RS, Nachtigal MW. Valproic acid inhibits leptin secretion and reduces leptin messenger ribonucleic acid levels in adipocytes. Endocrinology 2004;145:5493—503.

[264] Perseghin G, Petersen K, Shulman GI. Cellular mechanism of insulin resistance: potential links with inflammation. Int J Obes Relat Metab Disord 27 (Suppl.3)/ 2003(S6-11):S6—11.

[265] Shoelson SE, Lee J, Yuan M. Inflammation and the IKK beta/I kappa B/NF-kappa B axis in obesity- and diet-induced insulin resistance. Int J Obes Relat Metab Disord 2003;27:49—52.

[266] Zeender E, Maedler K, Bosco D, Berney T, Donath MY, Halban PA. Pioglitazone and sodium salicylate protect human beta-cells against apoptosis and impaired function induced by glucose and interleukin-1beta. J Clin Endocrinol Metab 2004;89:5059—66.

[267] Pinckard RN, Hawkins D, Farr RS. In vitro acetylation of plasma proteins, enzymes and DNA by aspirin. Nature 1968;219:68—9.

[268] Di Renzo F, Cappelletti G, Broccia ML, Giavini E, Menegola E. The inhibition of embryonic histone deacetylases as the possible mechanism accounting for axial skeletal malformations induced by sodium salicylate. Toxicol Sci 2008;104:397—404.

[269] Sun X, Han F, Yi J, Han L, Wang B. Effect of Aspirin on the Expression of Hepatocyte NF-kappaB and Serum TNF-alpha in Streptozotocin-Induced Type 2 Diabetic Rats. J Korean Med Sci 2011;26:765—70.

[270] Kim SN, Choi HY, Kim YK. Regulation of adipocyte differentiation by histone deacetylase inhibitors. Arch Pharm Res 2009;32:535—41.

[271] Yoo EJ, Chung JJ, Choe SS, Kim KH, Kim JB. Down-regulation of histone deacetylases stimulates adipocyte differentiation. J Biol Chem 2006;281:6608—15.

[272] Advani A, Huang Q, Thai K, Advani SL, White KE, Kelly DJ, et al. Long-term administration of the histone deacetylase inhibitor vorinostat attenuates renal injury in experimental diabetes through an endothelial nitric oxide synthase-dependent mechanism. Am J Pathol 2011;178:2205—14.

[273] Gilbert RE, Huang Q, Thai K, Advani SL, Lee K, Yuen DA, et al. Histone deacetylase inhibition attenuates diabetes-associated kidney growth: potential role for epigenetic modification of the epidermal growth factor receptor. Kidney Int 2011;79:1312—21.

[274] Schein PS, Cooney DA, Vernon ML. The use of nicotinamide to modify the toxicity of streptozotocin diabetes without loss of antitumor activity. Cancer Res 1967;27:2324—32.

[275] Stauffacher W, Burr I, Gutzeit A, Beaven D, Veleminsky J, Renold AE. Streptozotocin diabetes: time course of irreversible B-cell damage; further observations on prevention by nicotinamide. Proc Soc Exp Biol Med 1970;133:194—200.

[276] Sandler S, Andersson A, Korsgren O, Tollemar J, Petersson B, Groth CG, et al. Tissue culture of human fetal pancreas. Effects of nicotinamide on insulin production and formation of isletlike cell clusters. Diabetes 1989; 38(suppl.1):168—71 .

[277] Korsgren O, Andersson A, Sandler S. Pretreatment of fetal porcine pancreas in culture with nicotinamide accelerates reversal of diabetes after transplantation to nude mice. Surgery 1993;113:205–14.

[278] Gale EA, Bingley PJ, Emmett CL, Collier T. European Nicotinamide Diabetes Intervention Trial (ENDIT): a randomised controlled trial of intervention before the onset of type 1 diabetes. Lancet 2004;363: 925–31.

[279] Hedman M, Ludvigsson J, Faresjo MK. Nicotinamide reduces high secretion of IFN-gamma in high-risk relatives even though it does not prevent type 1 diabetes. J Interferon Cytokine Res 2006;26: 207–13.

[280] Liu HK, Green BD, Flatt PR, McClenaghan NH, McCluskey JT. Effects of long-term exposure to nicotinamide and sodium butyrate on growth, viability, and the function of clonal insulin secreting cells. Endocr Res 2004;30:61–8.

[281] Shin JS, Min BH, Lim JY, Kim BK, Han HJ, Yoon KH, et al. Novel culture technique involving an histone deacetylase inhibitor reduces the marginal islet mass to correct streptozotocin-induced diabetes. Cell Transplant 2011;20:1321–32.

[282] Honkanen J, Nieminen JK, Gao R, Luopajarvi K, Salo HM, Ilonen J, et al. IL-17 immunity in human type 1 diabetes. J Immunol 2010;185:1959–67.

[283] Jagannathan-Bogdan M, McDonnell ME, Shin H, Rehman Q, Hasturk H, Apovian CM, et al. Elevated proinflammatory cytokine production by a skewed T cell compartment requires monocytes and promotes inflammation in type 2 diabetes. J Immunol 2011;186:1162–72.

[284] Sutherland AP, Van Belle T, Wurster AL, Suto A, Michaud M, Zhang D, et al. Interleukin-21 is required for the development of type 1 diabetes in NOD mice. Diabetes 2009;58:1144–55.

[285] Spolski R, Kashyap M, Robinson C, Yu Z, Leonard WJ. IL-21 signaling is critical for the development of type I diabetes in the NOD mouse. Proc Natl Acad Sci USA 2008;105:14028–33.

[286] McGuire HM, Walters S, Vogelzang A, Lee CM, Webster KE, Sprent J, et al. Interleukin-21 is critically required in autoimmune and allogeneic responses to islet tissue in murine models. Diabetes 2011;60: 867–75.

[287] Arif S, Moore F, Marks K, Bouckenooghe T, Dayan CM, Planas R, et al. Peripheral and Islet Interleukin-17 Pathway Activation Characterizes Human Autoimmune Diabetes and Promotes Cytokine-Mediated {β }-Cell Death. Diabetes 2011;60:2112–9.

[288] Koenen HJ, Smeets RL, Vink PM, van Rijssen E, Boots AM, Joosten I. Human CD25highFoxp3pos regulatory T cells differentiate into IL-17-producing cells. Blood 2008;112:2340–52.

[289] Patel T, Patel V, Singh R, Jayaraman S. Chromatin remodeling resets the immune system to protect against autoimmune diabetes in mice. Immunol Cell Biol 2011;89:640–9.

[290] Gangaram-Panday ST, Faas MM, de Vos P. Towards stem-cell therapy in the endocrine pancreas. Trends Mol Med 2007;13:164–73.

[291] Lee JH, Hart SR, Skalnik DG. Histone deacetylase activity is required for embryonic stem cell differentiation. Genesis 2004;38:32–8.

[292] Romagnani P, Lasagni L, Mazzinghi B, Lazzeri E, Romagnani S. Pharmacological modulation of stem cell function. Curr Med Chem 2007;14:1129–39.

[293] Okazaki K, Maltepe E. Oxygen, epigenetics and stem cell fate. Regen Med 2006;1:71–83.

[294] Lumelsky N, Blondel O, Laeng P, Velasco I, Ravin R, McKay R. Differentiation of embryonic stem cells to insulin-secreting structures similar to pancreatic islets. Science 2001;292:1389–94.

[295] Calvanese V, Lara E, Suarez-Alvarez B, Abu Dawud R, Vazquez-Chantada M, Martinez-Chantar ML, et al. Sirtuin 1 regulation of developmental genes during differentiation of stem cells. Proc Natl Acad Sci USA 2011;107:13736–41.

[296] Tayaramma T, Ma B, Rohde M, Mayer H. Chromatin-remodeling factors allow differentiation of bone marrow cells into insulin-producing cells. Stem Cells 2006;24:2858–67.

[297] Sun Y, Chen L, Hou XG, Hou WK, Dong JJ, Sun L, et al. Differentiation of bone marrow-derived mesenchymal stem cells from diabetic patients into insulin-producing cells in vitro. Chin Med J (Engl) 2007;120:771–6.

[298] Goicoa S, Alvarez S, Ricordi C, Inverardi L, Dominguez-Bendala J. Sodium butyrate activates genes of early pancreatic development in embryonic stem cells. Cloning Stem Cells 2006;8:140–9.

[299] Jiang J, Au M, Lu K, Eshpeter A, Korbutt G, Fisk G, et al. Generation of Insulin-producing Islet-like Clusters from Human Embryonic Stem Cells. Stem Cells 2007;25:1940–53.

[300] Mutskov V, Raaka BM, Felsenfeld G, Gershengorn MC. The human insulin gene displays transcriptionally active epigenetic marks in islet-derived mesenchymal precursor cells in the absence of insulin expression. Stem Cells 2007;25:3223–33.

[301] Wilson LM, Wong SH, Yu N, Geras-Raaka E, Raaka BM, Gershengorn MC. Insulin but not glucagon gene is silenced in human pancreas-derived mesenchymal stem cells. Stem Cells 2009;27:2703–11.

[302] Davani B, Ikonomou L, Raaka BM, Geras-Raaka E, Morton RA, Marcus-Samuels B, et al. Human islet-derived precursor cells are mesenchymal stromal cells that differentiate and mature to hormone-expressing cells in vivo. Stem Cells 2007;25:3215−22.

[303] Rubenstein RC, Zeitlin PL. Sodium 4-phenylbutyrate downregulates Hsc70: implications for intracellular trafficking of DeltaF508-CFTR. Am J Physiol Cell Physiol 2000;278:C259−67.

[304] Vij N, Fang S, Zeitlin PL. Selective inhibition of endoplasmic reticulum-associated degradation rescues DeltaF508-cystic fibrosis transmembrane regulator and suppresses interleukin-8 levels: therapeutic implications. J Biol Chem 2006;281:17369−78.

[305] Burrows JA, Willis LK, Perlmutter DH. Chemical chaperones mediate increased secretion of mutant alpha 1-antitrypsin (alpha 1-AT) Z: A potential pharmacological strategy for prevention of liver injury and emphysema in alpha 1-AT deficiency. Proc Natl Acad Sci USA 2000;97:1796−801.

[306] Qi X, Hosoi T, Okuma Y, Kaneko M, Nomura Y. Sodium 4-phenylbutyrate protects against cerebral ischemic injury. Mol Pharmacol 2004;66:899−908.

[307] Vilatoba M, Eckstein C, Bilbao G, Smyth CA, Jenkins S, Thompson JA, et al. Sodium 4-phenylbutyrate protects against liver ischemia reperfusion injury by inhibition of endoplasmic reticulum-stress mediated apoptosis. Surgery 2005;138:342−51.

[308] Mulhern ML, Madson CJ, Kador PF, Randazzo J, Shinohara T. Cellular osmolytes reduce lens epithelial cell death and alleviate cataract formation in galactosemic rats. Mol Vis 2007;13:1397−405.

[309] Bonapace G, Waheed A, Shah GN, Sly WS. Chemical chaperones protect from effects of apoptosis-inducing mutation in carbonic anhydrase IV identified in retinitis pigmentosa 17. Proc Natl Acad Sci USA 2004;101:12300−5.

[310] de Almeida SF, Picarote G, Fleming JV, Carmo-Fonseca M, Azevedo JE,. de Sousa M.Chemical chaperones reduce endoplasmic reticulum stress and prevent mutant HFE aggregate formation. J Biol Chem 2007;282:27905−12.

[311] Kubota K, Niinuma Y, Kaneko M, Okuma Y, Sugai M, Omura T, et al. Suppressive effects of 4-phenylbutyrate on the aggregation of Pael receptors and endoplasmic reticulum stress. J Neurochem 2006;97:1259−68.

[312] Shao L, Sun X, Xu L, Young LT, Wang JF. Mood stabilizing drug lithium increases expression of endoplasmic reticulum stress proteins in primary cultured rat cerebral cortical cells. Life Sci 2006;78:1317−23.

[313] Wang JF, Bown C, Young LT. Differential display PCR reveals novel targets for the mood-stabilizing drug valproate including the molecular chaperone GRP78. Mol Pharmacol 1999;55:521−7.

[314] Cui J, Shao L, Young LT, Wang JF. Role of glutathione in neuroprotective effects of mood stabilizing drugs lithium and valproate. Neuroscience 2007;144:1447−53.

[315] Tveten K, Holla OL, Ranheim T, Berge KE, Leren TP, Kulseth MA. 4-Phenylbutyrate restores the functionality of a misfolded mutant low-density lipoprotein receptor. Febs J 2007;274:1881−93.

[316] Chiang DJ, Pritchard MT, Nagy LE. Obesity, diabetes mellitus, and liver fibrosis. Am J Physiol Gastrointest Liver Physiol 2011;300:G697−702.

[317] Morton GJ, Schwartz MW. Leptin and the central nervous system control of glucose metabolism. Physiol Rev 2011;91:389−411.

[318] Myers Jr MG, Leibel RL, Seeley RJ, Schwartz MW. Obesity and leptin resistance: distinguishing cause from effect. Trends Endocrinol Metab 2010;21:643−51.

[319] Hosoi T, Sasaki M, Miyahara T, Hashimoto C, Matsuo S, Yoshii M, et al. Endoplasmic reticulum stress induces leptin resistance. Mol Pharmacol 2008;74:1610−9.

[320] Ozcan L, Ergin AS, Lu A, Chung J, Sarkar S, Nie D, et al. Endoplasmic reticulum stress plays a central role in development of leptin resistance. Cell Metab 2009;9:35−51.

[321] Won JC, Jang PG, Namkoong C, Koh EH, Kim SK, Park JY, et al. Central administration of an endoplasmic reticulum stress inducer inhibits the anorexigenic effects of leptin and insulin. Obesity (Silver Spring) 2009;17:1861−5.

[322] Nogueira TC, Lellis-Santos C, Jesus DS, Taneda M, Rodrigues SC, Amaral FG, et al. Absence of melatonin induces night-time hepatic insulin resistance and increased gluconeogenesis due to stimulation of nocturnal unfolded protein response. Endocrinology 2011;152:1253−63.

[323] Ozcan U, Yilmaz E, Ozcan L, Furuhashi M, Vaillancourt E, Smith RO, et al. Chemical chaperones reduce ER stress and restore glucose homeostasis in a mouse model of type 2 diabetes. Science 2006;313:1137−40.

[324] Xiao C, Giacca A, Lewis GF. Sodium phenylbutyrate, a drug with known capacity to reduce endoplasmic reticulum stress, partially alleviates lipid-induced insulin resistance and beta-cell dysfunction in humans. Diabetes 2011;60:918−24.

[325] Hsu BR, Chen ST, Fu SH. Enhancing engraftment of islets using perioperative sodium 4-phenylbutyrate. Int Immunopharmacol 2006;6:1952−9.

[326] Zhou N, Moradei O, Raeppel S, Leit S, Frechette S, Gaudette F, et al. Discovery of N-(2-aminophenyl)-4-[(4-pyridin-3-ylpyrimidin-2-ylamino)methyl]benzamide (MGCD0103), an orally active histone deacetylase inhibitor. J Med Chem 2008;51:4072–5.

[327] Balasubramanian S, Ramos J, Luo W, Sirisawad M, Verner E, Buggy JJ. A novel histone deacetylase 8 (HDAC8)-specific inhibitor PCI-34051 induces apoptosis in T-cell lymphomas. Leukemia 2008;22:1026–34.

[328] Napper AD, Hixon J, McDonagh T, Keavey K, Pons JF, Barker J, et al. Discovery of indoles as potent and selective inhibitors of the deacetylase SIRT1. J Med Chem 2005;48:8045–54.

[329] Camins A, Sureda FX, Junyent F, Verdaguer E, Folch J, Pelegri C, et al. Sirtuin activators: designing molecules to extend life span. Biochim Biophys Acta 2010;1799:740–9.

[330] Miranda TB, Cortez CC, Yoo CB, Liang G, Abe M, Kelly TK, et al. DZNep is a global histone methylation inhibitor that reactivates developmental genes not silenced by DNA methylation. Mol Cancer Ther 2009;8:1579–88.

[331] Suzuki T, Miyata N. Lysine Demethylases Inhibitors. J Med Chem 2011;54:8236–50.

[332] Lizcano F, Romero C, Vargas D. Regulation of adipogenesis by nuclear receptor PPARgamma is modulated by the histone demethylase JMJD2C. Genet Mol Biol 2011;34:19–24.

[333] Huang J, Plass C, Gerhauser C. Cancer Chemoprevention by Targeting the Epigenome. Curr Drug Targets 2010;12:1925–56.

[334] Link A, Balaguer F, Goel A. Cancer chemoprevention by dietary polyphenols: promising role for epigenetics. Biochem Pharmacol 2010;80:1771–92.

[335] Meeran SM, Ahmed A, Tollefsbol TO. Epigenetic targets of bioactive dietary components for cancer prevention and therapy. Clin Epigenetics 2010;1:101–16.

[336] Spannhoff A, Kim YK, Raynal NJ, Gharibyan V, Su MB, Zhou YY, et al. Histone deacetylase inhibitor activity in royal jelly might facilitate caste switching in bees. EMBO Rep 2011;12:238–43.

[337] Song MY, Kim EK, Moon WS, Park JW, Kim HJ, So HS, et al. Sulforaphane protects against cytokine- and streptozotocin-induced beta-cell damage by suppressing the NF-kappaB pathway. Toxicol Appl Pharmacol 2009;235:57–67.

[338] Solowiej E, Solowiej J, Godlewski M, Motyl T, Perkowska-Ptasinska A, Jaskiewicz K, et al. Application of sulforaphane: histopathological study of intraportal transplanted pancreatic islets into livers of diabetic rats. Transplant Proc 2006;38:282–3.

[339] Dancik V, Seiler KP, Young DW, Schreiber SL, Clemons PA. Distinct biological network properties between the targets of natural products and disease genes. J Am Chem Soc 2010;132:9259–61.

[340] Zanichelli F, Capasso S, Cipollaro M, Pagnotta E, Carteni M, Casale F, et al. Dose-dependent effects of R-sulforaphane isothiocyanate on the biology of human mesenchymal stem cells, at dietary amounts, it promotes cell proliferation and reduces senescence and apoptosis, while at anti-cancer drug doses, it has a cytotoxic effect. Age (Dordr) 2012;34:281–93.

[341] Lee SJ, Krauthauser C, Maduskuie V, Fawcett PT, Olson JM, Rajasekaran SA. Curcumin-induced HDAC inhibition and attenuation of medulloblastoma growth in vitro and in vivo. BMC Cancer 2011;11:144.

[342] Chen Y, Shu W, Chen W, Wu Q, Liu H, Cui G. Curcumin, both histone deacetylase and p300/CBP-specific inhibitor, represses the activity of nuclear factor kappa B and Notch 1 in Raji cells. Basic Clin Pharmacol Toxicol 2007;101:427–33.

[343] Balasubramanyam K, Varier RA, Altaf M, Swaminathan V, Siddappa NB, Ranga U, et al. Curcumin, a novel p300/CREB-binding protein-specific inhibitor of acetyltransferase, represses the acetylation of histone/nonhistone proteins and histone acetyltransferase-dependent chromatin transcription. J Biol Chem 2004;279:51163–71.

[344] Marcu MG, Jung YJ, Lee S, Chung EJ, Lee MJ, Trepel J, et al. Curcumin is an inhibitor of p300 histone acetylatransferase. Med Chem 2006;2:169–74.

[345] Aggarwal BB. Targeting inflammation-induced obesity and metabolic diseases by curcumin and other nutraceuticals. Annu Rev Nutr 2010;30:173–99.

[346] Srinivasan M. Effect of curcumin on blood sugar as seen in a diabetic subject. Indian J Med Sci 1972;26:269–70.

[347] Wickenberg J, Ingemansson SL, Hlebowicz J. Effects of Curcuma longa (turmeric) on postprandial plasma glucose and insulin in healthy subjects. Nutr J 9:43

[348] Cuomo J, Appendino G, Dern AS, Schneider E, McKinnon TP, Brown MJ, et al. Comparative absorption of a standardized curcuminoid mixture and its lecithin formulation. J Nat Prod 2011;74:664–9.

[349] Nair HB, Sung B, Yadav VR, Kannappan R, Chaturvedi MM, Aggarwal BB. Delivery of antiinflammatory nutraceuticals by nanoparticles for the prevention and treatment of cancer. Biochem Pharmacol 2010;80:1833–43.

[350] Beher D, Wu J, Cumine S, Kim KW, Lu SC, Atangan L, et al. Resveratrol is not a direct activator of SIRT1 enzyme activity. Chem Biol Drug Des 2009;74:619–24.

[351] Szkudelski T, Szkudelska K. Anti-diabetic effects of resveratrol. Ann N Y Acad Sci 2011;1215:34—9.

[352] Czech MP, Aouadi M, Tesz GJ. RNAi-based therapeutic strategies for metabolic disease. Nat Rev Endocrinol 2011.

[353] Guay C, Roggli E, Nesca V, Jacovetti C, Regazzi R. Diabetes mellitus, a microRNA-related disease? Transl Res 2011;157:253—64.

[354] Kim N, Kim H, Jung I, Kim Y, Kim D, Han YM. Expression profiles of miRNAs in human embryonic stem cells during hepatocyte differentiation. Hepatol Res 2011;41:170—83.

[355] Kroon E, Martinson LA, Kadoya K, Bang AG, Kelly OG, Eliazer S, et al. Pancreatic endoderm derived from human embryonic stem cells generates glucose-responsive insulin-secreting cells in vivo. Nat Biotechnol 2008;26:443—52.

[356] Wittmann J, Jack HM. Serum microRNAs as powerful cancer biomarkers. Biochim Biophys Acta 2010;1806:200—7.

[357] Mattes J, Yang M, Foster PS. Regulation of microRNA by antagomirs: a new class of pharmacological antagonists for the specific regulation of gene function? Am J Respir Cell Mol Biol 2007;36:8—12.

CHAPTER 18

Epigenetic Aberrations in Human Allergic Diseases

Manori Amarasekera[1], David Martino[2], Meri K. Tulic[1], Richard Saffery[2], Susan Prescott[1]
[1]University of Western Australia, Subiaco, WA, Australia
[2]Murdoch Children's Research Institute, Parkville, VIC, Australia

369

18.1 INTRODUCTION AND CONTEXT: THE RISING PREVALENCE OF ALLERGIC DISEASES

The prevalence of allergic diseases such as asthma, allergic rhinitis, and eczema has risen at an alarming rate over the past 4–5 decades [1,2]. This has been clearly associated with the marked environmental changes associated with transition to more modern lifestyles. Moreover, the parallel rise in a wide range of other immune diseases during this short period provides additional strong evidence that the immune system is highly susceptible to these environmental changes [3]. Furthermore, there is mounting evidence that the effects of environmental change are potentially greatest during critical periods of life, when epigenetic modifications in immune gene expression can alter subsequent disease susceptibility.

T. Tollefsbol (Ed): Epigenetics in Human Disease. DOI: 10.1016/B978-0-12-388415-2.00018-4

The "allergy epidemic" was first evident in industrially developed countries initially as a surge of respiratory diseases such as asthma, rising at approximately 5% per year towards the new millennium [1]. The burden of these disorders is enormous, with more than 40% of the population in developed countries experiencing allergic symptoms [4,5]. While the prevalence of asthma and rhinitis may have reached a plateau, or may even have begun to decline in some regions [6–8], the global burden of these diseases continues to rise as the prevalence of respiratory allergies increases in developing countries as they undergo economic and lifestyle transition [6]. Of further concern, is an apparent "second wave" of allergic disease, manifest by a much more recent rise in food allergy, now looming as an epidemic in developed countries [9]. Food allergy was still uncommon at the time of the "first wave" of respiratory allergic disease, only emerging as a significant problem in the last 10–15 years. The reason for this earlier and more dramatic presentation of the allergic phenotype is not clear, but as this appears linked with ongoing environmental change, the same trends can also be anticipated in developing regions. This rise in disease burden is most evident in infants and children under 3 years of age, further highlighting the likely role of early environmental exposures. Of yet further concern, there is also evidence of increasing disease persistence. Food allergies (such as egg and milk allergy) which were previously transient in nature, are now more likely to persist into late childhood and adolescence [10]. Less common presentations of food allergy, for example eosinophilic eosophagitis, have also increased in recently [11]. Collectively, these changes in disease patterns are placing growing demands on healthcare systems globally.

While environmental change may be driving the recent rise in disease, differences in susceptibility and familial aggregation of allergic diseases also implies a genetic contribution to the risk of these diseases. Variations in genetic susceptibility cannot explain the rise in disease, but there was hope that identification of specific atopy/asthma genes could provide valuable insight into the causal pathways and disease pathogenesis. Although a large number of candidate genes have been associated with the asthma/allergy phenotype, the findings have been highly variable with poor reproducibility between populations [12,13]. A study carried out in early 1990s reported that monozygotic twins showed a greater concordance than dizygotic twins, whether reared apart or together, for asthma and rhinitis, indicating heritability as the major factor on expression of these disorders [14]. However, subsequent twin studies, with higher levels of within-pair discordance, revealed that environmental factors are equally or even more important in the development of disease [15,16]. This reflects the significant heterogeneity of these conditions that arise as a result of multiple and variable genetic and environmental influences. It is important to elucidate how environmental modifiers confer changes to gene expression to fully understand the gene–environment interactions.

18.2 MECHANISMS OF ALLERGIC RESPONSE

Differentiation of naive $CD4^+$ T cells into type 2 helper (Th2) cells to an "innocuous" environmental antigen (allergen) is a hallmark of the allergic response, and produces the propensity for IgE production (atopy) to that specific allergen. Once this pattern of response is established, subsequent exposure to the allergen leads to crosslinking of IgE on mast cells and an inflammatory cascade that culminates in the release of histamine and other mediators which produce the many signs and symptoms of allergic disease. The clinical pattern and severity vary according to the route of the allergen exposure, dose of the allergen, the pattern of tissues affected, and other individual factors. Although the culminating events in the IgE cascade and the underlying characteristics of the Th2 cellular response are well characterised [17][17], the factors initiating and driving this process are less clear. Naive $CD4^+$ T cells have the capacity to differentiate into a range of effector cells or regulatory cells depending on the local milieu at the time of allergen/antigen encounter (as discussed in more detail below). For example, the presence of interleukin 12 (IL-12) secreted from antigen-presenting cells (APC) promotes differentiation into Th1 subset. On the other hand, relative absence of IL-12 and the presence of IL-4 in the local microenvironment promote Th2 differentiation over the Th1 pathway. While the environmental changes favoring Th2 differentiation are not fully understood, declining microbial exposure (a potent stimulant for APC induced IL-12 production) has been a leading candidate. The range of other environmental factors which can also modify IL-12 production or T-cell differentiation, includes smoking [19], vitamin D [19], and antioxidants [20], which have all been implicated in the rise in allergic disease. Therefore, the processes regulating T-cell gene expression during differentiation and maturation are of central interest, particularly with emerging evidence that some of the environmental factors regulating the differentiation process have epigenetic effects which could influence CD4$^+$ T-cell lineage commitment and subsequent allergic propensity, as further discussed below.

18.3 FETAL LIFE: THE CRITICAL PERIOD OF IMMUNE DEVELOPMENT

As allergic disease is often first manifest in early childhood, it is clear that preceding events in development are important. Early life therefore represents a critical period when gene—environmental interactions play a determining role in specifying immune tolerance. Clear differences in the neonatal immune function of children who subsequently develop allergy strongly suggests that these disorders have their origins in fetal life [21—26]. The Developmental Origins of Health and Disease (DOHaD) hypothesis proposes that prenatal exposures have the potential to modify a range of developmental processes in the fetus, and these exposures may program susceptibility to many chronic diseases in later life [27,28]. There is some evidence that these changes in disease predisposition are the result of altered fetal gene expression induced through epigenetic changes. Although this has been best studied in the context of cardiovascular and metabolic diseases, epigenetic effects of environmental changes are also now being investigated as a mechanism of the dramatic rise in allergic diseases [29—31].

Complex immunological mechanisms have evolved to allow fetal and maternal immune systems to coexist during pregnancy. The maternal immune system adapts in a subtle way to a "Th2-state" in order to down-regulate Th1-mediated IFN-γ responses to fetal antigens which can adversely affect the developing fetus [32—34]. Reflecting this, neonatal cytokine production is dominated by Th2 cytokines, with relative suppression of IFN-γ production [35]. This down-regulation of Th1 response in the fetal environment is generally attributed to the production of Th1-antagonistic mediators produced by the placenta; however there is also evidence of direct epigenetic control of gene transcription (further discussed below). Regulatory T cells (Tregs) expand during pregnancy and are recruited to the feto—maternal interface where they orchestrate immune tolerance towards the fetus [36] which may also be under epigenetic control. Together, these observations suggest a role for epigenetic regulation in the establishment and maintenance of the fetal environment.

While the transition in early gene expression patterns from fetal to postnatal patterns is developmentally regulated, environmental forces, such as microbial exposure (which is known to promote Th1 and Treg differentiation), also appear to play a critical role in the success of this process [37]. A better understanding of these effects is important for developing strategies to prevent or suppress the allergic phenotype.

18.4 DEVELOPMENTAL DIFFERENCES IN GENE EXPRESSION IN ALLERGIC DISEASE

Many presymptomatic differences of immune maturation pathways have been observed between allergic and non-allergic children. Of these, relative immaturity of neonatal Th1 immune function has been one of the prominent antecedents of allergic disease [21,38]. Although Th1 responses are generally suppressed at birth, this appears to be more marked in individuals who develop subsequent allergic disease [21,38]. This is also coupled with a delayed postnatal maturation of Th1 immunity in high-risk children [22,26]. Differences in innate immunity [26] and Treg function [39,40] are also observed at birth between allergic and non-allergic children. These differences could reflect both genetic predisposition and environmental exposures in pregnancy at a time when the fetal immune system is potentially more vulnerable to epigenetic changes in gene expression, as discussed further below.

18.5 EPIGENETIC REGULATION OF IMMUNE DEVELOPMENT

Development of the immune system, like all other systems/organs, is under epigenetic control. Changes in epigenetic profile have been observed in developmental maturation of T cells with age [41]. In the fetus, there is low-level methylation of CpG sites within the promoter regions of the *IFNG* and Th2 cytokine loci (gene silencing) in naive CD4$^+$ T cells [42]. With age there is progressive demethylation of *IFNG* which is accompanied by an increased capacity of IFNγ production by adult naive CD4$^+$ T cells.

The best evidence of epigenetic regulation of immune pathways has been observed for T-cell differentiation [43–45]. Polarization of naive CD4$^+$ T cells by the cytokine milieu is mediated through activation of signaling pathways and transcription factors that are distinct for Th1 and Th2 subsets. IL-4 activation of STAT-6 signaling and the expression of the transcription factor GATA-3 promotes Th2 differentiation. On the other hand, IL-12-induced differentiation into IFN-γ producing Th1 cells is mediated through the activation of STAT-4 signaling and the transcription factor T-bet. Regulation of these Th cell lineages is governed by reciprocal inhibition; i.e. IL-4 activated GATA-3 inhibits Th1 cytokine expression, while inhibition of Th2 cytokine expression by T-bet ensures that once committed naive T helper cells generally differentiate in one dominant direction [46,47]. More recently another transcription factor, Runt-related transcriptional factor 3 (*RUNX3*), was found to be induced by T-bet, which together enhances expression of *IFNG* while repressing *IL4* expression [48].

In naive CD4$^+$ T cells, both *IFNG* and *IL4* are methylated, resulting in chromatin remodeling that is transcriptionally non-permissive [43,45]. During Th1 lineage commitment the promoter region of the *IFNG* gene undergoes progressive demethylation (activation), while the *IL4* is hypermethylated (silenced), thus inducing chromatin remodeling compatible with *IFNG* but not *IL4* expression. On the other hand, differentiation of naive CD4$^+$ cells to Th2 cells involves hypomethylation of the *IL4* and concomitant silencing (through methylation) of *IFNG* [42–44,46,49]. GATA-3-mediated chromatin remodeling at the Th2 cytokine loci (*IL4/IL5/IL13/RAD50*) is also essential for the Th2 lineage commitment and maintenance of Th2 phenotype through cell division [50]. Changes in histone modifications following T-cell receptor (TCR) signaling appear to be involved in this process. These modifications are orchestrated by site-specific enzymes including histone acetyltranferases (HATs) and histone deacetylases (HDACs) [46]. Acetylation of histone by HAT is generally associated with gene expression, whereas removal of acetyl group by HDAC is associated with more closed chromatin structure that makes the gene transcriptionally non-permissive. The importance of endogenous HDAC activity on Th cell differentiation was demonstrated by a shift in recall (memory) responses toward a Th2 phenotype when HDAC activity was inhibited by trichostatin-A (TSA). Bronchial biopsies from untreated asthmatics revealed a higher level of HAT and lower level of HDAC activity [51]. Of note, treatment with inhaled steroids was associated with a reverse in HAT/HDAC levels to that of controls, suggesting therapeutic pathways also involve epigenetic modulation.

IL-2, another important cytokine in maintaining survival and proliferation of activated T cells, provides another example of the epigenetic regulation of immune pathways. In naive T cells, the *IL2* promoter region is also methylated, undergoing demethylation following activation of the T cell with enhancing *IL2* expression [52,53].

Regulatory T-cell differentiation is also under epigenetic control [54,55]. Expression of the forkhead transcription factor 3 (FOXP3) is known to be critical for the development and regulatory functions of Tregs. The regulation of *FOXP3* expression is not completely understood, however, recent findings indicate a pivotal role for DNA methylation at both promoter and enhancer sequences [55]. In undifferentiated naive T cells, CpG sites within a CpG island associated with the *FOXP3* promoter are highly methylated and upon stimulation by TGF-β, these methylation marks are removed, ensuring stable expression of *FOXP3* and differentiation into Treg cell lineage.

A more recently recognized distinct helper-T cell subset, the Th17 lineage associated with a number of autoimmune diseases as well as with severe forms of allergic diseases [56–58] also appears to be regulated through changes in histone acetylation [59]. Although the developmental role of this IL-17-producing lineage is not clear, a recent study revealed that Th17 cells can differentiate from Treg cells, and that HDAC inhibitor TSA had a profound negative effect on the emergence of these IL-17-producing cells [60].

These observations have led to speculation that factors that increase methylation (i.e. of *IFNG* and *FOXP3*) may increase the risk of disease by silencing the Th1 and Treg pathways. While this notion may appear simplistic, there are now a number of studies pursuing this general concept that environmental changes can alter patterns of gene methylation.

While there is solid evidence that epigenetic machinery regulates genes/pathways directly linked to allergic response, epigenetic regulation of "master" transcription factors (such as NFκB) indirectly controls a wider range of downstream immune and inflammatory responses [61,62]. For example, glucocorticoid receptor (GR) function is regulated by HDAC2 which is sensitive to oxidative stress [63]. Reduced GR function up-regulates the NFκB activity [64] which in turn enhances the expression of inflammatory genes such as IL8 [63]. This highlights how disease risk may be modulated at multiple levels.

In addition to histone modifications and DNA methylation, other important gene regulatory networks contribute to the control of gene expression, including microRNAs (miRNAs), small interfering RNAs (siRNAs), and long non-coding RNAs (ncRNA). There is accumulating evidence that miRNAs are important for T-cell differentiation [65,66] and may be linked to development of inflammatory disease [67,68]. Since miRNAs often target hundreds of genes [69], they may be more important epigenetic regulators that mediate environmental assaults in specific inflammatory pathways.

Although early models of "lineage commitment" proposed distinct terminally differentiated Th subsets, more recent evidence argues against this more static view [70]. While CD4$^+$ cells differentiate according to their local cytokine milieu during stimulation, they retain a degree of cellular plasticity [70–72]. Epigenetic modifications are undoubtedly involved in regulating the switch that controls signature-cytokine genes determining naive CD4$^+$ cell fates, but may be equally important in contributing to the potential cellular flexibility.

Now there is a growing body of data in support of epigenetic regulation of cellular plasticity of Th cells [70]. While Foxp3 is essential for the maintenance of suppressive function in Treg cells, it has been shown that depletion of Foxp3 expression results in acquisition of "effector" functions by these cells with concomitant loss of regulatory properties [73,74]. These observations have led to the speculation that "Th-effector" functions in Treg cells are continuously maintained in a dormant state by an active Foxp3-mediated mechanism. Recently, Beyer et al. reported that repression of "Th-effector" functions of Treg cells are mostly mediated by SATB1, a chromatin organizer and a transcription factor, and appear to be under direct transcriptional control of Foxp3 (at *SATB1* locus) as well as indirectly by Foxp3-dependent miRNA [75].

Furthermore, release of *SATB1* from Foxp3 control was able to reprogram the Treg to gain "effector" function while losing its suppressive function. This shows that Treg population represents a differentiated cell lineage committed to a specific function but retains developmental plasticity that may be mediated through epigenetic mechanisms.

Challenging traditional models of epigenetic control of T-cell lineage commitment, Wei et al. mapped active and repressive histone marks genome-wide across a spectrum of Th1, Th2, Th17, natural and induced Treg phenotypes [76]. Modifications at signature cytokine genes conformed to previous models of T-cell commitment, however, master transcription factors, such as the gene encoding T-bet (*TBX21*) and GATA3, were found to exhibit a mixture of active and repressive (bivalent) chromatin states across these phenotypes. It is speculated that bivalent epigenetic marks in master regulators of the Th differentiation maintain these transcription factors at a "poised" state for expression in non-expressing cell lineages and under appropriate conditions they can be induced leading to an alternate cell fate [76,77]. This shows that epigenetic mechanisms play a dual role in Th differentiation: ensure a "committed" state of Th-cell response upon activation while conferring cellular plasticity.

18.6 FACTORS THAT MODULATE ALLERGIC DISEASE RISK THROUGH EPIGENETIC MECHANISMS

There is intense interest in the prenatal factors that may modify optimal patterns of immune gene activation or silencing. A range of exposures already implicated in the rise in allergic disease have been shown to have potential epigenetic effects on fetal immune function, including microbial exposure, maternal diet, and pollutants [37,38] (Figure 18.1). Table 18.1 summarizes the environmental factors that modulate disease risk possibly through epigenetic modifications.

18.6.1 Epigenetic Effects of Bacterial Exposure on Immune Development

Declining microbial exposure has long been implicated in the rise in allergic diseases [3], although the original "hygiene hypothesis" was focused more on the allergy-protective effects of infectious exposures rather than more general microbial burden [79]. The initial explanations for this protective effect centered around the ability of microbial components to up-regulate IFN-γ production and inhibit pro-allergic Th2 response [3]. This notion is supported

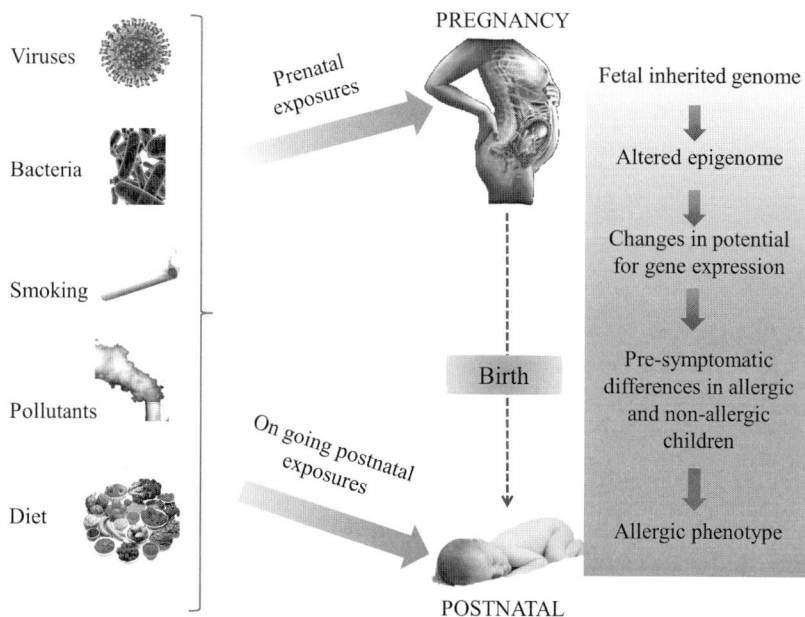

FIGURE 18.1

Environmental influences on developing immune system: This illustrates the environmental factors that have been shown to modify epigenetic profile and gene expression during early development. This figure is reproduced in the color plate section.

TABLE 18.1 Environmental Factors Known to be Associated with Epigenetic Modifications

Environmental Factor	Epigenetic Modification	Effect of Epigenetic Changes on Allergy	Reference
Prenatal exposure to microbial products	H4 acetylation of *IFNG* promoter in mice	Increase in *IFNG* expression	[80]
	Demethylation of Treg-specific demethylated region (TSDR) in neonates	Induce *FOXP3* expression	[87]
Maternal high-folate diet	Methylation changes in 82 gene loci including *RUNX3* in mice	Increase in airway hyperresponsiveness, airway eosinophilia, and production of inflammatory cytokines	[94]
Exposure to tobacco smoke	Suppress HDAC activity and overall HDAC activity in alveolar macrophages and bronchial biopsies in healthy smokers	Increase in expression of inflammatory mediators GM-CSF, IL-8, and TNF-α Reduced response to corticosteroids	[63]
	Hypomethyalation of *MAOB* promorter in circulating platelets and PBMCs in smokers	Significance to allergy is not yet known	[111]
Prenatal exposure to tobacco smoke	Global DNA hypomethylation, hypermethylation of *AXL* and *PTPRO*, varied pattern of *LINE1* methylation pattern by child's *GSPT1* haplotype	Relevance to pathogenesis of allergic disease is not yet known	[116]
Exposure to black carbon particles	Hypomethylation of *LINE1* of leukocyte DNA of elderly people	Relevance to allergy has not yet identified	[121]
Prenatal exposure to PAH	Hypermethylation of ACSL3 in cord blood mononuclear cells	Direct relevance to allergic disease pathogenesis has not been defined	[126]
Exposure to diesel exhaust particles	Hypermethylation of *IFNG* promoter and hypomethylation of *IL4* promorter in splenic CD4$^+$ cells in mice	Increase in the production of IgE upon intranasal administration of *Aspergillus fumigates*	[133]
Prenatal exposure to lead	Global DNA hypomethylation in neonates	Outcome on allergic disease is not yet known	[146]

by the impaired postnatal maturation of IFN-γ pathways in allergic disease [22,26]. Furthermore, an array of microbial exposures (such as enteric flora, farm animals, and house dust endotoxin) has been shown to both increase early IFN-γ production (Th1 response) and decrease the risk of allergic disease. Given the fact that ultimate regulation of IFN-γ expression is linked to methylation/demethylation events in the *IFNG* promoter in CD4$^+$ T cells, it is tempting to speculate that microbial exposures induce demethylation/acetylation of *IFNG* in CD4$^+$ T cells [31]. Preliminary results from an animal model provide some evidence that this is possible. Non-pathogenic microbial strains (*Acinetobacter lwoffi*) isolated from farming environments can induce epigenetic effects when administered to pregnant animals and protect the offspring from experimental postnatal asthma [80]. This effect depends on increased expression of IFN-γ mediated by an increase in H4 acetylation of the *IFNG* promoter. Notably, these effects were abolished by inhibition of histone acetylation following garcinol treatment. Even though there is no direct evidence for such association in humans, bacterial, viral, and parasitic agents have been shown to hold the capacity of inducing methylation events in host DNA [81–84]. This is an area for future research with clear implications on therapeutic as well as preventive strategies for allergic diseases.

While the postnatal microbial exposure has the most obvious implications for the developing immune system, there is emerging evidence that effects of microbial exposure may begin much earlier with maternal microbial exposure showing potential to modulate fetal immune function. Both human and animal studies clearly demonstrate that in utero exposure to both pathogenic and non-pathogenic microbial products can prevent allergic outcomes in the offspring, independent of postnatal exposure [85–88]. This protective effect has been associated with enhanced neonatal Treg numbers and function [87] along with increased *IFNG* expression [80]. Interestingly, exposure to microbial products in a farming environment appeared to stimulate *FOXP3* expression in neonates by demethylating an evolutionary conserved element within the *FOXP3* locus, Treg-specific demethylated region (TSDR) [87]. These findings together, are highly suggestive of microbial exposure during pregnancy can modify fetal immune responses through epigenetic mechanisms [81–83].

18.6.2 Epigenetic Effects of Maternal Diet on Immune Function

Modern diets differ in many aspects from more traditional diets with more processed and synthetic foods and less fresh fish, fruits, and vegetables. Of immediate relevance here, this dietary pattern in pregnancy appears to provide less tolerogenic conditions during early immune development promoting allergic outcomes in the offspring [89]. The specific nutritional changes implicated in the rising prevalence of asthma and other allergic diseases includes a decline in consumption of polyunsaturated fatty acids (PUFA) [90], soluble fiber [91], antioxidants, and other vitamins [92,93]. Indeed, diet and nutrition in pregnancy have been a dominant basis for notions of the "developmental origins" of many diseases [28]. The first evidence that maternal dietary changes in pregnancy can alter immune function and allergic outcomes through epigenetic modifications came from animal studies. A diet rich in methyl donors (folate) fed to pregnant mice induced allergic airway disease and a Th2 phenotype to in the offspring (F1 generation) [94]. This folate-rich maternal diet induced methylation changes in 82-gene loci in the offspring, resulting in increased airway hyperresponsiveness, airway eosinophilia, and production of inflammatory cytokines. This trait was then inherited to the subsequent F2 generation, demonstrating the transgenerational effects of environmental modification. Among these genes, *RUNX3*, a gene known to regulate T-lymphocyte development and to suppress airways inflammation [95] was hypomethylated with concordant transcriptional silencing of this gene in the progeny [94]. While some human studies reported that folic acid supplementation during pregnancy is associated with an increased risk of asthma and respiratory infections in infants [96], a recent Dutch study revealed no association between maternal folic acid supplementation and allergic outcomes in neonates [97]. However, until this is fully explored in human studies and the mechanistic pathways are clearly delineated, it is not appropriate to change the current practice of folate fortification to prevent neural tube defects.

The role of vitamin D as an immune-modulatory substance is currently under much debate. Epidemiological associations between vitamin D levels and allergic diseases remain inconclusive. Vitamin D intake during pregnancy has been associated with either increased risk [98] or decreased risk [99,100] of allergic disease in infants. At a cellular level, $1\alpha,25$-dihydroxyvitamin D_3 (active metabolite of vitamin D) also appears to have diverse actions on nuclear factor kappa B (NF-κB)-driven transcription of inflammatory genes [101,102]. However, it appears that HDAC activity is required for vitamin D-mediated NF-κB modulation [101].

Supplementation of fish oil (*n*-3 PUFA) is associated with effects on immune function of the offspring [90]. However, at this stage it is not clear whether it is related to epigenetic modulation. Similarly, antioxidants have the capacity to induce T-cell tolerance [103] and enhance the production of IL-12 by dendritic cells [104]. It can be postulated that through these effects antioxidants can favor the development of Th1 cells while suppressing the Th2 development. The effect of dietary antioxidants during pregnancy on fetal immune development is limited [105]. Evidence that oxidative stressors can modify the disease risk through epigenetic mechanisms suggests a role for these pathways [106].

Purified compounds isolated from garlic and broccoli have been reported to have epigenetic effects [107]. These bioactive compounds are found to be associated with HDAC inhibitory activity in animal models and may be related to increased cancer risk. Based on the immune modulatory property of these extracts, these common dietary components may be an additional source of epigenome modifiers in allergy risk and warrant further study.

18.6.3 Epigenetic Effects of Tobacco Smoke

Exposure to tobacco smoke represents a major risk factor for an array of diseases including asthma [108], chronic obstructive pulmonary disease (COPD) [109], and lung cancers [110]. There is mounting evidence that epigenetic modifications induced by tobacco smoke are associated with the development of these chronic diseases [78]. One possible epigenetic effect of tobacco smoke in the pathogenesis of respiratory inflammatory diseases is through perturbing the balance between the HAT/HDAC homeostasis of the airway immune cells. Bronchial biopsies and alveolar macrophages taken from healthy smokers and age-matched healthy non-smokers reveal that tobacco smoke suppresses the HDAC2 expression and overall HDAC activity and enhanced the expression of inflammatory mediators GM-CSF, IL-8, and TNF-α [63]. Of note, cigarette smoking reduces the response to corticosteroids by decreasing HDAC activity in key inflammatory cells such as alveolar macrophages, explaining the attenuated response to steroid therapy in patients with COPD (which also has a strong link with chronic exposure to tobacco smoke). The HDAC inhibitor TSA has been shown to reverse the proinflammatory changes and glucocorticoid responsiveness in the macrophages [63], implying their usefulness as an adjuvant drug for the treatment of asthma.

In addition to altering the HAT/HDAC balance, tobacco smoke can modulate the DNA methylation status of regulatory regions in a number of genes. Direct evidence for modified DNA methylation of epigenetic tags comes from a report that smoking leads to hypomethylation of monoamine oxidase (MAO) type B promoter in circulating platelets [111]. Furthermore, the authors describe similar results for PBMC of smokers and imply smoking-induced MAO gene deregulation could have a more general impact than vascular effects.

Exposure to cigarette smoke in pregnancy has many adverse effects on the fetus, including effects on lung function and asthma risk [112,113]. Smoking in the last trimester has been associated with early onset of airway hyperreactivity (likely asthma) by the age of 1 year [114]. Moreover, both maternal and grandmaternal smoking during pregnancy are associated with increased risk of childhood asthma, suggesting a persistent heritable effect [115]. Comparison of DNA methylation pattern of buccal cells from children born according to maternal smoking habits during pregnancy, revealed that exposure to tobacco smoke was associated with global DNA hypomethylation. Exposed children had significantly lower methylation of short interspersed nucleotide element AluYb8, a surrogate marker of global DNA methylation [116]. In addition, the study revealed that smoking affects epigenetic marks in gene specific manner. Using a CpG loci screen, eight genes were found differentially methylated in exposed children as opposed to unexposed children. Two genes, AXL and PTPRO, were validated by pyrosequencing and showed significant hypermethylation in exposed children but their significance in relation to asthma pathogenesis remains unclear. Moreover, methylation status of the DNA repetitive element LINE1 was observed only in children with common GSTM1-null genotype, while methylation pattern in exposed children varied with a common GSPT1 haplotype implying the genetic influence over the epigenetic−environmental interactions. It is interesting that maternal smoking had dual effects on fetal methylation profile; global DNA hypomethylation and hypermethylation of some specific genes [116]. Global hypomethylation could result from DNA damage caused by smoke-induced oxidative stress to DNA that interferes with the binding of DNA methyltransferase (DNMT) preventing DNA methylation [117]. The effect of tobacco smoke exposure on methylation of specific genes could possibly be due to de novo methylation in specific gene promoters, perhaps by incomplete erasure during methylation reprogramming that occurs in the embryo after fertilization [118].

18.6.4 Effects of Air Pollutants and Other Outdoor Pollutants on Epigenetics

Airborne pollutants such as particulate matter (PM) and noxious gases including benzene have been shown to be associated with asthma and other respiratory diseases [119,120]. These agents cause exacerbation of asthma symptoms in affected individuals but a causative link to asthma has not been well defined. Ambient level of black carbon particles, a marker of traffic pollution, has been consistently associated with a variety of adverse health outcomes and exposure even for a short duration was associated with hypomethylation of LINE1 but not Alu in blood DNA samples taken from a cohort of elderly people [121]. Polycyclic aromatic hydrocarbons (PAH) are one of the most widespread organic pollutants and also a major component of PM of urban aerosol. In addition to their presence in oil, coal, and tar deposits they are also formed by the incomplete combustion of carbon-containing fuels such as wood, coal, diesel, fat, tobacco, and incense. Grilled, smoked, or barbecued meats also appear to contain high levels of PAH [122,123]. In addition to its carcinogenic properties [124], it has been found not only to impair functions of airway cells and smooth muscle cells but also diminish responsiveness to standard therapy given to asthmatics [125]. Recently, a novel epigenetic marker for PAH-associated asthma has been identified and cord blood mono-nuclear cells (CBMC) of children born to mothers who had been exposed to considerable level of PAH showed hypermethylation of the acyl-CoA synthetase long-chian family member 3 (*ACSL3*) promoter [126]. Furthermore, the exposure level was highly correlated with increased risk of asthma symptoms in the offspring before age 5 years. *ACSL3* genes are expressed in lungs and thymus tissue and encode key enzymes in fatty acid metabolism [127,128]. Given the fact that *ACSL3* is located in 2q36.1 which is associated with regions of the asthma susceptibility loci in specific populations [129,130], it is tempting to speculate that hypermethylation of this gene in lung tissues can potentially influence the fatty acid metabolism and phospholipid composition of the membranes and lung function. The direct relevance of this finding to asthma pathogenesis has not been defined but epidemiological studies have revealed the alterations in fatty acid composition in the diet [131] and cord blood [132] are associated with the increased risk of asthma.

Diesel exhaust particles (DEP), in addition to a source of PAH, give rise to an array of chemicals dispersed in the air as ultrafine particles. Exposure of mice to inhaled DEP for 3 weeks can augment the production of IgE upon intranasal administration of *Aspergillus fumigates* [133]. Hypersensitization occurred through hypermethylation of *IFNG* promoter with concomitant hypomethylation of *IL4* promoter in DNA from splenic $CD4^+$ cells. The effects of PM could be mediated in the airways through induction of oxidative stress. Treatment of A549 cells (adenocarcinomic human alveolar basal epithelial cells) with either PM-10 or H_2O_2 increased the expression and release of IL-8, which increased the HAT activity, hence remodeled the *IL8* promoter region [134] suggesting PM exert their effects through chromatin remodeling of the susceptible genes.

Benzene, toluene, xylene, and other volatile organic compounds like PM are associated with adverse health effects including asthma. Children exposed to benzene have an increased risk of asthma symptoms [135] and this could possibly be mediated through changing the DNA methylation profile since exposure to benzene induced changes in DNA methylation in a global and gene-specific manner [136].

Although less is known about the immune effects of other pollutants released to the environment by agricultural and modern industrial processes, it has been shown that highly lipid soluble substances such as polychlorinated biphenyl compounds, organochlorine pesticide, dioxins, and phthalates accumulate in human tissue with age [137] possibly through contaminated water, food, and clothing. Of most concern, some of these products have been measured in breast milk, cord blood, and placental tissue [138–140] emphasizing the possible adverse outcome in early development and subsequent disease pathogenesis in offspring.

Higher levels of organic pollutants were associated with higher levels of cord blood IgE antibodies [141]. At higher levels these products can have immunosuppressive effects in humans [142], whereas at low levels some appear to selectively inhibit type 1 immune responses [143], leading to speculation that this could possibly favor allergic (type 2) immune responses. Many of these organic pollutants now have been associated with modified epigenetic tags in humans, as evidenced by variations in global DNA hypomethylation patterns with persistent low-dose exposure [144]. More evidence for epigenetic alterations induced by exposure to organic pollutants comes from rodent studies [145]. Prenatal exposure to lead was associated with global DNA hypomethylation in a human study suggesting epigenome of the developing fetus can be influenced by maternal cumulative lead burden [146]. This may influence long-term epigenetic programming and disease susceptibility throughout the life course. The levels of many pollutants are declining in some regions as a result of restrictions imposed on the use of pesticides and other toxic chemicals, and this is reflected in declining levels measured in adipose tissue [147]. Nevertheless, the effects of these factors should not be ignored, as epigenetic effects may potentially reflect exposure of subsequent several generations and this relationship may be obscured in cross-sectional epidemiological studies.

18.6.5 Other Maternal Factors that May Modulate Allergic Propensity in the Newborn Through Epigenetic Mechanisms

In the context of asthma and allergic diseases, the maternal phenotype appears to be a major factor determining the subsequent outcomes in the offspring. Maternal asthma and allergic status has a stronger effect than paternal allergy on both allergic diseases and Th1-IFNγ production in the neonate [148]. Lower IFNγ responses to HLA-DR-mismatched fetal antigens have also been observed in allergic mothers compared to non-allergic mothers [149]. This may affect the cytokine milieu at the feto–maternal interface and could be a mechanistic link of attenuated Th1 responses commonly observed in infants born to atopic mothers [38]. Rising rates of maternal allergy mean that the endogenous effects of the maternal allergic phenotype have the potential to amplify the effects of a proallergic exogenous environment. The underlying mechanisms, yet unclear, may involve epigenetic modifications of the specific immune genes.

The hypothalamic–pituitary–adrenal (HPA) axis is a major component of the neuroendocrine system. This axis controls many body processes and plays a major role in controlling stress responses. The immune system and the HPA axis are closely linked, particularly through the effects of glucocorticoids which connect these two critical systems. In pregnancy, the placental immune system is, at least in part, regulated by glucocorticoids. Under adverse (stressful) conditions HPA axis activation can induce up-regulation of placental Th1 cytokine production, resulting in poor fetal outcomes. In animal models, it has been shown that maternal stress can alter placental gene expression, in particular, genes involved in DNA methylation and histone modification and cell cycle regulation [150] and strongly suggest that maternal stress can induce epigenetic effects on fetal immune function with implications for the subsequent risk of disease in childhood. This is still poorly understood and an important area for ongoing research.

18.7 CONCLUSIONS

Exposure to a plethora of environmental factors (microbial exposure, maternal diet and smoking) during critical periods of early immune development, has the potential to modify the fetal immune development and the risk of subsequent disease. Notions of plasticity in gene expression that may be epigenetically modified by the early environment provide a new model to understand the gene–environmental interactions that contribute to the rising prevalence of asthma, allergy, and other immune diseases. Of greatest significance, this epigenetic plasticity may pave the way to develop novel early interventions to curb the epidemic of immune disease, ideally through primary prevention in early life.

379

The discovery of epigenetics as a key mechanism modulating immune machinery has profoundly changed perspectives and research approaches to allergy disease. However, many unanswered questions need to be addressed before these findings will be of any therapeutic value, including: can epigenetic profiles be used to accurately predict disease risk and susceptibility to treatment at the individual and population levels? How long will epigenetic memory last and can we reverse any events that occurred in early life at a later stage? Can we erase the epigenetic marks passed through generations by modulating the environment of the next generations or with therapeutic interventions?

References

[1] Devereux G. The increase in allergic disease: environment and susceptibility. Proceedings of a symposium held at the Royal Society of Edinburgh, 4th June 2002. Clin Exp Allergy 2003;33:394–406.

[2] Downs SH, Marks GB, Sporik R, Belosouva EG, Car NG, Peat JK. Continued increase in the prevalence of asthma and atopy. Arch Dis Child 2001;84:20–3.

[3] Bach JF. The effect of infections on susceptibility to autoimmune and allergic diseases. N Engl J Med 2002;347:911–20.

[4] Peat JK, van den Berg RH, Green WF, Mellis CM, Leeder SR, Woolcock AJ. Changing prevalence of asthma in Australian children. BMJ 1994;308:1591–6.

[5] Hopper JL, Jenkins MA, Carlin JB, Giles GG. Increase in the self-reported prevalence of asthma and hay fever in adults over the last generation: a matched parent-offspring study. Aust J Public Health 1995;19:120–4.

[6] Pearce N, Ait-Khaled N, Beasley R, Mallol J, Keil U, Mitchell E, et al. Worldwide trends in the prevalence of asthma symptoms: phase III of the International Study of Asthma and Allergies in Childhood (ISAAC). Thorax 2007;62:758–66.

[7] Robertson CF, Roberts MF, Kappers JH. Asthma prevalence in Melbourne schoolchildren: have we reached the peak? Med J Aust 2004;180:273–6.

[8] Bjorksten B, Clayton T, Ellwood P, Stewart A, Strachan D. Worldwide time trends for symptoms of rhinitis and conjunctivitis: Phase III of the International Study of Asthma and Allergies in Childhood. Pediatr Allergy Immunol 2008;19:110–24.

[9] Prescott S, Allen KJ. Food allergy: riding the second wave of the allergy epidemic. Pediatr Allergy Immunol 2011;22:155–60.

[10] Skripak JM, Matsui EC, Mudd K, Wood RA. The natural history of IgE-mediated cow's milk allergy. J Allergy Clin Immunol 2007;120:1172–7.

[11] Straumann A, Simon HU. Eosinophilic esophagitis: escalating epidemiology? J Allergy Clin Immunol 2005;115:418–9.

[12] Holloway JW, Yang IA, Holgate ST. Genetics of allergic disease. J Allergy Clin Immunol 2010;125:S81–94.

[13] Weiss ST, Raby BA, Rogers A. Asthma genetics and genomics 2009. Curr Opin Genet Dev 2009;19:279–82.

[14] Hanson B, McGue M, Roitman-Johnson B, Segal NL, Bouchard Jr TJ, Blumenthal MN. Atopic disease and immunoglobulin E in twins reared apart and together. Am J Hum Genet 1991;48:873–9.

[15] Skadhauge LR, Christensen K, Kyvik KO, Sigsgaard T. Genetic and environmental influence on asthma: a population-based study of 11,688 Danish twin pairs. Eur Respir J 1999;13:8–14.

[16] Liu X, Zhang S, Tsai HJ, Hong X, Wang B, Fang Y, et al. Genetic and environmental contributions to allergen sensitization in a Chinese twin study. Clin Exp Allergy 2009;39:991–8.

[17] Holgate ST. The epidemic of allergy and asthma. Nature 1999;402:B2–4.

[18] Noakes PS, Holt PG, Prescott SL. Maternal smoking in pregnancy alters neonatal cytokine responses. Allergy 2003;58:1053–8.

[19] D'Ambrosio D, Cippitelli M, Cocciolo MG, Mazzeo D, Di Lucia P, Lang R, et al. Inhibition of IL-12 production by 1,25-dihydroxyvitamin D3. Involvement of NF-kappaB downregulation in transcriptional repression of the p40 gene. J Clin Invest 1998;101:252–62.

[20] Murata Y, Shimamura T, Hamuro J. The polarization of T(h)1/T(h)2 balance is dependent on the intracellular thiol redox status of macrophages due to the distinctive cytokine production. Int Immunol 2002;14:201–12.

[21] Tang ML, Kemp AS, Thorburn J, Hill DJ. Reduced interferon-gamma secretion in neonates and subsequent atopy. Lancet 1994;344:983–5.

[22] Prescott SL, Macaubas C, Smallacombe T, Holt BJ, Sly PD, Holt PG. Development of allergen-specific T-cell memory in atopic and normal children. Lancet 1999;353:196–200.

[23] Amoudruz P, Holmlund U, Malmstrom V, Trollmo C, Bremme K, Scheynius A, et al. Neonatal immune responses to microbial stimuli: is there an influence of maternal allergy? J Allergy Clin Immunol 2005;115:1304—10.

[24] Schaub B, Campo M, He H, Perkins D, Gillman MW, Gold DR, et al. Neonatal immune responses to TLR2 stimulation: influence of maternal atopy on Foxp3 and IL-10 expression. Respir Res 2006;7:40.

[25] Prescott SL, Noakes P, Chow BW, Breckler L, Thornton CA, Hollams EM, et al. Presymptomatic differences in Toll-like receptor function in infants who have allergy. J Allergy Clin Immunol 2008;122:391—9:399e1—5.

[26] Tulic MK, Hodder M, Forsberg A, McCarthy S, Richman T, D'Vaz N, et al. Differences in innate immune function between allergic and nonallergic children: new insights into immune ontogeny. J Allergy Clin Immunol 2011;127:470—8:e1.

[27] Barker DJ, Fall CH. Fetal and infant origins of cardiovascular disease. Arch Dis Child 1993;68:797—9.

[28] Waterland RA, Michels KB. Epigenetic epidemiology of the developmental origins hypothesis. Annu Rev Nutr 2007;27:363—88.

[29] Miller RL, Ho SM. Environmental epigenetics and asthma: current concepts and call for studies. Am J Respir Crit Care Med 2008;177:567—73.

[30] Bousquet J, Jacot W, Yssel H, Vignola AM, Humbert M. Epigenetic inheritance of fetal genes in allergic asthma. Allergy 2004;59:138—47.

[31] Vuillermin PJ, Ponsonby AL, Saffery R, Tang ML, Ellis JA, Sly P, et al. Microbial exposure, interferon gamma gene demethylation in naive T-cells, and the risk of allergic disease. Allergy 2009;64:348—53.

[32] Breckler LA, Hale J, Jung W, Westcott L, Dunstan JA, Thornton CA, et al. Modulation of in vivo and in vitro cytokine production over the course of pregnancy in allergic and non-allergic mothers. Pediatr Allergy Immunol 2010;21:14—21.

[33] Wegmann TG, Lin H, Guilbert L, Mosmann TR. Bidirectional cytokine interactions in the maternal-fetal relationship: is successful pregnancy a TH2 phenomenon? Immunol Today 1993;14:353—6.

[34] Breckler LA, Hale J, Taylor A, Dunstan JA, Thornton CA, Prescott SL. Pregnancy IFN-gamma responses to foetal alloantigens are altered by maternal allergy and gravidity status. Allergy 2008;63:1473—80.

[35] Prescott SL, Macaubas C, Holt BJ, Smallacombe TB, Loh R, Sly PD, et al. Transplacental priming of the human immune system to environmental allergens: universal skewing of initial T cell responses toward the Th2 cytokine profile. J Immunol 1998;160:4730—7.

[36] Schumacher A, Brachwitz N, Sohr S, Engeland K, Langwisch S, Dolaptchieva M, et al. Human chorionic gonadotropin attracts regulatory T cells into the fetal-maternal interface during early human pregnancy. J Immunol 2009;182:5488—97.

[37] Martino DJ, Prescott SL. Silent mysteries: epigenetic paradigms could hold the key to conquering the epidemic of allergy and immune disease. Allergy 2010;65:7—15.

[38] Rinas U, Horneff G, Wahn V. Interferon-gamma production by cord-blood mononuclear cells is reduced in newborns with a family history of atopic disease and is independent from cord blood IgE-levels. Pediatr Allergy Immunol 1993;4:60—4.

[39] Schaub B, Liu J, Hoppler S, Haug S, Sattler C, Lluis A, et al. Impairment of T-regulatory cells in cord blood of atopic mothers. J Allergy Clin Immunol 2008;121:1491—9:1499e1—13.

[40] Smith M, Tourigny MR, Noakes P, Thornton CA, Tulic MK, Prescott SL. Children with egg allergy have evidence of reduced neonatal CD4(+)CD25(+)CD127(lo/-) regulatory T cell function. J Allergy Clin Immunol 2008;121:1460—6:1466e1—7.

[41] Martino DJ, Tulic MK, Gordon L, Hodder M, Richman T, Metcalfe J, et al. Evidence for age-related and individual-specific changes in DNA methylation profile of mononuclear cells during early immune development in humans. Epigenetics 2011;6.

[42] White GP, Watt PM, Holt BJ, Holt PG. Differential patterns of methylation of the IFN-gamma promoter at CpG and non-CpG sites underlie differences in IFN-gamma gene expression between human neonatal and adult CD45RO- T cells. J Immunol 2002;168:2820—7.

[43] Fields PE, Kim ST, Flavell RA. Cutting edge: changes in histone acetylation at the IL-4 and IFN-gamma loci accompany Th1/Th2 differentiation. J Immunol 2002;169:647—50.

[44] Lee DU, Agarwal S, Rao A. Th2 lineage commitment and efficient IL-4 production involves extended de-methylation of the IL-4 gene. Immunity 2002;16:649—60.

[45] White GP, Hollams EM, Yerkovich ST, Bosco A, Holt BJ, Bassami MR, et al. CpG methylation patterns in the IFNgamma promoter in naive T cells: variations during Th1 and Th2 differentiation and between atopics and non-atopics. Pediatr Allergy Immunol 2006;17:557—64.

[46] Sawalha AH. Epigenetics and T-cell immunity. Autoimmunity 2008;41:245—52.

[47] Zhu J. Transcriptional regulation of Th2 cell differentiation. Immunol Cell Biol 2010;88:244—9.

[48] Djuretic IM, Levanon D, Negreanu V, Groner Y, Rao A, Ansel KM. Transcription factors T-bet and Runx3 cooperate to activate Ifng and silence Il4 in T helper type 1 cells. Nat Immunol 2007;8:145—53.

381

[49] Shin HJ, Park HY, Jeong SJ, Park HW, Kim YK, Cho SH, et al. STAT4 expression in human T cells is regulated by DNA methylation but not by promoter polymorphism. J Immunol 2005;175:7143−50.

[50] Yamashita M, Ukai-Tadenuma M, Miyamoto T, Sugaya K, Hosokawa H, Hasegawa A, et al. Essential role of GATA3 for the maintenance of type 2 helper T (Th2) cytokine production and chromatin remodeling at the Th2 cytokine gene loci. J Biol Chem 2004;279:26983−90.

[51] Ito K, Caramori G, Lim S, Oates T, Chung KF, Barnes PJ, et al. Expression and activity of histone deacetylases in human asthmatic airways. Am J Respir Crit Care Med 2002;166:392−6.

[52] Bruniquel D, Schwartz RH. Selective, stable demethylation of the interleukin-2 gene enhances transcription by an active process. Nat Immunol 2003;4:235−40.

[53] Bird A. Il2 transcription unleashed by active DNA demethylation. Nat Immunol 2003;4:208−9.

[54] Polansky JK, Kretschmer K, Freyer J, Floess S, Garbe A, Baron U, et al. DNA methylation controls Foxp3 gene expression. Eur J Immunol 2008;38:1654−63.

[55] Janson PC, Winerdal ME, Marits P, Thorn M, Ohlsson R, Winqvist O. FOXP3 promoter demethylation reveals the committed Treg population in humans. PLoS One 2008;3:e1612.

[56] Hu Y, Shen F, Crellin NK, Ouyang W. The IL-17 pathway as a major therapeutic target in autoimmune diseases. Ann N Y Acad Sci 2011;1217:60−76.

[57] Toh ML, Miossec P. The role of T cells in rheumatoid arthritis: new subsets and new targets. Curr Opin Rheumatol 2007;19:284−8.

[58] Stockinger B, Veldhoen M. Differentiation and function of Th17 T cells. Curr Opin Immunol 2007;19:281−6.

[59] Akimzhanov AM, Yang XO, Dong C. Chromatin remodeling of interleukin-17 (IL-17)-IL-17F cytokine gene locus during inflammatory helper T cell differentiation. J Biol Chem 2007;282:5969−72.

[60] Koenen HJ, Smeets RL, Vink PM, van Rijssen E, Boots AM, Joosten I. Human CD25highFoxp3pos regulatory T cells differentiate into IL-17-producing cells. Blood 2008;112:2340−52.

[61] Bonizzi G, Karin M. The two NF-kappaB activation pathways and their role in innate and adaptive immunity. Trends Immunol 2004;25:280−8.

[62] Okamoto T. The epigenetic alteration of synovial cell gene expression in rheumatoid arthritis and the roles of nuclear factor kappaB and Notch signaling pathways. Mod Rheumatol 2005;15:79−86.

[63] Ito K, Lim S, Caramori G, Chung KF, Barnes PJ, Adcock IM. Cigarette smoking reduces histone deacetylase 2 expression, enhances cytokine expression, and inhibits glucocorticoid actions in alveolar macrophages. FASEB J 2001;15:1110−2.

[64] De Bosscher K, Vanden Berghe W, Haegeman G. The interplay between the glucocorticoid receptor and nuclear factor-kappaB or activator protein-1: molecular mechanisms for gene repression. Endocr Rev 2003;24:488−522.

[65] Lu TX, Hartner J, Lim EJ, Fabry V, Mingler MK, Cole ET, et al. MicroRNA-21 limits in vivo immune response-mediated activation of the IL-12/IFN-gamma pathway, Th1 polarization, and the severity of delayed-type hypersensitivity. J Immunol 2011;187:3362−73.

[66] Steiner DF, Thomas MF, Hu JK, Yang Z, Babiarz JE, Allen CD, et al. MicroRNA-29 regulates T-box transcription factors and interferon-gamma production in helper T cells. Immunity 2011;35:169−81.

[67] Lu TX, Munitz A, Rothenberg ME. MicroRNA-21 is up-regulated in allergic airway inflammation and regulates IL-12p35 expression. J Immunol 2009;182:4994−5002.

[68] Tsitsiou E, Williams AE, Moschos SA, Patel K, Rossios C, Jiang X, et al. Transcriptome analysis shows activation of circulating CD8(+) T cells in patients with severe asthma. J Allergy Clin Immunol 2011.

[69] Cannell IG, Kong YW, Bushell M. How do microRNAs regulate gene expression? Biochem Soc Trans 2008;36:1224−31.

[70] O'Shea JJ, Paul WE. Mechanisms underlying lineage commitment and plasticity of helper CD4+ T cells. Science 2010;327:1098−102.

[71] Yang XO, Nurieva R, Martinez GJ, Kang HS, Chung Y, Pappu BP, et al. Molecular antagonism and plasticity of regulatory and inflammatory T cell programs. Immunity 2008;29:44−56.

[72] Xu L, Kitani A, Fuss I, Strober W. Cutting edge: regulatory T cells induce CD4+CD25-Foxp3- T cells or are self-induced to become Th17 cells in the absence of exogenous TGF-beta. J Immunol 2007;178:6725−9.

[73] Williams LM, Rudensky AY. Maintenance of the Foxp3-dependent developmental program in mature regulatory T cells requires continued expression of Foxp3. Nat Immunol 2007;8:277−84.

[74] Lahl K, Mayer CT, Bopp T, Huehn J, Loddenkemper C, Eberl G, et al. Nonfunctional regulatory T cells and defective control of Th2 cytokine production in natural scurfy mutant mice. J Immunol 2009;183:5662−72.

[75] Beyer M, Thabet Y, Muller RU, Sadlon T, Classen S, Lahl K, et al. Repression of the genome organizer SATB1 in regulatory T cells is required for suppressive function and inhibition of effector differentiation. Nat Immunol 2011;12:898−907.

[76] Wei G, Wei L, Zhu J, Zang C, Hu-Li J, Yao Z, et al. Global mapping of H3K4me3 and H3K27me3 reveals specificity and plasticity in lineage fate determination of differentiating CD4+ T cells. Immunity 2009;30:155−67.

[77] Northrup DL, Zhao K. Application of ChIP-Seq and related techniques to the study of immune function. Immunity 2011;34:830−42.

[78] Ho SM. Environmental epigenetics of asthma: an update. J Allergy Clin Immunol 2010;126:453−65.

[79] Strachan DP. Hay fever, hygiene, and household size. BMJ 1989;299:1259−60.

[80] Brand S, Teich R, Dicke T, Harb H, Yildirim AO, Tost J, et al. Epigenetic regulation in murine offspring as a novel mechanism for transmaternal asthma protection induced by microbes. J Allergy Clin Immunol 2011;128:618−25.e1−7.

[81] Bobetsis YA, Barros SP, Lin DM, Weidman JR, Dolinoy DC, Jirtle RL, et al. Bacterial infection promotes DNA hypermethylation. J Dent Res 2007;86:169−74.

[82] Gutierrez MI, Siraj AK, Khaled H, Koon N, El-Rifai W, Bhatia K. CpG island methylation in Schistosoma- and non-Schistosoma-associated bladder cancer. Mod Pathol 2004;17:1268−74.

[83] Kitajima Y, Ohtaka K, Mitsuno M, Tanaka M, Sato S, Nakafusa Y, et al. Helicobacter pylori infection is an independent risk factor for Runx3 methylation in gastric cancer. Oncol Rep 2008;19:197−202.

[84] Foster SL, Hargreaves DC, Medzhitov R. Gene-specific control of inflammation by TLR-induced chromatin modifications. Nature 2007;447:972−8.

[85] Blumer N, Herz U, Wegmann M, Renz H. Prenatal lipopolysaccharide-exposure prevents allergic sensitization and airway inflammation, but not airway responsiveness in a murine model of experimental asthma. Clin Exp Allergy 2005;35:397−402.

[86] Blumer N, Sel S, Virna S, Patrascan CC, Zimmermann S, Herz U, et al. Perinatal maternal application of Lactobacillus rhamnosus GG suppresses allergic airway inflammation in mouse offspring. Clin Exp Allergy 2007;37:348−57.

[87] Schaub B, Liu J, Hoppler S, Schleich I, Huehn J, Olek S, et al. Maternal farm exposure modulates neonatal immune mechanisms through regulatory T cells. J Allergy Clin Immunol 2009;123:774−82:e5.

[88] Douwes J, Cheng S, Travier N, Cohet C, Niesink A, McKenzie J, et al. Farm exposure in utero may protect against asthma, hay fever and eczema. Eur Respir J 2008;32:603−11.

[89] West CE, Videky DJ, Prescott SL. Role of diet in the development of immune tolerance in the context of allergic disease. Curr Opin Pediatr 2010;22:635−41.

[90] Dunstan JA, Mori TA, Barden A, Beilin LJ, Taylor AL, Holt PG, et al. Fish oil supplementation in pregnancy modifies neonatal allergen-specific immune responses and clinical outcomes in infants at high risk of atopy: a randomized, controlled trial. J Allergy Clin Immunol 2003;112:1178−84.

[91] Maslowski KM, Vieira AT, Ng A, Kranich J, Sierro F, Yu D, et al. Regulation of inflammatory responses by gut microbiota and chemoattractant receptor GPR43. Nature 2009;461:1282−6.

[92] Miller RL. Prenatal maternal diet affects asthma risk in offspring. J Clin Invest 2008;118:3265−8.

[93] Devereux G, Litonjua AA, Turner SW, Craig LC, McNeill G, Martindale S, et al. Maternal vitamin D intake during pregnancy and early childhood wheezing. Am J Clin Nutr 2007;85:853−9.

[94] Hollingsworth JW, Maruoka S, Boon K, Garantziotis S, Li Z, Tomfohr J, et al. In utero supplementation with methyl donors enhances allergic airway disease in mice. J Clin Invest 2008;118:3462−9.

[95] Fainaru O, Shseyov D, Hantisteanu S, Groner Y. Accelerated chemokine receptor 7-mediated dendritic cell migration in Runx3 knockout mice and the spontaneous development of asthma-like disease. Proc Natl Acad Sci USA 2005;102:10598−603.

[96] Haberg SE, London SJ, Stigum H, Nafstad P, Nystad W. Folic acid supplements in pregnancy and early childhood respiratory health. Arch Dis Child 2009;94:180−4.

[97] Magdelijns FJ, Mommers M, Penders J, Smits L, Thijs C. Folic Acid use in pregnancy and the development of atopy, asthma, and lung function in childhood. Pediatrics 2011;128:e135−44.

[98] Gale CR, Robinson SM, Harvey NC, Javaid MK, Jiang B, Martyn CN, et al. Maternal vitamin D status during pregnancy and child outcomes. Eur J Clin Nutr 2008;62:68−77.

[99] Miyake Y, Sasaki S, Tanaka K, Hirota Y. Dairy food, calcium and vitamin D intake in pregnancy, and wheeze and eczema in infants. Eur Respir J 2010;35:1228−34.

[100] Erkkola M, Kaila M, Nwaru BI, Kronberg-Kippila C, Ahonen S, Nevalainen J, et al. Maternal vitamin D intake during pregnancy is inversely associated with asthma and allergic rhinitis in 5-year-old children. Clin Exp Allergy 2009;39:875−82.

[101] Tse AK, Zhu GY, Wan CK, Shen XL, Yu ZL, Fong WF. 1alpha,25-Dihydroxyvitamin D3 inhibits transcriptional potential of nuclear factor kappa B in breast cancer cells. Mol Immunol 2010;47:1728−38.

[102] Riis JL, Johansen C, Gesser B, Moller K, Larsen CG, Kragballe K, et al. 1alpha,25(OH)(2)D(3) regulates NF-kappaB DNA binding activity in cultured normal human keratinocytes through an increase in Ikappa-Balpha expression. Arch Dermatol Res 2004;296:195−202.

383

[103] Tan PH, Sagoo P, Chan C, Yates JB, Campbell J, Beutelspacher SC, et al. Inhibition of NF-kappa B and oxidative pathways in human dendritic cells by antioxidative vitamins generates regulatory T cells. J Immunol 2005;174:7633—44.

[104] Utsugi M, Dobashi K, Ishizuka T, Endou K, Hamuro J, Murata Y, et al. c-Jun N-terminal kinase negatively regulates lipopolysaccharide-induced IL-12 production in human macrophages: role of mitogen-activated protein kinase in glutathione redox regulation of IL-12 production. J Immunol 2003;171:628—35.

[105] Devereux G, Barker RN, Seaton A. Antenatal determinants of neonatal immune responses to allergens. Clin Exp Allergy 2002;32:43—50.

[106] Rahman I, Marwick J, Kirkham P. Redox modulation of chromatin remodeling: impact on histone acetylation and deacetylation, NF-kappaB and pro-inflammatory gene expression. Biochem Pharmacol 2004;68:1255—67.

[107] Nian H, Delage B, Ho E, Dashwood RH. Modulation of histone deacetylase activity by dietary isothiocyanates and allyl sulfides: studies with sulforaphane and garlic organosulfur compounds. Environ Mol Mutagen 2009;50:213—21.

[108] McLeish AC, Zvolensky MJ. Asthma and cigarette smoking: a review of the empirical literature. J Asthma 2010;47:345—61.

[109] Lin JL, Thomas PS. Current perspectives of oxidative stress and its measurement in chronic obstructive pulmonary disease. COPD 2010;7:291—306.

[110] Papadopoulos A, Guida F, Cenee S, Cyr D, Schmaus A, Radoi L, et al. Cigarette smoking and lung cancer in women: Results of the French ICARE case-control study. Lung Cancer 2011.

[111] Launay JM, Del Pino M, Chironi G, Callebert J, Peoc'h K, Megnien JL, et al. Smoking induces long-lasting effects through a monoamine-oxidase epigenetic regulation. PLoS One 2009;4:e7959.

[112] Magnusson LL, Olesen AB, Wennborg H, Olsen J. Wheezing, asthma, hayfever, and atopic eczema in childhood following exposure to tobacco smoke in fetal life. Clin Exp Allergy 2005;35:1550—6.

[113] Bisgaard H, Loland L, Holst KK, Pipper CB. Prenatal determinants of neonatal lung function in high-risk newborns. J Allergy Clin Immunol 2009;123:651—7:657e1—4.

[114] Sears MR, Holdaway MD, Flannery EM, Herbison GP, Silva PA. Parental and neonatal risk factors for atopy, airway hyper-responsiveness, and asthma. Arch Dis Child 1996;75:392—8.

[115] Li YF, Langholz B, Salam MT, Gilliland FD. Maternal and grandmaternal smoking patterns are associated with early childhood asthma. Chest 2005;127:1232—41.

[116] Breton CV, Byun HM, Wenten M, Pan F, Yang A, Gilliland FD. Prenatal tobacco smoke exposure affects global and gene-specific DNA methylation. Am J Respir Crit Care Med 2009;180:462—7.

[117] Franco R, Schoneveld O, Georgakilas AG, Panayiotidis MI. Oxidative stress, DNA methylation and carcinogenesis. Cancer Lett 2008;266:6—11.

[118] Chong S, Whitelaw E. Epigenetic germline inheritance. Curr Opin Genet Dev 2004;14:692—6.

[119] Peden DB, Bush RK. Advances in environmental and occupational respiratory diseases in 2009. J Allergy Clin Immunol 2010;125:559—62.

[120] D'Amato G, Cecchi L, D'Amato M, Liccardi G. Urban air pollution and climate change as environmental risk factors of respiratory allergy: an update. J Investig Allergol Clin Immunol 2010;20:95—102. quiz following 102.

[121] Baccarelli A, Wright RO, Bollati V, Tarantini L, Litonjua AA, Suh HH, et al. Rapid DNA methylation changes after exposure to traffic particles. Am J Respir Crit Care Med 2009;179:572—8.

[122] Liu G, Niu Z, Van Niekerk D, Xue J, Zheng L. Polycyclic aromatic hydrocarbons (PAHs) from coal combustion: emissions, analysis, and toxicology. Rev Environ Contam Toxicol 2008;192:1—28.

[123] Simko P. Factors affecting elimination of polycyclic aromatic hydrocarbons from smoked meat foods and liquid smoke flavorings. Mol Nutr Food Res 2005;49:637—47.

[124] Brody JG, Moysich KB, Humblet O, Attfield KR, Beehler GP, Rudel RA. Environmental pollutants and breast cancer: epidemiologic studies. Cancer 2007;109:2667—711.

[125] Factor P, Akhmedov AT, McDonald JD, Qu A, Wu J, Jiang H, et al. Polycyclic Aromatic Hydrocarbons Impair β 2AR Function in Airway Epithelial and Smooth Muscle Cells. Am J Respir Cell Mol Biol 2011;45:1045—9.

[126] Perera F, Tang WY, Herbstman J, Tang D, Levin L, Miller R, et al. Relation of DNA methylation of 5'-CpG island of ACSL3 to transplacental exposure to airborne polycyclic aromatic hydrocarbons and childhood asthma. PLoS One 2009;4:e4488.

[127] Mashek DG, Bornfeldt KE, Coleman RA, Berger J, Bernlohr DA, Black P, et al. Revised nomenclature for the mammalian long-chain acyl-CoA synthetase gene family. J Lipid Res 2004;45:1958—61.

[128] Minekura H, Kang MJ, Inagaki Y, Suzuki H, Sato H, Fujino T, et al. Genomic organization and transcription units of the human acyl-CoA synthetase 3 gene. Gene 2001;278:185—92.

[129] Bouzigon E, Siroux V, Dizier MH, Lemainque A, Pison C, Lathrop M, et al. Scores of asthma and asthma severity reveal new regions of linkage in EGEA study families. Eur Respir J 2007;30:253—9.

[130] Choudhry S, Taub M, Mei R, Rodriguez-Santana J, Rodriguez-Cintron W, Shriver MD, et al. Genome-wide screen for asthma in Puerto Ricans: evidence for association with 5q23 region. Hum Genet 2008;123:455−68.

[131] Woods RK, Raven JM, Walters EH, Abramson MJ, Thien FC. Fatty acid levels and risk of asthma in young adults. Thorax 2004;59:105−10.

[132] Beck M, Zelczak G, Lentze MJ. Abnormal fatty acid composition in umbilical cord blood of infants at high risk of atopic disease. Acta Paediatr 2000;89:279−84.

[133] Liu J, Ballaney M, Al-alem U, Quan C, Jin X, Perera F, et al. Combined inhaled diesel exhaust particles and allergen exposure alter methylation of T helper genes and IgE production in vivo. Toxicol Sci 2008;102:76−81.

[134] Gilmour PS, Rahman I, Donaldson K, MacNee W. Histone acetylation regulates epithelial IL-8 release mediated by oxidative stress from environmental particles. Am J Physiol Lung Cell Mol Physiol 2003;284:L533−40.

[135] Nicolai T, Carr D, Weiland SK, Duhme H, von Ehrenstein O, Wagner C, et al. Urban traffic and pollutant exposure related to respiratory outcomes and atopy in a large sample of children. Eur Respir J 2003;21:956−63.

[136] Bollati V, Baccarelli A, Hou L, Bonzini M, Fustinoni S, Cavallo D, et al. Changes in DNA methylation patterns in subjects exposed to low-dose benzene. Cancer Res 2007;67:876−80.

[137] Smeds A, Saukko P. Identification and quantification of polychlorinated biphenyls and some endocrine disrupting pesticides in human adipose tissue from Finland. Chemosphere 2001;44:1463−71.

[138] Stacey CI, Perriman WS, Whitney S. Organochlorine pesticide residue levels in human milk: Western Australia, 1979-1980. Arch Environ Health 1985;40:102−8.

[139] Rhainds M, Levallois P, Dewailly E, Ayotte P. Lead, mercury, and organochlorine compound levels in cord blood in Quebec, Canada. Arch Environ Health 1999;54:40−7.

[140] Ataniyazova OA, Baumann RA, Liem AK, Mukhopadhyay UA, Vogelaar EF, Boersma ER. Levels of certain metals, organochlorine pesticides and dioxins in cord blood, maternal blood, human milk and some commonly used nutrients in the surroundings of the Aral Sea (Karakalpakstan, Republic of Uzbekistan). Acta Paediatr 2001;90:801−8.

[141] Reichrtova E, Ciznar P, Prachar V, Palkovicova L, Veningerova M. Cord serum immunoglobulin E related to the environmental contamination of human placentas with organochlorine compounds. Environ Health Perspect 1999;107:895−9.

[142] Daniel V, Huber W, Bauer K, Opelz G. Impaired in-vitro lymphocyte responses in patients with elevated pentachlorophenol (PCP) blood levels. Arch Environ Health 1995;50:287−92.

[143] Daniel V, Huber W, Bauer K, Suesal C, Conradt C, Opelz G. Associations of blood levels of PCB, HCHS, and HCB with numbers of lymphocyte subpopulations, in vitro lymphocyte response, plasma cytokine levels, and immunoglobulin autoantibodies. Environ Health Perspect 2001;109:173−8.

[144] Kim KY, Kim DS, Lee SK, Lee IK, Kang JH, Chang YS, et al. Association of low-dose exposure to persistent organic pollutants with global DNA hypomethylation in healthy Koreans. Environ Health Perspect 2010;118:370−4.

[145] Herceg Z. Epigenetics and cancer: towards an evaluation of the impact of environmental and dietary factors. Mutagenesis 2007;22:91−103.

[146] Pilsner JR, Hu H, Ettinger A, Sanchez BN, Wright RO, Cantonwine D, et al. Influence of prenatal lead exposure on genomic methylation of cord blood DNA. Environ Health Perspect 2009;117:1466−71.

[147] Noakes PS, Taylor P, Wilkinson S, Prescott SL. The relationship between persistent organic pollutants in maternal and neonatal tissues and immune responses to allergens: A novel exploratory study. Chemosphere 2006;63:1304−11.

[148] Prescott SL, Holt PG, Jenmalm M, Bjorksten B. Effects of maternal allergen-specific IgG in cord blood on early postnatal development of allergen-specific T-cell immunity. Allergy 2000;55:470−5.

[149] Prescott SL, Breckler LA, Witt CS, Smith L, Dunstan JA, Christiansen FT. Allergic women show reduced T helper type 1 alloresponses to fetal human leucocyte antigen mismatch during pregnancy. Clin Exp Immunol 2010;159:65−72.

[150] Gheorghe CP, Goyal R, Mittal A, Longo LD. Gene expression in the placenta: maternal stress and epigenetic responses. Int J Dev Biol 2010;54:507−23.

CHAPTER 19

Therapy of Airway Disease: Epigenetic Potential

Peter J. Barnes

National Heart & Lung Institute, Imperial College, London, UK

19.1 INTRODUCTION

Both asthma and COPD involve the increased expression of multiple inflammatory genes [1–3]. There is increasing recognition that epigenetics may play an important role in the regulation of inflammatory genes in diseases. Since epigenetic changes can be longstanding and may be passed to the offspring this is likely to be important in understanding the chronicity of inflammation and how environmental factors which affect the mother (such as cigarette smoking and diet) may affect the progeny. It has also been recognized that currently used treatments, such as corticosteroids, also work, at least in part, through epigenetic mechanisms. Understanding these epigenetic pathways may identify novel targets for the development of future therapy [4–6]. Histone acetylation has been studied in some detail in relation to the expression of inflammatory genes [7]. Histones may also be influenced by other post-translational modifications, including phosphorylation through the action of various kinases, methylation via histone methyltransferases, tyrosine nitration, ubiquitination and SUMOylation (SUMO = small ubiquitin modifier protein), resulting in changes in gene expression (Figure 19.1). Indeed these modifications may take place sequentially so that one modification then makes it possible for the next occurring and this "histone code" may account for cell specificity in inflammatory gene regulation [8]. For example, regulation of the gene encoding transforming growth factor (TGF)-β1 involves successive acetylation and methylation at different lysine and arginine residues on histones-3 and -4 [9]. Various existing

T. Tollefsbol (Ed): Epigenetics in Human Disease. DOI: 10.1016/B978-0-12-388415-2.00019-6

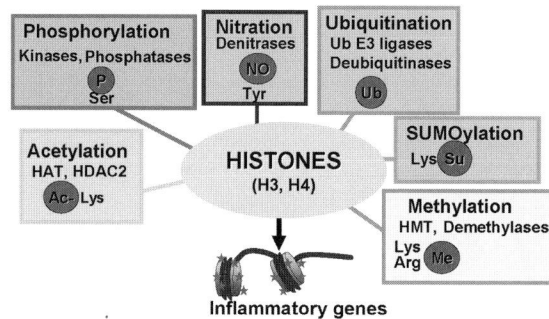

FIGURE 19.1

Post-translational histone modification. Histone-3 and -4 may be modified in several ways including acetylation (Ac) of lysine (Lys) residues by histone acetyltransferases (HAT) and deacetylation by histone deacetylases (HDAC); phosphorylation (P) of serine (Ser) residues by kinases and dephosphorylation by phosphatases; nitration (NO) of tyrosine (Tyr) residues by peroxynitrite and denitration by denitrases; ubiquitination (Ub) by E3 ubiquitin ligases and reversal by deubiquitinases; SUMOylation (Su) of Lys residues; methylation (Me) of Lys and arginine residues by histone methyltransferases (HMT) and demethylases. Drugs have the potential to interact with any of these modifications, particularly by effects on the signaling pathways that regulate the modifying enzymes. This figure is reproduced in the color plate section.

drugs may interfere with these histone modification signaling pathways and better understanding of these molecular mechanisms may identify novel molecular targets for discovering new anti-inflammatory therapies in the future.

19.2 HISTONE ACETYLATION AND INFLAMMATORY GENE REGULATION

Gene expression is regulated by various coactivator molecules, such as CREB-binding protein (CBP; CREB = cyclic AMP response element binding protein) and p300, all of which have intrinsic histone acetyltransferase (HAT) activity. Expression of inflammatory genes is regulated by increased acetylation of histone 4 [10]. In this way epigenetic factors play a critical role in chronic inflammation [11]. In asthma patients there is evidence for increased acetylation of histone-4, consistent with increased expression of multiple inflammatory genes [12]. Many of these genes are regulated by the transcription factor nuclear factor-κB (NF-κB), which may also be acetylated. In COPD peripheral lung, airway biopsies and alveolar macrophages there is an increase in the acetylation of histones associated with the promoter region of inflammatory genes, such as CXCL8 (interleukin-8) that are regulated by NF-κB and the degree of acetylation increases with disease severity [13]. This increased acetylation of histones associated with inflammatory genes is not due to an increase in HAT activity in lungs or macrophages as in asthma, but due to decreased histone deacetylase (HDAC) activity.

Histone acetylation is reversed by HDACs. There are 11 HDAC isoenzymes that deacetylate histones and other proteins within the nucleus and specific HDACs appear to be differentially regulated and to regulate different groups of genes [14]. HDACs play a critical role in the suppression of gene expression by reversing the hyperacetylation of core histones. For the regulation of inflammatory genes HDAC2 appears to be of critical importance [10,15]. The expression of inflammatory genes is determined by a balance between histone acetylation (which activates transcription) and deacetylation which switches off transcription.

19.3 ACETYLATION OF NON-HISTONE PROTEINS

It has been increasingly recognized that many regulatory proteins, particularly transcription factors and nuclear receptors, are also regulated by acetylation that is controlled by HATs and HDACs and may also play a key role in the regulation of inflammatory genes [16]. The proinflammatory transcription factor NF-κB may be acetylated at several Lys residues on p65 and this appears to enhance its ability to activate certain inflammatory genes, whereas other inflammatory genes may be less activated. There is a complex interplay between acetylation and ubiquitination which leads to loss of p65 protein [17]. Acetylation also plays a key role in the regulation of androgen and estrogen receptors, and this is also the case for glucocorticoid receptors (GR) [15]. GR is acetylated within the nucleus at specific lysine residues close to the hinge region of the receptor and only binds to its DNA-binding site in its acetylated form. However, in order to inhibit NF-κB-activated genes it is necessary to deacetylate the receptor and this is achieved by HDAC2 (Figure 19.2).

19.4 CORTICOSTEROIDS SUPPRESS INFLAMMATION VIA EPIGENETIC MECHANISMS

There have been major advances in understanding the molecular mechanisms whereby glucocorticoids suppress inflammation, which also provides important insights into how corticosteroid-resistance may arise [18, 19]. A major mechanism of action of corticosteroids involves changes in histone acetylation to regulate inflammatory and anti-inflammatory genes. Corticosteroids diffuse across the cell membrane and bind to glucocorticoid receptors (GR) in the cytoplasm. Upon ligand binding, GR are activated and released from chaperone proteins (heat shock protein 90 and others) and rapidly translocate to the nucleus where they exert their molecular effects through the activation and suppression of multiple genes involved in the inflammatory process. GR homodimerize and bind to glucocorticoid response elements (GRE) in the promoter region of glucocorticoid-responsive genes and this interaction switches on (or occasionally switches off) gene transcription. Genes that are switched on by glucocorticoids include genes encoding β_2-adrenergic receptors and the anti-inflammatory proteins secretory leukoprotease inhibitor and mitogen-activated protein kinase phosphatase-1 (MKP-1), which inhibits MAP kinase pathways. However, the major action of corticosteroids is to switch off multiple activated inflammatory genes that encode for cytokines, chemokines, adhesion molecules, inflammatory enzymes, and receptors, which are regulated by proinflammatory transcription factors, such as NF-κB and activator protein-1 (AP-1). These transcription factors activate inflammatory genes through histone acetylation, whereas activated GR reverse this process by interacting with corepressor molecules to attenuate NF-κB-associated coactivator activity, thus reducing histone acetylation [10, 18]. Reduction of histone

389

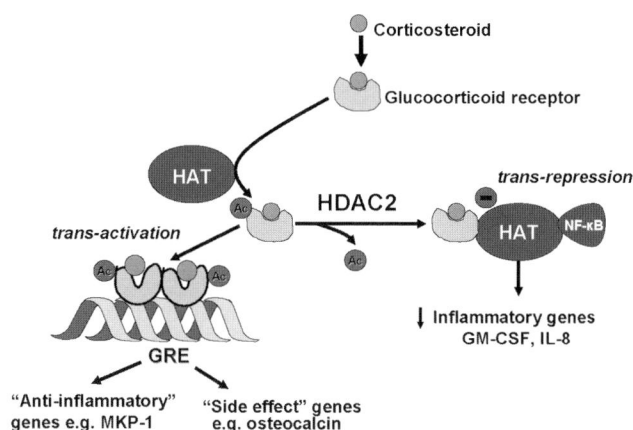

FIGURE 19.2
Acetylation of the glucocorticoid receptor (GR). After corticosteroid binding to GR the receptor translocates to the nucleus where it is acetylated by a histone acetyltransferase (HAT), which is necessary for GR to bind to its glucocorticoid receptor recognition element (GRE) in the promoter region of steroid-sensitive genes, which include the genes that mediate the side effects of corticosteroids, such as osteocalcin. It is necessary for the acetylated GR to be deacetylated by histone deacetylase 2 (HDAC2) in order to inhibit activated nuclear factor-κB (NF-κB) to suppress activated inflammatory genes. This figure is reproduced in the color plate section.

acetylation mainly occurs through the specific recruitment of HDAC2 to the activated inflammatory gene complex by activated GR, thereby resulting in effective suppression of activated inflammatory genes within the nucleus. GR becomes acetylated upon ligand binding, allowing it to bind to GREs and HDAC2 can target acetylated GR thereby allowing it to associate with the NF-κB complex [15] (Figure 19.2).

19.5 MOLECULAR MECHANISMS OF CORTICOSTEROID RESISTANCE

Several distinct molecular mechanisms contributing to decreased anti-inflammatory effects of glucocorticoids have now been identified, so that there is heterogeneity of mechanisms even within a single disease [20]. However, similar molecular mechanisms have also been identified in different inflammatory diseases, indicating that there may be common therapeutic approaches to these diseases in the future [19]. Histone acetylation plays a critical role in the regulation of inflammatory genes and corticosteroids switch on corticosteroid-responsive genes, such as MKP-1, via acetylation of specific lysine residues (K5 and K16) on histone-4 [21]. In a small proportion of patients with corticosteroid-resistant asthma, GR translocates normally to the nucleus after dexamethasone exposure but fails to acetylate K5 so that transactivation of genes does not occur [22].

As discussed above, recruitment of HDAC2 to activated inflammatory genes is a major mechanism of inflammatory gene repression by corticosteroids and reduced HDAC2 activity and expression is reduced in some diseases where patients respond poorly. For example, HDAC2 is markedly reduced in alveolar macrophages, airways, and peripheral lung in patients with COPD [13], and similar changes are found in peripheral blood mononuclear cells and alveolar macrophages of patients with refractory asthma [23] and in the airways of smoking asthmatics. The corticosteroid resistance of COPD bronchoalveolar macrophages is completely reversed by overexpressing HDAC2 (using a plasmid vector) to the level seen in control subjects [15]. The mechanisms for HDAC2 reduction in COPD are now being elucidated [24]. Oxidative and nitrative stress result in the formation of peroxynitrite, which nitrates tyrosine residues on HDAC2 resulting in its inactivation, ubiquitination, and degradation [25]. Oxidative stress also activates phosphoinositide-3-kinase (PI3K)-δ, which leads to phosphorylation and inactivation of HDAC2 (Figure 19.3) [26]. This suggests that oxidative stress may be an important mechanism of corticosteroid resistance and is increased in most severe and corticosteroid-resistant inflammatory diseases.

19.6 THEOPHYLLINE AS AN EPIGENETIC MODULATOR

An attractive therapeutic option is to reverse the cause of corticosteroid resistance by inhibiting the specific molecular pathways involved. Theophylline is a drug used to treat asthma and COPD for over 80 years and was previously used as a bronchodilator through inhibition of phosphodiesterases in airway smooth muscle. However, the use of theophylline has declined recently as side effects are common in the high doses needed for bronchodilatation and these are also mediated by phosphodiesterase inhibition, as well as through adenosine receptor antagonism. Recently, low concentrations of theophylline have been found to have a completely different pharmacological effect. Theophylline in low concentrations causes selective activation of HDAC2, and restores the reduced HDAC2 activity in COPD macrophages back to normal thereby reversing corticosteroid resistance [27,28]. In cigarette-smoke-exposed mice, which develop corticosteroid-resistant inflammation, oral theophylline is also effective in reversing resistance [26]. A clinical study has demonstrated that low doses of oral theophylline increase the anti-inflammatory effects of an inhaled corticosteroid in patients with COPD [29]. In smoking asthmatics, who are also corticosteroid-resistant, a low dose of theophylline appears to be effective in reversing resistance [30] and accelerates recovery of an

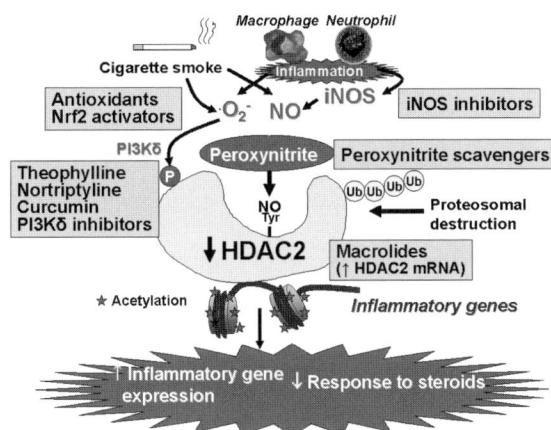

FIGURE 19.3

Mechanisms for decreased histone deacetylase(HDAC)2 in COPD and its reversal. Superoxide anions ($^{\cdot}O_2^-$) and nitric oxide (NO) generated by cigarette smoke and inflammatory cells combine to form peroxynitrite. NO production from inflammatory cells is derived from inducible NO synthase (iNOS) in response to inflammatory stimuli. Peroxynitrite nitrates HDAC2 at a tyrosine (Tyr) residue within the catalytic site, which inactivates HDAC2 and also leads to its ubiquitination (Ub) and proteasomal degradation. Oxidative stress also activates a phosphoinositide-3-kinase (PI3K)-δ pathway that phosphorylates (P) and inactivates HDAC2. Loss of HDAC function then results in enhanced inflammatory gene expression and blocks the anti-inflammatory action of corticosteroids. HDAC2 function may be restored by antioxidants (including Nrf2 activators), iNOS inhibitors or peroxynitrite scavengers, which reduce tyrosine nitration or by theophylline, nortriptyline, curcumin, or selective PI3K-δ inhibitors which restore HDAC function to normal through inhibition of PI3Kδ. HDAC2 mRNA may also be increased by macrolides. This figure is reproduced in the color plate section.

acute COPD exacerbation, an effect which is accompanied by increased HDAC activity ion sputum macrophages and reduced inflammatory mediators [31].

The molecular mechanism of action of theophylline in restoring HDAC2 appears to be via selective inhibition of PI3Kδ, which is activated by oxidative stress in COPD patients [26, 32, 33]. Selective PI3Kδ inhibitors are similarly effective and these drugs are currently in clinical development for other diseases.

19.7 OTHER DRUGS

The tricyclic antidepressant nortriptyline also increased HDAC2 by selectively inhibiting PI3Kδ and is an alternative therapeutic approach [34]. Since oxidative stress appears to me an important mechanism in reducing HDAC2 and leads to corticosteroid resistance, antioxidants should also be effective and the Nrf2 activator sulforaphane, which increases the expression of endogenous antioxidant genes, is also very effective in reversing corticosteroid resistance induced by oxidative stress. The naturally occurring compound curcumin is also effective in increasing HDAC activity and reversing corticosteroid resistance [35]. In the future, novel drugs which increase HDAC2 may be developed when the molecular signaling pathways that regulate HDAC2 are better understood [36].

It has long been recognized that macrolides have anti-inflammatory effects that may be independent of their antibiotic effects. Macrolides appear to inhibit inflammation by inhibiting NF-κB and other transcription factors. A non-antibiotic macrolide (EM-703) reverses corticosteroid resistance due to oxidative stress by increasing HDAC2 activity [37]. Several non-antibiotic macrolides are now in development as anti-inflammatory therapies.

19.8 FUTURE DIRECTIONS

The recognition that histone acetylation activates inflammatory genes and can be modified by anti-inflammatory therapies, such as corticosteroids, suggests that it may be possible to identify novel targets that may lead to the development of novel therapies. Epigenetic mechanisms may also account for the corticosteroid resistance of COPD, smoking asthma, and severe asthma, leading to the development of new treatments with the ability to reverse this corticosteroid resistance. This is already exemplified by the ability of low concentrations of theophylline to reverse corticosteroid resistance in COPD cells in vitro, smoking animals in vivo, and in patients with asthma and smoking asthmatics. The molecular mechanism of action of theophylline appears to be through inhibition of PI3Kδ and selective PI3Kδ inhibitors mimic the effects of theophylline. Several PI3Kδ inhibitors are now in clinical development. Other drugs may also interact with this pathway, opening up the possibility of a new therapeutic strategy for treating COPD and severe asthma.

While the role of histone acetylation and deacetylation in the regulation of inflammatory genes has been explored, the role of other histone modifications has hardly been investigated. DNA and histone methylation is associated with gene repression, so may be involved in switching off inflammatory genes. The methylation inhibitor 5-azacytidine increases expression of inflammatory genes in vitro and this is partially reversed by a corticosteroid, indicating that methylation may be involved [38]. Methylation of lysine (K) and arginine residues on histones may occur as a result of histone methyltransferases. Histone methylation is complex as methylation of K4 on histone-3 is associated with gene activation, whereas methylation of K9 is associated with gene repression. H3-K4 trimethylation is associated with increased expression of the anti-inflammatory gene-secretory leukocyte protease inhibitor and this is inhibited by 5-azacytidine [39]. Demethylase inhibitors may have anti-inflammatory potential as inhibitors of inflammatory gene expression. Several specific histone lysine demethylases have now been characterized, making it possible to now identify selective inhibitors [40]. Similarly, the therapeutic potential of affecting other histone modifications, such as phosphorylation, tyrosine nitration, ubiquitination, and SUMOylation, has hardly been explored.

References

[1] Barnes PJ. Pathophysiology of allergic inflammation. Immunol Rev 2011;242:31−50.

[2] Barnes PJ. Immunology of asthma and chronic obstructive pulmonary disease. Nat Immunol Rev 2008;8:183−92.

[3] Barnes PJ, Shapiro SD, Pauwels RA. Chronic obstructive pulmonary disease: molecular and cellular mechanisms. Eur Respir J 2003;22:672−88.

[4] Egger G, Liang G, Aparicio A, Jones PA. Epigenetics in human disease and prospects for epigenetic therapy. Nature 2004;429:457−63.

[5] Schwartz DA. Epigenetics and environmental lung disease. Proc Am Thorac Soc 2010;7:123−5.

[6] Barnes PJ. Targeting the epigenome in the treatment of asthma and chronic obstructive pulmonary disease. Proc Am Thorac Soc 2009;6:693−6.

[7] Barnes PJ, Adcock IM, Ito K. Histone acetylation and deacetylation: importance in inflammatory lung diseases. Eur Respir J 2005;25:552−63.

[8] Berger SL. The complex language of chromatin regulation during transcription. Nature 2007;447:407−12.

[9] Lee KY, Ito K, Hayashi R, Jazrawi EP, Barnes PJ, Adcock IM. NF-kB and activator protein 1 response elements and the role of histone modifications in IL-1b-Induced TGF-b1 gene transcription. J Immunol 2006;176:603−15.

[10] Ito K, Barnes PJ, Adcock IM. Glucocorticoid receptor recruitment of histone deacetylase 2 inhibits IL-1b-induced histone H4 acetylation on lysines 8 and 12. Mol Cell Biol 2000;20:6891−903.

[11] Adcock IM, Ford P, Barnes PJ, Ito K. Epigenetics and airways disease. Respir Res 2006;7:21.

[12] Ito K, Caramori G, Lim S, Oates T, Chung KF, Barnes PJ, et al. Expression and activity of histone deacetylases (HDACs) in human asthmatic airways. Am J Respir Crit Care Med 2002;166:392−6.

[13] Ito K, Ito M, Elliott WM, Cosio B, Caramori G, Kon OM, et al. Decreased histone deacetylase activity in chronic obstructive pulmonary disease. New Engl J Med 2005;352:1967−76.

[14] Thiagalingam S, Cheng KH, Lee HJ, Mineva N, Thiagalingam A, Ponte JF. Histone deacetylases: unique players in shaping the epigenetic histone code. Ann N Y Acad Sci 2003;983:84−100.

[15] Ito K, Yamamura S, Essilfie-Quaye S, Cosio B, Ito M, Barnes PJ, et al. Histone deacetylase 2-mediated deacetylation of the glucocorticoid receptor enables NF-kB suppression. J Exp Med 2006;203:7—13.

[16] Popov VM, Wang C, Shirley LA, Rosenberg A, Li S, Nevalainen M, et al. The functional significance of nuclear receptor acetylation. Steroids 2007;72:221—30.

[17] Li H, Wittwer T, Weber A, Schneider H, Moreno R, Maine GN, et al. Regulation of NF-kappaB activity by competition between RelA acetylation and ubiquitination. Oncogene 2011;10.

[18] Barnes PJ. Glucocorticosteroids: current and future directions. Br J Pharmacol 2011;163:29—43.

[19] Barnes PJ, Adcock IM. Glucocorticoid resistance in inflammatory diseases. Lancet 2009;342:1905—17.

[20] Adcock IM, Barnes PJ. Molecular mechanisms of corticosteroid resistance. Chest 2008;134:394—401.

[21] Barnes PJ. How corticosteroids control inflammation. Br J Pharmacol 2006;148:245—54.

[22] Matthews JG, Ito K, Barnes PJ, Adcock IM. Defective glucocorticoid receptor nuclear translocation and altered histone acetylation patterns in glucocorticoid-resistant patients. J Allergy Clin Immunol 2004;113:1100—8.

[23] Hew M, Bhavsar P, Torrego A, Meah S, Khorasani N, Barnes PJ, et al. Relative corticosteroid insensitivity of peripheral blood mononuclear cells in severe asthma. Am J Respir Crit Care Med 2006;174:134—41.

[24] Barnes PJ. Role of HDAC2 in the pathophysiology of COPD. Ann Rev Physiol 2009;71:451—64.

[25] Osoata G, Adcock IM, Barnes PJ, Ito K. Oxidative stress causes HDAC2 reduction by nitration, ubiquitnylation and proteasomall degradation. Proc Amer Thorac Soc 2005;2:A755.

[26] To Y, Ito K, Kizawa Y, Failla M, Ito M, Kusama T, et al. Targeting phosphoinositide-3-kinase-d with theophylline reverses corticosteroid insensitivity in COPD. Am J Resp Crit Care Med 2010;182:897—904.

[27] Ito K, Lim S, Caramori G, Cosio B, Chung KF, Adcock IM, et al. A molecular mechanism of action of theophylline: Induction of histone deacetylase activity to decrease inflammatory gene expression. Proc Natl Acad Sci USA 2002;99:8921—6.

[28] Cosio BG, Tsaprouni L, Ito K, Jazrawi E, Adcock IM, Barnes PJ. Theophylline restores histone deacetylase activity and steroid responses in COPD macrophages. J Exp Med 2004;200:689—95.

[29] Ford PA, Durham AL, Russell REK, Gordon F, Adcock IM, Barnes PJ. Treatment effects of low dose theophylline combined with an inhaled corticosteroid in COPD. Chest 2010;137:1338—44.

[30] Spears M, Donnelly I, Jolly L, Brannigan M, Ito K, McSharry C, et al. Effect of theophylline plus beclometasone on lung function in smokers with asthma-a pilot study. Eur Respir J 2009;33:1010—7.

[31] Cosio BG, Iglesias A, Rios A, Noguera A, Sala E, Ito K, et al. Low-dose theophylline enhances the anti-inflammatory effects of steroids during exacerbations of chronic obstructive pulmonary disease. Thorax 2009;64:424—9.

[32] Marwick JA, Caramori G, Stevenson CC, Casolari P, Jazrawi E, Barnes PJ, et al. Inhibition of PI3Kd restores glucocorticoid function in smoking-induced airway inflammation in mice. Am J Respir Crit Care Med 2009;179:542—8.

[33] Marwick JA, Wallis G, Meja K, Kuster B, Bouwmeester T, Chakravarty P, et al. Oxidative stress modulates theophylline effects on steroid responsiveness. Biochem Biophys Res Commun 2008;377:797—802.

[34] Mercado N, To Y, Ito K, Barnes PJ. Nortriptyline reverses corticosteroid insensitivity by inhibition of PI3K-d. J Pharmacol Exp Ther 2011;337:465—70.

[35] Meja KK, Rajendrasozhan S, Adenuga D, Biswas SK, Sundar IK, Spooner G, et al. Curcumin Restores Corticosteroid Function in Monocytes Exposed to Oxidants by Maintaining HDAC2. Am J Respir Cell Mol Biol 2008;39:312—23.

[36] Barnes PJ. Histone deacetylase-2 and airway disease. Ther Adv Respir Dis 2009;3:235—43.

[37] Charron C, Sumakuza T, Oomura S, Ito K. EM-703, a non-antibacterial erythromycin derivative, restores HDAC2 activition diminished by hypoxia and oxidative stress. Proc Am Thorac Soc 2007;175:A640.

[38] Kagoshima M, Wilcke T, Ito K, Tsaprouni L, Barnes PJ, Punchard N, et al. Glucocorticoid-mediated trans-repression is regulated by histone acetylation and DNA methylation. Eur J Pharmacol 2001;429:327—34.

[39] Wada H, Kagoshima M, Ito K, Barnes PJ, Adcock IM. 5-Azacytidine suppresses RNA polymerase II recruitment to the SLPI gene. Biochem Biophys Res Commun 2005;331:93—9.

[40] Anand R, Marmorstein R. Structure and mechanism of lysine-specific demethylase enzymes. J Biol Chem 2007;282:35425—9.

CHAPTER 20

The Role of Epigenetics in Cardiovascular Disease

Boda Zhou[1,2], Andriana Margariti[1], Qingbo Xu[1]
[1]King's College London BHF Centre, London, UK
[2]Peking University, Beijing, China

20.1 INTRODUCTION

Cardiovascular diseases, such as atherosclerosis and heart failure, are major public health problems and leading causes of mortality in Western countries [1]. Abundant research has been carried out to understand the factors associated with this disease. Although, there has been a substantial decline in the disease mortality in Europe and in the US over the past 30 years [1], this is mainly due to improvements in quality of care and to the prevention of the disease itself. As a result, understanding the mechanisms which lead to the development of cardiovascular disease keeps an important promise for the future.

T. Tollefsbol (Ed): Epigenetics in Human Disease. DOI: 10.1016/B978-0-12-388415-2.00020-2

Functions of endothelial (ECs) and smooth muscle cells (SMCs), such as proliferation, migration, and apoptosis have an indispensable role in atherosclerosis. Although the detailed mechanism of atherosclerosis is unclear, it is generally believed to be a multistep inflammatory disease [2,3]. The initial step is endothelial dysfunction involving EC proliferation, migration, and apoptosis, which leads to increased endothelial permeability to lipoproteins, increased leukocyte migration, and adhesion resulting in fatty streak formation [2]. In this process, the migration as well as proliferation of SMCs play critical role in the formation of fatty streak. Then, as fatty streaks progress to advanced lesions fibrous caps are formed [2]. At last, in most patients with myocardial infarction the final step is thinning and rupture of fibrous caps, in which the apoptosis of SMCs is considered as a possible cause [4]. Also, other cell types such as macrophages have a crucial role in this process [2].

Cardiomyocyte hypertrophy plays a part in the development of heart failure and left ventricular hypertrophy, which can be induced by hypertension, while myocardial hyper-trophy is initially an adaptive response. However, after sustained external load hearts can change into a state of decompensated hypertrophy resulting in dilation of the left ventricle (remodeling) and loss of contractile function. The molecular mechanisms responsible for myocardial remodeling and transition from compensated to decompensated hypertrophy are poorly defined, recent research showed the involvement of epigenetic modulations [5].

20.1.1 DNA Methylation

Epigenetics is the study of heritable alterations in phenotypes and gene expression that occur without changing the DNA sequence, providing a rapid and reversible regulation of the repertoire of expressed genes [6]. Epigenetic process includes DNA methylation as well as histone tails modification [7]. Methylation and acetylation are two major forms of histone modification. Histone methylation is modulated by two enzymes: histone methyltransferases (HMTs) and histone demethylases. The acetylation status of histone is fine tuned by histone acetyltransferases (HATs) and histone deacetylases (HDACs). The role of epigenetics in cancer and neurological diseases has been extensively examined [8]. However, the functions of DNA methylation and histone acetylation in cardiovascular disease are not clear yet. This chapter will discuss the impact of epigenetics on atherosclerosis and heart failure.

DNA methylation is a covalent modification that in mammals occurs predominantly at cytosines followed by guanines (i.e. CpG dinucleotides) to form 5-methylcytosines. CpG methylation generally is associated with gene silencing. DNA methylation is a form of epigenetic gene regulation that together with altered binding profile of transcription factors commonly leads to suppression of gene expression [9]. DNA methylation is catalyzed by three different DNA methyltransferases (DNMTs) encoded by different genes on distinct chromo-somes: DNMT1, DNMT3a, and DNMT3b. De novo methylation is catalyzed by DNMT3a and DNMT3b [10]. In contrast, DNMT1 is responsible for the propagation of DNA methylation patterns following DNA replication during mitotic cell division. Mechanisms responsible for the removal of DNA methylation marks remain poorly understood.

20.1.2 Histone Acetylation

Histone acetyltransferases (HAT) are enzymes that acetylate conserved lysine amino acids on histone proteins by transferring an acetyl group from acetyl CoA to form ε-N-acetyl lysine. Typical HATs are CREB-binding protein (CBP) and p300, while in most cases histone acetyl-ation is linked to transcriptional activation. HATs hyperacetylate histones that result in an incompact structure of the DNA, where transcription factors could potentially bind and promote gene expression. In contrast, HDACs act by removing the acetyl groups from hyperacetylated histones and lead to a compact chromatin structure and suppression of genes. HATs and HDACs are recruited to gene promoters by DNA-binding proteins that recognize certain DNA sequences, which results in modulation of specific gene expression [11].

There are 18 characterized members of human HDACs [12], which can be grouped into four classes based on function and DNA similarity (Table 20.1). The class I and class II HDACs are considered as the "classical" HDACs, whose activities could be inhibited by trichostatin A. Class III HDACs represent the silent information regulator 2 (Sir2) family of NAD^+-dependent HDACs (SIRT1-7) [12], which share structural and functional similarities with yeast Sir2 protein. Borradaile et al. [13] have summarized the role of Class III HDACs in cardiovascular disease. Finally, class IV HDAC is the recently discovered HDAC11. Although HDAC11 is most closely related to class I HDACs, it cannot be grouped into any of the three existing classes due to the low sequence similarities [14].

Class I HDACs (HDAC1, 2, 3, 8) are widely expressed, importantly HDAC1, 2, and 8 reside nearly exclusively in the nucleus, In contrast, HDAC3 is found to shuttle between nucleus and cytoplasm [12]. This indicates that the different localization of HDACs may be related with their distinct function. The members of the class II HDACs (HDAC4, 5, 6, 7, 9, 10) are shown to be involved in cell proliferation and cell survival [12]. In a similar manner Class II HDACs can shuttle between the nucleus and cytoplasm, which can be further divided into subclasses IIa (HDAC4, 5, 7, 9) and IIb (HDAC6, 10). Class IIa HDACs distinguish themselves by their extended N-terminal regulatory domain [15], whereas class IIb HDACs contain two catalytic domains [14]. Class II HDACs have been shown to possess limited HDAC activity compared with Class I HDACs, and they are found to form a complex with corepressors such as N-CoR (Nuclear receptor co-Repressor) and SMRT (Silencing Mediator for Retinoid and Thyroid receptors), as well as Class I HDACs such as HDAC3 [16,17].

Since it is generally believed that HDACs could suppress gene expression only when they are in the nucleus, modulation of cellular localization is considered to be critical in defining the functions of Class II HDACs. Timothy et al. [18] reported that calcium/calmodulin-dependent protein kinase (CaMK) signaling plays a central role in regulating class II HDAC nuclear export. They found that CaMK could phosphorylate class II HDACs and promote their nuclear exportation, thus provide derepression of HDAC-responsive genes. Ruijter et al. [19] summarized that after phosphorylation by CaMK, class II HDACs are exported to the cytoplasm with the help of cellular export factor CRM-1. 14-3-3 proteins (a cytoplasmic anchor protein family) then bind to and retain phosphorylated class II HDACs in the cytoplasm.

As HDACs have broad functions in the cell, various HDAC inhibitors are designed targeting on the catalytic sites of HDACs. They could be grouped into four classes based on their structural diversity, e.g. hydroxamic acids, short-chain fatty acids, cyclic tetrapeptides, and benzamides [20–43] (Table 20.2). Some inhibitors such as trichostatin A (TSA) can inhibit the HDAC activity of both class I and class II HDACs. However, specific inhibitors that could only inhibit the activity of one specific HDAC have been developed. An example is tubacin, a small

TABLE 20.1 Classes of HDACs

Class	Human HDAC	Subcellular Location
I	HDAC1	Nucleus
	HDAC2	Nucleus
	HDAC3	Nuc/Cyt
	HDAC8	Nucleus
IIa	HDAC4	Nuc/Cyt
	HDAC5	Nuc/Cyt
IIb	HDAC7	Nuc/Cyt
	HDAC9	Nuc/Cyt
	HDAC6	Mainly Cyt
	HDAC10	Nuc/Cyt
III	Sirt1- Sirt7	Nuc/Cyt
IV	HDAC11	Nuc/Cyt

TABLE 20.2 Structural Classes of HDAC Inhibitors

Class	Example	References
Hydroxamic acids	Trichostatin A(TSA)	[71,61]
	Suberoyl anilide bishydroxamide (SAHA)	[72,73]
	Pyroxamide	[74]
	M-carboxycinnamic acid bishydroxamide (CBHA)	[75,76]
	Scriptaid	[77,78]
Short-chain fatty acids	Valproic acid	[79,80]
	Phenylbutyrate	[81,82]
	Butyrate	[83,84]
Cyclic tetrapeptides	Trapoxin	[85,86]
	Depsipeptide (FK228)	[87,88]
	Apicidine	[89,90]
	Chlamydocin	[91]
Benzamides	MS-275	[92]

molecule that selectively inhibits HDAC6 activity [44]. More information about the selective HDAC inhibitors could be found in this review [45].

20.2 EPIGENETICS AND EC HOMEOSTASIS

EC proliferation has a major influence on EC turnover. Reports have shown that high EC turnover correlates with high EC permeability [46,47]. Therefore, EC proliferation plays a core part in regulating vascular homeostasis and atherosclerosis. DNA methylation and histone acetylation have been shown to regulate EC proliferation.

20.2.1 DNA Methylation and EC Proliferation

Jamaluddin et al. [48] showed that the cyclin A promoter contains a CpG island for DNA methyltransferase 1 (DNMT1). Homocysteine inhibited DNA methyltransferase 1 (DNMT1) activity by 30%, and reduced the binding of methyl CpG binding protein 2 (MeCP2) and increased the bindings of acetylated histone H3 and H4 in the cyclin A promoter. Finally, adenovirus-transduced DNMT1 gene expression reversed the inhibitory effect of homocysteine on cyclin A expression and EC growth inhibition. In conclusion, homocysteine inhibits cyclin A transcription and cell growth by inhibiting DNA methylation through suppression of DNMT1 in ECs. These results revealed the potential role of DNA methylation in EC proliferation.

20.2.2 Histone Acetylation and EC Proliferation

Our laboratory has found that HDAC7 controls EC growth through modulation of β-catenin translocation [49] (Figure 20.1). We found that overexpression of HDAC7 suppresses human umbilical vein EC proliferation by preventing nuclear translocation of β-catenin and down-regulating T-cell factor-1/Id2 (inhibitor of DNA binding 2) and cyclin D1, leading to G1 phase elongation. Knockdown of HDAC7 by shRNA induced β-catenin nuclear translocation but down-regulated cyclin D1, cyclin E1, and E2F2, causing EC hypertrophy. Further experiments showed that HDAC7 could retain β-catenin in the cytoplasm by direct binding. We also found that VEGF could induce HDAC7 degradation via PLCγ-IP3K signal pathway and partially rescued HDAC7-mediated suppression of proliferation. Our findings demonstrate a novel function of HDAC7 in the cytoplasm, indicating that class II HDACs are not only functional when they are localized in the nucleus as it was previously believed. Other researchers have also investigated HDAC7 in EC function. Importantly, Chang et al. [50] have shown that disruption of the HDAC7 gene in mice resulted in failure of endothelial cell–cell adhesion and enlargement of the branchial arteries. The enlargement of the branchial arteries may imply an increase in EC number or cell size, which is in agreement with our

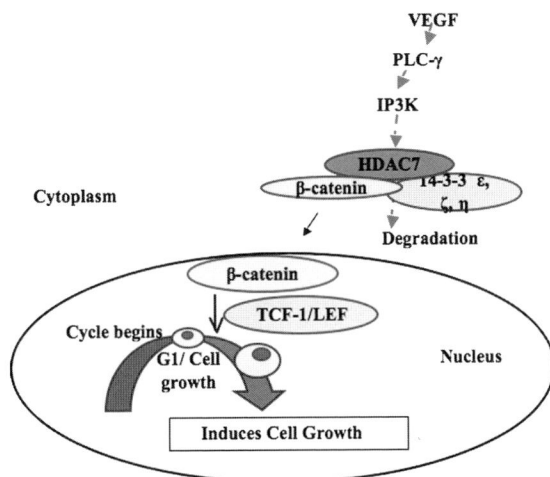

FIGURE 20.1
A schematic illustration shows the role of HDAC7 in controlling EC growth. HDAC7 acts as a bridge between 14-3-3 proteins and β-catenin, which stabilizes β-catenin in the cytoplasm, resulting in inhibition of EC growth and leading to G1 phase elongation. VEGF treatment increases HDAC7 degradation through a VEGF-PLC γ-IP3K pathway, releasing β-catenin from the HDAC7-β-catenin-14-3-3 complex, leading to β-catenin nuclear translocation. The overall effect is an increase in β-catenin target gene expression and EC growth. This figure is reproduced in the color plate section.

findings that knockdown of HDAC7 by shRNA increased EC size with a concomitant increase in cellular metabolism. These findings suggest that HDAC7 regulates EC cycle and growth [50]. However, Mottet et al. [51] reported that siRNA-mediated knockdown of HDAC7 does not influence EC proliferation. This discrepancy may result from different culture systems, different effects of gene-knockdown assays, or different methods of measuring cell proliferation and growth.

Injury of the lumen of vessel, e.g. percutaneous coronary intervention, or percutaneous transluminal coronary angioplasty, could cause endothelial denudation. As mentioned above, endothelial denudation induces local inflammation followed by the proliferation and migration of SMC towards the lesion, leading to neointima formation and restenosis. Re-endothelialization, which is promoted by EC migration and proliferation, can rescue this restenosis process [52]. Several studies have been performed to investigate the potential effect of epigenetics on EC migration.

20.2.3 Histone Acetylation and EC Migration

Urbich et al. [53] revealed that siRNA-mediated knockdown of HDAC5 could promote EC migration and sprouting, while knockdown of HDAC7 and HDAC9 decreased EC migration. Deletion and mutation study of HDAC5 revealed that the nuclear localization of HDAC5 is crucial for its function in EC migration, while its binding with MEF2 (myocyte enhancer factor 2) and deacetylase activity are dispensable. It seems that HDAC location is more important than HDAC activity. However, we could not exclude the role of deacetylation in this case, because class II HDACs can recruit other HDACs [16,17]. Microarray analysis indicated that HDAC5 silencing increased the expression of secretive protein fibroblast growth factor 2 (FGF2) [17]. In addition, the conditional medium from ECs transfected with HDAC5 siRNA attracted more migrated cells comparing with scramble siRNA. Chromatin immunoprecipitation assay showed that HDAC5 bound to the promoter region of FGF2, indicating HDAC5 functions as a repressor for FGF2 gene transcription. Their work suggests that HDAC5 is a repressor of EC migration and angiogenesis partially through modulation of FGF2 expression.

HDAC7 has also been reported to modulate EC migration. Mottet et al. [51] showed that siRNA-mediated knockdown of HDAC7 inhibited EC tube formation and migration.

Moreover, platelet-derived growth factor-B (PDGF-B) and its receptor (PDGFR-β) were the most up-regulated genes following HDAC7 silencing. The increased expression of PDGF-B and PDGFR-β are partially responsible for the inhibition of EC migration. Furthermore, treatment of ECs with phorbol 12-myristate 13-acetate resulted in the translocation of HDAC7 out of the nucleus through a protein kinase C/protein kinase D pathway and induced, similarly to HDAC7 silencing, an increase in PDGF-B expression, as well as a partial inhibition of EC migration. Collectively, their data identify HDAC7 as a key modulator of EC migration and hence angiogenesis, at least in part by regulating PDGF-B/PDGFR-β gene expression.

Ha et al. [54] demonstrated another way that the VEGF could modulate HDAC7 and EC migration. They found that VEGF stimulated phosphorylation of HDAC7 at the sites of Ser178, Ser344, and Ser479 in a dose- and time-dependent manner, leading to the cytoplasmic accumulation of HDAC7. The phosphorylation of HDAC7 has been proved to be mediated by phospholipase Cγ/protein kinase C/protein kinase D1 (PKD1)-dependent signal pathway. Infection of ECs with adenoviruses encoding a mutant of HDAC7 specifically deficient in PKD1-dependent phosphorylation inhibited VEGF-induced primary aortic EC migration.

Recent researches have shown that luminal EC apoptosis may be responsible for thrombus formation on eroded plaques without rupture [4,55]. So, the importance of EC apoptosis in atherosclerosis could not be neglected. Epigenetics have been shown to be involved in EC apoptosis.

20.2.4 DNA Methylation and EC Apoptosis

Mitra et al. [56] found that oxidized-LDL evoked a dose-dependent increase in apoptosis in the first passage ECs that was completely abrogated by LOX-1 (Lectin-like oxidized low-density-lipoprotein receptor-1) neutralizing antibody. Oxidized-LDL-induced apoptosis was associated with up-regulation of proapoptotic LOX-1, ANXA5, BAX, and CASP3 and inhibition of anti apoptotic BCL2 and cIAP-1 genes accompanied with reciprocal changes in the methylation of promoter regions of these genes. Based on these data, they conclude that exposure of ECs to oxidized-LDL induces epigenetic changes leading to resistance to apoptosis in subsequent generations, and this effect may be related to a LOX-1-mediated increase in DNA methylation.

Rao et al. [57] used methyl-CpG-binding domain protein 2 (MBD2), an interpreter for DNA methylome-encoded information, to dissect the impact of DNA methylation on endothelial function. They found that knockdown of MBD2 by siRNA significantly enhanced angiogenesis and provided protection against H_2O_2-induced apoptosis. Remarkably, Mbd2($-/-$) mice were protected against hindlimb ischemia evidenced by the significant improvement in perfusion recovery, along with increased capillary and arteriole formation. On ischemic insult, key endothelial genes such as eNOS and vascular endothelial growth factor receptor 2 undergo a DNA methylation turnover, and MBD2 interprets the changes of DNA methylation to suppress their expressions. Thus, Mbd2 could be a viable epigenetic target for modulating endothelial apoptosis in disease states.

20.2.5 Histone Acetylation and EC Survival

Our group has previously found that HDAC3 plays a crucial role in the differentiation of ECs from embryonic stem cells [58,59]. Zampetaki et al. [60] found that shRNA-mediated knockdown of HDAC3 resulted in an increase of cells showing extensive membrane blebs, reduced cell number and survival, enhanced presence of nucleosomes in cytosol, and more Annexin V staining. Ex vivo experiments showed loss of ECs in the aortic segments treated with ShRNA of HDAC3. Coimmunoprecipitation experiments resolved that HDAC3 forms a complex with Akt. Overexpression of HDAC3 resulted in increased phosphorylation of Akt and up-regulation of its kinease activity. Taken together, our findings demonstrated that HDAC3 plays a critical role in maintaining EC survival and prevents arteriosclerosis via Akt activation.

A class III HDAC SIRT1 has been shown to prevent EC apoptosis and senescence. Hou et al. [61] showed that EC SIRT1 is vital for the prevention of early membrane apoptotic phosphatidylserine externalization and subsequent DNA degradation, through a pathway involving Akt1 and FoxO3a. Zu et al. [62] demonstrated that SIRT1 could promote EC proliferation and prevent senescence by regulating a serine/threonine kinase and tumor suppressor LKB1.

HDAC inhibitor valproic acid can also affect EC apoptosis. Michaelis et al. [63] showed that valproic acid could increase extracellular signal-regulated kinase1/2 (ERK 1/2) phosphorylation in ECs. ERK 1/2 phosphorylation leads to phosphorylation of the antiapoptotic protein Bcl-2 and inhibits serum starvation-induced EC apoptosis and cytochrome C release from the mitochondria. Collectively, their results showed that valproic acid can prevent EC apoptosis through the phosphorylation of ERK1/2.

Apart from the findings of epigenetics in EC proliferation, migration, and apoptosis, Banerjee et al. [64] also found that specific DNA methyltransferase (DNMT) inhibitor 5'-aza-2'-deoxycytidine (aza-dC) could induce embryonic stem cell (ESC) differentiation towards endothelial cell (EC). Significant increase in angiogenesis and expression of the mediators of EC differentiation and EC-specific genes was only observed in aza-dC-treated cells. The DNMT inhibition-mediated increase in EC specification and marker gene expression was not associated with demethylation of these genes. These studies suggest the crucial role of epigenetics in EC differentiation.

20.3 EPIGENETICS AND SMC HOMEOSTASIS

SMC proliferation is a key event in neointima formation or arteriosclerosis. After EC injury and activation, various growth factors (e.g. PDGF, TGF-β) and cytokines (interferon-γ) are released by different cell types, including EC, platelets, and monocytes [2]. These cytokines and growth factors promote SMC proliferation, which exacerbates the formation of advanced lesion in arteriosclerosis.

20.3.1 DNA Methylation, Atherosclerosis, and SMC Phenotype Switch

Seppo and colleagues [65] have shown that significant genomic hypomethylation develops during the first replications of aortic SMCs in vitro and that hypomethylation occurs in some specific genes, such as 15-lipoxygenase and extracellular superoxide dismutase. It has also been shown that regional hypermethylation occurs in atherosclerosis. Estrogen receptor-α gene was found to have an increased methylation level in atheromas compared with normal aorta [66]. Estrogen receptor-α gene was also shown to be methylated in SMCs in vitro during the phenotypic switch [67]. These results indicate the methylation status of genes is closely related to atherosclerosis.

3-Deazaadenosine (c3Ado) is a potent inhibitor of S-adenosylhomocysteine hydrolase, which regulates cellular methyltransferase activity. Sedding et al. [68] showed that c3Ado dose-dependently prevented the proliferation and migration of human coronary SMCs in vitro. Mechanistically, c3Ado could reduce growth factor-induced extracellular signal-regulated kinase (ERK)1/2 and Akt phosphorylation. For in vivo study, the femoral artery of C57BL/6 mice was dilated and mice were fed a diet containing 150 μg of c3Ado per day. C3Ado prevented dilation-induced Ras activation, as well as ERK1/2 and Akt phosphorylation in vivo. At day 21, VSMC proliferation, as well as the neointima/media ratio, was significantly reduced. These findings highlight the importance of cellular methyltransferase activity in regulating SMC proliferation.

Diabetic patients continue to develop inflammation and vascular complications even after achieving glycemic control. This poorly understood "metabolic memory" phenomenon poses major challenges in treating diabetes; Villeneuve et al. [69] showed that in SMCs derived from type 2 diabetic db/db mice. These cells exhibit a persistent atherogenic and inflammatory phenotype even after culture in vitro. ChIP assays showed that H3K9me3 levels were significantly decreased at the promoters of key inflammatory genes in cultured db/db SMC relative to control db/+ cells. These results indicated the important role of DNA methylation in SMC phenotype switch in diabetes.

20.3.2 Histone Acetylation and SMC Proliferation

Okamoto et al. [70] have found that TSA at the concentrations of 0.1, 1, 10 μmol/l could time-dependently suppress proliferation of primary SMCs isolated from rat thoracic aorta. They have used cell count and [H^3]-thymidine incorporation methods to measure proliferation. Mechanistic studies revealed that TSA reduced the phosphorylation of Rb protein, and induced the expression of p21/WAF1 but not of p16INK4, p27KIP1, or p53. Finally, TSA inhibited HDAC activity of SMCs from p21/WAF1 knockout mice but did not influence the proliferation of these cells. Their work suggests that TSA inhibits SMC proliferation via the induction of p21/WAF1 and subsequent cell-cycle arrest with reduction of the phosphorylation of Rb.

Song et al. [71] argued that TSA at the concentration of 0.5 μmol/l could increase PDGF-BB-stimulated proliferation of primary SMC isolated from thoracic and abdominal aorta of rats. They pretreated SMC with TSA, and then stimulated with PDGF-BB, cell viability was measured and demonstrated that TSA treatment time-dependently decreased thioredoxin 1 expression in rat SMCs at both the mRNA and protein levels. Moreover, siRNA-mediated knockdown of thioredoxin 1 could potentiate Akt phosphorylation and enhance SMC proliferation in response to PDGF and serum. Collectively, these results indicate that TSA could enhance SMC proliferation by down-regulating thioredoxin 1, thus activating an Akt-dependent pathway.

This discrepancy may be attributed to different methods. Okamoto et al. [70] measured the proliferation of SMC in response to serum when treated with TSA, whilst Song et al. [71] measured the proliferation of SMC in response to PDGF-BB. Therefore, there may be a possibility that TSA activates or inhibits a certain pathway that is specifically involved in the PDGF stimulation. Also, based on the fact that TSA can inhibit both class I and class II HDACs, it may have different functions in the cell when different concentrations are applied.

Recently, studies of our group have revealed that two isoforms of HDAC7 are expressed in SMCs through alternative splicing. We sought to examine the effect of two HDAC7 isoforms on SMC proliferation and the function in neointima formation [72]. We found that over-expression of the unspliced HDAC7 isoform (HDAC7u) could suppress SMC proliferation through down-regulation of cyclin D1 and cell cycle arrest, while the spliced HDAC7 isoform (HDAC7s) did not have the same effect. siRNA-mediated knockdown of HDAC7 increased SMC proliferation and induced β-catenin nuclear translocation. Further experiments showed that only HDAC7u could bind with β-catenin and retain it in the cytoplasm. Reporter gene assay and reverse transcription PCR revealed less β-catenin activity in the cells overexpressing HDAC7u, but not HDAC7s. Deletion studies indicate that the C-terminal of HDAC7u is responsible for the binding with β-catenin. However, N-terminal additional amino acids disrupted the binding, which gives more strength to the fact that HDAC7s did not bind with β-catenin. Growth factor PDGF-BB increased the splicing of HDAC7, decreasing the expression of HDAC7u. Importantly, in an animal model of femoral artery wire injury we demonstrated that knockdown of HDAC7 by siRNA aggravates neointimal formation in comparison with control siRNA. Our findings demonstrate that splicing of HDAC7 modulates SMC proliferation and neointima formation through β-catenin translocation, which provide a potential therapeutic target in vascular disease (Figure 20.2).

SMC migrating from the media to the intima is an important step during atherogenesis, which turns fatty streak into advanced, complicated lesion [2]. Under physiological conditions SMCs are surrounded by extracellular matrix and kept in a low migratory activity. However, during the development of arteriosclerosis, matrix metalloproteinase could be released from various cell

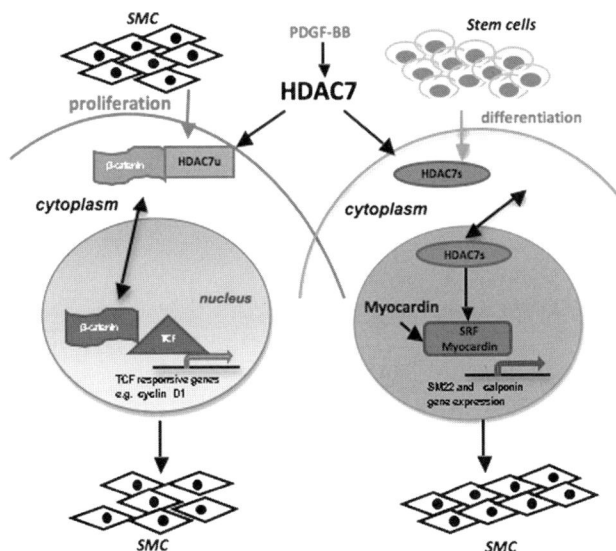

FIGURE 20.2

A schematic illustration shows the role of HDAC7 splicing in controlling SMC differentiation and proliferation. HDAC7 undergoes alternative splicing, producing spliced HDAC7 (HDAC7s) and unspliced HDAC7 (HDAC7u). HDAC7s, when it is in the nucleus, could promote stem cell differentiation towards SMC through HDAC7s-SRF-myocardin signaling. HDAC7u could inhibit this differentiation process. HDAC7u, on the other hand, could suppress SMC proliferation by sequestering β-catenin in the cytoplasm and preventing β-catenin nuclear translocation, thus, inhibiting the expression of β-catenin target genes such as cyclin D1. HDAC7s has no effect on SMC proliferation. PDGF-BB could increase the splicing of HDAC7. This figure is reproduced in the color plate section.

sources (e.g. activated macrophages) to degrade matrix proteins, thus promoting the migration of SMCs. Epigenetics is actively involved in the molecular mechanism controlling SMC migration.

20.3.3 Histone Acetylation and SMC Migration

SMCs respond to mechanical strain but the role of HDACs in modulating SMC migration induced by mechanical strain is not well elucidated. Yan et al. [73] stated that cyclic strain could significantly inhibit the migration of cultured SMCs. The cyclic strain up-regulated the levels of acetylased histone H3 and HDAC7 while it down-regulated the level of HDAC3/4 in SMCs. Furthermore, the mechanically induced SMC migration was diminished by treatment with tributyrin, an HDAC inhibitor. They also observed hyperacetylation of histone H3 and reduced expression of HDAC7 upon tributyrin treatment. These results provide evidence that HDACs are involved in the migration of SMCs induced by mechanical strain. Similarly, Song et al. [71] found that TSA could enhance SMC migration in response to PDGF-BB.

By comparing the SMC from normal human coronary arteries and from coronary plaques, Bennett el al. [74] observed higher rates of SMC apoptosis in plaque, which may ultimately contribute to plaque rupture. It was previously demonstrated that plaque-derived SMCs have reduced insulin-like growth factor 1 (IGF1) signaling, resulting from a decrease in the expression of IGF1 receptor compared with normal aortic SMCs. Furthermore, they have found that overexpression of IGF1 receptor could abolish oxidative-stress-induced apoptosis in SMCs [75], and oxidative stress repressed IGF1 receptor gene expression in turn, which needs HDAC1 [76]. These findings suggest that HDAC1 is a critical molecule in the signaling of oxidative-stress-induced SMC apoptosis.

20.3.4 Histone Acetylation and SMC Apoptosis

Class III HDACs have also been involved in SMC apoptosis. Veer et al. [77] reported that pre-B-cell colony-enhancing factor could reduce SMC apoptosis as revealed by TdT-mediated dUTP

Nick-End Labeling (TUNEL) analysis. They found that nicotinamide adenine dinucleotide (NAD^+)-dependent protein deacetylase activity was required for SMC maturation and that NAD^+-dependent HDAC activity was augmented by pre-B-cell colony-enhancing factor. These results provide a novel pathway that class III HDACs can influence SMC apoptosis and phenotype switch.

Traditionally, SMCs in the neointima of atherosclerosis were believed to be originated from the media of injured arteries. These media-originated SMCs then proliferate and migrate towards the intima in response to signals from inflammatory cells and ECs, which form neointima [2]. However, recent findings from different groups emphasized the importance of stem/progenitor cell-derived SMCs in neointima formation and atherosclerosis [78−84].

20.3.5 DNA Methylation and SMC Differentiation

Lockman et al. [85] identified the histone 3 lysine 9 (H3K9)-specific demethylase, Jmjd1a bound all three myocardin family members, and regulated SMC differentiation marker gene expression. Overexpression of Jmjd1a in multipotential 10T1/2 cells decreased global levels of dimethyl H3K9, stimulated the SM alpha-actin and SM22 promoters, and synergistically enhanced MRTF-A- and myocardin-dependent transactivation. Using chromatin immunoprecipitation assays, they also demonstrated that TGF-beta-mediated up-regulation of SMC differentiation marker gene expression in 10T1/2 cells was associated with decreased H3K9 dimethylation at the CArG-containing regions of the SMC differentiation marker gene promoters. Importantly, knockdown of Jmjd1a in 10T1/2 cells and primary rat aortic SMCs by retroviral delivery of siRNA attenuated TGF-beta-induced up-regulation of endogenous SM myosin heavy-chain expression. These results showed that histone methylation status could modulate SMC differentiation. Furthermore, Lee et al. [86] reported that DNA methylation could influence the transdifferentiation of myoblasts into smooth muscle cells. The DNA methylation inhibitor, zebularine, induced the morphological transformation of C2C12 myoblasts into smooth muscle cells accompanied by de novo synthesis of smooth muscle markers such as smooth muscle alpha-actin and transgelin.

20.3.6 Histone Acetylation and Smooth Muscle Cell-specific Gene Expression

Spin et al. [87] performed a subanalysis examining transcriptional time-course microarray data obtained using the A404 model of SMC differentiation. Ontology analysis indicated a high degree of p300 involvement in SMC differentiation. Knockdown of p300 expression accelerated SMC differentiation in A404 cells and human SMCs, while inhibition of p300 HAT activity blunted SMC differentiation. Also Qiu et al. [88] showed that the stimulation of the SM22 promoter by the coactivator CREB-binding protein (CBP) was dependent on HAT activity. These studies provides evidence that chromatin acetylation is involved in smooth muscle cell-specific gene regulation.

20.4 EPIGENETICS AND ATHEROSCLEROSIS

This section looks at epigenetics and atherosclerosis.

20.4.1 DNA Methylation in Atherosclerosis

Recent results from human and animal studies have shown that DNA methylation correlates well with atherosclerosis. Sharma et al. [89] observed that the genomic DNA methylation in peripheral lymphocytes in coronary artery disease (CAD) patients is significantly higher than in controls ($p < 0.05$). Moreover, a significant positive correlation of global DNA methylation with plasma homocysteine levels was seen in CAD patients ($p = 0.001$). Hiltunen et al. [90] evaluated the methylation status of genomic DNA from peripheral lymphocytes in a cohort of 287 individuals: 137 angiographically confirmed CAD patients and 150 controls. They found that (1) genomic hypomethylation occurs during atherogenesis in human, mouse, and rabbit lesions and that it correlates with increased transcriptional activity; (2) methyltransferase (MTase) is expressed in atherosclerotic lesions; and (3) hypomethylation is present in advanced lesions at the same level as in malignant tumors and may affect cellular proliferation and gene expression in atherosclerotic lesions. Kim et al. [91] investigated DNA methylation changes in cardiovascular atherosclerotic tissues and showed that coronary atherosclerotic tissues had higher methylation levels (28.7%) than normal-appearing arterial (6.7–10.1%) and venous tissues (18.2%).

For the methylation of specific genes, Post et al. [66] showed estrogen receptor alpha gene methylation appears to be increased in coronary atherosclerotic plaques when compared to normal proximal aorta. Lund et al. [92] showed that DNA methylation profiles, including both hyper- and hypomethylation were present in aortas and blood cells of mutant mice with no detectable atherosclerotic lesion. Castillo et al. [93] carried out a microarray-based survey of the methylation status of CpG islands (CGIs) in 45 human atherosclerotic arteries and 16 controls. Data from 10 367 CGIs revealed that a subset (151 or 1.4%) of these was hypermethylated in control arteries. The vast majority (142 or 94%) of this CGI subset was found to be unmethylated or partially methylated in atherosclerotic tissue, while only 17 of the normally unmethylated CGIs were hypermethylated in the diseased tissue. These results strongly indicate that atherosclerosis is closely related to methylation status (Table 20.3).

20.4.2 Histone Acetylation in Arteriosclerosis

Atherosclerosis develops at specific sites of the vasculature that experience disturbed blood flow [94]. Zampetaki et al. [60] found that HDAC3 expression was up-regulated in areas close to branch openings where disturbed flow occurs. In aortic isografts of apolipoprotein E-knockout mice treated with shHDAC3, a robust atherosclerotic lesion was formed. Surprisingly, three of the eight mice that received shHDAC3-infected grafts died within 2 days after the operation. Miller staining of the isografts revealed disruption of the basement membrane and rupture of the vessel. These findings indicate that HDAC3 in the endothelium of artery is crucial in preventing atherosclerosis. Moreover, Song et al. [71] used a mouse carotid artery ligation model to examine the effect of TSA on arteriosclerosis. They found that the mice injected with TSA showed 2.2 times more neointima/media ratio than controls 2 weeks after ligation. They attributed this effect to enhanced PDGF-stimulated SMC proliferation and migration after TSA treatment.

We recently used a mouse femoral artery wire injury model to investigate the role of HDAC7 in neointima formation [72]. SiRNA of HDAC7 or control siRNA was perivascularly delivered to the adventitial side of injured femoral arteries. Two weeks later the neointima area was compared. We found that HDAC7 siRNA aggravated neointima formation compared with control siRNA. These results highlight the crucial role of HDAC7 during the pathogenesis of restenosis.

Granger et al. [95] showed that utilizing a standard murine model of ischemia-reperfusion, chemical HDAC inhibitors significantly reduce infarct area even when delivered 1 hour after the ischemic insult. They demonstrate that HDAC inhibitors prevent ischemia-induced activation of gene programs that include hypoxia inducible factor-1alpha, cell death, and vascular permeability in vivo and in vitro. These results reveal that HDAC inhibitors can alter the response to ischemic injury in the heart and reduce infarct size.

20.5 EPIGENETICS AND HEART FAILURE

Recent genetic and biochemical analyses indicate that epigenetic changes play a crucial role in the development of cardiac hypertrophy and heart failure, with dysregulation in histone acetylation status. In particular it has been shown to be directly linked to an impaired contraction ability of cardiac myocytes. HATs and HDACs excert their role in this process (Table 20.3).

405

TABLE 20.3 Epigenetics and Cardiovascular Disease

Cell Type	Feature	Impact	Mechanism	References
EC	Proliferation	DNMT1↓	CyclinA	[47]
		HDAC7 ↓	β-catenin	[32]
		Knockdown HDAC7 no effect	Not specified	[50]
	Migration	Knockdown HDAC5 ↑	FGF2	[52]
		Knockdown HDAC7 ↓	Not specified	[52]
		Knockdown HDAC9 ↓	Not specified	[52]
		Knockdown HDAC7 ↓	PDGF-B/PDGFR-ββ	[50]
	Aapoptosis	Knockdown MBD2↓	VEGFR2, eNOS	[56]
		Knockdown HDAC3 ↑	Akt	[59]
		VPA ↓	ERK1/2	[62]
		SIRT1↓	Akt1, FoxO3a, LKB1	[60, 61]
	Differentiation	Aza-dC↑	VEGF-A, BMP4, EPAS-1, VE-cadherin	[63]
SMC	Proliferation	C3Ado↓	ERK1/2, Akt	[67]
		TSA ↓	P21 WAF1	[69]
		TSA ↑	Trx1	[70]
		HDAC7↓	β-catenin	[71]
	Migration	C3Ado↓	ERK1/2, Akt	[67]
		TSA ↑	Trx1	[70]
	Apoptosis	HDAC1 ↑ (indirect)	IGF1R	[74]
		NAD+ HDAC activity	Not specified	[76]
	Differentiation	Jmjd1a↑	SMA, SM22	[84]
		Zebularine↑	SMA, Transgelin	[85]
		Knockdown p300↑	Not specified	[86]
Atherosclerosis	Genomic DNA methylation↑ in CAD patients		Not specified	[88]
	Genomic hypomethylation↑ in CAD patients		Not specified	[89]
	Coronary atherosclerotic tissue had ↑ methylation		Not specified	[90]
	ERa gene methylation ↑ in atherosclerotic plaque		Not specified	[65]
	Knockdown HDAC3 ↑		Not specified	[59]
	TSA↑		Not specified	[70]
	Knockdown HDAC7↑		Not specified	[70]
Cardiomyocyte/heart failure	Phenylephrine↑ cardiomyocyte hypertrophy through p300		GATA4	[95]
			GATA4	[96]
	p300↑ ES differentiation into cardiomyocyte		Not specified	[97]
	p300↑ left ventricular remodeling and myocyte hypertrophy		GATA, MEF2	[98]
			Not specified	[99]
	p300↑ cardiac hyertrophy		Akt	[101]
	TSA↑ myocardial differentiation		Not specified	[102]
	HDAC2↑ cardiomyocyte hypertrophy		Not specified	[103]
	HDAC4 phosphorylation↑ hypertrophic growth		Not specified	[104]
	HDAC6 catalytic activity↑ by extracellular stimuli		Not specified	[105]
	HDAC9 knockout mice ↑ cariomegaly			
	Sir2a↑ size of cardiac myocyte			

20.5.1 HAT in Ventricular Remodeling

GATA-4 is a cardiac-specific transcription factor in primary cardiac myocytes derived from neonatal rats. Yanazume et al. [96] found that stimulation with phenylephrine increased an acetylated form of GATA-4 and its DNA-binding activity, as well as expression of p300. A dominant-negative mutant of p300 suppressed phenylephrine-induced nuclear acetylation, activation of GATA-4-dependent endothelin-1 promoters, and hypertrophic responses such as increase in cell size and sarcomere organization. These findings suggest that p300-mediated nuclear acetylation plays a critical role in the development of myocyte hypertrophy and represents a pathway that leads to decompensated heart failure.

In addition, Kawamura et al. [97] found the HAT activity of p300 is required for acetylation and DNA binding of GATA-4 and its full transcriptional activity as well as for promotion of a transcriptionally active chromatin configuration. However, the roles of HATs and HDACs in post-translational modification of GATA-4 during the differentiation of ES cells into cardiac myocytes remain unknown. In an ES cell model of developing embryoid bodies an acetylated form of GATA-4 and its DNA binding increased concomitantly with the expression of p300 during the stem cell differentiation into cardiac myocytes.

Left ventricular remodeling after myocardial infarction is associated with hypertrophy of surviving myocytes and represents a major process that leads to heart failure. Miyamoto et al. [98] generated transgenic mice overexpressing intact p300 or mutant p300 in the heart. As the result of its 2-amino acid substitution in the p300-histone acetyltransferase domain, this mutant lost its histone acetyltransferase activity and was unable to activate GATA-4-dependent transcription. The two kinds of transgenic mice and the wild-type mice were subjected to myocardial infarction or sham operation at the age of 12 weeks. Intact p300 transgenic mice showed significantly more progressive ventricular dilation and diminished systolic function after myocardial infarction than wild-type mice, whereas mutant p300 transgenic mice did not. These findings demonstrate that cardiac overexpression of p300 promotes ventricular remodeling after myocardial infarction in adult mice in vivo and that histone acetyltransferase activity of p300 is required for these processes.

Pressure overload induced by transverse aortic contraction, postnatal physiological growth and human heart failure were associated with large increases in p300. Wei et al. [99] reported that minimal transgenic overexpression of p300 (1.5- to 3.5-fold) induced striking myocyte and cardiac hypertrophy. Heterozygous loss of a single p300 allele reduced pressure overload-induced hypertrophy by approximately 50% and rescued the hypertrophic phenotype of p300 overexpression. Increased p300 expression enhanced acetylation of the p300 substrates histone 3 and GATA-4. Interestingly a twofold expression of p300 was associated with the de novo acetylation of MEF2. Consistent with this, genes specifically up-regulated in p300 transgenic hearts were highly enriched for MEF2 binding sites. Collectively, these results indicate that p300 has a direct function in regulating cardiomyocyte marker expression and cardiac hypertrophy through transcription factor GATA-4.

20.5.2 HDAC in Cardiac Hypertrophy

Hosseinkhani et al. [100] demonstrated that 24-h stimulation by a histone deacetylase inhibitor, TSA facilitated myocardial differentiation of monkey ES cells, indicating the relevance of HDAC activity with cardiomyocytes.

Kaneda et al. [101] used differential chromatin scanning to isolate genomic fragments associated with histones subject to differential acetylation. They applied DCS to H9C2 rat embryonic cardiomyocytes incubated with or without TSA, and found that 200 genomic fragments were readily isolated by differential chromatin scanning on the basis of the preferential acetylation of associated histones in TSA-treated cells. Their data establish a genome-wide profile of HDAC targets in cardiomyocytes, which should provide a basis for further investigations into the role of epigenetic modification in cardiac disorders.

Kee et al. [102] showed in cardiomyocytes that a forced expression of HDAC2 simulated hypertrophy in an Akt-dependent manner, whereas enzymatically inert HDAC2 H141A did not. Hypertrophic stimuli induced the expression of heat shock protein (Hsp) 70. The induced Hsp70 physically associated with and activated HDAC2. Hsp70 overexpression produced a hypertrophic phenotype, which was blocked either by siHDAC2 or by a dominant negative Hsp70DeltaABD. These results suggest that the induction of Hsp70 in response to diverse hypertrophic stresses and the ensuing activation of HDAC2 trigger cardiac hypertrophy.

Backs et al. [103] showed that calcium/calmodulin-dependent kinase II (CaMKII) signals specifically to HDAC4 by binding to a unique docking site that is absent in other class IIa HDACs. Phosphorylation of HDAC4 by CaMKII promotes nuclear export and prevents nuclear import of HDAC4, with consequent derepression of HDAC target genes. In cardiomyocytes CaMKII phosphorylation of HDAC4 results in hypertrophic growth, which can be blocked by a signal-resistant HDAC4 mutant. These findings reveal a central role for CaMKII-HDAC4 signaling pathways during cardiomyocyte hypertrophy.

Lemon et al. [104] developed assays to quantify catalytic activity of distinct HDAC classes in left and right ventricular cardiac tissue from animal models of hypertensive heart disease. Class I and IIa HDAC activity was elevated in some but not all diseased tissues. In contrast, catalytic activity of the class IIb HDAC, HDAC6, was consistently increased in stressed myocardium, but not in a model of physiologic hypertrophy. HDAC6 catalytic activity was also induced by diverse extracellular stimuli in cultured cardiac myocytes and fibroblasts. These findings suggest an important role for HDAC6 in chronic hypertension.

Importantly, Zhang et al. [105] showed that class II HDACs, which repress MEF2 activity, are substrates for a stress-responsive kinase specific for conserved serines that regulate MEF2−HDAC interactions. Signal-resistant HDAC mutants lacking these phosphorylation sites are refractory to hypertrophic signaling and inhibit cardiomyocyte hypertrophy. Conversely, mutant mice lacking the class II HDAC, HDAC9, are sensitized to hypertrophic signals and exhibit stress-dependent cardiomegaly. Thus, class II HDACs act as signal-responsive suppressors of the transcriptional program governing cardiac hypertrophy and heart failure.

Alcendor et al. [106] found that overexpression of Sir2 (silent information regulator 2) alpha protected cardiac myocytes from apoptosis in response to serum starvation and significantly increased the size of cardiac myocytes. Furthermore, Sir2 expression was increased significantly in hearts from dogs with heart failure induced by rapid pacing superimposed on stable, severe hypertrophy. These results suggest that endogenous Sir2α plays an essential role in mediating cell survival, whereas Sir2α overexpression protects myocytes from apoptosis and causes modest hypertrophy.

20.6 BIOMARKER AND MicroRNA

On the other hand, identifying early-stage cardiovascular disease biomarkers is of unparallel value. Peripheral blood leukocytes, which can be easily obtained from patients, have close relevance with inflammation, atherosclerosis, and cardiovascular disease etiology, could be targets for the development of novel epigenomic biomarkers. For instance, Castro et al. [107] found lower DNA methylation content in peripheral blood leukocytes from patients with cardiovascular disease. Results from the Normative Aging Study have also shown that lower LINE-1 methylation in peripheral blood leukocytes is a predictor of incidence and mortality from ischemic heart disease and stroke [108]. Elevated Alu methylation in peripheral blood leukocytes recently was related to prevalence of cardiovascular disease and obesity in Chinese individuals [109]. Thus, characterizing the methylation status of human peripheral blood leukocytes may be potentially beneficial for the early diagnosis of cardiovascular diseases.

MicroRNA is a recently discovered mechanism which plays an important role in cardiovascular disease. Ren et al. [110] performed miRNA arrays analysis to detect the expression pattern of miRNAs in murine hearts subjected to ischemia/reperfusion in vivo and ex vivo. Surprisingly, they found that only miR-320 expression was significantly decreased in the hearts on ischemia/reperfusion in vivo and ex vivo. Wang et al. [111] examined muscle-enriched miRNAs (miR-1, miR-133a, and miR-499) and cardiac-specific miR-208a in circulating blood. Evaluation of the miRNA levels in plasma from patients with heart disease demonstrated that all four miRNA levels were substantially higher than those from healthy people, patients with non-heart disease, or patients with other cardiovascular diseases. Notably, miR-208a remained undetectable in non-heart disease patients but it was easily detected in 90.9% of heart disease patients and in 100% patients within 4αh of the onset of symptoms. MiR-208a revealed the higher sensitivity and specificity for diagnosis in patients. Small et al. [112] reviewed recent progress of microRNA in cardiovascular research, which seems to be a promising direction in future research.

20.7 SUMMARY AND FUTURE PERSPECTIVES

In this chapter we summarized the role of epigenetics in the pathogenesis of cardiovascular disease. DNA methylation as well as histone acetylation has crucial functions in modulating SMC and EC homeostasis (proliferation, migration, apoptosis, and differentiation), atherosclerosis, cardiomyocyte hypertrophy, and heart failure. These results have been listed in Table 20.4, which provides more information and internet resources.

Epigenetic modifications, which are dynamic and reversible, could represent a way that organisms adapt to their environment. Thus understanding the relationship between environmental conditions and epigenetic changes is of potential value. Future research elucidating

TABLE 20.4 Internet Resources

Name	Description	Address
NIH Epigenomics	Deposite experiments and samples investigating the function of DNA methylation and histone modification	http://www.ncbi.nlm.nih.gov/epigenomics
NAME21	DNA methylation patterns were analyzed for 190 gene promoter regions on chromosome 21	http://biochem.jacobs-university.de/name21/
Human Epigenome Atlas	The Human Epigenome Atlas includes human reference epigenomes and the results of their integrative and comparative analyses	http://www.genboree.org/epigenomeatlas/index.rhtml
The Human Epigenome Project	Identifies methylation variable positions (MVPs) in the human genome, currently contains CpG methylation information for MHC	http://www.epigenome.org/
MethDB	The database for DNA methylation and environmental epigenetic effects	http://www.methdb.de/
HHMD	A comprehensive database for human histone modifications, which focuses on integrating useful histone modification information from experimental data that is essential for understanding these modifications at a systematic level	http://202.97.205.78/hhmd/
Geneimprint	Imprinted genes are monnoallelically expressed in a parent-of-origin-dependent manner because the same parental allele is always epigenetically silenced. This website is a collection of imprinted genes by species	http://www.geneimprint.com/site/home
Methprimer	Primer design for methylation PCR primers	www.urogene.org/methprimer/index1.html
Methblast	A sequence similarity search program designed to explore in silico bisulfite modified DNA (either or not methylated at its CpG dinucleotides)	http://medgen.ugent.be/methBLAST/
Methylator	An SVM-based method for DNA methylation prediction	bio.dfci.harvard.edu/Methylator/
Chromatin structure and function	Information regarding histones, histone modifications and their biological roles, and related links	http://www.chromatin.us/
Epigenetic Methylation station	Web source for information, reference, protocols, methods, techniques and links on epigenetics and DNA methylation	www.epigeneticstation.com/

Adapted from a table of the paper published in *Circulation Cardiovascular Genetics* 2010;3;567−573.

the mechanisms of how environment influences epigenetic modifications may help us to better treat cardiovascular diseases. Despite the outstanding progress on understanding the role of HDACs in arteriosclerosis, there are still several questions that need to be answered. One question is whether the effects of HDACs are tissue- or cell-type-specific. This is crucial because if we want to suppress the migration of SMC during atherogenesis, the impact on other cell types needs to be determined. Otherwise, suppression of SMC migration with HDACs (or HDACis) could lead to suppression of EC migration as well. To answer this question, more experiments need to be performed on the functions of HDACs in various cardiovascular cells.

In respect to a clinical prospect, several clinical trials of HDAC inhibitors have been completed in cancer and neurological patients [113,114]. Among these clinical trials, HDAC inhibitors have shown potent inhibition of HDAC activities in patients, and only limited side effects have been observed. Marks et al. [115] reviewed that HDAC inhibition potent to have effect on apoptosis of cancer versus normal human cells. This emphasizes the possibility of a tissue-specific effect of HDAC inhibition, which could explain the fact that HDAC inhibition had limited side effects in humans. Therefore, it seems HDAC inhibition could be a promising target in treating cardiovascular diseases. However, detailed attention should be paid to the application of such drugs, which may have unexpected effects due to the fact that the majority of HDAC inhibitors are non-specific. Since different HDACs possess distinct functions, non-selective inhibition of HDACs could trigger side effect responses. Additionally, as an epigenetic modifier of gene expression a single HDAC is usually involved in several signal pathways. Thus, development of highly selective and cell-type-specific HDAC inhibitors could be a promising field for the cardiovascular research in future. Along with the progress of the function of epigenetics in atherosclerosis and heart failure, our research on epigenetics will benefit patients with cardiovascular diseases.

Acknowledgments

This work was supported by grants from the British Heart Foundation and the Oak Foundation.

References

[1] Ordovas JM, Smith CE. Epigenetics and cardiovascular disease. Nat Rev Cardiol 2010;7:510—9.

[2] Ross R. Atherosclerosis—an inflammatory disease. N Engl J Med 1999;340:115—26.

[3] Wick G, Knoflach M, Xu Q. Autoimmune and inflammatory mechanisms in atherosclerosis. Annu Rev Immunol 2004;22:361—403.

[4] Fuster V. Lewis A. Conner Memorial Lecture. Mechanisms leading to myocardial infarction: insights from studies of vascular biology. Circulation 1994;90:2126—46.

[5] Serpi R, Tolonen AM, Huusko J, Rysa J, Tenhunen O, Yla-Herttuala S, et al. Vascular endothelial growth factor-B gene transfer prevents angiotensin II-induced diastolic dysfunction via proliferation and capillary dilatation in rats. Cardiovasc Res 2011;89:204—13.

[6] Pons D, de Vries FR, van den Elsen PJ, Heijmans BT, Quax PH, Jukema JW. Epigenetic histone acetylation modifiers in vascular remodelling: new targets for therapy in cardiovascular disease. Eur Heart J 2009;30:266—77.

[7] Bonasio R, Tu S, Reinberg D. Molecular signals of epigenetic states. Science 2010;330:612—6.

[8] Atadja PW. HDAC inhibitors and cancer therapy. Prog Drug Res 2011;67:175—95.

[9] Hiltunen MO, Yla-Herttuala S. DNA methylation, smooth muscle cells, and atherogenesis. Arterioscler Thromb Vasc Biol 2003;23:1750—3.

[10] Lei H, Oh SP, Okano M, Juttermann R, Goss KA, Jaenisch R, et al. De novo DNA cytosine methyltransferase activities in mouse embryonic stem cells. Development 1996;122:3195—205.

[11] Zhou B, Margariti A, Zeng L, Xu Q. Role of histone deacetylases in vascular cell homeostasis and arterio-sclerosis. Cardiovasc Res 2011;90:413—20.

[12] Gray SG, Ekstrom TJ. The human histone deacetylase family. Exp Cell Res 2001;262:75—83.

[13] Borradaile NM, Pickering JG. NAD(+), sirtuins, and cardiovascular disease. Curr Pharm Des 2009;15:110—7.

[14] Gao L, Cueto MA, Asselbergs F, Atadja P. Cloning and functional characterization of HDAC11, a novel member of the human histone deacetylase family. J Biol Chem 2002;277:25748—55.

[15] Lahm A, Paolini C, Pallaoro M, Nardi MC, Jones P, Neddermann P, et al. Unraveling the hidden catalytic activity of vertebrate class IIa histone deacetylases. Proc Natl Acad Sci USA 2007;104:17335—40.

[16] Fischle W, Dequiedt F, Hendzel MJ, Guenther MG, Lazar MA, Voelter W, et al. Enzymatic activity associated with class II HDACs is dependent on a multiprotein complex containing HDAC3 and SMRT/N-CoR. Mol Cell 2002;9:45—57.

[17] Fischle W, Dequiedt F, Fillion M, Hendzel MJ, Voelter W, Verdin E. Human HDAC7 histone deacetylase activity is associated with HDAC3 in vivo. J Biol Chem 2001;276:35826—35.

[18] McKinsey TA, Zhang CL, Lu J, Olson EN. Signal-dependent nuclear export of a histone deacetylase regulates muscle differentiation. Nature 2000;408:106—11.

[19] de Ruijter AJ, van Gennip AH, Caron HN, Kemp S, van Kuilenburg AB. Histone deacetylases (HDACs): characterization of the classical HDAC family. Biochem J 2003;370:737—49.

[20] Wanczyk M, Roszczenko K, Marcinkiewicz K, Bojarczuk K, Kowara M, Winiarska M. HDACi—going through the mechanisms. Front Biosci 2011;16:340—59.

[21] Laherty CD, Yang WM, Sun JM, Davie JR, Seto E, Eisenman RN. Histone deacetylases associated with the mSin3 corepressor mediate mad transcriptional repression. Cell 1997;89:349—56.

[22] Yoshida M, Nomura S, Beppu T. Effects of trichostatins on differentiation of murine erythroleukemia cells. Cancer Res 1987;47:3688—91.

[23] Richon VM, Emiliani S, Verdin E, Webb Y, Breslow R, Rifkind RA, et al. A class of hybrid polar inducers of transformed cell differentiation inhibits histone deacetylases. Proc Natl Acad Sci USA 1998;95:3003—7.

[24] Richon VM, Sandhoff TW, Rifkind RA, Marks PA. Histone deacetylase inhibitor selectively induces p21WAF1 expression and gene-associated histone acetylation. Proc Natl Acad Sci USA 2000;97:10014—9.

[25] Butler LM, Webb Y, Agus DB, Higgins B, Tolentino TR, Kutko MC, et al. Inhibition of transformed cell growth and induction of cellular differentiation by pyroxamide, an inhibitor of histone deacetylase. Clin Cancer Res 2001;7:962—70.

[26] Glick RD, Swendeman SL, Coffey DC, Rifkind RA, Marks PA, Richon VM, et al. Hybrid polar histone deacetylase inhibitor induces apoptosis and CD95/CD95 ligand expression in human neuroblastoma. Cancer Res 1999;59:4392—9.

[27] Zhang Y, Adachi M, Kawamura R, Imai K. Bmf is a possible mediator in histone deacetylase inhibitors FK228 and CBHA-induced apoptosis. Cell Death Differ 2006;13:129—40.

[28] Su GH, Sohn TA, Ryu B, Kern SE. A novel histone deacetylase inhibitor identified by high-throughput transcriptional screening of a compound library. Cancer Res 2000;60:3137—42.

[29] Cao H, Stamatoyannopoulos G, Jung M. Induction of human gamma globin gene expression by histone deacetylase inhibitors. Blood 2004;103:701—9.

[30] Gottlicher M, Minucci S, Zhu P, Kramer OH, Schimpf A, Giavara S, et al. Valproic acid defines a novel class of HDAC inhibitors inducing differentiation of transformed cells. EMBO J 2001;20:6969—78.

[31] Williams RS, Cheng L, Mudge AW, Harwood AJ. A common mechanism of action for three mood-stabilizing drugs. Nature 2002;417:292—5.

[32] Gondcaille C, Depreter M, Fourcade S, Lecca MR, Leclercq S, Martin PG, et al. Phenylbutyrate up-regulates the adrenoleukodystrophy-related gene as a nonclassical peroxisome proliferator. J Cell Biol 2005;169:93—104.

[33] Lu Q, Wang DS, Chen CS, Hu YD. Structure-based optimization of phenylbutyrate-derived histone deacetylase inhibitors. J Med Chem 2005;48:5530—5.

[34] Dangond F, Gullans SR. Differential expression of human histone deacetylase mRNAs in response to immune cell apoptosis induction by trichostatin A and butyrate. Biochem Biophys Res Commun 1998;247:833—7.

[35] Gibson PR. The intracellular target of butyrate's actions: HDAC or HDON'T? Gut 2000;46:447—8.

[36] Sambucetti LC, Fischer DD, Zabludoff S, Kwon PO, Chamberlin H, Trogani N, et al. Histone deacetylase inhibition selectively alters the activity and expression of cell cycle proteins leading to specific chromatin acetylation and antiproliferative effects. J Biol Chem 1999;274:34940—7.

[37] Furumai R, Komatsu Y, Nishino N, Khochbin S, Yoshida M, Horinouchi S. Potent histone deacetylase inhibitors built from trichostatin A and cyclic tetrapeptide antibiotics including trapoxin. Proc Natl Acad Sci USA 2001;98:87—92.

[38] Piekarz RL, Robey R, Sandor V, Bakke S, Wilson WH, Dahmoush L, et al. Inhibitor of histone deacetylation, depsipeptide (FR901228), in the treatment of peripheral and cutaneous T-cell lymphoma: a case report. Blood 2001;98:2865—8.

[39] Furumai R, Matsuyama A, Kobashi N, Lee KH, Nishiyama M, Nakajima H, et al. FK228 (depsipeptide) as a natural prodrug that inhibits class I histone deacetylases. Cancer Res 2002;62:4916—21.

[40] Kim MS, Son MW, Kim WB, In Park Y, Moon A. Apicidin, an inhibitor of histone deacetylase, prevents H-ras-induced invasive phenotype. Cancer Lett 2000;157:23—30.

411

[41] Han JW, Ahn SH, Park SH, Wang SY, Bae GU, Seo DW, et al. Apicidin, a histone deacetylase inhibitor, inhibits proliferation of tumor cells via induction of p21WAF1/Cip1 and gelsolin. Cancer Res 2000;60:6068−74.

[42] De Schepper S, Bruwiere H, Verhulst T, Steller U, Andries L, Wouters W, et al. Inhibition of histone deacetylases by chlamydocin induces apoptosis and proteasome-mediated degradation of survivin. J Pharmacol Exp Ther 2003;304:881−8.

[43] Rosato RR, Almenara JA, Grant S. The histone deacetylase inhibitor MS-275 promotes differentiation or apoptosis in human leukemia cells through a process regulated by generation of reactive oxygen species and induction of p21CIP1/WAF1 1. Cancer Res 2003;63:3637−45.

[44] Haggarty SJ, Koeller KM, Wong JC, Grozinger CM, Schreiber SL. Domain-selective small-molecule inhibitor of histone deacetylase 6 (HDAC6)-mediated tubulin deacetylation. Proc Natl Acad Sci USA 2003;100:4389−94.

[45] Itoh Y, Suzuki T, Miyata N. Isoform-selective histone deacetylase inhibitors. Curr Pharm Des 2008;14:529−44.

[46] Caplan BA, Schwartz CJ. Increased endothelial cell turnover in areas of in vivo Evans Blue uptake in the pig aorta. Atherosclerosis 1973;17:401−17.

[47] Somer JB, Schwartz CJ. Focal (3 H)cholesterol uptake in the pig aorta. 2. Distribution of (3 H)cholesterol across the aortic wall in areas of high and low uptake in vivo. Atherosclerosis 1972;16:377−88.

[48] Jamaluddin MD, Chen I, Yang F, Jiang X, Jan M, Liu X, et al. Homocysteine inhibits endothelial cell growth via DNA hypomethylation of the cyclin A gene. Blood 2007;110:3648−55.

[49] Margariti A, Zampetaki A, Xiao Q, Zhou B, Karamariti E, Martin D, et al. Histone deacetylase 7 controls endothelial cell growth through modulation of beta-catenin. Circ Res 2010;106:1202−11.

[50] Chang S, Young BD, Li S, Qi X, Richardson JA, Olson EN. Histone deacetylase 7 maintains vascular integrity by repressing matrix metalloproteinase 10. Cell 2006;126:321−34.

[51] Mottet D, Bellahcene A, Pirotte S, Waltregny D, Deroanne C, Lamour V, et al. Histone deacetylase 7 silencing alters endothelial cell migration, a key step in angiogenesis. Circ Res 2007;101:1237−46.

[52] Welt FG, Rogers C. Inflammation and restenosis in the stent era. Arterioscler Thromb Vasc Biol 2002;22:1769−76.

[53] Urbich C, Rossig L, Kaluza D, Potente M, Boeckel JN, Knau A, et al. HDAC5 is a repressor of angiogenesis and determines the angiogenic gene expression pattern of endothelial cells. Blood 2009;113:5669−79.

[54] Ha CH, Jhun BS, Kao HY, Jin ZG. VEGF stimulates HDAC7 phosphorylation and cytoplasmic accumulation modulating matrix metalloproteinase expression and angiogenesis. Arterioscler Thromb Vasc Biol 2008;28:1782−8.

[55] Tedgui A, Mallat Z. Apoptosis as a determinant of atherothrombosis. Thromb Haemost 2001;86:420−6.

[56] Mitra S, Khaidakov M, Lu J, Ayyadevara S, Szwedo J, Wang XW, et al. Prior Exposure to Oxidized Low Density Lipoprotein Limits Apoptosis in Subsequent Generations of Endothelial Cells by Altering Promoter Methylation. Am J Physiol Heart Circ Physiol 2011.

[57] Rao X, Zhong J, Zhang S, Zhang Y, Yu Q, Yang P, et al. Loss of Methyl-CpG-Binding Domain Protein 2 Enhances Endothelial Angiogenesis and Protects Mice Against Hind-Limb Ischemic Injury. Circulation 2011;123:2964−74.

[58] Xiao Q, Zeng L, Zhang Z, Margariti A, Ali ZA, Channon KM, et al. Sca-1+ progenitors derived from embryonic stem cells differentiate into endothelial cells capable of vascular repair after arterial injury. Arterioscler Thromb Vasc Biol 2006;26:2244−51.

[59] Zeng L, Xiao Q, Margariti A, Zhang Z, Zampetaki A, Patel S, et al. HDAC3 is crucial in shear- and VEGF-induced stem cell differentiation toward endothelial cells. J Cell Biol 2006;174:1059−69.

[60] Zampetaki A, Zeng L, Margariti A, Xiao Q, Li H, Zhang Z, et al. Histone deacetylase 3 is critical in endothelial survival and atherosclerosis development in response to disturbed flow. Circulation 2010;121:132−42.

[61] Hou J, Chong ZZ, Shang YC, Maiese K. Early apoptotic vascular signaling is determined by Sirt1 through nuclear shuttling, forkhead trafficking, bad, and mitochondrial caspase activation. Curr Neurovasc Res 2010;7:95−112.

[62] Zu Y, Liu L, Lee MY, Xu C, Liang Y, Man RY, et al. SIRT1 promotes proliferation and prevents senescence through targeting LKB1 in primary porcine aortic endothelial cells. Circ Res 2010;106:1384−93.

[63] Michaelis M, Suhan T, Michaelis UR, Beek K, Rothweiler F, Tausch L, et al. Valproic acid induces extracellular signal-regulated kinase 1/2 activation and inhibits apoptosis in endothelial cells. Cell Death Differ 2006;13:446−53.

[64] Banerjee S, Bacanamwo M. DNA methyltransferase inhibition induces mouse embryonic stem cell differentiation into endothelial cells. Exp Cell Res 2010;316:172−80.

[65] Laukkanen MO, Mannermaa S, Hiltunen MO, Aittomaki S, Airenne K, Janne J, et al. Local hypomethylation in atherosclerosis found in rabbit ec-sod gene. Arterioscler Thromb Vasc Biol 1999;19:2171−8.

[66] Post WS, Goldschmidt-Clermont PJ, Wilhide CC, Heldman AW, Sussman MS, Ouyang P, et al. Methylation of the estrogen receptor gene is associated with aging and atherosclerosis in the cardiovascular system. Cardiovasc Res 1999;43:985−91.

[67] Ying AK, Hassanain HH, Roos CM, Smiraglia DJ, Issa JJ, Michler RE, et al. Methylation of the estrogen receptor-alpha gene promoter is selectively increased in proliferating human aortic smooth muscle cells. Cardiovasc Res 2000;46:172−9.

[68] Sedding DG, Trobs M, Reich F, Walker G, Fink L, Haberbosch W, et al. 3-Deazaadenosine prevents smooth muscle cell proliferation and neointima formation by interfering with Ras signaling. Circ Res 2009;104:1192−200.

[69] Villeneuve LM, Reddy MA, Lanting LL, Wang M, Meng L, Natarajan R. Epigenetic histone H3 lysine 9 methylation in metabolic memory and inflammatory phenotype of vascular smooth muscle cells in diabetes. Proc Natl Acad Sci USA 2008;105:9047−52.

[70] Okamoto H, Fujioka Y, Takahashi A, Takahashi T, Taniguchi T, Ishikawa Y, et al. Trichostatin A, an inhibitor of histone deacetylase, inhibits smooth muscle cell proliferation via induction of p21(WAF1). J Atheroscler Thromb 2006;13:183−91.

[71] Song S, Kang SW, Choi C. Trichostatin A enhances proliferation and migration of vascular smooth muscle cells by downregulating thioredoxin 1. Cardiovasc Res 2010;85:241−9.

[72] Zhou B, Margariti A, Zeng L, Habi O, Xiao Q, Martin D, et al. Splicing of Histone Deacetylase 7 Modulates Smooth Muscle Cell Proliferation and Neointima Formation Through Nuclear β-Catenin Translocation. Arterioscler Thromb Vasc Biol 2011.

[73] Yan ZQ, Yao QP, Zhang ML, Qi YX, Guo ZY, Shen BR, et al. Histone deacetylases modulate vascular smooth muscle cell migration induced by cyclic mechanical strain. J Biomech 2009;42:945−8.

[74] Bennett MR, Evan GI, Schwartz SM. Apoptosis of human vascular smooth muscle cells derived from normal vessels and coronary atherosclerotic plaques. J Clin Invest 1995;95:2266−74.

[75] Patel VA, Zhang QJ, Siddle K, Soos MA, Goddard M, Weissberg PL, et al. Defect in insulin-like growth factor-1 survival mechanism in atherosclerotic plaque-derived vascular smooth muscle cells is mediated by reduced surface binding and signaling. Circ Res 2001;88:895−902.

[76] Kavurma MM, Figg N, Bennett MR, Mercer J, Khachigian LM, Littlewood TD. Oxidative stress regulates IGF1R expression in vascular smooth-muscle cells via p53 and HDAC recruitment. Biochem J 2007;407:79−87.

[77] van der Veer E, Nong Z, O'Neil C, Urquhart B, Freeman D, Pickering JG. Pre-B-cell colony-enhancing factor regulates NAD+-dependent protein deacetylase activity and promotes vascular smooth muscle cell maturation. Circ Res 2005;97:25−34.

[78] Han CI, Campbell GR, Campbell JH. Circulating bone marrow cells can contribute to neointimal formation. J Vasc Res 2001;38:113−9.

[79] Li J, Han X, Jiang J, Zhong R, Williams GM, Pickering JG, et al. Vascular smooth muscle cells of recipient origin mediate intimal expansion after aortic allotransplantation in mice. Am J Pathol 2001;158:1943−7.

[80] Sata M, Saiura A, Kunisato A, Tojo A, Okada S, Tokuhisa T, et al. Hematopoietic stem cells differentiate into vascular cells that participate in the pathogenesis of atherosclerosis. Nat Med 2002;8:403−9.

[81] Hu Y, Davison F, Ludewig B, Erdel M, Mayr M, Url M, et al. Smooth muscle cells in transplant atherosclerotic lesions are originated from recipients, but not bone marrow progenitor cells. Circulation 2002;106:1834−9.

[82] Hu Y, Zhang Z, Torsney E, Afzal AR, Davison F, Metzler B, et al. Abundant progenitor cells in the adventitia contribute to atherosclerosis of vein grafts in ApoE-deficient mice. J Clin Invest 2004;113:1258−65.

[83] Shimizu K, Sugiyama S, Aikawa M, Fukumoto Y, Rabkin E, Libby P, et al. Host bone-marrow cells are a source of donor intimal smooth- muscle-like cells in murine aortic transplant arteriopathy. Nat Med 2001;7:738−41.

[84] Hu Y, Mayr M, Metzler B, Erdel M, Davison F, Xu Q. Both donor and recipient origins of smooth muscle cells in vein graft atherosclerotic lesions. Circ Res 2002;91:e13−20.

[85] Lockman K, Taylor JM, Mack CP. The histone demethylase, Jmjd1a, interacts with the myocardin factors to regulate SMC differentiation marker gene expression. Circ Res 2007;101:e115−23.

[86] Lee WJ, Kim HJ. Inhibition of DNA methylation is involved in transdifferentiation of myoblasts into smooth muscle cells. Mol Cells 2007;24:441−4.

[87] Spin JM, Quertermous T, Tsao PS. Chromatin remodeling pathways in smooth muscle cell differentiation, and evidence for an integral role for p300. PLoS One 2010;5:e14301.

[88] Qiu P, Ritchie RP, Gong XQ, Hamamori Y, Li L. Dynamic changes in chromatin acetylation and the expression of histone acetyltransferases and histone deacetylases regulate the SM22alpha transcription in response to Smad3-mediated TGFbeta1 signaling. Biochem Biophys Res Commun 2006;348:351−8.

[89] Sharma P, Kumar J, Garg G, Kumar A, Patoway A, Karthikeyan G, et al. Detection of altered global DNA methylation in coronary artery disease patients. DNA Cell Biol 2008;27:357−65.

[90] Hiltunen MO, Turunen MP, Hakkinen TP, Rutanen J, Hedman M, Makinen K, et al. DNA hypomethylation and methyltransferase expression in atherosclerotic lesions. Vasc Med 2002;7:5−11.

[91] Kim J, Kim JY, Song KS, Lee YH, Seo JS, Jelinek J, et al. Epigenetic changes in estrogen receptor beta gene in atherosclerotic cardiovascular tissues and in-vitro vascular senescence. Biochim Biophys Acta 2007;1772:72−80.

[92] Lund G, Andersson L, Lauria M, Lindholm M, Fraga MF, Villar-Garea A, et al. DNA methylation polymorphisms precede any histological sign of atherosclerosis in mice lacking apolipoprotein E. J Biol Chem 2004;279:29147–54.

[93] Castillo-Diaz SA, Garay-Sevilla ME, Hernandez-Gonzalez MA, Solis-Martinez MO, Zaina S. Extensive demethylation of normally hypermethylated CpG islands occurs in human atherosclerotic arteries. Int J Mol Med 2010;26:691–700.

[94] Davies PF, Polacek DC, Handen JS, Helmke BP, DePaola N. A spatial approach to transcriptional profiling: mechanotransduction and the focal origin of atherosclerosis. Trends Biotechnol 1999;17:347–51.

[95] Granger A, Abdullah I, Huebner F, Stout A, Wang T, Huebner T, et al. Histone deacetylase inhibition reduces myocardial ischemia-reperfusion injury in mice. FASEB J 2008;22:3549–60.

[96] Yanazume T, Hasegawa K, Morimoto T, Kawamura T, Wada H, Matsumori A, et al. Cardiac p300 is involved in myocyte growth with decompensated heart failure. Mol Cell Biol 2003;23:3593–606.

[97] Kawamura T, Ono K, Morimoto T, Wada H, Hirai M, Hidaka K, et al. Acetylation of GATA-4 is involved in the differentiation of embryonic stem cells into cardiac myocytes. J Biol Chem 2005;280:19682–8.

[98] Miyamoto S, Kawamura T, Morimoto T, Ono K, Wada H, Kawase Y, et al. Histone acetyltransferase activity of p300 is required for the promotion of left ventricular remodeling after myocardial infarction in adult mice in vivo. Circulation 2006;113:679–90.

[99] Wei JQ, Shehadeh LA, Mitrani JM, Pessanha M, Slepak TI, Webster KA, et al. Quantitative control of adaptive cardiac hypertrophy by acetyltransferase p300. Circulation 2008;118:934–46.

[100] Hosseinkhani M, Hasegawa K, Ono K, Kawamura T, Takaya T, Morimoto T, et al. Trichostatin A induces myocardial differentiation of monkey ES cells. Biochem Biophys Res Commun 2007;356:386–91.

[101] Kaneda R, Ueno S, Yamashita Y, Choi YL, Koinuma K, Takada S, et al. Genome-wide screening for target regions of histone deacetylases in cardiomyocytes. Circ Res 2005;97:210–8.

[102] Kee HJ, Eom GH, Joung H, Shin S, Kim JR, Cho YK, et al. Activation of histone deacetylase 2 by inducible heat shock protein 70 in cardiac hypertrophy. Circ Res 2008;103:1259–69.

[103] Backs J, Song K, Bezprozvannaya S, Chang S, Olson EN. CaM kinase II selectively signals to histone deacetylase 4 during cardiomyocyte hypertrophy. J Clin Invest 2006;116:1853–64.

[104] Lemon DD, Horn TR, Cavasin MA, Jeong MY, Haubold KW, Long CS, et al. Cardiac HDAC6 catalytic activity is induced in response to chronic hypertension. J Mol Cell Cardiol 2011;51:41–50.

[105] Zhang CL, McKinsey TA, Chang S, Antos CL, Hill JA, Olson EN. Class II histone deacetylases act as signal-responsive repressors of cardiac hypertrophy. Cell 2002;110:479–88.

[106] Alcendor RR, Kirshenbaum LA, Imai S, Vatner SF, Sadoshima J. Silent information regulator 2alpha, a longevity factor and class III histone deacetylase, is an essential endogenous apoptosis inhibitor in cardiac myocytes. Circ Res 2004;95:971–80.

[107] Castro R, Rivera I, Struys EA, Jansen EE, Ravasco P, Camilo ME, et al. Increased homocysteine and S-adenosylhomocysteine concentrations and DNA hypomethylation in vascular disease. Clin Chem 2003;49:1292–6.

[108] Baccarelli A, Wright R, Bollati V, Litonjua A, Zanobetti A, Tarantini L, et al. Ischemic heart disease and stroke in relation to blood DNA methylation. Epidemiology 2010;21:819–28.

[109] Kim M, Long TI, Arakawa K, Wang R, Yu MC, Laird PW. DNA methylation as a biomarker for cardiovascular disease risk. PLoS One 2010;5:e9692.

[110] Ren XP, Wu J, Wang X, Sartor MA, Qian J, Jones K, et al. MicroRNA-320 is involved in the regulation of cardiac ischemia/reperfusion injury by targeting heat-shock protein 20. Circulation 2009;119:2357–66.

[111] Wang GK, Zhu JQ, Zhang JT, Li Q, Li Y, He J, et al. Circulating microRNA: a novel potential biomarker for early diagnosis of acute myocardial infarction in humans. Eur Heart J 2010;31:659–66.

[112] Small EM, Frost RJ, Olson EN. MicroRNAs add a new dimension to cardiovascular disease. Circulation 2010;121:1022–32.

[113] Muller S, Kramer OH. Inhibitors of HDACs—effective drugs against cancer? Curr Cancer Drug Targets 2010;10:210–28.

[114] Hirtz D, Iannaccone S, Heemskerk J, Gwinn-Hardy K, Moxley 3rd R, Rowland LP. Challenges and opportunities in clinical trials for spinal muscular atrophy. Neurology 2005;65:1352–7.

[115] Marks PA, Xu WS. Histone deacetylase inhibitors: Potential in cancer therapy. J Cell Biochem 2009;107:600–8.

414

CHAPTER 21

Epigenetics and Human Infectious Diseases

Hans Helmut Niller[1], Janos Minarovits[2]
[1]Institute for Medical Microbiology and Hygiene at the University of Regensburg, Regensburg, Germany
[2]National Center for Epidemiology, Budapest, Hungary

415

T. Tollefsbol (Ed): Epigenetics in Human Disease. DOI: 10.1016/B978-0-12-388415-2.00021-4

21.1 INTRODUCTION

Patho-epigenetics is a new discipline dealing with the pathological consequences of dysregulated epigenetic processes [1]. In recent years it became more and more obvious that — perhaps following the lead of cancer research, where epigenetic and genetic theories of neoplastic development complement each other peacefully — epigenetic ideas found their way to virtually all areas of biomedical research. Looking at the exponentially accumulating data one has the impression that epigenetic alterations, induced by certain pathogenic viruses and bacteria in the host cells they are interacting with, play an unexpected but most important role in disease initiation and progression. In this review we focus mainly on human pathogens eliciting epigenetic changes relevant, or at least potentially relevant, to disease initiation or progression. Although the idea that microbes do induce epigenetic changes in host cells is gaining more and more support, so far only a minority of infectious agents was analyzed in this respect. Thus, in this chapter we do not discuss, due to the lack of available data on their potential patho-epigenetic effects, even those pathogens, including protozoan parasites and fungi, that are known to use their own sophisticated epigenetic mechanisms to control the expression of their genomes [2—6]. For the very same reason, we omit alphaherpesviruses in spite of the fact that their latent genomes undergo epigenetic changes in their host cells [7,8], and oncogenic human adenoviruses that induce malignant tumors in experimental animals and elicit spectacular epigenetic alterations in tissue culture [9,10], but lack any association with human neoplasms. As to the bacterial pathogens discussed in this chapter, the epigenetic aspects of *Helicobacter pylori* infection were analyzed most intensively, due to its association with gastric carcinoma. Regarding viruses infecting humans, the epigenotypes of tumor-associated DNA viruses and proviral DNA copies of retroviral genomes were characterized in detail. These data have potential implications as to the therapy of neoplasms and elimination of latent reservoirs of human immunodeficiency virus and human T-cell lymphotropic virus. Deciphering how viral and other microbial proteins and non-translated RNAs alter the host cell epigenome may result in new tools permitting early and stage-specific detection of pathological alterations. Epigenetic studies may contribute to a better understanding of the diseases caused by infectious pathogens, first of all microbes, but also macroparasites, including helminths, fungi, and arthropods.

21.2 EPIGENETIC MODIFICATIONS ELICITED IN HOST CELLS DURING BACTERIAL INFECTIONS

Bacteria may affect the epigenetic regulatory mechanisms of their hosts via the metabolic products and toxins released into their environment. In addition, certain pathogens synthesize effector proteins capable of inducing epigenetic alterations. Such effector proteins are either injected into their target cells or enter the host cell nuclei as the products of bacteria that are capable of surviving within the cytoplasm of infected cells. In several cases the exact mechanism of epigenetic changes elicited by bacterial infections remains to be clarified.

21.2.1 Epigenetic Alterations Induced by Bacterial Infections in Periodontal Disease

Pathogenic Gram-negative anaerobic bacteria populating the tooth-associated biofilm may cause an inflammatory reaction (periodontitis). The resulting chronic inflammation, periodontal disease, affects the majority (50–90%) of the worldwide population [11]. In addition to its local consequences (occasional pain, impaired mastication, loosening and loss of teeth), subgingival infection may have a systemic impact including adverse pregnancy outcomes, atherosclerosis, rheumatoid arthritis, diabetes, and aspiration pneumonia. A critical step in periodontal disease progression is the disruption of the host innate immune system by anaerobic bacteria (reviewed by [12]). The cells of the innate immune system, including gingival epithelial cells (GECs), recognize pathogen-associated molecular patterns interacting with Toll-like receptors (TLRs) and other receptor classes, and respond by the production of antimicrobial peptides such as human β-defensins and chemokines that activate adaptive immune responses as well [13–15]. In addition to *Porphyromonas gingivalis, Tannerella forsythia, Treponema denticola,* and *Prevotella intermedia,* the pathogenic bacteria causing periodontitis, a series of non-pathogenic bacteria (e.g. *Fusobacterium nucleatum*) also inhabit the oral cavity of periodontitis patients, forming a microbial community (reviewed by [12]). Similarly to their pathogenic counterparts, non-pathogenic bacteria may also interact with GECs, modulating native immune respones to invading pathogens.

Porphyromonas gingivalis, one of the major etiological agents of periodontal disease, elicited complex changes when incubated with gingival epithelial cells: it suppressed the expression of histone deacetylase 1 and 2 (HDAC1 and HDAC2) and DNA methyltransferase 1 (DNMT1) [16]. In parallel, it stimulated human β-defensin and CCL20 (CC chemokine ligand 20) expression and increased promoter methylation of six genes, including the immune regulator *CD276, elastase 2, TLR2, IL-12A,* and two putative tumor-suppressor genes (TSG). The level of activating histone modification H3K4me3 decreased in GECs incubated with *Porphyromonas gingivalis,* but not in the presence of the non-pathogenic *Fusobacterium nucleatum* [16]. Short-chain fatty acids, the metabolic by-products of *Porphyromonas gingivalis,* are secreted extracellularly and may affect local immune responses, contributing thereby to the development of periodontal disease [17–19]. Butyric acid, the major short-chain fatty acid species produced by pathogenic *Porphyromonas gingivalis* strains, accumulated in periodontal pockets [18–20], and it was suggested that as a HDAC inhibitor [21,22] (reviewed by [23]), butyric acid may induce acetylation of histone H3 and H4 in neighboring cells. Indeed, supernatants of *Porphyromonas gingivalis* cultures, and butyric acid induced histone acetylation in T-cells and macrophages carrying latent proviral genomes of type 1 human immunodeficiency virus (HIV-1), and caused HIV-1 reactivation [20]. The role of periodontal disease in the progression of acquired immunodeficiency syndrome needs further studies.

Another bacterium involved in periodontal infections, *Campylobacter rectus* may also induce epigenetic alterations in human cells. In experimental mice *Campylobacter rectus* infection down-regulated the expression of the *Igf2* (insulin-like growth factor 2) gene via hypermethylation of the *Igf2* promoter in the murine placenta [24]. This epigenetic change resulted in reduced placental growth and fetal growth restriction, suggesting that a similar mechanism may be involved in preterm births associated with *Campylobacter rectus* infection in humans [25]. The potential epigenetic consequences of *Campylobacter rectus*–gingival epithelial cell interactions remain to be established.

21.2.2 Histone modifications Caused by *Listeria monocytogenes*

The Gram-positive bacterium *Listeria monocytogenes* appears to cause only little inflammation during infection. *L. monocytogenes* is associated with foodborne infections, causing usually self-limiting gastroenteritis, but also sepsis and meningitis in immunocompromised patients, perinatal and intrauterine infection, and abortion in pregnant women. It acquired the tools for survival within the cytoplasm of macrophages and endothelial cells (reviewed by [26]). *L. monocytogenes* appears to use its toxin listeriolysin O (LLO) not only as a pore-forming toxin to escape from the phagosome and enter the host cytoplasm, but also in an extracellular form, before entering the host cell, to induce histone modifications in the cells to be infected via an unexplored signaling pathway. Extracellular, pathogenic *Listeria* or purified LLO protein induced, at a subset of genes and within 3 hours or even within 20 minutes, dephosphory-lation of histone H3 at serine 10 and deacetylation of histone H4 [27]. Pore formation was dispensable for the induction of histone H3 dephosphorylation. These changes correlated with a decreased expression of the chemoattractant chemokine CXCL2 involved in the recruitment of polymorphonuclear cells during epithelial cell infection. Expression of the phosphatase DUSP4/MKP2 was also down-regulated, similarly to interferon regulatory factor-3 (IRF-3) and the transcription factor EGR1 (early growth response 1). Thus, *L. monocytogenes* and its toxin, LLO, down-regulates key mediators of innate immunity before invading the host cells, by eliciting targeted, local histone modifications at key immune response genes. The situation is complex, however, because transcriptome analysis revealed not only 47 repressed, but also 99 induced genes after 20 min of LLO incubation [27].

Pore-forming toxins of other bacteria, belonging to different genera and unrelated to *Listeria monocytogenes*, also induced dephosphorylation of histone H3. These cholesterol-dependent cytolysines of the LLO family, perfringolysin (PFO) produced by *Clostridium perfringens* and pneumolysin (PLY) secreted by *Streptococcus pneumoniae* also exerted their histone-modifying effect in a pore-forming-independent manner [27].

21.2.3 Histone Dephosphorylation by OspF, the Effector Protein of *Shigella flexneri*

Shigella flexneri, a Gram-negative bacterium, causes diarrhea that is usually self-limiting. Life-threatening disease may develop, however, in the absence of adequate medical care or in immunocompromised patients [28]. *Shigella flexneri* abrogated histone H3 phosphorylation at selected cellular promoters by injection of a phosphatase into epithelial cells [29]. The effector protein OspF entered the nucleus and dephosphorylated the mitogen-activated cellular protein kinases Erk and p38, thereby blocking histone H3 phosphorylation at Ser10 [29]. These events resulted in repression of a narrow set of genes including the NF-κB responsive *IL8* and *CCL20*. By down-regulating IL-8 expression, OspF blocked neutrophil recruitment in vivo in a rabbit model of infection [29].

21.2.4 Histone Modifications in *Anaplasma phagocytophilum*-Infected Cells

Anaplasma phagocytophilum, a tick-transmitted rickettsial pathogen causes an acute febrile disease [30]. It is capable of infecting and surviving in granulocytes by blocking or delaying key antimicrobial mechanisms including oxidative burst, apoptosis, and phagocytosis, and down-regulating defense gene expression in its host cells [31]. AnkA, the effector protein of *Anaplasma phagocytophilum* is translocated into the nucleus of the host where it binds to the host DNA and nuclear proteins and silences the expression of 19 defense genes arranged into three gene clusters. They code for antimicrobial peptides and enzymes or proteins involved in the generation of reactive oxygen intermediates. Silencing correlated with a transient increase of HDAC1 expression and a steady increase of HDAC2 mRNA levels in *Anaplasma phagocytophilum*-infected cells. In addition, at 48 hours following infection of the THP-1 acute monocytic leukemia cell line, an increased binding of HDAC1 and a parallel decrease in acetylated histone H3 was observed at most of the defense gene promoters, suggesting that *Anaplasma phagocytophilum* takes over the epigenetic control of host defense gene clusters [31].

21.2.5 Epigenetic Alterations Associated with *Helicobacter pylori* Infection

Helicobacter pylori (Hp), a Gram-negative spiral bacterium, colonizes more than half of the world's population in early childhood. Primary infection of the stomach regularly triggers an inflammation. Infection can lead to gastric or duodenal ulcer disease and different types of chronic gastritis (atrophic gastritis, enlarged fold gastritis, pangastritis). Chronic inflammation can progress to gastric adenocarcinoma in about 1—2% of the infected patients through the premalignant stages of gastric atrophy, intestinal metaplasia, and dysplasia. The risk of developing intestinal metaplasia or different types of malignant neoplasms (intestinal type gastric carcinoma (GC), sporadic diffuse type GC, MALT lymphoma) is significantly associated with Hp carrier status [32] (reviewed by [33,34]). Hp codes for the virulence factors vacuolating cytotoxin VacA and cytotoxicity-associated antigen CagA. CagA is encoded within the bacterial cag-pathogenicity island (cag-PAI). Both proteins are strongly associated with Hp pathogenesis. CagA, a bacterial oncoprotein can be injected into gastric epithelial cells by a type IV bacterial secretion system and interfere with multiple cellular signal transduction pathways (reviewed by [35—37]). Through activation of the tyrosine phosphatase SHP-2, CagA may elicit uncontrolled cell proliferation, and induce chromosomal instability by destabilizing microtubules during mitosis [38].

Epigenetic dysregulation is strongly associated with gastric carcinogenesis [39—41]. Both global genomic hypomethylation and CpG island (CGI) hypermethylation correlated with the presence of Hp in gastritis patients, and malignant progression in cancer patients [42—47]. Non-tumorous gastric tissue of Hp-negative GC patients also showed increased CGI hypermethylation [42,43,48]. Concordantly, Hp eradication led to a significant decline of CGI hypermethylation in gastritis patients, and in a small rodent model, but remained still higher than in non-infected individuals [44,49—51] (reviewed by [52]) or animals [53]. With the progress of Hp-induced chronic gastritis through the advanced stages of atrophy, metaplasia, dysplasia, and GC, Hp was frequently lost from the gastric mucosa [54]. This explains — through an epigenetic-based "hit-and-run" scenario — why the hypermethylation profile of gastric mucosa does not regularly correlate with the Hp-carrier status in the advanced stages [46].

CPG-METHYLATION PROFILES OF *HELICOBACTER PYLORI*-ASSOCIATED GASTRIC CARCINOMA

Tumor-suppressor genes in gastric carcinogenesis are more frequently inactivated by promoter methylation than by mutations [55]. One may wonder, however, whether epigenetic dysregulation is causative for carcinogenesis or a mere epiphenomenon. The study of the *E-cadherin* (*CDH1*) promoter may help to clarify this question. *CDH1* down-regulation decreases cellular adhesion and therefore increases cellular motility and the disposition of cancer cells for metastasis. *CDH1* germline mutation is regularly observed in families with hereditary diffuse GC. Thus, *CDH1* inactivation clearly predisposes to GC and is therefore co-causal [56]. Somatic *CDH1* mutation is also common in sporadic GC, while transcriptional silencing by promoter CGI hypermethylation represents a second hit for carcinogenesis [57,58]. The following observations support a causative role for epigenetic dysregulation of *CDH1* in gastric carcinogenesis: (1) *CDH1* promoter hypermethylation is especially frequent in enlarged fold gastritis, a high-risk factor for GC [59]; (2) in the gastric mucosa of Hp-infected patients *CDH1* expression is down-regulated [60], due to promoter methylation [42,49,50,61]; (3) epigenetic silencing of

CDH1 occurs early in gastric carcinogenesis [44] (reviewed by [33]). Overall, hypermethylation of *CDH1* is clearly co-causal in the development of sporadic GC (reviewed by [62]).

The patterns of hypermethylated CGIs in non-cancerous mucosa from GC patients presented a rather characteristic fingerprint of Hp infection [63,64]. Surprisingly, however, and unlike in other malignancies, DNMTs were not induced by Hp in human gastric mucosa [63], and they were down-regulated when Hp was cocultured with gastric epithelial cells from gerbils [53]. In contrast, another study found an up-regulation of DNMT1 and DNMT3A through coculturing Hp with two human GC cell lines [65]. However, cancer cell lines may react differently from primary cells. Thus, other mechanisms, like altered distribution of DNMTs or inflammation-induced disturbance of factors protecting CGIs against methylation, may explain Hp-associated CGI methylation [53,66,67]. The role of Hp-encoded 5'-CpG-DNMTs [68] remains to be explored.

Ushijima observed a clear association between the methylation levels in the histologically normal gastric mucosae and the risk of gastric cancer development. He suggested that increased CGI-hypermethylation predisposed to carcinogenesis and proposed that the epigenetically altered mucosal tissue corresponded to an *epigenetic field* for cancerization [55]. Methylation was found at the *p16, MLH1, ECAD, DAPK,* and *MTSS1* loci of the healthy mucosa of GC patients with and without Hp-infection [41,45,48,49]. The *MLH1* repair-gene methylation was not seen in non-neoplastic epithelia from healthy persons, but was frequent in non-neoplastic tissue from GC patients [44,48], and it was significantly associated with the intestinal type of GC [69]. The metastasis suppressor gene *MTSS1* is highly expressed in normal gastric mucosa and frequently affected by loss of heterozygosity in GC, suggesting a physiological role for *MTSS1* as a suppressor of gastric carcinogenesis [41].

A series of TSGs and many other genomic loci were methylated at a higher level in Hp-positive than Hp-negative cases [42,44,64,65,70–72]. The degree of methylation at the *CDH1, p16, MLH1, APC,* and *COX2* loci correlated to malignant progression, with *CDH1* methylation as an early, and that of *MLH1* as a late, event. *COX2* hypermethylation was associated with the Hp virulence factor VacA [69], whereas hypermethylation at *p16* and *THBD* (*thrombomodulin*) correlated significantly with inflammatory cell infiltration. Thus, chronic inflammation was potentially involved in methylation induction [45,51]. Hp-eradication led to a decreased methylation at *CDH1, p16, APC, FLNc, THBD, MGMT,* and to the complete disappearance of *COX2* (*cyclooxygenase 2*) methylation [44,51,72]. The specific methylation profiles associated with GC and Hp infection may allow the development of new diagnostic tools to detect precancerous stages, assess the risk of GC development, and estimate its prognosis [64].

EPIGENETIC SILENCING OF miRNA GENES IN *HELICOBACTER PYLORI*-ASSOCIATED NEOPLASMS

Ando et al. found that three putative tumor-suppressor miRNA genes, *miR-124a1, miR-124a2, miR-124a3,* were unmethylated in normal mucosa, but hypermethylated in tumor cell lines. In biopsies, hypermethylation was strongly associated with a positive Hp carrier status [73]. *miR-34b* and *miR-34c* were highly methylated in GC, but not in normal gastric mucosa from Hp-negative persons. Transfection of these miRNAs suppressed cell growth in vitro, suggesting a tumor-suppressor function [74]. The tumor suppressor *miR-203* was significantly down-regulated by hypermethylation in human low-grade MALT-lymphoma biopsies. Methylation of *miR-203* coincided with the increased expression of its direct target, the non-receptor tyrosine kinase and oncogene *ABL1* [75,76]. Overexpression of miR-203 in ABL1-expressing human lymphoma cells inhibited proliferation. Further, the ABL1 inhibitor imatinib caused tumor regression in a MALT lymphoma mouse model based on chronic *Helicobacter felis* infection [75]. Hypermethylation of the *miR-124* and *miR-34* groups and of *miR-203* may contribute to the *epigenetic field for cancerization* and may also serve as a predictive marker for GC risk [55,73,74].

THE ROLE OF POLYCOMB GROUP PROTEINS AND HISTONE MODIFICATIONS IN *HELICOBACTER PYLORI*-ASSOCIATED GASTRIC CANCER

Polycomb group (PcG) proteins play an important role in reversibly repressing transcription factor genes which are involved in development and differentiation in embryonic stem cells. This reversible PcG-mediated repression utilizes trimethylation on lysine 27 of histone 3 (H3K27me3). In cancer cells, but not in normal cells, the H3K27me3 chromatin mark recruited DNMTs, resulting in a permanent CpG-methylation [77]. Thus, in cancer cells, reversible repression was more frequently replaced by methylation and permanent silencing at stem-cell PcG-target promoters than at non-PcG target promoters [78] (reviewed by [79]). A glutathione peroxidase knockout mouse model for inflammatory bowel disease and intestinal cancer confirmed these observations. Healthy wild-type mice mostly kept the H3K27me3 chromatin mark at PcG target genes, while inflammation in KO mice led to aberrant methylation mostly of PcG targets which was later also observed in cancer tissue [80].

Changes in global histone modification profiles and HDAC expression patterns correlated with the progression of GC in general and may serve as prognostic markers (reviewed by [35]). Further, the epigenetic repression through both DNA methylation and histone modification of specific genes involved in chromatin remodeling, cell cycle control and tumor suppression has been linked with GC as well (reviewed by [35]). Coculture of Hp with mouse macrophages triggered, through a secreted bacterial factor, NF-κB and MAPK signaling and led to increased *IL6* transcription via phosphorylation of histone H3 at the *IL6* promoter [81]. In contrast, coculture of Hp with gastric epithelial cells caused a global decrease of histone H3 phosphorylation and deacetylation of H3 at lysine 23 [82,83]. Histone dephosphorylation was dependent on the presence of the Hp cag-PAI, but not on the CagA and VacA proteins. As a consequence of Hp-triggered epigenetic dysregulation, the expression of the *c-jun* proto-oncogene increased, and that of *hsp-70*, coding for a heat shock protein, decreased [83].

21.2.6 Uropathogenic *Escherichia coli* Infection Down-Regulates CDKN2A (p16^INK4A)

Coculture of uropathogenic *E. coli* with human uroepithelial cell lines strongly induced DNMT1 expression in comparison with non-pathogenic strains. In parallel, the tumor suppressor *CDKN2A* and the DNA repair gene *MGMT* were down-regulated, while a set of other genes (*CDH1*, *MLH1*, *DAPK1*, and *TLR4*) were not affected. Down-regulation of *CDKN2A* correlated to DNA methylation of its promoter. However, the MGMT gene was not methylated. Frequent UPEC infections might increase the risk for bladder cancer through increasing methylation of TSGs [84].

21.2.7 *Chlamydophila* spp. Encoded Histone Methyltransferases: Putative Inducers of Epigenetic Reprogramming in Host Cell Nuclei

Chlamydophila pneumoniae, an obligatory intracellular bacterium involved in acute respiratory diseases and implicated in chronic inflammatory processes, may modify, in principle, the epigenotype of its host cells. Similarly to *Anaplasma phagocytophilum*, *C. pneumoniae* introduces a putative bacterial effector, in this case a SET domain protein, into the host cell nucleus. Although histone-like proteins play an important role in the lifecycle of *Chlamydophila pneumoniae*, its putative effector protein, a histone methyltransferase (HMT), preferentially targets histone H3 in murine host cells [85] (Table 21.1).

Analogously, *Chlamydia trachomatis* codes for a SET domain-protein, called "nuclear effector" (NUE), which functions as a HMT. During infection, NUE associates with the host cell chromatin and methylates histones H2B, H3, and H4 [86]. Modification of host-encoded histones may reprogram the infected cell, perhaps by silencing host defense genes, as the *Anaplasma phagocytophilum* effector AnkA happens to do it.

421

TABLE 21.1 Epigenetic Alterations Elicited by Bacterial Infections in Host Cells

Microbe	Effector Molecule	Epigenetic Alteration	Effect
Porphyromonas gingivalis	Butyric acid	Histone acetylation	HIV-1 reactivation
Listeria monocytogenes	Listeriolysin O	Histone dephosphorylation Histone deacetylation	Blocking innate immunity
Clostridium perfringens	Perfringolysin	Histone dephosphorylation	?
Streptococcus pneumoniae	Pneumolysin	Histone dephosphorylation	?
Shigella flexneri	OspF phosphatase	Inhibition of histone phosphorylation	Blocking innate immunity
Anaplasma phagocytophilum	AnkA	Upregulation of HDAC	Blocking defence genes
Helicobacter pylori	Bacterial DNMTs	?	?
	?	Promoter hypermethylation	Gastric carcinogenesis
	?	Histone dephosphorylation Histone deacetylation	?
Campylobacter rectus	?	Methylation of the Igf2 promoter	Pre-term births
Uropathogenic E. coli	?	Promoter hypermethylation	CDKN2A silencing
Chlamydophila spp.	Histone methyltransferase	Histone methylation	Gene silencing

422

21.3 VIRUS-INDUCED EPIGENETIC ALTERATIONS

Similarly to certain bacteria, viruses harboring either RNA or DNA genomes can also elicit epigenetic changes in their host cells. Although most of the studies are related to "tumor viruses" that are associated with neoplasms, epigenetic dysregulation may contribute to other virus-induced pathological alterations as well.

21.3.1 Epigenetic Alterations in Cells Carrying Latent Gammaherpesvirus Genomes

Gammaherpesvirus genomes regularly carry viral oncogenes and are associated, consequently, with malignant tumors. Some of the viral oncoproteins encoded by human gamma-herpesviruses turned to be modifiers of the cellular epigenome.

EPIGENETIC REPROGRAMMING IN EPSTEIN–BARR VIRUS-ASSOCIATED NEOPLASMS

Epstein–Barr virus (EBV), a human gammaherpesvirus, is associated with a series of malignant neoplasms including lymphomas (Burkitt's lymphoma, Hodgkin's disease, T/NK-cell lymphoma, post-transplant lymphoproliferative disease, AIDS-associated lymphoma, X-linked lymphopro-liferative syndrome), carcinomas (nasopharyngeal carcinoma, gastric carcinoma, carcinomas of major salivary glands, thymic carcinoma, mammary carcinoma) and a sarcoma (leiomyo-sarcoma) (reviewed by [87]). In infected cells, the terminal repeats of the linear double-stranded DNA packaged into the virions fuse with each other. Accordingly, latent EBV genomes persist in the tumor cells as circular episomes. They attach to the nuclear matrix in interphase nuclei and coreplicate with the cellular DNA once per cell cycle using *oriP*, the latent origin of EBV replication.

The expression of latent EBV genes is cell type specific. Epigenetic regulatory mechanisms including DNA methylation and histone modifications, and binding of key cellular regulatory proteins control the activity of the alternative promoters for transcripts encoding the nuclear antigens EBNA1 to 6. Similar mechanisms affect the activity of promoters for transcripts encoding transmembrane proteins (LMP1, LMP2A, LMP2B), too. In addition to the genes transcribed by RNA polymerase II, there are also two RNA polIII transcribed genes in the EBV genome (EBER1, EBER2). The 5′ and internal regulatory sequences of the EBER1 and EBER2 transcription units are invariably unmethylated. The highly abundant EBER1 and EBER2 RNAs are not translated to protein. In addition, microRNAs are also generated by RNase-processing of latent EBV transcripts. EBV-derived microRNAs may interact both with cellular and viral mRNAs, thereby modulating their post-transcriptional level [88].

Based on the cell-type-specific epigenetic marks associated with latent EBV genomes one can distinguish between unique viral epigenotypes [7]. Although containing identical or nearly identical DNA sequences, each viral epigenotype is associated with a different pattern of gene expression. In addition, alternative conformations adopted by the viral episomes in a latency-type-specific manner may also affect the activity of latent EBV promoters [89].

Latent EBV genomes are regularly targeted by epigenetic control mechanisms in different cell types. EBV-encoded oncoproteins may, in turn, affect the activity of a set of cellular promoters. Thus, their interaction with the cellular epigenetic regulatory machinery results in epigenetic "reprogramming" of the host cells. The nuclear antigen EBNA2 transactivates both viral and cellular promoters. EBNA2 interacts with the cellular histone acetyltransferases p300, CBP, and PCAF, whereas the leader protein EBNA-LP (EBNA5) coactivates transcription by displacing histone deacetylase 4 from EBNA2-bound promoter sites [90,91]. At LMP1p, the EBNA2-associated histone acetyltransferases may counteract the silencing effect of histone deacetylases [90].

EBNA3C (EBNA6) was associated both with histone acetylases and deacetylases, although in separate complexes [92,94]. EBNA3C and EBNA3A interact with C terminal binding protein (CtBP) and repress the tumor-suppressor gene $p16^{INK4A}$ in lymphoid cells [95]. These interactions contribute to the maintenance of the H3K27me3 repressive chromatin mark at the silent $p16^{INK4A}$ promoter as well as the sustained growth of lymphoblastoid cells immortalized by EBV. EBNA3C and EBNA3A cooperated in repression of *Bim*, a gene coding for the proapoptotic protein Bcl-2-interacting mediator of cell death [96]. In cells carrying latent EBV genomes the *Bim* promoter was enriched in the repressive histone mark H3K27me3 and methylated CpGs [97].

LMP1, a transmembrane oncoprotein, affected both alternative systems of epigenetic memory, DNA methylation, and Polycomb group complexes. In epithelial cells LMP1 up-regulated DNMT1, DNMT3A, and DNMT3B, whereas in Hodgkin lymphoma cells it induced the Polycomb group protein Bmi-1 [98–100]. LMP1 activated the expression of DNMT1 via the c-jun NH(2)-terminal kinase/activator protein-1 (JNK-AP-1) signaling pathway [99]. The NF-κB pathway was involved in Bmi-1 induction [100]. Dutton et al. suggested that by up-regulating Bmi-1 via the NF-κB pathway, LMP1 may induce the loss of B-cell identity in EBV-positive Hodkin's lymphomas via down-regulating B-cell markers (CD21/MS4A1, BLK, LY9). In addition, Bmi-1 may mark a set of promoters for de novo methylation by DNMTs, thereby silencing tumor-suppressor genes, including *IGSF4* and *ATM*. Other LMP1 target genes are activated by Bmi-1, like *STAT*, *c-MET*, and *HK*, coding for signaling molecules and hexokinase (reviewed by [101]). Latent EBV infection was associated with hypermethylation of a set of cellular promoters in LMP1-positive tumors including nasopharyngeal carcinoma [102], Hodgkin lymphoma [103–105], and lymphomas developing in AIDS patients [106]. In contrast, aberrant methylation of *p15* and *p16* tumor-suppressor genes was infrequent in iatrogenic lymphomas developing in methotrexate-treated rheumatoid arthritis patients [107].

The transmembrane protein LMP2A is also involved in promoter silencing by CpG methylation [108]. LMP2A induces the phosphorylation of STAT3 in gastric carcinoma cell lines resulting in the up-regulation of DNMT1 transcription and methylation of the tumor-suppressor gene *PTEN* (*phosphatase and tensin homolog, deleted on chromosome ten*). PTEN functions as a phosphatidylinositol-3,4,5-triphosphate (PIP_3) phosphatase causing cell cycle arrest and inhibition of cell migration. PTEN contributes to the maintenance of apical−basal polarity of epithelial cells, too (reviewed by [109]). Because LMP1 is absent from EBV-associated gastric carcinomas, Hino et al. suggested that CpG methylation of cellular promoters [110,111], including the *PTEN* promoter, is due to LMP2A expression in these neoplasms [108]. One has to add, however, that in a significant portion of EBV-associated gastric carcinomas LMP2 expression can't be detected, either [112], suggesting that other latent EBV proteins or RNAs may also contribute to the development of the CGI methylator phenotype (CIMP).

EPIGENETIC ALTERATIONS IN NEOPLASMS CARRYING KAPOSI'S SARCOMA-ASSOCIATED HERPESVIRUS GENOMES

Human herpesvirus 8 (HHV-8) or Kaposi's sarcoma-associated herpesvirus (KSHV) belongs to the genus rhadinovirus of the human gammaherpesviruses, a subfamily of herpesviruses. HHV8 is causally associated with Kaposi's sarcoma (KS), a tumor of endothelial cell origin. KSHV is also associated with multicentric Castleman's disease (MCD) and primary effusion lymphoma (PEL, also called body cavity-based lymphoma) (reviewed by [113]). KSHV exists in two replication states, lytic or latent. The key lytic switch and transcriptional transactivator protein Rta is encoded by the viral immediate early *ORF50*. During viral latency in B-cells, Rta expression is down-regulated, together with most other viral proteins. The key regulator of viral latency, LANA (latency-associated nuclear antigen encoded by *ORF73*), suppresses both lytic activation and apoptosis (reviewed by [113]). The KSHV latent origin of DNA replication is located to the terminal repeats (TRs) of the genome where LANA has two binding sites. LANA recruits cellular replication proteins to the TRs. Histone acetylation at the TRs is high, while histone methylation changes throughout the cell cycle [114]. In addition to replication factors, LANA also recruits to the TRs a repressive complex consisting of heterochromatin protein (HP) 1 and the HMT SUV39H1 [115].

Two recent studies analyzed the chromatin structure of entire KSHV genomes in latently infected cells. Immunoprecipitation of methylated DNA and of covalently modified histones in combination with microarrays yielded distinct profiles of epigenetic modifications for latency in tumor cell lines and in newly infected epithelial cells. Characteristic DNA-methylation patterns were found all over the genome. Latency-specific histone modifications were rapidly established upon infection. Silencing of lytic viral promoters was not established by removal of the activating histone marks H3K9ac, H3K14ac, and H3K4me3, but by the deposition of H3K27me3 across the genome. This "bivalent" modification is transcriptionally repressive but enables rapid activation upon the induction of the lytic cycle [116]. A similar "bivalent" chromatin structure was found at the *IE* genes *ORF50* and *ORF48* in the second study. In addition, Enhancer of Zeste Homologue (EZH) 2, the HMT enzyme of PcG protein complexes colocalized with the H3K27me3 marks during latency, while the reactivation of the lytic cycle led to the dissociation of EZH2 and H3K27me3 from the viral genome and to the expression of the lytic protein cascade [117].

Binding of the chromatin insulator protein CTCF in association with cohesins to the KSHV episome may also influence viral gene expression by establishing alternative three-dimensional conformations during latency and lytic replication [118].

In addition to PcG proteins, viral microRNAs also contribute to KSHV-latency. A miRNA deletion mutant virus eliminating ten of the 12 viral miRNAs from the KSHV genome was more lytically active than the wild-type virus. This effect might be explained by the fact that two of the deleted viral miRNAs, miR-K12-5 and miR-K12-4-5p, have a seed sequence targeting and down-regulating transcripts of the viral transactivator *ORF50* and *retinoblastoma like protein (Rbl) 2*, a repressor of *DNMT3A* and *DNMT3B* transcription, respectively [119].

KSHV infection influences the epigenetic marks of the host genome. In a *LANA*-transduced cell line, the promoters for *CDH13* (*H-Cadherin*), *CCND2* (*cyclin D2*), *LDHB* (*lactate dehydrogenase B*), *FOXG1B_FKHL1* (*Forkhead box protein G1B_Forkhead-related protein 1*) and *CREG* (*cellular repressor of E1A stimulated genes*) were strongly down-regulated. LANA was associated with repressed cellular promoters and recruited DNMT3A, but also DNMT1 and DNMT3B, resulting in promoter methylation, as demonstrated in the case of the *CDH13* promoter [120]. At the TGF-β type II receptor (*TbetaRII*) promoter, deacetylation of promoter-associated histones was also demonstrated, in addition to promoter methylation and LANA binding to the transcriptional corepressor complex mSIN3A [121].

In PEL cells carrying both KSHV and EBV genomes, transfection of *LANA* decreased the activity of the EBV promoters Qp and Cp [122]. Upon transfection into immortalized epithelial cell lines, LANA was also associated with the repressive HP1 at the cellular chromatin [123]. Accordingly, LANA seems to work as an epigenetic modifier of the viral and host genomes. Transfection of LANA into several tumor cell lines led to a fundamental reorganization of the cellular chromatin. LANA-induced redistribution of DNA staining coincided with redistribution of the methyl-cytosin binding protein MeCP2. This effect was attributed to the interaction of LANA with the HMT enzyme SUV39H1 [124]. The interaction of LANA with MeCP2 and nucleosomes would enable LANA to direct regulatory complexes to chromosomal sites and thereby stably reprogram cellular and viral gene expression [125]. Accordingly, $p16^{INK4A}$ was frequently silenced through CpG-methylation in PEL cell lines and primary PEL cells [126]. Furthermore, CpG-hypermethylation at the *MGMT* promoter and a loss of the EBV genome in cell culture was observed in a PEL cell subclone derived from a PEL tumor, which was positive for KSHV and EBV, but unmethylated at *MGMT* [127] (Table 21.2).

21.3.2 Interactions of Lytic Cycle Proteins of Human Cytomegalovirus with Histone Deacetylases

Following primary infection, human cytomegalovirus (HCMV), a betaherpesvirus, persists or establishes latency in cells of the myeloid lineage. Although most infections are asymptomatic, life-threatening disease can develop in immunocompromised patients or if infection occurs in utero. Gönczöl et al. observed that HCMV latency and reactivation is connected to the undifferentiated or differentiated nature of the host cell [128]. In undifferentiated cells the major immediate-early (IE) promoter (MIEP) of HCMV was repressed and located to a hypoacetylated chromatin domain associated with HP1 and Ets-2 repressor factor (ERF), a transcriptional repressor recruiting HDAC1 to MIEP [129,130]. During lytic infection,

TABLE 21.2 Epigenetic Alterations Induced by Latent Human Gammaherpesvirus Proteins

Microbe	Effector Molecule	Epigenetic Alteration	Effect
EBV	EBNA2	Histone acetylation	Switching on promoters
	EBNALP (EBNA5)	HDAC displacement	Switching on promoters
	EBNA3C (EBNA6)	Histone acetylation, histone deacetylation	Modulation of promoter activity
	LMP1	Up-regulation of DNMTs	Promoter silencing
	LMP2A	Up-regulation of DNMT1	Promoter silencing
KSHV	LANA	Recruitment of DNMTs to cellular promoters	Promoter silencing
		Association with mSIN3A, recruitment of HDACs	Promoter silencing
		Association with HP1	Promoter silencing
		Interaction with SUV39H1, Redistribution of MeCP2	Reorganization of the nucleus

however, for example when monocytes differentiated to macrophages, MIEP was activated and associated with hypermethylated histones [129]. In addition, IE1 and IE2, the abundant protein products of the IE genes interacted with histone deacetylases, abolishing their repressive effect at the MIEP and the early promoter of the viral polymerase [131,132]. Thus, IE proteins modulate the host epigenetic regulation of viral promoters, thereby facilitating viral replication.

IE1 and IE2 activated cellular promoters either by direct binding to CCAAT box binding factor (CTF1) at TATA-less promoters, or TATA-binding protein (TBP), at promoters with a TATA motif, respectively [133,134]. In human fibroblasts, activation of the human telomerase reverse transcriptase gene (*hTERT*) by IE1 was accompanied by increased binding of the nuclear protein Sp1, acetylation of histone H3, and a reduction of HDAC binding at the *hTERT* promoter [135]. Since telomerase activation is crucial for immortalization and malignant transformation of cells, these data support the point that HCMV may contribute to tumorigenesis [136].

21.3.3 Epigenetic Dysregulation in Human Retrovirus-Infected Cells

The RNA genomes of human T-cell lymphotropic virus (HTLV) and human immunodeficiency virus (HIV) replicate via a DNA intermediate that becomes integrated into the host cell genome (provirus). Proviral genomes are frequently silenced by epigenetic mechanisms. In turn, retroviral proteins interacting with the epigenetic regulatory machinery modulate the gene expression pattern of their host cells.

CPG ISLAND METHYLATOR PHENOTYPE IN ADULT T-CELL LEUKEMIA/ LYMPHOMA (ATLL): A PUTATIVE ROLE FOR THE HTLV-1 PROTEIN TAX IN THE SILENCING OF KEY CELLULAR PROMOTERS?

The proviral genome of type I human T-lymhotropic virus (HTLV-1), a retrovirus associated with ATLL and tropical spastic paraparesis, frequently undergoes epigenetic silencing [137]. Therefore, in tropical spastic paraparesis patients the activation of viral gene expression using the HDAC inhibitor valproate seems to be a justified therapeutic approach, because the virus-positive cells may thereby be exposed to the host immune response, resulting in the collapse of the latent viral reservoir [138]. In ATLL patients the CGI methylation frequency of the *Src homology-2-containing protein tyrosine phosphatase (Shp1, PTPN6)* gene gradually increased during disease progression [139]. In parallel, the number of CGI methylated genes also increased. Because the viral oncoprotein Tax induced the dissociation of transcription factors from the *Shp1* promoter and subsequent promoter hypermethylation [140], Niller et al. speculated that Tax may act as a "hit-and-run" oncoprotein by initiating the down-regulation of Shp1 expression in an early stage of leukemogenesis [141]. This early step may be followed by silencing of additional cellular promoters and down-regulation of Tax expression itself, due to deletions of the proviral genome or its epigenetic silencing.

HUMAN IMMUNODEFICIENCY VIRUS: EPIGENETIC SILENCING OF LATENT VIRAL GENOMES AND HOST GENES

HIV causes the acquired immunodeficiency syndrome. The virus is transmitted parenterally and, with the exception of very few long-term non-progressors, ends fatally for the infected patients due to the severe course of opportunistic infections which would normally run less severely or even subclinically. Due to the HIV-associated severe immune suppression, HIV patients have an increased risk to develop virus-associated neoplasms during all stages of HIV disease [142–144].

Although highly active antiretroviral therapy (HAART) that combines several drugs with different viral targets significantly improved the life expectancy of HIV-infected individuals, there is still no cure for HIV infection. One reason for this failure is the existence of a dormant viral reservoir in resting memory $CD4^+$ T cells [145]. Upon cessation of HAART and due to stimulation of the memory cells, the latent viral genomes start transcription and replication again, leading to viral rebound. The half-life of the latent viral reservoir is approximately 40 months. In patients undergoing HAART, occasional short periods of asymptomatic viral replication, termed "blips", may allow the virus to replenish the latent reservoir [146]. Therefore, viral persistence in resting memory CD4 T cells has been recognized as a major obstacle against curative treatment (reviewed by [147,148]).

Mechanisms of HIV Latency: CpG Methylation, Repressive Chromatin, Transcriptional Interference, and miRNAs

Enzymatic methylation of CpG dinucleotides prevented binding of transcription factors to the HIV-LTR and led to a transcriptional block in transfected cells which could be overcome by demethylation through 5-aza-C or the presence of the HIV-transactivator Tat. This suggested that DNA-methylation might contribute to the silencing of viral transcription in the latent state [149–151]. HIV transcription and chromatin structure were examined in several latently infected cell lines. Induction of viral transcription through the NF-κB stimulator TNFα and HDAC inhibitors correlated with an overall increase in histone acetylation and the disruption of nucleosome 1 which blocks the transcriptional start site of the viral 5′-LTR [152]. Two transcriptional repressors, YY1 (Ying Yang 1) and LSF (late simian virus 40 transcription factor), were shown to associate with the HIV LTR, and recruited HDAC1 in vitro. YY1-dependent repression was reversed by the HDAC inhibitor trichostatin A (TSA), suggesting that repressive histone marks also contribute to the silencing of HIV-transcription [153].

In resting CD4$^+$ T cells isolated from aviremic HIV-infected cell donors on HAART, the HDAC inhibitor valproic acid (VPA) led to the acetylation of the integrated provirus-LTR and to an outgrowth of HIV from the resting cells without inducing T-cell activation markers [154]. The contribution of different histone methylation marks to LTR-silencing was examined in latently transfected T-cell lines. EZH2, a component of the polycomb repressive complex (PRC) 2 involved in H3K27me3 methylation, was enriched at the repressed LTR, while upon LTR activation it was rapidly displaced. A smaller number of proviruses were silenced through SUV39H1 which is involved in H3K9me3 methylation. Thus, specific inactivation of PRC2 may give a cue to future therapeutic attempts to selectively activate the LTR [155]. The LTRs in isolated resting T cells from aviremic HIV patients and stably transfected HeLa cells were repressed through a protein complex of transcription factors of c-Myc and Sp1, attracting HDAC1 [156]. CBF1, a transcriptional repressor and effector in the Notch signaling pathway, also silenced HIV transcription. CBF1 recruited HDAC1 and the corepressor proteins CIR (CBF1-interacting protein) and mSIN3A to the LTR. Knockdown of CBF1 led to transcriptional derepression, with a concurrent increase in histone acetylation levels, loss of HDAC1 and loss of co-repressor proteins [157].

HIV proviruses preferentially integrate within actively transcribed genes. Nevertheless, they are transcriptionally suppressed in resting CD4 T cells from patients under HAART. Proviruses, mostly within introns, were randomly inserted as to their transcriptional orientation, suggesting that proviral transcripts may partially end up as rapidly degraded intronic RNA upon transcription of the respective host gene. Thus, intronic integration represents a distinct molecular mechanism of HIV silencing in latency [158,159]. Targeted insertion of a HIV provirus by homologous recombination into the third intron of the *HPRT* gene was used as a model to study the effect of the transcriptional orientation of host and viral genes. *HPRT*-readthrough transcription enhanced HIV transcription, when running in the same direction, but inhibited it, when running in the opposite direction. In this model, repressive chromatin marks did not play a significant role in HIV latency [160].

Another mechanism of HIV latency in memory T cells may rely on miRNAs. A series of CD4$^+$ T-cell lines expressed high levels of Dicer which belongs to the miRNA processing machinery. In HIV-infected T cells, the viral TAR stem-loop RNA was bound by Dicer and processed to

427

a miRNA. Transfection of a corresponding siRNA-expressing construct decreased LTR-dependent expression of a reporter protein. Furthermore, the TAR-siRNA was able to recruit HDAC1 to the LTR, but not to downstream sequences [161]. Cellular miRNAs may also contribute to HIV latency. miR-28, miR-125b, miR-150, miR-223, and miR-382 were enriched in resting T cells compared to activated T cells. These miRNAs were recruited to the 3′-end of the HIV transcript and inhibited the translation of HIV proteins. miRNA antisense-constructs released the translational block and led to viral replication [162].

Silencing of Host Genes in HIV-Infected Cells

The chromatin structure in host cell genes is influenced by HIV infection of CD4$^+$ T cells. DNMT1 levels increased significantly in CD4$^+$ T-cell lines and primary CD4 T cells upon infection with HIV. Acute infection of primary T$_H$1 cells led to a decrease of IFNγ expression which coincided with the de novo methylation of the *IFNγ* promoter. An antisense-mediated knockdown of DNMT1 led to transcriptional derepression of the *IFNγ* promoter [163]. An increase in the level of DNMT1 was also observed in a T-cell line infected with an integration- and therefore replication-defective HIV mutant. Thus, the presence of viral nucleic acids was sufficient, but viral replication was not required for the increase of DNMT1 expression. The overall methylation level of the cellular DNA increased, the CGI at the *p16^{INK4A}* promoter became hypermethylated and repressed upon HIV infection, while 5-aza-C treatment led to the induction of p16^{INK4A} expression [164].

Another gene down-regulated in virus-infected immortalized T cells was UDP-N-acetylglucosamine 2-epimerase/N-acetylmannosamine kinase (GNE). GNE produces the sialyl-donor substrate for all cellular sialyl-transferases and thereby profoundly influences the receptor and homing functions of immune cells. In T-cell lines the CGI at the GNE promoter became de novo hypermethylated upon acute HIV infection, as demonstrated by bisulfite sequencing. Treatment with 5-aza-C released both the transcriptional repression and methylation, and restored sialylation function. GNE hypermethylation may to some degree explain the frequently persisting immune disorder of aviremic HIV patients on long-term successful HAART [165].

Infection of a T-cell line or transfection of HeLa cells with an integration-defective HIV increased DNMT1 expression. The effect was dependent on expression of HIV early proteins Tat, Rev, and Nef. The transcription factor AP1 was implicated in the HIV-dependent up-regulation of DNMT1 [166]. Furthermore, in a human neuroblastoma cell line and in human primary neuronal cells, Tat was shown to induce HDAC2. This led to the silencing of the *CREB* and *CaMKIIa* genes which contribute to synaptic plasticity and neuronal function. Thus, HIV-induced dysregulation of HDAC2 may be involved in HIV-associated neurocognitive disorders (HAND) [167] (Table 21.3).

21.3.4 Local Hypermethylation and Global Hypomethylation of the Host Cell DNA in Hepatocellular Carcinomas Associated with Hepatitis B Virus

Hepatocellular carcinoma (HCC) is the predominant form of human liver cancer. Hepatitis B virus (HBV) and hepatitis C virus (HCV) infections are the major etiological factors for

TABLE 21.3 Epigenetic Alterations Induced by Human Retroviruses

Microbe	Effector Molecule	Epigenetic Alteration	Effect
HTLV-I	Tax	*Shp1* promoter methylation	Promoter silencing
HIV-1	Tat, rev, nef	Upregulation of DNMT1	Promoter silencing
	Tat	Induction of HDAC2	Promoter silencing

HCC. HBV-related HCC develops through distinct stages (reviewed by [168]). Epigenetic alterations including aberrant CGI hypermethylation appear already in the preneoplastic lesions (cirrhotic nodules) and their frequency gradually increases during the early stages of liver carcinogenesis (low-grade dysplastic nodules, high-grade dysplastic nodules, early hepatocellular carcinoma), although the methylation level of certain genes may decrease in progressed HCCs. Shim et al. found that hypermethylation of the *p16* gene starts at an early stage of hepatocarcinogenesis. Sixty-two percent of cirrhotic nodules (putative preneoplastic lesions) and 70% of dysplastic nodules, surrounding HBV-positive hepato-cellular carcinoma lesions, showed *p16* hypermethylation [169]. Um et al. described that *APC*, *RASSF1A*, and *SOCS1* were methylated in a fraction of cirrhotic nodules but the methylation levels of *APC* and *RASSF1* increased significantly in low-grade dysplastic nodules. *SOCS1* methylation gradually increased during multistep carcinogenesis, peaked in early hepatocellular carcinoma, and decreased in progressed liver carcinomas [170].

The genome of hepatitis B virus codes for a pleiotropic regulator, called X protein (HBx or pX). pX interacts with multiple key signaling pathways and affects the regulators of cell cycle progression [171—175]. In vitro, pX up-regulates cyclin D1 and activates DNMT1 expression via the cyclin D1-CDK4/6-pRb-E2F1 pathway, resulting in DNA-methylation-mediated down-regulation of the tumor-suppressor protein p16^{INK4a} [176,177]. In addition to DNMT1, pX also up-regulated two variants of the de novo DNA methyltransferase DNMT3A in liver cells. DNMT3A1 and DNMT3A2 are translated from differentially spliced trancripts of the *DNMT3A* gene and induce *regional hypermethylation* of specific tumor-suppressor genes [178]. pX *down-regulated*, however, another de novo DNA methyl-transferase, DNMT3B, in the very same cells. Because satellite 2 repeat sequences are methylated by DNMT3B, pX-induced down-regulation of the enzyme led to a *global hypo-methylation* of these repeats [178]. This important observation connected the viral onco-protein pX both to global DNA hypomethylation, a phenomenon as widespread in neoplastic cells [179], and regional hypermethylation [180], which is another key feature of tumor cells. In hepatocellular carcinomas carrying integrated HBV genomes, CpG-rich sequences were also found to be *hypomethylated* in the pericentromeric regions of acro-centric chromosomes [181]. pX was also implicated in hypermethylation-mediated-silencing of the E-cadherin promoter [182—184]. The tumor-suppressor gene *RASSF1A* (Ras association domain family 1A) and the *GSTP1* gene, coding for the π-class glutathione S-transferase, an enzyme involved in protecting against electrophilic carcinogens, was also frequently inactivated by CGI methylation in HBV-positive hepatocellular carcinomas [185,186].

The epigenetic profile of hepatocellular carcinoma associated with HBV appears to be different from that of HCV-associated liver cancer. Feng et al. observed that certain genes (*HOXA9*, *RASSF1*, *SFRP1*) were methylated more frequently in HBV-positive hepatocellular carcinomas than in HCV-positive liver tumors. In contrast, *CDKN2A* was significantly more frequently methylated in HCV-positive than in HBV-positive liver carcinomas [187]. These data support the idea that hepatocellular carcinomas of different viral etiologies are associated with unique, *virus-specific epigenetic signatures*.

21.3.5 Hepatitis C Virus (HCV)-Induced Epigenetic Alterations

Hepatitis C virus belongs to the genus hepacivirus of the flaviviruses. It is transmitted through blood—blood contacts and causes inflammatory liver disease which turns chronic in about 80% of the infected patients. Chronic infection may lead to liver cirrhosis and hepatocellular carcinoma in the long term. Several large studies described epigenetic down-regulation of numerous tumor-suppressor genes in HCC tissue samples [187—190] (reviewed by [191]). The overall genomic methylation levels, reflected in the methylation of LINE1 and SAT2 repeats,

TABLE 21.4 Epigenetic Alterations Induced by Human Hepatitis Viruses			
Microbe	Effector Molecule	Epigenetic Alteration	Effect
HBV	pX	Up-regulation of DNMT1	Promoter silencing
		Up-regulation of DNMT3A1and DNMT3A2	Promoter silencing
		Down-regulation of DNMT3B	Satellite 2 repeat hypomethylation
HCV	Core protein	Induction of DNMT1, and DNMT3B	Promoter silencing
		Induction of HDAC (SIRT1)	Promoter silencing
		Induction of SMYD3, a histone methyltransferase	Silencing of RASSF1A

decreased, while methylation at CGIs of TSGs increased with the progression of hepatitis via cirrhosis to HCC [168,192,193] (Table 21.4).

DNA methylation at distinct tumor-suppressor gene loci is significantly associated with HBV, HCV, or alcoholism, respectively [190] (reviewed by [191]). Methylation at the tumor-suppressor genes CDKN2A, SOCS-1, GADD45B, STAT1, APC, and p15 is more prevalent in HCV-positive than in HCV-negative tumors, while HOXA9, RASSF1, and SFRP1 were more methylated in HBV-positive tumors [187,192,194,195] (reviewed by [191]). Methylation of the kallikrein (KLK) 10 promoter correlated with HCV infection, while methylation at KLK10 and GSTP1 inversely correlated with HBV infection [190,196]. Another study applied methylated-DNA immunoprecipitation on chip and mass spectrometry on HCC tissue samples to discriminate between HCV- and HBV-associated HCC. While hypermethylation of only the CYP7B1 locus was found in HBV-HCC, a multitude of 15 loci was hypermethylated in HCV-HCC: age-independent methylation markers included genes inhibitory to Ras/Raf/ERK signaling (NPR1, DUSP4, LOX, and RRAD) and inhibitory to Wnt/β-catenin signaling (SFRP4 and RUNX3) [197].

In a transgenic mouse and a cell line model, epigenetic down-regulation of SOCS1 was dependent on the expression of the HCV core protein [198]. This observation is in accordance with a clinical study showing that HCV-infection was associated with SOCS1 methylation, whereas HBV infection was inversely correlated with SOCS1 methylation [199]. In HepG2 cells the expression of the core protein induced DNMT1 and DNMT3B resulting in CDH1 promoter hypermethylation and a more aggressive growth behavior [200]. HCV core protein expression in the Huh cell line also led to hypermethylation and silencing of CDH1. In addition, the HDAC SIRT1 was induced through core protein expression. Inhibiting SIRT1 derepressed the CDH1 promoter [201]. A similar mechanism may silence the promoters of interferon-stimulated genes, too [202].

Alterations of the histone modification machinery have also been associated with the progression of HCC (reviewed by [191]). HCC is unusually low in H3K4me2, due to deregulation of the trithorax protein Ash2 and histone demethylase LSD1 [203]. Inducible HCV protein expression in cell lines led to the overexpression of protein phosphatase 2A (PP2A) which influenced the histone modification machinery through binding to protein arginine methyltransferase 1 (PRMT1), and the DNA repair machinery through dephosphorylating H2AX. The deregulation of both systems is considered critical in HCC development. Accordingly, the expression of a set of indicator genes involved in HCC tumorigenesis was changed concordantly by either the expression of HCV proteins or the overexpression of PP2A in cell culture [194] (reviewed by [191]). Transfection of the HCV core protein into cholangiocarcinoma cell lines induced the SET and MYND domain containing protein, SMYD3, a novel HMT. Overexpression of SMYD3 correlated with hypermethylation of the RASSF1A promoter [204].

21.3.6 Induction of Histone Methyltransferase and Histone Demethylase Enzymes by the Human Papillomavirus Oncoprotein E7

Infection with high-risk human papillomavirus (HPV) strains (especially type 16 and 18) is causally related to the development of cervical cancer. Progression of HPV-positive pre-malignant lesions to invasive cancer is a rare event, however, suggesting that additional steps including chromosomal alterations and epigenetic changes are necessary for neoplastic development [205]. In an in vitro model system of cervical carcinogenesis, Henken et al. observed an accumulation of frequent methylation events involving five tumor-suppressor genes when primary keratinocyte cell lines transfected with HPV genomes acquired anchorage independence [206]. In contrast, no methylation was evident in pre-immortal HPV-18 transfected cells. Certain cellular genes are also hypermethylated during in vivo cervical carcinogenesis in a histological type- or clinical stage-dependent manner (reviewed by [205]). The E7 oncoprotein of HPV-16 was shown to bind to DNMT1 and stimulate its methyl-transferase activity [207]. This may explain the hypermethylation of selected cellular promoters in HPV-infected cells, and Laurson et al. speculated that E7—DNMT1 interaction may target DNMT1 to specific sequences, similarly to the recruitment of DNMT3A to specific genomic regions by the KSHV oncoprotein LANA [208].

Consistent hypermethylation patterns were described in laryngeal papillomas associated with HPV-6 as well [209]. Laryngeal papillomas are usually benign lesions, although they may progress to squamous cell carcinoma in a fraction of cases. Promoter hypermethylation affecting *TIMP3* and *CDKN2B* occurred early during tumor progression and was maintained throughout neoplastic development [209].

HPV-16 E7 activated the expresssion of EZH2, a Polycomb protein with histone H3K27 methyltransferase activity, too. Continuous expression of EZH2 was indispensable for the proliferation of HPV-positive tumor cells [210]. Histone H3K27me3 marks may contribute to promoter silencing either in concert with DNMTs or independently of DNA methylation. E7 is a pleiotropic regulator protein involved both in silencing and activation of certain target gene sets. HPV-16 E7 increased histone H3 acetylation at *E2F1* and *CDC25A* promoters in human foreskin keratinocytes [211], and reduced the level of the repressive histone modification H3K27me3 in primary human epithelial cells [212]. The latter change was due to induction of the histone demethylases KDM6A and KDM6B targeting specifically histone H3K27me3 and disrupting Polycomb repressor complexes. E7-mediated reprogramming resulted in an increased expression of KDM6A- and KDM6B-responsive Homeobox genes (*HOXC5, HOXC8*) that are known to be up-regulated in cervical carcinomas.

21.3.7 Epigenetic Transcriptional Silencing in Merkel Cell Polyomavirus-Associated Carcinoma of the Skin

Merkel cell carcinoma (MCC), a neuroectodermal tumor arising from mechanoreceptor Merkel cells, frequently carries integrated genomes of a recently discovered human polyoma virus, called Merkel cell polyomavirus (MCPyV) [213]. MCCs are rare but highly aggressive neoplasms. Similarly to other polyomaviruses, the dsDNA genome of MCPyV encodes tumor antigens (large T antigen and small t antigen) implicated in oncogenesis. In MCCs the large T antigen is regularly truncated rendering the virus replication incompetent [214]. The proneural transcription factor Atonal Homologue 1 (Atoh1) is essential for cell fate commitment of multiple neuronal lineages and acts as a developmental regulator of the mechanoreceptive Merkel cells in the skin, too [215]. *ATOH1* expression was found to be reduced due to deletion or inactivation by CpG methylation in MCCs as well as colorectal carcinomas [216]. In addition, the promoter of the tumor-suppressor gene *RASSF1A* was also hypermethylated in about half of the MCC samples [217]. These observations indicate that epigenetic events may contribute to the pathogenesis of MCPyV-associated Merkel cell carcinoma.

21.4 EPIGENETIC ALTERATIONS ELICITED IN THE HOST TISSUE BY TREMATODE INFECTIONS

Trematode infections can lead to epigenetic alterations in the host tissue, which we will subsequently further explore.

21.4.1 Promoter Hypermethylation in Cholangiocarcinoma Associated with *Opistorchis viverrini* (Liver Fluke) Infection

Cholangiocarcinoma in certain geographical areas is related to liver fluke infection. In northeast Thailand most patients with the malignancy of the bile duct epithelium carry the macroparasite *Opistorchis viverrini* whereas in Cambodia, China, Japan, Korea, Laos, and Vietnam *Clonorchis sinensis* infection is spreading via the consumption of raw freshwater fish, causing cholangiocarcinoma [218,219]. In *Opistorchis viverrini*-associated cholangio-carcinomas down-regulation of *p14ARF* expression was predominantly related to inactivation by DNA methylation whereas *p16^{INK4A}*, another tumor-suppressor gene of the 9p21 gene cluster was affected by allelic loss. Both mechanisms contributed to down-regulation of *p15^{INK4B}* [220]. Sriraksa et al. analyzed the methylation pattern of 26 CGIs in *Opistorchis viverrini*-associated cholangiocarcinoma samples. There was an elevated frequency of hyper-methylation at five areas including the ovarian tumor-suppressor *OPCML (opioid binding protein/cell adhesion molecule-like gene)* and the immunomodulator *DcR1 (decoy receptor 1)* [221]. Thus, *OPCML* and DcR1may serve as methylation biomarkers for cholangiocarcinoma.

21.4.2 CpG Island Methylation in *Schistosoma haematobium*-Associated Bladder Carcinoma

Squamous cell carcinoma, and to some extent also transitional cell carcinoma of the bladder in East Africa and the Middle East and other subtropical areas is frequently associated with chronic urinary *Schistosoma haematobium* infection. Chronic mechanical irritation by calcified eggs deposited by the worms in the bladder epithelium and accumulation of carcinogenic compounds in the urine may be involved in the transformation of the uroepithelium. Although the genomic DNA isolated from adult worms of the related *Schistosoma mansoni* is unmethylated [222], in bladder cancer samples of most probably *Schistosoma haematobium*-infected patients from Egypt there was a greater degree of CGI methylation than in non-*Schistosoma*-associated tumors [223].

21.5 CONCLUSIONS

Although most of the work related to epigenetic alterations induced by infectious agents in host cells focused on tumor-associated microbes and macroparasites implicated in tumori-genesis, it is clear from the overview of the literature that other important bacterial and viral pathogens directly not involved in the initiation or maintenance of neoplasia leave their epigenetic marks on their target cells as well. There is no doubt in our mind that in addition to viruses and bacteria, other microparasites, i.e. protozoa that do have their own sophisticated epigenetic regulatory systems, may also cause epigenetic dysregulation in their hosts. Furthermore, tumor-associated viruses may have a role in other diseases as well, thus the knowledge as to the epigenetic control of their genomes and the epigenetic changes they elicit in neoplastic cells may help to decipher the patho-epigenetic mechanisms causing dysfunc-tions in non-neoplastic cells. In this respect it is worthy to mention that the epigenetic control of Epstein−Barr virus latency appears to be inadequate in certain autoimmune diseases, and was implicated in triggering and perpetuating the pathogenic processes [224]. Other impor-tant research topics are also emerging, including the potential role of microbial and other infections in the patho-epigenetics of allergic diseases [225] and the modification of host epigenetic processes by the microbial communities inhabiting mucosal surfaces and the skin, or even by probiotic bacteria [226]. The epigenetic changes induced by the vast majority of macroparasites (helminths, fungi, arthropods) infecting humans as well as animals remain to be explored as well.

References

[1] Minarovits J. Microbe-induced epigenetic alterations in host cells: the coming era of patho-epigenetics of microbial infections. A review. Acta Microbiol Immunol Hung 2009;56:1–19.

[2] Lopez-Rubio JJ, Riviere L, Scherf A. Shared epigenetic mechanisms control virulence factors in protozoan parasites. Curr Opin Microbiol 2007;10:560–8.

[3] Verstrepen KJ, Fink GR. Genetic and epigenetic mechanisms underlying cell-surface variability in protozoa and fungi. Annu Rev Genet 2009;43:1–24.

[4] Dixon SE, Stilger KL, Elias EV, Naguleswaran A, Sullivan WJ. A decade of epigenetic research in Toxoplasma gondii. Mol Biochem Parasitol 2010;173:1–9.

[5] Merrick CJ, Duraisingh MT. Epigenetics in Plasmodium: what do we really know? Eukaryot Cell 2010;9:1150–8.

[6] da Rosa JL, Kaufman PD. Chromatin-mediated Candida albicans virulence. Biochim Biophys Acta 2012;1819:349–55.

[7] Minarovits J. Epigenotypes of latent herpesvirus genomes. Curr Top Microbiol Immunol 2006;310:61–80.

[8] Bloom DC, Giordani NV, Kwiatkowski DL. Epigenetic regulation of latent HSV-1 gene expression. Biochim Biophys Acta 2010;1799:246–56.

[9] Horwitz GA, Zhang K, McBrian MA, Grunstein M, Kurdistani SK, Berk AJ. Adenovirus small e1a alters global patterns of histone modification. Science 2008;321:1084–5.

[10] Ferrari R, Pellegrini M, Horwitz GA, Xie W, Berk AJ, Kurdistani SK. Epigenetic reprogramming by adenovirus e1a. Science 2008;321:1086–8.

[11] Pihlstrom BL, Michalowicz BS, Johnson NW. Periodontal diseases. Lancet 2005;366:1809–20.

[12] Darveau RP. The oral microbial consortium's interaction with the periodontal innate defense system. DNA Cell Biol 2009;28:389–95.

[13] Pedra JH, Cassel SL, Sutterwala FS. Sensing pathogens and danger signals by the inflammasome. Curr Opin Immunol 2009;21:10–6.

[14] Manavalan B, Basith S, Choi S. Similar Structures but Different Roles - An Updated Perspective on TLR Structures. Front Physiol 2011;2:41.

[15] Philbin VJ, Levy O. Developmental biology of the innate immune response: implications for neonatal and infant vaccine development. Pediatr Res 2009;65:98R–105R.

[16] Yin L, Chung WO. Epigenetic regulation of human beta-defensin 2 and CC chemokine ligand 20 expression in gingival epithelial cells in response to oral bacteria. Mucosal Immunol 2011;4:409–19.

[17] Kurita-Ochiai T, Fukushima K, Ochiai K. Volatile fatty acids, metabolic by-products of periodontopathic bacteria, inhibit lymphocyte proliferation and cytokine production. J Dent Res 1995;74:1367–73.

[18] Niederman R, Buyle-Bodin Y, Lu BY, Robinson P, Naleway C. Short-chain carboxylic acid concentration in human gingival crevicular fluid. J Dent Res 1997;76:575–9.

[19] Tonetti M, Eftimiadi C, Damiani G, Buffa P, Buffa D, Botta GA. Short chain fatty acids present in periodontal pockets may play a role in human periodontal diseases. J Periodontal Res 1987;22:190–1.

[20] Imai K, Ochiai K, Okamoto T. Reactivation of latent HIV-1 infection by the periodontopathic bacterium Porphyromonas gingivalis involves histone modification. J Immunol 2009;182:3688–95.

[21] Riggs MG, Whittaker RG, Neumann JR, Ingram VM. n-Butyrate causes histone modification in HeLa and Friend erythroleukaemia cells. Nature 1977;268:462–4.

[22] Sealy L, Chalkley R. The effect of sodium butyrate on histone modification. Cell 1978;14:115–21.

[23] Myzak MC, Ho E, Dashwood RH. Dietary agents as histone deacetylase inhibitors. Mol Carcinog 2006;45:443–6.

[24] Bobetsis YA, Barros SP, Lin DM, Weidman JR, Dolinoy DC, Jirtle RL, et al. Bacterial infection promotes DNA hypermethylation. J Dent Res 2007;86:169–74.

[25] Paschos K, Allday MJ. Epigenetic reprogramming of host genes in viral and microbial pathogenesis. Trends Microbiol 2010;18:439–47.

[26] Freitag NE, Port GC, Miner MD. Listeria monocytogenes - from saprophyte to intracellular pathogen. Nat Rev Microbiol 2009;7:623–8.

[27] Hamon MA, Batsche E, Regnault B, Tham TN, Seveau S, Muchardt C, et al. Histone modifications induced by a family of bacterial toxins. Proc Natl Acad Sci USA 2007;104:13467–72.

[28] Schroeder GN, Hilbi H. Molecular pathogenesis of Shigella spp.: controlling host cell signaling, invasion, and death by type III secretion. Clin Microbiol Rev 2008;21:134–56.

[29] Arbibe L, Kim DW, Batsche E, Pedron T, Mateescu B, Muchardt C, et al. An injected bacterial effector targets chromatin access for transcription factor NF-kappaB to alter transcription of host genes involved in immune responses. Nat Immunol 2007;8:47—56.

[30] Ismail N, Bloch KC, McBride JW. Human ehrlichiosis and anaplasmosis. Clin Lab Med 2010;30:261—92.

[31] Garcia-Garcia JC, Barat NC, Trembley SJ, Dumler JS. Epigenetic silencing of host cell defense genes enhances intracellular survival of the rickettsial pathogen Anaplasma phagocytophilum. PLoS Pathog 2009;5:e1000488.

[32] Uemura N, Okamoto S, Yamamoto S, Matsumura N, Yamaguchi S, Yamakido M, et al. Helicobacter pylori infection and the development of gastric cancer. N Engl J Med 2001;345:784—9.

[33] Nardone G, Compare D, De Colibus P, de Nucci G, Rocco A. Helicobacter pylori and epigenetic mechanisms underlying gastric carcinogenesis. Dig Dis 2007;25:225—9.

[34] Yamamoto E, Suzuki H, Takamaru H, Yamamoto H, Toyota M, Shinomura Y. Role of DNA methylation in the development of diffuse-type gastric cancer. Digestion 2011;83:241—9.

[35] Ding SZ, Goldberg JB, Hatakeyama M. Helicobacter pylori infection, oncogenic pathways and epigenetic mechanisms in gastric carcinogenesis. Future Oncol 2010;6:851—62.

[36] Suerbaum S, Michetti P. Helicobacter pylori infection. N Engl J Med 2002;347:1175—86.

[37] Hatakeyama M. Oncogenic mechanisms of the Helicobacter pylori CagA protein. Nat Rev Cancer 2004;4:688—94.

[38] Umeda M, Murata-Kamiya N, Saito Y, Ohba Y, Takahashi M, Hatakeyama M. Helicobacter pylori CagA causes mitotic impairment and induces chromosomal instability. J Biol Chem 2009;284:22166—72.

[39] Kang GH, Shim YH, Jung HY, Kim WH, Ro JY, Rhyu MG. CpG island methylation in premalignant stages of gastric carcinoma. Cancer Res 2001;61:2847—51.

[40] Kang GH, Lee S, Kim JS, Jung HY. Profile of aberrant CpG island methylation along the multistep pathway of gastric carcinogenesis. Lab Invest 2003;83:635—41.

[41] Yamashita S, Tsujino Y, Moriguchi K, Tatematsu M, Ushijima T. Chemical genomic screening for methylation-silenced genes in gastric cancer cell lines using 5-aza-2'-deoxycytidine treatment and oligonucleotide microarray. Cancer Sci 2006;97:64—71.

[42] Maekita T, Nakazawa K, Mihara M, Nakajima T, Yanaoka K, Iguchi M, et al. High levels of aberrant DNA methylation in Helicobacter pylori-infected gastric mucosae and its possible association with gastric cancer risk. Clin Cancer Res 2006;12:989—95.

[43] Nakajima T, Maekita T, Oda I, Gotoda T, Yamamoto S, Umemura S, et al. Higher methylation levels in gastric mucosae significantly correlate with higher risk of gastric cancers. Cancer Epidemiol Biomarkers Prev 2006;15:2317—21.

[44] Perri F, Cotugno R, Piepoli A, Merla A, Quitadamo M, Gentile A, et al. Aberrant DNA methylation in non-neoplastic gastric mucosa of H. Pylori infected patients and effect of eradication. Am J Gastroenterol 2007;102:1361—71.

[45] Kaise M, Yamasaki T, Yonezawa J, Miwa J, Ohta Y, Tajiri H. CpG island hypermethylation of tumor-suppressor genes in H. pylori-infected non-neoplastic gastric mucosa is linked with gastric cancer risk. Helicobacter 2008;13:35—41.

[46] Park SY, Yoo EJ, Cho NY, Kim N, Kang GH. Comparison of CpG island hypermethylation and repetitive DNA hypomethylation in premalignant stages of gastric cancer, stratified for Helicobacter pylori infection. J Pathol 2009;219:410—6.

[47] Compare D, Rocco A, Liguori E, D'Armiento FP, Persico G, Masone S, et al. Global DNA hypomethylation is an early event in Helicobacter pylori-related gastric carcinogenesis. J Clin Pathol 2011;64:677—82.

[48] Waki T, Tamura G, Tsuchiya T, Sato K, Nishizuka S, Motoyama T. Promoter methylation status of E-cadherin, hMLH1, and p16 genes in nonneoplastic gastric epithelia. Am J Pathol 2002;161:399—403.

[49] Chan AO, Peng JZ, Lam SK, Lai KC, Yuen MF, Cheung HK, et al. Eradication of Helicobacter pylori infection reverses E-cadherin promoter hypermethylation. Gut 2006;55:463—8.

[50] Leung WK, Man EP, Yu J, Go MY, To KF, Yamaoka Y, et al. Effects of Helicobacter pylori eradication on methylation status of E-cadherin gene in noncancerous stomach. Clin Cancer Res 2006;12:3216—21.

[51] Nakajima T, Enomoto S, Yamashita S, Ando T, Nakanishi Y, Nakazawa K, et al. Persistence of a component of DNA methylation in gastric mucosae after Helicobacter pylori eradication. J Gastroenterol 2010;45:37—44.

[52] Kabir S. Effect of Helicobacter pylori eradication on incidence of gastric cancer in human and animal models: underlying biochemical and molecular events. Helicobacter 2009;14:159—71.

[53] Niwa T, Tsukamoto T, Toyoda T, Mori A, Tanaka H, Maekita T, et al. Inflammatory processes triggered by Helicobacter pylori infection cause aberrant DNA methylation in gastric epithelial cells. Cancer Res 2010;70:1430—40.

[54] Kang HY, Kim N, Park YS, Hwang JH, Kim JW, Jeong SH, et al. Progression of atrophic gastritis and intestinal metaplasia drives Helicobacter pylori out of the gastric mucosa. Dig Dis Sci 2006;51:2310—5.

[55] Ushijima T. Epigenetic field for cancerization. J Biochem Mol Biol 2007;40:142—50.

[56] Guilford P, Hopkins J, Harraway J, McLeod M, McLeod N, Harawira P, et al. E-cadherin germline mutations in familial gastric cancer. Nature 1998;392:402—5.

[57] Grady WM, Willis J, Guilford PJ, Dunbier AK, Toro TT, Lynch H, et al. Methylation of the CDH1 promoter as the second genetic hit in hereditary diffuse gastric cancer. Nat Genet 2000;26:16—7.

[58] Machado JC, Oliveira C, Carvalho R, Soares P, Berx G, Caldas C, et al. E-cadherin gene (CDH1) promoter methylation as the second hit in sporadic diffuse gastric carcinoma. Oncogene 2001;20:1525—8.

[59] Miyazaki T, Murayama Y, Shinomura Y, Yamamoto T, Watabe K, Tsutsui S, et al. E-cadherin gene promoter hypermethylation in H. pylori-induced enlarged fold gastritis. Helicobacter 2007;12:523—31.

[60] Terres AM, Pajares JM, O'Toole D, Ahern S, Kelleher D. H pylori infection is associated with downregulation of E-cadherin, a molecule involved in epithelial cell adhesion and proliferation control. J Clin Pathol 1998;51:410—2.

[61] Chan AO, Lam SK, Wong BC, Wong WM, Yuen MF, Yeung YH, et al. Promoter methylation of E-cadherin gene in gastric mucosa associated with Helicobacter pylori infection and in gastric cancer. Gut 2003;52:502—6.

[62] Yamashita K, Sakuramoto S, Watanabe M. Genomic and epigenetic profiles of gastric cancer: potential diagnostic and therapeutic applications. Surg Today 2011;41:24—38.

[63] Nakajima T, Yamashita S, Maekita T, Niwa T, Nakazawa K, Ushijima T. The presence of a methylation fingerprint of Helicobacter pylori infection in human gastric mucosae. Int J Cancer 2009;124:905—10.

[64] Shin CM, Kim N, Jung Y, Park JH, Kang GH, Park WY, et al. Genome-wide DNA methylation profiles in noncancerous gastric mucosae with regard to Helicobacter pylori infection and the presence of gastric cancer. Helicobacter 2011;16:179—88.

[65] Yan J, Zhang M, Zhang J, Chen X, Zhang X. Helicobacter pylori infection promotes methylation of WWOX gene in human gastric cancer. Biochem Biophys Res Commun 2011;408:99—102.

[66] Takeshima H, Yamashita S, Shimazu T, Niwa T, Ushijima T. The presence of RNA polymerase II, active or stalled, predicts epigenetic fate of promoter CpG islands. Genome Res 2009;19:1974—82.

[67] Takeshima H, Ushijima T. Methylation destiny: Moira takes account of histones and RNA polymerase II. Epigenetics 2010;5:89—95.

[68] Vitkute J, Stankevicius K, Tamulaitiene G, Maneliene Z, Timinskas A, Berg DE, et al. Specificities of eleven different DNA methyltransferases of Helicobacter pylori strain 26695. J Bacteriol 2001;183:443—50.

[69] Alves MK, Ferrasi AC, Lima VP, Ferreira MV, de Moura Campos Pardini MI, Rabenhorst SH. Inactivation of COX-2, HMLH1 and CDKN2A Gene by Promoter Methylation in Gastric Cancer: Relationship with Histological Subtype, Tumor Location and Helicobacter pylori Genotype. Pathobiology 2011;78:266—76.

[70] Bussiere FI, Michel V, Memet S, Ave P, Vivas JR, Huerre M, et al. H. pylori-induced promoter hypermethylation downregulates USF1 and USF2 transcription factor gene expression. Cell Microbiol 2010;12:1124—33.

[71] Touati E. When bacteria become mutagenic and carcinogenic: lessons from H. pylori. Mutat Res 2010;703:66—70.

[72] Sepulveda AR, Yao Y, Yan W, Park DI, Kim JJ, Gooding W, et al. CpG methylation and reduced expression of O6-methylguanine DNA methyltransferase is associated with Helicobacter pylori infection. Gastroenterology 2010;138:1836—44.

[73] Ando T, Yoshida T, Enomoto S, Asada K, Tatematsu M, Ichinose M, et al. DNA methylation of microRNA genes in gastric mucosae of gastric cancer patients: its possible involvement in the formation of epigenetic field defect. Int J Cancer 2009;124:2367—74.

[74] Suzuki H, Yamamoto E, Nojima M, Kai M, Yamano HO, Yoshikawa K, et al. Methylation-associated silencing of microRNA-34b/c in gastric cancer and its involvement in an epigenetic field defect. Carcinogenesis 2010;31:2066—73.

[75] Craig VJ, Cogliatti SB, Rehrauer H, Wundisch T, Muller A. Epigenetic silencing of microRNA-203 dysregulates ABL1 expression and drives Helicobacter-associated gastric lymphomagenesis. Cancer Res 2011;71:3616—24.

[76] Matsushima K, Isomoto H, Inoue N, Nakayama T, Hayashi T, Nakayama M, et al. MicroRNA signatures in Helicobacter pylori-infected gastric mucosa. Int J Cancer 2011;128:361—70.

[77] Schlesinger Y, Straussman R, Keshet I, Farkash S, Hecht M, Zimmerman J, et al. Polycomb-mediated methylation on Lys27 of histone H3 pre-marks genes for de novo methylation in cancer. Nat Genet 2007;39:232—6.

[78] Widschwendter M, Fiegl H, Egle D, Mueller-Holzner E, Spizzo G, Marth C, et al. Epigenetic stem cell signature in cancer. Nat Genet 2007;39:157—8.

[79] Ballestar E, Esteller M. Epigenetic gene regulation in cancer. Adv Genet 2008;61:247—67.

[80] Hahn MA, Hahn T, Lee DH, Esworthy RS, Kim BW, Riggs AD, et al. Methylation of polycomb target genes in intestinal cancer is mediated by inflammation. Cancer Res 2008;68:10280—9.

[81] Pathak SK, Basu S, Bhattacharyya A, Pathak S, Banerjee A, Basu J, et al. TLR4-dependent NF-kappaB activation and mitogen- and stress-activated protein kinase 1-triggered phosphorylation events are central to

Helicobacter pylori peptidyl prolyl cis-, trans-isomerase (HP0175)-mediated induction of IL-6 release from macrophages. J Immunol 2006;177:7950−8.

[82] Fehri LF, Rechner C, Janssen S, Mak TN, Holland C, Bartfeld S, et al. Helicobacter pylori-induced modification of the histone H3 phosphorylation status in gastric epithelial cells reflects its impact on cell cycle regulation. Epigenetics 2009;4:577−86.

[83] Ding SZ, Fischer W, Kaparakis-Liaskos M, Liechti G, Merrell DS, Grant PA, et al. Helicobacter pylori-induced histone modification, associated gene expression in gastric epithelial cells, and its implication in pathogenesis. PLoS ONE 2010;5:e9875.

[84] Tolg C, Sabha N, Cortese R, Panchal T, Ahsan A, Soliman A, et al. Uropathogenic E. coli infection provokes epigenetic downregulation of CDKN2A (p16INK4A) in uroepithelial cells. Lab Invest 2011;91:825−36.

[85] Murata M, Azuma Y, Miura K, Rahman MA, Matsutani M, Aoyama M, et al. Chlamydial SET domain protein functions as a histone methyltransferase. Microbiology 2007;153:585−92.

[86] Pennini ME, Perrinet S, Dautry-Varsat A, Subtil A. Histone methylation by NUE, a novel nuclear effector of the intracellular pathogen Chlamydia trachomatis. PLoS Pathog 2010;6:e1000995.

[87] Niller HH, Wolf H, Minarovits J. Epstein−Barr Virus. In: Minarovits J, Gonczol E, Valyi-Nagy T, editors. Latency Strategies of Herpesviruses. New York: Springer; 2007. p. 154−91.

[88] Swaminathan S. Noncoding RNAs produced by oncogenic human herpesviruses. J Cell Physiol 2008;216:321−6.

[89] Tempera I, Klichinsky M, Lieberman PM. EBV latency types adopt alternative chromatin conformations. PLoS Pathog 2011;7:e1002180.

[90] Wang L, Grossman SR, Kieff E. Epstein−Barr virus nuclear protein 2 interacts with p300, CBP, and PCAF histone acetyltransferases in activation of the LMP1 promoter. Proc Natl Acad Sci USA 2000;97:430−5.

[91] Portal D, Rosendorff A, Kieff E. Epstein−Barr nuclear antigen leader protein coactivates transcription through interaction with histone deacetylase 4. Proc Natl Acad Sci USA 2006;103:19278−83.

[92] Cotter MA, Robertson ES. Modulation of histone acetyltransferase activity through interaction of Epstein−Barr nuclear antigen 3C with prothymosin alpha. Mol Cell Biol 2000;20:5722−35.

[93] Subramanian C, Hasan S, Rowe M, Hottiger M, Orre R, Robertson ES. Epstein−Barr virus nuclear antigen 3C and prothymosin alpha interact with the p300 transcriptional coactivator at the CH1 and CH3/HAT domains and cooperate in regulation of transcription and histone acetylation. J Virol 2002;76:4699−708.

[94] Knight JS, Lan K, Subramanian C, Robertson ES. Epstein−Barr virus nuclear antigen 3C recruits histone deacetylase activity and associates with the corepressors mSin3A and NCoR in human B-cell lines. J Virol 2003;77:4261−72.

[95] Skalska L, White RE, Franz M, Ruhmann M, Allday MJ. Epigenetic repression of p16(INK4A) by latent Epstein−Barr virus requires the interaction of EBNA3A and EBNA3C with CtBP. PLoS Pathog 2010;6:e1000951.

[96] Anderton E, Yee J, Smith P, Crook T, White RE, Allday MJ. Two Epstein−Barr virus (EBV) oncoproteins cooperate to repress expression of the proapoptotic tumour-suppressor Bim: clues to the pathogenesis of Burkitt's lymphoma. Oncogene 2008;27:421−33.

[97] Paschos K, Smith P, Anderton E, Middeldorp JM, White RE, Allday MJ. Epstein−Barr virus latency in B cells leads to epigenetic repression and CpG methylation of the tumour suppressor gene Bim. PLoS Pathog 2009;5:e1000492.

[98] Tsai CN, Tsai CL, Tse KP, Chang HY, Chang YS. The Epstein−Barr virus oncogene product, latent membrane protein 1, induces the downregulation of E-cadherin gene expression via activation of DNA methyltransferases. Proc Natl Acad Sci USA 2002;99:10084−9.

[99] Tsai CL, Li HP, Lu YJ, Hsueh C, Liang Y, Chen CL, et al. Activation of DNA methyltransferase 1 by EBV LMP1 Involves c-Jun NH(2)-terminal kinase signaling. Cancer Res 2006;66:11668−76.

[100] Dutton A, Woodman CB, Chukwuma MB, Last JI, Wei W, Vockerodt M, et al. Bmi-1 is induced by the Epstein−Barr virus oncogene LMP1 and regulates the expression of viral target genes in Hodgkin lymphoma cells. Blood 2007;109:2597−603.

[101] Niller HH, Wolf H, Minarovits J. Epigenetic dysregulation of the host cell genome in Epstein−Barr virus-associated neoplasia. Semin Cancer Biol 2009;19:158−64.

[102] Kwong J, Lo KW, To KF, Teo PM, Johnson PJ, Huang DP. Promoter hypermethylation of multiple genes in nasopharyngeal carcinoma. Clin Cancer Res 2002;8:131−7.

[103] Murray PG, Qiu GH, Fu L, Waites ER, Srivastava G, Heys D, et al. Frequent epigenetic inactivation of the RASSF1A tumor suppressor gene in Hodgkin's lymphoma. Oncogene 2004;23:1326−31.

[104] Doerr JR, Malone CS, Fike FM, Gordon MS, Soghomonian SV, Thomas RK, et al. Patterned CpG methylation of silenced B cell gene promoters in classical Hodgkin lymphoma-derived and primary effusion lymphoma cell lines. J Mol Biol 2005;350:631−40.

[105] Ushmorov A, Leithauser F, Sakk O, Weinhausel A, Popov SW, Moller P, et al. Epigenetic processes play a major role in B-cell-specific gene silencing in classical Hodgkin lymphoma. Blood 2006;107:2493−500.

[106] Rossi D, Gaidano G, Gloghini A, Deambrogi C, Franceschetti S, Berra E, et al. Frequent aberrant promoter hypermethylation of O6-methylguanine-DNA methyltransferase and death-associated protein kinase genes in immunodeficiency-related lymphomas. Br J Haematol 2003;123:475—8.

[107] Au WY, Ma ES, Choy C, Chung LP, Fung TK, Liang R, et al. Therapy-related lymphomas in patients with autoimmune diseases after treatment with disease-modifying anti-rheumatic drugs. Am J Hematol 2006;81:5—11.

[108] Hino R, Uozaki H, Murakami N, Ushiku T, Shinozaki A, Ishikawa S, et al. Activation of DNA methyltransferase 1 by EBV latent membrane protein 2A leads to promoter hypermethylation of PTEN gene in gastric carcinoma. Cancer Res 2009;69:2766—74.

[109] Georgescu MM. PTEN Tumor Suppressor Network in PI3K-Akt Pathway Control. Genes Cancer 2010;1:1170—7.

[110] Kang GH, Lee S, Kim WH, Lee HW, Kim JC, Rhyu MG, et al. Epstein—Barr virus-positive gastric carcinoma demonstrates frequent aberrant methylation of multiple genes and constitutes CpG island methylator phenotype-positive gastric carcinoma. Am J Pathol 2002;160:787—94.

[111] Chang MS, Uozaki H, Chong JM, Ushiku T, Sakuma K, Ishikawa S, et al. CpG island methylation status in gastric carcinoma with and without infection of Epstein—Barr virus. Clin Cancer Res 2006;12:2995—3002.

[112] Luo B, Wang Y, Wang XF, Liang H, Yan LP, Huang BH, et al. Expression of Epstein—Barr virus genes in EBV-associated gastric carcinomas. World J Gastroenterol 2005;11:629—33.

[113] Pantry SN, Medveczky PG. Epigenetic regulation of Kaposi's sarcoma-associated herpesvirus replication. Semin Cancer Biol 2009;19:153—7.

[114] Stedman W, Deng Z, Lu F, Lieberman PM. ORC, MCM, and histone hyperacetylation at the Kaposi's sarcoma-associated herpesvirus latent replication origin. J Virol 2004;78:12566—75.

[115] Sakakibara S, Ueda K, Nishimura K, Do E, Ohsaki E, Okuno T, et al. Accumulation of heterochromatin components on the terminal repeat sequence of Kaposi's sarcoma-associated herpesvirus mediated by the latency-associated nuclear antigen. J Virol 2004;78:7299—310.

[116] Gunther T, Grundhoff A. The epigenetic landscape of latent Kaposi sarcoma-associated herpesvirus genomes. PLoS Pathog 2010;6:e1000935.

[117] Toth Z, Maglinte DT, Lee SH, Lee HR, Wong LY, Brulois KF, et al. Epigenetic analysis of KSHV latent and lytic genomes. PLoS Pathog 2010;6:e1001013.

[118] Kang H, Wiedmer A, Yuan Y, Robertson E, Lieberman PM. Coordination of KSHV latent and lytic gene control by CTCF-cohesin mediated chromosome conformation. PLoS Pathog 2011;7:e1002140.

[119] Lu F, Stedman W, Yousef M, Renne R, Lieberman PM. Epigenetic regulation of Kaposi's sarcoma-associated herpesvirus latency by virus-encoded microRNAs that target Rta and the cellular Rbl2-DNMT pathway. J Virol 2010;84:2697—706.

[120] Shamay M, Krithivas A, Zhang J, Hayward SD. Recruitment of the de novo DNA methyltransferase Dnmt3a by Kaposi's sarcoma-associated herpesvirus LANA. Proc Natl Acad Sci USA 2006;103:14554—9.

[121] Di Bartolo DL, Cannon M, Liu YF, Renne R, Chadburn A, et al. KSHV LANA inhibits TGF-beta signaling through epigenetic silencing of the TGF-beta type II receptor. Blood 2008;111:4731—40.

[122] Krithivas A, Young DB, Liao G, Greene D, Hayward SD. Human herpesvirus 8 LANA interacts with proteins of the mSin3 corepressor complex and negatively regulates Epstein—Barr virus gene expression in dually infected PEL cells. J Virol 2000;74:9637—45.

[123] Lim C, Lee D, Seo T, Choi C, Choe J. Latency-associated nuclear antigen of Kaposi's sarcoma-associated herpesvirus functionally interacts with heterochromatin protein 1. J Biol Chem 2003;278:7397—405.

[124] Stuber G, Mattsson K, Flaberg E, Kati E, Markasz L, Sheldon JA, et al. HHV-8 encoded LANA-1 alters the higher organization of the cell nucleus. Mol Cancer 2007;6:28.

[125] Matsumura S, Persson LM, Wong L, Wilson AC. The latency-associated nuclear antigen interacts with MeCP2 and nucleosomes through separate domains. J Virol 2010;84:2318—30.

[126] Platt G, Carbone A, Mittnacht S. p16INK4a loss and sensitivity in KSHV associated primary effusion lymphoma. Oncogene 2002;21:1823—31.

[127] Carbone A, Cilia AM, Gloghini A, Capello D, Fassone L, Perin T, et al. Characterization of a novel HHV-8-positive cell line reveals implications for the pathogenesis and cell cycle control of primary effusion lymphoma. Leukemia 2000;14:1301—9.

[128] Gonczol E, Andrews PW, Plotkin SA. Cytomegalovirus replicates in differentiated but not in undifferentiated human embryonal carcinoma cells. Science 1984;224:159—61.

[129] Murphy JC, Fischle W, Verdin E, Sinclair JH. Control of cytomegalovirus lytic gene expression by histone acetylation. EMBO J 2002;21:1112—20.

[130] Wright E, Bain M, Teague L, Murphy J, Sinclair J. Ets-2 repressor factor recruits histone deacetylase to silence human cytomegalovirus immediate-early gene expression in non-permissive cells. J Gen Virol 2005;86:535—44.

437

[131] Park JJ, Kim YE, Pham HT, Kim ET, Chung YH, Ahn JH. Functional interaction of the human cytomegalovirus IE2 protein with histone deacetylase 2 in infected human fibroblasts. J Gen Virol 2007;88:3214−23.

[132] Nevels M, Paulus C, Shenk T. Human cytomegalovirus immediate-early 1 protein facilitates viral replication by antagonizing histone deacetylation. Proc Natl Acad Sci USA 2004;101:17234−9.

[133] Hayhurst GP, Bryant LA, Caswell RC, Walker SM, Sinclair JH. CCAAT box-dependent activation of the TATA-less human DNA polymerase alpha promoter by the human cytomegalovirus 72-kilodalton major immediate-early protein. J Virol 1995;69:182−8.

[134] Caswell R, Bryant L, Sinclair J. Human cytomegalovirus immediate-early 2 (IE2) protein can transactivate the human hsp70 promoter by alleviation of Dr1-mediated repression. J Virol 1996;70:4028−37.

[135] Straat K, Liu C, Rahbar A, Zhu Q, Liu L, Wolmer-Solberg N, et al. Activation of telomerase by human cytomegalovirus. J Natl Cancer Inst 2009;101:488−97.

[136] Michaelis M, Doerr HW, Cinatl J. The story of human cytomegalovirus and cancer: increasing evidence and open questions. Neoplasia 2009;11:1−9.

[137] Matsuoka M. Human T-cell leukemia virus type I (HTLV-I) infection and the onset of adult T-cell leukemia (ATL). Retrovirology 2005;2:27.

[138] Lezin A, Gillet N, Olindo S, Signate A, Grandvaux N, Verlaeten O, et al. Histone deacetylase mediated transcriptional activation reduces proviral loads in HTLV-1 associated myelopathy/tropical spastic paraparesis patients. Blood 2007;110:3722−8.

[139] Sato H, Oka T, Shinnou Y, Kondo T, Washio K, Takano M, et al. Multi-step aberrant CpG island hyper-methylation is associated with the progression of adult T-cell leukemia/lymphoma. Am J Pathol 2010;176:402−15.

[140] Nakase K, Cheng J, Zhu Q, Marasco WA. Mechanisms of SHP-1 P2 promoter regulation in hematopoietic cells and its silencing in HTLV-1-transformed T cells. J Leukoc Biol 2009;85:165−74.

[141] Niller HH, Wolf H, Minarovits J. Viral hit and run-oncogenesis: Genetic and epigenetic scenarios. Cancer Lett 2011;305:200−17.

[142] Aoki Y, Tosato G. Neoplastic conditions in the context of HIV-1 infection. Curr HIV Res 2004;2:343−9.

[143] Grulich AE, van Leeuwen MT, Falster MO, Vajdic CM. Incidence of cancers in people with HIV/AIDS compared with immunosuppressed transplant recipients: a meta-analysis. Lancet 2007;370:59−67.

[144] Elgui de Oliveira D. DNA viruses in human cancer: an integrated overview on fundamental mechanisms of viral carcinogenesis. Cancer Lett 2007;247:182−96.

[145] Finzi D, Hermankova M, Pierson T, Carruth LM, Buck C, Chaisson RE, et al. Identification of a reservoir for HIV-1 in patients on highly active antiretroviral therapy. Science 1997;278:1295−300.

[146] Siliciano JD, Kajdas J, Finzi D, Quinn TC, Chadwick K, Margolick JB, et al. Long-term follow-up studies confirm the stability of the latent reservoir for HIV-1 in resting CD4+ T cells. Nat Med 2003;9:727−8.

[147] Richman DD, Margolis DM, Delaney M, Greene WC, Hazuda D, Pomerantz RJ. The challenge of finding a cure for HIV infection. Science 2009;323:1304−7.

[148] Margolis DM. Mechanisms of HIV latency: an emerging picture of complexity. Curr HIV/AIDS Rep 2010;7:37−43.

[149] Bednarik DP, Mosca JD, Raj NB. Methylation as a modulator of expression of human immunodeficiency virus. J Virol 1987;61:1253−7.

[150] Bednarik DP, Cook JA, Pitha PM. Inactivation of the HIV LTR by DNA CpG methylation: evidence for a role in latency. EMBO J 1990;9:1157−64.

[151] Schulze-Forster K, Gotz F, Wagner H, Kroger H, Simon D. Transcription of HIV1 is inhibited by DNA methylation. Biochem Biophys Res Commun 1990;168:141−7.

[152] Van Lint C, Emiliani S, Ott M, Verdin E. Transcriptional activation and chromatin remodeling of the HIV-1 promoter in response to histone acetylation. EMBO J 1996;15:1112−20.

[153] Coull JJ, Romerio F, Sun JM, Volker JL, Galvin KM, Davie JR, et al. The human factors YY1 and LSF repress the human immunodeficiency virus type 1 long terminal repeat via recruitment of histone deacetylase 1. J Virol 2000;74:6790−9.

[154] Ylisastigui L, Archin NM, Lehrman G, Bosch RJ, Margolis DM. Coaxing HIV-1 from resting CD4 T cells: histone deacetylase inhibition allows latent viral expression. AIDS 2004;18:1101−8.

[155] Friedman J, Cho WK, Chu CK, Keedy KS, Archin NM, Margolis DM, et al. Epigenetic Silencing of HIV-1 by the Histone H3 Lysine 27 Methyltransferase Enhancer of Zeste 2. J Virol 2011;85:9078−89.

[156] Jiang G, Espeseth A, Hazuda DJ, Margolis DM. c-Myc and Sp1 contribute to proviral latency by recruiting histone deacetylase 1 to the human immunodeficiency virus type 1 promoter. J Virol 2007;81:10914−23.

[157] Tyagi M, Karn J. CBF-1 promotes transcriptional silencing during the establishment of HIV-1 latency. EMBO J 2007;26:4985−95.

[158] Schroder AR, Shinn P, Chen H, Berry C, Ecker JR, Bushman F. HIV-1 integration in the human genome favors active genes and local hotspots. Cell 2002;110:521−9.

[159] Han Y, Lassen K, Monie D, Sedaghat AR, Shimoji S, Liu X, et al. Resting CD4+ T cells from human immunodeficiency virus type 1 (HIV-1)-infected individuals carry integrated HIV-1 genomes within actively transcribed host genes. J Virol 2004;78:6122–33.

[160] Han Y, Lin YB, An W, Xu J, Yang HC, O'Connell K, et al. Orientation-dependent regulation of integrated HIV-1 expression by host gene transcriptional readthrough. Cell Host Microbe 2008;4:134–46.

[161] Klase Z, Kale P, Winograd R, Gupta MV, Heydarian M, Berro R, et al. HIV-1 TAR element is processed by Dicer to yield a viral micro-RNA involved in chromatin remodeling of the viral LTR. BMC Mol Biol 2007;8:63.

[162] Huang J, Wang F, Argyris E, Chen K, Liang Z, Tian H, et al. Cellular microRNAs contribute to HIV-1 latency in resting primary CD4+ T lymphocytes. Nat Med 2007;13:1241–7.

[163] Mikovits JA, Young HA, Vertino P, Issa JP, Pitha PM, Turcoski-Corrales S, et al. Infection with human immunodeficiency virus type 1 upregulates DNA methyltransferase, resulting in de novo methylation of the gamma interferon (IFN-gamma) promoter and subsequent downregulation of IFN-gamma production. Mol Cell Biol 1998;18:5166–77.

[164] Fang JY, Mikovits JA, Bagni R, Petrow-Sadowski CL, Ruscetti FW. Infection of lymphoid cells by integration-defective human immunodeficiency virus type 1 increases de novo methylation. J Virol 2001;75:9753–61.

[165] Giordanengo V, Ollier L, Lanteri M, Lesimple J, March D, Thyss S, et al. Epigenetic reprogramming of UDP-N-acetylglucosamine 2-epimerase/N-acetylmannosamine kinase (GNE) in HIV-1-infected CEM T cells. FASEB J 2004;18:1961–3.

[166] Youngblood B, Reich NO. The early expressed HIV-1 genes regulate DNMT1 expression. Epigenetics 2008;3:149–56.

[167] Saiyed ZM, Gandhi N, Agudelo M, Napuri J, Samikkannu T, Reddy PV, et al. HIV-1 Tat upregulates expression of histone deacetylase-2 (HDAC2) in human neurons: implication for HIV-associated neurocognitive disorder (HAND). Neurochem Int 2011;58:656–64.

[168] Tischoff I, Tannapfel A. DNA methylation in hepatocellular carcinoma. World J Gastroenterol 2008;14:1741–8.

[169] Shim YH, Yoon GS, Choi HJ, Chung YH, Yu E. p16 Hypermethylation in the early stage of hepatitis B virus-associated hepatocarcinogenesis. Cancer Lett 2003;190:213–9.

[170] Um TH, Kim H, Oh BK, Kim MS, Kim KS, Jung G, et al. Aberrant CpG island hypermethylation in dysplastic nodules and early HCC of hepatitis B virus-related human multistep hepatocarcinogenesis. J Hepatol 2011;54:939–47.

[171] Doria M, Klein N, Lucito R, Schneider RJ. The hepatitis B virus HBx protein is a dual specificity cytoplasmic activator of Ras and nuclear activator of transcription factors. EMBO J 1995;14:4747–57.

[172] Haviv I, Shamay M, Doitsh G, Shaul Y. Hepatitis B virus pX targets TFIIB in transcription coactivation. Mol Cell Biol 1998;18:1562–9.

[173] Benn J, Schneider RJ. Hepatitis B virus HBx protein activates Ras-GTP complex formation and establishes a Ras, Raf, MAP kinase signaling cascade. Proc Natl Acad Sci USA 1994;91:10350–4.

[174] Choi BH, Choi M, Jeon HY, Rho HM. Hepatitis B viral X protein overcomes inhibition of E2F1 activity by pRb on the human Rb gene promoter. DNA Cell Biol 2001;20:75–80.

[175] Wang WH, Hullinger RL, Andrisani OM. Hepatitis B virus X protein via the p38MAPK pathway induces E2F1 release and ATR kinase activation mediating p53 apoptosis. J Biol Chem 2008;283:25455–67.

[176] Jung JK, Arora P, Pagano JS, Jang KL. Expression of DNA methyltransferase 1 is activated by hepatitis B virus X protein via a regulatory circuit involving the p16INK4a-cyclin D1-CDK 4/6-pRb-E2F1 pathway. Cancer Res 2007;67:5771–8.

[177] Zhu YZ, Zhu R, Shi LG, Mao Y, Zheng GJ, Chen Q, et al. Hepatitis B virus X protein promotes hypermethylation of p16(INK4A) promoter through upregulation of DNA methyltransferases in hepatocarcinogenesis. Exp Mol Pathol 2010;89:268–75.

[178] Park IY, Sohn BH, Yu E, Suh DJ, Chung YH, Lee JH, et al. Aberrant epigenetic modifications in hepatocarcinogenesis induced by hepatitis B virus X protein. Gastroenterology 2007;132:1476–94.

[179] Ehrlich M. DNA hypomethylation and cancer. In: Ehrlich M, editor. DNA Alterations in Cancer. Westborough: Eaton Publishing; 2000. p. 273–91.

[180] Baylin SB, Herman JG. Epigenetics and loss of gene function in cancer. In: Ehrlich M, editor. DNA Alterations in Cancer. Westborough: Eaton Publishing; 2000. p. 293–309.

[181] Nagai H, Baba M, Konishi N, Kim YS, Nogami M, Okumura K, et al. Isolation of NotI clusters hypomethylated in HBV-integrated hepatocellular carcinomas by two-dimensional electrophoresis. DNA Res 1999;6:219–25.

[182] Wei Y, Van Nhieu JT, Prigent S, Srivatanakul P, Tiollais P, et al. Altered expression of E-cadherin in hepatocellular carcinoma: correlations with genetic alterations, beta-catenin expression, and clinical features. Hepatology 2002;36:692–701.

[183] Lee JO, Kwun HJ, Jung JK, Choi KH, Min DS, Jang KL. Hepatitis B virus X protein represses E-cadherin expression via activation of DNA methyltransferase 1. Oncogene 2005;24:6617−25.

[184] Liu J, Lian Z, Han S, Waye MM, Wang H, Wu MC, et al. Downregulation of E-cadherin by hepatitis B virus X antigen in hepatocellullar carcinoma. Oncogene 2006;25:1008−17.

[185] Zhong S, Yeo W, Tang MW, Wong N, Lai PB, Johnson PJ. Intensive hypermethylation of the CpG island of Ras association domain family 1A in hepatitis B virus-associated hepatocellular carcinomas. Clin Cancer Res 2003;9:3376−82.

[186] Zhong S, Tang MW, Yeo W, Liu C, Lo YM, Johnson PJ. Silencing of GSTP1 gene by CpG island DNA hypermethylation in HBV-associated hepatocellular carcinomas. Clin Cancer Res 2002;8:1087−92.

[187] Feng Q, Stern JE, Hawes SE, Lu H, Jiang M, Kiviat NB. DNA methylation changes in normal liver tissues and hepatocellular carcinoma with different viral infection. Exp Mol Pathol 2010;88:287−92.

[188] Ahmed R, Salama H, Fouad A, Sabry D, AbdAlah e, Kamal M. Detection of aberrant p16INK4A methylation in sera of patients with HCV-related liver diseases: An Egyptian study. Med Sci Monit 2010;16:CR410−5.

[189] Formeister EJ, Tsuchiya M, Fujii H, Shpyleva S, Pogribny IP, Rusyn I. Comparative analysis of promoter methylation and gene expression endpoints between tumorous and non-tumorous tissues from HCV-positive patients with hepatocellular carcinoma. Mutat Res 2010;692:26−33.

[190] Lambert MP, Paliwal A, Vaissiere T, Chemin I, Zoulim F, Tommasino M, et al. Aberrant DNA methylation distinguishes hepatocellular carcinoma associated with HBV and HCV infection and alcohol intake. J Hepatol 2011;54:705−15.

[191] Herceg Z, Paliwal A. Epigenetic mechanisms in hepatocellular carcinoma: how environmental factors influence the epigenome. Mutat Res 2011;727:55−61.

[192] Yang B, Guo M, Herman JG, Clark DP. Aberrant promoter methylation profiles of tumor suppressor genes in hepatocellular carcinoma. Am J Pathol 2003;163:1101−7.

[193] Archer KJ, Mas VR, Maluf DG, Fisher RA. High-throughput assessment of CpG site methylation for distinguishing between HCV-cirrhosis and HCV-associated hepatocellular carcinoma. Mol Genet Genomics 2010;283:341−9.

[194] Duong FH, Christen V, Lin S, Heim MH. Hepatitis C virus-induced up-regulation of protein phosphatase 2A inhibits histone modification and DNA damage repair. Hepatology 2010;51:741−51.

[195] Higgs MR, Lerat H, Pawlotsky JM. Downregulation of Gadd45beta expression by hepatitis C virus leads to defective cell cycle arrest. Cancer Res 2010;70:4901−11.

[196] Lu CY, Hsieh SY, Lu YJ, Wu CS, Chen LC, Lo SJ, et al. Aberrant DNA methylation profile and frequent methylation of KLK10 and OXGR1 genes in hepatocellular carcinoma. Genes Chromosomes Cancer 2009;48:1057−68.

[197] Deng YB, Nagae G, Midorikawa Y, Yagi K, Tsutsumi S, Yamamoto S, et al. Identification of genes preferentially methylated in hepatitis C virus-related hepatocellular carcinoma. Cancer Sci 2010;101:1501−10.

[198] Miyoshi H, Fujie H, Shintani Y, Tsutsumi T, Shinzawa S, Makuuchi M, et al. Hepatitis C virus core protein exerts an inhibitory effect on suppressor of cytokine signaling (SOCS)-1 gene expression. J Hepatol 2005;43:757−63.

[199] Ko E, Kim SJ, Joh JW, Park CK, Park J, Kim DH. CpG island hypermethylation of SOCS-1 gene is inversely associated with HBV infection in hepatocellular carcinoma. Cancer Lett 2008;271:240−50.

[200] Arora P, Kim EO, Jung JK, Jang KL. Hepatitis C virus core protein downregulates E-cadherin expression via activation of DNA methyltransferase 1 and 3b. Cancer Lett 2008;261:244−52.

[201] Ripoli M, Barbano R, Balsamo T, Piccoli C, Brunetti V, Coco M, et al. Hypermethylated levels of E-cadherin promoter in Huh-7 cells expressing the HCV core protein. Virus Res 2011;160:74−81.

[202] Naka K, Abe K, Takemoto K, Dansako H, Ikeda M, Shimotohno K, et al. Epigenetic silencing of interferon-inducible genes is implicated in interferon resistance of hepatitis C virus replicon-harboring cells. J Hepatol 2006;44:869−78.

[203] Magerl C, Ellinger J, Braunschweig T, Kremmer E, Koch LK, Holler T, et al. H3K4 dimethylation in hepato-cellular carcinoma is rare compared with other hepatobiliary and gastrointestinal carcinomas and correlates with expression of the methylase Ash2 and the demethylase LSD1. Hum Pathol 2010;41:181−9.

[204] Guo N, Chen R, Li Z, Liu Y, Cheng D, Zhou Q, et al. Hepatitis C virus core upregulates the methylation status of the RASSF1A promoter through regulation of SMYD3 in hilar cholangiocarcinoma cells. Acta Biochim Biophys Sin (Shanghai) 2011;43:354−61.

[205] Szalmas A, Konya J. Epigenetic alterations in cervical carcinogenesis. Semin Cancer Biol 2009;19:144−52.

[206] Henken FE, Wilting SM, Overmeer RM, van Rietschoten JG, Nygren AO, et al. Sequential gene promoter methylation during HPV-induced cervical carcinogenesis. Br J Cancer 2007;97:1457−64.

[207] Burgers WA, Blanchon L, Pradhan S, de Launoit Y, Kouzarides T, Fuks F. Viral oncoproteins target the DNA methyltransferases. Oncogene 2007;26:1650−5.

[208] Laurson J, Khan S, Chung R, Cross K, Raj K. Epigenetic repression of E-cadherin by human papillomavirus 16 E7 protein. Carcinogenesis 2010;31:918−26.

[209] Stephen JK, Chen KM, Shah V, Schweitzer VG, Gardner G, Benninger MS, et al. Consistent DNA hyper-methylation patterns in laryngeal papillomas. Int J Head Neck Surg 2010;1:69−77.

[210] Holland D, Hoppe-Seyler K, Schuller B, Lohrey C, Maroldt J, Durst M, et al. Activation of the enhancer of zeste homologue 2 gene by the human papillomavirus E7 oncoprotein. Cancer Res 2008;68:9964−72.

[211] Zhang B, Laribee RN, Klemsz MJ, Roman A. Human papillomavirus type 16 E7 protein increases acetylation of histone H3 in human foreskin keratinocytes. Virology 2004;329:189−98.

[212] McLaughlin-Drubin ME, Crum CP, Munger K. Human papillomavirus E7 oncoprotein induces KDM6A and KDM6B histone demethylase expression and causes epigenetic reprogramming. Proc Natl Acad Sci USA 2011;108:2130−5.

[213] Feng H, Shuda M, Chang Y, Moore PS. Clonal integration of a polyomavirus in human Merkel cell carcinoma. Science 2008;319:1096−100.

[214] Shuda M, Feng H, Kwun HJ, Rosen ST, Gjoerup O, Moore PS, et al. T antigen mutations are a human tumor-specific signature for Merkel cell polyomavirus. Proc Natl Acad Sci USA 2008;105:16272−7.

[215] Ben Arie N, Hassan BA, Bermingham NA, Malicki DM, Armstrong D, Matzuk M, et al. Functional conser-vation of atonal and Math1 in the CNS and PNS. Development 2000;127:1039−48.

[216] Bossuyt W, Kazanjian A, De Geest N, Van Kelst S, De Hertogh G, Geboes K, et al. Atonal homolog 1 is a tumor suppressor gene. PLoS Biol 2009;7:e39.

[217] Helmbold P, Lahtz C, Enk A, Herrmann-Trost P, Marsch WC, Kutzner H, et al. Frequent occurrence of RASSF1A promoter hypermethylation and Merkel cell polyomavirus in Merkel cell carcinoma. Mol Carcinog 2009;48:903−9.

[218] Vatanasapt V, Uttaravichien T, Mairiang EO, Pairojkul C, Chartbanchachai W, Haswell-Elkins M. Chol-angiocarcinoma in north-east Thailand. Lancet 1990;335:116−7.

[219] Shin HR, Oh JK, Masuyer E, Curado MP, Bouvard V, Fang YY, et al. Epidemiology of cholangiocarcinoma: an update focusing on risk factors. Cancer Sci 2010;101:579−85.

[220] Chinnasri P, Pairojkul C, Jearanaikoon P, Sripa B, Bhudhisawasdi V, Tantimavanich S, et al. Preferentially different mechanisms of inactivation of 9p21 gene cluster in liver fluke-related cholangiocarcinoma. Hum Pathol 2009;40:817−26.

[221] Sriraksa R, Zeller C, El Bahrawy MA, Dai W, Daduang J, et al. CpG-island methylation study of liver fluke-related cholangiocarcinoma. Br J Cancer 2011;104:1313−8.

[222] Fantappie MR, Gimba ER, Rumjanek FD. Lack of DNA methylation in Schistosoma mansoni. Exp Parasitol 2001;98:162−6.

[223] Gutierrez MI, Siraj AK, Khaled H, Koon N, El Rifai W, Bhatia K. CpG island methylation in Schistosoma- and non-Schistosoma-associated bladder cancer. Mod Pathol 2004;17:1268−74.

[224] Niller HH, Wolf H, Minarovits J. Regulation and dysregulation of Epstein−Barr virus latency: implications for the development of autoimmune diseases. Autoimmunity 2008;41:298−328.

[225] North ML, Ellis AK. The role of epigenetics in the developmental origins of allergic disease. Ann Allergy Asthma Immunol 2011;106:355−61.

[226] Licciardi PV, Wong SS, Tang ML, Karagiannis TC. Epigenome targeting by probiotic metabolites. Gut Pathog 2010;2:24.

441

CHAPTER 22

The Epigenetics of Endometriosis

Sun-Wei Guo
Fudan University Shanghai College of Medicine, Shanghai, China

443

22.1 INTRODUCTION

Endometriosis, defined as the presence and growth of functional endometrial-like tissues outside the uterine cavity, is a common and benign gynecological disorder with a poorly understood and somewhat enigmatic etiopathogenesis and pathophysiology [1]. It is a leading cause of disability in women of reproductive age, responsible for dysmenorrhea, pelvic pain, and subfertility [2]. As such, it impacts negatively on patients' physical, mental, relational, and social wellbeing [3]. It also consumes tremendous healthcare resources. In the

T. Tollefsbol (Ed): Epigenetics in Human Disease. DOI: 10.1016/B978-0-12-388415-2.00022-6

United States, endometriosis is the third leading cause of gynecologic hospitalization [4,5]. In China, endometriosis-related surgeries constitute about one quarter of all gynecological surgeries.

The direct healthcare cost per patient in the US was estimated to range from $2801 to $12 644 in the early 2000s [6−8]. In Italy, more than 65% of patients with endometriosis had endometriosis-related surgical procedures, including hysterectomy, within 1 year of firm diagnosis [9]. In Belgium, the average non-healthcare costs associated with endometriosis incurred during the 6 months prior to and following surgical treatment are 1514 and 2496 Euros, respectively [10]. In Canada, the estimated mean annual societal cost of endometriosis was $5200 per patient (95% confidence interval: $3700−7100), 78% of which is due to the lost productivity and lost leisure time [11]. Thus, the economical burden of endometriosis to society, due to either healthcare costs or loss of productivity, is enormous because of its high prevalence, costly treatment, and debilitating nature [4,5,12]. Evidently, endometriosis poses a serious public health problem worldwide.

Various theories on the pathogenesis of endometriosis have been proposed, but none has been unequivocally proven [13]. These theories can be grouped into roughly three themes: in situ development (such as coelomic metaplasia or embryonic cell rests), implantation, or a combination of in situ development and implantation. The implantation theory of Sampson [14], or the retrograde menstruation theory, is the most widely accepted. This theory stipulates that viable endometrial cells regurgitate through the fallopian tubes during menstruation to implant and grow in peritoneum or other ectopic sites. Indeed, retrograde menstruation is reported to occur in over 95% of women of reproductive age with patent fallopian tubes [15]. However, far less women are actually inflicted with endometriosis. Thus, why there is such discrepancy remains unresolved.

22.1.1 Diagnosis and Classification

A definite diagnosis of endometriosis is based principally on direct visualization through laparoscopy and appropriate biopsies in conjunction with a thorough medical history, although ultrasound and magnetic resonance imaging may also be useful [16]. While tremendous effects have been devoted to the search for non-invasive diagnostic procedures such as serum biomarkers, so far no single biomarker or group of biomarkers have been proven to be unequivocally useful clinically [17].

Endometriosis is staged by the classification system of the revised American Fertility Society (rAFS) [18]. Yet the staging system does not correlate well with either the severity of pain or the extent of infertility, nor does it correlate well with the prognosis [19]. Therefore, the development of a better classification system is currently an active research area [20].

It has been generally regarded that endometriosis has at least three different subtypes, i.e. ovarian endometriomas, peritoneal endometriosis, and adenomyotic nodules of the recto-vaginal septum [21]. This view has been supported by different gene expression patterns between ovarian and peritoneal endometriosis based on large-scale gene expression profiling studies [22].

22.1.2 Treatment

In treating women with endometriosis, the efficacy has been measured by means of assessment of pains and/or pregnancy rate [23]. The current treatment modalities include medical, surgical, or a combination of both, with surgery being the treatment of choice. However, the recurrence risk after surgery is high: 7−30% of patients reported recurrences 3 years after laparoscopic surgery [24]. The risk increases to 40−50% 5 years after surgery [25,26]. Since repeated surgeries are positively associated with increased morbidity and healthcare costs and, in endometriosis, with damage to ovarian reserve [27−31], the risk for reoperation poses a serious challenge to the effective management of endometriosis. Therefore, non-surgical medical therapy, preferably with high safety and cost profiles, is sorely needed.

Non-surgical medical therapy is also used as a first-line therapy for treating endometriosis, and may be used in conjunction with those patients who undergo surgical therapy for pain. The current medical treatment for endometriosis has so far focused on the hormonal alteration of the menstrual cycle to produce a pseudo-pregnancy, pseudo-menopause, or chronic anovulation, creating an acyclic, hypoestrogenic environment [23]. This is achieved either by blocking ovarian estrogen production (GnRH agonists, GnRH-a), by inducing pseudo-pregnancy (progestins), or by locally inhibiting estrogenic stimulation of the ectopic endometrium (progestins, androgenic progestins) [16,23,32]. While all hormonal treatments are more or less equally effective in relieving pains [33], the relief, however, appears to be relatively short term [34]. Given the lack of long-term efficacious medical therapy for endometriosis-associated pelvic pains and for minimizing recurrence risk, and the lack of efficacious medical therapy for endometriosis-associated subfertility, there is a clear and pressing need for novel medical therapies with more tolerable side effects and cost profiles [35].

22.1.3 Unmet Medical Needs

In response to this unmet need, numerous encouraging preclinical studies of a vast array of potential therapeutics for endometriosis have been reported in the last two decades. A handful of these have undergone phase II/III clinical trials. Unfortunately, most of these completed trials were found to be unpublished [36]. For those trials that have been published, the efficacy turns out to be much less impressive than that found in preclinical studies [35]. Thus, there seems to be a bewildering lost in translation in the effort to turn discoveries in basic research in endometriosis into better patient care. In fact, there is a palpable disappointment over the drug research and development (R&D) in endometriosis: Vercellini and co-workers recently likened the process to the "waiting for Godot" [37].

All these unmet medical needs, i.e. a classification system that can better relate symptomology and/or prognosis, the development of better, more efficacious therapeutics, better non-invasive diagnostic procedures, and possible prevention, stem from the fact that our current understanding of the molecular mechanisms underlying endometriosis pathogenesis is woefully inadequate. In this paper, which is an updated version of the previous one [38], I shall review our current knowledge of the epigenetic aberrations in endometriosis, discuss their implications for delineating the molecular mechanisms, clinical diagnosis, therapeutics, and intervention. Aside from genes with known aberrant methylation, the author shall restrict my focus on proteins/enzymes known to be involved in DNA methylation and histone modifications — sort of "writers" and "erasers" of epigenetic codes or marks — in endometriosis. This author shall not review, however, work on "reader/effector" of histone modifications since, as of writing, no work has been published in the field of endometriosis per se. For the same reason, this author will confine myself on DNA methylation, histone acetylation/deacetylation, histone methylation/demethylation, since no work on other types of histone modifications has been published in this field as of now. For ease of exposition and also for coherence, the author will leave out the part on microRNA and its response to steroid hormones in endometriosis, which could be rightfully subsumed into the realm of endometriosis epigenetics and has received a great deal of attention recently. The interested readers should consult, for example, Toloubeydokhti et al. [39]. At the end of this review, areas in need of future research will be exposed.

22.2 METHODS

A systematic and comprehensive search of PUBMED was performed for all studies published up to September 30, 2011, using the following search terms: "endometriosis", "epigenetics", "histone", "methylation", "histone acetylation", "histone phosphorylation", "histone ubiquitylation", "histone sumoylation", "post-translational modifications", or a combination of these. The studies had to report epigenetic aberrations in endometriosis. The search was limited to publications written in English. Evidence for or against endometriosis epigenetics was presented, and its therapeutical, diagnostic, and prognostic implications were discussed.

22.3 ALL ROADS LEAD TO EPIGENETICS

Endometriosis has been regarded as an ultimate hormonal disease, owing much to its estrogen-dependency and aberrations in estrogen production and metabolism [40–42]. It also has been viewed as an immunological disease due to a myriad immunological aberrations in endometriosis [43,44]. In addition, it has been thought of a disease caused by exposure to environmental pollution and toxins [45,46] although so far there are no solid human data [47]. Finally, it has been regarded as a genetic disease [48,49], ostensibly due to its reported familial aggregation. Yet even the reported familial aggregation, when examined closely, may be debatable [50]. Incidentally, beyond reported associations with various polymorphisms, there has been little headway made so far into the identification of genetic variants that predispose women to endometriosis [50–52].

Endometriosis is undoubtedly a hormonal disease and certainly entails an array of immunological aberrations. While so far there is no solid evidence linking dioxin exposure to endometriosis, it may still be plausible that dioxin exposure, at the right time and dosage, might precipitate the initiation or progression of endometriosis through interaction with estrogen receptors [53] or suppressing expression of progesterone receptors [54]. So what is the common denominator for a disease that is hormonal, immunological, and possibly environmental and genetic?

In the last decade, numerous large-scale gene expression profiling studies have demonstrated, unequivocally, that many genes are deregulated in endometriosis [22,55–71]. It also has been shown that a single focus of endometriotic lesion originates from a single progenitor cell [72], forming a cellular lineage. During their development from single progenitor cells to endometriotic lesions leading to various symptoms, endometriotic cells presumably need to make a series of sequential, perhaps dichotomous, and irrevocable cell fate choices. These choices are likely to be made without any change in DNA sequences. This cellular lineage, or identity, inevitably requires that cells transcribe, or enable transcription of, specific sets of genes while at the same time repressing others. To maintain cellular identity, the gene expression program must be iterated through cell divisions in a heritable fashion by epigenetic processes.

Indeed, transcription is regulated, in part, by the assembly of a plethora of complexes of transcription factors on regulatory regions of genes, and can be regulated at various levels: DNA modifications (both chemical and structural), post-transcriptional modifications, and post-translational modifications. These involve chemical modification of DNA (methylation), histone modification, and various machineries, such as specific factors, repressors, activators, general transcriptional factors, enhancers, microRNAs (miRNAs) [73,74], and recently discovered, double-stranded, non-coding RNAs (ncRNAs) [75]. These levels are either part of the epigenetic regulation (DNA methylation, histone modifications, miRNA) or closely related. After the DNA is transcribed and mRNA formed, there are extra levels of regulation on how much the mRNA is translated into proteins. Post-translational modifications of protein products, localization and higher-order interactions with other transcription factors, coactivators or corepressors are one set of mechanisms through which transcription can be controlled at another level.

In light of these, epigenetics is very likely to be involved in maintaining cellular identity in ectopic endometrial cells. This is the view that was first expressed in 2005 by Wu et al. that "endometriosis, like neoplasia, may also be an epigenetic disease " [76], after realizing that epigenetic aberration, as a more general biological phenomenon and possibly one major mechanism for gene deregulation, should not be exclusively restricted in cancers or developmental diseases.

22.4 EVIDENCE IN SUPPORT THAT ENDOMETRIOSIS IS AN EPIGENETIC DISEASE

Evidence that supports the theory that endometriosis is an epigenetic disease is discussed below.

22.4.1 HOXA10 Hypermethylation

The very first piece of evidence suggesting that endometriosis may be an epigenetic disease came from a study showing that the putative promoter of HOXA10 in endometrium from women with endometriosis is hypermethylated as compared with that from women without endometriosis [76]. HOXA10 is a member of a family of homeobox genes that serve as transcription factors during development and has been shown to be important for uterine function. It is expressed in human endometrium, and its expression is dramatically increased during the midsecretory phase of the menstrual cycle, corresponding to the time of implantation and increase in circulating progesterone [77]. This suggests that HOXA10 may have an important function in regulating endometrial development during the menstrual cycle and in establishing conditions necessary for implantation [78].

In endometrium of women with endometriosis, however, HOXA10 gene expression is significantly reduced, indicating some defects in uterine receptivity [79,80], which may be responsible for reduced fertility in women with endometriosis. As promoter hypermethylation is generally associated with gene silencing, the observed HOXA10 promoter hypermethylation provides a plausible explanation as to why HOXA10 gene expression is reduced in endometrium of women with endometriosis [76]. In infertile women with minimal endometriosis, it was recently reported that HOXA10 expression in eutopic endometrium is reduced, and bisulfite sequencing confirmed that the HOXA10 promoter is hypermethylated [81].

The HOXA10 promoter hypermethylation also has been demonstrated in a baboon model of endometriosis, which coincides with reduced HOXA10 expression [82]. What is interesting in the baboon study is the time course of HOXA10 expression levels, which was reduced progressively after induction of endometriosis but only became significant 1 year after the induction [82]. In mouse, surgical induction of endometriosis also resulted in the down-regulation of Hoxa10 as well as hypermethylation [83]. Besides serving as a validation of the human observation, these two experimental studies also challenge the view that endometriosis may originate from eutopic endometrium that harbor certain, yet to be identified, molecular aberrations through retrograde menstruation. What is puzzling and remains unanswered is just how endometriotic lesions situated in the peritoneal cavity apparently result in molecular genetic changes in eutopic endometrium.

Hoxa10 hypermethylation, accompanied by overexpression of Dnmt1 and Dnmt3b, also has been reported recently in female mice prenatally exposed to diethylstilbestrol (DES) [84]. Interestingly, prenatal exposure to DES also results in Hoxa10 down-regulation in gubernaculum and inhibition of transabdominal testicular descent in male rat fetuses [85]. This aberrant methylation seems to be a novel mechanism of altered developmental programming induced by in utero DES exposure.

22.4.2 PR-B Hypermethylation

The second piece of evidence came from the study demonstrating that the promoter of PR-B is hypermethylated in endometriosis [86]. In addition, the PR-B promoter hypermethylation is concomitant with reduced PR-B gene expression, providing support for the role of epigenetic aberration in PR-B down-regulation. It is well-known that there is a general tendency of progesterone resistance in endometriosis [1]. It is also known that PR-B is down-regulated in endometriosis [87] and may be responsible for, at least in part,

447

progesterone resistance since progesterone is mediated through its receptors, including PR-B. Yet why there is a persistent PR-B down-regulation was a mystery. PR-B promoter hypermethylation provides a simple yet biologically plausible explanation as to why PR-B is persistently down-regulated in endometriosis. In addition, promoter hypermethylation as a cause for PR-B down-regulation appears to satisfy the Occam's razor principle, that is, the explanation of any phenomenon should make as few assumptions as possible, eliminating those that make no difference in the observable predictions of the explanatory hypothesis or theory.

22.4.3 Aberrant Expression of DNMT1, DNMT3A, and DNMT3B

Perhaps the most important piece of evidence showing that endometriosis is an epigenetic disease comes from a study demonstrating that DNMT1, DNMT3A, and DNMT3B, the three genes coding for DNA methyltransferases that are involved in genomic DNA methylation, are all overexpressed in endometriosis [88]. However, another study reports that DNMT1 and DNMT3B protein expression in ectopic endometrium was significantly lower than that in control endometrium [89]. The discrepancy is likely due to the use of different materials: the former study used endometriotic epithelial cells harvested through laser capture microdissection while the latter used tissue culture, which consists of several, mixed cell types. Regardless, the two studies both suggest aberrant expression of DNMTs in endometriotic tissues. Since these genes are involved in de novo as well as maintenance methylation, their aberrant expression suggests that aberrant methylation may be widespread in endometriosis. As methylation is closely linked with chromatin remodeling, the aberrant expression of these genes may also signal that there are aberrant epigenetic changes, other than DNA methylation, in endometriosis.

22.4.4 SF-1 and ERβ Hypomethylation

Consistent with the view that aberrant methylation may be widespread in endometriosis, several very recent studies provide further evidence for epigenetic changes in endometriosis. Steroidogenic factor-1 (SF-1), a transcriptional factor essential for activation of multiple steroidogenic genes for estrogen biosynthesis, is usually undetectable in normal endometrial stromal cells but is aberrantly expressed in endometriotic stromal cells. Xue et al. show that SF-1 promoter has increased methylation in endometrial cells yet in endometriotic cells it is hypomethylated [90]. They also find that ERβ promoter is hypomethylated in endometriotic cells, which accounts for its overexpression [91].

22.4.5 Epigenetic Aberrations in Other Genes

Izawa et al. show that the treatment of endometrial stromal cells, which normally do not express aromatase, with a demethylation agent (DMA), 5-aza-deoxycytidine, dramatically increased the aromatase mRNA expression [92]. Further study by the same group found that a stretch of CpG demethylation within a non-promoter CpG island of the aromatase gene in endometriotic cells while the same region is heavily methylated and associated with methyl-CpG-binding proteins in endometrial cells [93]. Thus, the increased expression of the aromatase gene in endometriosis is likely attributable to the epigenetic disorder associated with aberrant DNA hypomethylation in a non-promoter CpG island [93].

Endometriotic cells are found to lack the intercellular adhesion protein E-cadherin, a known metastasis-suppressor protein in epithelial tumor cells whose deregulation also seems to be associated with invasiveness of endometriotic cells [94,95]. In two immortalized endometriotic cell lines, E-cadherin was found to be hypermethylated, and the treatment with trichostatin A (TSA) resulted in its reactivated expression with concomitant attenuated invasion [96]. This seems to suggest that, at least in endometriotic cell lines, E-cadherin silenced by methylation is associated with invasiveness.

One in silico study based on large-scale gene expression profiling of paired ectopic and eutopic endometrium also suggests a theme of post-translational modification and histone deacetylation [97], again supporting the role of epigenetics in endometriosis. Consistent with this view, a recent study based on whole-genome scanning of methylation status in 25 500 promoters compared endometriotic cells in three subtypes of endometriosis (superficial endometriosis or SUP, ovarian endometriomas or OMA, and deep infiltrating endometriosis or DIE) with their respective eutopic endometrium, and report that, with a pre-determined threshold of 1.5 or 0.66, there are 153 (14), 29 (53), and 19 (20) hypermethylated (hypomethylated) promoters in SUP, OMA, and DIE, respectively, as compared with eutopic endometrium [98]. There are 11 hypermethylated and nine hypomethylated chromosomal regions common to all three subtypes of endometriosis. Hypermethylated regions appear to be located at the ends of chromosomes, while hypomethylated regions are found to be randomly distributed along the chromosomes [98]. While this high-thoughput technology can identify many aberrant methylations in a single study, caution should be made. First, not all aberrant methylations are associated with aberrant gene expression, as evidenced by the poor correlation coefficient between the gene expression induction ratio and methylation ratio ($r = 0.03$, $p = 0.90$, based on data in Table 1 in [98]). In fact, among 20 transcription factors for which expression data were also available, the agreement between expression levels (in direction and in terms of statistical significance) and methylation patterns is merely 20% (or four out of 20). Second, while the use of paired eutopic and ectopic endometrium can effectively minimize the between-individual variation and reveal difference between the two tissues, the study design cannot detect aberrant methylations that are *shared* by the two tissues.

Compared with the work on DNA methylation, there has been scanty report on aberrant histone modification in endometriosis. Kawano et al. recently reported decreased acetylated histone H3 and H4 in endometriotic stromal cells as compared with normal endometrial stromal cells [99]. In our lab, we found that immunoreactivity to lysine-specific demethylase 1 (LSD1, or KDM1A), which can demethylate mono- and di-methylated lysines, specifically histone 3, lysines 4 and 9 (H3K4 and H3K9), is elevated in endometriotic lesions as compared with normal endometrium [Ding et al., unpublished data]. In addition, we also found that SIRT1, a class III histone deacetylase, and EZH2, a histone methyltransferase that methylates H3K27, have increased immunoreactivity in endometriotic lesions as compared with normal endometrium [Ding et al., unpublished data].

Table 22.1 provides a complete list of epigenetic aberrations in endometriosis identified so far.

22.5 HISTONE MODIFICATIONS IN ENDOMETRIOSIS: AN UNEXPLORED FRONTIER

Histones can undergo various kinds of modifications which alter their interaction with DNA and nuclear proteins. The H3 and H4 histones, in particular, have long tails protruding from the nucleosome and can be covalently modified post-translationally at various residuals and in various ways. Histone modifications, primarily on the N-terminal tail, include acetylation, methylation, phosphorylation, ubiquitination, sumoylation, citrullination, and ADP-ribosylation. The core of the histones H2A and H3 can also be modified. Thus, histone modifications can take place in different histones (e.g. H3 or H4), histone variants (e.g. H3.3), and histone residuals (e.g. argine, lysine, and serine). The modification can involve different chemical groups (e.g. acetyl, methyl, and phosphate), and, for methylation, there can be different degrees (e.g. mono-, di-, and tri-methylation for lysines, mono-, symmetric and assymetric di-methylation for argines). These kaleidoscopic combinations of histone modifications influence the interaction of histones with DNA and nuclear proteins, and act concertedly in gene regulation. Thus, the combinations of modifications are proposed to constitute a code for gene expression, the so-called "histone code" [100,101].

TABLE 22.1 Epigenetic Aberrations in Endometriosis that have been Identified so far.
The Finding Reported in [98] is not Included.

Year of the First Report	Gene Name	Major Finding	Reference
2005	HOXA10	Hypermethylated in eutopic endometrium	[76,81,82,98]
2006	PR-B	Hypermethylated in ectopic endometrium	[86]
2007	Aromatase	Endometriotic cells secreted more aromatase than endometrial cells with added testosterone, yet when treated with a demethylation agent, endometrial cells increased the secretion	[92, 93]
2007	ERβ	Hypomethylated in ectopic endometrium	[91]
2007	SF-1	Hypomethylated in ectopic endometrium	[90]
2007	E-cadherin	Methylated and inactivated in an endometriotic epithelial-like cell line, and can be demethylated and reactivated by the treatment of TSA	[96]
2007	DNMT1, DNMT3A, DNMT3B	Overexpressed in ectopic endometriotic epithelial cells	[88]
2010	SRC-1	Reduced immunoreactivity in endometriotic epithelium	[103]
2010	HDAC2	Elevated immunoreactivity in ectopic endometrium and dorsal root ganglia in rats with induced endometriosis	[106]
2011	Acetylated H3 and H4	Decreased acetylated H3 and H4 in endometriotic stromal cells as compared with normal endometrial stromal cells	[99]
2011	DNMT1, DNMT3B, MBD1, MBD2	Reduced expression in ectopic endometrial tissues	[89]
2011	SRC-1	Elevated immunoreactivity in ovarian endometriosis	[104]

The investigation of aberrant histone modifications in endometriosis has been scarce. For ease of exposition, we shall review published work, albeit few, in "writers" and "erasers" of histone acetylation and methylation.

22.5.1 "Writers" and "Erasers" of Histone Modifications

One can view histone modifications of various types as some kind of "marking", and, as such, the enzymes that make the "marking" happen can be viewed as "writers", while those making the "marks" removed as "erasers".

HISTONE ACETYLTRANSFERASES

Histone acetyltransferases (HATs) are enzymes that acetylate conserved lysine residuals on histone proteins by transferring an acetyl group from acetyl CoA to form ε-N-acetyl lysine. This modification neutralizes the positive charge of lysine and may thus disrupt the interaction between DNA and histone tails. Acetylated histones are generally associated with euchromatin and transcriptional activation. In contrast to histone acetylation, deacetylation restricts DNA accessibility through revealing the positive charge of lysine, permitting interaction between DNA and the histone tail and thus chromatin compaction.

There are two general categories of HATs: type A and type B. Type A HATs are nuclear and acetylate nucleosomal histones and other chromatin-associated proteins, and type B HATs are located in the cytoplasm, acetylate newly synthesized histones, and have no direct impact on transcription [102]. Type A HATs can be further categorized into five families: nuclear receptor coactivators and general transcription factors, p300/CBP (CREB (cAMP-response element binding protein)-binding protein), GNAT (GCN5 (general control of nuclear-5)-related N-actyl transferases), MYST (MOZ (monocytic leukemia zinc-finger protein), YBF2 (yeast binding factor 2)/SAS3 (something about silencing 3), and SAS2, TIP60 (Tat interactive protein-60).

In endometriosis, several HATs are investigated in the context of steroid receptor coactivators, since many such coactivators are HATs. Suzuki et al. report that SRC-1 immunoreactivity is reduced [103]. Yet in ovarian endometriosis, Kumagami et al. report that SRC-1 colocalizes with ERα and may thus affect the transcriptional activity of ERα [104].

HDACS

HDACs are a class of enzymes that remove acetyl groups from an ε-N-acetyl lysine amino acid on a histone, which are opposite to the action of HATs. Depending on sequence identity and domain organization, HDACs can be classified into four groups: class I consists of HDAC1, 2, 3, and 8; class II, HDAC4, 5, 7A, 9, and 10; class III or sirtuins, consists of SIRT1-7; class IV has one member, HDAC11. Classes I and II are considered "classical" HDACs whose activities can be inhibited by trichostatin A (TSA).

So far, no report on HDACs activity in endometriotic lesions has been published. One recent study, however, found that an epithelial-like endometriotic cell line expresses class I HDAC1, HDAC2, HDAC3, and HADC6, and class II HDAC4 and HDAC5 [105]. Given this result, and in view of many encouraging results of HDAC inhibitors (HDACIs) on endometroitic cells (detailed below), it is very likely that some HDACs may be aberrantly overexpressed in endometriosis. In fact, our study on a rat model of endometriosis indicates that HDAC2 is aberrantly expressed in ectopic endometrium [106]. We also found that SIRT1, a class III HDAC, has increased immunoreactivity in endometriotic lesions as compared with normal endometrium [Ding et al., unpublished data]. We also found that class I HDAC1, HDAC2, and HDAC2 immunoreactivity is elevated in adenomyosis as compared with normal endometrium [212].

HISTONE METHYLTRANSFERASES

Histone methylation occurs on lysines (Ks) and arginines (Rs). Hisone lysines can be methylated in different forms: mono- (me1), di- (me2), or trimethylated (me3). Histone arginine methylation can be monomethylated, symmetrically or asymmetrically dimethylated. While histone methylation has been known since the early 1960s, it was generally thought that histone methylation, unlike acetylation and phorsphorylation, was biochemically stable and irreversible. Yet the identification of the first histone methyltransferases (HMT) in 2000 [107], and especially the identification of the first HDM in 2004 [108], challenged the notion that histone methylation is a permanent, irreversible mark. With more HMTs and HDMs being discovered, it is now held that histone methylation can be dynamic [109]. A list of known site-specific lysine and arginine HMTs is shown in Table 22.2.

Depending on which residual to catalyze on, there are two types of HMTs: histone lysine N-methyltransferase (KMT) and histone arginine N-methyltransferase. HMTs catalyze the transfer of one to three methyl groups from S-adenosyl methionine (SAM) to lysine and arginine

TABLE 22.2 List of Known Site-Specific Lysine and Arginine HMTs
Adapted from Table 1 in Nimura et al. [210] with modifications

Histone Residual	Methylation	Chromatin Status	HMT Names
H3K4	me1	Active	MLL, SET1A/B, ASH1L, ASH2L,
	me2	Active	Smyd1, Smyd3, SET7/9
	me3	Active	
H3K9	me1	Active	Suv39hl, Suv39h2, G9a,GLP,ESET,
	me2	Inactive	Riz1
	me3	Inactive	
H3K27	me1	Active	EZH2
	me2	Inactive	
	me3	Inactive	
H3K36	me1	Active	NSD1, NSD2/WHSC1, Smyd2,
	me2	No	HYPB/SETD2, NSD3/WHSCl1
	me3	Active	
H3K79	me1	No	DOT1L
	me2	No	
	me3	Inactive	
H4K20	me1	Active	PR-Set7, Suv4-20H1/2, NSD1,
	me2	Unknown	NSD2/WHSC1
	me3	Inactive	
H3R2	me1	Unknown	PRMT4/CARM1, PRMT6
	me2(as)	No	
	me2(s)	No	
H4R3	me1	Unknown	PRMT1, PRMT5
	me2(as)	Unknown	
	me2(s)	No	
H3R17		Unknown	PRMT4/CARM1
H3R26			

As, asymmetrical; s, symmetrical; Active, correlation of actively transcribed region; Inactive, correlation of inactively transcribed region; No, not significant correlation of transcription.

residuals. With few exceptions, HMTs contain a conserved SET (Su(var)3-9, Enhancer of Zeste, Trithorax) domain that methylates specific residues of histone as well as non-histone proteins.

As of the time of writing, there has been no published study on aberrant expression of any HMT in endometriosis. In our lab, we found that immunoreactivity to EZH2, an HMT that methylates H3K27, is increased in endometriotic lesions as compared with normal endometrium [Ding et al., unpublished data].

HISTONE DEMETHYLASES

Similar to HDACs, histone demethylases (HDMs) are "eraser" enzymes that site-specifically remove the methyl group(s) from histone lysine residuals. Two classes of HDMs have been identified thus far. One is the KDM1 family HDMs that are flavin-adenine dinucleotide (FAD)-dependent amine oxidases, which can act only on mono- and dimethylated lysines. The other is characterized by the Jumonji C (JmjC) domain. HDMs in the latter class are Fe(II) and 2-oxoglutarate-dependent enzymes, and, depending on sequence homology and the overall architecture of associated motifs, can be further classified into different subgroups (Table 22.3). The KDM1 family has two members, KDM1A (LSD1) and KDM1B (LSD2), while the JmjC domain-containing KDMs have over 30 members identified so far, each with exclusive histone- and residue-specific demthylating capabilities. Table 22.3 gives a list of all current KDM families in mammals.

Histone acetylation, methylation, and phosphorylation are the most investigated histone modifications. In general, histone acetylation is associated with transcriptional activation while

TABLE 22.3 The KDM Families in Mammals.
The Family Names are Arranged According to the Phylogenetic Relationship of DNA Sequences Displayed in Figure 2 of Pedersen et al. [211]

Family	Members	Synonyms	Specificity	Remarks
KDM1	KDM1A	LSD1/AOF2	H3K4me1/me2, H3K9me1/me2	First discovered KDM
	KDM1B	LSD2/AOF1	H3K4me1/me2	
KDM2	KDM2A	JHDM1A/FBXL11	H3K36me1/me2	
	KDM2B	JHDM1B/FBXL10	H3K36me1/me2, H3K4me3	
KDM3	KDM3A	JHDM2A/JMJD1A	H3K9me1/me2	
	KDM3B			
KDM4	KDM4A	JHDM3A/JMJD2A		
	KDM4B	JHDM3B/JMJD2B	H3K36me2/me3	
	KDM4C	JHDM3C/JMJD2C	H3K9me2/me3	
	KDM4D	JHDM3D/JMJD2D		
KDM5	KDM5A	JARID1A/RBP2	H3K4me2/me3	
	KDM5B	JARID1B/PLU1		
	KDM5C	JARID1C/SMCX		
	KDM5D	JARID1D/SMCY		
KDM6	KDM6A	UTX	H3K27me2/me3	
	KDM6B	JMJD3		
	HIF1AN			
	HSPBAP1			
	JMJD4			
	JMJD5			
	JMJD6	PSR/PTDSR	H3R2, H4R3	
	JMJD7			
	JMJD8			
	JHDM1D	KIAA1718	H3K9me1/me2, H3K27me1/me2	
	PHF2	JHDM1E		
	PHF8	JHDM1F	H3K9me1/me2	
	HR			Phylogenetically related with KDM3
	JMJD1C			
	UTY			Phylogenetically related with KDM6
	JARID2			
	MINA			
	NO66		H3K4me2/me3, H3K36me2/me3	

deacetylation is associated with transcriptional repression [100,110]. In contrast, the effect of histone methylation depends on the histone residuals, their positions, and degrees [10,110].

As at time of writing, there has been no published study on aberrant expression of any HDMs in endometriosis. In our lab, we found that immunoreactivity to LSD1 (KDM1A) that demethylates H3K4me1/me2 and H3K9me1/me2, is increased in endometriotic lesions as compared with normal endometrium [Ding et al., unpublished data].

22.5.2 "Reader/Effector" Modules

Histone modifications are recognized by "reader/effector" modules, which read and interpret modification codes or marks and then execute conformational changes in chromatins and provide signals to regulate chromatin dynamics. These modules include **P**lant **H**omeo **D**omain (PHD), chromo (for lysine methylation), bromo (for lysine acetylation), tudor, proline-tryptophan-tryptophan-proline (PWWP), **SWI**3p-**R**sc8p-**M**oira (SWIRM), SWI3-ADA2-N-CoR-TFIIIB (SANT), and **M**alignant **B**rain **T**umor (MBT) domains. By recruiting these reader/effector proteins, histone modifications lead to changes in chromatin structure as well as dynamics [111]. As at time of writing, there has been no published account on aberration of any "reader/effector" modules in histone modifications.

22.6 EPIGENETIC ABERRATION: CAUSE OR CONSEQUENCE?

Given the reported epigenetic aberrations in endometriosis, one question is whether these aberrations are the cause or merely the consequence of endometriosis. Since most, if not all, human studies reporting epigenetic aberrations in endometriosis are carried out cross-sectionally, the reported aberration may be a cause for, but also could be a consequence of, endometriosis. In a linearly causal relationship, the cause and consequence can be clearly defined, with temporal sequences, and necessary and sufficient cause distinguished. In a complex system, such as endometriosis which appears to be a system-wide disease [112,113], in which there are usually many interconnected parts, linearly causal relationship may be rare in the first place. In many ways, a complex transcription network often has a highly optimized tolerance featuring high efficiency, performance, and robustness to designed-for-uncertainties yet hypersensitive to design flaws and unanticipated perturbations [114]. In such a system, the demarcation of cause and consequence could be difficult since the removal of one part may affect other parts of the system, especially when the system is redundant. Such complex systems often display emergent properties. Therefore, it may be difficult to prove that in endometriosis aberrant methylation is a cause rather than a consequence.

Despite this challenge, it is known that methylation can be induced by various factors. Aging [115−117], diet [118], chronic inflammation [119,120], prolonged transcriptional suppression [121,122], maternal care [123], and prolonged use of intrauterine devices [124]. In endometriosis, it has been shown that prolonged stimulation of an endometriotic epithelial-like cell line by TNFα, which has been shown to have increased production in endometriosis, resulted in at least partial methylation in the PR-B promoter [125]. This provides evidence that certain phenotypic changes in endometriosis, such as increased production of proinflammatory cytokines, may also cause epigenetic aberrations, which in turn result in changes in gene expression and subsequently other phenotypic changes such as increased cellular proliferation [126] and perhaps some phenotypic changes.

In a baboon model of endometriosis, it is reported that the induction of endometriosis resulted in progressively decreased expression of HOXA10 and subsequent alteration in gene expression of its downstream target genes in *eutopic* endometrium [82]. More remarkably, the decreased HOXA10 expression is accompanied by the promoter hypermethylation [82]. This finding has also been replicated in a mouse model of endometriosis [83]. In this case, the aberrant methylation at HOXA10 promoter is apparently a consequence of endometriosis.

Remarkably, developmental exposures to chemicals can also result in aberrant methylation. Mice neonataly exposed to diethylstilbestrol (DES) are reported to have demethylation of estrogen-responsive gene lactoferrin in their uteri, along with uterine tumor [127,128]. Neonatal exposure to DES can also lead to the hypomethylation nucleosomal binding protein 1 (Nsbp1) in mice [129]. In mice exposed to DES in utero, Hoxa10 hyper-methylation has been reported very recently [84]. Nutritional factors and stress have also been reported to alter DNA methylation during early life [123,130−133]. While it is still unclear as to how much nutritional factors, stress, and exposure to certain chemicals in early life and thus aberrant epigenetic changes that they may cause contribute to the risk of endometriosis, it should be noted that the concept of "fetal origins of adult-onset diseases" is fairly new, and the human research in this area can be quite challenging for obvious reasons. Nevertheless, the developmental origins of many chronic diseases such as type 2 diabetes have now been demonstrated epidemiologically [134]. Incidentally, Missmer et al. reported that in utero exposure to DES nearly doubles the risk of developing endometriosis in women while low birthweight increases the risk by 30% [135]. Further research in this area is sorely needed, not just for the sake of understanding of endometriosis pathogenesis but also because proper nutritional intervention may reverse the aberrant epigenetic changes [136,137].

22.7 THERAPEUTIC IMPLICATIONS

Unlike DNA mutations or copy number changes, DNA methylation, histone and protein modifications are reversible. Hence, enzymes that regulate the epigenetic changes could be ideal targets for intervention by pharmacological means. Given the accumulating evidence that endometriosis may be an epigenetic disease, naturally one may wonder as to whether endometriosis can be treated by correcting epigenetic aberrations through pharmacological means. Indeed, encouraging in vitro and in vivo results on the use of HDAC inhibitors (HDACIs) as a potential therapeutics for endometriosis have been reported.

Treatment of an endometrial stromal cell line with trichostatin A (TSA) resulted in decreased proliferation [138]. Treatment with TSA or valproic acid (VPA) resulted in cell cycle arrest and induction of p21, a cell-cycle related gene [139]. The effect is likely through, perhaps in part, the up-regulation of PR-B by TSA [138], possibly through increased acetylation of histones in chromatins. These results have been replicated recently by Kawano et al. in primary endometriotic stromal cells using three different classes of HDACIs (VPA, suberoyl anilide bishydroxamine or SAHA, and apicidin) [99]. Kawano et al. also show that treatment with HDACIs induced expression of cell-cycle-related proteins such as p21, p16, p27, and chk2 as well as apoptosis-related proteins such as cleaved caspase 9 and Bcl-X_L, and also elevated acetylation levels in the promoter region of p21, p16, p27, and chk2 as well as acetylated H3 and H4 in endometriotic stromal cells [99]. What is remarkable is that, when it comes to inhibition of proliferation, endometriotic cells are more sensitive to treatment with HDACs than normal endometrial stromal cells [99,140], a fact that may be further exploited when considering dosing.

Treatment of TSA also inhibited IL-1β-induced COX-2 expression [141]. This is significant, since COX-2 overexpression has been observed in ectopic endometrium [142], found to correlate with endometriosis-associated pain [143,144], and reported to be a biomarker for recurrence [145]. TSA treatment up-regulated PPARγ expression in endometrial stromal cells [146]. PPARγ agonists have been reported to inhibit VEGF expression and angiogenesis in endometrial cells [147], inhibit TNF-induced IL-8 production in endometriotic cells [148], and repress ectopic implants in animal models of endometriosis [149–151]. TSA treatment can also attenuate constitutive and TNFα-induced NF-κB activation in endometriotic cells [140]. Since NF-κB plays pivotal roles in inflammation, proliferation, and angiogenesis [152] and is known to be constitutively activated in endometriosis [153,154], its attenuation by TSA strongly suggests that HDACIs may be a promising therapeutic for endometriosis.

In two endometriotic cell lines, TSA treatment resulted in attenuated invasion and reactivated E-cadherin expression [96]. This appears to suggest that some cellular phenotypes of endometriotic cells, such as invasiveness, may be mediated epigenetically and, as such, could be tamed by epigenetic reprogramming through pharmaceutical means.

In an endometriotic epithelial-like cell line, another HDACI, romidepsin, also known as FK-228 and depsipeptide that was originally isolated from a broth culture of *Chromobacterium violaceum*, has been shown to reduce HDAC activity, induce acetylation of H2A, H2B, H3, and H4, inhibit proliferation, and activate apoptosis through induction of p21, caspase 3, caspase 9, and PARP-1 as well as reduction of Cyclin B1 and Cyclin D1 [105]. Romidepsin also inhibits the transcription, expression, and secretion of VEGF, a known and major factor involved in angiogenesis in endometriosis [155].

In a preliminary study, TSA has been found to inhibit the expression of SLIT2 [Zhao et al., unpublished data], a member of the SLIT family of secretory glycoproteins that was recently found to attract vascular endothelial cells *in vitro* and promote tumor-induced angiogenesis [156], and, more recently, found to be a constituent biomarker for recurrence of endometriosis [157].

There are indications that show HDACIs may be analgesic when treating endometriosis. The first such indication comes from the report that three HDACIs, TSA, suberic bishydroxamate,

455

and VPA, suppress spontaneous and oxytoxin-induced uterine contractility [158]. It has been shown that women with endometriosis have aberrant uterine contractility during menses with increased frequency, amplitude, and basal pressure tone as compared with those without [159]. There is a sign that in the uterus of women with dysmenorrhea there is a lack of synchronization in fundal-cervical contraction [160]. Incidentally, progesterone, a traditional drug for treating endometriosis-associated dysmenorrhea, can also inhibit myometrial contraction [161].

Animal studies also show the protential of HDACIs in treating endometriosis. In mice with surgically induced endometriosis, treatment with TSA significantly reduced the average size of ectopic implants as compared with the controls [162]. This finding has been replicated in rats treated with VPA [163]. More remarkably, it was found that induced endometriosis resulted in hyperalgesia or "central sensitization", while TSA or VPA treatment significantly improved mice's or rats' perception of pain induced by noxious stimuli [162,163]. It should be added that the improvement in pain behavior in rats with induced endometriosis is endometriosis-specific, not due to the general analgesic property that VPA may have [163].

Taking advantage of an existing drug, VPA, that is an HDACI with known pharmacology, and the advantage that adenomyosis, once called endometriosis *interna*, can be diagnosed quite accurately by non-invasive imaging techniques and that adenomyosis shares with endometriosis many similarities, Liu and Guo tested VPA on three patients as a new therapeutic and found that it was well tolerated and, after 2 months of use, the pain symptoms were dramatically reduced [164]. In addition, the uterus size was reduced by an average of one third. Results from more patients show that VPA can effectively alleviate adenomyosis-associated pain and reduce uterus size [165]. These clinical observations corroborate well with the in vitro data that TSA treatment suppresses proliferation and cell cycle progression in ectopic endometrium in adenomyosis [166]. They are also consistent with the in vivo data that VPA treatment results in reduction in myometrial infiltration, uterine contractility, and contractile irregularity [167], along with alleviation of adenomyosis-associated pain [167,168].

Chronic administration of VPA has been shown to reduce brain N-methyl-D-asparate (NMDA) signaling in rats [169]. NMDA receptors (NMDARs), along with calcitonin gene-related peptide (CGRP), c-Fos, acid-sensing ion channel 3 (ASIC-3), are known to be expressed in sensory neurons in dorsal root ganglion (DRG) in the presence of central sensitization [170—173]. In particular NMDARs and CGRP are known to be synaptic triggers of central sensitization [174]. NGF and its high-affinity receptor, TrkA, are mediators of inflammatory pain [175]. NMDAR blockade and anti-NGF therapy have been shown to be effective in reducing central sensitization [170,176]. In rats with induced endometriosis, VPA treatment results in significantly decreased immunoreactivity of NMDAR1, c-Fos, ASIC3, TrkA, and CGRP in DRG, along with improved thermal latency [106], demonstrating that VPA, and perhaps other HDACIs as well, may be efficacious in reducing central sensitization induced by endometriosis and possibly in alleviating endometriosis-associated pain in humans.

On the surface, there may seem to be a mismatch between the reported aberrant methylation in endometriosis and the focus of current therapeutic approach, which has been so far confined to HDACIs. Yet this mismatch is merely a *trompe l'oeil*, since these exists a cross-talk between DNA methylation and histone modifications and they work in concert to control gene expression [177,178], although it is unclear as to whether DNA methylation or histone modification is the primary signal by which gene expression is determined. Hence the change in histone modification may result in change in DNA methylation, and vice versa. The inhibition of histone deacetylation can result in DNA demethylation, as evidenced by the demethylation of E-cadherin as a result of HDACI treatment [96].

22.7.1 Possible Mechanisms of Action in HDACIs as a Therapeutics

As alluded to above, quite extensive in vitro and in vivo studies have shown that certain HDACIs such as VPA and TSA can promote apoptosis, hinder cell cycle progression, inhibit proliferation, reduce inflammation and angiogenesis, and attenuate invasiveness in endometriosis. In other cell types, VPA has been shown to reduce basal and FSH-stimulated estrodiol secretion and FSH-induced aromatase activity in human ovarian follicular cells [179] and in forskolin-stimulated H295R cells [180]. Thus, VPA may potentially interfere with steroidogenesis in endometriosis.

TABLE 22.4 Summary of the Activities of HDACIs in Endometriotic Lesions and Other Cell/Tissues

Name of HDACI	Gene/Protein Name	Effect	Cell/Tissue Type	Reference
TSA	PR-B	↑	NESCL	[138]
TSA, VPA, SAHA, apicidin Romidepsin	p21	↑	NESCL, EESC, EECL	[99,105,139]
VPA, SAHA, apicidin	p16	↑	EESC	[99]
VPA, SAHA, apicidin	p27	↑	EESC	[99]
VPA, SAHA, apicidin	chk2	↑	EESC	[99]
Romidepsin	Cyclin B1	↓	EECL	[105]
Romidepsin	Cyclin D1	↓	EECL	[105]
Romidepsin	Caspase 3	↑	EECL	[105]
Romidepsin	Caspase 9	↑	EECL	[105]
Romidepsin	PARP-1	↑	EECL	[105]
VPA, SAHA, apicidin	Bcl-XL	↑	EESC	[99]
TSA	COX-2	↓	NESCL	[141]
TSA, VPA	NF-κB	↓	EECL, also in a rat model of endometriosis	[106,140]
TSA	PPARγ	↑	NESCL	[146]
TSA	E-cadherin	↑	EECL, ESCL	[96]
Romidepsin	VEGF	↓	EECL	[155]
Romidepsin	HIF-1α	↓	EECL	[105]
VPA	aromatase	↓	Forskolin-stimulated human adrenal carcinoma cell line (H295R)	
VPA	c-Fos	↓	DRG, in a rat model of endometriosis	[106]
VPA	CGRP	↓	DRG, in a rat model of endometriosis	[106]
VPA	TrkA	↓	DRG, in a rat model of endometriosis	[106]
VPA	ASIC3	↓	DRG, in a rat model of endometriosis	[106]
VPA	HDAC2	↓	DRG, in a rat model of endometriosis	[106]
VPA	NMDAR1	↓	DRG, in a rat model of endometriosis	[106]
TSA	TRPV1	↓	Eutopic endometrium, in a mouse model of endometriosis	[162]
TSA	PKCε	↓	Ectopic endometrium, in a mouse model of endometriosis	[162]
TSA	PGP9.5	↓	Vagina, in a mouse model of endometriosis	[162]
VPA	HDAC2	↓	Ectopic endometrium, in a rat model of endometriosis	[106]
VPA	TrkA	↓	Ectopic endometrium, in a rat model of endometriosis	[106]
VPA	CGRP	↓	Ectopic endometrium, in a rat model of endometriosis	[106]
VPA	OTR	↓	Primary myometrial smooth muscle cells	[Guo et al. unpublished data]

NESCL, normal endometrial stromal cell line; EESC, endometriotic stromal cells; EECL, epithelial-like endometriotic cell line; ESCL, stromal-like endometriotic cell line; DRG, dorsal root ganglia.

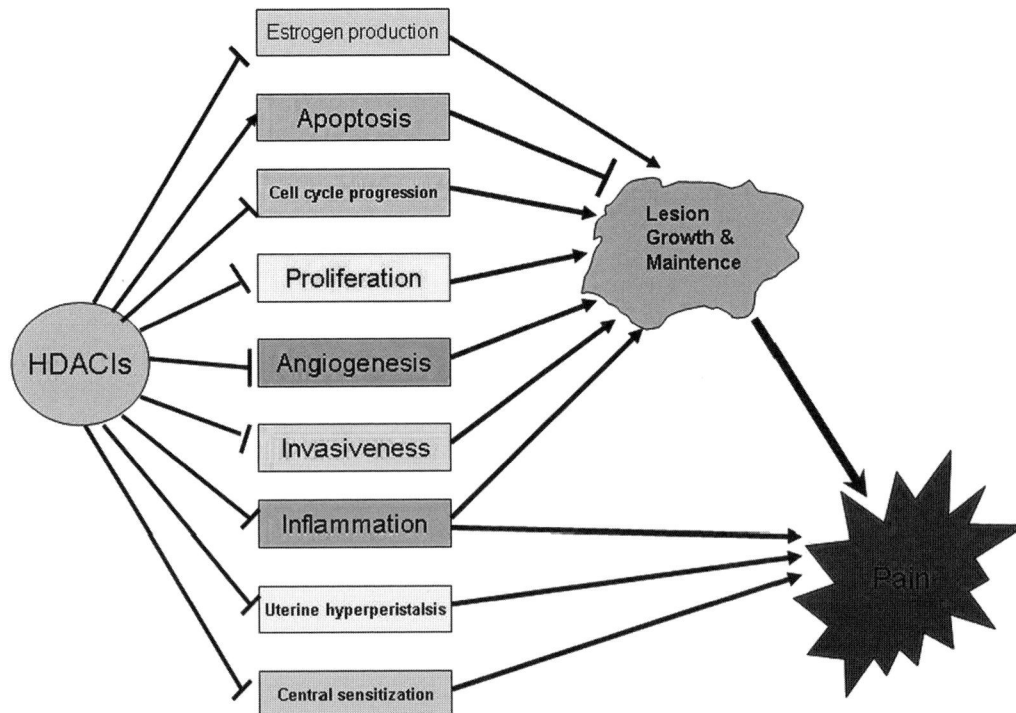

FIGURE 22.1
Schematic illustration of potential therapeutic effects of HDACIs in endometriosis. This figure is reproduced in the color plate section.

In human myometrial cells, it has been shown that long-term treatment resulted in repression of NF-κB DNA binding and inhibition of the expression of proinflammatory genes such as COX-2, IL-8, IL-6, and RANTES [181]. HDACIs have been reported to suppress TNFα-induced tissue factor expression [182] and also suppress the transcription, expression, and secretion of vascular endothelial growth factor (VEGF) in endometriotic cells [155] and other cell types [183]. Both tissue factor and VEGF (and its receptors) are known to be key players involved in angiogenesis in endometriosis [184–186]. Our unpublished data also indicate that VPA can inhibit the expression of oxytocin receptor (OTR) in primary myometrial smooth muscle cells [Guo et al., unpublished data]. The OTR expression has been found to be positively correlated with the amplitude of uterine contractility and also with the the severity of dysmenorrhea in women with adenomyosis [Guo et al., unpublished data].

Table 22.4 summarizes various effects of HDACIs on expression of some important genes that are known to be involved in endometriosis. Figure 22.1 depicts possible mechanisms of action as to how HDACIs such as VPA can be therapeutically valuable.

There is an indication that HDACIs appears to synergize with demethylation agents (DMAs), resulting in more potent antiproliferative effects than either used alone and more robust re-expression of methylation-silenced genes [166], as in cancer cells [187]. Clearly, future research should illuminate this further.

22.7.2 Potential Detrimental Effects of Epigenetic Therapies and Possible Ways to Circumvent Them

Since global hypomethylation is a notable feature of cancer and is reported to cause genomic instability [188,189], there may be a legitimate concern as whether the use of demethylation agents and/or HDACIs in treating endometriosis would increase the risk for cancer. After all, endometriosis is not a fatal disease even if left untreated, hence the demand for better safety and side effect profiles is higher than anticancer drugs [35].

Several studies have shown that only a small percentage (0.2–3%) of silenced genes are up-regulated by DMA treatment in cancer cells [190,191] and in normal fibroblast cell lines the number of genes affected is even lower (0.4%) [192]. Similarly, HDACIs also up-regulate a small subset of genes (0.4–2%) and are quite specific in their activation and repression of distinct genes [191,193]. While the percentage of affected genes is generally small, it is still possible that these affected genes may be important enough in causing unacceptable side effects, even though such data are lacking as of now. In addition, it has been shown that the withdrawal of methyltransferase inhibitors (such as DMAs) is followed by a rapid return of methylation [194], suggesting that achieving long-lasting epigenetic reprogramming may require continued drug treatment.

Even with these concerns, it should be noted that two DMAs, 5-azacitidine and 5-aza-2′-deoxycitidine (decitabine), appear to be well-tolerated, and no significant demethylation of repetitive elements or any indication of secondary malignancies was found [195]. In fact, chromosomal abnormalities were even found to be reversed in 31% of patients with myelo-dysplastic syndrome (MDS) who took decitabine [196]. Therefore, the two drugs have now been approved in the US for treating MDS and the only agents known to improve the natural history of MDS [197].

There is also a concern that VPA use may increase the risk of polycystic ovarian syndrome (PCOS) [198]. Yet from the very same study based on which the concern was raised, the authors actually stated explicitly that "[n]one of the tested AEDs influenced 3βHSDII or P450c17 activities *at concentrations normally used in AED therapy*" [199] (AEDs stands for antiepileptic drugs, including VPA; author's italics). As Paracelsus, considered to be the father of modern toxicology, said, *Sola dosis facit venenum* (only dose makes the poison). Hence the extrapolation of the observation under the high dosage to the situation of low dosage should be made with extreme caution.

Just as the dose, the duration of medication, perhaps to a lesser degree, can also make a difference. The two studies based on which the concern was raised actually examined women taking VPA for a period of time much longer than 3 months as we used. The women in one study [200] had taken VPA for ≥2 years while in the other [201] the average duration of taking VPA was 28 months. Even if VPA is proved to have an unfavorable risk to benefit ratio, it is still premature to throw the baby out with the bathwater, since VPA is just one of many HDACIs and some novel HDACIs may still hold the promise of being more efficacious while having less side effects.

Besides dose and treatment duration, one promising way to improve drug safety and to minimize side effects is to use the local route for administration, namely the drug-containing intrauterine system (IUS). There is an indication that levonorgestrel-releasing IUS (LNG-IUS) appears to be efficacious in treating endometriosis, adenomyosis, and their associated pain [102], but little work has been done in this area.

Regardless, the long-term safety of HDACIs and/or DMAs, when used to treat endometriosis, should be carefully evaluated even when they prove to be efficacious. Fortunately, such data may come from cancer clinical trials.

One final point is that many natural products contain weak HDACI activities. In particular, royal jelly, the material responsible for the making of honey bee queens, has weak HDACI activities [203], and may be responsible for the epigenetic changes leading to strikingly different phenotypes in the queen as compared with her workers who both share identical genomes [204]. The divergent developmental pathways of the queen and workers are associated with changes in subtle gene expression patterns in a particular group of genes encoding conserved physiometabolic proteins [204,205]. Thus, reduced methylation, or the change of epigenome, in young larvae can mimic the effects of royal jelly. This example demonstrates that HDACI activity, at least in weak form, does not necessarily cause detrimental health effect.

22.8 DIAGNOSTIC AND PROGNOSTIC IMPLICATIONS

Besides providing novel targets for drug therapy, epigenetic aberrations, once identified, may also provide promising prospects for diagnostic and/or prognostic purposes. One attractive approach is the identification of DNA methylation markers, which can be used for many specimens, such as menstrual blood.

Any biomarker, in order to be clinically useful, should ideally have high specificity and sensitivity. In addition, it should be easily detectable in specimens procured through a minimally invasive manner. DNA methylation biomarkers appear to fit the latter requirement quite well.

Since menstrual blood contains the same DNA (and thus methylation status) as that from endometrial cells, and since the endometrium from women with endometriosis is somewhat different from that of women without [113], menstrual blood could be a valuable, abundant, non-invasive, and convenient source for detection of methylation changes, as reported [206]. A recent preliminary study using menstrual blood provides the evidence that the frequency of ERβ hypermethylation in women with endometriosis is significantly lower than that in women without endometriosis [Shen et al. unpublished data]. This seems to echo the result by Xue et al. that ERβ is hypomethylated in endometriosis [91].

Of course, it is unclear as of now as to whether the DNA methylation markers based on menstrual blood are of any use for early diagnosis of endometriosis. It is also unclear as to whether they would be of value for the differential diagnosis of endometriosis, which could be more challenging. Much more work is warranted.

DNA methylation markers may also prove to be useful for prognostic purposes. The preliminary results in our lab seem to suggest that PR-B promoter hypermethylation found in tissue samples harvested at the time of surgery may be a biomarker for recurrence [Shen et al. unpublished data], which is consistent with the published findings [86,207]. In any case, very little has been published in this area, even though it is an area that is likely to be clinically most useful and could bring tangible results to better patient care. The identification of patients with high risk of recurrence should accord for further intervention. On the other hand, patients with low risk of recurrence may be advised not to take any medication, which often have side effects.

22.9 CONCLUSIONS AND FUTURE RESEARCH DIRECTIONS

Growing evidence now suggests that endometriosis is an epigenetic disorder, in the sense that epigenetics plays a *definite* role in the pathogenesis and pathophysiology of endometriosis. This is characterized, at least in part, by aberrant methylation, dysregulated microRNA, and very recently by deregulation of some ill-behaved epigenetic "writers" and "erasers" in eutopic as well as ectopic endometrium. Quite extensive data also have shown that HDACIs have many desirable effects and, as such, have great potential as a therapeutic for endometriosis. In addition, DNA methylation-based as well as miRNA-based biomarkers may hold potential in the diagnosis of endometriosis and predicting recurrence risks after surgery.

Despite these advances, however, our current knowledge on the epigenetics of endometriosis, and its pathophysiological significance is still in its infancy. This can be seen from the vast gap between a myriad of enzymes/proteins involved in DNA methylation and histone modifications and just a handful of them that just have been evaluated in endometriosis. Indeed, so far we have merely scratched the surface of the epigenetics of endometriosis.

While a complete understanding of the epigenetics of endometriosis holds keys to a full knowledge of how genes are dysregulated and coordinated in the genesis and development of endometriosis and the manifest of variable symptoms, such an enormous task is quite challenging, and somewhat daunting. This is because, first of all, epigenetics itself is evolving and rapidly developing field. Many issues, such as the existance of DNA demethylase(s) or not, are still unresolved. In addition, while aberrant DNA methylation in endometriosis has begun to be investigated, few histone modifications have, if any, been investigated in endometriosis. There are 17 HDACs, over 50 HMTs [208], and possibly a similar number of HDMs, yet few of them have been evaluated in endometriosis. Besides DNMTs, methyl-CpG binding proteins, HDACs, HATs, HMT, HDMs, and microRNAs, there are E3 ubiquitin ligases, kinases, small ubiquitin-related modifier (SUMO)-conjugated enzymes, and ADP-ribosyl transferases (ADPRTs) and other enzymes that may also play important roles in endometriosis epigenetics. Even for histone methylation, there is an added layer of complexity of having three forms of lysine methylation and two forms of arginine methylation. These complexities are further compounded by nearly an astronomical number of combinatorial assortment of various histone modifications, by crosstalks between and among different DNA and histone modifications, and by possible temporal and spatial dynamics of these modifications. Moreover, the epigenetic aberrations may exist not only in endometriotic cells, but also in other types of cells, such as endothelial cells and macrophages, that are within or surround the endometriotic lesions which are also intimately involved in the pathogenesis of endometriosis. As of the time of writing, no work in this area has been published.

Regardless, it is now clear that chromatin, once considered just a structural scaffold allowing the packaging of DNA, is actually a dynamic and key regulatory element of gene expression and actively participates in several cellular processes such as mitosis and differentiation [209]. The epigenetics of endometriosis is a rapidly growing field, and may likely transform our understanding of the pathogenesis and pathophysiology of endometriosis, opening new avenues for diagnosis, treatment, and prognostic prediction. So far we have only scratched its surface. With more research, we may come closer to the full understanding of the etiopathogenesis of endometriosis and will be in a better position to treat or perhaps prevent this unrelentingly painful and dreadful disease that is endometriosis.

Acknowledgment

This work was supported by grant 30872759 from the National Science Foundation of China, grants 09PJD015 and 10410700200 from the Shanghai Science and Technology Commission (SWG), and grant 09-11 from the State Key Laboratory of Medical Neurobiology of Fudan University, and financial support from Shanghai Key Laboratory of Female Reproductive Endocrine-Related Diseases and from the Key Specialty Project of the Ministry of Health, China.

References

[1] Giudice LC, Kao LC. Endometriosis. Lancet 2004;364:1789−99.

[2] Farquhar CM. Extracts from the clinical evidence. Endometriosis. Bmj 2000;320:1449−52.

[3] Fourquet J, Gao X, Zavala D, Orengo JC, Abac S, Ruiz A, Laboy J, Flores I. Patients' report on how endometriosis affects health, work, and daily life. Fertil Steril 93:2424−8.

[4] Boling RO, Abbasi R, Ackerman G, Schipul Jr AH, Chaney SA. Disability from endometriosis in the United States Army. J Reprod Med 1988;33:49−52.

[5] Kjerulff KH, Erickson BA, Langenberg PW. Chronic gynecological conditions reported by US women: findings from the National Health Interview Survey, 1984 to 1992. Am J Public Health 1996;86:195−9.

[6] Simoens S, Hummelshoj L, D'Hooghe T. Endometriosis: cost estimates and methodological perspective. Hum Reprod Update 2007;13:395−404.

[7] Zhao SZ, Wong JM, Davis MB, Gersh GE, Johnson KE. The cost of inpatient endometriosis treatment: an analysis based on the Healthcare Cost and Utilization Project Nationwide Inpatient Sample. Am J Manag Care 1998;4:1127−34.

[8] Gao X, Outley J, Botteman M, Spalding J, Simon JA, Pashos CL. Economic burden of endometriosis. Fertil Steril 2006;86:1561−72.

[9] Fuldeore M, Chwalisz K, Marx S, Wu N, Boulanger L, Ma L, Lamothe K. Surgical procedures and their cost estimates among women with newly diagnosed endometriosis: a US database study. J Med Econ 14:115−23.

[10] Simoens S, Meuleman C, D'Hooghe T. Non-health-care costs associated with endometriosis. Hum Reprod 26:2363−7.

[11] Levy AR, Osenenko KM, Lozano-Ortega G, Sambrook R, Jeddi M, Belisle S, Reid RL. Economic burden of surgically confirmed endometriosis in Canada. J Obstet Gynaecol Can 33:830–7.

[12] Luisi S, Lazzeri L, Ciani V, Petraglia F. Endometriosis in Italy: from cost estimates to new medical treatment. Gynecol Endocrinol 2009;25:734–40.

[13] Vinatier D, Orazi G, Cosson M, Dufour P. Theories of endometriosis. Eur J Obstet Gynecol Reprod Biol 2001;96:21–34.

[14] Sampson J. Peritoneal endometriosis due to the menstrual dissemination of endometrial tissue into the peritoneal cavity. Am J Obstet Gynecol 1927;14:422–69.

[15] Halme J, Hammond MG, Hulka JF, Raj SG, Talbert LM. Retrograde menstruation in healthy women and in patients with endometriosis. Obstet Gynecol 1984;64:151–4.

[16] Valle RF, Sciarra JJ. Endometriosis: treatment strategies. Ann N Y Acad Sci 2003;997:229–39.

[17] May KE, Conduit-Hulbert SA, Villar J, Kirtley S, Kennedy SH, Becker CM. Peripheral biomarkers of endometriosis: a systematic review. Hum Reprod Update 16:651–74.

[18] Revised American Society for Reproductive Medicine classification of endometriosis: 1996. Fertil Steril 1997;67:817–21.

[19] Canis M, Bouquet De Jolinieres J, Wattiez A, Pouly JL, Mage G, Manhes H, et al. Classification of endometriosis. Baillieres Clin Obstet Gynaecol 1993;7:759–74.

[20] Adamson GD. Endometriosis classification: an update. Curr Opin Obstet Gynecol 23:213–20.

[21] Nisolle M, Donnez J. Peritoneal endometriosis, ovarian endometriosis, and adenomyotic nodules of the rectovaginal septum are three different entities. Fertil Steril 1997;68:585–96.

[22] Wu Y, Kajdacsy-Balla A, Strawn E, Basir Z, Halverson G, Jailwala P, et al. Transcriptional characterizations of differences between eutopic and ectopic endometrium. Endocrinology 2006;147:232–46.

[23] Olive DL, Pritts EA. Treatment of endometriosis. N Engl J Med 2001;345:266–75.

[24] Weedon MN, Lango H, Lindgren CM, Wallace C, Evans DM, Mangino M, et al. Genome-wide association analysis identifies 20 loci that influence adult height. Nat Genet 2008;40:575–83.

[25] Guo SW. Recurrence of endometriosis and its control. Hum Reprod Update 2009.

[26] Wheeler JM, Malinak LR. Recurrent endometriosis: incidence, management, and prognosis. Am J Obstet Gynecol 1983;146:247–53.

[27] Ragni G, Somigliana E, Benedetti F, Paffoni A, Vegetti W, Restelli L, et al. Damage to ovarian reserve associated with laparoscopic excision of endometriomas: a quantitative rather than a qualitative injury. Am J Obstet Gynecol 2005;193:1908–14.

[28] Hachisuga T, Kawarabayashi T. Histopathological analysis of laparoscopically treated ovarian endometriotic cysts with special reference to loss of follicles. Hum Reprod 2002;17:432–5.

[29] Somigliana E, Ragni G, Benedetti F, Borroni R, Vegetti W, Crosignani PG. Does laparoscopic excision of endometriotic ovarian cysts significantly affect ovarian reserve? Insights from IVF cycles. Hum Reprod 2003;18:2450–3.

[30] Somigliana E, Ragni G, Infantino M, Benedetti F, Arnoldi M, Crosignani PG. Does laparoscopic removal of nonendometriotic benign ovarian cysts affect ovarian reserve? Acta Obstet Gynecol Scand 2006;85:74–7.

[31] Candiani M, Barbieri M, Bottani B, Bertulessi C, Vignali M, Agnoli B, et al. Ovarian recovery after laparoscopic enucleation of ovarian cysts: insights from echographic short-term postsurgical follow-up. J Minim Invasive Gynecol 2005;12:409–14.

[32] Lessey BA. Medical management of endometriosis and infertility. Fertil Steril 2000;73:1089–96.

[33] Kennedy S, Bergqvist A, Chapron C, D'Hooghe T, Dunselman G, Greb R, et al. ESHRE guideline for the diagnosis and treatment of endometriosis. Hum Reprod 2005;20:2698–704.

[34] Waller KG, Shaw RW. Gonadotropin-releasing hormone analogues for the treatment of endometriosis: long-term follow-up. Fertil Steril 1993;59:511–5.

[35] Guo SW. Emerging drugs for endometriosis. Expert Opin Emerging Drugs 2008;13:547–71.

[36] Guo SW, Hummelshoj L, Olive DL, Bulun SE, D'Hooghe TM, Evers JL. A call for more transparency of registered clinical trials on endometriosis. Hum Reprod 2009;24:1247–54.

[37] Vercellini P, Crosignani P, Somigliana E, Vigano P, Frattaruolo MP, Fedele L. 'Waiting for Godot': a common-sense approach to the medical treatment of endometriosis. Hum Reprod 26:3–13.

[38] Guo SW. Epigenetics of endometriosis. Mol Hum Reprod 2009;15:587–607.

[39] Toloubeydokhti T, Pan Q, Luo X, Bukulmez O, Chegini N. The expression and ovarian steroid regulation of endometrial micro-RNAs. Reprod Sci 2008;15:993–1001.

[40] Bulun SE, Yang S, Fang Z, Gurates B, Tamura M, Sebastian S. Estrogen production and metabolism in endometriosis. Ann N Y Acad Sci 2002;955:75–85. discussion 86–8, 396–406.

[41] Gurates B, Bulun SE. Endometriosis: the ultimate hormonal disease. Semin Reprod Med 2003;21:125–34.

[42] Kitawaki J, Kado N, Ishihara H, Koshiba H, Kitaoka Y, Honjo H. Endometriosis: the pathophysiology as an estrogen-dependent disease. J Steroid Biochem Mol Biol 2002;83:149−55.

[43] Paul Dmowski W, Braun DP. Immunology of endometriosis. Best Pract Res Clin Obstet Gynaecol 2004;18:245−63.

[44] Ulukus M, Arici A. Immunology of endometriosis. Minerva Ginecol 2005;57:237−48.

[45] Rier SE, Martin DC, Bowman RE, Dmowski WP, Becker JL. Endometriosis in rhesus monkeys (Macaca mulatta) following chronic exposure to 2,3,7,8-tetrachlorodibenzo-p-dioxin. Fundam Appl Toxicol 1993;21:433−41.

[46] Rier SE. The potential role of exposure to environmental toxicants in the pathophysiology of endometriosis. Ann N Y Acad Sci 2002;955:201−12. discussion 230-2, 396-406.

[47] Guo SW. The Link between Exposure to Dioxin and Endometriosis: A Critical Reappraisal of Primate Data. Gynecol Obstet Invest 2004;57:157−73.

[48] Simpson JL, Bischoff FZ, Kamat A, Buster JE, Carson SA. Genetics of endometriosis. Obstet Gynecol Clin North Am 2003;30:21−40. vii.

[49] Barlow DH, Kennedy S. Endometriosis: new genetic approaches and therapy. Annu Rev Med 2005;56:345−56.

[50] Di W, Guo SW. The search for genetic variants predisposing women to endometriosis. Curr Opin Obstet Gynecol 2007;19:395−401.

[51] Falconer H, D'Hooghe T, Fried G. Endometriosis and genetic polymorphisms. Obstet Gynecol Surv 2007;62:616−28.

[52] Guo SW. The relevance of genetics in endometriosis. In: Garcia-Velasco JA, Rizk B, editors. Endometriosis: Current management and future trends, Vol. aypee Medical Publishers; 2009. In press.

[53] Ohtake F, Takeyama K, Matsumoto T, Kitagawa H, Yamamoto Y, Nohara K, et al. Modulation of oestrogen receptor signalling by association with the activated dioxin receptor. Nature 2003;423:545−50.

[54] Igarashi TM, Bruner-Tran KL, Yeaman GR, Lessey BA, Edwards DP, Eisenberg E, et al. Reduced expression of progesterone receptor-B in the endometrium of women with endometriosis and in cocultures of endometrial cells exposed to 2,3,7,8-tetrachlorodibenzo-p-dioxin. Fertil Steril 2005;84:67−74.

[55] Carson DD, Lagow E, Thathiah A, Al-Shami R, Farach-Carson MC, Vernon M, et al. Changes in gene expression during the early to mid-luteal (receptive phase) transition in human endometrium detected by high-density microarray screening. Mol Hum Reprod 2002;8:871−9.

[56] Arimoto T, Katagiri T, Oda K, Tsunoda T, Yasugi T, Osuga Y, et al. Genome-wide cDNA microarray analysis of gene-expression profiles involved in ovarian endometriosis. Int J Oncol 2003;22:551−60.

[57] Kao LC, Germeyer A, Tulac S, Lobo S, Yang JP, Taylor RN, et al. Expression profiling of endometrium from women with endometriosis reveals candidate genes for disease-based implantation failure and infertility. Endocrinology 2003;144:2870−81.

[58] Matsuzaki S, Canis M, Vaurs-Barriere C, Boespflug-Tanguy O, Dastugue B, Mage G. DNA microarray analysis of gene expression in eutopic endometrium from patients with deep endometriosis using laser capture micro-dissection. Fertil Steril 2005;84(Suppl. 2):1180−90.

[59] Burney RO, Talbi S, Hamilton AE, Vo KC, Nyegaard M, Nezhat CR, et al. Gene expression analysis of endo-metrium reveals progesterone resistance and candidate susceptibility genes in women with endometriosis. Endocrinology 2007;148:3814−26.

[60] Eyster KM, Klinkova O, Kennedy V, Hansen KA. Whole genome deoxyribonucleic acid microarray analysis of gene expression in ectopic versus eutopic endometrium. Fertil Steril 2007;88:1505−33.

[61] Flores I, Rivera E, Ruiz LA, Santiago OI, Vernon MW, Appleyard CB. Molecular profiling of experimental endometriosis identified gene expression patterns in common with human disease. Fertil Steril 2007;87:1180−99.

[62] Hever A, Roth RB, Hevezi P, Marin ME, Acosta JA, Acosta H, et al. Human endometriosis is associated with plasma cells and overexpression of B lymphocyte stimulator. Proc Natl Acad Sci USA 2007;104:12451−6.

[63] Konno R, Fujiwara H, Netsu S, Odagiri K, Shimane M, Nomura H, et al. Gene expression profiling of the rat endometriosis model. Am J Reprod Immunol 2007;58:330−43.

[64] Mettler L, Salmassi A, Schollmeyer T, Schmutzler AG, Pungel F, Jonat W. Comparison of c-DNA microarray analysis of gene expression between eutopic endometrium and ectopic endometrium (endometriosis). J Assist Reprod Genet 2007;24:249−58.

[65] Chand AL, Murray AS, Jones RL, Hannan NJ, Salamonsen LA, Rombauts L. Laser capture microdissection and cDNA array analysis of endometrium identify CCL16 and CCL21 as epithelial-derived inflammatory mediators associated with endometriosis. Reprod Biol Endocrinol 2007;5:18.

[66] Van Langendonckt A, Punyadeera C, Kamps R, Dunselman G, Klein-Hitpass L, Schurgers LJ, et al. Identifi-cation of novel antigens in blood vessels in rectovaginal endometriosis. Mol Hum Reprod 2007;13:875−86.

[67] Hull ML, Escareno CR, Godsland JM, Doig JR, Johnson CM, Phillips SC, et al. Endometrial-peritoneal interactions during endometriotic lesion establishment. Am J Pathol 2008;173:700−15.

463

[68] Sherwin JR, Sharkey AM, Mihalyi A, Simsa P, Catalano RD, D'Hooghe TM. Global gene analysis of late secretory phase, eutopic endometrium does not provide the basis for a minimally invasive test of endometriosis. Hum Reprod 2008;23:1063—8.

[69] Zafrakas M, Tarlatzis BC, Streichert T, Pournaropoulos F, Wolfle U, Smeets SJ, et al. Genome-wide microarray gene expression, array-CGH analysis, and telomerase activity in advanced ovarian endometriosis: a high degree of differentiation rather than malignant potential. Int J Mol Med 2008;21:335—44.

[70] Pelch KE, Schroder AL, Kimball PA, Sharpe-Timms KL, Wade Davis J, Nagel SC. Aberrant gene expression profile in a mouse model of endometriosis mirrors that observed in women. Fertil Steril 2010;93:1615—27.

[71] Umezawa M, Tanaka N, Tainaka H, Takeda K, Ihara T, Sugamata M. Microarray analysis provides insight into the early steps of pathophysiology of mouse endometriosis model induced by autotransplantation of endometrium. Life Sci 2009;84:832—7.

[72] Wu Y, Basir Z, Kajdacsy-Balla A, Strawn E, Macias V, Montgomery K, et al. Resolution of clonal origins for endometriotic lesions using laser capture microdissection and the human androgen receptor (HUMARA) assay. Fertil Steril 2003;79(Suppl. 1):710—7.

[73] Ambros V. The functions of animal microRNAs. Nature 2004;431:350—5.

[74] Bartel DP. MicroRNAs: genomics, biogenesis, mechanism, and function. Cell 2004;116:281—97.

[75] Kurokawa R, Rosenfeld MG, Glass CK. Transcriptional regulation through noncoding RNAs and epigenetic modifications. RNA Biol 2009;6.

[76] Wu Y, Halverson G, Basir Z, Strawn E, Yan P, Guo SW. Aberrant methylation at HOXA10 may be responsible for its aberrant expression in the endometrium of patients with endometriosis. Am J Obstet Gynecol 2005;193:371—80.

[77] Troiano RN, Taylor KJ. Sonographically guided therapeutic aspiration of benign-appearing ovarian cysts and endometriomas. AJR Am J Roentgenol 1998;171:1601—5.

[78] Taylor HS, Arici A, Olive D, Igarashi P. HOXA10 is expressed in response to sex steroids at the time of implantation in the human endometrium. J Clin Invest 1998;101:1379—84.

[79] Taylor HS, Bagot C, Kardana A, Olive D, Arici A. HOX gene expression is altered in the endometrium of women with endometriosis. Hum Reprod 1999;14:1328—31.

[80] Gui Y, Zhang J, Yuan L, Lessey BA. Regulation of HOXA-10 and its expression in normal and abnormal endometrium. Mol Hum Reprod 1999;5:866—73.

[81] Szczepanska M, Wirstlein P, Luczak M, Jagodzinski PP, Skrzypczak J. Reduced expression of HOXA10 in the midluteal endometrium from infertile women with minimal endometriosis. Biomed Pharmacother 2010;64:697—705.

[82] Kim JJ, Taylor HS, Lu Z, Ladhani O, Hastings JM, Jackson KS, et al. Altered expression of HOXA10 in endometriosis: potential role in decidualization. Mol Hum Reprod 2007;13:323—32.

[83] Lee B, Du H, Taylor HS. Experimental murine endometriosis induces DNA methylation and altered gene expression in eutopic endometrium. Biol Reprod 2009;80:79—85.

[84] Bromer JG, Wu J, Zhou Y, Taylor HS. Hypermethylation of HOXA10 by in utero diethylstilbestrol exposure: an epigenetic mechanism for altered developmental programming. Endocrinology 2009;150:3376—82.

[85] Zhang L, Zheng XM, Hubert J, Zheng H, Yang ZW, Li SW. Prenatal exposure to diaethylstilbestrol in the rat inhibits transabdominal testicular descent with involvement of the INSL3/LGR8 system and HOXA10. Chin Med J (Engl) 2009;122:967—71.

[86] Wu Y, Strawn E, Basir Z, Halverson G, Guo SW. Promoter hypermethylation of progesterone receptor isoform B (PR-B) in endometriosis. Epigenetics 2006;1:106—11.

[87] Attia GR, Zeitoun K, Edwards D, Johns A, Carr BR, Bulun SE. Progesterone receptor isoform A but not B is expressed in endometriosis. J Clin Endocrinol Metab 2000;85:2897—902.

[88] Wu Y, Strawn E, Basir Z, Halverson G, Guo SW. Aberrant expression of deoxyribonucleic acid methyltransferases DNMT1, DNMT3A, and DNMT3B in women with endometriosis. Fertil Steril 2007;87:24—32.

[89] van Kaam KJ, Delvoux B, Romano A, D'Hooghe T, Dunselman GA, Groothuis PG. Deoxyribonucleic acid methyltransferases and methyl-CpG-binding domain proteins in human endometrium and endometriosis. Fertil Steril 2011;95:1421—7.

[90] Xue Q, Lin Z, Yin P, Milad MP, Cheng YH, Confino E, et al. Transcriptional activation of steroidogenic factor-1 by hypomethylation of the 5′ CpG island in endometriosis. J Clin Endocrinol Metab 2007;92:3261—7.

[91] Xue Q, Lin Z, Cheng YH, Huang CC, Marsh E, Yin P, et al. Promoter methylation regulates estrogen receptor 2 in human endometrium and endometriosis. Biol Reprod 2007;77:681—7.

[92] Izawa M, Harada T, Taniguchi F, Ohama Y, Takenaka Y, Terakawa N. An epigenetic disorder may cause aberrant expression of aromatase gene in endometriotic stromal cells. Fertil Steril 2008;89:1390—6.

[93] Izawa M, Taniguchi F, Uegaki T, Takai E, Iwabe T, Terakawa N, Harada T. Demethylation of a nonpromoter cytosine-phosphate-guanine island in the aromatase gene may cause the aberrant up-regulation in endometriotic tissues. Fertil Steril 95:33—9.

[94] Starzinski-Powitz A, Gaetje R, Zeitvogel A, Kotzian S, Handrow-Metzmacher H, Herrmann G, et al. Tracing cellular and molecular mechanisms involved in endometriosis. Hum Reprod Update 1998;4:724—9.

[95] Starzinski-Powitz A, Zeitvogel A, Schreiner A, Baumann R. In search of pathogenic mechanisms in endometriosis: the challenge for molecular cell biology. Curr Mol Med 2001;1:655—64.

[96] Wu Y, Starzinski-Powitz A, Guo SW. Trichostatin A, a histone deacetylase inhibitor, attenuates invasiveness and reactivates E-cadherin expression in immortalized endometriotic cells. Reprod Sci 2007;14:374—82.

[97] Wren JD, Wu Y, Guo SW. A system-wide analysis of differentially expressed genes in ectopic and eutopic endometrium. Hum Reprod 2007;22:2093—102.

[98] Borghese B, Barbaux S, Mondon F, Santulli P, Pierre G, Vinci G, Chapron C, Vaiman D. Research resource: genome-wide profiling of methylated promoters in endometriosis reveals a subtelomeric location of hypermethylation. Mol Endocrinol 24:1872—85.

[99] Kawano Y, Nasu K, Li H, Tsuno A, Abe W, Takai N, Narahara H. Application of the histone deacetylase inhibitors for the treatment of endometriosis: histone modifications as pathogenesis and novel therapeutic target. Hum Reprod 26:2486—98.

[100] Jenuwein T, Allis CD. Translating the histone code. Science 2001;293:1074—80.

[101] Strahl BD, Allis CD. The language of covalent histone modifications. Nature 2000;403:41—5.

[102] Keppler BR, Archer TK. Chromatin-modifying enzymes as therapeutic targets — Part 1. Expert Opin Ther Targets 2008;12:1301—12.

[103] Suzuki A, Horiuchi A, Oka K, Miyamoto T, Kashima H, Shiozawa T. Immunohistochemical detection of steroid receptor cofactors in ovarian endometriosis: involvement of down-regulated SRC-1 expression in the limited growth activity of the endometriotic epithelium. Virchows Arch 456:433—41.

[104] Kumagami A, Ito A, Yoshida-Komiya H, Fujimori K, Sato A. Expression patterns of the steroid receptor coactivator family in human ovarian endometriosis. J Obstet Gynaecol Res 2011;37:1269—76.

[105] Imesch P, Fink D, Fedier A. Romidepsin reduces histone deacetylase activity, induces acetylation of histones, inhibits proliferation, and activates apoptosis in immortalized epithelial endometriotic cells. Fertil Steril 94:2838—42.

[106] Zhao T, Liu X, Zhen X, Guo SW. Levo-tetrahydropalmatine retards the growth of ectopic endometrial implants and alleviates generalized hyperalgesia in experimentally induced endometriosis in rats. Reprod Sci 18:28—45.

[107] Rea S, Eisenhaber F, O'Carroll D, Strahl BD, Sun ZW, Schmid M, et al. Regulation of chromatin structure by site-specific histone H3 methyltransferases. Nature 2000;406:593—9.

[108] Shi Y, Lan F, Matson C, Mulligan P, Whetstine JR, Cole PA, et al. Histone demethylation mediated by the nuclear amine oxidase homolog LSD1. Cell 2004;119:941—53.

[109] Pedersen MT, Helin K. Histone demethylases in development and disease. Trends Cell Biol 2010;20:662—71.

[110] Bernstein BE, Meissner A, Lander ES. The mammalian epigenome. Cell 2007;128:669—81.

[111] Klose RJ, Zhang Y. Regulation of histone methylation by demethylimination and demethylation. Nat Rev Mol Cell Biol 2007;8:307—18.

[112] Leyendecker G. Redefining endometriosis: endometriosis is an entity with extreme pleiomorphism. Hum Reprod 2000;15:4—7.

[113] Vinatier D, Cosson M, Dufour P. Is endometriosis an endometrial disease? Eur J Obstet Gynecol Reprod Biol 2000;91:113—25.

[114] Carlson JM, Doyle J. Highly optimized tolerance: robustness and design in complex systems. Phys Rev Lett 2000;84:2529—32.

[115] Richardson BC. Role of DNA methylation in the regulation of cell function: autoimmunity, aging and cancer. J Nutr 2002;132:2401S—5S.

[116] Wilson VL, Jones PA. DNA methylation decreases in aging but not in immortal cells. Science 1983;220:1055—7.

[117] Toyota M, Issa JP. CpG island methylator phenotypes in aging and cancer. Semin Cancer Biol 1999;9:349—57.

[118] Jacob RA, Gretz DM, Taylor PC, James SJ, Pogribny IP, Miller BJ, et al. Moderate folate depletion increases plasma homocysteine and decreases lymphocyte DNA methylation in postmenopausal women. J Nutr 1998;128:1204—12.

[119] Hsieh CJ, Klump B, Holzmann K, Borchard F, Gregor M, Porschen R. Hypermethylation of the p16INK4a promoter in colectomy specimens of patients with long-standing and extensive ulcerative colitis. Cancer Res 1998;58:3942—5.

[120] Issa JP, Ahuja N, Toyota M, Bronner MP, Brentnall TA. Accelerated age-related CpG island methylation in ulcerative colitis. Cancer Res 2001;61:3573—7.

[121] Song JZ, Stirzaker C, Harrison J, Melki JR, Clark SJ. Hypermethylation trigger of the glutathione-S-transferase gene (GSTP1) in prostate cancer cells. Oncogene 2002;21:1048—61.

[122] Stirzaker C, Song JZ, Davidson B, Clark SJ. Transcriptional gene silencing promotes DNA hypermethylation through a sequential change in chromatin modifications in cancer cells. Cancer Res 2004;64:3871−7.

[123] Champagne FA, Weaver IC, Diorio J, Dymov S, Szyf M, Meaney MJ. Maternal care associated with methylation of the estrogen receptor-alpha1b promoter and estrogen receptor-alpha expression in the medial preoptic area of female offspring. Endocrinology 2006;147:2909−15.

[124] Lu Y, Nie J, Liu X, Guo SW. Reduced expression and concomitant promoter hypermethylation of HOXA10 in endometrium from women wearing intrauterine devices. Fertil Steril 94:1583−8.

[125] Wu Y, Starzinski-Powitz A, Guo SW. Prolonged stimulation with tumor necrosis factor-alpha induced partial methylation at PR-B promoter in immortalized epithelial-like endometriotic cells. Fertil Steril 2008;90:234−7.

[126] Wu Y, Shi X, Guo SW. The knockdown of progesterone receptor isoform B (PR-B) promotes proliferation in immortalized endometrial stromal cells. Fertil Steril 2008;90:1320−3.

[127] Li S, Washburn KA, Moore R, Uno T, Teng C, Newbold RR, et al. Developmental exposure to diethylstilbestrol elicits demethylation of estrogen-responsive lactoferrin gene in mouse uterus. Cancer Res 1997;57:4356−9.

[128] McLachlan JA, Simpson E, Martin M. Endocrine disrupters and female reproductive health. Best Pract Res Clin Endocrinol Metab 2006;20:63−75.

[129] Tang WY, Newbold R, Mardilovich K, Jefferson W, Cheng RY, Medvedovic M, et al. Persistent hypomethylation in the promoter of nucleosomal binding protein 1 (Nsbp1) correlates with overexpression of Nsbp1 in mouse uteri neonatally exposed to diethylstilbestrol or genistein. Endocrinology 2008;149:5922−31.

[130] Li S, Hursting SD, Davis BJ, McLachlan JA, Barrett JC. Environmental exposure, DNA methylation, and gene regulation: lessons from diethylstilbesterol-induced cancers. Ann N Y Acad Sci 2003;983:161−9.

[131] McGowan PO, Sasaki A, D'Alessio AC, Dymov S, Labonte B, Szyf M, et al. Epigenetic regulation of the glucocorticoid receptor in human brain associates with childhood abuse. Nat Neurosci 2009;12:342−8.

[132] Weaver IC, Cervoni N, Champagne FA, D'Alessio AC, Sharma S, Seckl JR, et al. Epigenetic programming by maternal behavior. Nat Neurosci 2004;7:847−54.

[133] Waterland RA, Jirtle RL. Early nutrition, epigenetic changes at transposons and imprinted genes, and enhanced susceptibility to adult chronic diseases. Nutrition 2004;20:63−8.

[134] Barker DJ. The developmental origins of adult disease. Eur J Epidemiol 2003;18:733−6.

[135] Missmer SA, Hankinson SE, Spiegelman D, Barbieri RL, Michels KB, Hunter DJ. In utero exposures and the incidence of endometriosis. Fertil Steril 2004;82:1501−8.

[136] Dolinoy DC, Huang D, Jirtle RL. Maternal nutrient supplementation counteracts bisphenol A-induced DNA hypomethylation in early development. Proc Natl Acad Sci USA 2007;104:13056−61.

[137] Dolinoy DC, Weidman JR, Waterland RA, Jirtle RL. Maternal genistein alters coat color and protects Avy mouse offspring from obesity by modifying the fetal epigenome. Environ Health Perspect 2006;114:567−72.

[138] Wu Y, Guo SW. Inhibition of proliferation of endometrial stromal cells by trichostatin A, RU486, CDB-2914, N-acetylcysteine, and ICI 182780. Gynecol Obstet Invest 2006;62:193−205.

[139] Wu Y, Guo SW. Histone deacetylase inhibitors trichostatin A and valproic acid induce cell cycle arrest and p21 expression in immortalized human endometrial stromal cells. Eur J Obstet Gynecol Reprod Biol 2008;137:198−203.

[140] Wu Y, Starzinski-Powitz A, Guo SW. Constitutive and tumor necrosis factor-alpha-stimulated activation of nuclear factor-kappaB in immortalized endometriotic cells and their suppression by trichostatin A. Gynecol Obstet Invest 70:23−33.

[141] Wu Y, Guo SW. Suppression of IL-1beta-induced COX-2 expression by trichostatin A (TSA) in human endometrial stromal cells. Eur J Obstet Gynecol Reprod Biol 2007;135:88−93.

[142] Ota H, Igarashi S, Sasaki M, Tanaka T. Distribution of cyclooxygenase-2 in eutopic and ectopic endometrium in endometriosis and adenomyosis. Hum Reprod 2001;16:561−6.

[143] Matsuzaki S, Canis M, Pouly JL, Wattiez A, Okamura K, Mage G. Cyclooxygenase-2 expression in deep endometriosis and matched eutopic endometrium. Fertil Steril 2004;82:1309−15.

[144] Buchweitz O, Staebler A, Wulfing P, Hauzman E, Greb R, Kiesel L. COX-2 overexpression in peritoneal lesions is correlated with nonmenstrual chronic pelvic pain. Eur J Obstet Gynecol Reprod Biol 2006;124:216−21.

[145] Yuan L, Shen F, Lu Y, Liu X, Guo SW. Cyclooxygenase-2 overexpression in ovarian endometriomas is associated with higher risk of recurrence. Fertil Steril 2008;91:1303−6.

[146] Wu Y, Guo SW. Peroxisome proliferator-activated receptor-gamma and retinoid X receptor agonists synergistically suppress proliferation of immortalized endometrial stromal cells. Fertil Steril 2009;91:2142−7.

[147] Peeters LL, Vigne JL, Tee MK, Zhao D, Waite LL, Taylor RN. PPARgamma represses VEGF expression in human endometrial cells: implications for uterine angiogenesis. Angiogenesis 2005;8:373−9.

466

[148] Ohama Y, Harada T, Iwabe T, Taniguchi F, Takenaka Y, Terakawa N. Peroxisome proliferator-activated receptor-gamma ligand reduced tumor necrosis factor-alpha-induced interleukin-8 production and growth in endometriotic stromal cells. Fertil Steril 2008;89:311—7.

[149] Aytan H, Caliskan AC, Demirturk F, Aytan P, Koseoglu DR. Peroxisome proliferator-activated receptor-gamma agonist rosiglitazone reduces the size of experimental endometriosis in the rat model. Aust N Z J Obstet Gynaecol 2007;47:321—5.

[150] Lebovic DI, Kir M, Casey CL. Peroxisome proliferator-activated receptor-gamma induces regression of endometrial explants in a rat model of endometriosis. Fertil Steril 2004;82(Suppl. 3): 1008—13.

[151] Lebovic DI, Mwenda JM, Chai DC, Mueller MD, Santi A, Fisseha S, et al. PPAR-gamma receptor ligand induces regression of endometrial explants in baboons: a prospective, randomized, placebo- and drug-controlled study. Fertil Steril 2007;88:1108—19.

[152] Guo SW. Nuclear Factor-kappaB (NF-kappaB): An Unsuspected Major Culprit in the Pathogenesis of Endometriosis That is Still At Large? Gynecol Obestet Invest 2006;63:71—97.

[153] Gonzalez-Ramos R, Donnez J, Defrere S, Leclercq I, Squifflet J, Lousse JC, et al. Nuclear factor-kappa B is constitutively activated in peritoneal endometriosis. Mol Hum Reprod 2007;13:503—9.

[154] Gonzalez-Ramos R, Van Langendonckt A, Defrere S, Lousse JC, Colette S, Devoto L, et al. Involvement of the nuclear factor-kappaB pathway in the pathogenesis of endometriosis. Fertil Steril 2010;94:1985—94.

[155] Imesch P, Samartzis P, Schneider M, Fink D, Fedier A. Inhibition of transcription, expression, and secretion of the vascular epithelial growth factor in human epithelial endometriotic cells by romidepsin. Fertil Steril 95:1579—83.

[156] Wang B, Xiao Y, Ding BB, Zhang N, Yuan X, Gui L, et al. Induction of tumor angiogenesis by Slit-Robo signaling and inhibition of cancer growth by blocking Robo activity. Cancer Cell 2003;4:19—29.

[157] Shen FH, Liu XS, Geng JG, Guo SW. Increased immunoreactivity to SLIT/ROBO1 in ovarian endometriomas and as a likely constituent biomarker for recurrence. Amer J Pathol 2009;175:479—88.

[158] Moynihan AT, Hehir MP, Sharkey AM, Robson SC, Europe-Finner GN, Morrison JJ. Histone deacetylase inhibitors and a functional potent inhibitory effect on human uterine contractility. Am J Obstet Gynecol 2008;199(167):e1—7.

[159] Bulletti C, D, D. E. Z., Setti PL, Cicinelli E, Polli V, Flamigni C. The patterns of uterine contractility in normal menstruating women: from physiology to pathology. Ann N Y Acad Sci 2004;1034:64—83.

[160] Kitlas A, Oczeretko E, Swiatecka J, Borowska M, Laudanski T. Uterine contraction signals—application of the linear synchronization measures. Eur J Obstet Gynecol Reprod Biol 2009;144(Suppl. 1):S61—4.

[161] Ruddock NK, Shi SQ, Jain S, Moore G, Hankins GD, Romero R, et al. Progesterone, but not 17-alpha-hydroxyprogesterone caproate, inhibits human myometrial contractions. Am J Obstet Gynecol 2008;199(391):e1—7.

[162] Lu Y, Nie J, Liu X, Zheng Y, Guo SW. Trichostatin A, a histone deacetylase inhibitor, reduces lesion growth and hyperalgesia in experimentally induced endometriosis in mice. Hum Reprod 25:1014—25.

[163] Liu M, Liu X, Zhang Y, Guo SW. Valproic acid and progestin inhibit lesion growth and reduce hyperalgesia in experimentally induced endometriosis in rats. Reprod Sci 2012; Feb 16. [Epub ahead of print].

[164] Liu X, Guo SW. A pilot study on the off-label use of valproic acid to treat adenomyosis. Fertil Steril 2008;89:246—50.

[165] Liu XS, Guo SW. Valproic acid alleviates generalized hyperalgesia in mice with induced adenomyosis. J Gynaecol Obstet Res 2011;37:696—708.

[166] Jichan N, Xishi L, Guo SW. Promoter hypermethylation of progesterone receptor isoform B (PR-B) in adenomyosis and its rectification by a histone deacetylase inhibitor and a demethylation agent. Reprod Sci 17:995—1005.

[167] Mao X, Wang Y, Carter AV, Zhen X, Guo SW. The Retardation of Myometrial Infiltration, Reduction of Uterine Contractility, and Alleviation of Generalized Hyperalgesia in Mice With Induced Adenomyosis by Levo-Tetrahydropalmatine (l-thp) and Andrographolide. Reprod Sci 2011;18:1025—37.

[168] Liu X, Guo SW. Valproic acid alleviates generalized hyperalgesia in mice with induced adenomyosis. J Obstet Gynaecol Res 37:696—708.

[169] Basselin M, Chang L, Chen M, Bell JM, Rapoport SI. Chronic administration of valproic acid reduces brain NMDA signaling via arachidonic acid in unanesthetized rats. Neurochem Res 2008;33:2229—40.

[170] Woolf CJ, Thompson SW. The induction and maintenance of central sensitization is dependent on N-methyl-D-aspartic acid receptor activation; implications for the treatment of post-injury pain hypersensitivity states. Pain 1991;44:293—9.

[171] Gao YJ, Ji RR. c-Fos and pERK, which is a better marker for neuronal activation and central sensitization after noxious stimulation and tissue injury? Open Pain J 2009;2:11—7.

[172] Sun RQ, Lawand NB, Willis WD. The role of calcitonin gene-related peptide (CGRP) in the generation and maintenance of mechanical allodynia and hyperalgesia in rats after intradermal injection of capsaicin. Pain 2003;104:201—8.

[173] Ikeuchi M, Kolker SJ, Burnes LA, Walder RY, Sluka KA. Role of ASIC3 in the primary and secondary hyperalgesia produced by joint inflammation in mice. Pain 2008;137:662—9.

[174] Latremoliere A, Woolf CJ. Central sensitization: a generator of pain hypersensitivity by central neural plasticity. J Pain 2009;10:895—926.

[175] McMahon SB. NGF as a mediator of inflammatory pain. Philos Trans R Soc Lond B Biol Sci 1996;351:431—40.

[176] Sevcik MA, Ghilardi JR, Peters CM, Lindsay TH, Halvorson KG, Jonas BM, et al. Anti-NGF therapy profoundly reduces bone cancer pain and the accompanying increase in markers of peripheral and central sensitization. Pain 2005;115:128—41.

[177] Fuks F. DNA methylation and histone modifications: teaming up to silence genes. Curr Opin Genet Dev 2005;15:490—5.

[178] Fuks F, Burgers WA, Brehm A, Hughes-Davies L, Kouzarides T. DNA methyltransferase Dnmt1 associates with histone deacetylase activity. Nat Genet 2000;24:88—91.

[179] Tauboll E, Gregoraszczuk EL, Wojtowicz AK, Milewicz T. Effects of levetiracetam and valproate on reproductive endocrine function studied in human ovarian follicular cells. Epilepsia 2009;50:1868—74.

[180] von Krogh K, Harjen H, Almas C, Zimmer KE, Dahl E, Olsaker I, Tauboll E, Ropstad E, Verhaegen S. The effect of valproate and levetiracetam on steroidogenesis in forskolin-stimulated H295R cells. Epilepsia 51:2280—8.

[181] Lindstrom TM, Mohan AR, Johnson MR, Bennett PR. Histone deacetylase inhibitors exert time-dependent effects on nuclear factor-kappaB but consistently suppress the expression of proinflammatory genes in human myometrial cells. Mol Pharmacol 2008;74:109—21.

[182] Wang J, Mahmud SA, Bitterman PB, Huo Y, Slungaard A. Histone deacetylase inhibitors suppress TF-kappaB-dependent agonist-driven tissue factor expression in endothelial cells and monocytes. J Biol Chem 2007;282:28408—18.

[183] Dong XF, Song Q, Li LZ, Zhao CL, Wang LQ. Histone deacetylase inhibitor valproic acid inhibits proliferation and induces apoptosis in KM3 cells via downregulating VEGF receptor. Neuro Endocrinol Lett 2007;28:775—80.

[184] Krikun G, Hu Z, Osteen K, Bruner-Tran KL, Schatz F, Taylor HS, Toti P, Arcuri F, Konigsberg W, Garen A, Booth CJ, Lockwood CJ. The immunoconjugate "icon" targets aberrantly expressed endothelial tissue factor causing regression of endometriosis. Am J Pathol 176:1050—6.

[185] Krikun G, Schatz F, Taylor H, Lockwood CJ. Endometriosis and tissue factor. Ann N Y Acad Sci 2008;1127:101—5.

[186] Rogers PA, Donoghue JF, Walter LM, Girling JE. Endometrial angiogenesis, vascular maturation, and lymphangiogenesis. Reprod Sci 2009;16:147—51.

[187] Cameron EE, Bachman KE, Myohanen S, Herman JG, Baylin SB. Synergy of demethylation and histone deacetylase inhibition in the re-expression of genes silenced in cancer. Nat Genet 1999;21:103—7.

[188] Eden A, Gaudet F, Waghmare A, Jaenisch R. Chromosomal instability and tumors promoted by DNA hypomethylation. Science 2003;300:455.

[189] Gaudet F, Hodgson JG, Eden A, Jackson-Grusby L, Dausman J, Gray JW, et al. Induction of tumors in mice by genomic hypomethylation. Science 2003;300:489—92.

[190] Suzuki H, Gabrielson E, Chen W, Anbazhagan R, van Engeland M, Weijenberg MP, et al. A genomic screen for genes upregulated by demethylation and histone deacetylase inhibition in human colorectal cancer. Nat Genet 2002;31:141—9.

[191] Heller G, Schmidt WM, Ziegler B, Holzer S, Mullauer L, Bilban M, et al. Genome-wide transcriptional response to 5-aza-2'-deoxycytidine and trichostatin a in multiple myeloma cells. Cancer Res 2008;68:44—54.

[192] Liang G, Gonzales FA, Jones PA, Orntoft TF, Thykjaer T. Analysis of gene induction in human fibroblasts and bladder cancer cells exposed to the methylation inhibitor 5-aza-2'-deoxycytidine. Cancer Res 2002;62:961—6.

[193] Van Lint C, Emiliani S, Verdin E. The expression of a small fraction of cellular genes is changed in response to histone hyperacetylation. Gene Expr 1996;5:245—53.

[194] Bodden-Heidrich R, Hilberink M, Frommer J, Stratkotter A, Rechenberger I, Bender HG, et al. [Qualitative research on psychosomatic aspects of endometriosis]. Z Psychosom Med Psychother 1999;45:372—89.

[195] Yang AS, Estecio MR, Garcia-Manero G, Kantarjian HM, Issa JP. Comment on "Chromosomal instability and tumors promoted by DNA hypomethylation" and "Induction of tumors in nice by genomic hypomethylation". Science 2003;302:1153. author reply 1153.

468

[196] Lubbert M, Wijermans P, Kunzmann R, Verhoef G, Bosly A, Ravoet C, et al. Cytogenetic responses in high-risk myelodysplastic syndrome following low-dose treatment with the DNA methylation inhibitor 5-aza-2′-deoxycytidine. Br J Haematol 2001;114:349–57.

[197] Garcia-Manero M, Santana GT, Alcazar JL. Relationship between Microvascular Density and Expression of Vascular Endothelial Growth Factor in Patients with Ovarian Endometriosis. J Womens Health (Larchmt) 2008;17:777–82.

[198] Chandrareddy A, Muneyyirci-Delale O. Risks versus benefits of valproic acid? Fertil Steril 2008;90:238. author reply 238–9.

[199] Fluck CE, Yaworsky DC, Miller WL. Effects of anticonvulsants on human p450c17 (17alpha-hydroxylase/17,20 lyase) and 3beta-hydroxysteroid dehydrogenase type 2. Epilepsia 2005;46:444–8.

[200] Bofinger DP, Feng L, Chi LH, Love J, Stephen FD, Sutter TR, et al. Effect of TCDD exposure on CYP1A1 and CYP1B1 expression in explant cultures of human endometrium. Toxicol Sci 2001;62:299–314.

[201] McIntyre RS, Mancini DA, McCann S, Srinivasan J, Kennedy SH. Valproate, bipolar disorder and polycystic ovarian syndrome. Bipolar Disord 2003;5:28–35.

[202] Bahamondes L, Petta CA, Fernandes A, Monteiro I. Use of the levonorgestrel-releasing intrauterine system in women with endometriosis, chronic pelvic pain and dysmenorrhea. Contraception 2007;75:S134–9.

[203] Spannhoff A, Kim YK, Raynal NJ, Gharibyan V, Su MB, Zhou YY, et al. Histone deacetylase inhibitor activity in royal jelly might facilitate caste switching in bees. EMBO Rep 2011;12:238–43.

[204] Kucharski R, Maleszka J, Foret S, Maleszka R. Nutritional control of reproductive status in honeybees via DNA methylation. Science 2008;319:1827–30.

[205] Barchuk AR, Cristino AS, Kucharski R, Costa LF, Simoes ZL, Maleszka R. Molecular determinants of caste differentiation in the highly eusocial honeybee Apis mellifera. BMC Dev Biol 2007;7:70.

[206] Fiegl H, Gattringer C, Widschwendter A, Schneitter A, Ramoni A, Sarlay D, et al. Methylated DNA collected by tampons—a new tool to detect endometrial cancer. Cancer Epidemiol Biomarkers Prev 2004;13:882–8.

[207] Shen F, Wang Y, Lu Y, Yuan L, Liu X, Guo SW. Immunoreactivity of progesterone receptor isoform B and nuclear factor kappa-B as biomarkers for recurrence of ovarian endometriomas. Am J Obstet Gynecol 2008;199:e1–486.e10.

[208] Jenuwein T. The epigenetic magic of histone lysine methylation. FEBS J 2006;273:3121–35.

[209] Wolffe AP. Chromatin remodeling: why it is important in cancer. Oncogene 2001;20:2988–90.

[210] Nimura K, Ura K, Kaneda Y. Histone methyltransferases: regulation of transcription and contribution to human disease. J Mol Med 88:1213–20.

[211] Pedersen MT, Helin K. Histone demethylases in development and disease. Trends Cell Biol 20:662–71.

[212] Liu XS, Nie J, Guo SW. Elevated immunoreactivity against class I histone deacetylases (HDACs) in adenomyosis. Gynecol Obstet Invest 2012. in press.

CHAPTER 23

Aberrant DNA Methylation in Endometrial Cancer

Kenta Masuda, Kouji Banno, Megumi Yanokura, Kosuke Tsuji, Iori Kisu, Arisa Ueki, Yusuke Kobayashi, Hiroyuki Nomura, Akira Hirasawa, Nobuyuki Susumu, Daisuke Aoki
Keio University School of Medicine, Tokyo, Japan

23.1 INTRODUCTION

Epigenetics refers to the information stored after somatic cell division that is not contained within the DNA base sequence. Recent findings have shown that epigenetic changes — selective abnormalities in gene function that are not due to DNA base sequence abnormalities — play a significant role in carcinogenesis in various organs [1,2]. In particular, the relationship between cancer and aberrant hypermethylation of specific genome regions has attracted attention. A completely new model for the mechanism of carcinogenesis has been proposed in which hypermethylation of unmethylated CpG islands in the promoter regions of cancer-related genes in normal cells silences these genes and leads to the cell becoming cancerous (Figure 23.1). Both genetic and epigenetic changes are intricately involved in the process through which cells become cancerous, and hypermethylation of cancer-related genes such as *p16*, *APC*, and *hMLH1* has been associated with several types of cancer [3,4]. The main difference between epigenetic abnormalities and genetic abnormalities, such as gene mutations, is that epigenetic changes are reversible and do not involve changes in base sequence, which suggests that gene re-expression is possible and that epigenetic data may lead to important molecular targets for treatment. Attempts have begun to detect aberrant DNA methylation of cancer cells present in minute quantities in biological samples and to apply the results to cancer diagnosis, prediction of the risk of carcinogenesis, and definition of the properties of a particular cancer [5].

In Japan, the number of women with endometrial cancer and the prevalence and mortality rate of this cancer continue to increase due to westernization of lifestyles and environmental

T. Tollefsbol (Ed): Epigenetics in Human Disease. DOI: 10.1016/B978-0-12-388415-2.00023-8

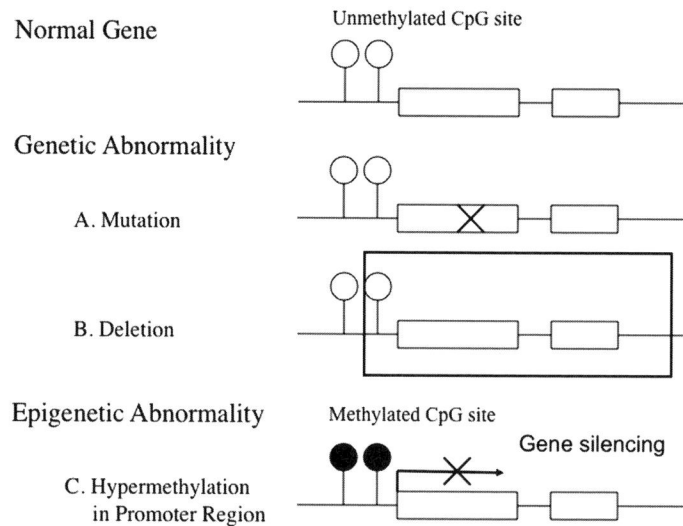

FIGURE 23.1
Mechanism of inactivation of cancer-related genes.

changes. Endometrial cancer is positively associated with higher BMI and obesity at age 20 and weight gain during adulthood among Japanese women [6,7]. Endometrial cancer currently accounts for approximately 40% of all cancers of the uterus and an increase in the total number of patients and the number of young women with this condition has been forecast. Elucidation of the pathogenesis and establishment of effective treatment for endometrial cancer are significant challenges in gynecological oncology, but many aspects of the carcinogenic mechanism are still poorly understood. The conventional explanation of the mechanism involving genetic changes — mutations of cancer-related genes — is inadequate and epigenetic changes in endometrial cancer are now being examined. In particular, aberrant DNA methylation is thought to play a key role in endometrial carcinogenesis. Breakdown of the DNA mismatch repair mechanism plays a particularly important role in the development of type I endometrial cancer, and inhibition of *hMLH1* expression due to DNA methylation may contribute significantly to this mechanism. Therefore, an understanding of the epigenetics of DNA methylation may shed light on the mechanism of carcinogenesis and improve diagnosis, risk evaluation, treatment, and prevention of endometrial cancer.

23.2 EPIGENETIC DNA HYPERMETHYLATION IN CANCER CELLS

Mechanisms involved in epigenetic regulation of gene expression involve DNA methylation, histone modification, and polycomb-group proteins [8]. DNA methylation patterns are faithfully stored after cell division and DNA methylation is one of the most common and best-studied epigenetic modifications in mammals. Genomic DNA methylation in vertebrates occurs at the cytosine in CpG sites; that is, where a cytosine is directly followed by a guanine in the DNA sequence. Transfer of the methyl group from S-adenosyl-L-methionine is catalyzed by DNA methyltransferase enzymes in two distinct processes referred to as maintenance methylation and de novo methylation. Maintenance methylation takes place after DNA replication, which produces hemimethylated DNA in which only one of the strands is methylated. CpG sites on the daughter strand are then methylated in exactly the same way as the parent strand. In maintenance methylation, the methylation pattern of the parent strand is copied onto the daughter strand produced by DNA replication during cell division, thus allowing maintenance of the pattern. De novo methylation involves methylation of a completely unmethylated CpG, allowing new methylation in the course of the generation or differentiation of cells, aging, or neoplastic transformation. The known DNA methyltransferases Dnmt1, Dnmt2, Dnmt3a,

Dnmt3b, and Dnmt3L are classified according to whether they catalyze maintenance or de novo methylation. Dnmt1 maintains attachment of methyl groups to hemimethylated DNA during replication, whereas Dnmt3a and Dnmt3b can catalyze de novo methylation of DNA.

DNA methylation in a region with a dense concentration of CpG sites (CpG islands) upstream from the transcription initiation site has a critical effect on gene expression [9]. It has also been shown that a region of unmethylated DNA tends not to form nucleosomes, allowing transcription to occur, whereas methylated DNA induces nucleosome formation that renders transcription impossible [10,11].

Tumor suppressor genes such as *CDKN2A*, *CDH1* (*E-cadherin*), and *hMLH1* are silenced as a result of aberrant DNA methylation of CpG islands in their promoter regions. Aberrant DNA methylation may play a significant role in carcinogenesis, as for gene mutations [12].

Decreased DNA methylation (hypomethylation) is an early event in carcinogenesis, and one of the first epigenetic alterations [13]. It is associated with early-stage genetic instability and up-regulation of gene expression [14].

23.3 ABERRANT DNA METHYLATION IN ENDOMETRIAL CANCER

Endometrial cancer is classified into types I and II according to clinicopathological characteristics. Type I endometrial cancer mainly occurs in pre- or perimenopausal women; is estrogen-dependent and positive for both estrogen and progesterone receptors; and develops from endometrial hyperplasia. Pathologically, type I endometrial cancer is a well-differentiated endometrioid adenocarcinoma with a low incidence of lymph node metastasis and myometrial invasion. The prognosis is usually comparatively favorable. Type II endometrial cancer mainly occurs in postmenopausal women; is estrogen-independent; and is thought to develop from a normal endometrium directly or from unspecified precancerous lesions, but not from endometrial hyperplasia. Histologically, it is an unusual type with poorly differentiated endometrioid adenocarcinoma or serous adenocarcinoma, and the prognosis is usually poor. Different molecular mechanisms are thought to be involved in the development of the two types of endometrial cancer. An epigenetic mechanism has been proposed for development of type I endometrial cancer.

Aberrant DNA methylation is common in type I but not type II endometrial cancer [15]. This difference in the methylation levels between types I and II cancers could be explained by the increased expression levels in DNA methyltransferase1 (DNMT1) and DNMT3B in type I endometrial cancer in contrast to the reduced DNMT1 and DNMT3B expression levels in type II endometrial cancer [16].

In type I endometrial cancer, DNA mismatch repair (MMR) deficiency is a typical genetic defect. The DNA mismatch repair system corrects errors in bases that arise when genes are replicated during cell division and silencing of DNA mismatch repair genes reduces the ability to repair gene mutations. This results in an accumulation of cancer-related gene mutations, leading to carcinogenesis. The MMR gene *hMLH1* is a typical gene that is silenced by promoter DNA methylation. In endometrial cancer, *hMLH1* silencing is found in approximately 40% of cases and is an important step in the early stages of carcinogenesis, with the loss of DNA mismatch repair function proposed to lead to mutation of genes such as *PTEN* [17,18].

Microsatellite instability (MSI) occurs when the mismatch repair system is damaged. Microsatellites are DNA sequences of repeating units of between one and five base-pairs. Abnormalities in the mismatch repair system may cause replication errors in the repeating unit, leading to changes in length that are referred to as microsatellite instability. MSI is observed in certain types of cancer, including approximately 20—30% of cases of endometrial cancer [19]. These results suggest that MMR gene abnormalities occur frequently in endometrial cancer.

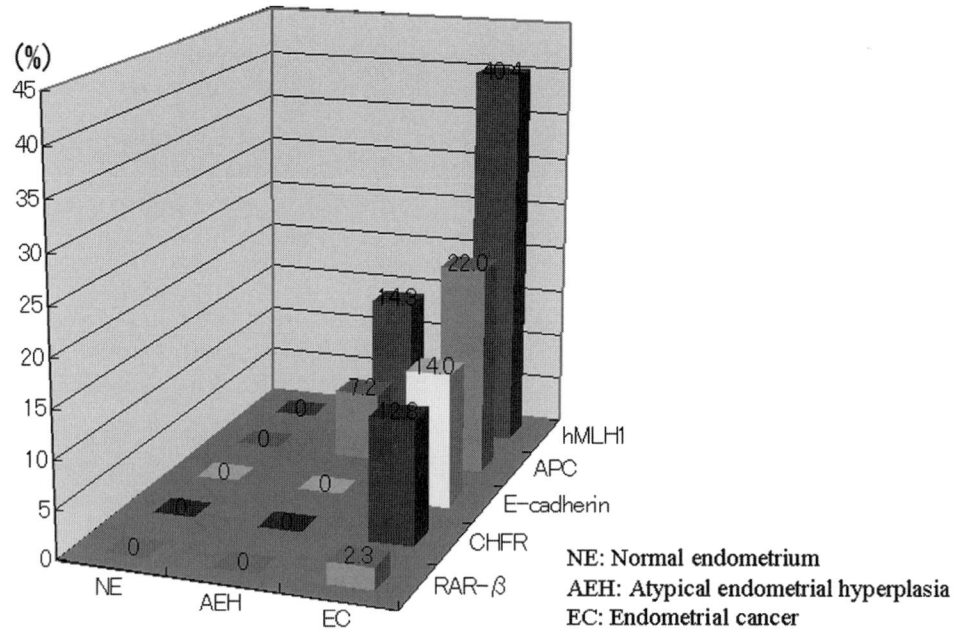

FIGURE 23.2
Frequencies of aberrant methylation of cancer-related genes in specimens from normal endometria, atypical endometrial hyperplasia and endometrial cancer. NE, normal endometrium; AEH, atypical endometrial hyperplasia; EC, endometrial cancer. This figure is reproduced in the color plate section.
Source: Reproduced from [17], with permission.

474

In patients with endometrial cancer, aberrant DNA hypermethylation was found in the promoter CpG islands of *hMLH1, APC, E-cadherin*, and *CHFR*. The frequencies of aberrant hypermethylation were 40.4% in *hMLH1*, 22.0% in *APC*, 14.0% in *E-cadherin*, and 13.3% in *CHFR*, respectively. A significant decrease in protein expression was found in patients with aberrant methylation of *hMLH1* ($p < 0.01$) and *E-cadherin* ($p < 0.05$), and aberrant methylation of *hMLH1* was also found in 14.3% of patients with atypical endometrial hyperplasia. However, no aberrant methylation of the four cancer-related genes was found in patients with a normal endometrium. These results indicate that aberrant methylation of specific genes associated with carcinogenesis in endometrial cancer does not occur in a normal endometrium, with aberrant methylation of the *hMLH1* gene being most frequent [17]. The aberrant methylation of *hMLH1* in atypical endometrial hyperplasia, which is found in the first stage of endometrial cancer, supports the hypothesis that *hMLH1* aberrant methylation is an important event in carcinogenesis in endometrial cancer [17] (Figure 23.2).

23.4 METHYLATION OF microRNA IN ENDOMETRIAL CANCER

MicroRNAs (miRNAs) are small non-coding RNAs of approximately 22 base-pairs that regulate the expression of genes by targeting mRNA with complementarity with the miRNA base sequence. Regulation of gene expression by miRNAs is important in cellular development and differentiation, and recent studies suggest a relationship between human diseases and the breakdown of gene silencing mechanisms induced by miRNA abnormalities. In particular, abnormal miRNA expression has been detected in various cancers and the target genes have been identified.

The first evidence of a correlation between miRNAs and cancer was reported by Calin et al., who observed knockdown or knockout of *miR-15a* and *miR-16-1* in approximatively 69% of CLL patients [20]. It is now known that many miRNAs have actions that make them equivalent to oncogenes or tumor-suppressor genes in cancer development and progression.

MiRNA expression can be regulated by several mechanisms, chromosomal abnormalities, mutations, polymorphisms (SNPs) [21,22]. In addition, also epigenetic mechanisms, such as promoter methylation or histone acetylation, can modulate miRNA expression, and an aberrant regulation at this level is found in different diseases, including cancer. Several evidences have indeed proved that an altered methylation status can be responsible for the deregulated expression of miRNAs in cancer. In endometrial cancer, Huang et al. reported that *SOX4* oncogene was overexpressed, and miRNA, *miR-129-2*, was validated to be an upstream regulator of SOX4. The hypermethylation of the miR-129-2 CpG island, which was observed in endometrial cancer cell lines and primary tumors, silences *miR-129-2* and depresses its oncogenic target, *SOX4* [23]. Tsuruta et al. found *miR-152* was a tumor-suppressor miRNA gene in endometrial cancer, and *miR-152* was silenced by DNA hypermethylation. They also identified *E2F3*, *MET*, and *Rictor* as candidate targets of miR-152 [24]. MiRNAs themselves can regulate the expression of components of the epigenetic machinery, creating a highly controlled feedback mechanism.

An aberrant expression of these miRNAs, called "epi-miRNAs", has been often related to development or progression of human cancer. The first evidence of the existence of epi-miRNAs was reported in lung cancer, where *miR-29* family has been shown to directly target the de novo DNA methyltransferases DNMT-3A and -3B, and more recently the *maintenance* DNA methyl transferase DNMT [25].

23.5 APPLICATION OF ABERRANT DNA HYPERMETHYLATION TO DIAGNOSTICS

Aberrant DNA methylation can be analyzed using the polymerase chain reaction (PCR), which shows a high degree of sensitivity for minute quantities of DNA in biological samples. However, it is important to ensure that the DNA hypermethylation is specific to cancer cells, since aberrant DNA hypermethylation can also occur in non-cancerous cells. Specific aberrant DNA hypermethylation may be applicable to cancer diagnosis, but use of this method for cancer screening requires detection of cancer-cell genes with aberrant hypermethylation in clinical samples that also contain normal cells. Methylation-specific PCR (MSP), which combines bisulfite sequencing and PCR, can detect aberrant hypermethylation with a high degree of sensitivity using small quantities of DNA. In bisulfite sequencing, cytosine is converted to uracil, but methylated cytosine is not converted. The PCR is thus set up with primers for sequences containing cytosine or uracil and amplification is performed to detect aberrant hypermethylation (Figure 23.3). In several types of cancer, this procedure can be used with various biological samples including sputum, plasma, and urine. In endometrial cancer, MSP has been used with endometrial cell samples to detect aberrant methylation of cancer-associated genes [17].

The methylation status of common tumor-suppressor genes could be useful to distinguish between histological subtypes of endometrial cancer. Seeber et al. compared methylation

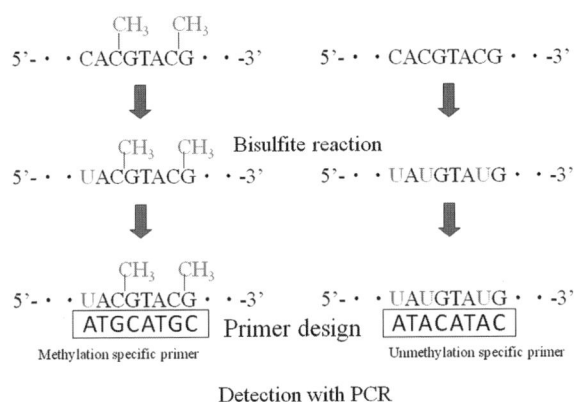

FIGURE 23.3
Principles of bisulfite treatment and methylation-specific PCR.

status of a set of common tumor-suppressor genes, previously studied by Joensuu et al. [26], in endometrioid endometrial carcinoma and uterine serous carcinoma. Promoter methylation of *CDH13* (Cadherin 13, H-cadherin) and *MLH1* was more frequently present in endometrioid carcinoma, while *CDKN2B* and *TP73* were more frequently methylated in serous carcinoma. Almost 90% of endometrioid carcinoma and 70% of serous carcinoma could be predicted by *CDH13* and *TP73* [15].

Aberrant DNA hypermethylation has been reported to affect several genes in endometrial cancer, in addition to *hMLH1*. Such genes include *CASP8* (27), an apoptosis-related gene; *TGF-βRII* (28), a TGF-β receptor with a tumor-suppressor effect; *p73* (27), a tumor-suppressor gene; *HOXA11* (29), which is important in uterine development; and *COMT* (30), which codes for the catechol-O-methyltransferase that metabolizes catechol, an estrogen metabolite that plays a role in carcinogenesis. Methylation of each of these genes results in a loss of protein expression that promotes cancer, and the degree of methylation of the genes differs significantly between normal tissues and endometrial cancer tissues [31].

Aberrant DNA methylation of some tumor-suppressor genes was evident before endometrial carcinoma diagnosis in women with the DNA mismatch repair gene mutation. That means the timing and molecular alterations of the critical events in endometrial carcinogenesis may be useful to identify DNA methylation profile for early detection of endometrial cancer [32].

The properties of cancer cells can be significantly affected by aberrant DNA hypermethylation. For example, in colorectal cancer, methylation of *CHFR* is observed in 40% of cases and is closely related to sensitivity to the microtubule inhibitor docetaxel [33]. Other examples of similar effects include the relationship between aberrant hypermethylation of the DNA repair enzyme *MGMT* and alkylating agent sensitivity [34], the link between simultaneous hypermethylation of *CDKN2A* and *FHIT* and recurrence of lung cancer [35], and the link between methylation of ER-α and prognosis during treatment with tamoxifen [36]. This last example indicates that detection of aberrant hypermethylation can be used for prognostic evaluation. Aberrant DNA hypermethylation is sometimes seen in the elderly and in non-cancerous regions in cancer patients [37−39]. In particular, it has been shown that exposure to *H. pylori*, a carcinogenic factor for gastric cancer, induces aberrant DNA hypermethylation and that the degree of aberration is correlated with carcinogenic risk. This leads to the concept of an "epigenetic field for cancerization", with similar results reported for colorectal and breast cancer [39]. Accumulation of aberrant methylation in normal tissue may trigger carcinogenesis and such information may be useful to evaluate carcinogenic risk.

To analyze DNA methylation patterns on a genome-wide scale, several techniques have been developed. Immunoprecipitation approach by an antibody against 5-methylcytosine combined with DNA microarrays was developed [40]. As a similar method, the MIRA (methylated-CpG island recovery assay) microarray approach was devised. It makes use of the high affinity of the MBD (methyl-CpG binding proteins) complex for methylated DNA [41]. One of quantitative approaches currently used as standard methods employs base-specific cleavage and MALDI-TOF MS (matrix-assisted laser desorption ionization time-of-flight mass spectrometry) analysis [42,43].

23.6 APPLICATION OF ABERRANT DNA HYPERMETHYLATION TO TREATMENT

Unlike irreversible genetic changes, epigenetic patterns observed in cancer can be partly or fully reversed pharmacologically. Indeed, this is the main reason for development of anticancer drugs for epigenetic modification. Treatment of cancer using demethylating agents to restore expression of cancer-suppressor genes silenced through methylation has been attempted for some time and use of methylation inhibitors to treat cancer has a long history. Development of an antileukemic agent based on this principle was performed in the 1960s, but the drug could not be used clinically due to toxicity problems at high concentrations. More recent reports have shown antitumor effects at lower drug concentrations with a lower incidence of adverse drug reactions and concomitant use with other chemotherapeutic agents may further improve efficacy. Lubbert et al. reported a response rate of 60% in patients with myelodysplastic syndrome (MDS) following administration of the methylation inhibitor 5-aza-2'-deoxy-cytidine (5-aza-dC) [44], with induction of expression of *p15INK4A* following demethylation in patients who were responsive to 5-aza-dC. A group from MD Anderson Cancer Center reported that repeated administration of low-concentration 5-aza-dC produced a response rate of approximately 60% in patients with acute myeloid leukemia [45]. However, the disadvantage of this methylation inhibitor is that it is not sequence-specific, which may lead to adverse effects through demethylation of physiologically important genes and reactivation of cancer genes silenced by methylation. Therefore, development of sequence-specific demethylating agents based on binding sequence of transcription factors is a current area of research.

Epigenetic abnormalities have also been examined as markers of anticancer drug sensitivity. Esteller et al. discovered that the DNA repair enzyme *MGMT* gene is silenced by methylation and showed that tumor cell lines in which *MGMT* is methylated are highly responsive to alkylating agents, with a greater antitumor effect of BNCU observed in patients in whom *MGMT* methylation was detected in the tumor compared to those with no *MGMT* methylation [46]. Sato et al. reported that aberrant hypermethylation of *CHFR*, a mitotic checkpoint gene, is strongly correlated with responsiveness to taxanes, which are microtubule inhibitors [47] (Figure 23.4). These findings suggest that it may be possible to select a treatment based on methylation as an indicator of the biological characteristics of tumor cells. *CHFR* methylation may be a particularly sensitive marker in endometrial cancer and analysis of this gene may play an important role in treatment of this type of cancer. *CHFR* is an M-phase checkpoint gene with folk-head associated (FHA) and ring finger domains that was first identified as a yeast DMA 1 gene homolog [48]. If the cell is subjected to mitotic stress in the M-phase, CHFR delays progression from the start of prophase to the later part of prophase in mitosis. CHFR is an ubiquitin ligase that includes Aurora-A and PLK1 among its substrates, and degradation of these proteins is thought to stop the cell cycle following *CHFR* activation by microtubular stress [49,50]. The FHA domain of CHFR is involved in binding of phosphorylated proteins and is important in the checkpoint function, but further studies are required to identify molecules that interact with CHFR.

CHFR methylation is thought to occur frequently in cancer [33,47] and is related to the mitotic index as follows. In normally functioning cells with no *CHFR* methylation, cell-cycle arrest occurs during the G2/M phase following administration of docetaxel and the mitotic index is low. Conversely, in cells in which *CHFR* hypermethylation has reduced expression of the gene,

477

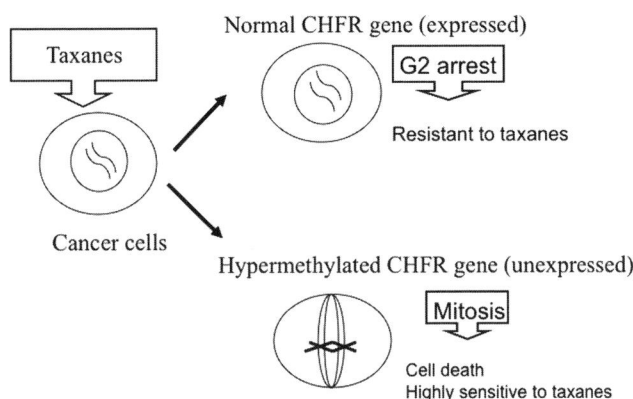

FIGURE 23.4
Hypothesis for the relationship of expression of CHFR and sensitivity of cancer cells to taxanes.

the mitotic index is high following administration of docetaxel. These results show that the expression level of *CHFR* is negatively correlated with the mitotic index [33]. In cells in which the *CHFR* checkpoint does not function, translocation of cyclin B1 to the nucleus during mitotic stress cannot be prevented. Treatment of cells with *CHFR* hypermethylation with the methylation inhibitor 5-aza-dC restores the checkpoint function and decreases the mitotic index. *CHFR* methylation and microtubule inhibitor sensitivity are useful molecular markers in gastric and cervical cancer, as well as in endometrial cancer, and methylation may be a useful predictor of anticancer drug responsiveness [51—53].

23.7 FUTURE DIRECTIONS AND CONCLUSION

Progress made in epigenetics in recent years has suggested that aberrant DNA hypermethylation plays a role in carcinogenesis in several types of cancer. Various genes are silenced as a result of aberrant hypermethylation, including cell cycle regulatory genes, apoptosis-related genes, and DNA repair enzymes.

Epigenetic research in endometrial cancer suggests that damage to the mismatch repair system plays a significant role in development of type I endometrial cancer and that *hMLH1* hypermethylation is important in this mechanism. Such research has potential for prevention, diagnosis, risk assessment, and treatment of endometrial cancer. Cancer-specific DNA methylation may be useful for diagnosis using methods such as MSP for detection of such abnormalities. Aberrant DNA hypermethylation can be detected with a high level of sensitivity and cancer cells can be detected in minute quantities of endometrial samples. Treatment with methylation inhibitors such as 5-aza-dC may also be effective, since a low concentration of this drug has an antitumor effect with a reduced incidence of adverse drug reactions, and concomitant use with other chemotherapy drugs may show even greater efficacy. Attempts are also being made to use epigenetic abnormalities as indicators of anticancer drug sensitivity, which may allow selection of the most appropriate treatment based on the biological characteristics of tumor cells. Aberrant *CHFR* hypermethylation is strongly correlated with sensitivity to microtubule inhibitors and these findings may be applicable in treatment of endometrial cancer. The main objective of epigenetics in oncology research is to identify aberrant gene hypermethylation associated with carcinogenesis. These findings may lead to new methods of diagnosis and treatment based on control of methylation, including new approaches to treatment of endometrial cancer.

References

[1] Chan MW, Chan LW, Tang NL, Tong JH, Lo KW, Lee TL, et al. Hypermethylation of multiple genes in tumor tissues and voided urine in urinary bladder cancer patients. Clin Cancer Res 2002;8(2):464—70.

[2] Palmisano WA, Divine KK, Saccomanno G, Gilliland FD, Baylin SB, Herman JG, et al. Predicting lung cancer by detecting aberrant promoter methylation in sputum. Cancer Res 2000;60(21):5954—8.

[3] Baylin SB, Herman JG. DNA hypermethylation in tumorigenesis: epigenetics joins genetics. Trends Genet 2000;16:168—74.

[4] Esteller M, Corn PG, Baylin SB, Herman JG. A gene hypermethylation profile of human cancer. Cancer Res 2001;61:3225—9.

[5] Fiegl H, Gattringer C, Widschwendter A, Schneitter A, Ramoni A, Sarlay D, et al. Methylated DNA collected by tampons—A new tool to detect endometrial cancer. Cancer Epidemiol. Biomarkers Prev 2004;13(5):882—8.

[6] Kuriyama S, Tsubono Y, Hozawa A, Shimazu T, Suzuki Y, Koizumi Y, et al. Obesity and risk of cancer in Japan. Inter J Cancer 2005;113:148—57.

[7] Satoyo H, Keitaro M, Kaoru H, Hidemi I, Takeshi S, Takakazu K, et al. Weight gain during adulthood and body weight at age 20 are associated with the risk of endometrial cancer in Japanese Women. J Epidemiol 2011;21(6):466—73.

[8] Goldberg AD, Allis CD, Bernstein E. Epigenetics: a landscape takes shape. Cell 2007;128(4):635—8.

[9] Ushijima T. Detection and interpretation of altered methylation patterns in cancer cells. Nat Rev Cancer 2005;5:223—31.

[10] Gal-Yam EN, Jeong S, Tanay A, Egger G. Lee AS and Jones PA: Constitutive nucleosome depletion and ordered factor assembly at the GRP78 promoter revealed by single molecule footprinting. PLoS Genet 2006;2:160.

[11] Appanah R, Dickerson DR, Goyal P, Groudine M, Lorincz MC. An unmethylated 3′ promoter-proximal region is required for efficient transcription initiation. PLoS Genet 2007;3:27.

[12] Jones PA, Baylin SB. The epigenomics of cancer. Cell 2007;128:683—92.

[13] Feinberg AP, Vogelstein B. Hypomethylation distinguishes genes of some human cancers from their normal counterparts. Nature 1983;301:89—92.

[14] Wilson AS, Power BE, Molloy PL. DNA hypomethylation and human diseases. Biochim Biophs Acta 2007;1775:138—62.

[15] Seeber LM, Zweemer RP, Marchionni L, Massuger LF, Smit VT, van Baal WM, et al. Methylation profiles of endometrioid and serous endometrial cancers. Endocr Relat Cancer 2010;17(3):663—73.

[16] Zhou XC, Dowdy SC, Podratz KC, Jiang SW. Epigenetic considerations for endometrial cancer prevention, diagnosis and treatment. Gynecol Oncol 2007;107:143—53.

[17] Banno K, Yanokura M, Susumu N, Kawaguchi M, Hirao N, Hirasawa A, et al. Relationship of aberrant DNA hypermethylation of cancer-related genes with carcinogenesis of endometrial cancer. Oncol Rep 2006;16:1189—96.

[18] Kanaya T, Kyo S, Sakaguchi J, Maida Y, Nakamura M, Takakura M, et al. Association of mismatch repair deficiency with PTEN frameshift mutations in endometrial cancers and the precursors in a Japanese population. Am J Clin Pathol 2005;124:89—96.

[19] Kanaya T, Kyo S, Maida Y, Yatabe N, Tanaka M, Nakamura M, et al. Frequent hypermethylation of MLH1 promoter in normal endometrium of patients with endometrial cancers. Oncogene 2003;22:2352—60.

[20] Calin GA, Dumitru CD, Shimizu M, Bichi R, Zupo S, Noch E, et al. Frequent deletions and down-regulation of microRNA genes miR15 and miR16 at 13q14 in chronic lymphocytic leukemia. Proc Natl Acad Sci USA 2002;99:15524—9.

[21] Zhang L, Huang J, Yang N, Greshock J, Megraw MS, Giannakakis A, et al. MicroRNAs exhibit high frequency genomic alterations in human cancer. Proc Natl Acad Sci USA 2006;103:9136—41.

[22] Hu Z, Chen J, Tian T, Zhou X, Gu H, Xu L, et al. Genetic variants of miRNA sequences and non-small cell lung cancer survival. J Clin Invest 2008;118:2600—8.

[23] Huang YW, Liu JC, Deatherage DE, Luo J, Mutch DG, Goodfellow PJ, et al. Epigenetic repression of microRNA-129-2 leads to overexpression of SOX4 oncogene in endometrial cancer. Cancer Res 2009;69:9038—46.

[24] Tsuruta T, Kozaki KI, Uesugi A, Furuta M, Hirasawa A, Imoto I, et al. miR-152 is a tumor suppressor microRNA that is silenced by DNA hypermethylation in endometrial cancer. Cancer Res 2011;71(20):6450—62.

[25] Fabbri M, Garzon R, Cimmino A, Liu Z, Zanesi N, Callegari E, et al. MicroRNA-29 family reverts aberrant methylation in lung cancer by targeting DNA methyltransferases3A and 3B. Proc Natl Acad Sci USA 2007;104:15805—10.

[26] Joensuu EI, Abdel-Rahman WM, Ollikainen M, Ruosaari S, Knuutila S, Peltomaki P. Epigenetic signatures of familial cancer are characteristic of tumor type and family category. Cancer Res 2008;68:4597—605.

[27] Yang HJ, Liu VW, Wang Y, Tsang PC, Ngan HY. Differential DNA methylation profiles in gynecological cancers and correlation with clinico-pathological data. BMC Cancer 2006;6:212.

[28] Sakaguchi J, Kyo S, Kanaya T, Maida Y, Hashimoto M, Nakamura M, et al. Aberrant expression and mutations of TGF-beta receptor type II gene in endometrial cancer. Gyneol Oncol 2005;98:427—33.

[29] Whitcomb BP, Mutch DG, Herzog TJ, Rader JS, Gibb RK, Goodfellow PJ. Frequent HOXA11 and THBS2 promoter methylation, and a methylator phenotype in endometrial adenocarcinoma. Clin Cancer Res 2003;9:2277—87.

[30] Sasaki M, Kaneuchi M, Sakuragi N, Dahiya R. Multiple promoters of catechol-O-methyltransferase gene are selectively inactivated by CpG hypermethylation in endometrial cancer. Cancer Res 2003;63:3101—6.

[31] Fiegl H, Gattringer C, Widschwendter A, Schneitter A, Ramoni A, Sarlay D, et al. Methylated DNA collected by tampons—a new tool to detect endometrial cancer. Cancer Epidemiol Biomarkers Prev 2004;13:882—8.

[32] Nieminen TT, Gylling A, Abdel-Rahman WM, Nuorva K, Aarnio M, Renkonen-Sinisalo L, et al. Molecular analysis of endometrial tumorigenesis: importance of complex hyperplasia regardless of atypia. Clin Cancer Res 2009;83(15):5772—83.

[33] Toyota M, Sasaki Y, Satoh A, Ogi K, Kikuchi T, Suzuki H, et al. Epigenetic inactivation of CHFR in human tumors. Proc Natl Acad Sci USA 2003;100:7818—23.

[34] Esteller M, Herman JG. Generating mutations but providing chemosensitivity: the role of O6-methylguanine DNA methyltransferase in human cancer. Oncogene 2004;23:1—8.

[35] Kim JS, Kim JW, Han J, Shim YM, Park J, Kim DH. Cohypermethylation of p16 and FHIT promoters as a prognostic factor of recurrence in surgically resected stage I non-small cell lung cancer. Cancer Res 2006;66:4049—54.

[36] Mori T, Martinez SR, O'Day SJ, Morton DL, Umetani N, Kitago M, et al. Estrogen receptor-alpha methylation predicts melanoma progression. Cancer Res 2006;66:6692—8.

[37] Waki T, Tamura G, Sato M, Motoyama T. Age-related methylation of tumor suppressor and tumor-related genes: an analysis of autopsy samples. Oncogene 2003;22:4128—33.

[38] Maekita T, Nakazawa K, Mihara M, Nakajima T, Yanaoka K, Iguchi M, et al. High levels of aberrant DNA methylation in Helicobacter pylori-infected gastric mucosae and its possible association with gastric cancer risk. Clin Cancer Res 2006;12:989—95.

[39] Ushijima T. Epigenetic field for cancerization. J Biochem Mol Biol 2007;40:142—50.

[40] Weber M, Davies JJ, Wittig D, Oakeley EJ, Haase M, Lam WL, et al. Chromosome-wide and promoter-specific analyses identify sites of differential DNA methylation in normal and transformed human cells. Nat Genet 2005;37(8):853—62.

[41] Rauch T, Li H, Wu X, Pfeifer GP. MIRA-assisted microarray analysis, a new technology for the determination of DNA methylation patterns, identifies frequent methylation of homeodomain-containing genes in lung cancer cells. Cancer Res 2006;66(16):7939—47.

[42] Ehrich M, Nelson MR, Stanssens P, Zabeau M, Liloglou T, Xinarianos G, et al. Quantitative high-throughput analysis of DNA methylation patterns by base-specific cleavage and mass spectrometry. Proc Natl Acad Sci USA 2005;102(44):15785—90.

[43] Schatz P, Distler J, Berlin K, Schuster M. Novel method for high throughput DNA methylation marker evaluation using PNA-probe library hybridization and MALDI-TOF detection. Nucleic Acids Res 2006;34(8):e59.

[44] Daskalakis M, Nguyen TT, Nguyen C, Guldberg P, Köhler G, Wijermans P, et al. Demethylation of a hyper-methylated P15/INK4B gene in patients with myelodysplastic syndrome by 5-Aza-2'-deoxycytidine (decitabine) treatment. Blood 2002;100:2957—64.

[45] Issa JP, Garcia-Manero G, Giles FJ, Mannari R, Thomas D, Faderl S, et al. Phase 1 study of low-dose prolonged exposure schedules of the hypomethylating agent 5-aza-2'-deoxycytidine (decitabine) in hematopoietic malignancies. Blood 2003;103:1635—40.

[46] Esteller M, Garcia-Foncillas J, Andion E, Goodman SN, Hidalgo OF, Vanaclocha V, et al. Inactivation of the DNA-repair gene MGMT and the clinical response of gliomas to alkylating agents. N Engl J Med 2000;343:1350—4.

[47] Satoh A, Toyota M, Itoh F, Sasaki Y, Suzuki H, Ogi K, et al. Epigenetic inactivation of CHFR and sensitivity to microtubule inhibitors in gastric cancer. Cancer Res 2003;63:8606—13.

[48] Scolnick DM, Halazonetis TD. CHFR defines a mitotic stress checkpoint that delays entry into metaphase. Nature 2000;406:430—5.

[49] Kang D, Chen J, Wong J, Fang G. The checkpoint protein CHFR is a ligase that ubiquitinates Plk1 and inhibits Cdc2 at the G2 to M transition. J Cell Biol 2002;156:249—59.

[50] Yu X, Minter-Dykhouse K, Malureanu L, Zhao WM, Zhang D, Merkle CJ, et al. CHFR is required for tumor suppression and Aurora A regulation. Nat Genet 2005;37:401—6.

[51] Ogi K, Toyota M, Mita H, Satoh A, Kashima L, Sasaki Y, et al. Small interfering RNA-induced CHFR silencing sensitizes oral squamous cell cancer cells to microtubule inhibitors. Cancer Biol Ther 2005;4:773—80.

[52] Banno K, Yanokura M, Kawaguchi M, Kuwabara Y, Akiyoshi J, Kobayashi Y, et al. Epigenetic inactivation of the CHFR gene in cervical cancer contributes to sensitivity to taxanes. Int J Oncol 2007;31:713—20.

[53] Koga Y, Kitajima Y, Miyoshi A, Sato K, Sato S, Miyazaki K. The significance of aberrant CHFR methylation for clinical response to microtubule inhibitors in gastric cancer. J Gastroenterol 2006;41:133—9.

CHAPTER 24

Stem Cell Epigenetics and Human Disease

Mehdi Shafa, Derrick E. Rancourt
University of Calgary, Calgary, AB, Canada

481

24.1 INTRODUCTION

Recent analysis of the mechanisms by which chromatin remodeling regulates embryonic and adult stem cell pluripotency/multipotency and differentiation have provided important insights into our understanding of human and mouse developmental biology. There is now a growing body of information, which shows the crucial key role of epigenetic changes and chromatin organization in the activation or repression of genes during embryogenesis and in maintaining pluripotency in stem cells. Epigenetic changes include histone acetylation, methylation and phosphorylation, and DNA methylation/demethylation. Higher-order chromatin architecture integrity is also crucial for proper gene activity in stem cells. Pluripotent stem cells, including embryonic stem cells (ESCs) and primordial germ cells, have the potential to differentiate into any cell type in an organism, whereas multipotent or unipotent stem cells have limited differentiation capacity, giving rise to defined progenies. Normally, progression from a stem state to a more differentiated cell lineage needs distinguished changes in cell function, gene expression patterns, and morphology. Recent evidence reveals the role of epigenetic modifications during a stem cell's maintenance of self-renewal, as well as during differentiation. Alterations in the epigenetic status of cells are associated with different types of human cancer and congenital disease. Induced pluripotent stem cells (iPSCs) have recently been developed by reprogramming of mouse and human dermal fibroblasts following the transduction of defined transcription factors using retroviral vectors. The reprogramming of somatic cells through this technology is a valuable tool to identify the mechanisms of pluripotency. It also provides the potential of modeling human degenerative disorders and producing patient-specific pluripotent stem cells. In this chapter, the fundamental impact of

T. Tollefsbol (Ed): Epigenetics in Human Disease. DOI: 10.1016/B978-0-12-388415-2.00024-X

chromatin dynamics in stem cells, as well as the critical role of epigenetic changes in the generation of human diseases, will be discussed. Furthermore, the relationship between stem cell epigenetics and human diseases such as cancer and the application of iPSCs in modeling human epigenetic disorders will be discussed.

24.2 EPIGENETICS

The functional properties of a specific cell are not only dependent on DNA sequence, but also on the combination of active and silent genes in a particular time during development. Although all of the cells in the body have the same identical DNA sequence, every cell has its own exclusive phenotype and gene expression pattern, which shows that the DNA is more complex and dynamic than has previously been known. There are other important mechanisms of gene regulation that rely mostly on the cell's epigenetic status, which controls the timing and degree of gene expression in a particular time. In eukaryotes, genomic DNA is organized into DNA/protein complexes known as chromatin. The central unit of chromatin is the nucleosome, which is comprised of a family of small, basic proteins called histones. The nucleosome consists of two copies of each histone protein termed H2A, H2B, H3, and H4 with a ~146 base-pair of DNA wound around its surface [1]. These bead-like core proteins are often associated with a 15—55-bp sequence of linker DNA and a fifth histone, H1, which is believed to be responsible for organizing higher-order structure between the beads.

Epigenetics is a term initially used by Waddington in 1942 describing the relationship between genotype and phenotype through different gene interactions [2]. The term epigenetic is used today to describe transcriptional memory without affecting DNA sequence and is mediated mostly by variation in DNA methylation and chromatin structure. For example, a chromatin change in the brain, influencing gene expression, can be "epigenetic" if it lasts regardless of cell division.

Specific residues in the N-terminal ends of histones, which protrude from the nucleosome surface, are susceptible to numerous reversible post-translational modifications including methylation, acetylation, and phosphorylation [3]. These alterations are obtained via various chromatin-modifying enzyme complexes with different antagonized functions, which are accountable for the dynamic performance of chromatin. The "histone code" theory is assigned to the two different states of chromatin acquired by these modifying complexes [4]. Generally, regions of transcriptional activity are distinguished by lysine acetylation alleviated by histone acetyltranferases (HATs). Alternatively, histone deacetylases (HDACs), which mediate lysine deacetylation, are related with regions of transcriptional quiescence.

Higher-order chromatin organization is distinguished by highly packed heterochromatin and relatively broadened euchromatin regions within the genome [5]. It has long been assumed that heterochromatin is transcriptionally inert compared to euchromatin. However, many recent studies have questioned this model of transcriptionally silent heterochromatin. Some studies have shown that transcription of heterochromatin is indispensable for its own repression [6]. Furthermore, a few protein-coding genes in heterochromatic domain [7,8] as well as RNAs expressed in telomeric and pericentric regions have been also reported in different species [9,10]. The crucial role of epigenetics in the regulation of stem cells and the etiology of human disorders is increasingly acknowledged and will be discussed in detail in the following sections.

24.3 STEM CELL EPIGENETICS

The important characteristics of stem cells (SCs), including self-renewal, maintenance of pluripotency and potency — the capacity to differentiate along specialized cell lineages — require the presence of specific molecular mechanisms during early development. The mechanism by which SCs select between self-renewal and differentiation seems to not only rely on specific molecular cues, but also more importantly on prominent modifications in their chromatin state. Progression from the pluripotent state to a differentiated phenotype is typically highlighted by distinguished changes in cellular function, which are predetermined by global gene expression patterns during early development. Recent understanding of the mechanisms by which chromatin remodeling controls stem cell pluripotency and differentiation have had significant influence on our knowledge about developmental biology. There is currently a growing body of evidence, which reveals the decisive role of epigenetic variations and chromatin reshaping in the repression or activation of genes during early development and also maintaining the identity and self-renewal of stem cells (Figure 24.1).

The regulatory mechanisms that regulate stem cell self-renewal are yet to be understood; however, the important function of transcription factors like Oct4, Sox2, and Nanog has been elucidated. Several recent lines of evidence underscore that stem cell differentiation and early mammalian development largely depend on the elasticity of epigenetic alterations [11]. The mechanism of gene regulation during early development does not only rely upon the interactions among different transcription factors and signaling pathways, but also on epigenetic modifications, such as the ATP-dependent chromatin remodeling [12], covalent alterations of histones [13], exchange of histones and histone variant [14], DNA methylation at CpG islands [15], RNA-mediated gene regulation including RNAi pathways and non-protein-coding RNAs (ncRNA) [16,17]. All of these have been found to play critical roles in maintenance of stem cell pluripotency/multipotency, blocking differentiation in stem cells, and controlling the inherited characteristics of cellular memories during early development.

One of the significant determining factors of gene expression in stem cells is DNA and histone methylation. DNA methylation occurs mostly at the 5′ end of the cytosine nucleotide of CpGs dinucleotide islands. However, non-CpG methylation like in CpT and CpA motifs can also take place in ESCs mediated by methyltransferase 3a [18]. Particularly, unmethylated clusters of CpG islands are located at the promoters of tissue-specific and housekeeping genes, which are required to be expressed for maintenance purposes. These unmodified CpG pairs recruit other transcription factors to start transcription. Conversely, methylated CpGs are normally located at the promoters of silent genes. Although DNA methylation at promoter regions conversely associates with gene activity, this inverse relationship is reliant upon the amount of CpG motifs within a specific promoter. Additionally, there is no apparent relationship between gene expression and methylation in promoters without distinct CpG content [19]. Methylated CpG islands are detected within promoters of definite tissue-specific genes [20], but they are absent in other regions of the genome.

DNA methylation patterns are directed and conserved by the DNA methyltransferase (DNMT) family of proteins, whereas the effects of DNA methylation are mediated by methyl-CpG-binding domain (MBD) familys, which distinguish DNA sequences with high CpG content [21]. There are three main functional DNA methytransferases in mammals including Dnmt1, Dnmt3a, Dnmt3b, and Dnmt3L. Dnmt1 is ubiquitously expressed in dividing somatic cells throughout mammalian development. Dnmt1 was the first DNA methyltransferase enzyme to be cloned [22] and shows a preference toward hemimethylated DNA over unmethylated DNA strands [23]. Dnmt3a and Dnmt3b are recognized as de novo methyltransferases and have the responsibility of establishing methylation pattern during mammalian early development and in germ cells. Both of the enzymes are highly expressed in ESCs and down-regulated upon differentiation. Dnmt3L is enzymaticaly inactive and has some regulatory functions in germ cells. It has been shown that the DNA methylation pattern in differentiated cells is persistent and can be inherited [24]. Cells achieve this pattern progressively as they move towards their ultimate specific cell lineage during development. As the zygote divides, methylation of paternal and maternal alleles are removed through down-regulation of Dnmt1 gene expression within the nucleus [25]. This phenomenon is followed by reacquirement of a new DNA methylation pattern during embryogenesis and germ layer formation. The epigenomic status of ESCs inside the blastocyst's inner cell mass is restored during this period, to re-establish pluripotency.

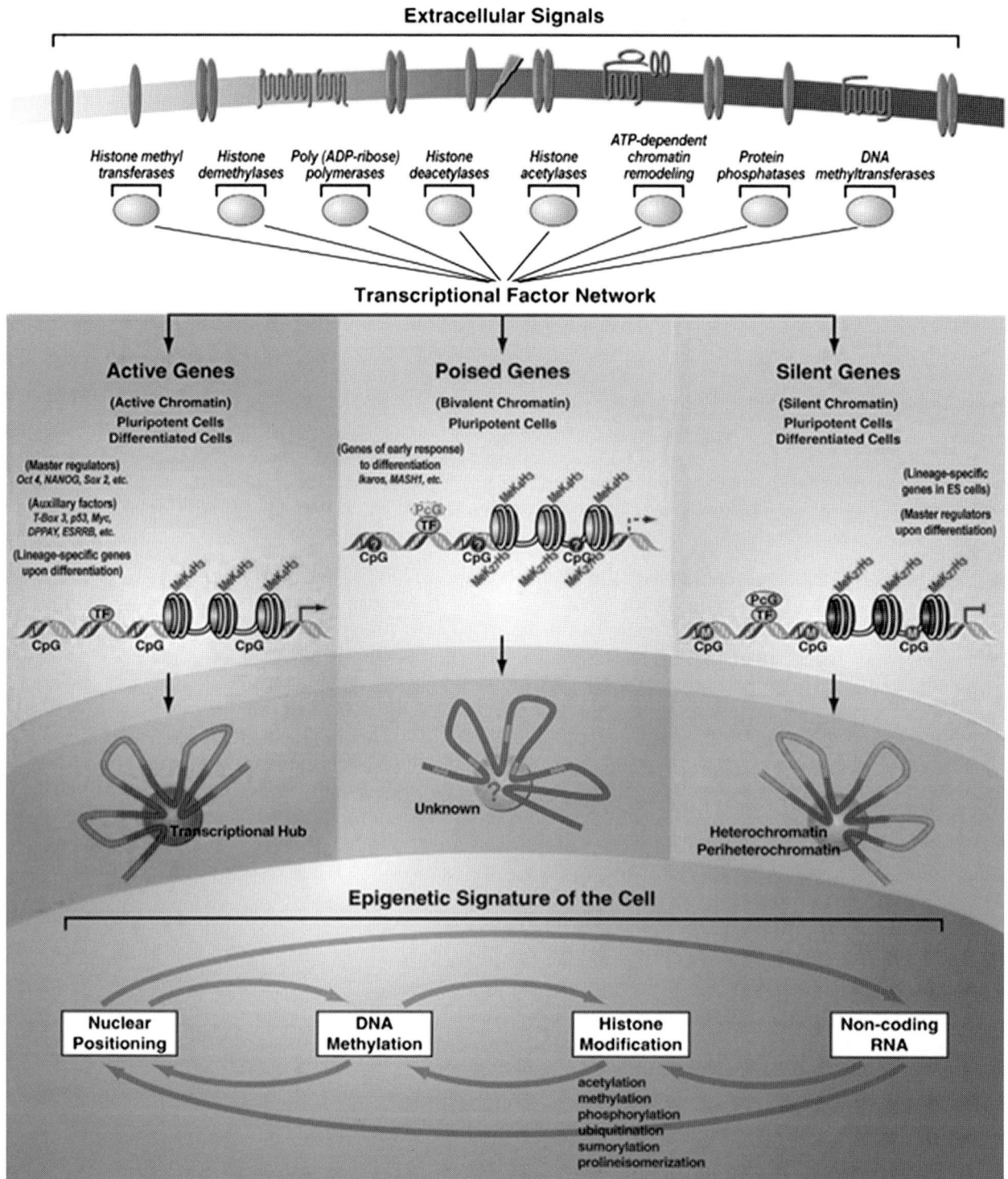

FIGURE 24.1

The pluripotency of ES cells and their differentiation rely largely on transcription factor circuitry and chromatin modifications. In this model, there is a multidirectional interaction between transcription factors and the epigenetic signature of the cell. Within this scenario, the extracellular matrix (ECM) of ES cells play an important role by providing necessary environmental signals. This figure is reproduced in the color plate section. *Source: Adapted from [132]*

The role of various DNMTs in mouse ESCs has been explained using homozygous mutants. Dnmt1$^{-/-}$ ESCs can multiply when cultured as undifferentiated cells, suggesting that the hypomethylated state can preserve self-renewal. The induction of apoptosis in these ESCs upon differentiation is highly correlated to their inability to suppress pluripotency factors like Oct4 and Nanog by DNA methylation [26]. Recent advances suggest that the CpG motif within the Oct4 promoter is hypomethylated in pluripotent ESCs yet is highly methylated in different kinds of somatic cells. In a study on a neuronally committed human teratocarcinoma cell line (NT2), down-regulation of Oct4 and Nanog expression during differentiation is corelated to methylation of 5′-flanking regions of both these genes [27]. ESCs have the potential capability to impose their pluripotent epigenetic status to differentiated cells upon fusion. For instance, when terminally differentiated cells are fused with ESCs, resulting heterokaryons acquire the pluripotent state [28]. It has been believed that this capacity is largely restricted to early embryonic cells including ESCs and that lineage-restricted cells loose this capacity to revert the epigenetic status of other nuclei.

Global chromatin dynamism is also of interest in ESCs, as recent studies stress its fundamental role in the preservation of pluripotency and regulation of gene expression [29]. Several studies point out that ESCs are delineated by less-condensed chromatin and higher transcription activity compared to differentiated cells. During the advancement of stem cell differentiation, active chromatin is substituted with a repressed and inactive state [30]. For example, ESCs accumulate highly compacted heterochromatin in the pericentric domain of some chromosomes during differentiation, which is not detected in the pluripotent state [31,32]. Likewise, the number, size, and distribution of highly compact heterochromatin foci are changed during ESC neural differentiation [33,34].

Epigenetic mechanisms have profound effects on mammalian stem cell regulation and their dysfunction can give rise to several human diseases such as neurodegenerative disorders and cancer. Nevertheless, the extents to which these modifications can determine stem cell fate are still mainly unrecognized and many questions have to be answered.

24.4 HISTONE VARIANTS AND EXCHANGE OF HISTONES

Several other replacement variants for H2A (H2A.X, H2A.Z, H2A-Bbd, MacroH2A) and H3 (H3.2, H3.3, CenpA) have been identified in addition to the four conserved histone proteins (H2A, H2B, H3, and H4). Unlike core proteins, these histone variants are transcribed from a polyA mRNA and their assembly into chromatin is controlled separately from DNA replication. The vast majority of the histone variants are found at specific chromatin conformations and at defined developmental stages. For example, MacroH2A is found at the inactive X-chromosome [35] and acts like a suppressor variant [36], whereas H2A.Z chaperons euchromatin from the ectopic distribution of inactive heterochromatin by mediating stabilization of the nucleosome [37]. Conversely, Histone H2A-Bbd has the opposite effect of destabilizing the nucleosome components.

The role of histone variants in the regulation of ESC differentiation has been documented in several studies. For instance, H2A.Z is highly expressed in human embryonic carcinoma cell lines, but its expression is diminished upon differentiation [38]. Tip60-p400, a chromatin remodeling protein, behaves as a transcriptional enhancer by including the H2A.Z to target promoters in ESCs [39]. It can be hypothesized that H2A.Z helps maintain ESC's open chromatin state, which is indispensable for pluripotency. It is important to verify the hypothesis by understanding how different histone variants like H2A.Z target specific genes in ESCs and how they may correlate to the suppression or expression of particular genes during ESCs self-renewal and differentiation.

To better understand how the histone variants contribute to pluripotency in ESCs, Meshorer et al. looked for the exchange rate of different chromatin proteins in mouse ESCs and their differentiated cell progeny using fluorescent recovery after photobleaching (FRAP) [34]. They suggested that the histone structure in ESCs is highly dynamic and various histone variants interact with chromatin rapidly and transiently, ranging from a few seconds to a few minutes. These findings demonstrated that a large number of specific histone variants (25%) attach loosely to chromatin in ESC, as compared to differentiated cells.

Recently, Ng and Gurdon showed the significant role of histone H3.3 variants in cellular memory in *Xenopus* [40]. They provide evidence that the epigenetic memory of a transcriptionally active state relies upon histone H3.3 in chromatin. So, DNA methylation is not the sole mechanism involved in cellular memory. This new discovery may define how ESCs can pass their identity through their daughter cells after division, thereby maintaining self-renewal.

These findings have resulted in the hyperdynamic ("breathing") theory of chromatin architecture in ESCs. Chromatin hyperdynamics is an exclusive characteristic of ESCs and is important in maintaining ESC self-renewal and pluripotency, as differentiated and non-pluripotent cells do not show this property [34]. This poised chromatin structure, which is partly dependent upon histone variants, gives an opportunity to ESCs to rapidly differentiate into any lineage-specific cell.

24.5 CHROMATIN BIVALENCY IN ESCS

Recent studies propose that several types of protein complexes mediate the histone methylation in eukaryotes. Several studies in *Drosophila* have emphasized the crucial role of Polycomb (PcG) and Thrithorax multiprotein complexes in regulating lineage-specific gene expression [41]. Polycomb proteins (PcG) repress gene activity by regulating trimethylation of lysine 27 at histone H3 (H3K27me3) and imparting a silent state of gene expression. PcG proteins downregulate many signaling and developmental genes by inducing methylation of H3K27. PRC1 and PRC2, as two major PcG complexes, have been found to mediate gene silencing in mammals. PRC2 mediates both H3K27me3 and H3K27me2 methylation through its Ezh2, Suz12, and EeD subunits [41,42]. PRC2 has been suggested to be responsible for gene silencing by ubiquitination of H2A at Lysine-119, which accelerate the induction of chromatin compaction [43]. Thrithorax proteins have the opposite effect by acting to control gene expression through regulation of chromatin structure by catalyzing or interacting with H3K4 methylation (H3K4me3). Other members of this family can provoke ATP-dependent chromatin remodeling.

In 2006, two studies suggested that PcG protein complexes play a regulatory function in maintaining ESC pluripotency [44–46]. Mouse development and ESC self-renewal are dependent upon the expression of PcG genes, as no ESC lines can be derived from Ezh2-deficient mice [47]. It should be noted that PRC2 was found to be dispensable for ESCs pluripotency [48]. Some of the developmentally important genes in ESCs are regulated by PcG repressor complex [49,50]. Interestingly, a significant number of these sites are also occupied by Oct4, Nanog, and Sox2 [49]. This is interesting since it appears that the promoters of some of the developmental regulatory genes are controlled simultaneously by activators and repressors. This may point out that ESCs modulate their chromatin organization by an unusual and complex mechanism compared to somatic cells.

Recent studies indicate that ESCs engage an uncommon and novel mechanism for lineage-specific gene regulation. These genes are normally silent in pluripotent stem cells, but may be expressed upon differentiation. Recently, a new "histone bivalent" model (active and inactive histone modifications) has been suggested to explain the mechanism of gene expression regulation in ESCs. According to this model, some lineage-specific genes are marked with both repressive and active modifications at the same time. By applying a quantitative sequential chromatin immune-precipitation (ChIP) approach, Azuara et al. found that several important inactive developmental genes in mouse ESCs such as Pax3, Irx3, Sox1, and Nkx2.2 were simultaneously enriched by repressive modifications (H3K27met3) and activating marks (H3K4me3 and H3K9ac3) in their promoters [44]. Their results also suggested that ESCs, which are defective in the Eed-dependent methytransferase enzyme, expressed lineage-specific genes. They concluded that H3K27 methylation in ESCs assists to maintain pluripotency and suppress developmental gene expression.

Bernstein et al. also supported this bivalent model by demonstrating that promoter regions of lineage-control genes including POU, Pax, Sox, and Hox are distinguished by both suppressive (H3K27me3) and activating (H3K4me3) marks [45]. They found that these opposite histone modifications take place at the same chromosomal locations within the same ESC population. Most of these uncommon bivalent patterns of histone changes are erased upon ESC differentiation into neural progenitor cells. Specifically, some of the neural-specific genes continue to be expressed (H3K4me3), but lose their repressing (H3K27me) histone modifications. In contrast, inactive pluripotent genes preserved the repressor mark (H3K27me3), while losing the activator (H3K4me3) changes. These findings suggest that key developmental regulating genes exhibit a "primed or poised status" in ESCs as defined by opposite combinations of histone modifications.

Other studies revealed that this interesting histone mark can also be found at later stages in development, indicating that differentiated cells have the capacity to gain some characteristics of ESCs [51,52]. Fewer bivalent domains are observed in differentiated cells, whereas the number of these domains in ESCs is about 2000−3000 genes [53−55]. These studies pointed out that this chromatin mark might, at least in part, be responsible for the pluripotency of ESCs, since these changes down-regulate the expression of genes in ESCs due to the dominant influence of H3K27me3 over H3K4me3. Lineage-specific genes are maintained inactive by chromatin modifications, but may be poised for subsequent induction as ESCs select to differentiate to three embryonic lineages. This bivalent model supports the idea that ESC pluripotency and self-renewal is preserved by the expression of differentiation genes, which are in a dormant but poised state.

Histone demethylases may also play an important role in ESC pluripotency and epigenetics. The exact mechanism of histone demethylation in ESCs has only recently been revealed [3,56,57]. Several studies have implicated specific H3K27 trimethylation removal by two enzyme families: UTX and jumonji (JMJD) [51,58,59]. These two groups have the opposite functions in modifying gene expression within PcG complexes and are crucial in early development and differentiation. Oct4 can up-regulate Jmjd1a and Jmjd2c genes, which are responsible for H3K9me2 and H3K9me3 demethylation. While a corresponding decrease in lineage-specific gene expression occurs upon depletion of either Jmjd1a or Jmjd2c, an increase in the expression of ESC-specific genes happens. Up-regulation of other key pluripotency factors, such as Nanog, is also induced by Jmjd2a, whereas Jmjd1a affects the demethylation of activating mark H3K9me2 at the promoter region of Tc11, Zfp57, and Tcfcp211, among others [60]. Predominantly, histone demethylases play a prominent role in self-renewal and pluripotency. They are an inseparable portion of the ESC transcriptome network regulating the transcription circuitry to chromatin structure during early development, and later within tissue-specific differentiation.

24.6 CHANGING THE EPIGENETIC LANDSCAPE DURING CELLULAR REPROGRAMMING

In 2006, a milestone was achieved by artificially converting mouse embryonic and adult fibroblasts to induced pluripotent stem cells (iPSCs) using Oct4, Klf4, c-myc, and Sox2 transcription factors [61]. In the subsequent year and in two independent studies led by Thomson and Yamanaka, the same process reverted adult human somatic cells to ES-like cells by viral transduction of four defined pluripotency transcription factor genes [62,63]. Yamanaka's group used the same cocktail in mouse, while Thomson's group replaced c-Myc and Klf4 with Nanog and Lin28. These iPSCs are morphologically very similar to that of ESCs; they express cell surface markers characteristic of ESCs, demonstrate multilineage differentiation both in vivo and in vitro and give rise to live chimeras. Since 2006, hundreds of studies have tried to derive iPSCs from different mouse and human sources using various approaches [64−71].

Although some of the molecular highlights that happen during the course of iPSCs derivation have been elucidated, the mechanism and dynamics of the process by which ectopic expression of a few transcription factors may change cell destiny remains to be largely unknown. Several lines of evidence point out to the role of epigenetic marks during iPSC generation and the concomitant overturning of cell state [72,73]. It has been suggested that cell fate can be reset through modification of lineage-specific epigenetic marks such as DNA methylation, histone methylation/acetylation, and nucleosome spacing. Experimental data propose that mouse-derived iPSCs have the ability to gain the chromatin signature of pluripotent ESCs and can transmit these epigenetic patterns to their progenies through the germline [74].

There are several important questions to be answered: what kinds of chromatin remodeling are really taking place during the immediate expression of reprogramming factors until the appearance of the first pluripotency markers such as E-Cadherin, SSEA1 (mouse), or SSEA3 (human) and how do the reprogramming factors induce the reversion of epigenetic marks? The role of each reprogramming factor has been investigated in more detail since the initial iPSC studies. In differentiated cells, Oct4 and Sox2 cannot find their target genes, possibly due to large differences in epigenetic status between ESCs and their somatic cell progenies. Although, the application of Oct4 and Sox2 as part of an autoregulatory loop are indispensable for iPSC generation, c-Myc and Klf4 are not necessary for reprogramming, but they significantly increase the efficiency, possibly through interaction with the chromatin [75] or suppression of genes related to differentiation [76]. It has been postulated that c-Myc and Klf4 modify the organization of chromatin enabling Oct4 and Sox2 access to their targets, thereby increasing the expression of downstream genes [61]. For instance, c-Myc is a well-characterized oncogene transcription activator and a modulator of DNA replication. It has been proposed that c-Myc induces the up-regulation of histone acetyltransferase gene (Gcn5), which is a key factor in histone structure, there by improving the accessibility of target genes to Oct4. Klf4 is also acetylated by p300 (acetyl transferase protein) and has the capacity to control gene transcription through regulation of histone acetylation [77]. These results are in agreement with the open chromatin theory of ESC self-renewal and pluripotency.

Using ChIP and cDNA microarray approaches, it is well established that Oct4 regulates the expression of over 350 genes in ESCs including several epigenetic regulators [78]. As mentioned earlier, two histone demethylases, Jmjd1a and Jmjd2c have been identified to be part of the groups of the genes regulated by Oct4 [60]. This study also demonstrated that Jmdj2c is recruited to the Nanog promoter (an important regulator of the ESC self renewal machinery) and found that upon depletion of Jmdj2c, subsequent differentiation could only be rescued by ectopic expression of Nanog. These results clearly demonstrate that Oct4 both directly and indirectly regulates the genes necessary for maintaining the accessible and open chromatin state needed for ESC self-renewal and pluripotency. A positive feedback loop between transcriptional circuitry and epigenetic modification has also been found. In this way, ESC transcription factors control the expression of chromatin remodeling genes and, in turn, assist to open up chromatin structure in the promoter regions of target genes allowing for self-regulation of the epigenetic network.

In 1957, Waddington proposed his famous "epigenetic landscape" model by comparing the early developmental differentiation with a ball travelling downward a canal. This journey starts from a fertilized totipotent embryo and ends up as different lineage-committed cells. According to this developmental model, cells inside the canal move through various one-way branched valleys and select their final irreversible cellular fates during this trip [79]. As they reach the end of each valley (cell lineage), they are urged to stay in that valley and cannot jump over boundaries into other branches or return to their starting point. With the discovery of cellular reprogramming, now this trip is not irreversible. The cells can now move back within the valley from somatic cell to pluripotent state or even transdifferentiate from one lineage to the other without returning back to the pluripotent state [80]. With regard to iPSC generation, it has been postulated that the four reprogramming transcription factors push cells backward in this canal by removing specific epigenetic barriers, which stabilize cells in their differentiated status/valley under normal conditions [81].

Recently, based on Waddington's epigenetic model, a stochastic model for iPSC generation has been proposed [81]. This model is built on an opposite concept suggested by Waddington. Specially, reprogrammed cells undergo four various events. As they begin their journey backwards along the slope toward pluripotency, some epigenetic obstacles hinder the cells from rolling back and consequently they will achieve self-renewal capability. The second group of cells will be partially reprogrammed and lose their pluripotency state and differentiate into specific cell lineage without constant expression of reprogramming factors. The third category may trans-differentiate as a result of insufficient and improper expression of ectopic factors. The fourth group doesn't even begin the reprogramming and experience apoptosis or cellular senescence (Figure 24.2). In agreement with these series of events and based on a stochastic model of ESCs, any gene or molecule, with the ability to enhance this movement up the slope and avert the cell from returning back would facilitate the derivation of iPSCs.

The endogenous loci of all reprogramming factors are excessively methylated in somatic cells, though their loci are hypomethylated in ESCs and iPSCs [82]. For iPSC derivation, all of these promoters require to be reactivated by enzymes responsible for DNA demethylation. This event requires other downstream activated epigenetic modifiers, since the factors used in direct reprogramming do not have known demethylating activity. Based on CpG methylation differences between lineage-specific cells and ESCs, one group demonstrated an essential role for CpG methylation as an epigenetic modifier in iPSC generation. Doi et al. found that CpG

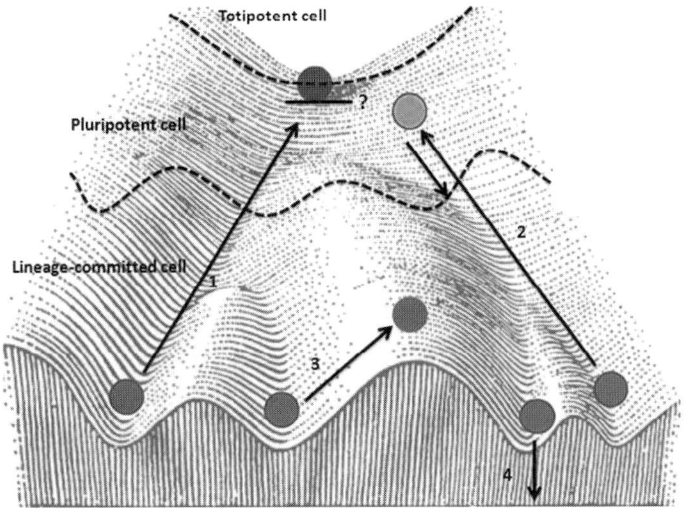

FIGURE 24.2

The stochastic model proposed by Yamanaka for iPS cell generation is based on the Waddington model. The reprogrammed cells may encounter four possible scenarios. One group of cells is blocked by epigenetic barriers and begins self-renewal (1). Other cells are trapped inside a semi-reprogrammed state due to inefficient epigenetic modification and travel back down their valley in the absence of ectopic expression of pluripotency factors (2). Some of the cells may move to other neighboring valleys and transdifferentiate into other cell types due to inefficient expression of reprogramming factors (3). The fourth group of the cells experiences apoptosis or cellular senescence (4). This figure is reproduced in the color plate section. *Source: Adapted from [133]* methylation can distinguish iPSCs, ESCs, and fibroblasts. They proposed that epigenetic reprogramming incorporates the same differentially methylated regions (DMRs) of CpG island "shores" that mark normal differentiation [83]. Specific loci in iPSCs remain semi-reprogrammed, which means that the methylation pattern of iPSCs differs from ESCs.

Another obstacle toward iPSC generation is the requirement for histone modification and remodeling during reprogramming. Although, it has been shown that histone H4 within the promoters of reprogramming genes is deacetylated in somatic cells, it is hyperacetylated in iPSCs and ESCs. Several studies indicate that the H3 and H4 histones within the Oct4 and Nanog promoters are hyperacetylated [84,85]. Among the four factors, only c-Myc has the ability to modify chromatin. It mediates chromatin modification by increasing the expression of Gcn5 recruiting it for the modification of target genes [86].

Another factor in the derivation of iPSCs is the state of histone methylation. ESCs and iPSCs are tagged by activator histone modifications, which are marked by H3K4me3 and deme-thylation of lysine 9 within pluripotency genes. These two histone modifications show the bivalent chromatin characteristics of pluripotent genes, which are accomplished by simulta-neous methylation at H3K27me3 and H3K4me3. For efficient generation of iPSCs, it is required that this histone modification be achieved either through genetic approaches or with the aid of some small molecules.

Reflecting on the role of chromatin remodeling during reprogramming, researchers have recently applied small molecules to circumvent these epigenetic blocks and enhance the generation of iPSCs. Several different chemical inhibitors for histone deacetylases, as well as DNA and histone methyltransferases, have been used in combination with genetic factors [87,88] (Table 24.1). Kubicek et al. found a small-molecule inhibitor of G9a histone methyl-transferase, BIX-01294, could enhance the induction of reprogramming in neural stem cells while replacing Oct4 [89]. Since G9a is a down-regulator of Oct4 during early development, they suggested that BIX-01294 enhances iPSC formation by inhibiting G9a and subsequently releasing Oct4 from negative regulation. This group also generated iPSCs from mouse embryonic fibroblasts (MEFs) with only two factors; Oct4 and Klf4 in the presence of BIX-

TABLE 24.1 Chemicals with Epigenetic Modification Properties Used to Enhance iPSC Generation

Name	Function	Effects	Reference
RG-108	DNA methyltransferase inhibitor	Promote MEFs reprogramming	[134]
BIX-01294	G9 histone methyltransferase inhibitor	Reprogramming of neural progenitor cells and MEFs	[134,135]
VPA	Histone deacetylase inhibitor	Enhanced reprogramming of mouse and human fibroblasts	[134,135]
SAHA	Histone deacetylase inhibitor	Enhances reprogramming of mouse fibroblasts	[135]
Trichostatin-A	Histone deacetylase inhibitor	Enhances reprogramming of mouse fibroblasts	[135]
5-azacytidine	Methyltransferase inhibitor	Enhanced reprogramming of mouse fibroblasts	[135,136]

01294. Interestingly, valproic acid (VPA) increased the efficiency and kinetics of reprogramming by 100-fold in a four-factor system. VPA could also replace either Klf4 or c-Myc in reprogramming and enabled iPSC generation with only Oct4 and Sox2 from human fibroblasts [90]. These results imply that DNA methylation, histone methylation, and histone deacetylation contribute to epigenetic hurdles, which have to be overcome for successful iPSC generation.

24.7 STEM CELL EPIGENETICS AND HUMAN DISEASE

Studies in the past two decades have revealed a great deal about the basis of known epigenetic defects causing disease. There are two major categories of disease-related epigenetic defects: histone modification and DNA methylation. Enzymes such as histone deacetylases (HDACs), histone methyltransferases, histone acetylases, DNA methyltransferases, and methyl-CpG binding domain protein (MECP2) are among the best-known epigenetic modifiers prone to abnormality. Any variation in these epigenetic apparatus may alter gene expression, which in turn has enormous and critical clinical consequences. One of the best-known epigenetic causes of disease is the loss of DNA methylation patterns. Alterations in DNA methylation have been observed in various disorders involving genomic imprinting, X chromosome inactivation, and cancer.

Cancer comprises a multistep process wherein both genetic and epigenetic abnormalities work together to transform a normal cell into an abnormal malignant tumor cell. Cancer is generally characterized by global hypomethylation and gene-specific hypermethylation [91]. Modified DNA methylation in a cell alters the expression of cancer-related genes. DNA hypermethylation commences the down-regulation of tumor-suppressor genes, whereas DNA hypomethylation activates proto-oncogenes and promotes cell instability via chromosomal aberrations [92]. DNA methylation patterns are created and preserved by a group of enzymes called DNMTs, which are important for appropriate gene expression. Knocking out of DNMT genes in animal models is detrimental and the overexpression of these enzymes in human may lead to various kinds of cancers. In 1983, hypomethylation patterns were found to distinguish human cancer cells from their normal counterparts [93] and now there is a growing body of evidence that hypomethylation of certain proto-oncogenes, such as HRAS and cyclin D2, cause tumor formation. To the contrary, it has been revealed that hypermethylation of promoter regions of tumor suppressors such as RB, P16, VHL, APC, and E-Cadherin induce malignancy and are mostly found in sporadic cancer [91,94—96].

Recent studies have shed light over the role of stem cell chromatin marks in cancer development. They report that there is an epigenetic stem cell signature in cancer and this stem-cell-like chromatin pattern makes the tumor-suppressor genes prone to DNA hypermethylation and subsequent silencing [97,98]. It is interesting that several numbers of these tumor-related genes are associated with the bivalent chromatin state in ESCs. For example, two suppressive marks including H3K9me3 and H3K9me2 are linked to DNA hypermethylation in cancer [98]. Also, methylation of H3K27 by polycomb protein, along with the previously mentioned repressive marks, targets some genes for de novo methylation by methyltransferases in cancer [99].

Understanding the exact epigenetic mechanism governing cancer cells can have significant therapeutic consequences. For instance, it has been revealed that any modification in chromatin organization can affect normal development and cellular tumorigenic transformation [100]. These observations have resulted in the development of new drugs such as HDAC and DNMT inhibitors for the treatment of different cancers [101,102]. Elucidation of chromatin defects, which transform a normal cell to a malignant tumor, may lead to the development of new designs for cancer treatment as well as its early diagnosis.

Recent studies have evaluated the role of epigenetic defects in the onset of various pediatric and adult neurodegenerative disorders [103] (Table 24.2). Some mood disorders such as

491

TABLE 24.2 Relationship Between Human Diseases and Epigenetic Modifications

Disease	Gene	Function	Reference
Neurodegenerative Disorders			
Schizophrenia	*RELN*	Hypermethylation	[137]
Bipolar disorders, memory formation	*RELN*	Hypermethylation	[138]
Psychosis	*DNMT1*	Overexpression	[137]
Alzheimer's diseases	*Presenilin I*	Hypomethylation	[139]
Imprinting Diseases			
BWS	11p15	Imprinting	[140]
ICF syndrome	*DNMT3B*	Mutation	[141]
Rett syndrome	*MECP2*	Mutation	[142]
ATRX	*ATRX*	Chromatin conformation	[143]
PWS/AS	15q11-q13	Imprinting	[144]
FraX	*FMR1*	Gene suppression	[145]
Cancer			
Brain	*MGMT*	Hypermethylation	[416]
Colon	Different genes	Hypermethylation	[147]
Esophagus	*CDH1*	Hypermethylation	[147]
Kidney	*TIMP-3*	Hypermethylation	[147]
Ovary	*SAT2*	Hypomethylation	[148]
Pancreas	*APC*	Hypermethylation	[147]
Prostate	*BRCA2*	Hypermethylation	[149]
Uterus	*hMLH1*	Hypermethylation	[147]
Aging	DNA and Chromatin	Hypo-, hypermethylation	[150]
Immunity and Related Disorders			
Lupus	*LFAr*	Hypomethylation	[151]

schizophrenia have been reported to be associated with DNA methyltransferase gene mutations. For example, the overexpression of DNMTI has been found in the gamma-aminobutyric acid (GABA)-ergic neurons of schizophrenic individuals. This overexpression inhibits the activity of Reelin protein in patients with bipolar disorders and psychosis, which is normally needed for proper memory composition and normal neurotransmission [104,105]. Furthermore, it has been suggested that valproic acid, an HDAC inhibitor, corrects the schizophrenia-like behavior of a mouse model of schizophrenia by preventing hypermethylation of Reelin promoter [106].

New findings in Alzheimer's disease (AD) have shed light over the connection between methylation pattern in specific genes and neural survival, as well as memory loss. Some studies revealed the connection between the hypomethylation of presenilin I gene and its up-regulation and beta-amylois production. Also, a role for folate-mediated methylation in Alzheimer's disease has been suggested [107,108].

Recently, a connection between genetic/epigenetic changes and autism has been proposed [109]. Autism is a disease of neural development in the brain which impairs an individual's normal development of social and communication interactions. The main defect involved in the onset of autism is known to be mutation or changes in the activity of "methyl CpG binding protein 2" (*MECP2*), which is believed to be essential for normal brain development. *MECP2* is an X-linked gene and can bind specifically to the region of methylated DNA. It is also capable of repressing transcription from methylated gene promoters. Once bound, MeCP2 forms a complex with histone deacetylases (HDAC) to condense chromatin structure. If the gene undergoes any defect, cells do not have enough MeCP2 protein for silencing other genes. Interestingly, mutations in the *MECP2* gene cause other neurodevelopmental disorders including Rett syndrome, some cases of X-linked mental retardation, and systemic lupus erythematosus. Based upon all of these findings, there may be a potential future role for epigenetics in the diagnosis and treatment of neurological diseases.

24.8 MODELING OF HUMAN EPIGENETIC DISORDERS USING IPSCS

Studies on the mechanism and pathophysiology of human diseases have been impeded by the lack of suitable in vitro practical models and the inability to obtain sufficient numbers of primary cells. iPSCs may provide potent models of human genetic and multifactoral diseases due to their enormous similarities to ESCs. Reprogrammed cells can be differentiated to desired cell types using established protocols for hESCs. With this tool, the molecular mechanism of disease, as well as drug and therapeutic screening, can be performed using derived affected cells. So far, iPSCs have been generated from patients with different types of diseases including neurological and neurodegenerative disorders (HD, PD, AD, ALS, SMA, Rett syndrome), muscular dystrophy (DMD, BMD), hematopoietic disorders (beta-thalassemia, sickle cell anemia, Fanconi anemia), cardiovascular diseases (long QT, LEOPARD syndrome), diabetes, Down syndrome, and dyskeratosis congenita (DC) [110]. In this review, we will only focus on diseases with an epigenetic component.

Autism spectrum disorders (ASDs) are complex neurodevelopmental disorders wherein the phenotype of patients is the result of various combinations of genetic mutations. Rett syndrome (RTT) is an X-linked neurodevelopment disorder caused by a heterozygous mutation in the *MECP2* gene, which can arise sporadically or from germline mutations. It affects females more than males. In about 10% of cases, mutation in the genes *FOXG1* or *CDKL5* have also been found to cause the syndrome. The mutation affects postnatal neural development and results in communication defects and mental retardation in affected children. As mentioned earlier, with the mutant gene, cells do not have enough MeCP2 protein for transcriptional silencing or activating of other genes.

For the first time, fibroblasts of an 8-year-old girl with R306C missense mutation in *MECP2* have been reprogrammed using four retroviral factors including Oct4, Sox2, Klf4, and c-Myc [111]. The pluripotency of derived cells was confirmed by in vivo teratoma assay and in vitro differentiation through EB formation. The authors could differentiate Rett-derived iPSCs to neurons, but further characterization was not reported. In the other study, iPSCs generated from a Rett syndrome patient induced neural differentiation. Differentiated cells demonstrated abnormalities in neural characteristics including smaller size, modified calcium signaling, diminished synapse, and electrophysiological abnormalities [112].

Recently, MECP2-null Rett syndrome hiPSCs and isogenic controls have been derived through X-chromosome inactivation. Phenotypic evaluation of mutant RTT-hiPSC-derived neurons revealed a reduction in soma size compared with the isogenic control neurons from the same patient [113]. Such state-of-the-art cell characterization is going to open a new era in molecular medicine for revealing mechanisms of disease and new approaches for drug screening.

iPSCs have also been recently derived from the genomic imprinting diseases Prader–Willi and Angelman syndromes. Genomic imprinting is a genetic event by which particular gene loci become transcribed in a parent-of-origin-determined way. This means that the phenotype triggered from a certain locus is differentially altered by the sex of the parent providing that specific allele. Genomic imprinting is an epigenetic event that applies histone and DNA methylation patterns in order to acquire monoallelic gene transcription without altering the

DNA sequence. These epigenetic hallmarks are placed during germline establishment and are preserved throughout the lifetime of an organism. Proper expression of imprinted genes is crucial for normal development. Multiple genetic diseases are associated with defects in imprinting loci such as Angelman and Prader–Willi syndromes, Beckwith–Wiedemann syndrome and Silver–Russell syndrome [114]. Among these disorders, iPSCs have been only derived from dermal fibroblasts of patients with Angelman syndrome (AS) and Prader–Willi syndrome (PWS) [115,116]. AS is a complex neurogenetic disorder which initially affects the nervous system. It is characterized by speech impairment, frequent seizures, intellectual disability, and ataxia and affected children have typically a happy and excitable demeanor. PWS is characterized by speech delay, obesity, small stature, and behavior problems. AS is usually caused by deletion or inactivation of genes on chromosome 15q11-q13 region; this region is inherited from the mother while the paternal allele is imprinted and hence silenced. Many symptoms of AS result from the loss of function of E3 ubiquitin ligase UBE3. PWS is caused by the deletion of paternally expressed genes on the chromosome 15q11-q13, but its genetic basis is largely unknown [117,118].

Fibroblasts from AS and PWS patients were reprogrammed using retroviral vectors encoding Oct4, Sox2, Klf4, c-Myc, and LIN28 transcription factors. iPSCs derived from AS and PWS patients were induced to differentiate to functional neurons [115]. It is interesting that the iPSCs showed no sign of DNA methylation removal at the imprinting center (IC) of chromosome 15. Furthermore, in normal brain neurons, the imprinting of *UBE3A* is established during neural differentiation of AS iPSCs. This study highlights the importance of the iPSC model of epigenetic disorders in elucidating the disease mechanism and developmental timing of gene activation or silencing in the affected cells of patients.

There are growing bodies of evidence that show a relationship between premature aging and adult stem cell malfunction. Aging is an exceedingly complex trend whose molecular mechanism is still mainly unknown. Delineation of many molecular aspects of aging has been facilitated by investigations on premature aging syndromes [119]. The premature-ageing disease Hutchinson–Gilford progeria syndrome (HGPS) is one of the rare fatal human genetic disorders that the role of both epigenetic changes and stem cell misregulation have been elucidated. HGPS is caused by a single point mutation in the *LMNA* gene resulting in constitutive production of progerin, a truncated splicing mutant form of the nuclear structural protein lamin A [120]. Accumulation of progerin in several tissues leads to diverse aging-related nuclear defects such as structural disorganization of nuclear lamina and function of the nucleus and chromatin [121]. It also causes epigenetic changes by histone modifications and increased DNA damage [122].

Interestingly, recent studies demonstrated that Lamin A-dependent dysfunction of adult stem cells is associated with accelerated aging in humans [123]. The authors could provide evidence that the induction of progerin hinders the proper function of human mesenchymal stem cells (hMSCs) by activating downstream effector proteins of the Notch signaling pathway and modified cellular characteristics and differentiation capacity. These findings are interesting since the Notch signaling pathway regulates stem cell differentiation [124] and many of the affected tissues in HGPS patients are of mesenchymal origin [125]. Alternatively, there is a possibility that the Notch effector genes are regulated by their direct association with lamina or misregulation of the pathway is due to the epigenetic alterations normally observed in normal aging and specifically in HGPS patients [126]. Malfunction of adult stem cells may also have implications for the normal aging process since the progerin is present at very low levels in cells from normal individuals [126].

In another study, Zmpste24-null progeroid mice (with nuclear lamina defects and accelerated aging) were evaluated for the number and functional capacity of stem cells [127]. The authors used telogen hair follicles, which contain multipotent stem cells of both epidermal and neural origin. They showed that Zmpste24 deficiency causes a change in the number and proliferative capacity of epidermal stem cells characteristic with nuclear architecture disorganization and increased apoptosis. These changes are associated with some signaling pathways such as Wnt and microphthalmia transcription factor. These results confirmed the existence of a relationship between stem cell misregulation and age-related nuclear envelope deformity.

Recently, several studies investigated the possibility of reprogramming cells from premature aging Hutchinson—Gilford progeria syndrome to the iPSC state [128,129]. Adult dermal fibroblasts from HGPS patients were transfected with Oct4, Sox2, Klf4, and c-Myc. HGPS-iPSCs show absence of progerin, and lack the nuclear envelope and epigenetic alterations that are normally associated with premature aging. Furthermore, HGPS-iPSCs revealed vascular smooth muscle and mesenchymal stem cell defects. These studies clearly provide an in vitro iPSC-based model to study the pathogenesis of human premature and physiological vascular aging.

24.9 FUTURE STUDIES

Early development and ESC differentiation necessitate complex interactions between various transcription factors and epigenetic regulators such as histone-modifying enzymes and chromatin-remodeling proteins. Based on what is currently known about this interaction, it seems that chromatin dynamics is one of the most critical determinants of ESC pluripotency and self-renewal. iPSC derivation has triggered a paradigm shift in developmental biology. However, this revolutionary discovery has raised several discussions regarding the exact mechanism of reprogramming and the function of epigenetic changes. At the present time, there is uncertainty about the epigenetic status of cells during the iPSC procedure. For instance, do the iPSCs have any "memory" of their original epigenetic state? If so, is this memory favorable or disadvantageous for their future clinical applications? Recently, several studies have demonstrated the existence of an epigenetic memory during iPSC generation. It was revealed that reprogramming leaves an epigenetic memory of the tissue of origin which may affect their differentiation and application in disease modeling [130,131].

Advancements in understanding the role of epigenetic obstacles will definitely move this field forward, establishing straightforward and more efficient methods. Furthermore, iPSCs may provide an important model of human genetic and multifactorial disorders. Specially, with the lack of a proper in vitro cellular models of epigenetic diseases to study the mechanism, the field will move toward the establishment of iPSCs from different epigenetic disorders such as Beckwith—Wiedemann syndrome, Silver—Russell syndrome, Albright hereditary osteodystrophy, pseudohypoparathyroidism type IA (PHPIA), immunodeficiency/centromeric instability/facial anomalies (ICF), alpha-thalassemia/mental retardation, X-linked (ATRX), Rubinstein-Taybi syndrome, asthma, multiple sclerosis, obesity, and different types of cancer cells. A common characteristic of these disorders is that mutations in the components of chromatin regulators and epigenetic machinery cause the pathophysiological symptoms. As the mechanisms of human epigenetic disorders are recognized (for example through disease-derived iPSCs modeling), there will be even more developments in epigenetic therapies. Because epigenetic changes are the key factors in human health and disease, there is hope that understanding the mechanism of epigenome regulation will aid in the treatment of human sickness that may ultimately be beneficial for the health and wellbeing of mankind. Finally, if we can modify the epigenetic status of adult somatic cells toward pluripotency without intervening in their genetic integrity, we will be one step closer toward the clinical application of these cells in the near future.

References

[1] Luger K, Mader AW, Richmond RK, Sargent DF, Richmond TJ. Crystal structure of the nucleosome core particle at 2.8 A resolution. Nature 1997;389:251—60.

[2] Waddington CH. L'epigenotype. Endeavour 1942;1:18—20.

[3] Kouzarides T. Chromatin modifications and their function. Cell 2007;128:693–705.

[4] Jenuwein T, Allis CD. Translating the histone code. Science 2001;293:1074–80.

[5] Horn PJ, Peterson CL. Heterochromatin assembly: a new twist on an old model. Chromosome Res 2006;14:83–94.

[6] Grewal SI, Jia S. Heterochromatin revisited. Nat Rev Genet 2007;8:35–46.

[7] Yasuhara JC, Wakimoto BT. Oxymoron no more: the expanding world of heterochromatic genes. Trends Genet 2006;22:330–8.

[8] Smith CD, Shu S, Mungall CJ, Karpen GH. The Release 5.1 annotation of Drosophila melanogaster heterochromatin. Science 2007;316:1586–91.

[9] Pezer Z, Ugarkovic D. RNA Pol II promotes transcription of centromeric satellite DNA in beetles. PLoS One 2008:3:e1594.

[10] Li F, Sonbuchner L, Kyes SA, Epp C, Deitsch KW. Nuclear non-coding RNAs are transcribed from the centromeres of Plasmodium falciparum and are associated with centromeric chromatin. J Biol Chem 2008;283:5692–8.

[11] Reik W. Stability and flexibility of epigenetic gene regulation in mammalian development. Nature 2007;447:425–32.

[12] Lusser A, Kadonaga JT. Chromatin remodeling by ATP-dependent molecular machines. Bioessays 2003;25:1192–200.

[13] Iizuka M, Smith MM. Functional consequences of histone modifications. Curr Opin Genet Dev 2003;13:154–60.

[14] Ahmad K, Henikoff S. Histone H3 variants specify modes of chromatin assembly. Proc Natl Acad Sci USA 2002;99(Suppl. 4):16477–84.

[15] Bird AP, Wolffe AP. Methylation-induced repression—belts, braces, and chromatin. Cell 1999;99:451–4.

[16] Almeida R, Allshire RC. RNA silencing and genome regulation. Trends Cell Biol 2005;15:251–8.

[17] Bartel DP. MicroRNAs: Genomics, biogenesis, mechanism, and function. Cell 2004;116:281–97.

[18] Ramsahoye BH, Biniszkiewicz D, Lyko F, Clark V, Bird AP, Jaenisch R. Non-CpG methylation is prevalent in embryonic stem cells and may be mediated by DNA methyltransferase 3a. Proc Natl Acad Sci USA 2000;97:5237–42.

[19] Weber M, Hellmann I, Stadler MB, Ramos L, Paabo S, Rebhan M, et al. Distribution, silencing potential and evolutionary impact of promoter DNA methylation in the human genome. Nat Genet 2007;39:457–66.

[20] Suzuki Y, Tsunoda T, Sese J, Taira H, Mizushima-Sugano J, Hata H, et al. Identification and characterization of the potential promoter regions of 1031 kinds of human genes. Genome Res 2001;11:677–84.

[21] Hendrich B, Tweedie S. The methyl-CpG binding domain and the evolving role of DNA methylation in animals. Trends Genet 2003;19:269–77.

[22] Fatemi M, Hermann A, Pradhan S, Jeltsch A. The activity of the murine DNA methyltransferase Dnmt1 is controlled by interaction of the catalytic domain with the N-terminal part of the enzyme leading to an allosteric activation of the enzyme after binding to methylated DNA. J Mol Biol 2001;309:1189–99.

[23] Bestor T, Laudano A, Mattaliano R, Ingram V. Cloning and sequencing of a cDNA encoding DNA methyltransferase of mouse cells. The carboxyl-terminal domain of the mammalian enzymes is related to bacterial restriction methyltransferases. J Mol Biol 1988;203:971–83.

[24] Probst AV, Dunleavy E, Almouzni G. Epigenetic inheritance during the cell cycle. Nat Rev Mol Cell Bio 2009;10:192–206.

[25] Reik W, Dean W, Walter J. Epigenetic reprogramming in mammalian development. Science 2001;293:1089–93.

[26] Jackson M, Krassowska A, Gilbert N, Chevassut T, Forrester L, Ansell J, et al. Severe global DNA hypomethylation blocks differentiation and induces histone hyperacetylation in embryonic stem cells. Mol Cell Biol 2004;24:8862–71.

[27] Deb-Rinker P, Ly D, Jezierski A, Sikorska M, Walker PR. Sequential DNA methylation of the Nanog and Oct-4 upstream regions in human NT2 cells during neuronal differentiation. J Biol Chem 2005;280:6257–60.

[28] Do JT, Han DW, Scholer HR. Reprogramming somatic gene activity by fusion with pluripotent cells. Stem Cell Rev 2006;2:257–64.

[29] Boyer LA, Mathur D, Jaenisch R. Molecular control of pluripotency. Curr Opin Genet Dev 2006;16:455–62.

[30] Kashyap V, Rezende NC, Scotland KB, Shaffer SM, Persson JL, Gudas LJ, et al. Regulation of stem cell pluripotency and differentiation involves a mutual regulatory circuit of the NANOG, OCT4, and SOX2 pluripotency transcription factors with polycomb repressive complexes and stem cell microRNAs. Stem Cells Dev 2009;18:1093–108.

[31] Wiblin AE, Cui W, Clark AJ, Bickmore WA. Distinctive nuclear organisation of centromeres and regions involved in pluripotency in human embryonic stem cells. J Cell Sci 2005;118:3861–8.

[32] Williams RR, Azuara V, Perry P, Sauer S, Dvorkina M, Jorgensen H, et al. Neural induction promotes large-scale chromatin reorganisation of the Mash1 locus. J Cell Sci 2006;119:132—40.

[33] Meshorer E, Misteli T. Chromatin in pluripotent embryonic stem cells and differentiation. Nat Rev Mol Cell Biol 2006;7:540—6.

[34] Meshorer E, Yellajoshula D, George E, Scambler PJ, Brown DT, Misteli T. Hyperdynamic plasticity of chromatin proteins in pluripotent embryonic stem cells. Dev Cell 2006;10:105—16.

[35] Csankovszki G, Panning B, Bates B, Pehrson JR, Jaenisch R. Conditional deletion of Xist disrupts histone macroH2A localization but not maintenance of X inactivation. Nat Genet 1999;22:323—4.

[36] Angelov D, Molla A, Perche PY, Hans F, Cote J, Khochbin S, et al. The histone variant macroH2A interferes with transcription factor binding and SWI/SNF nucleosome remodeling. Mol Cell 2003;11:1033—41.

[37] Meneghini MD, Wu M, Madhani HD. Conserved histone variant H2A.Z protects euchromatin from the ectopic spread of silent heterochromatin. Cell 2003;112:725—36.

[38] Hatch CL, Bonner WM. An upstream region of the H2AZ gene promoter modulates promoter activity in different cell types. Biochim Biophys Acta 1996;1305:59—62.

[39] Squatrito M, Gorrini C, Amati B. Tip60 in DNA damage response and growth control: many tricks in one HAT. Trends Cell Biol 2006;16:433—42.

[40] Ng RK, Gurdon JB. Epigenetic memory of an active gene state depends on histone H3.3 incorporation into chromatin in the absence of transcription. Nat Cell Biol 2008;10:102—9.

[41] Schuettengruber B, Chourrout D, Vervoort M, Leblanc B, Cavalli G. Genome regulation by polycomb and trithorax proteins. Cell 2007;128:735—45.

[42] Kirmizis A, Bartley SM, Kuzmichev A, Margueron R, Reinberg D, Green R, et al. Silencing of human polycomb target genes is associated with methylation of histone H3 Lys 27. Genes Dev 2004;18:1592—605.

[43] Wang H, Wang L, Erdjument-Bromage H, Vidal M, Tempst P, Jones RS, et al. Role of histone H2A ubiquitination in Polycomb silencing. Nature 2004;431:873—8.

[44] Azuara V, Perry P, Sauer S, Spivakov M, Jorgensen HF, John RM, et al. Chromatin signatures of pluripotent cell lines. Nat Cell Biol 2006;8:532—8.

[45] Bernstein BE, Mikkelsen TS, Xie X, Kamal M, Huebert DJ, Cuff J, et al. A bivalent chromatin structure marks key developmental genes in embryonic stem cells. Cell 2006;125:315—26.

[46] Jorgensen HF, Giadrossi S, Casanova M, Endoh M, Koseki H, Brockdorff N, et al. Stem cells primed for action: polycomb repressive complexes restrain the expression of lineage-specific regulators in embryonic stem cells. Cell Cycle 2006;5:1411—4.

[47] O'Carroll D, Erhardt S, Pagani M, Barton SC, Surani MA, Jenuwein T. The polycomb-group gene Ezh2 is required for early mouse development. Mol Cell Biol 2001;21:4330—6.

[48] Chamberlain SJ, Yee D, Magnuson T. Polycomb repressive complex 2 is dispensable for maintenance of embryonic stem cell pluripotency. Stem Cells 2008;26:1496—505.

[49] Lee TI, Jenner RG, Boyer LA, Guenther MG, Levine SS, Kumar RM, et al. Control of developmental regulators by Polycomb in human embryonic stem cells. Cell 2006;125:301—13.

[50] Boyer LA, Plath K, Zeitlinger J, Brambrink T, Medeiros LA, Lee TI, et al. Polycomb complexes repress developmental regulators in murine embryonic stem cells. Nature 2006;441:349—53.

[51] De Santa F, Totaro MG, Prosperini E, Notarbartolo S, Testa G, Natoli G. The histone H3 lysine-27 demethylase Jmjd3 links inflammation to inhibition of polycomb-mediated gene silencing. Cell 2007;130:1083—94.

[52] Roh TY, Cuddapah S, Cui K, Zhao K. The genomic landscape of histone modifications in human T cells. Proc Natl Acad Sci USA 2006;103:15782—7.

[53] Mikkelsen TS, Ku M, Jaffe DB, Issac B, Lieberman E, Giannoukos G, et al. Genome-wide maps of chromatin state in pluripotent and lineage-committed cells. Nature 2007;448:553—60.

[54] Pan G, Tian S, Nie J, Yang C, Ruotti V, Wei H, et al. Whole-genome analysis of histone H3 lysine 4 and lysine 27 methylation in human embryonic stem cells. Cell Stem Cell 2007;1:299—312.

[55] Barski A, Cuddapah S, Cui K, Roh TY, Schones DE, Wang Z, et al. High-resolution profiling of histone methylations in the human genome. Cell 2007;129:823—37.

[56] Shi Y, Whetstine JR. Dynamic regulation of histone lysine methylation by demethylases. Mol Cell 2007;25:1—14.

[57] Seward DJ, Cubberley G, Kim S, Schonewald M, Zhang L, Tripet B, et al. Demethylation of trimethylated histone H3 Lys4 in vivo by JARID1 JmjC proteins. Nat Struct Mol Biol 2007;14:240—2.

[58] Agger K, Cloos PA, Christensen J, Pasini D, Rose S, Rappsilber J, et al. UTX and JMJD3 are histone H3K27 demethylases involved in HOX gene regulation and development. Nature 2007;449:731—4.

[59] Lee MG, Villa R, Trojer P, Norman J, Yan KP, Reinberg D, et al. Demethylation of H3K27 regulates polycomb recruitment and H2A ubiquitination. Science 2007;318:447—50.

[60] Loh YH, Zhang W, Chen X, George J, Ng HH. Jmjd1a and Jmjd2c histone H3 Lys 9 demethylases regulate self-renewal in embryonic stem cells. Genes Dev 2007;21:2545–57.

[61] Takahashi K, Yamanaka S. Induction of pluripotent stem cells from mouse embryonic and adult fibroblast cultures by defined factors. Cell 2006;126:663–76.

[62] Takahashi K, Tanabe K, Ohnuki M, Narita M, Ichisaka T, Tomoda K, et al. Induction of pluripotent stem cells from adult human fibroblasts by defined factors. Cell 2007;131:861–72.

[63] Yu J, Vodyanik MA, Smuga-Otto K, Antosiewicz-Bourget J, Frane JL, Tian S, et al. Induced pluripotent stem cell lines derived from human somatic cells. Science 2007;318:1917–20.

[64] Stadtfeld M, Nagaya M, Utikal J, Weir G, Hochedlinger K. Induced pluripotent stem cells generated without viral integration. Science 2008;322:945–9.

[65] Okita K, Nakagawa M, Hyenjong H, Ichisaka T, Yamanaka S. Generation of mouse induced pluripotent stem cells without viral vectors. Science 2008;322:949–53.

[66] Soldner F, Hockemeyer D, Beard C, Gao Q, Bell GW, Cook EG, et al. Parkinson's disease patient-derived induced pluripotent stem cells free of viral reprogramming factors. Cell 2009;136:964–77.

[67] Kaji K, Norrby K, Paca A, Mileikovsky M, Mohseni P, Woltjen K. Virus-free induction of pluripotency and subsequent excision of reprogramming factors. Nature 2009;458:771–5.

[68] Woltjen K, Michael IP, Mohseni P, Desai R, Mileikovsky M, Hamalainen R, et al. piggyBac transposition reprograms fibroblasts to induced pluripotent stem cells. Nature 2009;458:766–70.

[69] Zhou H, Wu S, Joo JY, Zhu S, Han DW, Lin T, et al. Generation of induced pluripotent stem cells using recombinant proteins. Cell Stem Cell 2009;4:381–4.

[70] Yakubov E, Rechavi G, Rozenblatt S, Givol D. Reprogramming of human fibroblasts to pluripotent stem cells using mRNA of four transcription factors. Biochem Biophys Res Commun 394:189–193

[71] Miyoshi N, Ishii H, Nagano H, Haraguchi N, Dewi DL, Kano Y, et al. Reprogramming of Mouse and Human Cells to Pluripotency Using Mature MicroRNAs. Cell Stem Cell 2011;8:633–8.

[72] Zaehres H, Scholer HR. Induction of pluripotency: from mouse to human. Cell 2007;131:834–5.

[73] Rodriguez RT, Velkey JM, Lutzko C, Seerke R, Kohn DB, O'Shea KS, et al. Manipulation of OCT4 levels in human embryonic stem cells results in induction of differential cell types. Exp Biol Med (Maywood) 2007;232:1368–80.

[74] Okita K, Ichisaka T, Yamanaka S. Generation of germline-competent induced pluripotent stem cells. Nature 2007;448:313–7.

[75] Knoepfler PS. Why myc? An unexpected ingredient in the stem cell cocktail. Cell Stem Cell 2008;2:18–21.

[76] Sridharan R, Tchieu J, Mason MJ, Yachechko R, Kuoy E, Horvath S, et al. Role of the Murine Reprogramming Factors in the Induction of Pluripotency. Cell 2009;136:364–77.

[77] Evans PM, Zhang W, Chen X, Yang J, Bhakat KK, Liu C. Kruppel-like factor 4 is acetylated by p300 and regulates gene transcription via modulation of histone acetylation. J Biol Chem 2007;282:33994–4002.

[78] Matoba R, Niwa H, Masui S, Ohtsuka S, Carter MG, Sharov AA, et al. Dissecting Oct3/4-regulated gene networks in embryonic stem cells by expression profiling. PLoS One 2006;1:e26.

[79] Waddington C. The strategy of the genes. london: Allen and Unwin; 1957.

[80] Ieda M, Fu JD, Delgado-Olguin P, Vedantham V, Hayashi Y, Bruneau BG, et al. Direct Reprogramming of Fibroblasts into Functional Cardiomyocytes by Defined Factors. Cell 2010;142:375–86.

[81] Yamanaka S. Elite and stochastic models for induced pluripotent stem cell generation. Nature 2009;460:49–52.

[82] Imamura M, Miura K, Iwabuchi K, Ichisaka T, Nakagawa M, Lee J, et al. Transcriptional repression and DNA hypermethylation of a small set of ES cell marker genes in male germline stem cells. BMC Dev Biol 2006;6:34.

[83] Doi A, Park IH, Wen B, Murakami P, Aryee MJ, Irizarry R, et al. Differential methylation of tissue- and cancer-specific CpG island shores distinguishes human induced pluripotent stem cells, embryonic stem cells and fibroblasts. Nat Genet 2009;41:1350–3.

[84] Hattori N, Nishino K, Ko YG, Hattori N, Ohgane J, Tanaka S, et al. Epigenetic control of mouse Oct-4 gene expression in embryonic stem cells and trophoblast stem cells. J Biol Chem 2004;279:17063–9.

[85] Kimura H, Tada M, Nakatsuji N, Tada T. Histone code modifications on pluripotential nuclei of reprogrammed somatic cells. Mol Cell Biol 2004;24:5710–20.

[86] Knoepfler PS, Zhang XY, Cheng PF, Gafken PR, McMahon SB, Eisenman RN. Myc influences global chromatin structure. Embo J 2006;25:2723–34.

[87] Shi Y, Do JT, Desponts C, Hahm HS, Scholer HR, Ding S. A combined chemical and genetic approach for the generation of induced pluripotent stem cells. Cell Stem Cell 2008;2:525–8.

[88] Huangfu D, Maehr R, Guo W, Eijkelenboom A, Snitow M, Chen AE, et al. Induction of pluripotent stem cells by defined factors is greatly improved by small-molecule compounds. Nat Biotechnol 2008;26:795–7.

[89] Kubicek S, O'Sullivan RJ, August EM, Hickey ER, Zhang Q, Teodoro ML, et al. Reversal of H3K9me2 by a small-molecule inhibitor for the G9a histone methyltransferase. Mol Cell 2007;25:473–81.

[90] Huangfu D, Osafune K, Maehr R, Guo W, Eijkelenboom A, Chen S, et al. Induction of pluripotent stem cells from primary human fibroblasts with only Oct4 and Sox2. Nat Biotechnol 2008;26:1269—75.

[91] Jones PA, Baylin SB. The fundamental role of epigenetic events in cancer. Nat Rev Genet 2002;3:415—28.

[92] Gaudet F, Hodgson JG, Eden A, Jackson-Grusby L, Dausman J, Gray JW, et al. Induction of tumors in mice by genomic hypomethylation. Science 2003;300:489—92.

[93] Feinberg AP, Vogelstein B. Hypomethylation Distinguishes Genes of Some Human Cancers from Their Normal Counterparts. Nature 1983;301:89—92.

[94] Greger V, Passarge E, Hopping W, Messmer E, Horsthemke B. Epigenetic changes may contribute to the formation and spontaneous regression of retinoblastoma. Hum Genet 1989;83:155—8.

[95] Jones PA, Baylin SB. The epigenomics of cancer. Cell 2007;128:683—92.

[96] Baylin SB. DNA methylation and gene silencing in cancer. Nat Clin Pract Oncol 2005;2(Suppl. 1):S4—11.

[97] Widschwendter M, Fiegl H, Egle D, Mueller-Holzner E, Spizzo G, Marth C, et al. Epigenetic stem cell signature in cancer. Nat Genet 2007;39:157—8.

[98] Ohm JE, McGarvey KM, Yu X, Cheng L, Schuebel KE, Cope L, et al. A stem cell-like chromatin pattern may predispose tumor suppressor genes to DNA hypermethylation and heritable silencing. Nat Genet 2007;39:237—42.

[99] Schlesinger Y, Straussman R, Keshet I, Farkash S, Hecht M, Zimmerman J, et al. Polycomb-mediated methylation on Lys27 of histone H3 pre-marks genes for de novo methylation in cancer. Nat Genet 2007;39:232—6.

[100] Shukla V, Vaissiere T, Herceg Z. Histone acetylation and chromatin signature in stem cell identity and cancer. Mutat Res 2008;637:1—15.

[101] Moradei O, Vaisburg A, Martell RE. Histone deacetylase inhibitors in cancer therapy: new compounds and clinical update of benzamide-type agents. Curr Top Med Chem 2008;8:841—58.

[102] Oki Y, Issa JP. Review: recent clinical trials in epigenetic therapy. Rev Recent Clin Trials 2006;1:169—82.

[103] Abdolmaleky HM, Smith CL, Faraone SV, Shafa R, Stone W, Glatt SJ, et al. Methylomics in psychiatry: Modulation of gene-environment interactions may be through DNA methylation. Am J Med Genet B Neuropsychiatr Genet 2004;127B:51—9.

[104] Veldic M, Guidotti A, Maloku E, Davis JM, Costa E. In psychosis, cortical interneurons overexpress DNA-methyltransferase 1. Proc Natl Acad Sci USA 2005;102:2152—7.

[105] Chen Y, Sharma RP, Costa RH, Costa E, Grayson DR. On the epigenetic regulation of the human reelin promoter. Nucleic Acids Res 2002;30:2930—9.

[106] Tremolizzo L, Doueiri MS, Dong E, Grayson DR, Davis J, Pinna G, et al. Valproate corrects the schizophrenia-like epigenetic behavioral modifications induced by methionine in mice. Biol Psychiatry 2005;57:500—9.

[107] Fuso A, Seminara L, Cavallaro RA, D'Anselmi F, Scarpa S. S-adenosylmethionine/homocysteine cycle alterations modify DNA methylation status with consequent deregulation of PS1 and BACE and beta-amyloid production. Mol Cell Neurosci 2005;28:195—204.

[108] Mulder C, Schoonenboom NS, Jansen EE, Verhoeven NM, van Kamp GJ, Jakobs C, et al. The trans-methylation cycle in the brain of Alzheimer patients. Neurosci Lett 2005;386:69—71.

[109] Jiang YH, Sahoo T, Michaelis RC, Bercovich D, Bressler J, Kashork CD, et al. A mixed epigenetic/genetic model for oligogenic inheritance of autism with a limited role for UBE3A. Am J Med Genet A 2004;131:1—10.

[110] Unternaehrer JJ, Daley GQ. Induced pluripotent stem cells for modelling human diseases. Philos Trans R Soc Lond B Biol Sci 2011;366:2274—85.

[111] Hotta A, Cheung AY, Farra N, Vijayaragavan K, Seguin CA, Draper JS, et al. Isolation of human iPS cells using EOS lentiviral vectors to select for pluripotency. Nat Methods 2009;6:370—6.

[112] Marchetto MC, Carromeu C, Acab A, Yu D, Yeo GW, Mu Y, et al. A model for neural development and treatment of Rett syndrome using human induced pluripotent stem cells. Cell 2010;143:527—39.

[113] Cheung AY, Horvath LM, Grafodatskaya D, Pasceri P, Weksberg R, Hotta A, et al. Isolation of MECP2-null Rett Syndrome patient hiPS cells and isogenic controls through X-chromosome inactivation. Hum Mol Genet 2011;20:2103—15.

[114] Butler MG. Genomic imprinting disorders in humans: a mini-review. J Assist Reprod Genet 2009;26:477—86.

[115] Chamberlain SJ, Chen PF, Ng KY, Bourgois-Rocha F, Lemtiri-Chlieh F, Levine ES, et al. Induced pluripotent stem cell models of the genomic imprinting disorders Angelman and Prader—Willi syndromes. Proc Natl Acad Sci USA 2010;107:17668—73.

[116] Yang J, Cai J, Zhang Y, Wang X, Li W, Xu J, et al. Induced pluripotent stem cells can be used to model the genomic imprinting disorder Prader—Willi syndrome. J Biol Chem 2010;285:40303—11.

[117] Rougeulle C, Glatt H, Lalande M. The Angelman syndrome candidate gene, UBE3A/E6-AP, is imprinted in brain. Nat Genet 1997;17:14—5.

[118] Chamberlain SJ, Lalande M. Neurodevelopmental disorders involving genomic imprinting at human chromosome 15q11-q13. Neurobiol Dis 2010;39:13—20.

499

[119] Kipling D, Davis T, Ostler EL, Faragher RG. What can progeroid syndromes tell us about human aging? Science 2004;305:1426—31.

[120] Eriksson M, Brown WT, Gordon LB, Glynn MW, Singer J, Scott L, et al. Recurrent de novo point mutations in lamin A cause Hutchinson—Gilford progeria syndrome. Nature 2003;423:293—8.

[121] Dechat T, Pfleghaar K, Sengupta K, Shimi T, Shumaker DK, Solimando L, et al. Nuclear lamins: major factors in the structural organization and function of the nucleus and chromatin. Genes Dev 2008;22:832—53.

[122] Scaffidi P, Misteli T. Reversal of the cellular phenotype in the premature aging disease Hutchinson—Gilford progeria syndrome. Nat Med 2005;11:440—5.

[123] Scaffidi P, Misteli T. Lamin A-dependent misregulation of adult stem cells associated with accelerated ageing. Nat Cell Biol 2008;10:452—9.

[124] Chiba S. Notch signaling in stem cell systems. Stem Cells 2006;24:2437—47.

[125] Hennekam RC. Hutchinson—Gilford progeria syndrome: review of the phenotype. Am J Med Genet A 2006;140:2603—24.

[126] Scaffidi P, Misteli T. Lamin A-dependent nuclear defects in human aging. Science 2006;312:1059—63.

[127] Espada J, Varela I, Flores I, Ugalde AP, Cadinanos J, Pendas AM, et al. Nuclear envelope defects cause stem cell dysfunction in premature-aging mice. J Cell Biol 2008;181:27—35.

[128] Liu GH, Barkho BZ, Ruiz S, Diep D, Qu J, Yang SL, et al. Recapitulation of premature ageing with iPSCs from Hutchinson—Gilford progeria syndrome. Nature 2011;472:221—5.

[129] Zhang J, Lian Q, Zhu G, Zhou F, Sui L, Tan C, et al. A human iPSC model of Hutchinson—Gilford Progeria reveals vascular smooth muscle and mesenchymal stem cell defects. Cell Stem Cell 2011;8:31—45.

[130] Bar-Nur O, Russ HA, Efrat S, Benvenisty N. Epigenetic memory and preferential lineage-specific differenti-ation in induced pluripotent stem cells derived from human pancreatic islet Beta cells. Cell Stem Cell 2011;9:17—23.

[131] Kim K, Doi A, Wen B, Ng K, Zhao R, Cahan P, et al. Epigenetic memory in induced pluripotent stem cells. Nature 2010;467:285—90.

[132] Lunyak VV, Rosenfeld MG. Epigenetic regulation of stem cell fate. Hum Mol Genet 2008;17:R28—36.

[133] Shafa M, Krawetz R, Rancourt DE. Returning to the stem state: epigenetics of recapitulating pre-differentiation chromatin structure. Bioessays 2010;32:791—9.

[134] Shi Y, Desponts C, Do JT, Hahm HS, Scholer HR, Ding S. Induction of pluripotent stem cells from mouse embryonic fibroblasts by Oct4 and Klf4 with small-molecule compounds. Cell Stem Cell 2008;3:568—74.

[135] Huangfu D, Maehr R, Guo W, Eijkelenboom A, Snitow M, Chen AE, et al. Induction of pluripotent stem cells by defined factors is greatly improved by small-molecule compounds. Nat Biotechnol 2008;26:795—7.

[136] Mikkelsen TS, Hanna J, Zhang XL, Ku MC, Wernig M, Schorderet P, et al. Dissecting direct reprogramming through integrative genomic analysis (vol 454, pg 49, 2008). Nature 2008;454. 794-794.

[137] Veldic M, Guidotti A, Maloku E, Davis JM, Costa E. In psychosis, cortical interneurons overexpress DNA-methyltransferase 1. Proc Natl Acad Sci USA 2005;102:2152—7.

[138] Chen Y, Sharma RP, Costa RH, Costa E, Grayson DR. On the epigenetic regulation of the human reelin promoter. Nucleic Acids Res 2002;30:2930—9.

[139] Fuso A, Seminara L, Cavallaro RA, D'Anselmi F, Scarpa S. S-adenosylmethionine/homocysteine cycle alter-ations modify DNA methylation status with consequent deregulation of PS1 and BACE and beta-amyloid production. Mol Cell Neurosci 2005;28:195—204.

[140] Weksberg R, Smith AC, Squire J, Sadowski P. Beckwith—Wiedemann syndrome demonstrates a role for epigenetic control of normal development. Hum Mol Genet 2003;12. Spec No 1:R61—8.

[141] Ehrlich M. The ICF syndrome, a DNA methyltransferase 3B deficiency and immunodeficiency disease. Clin Immunol 2003;109:17—28.

[142] Van den Veyver IB, Zoghbi HY. Mutations in the gene encoding methyl-CpG-binding protein 2 cause Rett syndrome. Brain Dev 2001;23(Suppl. 1):S147—51.

[143] Gibbons RJ, McDowell TL, Raman S, O'Rourke DM, Garrick D, Ayyub H, et al. Mutations in ATRX, encoding a SWI/SNF-like protein, cause diverse changes in the pattern of DNA methylation. Nat Genet 2000;24:368—71.

[144] Nicholls RD, Knepper JL. Genome organization, function, and imprinting in Prader—Willi and Angelman syndromes. Annu Rev Genomics Hum Genet 2001;2:153—75.

[145] Tassone F, Hagerman PJ. Expression of the FMR1 gene. Cytogenet Genome Res 2003;100:124—8.

[146] Bello MJ, Alonso ME, Aminoso C, Anselmo NP, Arjona D, Gonzalez-Gomez P, et al. Hypermethylation of the DNA repair gene MGMT: association with TP53 G: C to A: T transitions in a series of 469 nervous system tumors. Mutat Res 2004;554:23—32.

[147] Esteller M, Corn PG, Baylin SB, Herman JG. A gene hypermethylation profile of human cancer. Cancer Res 2001;61:3225—9.

[148] Widschwendter M, Jiang G, Woods C, Muller HM, Fiegl H, Goebel G, et al. DNA hypomethylation and ovarian cancer biology. Cancer Res 2004;64:4472—80.

[149] Li LC, Okino ST, Dahiya R. DNA methylation in prostate cancer. Biochim Biophys Acta 2004;1704:87—102.

[150] Richardson B. Impact of aging on DNA methylation. Ageing Res Rev 2003;2:245—61.

[151] Sekigawa I, Okada M, Ogasawara H, Kaneko H, Hishikawa T, Hashimoto H. DNA methylation in systemic lupus erythematosus. Lupus 2003;12:79—85.

CHAPTER 25

Non-Coding RNA Regulatory Networks, Epigenetics, and Programming Stem Cell Renewal and Differentiation: Implications for Stem Cell Therapy

503

Rajesh C. Miranda
Texas A&M Health Science Center, Bryan, TX, USA

T. Tollefsbol (Ed): Epigenetics in Human Disease. DOI: 10.1016/B978-0-12-388415-2.00025-1

25.1 MAJOR TYPES OF STEM CELLS

Stem cells have a near infinite capacity for self-renewal and can be guided by specific signaling mechanisms to differentiate into selected cell lineages. The capacity for self-renewal, i.e. for parent stem cells to produce daughter cells that are functionally and phenotypically identical to the parent cell, means that at least theoretically, every tissue and organ with a resident population of stem cells has a nearly indefinite capacity for repair and regeneration. Moreover, the capacity of restricted-potential adult stem cells to replenish lost and damaged cells and to repair tissues, with programming by growth factors, makes these cells useful to harness for therapeutic purposes.

An extensive review of the types of stem cells and their properties is beyond the scope of this chapter. (The reader is referred to several well-annotated public resources (e.g. at the National Institutes of Health http://stemcells.nih.gov/info/basics/; The Howard Hughes Medical Institute http://www.hhmi.org/biointeractive/stemcells/lectures.html).)

Briefly, major classes of indigenous stem cells include embryonic (ES), fetal, and adult (or tissue) stem cells. These cells range in their capacity for differentiation. "Totipotent" stem cells, which retain a capacity to generate an entire organism, can be derived from nuclear transfer into an oocyte, or following fertilization, from early divisions of the zygote and resulting morula (Figure 25.1). (For an interesting essay into the history of nuclear transfer and cloning, see [1].) The cells of the morula subsequently form a blastocyst, where they differentiate into two cell layers, an outer trophoectoderm and an inner cell mass. "Pluripotent" stem cells can be isolated from the inner cell mass, and exhibit the possibility of differentiation into ectoderm, mesoderm, and endoderm cell lineages, but are not, by themselves, capable of forming an entire organism. Finally, within each fetal and adult tissue, resident stem cells continue to generate, with variable efficiency, replacement cells through the life of the tissue and organism. These cells normally exhibit restricted potential, ranging from multipotent, to bi- or even unipotency, from tissue-specific cell types. For example, hematopoietic stem cells are multipotent because they can generate a wide variety of lymphoid, erythroid, and myeloid cell

504

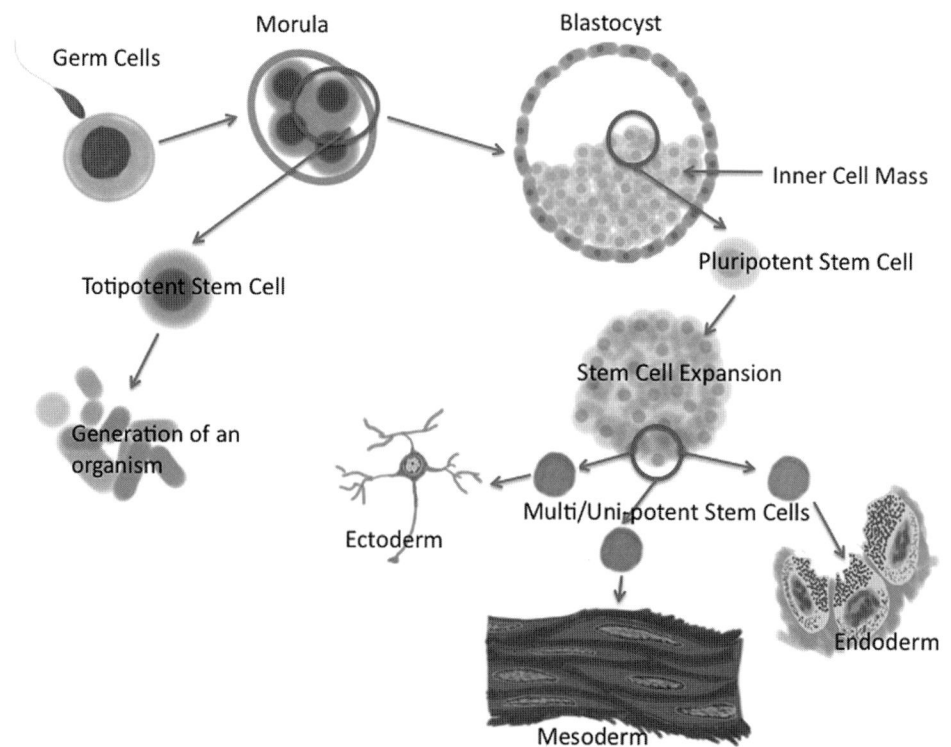

FIGURE 25.1
Schematic for the general maturation of stem cells and loss of differentiation potential. This figure is reproduced in the color plate section.

lineages. Mesenchymal stem cells also exhibit multipotency as they can generate osteoblasts, chondroblasts, adipocytes, and fibroblasts among a wide variety of cells. In contrast, neural stem cells exhibit a more restricted capacity to differentiate into neurons and glia, and epidermal stem cells exhibit unipotential capacity to differentiate into keratinocytes.

Finally, increasing evidence suggests that adult multipotential stem cells, even within a single tissue are heterogeneous and exhibit a hierarchy of "stemness" that may make them more or less suited as therapeutic targets. For example, the G-protein-coupled receptor Lgr5 (Gpr49) marks a population of rapidly cycling cells within intestinal crypts that have the capacity to generate every other cell of the intestinal epithelium [2]. However, when crypt Lgr5+ cells were depleted, a rare population of slowly proliferating cells that are positive for Bmi1 (a member of the polycomb transcription regulatory complex), exhibit a parallel capacity to regenerate the intestinal epithelium including the Lgr5+ population [3]. Bmi1+ and Lgr5+ cells may represent long- vs. short-term enteric stem cell populations that may be selectively targeted to promote long-range tissue re-engineering vs. shorter-term repair.

25.1.1 Stem Cell Renewal and the Pluripotency Transcription Factor Network: Lessons from Induced Pluripotent Stem Cell (IPSC) Biology

Non-protein-coding RNA networks and epigenetics have particular relevance for stem cell biology in the context of recent successful efforts to transform somatic cells into stem cells, i.e. induced pluripotency. In 2007, two independent research groups published evidence for the existence of a minimal network of transcription factors that could be coordinately activated, to induce differentiated *human* cells to revert to a pluripotent state [4,5]. The research group led by Shinya Yamanaka used a combination of four transcription factors, Oct3/4 (POU5F1), Sox2, Klf4, and c-Myc to reprogram human dermal fibroblasts to express properties of embryonic stem cells (Figure 25.2). The research group lead by James Thompson in contrast, used an overlapping combination of transcription factors, OCT3/4, SOX2, NANOG, and LIN28, to accomplish the same purpose. Both research groups showed that the resultant IPSCs had attained the capacity to differentiate into cells belonging to all three germ layers, suggesting that these cells were pluripotent. A large amount of subsequent research indicates that Sox2, c-Myc, Oct3/4, Klf4, and NANOG constitute a core network of transcription factors that maintains pluripotency of embryonic stem cells and confers pluripotency characteristics

505

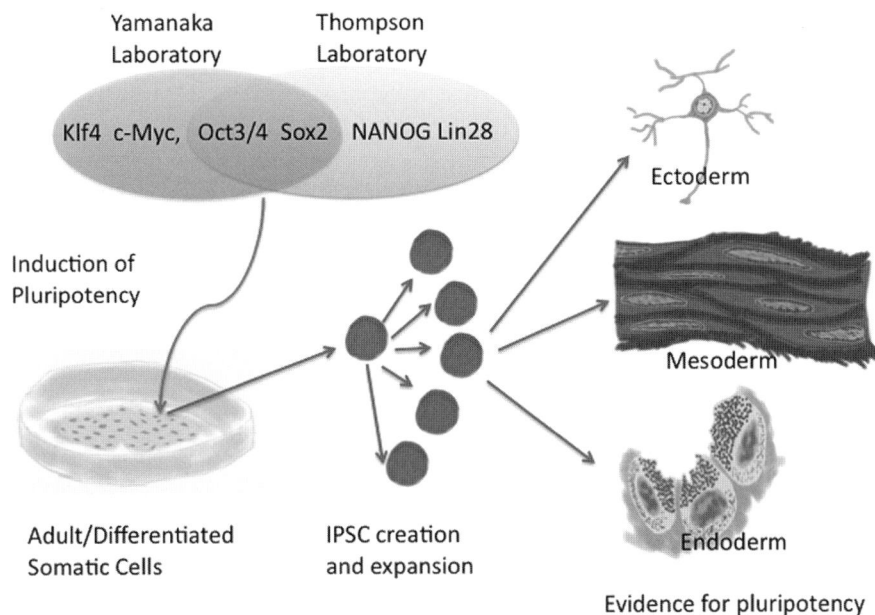

FIGURE 25.2
Schematic for the induction of pluripotency in somatic cells by the pluripotency network of transcription factors. This figure is reproduced in the color plate section.

to a wide variety of differentiated somatic cell types derived from diverse tissue sources (reviewed in [6,7]).

The pluripotency factor network has evolved functionally through vertebrate evolution, to target new downstream genes, and presumably new tissue- and species-specific stem cell renewal and differentiation programs. For example, the Fibroblast Growth Factor-4 (FGF4) gene is present in vertebrates, but only emerged in recent eutherian mammal evolution, as a positively regulated target of Oct3/4 [8]. FGF4 in turn mediates the transition between stem cell renewal and cell differentiation [9], implying that Oct3/4-induced, FGF4 autocrine/ paracrine signaling may have evolved to specifically limit the renewal capacity of the pluripotency transcription factor network in mammals. Interestingly, the FGF family itself has expanded in numbers significantly during vertebrate and even recent mammalian development [10], and has come to play a critical role in mediating lineage specification of stem cells [11]. Though the core pluripotency network has remained relatively evolutionarily conserved (stem cells in amphibian [12] and mammalian species [4,5] make use of an overlapping complement of pluripotency factors), the integration of the pluripotency network with lineage specification networks exhibits a great deal of evolutionarily diversity. As we will discuss below, some of this evolutionary recent regulatory complexity is due to the emergence of novel epigenetic programs.

25.1.2 Regulatory Networks and Control of Stem Cell Differentiation

Growth factors constitute an effective means for guiding multipotent stem cells down selected differentiation pathways because of their capacity to activate specific transcription factor networks (e.g. Figure 25.3A). FGF4 is an important example in this regard. FGF4 directs mouse embryonic stem cells towards a neural (ectoderm) lineage by activating ERK1/2 [13]. The concurrent overexpression of Sprouty1 (Spry1) on the other hand, prevents neural lineage progression, and promotes mesoderm lineage differentiation instead [13]. Further, within the mesoderm lineage, other growth factors, the Bone Morphogenetic Proteins (BMP) 4 and 7 for example, promote osteoblast differentiation by inducing the expression of the transcription factor DLX5 [14,15]. Similarly, Wnt family members, Wnt1 and Wnt10b, direct differentiation of mesenchymal stem cells towards an osteoblast lineage, while inhibiting adipocyte lineage differentiation [16], in part, by promoting expression of osteoblast transcription factors like Runx2 and Dlx5 while inhibiting adipocyte-associated transcription factors like PPARγ, C/EBPα, and C/EBPβ. In contrast, BMP2 in cooperation with other signaling molecules like TGFβ, drives mesenchymal stem cells into a chrondrocyte lineage [17]. Interestingly, in portions of the cranial ectomesenchyme, Dlx5/6 induces expression of the transcription factor Hand2, which acts as a negative regulator of Dlx5/6. The expression of Hand2 results in the transformation of primitive stem cells into tissues of the tongue, whereas its absence results in a Dlx5/6-driven transformation to bone [18]. Specific lineage commitment is, therefore, dependent on the contextual cues provided by competing signaling molecules. Moreover, a limited set of interacting secreted signaling factors and a related network of intracellular signaling cascades and transcription factors can clearly drive divergent differentiation from a common cohort of stem cells.

The core pluripotency network arguably emerged early in vertebrate evolution to support two main functions, to promote stem cell renewal while simultaneously suppressing differentiation, and appears to be largely similar from one stem cell compartment to the next. Differentiation, on the other hand, is likely to involve a diversity of maturation genes that adapt differentiating cells to specific tissues and organs. It is likely therefore that there is a layer of cellular regulation that adapts the common pluripotency network to cell-, tissue-, and even organism-specific differentiation. Several pieces of evidence indirectly point to an intervening regulatory layer between stem cell renewal and differentiation. Firstly, though stem cells reside in every tissue and organ, not all stem cells have an equivalent capacity to regenerate or repair

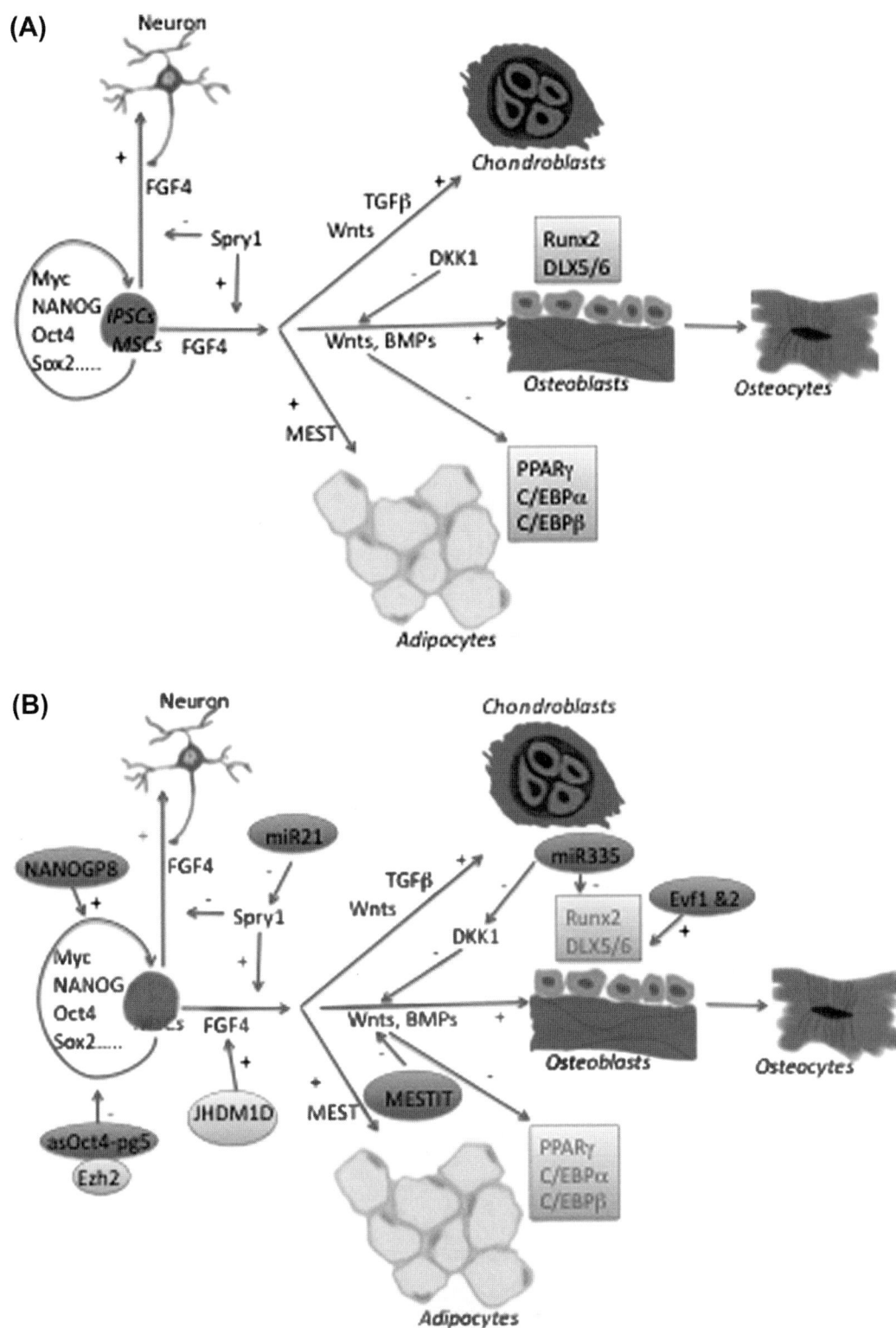

FIGURE 25.3

(A) Schematic of the competing network of transcription factors and signaling molecules that direct stem cells down specific mesoderm or ectoderm lineages. Green boxes indicate core transcription factors necessary for osteoblast and adipocyte lineage specific differentiation. (B) Overlay of the epigenetic/ncRNA regulatory network. This schematic depicts examples of ncRNAs (blue boxes) and chromatin modification factors (pink boxes) that control the balance between signaling molecules and transcription factors, to direct stem cell differentiation. This figure is reproduced in the color plate section.

tissues and organs. Even cells that share a common developmental lineage, i.e. mesenchymal stem cells, can exhibit varying differentiation potential related to their tissues of origin [19,20], suggesting the emergence of non-genetically programmed regulatory mechanisms. Secondly, the biology of stem cells is intimately associated with evolution and speciation. Similar types of stem cells in different organisms produce diverse body plans and exhibit divergent regenerative capacities. For example, in the python but not humans, an episode of feeding produces a large and rapid increase in enterocytes in the small intestine [21], suggesting that stem cell renewal and maturation is remarkably attuned to the environment, but in a species-specific manner. Stem cells in amphibian [12] and mammalian species [4,5] make use of an overlapping complement of pluripotency factors, yet amphibian, but not mammalian stem cells have the capacity to regenerate complex tissues like limbs. Finally, not only do stem cells in older organisms exhibit diminished regenerative capacities [22], but stem cells can exhibit altered patterns of lineage commitment with age; i.e. aging mesenchymal stem cells favor adipocyte rather than osteoblast differentiation [23]. The question is why is there such diversity in stem cell differentiation potential from tissue type, speciation, and age? One answer to this question might lie in the existence of a new and relatively poorly understood network of regulatory mechanisms collectively termed, "epigenetics".

25.2 A BRIEF OVERVIEW OF EPIGENETICS

Epigenetics, literally "over the genome", represents a newly discovered and dynamic regulatory network that is superimposed over, and therefore influences, the interpretation of genetic information contained within a cell. At the level of the organism, epigenetics serves to promote adaptation and is increasingly thought to be a major mechanism for speciation, and at the molecular level, a mechanism to control cellular differentiation and homeostasis. Epigenetic regulatory networks are increasingly being found to be critical facilitators of the successful transformation of stem cells into tissues and organs, but may also serve the aberrant transformation of stem cells in cancer. A detailed and comprehensive overview of the field of epigenetics is well beyond the scope of this chapter. Excellent recent reviews have outlined the history and basic mechanisms underlying epigenetics [24], and detailed their relevance to tissue and organism development [25,26] and to cancer mechanisms [27]. Briefly, the term "epigentics" was coined by the developmental biologist, Conrad Waddington, based on early observations that an environmental stimulus (heat) administered to a cohort of fruitflies, produced multigenerational alterations in wing structure without any apparent change in the fly genome [28]. A variety of cellular mechanisms that regulate nuclear chromatin structure and control gene transcription and translation are collectively classified as "epigenetic mechanisms", if these mechanisms result in relatively irreversible changes in the function of cells and tissues.

25.2.1 DNA Methylation, Histone Modifications, and Gene Imprinting

Activity of DNA methylases (DNMTs) results in methylation of genes at chromosomal loci that are GC-rich (extensive regions of GC enrichment are termed CpG islands), and can result in the gene inactivation by the recruitment of methyl-CpG-binding proteins and the segregation of chromosome regions containing those genes into the nuclear heterochromatin. Similarly, post-translational histone modifications can also alter the compactness of nucleosomes to regulate gene expression. The methylation of histones, such as di- or trimethylation of histone H3 on lysine-4 (H3K4me2 and H3K4me3), result in increased activation, whereas di- and trimethylation on H3K9 and histone acetylation are associated with repression [29]. The methylation and demethylation of chromatin is an important component of the stem cell differentiation process. For example, adipose-derived mesenchymal stem cells exhibit demethylation at Dlx5 and other osteoblast-specific transcription factors during the process of transformation into osteoblasts [30]. The process of demethylation results in permissive activation of the Dlx5 locus and presumably plays an important and contextual role in enabling growth-factor-mediated transformation of stem cells.

During vertebrate evolution the emergence of mammals, particularly eutherian mammals, was accompanied by an important variation in the epigenetic program, the phenomenon of gene imprinting, where one allele (either the paternal or the maternal allele) of a gene undergoes DNA methylation, and is silenced. The dominant model for transcription at these loci is that it proceeds from the remaining active allele. Frequently, the non-silenced allele exhibits post-translational histone modifications like trimethylation of lysine 4 (H3K4me3) that are known to facilitate transcription activation [31]. The human genome is predicted to contain as many as 156 imprinted genes [32], and many of these do not overlap with the cohort of imprinted genes in the mouse [33], suggesting the likelihood of shifts in imprinting with mammalian speciation. The net effect is to decrease the gene dosage in tissues and the emergence of this phenomenon with mammalian evolution is thought to be a mechanism for the control of fetal size. Paternal alleles are thought to promote, while maternal alleles are thought to constrain, fetal growth (reviewed in [34]). The implication of imprinting as an epigenetic phenomenon that regulates stem cells is enormous. Because of their capacity to control tissue growth [35], it is likely that imprinted genes play an important role in stem cell maturation [36]. The species variation in gene imprinting suggests that the epigenetic controls over stem cell renewal and maturation are likely to be species-specific. Moreover, gene imprinting may vary as a function of the state of cellular differentiation. For example, neurons of the adult brain have recently been shown to exhibit monoallelic expression of the Angelman's syndrome locus, Ube3a, whereas Glial Fibrillary Acidic Protein (GFAP)-positive cells of the ventricular zone have been found to exhibit biallelic expression from this locus [37]. Since GFAP-positive cells adjacent to the lateral ventricles of the brain are presumptive neural stem cells [38] these data suggest that the gene dosage of Ube3a is decreased during the differentiation of neural stem cells by epigenetic mechanisms. These data suggest that the epigenetic programming of stem cells may vary as a function of both species and tissue of origin, and that the replication of tissue- and species-specific epigenetic programs will be critical for the successful therapeutic manipulation of stem cells.

25.2.2 Epigenetics of Non-Coding RNAs

In the abstract to a 2007 article [39], John Mattick proposed that, *"the epigenetic trajectories of differentiation and development are primarily programmed by feed-forward* (non-coding) *RNA regulatory networks and that most of the information required for multicellular development is embedded in these networks"* (author inserted text in parenthses). Since "differentiation" and "development" programs implicitly involve the maturation of stem cells, the hypothesis advanced by Mattick implies that non-coding RNAs (ncRNAs) are a significant locus of epigenetic control of stem cell maturation programs.

Sequencing the human genome has shown unexpectedly that the human genome contains a surprisingly small number of protein-coding genes [40]. The Ensembl database (http://www.ensembl.org/, Genebuild 2011) counts 20 599 protein-coding genes, compared to the mouse at 21 873, the zebrafish at 18 572, and the invertebrate, *Caenorhabditis elegans* at 20 389 known protein-coding genes. Clearly the protein coding gene content of animal chromosomes does not change dramatically with vertebrate and mammalian evolution. The approximately 2000 additional protein-coding genes that separate mammals, including humans, from zebrafish, do not account for the evolutionary complexity of humans, and consequently, cannot readily explain the complexity of stem cell programs that generate the human body plan. However, recent evidence indicates that other products of genome transcription, namely non-coding RNAs, do exhibit increased representation within genomes with increasing evolutionary complexity [39]. A 2007 report of the ENCODE (Encyclopedia of DNA Elements) project, suggested that the human genome is *"pervasively transcribed, such that the majority of its bases are associated with at least one primary transcript*" [41]. This paper showed that the human genome is filled with large numbers of transcription start sites that code for ncRNAs and that regions of the genome that had hitherto been thought of as "silent", were in fact, transcriptionally active. It is likely that, in terms of sheer numbers, ncRNAs will dwarf the numbers of protein-coding genes. RNA molecules can participate in a variety of cell functions because they exhibit two important functions. Firstly, their primary nucleotide sequence complementarity can give RNA molecules the capacity to bind other RNA and DNA molecules with exquisite specificity. Secondly, RNAs can fold into a variety of functional shapes in the presence of helicase chaperones to participate in RNA–protein complexes [42]. Therefore, ncRNAs can effectively serve as receptors or adapter molecules to target regulatory protein complexes to DNA or other RNAs, with important consequences for stem cell renewal.

The family of ncRNAs ranges in size from small, miRNAs and siRNAs (17–24 nucleotides), and piRNAs (25–30 nucleotides), to extremely long ncRNAs like the 19.3-kb X-chromosome inactivation factor, XIST [43]. Both long ncRNAs (reviewed in [44]) and short ncRNAs like piRNAs [45] have roles in chromatin reorganization, and consequently, epigenetic programming. NcRNAs like miR9/miR9* and miR124 can control chromatin remodeling complexes [46] to promote the renewal or maturation of neural stem cells. However, recent deep-sequencing data suggest an additional layer of regulation, in that substantial numbers of processed miRNAs traffic back into the nucleus [47] where they may also directly act as epigenetic factors.

25.3 EPIGENETIC PROGRAMMING OF STEM CELLS

We will now look at the epigenetic programming of stem cells.

25.3.1 Epigenetic Programming and the Core Pluripotency Transcription Factor Network

Recent evidence points to a complex interplay between core pluripotency factors, epigenetics, and non-coding RNAs. Several miRNA families can serve as substitutes for or enhance the activity of core pluripotency factors to transform somatic cells into stem cells [48,49].

The miR17-92 miRNA cluster, among others, for example, has been recently found to be highly expressed during early stages of induced stem cell programming [49] and therefore, it is instructive to examine the interactions of members of this cluster with the core pluripotency factor network. The pluripotency factor c-Myc, for example, is thought to promote cell proliferation in part, by binding to the common promoter for miR17-5p and miR20a, in the oncogenic miR17-92 microRNA cluster (Figure 25.4) to promote the expression of miR20a and therefore repress translation of the retinoblastoma-like protein-2 (Rbl2) [50]). Rbl2 and other members of the retinoblastoma family promote and stabilize nuclear heterochromatin by recruiting histone deacetylases and increasing DNA methylation [51,52], suggesting that miR20a might indirectly interfere with DNA methylation. However, miR20a and miR17-5p (another member of the miR17-92 cluster), also translocate to the nucleus, and recruit Argonaut complexes to promoters of target genes (in this case, a cyclin-dependent kinase inhibitor CDKN2B/INK4B), resulting in increased trimethylation of histone H3 at Lysine 9 (H3K9me3, a repressive epigenetic marks), and, like Rbl2, promote the formation of heterochromatin [53]. These apparently contradictory data suggest that Myc-mediated epigenetic programming is complex, but taken as a whole, prevents cell cycle arrest. Recent observations [54] of genome-wide DNA methylation patterns have reported a diversification in the patterns of moderately methylated genes during stem cell maturation. Some genes that are moderately methylated during stem cell renewal, become hypomethylated, while others exhibit increased methylation. The association between Myc and the miR17-92 cluster provides a mechanistic hypothesis that explains the observed diversification in DNA methylation patterns that reshape the nuclear heterochromatin landscape during transitions between stem cell renewal and differentiation.

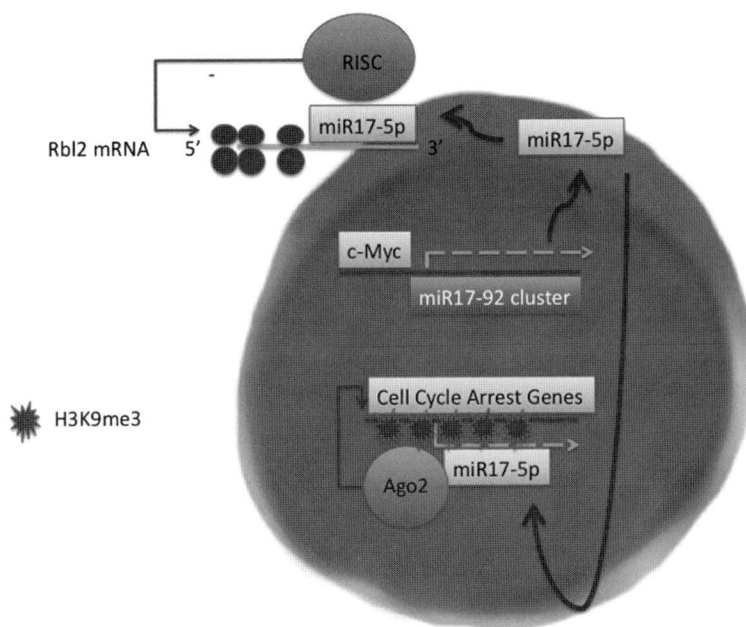

FIGURE 25.4

Example of interactions between a pluripotency factor, Myc, and the ncRNA/epigenetics network. Myc promotes the expression of the miR17-92 miRNA cluster, which is processed, transported to the cytoplasm and directs translational repression of cell cycle arrest genes by the RNA-induced silencing complex (RISC). However, miRNAs are also localized to the nucleus, where they can bind to nascent mRNA transcripts and recruit Argonaut proteins (Ago2) and promote histone methylation marks (e.g. H3K9me3) that result in transcription suppression of cell cycle arrest proteins. Collectively, these factors contribute to Myc-mediated epigenetic control over stem cell renewal and maintenance of pluripotency. This figure is reproduced in the color plate section.

Following expression of the pluripotency factor network, Myc-mediated transcription at the miR17-92 cluster may favor direct miRNA-mediated chromatin remodeling and indirectly, by translation repression, prevent inactivation of cell cycle genes.

Myc's role in pluripotency is complex. Myc normally promotes or represses gene transcription, but this function does not appear to be an important factor in the phase of DNA replication to promote cell cycle. During this phase, Orc1, a member of the DNA "origin of replication (ORC)" complex, sequesters and inactivates Myc control of transcription [55], and Myc itself localizes to early DNA replication complexes to promote DNA replication [56]. Interestingly, the Myc gene itself serves as a site for ORC assembly and initiation of DNA replication. A network of small ncRNAs termed "yRNAs" [57,58] promote DNA replication by stabilizing ORCs [59] to chromatin during periods of DNA replication. However, ORCs also recognize repressive epigenetic heterochromatin marks and localize to heterochromatic regions of the genome during other periods of the cell cycle [60]. The oscillation between Myc control of transcription and Myc-ORC control of DNA replication recruits layers of epigenetic mechanisms and is critical for the processes of stem cell renewal and maturation.

A further analysis of Myc function is also instructive in terms of the assessment of the contribution of long ncRNAs to stem cell programming. Myc also directly binds to, and strongly represses, the transcription of Gata6, a transcription factor that promotes endoderm differentiation of stem cells. However, while the expression of the Gata6 mRNA transcript is repressed during stem cell renewal, not all transcription at the Gata6 gene locus is suppressed. Recent evidence shows that the Gata6 gene locus is also the source of a long ncRNA [61], termed Gata6bt (for Gata6 bidirectional transcript) that is transcribed in the antisense orientation relative to Gata6. The expression of Gata6bt is inversely correlated with the mRNA transcript for Gata6, being highly expressed in stem cells and suppressed during differentiation. Gata6bt, like Myc, may serve as a negative regulator of Gata6 transcription [61], to maintain stem cell pluripotency, and like other large antisense ncRNAs (HOTAIR for example) [62]), may regulate chromatin state by serving as scaffolds for the assembly of histone modification complexes.

Other members of the pluripotency network are also subject to epigenetic regulatory programs. Both Oct4 and NANOG are subject to extensive regulation by DNA methylation and histone modifications [63]. Moreover, in silico screens have found that mammalian genomes are specifically enriched in pseudogenes for Oct4 and NANOG. The human genome contains six pseudogenes for Oct3/4 and ten pseudogenes for Nanog, compared to a relative paucity of psuedogenes for other non-pluripotency-related transcription factors [64]. The Oct4 pseudogene family has been recently found to exert complex and mutually interdependent epigenetic regulation of the Oct4 promoter. Oct4-pseudogene-5 for example, generates an antisense transcript (asOct4-pg5), that serves as a negative regulator of Oct4, in part by recruiting the histone methyl transferase, Ezh2 to the Oct3/4 promoter and increasing H3K9me2 and H3K9me3 histone methylation marks (Figure 25.3B) [65]). The more extensive NANOG pseudogene family may serve a similar epigenetic regulatory purpose. Interestingly, the evolutionarily recent appearance of NANOG pseudogene-8 (NANOGP8), following the divergence between humans and chimpanzees [66], suggests that human stem cells may have acquired novel epigenetic regulatory mechanisms that are not present in stem cells of other mammals or vertebrates. An analysis of the EST database for expressed sequences ascribed to transcription from the human NANOGP8 genomic locus (http://genome.ucsc.edu; chr15:35,371,776-35,450,927), indicates that this locus is the origin of multiple non-coding RNA transcripts. NANOGP8 has been found to promote cell proliferation and is also the origin of a NANOG-like protein that is expressed in cancers [67], opening up the possibility that this evolutionarily recent pseudogene may have the capacity to promote the oncogenic transformation stem cells.

Finally, a recent study by Guttman et al., showed that a large network of long intergenic ncRNAs (termed lincRNA), maintain the pluripotency of stem cells [68]. Guttman et al. were able to identify at least 26 lincRNAs that positively regulated the activity of Oct4 and NANOG. Using Oct4 and Nanog promoter linked to a Luciferase reporter, the authors were able to show that a suppression of any one of these lincRNAs resulted in significant suppression of promoter activity for these pluripotency factors. Interestingly, these data imply that each of the 26 lincRNAs exhibits independent, non-redundant control over the Oct4 and Nanog promoters and suggests that all of the lincRNAs are required to maintain at least these components of the minimal pluripotency transcription factor network. These data lend further support to Mattick's hypothesis that extensive ncRNA regulatory networks [39] are likely to control stem cell renewal.

25.3.2 Epigenetic Programming, Non-Coding RNAs, and Early Stem Cell Differentiation Programs

We previously presented evidence for FGF4 as an example of an early differentiation gene that directed stem cells towards a neural ectoderm lineage in preference to mesoderm or endoderm lineages. Recent evidence indicates that the FGF4 locus is inactivated epigenetically, by dimethyation of histone H3 (H3K9me2 and H3K27me2), and that a histone demethylase, JHDM1D/KIAA1718, results in the elimination of these repressive methylation marks, induction of FGF4 expression and consequently, neural lineage commitment [69]. Spry1, the negative regulator of FGF4 signaling, is a target of the miRNA, miR21 [70], which is highly expressed in neural lineage-committed stem cells [71]. Therefore, in the presence of miR21, FGF4-programmed stem cells will preferentially commit to neural lineages (Figure 25.3B).

Imprinted gene loci play an important role in tissue growth in mammals and therefore an analysis of how they control stem cell differentiation is particularly important for the therapeutic use of stem cells. The Mest/Peg (Paternally-Expressed Gene)-1 locus is a good example of the role of epigenetics in stem cell maturation. An analysis of the Peg-1 locus (human chromosome 7:130,125,573-130,146,478) shows the presence of two CpG islands, coinciding with transcription factor binding sites and constituting two presumptive promoters. Interestingly, these regions, particularly at the second CpG island also coincide with a high density of activation acetylation (H3K27Ac) and methylation (H3K4me3 and H3K4Me1) marks on histones, suggesting differential activation of maternal and paternal alleles. While the exon structure of MEST appears to be conserved through vertebrate evolution, two ncRNA transcripts, a miRNA (miR335) and a long antisense transcript overlapping the presumptive promoter region (MESTIT1) late in evolution, in eutherian mammals and primates specifically. Like MEST, MESTIT1 also exhibits monoallelic expression from the paternal allele and like other long ncRNAs [62] may recruit heterochromatin marks to the parental gene. We previously observed that the MEST intronic miRNA, miR335 is expressed at moderate levels in neural stem cells and that miR335 knockdown results in increased neural stem cell proliferation [71]. The MEST/miR335 locus appears to cooperatively specific differentiation of mesenchymal stem cells, by interacting with core differentiation signaling programs. We previously discussed evidence, for example, that Wnt signaling directs mesenchymal stem cells towards osteoblast-specific differentiation and inhibits adipocyte differentiation. MEST has recently been reported to block Wnt signaling by inhibiting post-translational processing of the Wnt co-receptor LRP6, and knockdown of MEST prevents adipogenesis and results in reduced expression of the adipogenic factors C/EBPα and PPARγ (Figure 25.3B) [72]. The MEST intronic miRNA on the other hand, is highly expressed in bone-marrow-derived mesenchymal stem cells compared to differentiated cells belonging to the mesoderm lineage, and its expression decreases during osteoblast and adipocyte differentiation [73]. MiR335 promotes Wnt signaling by suppressing the expression of the Wnt antagonist DKK1 [74]. However, miR335 also acts as a direct negative regulator of Runx2, a factor required for osteogenic differentiation [73]. Collectively, these data suggest that osteoblast differentiation from mesenchymal stem cells is facilitated by careful temporal choreography of gene expression from the imprinted MEST/miR335 locus. Runx2 and other important components of the osteoblastic transformation of mesenchymal stem cells, like DLX5 [14,15] are also loci for the interactions between epigenetics and ncRNA networks. The human Runx2 (human chr6:45,296,054-45,518,818) and DLX5 (human chr7:96,649,667-96,654,230) gene loci for example, contain CpG islands, and demethylation leads to increased expression of these factors and supports osteoblast differentiation [75]. DLX2, 5, and 6 are coexpressed strongly in immature osteoblasts [76], but this cluster is also important for neural stem cell differentiation [77], and may provide redundant control of early stem cell maturation. The DLX5/6 gene loci are the site for transcription of two long ncRNAs, generated by alternate splicing, the 2.7-kb and 3.8-kb Evf1 and Evf2 RNAs. These ncRNAs expand activation of the DLX network by serving as transcription coactivators for DLX2, further increasing transcription of DLX5/6 [78]. The Evf ncRNAs have evolved to coordinate activity of the DLX gene family, and in coordination with other epigenetic mechanisms that regulate the methylation state of DLX5, can promote specific patterns of stem cell differentiation. The emerging story is that pluripotency and differentiation-related transcription factors, and many of their associated signaling pathways are subject to a layer of additional regulation from a network of proteins and ncRNAs that guide the renewal and differentiation potential of stem cells. Finally, ncRNAs like miR-9/9* and miR-124 have reportedly been used to transform cells from one lineage, mesoderm, to another, i.e. neural [79], without the requirement for passing through an intermediate "stem cell" stage. This suggests that the pluripotency factor network may be dispensable, and that the direct ncRNA-mediated transformation of somatic cells may be a viable strategy for therapeutic purposes.

25.3.3 Epigenetic Mutations and Somatic Cell Mosaicism

A major barrier to the adoption of stem cell therapy is that stem cells develop epigenetic mutations, and that these may persist and influence differentiation patterns, cancer transformation, and premature stem cell senescence. For example, researchers have reported the loss of X-chromosome inactivation in well-established human embryonic stem cell lines [80] suggesting that stem cells can experience epigenetic drift. Loss of expression of the ncRNA, XIST, is related to the loss of X-chromosome inactivation and to the resistance of cancer to chemotherapy [81] suggesting that disruptions in epigenetic programs may promote the emergence of cancer stem cells. We [71] and others [54] have shown that ncRNAs coded within imprinted genomic regions, and methylation patterns in stem cells are both sensitive to teratogens like ethanol. This suggests that the environment can reprogram epigenetic controls over stem cell renewal and maturation.

Most epigenetic changes do not lead to alterations in the primary sequence of genes and are potentially reversible. However, some epigenetic mutations do lead to genetic mosaicism in somatic stem cells, potentially leading to permanent alterations in differentiation. The retro-transposon genes, which constitute approximately 45% of the sequence of the human genome [82] are a good example of how mutations in the epigenome may produce genetic drift among somatic cells, and perhaps even among stem cells. Long interspersed element 1 (LINE-1) retro-transposons code for RNA molecules that can be reverse-transcribed and reinserted into genes to alter their function either subtly, by altering promoter regions for example, or fundamentally, by shuffling exon structure of protein-coding genes [82]. Retrotransposons are normally silenced in stem cell genome by piRNAs [45] and by recruitment of DNA methylation mechanisms [83]. Recent research has shown that LINE-1 retrotransposons are activated during neural stem cell differentiation [84] and IPSCs obtained from patients with diagnosed mutations in Methyl CpG binding protein-2 (MeCP2) exhibited increase LINE-1 activity during neural differentiation [85].

Collectively these data show that not only are epigenetic mechanisms and RNA networks needed to maintain chromosome stability in stem cells, but that evolution may have equipped mammalian and primate stem cells to actually utilize chromatin destabilization mechanisms to drive genetic diversification of somatic cells during the process of stem cell differentiation. However, epigenetic mutations may contribute to senescence of adult tissue stem cells, compromising their regenerative capacity [22]. Moreover, epigenetically driven genetic diversi-fication of somatic cells means that these cells may not be equipped to recapitulate native pluripotency states of embryonic stem cells derived from the blastocyst. For example, a recent study showed that IPSC cells derived from muscle-derived pericytes retained their epigenetic programming, and preferentially differentiated into muscle tissue [86]. Similar fate preference was shown for IPSCs derived from pancreatic beta cells [87]. Therefore IPSC-type stem cells may have limited functional therapeutic applicability to replacement of their tissues of origin.

25.4 FINAL COMMENTS

This real potential for stem differentiation for tissue repair and engineering in human populations was emphasized by an early clinical report [88], outlining the steps towards engineering a functional mandible in a human patient. In this study, the authors impregnated an artificial mandible-shaped, titanium-encased bone mineral matrix, with patient-derived bone marrow (rich in adult mesenchymal stem cells) treated with bone morphogenetic protein-7 (BMP7). The authors were able to show vascularization, osteogenesis, and successful functional engraftment of the engineered mandible into a patient. BMPs are now used clini-cally to induce bone formation and agents, by programming adult mesenchymal stem cells [89], showing that the guided differentiation of restricted-potential adult stem cells has practical value. The generation of embryonic stem cells and patient-specific stem cells by IPSC technology also bears significant promise, at least in animal studies of a variety of disease model systems (for example [90–93]). A clearer understanding of the epigenetic landscape of the stem cell during renewal and through successive stages of maturation will be a critical requirement for the development of effective stem cell therapy.

References

[1] McLaren A. Cloning: pathways to a pluripotent future. Science 2000;288(5472):1775–80.

[2] Barker N, van Es JH, Kuipers J, et al. Identification of stem cells in small intestine and colon by marker gene Lgr5. Nature 2007;449(7165):1003–7.

[3] Tian H, Biehs B, Warming S, et al. A reserve stem cell population in small intestine renders Lgr5-positive cells dispensable. Nature 2011;478(7368):255–9.

[4] Takahashi K, Tanabe K, Ohnuki M, et al. Induction of pluripotent stem cells from adult human fibroblasts by defined factors. Cell 2007;131(5):861–72.

[5] Yu J, Vodyanik MA, Smuga-Otto K, et al. Induced pluripotent stem cell lines derived from human somatic cells. Science 2007;318(5858):1917–20.

[6] Masip M, Veiga A, Izpisua Belmonte JC, Simon C. Reprogramming with defined factors: from induced pluripotency to induced transdifferentiation. Mol Hum Reprod 2010;16(11):856–68.

[7] Pan G, Thomson JA. Nanog and transcriptional networks in embryonic stem cell pluripotency. Cell Res 2007;17(1):42–9.

[8] Fernandez-Tresguerres B, Canon S, Rayon T, et al. Evolution of the mammalian embryonic pluripotency gene regulatory network. Proc Natl Acad Sci USA 2010;107(46):19955–60.

[9] Kunath T, Saba-El-Leil MK, Almousailleakh M, et al. FGF stimulation of the Erk1/2 signalling cascade triggers transition of pluripotent embryonic stem cells from self-renewal to lineage commitment. Development 2007;134(16):2895–902.

[10] Itoh N, Ornitz DM. Functional evolutionary history of the mouse Fgf gene family. Dev Dyn 2008;237(1):18–27.

[11] Villegas SN, Canham M, Brickman JM. FGF signalling as a mediator of lineage transitions–evidence from embryonic stem cell differentiation. J Cell Biochem 2010;110(1):10–20.

[12] Maki N, Suetsugu-Maki R, Tarui H, et al. Expression of stem cell pluripotency factors during regeneration in newts. Dev Dyn 2009;238(6):1613–6.

[13] Jung JE, Moon SH, Kim DK, et al. Sprouty1 Regulates Neural and Endothelial Differentiation of Mouse Embryonic Stem Cells. Stem Cells Dev 2011;21(4):554–61.

[14] Holleville N, Quilhac A, Bontoux M, Monsoro-Burq AH. BMP signals regulate Dlx5 during early avian skull development. Dev Biol 2003;257(1):177–89.

[15] Miyama K, Yamada G, Yamamoto TS, et al. A BMP-inducible gene, dlx5, regulates osteoblast differentiation and mesoderm induction. Dev Biol 1999;208(1):123–33.

[16] Bennett CN, Longo KA, Wright WS, et al. Regulation of osteoblastogenesis and bone mass by Wnt10b. Proc Natl Acad Sci USA 2005;102(9):3324–9.

[17] zur Nieden NI, Kempka G, Rancourt DE, Ahr HJ. Induction of chondro-, osteo- and adipogenesis in embryonic stem cells by bone morphogenetic protein-2: effect of cofactors on differentiating lineages. BMC Dev Biol 2005;5:1.

[18] Barron F, Woods C, Kuhn K, et al. Downregulation of Dlx5 and Dlx6 expression by Hand2 is essential for initiation of tongue morphogenesis. Development 2011;138(11):2249–59.

[19] Sakaguchi Y, Sekiya I, Yagishita K, Muneta T. Comparison of human stem cells derived from various mesenchymal tissues: superiority of synovium as a cell source. Arthritis Rheum 2005;52(8):2521–9.

[20] Yoshimura H, Muneta T, Nimura A, et al. Comparison of rat mesenchymal stem cells derived from bone marrow, synovium, periosteum, adipose tissue, and muscle. Cell Tissue Res 2007;327(3):449–62.

[21] Secor SM, Whang EE, Lane JS, Ashley SW, Diamond J. Luminal and systemic signals trigger intestinal adaptation in the juvenile python. Am J Physiol Gastrointest Liver Physiol 2000;279(6):G1177–87.

[22] Rando TA. Stem cells, ageing and the quest for immortality. Nature 2006;441(7097):1080–6.

[23] Moerman EJ, Teng K, Lipschitz DA, Lecka-Czernik B. Aging activates adipogenic and suppresses osteogenic programs in mesenchymal marrow stroma/stem cells: the role of PPAR-gamma2 transcription factor and TGF-beta/BMP signaling pathways. Aging Cell 2004;3(6):379–89.

[24] Holliday R. Epigenetics: a historical overview. Epigenetics 2006;1(2):76–80.

[25] Bernal AJ, Jirtle RL. Epigenomic disruption: the effects of early developmental exposures. Birth Defects Res A Clin Mol Teratol 2010;88(10):938–44.

[26] Skinner MK. Role of epigenetics in developmental biology and transgenerational inheritance. Birth Defects Res C Embryo Today 2011;93(1):51–5.

[27] Sharma S, Kelly TK, Jones PA. Epigenetics in cancer. Carcinogenesis 2010;31(1):27–36.

[28] Waddington CH. Genetic Assimilation of an acquired character. Evolution 1953;7:118–26.

[29] Wang Z, Zang C, Rosenfeld JA, et al. Combinatorial patterns of histone acetylations and methylations in the human genome. Nat Genet 2008;40(7):897–903.

[30] Zhang RP, Shao JZ, Xiang LX. Gadd45a plays an essential role in active DNA demethylation during terminal osteogenic differentiation of adipose-derived mesenchymal stem cells. J Biol Chem 2011;286(47):41083–94.

[31] Dindot SV, Person R, Strivens M, Garcia R, Beaudet AL. Epigenetic profiling at mouse imprinted gene clusters reveals novel epigenetic and genetic features at differentially methylated regions. Genome Res 2009;19(8):1374–83.

515

[32] Luedi PP, Dietrich FS, Weidman JR, et al. Computational and experimental identification of novel human imprinted genes. Genome Res 2007;17(12):1723−30.

[33] Morison IM, Ramsay JP, Spencer HG. A census of mammalian imprinting. Trends Genet 2005;21(8):457−65.

[34] Abu-Amero S, Monk D, Apostolidou S, Stanier P, Moore G. Imprinted genes and their role in human fetal growth. Cytogenet Genome Res 2006;113(1-4):262−70.

[35] Constancia M, Hemberger M, Hughes J, et al. Placental-specific IGF-II is a major modulator of placental and fetal growth. Nature 2002;417(6892):945−8.

[36] Hamidouche Z, Fromigue O, Ringe J, Haupl T, Marie PJ. Crosstalks between integrin alpha 5 and IGF2/IGFBP2 signalling trigger human bone marrow-derived mesenchymal stromal osteogenic differentiation. BMC Cell Biol 2010;11:44.

[37] Dindot SV, Antalffy BA, Bhattacharjee MB, Beaudet AL. The Angelman syndrome ubiquitin ligase localizes to the synapse and nucleus, and maternal deficiency results in abnormal dendritic spine morphology. Hum Mol Genet 2008;17(1):111−8.

[38] van den Berge SA, Middeldorp J, Zhang CE, et al. Longterm quiescent cells in the aged human subventricular neurogenic system specifically express GFAP-delta. Aging Cell 2010;9(3):313−26.

[39] Mattick JS. A new paradigm for developmental biology. J Exp Biol 2007;210(Pt 9):1526−47.

[40] International Human Genome Sequencing Consortium. Finishing the euchromatic sequence of the human genome. Nature 2004;431(7011):931−45.

[41] Birney E, Stamatoyannopoulos JA, Dutta A, et al. Identification and analysis of functional elements in 1% of the human genome by the ENCODE pilot project. Nature 2007;447(7146):799−816.

[42] Pan C, Russell R. Roles of DEAD-box proteins in RNA and RNP Folding. RNA Biol 2010;7(6):667−76.

[43] Hong YK, Ontiveros SD, Strauss WM. A revision of the human XIST gene organization and structural comparison with mouse Xist. Mamm Genome 2000;11(3):220−4.

[44] Spitale RC, Tsai MC, Chang HY. RNA templating the epigenome: long noncoding RNAs as molecular scaffolds. Epigenetics 2011;6(5):539−43.

[45] Khurana JS, Theurkauf W. piRNAs, transposon silencing, and Drosophila germline development. J Cell Biol 2010;191(5):905−13.

[46] Yoo AS, Staahl BT, Chen L, Crabtree GR. MicroRNA-mediated switching of chromatin-remodelling complexes in neural development. Nature 2009;460(7255):642−6.

[47] Liao JY, Ma LM, Guo YH, et al. Deep sequencing of human nuclear and cytoplasmic small RNAs reveals an unexpectedly complex subcellular distribution of miRNAs and tRNA 3′ trailers. PLoS One 2010;5(5):e10563.

[48] Anokye-Danso F, Trivedi CM, Juhr D, et al. Highly efficient miRNA-mediated reprogramming of mouse and human somatic cells to pluripotency. Cell Stem Cell 2011;8(4):376−88.

[49] Li Z, Yang CS, Nakashima K, Rana TM. Small RNA-mediated regulation of iPS cell generation. EMBO J 2011;30(5):823−34.

[50] Smith KN, Singh AM, Dalton S. Myc represses primitive endoderm differentiation in pluripotent stem cells. Cell Stem Cell 2010;7(3):343−54.

[51] Gonzalo S, Garcia-Cao M, Fraga MF, et al. Role of the RB1 family in stabilizing histone methylation at constitutive heterochromatin. Nat Cell Biol 2005;7(4):420−8.

[52] Montoya-Durango DE, Liu Y, Teneng I, et al. Epigenetic control of mammalian LINE-1 retrotransposon by retinoblastoma proteins. Mutat Res 2009;665(1−2):20−8.

[53] Gonzalez S, Pisano DG, Serrano M. Mechanistic principles of chromatin remodeling guided by siRNAs and miRNAs. Cell Cycle 2008;7(16):2601−8.

[54] Zhou FC, Balaraman Y, Teng M, et al. Alcohol Alters DNA Methylation Patterns and Inhibits Neural Stem Cell Differentiation. Alcohol Clin Exp Res 2011;35(4):1−12.

[55] Takayama MA, Taira T, Tamai K, Iguchi-Ariga SM, Ariga H. ORC1 interacts with c-Myc to inhibit E-box-dependent transcription by abrogating c-Myc-SNF5/INI1 interaction. Genes Cells 2000;5(6):481−90.

[56] Dominguez-Sola D, Ying CY, Grandori C, et al. Non-transcriptional control of DNA replication by c-Myc. Nature 2007;448(7152):445−51.

[57] Christov CP, Gardiner TJ, Szuts D, Krude T. Functional requirement of noncoding Y RNAs for human chromosomal DNA replication. Mol Cell Biol 2006;26(18):6993−7004.

[58] Gardiner TJ, Christov CP, Langley AR, Krude T. A conserved motif of vertebrate Y RNAs essential for chromosomal DNA replication. RNA 2009;15(7):1375−85.

[59] Norseen J, Thomae A, Sridharan V, et al. RNA-dependent recruitment of the origin recognition complex. EMBO J 2008;27(22):3024−35.

[60] Chakraborty A, Shen Z, Prasanth SG. "ORCanization" on heterochromatin: linking DNA replication initiation to chromatin organization. Epigenetics 2011;6(6):665−70.

[61] Dinger ME, Amaral PP, Mercer TR, et al. Long noncoding RNAs in mouse embryonic stem cell pluripotency and differentiation. Genome Res 2008;18(9):1433–45.

[62] Tsai MC, Manor O, Wan Y, et al. Long noncoding RNA as modular scaffold of histone modification complexes. Science 2010;329(5992):689–93.

[63] Freberg CT, Dahl JA, Timoskainen S, Collas P. Epigenetic reprogramming of OCT4 and NANOG regulatory regions by embryonal carcinoma cell extract. Mol Biol Cell 2007;18(5):1543–53.

[64] Pain D, Chirn GW, Strassel C, Kemp DM. Multiple retropseudogenes from pluripotent cell-specific gene expression indicates a potential signature for novel gene identification. J Biol Chem 2005;280(8):6265–8.

[65] Hawkins PG, Morris KV. Transcriptional regulation of Oct4 by a long non-coding RNA antisense to Oct4-pseudogene 5. Transcription 2010;1(3):165–75.

[66] Fairbanks DJ, Maughan PJ. Evolution of the NANOG pseudogene family in the human and chimpanzee genomes. BMC Evol Biol 2006;6:12.

[67] Jeter CR, Badeaux M, Choy G, et al. Functional evidence that the self-renewal gene NANOG regulates human tumor development. Stem Cells 2009;27(5):993–1005.

[68] Guttman M, Donaghey J, Carey BW, et al. lincRNAs act in the circuitry controlling pluripotency and differentiation. Nature 2011;477(7364):295–300.

[69] Huang C, Xiang Y, Wang Y, et al. Dual-specificity histone demethylase KIAA1718 (KDM7A) regulates neural differentiation through FGF4. Cell Res 2010;20(2):154–65.

[70] Darimipourain M, Wang S, Ittmann M, Kwabi-Addo B. Transcriptional and post-transcriptional regulation of Sprouty1, a receptor tyrosine kinase inhibitor in prostate cancer. Prostate Cancer Prostatic Dis 2011;14(4):279–85.

[71] Sathyan P, Golden HB, Miranda RC. Competing interactions between micro-RNAs determine neural progenitor survival and proliferation after ethanol exposure: evidence from an ex vivo model of the fetal cerebral cortical neuroepithelium. J Neurosci 2007;27(32):8546–57.

[72] Jung H, Lee SK, Jho EH. Mest/Peg1 inhibits Wnt signalling through regulation of LRP6 glycosylation. Biochem J 2011;436(2):263–9.

[73] Tome M, Lopez-Romero P, Albo C, et al. miR-335 orchestrates cell proliferation, migration and differentiation in human mesenchymal stem cells. Cell Death Differ 2011;18(6):985–95.

[74] Zhang J, Tu Q, Bonewald LF, et al. Effects of miR-335-5p in modulating osteogenic differentiation by specifically downregulating Wnt antagonist DKK1. J Bone Miner Res 2011;26(8):1953–63.

[75] Zhou GS, Zhang XL, Wu JP, et al. 5-Azacytidine facilitates osteogenic gene expression and differentiation of mesenchymal stem cells by alteration in DNA methylation. Cytotechnology 2009;60:11–22.

[76] Li H, Marijanovic I, Kronenberg MS, et al. Expression and function of Dlx genes in the osteoblast lineage. Dev Biol 2008;316(2):458–70.

[77] Kohwi M, Petryniak MA, Long JE, et al. A subpopulation of olfactory bulb GABAergic interneurons is derived from Emx1- and Dlx5/6-expressing progenitors. J Neurosci 2007;27(26):6878–91.

[78] Feng J, Bi C, Clark BS, et al. The Evf-2 noncoding RNA is transcribed from the Dlx-5/6 ultraconserved region and functions as a Dlx-2 transcriptional coactivator. Genes Dev 2006;20(11):1470–84.

[79] Yoo AS, Sun AX, Li L, et al. MicroRNA-mediated conversion of human fibroblasts to neurons. Nature 2011;476(7359):228–31.

[80] Tanasijevic B, Dai B, Ezashi T, et al. Progressive accumulation of epigenetic heterogeneity during human ES cell culture. Epigenetics 2009;4(5):330–8.

[81] Huang KC, Rao PH, Lau CC, et al. Relationship of XIST expression and responses of ovarian cancer to chemotherapy. Mol Cancer Ther 2002;1(10):769–76.

[82] Callinan PA, Batzer MA. Retrotransposable elements and human disease. Genome Dyn 2006;1:104–15.

[83] Yu F, Zingler N, Schumann G, Stratling WH. Methyl-CpG-binding protein 2 represses LINE-1 expression and retrotransposition but not Alu transcription. Nucleic Acids Res 2001;29(21):4493–501.

[84] Muotri AR, Chu VT, Marchetto MC, et al. Somatic mosaicism in neuronal precursor cells mediated by L1 retrotransposition. Nature 2005;435(7044):903–10.

[85] Muotri AR, Marchetto MC, Coufal NG, et al. L1 retrotransposition in neurons is modulated by MeCP2. Nature 2010;468(7322):443–6.

[86] Quattrocelli M, Palazzolo G, Floris G, et al. Intrinsic cell memory reinforces myogenic commitment of pericyte-derived iPSCs. J Pathol 2011;223(5):593–603.

[87] Bar-Nur O, Russ HA, Efrat S, Benvenisty N. Epigenetic memory and preferential lineage-specific differentiation in induced pluripotent stem cells derived from human pancreatic islet Beta cells. Cell Stem Cell 2011;9(1):17–23.

[88] Warnke PH, Springer IN, Wiltfang J, et al. Growth and transplantation of a custom vascularised bone graft in a man. Lancet 2004;364(9436):766–70.

517

[89] Lavery K, Swain P, Falb D, Alaoui-Ismaili MH. BMP-2/4 and BMP-6/7 differentially utilize cell surface receptors to induce osteoblastic differentiation of human bone marrow-derived mesenchymal stem cells. J Biol Chem 2008;283(30):20948−58.

[90] Hanna J, Wernig M, Markoulaki S, et al. Treatment of sickle cell anemia mouse model with iPS cells generated from autologous skin. Science 2007;318(5858):1920−3.

[91] Lamba DA, McUsic A, Hirata RK, et al. Generation, purification and transplantation of photoreceptors derived from human induced pluripotent stem cells. PLoS One 2010;5(1):e8763.

[92] Lu SJ, Li F, Yin H, et al. Platelets generated from human embryonic stem cells are functional in vitro and in the microcirculation of living mice. Cell Res 2011;21(3):530−45.

[93] Mauritz C, Martens A, Rojas SV, et al. Induced pluripotent stem cell (iPSC)-derived Flk-1 progenitor cells engraft, differentiate, and improve heart function in a mouse model of acute myocardial infarction. Eur Heart J 2011;32:2634−41.

Aging and Disease: The Epigenetic Bridge

Andrea Fuso
Sapienza University of Rome, Rome, Italy

519

26.1 INTRODUCTION

Conrad Hal Waddington, in 1942, was the first to define epigenetics as "the branch of biology which studies the causal interactions between genes and their products, which bring the phenotype into being". At that time, the biochemical nature of genes was unknown as well as their role as repositories and transmitters of the genetic information. Waddington imagined the epigenetics as a conceptual model to explain his theory sustaining that different interactions between the genes and their surroundings (or, we could say their "environment") could result in different phenotypes, starting from the same genetic material. He used the metaphor of the "epigenetic landscape" to explain the biological development. Waddington stated that cell fates were established during the development similarly to a stone (a marble) that rolls down from high places to the point of lowest local elevation; the increasing irreversibility associated with cell-type differentiation was imagined as due to ridges, rising along the slope where the stone is rolling down, directing the marble into different valleys [1].

More recently, Holliday defined epigenetics in a more formal way as "the study of the mechanisms of temporal and spatial control of gene activity during the development of complex organisms" [2]. According to this definition, the term "epigenetic" could be used to describe anything other than a DNA sequence itself that is able to influence differentiation, physiology,

T. Tollefsbol (Ed): Epigenetics in Human Disease. DOI: 10.1016/B978-0-12-388415-2.00026-3

and the fate of a cell or of a whole organism. Actually, the present use of the term is more narrow, referring to the combined epigenetic modifications of a given domain of DNA sequence, characterized by heritability (over rounds of cell division and sometimes trans-generationally), that do not involve changes to the underlying DNA sequence. On this basis, epigenetics was recently defined as "the study of any potentially stable and, ideally, heritable change in gene expression or cellular phenotype that occurs without changes in Watson–Crick base pairing of DNA" [3]. Epigenetic modifications include (among others) methylation of cytosine residues of DNA and post-translational modifications of the histone proteins. Specific combinations of epigenetic modifications determine the conformation of the chromatin fiber, thereby having the possibility to regulate the transcriptional potential of the associated genes.

Despite the advances in our knowledge about cell differentiation and epigenetic phenomena, and with the unavoidable adjustments and corrections, Waddington's model still represents a nice visualization of the epigenetics. As a matter of fact, it appears really useful to suggest that aging processes are particularly prone to epigenetic mechanisms. The notion Waddington could not know, indeed, was that once differentiation has been completed (i.e. the "lowest local elevation" has been reached) cell fate could be not definitive; on the contrary, the phenotype of differentiated cells still evolves either with the normal aging processes or as a consequence of external (pathogenic) stimuli. To resume and apply Waddington's model to the aging, we can imagine that erosive processes can change the shape of the slope and of the surroundings of the stone, causing the reprise of its rolling down through new ridges and valleys. According to this view, the terminally differentiated cell is subjected to "environmental" stimuli (originated either from the organism itself or from the external environment) able to induce changes in gene expression through epigenetic mechanisms. The higher the mountain, the longer the slope; consequently, the stone encounters many more possibilities to be subjected to changes of directions and shape. This view recalls the idea that a longer life (of a cell or organism) is associated with a more frequent probability that epigenetic changes arise, possibly causing aging-associated dysregulation. On the basis of this metaphoric view, aging (and aging-associated diseases) represents the inevitable companion of a long life. Fortunately, epigenetic modifications are, by definition, reversible; consequently, they are potential targets for pharmacological interventions aimed at re-establishing the correct epigenome. Interestingly, recent evidences indicate that the epigenome can be also modulated by non-pharmacological interventions like diet, physical exercise, lifestyle, and behavioral stresses.

In the present chapter, evidences related to the connection between epigenetics and aging are presented and discussed in the light of the most recent advances in this field of biomedical research. Particular attention is devoted to the aging brain, which appears to be the organ most interesting in normal and pathological aging processes, due to the relevance of neurodegeneration among the age-associated diseases and to the recent scientific evidences indicating substantial involvement of epigenetic phenomena in brain aging.

26.2 GENES AND AGING

A common saying among geriatricians sounds like "Life is a chronic disease with unpropitious outcome". Probably, being faced daily with aged and diseased patients negatively influences the humor and mood of clinicians and researchers working on aging, but the sentence is undeniably correct. Indeed, humans cannot escape (as far as we know) aging and, in that case, aging-related diseases [4]. Aging, although not considered a disease, is itself associated with progressive, cognitive, physical, and physiological impairment; moreover, many diseases are age-related. The main risk factor for late-onset Alzheimer's disease (LOAD), for instance, is aging, since the incidence of the disease is constantly increasing with age [5].

As a matter of fact, the evolution of the human species has allowed the possibility to constantly improve life quality, particularly in terms of better hygiene, nutrition, protection, and medical care. These improvements contributed not just to the increase in average life expectancy but also, in many cases, to reach and spend the oldest age in better physical and cognitive condition than in the past. Despite this progress in life expectancy, it is interesting to note that almost no progress was observed for the oldest age that it is possible to reach (the "maximum lifespan potential"); moreover, in association with the increased life expectancy, many (and sometimes new) diseases show an increased morbidity dependent on aging [6].

The existence of a genetic determinant of life duration is supported by the apparent impossibility of going beyond a certain maximum lifespan potential and also by the observations indicating that this potential seems to be determined and characteristic for each species. This information induced the theory that even if we could cure or prevent the diseases most responsible for human death, we will be able to just further extend life expectancy, but won't be able to significantly overcome the maximum lifespan potential determined by the advent of fatal age-associated physiological impairment [7]. This hypothesis depicts the age-related diseases, mainly the neurodegenerative diseases, as the price to pay to survive cancer; in other words, the organism has two possible fates during aging: facing cells prone to errors that would shift toward a relatively longer (cancerous) life, or coping with the progressive cell impairment and death due to physiological cellular degeneration.

Molecular and cellular degeneration, resulting in the aging of the whole organism, are often evident and clearly observable, although the initial and causal molecular mechanisms are difficult to study because of their extreme complexity, interconnections, presence of side effects, and concurrent external and environmental factors [8]. The study of the picture representing the age-associated diseases is complicated by the possible early start of the pathological mechanisms, possibly initiating in early age, and also by the above-cited difference in the regulation of aging mechanisms in different organisms, which makes it difficult to use surrogated animal models to study human aging.

In this context, it is not easy to hypothesize a unifying and comprehensive theory of aging. Initially, it was postulated that biological aging was dependent on changes residing in the information-containing molecules, leading to formulate the "error theory", the "redundant message theory", the "codon restriction theory", and the "transcriptional event theory" [9]. In a second moment, the attention of the researchers has been shifted to the hypothesis that possible "gerontogenes", mainly involved in DNA repair, could be responsible for cellular senescence [10]. To summarize, it is possible to differentiate two major groups of theories: stochastic and genetic; however, it becomes evident that these groups could be, in some cases, not mutually exclusive.

A list of the principal theories explaining causes and possible mechanisms of aging is reported here [8,11]:

1. Evolutionary: evolution presses the organisms to reach the reproductive age, procreate, and care for the offspring. According to this point of view, the physiology of an organism after the end of the reproductive period could be the manifestation of the epigenetic events occurring on the basis of the genetic development during the previous stage of the life. The conclusion is that cellular senescence could be the price to pay in order to avoid other damage, like tumorigenesis, potentially caused by the prolonged expression of the genes involved in the reaching of reproductive fitness [12].
2. Stochastic: the stochastic theories are generally based on the idea that aging is caused by casual damage to cell molecules, particularly to DNA.
    2.1. Mutation and repair: DNA is subjected to continuous damage that induces mutations. The effect is cumulative and becomes increasingly evident because of the loss of the DNA repair ability [9].
    2.2. Error-catastrophe: casual errors in the synthesized proteins are normally bypassed by the normal molecular turnover, but errors in the molecules involved in protein or DNA synthesis or in postsynthetic modification could represent the start of a process

of error amplification, resulting in the accumulation of the wrong molecules able to interfere with normal cell physiology [13].

2.3. Protein modification: the worsening of the enzymatic activities in aging could be a consequence of the altered postsynthetic modifications, altered turnover and proteins cross-linking [13].

2.4. Oxidative stress: this is one of the most investigated areas of cellular senescence; the involvement of free radicals and the alteration of the oxidative status in aging has been characterized in several models and organisms and in different pathologies associated with older age, like Alzheimer's disease and Parkinson's disease. The balance between pro- and antioxidants in the cell is finely and complexly regulated and the impairment of this regulation is critical to mitochondrial, cellular, and tissue physiology during aging [14].

3. Genetic: in the genetic (or developmental) theories, aging is considered as a programmed and genetically controlled process of maturation, successive to the development of the organism or cell. These theories are supported by the elevated species-specificity of the maximum lifespan but are in contrast with the variable control and manifestation of aging in different individuals of the same species.

3.1. Longevity genes: there are several evidences about the existence of genetic elements able to regulate senescence, in particular responsible for the regulation of the maximum lifespan. Studies regarding the role of genes involved in the increment of lifespan were primarily performed on "simple" eukaryotes like yeast and *C. elegans*, but significant data also arise from studies on *D. melanogaster* and rodents [15,16].

3.2. Aging syndromes: the existence of a number of human genetic diseases (Hutchinson's syndrome, Werner's syndrome, Down's syndrome) displaying some characteristics distinctive of accelerated aging, lead to hypothesize that aging could be a kind of disease itself, regulated by specific genes. Recently, different transgenic mouse models showing aging phenotypes similar to those observed in humans were also settled [17].

3.3. Neuroendocrine theory: this is based on the importance of the hormones secreted in the brain (hypothalamic, pituitary, and adrenal hormones) in the regulation of organismic aging and on the decrement in brain neurons [18].

3.4. Immunologic theory: this is based on the decreased T-cell response and increased autoimmune reactions during aging [19]. As for the neuroendocrine theory, the weak point is that complex immune and neuronal systems are not present in simple eukaryotes although they show characteristics of aging comparable to higher organisms.

3.5. Cellular senescence: cellular cultures were used as a model for the comprehension of senescence processes due to their usefulness in studying the basic molecular mechanisms, unlike the whole organisms. Data on the genetic effectors responsible for the regulation of cell senescence sustain the hypothesis that organismic aging reflects the senescence of single cell lines or tissues. Cellular senescence is often indicated as "replicative senescence", since the genes involved in this phenomenon are mainly genes related to the replication machinery and since the cellular senescence becomes evident through decline in growth rate and proliferative activity and alterations in the signal transduction and adaptive response pathways. All these alterations characterize a senescent cell growth status, which is quite different from the young cells [20]. The first event characterized as a potential cellular clock was the mechanism of telomere shortening [21]. The function of telomeres (sequences of non-coding DNA at the end of chromosomes) is to avoid the degradation of coding regions and fusion with other DNA sequences; the reduction of the telomeric length with cell aging is the main evidence in cellular senescence. Another two genes of the replicative machinery, retinoblastoma and p53, are well known to be involved in cell senescence; their activity is generally increased in senescent cells [22]. The complexity of the mechanisms of cell cycle progression and the balance between positive and negative regulators is due to the importance of this regulation in order to avoid the onset of tumorigenic events in the cell. These data are consistent with the hypothesis that cellular senescence has evolved as a mechanism of tumor suppression [8].

3.6. Cell death: strictly linked to the mechanisms of cellular replication and senescence, the mechanism of apoptosis is considered as a cause of aging since it consists of a process of active, gene-dependent and injury-independent cell death [23].

More recently, evidence that epigenetic mechanisms could have a role in cellular degeneration and aging has been supported by technical advances allowing a detailed study of the epigenome and of the epigenetic mechanisms and by the discovery of a complex, non-Mendelian, nature of many age-associated disorders.

In yeast and mice, significant changes in gene expression during cellular degeneration are related to significant and net loss of heterochromatin, with consequent overexpression of heterochromatin-associated silenced genes [24]. Changes in chromatin structure are associated with epigenetic modifications that consist of DNA methylation, DNA hydroxymethylation, histone post-transcriptional modifications (methylation, acetylation, and phosphorylation), and ATP-mediated chromatin modifications. DNA is associated with euchromatin in actively transcribed loci in proliferating cells; these loci mainly contain genes involved in cell growth regulation and in basal cell metabolism [25]. For this reason, it has been proposed that loss of repressive chromatin domains (heterochromatin) may contribute to cellular degeneration and aging processes.

The observed heterochromatin loss is in agreement with, and apparently linked to, the generally observed loss of DNA methylation in the elderly [26,27]. Other experimental evidence also points out the spreading of DNA methylation and the presence of hypermethylated DNA in senescence and cancer [28,29]; on the one hand, these contrasting data stress the complexity of the epigenetic phenomena and contribute to confound the picture but, on the other, the contrast can be clarified when DNA methylation is analyzed at a sequence-specific level in regions and sequences undergoing changes of methylation during aging. Observed global changes in DNA methylation, at genomic level, could indeed be due to changes in methylation of bulk or non-regulatory DNA regions. The role of these changes is not yet ascertained and we cannot conclude it is directly associated with changes in gene expression, unlike methylation changes in promoters and other regulatory DNA sequences. However, the evidence about the general DNA methylation decreasing in proliferating cell cultures and in normal organismic aging, but not in immortal cell lines, and the concurrent impairment of the DNA methyltransferase activity, stress the hypothesis that DNA methylation is susceptible to aging-linked variations and that these variations have a specific role in the aging processes [30]. One of the main questions to be ascertained in this sense involves the "cause or consequence" role of DNA methylation in aging.

DNA methylation is strictly related to histone modification and "packing"; in turn, histone modifications can influence gene transcription by modulating chromatin assembly. Histone acetylation is the most studied modification; it is regulated by the classical histone deacetylases (HDAC) family and by the SIR2 (silent information regulator) family. It is worthwhile to underline that the SIR2 family was also identified as being involved in the regulation of life-span in other eukaryotes [31,32].

Other epigenetic modifications, as histone methylation and ATP-mediated chromatin remodeling, are not yet well characterized for their role in cellular degeneration and aging, but, as a matter of fact, it was shown that euchromatinization of specific domains or, vice versa, heterochromatinization of specific growth genes, due to alterations in balance of chromatin modifying molecules, is modulated during aging [33].

Besides their role in normal cellular senescence, epigenetic mechanisms are also to be considered in the light of their association (DNA methylation in particular) to cancer onset and development, since it is well known that cancer onset increases with aging.

523

Methylation, aging, and cancer are representatively connected in the observed CpG island methylation increasing associated to the estrogen receptor (ER) gene, which resulted age-associatedly in the normal human colonic mucosa. ER gene hypermethylation was also found in all colonic tumors examined including small adenomas, suggesting that methylation-dependent inactivation of the ER gene in the aging colorectal mucosa is one early event in colorectal tumorigenesis. Other CpG islands in promoter regions of several genes exhibit age-related hypermethylation in colon mucosa [29,34]. Most of the CpG islands found hyper-methylated in primary colon tumors were hypermethylated to a lesser extent in the aging colon, but a minor number of islands were hypermethylated only in subsets of colon cancers. These findings stress the hypothesis that two kinds of methylation exist: (1) one age-related methylation, presents in the normal mucosa as a function of the age and (2) a cancer-related methylation, not observed in normal colon. The mechanisms responsible for the concurrent induction of global DNA hypomethylation and sequence-specific increase in de novo methyl-ation with aging are still an open field of research. It is well known that different DNA-methyltransferase (DNMT) enzymes with peculiar functions exist. Maintenance methylation of hemimethylated DNA during cellular replication is guarantee by DNMT1, which is the most abundant. Other DNMTs (namely, DNMT3a and DNMT3b) have a high affinity for un-methylated DNA and were, for this reason, indicated as de novo methyltransferases, able to methylate DNA without pre-existing hemimethylation. These methylation mechanisms are schematized in Figure 26.1. Since the levels of DNMT1 decrease with aging, it can be hypothesized that the overall decrease in DNA methylation can be referred to maintenance methylation, whereas the sequence-specific increase in de novo methylation depends on the increased activity of the other DNMTs [35]. The observed general decrease in total genomic methylation with aging in different organisms and the apparently finite number of cell divisions characteristic of most somatic cells stress the view of overall DNA methylation (or even methyl-cytosine residues) loss as a cellular countdown mechanism to trigger normal cellular senescence or degeneration.

More recently, thanks to the power of the genome-wide studies comparing younger to older subjects, it was possible to confirm on a large-scale basis that methylation changes (both in the direction of hyper- and hypo-methylation) are associated with aging, both in humans [36–38] and in animal models [39]. Even after these recent results, the idea that the methylation status of a larger part of the examined genes and sequences seem unchanged during aging [35] is, so far, still preserved. This is not at all, of course, a "negative" or controversial result for the disciples of the epigenetic theory of aging, but it just points out the idea that the age-associated epigenetic drift targets specific genes involved in aging processes. The alterations in CpG island methylation are crucial to modulate the binding of transcription factors and methyl-DNA binding proteins. Aberrant methylation of CpG islands in the promoter region may contribute to the progressive inactivation of growth-inhibitory genes during aging, resulting in the clonal selection of cells with growth advantage towards cancer development.

DNA methylation also presents a link with the regulation of telomerase activity, another key factor in the regulation of senescence. Human telomerase consists of an RNA component (hTR) and a catalytic subunit with reverse transcriptase activity (hTERT). Most differentiated somatic cells lack detectable telomerase activity, in contrast to the high activities found in immortalized cancer cells, germline cells, and stem cells and the high GC content of the hTERT promoter indicates that DNA methylation may be important in controlling hTERT expression. One emerging theory indicates that telomere control and DNA methylation may be strictly involved in the aging processes; in fact DNMT1 expression results decreased in aging human fibroblasts, with consequent decreasing of methylation activity in the cells, but hTERT was able to activate DNMT1 expression avoiding the loss of methylation [40].

All the above-reported data evidence that the methylation pattern established during the development is not stable or definitive in adult life and, in particular, during aging. Therefore,

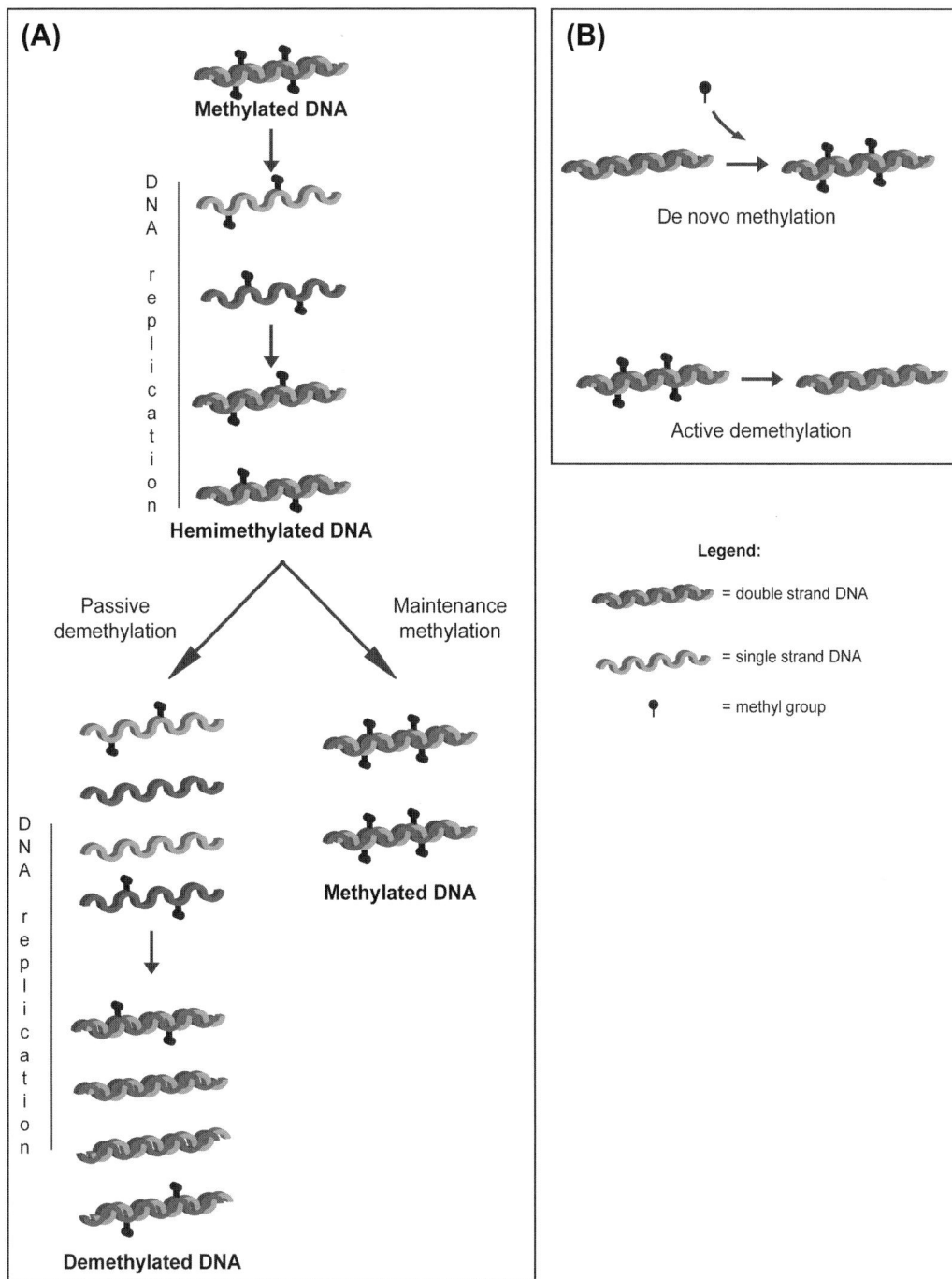

FIGURE 26.1

Schematic representation of the DNA methylation reactions: maintenance methylation, passive demethylation, de novo methylation, and active demethylation. (A) After rounds of cell divisions, methylated sequences result in demethylated neosynthesized DNA molecules (passive methylation) unless the maintenance methylase activity guarantees the transmission of the methylation pattern retained by the parental DNA strand. (B) De novo methylase activity has the ability to methylate DNA sequences independently on the presence of methyl groups in one of the two strands, whereas active demethylase activity has the ability to remove methyl groups; both these activities can be exerted in absence of cellular replication. This figure is reproduced in the color plate section.

the epigenetic approach seems promising for the understanding of the mechanisms that regulate aging and age-related diseases, whereas the consideration that the preservation of epigenetic patterns may help the maintaining of a healthy status could provide the basis for a therapeutic and preventive action.

26.3 THE DYNAMIC METHYLOME

The idea that epigenetic changes are responsible for aging and age-related diseases directly implies that the epigenome is a dynamic entity, undergoing modulation after particular stimuli. DNA methylation was long considered as the repository of the epigenetic "message" since the methylation pattern could be transmitted to the offspring and, once established at the end of the developmental processes, could be maintained during cell replication in a semiconservative manner. This feature is permitted by the ability of the maintenance DNMT(s) to replicate in the newly synthesized DNA strand the methylation pattern of the parent strand. For the last two decades it was supposed that the only mechanism leading to DNA demethylation could be the passive, replication-dependent, loss of methyl moieties occurring when the maintenance DNMT activity was defective during cell replication. This hypothesis stressed the impossibility that DNA demethylation could occur in adult organisms, at least in non-proliferating tissues, depicting a DNA methylation pattern that is, once established, relatively fixed and stable [41]. These assumptions were initially contrasted by the identification of putative DNA demethylase activity (Figure 26.1) in different candidate proteins like 5-methylcytosine-DNA glycosilase, Gadd45a, MBD2, MBD2b, MBD3, MBD4 [42–48]. It was even evidenced that DNMTs could be associated with demethylase activity [45,49,50]. Among these putative demethylases, MBD2 is the best characterized thanks to the pioneering work by the group of M. Szyf: they hypothesized that MBD2 can directly bind to methylated CpGs removing only the methyl group without the cytosine excision postulated by others [46]. Data are still not definitive and the possibility that MBD2 could interact with other (effectors) proteins could not be excluded; if the real nature of the active DNA demethylase is not yet definitely revealed, the existence of such enzymatic activity cannot be ignored any longer [51]. In our laboratory, we obtained earlier indirect indications that rapid demethylation, not compatible with the time necessary for cellular replication, occurred at a specific CpG site of myogenin gene promoter during myogenic differentiation in vitro [52]. More recently, we obtained evidence, in the same experimental system, that DNA demethylation pattern of myogenin gene promoter is regulated by a dynamic balance between DNA methylase and demethylase activities and that inhibition of DNMTs results in improved active demethylation, non necessarily dependent on DNA replication [53]. In a different experimental model, more strictly connected to age-related disease processes, we were able to demonstrate that DNA methylase and demethylase activities could be modulated in the adult brain. Using a transgenic mouse model of Alzheimer's disease we showed that the inhibition of the metabolic pathway that generates the methyl donor S-adenosylmethionine resulted in the impairment of the methylase activities and the improvement of the demethylase activity in mice brain — a tissue known for its scarce cellular proliferation. On the contrary, the supplementation of the methyl donor resulted in increased DNA methylation and decreased DNA demethylation activities. These modulations had a direct effect on the methylation pattern of Presenilin1 (PSEN1) gene, which is involved in amyloid processing in Alzheimer's disease [54,55]. We can therefore conclude that the methylation pattern of specific genes is not fixed in adults and non-proliferating tissues but undergoes dynamic regulation under appropriate stimuli.

26.4 EPIGENETIC DYNAMICS IN THE AGING BRAIN

The brain is certainly the human organ most hit by aging: it shows sensitive impairment in normal aging, and appears more prone than other organs to the occurrence of aging-related diseases. Neurodegenerative disorders represent the main class of age-associated diseases and, among these, Alzheimer's disease represents the most prevalent form of neurodegenerative disease. Moreover, many of these disorders have been recently associated with epigenetic events. For this reasons, it seems of particular relevance in the discussion of the epigenetic changes occurring in the brain and observed in adulthood and aging; an excellent and comprehensive review of these mechanisms was recently published by J. Rogers and colleagues [56].

As previously discussed, the knowledge that postmitotic cells could present functional alterations of the methylation patterns is a quite recent acquisition. In recent years, it was demonstrated that dynamic and rapid (active) DNA methylation and demethylation are present in postmitotic cells and, in particular, in neurons [45,50,57] stressing the involvement of epigenetic processes in neuronal plasticity. The paper by Murgatroyd and colleagues [50] has to be considered as a milestone in this area. In their work, they demonstrated that early life stress in mice was associated with behavioral changes in adult life via a mechanism involving epigenetic modifications of hypothalamic neurons. Early exposure of mice to environmental stress during the first 10 days of life resulted in impaired avoidance learning, sustained hyperactivity of the hypothalamic–pituitary–adrenal axis, and corticosterone and pituitary adrenocorticotropin prohormone hypersecretion. These effects were linked to arginine-vasopressin (AVP) overexpression by parvocellular neurons of the hypothalamic paraventricular nucleus, and to hypomethylation of specific CpG sites within the AVP gene. The authors also showed the age-dependent hypomethylation of AVP gene at different CpG sites in control mice, although hypomethylation of CpGs within a CpG island included in the AVP enhancer region were responsible for AVP overexpression in mice that underwent the early-life stress. Hypomethylation of this CpG island was specific to the paraventricular nucleus, suggesting a possible mechanism for the observed epigenetic change that involves the binding of MeCP2 protein to the CpG sites in this island. The authors finally demonstrated that phosphorylation of MeCP2 by calmodulin-dependent protein kinase II decreased MeCP2 occupancy of CpGs in the CpG island, enhancing gene expression. Experimental models taking advantage of early-life stresses represent promising approaches to study the modification of adult stress response, cognition and behavior, induced by epigenetic modifications [58,59].

A different approach was used to study the age-dependent decrease of caspase-3 in rat brain, associated with alterations of the methylation pattern of specific CpG sites in the promoter of the gene [60]. The promoter sequence interested by methylation alteration lies in a region of the promoter necessary for its activity. This region is predicted to bind the transcription factors ETS-1 and -2, which are important, besides in controlling caspase-3 transcription, in regulating neuronal differentiation and death. Since these two factors seem not to be altered during aging, it is possible to appreciate the relevance of methylation status, mediating transcription factors binding and activity.

Dynamic changes of methylation patterns even showed cyclical regulation associated with cyclical activation/inactivation of transcription [45,57]. In this model, DNMT3a and 3b are responsible for deamination of methylated CpG sites, leading to a process of active de-methylation mediated by an excision/repair mechanism; DNA demethylation induces gene transcription, followed by MeCP2 and DNMTs new recruitment responsible for new DNA methylation and gene silencing.

While much is known about DNA methylation mechanisms and the possible role of DNA methylation dysregulation has become an intense field of research in respect to aging and age-related neurodegeneration [61], little is known about the relevance to the aging processes of DNA hydroxymethylation, a relatively new epigenetic modification recently identified [62]. 5-Hydroxymethylcytosine (5-hmC) is generated by the action of the ten-eleven translocation enzymes (TET), which are able to hydroxylate methylated cytosines [63,64]. 5-hmCs is then formed after a previous cytosine methylation and is not efficiently recognized by methyl-binding proteins [65,66]. Unlike DNA methylation, hydroxymethylation is generally associated with euchromatinization and gene expression [67,68]. A role for this epigenetic modification in aging could be hypothesized on the basis of the work by Chouliaras and colleagues that recently showed an increase in hydroxymethylation in aging mouse hippocampus associated with increases in Dnmt3a levels [69]; interestingly, this increase was prevented by caloric restriction (CR) [70].

As for DNA methylation and demethylation, a recent study suggested that also histone acetylation and deacetylation are subjected to dynamic and rapid changes that can influence gene transcription [71]. Histone modifications are involved in several neurobiological processes in the differentiated brain, operating dynamic regulatory mechanisms in postmitotic neurons: post-traumatic stress disorders [72], addiction [73], choline acetyltransferase activity regulation [74], GDNF and BDNF transcription [75], and microglial apoptosis [76]. Moreover, histone modifications are known to be associated with different neurological and neurodegenerative disorders, including Parkinson's disease [75], motor neuron disease [77], multiple sclerosis [78], and X-linked mental retardation [79].

Another class of molecules with transcriptional regulatory functions, today included in the list of the epigenetic mechanisms, is represented by the microRNAs (miRNAs). The study of these regulatory molecules is facing an increasing boost in the very recent times since miRNAs were found to be associated with different physiological and pathological processes, also involving the brain, and appear modulated in postmitotic neurons [80]. For example, miRNA-329, miRNA-134, and miRNA-381, which are induced by neuronal activity, seem to be essential for the dendritic outgrowth of hippocampal neurons [81] and were also investigated in various neurological disorders [82].

26.5 THE COMPLEXITY OF THE AGE-ASSOCIATED EPIGENETIC CHANGES

Both normal and pathological aging processes in the brain are characterized by increased DNA damage, synaptic dysfunction, brain shrinkage, structural brain changes, and cognitive decline [83–85]. As a matter of fact, specific brain regions such as cortex and hippocampus appear more prone to be interested by these aging-related changes [86–88]. This region-specificity could be also observed when alterations of gene transcription are analyzed. These alterations can be summarized, on the one hand, in down-regulation of genes devoted to maintain synaptic plasticity, neurotrophic support to neurons, neurotransmitter synthesis, and DNA repair machinery and, on the other, in up-regulation of immune-related genes [61,89,90]. A role for epigenetic mechanisms, which are responsible for regulating gene expression, is now clearly claimed in these aging-related changes [91–93].

As previously stated for Alzheimer's disease, aging is the most evident risk factor associated with aging-related diseases [94]. Unfortunately, the causal or consequential nature of this association is not at all yet clear, although a number of possible concurrent alterations were identified both in normal and pathological aging; these include oxidative stress, sexual hormone effects, calcium dyshomeostasis, neurotransmitter and glucocorticoid deregulation, neuroinflammation, diabetes-associated alterations, neurovascular deficits, toxic protein deposition, and transcriptional alterations of many genes [91,95–103]. Epigenetic modifications, being involved in aging-associated changes in different experimental models and organisms, could represent a link between normal and pathological aging [104].

As in other organisms [105], progressive, age-related and genome-wide DNA hypomethylation was observed in humans both in vivo and in in vitro models [106,107] and was associated with the concurrent DNMT1 impairment [35,108]. Age-associated loss of DNA methylation was also associated with the parallel increase in S-adenosylhomocysteine [109] causing the inhibition of methyltransferase reactions (as discussed in depth in the prosecution of this chapter). Besides this general loss of methylation, specific hypomethylation of both coding and non-coding regions was observed during aging.

As for non-coding regions, since these are normally repressed by methylation, it seems possible that hypomethylation processes activating repetitive sequences, retrotransposons, and endogenous retroviruses occurring with aging, could induce chromosome instability and retroviruses activation [110]. Moreover, a very recent paper evidenced that hypomethylation of non-coding sequences like LINE-1 is possibly associated with neurodegeneration and, in particular, with LOAD [111].

The effect of the age-related hypomethylation on coding regions and specific genes in humans was also particularly studied in LOAD, probably because it is the most frequent pathology in older people. The work by Tohgi and his group, for example, showed hypomethylation of Amyloid Precursor Protein (APP) gene promoter in the brain of older people and also complex variations in methylation patterns of Tau and RAGE (Receptor for Advanced Glycation End-products) gene promoters [112–114]. Other researchers evidenced the age-dependent hypo-methylation of regulatory regions of the immune/inflammatory antigen CD11a [115].

Examples of gene-specific hypermethylation related to aging were also evidenced, together with the silencing of the associated genes. This was particularly observed for genes associated with CpG islands like tumor-suppressor genes, ER genes, and insulin-like growth factor 2 [116,117].

The very complex draft of the age-dependent changes in DNA methylation patterns, representing both hypo- and hypermethylation events at specific DNA sequences, is further complicated by the presence of tissue-specific patterns. Once again, it is the brain that offers a clear example of this complexity, since it was demonstrated that methylation patterns are region-specific [118].

Finally, a further degree of complexity is added by the possibility that many of the above-discussed changes in the epigenome of the aging brain could have an unsuspected early (even developmental) origin. The theory that early-life events could induce epigenetic modifications that are phenotypically silent until middle or old age, accumulating potential toxic features (amyloid deposition, for example) during many years of asymptomatic state, is well summarized by the LEARn (latent early-life associated regulation) model of age-related neurological disorders [119]. Experimental evidence of this model was given by the above-mentioned effect of early-life stress that results in epigenetic modifications and behavioral deficits in adult age [50,59] and by the finding that early exposure of monkeys to Pb decreased DNMT activity and had effects on amyloid-beta (Aβ) deposition in late life [87]. Also, the work performed in our laboratory so far, seems to support the LEARn theory since we observed that early interventions on methylation machinery in AD transgenic mice are able to modulate PSEN1 expression and Aβ deposition [54,55,120].

26.6 HEALTHY AND PATHOLOGICAL AGING

The variability of age-associated cognitive change is related to many factors influenced by demographic, social, educational, medical, nutritional, and biological stimuli. Biological factors include both genetic and epigenetic mechanisms. Epigenetic modifications, such as DNA methylation and histone acetylation regulate replication and transcription and are responsible for chromatin (re)modeling, chromosome stability, and DNA imprinting. Among the different epigenetic modifications, DNA methylation plays a pivotal role; methylation homeostasis is fundamental for normal brain physiology, whereas the impairment of methylation reactions and particularly DNA methylation are associated with markers related to neurodegeneration. Moreover, epigenetic mechanisms are greatly involved in brain development during the early life and the adult neurogenesis and are also involved in the onset and development of AD and other neurodegenerative disorders [121]. The hypothesis that gene transcription in brain is epigenetically regulated by DNA methylation much more than tissue-specific expression in other organs is now largely accepted. This hypothesis is supported by the strong association between age-related low methylation status and cognitive deficits or neurological and neurodegenerative pathologies [4,26,27]. However, it is still not completely clear whether epigenetic changes actually represent a cause or a consequence of the disease. This gap is due to the high complexity of the epigenetic mechanisms and of their regulation

529

during aging; moreover it is possible that epigenetic studies on pathological aging could be biased by the involvement of subjects in an advanced stage of disease or, in any case, subjects in which the epigenetic changes started even many years before the comparison of the first symptoms. An active role for epigenetics in normal and pathological aging must meet two conditions: specific epigenetic changes must occur during aging and they must be functionally associated with the aged and/or the diseased phenotype. Assuming that specific epigenetic modifications can have a direct functional outcome in aging or age-related diseases, it is also essential to establish whether they depend on genetic, environmental, or stochastic factors [61]. Few objections could be moved to the statement that the two cited conditions (the specificity of the epigenetic changes and the functional association to a phenotype) are demonstrated in the relationship between aging and cancer. As a matter of fact, epigenetic modifications play a major role in cancer, influencing tumor outcome by interfering with key senescence pathways [122]. As for other physiopathological mechanisms, a direct and causal role can be, so far, just supposed; the complexity of the immune system, for example, opens the possibility to predict that a higher-order, supragenetic regulation is indispensable for generation of its constituents and control of its functions during aging [123].

In human brain, a recent study attempted to quantify the extent and the identity of epigenetic changes in the aging process. DNA methylation at >27 000-CpG sites throughout the human genome, was examined in frontal cortex, temporal cortex, pons, and cerebellum from 387 subjects between 1 and 102 years of age. The authors were able to demonstrate the presence of CpG loci, mainly associated to CpG islands, which showed highly significant correlation between DNA methylation and age; they also confirmed the positive correlation between age and DNA methylation level. Moreover, the loci showing the most significant association with age were physically close to genes and involved in DNA binding and regulation of transcription, suggesting that specific age-dependent DNA methylation changes could be responsible for regulating gene expression in the human brain [124].

A very intriguing and useful model to study normal and pathological aging is represented by the monozygotic twins discordant for the occurrence of age-related diseases. Monozygotic twin siblings share the same genotype because they are derived from the same zygote. Despite the appearance, they frequently present phenotypic differences, such as their susceptibility to disease. Recent studies suggest that phenotypic discordance between monozygotic twins could be at least in part due to epigenetic differences and factors changing over their lifetime. The epigenetic drift occurring during the development is probably resulting by a combination of stochastic and environmental factors [125]. One example is represented by a study on twins discordant for Lewy body dementia that allowed postulating that epigenetic factors could play a role in Lewy body pathology [126]. In another elegant work taking advantage of a rare set of monozygotic twins discordant for AD, it was possible to observe significantly reduced levels of DNA methylation in temporal neocortex neuronal nuclei of the AD twin. This result stressed the hypothesis that the effects of life events on AD risk could be mediated by epigenetic mechanisms, providing a more general potential explanation for AD discordance despite genetic similarities [127].

We can conclude that aging is a process characterized by genetic and epigenetic interactions, where epigenetics has an important function in determining phenotypic differences. Epigenetics also plays a key role in the development of diseases associated with aging and explains the relationship between individual genetic background, environment, aging, and disease [128].

26.7 ENVIRONMENT, EPIGENETICS, AND AGING

The discussion about the studies on monozygotic twins also helps to introduce another fundamental concept regarding epigenetic changes occurring with aging: it is the idea that epigenetic mechanisms could be triggered by environmental factors or, in a slightly different point of view, that epigenetics exerts the role of mediator of environmental stimuli. The term "environmental" encompasses, in this case, many different processes and conditions occurring outside but entering in contact with the organism. These environmental factors include, among others: stresses (physical and behavioral), nutritional factors, pollutants and pesticides, chemicals, metals, physical exercise, and lifestyle. All these factors are able to cause, in the organism, biological effects that could determine (or contribute to) the onset and progression of the disease; many of these factors were found associated with the induction of epigenetic changes, as depicted in Figure 26.2. Obviously, these factors become increasingly relevant with aging to the healthy or pathological status of an individual, due to the increased possibility to encounter different environmental "hits" or to cumulate the reiterated effects of one of these factors.

Since monozygotic twins are genetically identical, they are considered as ideal "experimental models" to study the role of environmental factors as determinants of complex diseases and phenotypes. As in the case of the above-mentioned study on twins discordant for AD, for example, the AD twin presented a history of contact with chemical species [127]. Another example of association between acute environmental stimuli and epigenetic-dependent disease phenotypes is given by the observed increase of CpG-island promoter hyper-methylation in tumor-suppressor genes in the oral mucosa of smokers (reviewed in [125]). Environmental factors are increasingly claimed as responsible for neurodegeneration-related modifications; a link between environmental-induced epigenetic modification, oxidation, and repair of AD-related genes, was discussed in depth in a recent review [129].

One of the most intriguing interactions between environment and epigenome is represented by the discussed results evidencing that even an apparently "mild" environmental factor as the

531

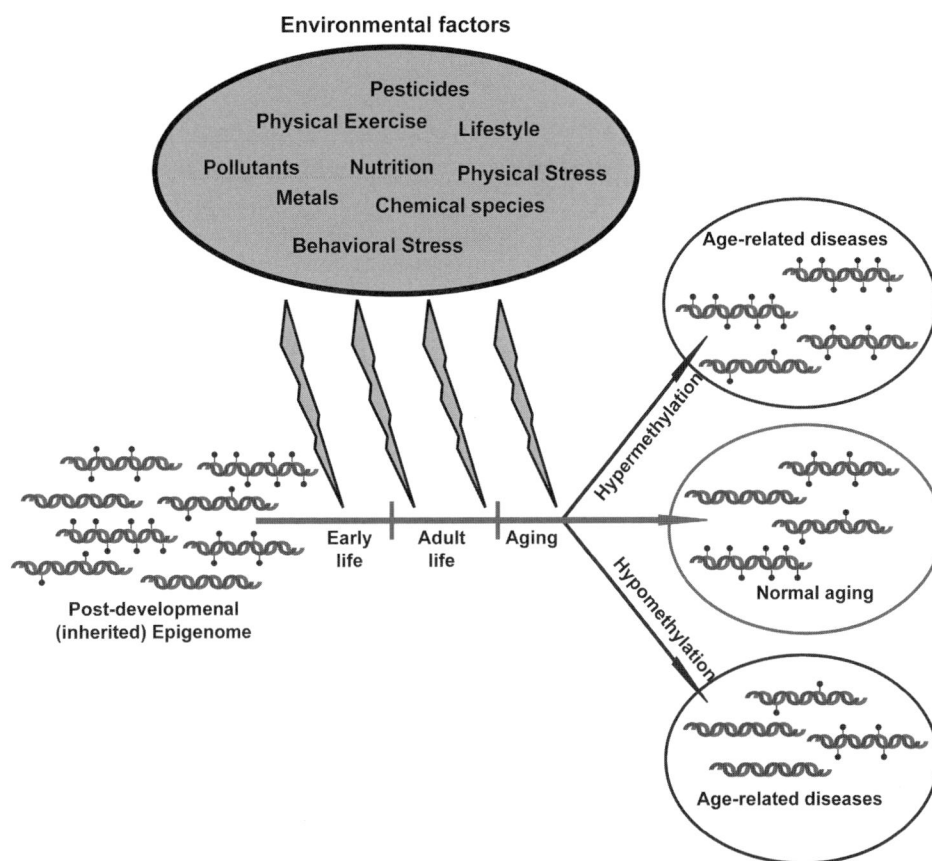

FIGURE 26.2

Many environmental factors can interfere with the organism, inducing epigenetic modifications. DNA methylation patterns established after developmental processes, for example, can be modified in sense of local and sequence-specific hyper- or (more frequently) hypomethylation. These alterations can be responsible for deviations from the normal aging processes, resulting in higher susceptibility to age-associated disease. This figure is reproduced in the color plate section.

behavioral stress (since it does not involve a physical contact of the individual with any chemical species), could result in a long-lasting alteration of epigenetic markers, leading to functional alterations. This surprising effect was very well demonstrated by the above-discussed study showing that perinatal behavioral stress in rats was responsible for altered AVP or BDNF expression, mediated by DNA methylation [50,59]. Another "mild" environmental factor that seems involved in the development of healthy or pathological aging is represented by the physical exercise. Physical exercise improves the efficiency of the capillary system and increases the oxygen supply to the brain, thus enhancing metabolic activity and oxygen intake in neurons, and increases neurotrophin levels and resistance to stress. Regular exercise and active lifestyle during adulthood have been associated with reduced risk and protective effects for mild cognitive impairment and AD. Similarly, studies in animal models show that physical activity has positive physiological and cognitive effects that correlate with changes in transcriptional profiles possibly mediated by epigenetic modifications [130].

Nutrition and diet represent another environmental factor that can exert its influence on aging. Dietary exposures can have consequences even many years later and this observation raises questions about the mechanisms through which such exposures are "remembered" and how they can result in altered disease risk. There is growing evidence that epigenetic mechanisms may mediate the effects of nutrients, micronutrients, and even non-nutrient dietary factors may be causal for the development of complex diseases [131]. Alterations in DNA methylation during aging can depend on alterations in dietary status and the great influence of nutritional components on health and lifespan it is largely accepted. Among the various mechanisms by which nutritional elements could affect the progress of aging, two pathways involve DNA methylation: the first involves the supply of metabolites of the S-adenosylmethionine cycle (and will be extensively discussed in the next paragraph), whereas the second is referred to elements able to directly modify the DNMT activity (selenium, cadmium, and nickel). However, other nutritional factors seem able to determine epigenetic modifications without directly perturbing the core of the methylation reactions. One example is given by the link between under- and overnutrition during pregnancy and the consequent (later in life) development of diseases such as diabetes and obesity. Epigenetic modifications may be one mechanism by which exposure to an altered intrauterine milieu may influence the onset of these disturbances much later in life. As a matter of fact, it was demonstrated that epigenetic modifications affecting processes important to glucose regulation and insulin secretion are present in the pancreatic β-cells and muscle of the intrauterine growth-retarded offspring, characteristics essential to the pathophysiology of type 2 diabetes. Moreover, epigenetic regulation of gene expression contributes to both adipocyte determination and differentiation in in vitro models [132]. The epigenetic connection between nutrition and age-related diseases was well presented by Tollefsbol and colleagues in relation to cancer onset [35], whereas the work performed in my laboratory in relation to the connection between nutrition and epigenetics in LOAD and the role of nutrition in the modulation of methylation reaction were summarized in a recent commentary [133] and will be here exposed more in detail.

26.8 EPIGENETICS AND AGE-ASSOCIATED DISEASES

Epigenetic modifications appear to be causative of, or at least involved in, an increasing number of human diseases. As previously discussed, a modern and developing concept points out the fetal or perinatal origin of adult diseases and the adaptation response to environmental stimuli leading to increased susceptibility to age-associated diseases [134]. Although the mechanisms mediating and expressing this "memory" of the early life throughout aging are not clearly unraveled, it is clear that an epigenetic basis exists. Apparently, the consequent increased susceptibility to the disease recapitulates as well the mechanisms typical of the decline observed in normal aging. The involvement of multiple organ systems in the pathological aging phenotype can be assimilated to the "frail syndrome". Frailty is defined as a non-specific state of global vulnerability reflecting multisystem pathology due to decreased adaptation to stressors and reduced functional reserve that becomes not sufficient to maintain and repair the aging body [135,136]. Clinically, frailty is characterized by low physical activity, global weakness, low muscle strength, fatigability, general slowness, and loss of weight [137]. As a matter of fact, frailty is strictly related to epigenetic mechanisms. It was demonstrated that global DNA methylation levels were correlated to the frailty status and that the worsening of frailty was significantly associated with DNA demethylation [138]. Identification of the role of epigenetic drift in the onset of frail status also represents the opportunity to underline the connection between epigenetics and other age-associated diseases. Part of the frail phenotype is, in fact, connected to other diseases typical of old age and characterized by evident epigenetic bases. As a matter of fact, significant high levels of inflammatory markers and activation of the clotting cascade were found in frail patients; these markers are, as is well known, also risk factors for cardiovascular diseases (CVD) [139]. An emerging theory identifies an epigenetic basis also for the chronic low-grade inflammation typical of aging, generated by the increase in the production of proinflammatory cytokines and other markers that lead to the definition of "inflamm-aging" status. This status is, in turn, comparable to frailty due to the multiorgan (brain, liver) and different tissue (adipose, muscle) contribution [140]. Finally, this complex picture involving inflammation and multi-organ contribution to the aged phenotype, finds a further piece of the jigsaw in the epigenetic basis of another complex disease like diabetes. Diabetes also involves alterations of the inflammatory markers and undergoes environmental influences mediated mainly by the nutrient intake; thus aging, frailty, nutrition, early life-events, and epigenetics are, once again, connected [134,141–143]. A very important concept emerging from these studies is that malnutrition is often associated with aging but that this deficit should be seen in terms of quality and variety of foods rather than in terms of quantity [141].

Epigenetics, disease, and aging are connected also in another complex relationship represented by the telomere attrition and the onset of cancer. Telomere attrition and mutations accumulate due to the deficit of the DNA repair machinery occurring with age is one of the main causes of age-associated genome instability. Telomeres are composed by G-rich repetitive DNA sequences (TTAGGG); these sequences are bound to complexes and specialized proteins, localized at the ends of linear chromosomes. Telomere length shortens with age as a consequence of decreased DNA methylation, until the minimal required length necessary to maintain telomere structure and function is reached [144–146]. The activation of the mechanism responsible for the maintenance of telomeric length (telomerase) is considered a hallmark of cancer while attrition is associated with the aging phenotype [147] depending on the response of the DNA damage checkpoint [148]; as previously discussed, the human telomerase reverse transcriptase gene (hTERT) is regulated by epigenetic mechanisms [149]. Recently, different models of transgenic mice deleted for the shelterin proteins (the major complex bound to telomeres) have been generated and could help the future study on the role of telomeric attrition and instability in aging and cancer [146]. Besides the epigenetic regulation of the telomeric length, cancer onset is clearly associated with epigenetics via the transcriptional repression of tumor-suppressor genes by CpG island promoter hypermethylation (reviewed in [150]); in addition to classical tumor-suppressor and DNA repair genes, cancer-associated hypermethylation also affects genes involved in premature aging and miRNAs with growth-inhibitory functions.

Epigenetic changes associated with aging and very often induced by environmental stimuli, seem therefore responsible for the possible onset of different, although strictly interconnected, pathologies typical of the elderly.

26.9 ONE CARBON METABOLISM

Studies on epigenetics exponentially increased, as results evident by the raise, from a scarce dozen in 1980 to over 5000 papers published in 2010, of PubMed citations involving epigenetics. Despite this great increase, one could suspect that a real knowledge of the biochemical basis regulating epigenetic mechanisms (DNA methylation, for instance) is often missing in these works; as a matter of fact, we can see that PubMed citations in which "epigenetics" and "homocysteine" are both present are only 96 in the period 1980–2010. This observation can be considered as a clue that the biochemical pathway underneath the transfer of methyl groups (the most relevant epigenetic modification) is quite unknown, even though this pathway (known as "one-carbon metabolism", being characterized by the transfer of monocarbonic units like the methyl group $-CH_3$) has a number of biochemical reactions comparable to ATP pathway.

One carbon metabolism, also known as the homocysteine (HCY) cycle, is a complex biochemical pathway regulated by the presence of folate, vitamin B12, and vitamin B6 (among other metabolites) and leading to the production of methyl donor molecule S-adenosylme-thionine (SAM). A schematic and partial (focused on SAM and B vitamins) representation of this metabolic cycle is reported in Figure 26.3. SAM can donate the methyl group to different substrates (lipids, proteins, and DNA) being converted in S-adenosylhomocysteine (SAH), a strong competitive inhibitor of methyltransferases. SAH is then hydrolyzed to adenosine and HCY, a thiol-containing amino acid produced during the methionine metabolism via the adenosylated compound SAM, once formed is either converted to cysteine by transsulfuration (further leading to glutathione synthesis), remethylated to form methionine, or exported to blood [151–155]. In the remethylation pathway HCY is remethylated by the vitamin B12-dependent enzyme methionine synthase (MS) using 5-methyltetrahydrofolate as cosub-strate. Alternatively, mainly in the liver, betaine can donate a methyl group in a vitamin B12-independent reaction, catalyzed by betaine-homocysteine methyltransferase (BHMT). In the transsulfuration pathway, HCY can condense with serine to form cystathionine in a reac-tion catalyzed by the cystathionine beta synthase (CBS), a vitamin B6-dependent enzyme, and the cystathionine is hydrolyzed to cysteine (Cys); it is not known for sure whether this pathway

534

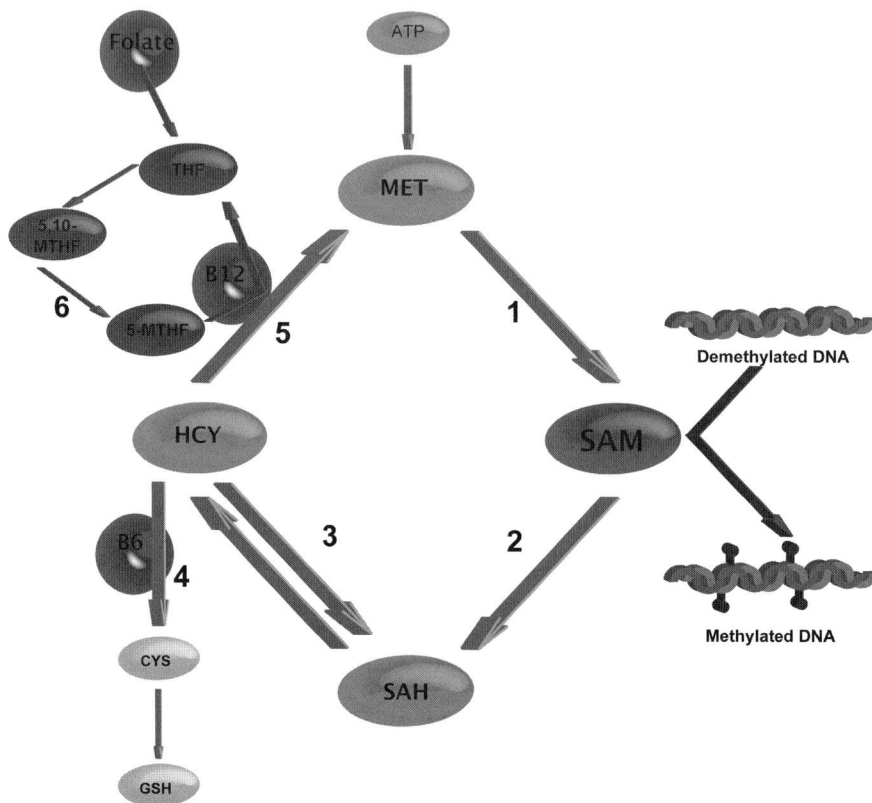

FIGURE 26.3
Schematic representation of the most relevant reactions comprises of the "one-carbon metabolism", emphasizing a vitamin cofactors cycle (remethylation) and SAM-dependent DNA methylation (transmethylation). Met, methionine; SAM, S-adenosylmethionine; SAH, S-adenosylhomocysteine; HCY, homocysteine; CYS, cystathionine, GSH, glutathione; THF, tetrahydrofolate; MTHF, methyltetrahydrofolate; B12, vitamin B12; B6, vitamin B6; 1, methionine adenosyltransferase (MAT); 2, methyltransferase(s); 3, SAH hydrolase; 4, cystathionine–β–synthase (CBS); 5, methionine synthase; 6, methylenetetrahydrofolate reductase (MTHFR). This figure is reproduced in the color plate section.

is normally active in the brain, due to the lack of the enzyme γ-cystathionase, but a recent study demonstrated the presence of a functional (maybe alternative) transsulfuration pathway also in this tissue [156]. Cys is used for protein synthesis, metabolized to sulfate or used for glutathione (GSH) synthesis. The tripeptide GSH is the most abundant intracellular non-protein thiol, and it is a versatile reductant, serving multiple biological functions, acting, among others, as a quencher of free radicals and a cosubstrate in the enzymatic reduction of peroxides [157].

HCY accumulation causes the accumulation of SAH because of the reversibility of the reaction converting SAH to HCY and adenosine (Ado); the equilibrium dynamics favors SAH synthesis. The reaction proceeds in the hydrolytic direction only if HCY and Ado are efficiently removed. At least in brain and other tissues, SAH is a DNA methyltransferase inhibitor, which reinforces DNA hypomethylation [158]. Therefore, an alteration of the metabolism through either remethylation or transsulfuration pathways can lead to hyperhomocysteinemia, decreasing of SAM/SAH ratio (often indicated as the "methylation potential", MP) and alteration of GSH levels, suggesting that hypomethylation is a mechanism through which HCY is involved in vascular disease and AD, together with the oxidative damage [4,159,160].

26.10 ONE-CARBON METABOLISM IN AGING AND NEURODEGENERATION

Methylation of CpG dinucleotides plays an important role in regulation of gene expression in the brain. It was discovered that brain-specific promoter-related sequences are surprisingly enriched in CpG sites. This leads to the conclusion that it is likely that brain-specific transcription is regulated by methylation at an epigenetic level much more frequently than tissue-specific expression in other organs. Low methylation status is strongly associated with neurological and cognitive deficits. Many epidemiological studies have shown that factors connected to low methylation status such as elevated total homocysteine, low folate or low vitamin B12 levels are associated with increased risk of cognitive decline, dementia, and brain atrophy. In recent years, hyperhomocysteinemia has begun to be widely considered a risk factor in AD and this may be ascribed to alteration of the S-adenosylmethionine/homocysteine (SAM/HCY) metabolism [161]. Indeed, a frequently observed condition in AD-affected people is the increase in plasmatic HCY, sometimes along with a decrease in vitamin B12, B6, and folate uptake [162]. In AD, the loss of precise control through gene methylation may alter a delicate equilibrium among the three enzymes (α-, β-, and γ-secretases) known to be involved in the production of either amyloid-beta (Aβ) or other non-dangerous catabolites [163].

It is well established that α-secretase cleavage of APP does not produce the amyloidogenic peptides and that, on the contrary, they are produced by the activity of β-secretases that generate an N-terminal soluble fragment and a C-terminal fragment that is sequentially cleaved by γ-secretase to produce Aβ peptides [164]. The alteration of SAM/SAH ratio is tightly related to the altered expression of two genes involved in APP metabolism, finally producing the accumulation of Aβ peptide in the senile plaque. In the last 50 years, SAM has been shown to be perhaps the most frequently used substrate, after ATP, and therefore occupies a central position in human health, disease, and aging. SAM is known to be the primary methyl-donor present in eukaryotes and it is involved in methylation of target molecules as DNA, RNA, proteins, lipids, and polyamines synthesis [165]. SAM appears to be altered in some neurological disorders, including LOAD [26]. About 95% of SAM is engaged in methylation reactions. These metabolic alterations may be responsible for the generalized reduction of DNA methylation observed in aging, which can lead to overactivation of methylation-controlled genes. DNA methylation represents an important mechanism for epigenetic control of gene expression and the maintenance of genome integrity. Methyl deficiency results in global hypomethylation of the genome. Hence, methyl deficiency leads to disturbances in gene expression. Interestingly, the aging process leads to similar changes in the methylation pattern.

26.11 EPIGENETICS AND NEURODEGENERATION: THE ALZHEIMER'S DISEASE PARADIGM

The sporadic form of Alzheimer's disease (LOAD) is considered a multifactorial disease. Genetic, nutritional, metabolic, environmental, and social factors are associated with onset and progression of the disease [75,120,166] but, despite the extent of researches in this area, most causes of LOAD remain obscure. The complex, interconnected, non-Mendelian, etiology of LOAD suggests that epigenetic components could be involved and epigenetic modifications could represent candidate effector mechanisms mediating the above-mentioned risk factors in AD onset and progression. Despite the fact that it is already well known that epigenetic changes could act in several physiological and pathological processes, few papers pointed the attention on epigenetic regulation of aging and neurodegeneration. Nevertheless, it is evident that whereas genetic factors are clearly associated with the early-onset form of AD, epigenetic factors could be more easily linked to LOAD, since the epigenome is prone to changes during development and also aging [167–169]. As a matter of fact, the epigenetic mechanisms can be considered as a link between environmental stimuli and their effect on the genome and on the pathologies. For these reasons, LOAD research is facing an increasing interest in the study of epigenetic mechanisms possibly involved in the pathology [56,133,170]. Due to the large development of wide-scale analysis techniques and their application to the study of epigenome, it has been possible to demonstrate that DNA methylation in individuals changes over time [171]. Moreover, the DNA methylation pattern in human brain is different in different brain areas [120] and can dynamically change during the lifespan [172]. Finally, it has been demonstrated that the existence of epigenetic changes in LOAD patients can easily contribute to the onset and development of disease [173]. In particular, Wang and colleagues [173] evidenced that, among others, PS1 gene was differentially methylated and that there was a correlation between gene demethylation and LOAD. In agreement with this result, we also demonstrated that PSEN1 promoter demethylation is responsible for the overexpression of the gene [54,55,120].

As previously described, methylation metabolism is dependent on the "one-carbon" metabolism (Figure 26.3). High HCY and low B vitamin levels are positively associated with LOAD [174–176], even if the cause–effect relationship is still not completely demonstrated. In this scenario, epigenome modulation could represent the molecular link between hyperhomocysteinemia and LOAD course and onset [177–179]. Moreover, one-carbon metabolism alteration and consequent methylation reactions unbalance (i.e. loss of methyl groups) during aging represents one of the mechanisms by which environmental and dietary factors can promote LOAD [75,158,180–182]. Finally, the finding that one-carbon metabolism alterations can affect also tau processing [183] stress the hypothesis that this metabolism, and methylation reactions in particular, could retain a central role in LOAD, connecting both amyloidogenic and neurofibrillary pathways.

Two genes directly involved in AD showed changes in methylation pattern. It was demonstrated that the APP gene promoter was differentially modulated [184] and it was found hypomethylated in postmortem brains of AD patients [112,185]. Besides APP, also PSEN1 has been found hypomethylated both in AD patients [173] and in our experimental models characterized by unbalanced methylation pathway [54,55,120].

26.12 AGED AD MICE AND EPIGENETICS

These studies, spacing from specific gene to large-scale studies, both in LOAD patients and in experimental models, are very well summarized in two recent reviews [56,186]. Studies on

LOAD subjects evidenced that DNA methylation, the main epigenetic modification, is associated with Alzheimer's mechanisms. Studies on experimental models, although not supported by a final proof in AD patients, had the fundamental value of investigating the molecular mechanisms linking DNA methylation to the onset and the progression of the AD-like features observed in these models. Several groups investigated the role of epigenetics in AD in different areas of cell biology as Aβ processing and scavenging, tau phosphorylation, inflammation, apoptosis, cell cycle dysregulation, and ApoE expression (summarized in [56]).

In our laboratory, during the last 10 years, we addressed our studies to the area of DNA methylation in human neuroblastoma cells [187,188] and then in a murine model of Alzheimer's disease [120], developing an experimental model of nutritionally induced hyperhomocysteinemia. Taking advantage of the massive and rapid amyloid deposition of the TgCRND8 mice, carrying a double-mutated human APP gene, we were able to demonstrate that hyperhomocysteinemia induced by B vitamins deficiency (folate, B12, and B6) caused an alteration in SAM and SAH levels, inhibiting SAM-dependent methyltransferases. This alteration was responsible for dysequilibrium in the DNA methylation machinery consisting of the decrease of the methyltransferase and increase of the demethylase activities [54]. The DNA methylation impairment was correlated to the site-specific demethylation of PSEN1 promoter, with consequent gene overexpression [55]. Other studies found similar results in a different cellular model [189], also suggesting the role of DNA hypomethylation in the increase of PSEN1 expression and amyloid production.

If the involvement of DNA methylation in LOAD is now largely accepted, its causal or consequential role still remains to be ascertained [121,190]. One hypothesis is that alteration of methylation reactions could be dependent on other major pathways involved in LOAD onset, representing just a marker of the disease. Experimental evidences in this sense are given by the finding that exogenous Aβ 1-40 seems to induce the hypermethylation of the Neprilysin gene [191]. To prove that PSEN1 hypomethylation is a cause and not a consequence of the AD-like phenotype in our mouse model, we analyzed the methylation pattern in old mice (presenting increased Aβ pathology) versus young mice and in transgenic versus wild-type mice (which are not affected at all by Aβ pathology); our results (submitted at the moment this chapter was written) clearly indicate that PSEN1 methylation is not dependent on Aβ levels, since similar methylation patterns were found in young and old mice, both transgenic and wild-type. This result further stresses the idea that epigenetics plays a pivotal and fundamental role in the onset and progression of age-related diseases like LOAD.

26.13 CONCLUSION

The great technological advance in biomedical research makes possible the constant increase in our ability to study the extremely complex phenomena encompassing the different, but highly interconnected, epigenetic modifications. This ability opened the window, in recent years, on a previously hidden scenario where epigenetics retains a causal role in mediating the effects that environmental stimuli exert in the organism. This growing area of the science is particularly relevant to the study of aging-associated processes, because the aging organism is increasingly exposed to continuous and different external stimuli. Now that the window is open on the mechanisms possibly responsible for the shifting from healthy to pathological aging, new questions rise together with the first results. How many, and how long, stimuli are necessary to induce changes in the normal aging processes? In which manner does the organism translate these stimuli in processes that alter epigenetic modifications? Why do environmental-induced epigenetic changes have different tissue, cell population, genomic sequence, or even DNA site specificity? Alternatively, is this different susceptibility an endogenous characteristic of these tissue, cells, or sequences? Are these "toxic" epigenetic modifications really reversible? And, in this case, are these modifications "druggable", in order to hypothesize epigenetic intervention and therapies? Deciphering the epigenome and its relevance to the aging processes is probably one of the most promising challenges for the researchers in the coming years and, due to its extreme complexity, it surely requires the interaction of different skills and knowledge in an interdisciplinary effort.

References

[1] Waddington CH. Endeavour 1942;1:18—20.

[2] Holliday R. Mechanisms for the control of gene activity during development. Biol Rev Cambr Philos Soc 1990;65:431—71.

[3] Goldberg AD, Allis CD, Bernstein E. Epigenetics: a landscape takes shape. Cell 2007;128:635—8.

[4] Scarpa S, Cavallaro RA, D'Anselmi F, Fuso A. Gene silencing through methylation: an epigenetic intervention on Alzheimer disease. J. Alzheimer's Dis 2006;9:407—14.

[5] Reitz C, Brayne C, Mayeux R. Epidemiology of Alzheimer disease. Nat Rev Neurol 2011;7:137—52.

[6] Cutler RG. Evolutionary perspective of human longevity. In: Hazzard WR, Andres R, Bierman EL, Blass JP, editors. Principles of geriatric medicine and gerontology. New York: McGraw-Hill; 1990. p. 15—21.

[7] Roush W. Live long and prosper? Science 1996;273:42—6.

[8] Troen BR. The biology of aging. Mt Sinai J Med 2003;70:3—22.

[9] Hayflick L. Current theories of biological ageing. Federation Proceedings 1975;34:9—13.

[10] Rattan SI. Cellular and molecular determinants of ageing. Indian J Exp Biol 1996;34:1—6.

[11] Hamet P, Tremblay J. Genes of Aging. Metab 2003;52:5—9.

[12] Kirkwood TB. Human senescence. Bioassays 1996;18:1009—16.

[13] Orgel LE. The maintenance of the accuracy of protein synthesis and its relevance to aging. Proc Natl Acad Sci USA 1963;49:517—21.

[14] Suzuki YJ, Forman HJ, Sevanian A. Oxidants as stimulators of signal transduction. Free Radic Biol Med 1997;22:269—85.

[15] Finch CE, Tanzi RE. Genetics of aging. Science 1997;278:407—11.

[16] Perls T, Levenson R, Regan M, Puca A. What does it take to live to 100? Mech Ageing Dev 2002;123:231—42.

[17] Kuro-o M. Disease model: human aging. Trends Mol Med 2001;7:179—81.

[18] Denckla WD. A time to die. Life Sci 1975;16:31—44.

[19] Walford RL. Immunologic theory of aging: current status. Federation Proc 1974;33:2020—7.

[20] Rubin H. Cell aging in vivo and in vitro. Mech Ageing Dev 1997;98:1—35.

[21] Harley CB. Telomere loss: mitotic clock or genetic time bomb? Mutation Res 1991;256:271—82.

[22] Campisi J. Aging and cancer: the double-edged sword of replicative senescence. J Am Geriatric Soc 1997;45:482—8.

[23] Jacobson MD, Weil M, Raff MC. Programmed cell death in animal development. Cell 1997;88:347—54.

[24] Villeponteau B. The heterochromatin loss model of aging. Exp Gerontol 1997;32:383—94.

[25] Howard BH. Replicative senescence: considerations relating to the stability of heterochromatin domains. Exp Gerontol 1996;31:281—93.

[26] Bottiglieri T, Hyland K. S-adenosylmethionine levels in psychiatric and neurological disorders: a review. Acta Neurologica Scandinava: Supplements 1994;154:19—26.

[27] Morrison LD, Smith DD, Kish SJ. Brain S-adenosylmethionine levels are severely decreased in Alzheimer's disease. J Neurochem 1996;67:1328—31.

[28] Tollefsbol TO, Andrews LG. Mechanisms for methylation-mediated gene silencing and aging. Med Hypotheses 1993;41:83—92.

[29] Ahuja N, Issa JP. Aging, methylation and cancer. Histol Histopathol 2000;15:835—42.

[30] Young JI, Smith JR. DNA methyltransferase inhibition in normal human fibroblasts induces a p21-dependent cell cycle withdrawal. J Biol Chem 2001;276:19610—6.

[31] Razin A. CpG methylation, chromatin structure and gene silencing-a three-way connection. EMBO J 1998;17:4905—8.

[32] Guarente L. Sir2 links chromatin silencing, metabolism, and aging. Gene Dev 2000;14:1021—6.

[33] Bandyopadhyay D, Medrano EE. The emerging role of epigenetics in cellular and organismal aging. Exp Gerontol 2003;38:1299—307.

[34] Yuasa Y. DNA methylation in cancer and ageing. Mech Ageing Dev 2002;123:1649—54.

[35] Liu L, Wylie RC, Andrews LG, Tollefsbol TO. Aging, cancer and nutrition: the DNA methylation connection. Mech Ageing Dev 2003;124:989—98.

[36] Teschendorff AE, Menon U, Gentry-Maharaj A, Ramus SJ, Weisenberger DJ, Shen H, et al. Age- dependent DNA methylation of genes that are suppressed in stem cells is a hallmark of cancer. Genome Res 2010;20:440—6.

[37] Rakyan VK, Down TA, Maslau S, Andrew T, Yang TP, Beyan H, et al. Human aging-associated DNA hyper-methylation occurs preferentially at bivalent chromatin domains. Genome Res 2010;20:434—9.

[38] Christensen BC, Houseman EA, Marsit CJ, Zheng S, Wrensch MR, Wiemels JL, et al. Aging and environmental exposures alter tissue-specific DNA methylation dependent upon CpG island context. PLoS Genet 2009;5 e1000602:

[39] Maegawa S, Hinkal G, Kim HS, Shen L, Zhang L, Zhang J, et al. Widespread and tissue specific age-related DNA methylation changes in mice. Genome Res 2010;20:332—40.

[40] Young JI, Sedivy JM, Smith JR. Telomerase expression in normal human fibroblasts stabilizes DNA 5-methylcytosine transferase I. J Biol Chem 2003;278:19904—8.

[41] Szyf M. A dynamic methylome; implications of non-CpG methylation/demethylation. Cell Cycle 2010;9:3846—7.

[42] Zhu B, Zheng Y, Hess D, Angliker H, Schwarz S, Siegmann M, et al. 5-methylcytosine-DNA glycosylase activity is present in a cloned G/T mismatch DNA glycosylase associated with the chicken embryo DNA demethylation complex. Proc Natl Acad Sci USA 2000;97:5135—9.

[43] Barreto G, Schäfer A, Marhold J, Stach D, Swaminathan SK, Handa V, et al. Gadd45a promotes epigenetic gene activation by repair-mediated DNA demethylation. Nature 2007;445:671—5.

[44] Brown SE, Suderman MJ, Hallett M, Szyf M. DNA demethylation induced by the methyl—CpG—binding domain protein MBD3. Gene 2008;420:99—106.

[45] Métivier R, Gallais R, Tiffoche C, Le Péron C, Jurkowska RZ, Carmouche RP, et al. Cyclical DNA methylation of a transcriptionally active promoter. Nature 2008;452:45—50.

[46] Bhattacharya SK, Ramchandani S, Cervoni N, Szyf M. A mammalian protein with specific demethylase activity for mCpG DNA. Nature 1999;397:579—83.

[47] Cervoni N, Bhattacharya S, Szyf M. DNA demethylase is a processive enzyme. J Biol Chem 1999;274:8363—6.

[48] Detich N, Hamm S, Just G, Knox JD, Szyf M. The methyl donor S-Adenosylmethionine inhibits active demethylation of DNA: a candidate novel mechanism for the pharmacological effects of S-Adenosylmethionine. J Biol Chem 2003;278:20812—20.

[49] Levenson JM, Roth TL, Lubin FD, Miller CA, Huang IC, Desai P, et al. Evidence that DNA (cytosine-5) methyltransferase regulates synaptic plasticity in the hippocampus. J Biol Chem 2006;281:15763—73.

[50] Murgatroyd C, Patchev AV, Wu Y, Micale V, Bockmühl Y, Fischer D, et al. Dynamic DNA methylation programs persistent adverse effects of early-life stress. Nat Neurosci 2009;12:1559—66.

[51] Szyf M. Epigenetics, DNA methylation, and chromatin modifying drugs. Annu Rev Pharmacol Toxicol 2009;49:243—63.

[52] Lucarelli M, Fuso A, Strom R, Scarpa S. The dynamics of myogenin site-specific demethylation is strongly correlated with its expression and with muscle differentiation. J Biol Chem 2001;276:7500—6.

[53] Fuso A, Ferraguti G, Grandoni F, Ruggeri R, Scarpa S, Strom R, et al. Early demethylation of non-CpG, CpC-rich, elements in the myogenin 5'-flanking region: a priming effect on the spreading of active demethylation. Cell Cycle 2010;9:3965—76.

[54] Fuso A, Nicolia V, Cavallaro RA, Scarpa S. DNA methylase and demethylase activities are modulated by one-carbon metabolism in Alzheimer's disease models. J Nutr Biochem 2011;122:242—51.

[55] Fuso A, Nicolia V, Pasqualato A, Fiorenza MT, Cavallaro RA, Scarpa S. Changes in Presenilin 1 gene methylation pattern in diet-induced B vitamin deficiency. Neurobiol Aging 2011;32:187—99.

[56] Mastroeni D, Grover A, Delvaux E, Whiteside C, Coleman PD, Rogers J. Epigenetic mechanisms in Alzheimer's disease. Neurobiol Aging 2011;7:1161—80.

[57] Kangaspeska S, Stride B, Métivier R, Polycarpou-Schwarz M, Ibberson D, Carmouche RP, et al. Transient cyclical methylation of promoter DNA. Nature 2008;452:12—5.

[58] Weaver IC, Champagne FA, Brown SE, Dymov S, Sharma S, Meaney MJ, et al. Reversal of maternal programming of stress responses in adult offspring through methyl supplementation: altering epigenetic marking later in life. J Neurosci 2005;25:11045—54.

[59] Roth TL, Sweatt JD. Epigenetic marking of the BDNF gene by early-life adverse experiences. Horm Behav 2011;59:315—20.

[60] Yakovlev A, Khafizova M, Abdullaev Z, Loukinov D, Kondratyev A. Epigenetic regulation of caspase-3 gene expression in rat brain development. Gene 2010;450:103—8.

539

[61] Calvanese V, Lara E, Kahn A, Fraga MF. The role of epigenetics in aging and age-related diseases. Ageing Res Rev 2009;8:268—76.

[62] Kriaucionis S, Heintz N. The nuclear DNA base 5-hydroxymethylcytosine is present in Purkinje neurons and the brain. Science 2009;324:929—30.

[63] Tahiliani M, Koh KP, Shen Y, Pastor WA, Bandukwala H, Brudno Y, et al. Conversion of 5-methylcytosine to 5- hydroxymethylcytosine in mammalian DNA by MLL partner TET1. Science 2009;324:930—5.

[64] Ito S, D'Alessio AC, Taranova OV, Hong K, Sowers LC, Zhang Y. Role of Tet proteins in 5mC to 5hmC conversion, ES-cell self-renewal and inner cell mass specification. Nature 2010;466:1129—33.

[65] Williams K, Christensen J, Pedersen MT, Johansen JV, Cloos PA, Rappsilber J, et al. TET1 and hydroxy-methylcytosine in transcription and DNA methylation fidelity. Nature 2011;473:343—8.

[66] Valinluck V, Tsai HH, Rogstad DK, Burdzy A, Bird A, Sowers LC. Oxidative damage to methyl-CpG sequences inhibits the binding of the methyl-CpG binding domain (MBD) of methyl-CpG binding protein 2 (MeCP2). Nucleic Acids Res 2004;32:4100—8.

[67] Ficz G, Branco MR, Seisenberger S, Santos F, Krueger F, Hore TA, et al. Dynamic regulation of 5-hydroxymethylcytosine in mouse ES cells and during differentiation. Nature 2011;473:398—402.

[68] Song CX, Szulwach KE, Fu Y, Dai Q, Yi C, Li X, et al. Selective chemical labeling reveals the genome-wide distribution of 5-hydroxymethylcytosine. Nat Biotechnol 2011;29:68—72.

[69] Chouliaras L, van den Hove DL, Kenis G, Dela Cruz J, Lemmens MA, van Os J, et al. Caloric restriction attenuates age-related changes of DNA methyltransferase 3a in mouse hippocampus. Brain Behav Immun 2011;25:616—23.

[70] Chouliaras L, van den Hove DL, Kenis G, Keitel S, Hof PR, van Os J, et al. Prevention of age-related changes in hippocampal levels of 5-methylcytidine by caloric restriction. Neurobiol Aging 2011. DOI: http://dx.doi.org/10.1016/j.neurobiolaging.2011.06.003.

[71] Clayton AL, Hazzalin CA, Mahadevan LC. Enhanced histone acetylation and transcription: a dynamic perspective. Mol Cell 2006;23:289—96.

[72] Sokolova NE, Shiryaeva NV, Dyuzhikova NA, Savenko YN, Vaido AI. Effect of long-term mental and pain stress on the dynamics of H4 histone acetylation in hippocampal neurons of rats with different levels of nervous system excitability. Bull Exp Biol Med 2006;142:341—3.

[73] Impey S. A histone deacetylase regulates addiction. Neuron 2007;56:415—7.

[74] Aizawa S, Yamamuro Y. Involvement of histone acetylation in the regulation of choline acetyltransferase gene in NG108-15 neuronal cells. Neurochem Int 2010;56:627—33.

[75] Wu J, Basha MR, Zawia NH. The environment, epigenetics and amyloidogenesis. J Mol Neurosci 2008;34:1—7.

[76] Chen PS, Wang CC, Bortner CD, Peng GS, Wu X, Pang H, et al. Valproic acid and other histone deacetylase inhibitors induce microglial apoptosis and attenuate lipopolysaccharide-induced dopaminergic neurotoxicity. Neuroscience 2007;149:203—12.

[77] Echaniz-Laguna A, Bousiges O, Loeffler JP, Boutillier AL. Histone deacetylase inhibitors: therapeutic agents and research tools for deciphering motor neuron diseases. Curr Med Chem 2008;15:1263—73.

[78] Gray SG, Dangond F. Rationale for the use of histone deacetylase inhibitors as a dual therapeutic modality in multiple sclerosis. Epigenetics 2006;1:67—75.

[79] Tahiliani M, Mei P, Fang R, Leonor T, Rutenberg M, Shimizu F, et al. The histone H3K4 demethylase SMCX links REST target genes to X-linked mental retardation. Nature 2007;447:601—5.

[80] Yoo AS, Staahl BT, Chen L, Crabtree GR. MicroRNA-mediated switching of chromatin-remodelling complexes in neural development. Nature 2009;460:642—6.

[81] Khudayberdiev S, Fiore R, Schratt G. MicroRNA as modulators of neuronal responses. Commun Integr Biol 2009;2:411—3.

[82] Maes OC, Chertkow HM, Wang E, Schipper HM. MicroRNA: Implications for Alzheimer Disease and other Human CNS. Disorders. Curr Genomics 2009;10:154—68.

[83] Morrison JH, Hof PR. Life and death of neurons in the aging brain. Science 1997;278:412—9.

[84] Hof PR, Morrison JH. The aging brain: morphomolecular senescence of cortical circuits. Trends Neurosci 2004;27:607—13.

[85] Dickstein DL, Kabaso D, Rocher AB, Luebke JI, Wearne SL, Hof PR. Changes in the structural complexity of the aged brain. Aging Cell 2007;6:275—84.

[86] Rutten BP, Schmitz C, Gerlach OH, Oyen HM, de Mesquita EB, et al. The aging brain: accumulation of DNA damage or neuron loss? Neurobiol Aging 2007;28:91—8.

[87] Wu W, Brickman AM, Luchsinger J, Ferrazzano P, Pichiule P, Yoshita M, et al. The brain in the age of old: the hippocampal formation is targeted differentially by diseases of late life. Ann Neurol 2008;64:698—706.

[88] Luebke JI, Weaver CM, Rocher AB, Rodriguez A, Crimins JL, Dickstein DL, et al. Dendritic vulnerability in neurodegenerative disease: insights from analyses of cortical pyramidal neurons in transgenic mouse models. Brain Struct Funct 2010;214:181–99.

[89] Lee CK, Weindruch R, Prolla TA. Gene-expression profile of the ageing brain in mice. Nat Genet 2000;25:294–7.

[90] Lu T, Pan Y, Kao SY, Li C, Kohane I, Chan J, et al, B. A. Gene regulation and DNA damage in the ageing human brain. Nature 2004;429:883–91.

[91] Berchtold NC, Cribbs DH, Coleman PD, Rogers J, Head E, Kim R, et al. Gene expression changes in the course of normal brain aging are sexually dimorphic. Proc Natl Acad Sci USA 2008;105:15605–10.

[92] Penner MR, Roth TL, Barnes CA, Sweatt JD. An epigenetic hypothesis of aging-related cognitive dysfunction. Front Aging Neurosci 2010;2:9.

[93] Peleg S, Sananbenesi F, Zovoilis A, Burkhardt S, Bahari-Javan S, Agis-Balboa RC, et al. Altered histone acetylation is associated with age-dependent memory impairment in mice. Science 2010;328:753–6.

[94] Kukull WA, Higdon R, Bowen JD, McCormick WC, Teri L, Schellenberg GD, et al. Dementia and Alzheimer disease incidence: a prospective cohort study. Arch Neurol 2002;59:1737–46.

[95] Crouch PJ, Cimdins K, Duce JA, Bush AI, Trounce IA. Mitochondria in aging and Alzheimer's disease. Rejuvenation Res 2007;10:349–57.

[96] Fuller SJ, Tan RS, Martins RN. Androgens in the etiology of Alzheimer's disease in aging men and possible therapeutic interventions. J Alzheimer's Dis 2007;12:129–42.

[97] Thibault O, Gant JC, Landfield PW. Expansion of the calcium hypothesis of brain aging and Alzheimer's disease: minding the store. Aging Cell 2007;6:307–17.

[98] Landfield PW, Blalock EM, Chen KC, Porter NM. A new glucocorticoid hypothesis of brain aging: implications for Alzheimer's disease. Curr Alzheimer Res 2007;4:205–12.

[99] Duenas-Gonzalez A, Candelaria M, Perez-Plascencia C, Perez-Cardenas E, de la Cruz-Hernandez E, Herrera LA. Valproic acid as epigenetic cancer drug: preclinical, clinical and transcriptional effects on solid tumors. Cancer Treat Rev 2008;34:206–22.

[100] Craft S. Insulin resistance syndrome and Alzheimer's disease: age- and obesity-related effects on memory, amyloid, and inflammation. Neurobiol Aging 2005;26:65–9.

[101] Bailey TL, Rivara CB, Rocher AB, Hof PR. The nature and effects of cortical microvascular pathology in aging and Alzheimer's disease. Neurol Res 2004;26:57357–8.

[102] Selkoe DJ. Aging, amyloid, and Alzheimer's disease: a perspective in honor of Carl Cotman. Neurochem Res 2003;28:1705–13.

[103] Maeda S, Sahara N, Saito Y, Murayama S, Ikai A, Takashima A. Increased levels of granular tau oligomers: an early sign of brain aging and Alzheimer's disease. Neurosci Res 2006;54:197–201.

[104] Fraga MF, Esteller M. Epigenetics and aging: the targets and the marks. Trends Genet 2007;23:413–8.

[105] Mays-Hoopes LL. DNA methylation in aging and cancer. J Gerontol 1989;44:35–6.

[106] Yu MK. Epigenetics and chronic lymphocytic leukemia. Am J Hematol 2006;81:864–9.

[107] Wilson VL, Jones PA. DNA methylation decreases in aging but not in immortal cells. Science 1983;220:1055–10057.

[108] Lopatina N, Haskell JF, Andrews LG, Poole JC, Saldanha S, et al. Differential maintenance and de novo methylating activity by three DNA methyltransferases in aging and immortalized fibroblasts. J Cell Biochem 2002;84:324–34.

[109] Varela-Moreiras G, Pérez-Olleros L, García-Cuevas M, Ruiz-Roso B. Effects of ageing on folate metabolism in rats fed a long-term folate deficient diet. Int J Vitam Nutr Res 1994;64:294–9.

[110] Richardson BC. Role of DNA methylation in the regulation of cell function: autoimmunity, aging and cancer. J Nutr 2002;132:2401S–5S.

[111] Bollati V, Galimberti D, Pergoli L, Dalla Valle E, Barretta F, Cortini F, et al. DNA methylation in repetitive elements and Alzheimer disease. Brain Behav Immun 2011;25:1078–83.

[112] Tohgi H, Utsugisawa K, Nagane Y, Yoshimura M, Genda Y, Ukitsu M. Reduction with age in methylcytosine in the promoter region -224 approximately -101 of the amyloid precursor protein gene in autopsy human cortex. Mol. Brain Res 1999;70:288–92.

[113] Tohgi H, Utsugisawa K, Nagane Y, Yoshimura M, Ukitsu M, Genda Y. The methylation status of cytosines in a tau gene promoter region alters with age to downregulate transcriptional activity in human cerebral cortex. Neurosci Lett 1999;275:89–92.

[114] Tohgi H, Utsugisawa K, Nagane Y, Yoshimura M, Ukitsu M, Genda Y. Decrease with age in methylcytosines in the promoter region of receptor for advanced glycated end products (RAGE) gene in autopsy human cortex. Brain Res Mol Brain Res 1999;65:124–8.

541

[115] Zhang Z, Deng C, Lu Q, Richardson B. Age-dependent DNA methylation changes in the ITGAL (CD11a) promoter. Mech Ageing Dev 2002;123:1257–68.

[116] Issa JP, Ottaviano YL, Celano P, Hamilton SR, Davidson NE, Baylin SB. Methylation of the oestrogen receptor CpG island links ageing and neoplasia in human colon. Nat Genet 1994;7:536–40.

[117] Issa JP, Vertino PM, Boehm CD, Newsham IF, Baylin SB. Switch from monoallelic to biallelic human IGF2 promoter methylation during aging and carcinogenesis. Proc Natl Acad Sci USA 1996;93:11757–62.

[118] Ladd-Acosta C, Pevsner J, Sabunciyan S, Yolken RH, Webster MJ, Dinkins T, et al. DNA methylation signatures within the human brain. Am J Hum Genet 2007;81:1304–15.

[119] Lahiri DK, Maloney B, Basha MR, Ge YW, Zawia NH. How and when environmental agents and dietary factors affect the course of Alzheimer's disease: the "LEARn" model (latent early-life associated regulation) may explain the triggering of AD. Curr Alzheimer Res 2007;4:219–28.

[120] Fuso A, Nicolia V, Cavallaro RA, Ricceri L, D'Anselmi F, Coluccia P, et al. B-Vitamin deprivation induces hyperhomocysteinemia and brain S-adenosylhomocsyteine, depletes brain S-adenosylmethionine, and enhances PSEN1 and BACE expression and Amyloid-β deposition in mice. Mol Cell Neurosci 2008;37:731–46.

[121] Chouliaras L, Rutten BP, Kenis G, Peerbooms O, Visser PJ, Verhey F, et al. Epigenetic regulation in the pathophysiology of Alzheimer's disease. Prog Neurobiol 2010;90:498–510.

[122] Carnero A, Lleonart ME. Epigenetic mechanisms in senescence, immortalisation and cancer. Biol Rev Camb Philos Soc 2011;86:443–55.

[123] Fernández-Morera JL, Calvanese V, Rodríguez-Rodero S, Menéndez-Torre E, Fraga MF. Epigenetic regulation of the immune system in health and disease. Tissue Antigens 2010;76:431–9.

[124] Hernandez DG, Nalls MA, Gibbs JR, Arepalli S, van der Brug M, Chong S, et al. Distinct DNA methylation changes highly correlated with chronological age in the human brain. Hum Mol Genet 2011;20:1164–72.

[125] Poulsen P, Esteller M, Vaag A, Fraga MF. The epigenetic basis of twin discordance in age-related diseases. Pediatr Res 2007;61:38R–42R.

[126] Wang CS, Burke JR, Steffens DC, Hulette CM, Breitner JC, Plassman BL. Twin pairs discordant for neuropathologically confirmed Lewy body dementia. J Neurol Neurosurg Psychiatry 2009;80:562–5.

[127] Mastroeni D, McKee A, Grover A, Rogers J, Coleman PD. Epigenetic differences in cortical neurons from a pair of monozygotic twins discordant for Alzheimer's disease. PLoS One 2009;4e6617:

[128] Rodríguez-Rodero S, Fernández-Morera JL, Fernandez AF, Menéndez-Torre E, Fraga MF. Epigenetic regulation of aging. Discov Med 2010;10:225–33.

[129] Coppedè F, Migliore L. Evidence linking genetics, environment, and epigenetics to impaired DNA repair in Alzheimer's disease. J Alzheimer's Dis 2010;20:953–66.

[130] Kaliman P, Párrizas M, Lalanza JF, Camins A, Escorihuela RM, Pallàs M. Neurophysiological and epigenetic effects of physical exercise on the aging process. Ageing Res Rev 2011;10(4):475–86.

[131] McKay JA, Mathers JC. Diet induced epigenetic changes and their implications for health. Acta Physiol 2011;202:103–18.

[132] Simmons R. Epigenetics and maternal nutrition: nature v. nurture. Proc Nutr Soc 2011;70:73–81.

[133] Fuso A, Scarpa S. One-carbon metabolism and Alzheimer's disease: is it all a methylation matter? Neurobiol Aging 2011;32:1192–5.

[134] Thompson RF, Einstein FH. Epigenetic basis for fetal origins of age-related disease. J Womens Health (Larchmt) 2010;19:581–7.

[135] Fulop T, Larbi A, Witkowski JM, McElhaney J, Loeb M, Mitnitski A, et al. Aging, frailty and age-related diseases. Biogerontology 2010;11:547–63.

[136] Lang PO, Michel JP, Zekry D. Frailty syndrome: a transitional state in a dynamic process. Gerontology 2009;55:539–49.

[137] Topinková E. Aging, disability and frailty. Ann Nutr Metab 2008;52:6–11.

[138] Bellizzi D, D'Aquila P, Montesanto A, Corsonello A, Mari V, Mazzei B, et al. Global DNA methylation in old subjects is correlated with frailty. Age 2012;34:169–79.

[139] Phan HM, Alpert JS, Fain M. Frailty, inflammation, and cardiovascular disease: evidence of a connection. Am J Geriatr Cardiol 2008;17:101–7.

[140] Cevenini E, Caruso C, Candore G, Capri M, Nuzzo D, Duro G, et al. Age-related inflammation: the contribution of different organs, tissues and systems. How to face it for therapeutic approaches. Curr Pharm Des 2010;16:609–18.

[141] Kaiser M, Bandinelli S, Lunenfeld B. Frailty and the role of nutrition in older people. A review of the current literature. Acta Biomed 2010;81:37–45.

[142] Choi SW, Friso S. Epigenetics: A New Bridge between Nutrition and Health. Adv Nutr 2010;1:8–16.

[143] Ozanne SE, Sandovici I, Constância M. Maternal diet, aging and diabetes meet at a chromatin loop. Aging 2011;3:548−54.

[144] Gonzalo S. Epigenetic alterations in aging. J Appl Physiol 2010;109:586−97.

[145] Kaszubowska L. Telomere shortening and ageing of the immune system. J Physiol Pharmacol 2008;59:169−86.

[146] Donate LE, Blasco MA. Telomeres in cancer and ageing. Philos Trans R Soc Lond B Biol Sci 2011;366:76−84.

[147] Gonzalez-Suarez I, Gonzalo S. Crosstalk between chromatin structure, nuclear compartmentalization, and telomere biology. Cytogenet Genome Res 2008;122:202−10.

[148] Artandi SE, De Pinho RA. Telomeres and telomerase in cancer. Carcinogenesis 2010;31:9−18.

[149] Zhu J, Zhao Y, Wang S. Chromatin and epigenetic regulation of the telomerase reverse transcriptase gene. Protein Cell 2010;1:22−32.

[150] Esteller M. Epigenetic gene silencing in cancer: the DNA hypermethylome. Hum Mol Genet 2007;16:R50−9.

[151] Chiang PK. Biological effects of inhibitors of S-adenosylhomocysteine hydrolase. Pharmacol Ther 1998;77:115−34.

[152] Chiang PK, Gordon RK, Tal J, Zeng GC, Doctor BP, Pardhasaradhi K, et al. S-Adenosylmethionine and methylation. FASEB J 1996;10:471−80.

[153] Medina MA, Urdiales JL, Amores-Sanchez MI. Roles of homocysteine in cell metabolism: old and new functions. Eur J Biochem 2001;268:3871−82.

[154] Finkelstein JD. Homocysteine: a History in progress. Nutr Rev 2000;58:193−204.

[155] Williams KT, Schalinke KL. New insight into the regulation of methyl group and homocysteine metabolism. J Nutr 2007;137:311−4.

[156] Vitvitsky V, Thomas M, Ghorpade A, Gendelman HE, Banerjee R. A functional transsulfuration pathway in the brain links to glutathione homeostasis. J Biol Chem 2006;281:35785−93.

[157] Liu H, Wang H, Shenvi S, Hagen TM, Liu RM. Glutathione metabolism during aging and in Alzheimer disease. Ann N Y Acad Sci 2004;1019:346−9.

[158] Caudill MA, Wang JC, Melnyk S, Pogribny IP, Jernigan S, Collins MD, et al. Intracellular S-adenosylhomo-cysteine concentrations predict global DNA hypomethylation in tissues of methyl-deficient cystathionine beta-synthase heterozygous mice. J Nutr 2001;131:2811−8.

[159] Ho PI, Collins SC, Dhitavat S, Ortiz D, Ashline D, Rogers E, et al. Homocysteine potentiates beta-amyloid neurotoxicity: role of oxidative stress. J Neurochem 2001;78:249−53.

[160] Ulrey CL, Liu L, Andrews LG, Tollefsbol TO. The impact of metabolism on DNA methylation. Hum. Mol. Genet 2005;14:R139−47.

[161] Seshadri S, Beiser A, Selhub J, Jacques PF, Rosenberg IH, D'Agostino RB, et al. Plasma homocysteine as a risk factor for dementia and Alzheimer's disease. N Engl J Med 2002;346:476−83.

[162] Quadri P, Fragiacomo C, Pezzati R, Zanda E, Forloni G, Tettamanti M, et al. Homocysteine, folate, and vitamin B-12 in mild cognitive impairment, Alzheimer disease, and vascular dementia. Am J Clin Nutr 2004;80:114−22.

[163] De Strooper B. Alzheimer's disease. Closing in on gamma-secretase. Nature 2000;405:627−9.

[164] Vassar R, Bennett BD, Babu-Khan S, Kahn S, Mendiaz EA, Denis P, et al. Beta-secretase cleavage of Alzheimer's amyloid precursor protein by the transmembrane aspartic protease BACE. Science 1999;286:735−41.

[165] Fontecave M, Atta M, Mulliez E. S-adenosylmethionine: nothing goes to waste. Trends Biochem Sci 2004;29:243−9.

[166] Cacabelos R. Influence of pharmacogenetic factors on Alzheimer's disease therapeutics. Neurodegener Dis 2008;5:176−8.

[167] Liddell MB, Lovestone S, Owen MJ. Genetic risk of Alzheimer's disease: advising relatives. Br J Psychiatry 2001;178:7−11.

[168] Kaminsky Z, Wang SC, Petronis A. Complex disease, gender and epigenetics. Ann Med 2006;38:530−44.

[169] Dolinoy DC, Das R, Weidman JR, Jirtle RL. Metastable epialleles, imprinting, and the fetal origins of adult diseases. Pediatr Res 2007;61:30R−7R.

[170] Liu L, van Groen T, Kadish I, Tollefsbol TO. DNA methylation impacts on learning and memory in aging. Neurobiol Aging 2009;30:549−60.

[171] Bjornsson HT, Sigurdsson MI, Fallin MD, Irizarry RA, Aspelund T, Cui H, et al. Intra-individual change over time in DNA methylation with familial clustering. J Am Med Ass 2008;299:2877−83.

[172] Siegmund KD, Connor CM, Campan M, Long TI, Weisenberger DJ, Biniszkiewicz D, et al. DNA methylation in the human cerebral cortex is dynamically regulated throughout the life span and involves differentiated neurons. PLoS ONE 2007;2 e895:

[173] Wang SC, Oelze B, Schumacher A. Age-specific epigenetic drift in late-onset Alzheimer's disease. PLoS ONE 2008;3 e2698:

543

[174] Gillette Guyonnet S, Abellan Van Kan G, Andrieu S, Barberger Gateau P, Berr C, Bonnefoy M, et al. IANA task force on nutrition and cognitive decline with aging. J Nutr Health Aging 2007;11:132−52.

[175] Haan MN, Miller JW, Aiello AE, Whitmer RA, Jagust WJ, Mungas DM, et al. Homocysteine, B vitamins, and the incidence of dementia and cognitive impairment: results from the Sacramento Area Latino Study on Aging. Am J Clin Nutr 2007;85:511−7.

[176] Luchsinger JA, Tang MX, Miller J, Green R, Mehta PD, Mayeux R. Relation of plasma homocysteine to plasma amyloid beta levels. Neurochem Res 2007;32:775−81.

[177] Lee ME, Wang H. Homocysteine and hypomethylation. A novel link to vascular disease. Trends Cardiovasc. Med 1999;9:49−54.

[178] Miller AL. The methionine-homocysteine cycle and its effects on cognitive diseases. Altern Med Rev 2003;8:7−19.

[179] Kennedy BP, Bottiglieri T, Arning E, Ziegler MG, Hansen LA, Masliah EJ. Elevated S-adenosylhomocysteine in Alzheimer brain: influence on methyltransferases and cognitive function. Neural Transm 2004;111:547−67.

[180] Grant WB, Campbell A, Itzhaki RF, Savory J. The significance of environmental factors in the etiology of Alzheimer's disease. J Alz Dis 2002;4:179−89.

[181] Dosunmu R, Wu J, Basha MR, Zawia NH. Environmental and dietary risk factors in Alzheimer's disease. Expert Rev Neurother 2007;7:887−900.

[182] Liu L, Li Y, Tollefsbol TO. Gene-environment interactions and epigenetic basis of human diseases. Curr Issues Mol Biol 2008;10:25−36.

[183] Sontag E, Nunbhakdi-Craig V, Sontag JM, Diaz-Arrastia R, Ogris E, Dayal S, et al. Protein phosphatase 2A methyltransferase links homocysteine metabolism with tau and amyloid precursor protein regulation. J Neurosci 2007;27:2751−9.

[184] Rogaev EI, Lukiw WJ, Lavrushina O, Rogaeva EA, St George-Hyslop PH. The upstream promoter of the beta-amyloid precursor protein gene (APP) shows differential patterns of methylation in human brain. Genomics 1994;22:340−7.

[185] West RL, Lee JM, Maroun LE. Hypomethylation of the amyloid precursor protein gene in the brain of an Alzheimer's disease patient. J Mol Neurosci 1995;6:141−6.

[186] Fleming JL, Phiel CJ, Toland AE. The Role for Oxidative Stress in Aberrant DNA Methylation in Alzheimer's Disease. Curr Alzheimer Res 2011. [In Press].

[187] Fuso A, Seminara L, Cavallaro RA, D'Anselmi F, Scarpa. S. S-adenosylmethionine/homocysteine cycle alterations modify DNA methylation status with consequent deregulation of PS1 and BACE and beta-amyloid production. Mol Cell Neurosci 2005;28:195−204.

[188] Fuso A, Cavallaro RA, Zampelli A, D'Anselmi F, Piscopo P, Confaloni A, et al. Gamma-Secretase is differentially modulated by alterations of homocysteine cycle in neuroblastoma and glioblastoma cells. J Alzheimer's Dis 2007;11:275−90.

[189] Lin HC, Hsieh HM, Chen YH, Hu ML. S-Adenosylhomocysteine increases beta-amyloid formation in BV-2 microglial cells by increased expressions of beta-amyloid precursor protein and presenilin 1 and by hypomethylation of these gene promoters. Neurotoxicology 2009;30:622−7.

[190] Lee J, Ryu H. Epigenetic modification is linked to Alzheimer's disease: is it a maker or a marker? BMB Rep 2010;43:649−55.

[191] Chen KL, Wang SS, Yang YY, Yuan RY, Chen RM, Hu CJ. The epigenetic effects of amyloid-beta(1-40) on global DNA and neprilysin genes in murine cerebral endothelial cells. Biochem Biophys Res Commun 2009;378:57−61.

CHAPTER 27

Early-Life Epigenetic Programming of Human Disease and Aging

Alexander M. Vaiserman
Institute of Gerontology, Kiev, Ukraine

27.1 INTRODUCTION

Aging is a complex process resulting from the progressive reduction of an individual's ability to maintain homeostasis and involving numerous factors centered on transcriptional changes with advanced age [1]. The genetic component of aging received initially all of the attention. However, epigenetic mechanisms including DNA methylation, histone modifications, and non-coding RNAs have now emerged as key contributors to the alterations of genome structure and function that accompany aging [2]. Aging is associated with increased stochastic deregulation of gene expression caused by errors in maintaining the established epigenetic patterns. Such stochastic changes in the epigenome were called "epimutations" by Robin Holliday [3].

T. Tollefsbol (Ed): Epigenetics in Human Disease. DOI: 10.1016/B978-0-12-388415-2.00027-5

Epimutations have been found to be crucially important as causal factors in the age-related increase in incidence of cancer [4], but can also play a pivotal role in driving other aging-associated diseases. It has been suggested that accumulation of epimutations over a lifetime is a major contributor to age-related decline of gene function [5]. However, the available data indicate that most of the detectable epigenetic modifications are systematic, and DNA responds to the environmental stimuli by modifying its epigenetic status in an adaptive manner, in order to maintain a proper functionality [6,7]. Gravina and Vijg [7] suggested that aging in part is driven by an epigenetic-mediated loss of phenotypic plasticity. Particularly, age-associated DNA hypomethylation can initiate chromosome instability while DNA hypermethylation in promoter regions suppresses expression of normal genes (e.g. tumor-suppressor genes).

Thus, theoretically, age-related hypo- or hypermethylation can impair or enhance normal gene responsiveness to environmental signals, in turn contributing to generalized functional decline and failure of homeostasis [7]. Such changes may be a result of an "epigenetic drift" caused by insufficient maintenance of epigenetic marks, but can also be induced by environmental factors. Fraga et al. [8] found that patterns of DNA methylation and histone acetylation diverge with age in all sets of monozygotic twins studied, but those twins who had different lifestyles, and had spent less of their lives together, were found to be the most epigenetically dissimilar. These results suggest that epigenetic divergence that occurs over a lifetime is not solely due to intrinsic epigenetic drift, but can be at least partially linked to environmental factors. Thus, age-associated changes in the epigenome could be seen as a process of laying down memories of the environments encountered throughout life. There are, however, specific stages of development during which the epigenetic landscape is more labile than it is during adulthood, therefore the environment with the potential for the most profound influence on epigenetic states is that encountered in prenatal life [9,10].

Recently, several new hypotheses have been proposed postulating the important role of early-life events in determining late-life diseases including cardiovascular disease, type 2 diabetes, osteoporosis, depression, cognitive impairments, and cancer [11–16]. According to the "developmental programming" concept proposed by Alan Lucas 20 years ago, events during critical or sensitive periods of development may program long-term or life-time structure or function of the organism [17]. The developmental origins of adult health and disease (DOHaD) hypothesis states that adverse influences early in development, and particularly during intrauterine life, can program the risks for adverse health outcomes in adult life [18,19]. This hypothesis has since been confirmed in a number of animal and human studies [20–23].

Increasing evidence has been accumulated indicating the important role of epigenetic regulation in developmental programming. The genome undergoes major epigenetic alterations during early development, when genome-wide changes in epigenetic marks orchestrate chromatin in a way destined to form different organs and tissues in the body. Once established, the epigenetic marks are stably maintained through somatic cell divisions and create unique, lineage-specific patterns of gene expression. In mammalian development, there are two main periods of epigenetic modification: gametogenesis and early embryogenesis [24]. During germ cell development, genome-wide DNA demethylation occurs followed by remethylation before fertilization. Early embryogenesis is then characterized by a second genome-wide demethylation wave, and patterns of methylation are re-established after implantation. The postfertilization demethylation and remethylation phases are likely to play a role in the removal of acquired epigenetic modifications, which can be influenced by individual genetic and environmental factors [25]. The epigenome is therefore likely to be particularly vulnerable to the adverse influences during gametogenesis and early embryogenesis [24]. Nutritional and endocrine factors have been repeatedly shown to be able to reprogram the epigenotype of the embryo [26,27]. In human beings, the window of epigenetic developmental plasticity extends from preconception to early childhood and involves epigenetic responses to environmental changes, which exert their effects during life-history phase transitions [28].

Presently, alteration in epigenetic pathways is considered to be a key mechanism connecting the early-life events with age-associated diseases including cancer, neurodegenerative diseases, and type 2 diabetes [29–32]. The accelerated adult atherogenesis associated with maternal hyperlipidemia is another example of the long-term epigenetic programming [33,34]. Dysregulation in epigenetic pathways can contribute to aging in general as well [2,28,35]. The role of early-life epigenetic events in developmental programming of adult disease and aging has been repeatedly reported in animal models. However, human data are comparably scarce. The purpose of this chapter is to provide a summary of theoretical models and recent research findings which indicate that early-life conditions can program human adult health and aging via epigenetic mechanisms.

27.2 INTRAUTERINE GROWTH RESTRICTION

It is well-established that intrauterine growth restriction (IUGR) induced by dietary manipulation, hypoxia, prenatal glucocorticoid administration, placental insufficiency, maternal stress, smoking, alcohol consumption, etc., can lead to postnatal abnormalities in cardiovascular, metabolic, and endocrine function in rats, guinea pigs, sheep, pigs, horses, and primates [27,36]. In humans, IUGR resulting in low birth weight and subsequent rapid catch-up growth has been repeatedly implicated as an important risk factor for the development of a variety of disorders including postnatal hypertension, glucose intolerance, obesity, type 2 diabetes, and cardiovascular disease, manifesting as late as decades later [27,37]. Multiple cohort studies have provided evidence that being born small for gestational age (SGA) increases the risk for adult diseases through various pathways of metabolic dysregulation. SGA infants can exhibit rapid postnatal weight gain (catch-up growth), altered body composition, increased visceral adiposity and low adiponectin levels which predispose to metabolic syndrome (a prediabetic condition associated with abdominal obesity, arterial hypertension, and insulin resistance), as well as to cardiovascular disease and type 2 diabetes mellitus in adulthood [38]. The precise mechanisms of IUGR-related metabolic programming are poorly understood. Recently, epigenetics has been proposed as an important mechanism for IUGR-associated programmed changes through environmentally induced changes in gene expression [39–42].

The DOHaD hypothesis has been investigated in a wide range of epidemiological and animal studies; these investigations highlight adaptations made by the nutritionally manipulated fetus that aim to maintain energy homeostasis to ensure survival. One consequence of such developmental adaptation may be a long-term resetting of cellular energy homeostasis via epigenetic modification of genes involved in a number of key regulatory pathways. For example, reduced maternal-fetal nutrition during early and mid gestation affects adipose tissue development and adiposity of the fetus by setting an increased number of adipocyte precursor cells [43]. The changes in the development of key endocrine axes including the reprogramming of the hypothalamic-pituitary-adrenal (HPA) axis and insulin-signaling pathways could be involved [44].

In adults born with IUGR, abnormalities in the circulating concentrations of insulin, insulin-like growth factors (IGFs), catecholamines, cortisol, and growth hormone (GH) were repeatedly observed [44,45]. Most of the present data regarding IUGR-related epigenetic modifications have been generated in animal experiments [40,41], while only few human studies have been carried out to date. Recently, Lee et al. [46] identified differentially expressed genes related to the glycolytic pathway between uncomplicated pregnancies and pregnancies with IUGR. They also compared the concentrations of insulin and IGFs in cord blood between the two groups. Microarray experiments identified increased expression of glycolytic enzyme-related genes, including lactate dehydrogenase C, dihydrolipoamide S-acetyltransferase, 6 phosphofructo-2-kinase/fructose-2, 6-biphosphatase 2, oxoglutarate dehydrogenase, phosphorylase, and *IGF-2* and decreased expression of *IGF-1* in placentas from pregnancies with

IUGR. There were significantly lower concentrations of glucose, insulin, IGF-1, and IGF-2 in the fetal cord blood of pregnancies with IUGR. Microarray analysis revealed increased expression of enzyme genes related to the tricarboxylic acid cycle pathway in placentas from pregnancies with IUGR. In the McCarthy et al. [47] study, microarray technology was used to identify genes, which may impair placentation resulting in IUGR. The RNA was isolated from both IUGR term placentas and normal term placentas. Microarray experiments identified increased expression of certain genes including leptin, soluble vascular endothelial growth factor receptor, human chorionic gonadotropin, follistatin-like 3, and hypoxia-inducible factor 2alpha in the IUGR. These results were confirmed by real-time PCR. The up-regulation of soluble vascular endothelial growth factor receptor and hypoxia-inducible factor 2alpha at this period in pregnancy indicate that placental angiogenesis is altered in IUGR and that hypoxia is a major contributor to maldevelopment of the placental vasculature. IUGR is known to be associated with increased expression of genes regulating both cell proliferation and the intrinsic pathway of apoptosis. Specifically, IUGR is associated with altered cell turnover in the villous trophoblast, an essential functional cell type of the human placenta. The intrinsic pathway of apoptosis, particularly p53, is important in regulating placental cell turnover in response to damage. In the recent Heazell et al. [48] study, *p53* mRNA and protein expression were increased in IUGR, which localized to the syncytiotrophoblast.

Currently, a genome-wide epigenetic profiling has become feasible, and a recent study by Einstein et al. [49] employing such a strategy found that the umbilical cord blood cells of infants with IUGR demonstrate subtle but widespread DNA methylation changes across the genome. Einstein et al. [49] hypothesized that IUGR-related programming is mediated by permanent epigenetic alterations in stem cell populations, and focused their study specifically on DNA methylation in CD34$^+$ hematopoietic stem and progenitor cells from cord blood from neonates with IUGR and control subjects. They found that changes in cytosine methylation occur in response to IUGR of moderate degree and involving a restricted number of loci. They also identify specific loci that are targeted for dysregulation of DNA methylation, in particular the hepatocyte nuclear factor 4alpha (*HNF4A*) gene, a well-known diabetes candidate gene, and other loci encoding *HNF4A*-interacting proteins. This genome-wide study suggests that many genes are epigenetically susceptible to alterations in maternal nutrition, and that comprehensive effects on the epigenome can be induced by mild as well as severe intrauterine insults. It gives the possibility that the epigenetic alterations underlying developmental programming are not restricted to a few specific genes. It is also possible that small but widespread epigenetic alterations induced by a poor intrauterine environment can persisted over a lifetime and hence can lead to the acceleration of an age-associated epigenetic decline [10]. A schematic diagram of hypothetical mechanisms potentially underlying the link between IUGR and its long-term adverse health outcomes is presented in Figure 27.1

27.3 FETAL MACROSOMIA

While the DOHaD hypothesis has focused on the adverse effects of maternal undernutrition, exposure to overnutrition in fetal life also results in a series of central and peripheral neuro-endocrine responses that in turn program development of the fat cell and central appetite regulatory system [19]. Epidemiologic studies have found that higher maternal gestational weight gain is associated with fetal macrosomia (arbitrarily defined as a birth weight of more than 4000 g) and consequent risk for obesity and its cardiometabolic complications among offspring. There is also some evidence that epigenetic changes might occur in response to maternal overnutrition [50,51]. Altered epigenetic regulation can be induced by both maternal under- and overnutrition within genes that control lipid and carbohydrate metabolism and within genes involved in the central appetite—energy balance neural network [51].

In the context of the "small baby syndrome hypothesis," it has usually been suggested that the relationship between birth weight and type 2 diabetes is inversely linear, implying that high

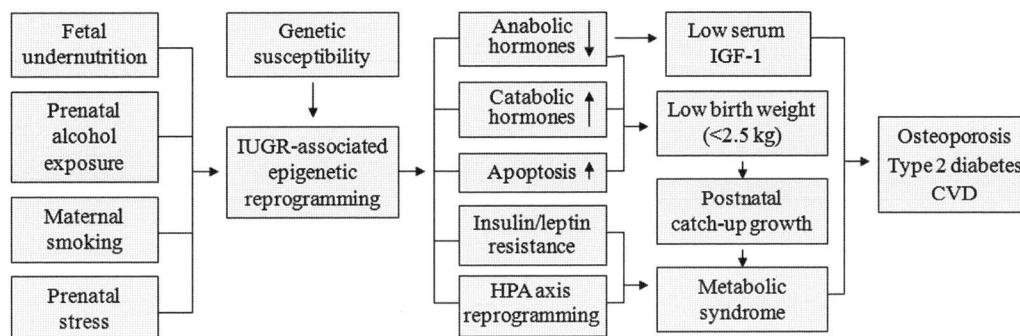

FIGURE 27.1

Hypothetical model of causes and consequences of IUGR. Exposure to adverse environmental/lifestyle factors (maternal stress, undernutrition, smoking, alcohol consumption, etc.) during fetal life can result in IUGR and cause epigenetic reprogramming of gene expression. IUGR conditions are associated with decreased anabolic hormone levels, increased catabolic hormone concentrations, higher apoptosis rate, and reprogramming of the HPA axis and GH—IGF axis in the fetus. IUGR leads to low birth weight (<2.5 kg) and subsequent rapid catch-up growth, as well as to reduced serum IGF-1 concentrations and metabolic syndrome in adulthood. IUGR is associated with subsequent risk of cardiovascular disease (CVD), type 2 diabetes, and osteoporosis.

birth weight leads to decreased risk [52]. Applying different meta-analytic techniques, however, Harder et al. [53] found that the relation between birth weight and type 2 diabetes is not linearly inverse but U-shaped, and high birth weight was found to be associated with increased risk of type 2 diabetes in later life to the same extent as low birth weight. Perinatally acquired microstructural and epigenomic alterations in regulatory systems of metabolism and body weight seem to be critical, leading to a cardiometabolic risk disposition throughout life [54]. People with high birth weight also were shown to have higher death rates from both prostate cancer and breast cancer in adulthood [55–57]. According to the Trichopoulos [58] hypothesis about the fetal origins of cancer, various perinatal factors including birth weight, birth order, maternal age, gestational age, twin status, and parental smoking can affect breast cancer risk in daughters by altering the hormonal environment of the developing mammary glands. Intrauterine exposure to the high levels of growth hormones was initially proposed as an underlying mechanism, increasing both cell proliferation and birth weight and predisposing to cancer in later life [57]. Both human and animal evidence suggest that exposure to obese intrauterine environment can epigenetically program the offspring obesity risk by influencing appetite, metabolism, and activity levels [59,60]. In the study by Gemma et al. [50], a positive correlation between maternal body mass index and the extent to which the peroxisome proliferator-activated receptor g coactivator 1-a (*PPARGC1A*) gene is methylated in umbilical cord blood of offspring was reported. Given that lipids act as both transcriptional activators and signaling molecules, excess fetal lipid exposure may regulate genes involved in lipid sensing and metabolism through epigenetic mechanisms [61]. A hypothetical model of relationships between fetal macrosomia and its adverse health outcomes is shown in Figure 27.2.

27.4 ENDOCRINE PROGRAMMING DURING INTRAUTERINE DEVELOPMENT

Hormones play a key role in regulating intrauterine development, and their concentrations and bioactivity change in response to the environmental challenges known to cause intrauterine programming. Undernutrition, hypoxemia, and other fetal stresses can alter both maternal and fetal concentrations of many hormones including glucocorticoids, catecholamines, insulin, GH, IGFs, leptin, thyroid hormones, and placental hormones such as the eicosanoids, sex steroids, and placental lactogen [27,28,36]. In general, IUGR-associated conditions lower anabolic hormone levels and increase catabolic hormone concentrations in the fetus [27,36]. These endocrine changes then affect fetal growth either directly or indirectly

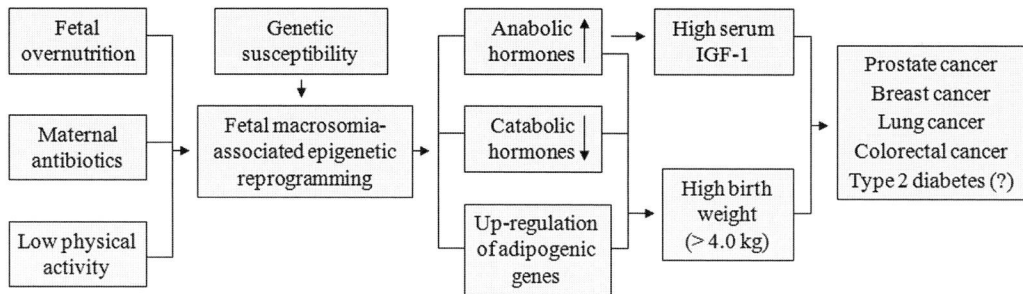

FIGURE 27.2

Hypothetical model of causes and consequences of fetal macrosomia. Prenatal overnutrition and maternal antibiotics, as well as low physical activity during pregnancy can result in fetal macrosomia (larger than normal fetal size and weight). Fetal macrosomia-related epigenetic reprogramming causes increased anabolic hormone levels, decreased catabolic hormone concentrations, and up-regulation of adipogenic genes. It results in high birth weight (>4.0 kg) of the baby and high serum IGF-1 levels in adulthood. Fetal macrosomia is associated with long-term health problems including some cancers and probably type 2 diabetes.

by altering the delivery, uptake, and metabolic fate of nutrients in the fetoplacental tissues. In utero manipulation of glucocorticoid, androgen, and thyroid hormone levels alters fetal development and has long-term consequences for cardiovascular, reproductive, and metabolic function [27].

27.5 INTRAUTERINE GROWTH RESTRICTION AND REPROGRAMMING OF THE HYPOTHALAMIC—PITUITARY—ADRENAL AXIS

Several experimental and human studies suggest that IUGR permanently resets the HPA axis that plays an essential role in the body's response to stressful events. Programming of the HPA axis involves epigenetic remodeling of chromatin, leading to alterations in the expression of genes in many organs and tissues involved in HPA activation and response, including the hippocampus and peripheral tissues [62]. HPA axis reprogramming may involve persistently altered expression of the hippocampal glucocorticoid receptor (GR), an important regulator of HPA axis reactivity, and postnatal changes in hippocampal GR variant and total mRNA expression may underlie IUGR-associated HPA axis reprogramming [63]. Crucial to proper infant growth and development is the placenta, and alterations to placental gene function may reflect differences in the intrauterine environment which functionally contribute to infant growth and may affect the consequent health outcomes. To examine whether epigenetic alterations of the GR gene are linked to infant growth, Filiberto et al. [64] analyzed 480 human placentas, and examined how differential methylation of the GR gene exon 1F is associated with fetal growth. A significant association between differential methylation of the GR gene and large-for-gestational-age (LGA) status was revealed. This work is one of the first to link infant growth as a measure of the intrauterine environment and epigenetic alterations of the GR gene and suggests that DNA methylation may be a critical determinant of placental function.

27.6 EARLY-LIFE PROGRAMMING OF THE GROWTH HORMONE/ INSULIN-LIKE GROWTH FACTORS AXIS

The GH/IGF axis plays a fundamental role in somatic growth and cellular differentiation, as well as in metabolism and survival. The processes linking nutrition, metabolism, and growth are thought to involve a complex interrelationship among insulin, GH, IGFs, and insulin-like growth factor binding proteins (IGFBPs). Fetal insulin and IGFs are thought to have a central role in the regulation of fetal growth, and the IGFs and IGFBPs are shown to be nutritionally regulated in the fetus [65]. Fetal growth retardation leads to long-term abnormalities in the GH/IGF axis [27,36,65], e.g., subjects born SGA had significantly lower mean serum IGF-I,

IGF-binding protein-3 (IGFBP-3) and IGF-I/IGFBP-3 ratio in adulthood than those born appropriate for gestational age [66].

Intrauterine programming of the GH—IGF axis has been proposed as a potential candidate mechanism to explain the link between low birth weight and adult disease [13,65,67]. Accumulating data from experimental and molecular epidemiological studies indicate that growth factors may be important in the pathophysiological processes underlying adult-onset chronic disease, including type 2 diabetes, coronary heart disease, and cancer, and IGF-1 signaling is one of the key pathways implicated in aging and longevity [68]. Abnormalities in the insulin signaling pathway generate age-related diseases and increased mortality, whereas the GH/IGF-1 axis could potentially modulate human longevity [69].

A number of experimental and observational studies suggest that higher levels of circulating IGF-1 may increase risk of several cancers during adulthood including breast, colorectal, lung, and prostate cancers [70—73], and reduce risk of type 2 diabetes and coronary heart disease [67,74]. Overexpression of IGF-1 occurs in tumors diagnosed in childhood (osteosarcoma, Wilms tumor, neuroblastoma, etc.) and in adults (breast, ovaries, colon, and prostate cancer), and congenital IGF-1 deficiency acts as a protecting factor for the development of cancer [75]. Low-birth-weight infants, on the contrary, have low levels of IGF-1 [76] and high risk of insulin resistance and type 2 diabetes [77], ischemic heart disease [78], cognitive decline [79], and osteoporosis [80] in adulthood. Ben-Shlomo et al. [81] suggest that low birth weight followed by accelerated catch-up growth during infancy and childhood is associated with lower life course IGF-1 levels and this may act as one of the biological pathways linking early growth with adult insulin resistance, hypertension, and cardiovascular disease. Similarly, this pattern should be associated with decreased cancer risk and is consistent with the observational data on breast-feeding, which has been shown to be protective for IGF-related premenopausal breast cancers [73].

27.7 EARLY INTERVENTIONS TO PREVENT AND TREAT ENDOCRINE—METABOLIC DISTURBANCES

Several nutritional and hormonal interventions have been proposed to prevent and treat the endocrine—metabolic disturbances. Milk intake in childhood and in adulthood is positively associated with higher levels of circulating IGF-1, and higher circulating IGF-1 promotes linear growth. A recent study conducted by Martin et al. [82] indicated that milk consumption in childhood appears to have long-term, programming effects. Specifically, some studies suggest that the long-term effect of higher levels of milk intake in early childhood is opposite to the expected short-term effect, because milk intake in early-life is inversely associated with IGF-1 levels throughout adult life. The authors hypothesized that this long-term programming effect is via a resetting of pituitary control in response to raised levels of IGF-1 in childhood. Such a programming effect of milk intake in early life could potentially have implications for cancer and ischemic heart disease risk many years later. Early-life diet has been found to affect the risk of childhood leukemia. In the study by Tower and Spector [83], maternal dietary DNA topoisomerase II (DNAt2) inhibitor intake was associated with infant acute myeloid leukemia with the *MLL* gene translocation. Increased intake of fruits and vegetables has been associated with decreased leukemia risk, and lack of maternal folate supplementation has been associated with increased childhood leukemia risk, possibly by causing DNA hypomethylation and increased DNA strand breaks. Administration of low-dose GH therapy, at a dose that minimizes the lipolytic effects of GH and has the ability to increase IGF-1 levels, were shown to enhance insulin sensitivity in young healthy adults and in GH-deficient adults and increases insulin secretion in individuals with impaired glucose tolerance [84]. IGF system is emerging as a promising new target in cancer therapy. Different strategies are being pursued to target this pathway. Several monoclonal antibodies and tyrosine kinase inhibitors targeting the IGF-1 receptor are in clinical development [85,86].

27.8 EARLY-LIFE NUTRITIONAL PROGRAMMING OF ADULT HEALTH AND AGING

There is increasing epidemiological evidence linking prenatal and early-life nutritional status to later adiposity and metabolic disease risk [87,88]. Different gestational dietary stressors (undernutrition, overnutrition, or a modified supply of key nutrients) can elicit similar metabolic responses in offspring [9]. High carbohydrate/protein ratio in the maternal diet was shown to be linked to impaired glucose homeostasis and raised blood pressure in offspring [89,90]. In the longitudinal studies conducted in the United Kingdom and Finland, it has been shown that subjects that were underweight at birth had high rates of coronary heart disease, high blood pressure, high cholesterol concentrations, and abnormal glucose–insulin metabolism in adulthood [52,91,92]. The authors suggested that a high carbohydrate intake in early pregnancy suppresses placental growth, especially if combined with a low dairy protein intake in late pregnancy, and such an effect could have long-term consequences for the offspring's risk of cardiovascular disease [91].

While the molecular basis of prenatal nutritional programming is unknown, available animal and human data suggest that epigenetic changes in gene expression play a substantial role in the link between the maternal diet, and altered metabolism and body composition in the adult offspring [93–95]. According to the "epigenetic programming" hypothesis, suboptimal maternal diet induces epimutations in offspring during early embryonic development, and that altered expression of affected genes is maintained into adulthood, eventually affecting health [9]. Epigenetic mechanisms play a key role in mediating between the early-life nutrient inputs and the ensuing phenotypic changes throughout the entire life and seem to be responsible, in part, for the biological changes that occur during aging [96].

In animal models, maternal diet alters offspring body composition, accompanied by epigenetic changes in metabolic control genes. To study whether such processes operate in humans, Godfrey et al. [95] analyzed the associations of methylation status of 68 CpGs 5′ from five candidate genes in umbilical cord tissue DNA from healthy neonates with maternal pregnancy diet and with child's adiposity at age 9 years. Replication was sought in a second independent cohort. In cohort 1, retinoid X receptor-α (RXRA) chr9:136355885+ and endothelial nitric oxide synthase (eNOS) chr7:150315553+ methylation had independent associations with sex-adjusted childhood fat mass and fat mass percentage. Regression analyses including sex and neonatal epigenetic marks explained >25% of the variance in childhood adiposity. Higher methylation of RXRA chr9:136355885+, but not of eNOS chr7:150315553+, was associated with lower maternal carbohydrate intake in early pregnancy, previously linked with higher neonatal adiposity in this population. In cohort 2, cord eNOS chr7:150315553+ methylation showed no association with adiposity, but RXRA chr9:136355885+ methylation showed similar associations with fat mass and fat mass percentage.

27.9 THE THRIFTY PHENOTYPE AND THRIFTY EPIGENOTYPE CONCEPTS

In 1962, James Neel proposed the "thrifty genotype" hypothesis, the idea that the same genes that helped our ancestors survive occasional famines are now being challenged by modern life conditions in which food is plentiful [97]. Thirty years ago, in 1992, Hales and Barker suggested the "thrifty phenotype" hypothesis, the concept that environmental factors acting in early life, in particular undernutrition, might influence later risk of type 2 diabetes [98]. According to this hypothesis, undernutrition during in utero development results in long-term adaptive changes in glucose–insulin metabolism (including reduced capacity for insulin secretion and insulin resistance) that, due to an enhanced ability to store fat, improves survival under postnatal conditions of nutritional deprivation. However, windows of plasticity close early in life, and postnatal environmental exposures may result in the selected trajectory becoming inappropriate, resulting in adverse effects on adult health. If mismatch exists between the environment predicted in utero and the actual environment experienced in subsequent life (e.g. excess food consumption), diabetes and other features of the metabolic syndrome will result. Fetal growth restriction followed by rapid weight gain during early infancy ("catch-up growth") has also been proposed to play an important role in promoting central adiposity and insulin resistance [99].

Epigenetic regulation of gene expression is one mechanism by which genetic susceptibility and environmental insults can lead to type 2 diabetes. Several studies clearly show that environmental effects can induce epigenetic alterations, ultimately affecting expression of key genes linked to the development of type 2 diabetes including genes critical for pancreatic development and beta-cell function, peripheral glucose uptake, and insulin resistance, as well as atherosclerosis [100]. Recently, Reinhard Stöger used elements of the thrifty phenotype and thrifty genotype concepts to synthesize a "thrifty epigenotype" hypothesis [101]. According to this hypothesis, it has been postulated that: (1) metabolic thrift, the capacity for efficient acquisition, storage, and use of energy, is an ancient, complex trait; (2) the environmentally responsive gene network encoding this trait is subject to genetic canalization and thereby has become robust against mutational perturbations; (3) DNA sequence polymorphisms play a minor role in the etiology of obesity and type 2 diabetes — instead, disease susceptibility is predominantly determined by epigenetic variations; (4) corresponding epigenotypes have the potential to be inherited across generations; and (5) leptin is a candidate gene for the acquisition of a thrifty epigenotype. According to Stöger's visual metaphor, the efficiency of anabolic metabolism ("metabolic thrift") is built upon large gene networks that form a rigid canal. In the absence of nutritional extremes (either over- or undernutrition), metabolism develops into the healthy norm. However, under conditions of intrauterine malnutrition, compensatory epigenetic changes can be induced in adipogenic and energy metabolism gene networks, and this can change the shape of the canal in such a way that metabolic phenotype is optimized for survival in these conditions. If the "thrifty epigenotype" hypothesis is correct, then the "thrifty epigenotype" is anticipated to be present at significantly higher frequencies in human populations experiencing recurrent food shortages [101]. Individuals exposed to these conditions will have a characteristic epigenetic profile, which could differ markedly from those for residents of developed countries.

Leptin is thought to be one of the best thrifty gene candidates since it encodes a hormone regulating appetite and energy homeostasis [101]. Leptin is secreted by adipocytes, and serum leptin level is thought to signal nutritional status to the hypothalamus and thus help govern appetite and energy expenditure. Leptin has been shown to be implicated in nutritional programming during fetal and neonatal growth with long-term effects on susceptibility to obesity, diabetes, and coronary heart disease [102]. The failure of elevated leptin levels to suppress feeding and mediate weight loss in common forms of obesity defines a state of so-called leptin resistance. The mechanisms underlying leptin resistance remain a matter of debate, but there is increasing evidence that it may be programmed during the fetal and neonatal life [103]. The promoter region of the leptin gene is methylated in somatic tissues of human and mouse and displays epigenetic variation [104], and it is the gene for which proximal promoter demethylation has been shown to induce its transcription in mature adipocytes [105]. Recently, new evidence of the key role of leptin in epigenetic programming of human metabolic disorders was obtained. Bouchard et al. [106] examined the possibility of leptin gene epigenetic adaptation to impaired glucose metabolism during pregnancy by studying whether the DNA methylation profile of leptin gene is altered in the offspring of mothers with gestational impaired glucose tolerance. They have shown that placental leptin gene DNA methylation levels were correlated with glucose levels in women with impaired glucose tolerance, and with decreased leptin gene expression in the whole cohort studied. The authors hypothesized that increasing maternal glycemia results in fetal leptin gene DNA demethylation, which leads to higher mRNA levels and subsequently higher leptin levels, possibly promoting leptin resistance and obesity development.

553

27.10 PRENATAL FAMINE AND ADULT HEALTH OUTCOMES

The opportunities to directly examine the relationship between early-life nutrition and late-life disease are limited because few cohort studies have information from birth until old age. The strong evidence linking early-life conditions with adult disease risk has been accumulated from natural experiments, i.e. naturally occurring circumstances in which subsets of the population have different levels of exposure to a supposed causal factor [23]. The famine has multiple beneficial features for use as a natural experiment. A relationship between the prenatal exposure to the famine and various adverse metabolic and mental phenotypes later in life including a higher body mass index, elevated plasma lipids, increased risks of obesity, cardiovascular disease, and psychiatric disorders was found in a large number of epidemiological studies. The bulk of these data were obtained in observational cohort studies of the long-term health consequences of the prenatal exposure to the Dutch famine of 1944–45 and to the Chinese famine of 1959–61 [107–109]. These associations were dependent on the timing of the exposure during gestation and lactation periods. Early gestation seems to be an especially vulnerable period [110]. The mechanisms contributing to associations between the prenatal exposure to famine and adult health outcomes are still unknown but may involve the persistent epigenetic alterations [108]. In the Heijmans et al. [110] study, the level of methylation of the IGF-2 gene, a key factor in human growth and development, was estimated to examine whether the prenatal exposure to famine can lead to persistent changes in the epigenome. It has been found that individuals who were exposed to famine in early gestation had a much lower level of methylation of IGF-2 than controls 60 years after the exposure [110]. More recently, this observation was extended by the study a set of 15 additional candidate loci implicated in growth, metabolic, and cardiovascular disorders [111]. Methylation of six of these loci has been shown to be associated with in utero exposure to famine. *IGF-2* was hypomethylated in individuals whose mothers were exposed to famine periconceptually, whereas interleukin-10, guanine nucleotide-binding protein, leptin, ATP-binding cassette A1, and maternally expressed 3(meg3) were hypermethylated. Taken together, these data indicate that differences in DNA methylation induced by exposure to prenatal famine may persist during the human life course. Moreover, Tobi et al. [112] investigated whether prenatal growth restriction early in pregnancy can be associated with changes in DNA methylation at loci that were previously shown by them to be sensitive to early gestational famine exposure. They compared 38 individuals born preterm (<32 weeks) and with a birth weight too small for gestational age (SGA, a proxy for intrauterine growth restriction), with individuals with normal postnatal growth. The DNA methylation levels of IGF-2, GNASAS, INSIGF, and LEP were 48.5%, 47.5%, 79.4%, and 25.7% respectively. Nevertheless, this was not significantly different between SGA and individuals with normal postnatal growth. Risk factors for being born SGA, including pre-eclampsia and maternal smoking, were also not associated with DNA methylation at these loci. Thus, growth restriction early in development was not associated with DNA methylation at loci shown earlier to be affected by prenatal famine exposure. Exposure to energy restriction during childhood and adolescence was also found to be associated with a lower risk of developing colorectal cancer. Within the Netherlands Cohort Study on diet and cancer, Hughes et al. [113] found that individuals exposed to severe famine during the Hunger Winter had a decreased risk of developing a tumor characterized by the (promoter) CpG island methylation phenotype compared to those not exposed.

27.11 EFFECT OF PRENATAL EXPOSURE TO METHYL DONORS ON DEVELOPMENTAL PROGRAMMING

It has been shown that organ-specific DNA methylation patterns established during the fetal period can be modified during the lifetime by the environment. Phenotypic changes through DNA methylation can be related to folate metabolism. One-carbon metabolism, using folate as a coenzyme, is important for methyl group transfer for DNA methylation. Whereas the aging process is accompanied by alterations in DNA methylation, the diminished activity of DNA methyltransferases (DNMTs) can be a potential mechanism for the decreased genomic DNA methylation during aging, along with reduced folate intake and altered folate metabolism [114].

Current research suggests that compounds that serve as metabolic methyl group donors and cofactors (choline, betaine, methionine, folic acid, and vitamin B_{12}) play a key role among the nutrients that alter the epigenetic status during the fetal development [115–118]. For example, Yajnik et al. [119] studied the association between maternal vitamin B_{12}, folate, and total homocysteine (tHcy) status during pregnancy, and offspring adiposity and insulin resistance at 6 years in the Pune (India). They measured maternal nutritional intake and circulating concentrations of folate, vitamin B_{12}, tHcy, and methylmalonic acid at 18 and 28 weeks of gestation. These parameters were correlated with offspring anthropometry, body composition, and insulin resistance at 6 years. Although short and thin, the 6-year-old children were relatively adipose compared with the UK standards. Higher maternal erythrocyte folate concentrations at 28 weeks predicted higher offspring adiposity. Low maternal vitamin B12 predicted higher risk of insulin resistance in children. The offspring of mothers with a combination of high folate and low vitamin B12 concentrations were the most insulin resistant. Consistent data were obtained in the Krishnaveni et al. [120] study, where low plasma vitamin B_{12} concentrations combined with high folate concentrations in pregnancy were associated with a higher incidence of gestational diabetes and later diabetes in Mysore (India).

The early-life dietary manipulation of methyl group donors (either deficiency or supplementation) can have a profound impact on the gene expression profile and, consequently, on the homeostatic mechanisms that ensure the normal course of physiological processes [117]. Recently, Steegers-Theunissen et al. [121] examined whether periconceptional maternal folic acid use and markers of global DNA methylation potential (S-adenosylmethionine and S-adenosylhomocysteine blood levels) in mothers and children affect methylation of the *IGF-2* gene differentially methylated region (DMR) in the child. Children of mothers who used folic acid had a 4.5% higher methylation of the *IGF-2* DMR than children who were not exposed to folic acid. An inverse association between *IGF-2* DMR methylation and birth weight was observed. These results indicate plasticity of *IGF-2* methylation by periconceptional folic acid use. The authors concluded that periconceptional folic acid administration is associated with epigenetic changes in *IGF-2* in the child that may affect intrauterine programming of growth and development with consequences for adult health and disease. Hoyo et al. [122] evaluated exposure to maternal folic acid supplementation before and during pregnancy in relation to aberrant DNA methylation at two DMRs regulating *IGF-2* expression in infants. Aberrant methylation at these regions has been associated with *IGF-2* deregulation and increased susceptibility to several chronic diseases. Compared to infants born to women reporting no folic acid intake before or during pregnancy, methylation levels at the *H19* DMR decreased with increasing folic acid intake. Ba et al. [123] investigated the relationship between folate, vitamin B_{12}, and methylation of the *IGF-2* gene in maternal and cord blood. The methylation patterns of *IGF-2* in promoter 2 (*P2*) and promoter 3 (*P3*) in cord blood were not associated with serum folate levels in either cord or maternal blood, whereas the *P3* methylation patterns were associated with serum levels of vitamin B_{12} in mother's blood. Methylation patterns in *P2* of maternal blood were associated with serum levels of vitamin B_{12} in mother's blood, exposure to passive smoking, and mother's weight gain during pregnancy.

Choline intake exceeding current dietary recommendations has also been shown to be able to preserve markers of cellular methylation and attenuate DNA damage in a genetic subgroup of folate-compromised men [124], indicating the epigenetic contribution to the link between choline and folate pathways. Although the molecular mechanisms by which methyl donors alter the epigenetic profile are still unclear, it was commonly considered that the primary mechanisms could be S-adenosylmethionine (SAM) availability. However, recent animal studies have demonstrated that SAM concentrations in rat liver are very stable, indicating that gene-specific epigenetic mechanisms are involved rather than simply SAM availability for methylation reactions [125,126].

27.12 LONG-TERM PROGRAMMING EFFECTS OF PRENATAL STRESS

A large number of animal and epidemiological studies have revealed that prenatal stress as well as excess exogenous glucocorticoids or inhibition of 11β-hydroxysteroid dehydrogenase type 2 (the placental barrier to maternal glucocorticoids) are linked with adverse health outcomes including low birth weight, risk of cardio-metabolic disorders, neuroendocrine dysfunction, and psychiatric diseases in later life [127–130], and with an increased risk of infectious diseases [131] in the offspring. Molecular mechanisms underlying the programming effects of early life stress and glucocorticoids have been suggested to include epigenetic changes in target gene promoters, notably affecting tissue-specific expression of the intracellular GR [132]. The long-term effects of maternal behavior on the stress responsiveness and behavior of the offspring during adulthood are well documented in animal models, and these experimental findings have been extended to humans by identifying an association between early-life adversity and epigenetic marks in adult life [133,134]. In humans, childhood abuse was found to alter HPA stress responses and to increase the risk of suicide. To test the hypothesis that epigenetic differences in critical loci in the brain are involved in the pathophysiology of suicide, McGowan et al. [133] investigated the extent of DNA methylation in the promoter of genes that encoded rRNA genes in the brains of suicide victims. Suicide subjects were selected for a history of early childhood neglect/abuse, which is associated with decreased hippocampal volume and cognitive impairments. rRNA was significantly hypermethylated throughout the promoter and 5′ regulatory region in the brain of suicide subjects, consistent with reduced rRNA expression in the hippocampus. More recently, McGowan et al. [134] examined epigenetic differences in a neuron-specific glucocorticoid receptor (*NR3C1*) promoter between postmortem hippocampus obtained from suicide victims, who were or were not abused as children. In this study, expression of total glucocorticoid receptor mRNA was significantly reduced in suicide victims with a history of childhood abuse relative to non-abused suicide victims or controls; there was no difference between non-abused suicide victims and controls. There was also a significant effect on the expression of transcripts containing the exon 1_F *NR3C1* promoter. The glucocorticoid receptor 1_F expression was significantly lower in samples from suicide victims with a history of childhood abuse compared with suicide victims without childhood abuse or controls. Similar to the findings with total glucocorticoid receptor mRNA expression, there was no difference between non-abused suicide victims and controls subjects. These findings suggest a common effect of parental care on the epigenetic regulation of hippocampal GR expression.

It is known that cesarean section can cause more severe stress in newborn infants compared with that of those born by vaginal delivery, who adapt to the new conditions better. Cesarean section is associated with significant risks for a baby's short- and long-term physical and emotional health compared to vaginal birth, including diabetes, leukemia, respiratory problems, and asthma later in life [135]. To study whether the mode of delivery affects epigenetic activity in newborn infants, Schlinzig et al. [135] analyzed the level of DNA methylation in leukocytes of 21 newborn infants delivered by cesarean section and 16 infants born by vaginal delivery. In this study, infants born by cesarean section exhibited significantly higher DNA methylation level in leukocytes compared with that of those born by vaginal delivery.

The secretion of glucocorticoids is a classic endocrine response to stress. The exposure to excess glucocorticoids in early life can permanently alter tissue glucocorticoid signaling, and these effects may have short-term adaptive benefits but increase the risk of later disease [129]. Currently, multiple courses of synthetic glucocorticoids are recommended for various conditions. Prenatal exposure to glucocorticoids is very efficient in reducing the incidence of respiratory distress syndrome in preterm babies. However, despite the beneficial therapeutic effect of antenatally administered glucocorticoids, their prenatal administration can result in transgenerational effects with respect to the risk of developing several metabolic and cardio-vascular disorders in later life which implies that these epigenetic effects can persist across generations [132,136].

27.13 LONG-TERM IMPACTS OF MATERNAL SUBSTANCE USE DURING PREGNANCY

It is well known that heavy prenatal alcohol exposure can severely affect human physical and neurobehavioral development. Epidemiological data offer some evidence that paternal alcohol consumption can affect birth weight, congenital heart defects, and mild cognitive impairments [137–139]. The detrimental effects following in utero alcohol exposure (fetal alcohol spectrum disorders, FASD) have been recognized for centuries, but only recently new data were obtained providing a molecular insight into the mechanisms involved in FASD. A substantial amount of data have been accumulated to support the role of environmentally induced epigenetic remodeling during gametogenesis and after conception as a key mechanism for the deleterious effects of prenatal alcohol exposure that persist into adulthood [139]. For example, the Ouko et al. [140] study has demonstrated a link between chronic alcohol use in men and hypomethylation of paternally imprinted loci in sperm DNA in genomic regions critical for embryonic development (*H19* and *IG-DMR*) thus providing a mechanism for paternal effects in the etiology of FASDs. There was a pattern of increased demethylation with alcohol consumption at the two imprinted loci with a significant difference observed at the *IG-DMR* between the non-drinking and heavy-alcohol-consuming groups. Greater interindividual variation in average methylation was observed at the *H19* DMR and individual clones were more extensively demethylated than those of the *IG-DMR*. CpG site #4 in the *IG-DMR* was preferentially demethylated among all individuals and, along with the *H19* DMR CpG site #7 located within the *CTCF* binding site 6, showed significant demethylation in the alcohol-consuming groups compared with the control group. The authors hypothesized that, should these epigenetic changes in imprinted genes be transmitted through fertilization, they would alter the critical gene expression dosages required for normal prenatal development resulting in offspring with features of FASD.

Three developmental periods are particularly vulnerable: preconception, preimplantation, and gastrulation. These periods of teratogenesis correlate with peak periods of epigenetic reprogramming which, together with the evidence that ethanol interferes with one-carbon metabolism, DNA methylation, histone modifications, and non-coding RNA, suggests an important role for epigenetic mechanisms in the etiology of FASDs [141]. Moreover, alcohol is known to lead to a variety of nutritional disturbances including nutrient intake, absorption, utilization, and excretion. Whereas some nutrients can essentially affect gene expression, the alcohol-induced nutrient disbalance may be a major contributor to impaired gene expression in FASD. Several metabolites, namely, acetate, S-adenosylmethionine, nicotinamide adenine dinucleotide, and zinc are supposed to be especially relevant to alcohol metabolism and ALD [142]. A wide range of fetal abnormalities and birth defects have been repeatedly reported in animals and humans after preconceptional alcohol exposure. The mechanisms, particularly in males, are likely to involve alcohol-induced epigenetic changes in the gametes or, alternatively, selection effects within the germline, resulting in the ontogenesis of "FASD-like" phenotypes in unexposed generations [137].

Maternal smoking can also result in intrauterine growth restriction. Children born to mothers who smoke are at an increased risk of obesity, hypertension, and diabetes [143,144]. Maternal smoking may be involved in fetal programming [145], and in utero tobacco exposure was shown to be associated with epigenetic changes in the offspring [146]. To demonstrate that differences in DNA methylation patterns occur in children exposed to prenatal tobacco smoking and that variation in detoxification genes may alter these associations Breton et al. [147] measured methylation of DNA repetitive elements *LINE1* and *AluYb8* in buccal cells of 348 children. DNA methylation patterns were associated with in utero exposure to maternal smoking. Exposed children had significantly lower methylation of *AluYb8*. Differences in smoking-related effects on *LINE1* methylation were observed in children with the common *GSTM1*-null genotype. Differential methylation of CpG loci in eight genes was identified through the screen. Two genes, *AXL* and *PTPRO*, showed significant increases in methylation and in exposed children. The authors concluded that life-long effects of in utero exposures may be mediated through alterations in DNA methylation, and variants in detoxification genes may modulate the effects of in utero exposure through epigenetic mechanisms. In a more recent study, Breton et al. [148] have shown that in utero tobacco smoke exposure negatively impacts respiratory health in later life and some of the detrimental effects of this exposure are likely due to epigenetic modifications. In a genome-wide survey of site-specific CPG methylation in asthmatics, they have demonstrated evidence of widespread epigenetic consequences of in utero smoke exposure, with differential methylation of more than 25% of all genes tested, including numerous cancer- and asthma-related genes.

It is known that prenatal exposure to maternal cigarette smoking (PEMCS) is associated with variations in brain and behavior in adolescence. Toledo-Rodriguez et al. [149] have found that epigenetic mechanisms may mediate some of the consequences of PEMCS through methylation of DNA in genes important for brain development, such as the brain-derived neurotrophic factor (*BDNF*). They used bisulfite sequencing to assess DNA methylation of the *BDNF* promoter in the blood of adolescents whose mothers smoked during pregnancy. PEMCS was associated with higher rates of DNA methylation in the *BDNF-6* exon. These results suggest that PEMCS may lead to long-term down-regulation of *BDNF* expression via the increase in DNA methylation in its promoter region. Such mechanisms could, in turn, lead to modifications in both development and plasticity of the brain exposed in utero to maternal cigarette smoking.

27.14 PROGRAMMING EFFECT OF EARLY-LIFE EXPOSURE TO ENVIRONMENTAL TOXICANTS

Prenatal and early postnatal exposure to environmental toxicantswas repeatedly found to be associated with aberrant DNA methylation of regulatory sequences in susceptible genes, leading to inappropriate gene expression and disease pathogenesis in later life [150,151]. The endocrine-disrupting chemicals (EDCs) are of specific concern among the detrimental environmental factors because they are widespread in the environment. Human epidemiological studies have shown that women exposed to EDCs such as agricultural pesticides have prolonged/irregular estrous cycles and difficulty in achieving pregnancy, failed assisted reproductive technology attempts, and loss of pregnancies [152,153]. An exposure to EDCs during early development could alter the epigenetic programming of the genome and thereby result in adult-onset disease [154].

Perinatal diethylstilbesterol (DES) exposure, which is associated with several reproductive tract abnormalities and increased vaginal and cervical cancer risk in women, provides a clear example of how estrogenic xenobiotic exposure during a critical period of development can abnormally demethylate DNA sequences during organ development and possibly increase cancer risk later in life [150,155]. Importantly, these effects can be epigenetically transmitted to the next generation [155,156]. Increasing epidemiologic evidence links specific pesticides, polychlorinated biphenyls (PCBs), and inorganic arsenic exposures to elevated prostate cancer risk in adulthood [157]. Importantly, the prostate seems to be particularly sensitive to these endocrine disruptors during the critical developmental windows including in utero and neonatal time points as well as during puberty.

In utero and early postnatal exposure to bisphenol A (BPA), an estrogenic environmental toxin widely used in the production of plastics, was shown to be able to produce a broad range of adverse health effects, including impaired brain development, sexual differentiation, behavior, and immune function, which could extend to future generations [158]. Molecular mechanisms that underlie the longlasting effects of BPA likely involve disruption of epigenetic programming of gene expression during development. Low-dose exposures to BPA were shown to be able to affect the prostate epigenome during development and, as a consequence, promote prostate disease with aging [151]. The alterations in DNA methylation patterns in multiple cell signaling genes in BPA-exposed prostates were obtained suggesting that exposure to the environmentally relevant doses of BPA can be imprinted in the developing prostate through epigenetic alterations [151,157]. Substantial evidence indicates that exposure to BPA during early development may also increase breast cancer risk later in life. Recently, Weng et al. [159] examined epigenetic changes in breast epithelial cells treated with low-dose BPA. They identified 170 genes with similar expression changes in response to BPA. Functional analysis confirms that gene suppression was mediated in part through an estrogen receptor α (ERα)-dependent pathway. As a result of exposure to BPA or other estrogen-like chemicals, the expression of lysosomal-associated membrane protein 3 (*LAMP3*) became epigenetically silenced in breast epithelial cells. Furthermore, increased DNA methylation in the promoter CpG island regions of LAMP3 loci was observed in ERα-positive breast tumors.

There is also convincing evidence that prenatal environmental exposures can influence the risk for subsequent asthma. Martino and Prescott [160] examined the epigenetic regulation of immune development and the early immune profiles that contribute to allergic risk. They reviewed new evidence that key environmental exposures, such as microbial exposure, dietary changes, tobacco smoke, and pollutants, can induce epigenetic changes in gene expression and alter disease risk. In particular, transplacental exposure to high levels of airborne traffic-related polycyclic aromatic hydrocarbons (PAHs) can cause aberrant DNA methylation changes leading to dysregulation of gene expression and childhood asthma [161]. In the Perera et al. study [162] it has been reported that methylation of ACSL3, a gene expressed in lung and thymus tissue, may be a possible biomarker linking prenatal exposure to PAHs to childhood asthma. In this study, methylation of the *ACSL3* 5′-CGI was found to be significantly associated with maternal airborne PAH exposure and with a parental report of asthma symptoms in children prior to age 5.

27.15 EPIGENETIC RISKS OF ASSISTED REPRODUCTIVE TECHNOLOGIES

Assisted reproductive technologies (ARTs) are methods of treating infertility in which pregnancy is achieved by artificial or partially artificial means. They generally include a stage of embryo culture that precisely coincides with zygotic epigenetic resetting. In vitro studies on the human embryos have suggested that imprinted genes in humans may be susceptible to ART conditions. Genomic imprinting is an epigenetic phenomenon by which certain genes are expressed in a parent-of-origin-dependent manner, i.e. primarily or exclusively from either the maternal or paternal allele [163]. Imprinted expression is a clear example of epigenetic inheritance, because genetically identical sequences are differentially transcribed depending on the sex of the parent from which the gene originates [164]. Most imprinted genes contain differentially methylated regions, where the methylation state of the parental alleles differs [165]. This variation allows for differential regulation of these alleles dependent on parental origin of the allele and leads to preferential expression of a specific allele, depending on its parental origin [25]. The underlying mechanisms by which culture media induce abnormal epigenetic modifications are still not clear but it has been suggested that embryonic developmental timing can be disturbed by the synthetic media and that this interferes with epigenetic reprogramming and gene expression [166].

Several reports have raised concerns that children conceived by ART are at increased risk of having imprinting disorders. Among them, Beckwith–Wiedemann syndrome and Angelman syndrome are the most extensively studied [167]. It was found the association of in vitro fertilization with Beckwith–Wiedemann syndrome, and with epigenetic alterations of *LIT1* and *H19* imprinted genes [168] and abnormal imprinting of the *KCNQ1OT* gene [169,170]. Intracytoplasmic sperm injection was shown to increase the risk of Angelman syndrome and some imprinting defects [171,172]. Although these studies examined only few cases, almost every case showed loss of methylation at imprinting control regions rather than the genetic defects generally responsible for these syndromes. Later, the conclusion that infertility and ovulation induction are risk factors for ART-related Angelman syndrome was supported by the findings from the surveys conducted in Great Britain [173] and Holland [174]. More recent studies on global methylation changes in ART-conceived patients with other imprinting disorders (including Prader–Willi syndrome, Silver–Russell syndrome, transient neonatal diabetes mellitus, and maternal hypomethylation syndrome) reveal conflicting results. Tierling et al. [175] reported no association with ART and imprinting in a study of ten loci known to be imprinted, whereas Katari et al. [176] observed altered expression in several genes implicated in metabolic disorders, such as obesity and diabetes. Evidence from clinical reports suggests that the association between imprinting syndromes and ART may be restricted to syndromes where the imprinting change takes the form of hypomethylation on the maternal allele [177]. Further studies are required to establish the level of risk of imprinting disorders in patients conceived by ART.

27.16 CONCLUSIONS AND FUTURE DIRECTIONS

The relationship between adverse events during prenatal and/or early postnatal life and development of chronic diseases later in life has been reported in a number of recent experimental and epidemiological studies. In these studies, it has been highlighted as the key role of epigenetic mechanisms in mediating the link between nutritional, hormonal, and metabolic environment early in life and lifelong health outcomes. Over recent years, there have been conducted numerous animal studies and limited human studies aimed at understanding the specific epigenetic mechanisms underlying developmental programming of later life pathology and aging.

Epigenetics has substantial potential for developing biological markers to predict which exposures would put exposed subjects at risk and which individuals will be more susceptible to develop disease. In human studies, this will require the use of highly sensitive laboratory methods, so that epigenetic alterations can be detected well ahead of disease diagnosis [178]. Given the reversibility of epigenetic modifications, the understanding of epigenetic mechanisms may represent a promising novel therapeutic target for prevention or reversion of human age-related disorders and healthy life extension. These therapeutic strategies may include changes in nutrition and lifestyle as well as pharmacological treatments. Several epigenetic drugs, such as DNA demethylating agents and histone deacetylase inhibitors have already been successfully tested in clinical trials [179]. However, all these drug candidates are very unspecific and, therefore, can cause large-scale epigenetic deregulation. In the future, it will be essential to develop therapies that target only specific elements of the epigenome. Such preventive approaches initiated in pre- and early postnatal periods of human development seem to be particularly promising. If one could modify the incorrect or deleterious epigenetic patterns through specific nutritional or pharmacological interventions during early ontogenesis, then it would be possible to correct the disrupted gene expression programs to treat age-related diseases and to achieve better health and longevity.

References

[1] Brink TC, Regenbrecht C, Demetrius L, Lehrach H, Adjaye J. Activation of the immune response is a key feature of aging in mice. Biogerontology 2009;10:721—34.

[2] Gonzalo S. Epigenetic alterations in aging. J Appl Physiol 2010;109:586—97.

[3] Holliday R. The inheritance of epigenetic defects. Science 1987;238:163—70.

[4] Feinberg AP, Ohlsson R, Henikoff S. The epigenetic progenitor origin of human cancer. Nat Rev Genet 2006;7:21—33.

[5] Calvanese V, Lara E, Kahn A, Fraga MF. The role of epigenetics in ageing and in age-related diseases. Ageing Res Rev 2009;8:268—76.

[6] Tremblay J, Hamet P. Impact of genetic and epigenetic factors from early life to later disease. Metab 2008;57:S27—31.

[7] Gravina S, Vijg J. Epigenetic factors in aging and longevity. Pflugers Arch 2010;459:247—58.

[8] Fraga MF, Ballestar E, Paz MF, Ropero S, Setien F, Ballestar ML, et al. Epigenetic differences arise during the lifetime of monozygotic twins. Proc Natl Acad Sci USA 2005;102:10604—9.

[9] Li CC, Maloney CA, Cropley JE, Suter CM. Epigenetic programming by maternal nutrition: shaping future generations. Epigenomics 2010;2:539—49.

[10] Thompson RF, Einstein FH. Epigenetic basis for fetal origins of age-related disease. J Women's Health 2010;19:581—7.

[11] Zwaan BJ. Linking development and aging. Sci Aging Knowledge Environ 2003;47:32.

[12] Gavrilov LA, Gavrilova NS. Early-life programming of aging and longevity: the idea of high initial damage load (the HIDL hypothesis). Ann N Y Acad Sci 2004;1019:496—501.

[13] Kajantie E. Early-life events. Effects on aging. Hormones 2008;7:101—13.

[14] Chen JH, Cottrell EC, Ozanne SE. Early growth and ageing. Nestle Nutr Workshop Ser Pediatr Program 2010;65:41—50.

[15] Fernandez-Capetillo O. Intrauterine programming of ageing. EMBO Rep 2010;11:32—6.

[16] Mentis AF, Kararizou E. Does ageing originate in utero? Biogerontology 2010;11:725—9.

[17] Lucas A. Programming by early nutrition in man. Ciba Found Symp 1991;156:38—50.

[18] Barker DJ. The developmental origins of adult disease. J Am Coll Nutr 2004;23:588S—95S.

[19] McMillen IC, MacLaughlin SM, Muhlhausler BS, Gentili S, Duffield JL, Morrison JL. Developmental origins of adult health and disease: the role of periconceptional and foetal nutrition. Basic Clin Pharmacol Toxicol 2008;102:82—9.

[20] Bateson P, Barker D, Clutton-Brock T, Deb D, D'Udine B, Foley RA, et al. Developmental plasticity and human health. Nature 2004;430:419—21.

[21] Gluckman PD, Hanson MA, Mitchell MD. Developmental origins of health and disease: reducing the burden of chronic disease in the next generation. Genome Med 2010;2:14.

[22] Hanson M, Godfrey KM, Lillycrop KA, Burdge GC, Gluckman PD. Developmental plasticity and developmental origins of non-communicable disease: Theoretical considerations and epigenetic mechanisms. Prog Biophys Mol Biol 2011;106:272—80.

[23] Vaiserman A. Early-life origin of adult disease: evidence from natural experiments. Exp Gerontol 2011;46:189—92.

[24] Vickaryous N, Whitelaw E. The role of the early embryonic environment on epigenotype and phenotype. Reprod Fertil Dev 2005;17:335—40.

[25] Reik W, Dean W, Walter J. Epigenetic reprogramming in mammalian development. Science 2001;293:1089—93.

[26] Waterland RA, Garza C. Potential mechanisms of metabolic imprinting that lead to chronic disease. Am J Clin Nutr 1999;69:179—97.

[27] Fowden AL, Giussani DA, Forhead AJ. Intrauterine programming of physiological systems: causes and consequences. Physiology (Bethesda) 2006;21:29—37.

[28] Hochberg Z, Feil R, Constancia M, Fraga M, Junien C, Carel JC, et al. Child health, developmental plasticity, and epigenetic programming. Endocr Rev 2011;32:159—224.

[29] Wolfson M, Tacutu R, Budovsky A, Aizenberg N, Fraifeld VE. MicroRNAs: relevance to aging and age-associated diseases. Open Longevity Science 2008;2:66—75.

[30] Szyf M. The early life environment and the epigenome. Biochim Biophys Acta 2009;1790:878—85.

[31] Gluckman PD, Hanson MA, Buklijas T, Low FM, Beedle AS. Epigenetic mechanisms that underpin metabolic and cardiovascular diseases. Nat Rev Endocrinol 2009;5:401—8.

[32] Gluckman PD, Hanson MA, Low FM. The role of developmental plasticity and epigenetics in human health. Birth Defects Res C Embryo Today 2011;93:12–8.

[33] Mukherjee S. Adult onset atherosclerosis may have its origin in the foetal state in utero. Curr Opin Lipidol 2009;20:155–6.

[34] Lund G, Zaina S. Atherosclerosis: an epigenetic balancing act that goes wrong. Curr Atheroscler Rep 2011;13:208–14.

[35] Rodríguez-Rodero S, Fernández-Morera JL, Fernandez AF, Menéndez-Torre E, Fraga MF. Epigenetic regulation of aging. Discov Med 2010;10:225–33.

[36] Fowden AL, Giussani DA, Forhead AJ. Endocrine and metabolic programming during intrauterine development. Early Hum Dev 2005;81:723–34.

[37] Cota BM, Allen PJ. The developmental origins of health and disease hypothesis. Pediatr Nurs 2010;36:157–67.

[38] Varvarigou AA. Intrauterine growth restriction as a potential risk factor for disease onset in adulthood. J Pediatr Endocrinol Metab 2010;23:215–24.

[39] Cutfield WS, Hofman PL, Mitchell M, Morison IM. Could epigenetics play a role in the developmental origins of health and disease? Pediatr Res 2007;61:68R–75R.

[40] Hall JG. Review and hypothesis: syndromes with severe intrauterine growth restriction and very short stature — are they related to the epigenetic mechanism(s) of fetal survival involved in the developmental origins of adult health and disease? Am J Med Genet A 2010;152A:512–27.

[41] Wadhwa PD, Buss C, Entringer S, Swanson JM. Developmental origins of health and disease: brief history of the approach and current focus on epigenetic mechanisms. Semin Reprod Med 2009;27:358–68.

[42] Joss-Moore LA, Metcalfe DB, Albertine KH, McKnight RA, Lane RH. Epigenetics and fetal adaptation to perinatal events: Diversity through fidelity. J Anim Sci 2010;88:E216–222.

[43] Symonds ME, Sebert SP, Hyatt MA, Budge H. Nutritional programming of the metabolic syndrome. Nat Rev Endocrinol 2009;5:604–10.

[44] Fowden AL, Forhead AJ. Endocrine mechanisms of intrauterine programming. Reproduction 2004;127:515–26.

[45] Phillips DI. Fetal growth and programming of the hypothalamic–pituitary–adrenal axis. Clin Exp Pharmacol Physiol 2001;28:967–70.

[46] Lee MH, Jeon YJ, Lee SM, Park MH, Jung SC, Kim YJ. Placental gene expression is related to glucose metabolism and fetal cord blood levels of insulin and insulin-like growth factors in intrauterine growth restriction. Early Hum Dev 2010;86:45–50.

[47] McCarthy C, Cotter FE, McElwaine S, Twomey A, Mooney EE, Ryan F, et al. Altered gene expression patterns in intrauterine growth restriction: potential role of hypoxia. Am J Obstet Gynecol 2007;196:70:e1–6.

[48] Heazell AE, Sharp AN, Baker PN, Crocker IP. Intra-uterine growth restriction is associated with increased apoptosis and altered expression of proteins in the p53 pathway in villous trophoblast. Apoptosis 2011;16:135–44.

[49] Einstein F, Thompson RF, Bhagat TD, Fazzari MJ, Verma A, Barzilai N, et al. Cytosine methylation dysregulation in neonates following intrauterine growth restriction. PLoS One 2010;5:e8887.

[50] Gemma C, Sookoian S, Alvariñas J, García SI, Quintana L, Kanevsky D, et al. Maternal pregestational BMI is associated with methylation of the PPARGC1A promoter in newborns. Obesity 2009;17:1032–9.

[51] Lillycrop KA, Burdge GC. Epigenetic changes in early life and future risk of obesity. Int J Obes (Lond) 2011;35:72–83.

[52] Godfrey KM, Barker DJ. Fetal nutrition and adult disease. Am J Clin Nutr 2000;71:1344S–52S.

[53] Harder T, Rodekamp E, Schellong K, Dudenhausen JW, Plagemann A. Birth weight and subsequent risk of type 2 diabetes: a meta-analysis. Am J Epidemiol 2007;165:849–57.

[54] Neitzke U, Harder T, Plagemann A. Intrauterine growth restriction and developmental programming of the metabolic syndrome: a critical appraisal. Microcirculation 2011;18:304–11.

[55] Ekbom A, Hsieh CC, Lipworth L, Wolk A, Pontén J, Adami HO, et al. Perinatal characteristics in relation to incidence of and mortality from prostate cancers. BMJ 1996;313:337–41.

[56] Vatten L. Can prenatal factors influence future breast cancer risk? Lancet 1996;348:1531.

[57] Park SK, Kang D, McGlynn KA, Garcia-Closas M, Kim Y, Yoo KY, et al. Intrauterine environments and breast cancer risk: meta-analysis and systematic review. Breast Cancer Res 2008;10. R8.

[58] Trichopoulos D. Hypothesis: Does breast cancer originate in utero? Lancet 1990;335. 9390-9340.

[59] Oken E. Maternal and child obesity: the causal link. Obstet Gynecol Clin North Am 2009;36:361–77.

[60] Ornoy A. Prenatal origin of obesity and their complications: Gestational diabetes, maternal overweight and the paradoxical effects of fetal growth restriction and macrosomia. Reprod Toxicol 2011;32:205–12.

[61] Heerwagen MJ, Miller MR, Barbour LA, Friedman JE. Maternal obesity and fetal metabolic programming: a fertile epigenetic soil. Am J Physiol Regul Integr Comp Physiol 2010;299:R711–722.

[62] Grace CE, Kim SJ, Rogers JM. Maternal influences on epigenetic programming of the developing hypothalamic-pituitary-adrenal axis. Birth Defects Res A Clin Mol Teratol 2011;91:797–805.

[63] Ke X, Schober ME, McKnight RA, O'Grady S, Caprau D, Yu X, et al. Intrauterine growth retardation affects expression and epigenetic characteristics of the rat hippocampal glucocorticoid receptor gene. Physiol Genomics 2010;42:177–89.

[64] Filiberto AC, Maccani MA, Koestler DC, Wilhelm-Benartzi C, Avissar-Whiting M, Banister CE, et al. Birthweight is associated with DNA promoter methylation of the glucocorticoid receptor in human placenta. Epigenetics 2011;6:566–72.

[65] Holt RI. Fetal programming of the growth hormone-insulin-like growth factor axis. Trends Endocrinol Metab 2002;13:392–7.

[66] Verkauskiene R, Jaquet D, Deghmoun S, Chevenne D, Czernichow P, Lévy-Marchal C. Smallness for gestational age is associated with persistent change in insulin-like growth factor I (IGF-I) and the ratio of IGF-I/IGF-binding protein-3 in adulthood. J Clin Endocrinol Metab 2005;10:5672–6.

[67] Sandhu M. Insulin-like growth factor-I and risk of type 2 diabetes and coronary heart disease: molecular epidemiology. Endocr Dev 2005;9:44–54.

[68] Rincon M, Muzumdar R, Atzmon G, Barzilai N. The paradox of the insulin/IGF-1 signaling pathway in longevity. Mech Ageing Dev 2004;125:397–403.

[69] Rincon M, Rudin E, Barzilai N. The insulin/IGF-1 signaling in mammals and its relevance to human longevity. Exp Gerontol 2005;40:873–7.

[70] Holly JMP, Gunnell DJ, Davey, Smith G. Growth hormone, IGF-I and cancer. Less intervention to avoid cancer? More intervention to prevent cancer? J Endocrinol 1999;162:321–30.

[71] Jenkins PJ, Bustin SA. Evidence for a link between IGF-I and cancer. Eur J Endocrinol 2004;151(Suppl 1):S17–22.

[72] Moschos SJ, Mantzoros CS. The role of the IGF system in cancer: from basic to clinical studies and clinical applications. Oncology 2002;63:317–32.

[73] Renehan AG, Zwahlen M, Minder C, O'Dwyer ST, Shalet SM, Egger M. Insulin-like growth factor (IGF)-1, IGF binding protein-3, and cancer risk: systematic review and meta-regression analysis. Lancet 2004;363:1346–53.

[74] Martin RM, Holly JM, Middleton N, Davey, Smith G, Gunnell D. Childhood diet and insulin-like growth factors in adulthood: 65-year follow-up of the Boyd Orr Cohort. Eur J Clin Nutr 2007;61:1281–92.

[75] Shevah O, Laron Z. Patients with congenital deficiency of IGF-I seem protected from the development of malignancies: a preliminary report. Growth Horm IGF Res 2007;17:54–7.

[76] Ong K, Kratzsch J, Kiess W, Costello M, Scott C, Dunger D. Size at birth and cord blood levels of insulin, insulin-like growth factor I (IGF-I), IGF-II, IGF-binding protein-1 (IGFBP-1), IGFBP-3, and the soluble IGF-II/mannose-6-phosphate receptor in term human infants. The ALSPAC Study Team. Avon Longitudinal Study of Pregnancy and Childhood. J Clin Endocrinol Metab 2000;85:4266–9.

[77] Sandhu MS, Heald AH, Gibson JM, Cruickshank JK, Dunger DB, Wareham NJ. Circulating concentrations of insulin-like growth factor-I and development of glucose intolerance: a prospective observational study. Lancet 2002;359:1740–5.

[78] Juul A, Scheike T, Davidsen M, Gyllenborg J, Jørgensen T. Low serum insulin-like growth factor I is associated with increased risk of ischemic heart disease: a population-based case-control study. Circulation 2002;106:939–44.

[79] Kalmijn S, Janssen JA, Pols HA, Lamberts SW, Breteler MM. A prospective study on circulating insulin-like growth factor I (IGF-I), IGF-binding proteins, and cognitive function in the elderly. J Clin Endocrinol Metab 2000;85:4551–5.

[80] Liu JM, Zhao HY, Ning G, Chen Y, Zhang LZ, Sun LH, et al. IGF-1 as an early marker for low bone mass or osteoporosis in premenopausal and postmenopausal women. J Bone Miner Metab 2008;26:159–64.

[81] Ben-Shlomo Y, Holly J, McCarthy A, Savage P, Davies D, Davey Smith G. Prenatal and postnatal milk supplementation and adult insulin-like growth factor I: long-term follow-up of a randomized controlled trial. Cancer Epidemiol Biomarkers Prev 2005;14:1336–9.

[82] Martin RM, Holly JM, Gunnell D. Milk and linear growth: programming of the IGF-I axis and implication for health in adulthood. Nestle Nutr Workshop Ser Pediatr Program 2011;67:79–97.

[83] Tower RL, Spector LG. The epidemiology of childhood leukemia with a focus on birth weight and diet. Crit Rev Clin Lab Sci 2007;44:203–42.

[84] Dunger D, Yuen K, Ong K. Insulin-like growth factor I and impaired glucose tolerance. Horm Res 2004;62:101–7.

[85] López-Calderero I, Sánchez Chávez E, García-Carbonero R. The insulin-like growth factor pathway as a target for cancer therapy. Clin Transl Oncol 2010;12:326–38.

563

[86] Rosenzweig SA, Atreya HS. Defining the pathway to insulin-like growth factor system targeting in cancer. Biochem Pharmacol 2010;80:1115—24.

[87] Gluckman PD, Hanson MA, Cooper C, Thornburg KL. Effect of in utero and early-life conditions on adult health and disease. N Engl J Med 2008;359:61—73.

[88] Hoffman DJ. Early nutrition and adult health: Perspectives for international and community nutrition programs and policies. Nutr Res Pract 2010;4:449—54.

[89] Campbell DM, Hall MH, Barker DJ, Cross J, Shiell AW, Godfrey KM. Diet in pregnancy and the offspring's blood pressure 40 years later. Br J Obstet Gynaecol 1996;103:273—80.

[90] Siemelink M, Verhoef A, Dormans JA, Span PN, Piersma AH. Dietary fatty acid composition during pregnancy and lactation in the rat programs growth and glucose metabolism in the offspring. Diabetologia 2002;45:1397—403.

[91] Godfrey K, Robinson S, Barker DJP, Osmond C, Cox V. Maternal nutrition in early and late pregnancy in relation to placental and fetal growth. BMJ 1996;312:410—4.

[92] Forsen T, Osmond C, Eriksson JG, Barker DJ. Growth of girls who later develop coronary heart disease. Heart 2004;90:20—4.

[93] Burdge GC, Slater-Jefferies J, Torrens C, Phillips ES, Hanson MA, Lillycrop KA. Dietary protein restriction of pregnant rats in the F0 generation induces altered methylation of hepatic gene promoters in the adult male offspring in the F1 and F2 generations. Br J Nutr 2007;97:435—9.

[94] Lillycrop KA, Hanson MA, Burdge GC. Epigenetics and the influence of maternal diet. In: Newnham JP, Ross MG, editors. Early Life Origins of Human Health and Disease. Basel: Karger; 2009. p. 11—20.

[95] Godfrey KM, Sheppard A, Gluckman PD, Lillycrop KA, Burdge GC, McLean C, et al. Epigenetic gene promoter methylation at birth is associated with child's later adiposity. Diabetes 2011;60:1528—34.

[96] Niculescu MD, Lupu DS. Nutritional influence on epigenetics and effects on longevity. Curr Opin Clin Nutr Metab Care 2011;14:35—40.

[97] Neel JV. Diabetes mellitus: a "thrifty" genotype rendered detrimental by "progress"? Am J Hum Genet 1962;14:353—62.

[98] Hales CN, Barker DJ. Type 2 (non-insulin-dependent) diabetes mellitus: the thrifty phenotype hypothesis. Diabetologia 1992;35:595—601.

[99] Dulloo AG. Thrifty energy metabolism in catch-up growth trajectories to insulin and leptin resistance. Best Pract Res Clin Endocrinol Metab 2008;22:155—71.

[100] Pinney SE, Simmons RA. Epigenetic mechanisms in the development of type 2 diabetes. Trends Endocr Metab 2010;21:223—9.

[101] Stöger R. The thrifty epigenotype: an acquired and heritable predisposition for obesity and diabetes? Bioessays 2008;30:156—66.

[102] Houseknecht KL, Spurlock ME. Leptin regulation of lipid homeostasis: dietary and metabolic implications. Nutr Res Rev 2003;16:83—96.

[103] Djiane J, Attig L. Role of leptin during perinatal metabolic programming and obesity. J Physiol Pharmacol 2008;59:55—63.

[104] Stöger R. In vivo methylation patterns of the Leptin promoter in human and mouse. Epigenetics 2006;1:155—62.

[105] Melzner I, Scott V, Dorsch K, Fischer P, Wabitsch M, Brüderlein S, et al. Leptin gene expression in human preadipocytes is switched on by maturation-induced demethylation of distinct CpGs in its proximal promoter. J Biol Chem 2002;277:45420—7.

[106] Bouchard L, Thibault S, Guay SP, Santure M, Monpetit A, St-Pierre J, et al. Leptin gene epigenetic adaptation to impaired glucose metabolism during pregnancy. Diabetes Care 2010;33:2436—41.

[107] Painter RC, Roseboom TJ, Bleker OP. Prenatal exposure to the Dutch famine and disease in later life: an overview. Reprod Toxicol 2005;20:345—52.

[108] Heijmans BT, Tobi EW, Lumey LH, Slagboom PE. The epigenome: archive of the prenatal environment. Epigenetics 2009;4:526—31.

[109] Lumey LH, Stein AD, Susser E. Prenatal famine and adult health. Annu Rev Public Health 2011;32:237—62.

[110] Heijmans BT, Tobi EW, Stein AD, Putter H, Blauw GJ, Susser ES, et al. Persistent epigenetic differences associated with prenatal exposure to famine in humans. Proc Natl Acad Sci USA 2008;105:17046—9.

[111] Tobi EW, Lumey LH, Talens RP, Kremer D, Putter H, Stein AD, et al. DNA methylation differences after exposure to prenatal famine are common and timing- and sex-specific. Hum Mol Genet 2009;18:4046—53.

[112] Tobi EW, Heijmans BT, Kremer D, Putter H, Delemarre-van de Waal HA, Finken MJ, et al. DNA methylation of IGF2, GNASAS, INSIGF and LEP and being born small for gestational age. Epigenetics 2011;6:171—6.

[113] Hughes LA, van den Brandt PA, de Bruïne AP, Wouters KA, Hulsmans S, Spiertz A, et al. Early life exposure to famine and colorectal cancer risk: a role for epigenetic mechanisms. PLoS One 2009;4:e7951.

[114] Kim KC, Friso S, Choi SW. DNA methylation, an epigenetic mechanism connecting folate to healthy embryonic development and aging. J Nutr Biochem 2009;20:917—26.

[115] Zeisel SH. Gene response elements, genetic polymorphisms and epigenetics influence the human dietary requirement for choline. IUBMB Life 2007;59:380—7.

[116] Zeisel SH. Is maternal diet supplementation beneficial? Optimal development of infant depends on mother's diet. Am J Clin Nutr 2009;89:685S—7S.

[117] Zeisel SH. Nutritional genomics: defining the dietary requirement and effects of choline. J Nutr 2011;141:531—4.

[118] Zeisel SH, da Costa KA. Choline: an essential nutrient for public health. Nutr Rev 2009;67:615—23.

[119] Yajnik CS, Deshpande SS, Jackson AA, Refsum H, Rao S, Fisher DJ, et al. Vitamin B$_{12}$ and folate concentrations during pregnancy and insulin resistance in the offspring: the Pune Maternal Nutrition Study. Diabetologia 2008;51:29—38.

[120] Krishnaveni GV, Hill JC, Veena SR, Bhat DS, Wills AK, Karat CL, et al. Low plasma vitamin B12 in pregnancy is associated with gestational 'diabesity' and later diabetes. Diabetologia 2009;52:2350—8.

[121] Steegers-Theunissen RP, Obermann-Borst SA, Kremer D, Lindemans J, Siebel C, Steegers EA, et al. Periconceptional maternal folic acid use of 400 microg per day is related to increased methylation of the IGF2 gene in the very young child. PLoS One 2009;4:e7845.

[122] Hoyo C, Murtha AP, Schildkraut JM, Jirtle RL, Demark-Wahnefried W, Forman MR, et al. Methylation variation at IGF2 differentially methylated regions and maternal folic acid use before and during pregnancy. Epigenetics 2011;6:928—36.

[123] Ba Y, Yu H, Liu F, Geng X, Zhu C, Zhu Q, et al. Relationship of folate, vitamin B12 and methylation of insulin-like growth factor-II in maternal and cord blood. Eur J Clin Nutr 2011;65:480—5.

[124] Shin W, Yan J, Abratte CM, Vermeylen F, Caudill MA. Choline intake exceeding current dietary recommendations preserves markers of cellular methylation in a genetic subgroup of folate-compromised men. J Nutr 2010;140:975—80.

[125] Min H. Effects of dietary supplementation of high-dose folic acid on biomarkers of methylating reaction in vitamin B(12)-deficient rats. Nutr Res Pract 2009;3:122—7.

[126] Engeham SF, Haase A, Langley-Evans SC. Supplementation of a maternal low-protein diet in rat pregnancy with folic acid ameliorates programming effects upon feeding behaviour in the absence of disturbances to the methionine-homocysteine cycle. Br J Nutr 2010;103:996—1007.

[127] Kajantie E. Fetal origins of stress-related adult disease. Ann NY Acad Sci 2006;1083:11—27.

[128] Beydoun H, Saftlas AF. Physical and mental health outcomes of prenatal maternal stress in human and animal studies: a review of recent evidence. Paediatr Perinat Epidemiol 2008;22:438—66.

[129] Cottrell EC, Seckl JR. Prenatal stress, glucocorticoids and the programming of adult disease. Front Behav Neurosci 2009;3:19.

[130] Virk J, Li J, Vestergaard M, Obel C, Lu M, Olsen J. Early life disease programming during the preconception and prenatal period: making the link between stressful life events and type-1 diabetes. PLoS One 2010;5:e11523.

[131] Nielsen NM, Hansen AV, Simonsen J, Hviid A. Prenatal stress and risk of infectious diseases in offspring. Am J Epidemiol 2011;173:990—7.

[132] Harris A, Seckl J. Glucocorticoids, prenatal stress and the programming of disease. Horm Behav 2011;59:279—89.

[133] McGowan PO, Sasaki A, Huang TC, Unterberger A, Suderman M, Ernst C, et al. Promoter-wide hypermethylation of the ribosomal RNA gene promoter in the suicide brain. PLoS One 2008;3:e2085.

[134] McGowan PO, Sasaki A, D'Alessio AC, Dymov S, Labonté B, Szyf M, et al. Epigenetic regulation of the glucocorticoid receptor in human brain associates with childhood abuse. Nat Neurosci 2009;12:342—8.

[135] Schlinzig T, Johansson S, Gunnar A, Ekström TJ, Norman M. Epigenetic modulation at birth — altered DNA-methylation in white blood cells after Caesarean section. Acta Paediatr 2009;98:1096—9.

[136] Seckl JR, Holmes MC. Mechanisms of disease: glucocorticoids, their placental metabolism and fetal 'programming' of adult pathophysiology. Nat Clin Pract Endocrinol Metab 2007;3:479—88.

[137] Abel E. Paternal contribution to fetal alcohol syndrome. Addict Biol 2004;9:127—33.

[138] Riley EP, McGee CL. Fetal alcohol spectrum disorders: an overview with emphasis on changes in brain and behavior. Exp Biol Med (Maywood) 2005;230:357—65.

[139] Ramsay M. Genetic and epigenetic insights into fetal alcohol spectrum disorders. Genome Med 2010;2:27—34.

[140] Ouko LA, Shantikumar K, Knezovich J, Haycock P, Schnugh DJ, Ramsay M. Effect of alcohol consumption on CpG methylation in the differentially methylated regions of H19 and IG-DMR in male gametes: implications for fetal alcohol spectrum disorders. Alcohol Clin Exp Res 2009;33:1615—27.

[141] Haycock PC. Fetal alcohol spectrum disorders: the epigenetic perspective. Biol Reprod 2009;81:607–17.

[142] Moghe A, Joshi-Barve S, Ghare S, Gobejishvili L, Kirpich I, McClain CJ, et al. Histone modifications and alcohol-induced liver disease: Are altered nutrients the missing link? World J Gastroenterol 2011;17:2465–72.

[143] Montgomery SM, Ekbom A. Smoking during pregnancy and diabetesmellitus in a British longitudinal birth cohort. BMJ 2002;324:26–7.

[144] Somm E, Schwitzgebel VM, Vauthay DM, Aubert ML, Hüppi PS. Prenatal nicotine exposure and the programming of metabolic and cardiovascular disorders. Mol Cell Endocrinol 2009;304:69–77.

[145] Rogers JM. Tobacco and pregnancy: overview of exposures and effects. Birth Defects Res C Embryo Today 2008;84:1–15.

[146] Suter M, Abramovici A, Aagaard-Tillery K. Genetic and epigenetic influences associated with intrauterine growth restriction due to in utero tobacco exposure. Pediatr Endocrinol Rev 2010;8:94–102.

[147] Breton CV, Byun HM, Wenten M, Pan F, Yang A, Gilliland FD. Prenatal tobacco smoke exposure affects global and gene-specific DNA methylation. Am J Respir Crit Care Med 2009;180:462–7.

[148] Breton CV, Siegmund KD, Kong H, Qiu W, Islam KTS, Salam MT, et al. The epigenetic impact of in utero smoke exposure in later life: a genome-wide survey of site-specific CPG methylation in asthmatics. Am J Respir Crit Care Med 2011;183:A1005.

[149] Toledo-Rodriguez M, Lotfipour S, Leonard G, Perron M, Richer L, Veillette S, et al. T. Maternal smoking during pregnancy is associated with epigenetic modifications of the brain-derived neurotrophic factor-6 exon in adolescent offspring. Am J Med Genet B Neuropsychiatr Genet 2010;153B:1350–4.

[150] Li S, Hursting SD, Davis BJ, McLachlan JA, Barrett JC. Environmental exposure, DNA methylation, and gene regulation: lessons from diethylstilbesterol-induced cancers. Ann NY Acad Sci 2003;983:161–9.

[151] Ho SM, Tang WY, Belmonte DF, Prins GS. Developmental exposure to estradiol and bisphenol A increases susceptibility to prostate carcinogenesis and epigenetically regulates phosphodiesterase type 4 variant 4. Cancer Res 2006;66:5624–32.

[152] Fuortes L, Clark MK, Kirchner HL, Smith EM. Association between female infertility and agricultural work history. Am J Ind Med 1997;31:445–51.

[153] Younglai EV, Holloway AC, Foster WG. Environmental and occupational factors affecting fertility and IVF success. Hum Reprod Update 2005;11:43–57.

[154] Zama AM, Uzumcu M. Fetal and neonatal exposure to the endocrine disruptor methoxychlor causes epigenetic alterations in adult ovarian genes. Endocrinology 2009;150:4681–91.

[155] Newbold RR. Lessons learned from perinatal exposure to diethylstilbestrol. Toxicol Appl Pharmacol 2004;199:142–50.

[156] Skinner MK. Environmental epigenetic transgenerational inheritance and somatic epigenetic mitotic stability. Epigenetics 2011;6:838–42.

[157] Prins GS. Endocrine disruptors and prostate cancer risk. Endocr Relat Cancer 2008;15:649–56.

[158] Kundakovic M, Champagne FA. Epigenetic perspective on the developmental effects of bisphenol A. Brain Behav Immun 2011;25:1084–93.

[159] Weng YI, Hsu PY, Liyanarachchi S, Liu J, Deatherage DE, Huang YW, et al. Epigenetic influences of low-dose bisphenol A in primary human breast epithelial cells. Toxicol Appl Pharmacol 2010;248:111–21.

[160] Martino D, Prescott S. Epigenetics and prenatal influences on asthma and allergic airways disease. Chest 2011;139:640–7.

[161] Miller RL, Ho SM. Environmental epigenetics and asthma: current concepts and call for studies. Am J Respir Crit Care Med 2008;177:567–73.

[162] Perera F, Tang WY, Herbstman J, Tang D, Levin L, Miller R, et al. Relation of DNA methylation of 5′-CPG island of ACSL3 to transplacental exposure to airborne polycyclic aromatic hydrocarbons and childhood asthma. PLoS One 2009;4:e4488.

[163] Koerner MV, Barlow DP. Genomic imprinting — an epigenetic gene-regulatory model. Curr Opin Genet Dev 2010;20:164–70.

[164] Uribe-Lewis S, Woodfine K, Stojic L, Murrell A. Molecular mechanisms of genomic imprinting and clinical implications for cancer. Expert Rev Mol Med 2011;13:e2.

[165] Neumann B, Kubicka P, Barlow DP. Characteristics of imprinted genes. Nat Genet 1995;9:12–3.

[166] De Rycke M, Liebaers I, Van Steirteghem A. Epigenetic risks related to assisted reproductive technologies. Hum Reprod 2002;17:2487–94.

[167] Odom LN, Segars J. Imprinting disorders and assisted reproductive technology. Curr Opin Endocrinol Diabetes Obes 2010;17:517–22.

[168] De Baun MR, Niemitz EL, Feinberg AP. Association of in vitro fertilisation with Beckwith–Wiedemann syndrome and epigenetic alterations of LIT1 and H19. Am J Hum Genet 2003;72:156–60.

[169] Gicquel C, Gaston V, Mandelbaum J, Siffroi JP, Flahault A, Le Bouc Y. In vitro fertilisation may increase the risk of Beckith–Wiedemann Syndrome related to the abnormal imprinting of the KCNQ1OT gene. Am J Hum Genet 2003;72:1338–41.

[170] Lim D, Bowdin SC, Tee L, Kirby GA, Blair E, Fryer A, et al. Clinical and molecular genetic features of Beckwith–Wiedemann syndrome associated with assisted reproductive technologies. Hum Reprod 2009;24:741–7.

[171] Cox GF, Bürger J, Lip V, Mau UA, Sperling K, Wu BL, et al. Intracytoplasmic sperm injection may increase the risk of imprinting defects. Am J Hum Genet 2002;71:162–4.

[172] Orstavik KH, Eiklid K, van der Hagen CB, Spetalen S, Kierulf K, Skjeldal O, et al. Another case of imprinting defect in a girl with Angelman Syndrome who was conceived by intracytoplasmic sperm injection. Am J Hum Genet 2003;72:218–9.

[173] Sutcliffe AG, Peters CJ, Bowdin S, Temple K, Reardon W, Wilson L, et al. Assisted reproductive therapies and imprinting disorders: a preliminary British survey. Hum Reprod 2006;21:1009–11.

[174] Doornbos ME, Maas SM, McDonnell J, Vermeiden JP, Hennekam RC. Infertility, assisted reproduction technologies and imprinting disturbances: a Dutch study. Hum Reprod 2007;22:2476–80.

[175] Tierling S, Souren NY, Gries J, Loporto C, Groth M, Lutsik P, et al. Assisted reproductive technologies do not enhance the variability of DNA methylation imprints in human. J Med Genet 2010;47:371–6.

[176] Katari S, Turan N, Bibikova M, Erinle O, Chalian R, Foster M, et al. DNA methylation and gene expression differences in children conceived in vitro or in vivo. Hum Mol Genet 2009;18:3769–78.

[177] Amor DJ, Halliday J. A review of known imprinting syndromes and their association with assisted reproduction technologies. Hum Reprod 2008;23:2826–34.

[178] Baccarelli A, Bollati V. Epigenetics and environmental chemicals. Curr Opin Pediatr 2009;21:243–51.

[179] Mani S, Herceg Z. DNA demethylating agents and epigenetic therapy of cancer. Adv Genet 2010;70:327–40.

Page references followed by *t*, *f*, and *b* denote tables, figures, and boxes, respectively.

570

environmental factors, 326, 553
epidemiology of, 321, 332
epigenetic modifications, 322, 332–334, 333f
  chromatin regulation, 334–339
  DNA methylation, 332–333, 334
  insulin resistance, 325–326, 342
  insulin secretion, 322–324
  microRNA, 339–341
  protein function, 334
epigenetic therapies, 346–353
  microRNA-based, 352–353
  nutrition-based, 350–352
gestational, 284–285
juvenile (type 1), 207t, 217, 321
pathogenesis of, 342–346, 345b
Diabetogenes, 334
Diagnosis. *See also* Biomarkers
  autoimmune disorders, 243, 246
  cancer, 40–41, 42f, 102
  diabetes, 321
  endometrial cancer, 475–476
  endometriosis, 444, 460
  neurobehavioral diseases, 132–136, 142
Diagnostic and Statistical Manual of Mental Disorders (DSM), 132
Diesel exhaust particles (DEP), 378
Diet. *See* Nutrition
Diethylstilbestrol (DES), 447, 454, 558
Differentially methylated regions (DMRs), 258, 261–263, 261f
Disease-modifying antirheumatic drugs (DMARDs), 226, 233, 242t
DNA demethylation, 30, 130, 255–256
DNA methylation, 508–509
  alterations
    aging, 281, 523–529, 525f, 536–537, 546
    Alzheimer's disease, 177–183, 179f, 180t, 186–187, 536–537
    cancer. *See* Cancer
    cardiovascular disease, 396, 398, 400–405
    diabetes, 322–324, 332–333, 334
    endometriosis, 446–449, 450t
    infectious disease, 31–33, 428–430
    rheumatoid arthritis, 210–212, 211f
    smoking, 377
    systemic lupus erythematosus, 208–210, 209f, 228, 236–237
  analysis methods, 8–11, 17t
  biological roles of, 29–30, 129–130, 255
  cross-talk with other systems, 257
  during development, 255–257, 281–282, 554–556
  histone modifications and, 36–37
  imprinted genes, 154, 195, 260
  obesity, 275, 279–280, 306
  processes, 154–155, 155f, 156f
  stem cells and, 483
  transcriptional silencing by, 255
DNA methyltransferases (DNMTs), 30, 111–112, 129, 154, 508
  aging, 524, 526, 555
  Alzheimer's disease, 179–180, 179f, 186–187

autoimmune disorders, 211–212, 236–237
cancer, 33–35, 117–118
  endometrial, 472–473
cardiovascular disease, 396, 398, 401
during development, 255–257, 256f
diabetes, 339–340
endometriosis, 448
neurodevelopmental disorders, 157–158
obesity, 306
stem cells and, 483, 485
DNA methyltransferase inhibitors (DNMTi), 227b, 234b
  autoimmune disorders, 236–237
  cancer, 117–118, 119t–121t
  drug interactions, 244–245
  side effects, 243–244, 244b
DNA replication timing, 14–15, 16f, 17t
  early/late-switch regions, 17–18, 18f
DNA tissue samples, 8
Dopamine metabolism, 145
Drug-induced lupus, 236–237
Drugs. *See also specific drug*
  autoimmune disorders, 226
  epigenetic. *See* Epigenetic therapies
  maternal use of, 557–558
  neurobehavioral diseases, 139, 145–146
  neurodevelopmental disorders, 196
Dutch Hunger Winter, 282–283, 298, 302, 326, 554

E

Early-life programming. *See* Development
E-cadherin, 419, 448
E2 conjugating enzymes, 62t, 63
Embryogenesis. *See* Development
Embryonic stem cells (ESCs). *See* Stem cells
ENCODE (Encyclopedia of DNA Elements), 509–510
Endocrine-disrupting chemicals (EDCs), 558
Endocrine programming, 549–551
Endometrial cancer, 5, 471–478
  carcinogenesis in, 471, 471f
  diagnosis, 475–476
  epidemiology of, 471–472
  epigenetic modifications
    DNA methylation, 472–474, 474f
    microRNA methylation, 474–475
  epigenetic therapy, 476–478
Endometriosis, 5, 443–461
  definition of, 443
  diagnosis and classification of, 444, 460
  epidemiology of, 444
  epigenetic modifications, 446, 450t
    cause *versus* consequence, 454
    DNA methylation, 446–449
    histone modifications, 446, 449–453
  future research, 460–461
  review methods, 445–446
  treatment of, 444–445, 455–459
Endophenotypes, 133, 134b
Endoplasmic reticulum (ER) stress, 343–346
Endothelial cells (ECs), 396
  homeostasis, 398–401, 399f, 406t

Ensembl database, 509
Environmental factors, 520. *See also specific factor*
  aging process, 530–532, 531f, 546
  allergic diseases, 370–371, 374–379, 375t, 559
  Alzheimer's disease, 176–177
  diabetes, 326, 553
  endometriosis, 454
  intrauterine. *See* Development; Maternal nutrition
  neurobehavioral diseases, 138–139
  neurobiological disorders, 194–196, 197f, 199–200
  obesity, 273–274, 310
Epialleles, 286–287
Epigenetic code, 75
Epigenetic drift, 182, 281, 546
Epigenetic drugs. *See* Epigenetic therapies; *specific drug*
Epigenetic field for cancerization, 476
Epigenetic inheritance, transgenerational, 197–198, 309–310
Epigenetic landscape model (Waddington), 128, 128f, 488–489, 519–520
Epigenetic modifications
  chromatin structure, 254–255, 254t
  DNA methylation, 508–509. *See also* DNA methylation
  histone modifications, 508–509. *See also* Histone modifications
  microRNA, 97–98. *See also* MicroRNA
  reversibility of, 47, 75, 199–200, 200f, 228, 514, 560
  types of, 254–257, 520
"Epigenetic programming" hypothesis, 552
Epigenetics
  overview of, 1–6, 508–510
  definition of, 128, 193–194, 206, 254, 482, 508, 519–520
  variation methods, 2, 7–22. *See also specific method*
Epigenetic therapies, 75, 560
  airway disease, 389–392, 391f
  Alzheimer's disease, 184–185, 187
  autoimmune disorders. *See* Autoimmune disorders
  balancing conventional therapy with, 244–245
  bipolar disease, 130–131
  cancer. *See* Cancer
  cellular reprogramming, 490, 490t
  diabetes, 346–353
  drug categories, 226, 227b. *See also specific drug*
  endometriosis, 455–459
  future research, 245
  neurobehavioral diseases, 145–146
  neurobiological disorders, 198–199
  obesity. *See* Obesity
  side effects, 243–244, 244b, 458–459
Epigenome, 128–129
  reference, 47

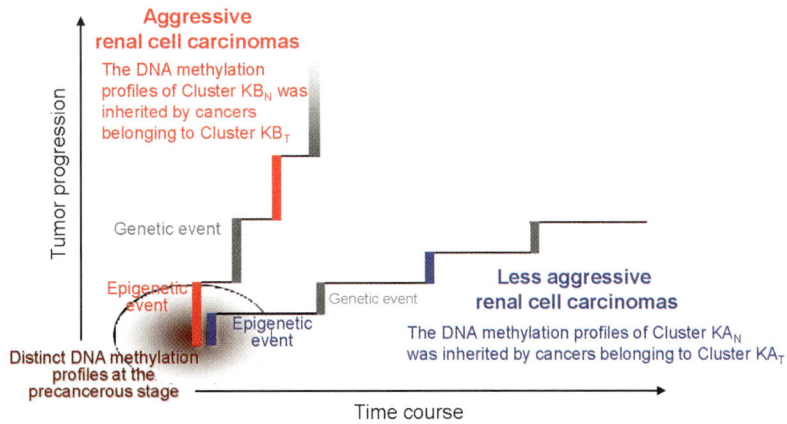

FIGURE 3.1

DNA methylation profiles in precancerous conditions and renal cell carcinomas (RCCs). See p. 39 for details.

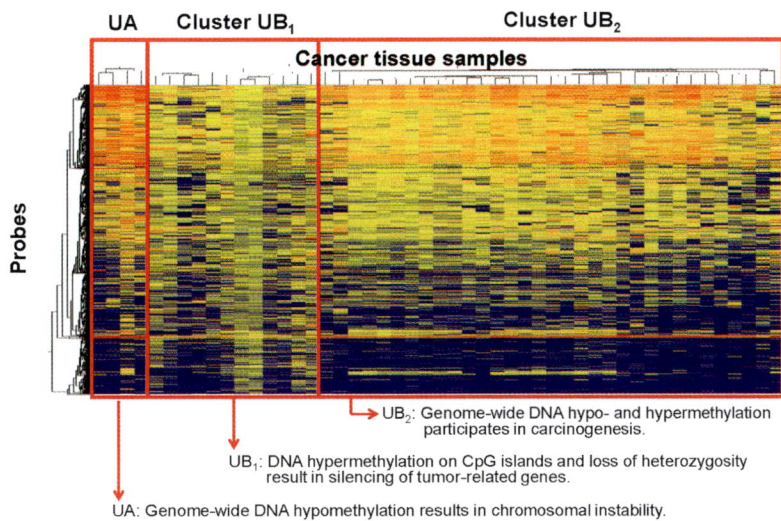

FIGURE 3.2

Hierarchical clustering analysis of urothelial carcinomas (UCs) based on array comparative genomic hybridization (CGH) data. See p. 40 for details.

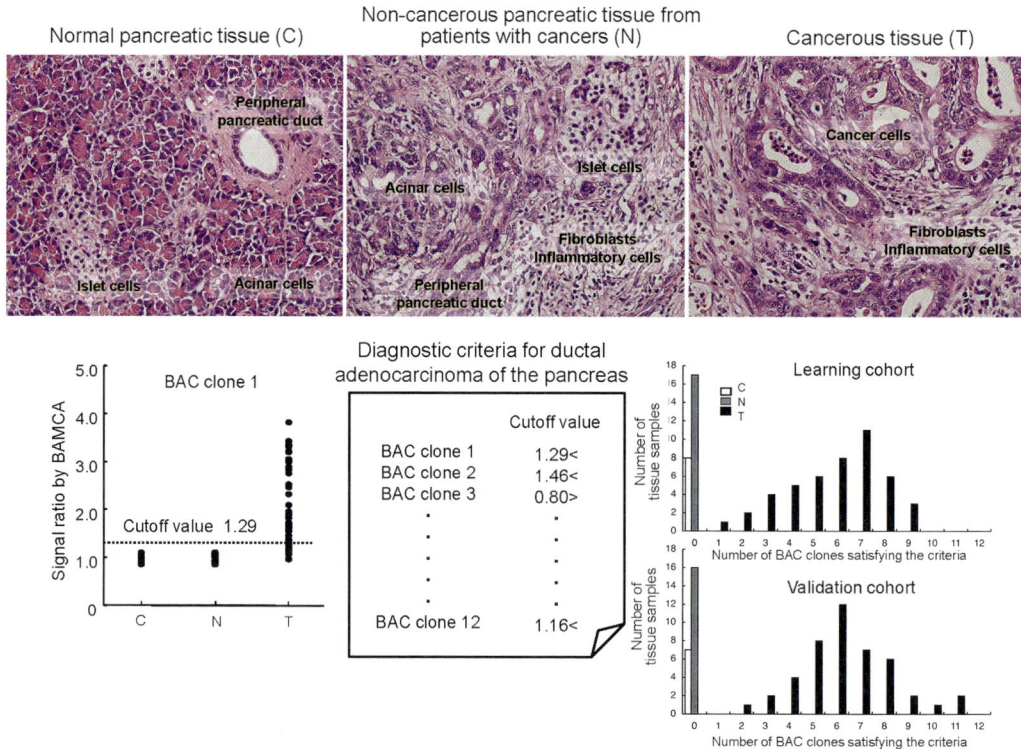

FIGURE 3.3

Diagnostic criteria based on DNA methylation profiles for ductal adenocarcinomas of the pancreas. See p. 42 for details.

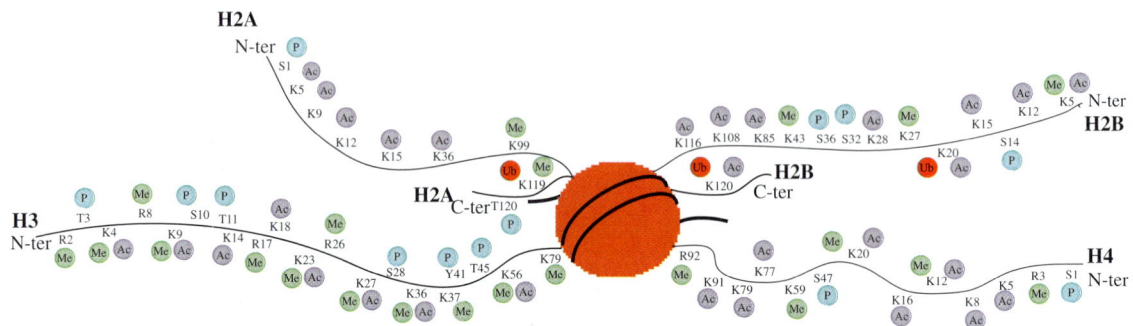

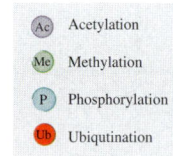

Different classes of histone modifications

FIGURE 4.1

Mammalian core histone modifications. See p. 55 for details.

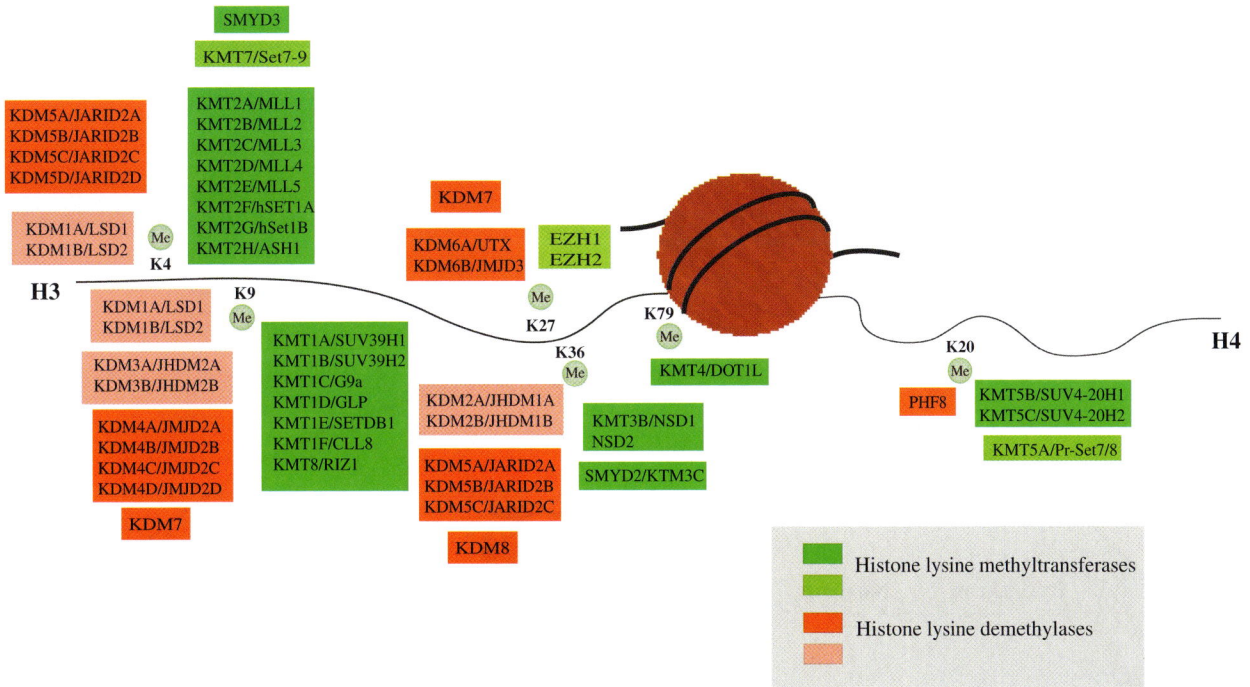

FIGURE 4.2

Histone lysine methylases and demethylases. See p. 60 for details.

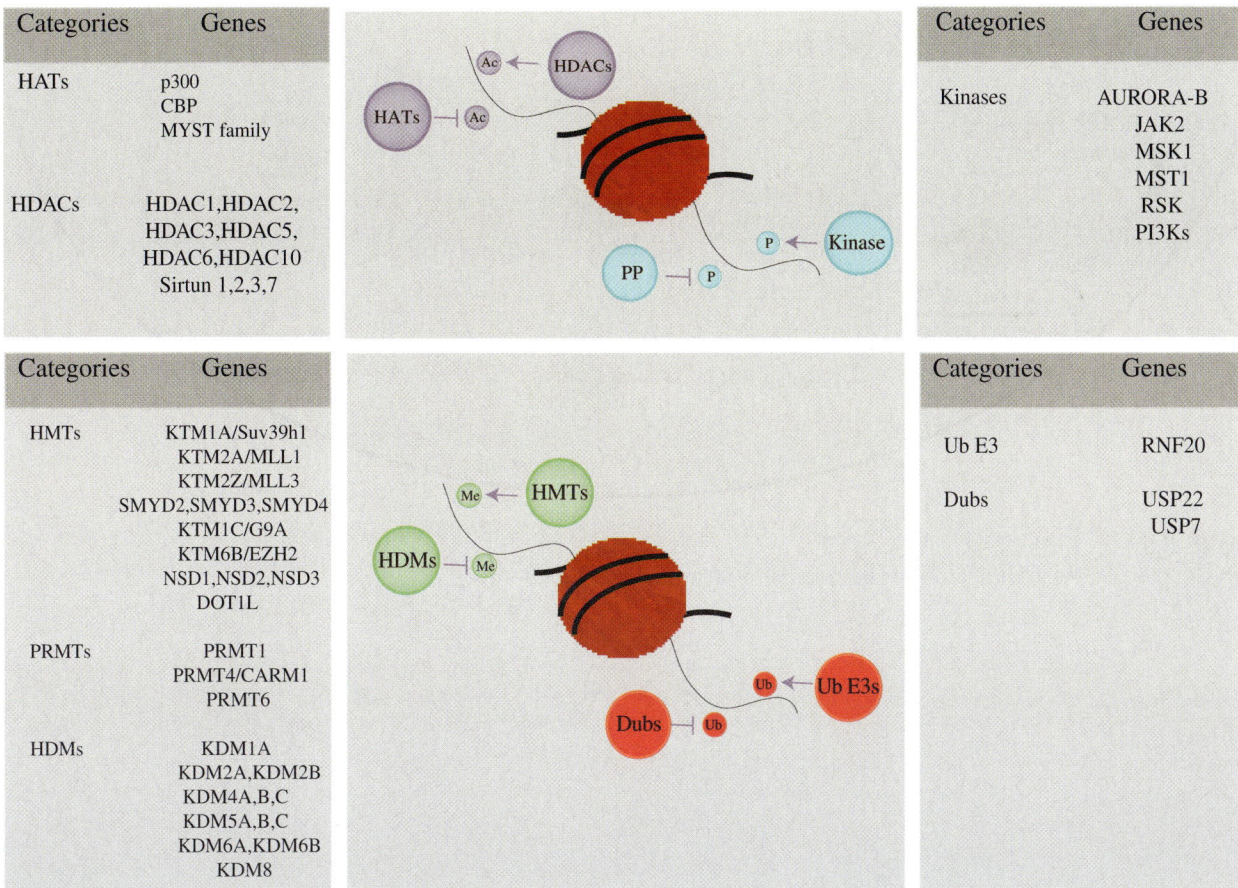

FIGURE 4.3

Histone-modifying enzymes and cancer. Selection of histone modifying enzymes altered in human cancer.

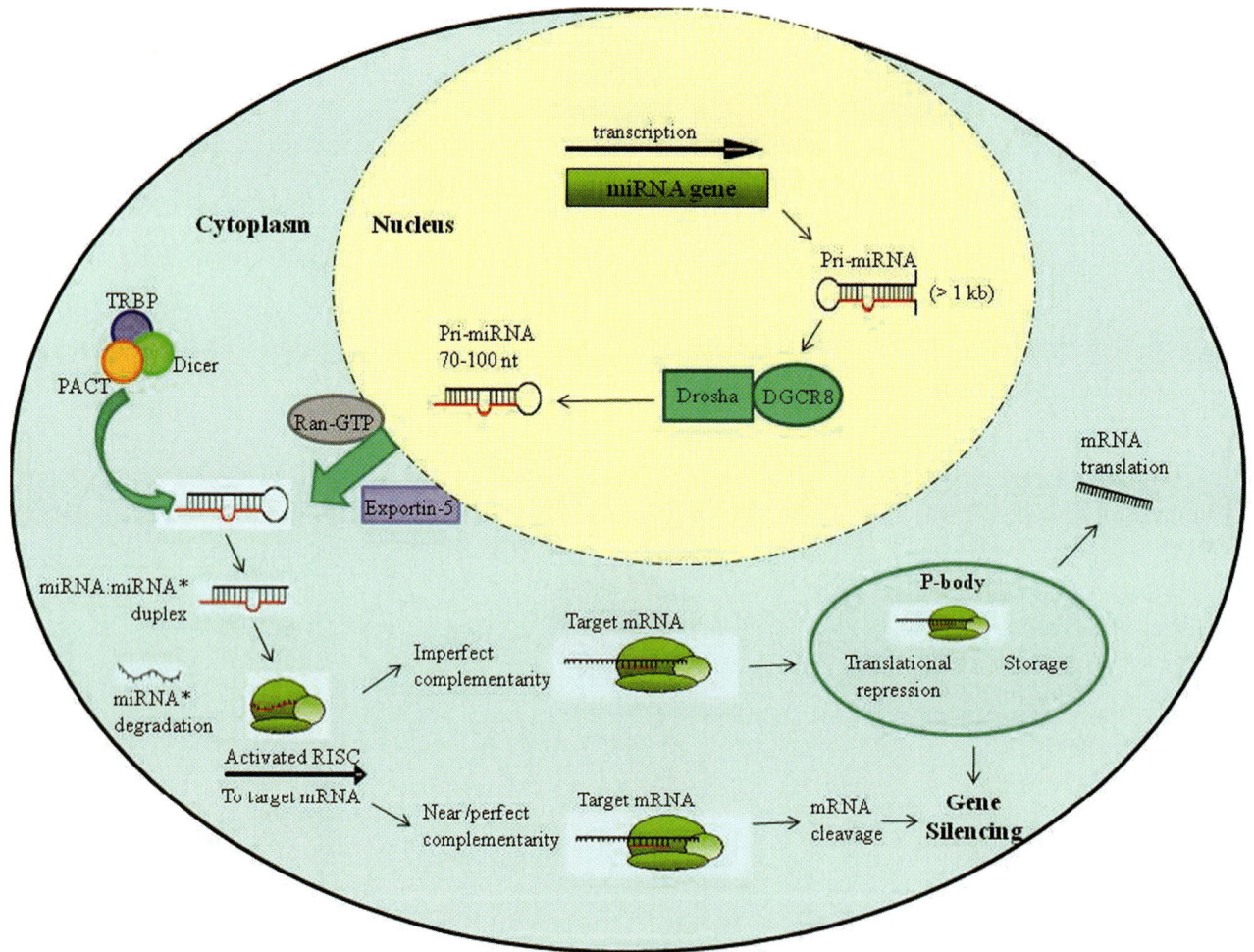

FIGURE 5.1

miRNA biogenesis and the mechanism of gene silencing. See p. 90 for details.

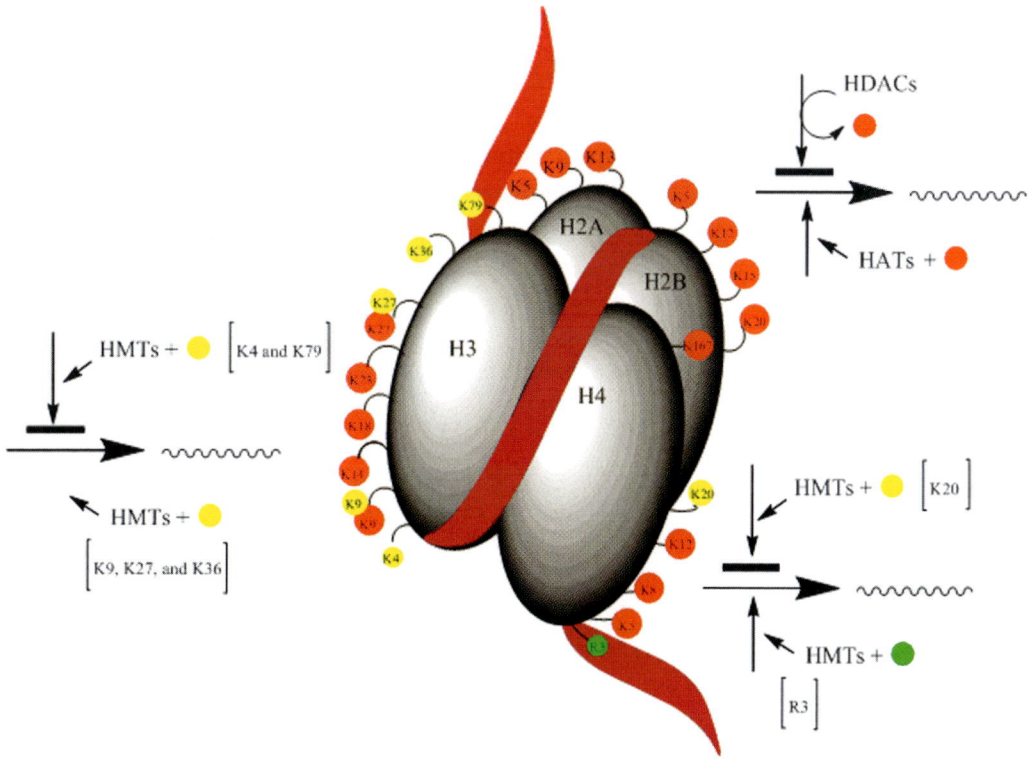

FIGURE 6.1

Effects of acetylation and methylation on histone residues. The red circle represents acetyl groups, the yellow circle symbolizes methylation and the green symbolizes methylation of arginine residue. See p. 112 for details.

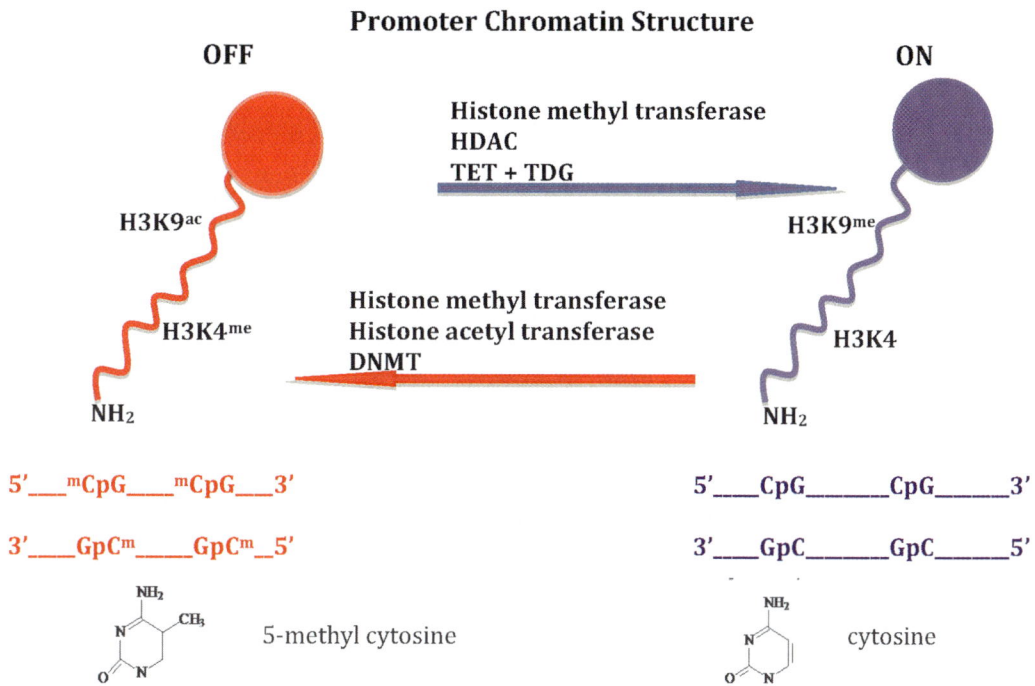

Promoter Chromatin Structure

OFF

Histone methyl transferase
HDAC
TET + TDG

ON

H3K9ᵃᶜ

H3K9ᵐᵉ

Histone methyl transferase
Histone acetyl transferase
DNMT

H3K4ᵐᵉ

H3K4

NH₂

NH₂

5'___ᵐCpG_____ᵐCpG___3'

5'____CpG_____CpG_____3'

3'___GpCᵐ_____GpCᵐ_5'

3'___GpC_____GpC_____5'

5-methyl cytosine

cytosine

FIGURE 7.2

Epigenomic programming of chromatin.

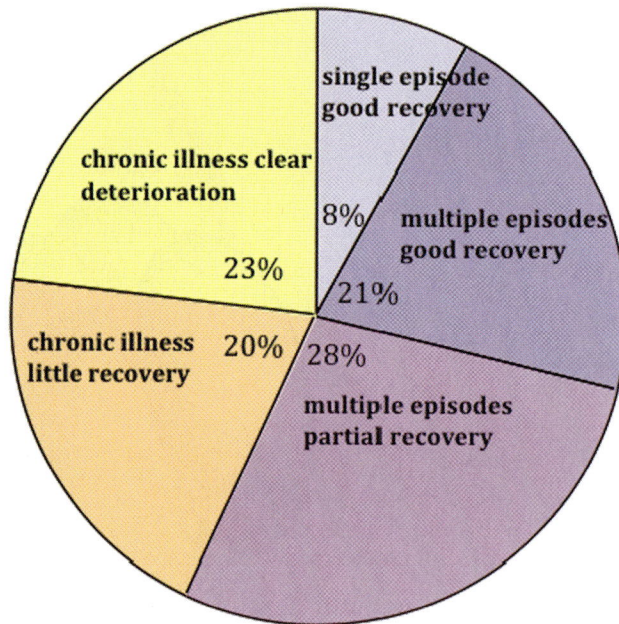

FIGURE 7.3
Outcome of schizophrenia diagnosis. *(adapted from http://www.science.org.au).*

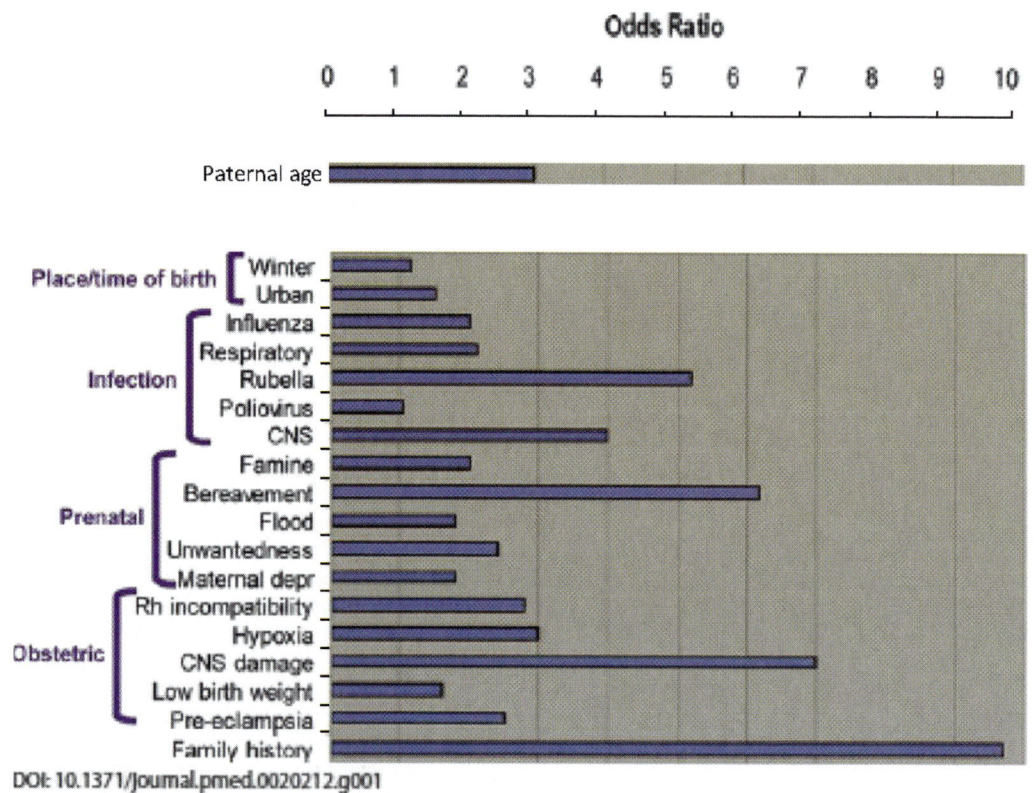

DOI: 10.1371/journal.pmed.0020212.g001

FIGURE 7.4
Odds ratio for schizophrenia as a function of genetic and environmental risk factors *(adapted from [73]).*

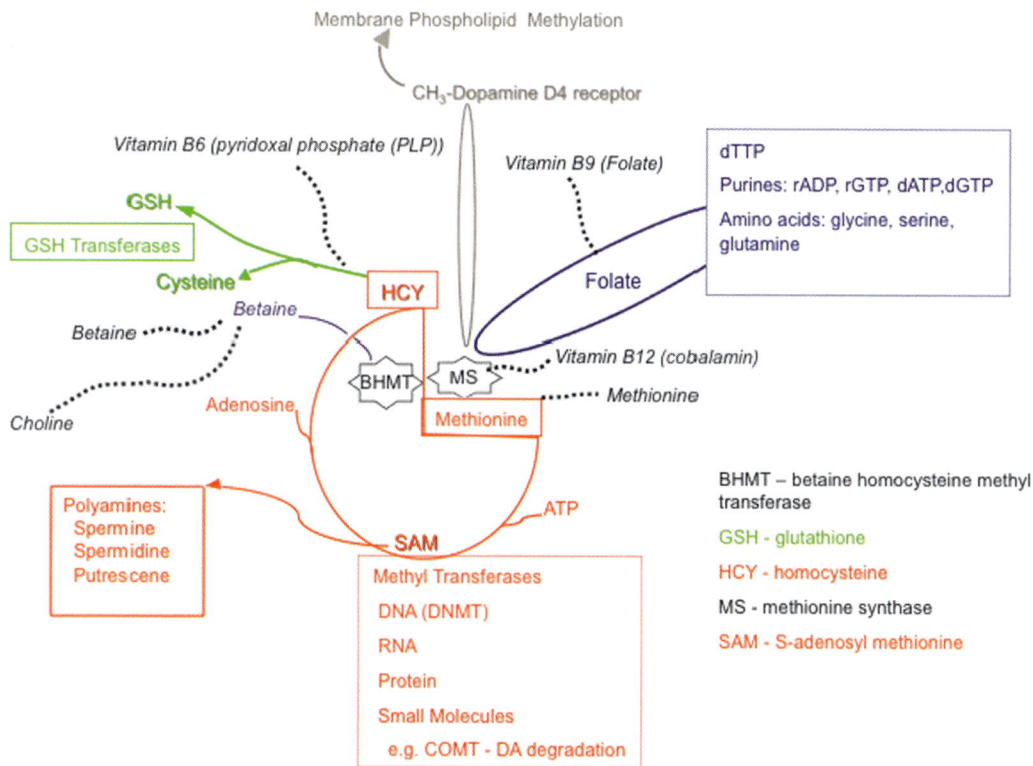

FIGURE 7.6

The FMTD hub: Metabolic link between DNA and RNA synthesis, energy production, epigenomic programming, oxidative stress responses, inflammation, and dopamine metabolism (not all metabolic steps are shown).

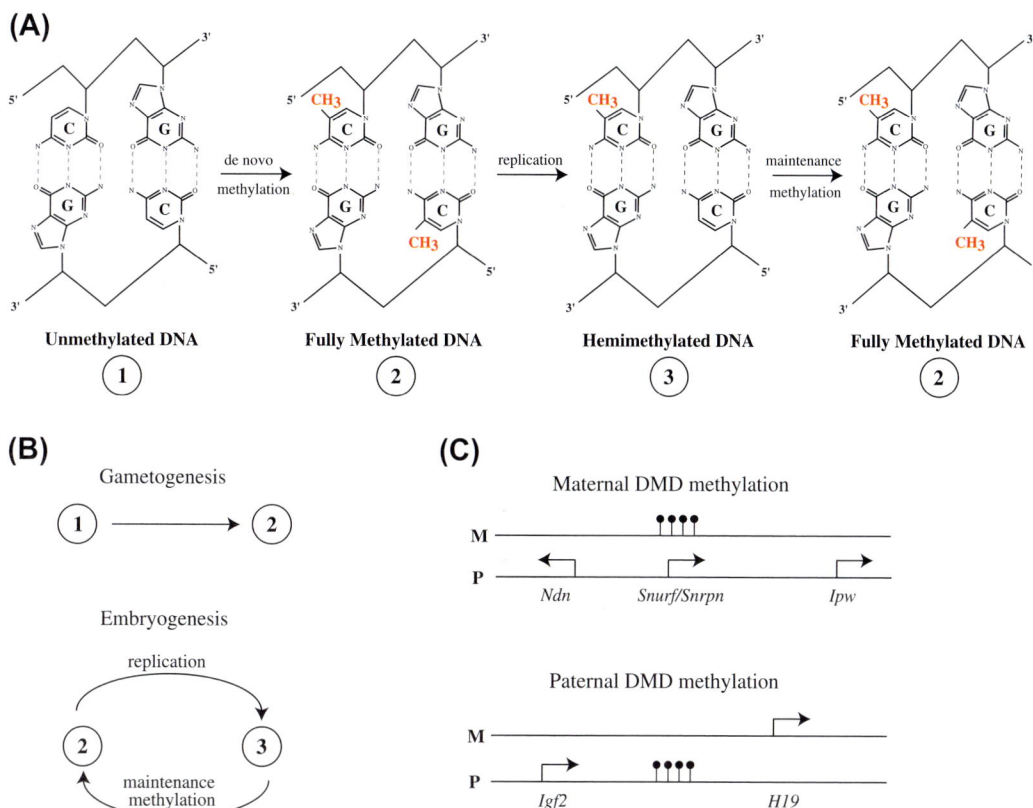

FIGURE 8.1

Establishing and maintaining DNA methylation in mammals. (A) Stepwise progression from unmethylated based-paired CpG dinucleotides (1) to fully methylated DNA (2), and its maintenance following DNA replication. Transition between (1) and (2) is de novo methylation. Maintenance requires an obligate hemimethylated intermediate (3). (B) De novo methylation of DNA occurs largely in the early stages of gametogenesis and maintenance methylation is primarily a post-fertilization process. (C) Differentially methylated domains (DMDs) of imprinted genes are created through a process in which only one parental allele undergoes de novo methylation in the germ lineage, and following its maintenance in the embryo, a difference in methylation of the two parental alleles is seen. *Snurf/Snrpn* is a maternally methylated DMD and *Igf2/H19* is a paternally methylated DMD.

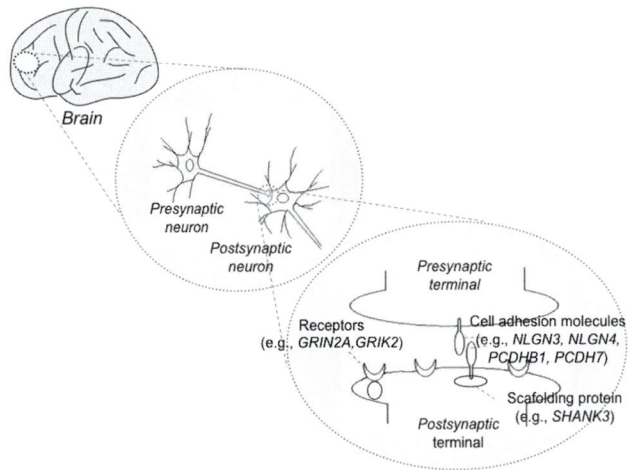

FIGURE 10.1

Location of the molecules in the synapse, which are associated with the pathogenesis of autism *(references [25] and [45]).*

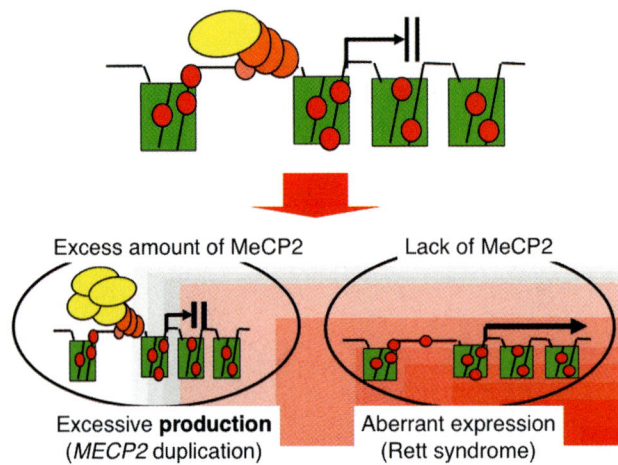

FIGURE 10.2

Schematic representation of epigenetic control of the gene expression. Yellow circle: MeCP2, orange circle: protein associated with histone modification, red circle: methyl-residue, green square: chromosomal histone protein, arrow: transcription start site. See p. 196 for details.

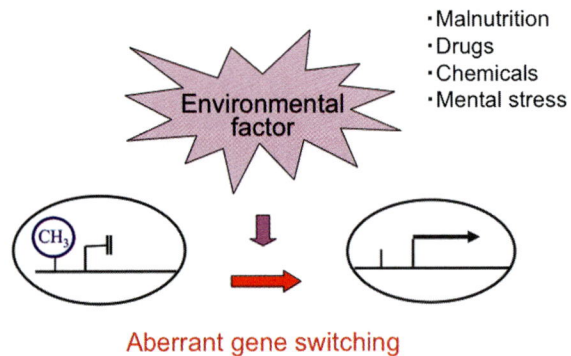

FIGURE 10.3

DNA methylation mechanisms can be affected by environmental factors.

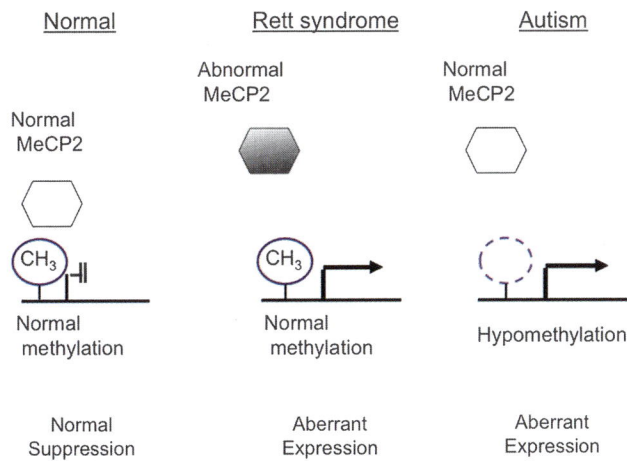

FIGURE 10.4
Pathogenesis of Rett syndrome and putative pathogenesis of autism.

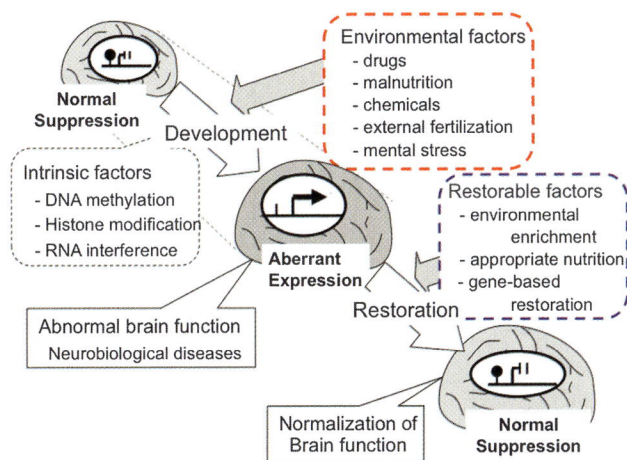

FIGURE 10.5
Overview of the DNA methylation changes and environmental factors. See p. 200 for details.

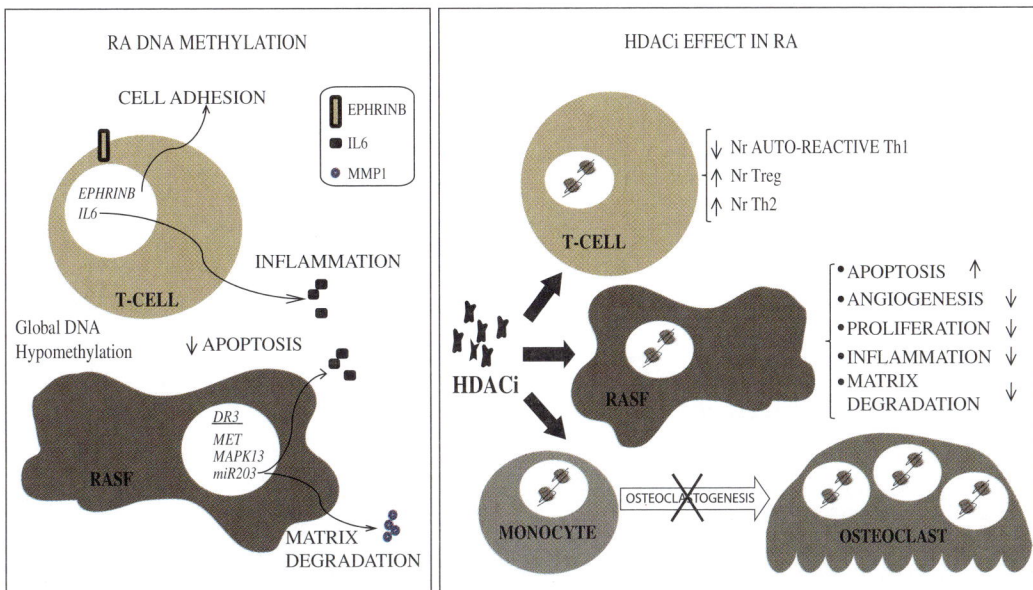

FIGURE 11.2
Epigenetic changes in RA. See p. 211 for details.

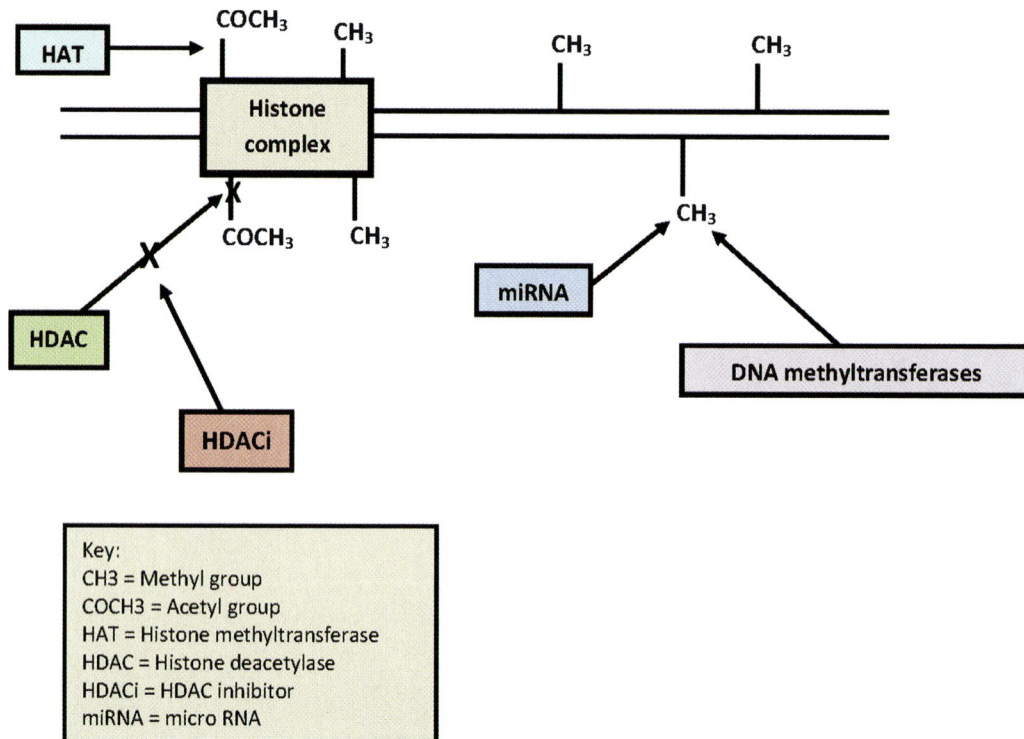

FIGURE 12.2
Molecular targets for epigenetic treatment of autoimmune diseases.

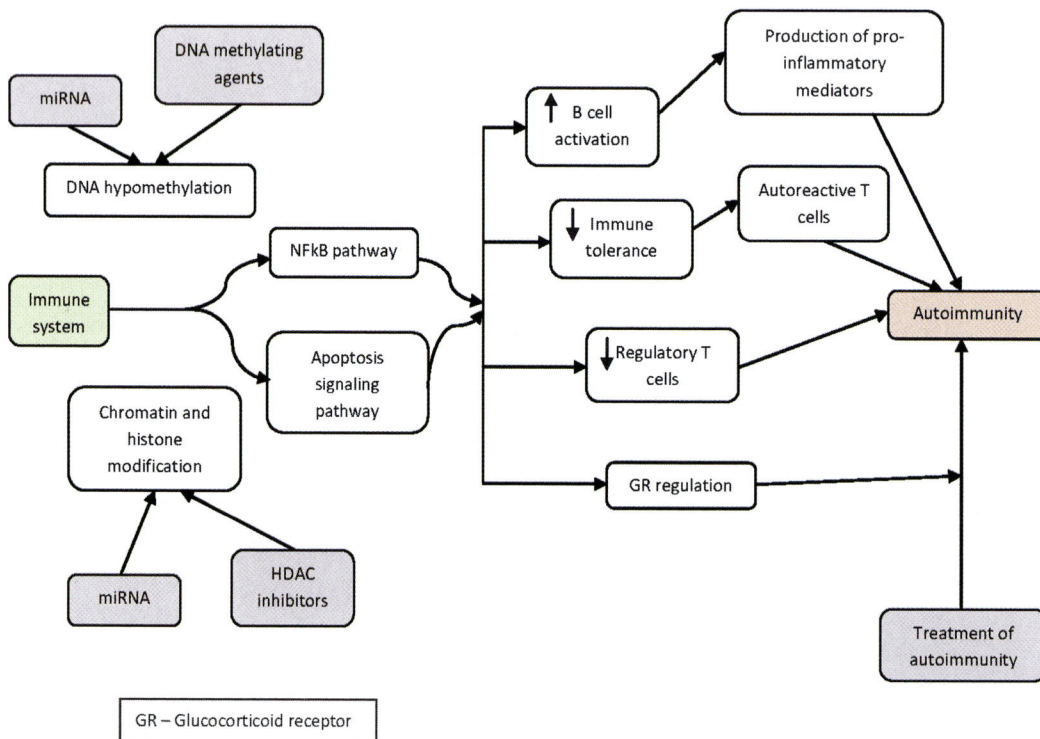

FIGURE 12.3
Potential effects of epigenetic changes in autoimmune diseases.

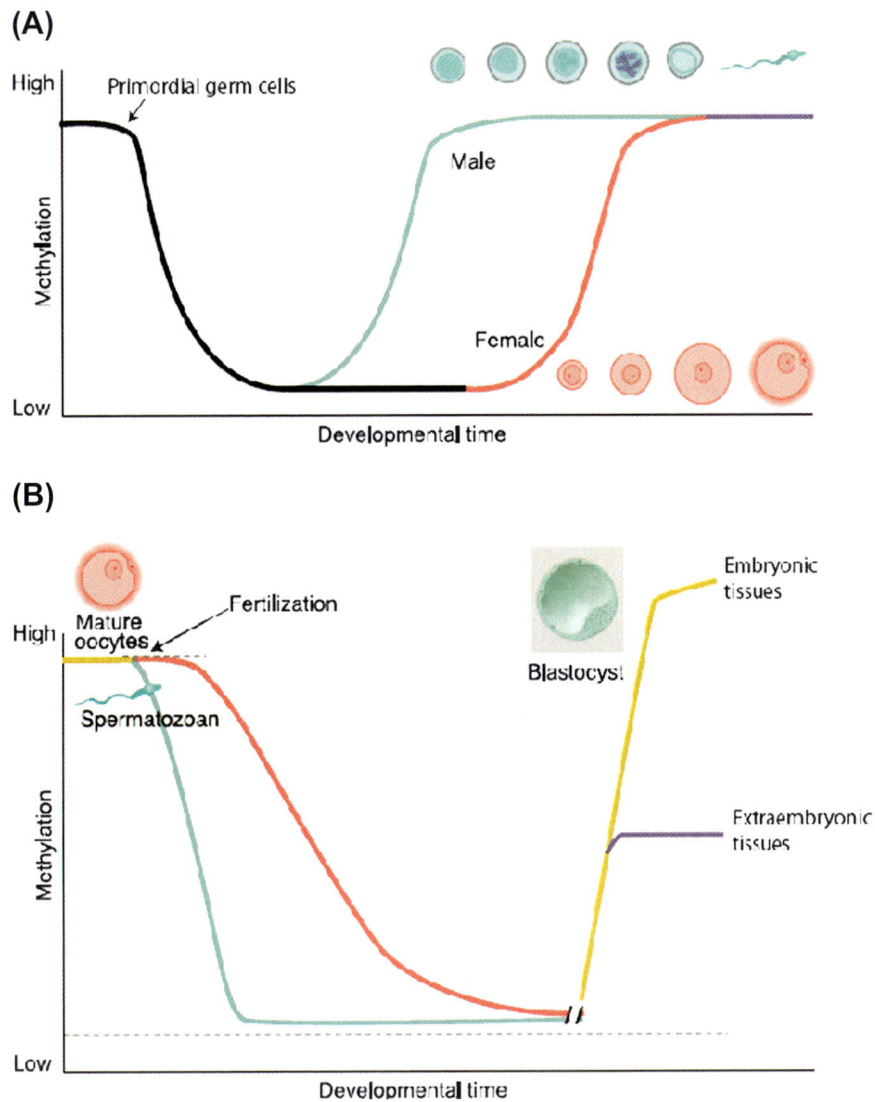

FIGURE 13.1
Changes in the overall level of DNA methylation during mammalian development. (A) Changes in DNA methylation in germ cells. (B) Changes in DNA methylation following fertilization. In both panels the level of DNA methylation is shown on the vertical axis and developmental time on the horizontal axis. *Adapted from [16].*

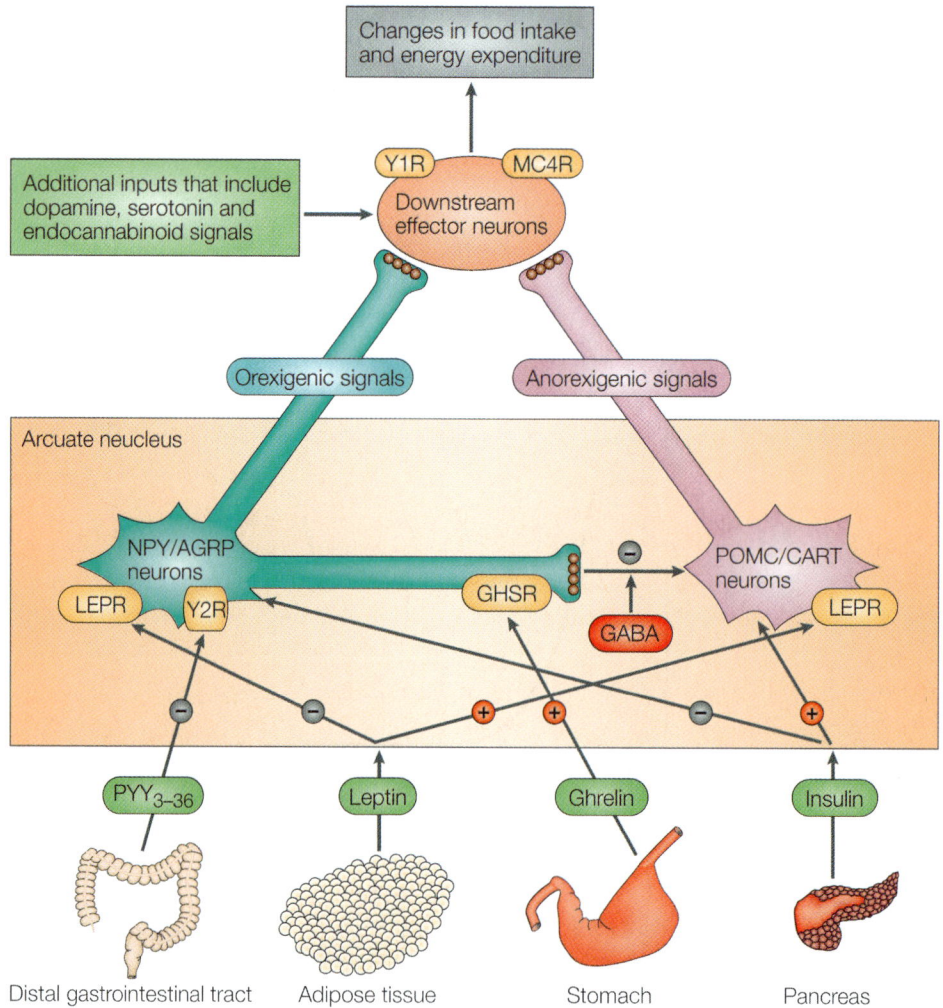

FIGURE 14.1

Arcuate nucleus control of central energy balance between food intake and energy expenditure. See p. 274 for details.

Distal gastrointestinal tract   Adipose tissue   Stomach   Pancreas

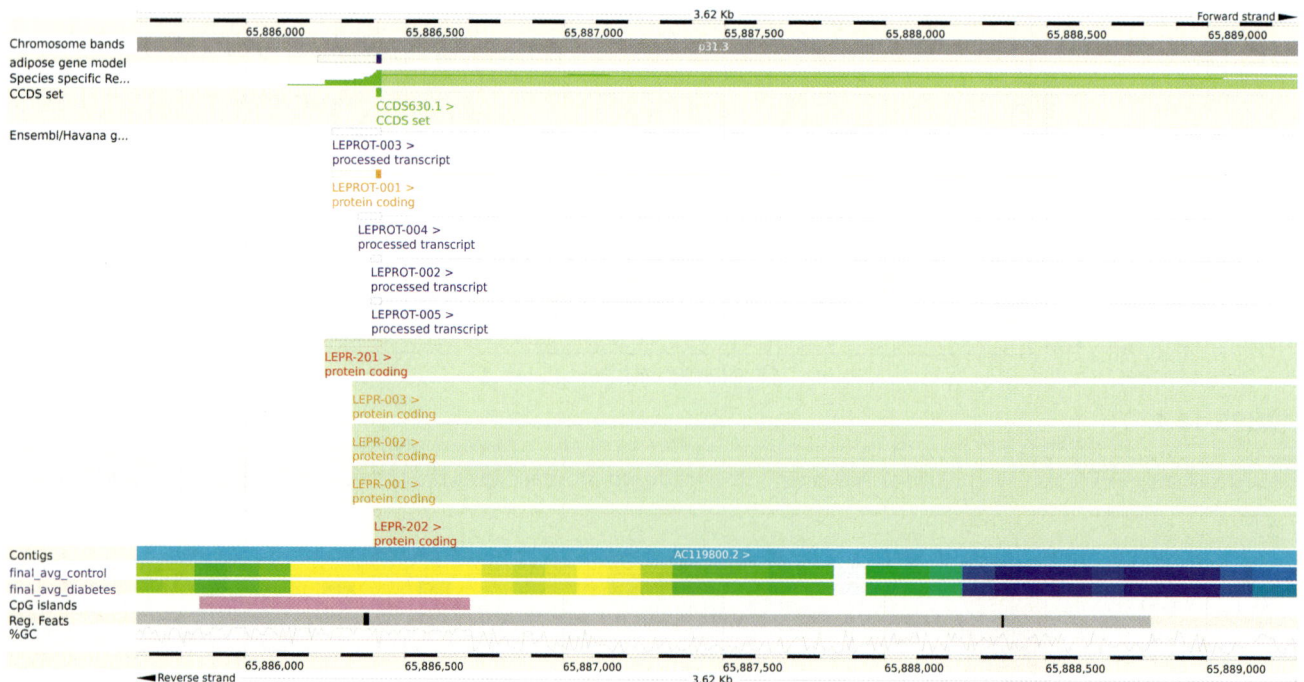

FIGURE 14.3

Methylation DNA immunoprecipitation (MeDIP) graphical result for the Leptin Promoter CpG Island. Sliding scale for methylation level: yellow = low, green = mid, blue = high. Hypomethylation seen over the CpG Island and no significant difference seen between T2D cases (final_avg_diabetes) versus controls (final_ave_control).

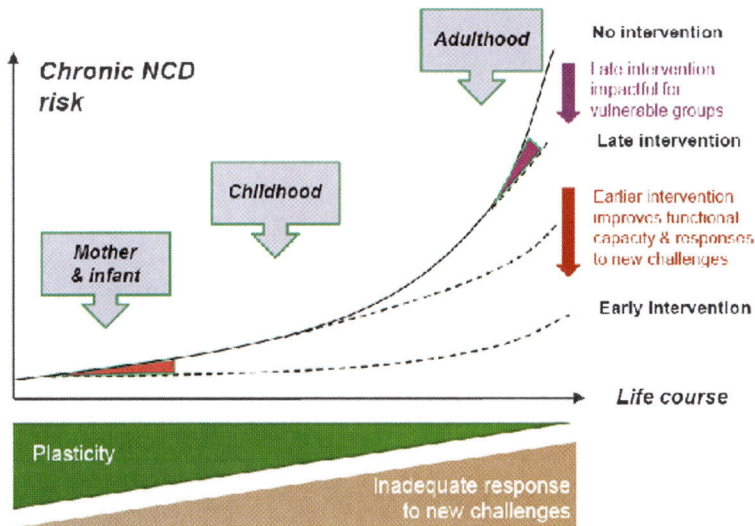

FIGURE 15.1
Non-communicable diseases (NCDs) do not fit the medical model in which an individual is healthy until they contract the disease. See p. 313 for details.

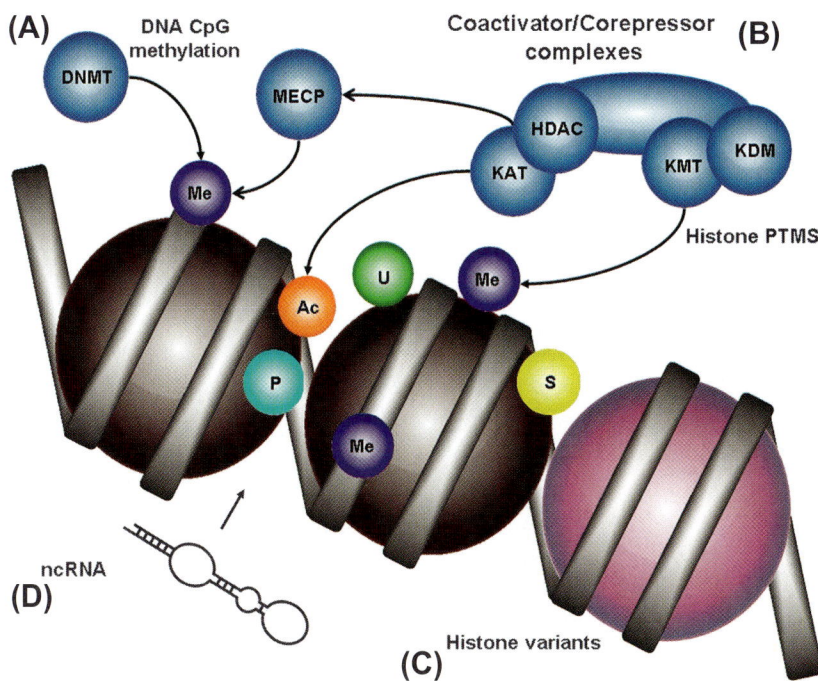

FIGURE 17.1
Overview of epigenetic mechanisms. See p. 333 for details.

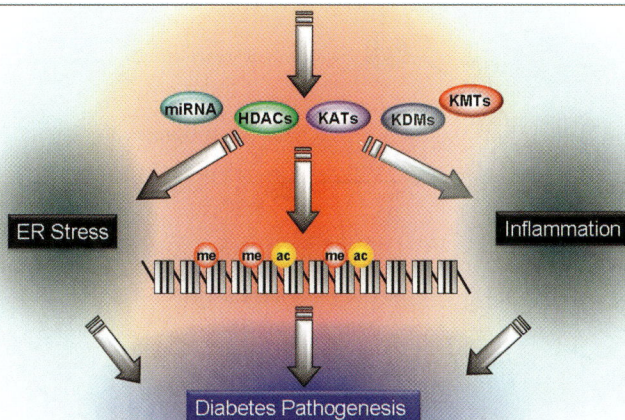

FIGURE 17.3
Simplified overview of how HDACi could target diabetes. See p. 343 for details.

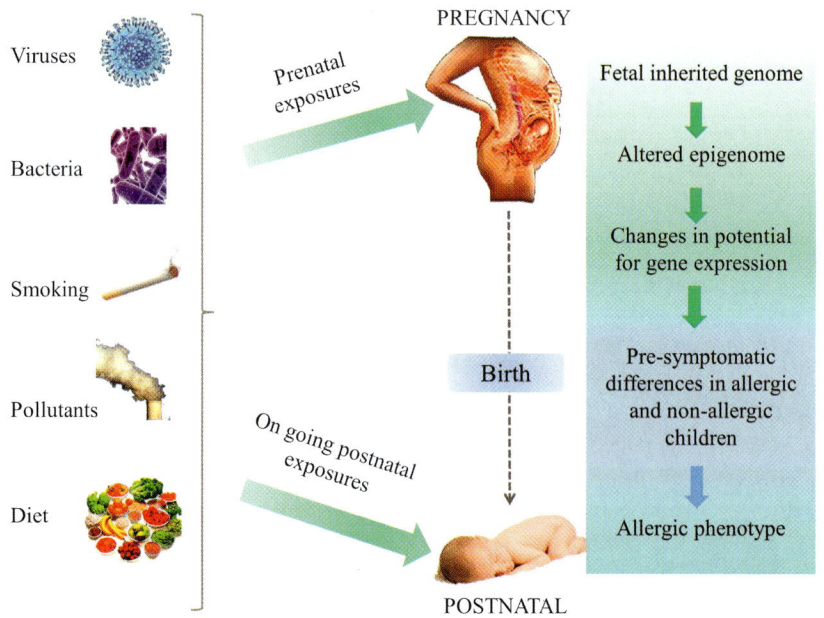

FIGURE 18.1
Environmental influences on developing immune system
See p. 374 for details.

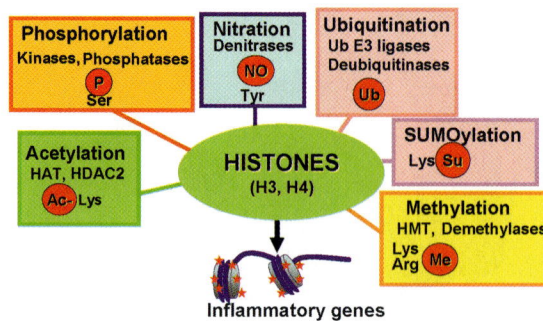

FIGURE 19.1
Post-translational histone modification. See p. 388 for details.

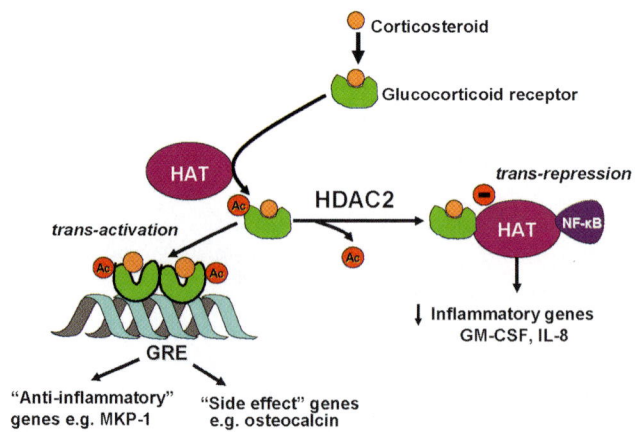

FIGURE 19.2
Acetylation of the glucocorticoid receptor (GR). See p. 389 for details.

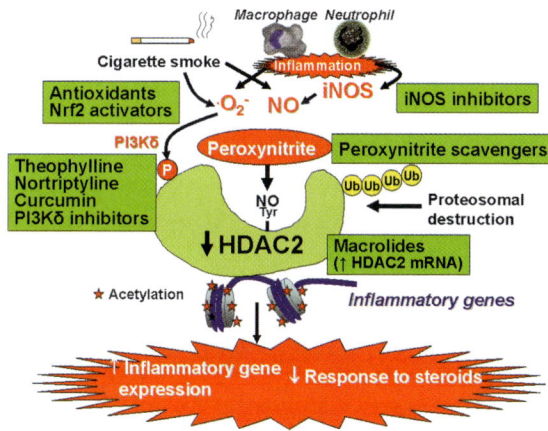

FIGURE 19.3
Mechanisms for decreased histone deacetylase(HDAC)2 in COPD and its reversal. See p. 391 for details.

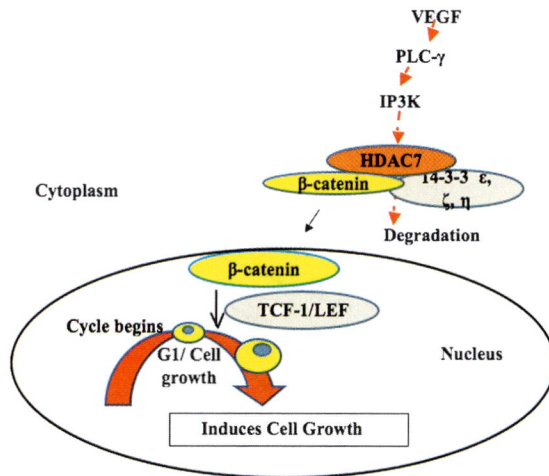

FIGURE 20.1
A schematic illustration shows the role of HDAC7 in controlling EC growth. See p. 399 for details.

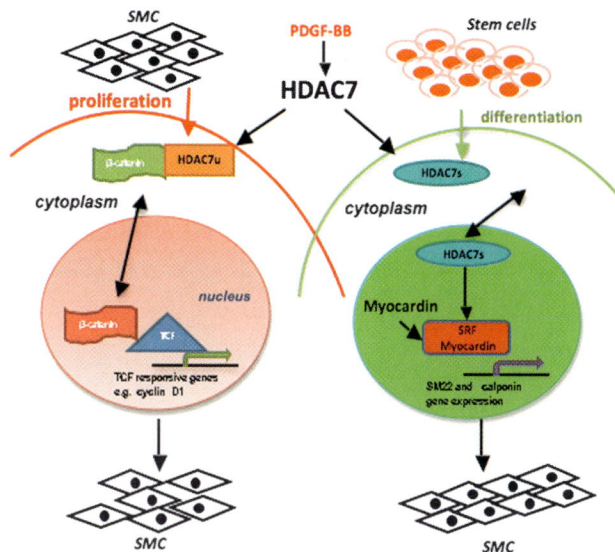

FIGURE 20.2
A schematic illustration shows the role of HDAC7 splicing in controlling SMC differentiation and proliferation. See p. 403 for details.

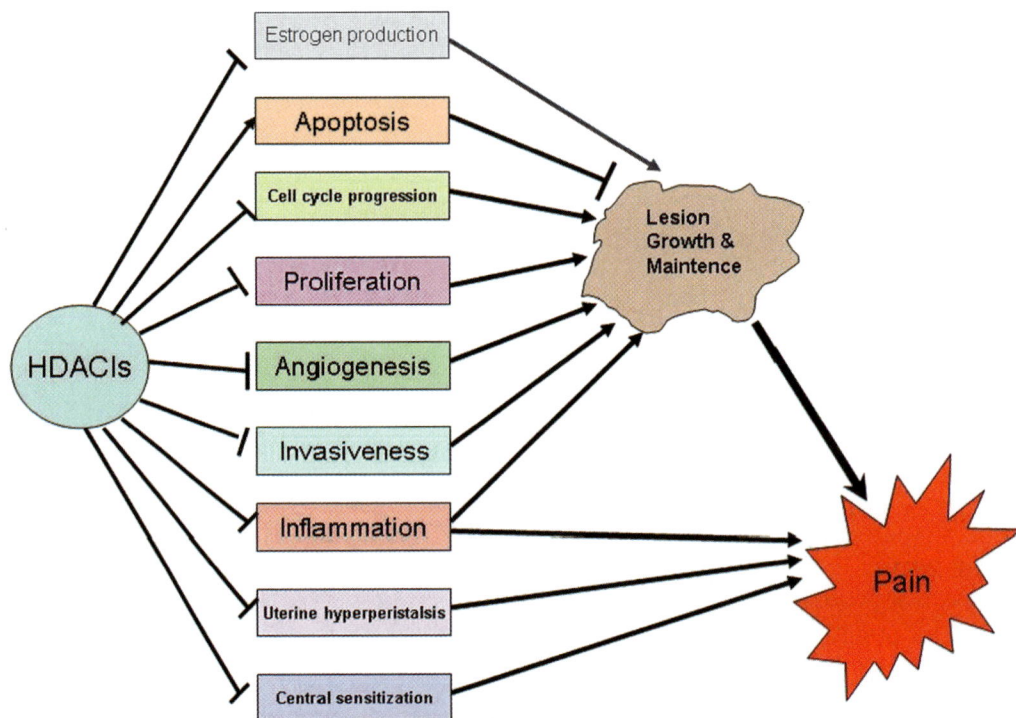

FIGURE 22.1
Schematic illustration of potential therapeutic effects of HDACIs in endometriosis. See p. 458 for details.

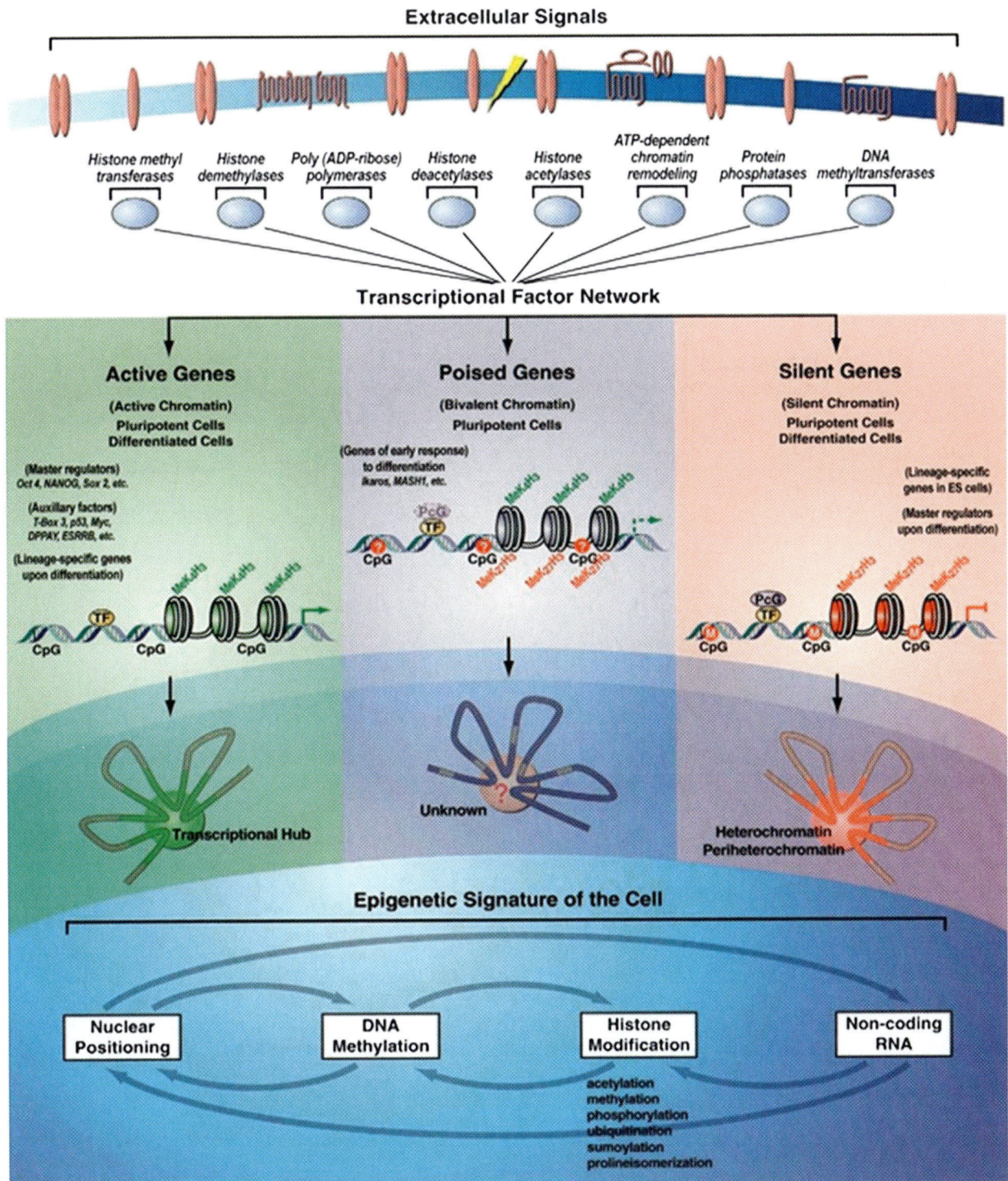

FIGURE 24.1

The pluripotency of ES cells and their differentiation rely largely on transcription factor circuitry and chromatin modifications. See p. 484 for details.

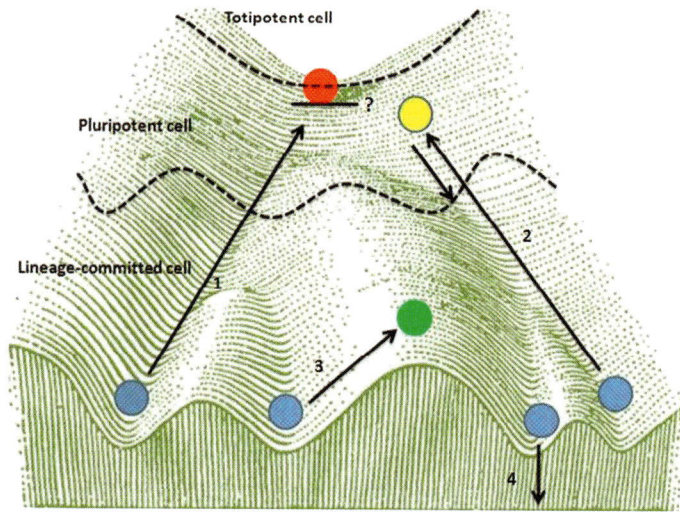

FIGURE 24.2
The stochastic model proposed by Yamanaka for iPS cell generation is based on the Waddington model. See p. 489 for details.

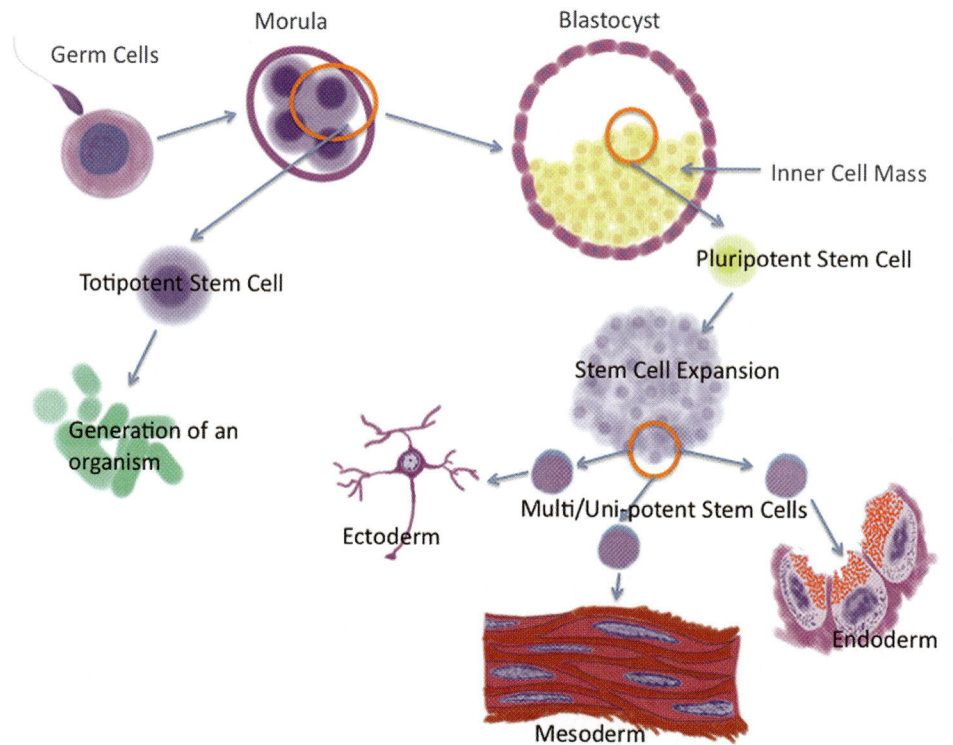

FIGURE 25.1
Schematic for the general maturation of stem cells and loss of differentiation potential.

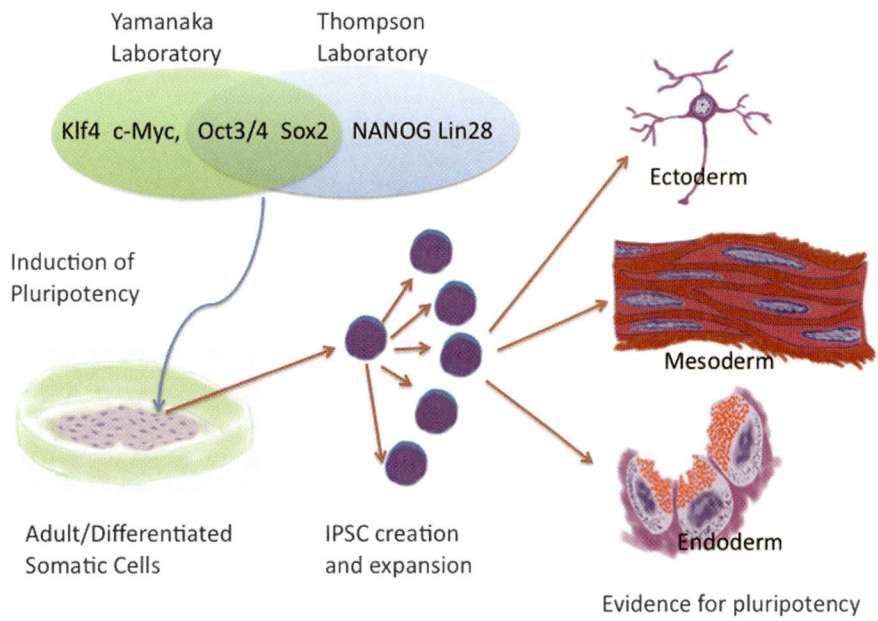

FIGURE 25.2

Schematic for the induction of pluripotency in somatic cells by the pluripotency network of transcription factors.

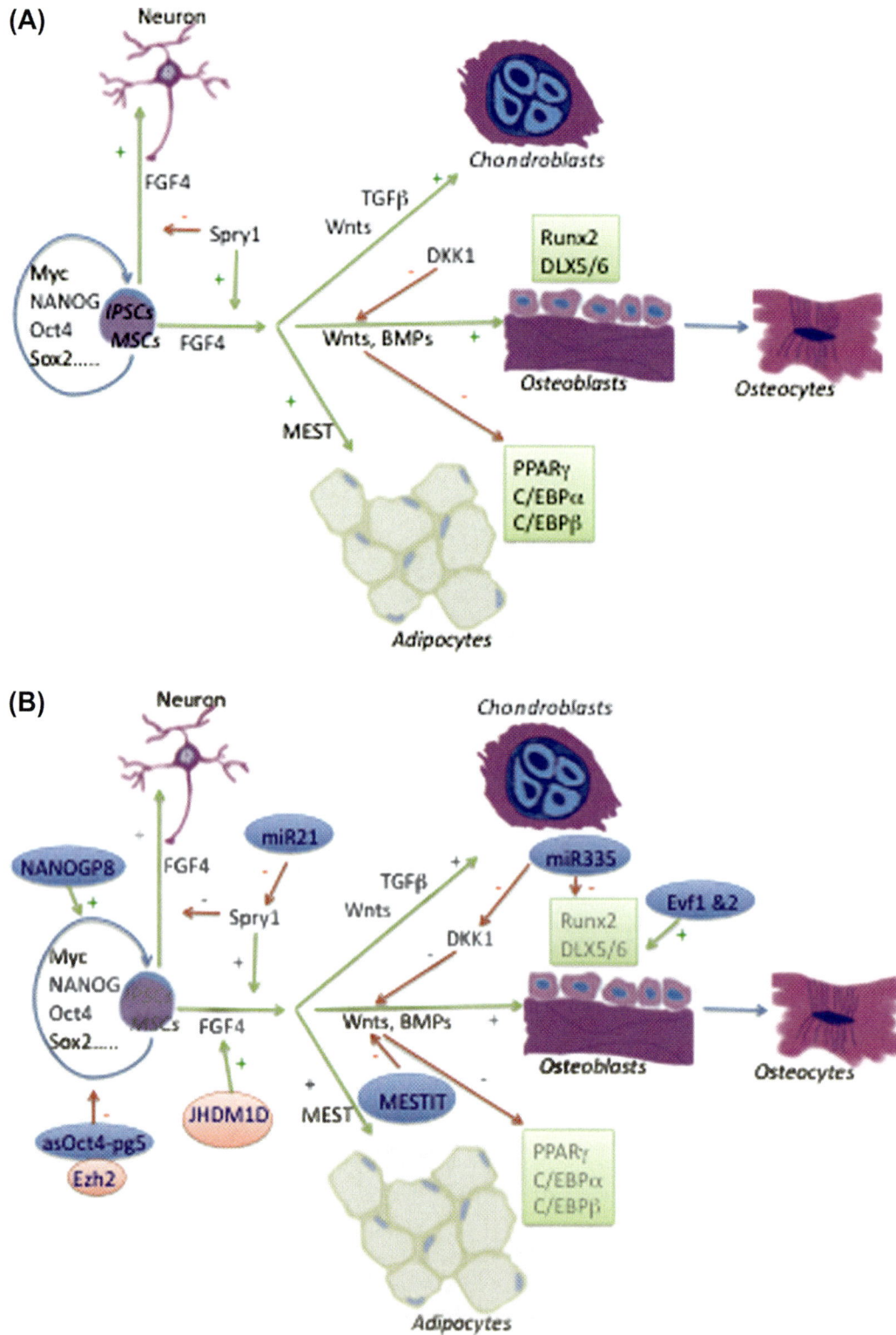

FIGURE 25.3

(A) Schematic of the competing network of transcription factors and signaling molecules that direct stem cells down specific mesoderm or ectoderm lineages. Green boxes indicate core transcription factors necessary for osteoblast and adipocyte lineage specific differentiation. (B) Overlay of the epigenetic/ncRNA regulatory network. This schematic depicts examples of ncRNAs (blue boxes) and chromatin modification factors (pink boxes) that control the balance between signaling molecules and transcription factors, to direct stem cell differentiation.

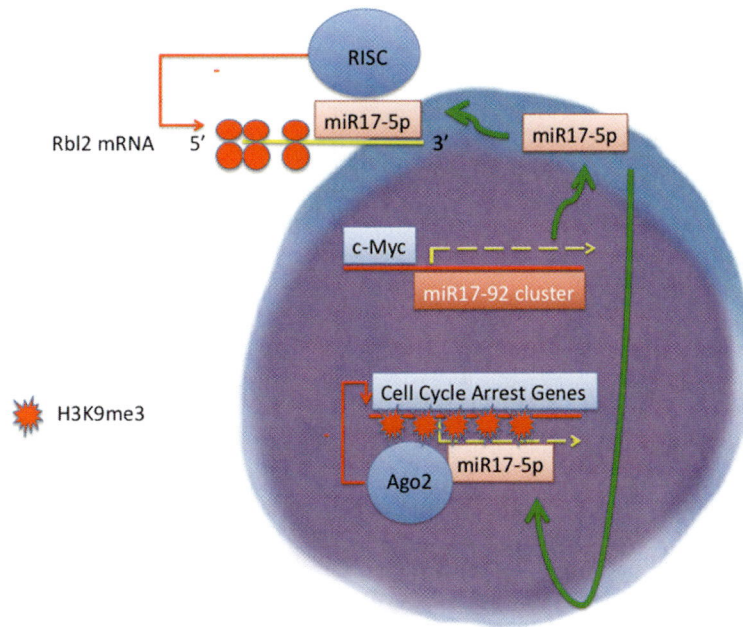

FIGURE 25.4

Example of interactions between a pluripotency factor, Myc, and the ncRNA/epigenetics network. See p. 511 for details.

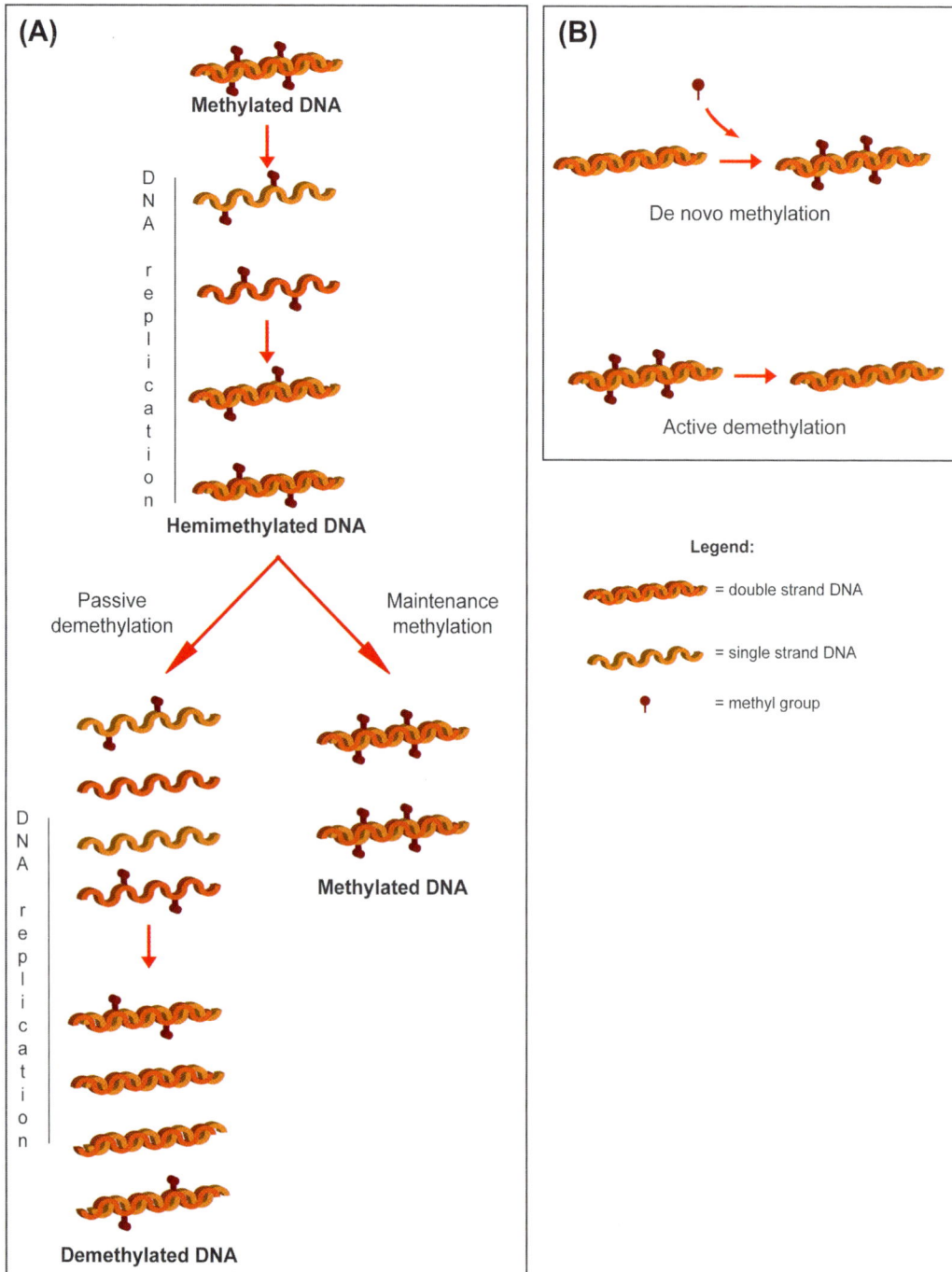

FIGURE 26.1

Schematic representation of the DNA methylation reactions: maintenance methylation, passive demethylation, de novo methylation, and active demethylation. See p. 525 for details.

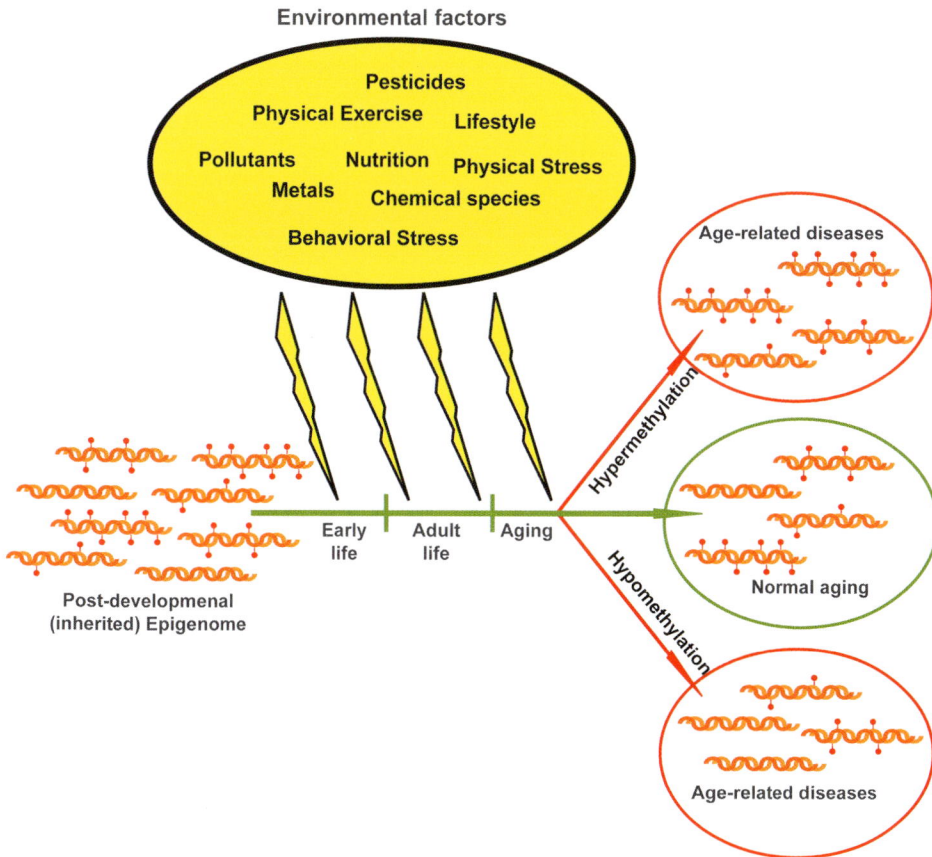

FIGURE 26.2

Many environmental factors can interfere with the organism, inducing epigenetic modifications. See p. 531 for details.

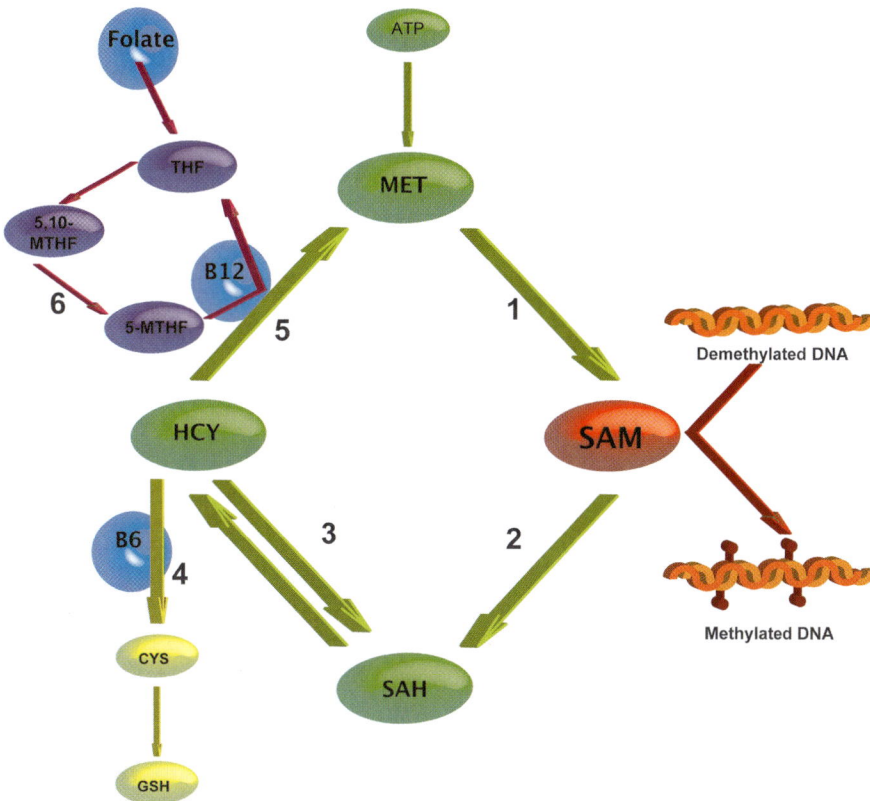

FIGURE 26.3

Schematic representation of the most relevant reactions comprises of the "one-carbon metabolism", emphasizing a vitamin cofactors cycle (remethylation) and SAM-dependent DNA methylation (transmethylation). See p. 534 for details.